KU-624-613

Handbook of
Water Analysis

edited by LEO M. L. NOLLET

Hogeschool Gent
Ghent, Belgium

MARCEL DEKKER, INC.

NEW YORK · BASEL

Library of Congress Cataloging-in-Publication Data

Handbook of water analysis / edited by Leo M. L. Nollet.
    p.   cm.—(Food science and technology)
Includes bibliographical references and index.
ISBN 0-8247-8433-2 (alk. paper)
    1. Water—Analysis—Handbooks, manuals, etc.   I. Nollet, Leo M. L., 1948–.   II. Food
science and technology (Marcel Dekker, Inc.) ; 102.

QD142 .H36 2000
628.1′61—dc21   00-037684

This book is printed on acid-free paper.

Headquarters
Marcel Dekker, Inc.
270 Madison Avenue, New York, NY 10016
tel: 212-696-9000; fax: 212-685-4540

Eastern Hemisphere Distribution
Marcel Dekker AG
Hutgasse 4, Postfach 812, CH-4001 Basel, Switzerland
tel: 41-61-261-8482; fax: 41-61-261-8896

World Wide Web
http://www.dekker.com

The publisher offers discounts on this book when ordered in bulk quantities. For more information, write to Special Sales/Professional Marketing at the headquarters address above.

Copyright © 2000 by Marcel Dekker, Inc. All Rights Reserved.

Neither this book nor any part may be reproduced or transmitted in any form or by any means, electronic or mechanical, including photocopying, microfilming, and recording, or by any information storage and retrieval system, without permission in writing from the publisher.

Current printing (last digit):
10 9 8 7 6 5 4 3 2 1

PRINTED IN THE UNITED STATES OF AMERICA

**Books are to be returned on or before
the last date below.**

2 1 OCT 2004

1 8 NOV 2004

LIBREX —

LIVERPOOL
JOHN MOORES UNIVERSITY
AVRIL ROBARTS LRC
TITHEBARN STREET
LIVERPOOL L2 2ER
TEL. 0151 231 4022

LIVERPOOL JMU LIBRARY

3 1111 00914 1803

Handbook of
Water Analysis

Preface

Handbook of Water Analysis discusses analysis techniques of all types of water: freshwater from glaciers, rivers, lakes, canals, and seawater, as well as groundwater from springs, ditches, drains, and brooks.

Each chapter describes the physical, chemical, and other relevant properties of water components, and covers sampling, cleanup, extraction, and derivatization procedures. Older techniques that are still in use are compared to recently developed techniques. The reader is also directed to future trends. A similar strategy is followed for discussion of detection methods. In addition, the applications of analysis of water types (potable water, tap water, wastewater, seawater) are reviewed. Information is summarized in graphs, tables, examples, and references.

Because water is an excellent solvent, it dissolves many substances. To get correct results and values, analysts have to follow sample strategies. Sampling has become a quality-determining step (Chapter 1). If samples can't be analyzed directly they have to be stored and preserved. Physical, chemical, or biological activities in a water sample can distort the chemical composition in water (Chapter 2). Statistical treatment of data ensures the reliability of the results. Statistical methods are discussed in Chapter 3.

Chapters 4, 5, and 6 deal with the water characteristics (physical, chemical, and organoleptical) and their analysis methods. Physical characteristics of water, such as temperature, color, turbidity, etc., are discussed, in addition to hardness, acidity, alkalinity, antioxidant demand, and how dissolved oxygen is detected.

Water is a living element housing a lot of organisms, wanted or unwanted, harmful or harmless. Some of these organisms produce toxic substances. Chapters 8 and 9 discuss bacteriological and algal analysis.

Humans both consume and pollute large quantities of water. Chapters 10 through 36 deal with injurious or toxic substances of domestic, agricultural, and industrial sources: sulfuric compounds, ammonia, nitrites, nitrates, organic nitrogen, phosphates, organic acids, phenolic compounds, cyanides, metals, pesticides, PCBs, dioxins, PAHs, BTEX compounds, oils, greases, petroleum hydrocarbons, asbestos, silicates, and surfactants. Chapter 7 discusses new technologies on radionuclides and their possible health hazards in water and the whole environment.

The last chapter gives detailed information on most of the cited techniques, sample preparation, separation, and detection methods. Separation techniques discussed are gas and liquid chromatography, supercritical fluid chromatography, and capillary electrophoresis.

This book may be used as a primary textbook for undergraduate students in the techniques of water analysis. Furthermore, it is intended for use by graduate students and scientists involved in the analysis of water.

All contributors are international experts in their field of water analysis. I thank them for their excellent efforts. This work is dedicated to my son Gerrit and his beloved girlfriend Veerle.

Leo M. L. Nollet

Contents

LIVERPOOL
JOHN MOORES UNIVERSITY
AVRIL ROBARTS LRC
TITHEBARN STREET
LIVERPOOL L2 2ER
TEL. 0151 231 4022

Preface   *iii*
Contributors   *ix*

1. Sampling Methods in Surface Waters   1
   Mathias Liess and Ralf Schulz

2. Methods of Water Preservation   25
   Bieluonwu Augustus Uzoukwu

3. Statistical Treatment of Data   29
   Bieluonwu Augustus Uzoukwu

4. Physical Properties of Water   41
   Jean-Yves Bottero

5. Chemical Parameters   51
   Hilda Ledo de Medina

6. Organoleptical Properties of Water   75
   Maria Anita Mendes and Marcos N. Eberlin

7. Radioanalytical Methodology for Water Analysis   101
   Jorge S. Alvarado

8. Bacteriological Analysis   115
   Chris W. Michiels and Els L. D. Moyson

9. Algal Analysis—Organisms and Toxins   143
   Glen Shaw and Maree Smith

v

10. Halogens   169
     Geza Nagy and Livia Nagy

11. Sulfate, Sulfite, and Sulfide   195
     Shreekant V. Karmarkar and M. A. Tabatabai

12. Analysis of Nitrates and Nitrites   201
     Maria Teresa Oms, Amalia Cerdà, and Victor Cerdà

13. Ammonia   223
     Stuart W. Gibb

14. Determination of Organic Nitrogen   261
     Amalia Cerdà, Maria Teresa Oms, and Victor Cerdà

15. Phosphates   273
     Ian D. McKelvie

16. Main Parameters and Assays Involved with the Organic Pollution of Water   297
     Angel Cuesta, Antonio Canals, and José Luis Todolí

17. Organic Acid Analysis   313
     Sigrid Peldszus

18. Determination of Phenolic Compounds in Water   347
     Magnus Knutsson and Jan Åke Jönsson

19. Cyanides   367
     Kazunori Ikebukuro, Hideaki Nakamura, and Isao Karube

20. Humic Substances in Water   387
     Juhani Peuravuori and Kalevi Pihlaja

21. Major Metals   409
     Payman Hashemi

22. Heavy Metals   439
     Elena González-Soto, Elia Alonso-Rodríguez, and Darío Prada Rodríguez

23. Trace Elements: Li, Be, B, Al, V, Cr, Co, Ni, Se, Sr, Ag, Sn, Sb, Ba, and Tl   459
     Rosa Cidu

24. Determination of Silicon and Silicates   483
     Bieluonwu Augustus Uzoukwu and Leo M. L. Nollet

25. Analysis of Urea Herbicides in Water   487
     Antonio Di Corcia

26. Analysis of Organochlorinated Pesticides in Water   517
     Filippo Mangani, Michela Maione, and Pierangela Palma

27. Residue Analysis of Carbamate Pesticides in Water 537
    Evaristo Ballesteros Tribaldo

28. Organophosphates 571
    Monica Culea and Simion Gocan

29. Fungicide and Herbicide Residues in Water 609
    H. S. Rathore and A. A. Khan

30. Methods for the Determination of Polychlorobiphenyls (PBCs) in Water 655
    Roger Fuoco and Alessio Ceccarini

31. Determination of PCDDs and PCDFs in Water 673
    Anna Laura Iamiceli, Luigi Turrio-Baldassarri, and Alessandro di Domenico

32. Polycyclic Aromatic Hydrocarbons 687
    Miren López de Alda-Villaizán

33. Analysis of BTEX Compounds in Water 721
    Ignacio Valor, Mónica Pérez, Carol Cortada, David Apraiz, and Juan Carlos Moltó

34. Oil and Greases and Petroleum Hydrocarbon Analysis 753
    Rossiza Belcheva

35. Asbestos in Water 765
    Leo M. L. Nollet

36. Analysis of Surfactants 767
    Bieluonwu Augustus Uzoukwu and Leo M. L. Nollet

37. Instruments and Techniques 785
    A-M Siouffi

Index 887

Contributors

Elia Alonso-Rodríguez Analytical Chemistry Department, University of A Coruña, A Coruña, Spain

Jorge S. Alvarado Environmental Research Division, Argonne National Laboratory, Argonne, Illinois

David Apraiz LABAQUA, Alicante, Spain

Evaristo Ballesteros Tribaldo Department of Physical and Analytical Chemistry, E. U. P. of Linares, University of Jaén, Jaén, Spain

Rossiza Belcheva Analytical Department, Refinery and Petrochemical Research Institute, Bourgas, Bulgaria

Jean-Yves Bottero Interfacial Physical Chemistry, CNRS and University of Aix-Marseille, Aix-en-Provence, France

Antonio Canals Department of Analytical Chemistry, University of Alicante, Alicante, Spain

Alessio Ceccarini Department of Chemistry and Industrial Chemistry, University of Pisa, Pisa, Italy

Amalia Cerdà University of Balearic Islands, Palma de Mallorca, Spain

Victor Cerdà Department of Chemistry, University of Balearic Islands, Palma de Mallorca, Spain

Rosa Cidu Department of Earth Sciences, University of Cagliari, Cagliari, Italy

Carol Cortada Department of Chromatography, LABAQUA, Alicante, Spain

Angel Cuesta Department of Analytical Chemistry, University of Alicante, Alicante, Spain

Monica Culea NATEX s.r.l., Cluj-Napoca, Romania

Antonio Di Corcia Department of Chemistry, University "La Sapienza," Rome, Italy

Alessandro di Domenico Department of Comparative Toxicology and Ecotoxicology, National Institute of Health, Rome, Italy

Marcos N. Eberlin   Institute of Chemistry, State University of Campinas—UNICAMP, Campinas, São Paulo, Brazil

Roger Fuoco   Department of Chemistry and Industrial Chemistry, University of Pisa, Pisa, Italy

Stuart W. Gibb   The University of the Highlands and Islands Project, Environmental Research Institute, Thurso, Caithness, Scotland, United Kingdom

Simion Gocan   Department of Analytical Chemistry, University "Babes-Bolyai," Cluj-Napoca, Romania

Elena González-Soto   Analytical Chemistry Department, University of A Coruña, A Coruña, Spain

Payman Hashemi   Department of Chemistry, Lorestan University, Khoramabad, Iran

Anna Laura Iamiceli   Department of Comparative Toxicology and Ecotoxicology, National Institute of Health, Rome, Italy

Kazunori Ikebukuro   Department of Biotechnology and Bioelectronics, Research Center for Science and Technology, University of Tokyo, Tokyo, Japan

Jan Åke Jönsson   Analytical Chemistry, Lund University, Lund, Sweden

Shreekant V. Karmarkar   Research and Development, Lachat Instruments Division, Zellweger Analytics, Milwaukee, Wisconsin

Isao Karube   Department of Biotechnology and Bioelectronics, Research Center for Science and Technology, University of Tokyo, Tokyo, Japan

A. A. Khan   Applied Chemistry Department, Zakir Husain College of Engineering and Technology, Aligarh Muslim University, Aligarh, India

Magnus Knutsson   Analytical Chemistry, Lund University, Lund, Sweden

Hilda Ledo de Medina   Environmental Chemistry Laboratory, Chemical Department, Faculty of Science, University of Zulia, Maracaibo, Zulia State, Venezuela

Mathias Liess   Zoological Institute, Technical University, Braunschweig, Germany

Miren López de Alda-Villaizán   Institute of Chemistry and Environmental Research, Spanish Council for Scientific Research, Barcelona, Spain

Ian D. McKelvie   Water Studies Centre, Chemistry Department, Monash University, Clayton, Victoria, Australia

Michela Maione   Department of Chemical Sciences, University of Urbino, Urbino, Italy

Filippo Mangani   Department of Chemical Sciences, University of Urbino, Urbino, Italy

Maria Anita Mendes   Institute of Chemistry, State University of Campinas—UNICAMP, Campinas, São Paulo, Brazil

Chris W. Michiels   Laboratory of Food Microbiology, Department of Food and Microbial Technology, Catholic University of Leuven, Leuven, Belgium

Juan Carlos Moltó   Laboratory of Food Chemistry and Toxicology, Faculty of Pharmacy, University of Valencia, Valencia, Spain

Els L. D. Moyson   Faculty of Agricultural and Applied Biological Sciences, Catholic University of Leuven, Leuven, Belgium

Geza Nagy   Department of General and Physical Chemistry, Janus Pannonius University, Pecs, Hungary

Livia Nagy   Research Group of Hungarian Academy of Sciences, Budapest, Hungary

Hideaki Nakamura   Department of Biotechnology and Bioelectronics, Research Center for Science and Technology, University of Tokyo, Tokyo, Japan

Leo M. L. Nollet   Department of Industrial Sciences, Hogeschool Gent, Ghent, Belgium

Maria Teresa Oms*   University of Balearic Islands, Palma de Mallorca, Spain

Pierangela Palma   Department of Chemical Sciences, University of Urbino, Urbino, Italy

Sigrid Peldszus   Department of Civil Engineering, University of Waterloo, Waterloo, Ontario, Canada

Mónica Pérez   LABAQUA, Alicante, Spain

Juhani Peuravuori   Physical/Environmental Chemistry, University of Turku, Turku, Finland

Kalevi Pihlaja   Physical Chemistry, University of Turku, Turku, Finland

H. S. Rathore   Applied Chemistry Department, Zakir Husain College of Engineering and Technology, Aligarh Muslim University, Aligarh, India

Darío Prada Rodríguez   Analytical Chemistry Department, University of A Coruña, A Coruña, Spain

Ralf Schulz   Zoological Institute, Technical University, Braunschweig, Germany

Glen Shaw   National Research Centre for Environmental Toxicology (NRCET), The University of Queensland, Brisbane, Queensland, Australia

A-M Siouffi   University of Aix-Marseille, Marseille, France

Maree Smith   Queensland Health Scientific Services, Brisbane, Queensland, Australia

M. A. Tabatabai   Department of Agronomy, Iowa State University, Ames, Iowa

José Luis Todolí   Department of Analytical Chemistry, University of Alicante, Alicante, Spain

Luigi Turrio-Baldassarri   Department of Comparative Toxicology and Ecotoxicology, National Institute of Health, Rome, Italy

Bieluonwu Augustus Uzoukwu   Department of Pure and Industrial Chemistry, University of Port Harcourt, Port Harcourt, Nigeria

Ignacio Valor   LABAQUA, Alicante, Spain

*Currently affiliated with the Department of Environment and Quality, TIRME, S.A., Palma de Mallorca, Spain.

1

Sampling Methods in Surface Waters

Mathias Liess and Ralf Schulz
Technical University, Braunschweig, Germany

I. INTRODUCTION

The first and usually the most important step in any analytical process is the sampling itself. Mistakes during the sampling process inevitably lead to erroneous results, which cannot be corrected afterwards (1–5). Especially because of progress in analytical protocols, including new and more sophisticated instrumental features that are described later on in this handbook, sampling is increasingly becoming the quality-determining step (6).

The main objective of this chapter is to describe and discuss methods for environmental sampling in surface waters (lakes, rivers, and the marine environment). This aspect of sampling is of major importance in view of the increasing concern about environmental contamination and its correct description and monitoring. Furthermore, the focus of this chapter is in accordance with the focus of the whole handbook on the sampling of water. The methods conventionally used for sampling solid material differ considerably and are not covered by this chapter. Where appropriate, short discussions of the sampling of suspended particulates (mineral or organic sediments) will be included. This portion is of great importance, for the less water-soluble chemicals (such as many insecticides) and similar chemicals are dynamically distributed between small suspended particles and the water phase.

One of the basic problems of environmental water analysis is that it must generally be carried out with selected portions (i.e., samples) of the water of interest, and the quality of this water of interest must then be inferred from that of the samples. If the quality is essentially constant in time and space, this inference would present no problem. Such constancy is, however, rarely if ever observed; in most circumstances virtually all waters show both spatial and temporal variations in quality. It follows that the times and positions of sampling must generally be chosen with great care. Since an increase in the number of sampling positions and sampling occasions increases the cost of the measurement program, there is a general need to define the minimal number of sampling positions and occasions that will provide the desired information.

The whole process of analyzing a material consists of several steps: sampling, sample storage, sample preparation, measurement, evaluation of results, comparison with standards or threshold values, assessment of results. This chapter is concerned with sampling strategy, storage of samples, and sampling equipment. The other steps will be described and discussed in the following chapters on specific chemical groups.

The second section of this chapter focuses on some general aspects of sampling design and some characteristics of the substances to be sampled and analyzed, since their properties, such as degradation and sorption, can substantially affect the results. In the third section an overview of sampling strategies in different ecosystems is given. The temporal and spatial scalings of sampling especially depend on the ecosystem under study and on the question to be answered. Finally, in the fourth section some types of sampling equipment will be presented and their specific properties will be discussed. This section covers general methods as well as specific methods, such as deep-water sampling and event-controlled sampling.

II. GENERAL ASPECTS AND SUBSTANCES

A. Initial Considerations

There are as many possibilities for sampling as there are possible moves in a chess game. The situation to be analyzed has to be defined first. Then the appropriate sampling design should be chosen, on the basis of the temporal and spatial processes of the investigated part of the ecosystem. Handling and storage of the samples should be adapted to the properties of the chemical

of interest, and the effort invested should be optimized in order to obtain the necessary information with such resources as are available. In order to achieve these objectives, the following considerations are useful (Fig. 1).

B. Spatial Aspects

Sampling for the quality control of material in the metal or food industry normally follows statistical approaches to ensure that relatively small subsamples will be representative of the material as a whole. Although similar requirements exist for environmental sampling, the principal difference is that the spatial variation is generally much greater in the case of environmental

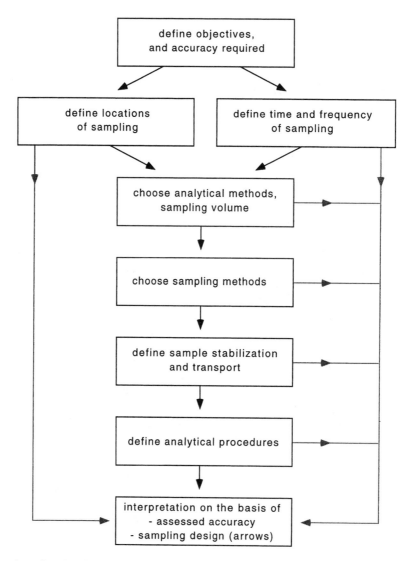

Fig. 1 Initial considerations for planning and carrying out sampling procedures.

contamination. Currents in flowing water and marine ecosystems must be considered. Especially in lakes, very often stratification crucially affects the distribution of substances (Sec. III.A). The locations for environmental sampling must be related to the expected sources of contamination, e.g., different distances downstream of a sewage effluent. A detailed description of the exact sampling site (longitudinal gradient, lateral gradient, depth, water level, distance to possible source of contamination, coordinates) is a basic requirement.

C. Temporal Aspects

The temporal pattern of sampling is of great importance if the environment to be sampled shows changes over time, e.g., river systems within minutes or hours or lakes within days or weeks. The schedule of the sampling program depends mainly on the expected temporal resolution of changes in the environment. In governmental programs for monitoring wastewater treatment effluents, sampling around the clock is necessary to determine whether control variables have been met or exceeded.

A single sample gives only a snapshot of the situation. The advantage is that the equipment necessary for this type of sampling usually is very simple and inexpensive. The power and reliability of the results are normally low and depend strongly on the background data and additional information available.

If many samples are taken over time, it is appropriate in many circumstances to match the sampling rate to the expected variations of the environment. To detect peak concentrations during short-term changes of water quality, for example, event-controlled samplers are useful. When it is necessary to quantify a contaminant load, discontinuous sampling systems may be needed. Various types of discontinuous sampling that are of special importance for quality control purposes and for automatic wastewater sampling in accordance with ISO 5667-10 are diagrammed in Fig. 2. *Time-proportional* sampling means that samples containing identical volumes are taken at constant time intervals. In *discharge-proportional* sampling the time intervals are constant but the volume of each sample is proportional to the discharge. In *quantity-proportional* sampling (or flow-weighted sampling) the volume of each sample is constant but the temporal resolution of sampling is proportional to the discharge. A fourth type is *event-controlled* sampling, which depends on a trigger signal (e.g., discharge threshold) and which will be discussed in detail later (see Sec. IV.E).

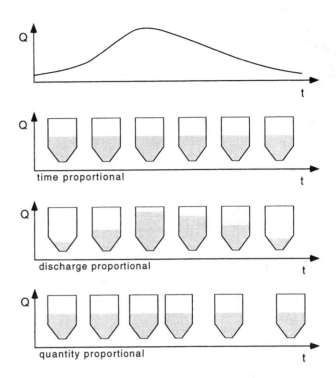

Fig. 2 Different types of discontinuous sampling.

In addition to single and discontinuous sampling, a continuous sampling and determination of analytical values is desirable in some cases. An example is the quality control for a very complex effluent with unpredictable temporal changes in composition that are not linked to possible trigger variables such as discharge and temperature. For this purpose automatic sampling and in some cases automatic analyzing units are useful. The expenditure of time and money is in general considerably higher for this type of sampling and cannot always be justified.

Another important type of sample is the composite sample that can result from mixing several single samples or from an automatic sampling program. The aforementioned types of discontinuous samples can also be mixed to produce composite samples representing the respective features.

D. Number of Samples

The number of samples depends of course on the problem to be addressed. If an average concentration is to be obtained from several samples, a general calculation of the necessary number of samples N can be done using the following equation:

$$N = 4\left(\frac{S}{\bar{x}d}\right)^2$$

where S = estimate of standard deviation of the arithmetic mean of all single samples, $\bar{x}$ = estimate of arithmetic mean of all single samples, and d = tolerable uncertainty of the result, e.g., 20% ($d = 0.2$).

If peak concentrations are to be quantified, the number of samples depends on the specific problem. Some examples are given in Chapter 4.

E. Sample Volume

The sample volume depends on the elements or substances to be analyzed, on their expected concentration in the sample, and on the required quantification limits. For heavy-metal analyses, sample volumes of about 100 ml are sufficient in most cases. For the analysis of organic chemicals (e.g., pesticides) 1–1 samples are commonly used. A 3–1 sample volume has been suggested for both first-flush and flow-weighted composite samples in the monitoring of storm water runoff from industries and municipalities (7). Fox (8) described an apparatus and procedure for the collection, filtration, and subsequent extraction of 20–1 water and suspended solid samples using readily available, inexpensive, and sturdy equipment. With this equipment he obtained quantification limits for several organochlorine substances at ng/L levels.

F. Storage and Conservation

Samples that are not analyzed immediately must be protected from further contamination, loss, sorption, or other unintended changes. For this purpose sampling bottles should be designed for long-term storage with as few changes as possible.

1. Contamination

An unintended contamination of samples can occur during the sampling process, either from external sources or as a result of contaminated sampling or storage equipment. In organic or inorganic trace analysis highly purified water must be used (filtered, distilled several times, cleanup with ion exchanger and sub-boiling cleanup). Polyethylene or Teflon bottles are normally used in inorganic trace analysis, and glass or quartz bottles in organic trace analysis. Organic compounds have been known to leach from the bottle material into the sample, react with the trace elements un-

der study, and cause systematic mistakes. Such problems become very important at detection limits below the μg/g level.

Many publications recommend that each sample container be rinsed two or three times with sample before finally being filled. This may lead to errors when undissolved materials, and perhaps also readily adsorbed substances, are of interest. It has been suggested not to rinse containers with the sample when trace organic compounds are of interest (1).

Empirical studies have shown that PTFE (polytetrafluoroethylene) and PVDF (polyvinyilidene fluoride) are of varying purity, often resulting in unexpected contamination problems in ultratrace analysis, whereas PFA (perfluoroalkoxy-fluorocarbon) proved to be cleaner by origin; consequently, acidic washing processes could be successfully applied. These different fluorinated polymers have been compared regarding their suitability for container or sampler material (9). It has been found that PFA exhibits the lowest nano-roughness and hence seems best suited as a container material.

2. Loss

Loss during storage can result from biological processes, hydrolysis, or evaporation. These processes can be reduced or prevented using the following procedures:

Acidification to pH 1.5: prevention of metabolism by microorganisms and of hydrolysis and precipitation

Cooling and freezing: reduction of bacterial activity

Addition of complexing substances: reduction of evaporation

UV irradiation (together with the addition of H_2O_2): destruction of biological and organic compounds to prevent complexation reactions

Loss of target elements or substances can also occur due to evaporation or volatilization. When contact of the sample with air is to be avoided (because it contains dissolved gases or volatile substances), sample containers or sample bottles should be completely filled. Evaporation is a problem during storage of mercury under reducing conditions; other elements evaporate as oxides (e.g., As, Sb), halogenides (e.g., Ti, Cr, Mo), or hydrides (e.g., As, Sb, Se), or they are able to diffuse through the walls of plastic bottles. Volatilization is a special problem in the case of organic compounds such as hydrocarbons and halogenated hydrocarbons.

3. Sorption

Sorption to the walls of sample bottles can reduce the concentration in the water phase considerably. Depending on the target substances, plastic or quartz bottles show the lowest adsorption and can therefore be used for the storage of samples in aqueous solution. In general the wall material of storage bottles can change over time and the potential for adsorption of target substances can increase considerably. In the case of heavy metals this problem can be reduced by acidifying the sample.

The affinity of selected organochlorine, pyrethroid, and triazine pesticides at concentrations of 0.25 ppb or less to glass and PTFE has been described (10). For the organochlorine pesticides, the adsorption behavior correlates well with the octanol–water partition coefficients. The triazines are not adsorbed to glass or PTFE, whereas α-BHC, lindane, dieldrin, and endrin are weakly adsorbed compared with DDT, DDE, TDE, permethrin, cypermethrin, and fenvalerate. Adsorption constants K_a (amount of adsorbed pesticide per unit area of surface) have been calculated (Table 1) by these authors to quantify the sorption affinity of the compounds on glass and PTFE:

$$K_a = (\text{amount of adsorbed pesticide per unit area of surface,}$$
$$\text{ng/cm}^2)/(\text{concentration aqueous solution, ng/cm}^3)$$

As an example, the adsorption of fenvalerate on a duran glass surface is calculated as follows using the equation: A bottle with a surface area of 325 cm^2 contains 500 ml of an aqueous solution of fenvalerate. Under these circumstances about 84% of the fenvalerate is adsorbed to the glass surface, and only about 16% remains in the solution; the concentration in water is reduced accordingly (e.g., given an initial concentration of 10 ng/L in a 500-ml bottle, 4.2 ng are adsorbed and 0.8 ng stays in solution, which equals a concentration of 1.6 ng/L after 48 h). For lindane and permethrin, 0.32% and 96%, respectively, of the chemical are absorbed to the glass wall after 48 h.

4. Recommended Storage

Even if the already-mentioned conservation methods are used, the storage time of water samples should not be longer than about 1 month, in some cases not longer than 6 hours (11). Table 2 gives an overview of sampling and storage bottles as well as conservation methods for different determinants in the sample.

For quality control and for the use of analytical results in forensic chemistry, national and international standardizations are necessary. Several international standards (ISO) have been defined for the design of sampling programs, the sampling techniques, and the preservation and handling of samples (12–14). A compilation of the EPA's sampling and analysis methods is available, which also covers sample preservation, sample preparation, quality control, and analytical instrumentation (15).

III. SAMPLING STRATEGIES FOR DIFFERENT ECOSYSTEMS

The strategy to be used in environmental sampling differs considerably depending on the details of the investigated ecosystem and the problems at issue. Hence strategies can be described in relation to either the

Table 1 Mean K_a Values for Duran Glass and PTFE Containers (48 h at 25°C) Appropriate to the Range of Concentrations in the Solution

Pesticide	Duran glass surface		PTFE surface	
	K_a (cm)[a]	Concentration range (ng/ml)	K_a (cm)	Concentration range (ng/ml)
α-BHC	0.014 (0.007)	0.05	0.036 (0.011)	0.01–0.04
Lindane	0.005	0.04–0.12	0.048	0.04–0.07
Dieldrin	0.027 (0.009)	0.17–0.19	0.093 (0.009)	0.11–0.15
Endrin	0.019 (0.006)	0.19–0.21	0.059 (0.005)	0.12–0.18
DDT	0.87 (0.25)	0.04–0.07	2.028 (0.116)	0.008–0.04
Permethrin	1.44 (0.30)	0.01–0.07	3.32 (1.68)	0.001–0.01
Cypermethrin	43.3 (16.8)	0.002–0.007	11.61 (5.97)	0.002–0.007
Fenvalerate	8.15 (2.48)	0.002–0.03	11.8 (3.99)	0.002–0.01

[a]The associated deviations are in parentheses.
Source: Ref. 10.

Table 2 Recommended Materials for Sample Containers and Sample Stabilization Procedures

Determinant	Container material[a]	Maximum storage time[b]	Stabilization procedure
Alkalinity	P or BG	mi	—
Alkaline ions	P or G	d	Not possible
Alkaline earth ions	P or G	mo	Add HCl to pH 2
Arsenic	P or G	mo	Add 2 ml 6M HCl per liter
Ammonium[c]	P or G	mo	Refrigerate or add HCl to pH 2
Biochemical oxygen demand BOD*	G	mo	Freeze, if storage unavoidable
Boron	P	mo	Not essential
Bromide	P or G	mo	Not essential
Carbon dioxide, free[c]	BG	mi	Refrigerate, if storage unavoidable
Chemical oxygen demand COD	G or P	mo	Freeze, if storage unavoidable
Chloride	P or G	mo	Not essential
Chlorine[c]	G	mi	—
Chromate	G	mi	—
Conductivity	P or G	d	Refrigerate
Cyanide	P or BG	d	Add NaOH to pH 12 and refrigerate
Fluoride	P	d	Not possible
Halogenated methanes	Special G	d	Add $Na_2S_2O_3$ and refrigerate
Hardness	P or G	mo	Not essential
Iodide	P or G	mo	Add NaOH to pH 12 and refrigerate
Mercury	G		Add $K_2Cr_2O_7$ and HNO_3 to pH 1
Metals	P	mo	Add 20 ml 5M HCl L^{-1} and refrigerate
Nitrate[c]	G or P	d	Refrigerate
Nitrite[c]	G or P	h	Refrigerate
Nitrogen, organic[c]	P or G	d	Add H_2SO_4 to pH 2
Nitrogen, total	P or G	d	Add H_2SO_4 to pH 2
Organic insecticides	G	mo	Freeze, after extraction
Oxygen[c]	G	mi	—
pH value[c]	P or G	h	Refrigerate
Phenolic compounds[c]	P or BG	d	Add H_3PO_4 to pH 4 and refrigerate
Phosphorus, ortho[c]	G	d	Refrigerate
Phosphorus, total	G	mo	Add H_2SO_4 to pH 2
Polycyclic aromatic hydrocarbons	Special G	d	Refrigerate
Silica	P	d	Refrigerate
Solids	BG	mo	Refrigerate
Sulphate	P or G	d	Refrigerate
Sulphide	P or BG	d	Add $Zn(CH_3COO)_2$ and NaOH to pH 12
Surfactants	BG	d	Add formaldehyde
Total organic carbon TOC	G	mo	Freeze
Turbidity	G	mi	—

[a]P = polyethylene or equivalent, G = glass, BG = borosilicate glass.
[b]mi = minutes (immediate measurement), h = hours (up to a few hours), d = days (up to a few days), mo = month (up to a few months). These values refer to approximate storage times.
[c]Determinant particularly liable to instability.
Source: Refs. 3, 11, 14, and 49.

goals of the study or the ecosystems involved. In the following section, the different sampling strategies appropriate to the main types of ecosystem (stagnant water, flowing water, marine environment) and their temporal and spatial scaling will be discussed. In Sec. III.D, considerations for sampling storm water runoff will be addressed as an example of a sampling design specific to urban areas.

A. Lakes and Reservoirs

Often a number of physical, chemical, and biological processes have to be considered, for they may markedly affect water quality and its spatial variations. The sources of heterogeneity within a body of water that need careful consideration in selecting sampling sites are as follows: thermal stratification, which leads to variations of quality in depth, and the effects of influent streams, lake morphology, and wind, which together may produce both lateral and vertical heterogeneity. Some important topics relevant to the choice of positions for sampling are noted, e.g., in Refs. 1 and 16. It must be kept in mind that shallow and/or relatively isolated embayments of lakes and reservoirs may show marked differences in quality from the main body of water.

In water bodies of sufficient depth in temperate climates, thermal stratification is often the most important source of vertical heterogeneity from spring to autumn (Fig. 3). Measurement of dissolved oxygen and temperature is a convenient means of following the development of such stratification, and has the advantage that both measurements can be made automatically and continuously in situ.

Thermal stratification may retard the mixing of streams entering lakes or reservoirs. Consequently, this important source of materials derived from the surrounding land has to be sampled with due consideration of its spatial variability.

The number of algae in the surface layer of a water body may have a marked effect on the concentrations of nutrients and other substances: It is often not possible to measure any dissolved nutrients during algal blooms, because all nutrients are bound in the algae. Therefore the trophic status and/or spatial heterogeneity in the distribution of algae should be considered while choosing a sampling site.

The choice of the correct sampling point can depend on the depth of a lake (17). These authors have compared different water sampling techniques in a series

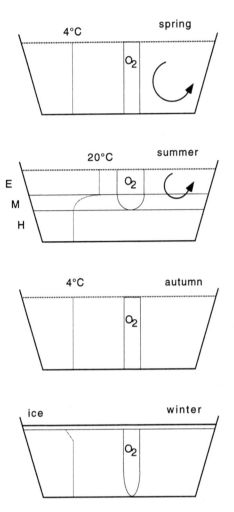

Fig. 3 Seasonal variation of oxygen and temperature within the different layers of a meso/eutrophic lake in temperate latitudes (E = epilimnion, M = metalimnion, H = hypolimnion).

of lakes. In deep lakes, they observed no significant differences between mean summer nutrient concentrations measured in a tube sample integrating over the photic zone, taken from the deepest point, and a surface dip sample taken by wading into the water's edge. In shallower lakes the integrating tube sampler gave significantly higher estimates of mean concentrations than the other method due to the increase in volume of the unmixed hypolimnion, which reduced the depth of the well-mixed epilimnion to less than the tube length. For national survey purposes they suggested samples taken from the edge of the lake as the most cost effective.

B. Streams and Rivers

1. Location of Sampling Within the Stream

Sampling locations, especially in larger streams and rivers, should, whenever possible, be at cross sections where vertical and lateral mixing of any effluents or tributaries are complete. To avoid nonrepresentative samples caused by surface films and/or the entrainment of bottom deposits, it has often been recommended that samples should, whenever possible, be collected no closer than 30 cm from the surface or the bottom (3).

Simple surface-grab procedures have been compared with more involved, cross-sectionally integrated techniques in streams (18). Paired samples for the analysis of selected constituents were collected over various flow conditions at four sites to evaluate differences between the two sampling methods. Concentrations of dissolved constituents were not consistently different. However, concentrations of suspended sediment and the total forms of some sediment-associated constituents, such as phosphorus, iron, and manganese, were significantly lower in the surface-grab samples than in the cross-sectionally integrated samples. The largest median percentage difference in concentration for a site was 60% (total recoverable manganese). Median percentage differences in concentration for sediment-associated constituents considering all sites grouped were in the range of 20–25%. The surface-grab samples underrepresented concentrations of suspended sediment and some sediment-associated constituents, thus limiting the applicability of such data for certain purposes.

When the quality of river water abstracted for a particular purpose (e.g., the production of drinking water) is of interest, the sampling point should, in general, be at or near the point of abstraction. It must be noted, however, that changes of quality may occur between the actual point of abstraction and the inlet to the treatment plant. If the amount or time of abstraction is to be controlled on the basis of the water quality, an additional sampling location upstream of the abstraction point will usually be needed, the distance upstream being dependent on the travel time of the river, the speed with which the relevant analysis can be made, and the upstream locations of the sources of the determinants. This is of course difficult to achieve. The need of an early warning system for drinking-water purposes was emphasized by the SANDOZ accident, since which online biomonitors have been in place in the Rhine River.

2. Description of the Longitudinal Gradient

When the aim is to assess the quality of a complete stream, river, or river basin, the number of potentially relevant sampling locations is usually extremely large. It is, therefore, usually necessary to assign different priorities to the various locations in order to arrive at a practicable sampling program (3,16). Such considerations are very closely connected to the question of sampling frequency, and a number of approaches have been described dealing with the overall design of sampling programs for river systems (1). The value of, and need for, the identification of locations where quality problems are or may be most acute have been stressed several times. These questions can be addressed with fixed-location monitoring and intensive, short-term surveys at selected locations for the routine assessment of rivers.

Water quality is usually monitored on a regular basis at only a small number of locations in a catchment, generally concentrated at the catchment outlet. This integrates the effect of all the point- and nonpoint-source processes occurring throughout the catchment. However, effective catchment management requires data that identify major sources and processes. As part of a wider study aimed at providing technical information for the development of integrated catchment management plans for a 5000-km^2 catchment in southeastern Australia, a "snapshot" of water quality was undertaken during stable summer-flow conditions (19). These low-flow conditions exist for long periods, so water quality at these flow levels is an important constraint on the health of in-stream biological communities. Over a 4-day period, a study of the low-flow water quality characteristics throughout the Latrobe River catchment was undertaken. Sixty-four sites were chosen to enable a longitudinal profile of water quality to be established. All tributary junctions and sites along major tributaries, as well as all major industrial inputs, were included. Samples were analyzed for a range of parameters, including total suspended solids concentration, pH, dissolved oxygen, electrical conductivity, turbidity, flow rate, and water temperature. Filtered and unfiltered samples were taken from 27 sites along the main stream and tributary confluences for analysis of total N, NH_4, oxidized N, total P, and dissolved reactive P concentrations. The data are used to illustrate the utility of this sampling methodology for establishing specific sources and estimating nonpoint-source loads of phosphorus, total suspended solids, and total dissolved solids. The methodology enabled several new insights into system behavior, including quantification of unknown point discharges, identification of key in-stream sources of suspended material, and the extent to which biological activity (phytoplankton growth) affects water quality. The costs and benefits of the sampling exercise are reviewed.

3. Temporal Changes of Water Quality

The discharge of streams in comparison with larger rivers is highly dynamic, depending mainly on local rainfall conditions and/or groundwater level (20). It follows that the chemical composition of the stream water is profoundly influenced by the allochthonous input of water, nutrients, sediments, and pesticides. Two-thirds of the contamination of headwater streams with sediments, nutrients, and pesticides is caused by those nonpoint sources (21). Substances with a high water solubility are introduced through soil filtration. Less water-soluble substances enter by way of the surface water runoff during heavy rains (22). The total loss of pesticides depends on the time period between the application and the rain event, the maximum precipitation, and various soil parameters. Consequently, streams with an agricultural catchment area are susceptible to unpredictable, brief pesticide inputs following precipitation (23,24).

To determine the influence of sampling frequency on the reliability of water quality estimates in small streams, a cultivated basin (0.12 km^2) and a forested basin (0.07 km^2) were studied in spring and autumn (25). During the 2-month spring season and 3-month autumn season, 97–99% of the annual loads of total nitrogen, total phosphorus, and suspended solids were acquired from the cultivated basin and 89–91% from the forested basin. During the same seasons, 99% and 87% of the total annual runoffs were recorded in the cultivated and forested basins, respectively. This means that in only 5 months of the year, more than 95% of the nutrient and water runoff occurred in the cultivated catchment, and about 90% in the forested catchment. Thus the values of nonpoint loads, normally presented as annual means, give a highly incorrect impression of the effects of nonpoint loading on watercourses, particularly in the case of relatively small streams.

The same author estimated the number of samples needed to calculate the load of various substances (25) by varying the sampling frequency at the two sites. In the cultivated basin the means of concentration data would be within ±20% of the mean of the whole data set in spring, if nitrogen and phosphorus samples were taken at least five times monthly and suspended-solids

samples at least three times monthly. In the forested basin the corresponding sampling frequencies were twice monthly for nitrogen samples and four times monthly for phosphorus and suspended solids. In autumn the concentration means in runoff waters would be within ±20% of the mean obtained using the whole data set if three samples per month were taken for nitrogen and phosphorus and five samples for suspended solids. In the forested basin the same deviation of the mean would be obtained with one nitrogen sample, five phosphorus samples, and 16 suspended-solids samples.

When intending to measure the peak concentration of slightly soluble substances in streams within a cultivated watershed, it is necessary to use runoff-triggered sampling methods. A headwater stream in an agricultural catchment in northern Germany was intensively monitored for insecticide occurrence (lindane, parathion-ethyl, fenvalerate). Brief insecticide inputs following precipitation with subsequent surface runoff result in high concentrations in water and suspended matter (e.g., fenvalerate: 6.2 μg L^{-1}, 302 μg kg^{-1}). These transient insecticide contaminations are typical of headwater streams with an agricultural catchment area, but have been rarely reported. Event-controlled sampling methods for the determination of this runoff-related contamination with a time resolution of as little as 1 hour make it possible to detect such events (26).

Within monitoring programs, loading errors are generally associated with an inadequate specification of the temporal variance of discharge and of the parameters of interest. Often little consideration is given to the impact of additional transport characteristics on contaminant sampling error and design. Detailed examination of five transport characteristics at a single river cross section emphasizes the importance of understanding the complete transport/loading regime at a sampling station, defining the required end products of the monitoring program, and defining the accuracy required to meet specific program needs before implementing or evaluating a monitoring program. River transport characteristics are (1) contaminant transport modes, (b) short-term temporal and seasonal variability, (c) the relationship between dissolved and particulate contaminant concentrations and discharge, (d) load distribution with sediment particle size, and (e) spatial variability in a cross section (27).

4. Using Sediments to Integrate over Time

It is also worth noting that the procedure known as catchment quality control, though intended for a different purpose, includes the identification of most important effluents entering a river system from the viewpoint of water quality (28). Finally, the analysis of river sediments has been suggested as a convenient means of reconnaissance of river systems to decide the locations where water quality is of particular interest with respect to pollutants.

C.   Estuarine and Marine Environments

The potential spatial heterogeneity (lateral and vertical—both time dependent) of these bodies of water again makes it essential that sampling locations be chosen with reference to the relevant basic processes (16). Sampling of ocean waters and the handling of such samples have been described in general (2).

A well-known practical problem is the unintended contamination of samples by material released from the research vessel or by the sampling apparatus. A sampling apparatus for the collection and filtration of up to 28 L of water at sea has been designed to minimize possible contamination from both the equipment and the ship's surroundings (29). It was used in the analysis of chlorinated biphenyls (CBs), persistent organochlorine pesticides (OCs), and pentachlorophenol (PCP), in both the aqueous and particulate phases. The system is suitable for the collection of estuarine and coastal waters where the levels of dissolved CBs, OCs, and PCP are above the limit of determination of 15 pg L^{-1}. The efficiency of the recovery of these compounds and the variance of the extraction and analysis have been estimated by analysis of filtered seawater spiked at a range of concentrations from pg L^{-1} to ng L^{-1}. Recoveries ranged from 66.5 to 97.3%, with coefficients of variation for the complete method from 7.2 to 29.9%.

The procedure of using small boats provided with sample bottles attached to a telescopic bar is recommended as a means to minimize contamination from the research vessel in coastal water sampling. Of the wires used to suspend samplers, plastic-coated steel gave negligible, and Kevlar and stainless steel only slight, contamination for some metals (5).

The high spatial and temporal variability of estuaries poses a challenge for characterizing estuarine water quality. This problem was examined by conducting monthly high-resolution transects for several water quality variables (chlorophyll a, suspended particulate matter, and salinity) in San Francisco Bay (30). Using these data, six different ways of choosing station locations along a transect, in order to estimate mean conditions, were compared. In addition, 11 approaches to estimating the variance of the transect mean when sta-

tions are equally spaced were compared, and the relationship between the variance of the estimated transect mean and the number of stations was determined. These results were used to derive guidelines for sampling along the axis of an estuary. In addition, the changes in the concentration of various substances due to the tide seem to be extremely important. Seawater with a low concentration of substances becomes mixed with the highly loaded water in the estuaries and along the shores. Automatic samplers can be used to integrate the concentrations of materials over time (see Sec. IV).

An overview of the analysis of polar pesticides in water samples has been presented (31). The sampling plans and strategies for different types of waters, such as rivers, wells, and seawater, are discussed. In situ preconcentration methods, involving online techniques or direct measurement, are mentioned as alternatives to conventional techniques. Attention is devoted to the influence of organic matter and its interaction with polar pesticides. The use of various types of filtration steps prior to the preconcentration of the analytes from water samples is also reviewed.

D.   Urban Areas

With respect to urban areas, sampling strategies for storm water runoff from industries and municipalities are of specific importance. The U.S. Federal Storm Water Regulations of 1990 specify protocols for such storm water runoff sampling. These regulations define two separate samples that must be collected when a storm occurs. A first-flush sample is to be collected during the first 30 minutes of the storm event. A flow-weighted composite sample must be collected for the entire storm event or for at least the first 3 hours of the event.

The first-flush sample and the flow-weighted composite sample must be analyzed for the pollutants listed in Table 3. In general, the sample volume required for laboratory analysis depends on the particular pollutants being monitored and varies for each application. As a general rule, a 3–1 sample volume for both first-flush and flow-weighted composite samples seems to be sufficient for the majority of applications (7).

Both manual and automatic methods can be used to collect samples for the required analysis (7). For manual sampling, the samples can be taken at fixed time intervals in individual bottles. After collection, a specific volume must be poured out of each bottle to form a flow-weighted composite. The exact volume must be calculated using the flow data taken when each bottle was filled. The advantage of manual sample collection

Table 3 Storm Water Analysis Requirements According to the U.S. Federal Storm Water Regulations

Pollutant	Industrial		Municipal
	First-flush grab sample	Flow-weighted composite sample	Flow-weighted composite sample
Oil and grease	X		X
pH	X		X
Biological oxygen demand (BOD)	X	X	X
Chemical oxygen demand (COD)	X	X	X
Total suspended solids (TSS)	X	X	X
Total phosphorus	X	X	X
Nitrate and nitrite nitrogen	X	X	X
Total Kjeldahl nitrogen	X	X	X
Any pollutant in the facility's effluent guideline	X	X	
Any pollutant in the facility's NPDES[a] permit	X	X	
Any pollutant in the EPA Form 2F tables believed to be present	X	X	
Total dissolved solids			X
Fecal streptococcus			X
Fecal coliforms			X
Dissolved phosphorus			X
Total ammonia plus organic nitrogen			X
13 metals, total cyanide, and total phenol			X
28 volatile compounds			X
11 acid compounds			X
46 base/neutral compounds			X
25 pesticides			X

[a]National pollutant discharge elimination system.
Source: Ref. 7.

is that, regardless of runoff amount, a fairly constant volume of sample is collected. This is because the flow-weighted composite is formed after the event and does not depend on calculations for runoff volume.

Automatic storm water monitoring systems consist of a rain gauge, a flow meter, an automatic sampler, and a power source. The rain gauge measures on-site rainfall. The flow meter measures the runoff water level and converts this level to a flow rate. In many systems, the flow meter activates the sampler when user-specified conditions of rainfall and water level have been reached. Once activated, the sampler collects water samples by pumping the runoff water into bottles inside the sampler.

Automatic storm water monitoring systems can form the flow-weighted composite sample automatically during the storm event, if there is sufficient storage capacity to accommodate variations in the runoff amount. Automatic samplers that fulfill the preceding requirements have been described (7).

IV. SAMPLING EQUIPMENT

A. General Comments

Now that the employment of sampling systems has been reviewed, we turn to the equipment itself. Systems designed for the various purposes already discussed will be introduced, and their advantages and disadvantages will be presented.

Any component of a sampling device that is not normally present in the water body may affect the concentrations of determinants in the water of interest through three main effects: (a) by disturbance of physical, chemical, and biological processes and equilibria, (b) by contaminating the water with some parts of the sampler (e.g., organic compounds may leach out of plastic materials), and (c) by direct reactions between determinants and the materials of which the device is constructed (e.g., dissolved oxygen can react with copper, decreasing the oxygen concentration in the water). Another source of error may lie in certain processes that occur within sample devices, such as the deposition of undissolved solid materials onto the walls of a device.

B. Manual Sampling Systems

1. Simple Sampler for Little Depth

For many purposes, specially designed and installed sampling devices are not required. It often suffices simply to immerse a bottle in the water of interest, and this technique may also be applicable for some purposes in water treatment plants.

2. Sampler for Large Quantities in Water with Little Depth

A system for the sampling and filtering of large quantities of surface seawater that is suitable for trace-metal analysis has been introduced (32). The water is brought onto the ship via an all-Teflon pump and PFA tubing from a buoy deployed away from the vessel. The sample is delivered directly into a polycarbonate pressure reservoir and is subsequently filtered through a polycarbonate filter and in-line holder.

Sampling systems based not on sample containers but on inlet tubes are commonly required in water treatment and other plants, and are also employed in a number of applications for natural waters (3). In some systems, the flow of sample through the inlet tube is achieved by the natural pressure differential, whereas in others the sample must be pumped, sucked (by vacuum), or pressurized (by a gas) through the tube. When dissolved gases and volatile organic compounds and possibly other determinants whose chemical forms and concentrations may be affected by dissolved gases are of interest, it is generally desirable to ensure that the sample is slightly pressurized to prevent gases from coming out of solution.

3. Simple Sampler for Deep Water

When it is necessary to sample from a particular depth in waters where this simple technique cannot be used, special sample collection containers are available that can be lowered into the water on a cable and that collect a sealed sample at the required depth.

One of the simplest kinds of equipment with which to obtain samples from various depths is a weighted bottle closed with a cork. This cork is connected to the bottleneck by a rope that can be used for releasing the cork and opening the bottle at the desired depth (scoop bottle according to MEYER, Fig. 4). For some purposes it could be a problem that the water flowing into the bottle will be mixed with the oxygen inside the bottle.

4. Deep Water Sampler (Not Adding Air to the Sample)

A common tool for taking water samples from different depths is the standard water sampler according to

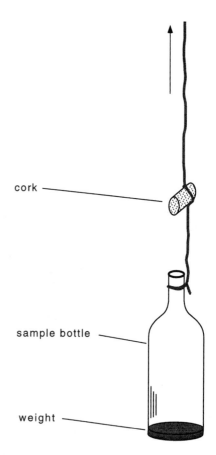

Fig. 4 Scoop bottle according to MEYER, a very simple sampler for various depths.

RUTTNER. The sampler, still open, is lowered by rope into the water. When it reaches the desired depth, a messenger is let down on the rope. Upon striking the standard water sampler, it releases the closing mechanism and the lids of the sampling tube close. In some versions another rope is used to close the sampling bottle. The advantage is that no mixing of air with the sample will occur. But this system has the major disadvantage that the bottle is in contact with water at all the depths through which the sampler travels on its way to the desired depth.

Figure 5 shows an apparatus from HYDROBIOS as an example of a sampler according to RUTTNER. This version contains a thermometer ranging from −2° to

+30°C, indicating the temperature of the sample; the temperature can easily be read through the plastic tube of the sampler. The water sample can be drawn off through the discharge cock in the lower lid for the various analyses. A similar version with a metal-free interior of the sampling tube for the determination of trace metals is also available.

5. Deep Water Sampler for Trace Elements (Adding Air to the Sample)

The MERCOS water sampler from HYDROBIOS (Fig. 6) is suitable for ultratrace-metal analysis. It consists of a holder device and exchangeable 500-ml Teflon bottles as sampling vessels that can be used down to 100-m water depth. All fittings are made of titanium. The sampler is attached to a plastic-coated steel hydrographic wire and lowered into the water in closed configuration in order to prevent sample contamination by surface water. The sampler is opened by means of plastic-covered messengers upon reaching the desired water depth. When the messenger hits the anvil, the silicone tubings spring up to allow water to move in and the air to leave the bottle. In case of serial operation, a second messenger for release of the next sampler is set free at the same time. A disadvantage of the system is the contact between air and the water sample within the bottle.

For the determination of trace elements in seawater, the system of sampling bottle = storage bottle = reaction vessel is used so that the samples cannot be falsified by pouring from one vessel to another. In order to sterilize the bottles for microbiological investigations, the Teflon bottles, along with the coupling pieces and silicone tubings, can easily be taken from the holder.

A modification of an inexpensive and easy-to-handle let-go system, a semiautomatic apparatus for primary production incubations at depths between 0 and 200 m, has been suggested (33). The system is composed of a buoy, a nylon line, a fiberglass ballast weight and about 15 sampling chambers. The entire volume of this apparatus is less than 80 dm³, and it weights about 7−8 kg. The sampling chambers sink with the ballast in an open position. When the line is stretched between the buoy at the surface and the ballast at the bottom, the chambers automatically enclose the water sample at the predetermined depth. The complete deployment of the apparatus takes less than 10 minutes. By means of an easy modification of the length of the line and/or the position of the chambers along it, the sampling depth can be varied for repeated deployment over variable

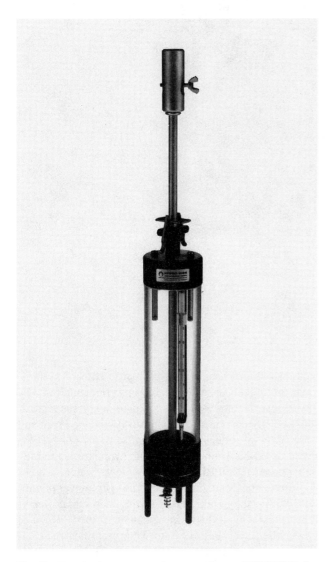

Fig. 5 Standard water sampler according to RUTTNER for various depths.

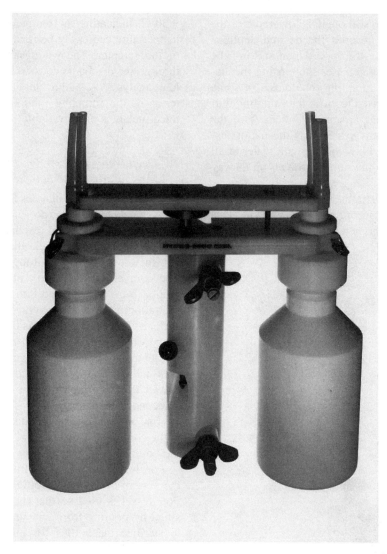

Fig. 6   MERCOS water sampler with two bottles that are lowered while closed and are opened at the desired depth.

depth. The advantage of this system is that parallel samples from different depths can be obtained at relatively low cost and with low technical complexity.

C. Systems for Sampling the Benthic Boundary Layer at Different Depths

1. Deep Water (>50 m)

Oceanographic studies of flow conditions and suspended-particle movements in the bottom nepheloid layer have been significantly more numerous in recent years. Instrumented tripods with flow meters, transmissiometers, optical backscatter sensors (OBSs), in situ

settling cylinders, and programmable camera systems have often been used in marine environments (34,35). These instruments were deployed to study suspended-sediment dynamics in the benthic boundary layer and were able to collect small water samples (1–2 L) at given distances from the seafloor. An instrumented tripod system (BIOPROBE) that collects water samples and time-series data on physical and geological parameters within the benthic layer in the deep sea at a maximum depth of 4000 m has been described (36). For biogeochemical studies, four water samples of 15 L each can be collected between 5 and 60 cm above the seafloor. BIOPROBE contains three thermistor flow meters, three temperature sensors, a transmissiometer, a compass with current direction indicator, and a bottom camera system.

2. Shallow Water (<50 m)

An instrument that collects water from the benthic boundary layer in more shallow waters with a maximum depth of 50 m is described in Ref. 37. Four water samples of 7 L each can be collected between 5 cm and 40 cm above the sediment. Handling is easy, and the sampling operation is brief enough to allow repeated employment, even on time-limited, routine investigations.

D. Automatic Sampling Systems

As already pointed out, the concentrations of substances change over time. To describe this situation in the field it might be necessary either to take an average of the contamination or to obtain information about the short-term peak concentration. To estimate the average contamination in the simplest way is to take a continuous sample or to generate a composite sample by sampling with constant time intervals. In this case the frequency of sampling should be at least as high as the frequency with which the concentration changes. If unpredictable peak concentrations are to be sampled, it is necessary to use a trigger, some easy-to-measure variable that specifies the optimum sampling time. A wide range of triggers is available (e.g., water level, conductivity, turbidity, temperature).

1. Sampling Average Concentrations

These systems consist, in general, of a sampling device and a unit that automatically controls the timing of the collection of a series of samples and houses the appropriate number of sampling containers (1,3). The control unit usually provides the ability to vary several factors, such as the number of samples in a given time period, the length of that period, and the time period over which each sample is collected. Some units also allow sample collection to be based on the flow rate of the water of interest rather than time. One common example of the application of such a system is the collection of 12 or 24 samples in a period of 24 h. The individual samples can be analyzed separately or subsequently can be used to prepare one composite sample for analysis. Such automatic sampler units, which allow composite samples to be taken, are available, for example, from ISCO Inc.

2. Sampling Average Concentrations— Sampling Buoy

An automatic suspensions and water sampler (SWS) that can be used in marine systems, lakes, and rivers is shown in Fig. 7. Pumps and bottles are situated in the underwater part of the floating device so that the sampler can be used in polar regions (no freezing of water samples) and tropical regions (no overheating of electronic circuits). Because the sampling time can be programmed, the data provided by this sampler reflect the average level of contamination (e.g., metals and pesticides) better than do single or random samples.

3. Event-Controlled Sampling of Industrial Short-Term Contamination

Short-term loading of toxic substances into natural waterways is a common phenomenon, with substantial impacts on biota. The effects of shock (pulse) pollution loading from two major industries on a river and wetland system in southern Ontario, Canada, has been described (38). The assessment of shock loading frequency indicated that sporadic discharges of polluted water occurred on average once every other day during the 38 days of monitoring in the period April 1986 to November 1987. To estimate the frequency and intensity of the shock loads, an automatic pump sampler that was triggered by a threshold conductivity was used. Samples were withdrawn from the river when the specific conductivity of the stream exceeded a threshold value of twice the background.

4. Event-Controlled Sampling: Surface Water Runoff from Arable Land

Streams in intensively cultivated areas are characterized by a high input of material from the surroundings. The characteristics of this contamination were detailed in Sec. III.B (Temporal Changes of Water Quality); most important are the unpredictability and the brevity of such inputs.

Because runoff water has low conductivity, the conductivity of a stream is lowered when it receives edge-of-field runoff. This conductivity decrease was used to trigger an automatic sampler that provides an accurate measurement of the maximum pesticide contamination of suspended matter and water, by sampling only the brief peak contamination levels during runoff events (26). The conductivity was measured every 4 min. During runoff events, a 500-ml water sample was taken every 8 min. For each event a mixture of subsamples was analyzed for insecticide content. An example of

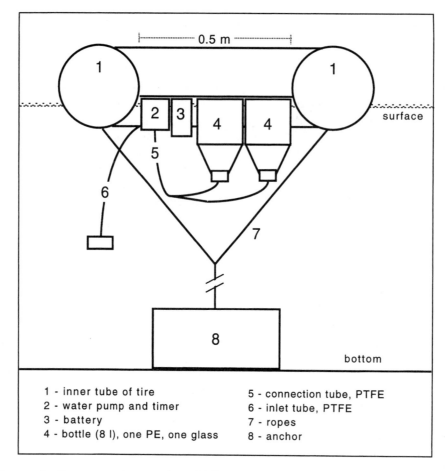

Fig. 7 Design of a sampler for water and suspensions (SWS) in marine systems including polar and tropical regions and in large rivers or lakes (Liess, Schulz, Duquesne in prep.).

the high variability of the measured concentrations of insecticides is shown in Fig. 8 and Table 4.

A simple qualitative method for the detection of pesticide contamination in the rainfall-induced surface runoff from agricultural fields into surface waters is described in Ref. 39. At each runoff site in an agricultural catchment a passively collecting glass bottle (2.5 L) was installed in the embankment. The glass bottle was placed in a plastic container with a lid, the upper surface of which was flush with the soil surface. The neck of the bottle passed through a bore (diameter: 4 cm) in the lid. The opening of the bottle (diameter: 3 cm) was 3 cm above the soil surface. This value was chosen to prevent animals (e.g., carabid beetles) from falling into the bottles and to sample runoff water from a particular amount onwards. A metal roof prevented rainfall from entering the collecting bottle (Fig. 9).

5. Final Considerations Regarding Automatic Sampling Equipment

Since automatic samplers generally include an inlet tube, problems like the development of biological films within the tube and control unit can arise. The influence of automatic sampling equipment on BOD test nitrification in nonnitrified final effluent has been evaluated (40). Samples were tested for BOD, carbonaceous BOD, nitrogenous oxygen demand, and concentration of nitrifying bacteria. Biofilms inside the equipment were tested for nitrification potential. A sampler utilizing continuous circulation of final effluent was found to support the attached growth of nitrifying bacteria and was associated with relatively high effluent nitrogenous oxygen demand. The effluent nitrogenous oxygen demand and nitrification potential of attached growth were significantly less with

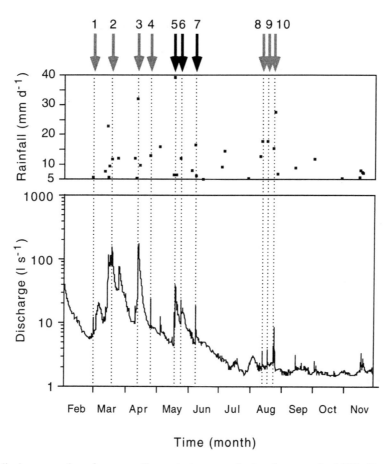

Fig. 8 Daily rainfall, discharge, and surface runoff events in an agricultural stream in 1994. (Arrows indicate the time of surface runoff. Black arrows point to insecticide contamination.)

a unit that aspirated effluent on an intermittent basis, purging the sample line with air before and after sampling. Peak nitrifier counts in samples from the continuous-flow equipment exceeded those in samples from the intermittent-flow equipment.

E. Extraction Techniques

All the sampling systems already mentioned are based on the principle of extracting the contaminants from the samples in the laboratory. There have also been attempts to concentrate and/or extract contaminants directly in the field. These systems take advantage of the tendency of determinants with low water solubility to bind to specific solid materials (solid-phase extraction) or the fact that a determinant is more soluble in specific solvents (liquid–liquid extraction) than in the water phase.

One main advantage of in situ concentration is that huge amounts of water can be extracted and need not

be transported. Because these systems stay mainly in the water for relatively long periods, there could be a great advantage of the increase in extracted water volume as well. At the same time, an integration of contaminants over time is possible with cheap systems. The disadvantage of the approach is that in most cases the contamination cannot be quantified. With in situ systems it is generally difficult to assess the amount of water from which the contaminants were extracted.

1. Semipermeable Membranes Filled with Solvents

A very common design employs semipermeable membrane devices [SPMDs (41)], which generally consist of a strip of nonporous polyethylene tubing filled with a small volume (1 ml) of a solvent (e.g., hexane) or a purified lipid, such as triolein. The combination of small solvent or lipid volume with a large

Table 4 Insecticide Contamination During Runoff Events in an Agricultural Stream, 1994 and 1995

Date	Number of event	Parathion-ethyl (μg L^{-1})	Lindane (μg L^{-1})	Fenvalerate (μg L^{-1})
1994				
March 1	1	—	—	—
March 17	2	—	—	—
April 13	3	—	—	—
April 25	4	0.04	0.02	nd
May 19	5	6.0	0.05	nd
May 25	6	0.9	0.05	nd
June 8	7	0.2	0.07	nd
August 13	8	nd	nd	nd
August 18	9	nd	nd	nd
August 24	10	nd	nd	nd
1995				
April 19/20	11	nd	nd	nd
May 27	12	0.6	0.2	6.2
June 1	13	0.15	nd	3.3
July 2	14	0.08	0.02	0.85
July 17	15	0.22	0.05	3.18
July 27	16	nd	nd	0.2
August 25	17	0.9	0.02	0.9
August 29	18	2.0	0.1	6.0
November 2	19	nd	nd	nd

— = not analyzed; nd = not determined.

membrane simulates the high surface-area-to-volume ratios characteristic of gill structures. Trace levels of contaminants that cannot be detected in conventional water samples are often concentrated to detectable levels by SPMDs placed in water for a controlled exposure period. An XAD-4 resin accumulative sampling method was tested (42) as a means of on-site extraction of surface waters. Recoveries for most organochlorine, organophosphorus, organonitrogen, chlorophenol, and chlorophenoxy acid pesticides and related pollutants were acceptable ($\geq$50%) when samples were spiked at the 10- and 0.1-ppb levels.

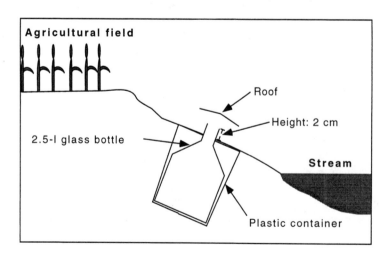

Fig. 9 Construction of a runoff-sampling apparatus, which is installed in erosion rills in the embankment.

Detection limits of 1–100 ppt were attainable for most compounds in river water, although lower levels required the use of an HPLC cleanup/fractionation step prior to those gas chromatography (GC) determinations using electron-capture detection. The XAD-4 accumulative sampling was competitive with the solvent extraction of grab samples in terms of recoveries, and it offered advantages in the volume of water sampled, the detection limits, and sample handling. The wide range of solute applicability combined with the ease of constructing and operating the accumulative sampler recommends it over grab sampling for many types of monitoring applications. The use of a similar system, supported liquid membranes for sampling and sample preparation of pesticides in water, has been described (43). A porous PTFE membrane is impregnated with a water-immiscible organic solvent, producing the supported liquid membrane. One general problem with membrane devices can be that they become covered with a biofilm that reduces the uptake of organic chemicals. Furthermore, the uptake also depends on temperature and on the boundary layer between membrane and water. All these parameters are difficult to assess over a long exposure period (weeks).

2. Solid-Phase Extraction Using Floating Complexing Granules

A technique for solid-phase preconcentration of contaminants dissolved in surface waters has been presented in (44). The method is based on the application of an extraction material that is not packed, as in con-

ventional solid-phase extraction systems, but instead is present in a freely floating form, as in a fluidized-bed reactor. The feasibility of the fluidized-bed extraction approach has been demonstrated for the determination of heavy metals in surface waters using 8-hydroxyquinoline attached to solid supports as complexing agent. Recoveries, repeatability, and sensitivity appear satisfactory for this application, even when no filtration of the sample is done. Because fluidized-bed extraction is based on free-floating, unpacked extraction material, the pressure drop over the column is minimal and filtration is not required. Hence the technique seems eminently suited for employment as an in situ long-term sampling method. As such it will provide time-integrated contamination levels that are not biased by biological variability or filtration artefacts, disadvantages of the commonly used methods for monitoring of contaminants in surface water.

3. Liquid–Liquid Extraction of Large Volumes

An accurate extraction and measurement procedure to determine chlorinated biphenyls (CBs) in surface waters is presented in Ref. 45. The procedure involved a 10–1 batch of liquid–liquid extraction directly from the sample bottle to prevent loss due to adsorption to the wall. Exhaustive extraction for recovery measurements was proposed, resulting in an extraction time of 10–45 h. A detection limit lower than 10 pg/kg and a coefficient of variation of 3–9% were obtained.

The equipment and procedure for the collection, filtration, and subsequent extraction of 20 L of water and

Table 5 Recovery of Spiked Organochlorine Contaminants from Lake Ontario Water and Suspended Solids

| | | Percentage recovery | | | | |
| | | Filtrate | | Suspended solids | | |
Compound	Amount spiked (ng/L)	Spike 1	Spike 2	Spike 1	Spike 2	Mean total
1,2,4-Trichlorobenzene	0.83	104	101	ND	ND	103
2,4,5-Trichlorotoluene	0.94	78	92	ND	ND	85
1,2,4,5-Tetrachlorobenzene	0.59	99	94	ND	ND	97
Pentachlorobenzene	0.12	68	87	ND	ND	78
Hexachlorobenzene	0.14	88	62	<1	ND	75
2,2′,3,3′-Tetrachlorobiphenyl	0.34	93	80	20	16	105
2,2′,4,4′,5,5′-Pentachlorobiphenyl	0.42	54	35	19	11	60
Overall mean recovery						86

ND = not detected.
Source: Ref. 8.

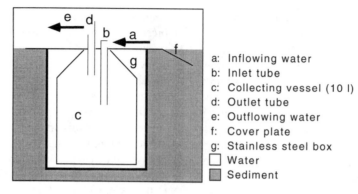

Fig. 10 Construction of the suspended-particle sampler (SPS).

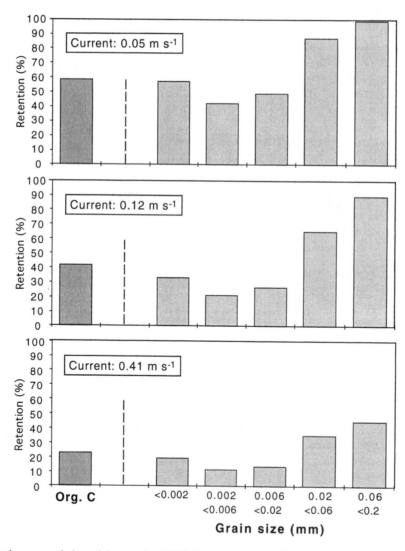

Fig. 11 Retention by the suspended-particle sampler (SPS) for current velocities between 0.05 and 0.41 m s⁻¹.

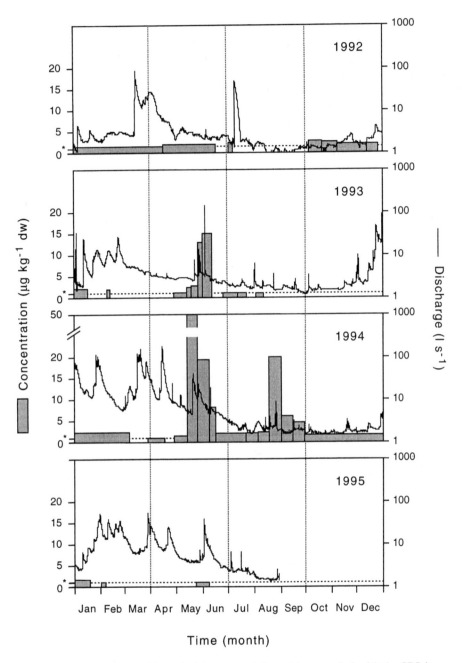

Fig. 12 Discharge and concentration of parathion-ethyl in suspended particles sampled with the SPS in an agricultural stream.

suspended-solid samples to be analyzed for organic chemicals, using readily available, inexpensive and sturdy apparatus, has been described (8). Water collection, filtration, and extraction of both phases can be accomplished in less than 1 hour. Recoveries of selected representative organochlorine contaminants spiked into "organic-free" water and Lake Ontario water at environmentally realistic levels are presented (Table 5).

F. Concentration of Contaminants in Suspensions and Sediment

It is generally accepted that heavy metals both of geogenic (i.e., natural) origin, like the so-called reference elements, and those introduced anthropogenically are preferentially found (by 85–95%) in the fine-grain fraction (≤20-μm fraction) of aquatic sediments and suspended solids. Similar conditions are assumed for pesticides with low water solubilities (46). For this reason, sampling of contaminants in the suspended particulates can sometimes be much more effective than sampling in the water phase. In a study in the North Saskatchewan River (47), the pathways and distribution of anthropogenic contaminants in water and in suspended and bottom sediments were examined. Synoptic sampling during stable low flow allows maximum inference of point sources and, therefore, of chemical complexity. After chemical extraction of contaminants into two fractions for aqueous samples and into five fractions for sediment samples, the authors were able to assess the toxicity of the fractions using two independent bioassays. The results called into question the efficacy of a standard menu of priority chemicals applied to water samples for ambient (i.e., not end-of-pipe) aquatic quality sensing (monitoring) purposes. They concluded that water may be an inappropriate medium upon which to base toxic chemical criteria for toxic chemical sensing purposes in aquatic systems.

1. Suspended-Particle Sampler for Small Streams

An economical alternative to event-controlled sampling for sampling potentially contaminated suspended particles in small streams is the employment of sedimentation vessels (suspended-particle sampler: SPS) through which the water flows continuously (48). The SPS (Fig. 10) is positioned in the stream so that the water flows through it at (a) and (e). It includes a 10-L sedimentation vessel (c) made of glass. This vessel is seated in a stainless steel box (g) that is lowered into the sediment at the bottom of the stream. The box, closed by a cover plate (f) made of stainless steel, comes to rest with the cover plate in the same plane as the stream bed, to prevent mobilization of the stream sediment in the region of the inlet tube (b). In addition to the inlet tube, an outlet tube (d) is mounted in the cover plate. Both tubes are 34 mm in diameter. The inlet tube is closed at its upper end and has a slit measuring 25×5 mm in the wall facing upstream. As a result, a flow pressure is directed into the collecting vessel. The outlet tube is cut at an angle of $45°$ at its upper end, forming an opening that faces downstream (e); this arrangement exerts suction on the collecting vessel. By rotating the outlet tube, the magnitude of the suction effect can be changed so that the velocity of flow into the inlet tube can be matched to the current of the stream. The height of the inlet opening can be adjusted to enable sampling at any desired level in the water. One implication of this method is that it cannot

be used to measure the exact maximal concentration, but instead gives the mean sediment contamination during the period from one emptying of the vessel to the next.

The grain-size-specific retention of the SPS was determined in laboratory experiments for the different flow velocities (Fig. 11). As expected, the amount retained by the SPS decreased with an increasing rate of flow. Because pollutants become attached preferentially to the grain-size fraction below 0.02-mm grain diameter (22), in evaluating retention properties particular weight was placed on the small grain sizes. At the highest flow velocity, 0.41 m s^{-1}, the retention rate was 15% for the grain-size fraction smaller than 0.02 mm. However, such rapid flow is rare in flatland streams. With the lower flow velocities tested here, the retention rate for this fraction was between 27% and 50%. The proportion of organic carbon retained, another parameter that is important regarding pesticide content, was between about 40% and 60% for these currents.

An example of long-term measurements of the concentrations of parathion-ethyl in an agricultural stream is shown in Fig. 12.

REFERENCES

1. EL Berg, eds. Handbook for Sampling and Sample Preservation of Water and Wastewater (Report PB 83-124503). Springfield, VA: U.S. National Technical Information Service, 1982.
2. K Grasshoff, M Ehrhardt, T Almgren, eds. Methods of Seawater Analysis. Weinheim, Germany: Verlag Chemie, 1983.
3. DTE Hunt, AL Wilson. The Chemical Analysis of Water—General Principles and Techniques. London: Royal Society of Chemistry, 1986.
4. B Kratochvil, JK Taylor. Sampling for chemical analysis. Anal Chem 53:924A–938A, 1981.
5. B Kratochvil, D Wallace, JK Taylor. Sampling for chemical analysis. Anal Chem 56:113R–129R, 1984.
6. P Hoffmann. Probenahme. Nachrichten aus Chemie, Technik und Laboratorium 40:M1–M32, 1992.
7. L Friling. Monitoring storm water runoff. Pollution Engineering 25:36–39, 1993.
8. ME Fox. A practical sampling and extraction system for the quantitative analysis of sub-ng/L organochlorine contaminants in filtered water and suspended solids. In: WC Sonzogni, DJ Dube, eds. Methods for Analysis of Organic Compounds in the Great Lakes. Madison, WI: University of Wisconsin Sea Grant Institute, 1986, pp 27–34.
9. HM Ortner, HH Xu, J Dahmen, K Englert, H Opfermann, W Görtz. Surface characterization of fluorinated

polymers (PTFE, PVDF, PFA) for use in ultratrace analysis. Fresenius J Anal Chem 355:657–664, 1996.

10. WA House. Determination of pesticides on suspended solids and sediments: investigations on the handling and separation. Chemosphere 24:819–832, 1992.

11. H Gudernatsch. Probenahme und Probeaufbereitung von Wässern. In: R Bock, ed. Analytiker-Taschenbuch. Berlin: Springer, 1983, pp 23–35.

12. ISO. Water quality—Sampling—Part 1: Guidance on the design of sampling programs. ISO 5667/1 1980.

13. ISO. Water quality—Sampling—Part 2: Guidance on sampling techniques. ISO 5667/2 1982.

14. ISO. Water quality—Sampling—Part 3: Guidance on the preservation and handling of samples. ISO 5667/3 1985.

15. LH Keith. Compilation of EPA's sampling and Analysis Methods. Boca Raton, FL: CRC Press, 1996, pp 1–1696.

16. Standing Committee of Analysts. General Principles of Sampling and Accuracy of the Results. London: H.M.S.O., 1980, pp 15–48.

17. J Hilton, T Carrick, E Rigg, JP Lishman. Sampling strategies for water quality monitoring in lakes: the effect of sampling method. Environ Pollut 57:223–234, 1989.

18. GR Martin, JL Smoot, KD White. A comparison of surface-grab and cross sectionally integrated stream-water-quality sampling methods. Water Environment Res 64:866–876, 1992.

19. RB Grayson, CJ Gippel, BL Finlayson, BT Hart. Catchment-wide impacts on water quality: the use of "snapshot" sampling during stable flow. J Hydrol 199:121–134, 1997.

20. LWG Higler. Caddis larvae in a Dutch lowland stream. Proceedings of the 3rd International Symposium on Trichoptera. 1981, pp 127–128.

21. CM Cooper. Biological effects of agriculturally derived surface-water pollutants on aquatic systems—a review. J Environ Qual 22:402–408, 1993.

22. H Ghadiri, CW Rose. Sorbed chemical transport in overland flow. 2. Enrichment ratio variation with erosion processes. J Environ Qual 20:634–642, 1991.

23. J Kreuger. Monitoring of pesticides in subsurface and surface water within an agricultural catchment in southern Sweden. British Crop Protection Council Monograph No 62: Pesticide Movement to Water 81–86, 1995.

24. RD Wauchope. The pesticide content of surface water draining from agricultural fields—a review. J Environ Qual 7:459–472, 1978.

25. T Kohonen. Influence of Sampling Frequency on the Estimates of Runoff Water Quality. Helsinki: Water Research Institute, 1982.

26. M Liess, R Schulz, B Rother, R Kreuzig. Quantification of insecticide contamination in agricultural headwater streams. Wat Res 23, 1:239–247, 1998.

27. IG Droppo, C Jaskot. Impact of river transport characteristics on contaminant sampling error and design. Environ Sci Technol 29:161–170, 1995.

28. ML Richardson, JM Bowron. Notes on Water Research No. 32. Medmenham, Bucks: Water Research Centre, 1983, pp 241.

29. AG Kelly, I Cruz, DE Wells. Polychlorobiphenyls and persistent organochlorine pesticides in sea water at the pg 1-1 level. Sampling apparatus and analytical methodology. Anal Chim Acta 276:3–13, 1993.

30. AD Jassby, BE Cole, JE Cloern. The design of sampling transects for characterizing water quality in estuaries. Estuarine Coastal Shelf Sci 45:285–302, 1997.

31. D Barcelo, MC Hennion. Sampling of polar pesticides from water matrices. Anal Chim Acta 338:3–18, 1997.

32. DJ Harper. A new trace metal–free surface water sampling device. Marine Chem 21:183–188, 1987.

33. P Conan, M Pujo-Pay, M Leveau, P Raimbault. Une Autre Utilisation du Systeme Let-Go: Echantillonnage Rapide d'une Colonne d'Eau Entre 0 et 200 M. Annales de L'Institut Oceanographique 72:221–227, 1996.

34. GC Kineke, RW Sternberg. The effect of particle settling velocity on computed suspended sediment concentration profiles. Mar Geol 90:159–174, 1989.

35. IN McCave, TJ Gross. The effect of particle settling velocity on computed suspended sediment concentration profiles. Mar Geol 99:403–413, 1991.

36. L Thomsen, G Graf, V Martens, E Steen. An instrument for sampling water from the benthic boundary layer. Continental Shelf Research 14:871–882, 1994.

37. U Eversberg. A new device for sampling water from the benthic boundary layer. Helgolaender Meeresuntersuchungen 44:329–334, 1990.

38. M Dickman, F Johnson. Deployment of a threshold activated pump sampler in an industrial shock load impact study. Hydrobiologia 344:181–193, 1997.

39. R Schulz, M Hauschild, M Ebeling, J Nanko-Drees, J Wogram, M Liess. A qualitative field method for monitoring pesticides in the edge-of-field runoff. Chemosphere 36:3071–3082, 1998.

40. B Koopman, CM Stevens, CL Logue, P Karney, G Bitton. Automatic sampling equipment and BOD test nitrification. Wat Res 23:1555–1562, 1989.

41. A Södergren. Solvent filled dialysis membranes simulate uptake of pollutants by aquatic organisms. Environ Sci Technol 21:855–859, 1987.

42. JE Woodrow, MS Majewski, JN Seiber. Accumulative sampling of trace pesticides and other organics in surface water using XAD-4 resin. J Environ Sci Health B, 21:143–164, 1986.

43. M Knutsson, G Nilve, L Mathiasson, JA Jonsson. Supported liquid membranes for sampling and sample preparation of pesticides in water. J Chromatogr A 754:197–205, 1996.

44. JW Hofstraat, JA Tierlrooij, H Compaan, WH Mulder. Fluidized-bed solid-phase extraction: a novel approach

to time-integrated sampling of trace metals in surface water. Environ Sci Technol 25:1722–1727, 1991.

45. JH Hermans, F Smedes, JW Hofstratt, WP Cofino. A method for estimation of chlorinated biphenyls in surface waters: influence of sampling method on analytical results. Environ Sci Technol 26:2028–2035, 1992.

46. WJ Adams, RA Kimerle, JW Barnett Jr. Sediment quality and aquatic life assessment. Environ Sci Technol 26: 1865–1875, 1992.

47. ED Ongley, DA Birkholz, JH Carey, MR Samoiloff. Is water a relevant sampling medium for toxic chemicals? An alternative environmental sensing strategy. J Environ Qual 17:391–401, 1988.

48. M Liess, R Schulz, M Neumann. A method for monitoring pesticides bound to suspended particles in small streams. Chemosphere 32:1963–1969, 1996.

49. R Wagner. Sampling and sample preparation. Z Anal Chem 282:315–328, 1976.

2

Methods of Water Preservation

Bieluonwu Augustus Uzoukwu

University of Port Harcourt, Port Harcourt, Nigeria

I. SCIENTIFIC METHODS OF WATER PRESERVATION AND STORAGE

Physical, chemical, or biological activities in a water sample can produce distortions in the microbiological or chemical composition of the sample. This is especially true when microbiological or interchemical reactions occur in a water sample, subjecting it to changes in chemical composition. The consequences of such activities may include the conversion of previously soluble species into insoluble species; change in temperature, pH, or conductivity of the water sample; and change in the concentration of some species in the water sample. Due to these activities it is known that polluted water samples can undergo changes in composition within a space of time after collection, and the most heavily polluted waters undergo significant changes in chemical composition within a few hours of collection. Water samples from sources that are less likely to be polluted, such as rainwater, stream, and other freshwater samples are also affected to some degree by these activities. For instance, a water sample that changed from being acidic to being basic due to microbiological and interchemical reactions can suddenly become turbid due to the precipitation of ions that form precipitates at high pH levels (e.g., Fe^{3+}, Fe^{2+}, Ca^{2+}, Pb^{2+}, Al^{3+}) if they are present in the water sample. Inorganic nitrogen and sulphur compounds in a water sample are subject to changes through microbiological activities. The phosphorus content of a water sample can also be affected by microbiological activities. Generally, organic substances may undergo microbiological degradation during storage, thereby changing the chemical oxygen demand (COD) or the biological oxygen demand (BOD). Hence, the concentrations of ions of nitrogen, sulphur, and phosphorus in a water sample, particularly those that are heavily polluted, can be affected significantly within a few hours of collection.

Some of the scientific methods for preserving and storing water can be classified as physical or as chemical methods for water preservation.

A. Physical Methods

1. Sterilization with Radiation

In this method a water sample is sterilized by the use of radiation to inactivate the activities of micro-organisms in the water body. This method is expensive and is of environmental concern. Many countries do not advocate the use of this method in water preservation and storage.

2. Temperature Reduction

This is a less expensive method that is commonly used for water preservation, and it involves the lowering of the temperature of a water sample down to 4°C immediately after collection, using solid carbon dioxide, or refrigerating and storing the sample in a chilled vacuum container in a dark place before transportation to a laboratory for analysis. Hence, the process is carried out in the field as an essential part of the field process

25

Table 1   Recommended Methods for Water Preservation

Parameter	Suitable container(s)	Remarks, and stabilization and preservation methods
pH	Plastic, glass	Analyze immediately, and avoid filtration of sample. If delays occur, fill container completely to exclude air, and store near 4°C in the dark.
Color	Plastic, glass	Cool and store in a dark place.
Dissolved gases	Glass	Fill container completely to exclude air, and analyze as soon as possible. Avoid the use of plastic containers and filtration of sample. Store near 4°C in the dark, and avoid freezing storage.
Total hardness	Plastic, glass	Fill container completely to exclude air. Special conditions are not required.
Sulphate	Plastic, glass	Fill container completely to exclude air. Special conditions are not required.
Sulphate	Plastic, glass	Store near 4°C in the dark.
Silica	Plastic	Add 10 ml chloroform or 10 ml toluene per liter of sample. Store near 4°C in the dark.
Nitrates and nitrites	Plastic, glass	Store near 4°C in the dark. Add phenyl mercuric acetate or 1 ml of $CHCl_3$ per liter of sample. Analyze as soon as possible.
Sulphide	Glass	Fill container completely to exclude air, and analyze as soon as possible.
Ammoniacal nitrogen	Plastic, glass	Store near 4°C in the dark. Acidify with HCl or H_2SO_4 to pH 2, or add 1 ml $CHCl_3$ per liter of sample. Analyze as soon as possible.
Total organic carbon	Glass	Analyze within 2 h. If delays occur, fill container completely to exclude air, and store near 4°C in the dark.
Phenolic compounds	Plastic, glass	Store near 4°C in the dark. Add Na_2AsO_3 and acidify with HCl to pH <2.
Surfactants	Plastic, glass	Analyze immediately. If delays occur, fill container completely, add formaldehyde, and store near 4°C in the dark.
Pesticides	Glass	The glass should be specially cleansed. Analyze within 3 days. If delays occur, fill container completely to exclude air, and store near 4°C in the dark.
Phosphates	Plastic, glass	The plastic container should be treated before use. Filter immediately and analyze as soon as possible. If delays occur, store near 4°C in the dark or freeze at −10°C.
Fluoride	Plastic	Special conditions are not required.
Arsenic	Plastic, glass	Special conditions are not required.
Arsenic	Plastic, glass	Add 2 ml 6 M HCl per liter of sample.
Antimony	Plastic, glass	Add 3 ml 6 M HCl per liter of sample.
Cyanide	Plastic, glass	Add NaOH to bring pH to over 12, and store near 4°C in the dark.
Polycyclic aromatic hydrocarbons	Glass	The glass should be specially cleansed. Store near 4°C in the dark.
Conductivity	Plastic, glass	Fill container completely to exclude air, and store near 4°C in the dark.
Boron	Plastic, glass	Store near 4°C in the dark.
Chlorinated hydrocarbons	Glass	The glass should be specially cleansed. Fill container completely to exclude air, and add $Na_2S_2O_3$ immediately. Store near 4°C in the dark.

Table 1 Continued

Parameter	Suitable container(s)	Remarks, and stabilization and preservation methods
Alkalinity	Plastic, glass	Analyze immediately. If storage is necessary, fill container completely to exclude air, and store near 4°C in the dark.
Metals	Plastic, glass	Add 2 ml of 5 M HCl or 2 ml concentrated HNO_3 per liter of sample to get pH under 2. Store near 4°C in the dark.
Sodium	Plastic	Special conditions are not required.
Potassium	Plastic, glass	Special conditions are not required.
Iron(II)	Plastic, glass	Fix in the field by adding some amount of 2,2′-bipyridyl to sample.
Hg(II)	Plastic, glass	Add HNO_3 and K_2CrO_7 to bring sample to pH = 1, and store near 4°C in the dark.

in water sample collection. The temperature-lowering technique is applied mainly to effect a reduction of the activities of micro-organisms in depleting the nitrogen, phosphorus, and sulphur content of a sample or depletion of any other species in the sample that is susceptible to microbial activities. Attempts should be made to ensure that the preserved samples are taken straight to the laboratory for analysis. And if a situation arises that indicates that the samples cannot be analyzed immediately on arrival in the laboratory, then the water samples should be kept in cold storage at around 4°C in a dark place until ready for analysis.

B. Chemical Methods

These methods involve the use of either inorganic or organic compounds as preservatives. The preservative ability of these compounds rests on their toxicity to micro-organisms such as bacteria. Hence, they are bacteriostatic agents. Some of the inorganic compounds used in water storage include mercury chloride, iodine, sulphuric acid, nitric acid, phosphoric acid, and hydrochloric acid.

Mercury as mercury chloride is very toxic to bacteria; this is the property of the element that is used in preventing the activities of bacteria in a stored water sample. Due to environmental concern, the use of mercury chloride is being discouraged because it may serve as a route for the entry of mercury into the environment. Iodine is used in water preservation because of its ability to inhibit microbial activities on phosphorus. Sulphuric and hydrochloric acids are used in bringing the pH of water samples down to pH <1. This is be-

cause microbial activities are seriously inhibited when the pH of a water medium is lower than 1. The use of these acids in water preservation also prevents the precipitation of some ions that may be present in a water sample. They include Fe, Ni, Cr, Ca, Al, Ba, Sr, Mn, V, Mo, Pb, Cd, and Co. Acidification of a water sample also prevents adsorption of metal ions on the body of the storage container.

Some of the organic compounds used for water sample preservation include dichloroethane, toluene, and chloroform. The main problem in the use of organic compounds in water sample preservation is that of solubility. The solubility in water of the organic compounds mentioned is very low. Another general problem associated with the use of chemical compounds in the preservation of water samples is the interference of the chemical compounds in the analysis of ionic species similar to those of the chemical preservative. For instance, a water sample acidified with hydrochloric acid cannot be used for the analysis of the chloride content of the water sample, hence necessitating the collection of separate samples for specific analysis. In addition, acidification can unduly interfere in the analysis of the sample if techniques such as atomic absorption or flame emission spectrophotometry are used for analysis. In such a case, acidification of the standards should be carried out in an identical way to give reproducible results.

The type of container used for preservation or storage is also vital. This is because some containers are capable of interfering in the chemistry of some of the water quality parameters and, hence, introducing error into the data obtained for these parameters during analysis. For instance, a glass container is not suitable for use in storing a water sample for which the silica content was to be determined. This is because glass itself is made of a silicate compound. A glass container, however, is very suitable for the storage of samples intended for use in the analysis of the pH, conductivity, dissolved gases, or organic content of a water body. These parameters are preserved in glass containers. Borosilicate glass containers are often the type of glass containers recommended. Plastics and other polythene containers are not suitable for use in storing samples meant for the analysis of odor, dissolved gases, or the organic content of a water body. Polythenes are made of organic compounds and can be porous to gaseous

constituents. High-density polythene containers are, however, suitable for use in preserving samples intended for use in the analysis of pH and the cation and anion contents of a water body. Air should be excluded from the storage container, particularly for those samples that will be used for phosphate and sulphide analysis. Hence, sufficient sample should be introduced into the storage container to fill it to the brim. Listed in Table 1 are containers suitable for preserving or storing water samples meant for the determination of the parameters listed along with them. Also indicated in the Table are recommended methods for fixing certain species in water, after sample collection.

3

Statistical Treatment of Data

Bieluonwu Augustus Uzoukwu

University of Port Harcourt, Port Harcourt, Nigeria

I. INTRODUCTION

There is a general need to subject data acquired from water analysis to regular statistical assessment in order to ensure the reliability of the results. In every application of statistics in the treatment and assessment of acquired data, the aim is primarily at making scientific meaning of the results. In water analysis the treatment of data this way helps to achieve specified aims and objectives. Some of the major aims and objectives are:

To provide a scientific basis for the level of confidence and reliability of results

To understand the trend of environmental processes associated with a water body

To provide a scientific basis for the utilization and planning programs where necessary

To properly characterize the quality parameters of a water sample, for comparison at known confidence levels with acceptable standards and for a thorough understanding of the changes in the physicochemical properties of the water body in space and time

These aims are frequently applied in many areas of water research, water resources, and development. Subsequently, measurement programs are designed to ensure that they are properly directed toward these objectives. A proper understanding of the methods being used and a sound knowledge of the appropriate statistical techniques required for the assessment and interpretation of the experimental data are very important

factors. Treatment of the data could be carried out to achieve a simple presentation and description of the data. Alternatively, it could be done to render an assessment and interpretation of the data and their limitations so that a sound conclusion can be made and professional advice rendered when utilizing the results of the analysis.

II. DEFINITIONS OF TERMS AND CONCEPTS

A. True, or Absolute, Result

In reality, there is a correct value for any measurement, which remains largely unknown. This correct result is called the *true*, or *absolute, result*. Hence, all results of measurements are subject to some degree of uncertainty. This uncertainty can be estimated with varying degrees of precision by experimental measurement and statistical analysis.

B. Accuracy

This is the closeness of an experimental result to the true, or absolute, result, estimated and expressed in terms of error. Since the determination of the true result of an experimental measurement is impossible, the estimated value of accuracy is, hence, a very important quantity.

C. Error

Generally, *error* is defined as the difference between the result from an experimental measurement and that of the true result. The numerical difference expressed in the same units as the measurement is known as the absolute error. If the error is expressed as a percentage or a proportion of the measured value, it is called relative error. The various types of error that are encountered in water analysis include gross errors, systematic (determinate) errors, and Random (indeterminate) errors.

D. Mean

This is the arithmetic average of a set of replicate results. This can be defined by the following equation:

$$\bar{x} = \frac{\sum_{i=1}^{n} x_i}{n}$$

where $\bar{x}$ represents the mean, x_i is the individual results, and n is the number of results.

E. Median

This is the middle value of a set of replicate results. The median, therefore, represents the central tendency of a set of replicate results.

F. Mode

This is the most frequently occurring value in a set of unsymmetrical distributed data. See Fig. 1 for a diagrammatic representation of the mean, the median, and the mode.

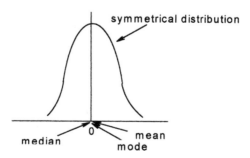

Fig. 1   Diagram of a normal distribution curve.

G. Degrees of Freedom

This is equal to the number of results in a set. However, when a quantity such as the mean is derived from the set of results, the degrees of freedom is reduced by 1.

H. Precision

This is a measure of the degree of agreement between a set of replicate results made at any one time under similar conditions. As in the case of error, it can be expressed as the standard deviation of results from their mean.

I. Spread

This is the numerical difference between the highest recorded value and the lowest recorded value in a set of replicate results. It offers one means of estimating precision, although it has some limitations.

J. Deviation

This is the numerical difference between an individual result in a set of replicate results and the mean or median of that set of results. Hence, the sign of the value could be positive or negative.

K. Standard Deviation

The standard deviation is a quantity that estimates the degree of precision in a set of replicate results. It essentially shows the degree of distribution of data around their mean. It can be expressed by the equation

$$\sigma = \sqrt{\frac{\sum_{i=1}^{i=n} (x_i - \mu)^2}{n}}$$

where x_i is the individual result, μ is the true mean, and n is the number of results in the set. A more practical value is $\bar{x}$ (the mean), which can be determined from the set of results, because μ can never be determined. Hence, in terms of the mean of the results, the equation for the true standard deviation, s, can be expressed as

$$s = \sqrt{\frac{\sum_{i=1}^{i=n} (x_i - \bar{x})^2}{n - 1}}$$

In this case the degree of freedom is reduced by 1. The

square of the standard deviation (σ^2 or s^2) gives the variance. The mean deviation is defined by the following equation:

$$\text{mean deviation} = \frac{\sum\limits_{i=1}^{i=n} (x_i - \bar{x})}{n}$$

L. Significant Digits

These are the digits in a number that are known with certainty. If the number is a zero it is taken as a significant figure when it is part of the number but not when it comes before the number as in the case of decimal numbers (e.g., 0.003).

III. TYPES OF ERRORS

Errors can occur during water analysis. Depending on the basis of their origin, errors can be classified as gross, systematic (determinate), or random (indeterminate). These three constitute the major types of errors identifiable in water analysis.

A. Gross Error

Gross error occurs when a measurement is invalidated by a major event, such as the loss of the sample or the failure of equipment, thus necessitating the complete repetition of the experiment. For instance, during analysis of the total suspended solids (TSS) of a water sample by gravimetry, a spill of part of the sample during transfer to the oven for drying will necessitate a repetition of the experiment, using fresh samples. In the analysis of the acidity of a water sample, if two drops of an acid–base indicator are required for titration and five drops of the indicator are mistakenly added, the results will have to be discarded and the experiment repeated. In the determination of the pH of a water sample, gross error could exist in the results if after restoration of power following an electric power failure, the measurement was carried out without first recalibrating the pH meter.

B. Systematic Errors

These are errors that can be measured and, thus, can be corrected. If the true value of a measurement is known, the systematic error can be determined from the difference between the mean of the results and the true value of the measurements. Systematic errors could be constant throughout the range of measurements and be discovered only when comparing them with the results of measuring the same quantity using different analytical methods.

Systematic errors can be either *constant* or *proportional*. When the systematic error is constant, the error will have a fixed value throughout the range of measurements made. This is not the same with proportional error, because it increases in proportion to the magnitude of the measurement and is independent of concentration. Since constant error has a fixed value, it follows that as the magnitude of the measurement increases, the value of the error decreases in significance. Subsequently, the effect of this error can be minimized by taking a larger volume of sample. For instance, in the determination of total hardness of water by the complexometric method, an error of 0.15 ml recorded during the analysis of a 50-ml water sample will amount to an error of 0.3%. However, if a 250-ml sample of water was analyzed, and given that the error is constant at 0.15 ml, the error this time would amount to 0.06%. Analysis of a 500-ml volume of the sample of water will amount to an error of 0.03%.

Some of the sources of systematic error include improper sampling techniques and handling of sample, mistakes by operators, and inadequate knowledge of a particular experimental procedure, faulty instrumentation and incorrect calibration of equipment, and the use of incorrect standards or contaminated reagents.

C. Random Errors

These are errors associated with the unpredictable inaccuracies of the individual worker manipulating an experimental procedure and with other external factors operating in the laboratory. They may arise from incorrect or poor knowledge of the use of an instrument, poor laboratory skills in the preparation of solutions, and spurious contamination of glassware and solutions. Hence, at the end of a measurement a degree of uncertainty is introduced into the result, leading to deviations of the measurements from the mean. For this reason the experimental data may appear scattered and can be assessed only by statistical tests and analysis. The deviations of a number of measurements from the mean of the measurement shows a *Gaussian distribution* about that mean. The distribution of the errors can be represented graphically by a curve known as *normal error curve*. The normal curve consists of a symmetrical distribution about the mean, as shown in Fig. 1, and can be described by the following equation:

$$y = \frac{e^{-(x-\mu)^2/2\sigma^2}}{\sigma\sqrt{2\pi}}$$

The term σ is the standard deviation, useful for estimating the width of the curve and the precision of the set of replicate results. Thus, precision may be expressed as the standard deviation of results from their mean. μ is the mean of the results in the set. In a normal distribution, the mode, the median, and the mean are numerically equal to one another. Random errors may also have an asymmetrical distribution about the mean, thereby giving rise to a skewed distribution, as shown in Fig. 2. In skewed distributions, the mode, the median, and the mean are frequently not equal to each other numerically.

Some of the sources of random error include experimental techniques that require long procedures, fluctuations in conditions at various stages of an experiment, such as changes in electric current or temperature, change in the pH of a water sample, and the unexpected formation of precipitates or particles.

IV. GRAPHS AND TABLES

The results of measurements from water analysis can be presented in graphical form. This enables the data to be presented in such a way that the distribution of the physicochemical parameters determined are clearer and the features of the data better understood. Collec-

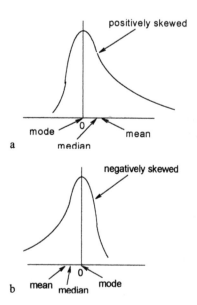

Fig. 2 Diagram of a skewed distribution curve: (a) positively skewed; (b) negatively skewed.

tion of water samples could be carried out through either a batch or a continuous process. Collection could also be done at various levels of a water body. Data acquired from these methods of water sampling and analysis could be presented graphically, and graphs have proved to be very valuable in the presentation and interpretation of analytical data of these kinds, because they enable a worker to have an overview of the trend of the experimental results. Graphs have proved to be very valuable in graphical analysis of experimental results through calibration with standards. Thus, during calibration they provide a worker with the means of determining the range over which an experimental measurement could be applied. For instance, in the colorimetric analysis of ions in water, the calibration curves at lower concentrations (from the Beer–Lambert law) should be a straight line. However, at higher concentrations of the analyte, the lines may show a curvature due to deviation from the law. This is illustrated by the calibration curve presented in Fig. 3 for the colorimetric analysis of the molybdenum content of a solution using the potassium thiocyanate–ascorbic acid method. Calibration curves of this nature are routinely used in quantitative analysis. Within the range of the limit of the straight-line portion of the graph, the line can be described by the following linear equation:

$$A = \epsilon c l$$

where A is the absorbance, l is a constant part length, ϵ is the molar absorptivity, and c is the concentration of the analyte being determined. This is the basis for the quantitative application of the calibration curve. Graphs show the nature of the relationship between variables within the range of study, thus establishing the experimental points where one or more variables can be properly correlated.

In the course of water analysis, particularly in continuous sampling and analysis of a water body, it may be important to monitor a particular parameter of water quality in terms of space and time. The purpose, essentially, is to determine the extent to which the parameter of interest has changed in value. Such parameters include pH, dissolved oxygen, temperature, and level of anionic and cationic contents, particularly for effluent water bodies emerging from an industrial area or places subjected to activities that have the potential for polluting the surrounding area. Monitoring of this sort helps to establish the trend of change of the parameter of interest and, where possible, to show if the change had occurred significantly and to show also when the change had begun. Problems in statistical presentation and assessment of time-dependent data obtained from

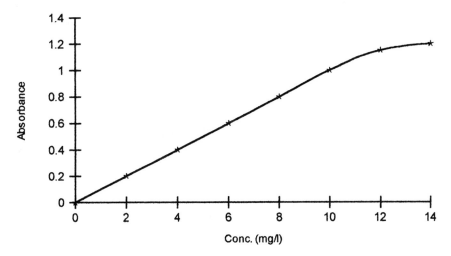

Fig. 3 Calibration curve for the determination of the Mo(VI) ion in an aqueous phase.

the analysis of water can be taken care of effectively by this type of time-dependent graphical presentation. Figure 4 shows a time-dependent graphical presentation of the temperature of a 100-L water body in a black plastic container recorded between the eighth and nineteenth hour in a particular day.

Other useful forms of graphs used in the presentation of data on water quality are *control charts*. There are two types of control chart in use: the shewhart chart and the cusum chart. *Shewhart charts* are used in quality assurance if an analysis requires that the performance of an analytical method using a reference sample be monitored at regular intervals. The chart is used in pairs of averages and ranges charts. In the *averages chart*, the results are plotted against the sequential

number of observations to be made so that any deviation from an expected value can be observed. The chart is marked with warning and action limits, done essentially to indicate the level of confidence to be placed on an observed deviation. In the *ranges chart*, the results are plotted against observation number. An increase in the range or a movement of the mean indicates a loss of precision in the measurement or the existence of bias in the methods employed, respectively. The *cusum chart* is derived from plotting the computed cumulative sum of the deviations of the observations from the expected value against the number of observations. The chart is very useful in monitoring the average changes in the parameters of water quality obtained for a stream, a river, or industrial effluent.

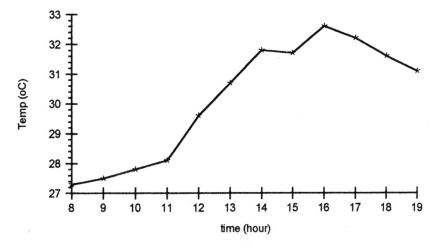

Fig. 4 Temperature- and- time-dependent plot of the surface temperature of a 100-ml water body in a black plastic container. (From BA Uzoukwu, unpublished results, 1998.)

Table 1 Results of the Analysis of the Cation Content of a Water Sample

Parameter	Concentration (mg L^{-1})
Zn	25.05 ± 0.15
Cu	12.57 ± 0.21
Fe	45.76 ± 0.06
Ni	4.66 ± 0.03
Mg	37.04 ± 0.11
Ca	69.98 ± 0.02
Mn	7.90 ± 0.01

Source: BA Uzoukwu, Personal communication, 1994.

The use of tables in the presentation of data from water analysis is very common. This is due to the simplicity in data presentation and the clarity in the details of data computation, for instance, when means and standard deviations are involved, as shown in Table 1.

Frequently, a table may consist of columns showing the water samples analyzed, indicated by their sampling codes, and rows listing the water quality parameters that have been determined. The main body of the table contains the accumulated data for the various water samples analyzed. Tables have also been found very useful in the presentation, classification, and interpretation of analytical data, particularly where water quality parameters are to be compared with accepted standards. Table 2 shows a presentation of the water quality parameters for water samples collected from some pits of water in the river Niger delta.

V. CALIBRATION

This is a very important experimental process in water analysis and chemical analysis in general. Calibration is the process of preparing data from standard preparations by adopting standardized analytical methods in order to provide a basis for comparison with a test solution to enable the concentration or other parameter of the test solution to be determined with a high degree of certainty. The process also offers a means by which analytical results based on similar methods can be easily compared. Examples include the calibration of instruments for pH measurement and for dissolved oxygen. Frequently, calibration is done using graphical methods and linear regression analysis, for which the linear calibration function can be expressed as follows:

$$y = mx + c_o$$

where y is the measured quantity, x is the standard concentration or quantity, m is the slope, which could be used qualitatively in certain cases, and c_o is a constant derived from the intercept on the ordinate and may be

Table 2 Physicochemical Data for Water Samples Collected from Pits W1–W4

Parameter	Sample (in mg L^{-1}) W1	W2	W3	W4	WHO standard
pH	7.4 ± 0.1	6.8 ± 0.1	6.7 ± 0.2	7.7 ± 0.2	6.5–8.5
Temperature (°C)	32.5	32.0	32.3	32.5	No guideline value set
Electrical conductivity (μScm^{-1})	68 ± 2	33 ± 2	308 ± 4	634 ± 7	—
Biochemical oxygen demand	2.2 ± 0.1	5.3 ± 0.5	1.8 ± 0.3	3.8 ± 0.5	—
Total suspended solids	5.9 ± 0.3	5.5 ± 0.2	104.4 ± 0.4	8.7 ± 0.3	—
Total dissolved solids	63.1 ± 0.8	22.2 ± 0.3	260.0 ± 0.9	467.4 ± 0.5	1000
Chloride	10.4 ± 0.9	5.1 ± 0.2	20.4 ± 0.6	24.8 ± 0.5	250
Sulphate	6.3 ± 0.7	2.3 ± 0.8	26.3 ± 0.6	39.1 ± 0.4	400
Phosphate	0.08 ± 0.01	0.70 ± 0.05	0.39 ± 0.06	0.15 ± 0.05	—
Nitrate	2.2 ± 0.2	4.7 ± 0.3	16.7 ± 0.2	24.7 ± 0.4	10
Iron	0.03 ± 0.01	0.34 ± 0.03	5.07 ± 0.02	0.02 ± 0.01	0.3
Zinc	0.04 ± 0.02	0.28 ± 0.04	0.18 ± 0.02	0.2 ± 0.01	5.0

Source: BA Uzoukwu, Environmental Impact Assessment Study, 1994.

indicative of the blank value. When the calibration graph or function is established from given standard quantities x and measured quantities y, the data presented graphically or in tables can then be used for converting measured quantities of the test sample into the desired parameter, for instance, for the conversion of an absorbance measurement from an atomic spectrophotometer into concentration. When a new method of analysis is involved, calibration may have to be carried out extensively to ascertain the linearity of the calibration within the working range, to enable intended parameters to be determined by the given method. During calibration, the distances between the individual standard quantities should be made equal, to ensure a normal distribution across the working range for the detection of the limits of the working range of the analytical method employed.

A. Blank Value

The blank value is a value that sets the lowest limit attainable when the measured quantity in solution does not exist in solution. Hence, through blank values some form of correction can be made for background concentration. The blank value is determined by extrapolating the calibration curve and taking the intercept on the ordinate. The dispersion of blank values may give an indication of the existence of systematic and random errors in a measurement, for which examples exist in the colorimetric analysis of ions in solution. It may indicate the existence of a deviation from a baseline value or uncover carryover errors due to unexpected changes in the physical properties of solution during measurement. It may also indicate the existence of a deviation of the working range at lower concentrations, for instance, in the application of the Beer–Lambert law in water analysis, or in the drift caused by the instrument or background noise.

B. Detection and Determination Limit

The *detection limit* can be defined as the smallest measurable quantity of a substance that can be qualitatively detected with the required statistical certainty using the procedure given. On the other hand, the *determination limit* can be defined as the smallest quantity of a substance that can be quantitatively detected with the required statistical certainty using the procedure given. In essence, both limits try to define the point at which a signal from a sample can be distinguished from those due to the background or the blank. Every measurement is subject to random error, and there is a chance

of erroneously identifying the presence of a substance in a sample when it is not there, or vice versa. Usually, the distribution of the errors should produce a normal error curve. The spread of replicate measurements from the blank and from the sample can be assessed statistically from the inherent normal distribution curve, which will enable the detection and determination limits to be defined statistically. Figure 5 shows an overlap of the normal distribution curves for both the blank and the sample measurements made, with equal standard deviations. Usually a 95% confidence level is used for deciding if a given measurement arises from the presence of a sample or blank, and vice versa. Thus, point L in Fig. 5 represents an upper limit on the blank distribution curve above which only 5% of blank measurements with true mean will lie, while on the sample distribution curve it represents a lower limit below which only 5% of sample measurements with true mean will lie. Note: The lowest measurable concentration of a sample is only partly related to the detection limit of the procedure employed. The relation between the true mean, μ_s, of the sample and L is given by the equation

$$\mu_s = 2L$$

Hence, for a single measurement that falls outside L and to the right, the chances that it arose from signals due to the sample has a 95% probability. Measurements that fall outside L but to the left have a 95% probability that they arose from signals due to background sources. Thus, measurements that gave results that fell between L and μ_s can be regarded as having been detected. Subsequently, the position of L on the curves represents the practical detection limit, while μ_s represents the theoretical detection limit for a measurement. In terms of the standard deviation, the practical detection limit L is defined as follows:

$$L = \begin{cases} \mu_b + 1.64\sigma & \text{for a large number of samples } (>20) \\ \mu_b + 2.33\sigma & \text{for a small number of samples } (<20) \end{cases}$$

where μ_b is the true mean for the blank.

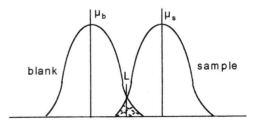

Fig. 5 Normal distribution curves for error from blank and sample measurements.

The theoretical detection limit is defined as follows:

$$\mu_s = \begin{cases} 2\mu_b + 3.28\sigma & \text{for a large number of samples} \\ 2\mu_b + 4.66\sigma & \text{for a small number of samples} \end{cases}$$

C. Regression and Correlation

Regression is the process of fitting a straight line to fit experimental data best. Thus, there must exist a relationship between two variable quantities that are measurable, in which case one quantity shall be dependent on the other. For instance, here is an equation describing a linear relationship:

$$y = mx + c_o$$

where y is dependent on the value of the quantity x, m is the slope of the plot, and c_o is the intercept on the ordinate.

In water analysis there exist many parameters that are determined by laws that involve a linear relationship between the measurable quantities. An example is the Beer law, applied in the spectrophotometric determination of species in solution based on the linear relationship between the absorbance and concentration of species in solution:

$$A = \epsilon c l + c_o$$

where c is the concentration, ϵ (the absorptivity and l the path length) are constant quantities; $c_0 = 0$. The variable quantity in the equation is c. The equation is, thus, similar to the general straight-line equation given earlier in which y and x are the variables in the experimental procedure while m is the regression coefficient and c_o is the regression constant. Through the method of least squares it can be shown that

$$m = \frac{\sum (x - \bar{x})(y - \bar{y})}{\sum (x - \bar{x})^2}$$

and

$$c_o = \bar{y} - m\bar{x}$$

Subsequently, the extent to which the values of the variables are correlated can be represented by the linear correlation coefficient, r:

$$r = \frac{\sum (x - \bar{x})(y - \bar{y})}{\sqrt{\sum (x - \bar{x})^2 \sum (y - \bar{y})^2}}$$

This is a measure of the linear correlation between the two randomly distributed quantities x and y.

The correlation coefficient r always lies between -1 and $+1$. If $r = +1$, this indicates a strong positive correlation. If $r = 0$, this indicates that there is no correlation, in which case the lines are perpendicular. $r = -1$ indicates a strong negative correlation. Numerical values of r related to different confidence levels and the number of data are presented in Table 3. If the pairs of the quantities x and y do not fit satisfactorily on a straight line or give a nonlinear regression line, this is an indication that other influencing factors are involved in a given process. In water analysis such factors may include influences from concentrations of different substances contained in a water sample with a large matrix of solutes, changes in the temperature and pH of a sample solution, the quantity of waste water, the length of storage, and the level of microbial degradation of nitrate or phosphate content of a water sample. Hence, if one of the quantities, y, is dependent upon a number of other quantities, such as $x_1, x_1, \ldots x_n$, this forms a linear multiple regression, which can be expressed as

Table 3 Correlation Coefficient r for Different Confidence Levels

Number of data	Confidence level 95%	Confidence level 99%	Number of data	Confidence level 95%	Confidence level 99%
5	0.75	0.87	18	0.44	0.56
6	0.71	0.83	20	0.42	0.54
7	0.67	0.80	25	0.38	0.49
8	0.63	0.77	30	0.35	0.45
9	0.60	0.74	40	0.30	0.39
10	0.58	0.71	50	0.27	0.35
12	0.53	0.66	60	0.25	0.33
14	0.50	0.62	80	0.22	0.23
16	0.47	0.59	100	0.20	0.25

$$y = c_o + m_1 x_1 + m_2 x_2 + \cdots m_n x_n$$

for which the nth regression coefficient is

$$m_n = \frac{\sum (x_n - \bar{x}_n)(y - \bar{y})}{\sum (x_n - \bar{x}_n)^2}$$

VI. INTERPRETATION AND ASSESSMENT OF DATA

At the end of water analysis, the data accumulated are examined to assess the reliability of the result and the interpretation of the meaning of the result. Interpretation of the physicochemical, chemical, and microbiological parameters is carried out to assess the data in relation to the use to which the water is to be put and to see if there is any pattern in the results that will support a given hypothesis sufficiently. In some instances, this is to see whether a purified water sample intended for drinking water has attained the quality specifications of a particular country for water meant for drinking (or potability), and, in the case of wastewater, whether the physicochemical, chemical, and microbiological parameters of the effluent wastewater are within environmentally acceptable limits for that country.

Assessment of data assembled during water analysis can be carried out statistically and the results compared with accepted standards or used in testing the fitness of a given hypothesis. Usually these results are assessed in terms of their precision, accuracy, and reliability. Some of the various approaches to the assessment and interpretation of data follow.

A. Reliability of Data

Replicate results obtained from water analysis, particularly those obtained from a certain level of the same water body, may follow a pattern that can be assessed statistically, since water is homogenous. Examples include temperature, total dissolved solids, dissolved oxygen, pH, and metal-ion and anion contents. On inspection, it is not unusual for one or more of the acquired results, called *outliers*, to appear higher or lower than the expected mean, thus throwing some doubt on their reliability due to one form of experimental error or another. In this circumstance questionable results should be subjected to proper statistical test. A statistical check is carried out starting with those results, or outliers, that deviate rather widely from the mean. This is a very important assessment of results,

in which the number of measurements in a set of replicate results is few. Note that a result that is significantly lower or higher than the true mean can produce a mean with gross error. Questionable results are examined for their reliability and rejected where the appropriate statistical test indicates so, particularly when the result is grossly different from the rest of the results in a set. Usually, the decision to reject a result has to be taken at a specified level of confidence. There are a number of methods for assessing the acceptance or rejection of questionable results; the most commonly used method is based on the rejection quotient (Q), known as the Q-test. Here a set of data are arranged in order of increasing magnitude; the potential questionable results will be those with the highest or the lowest values in the set. The Q-test is performed at 90% confidence level, which is considered an appropriate limit for the test. The rejection quotient Q is given as follows:

$$Q = \frac{x_n - x_{n-1}}{x_n - x_1}$$

where x_n is the questionable result in a set of results $x_1, x_2, \ldots, x_n$. The value obtained from the calculation of Q(Q-experimental) is compared with a table of critical values of Q(Q-critical). The critical values at the 90% confidence level are listed in Table 4. If Q-experimental exceeds Q-critical, then the result is rejected at the confidence level set.

B. Initial Statistical Assessment

At the end of water analysis the most useful assessment of the data acquired is through the determination of the arithmetic mean and the standard deviation of the data in the set once the reliability of the replicate set of

Table 4 Critical Values of Q at the 90% Confidence Level

Number of results	$Q_{critical}$ (90% confidence)
2	—
3	0.94
4	0.76
5	0.64
6	0.56
7	0.51
8	0.47
9	0.44
10	0.41

results has been established. A limit may have to be set for the confidence level for the experimental mean within which there is a known confidence of determining the true mean. For a large number of samples determined, the experimental mean will follow a normal distribution. For small samples ($n < 30$), the experimental mean will follow a t-distribution; hence the t-distribution can be used for calculating a confidence interval for an experimental mean using the relationship

$$\bar{x} \pm \frac{ts}{\sqrt{n}}$$

where $\bar{x}$ is the experimental mean, t is the statistical factor derived from the normal error at various confidence levels, s is the calculated standard deviation, and n is the number of results. The t-distribution also offers a means of comparing the experimental mean with a specified standard mean or standard value.

Circumstances may arise in which this type of comparison is needed to validate an analysis by comparing the mean with that of a reference sample, a control sample, or an accepted standard. Before carrying out the comparison of two separate sets of data or the comparison of an experimental mean with a control or an accepted standard, their respective precisions are first compared. For this purpose the F-test is employed. In the test, F is the ratio of the variances of the two sets of data, as shown by the following equation:

$$F = \frac{s_x^2}{s_y^2}$$

The larger of the two variances is often taken as the numerator in this equation. The calculated value obtained is used in establishing statistically if there is any significant difference between the precisions of the two sets of data by comparing the calculated value with those of the tabulated F-critical values listed in Table 5. The number of degrees of freedom is always ($n - 1$) when using the table. The usual assumption is that they can be exceeded on the probability of 5% of cases listed in the table. Subsequently, if the experimental value calculated for F exceeds that of F-critical, the difference between the precisions of the two sets of data being compared is said to be statistically significant. When it has been established that the standard deviation of the two sets of data agree to a reasonable level of confidence, the appropriate t-test can then be used for comparison of the experimental means with other means using any of the appropriate equations relating t to the standard deviation s.

The t-test will indicate whether the numerical difference between the experimental mean and the set standard value is significant or not. This could be done using the following equation, in which the value of t is calculated and compared with those of t-critical tabulated in Table 6 at various confidence levels:

$$t = \frac{(\bar{x} - \mu)\sqrt{n}}{s}$$

In this case μ is the true mean or the accepted standard. For a given number of degrees of freedom, if the calculated value of t exceeds t-critical derived from the tabulated values in the distribution table, the difference between the experimental mean and the accepted standard is considered significant. Comparison of two sets of data obtained for the same type of parameter for two different water samples can be carried out to see if there is any significant difference between the two experimental means. This is important if one of the water samples was used as the control while a set of data was also obtained for the other. The following equation can be used for the purpose of comparison:

$$t = \frac{\bar{x}_1 - \bar{x}_2}{\sqrt{\dfrac{s_1^2}{n_1} + \dfrac{s_2^2}{n_2}}}$$

C. Application of Statistical Tests to Results: An Example

Assessment of eight replicate results from an analysis of the calcium content of a water sample produced the following, arranged in increasing order of magnitude:

50.23, 50.23, 50.24, 50.24, 50.26, 50.26, 50.27, 50.29 mg L^{-1}

1. Test of the Reliability of the Results

An examination of the results shows that 50.29 mg L^{-1} appears to be the outlier and should be tested for reliability.

Solution   The rejection quotient Q is given as follows:

$$Q = \frac{x_n - x_{n-1}}{x_n - x_1}$$

$$Q = \frac{50.29 - 50.27}{50.29 - 50.23} = 0.333$$

The value of Q-critical for eight replicate results at the

Table 5 Values of *F*-Critical at the 5% Level

	Degrees of freedom (numerator)											
Denominator	2	3	4	5	6	7	8	9	10	12	20	∞
2	19.00	19.16	19.25	19.30	19.33	19.35	19.37	19.38	19.40	19.41	19.45	19.50
3	9.55	9.28	9.12	9.01	8.94	8.89	8.85	8.81	8.79	8.74	8.64	8.53
4	6.94	6.59	6.39	6.26	6.16	6.09	6.04	6.00	5.96	5.91	5.80	5.63
5	5.79	5.41	5.19	5.05	4.95	4.88	4.82	4.77	4.74	4.68	4.56	4.36
6	5.14	4.76	4.53	4.39	4.28	4.21	4.15	4.10	4.06	4.00	3.87	3.67
7	4.74	4.35	4.12	3.97	3.87	3.79	3.73	3.68	3.64	3.57	3.44	3.23
8	4.46	4.07	3.84	3.69	3.58	3.50	3.44	3.39	3.35	3.28	3.15	2.93
9	4.26	3.86	3.63	3.48	3.37	3.29	3.23	3.18	3.14	3.07	2.94	2.71
10	4.10	3.71	3.48	3.33	3.22	3.14	3.07	3.02	2.98	2.91	2.77	2.54
12	3.89	3.49	3.26	3.11	3.00	2.91	2.85	2.80	2.75	2.69	2.54	2.30
15	3.68	3.29	3.06	2.90	2.79	2.71	2.64	2.59	2.54	2.48	2.33	2.07
20	3.49	3.10	2.87	2.71	2.60	2.51	2.45	2.39	2.35	2.28	2.12	1.84
∞	3.00	2.60	2.37	2.21	2.10	2.01	1.94	1.88	1.83	1.75	1.57	1.00

Table 6 Values of *t*-Critical at Various Confidence Levels

Degrees of freedom	Confidence level (%)			
	90	95	99	99.9
1	6.31	12.7	63.7	636.62
2	2.92	4.30	9.92	31.60
3	2.35	3.18	5.84	12.94
4	2.13	2.78	4.60	8.61
5	2.02	2.57	4.03	6.87
6	1.94	2.45	3.71	5.96
7	1.90	2.36	3.50	5.41
8	1.86	2.31	3.36	5.04
9	1.83	2.26	3.25	4.78
10	1.81	2.23	3.17	4.59
11	1.80	2.20	3.11	4.44
12	1.78	2.18	3.06	4.32
13	1.77	2.16	3.01	4.22
14	1.76	2.14	2.98	4.14
15	1.75	2.13	2.95	4.07
16	1.75	2.12	2.92	4.02
17	1.74	2.11	2.90	3.97
18	1.73	2.10	2.88	3.92
19	1.73	2.09	2.88	3.88
20	1.72	2.09	2.85	3.85
25	1.71	2.06	2.79	3.73
30	1.70	2.04	2.75	3.65
40	1.68	2.02	2.70	3.55
∞	1.64	1.96	2.58	3.29

90% confidence level is 0.47. Since the Q-experimental value of 0.333 did not exceed the value of Q-critical, the result is retained in the set of data.

2. Determination of the Arithmetic Mean of the Data

Solution The arithmetic mean is given by

$$\bar{x} = \frac{\sum\limits_{i=1}^{n} x_i}{n}$$

$$\bar{x} = (50.23 + 50.23 + 50.24 + 50.24 + 50.26 + 50.26 + 50.27 + 50.29)/(8)$$

$$= 50.25 \text{ mg L}^{-1}$$

3. Determination of the Standard Deviation of the Data

Solution The standard deviation is determined using the following equation:

$$s = \sqrt{\frac{\sum\limits_{i=1}^{i=n} (x_i - \bar{x})^2}{n-1}}$$

The calculation can be done using the following table:

x	$(x - \bar{x})$	$(x - \bar{x})^2$	$\Sigma(x - \bar{x})^2$	s
50.23	−0.02	0.0004		
50.23	−0.02	0.0004		
50.24	−0.01	0.0001		
50.24	−0.01	0.0001		
50.26	0.01	0.0001		
50.26	0.01	0.0001		
50.27	0.02	0.0004		
50.29	0.04	0.0016		
			0.0032	0.02 mg L^{-1}

Hence, the experimental mean will be reported as 50.25 ± 0.02 mg L^{-1}.

4. Related Question

If the accepted value for the calcium content of a water sample intended for a certain industrial application, determined from previous analysis, was reported as 50.21 ± 0.03 mg L^{-1} for 10 determinations, determine if there is any agreement between the two standard deviations.

Solution This could be done using the F-test:

$$F = \frac{s_x^2}{s_y^2}$$

$$F = \frac{0.0009}{0.0004} = 2.25$$

The value of F-critical from Table 5 for 7 degrees of freedom for the denominator and 9 degrees of freedom for the numerator is 3.68. Hence, there is no significant difference between the two standard deviations.

5. Comparison of the Experimental Mean with the Accepted Mean

Solution This purpose is accomplished using the t-test:

$$t = \frac{\bar{x}_1 - \bar{x}_2}{\sqrt{\dfrac{s_1^2}{n_1} + \dfrac{s_2^2}{n_2}}}$$

$$t = \frac{50.25 - 50.21}{\sqrt{\dfrac{0.02^2}{8} + \dfrac{0.03^2}{10}}} = 3.38$$

Comparing the calculated t-experimental with the tabulated values of t-critical in Table 6 for $(8 + 10 - 2) = 16$ degrees of freedom shows that the calculated value exceeded t-critical at the 99% confidence level. Hence, the difference between the two means is significant at the 99% confidence level. That means that there is less than 1 chance out of 100 that a difference this large could occur between the two water samples.

4

Physical Properties of Water

Jean-Yves Bottero

CNRS and University of Aix-Marseille, Aix-en-Provence, France

I. INTRODUCTION

Water is essential to life, to all human activities, and to industry. The evolution of society and the growth of cities and economies led to an increase in the pollution of our waters. The physical properties of water, including its impurities, need to be known. The impurities must be eliminated in order to obtain environmentally acceptable drinking water for people and ultrapure water for silicon industries and food industries. Those three levels of quality can be obtained only if the analytical methods are adequate.

II. PHYSICAL PROPERTIES OF PURE WATER

A. Molecular Structure

A water molecule is formed of one H atom and two O atoms linked by covalent bonds. Water is an unusual liquid. It has a very high boiling point and a high heat of vaporization. The maximum density is at 4°C, and the water expands upon freezing. It has a very high surface tension and is a very good solvent for salts and polar molecules. These properties are a consequence of the dipolar character of H_2O. The electron cloud around H_2O results from the hybridization of s and p electrons, yielding two bonding orbitals between the O and the two H atoms and two nonbonding sp^3 orbitals on the O. The molecule has a high negative charge density near O and high positive charge density near H (1) (Fig. 1).

In the vapor phase the equivalent size of a water molecule is 3.3 Å, or 0.33 nm (1 Å = 10^{-10} m). The molecules move at high speed, and their translation energy is so high that during collisions the van der Waals forces are inefficient for creating bonds. The vapor expands when the temperature increases.

In the liquid phase the molecules are close together and occupy a volume of 29.7 $Å^3$, indicating a porosity of 36.7%. Liquid water forms a heterogeneous fluid structure of water molecules, clusters of molecules, and H^+ and OH^- molecules. The molecular structure influences the density and the viscosity. Liquid water consists of an ice-tridymite structure, a quartzlike structure, and a close-packed ammonia-like structure (2).

In the ice structure each O atom is bonded to four others by hydrogen bonds in a tetrahedral configuration (Fig. 2). The H atoms in the O–H $\cdots$ O bond are no longer 0.96 Å from the O atom but may be either 0.99 Å or 1.77 Å away. The volume per H_2O molecule in ice is 32.3 $Å^3$. The volume per gram-molecule is 19.56 ml, and the resulting density is 0.92.

The number of broken H bonds increases with temperature up to 50% at 40°C. Broken hydrogen bonds are responsible to a considerable extent for its high dielectric constant ($D = 78.55$ at 25°C), contributing to the fact that water is the best solvent for polar compounds.

B. Density and Elasticity

The density of water at 4°C is 1 g per ml. Table 1 gives density values at different temperatures.

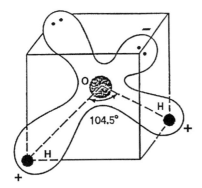

Fig. 1 Electron clouds of the H$_2$O molecule.

Water is assumed to be an incompressible fluid. Nevertheless it has a modulus of elasticity of about 300,000 psi, meaning a volumetric decrease of about 0.000048 for each added atmosphere of pressure.

C. Viscosity

The viscosity of a fluid is the proportionality factor in the expression for the intensity of viscous shear at a point in the moving fluid:

$$\tau = \mu \, \frac{d\nu}{ds}$$

where τ is the shear per unit area of surface normal to the s-direction, $d\nu/ds$ is the maximum velocity gradient at the point, with the s-direction representing the direction in which the maximum occurs, ν is the kinematic viscosity ($=\mu/\rho$), and μ is the absolute viscosity (force · time)/length^2.

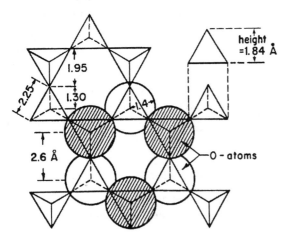

Fig. 2 Tetrahedral configuration of the ice crystal.

The unit of viscosity is the poise (dyne = s/cm^2). The viscosity of pure water is a function of the temperature. The viscosity of pure water at atmospheric pressure, as a function of the temperature, is presented in Table 2. The intensity of viscous shear corresponds to the internal energy loss. The velocity gradient and the shear intensity are important in flocculation, settling, and filtration processes.

D. Vapor Pressure and Relative Humidity

The vapor pressure of a liquid is the pressure of the liquid vapor in contact with the liquid at which vapor molecules condense as fast as they evaporate from it. Vapor pressure is a function of temperature. Table 3 shows the variation of the vapor pressure of pure water.

E. Surface Tension

Water molecules are held together by attractive forces. Beyond a certain radius, $R_{critical}$, the attractive forces become negligible. Molecules closer than $R_{critical}$ to a free surface are attracted to the interior of the liquid by the resultant force. The potential energy per unit surface area is the surface energy. The numerical value of the surface energy is equal to the surface tension of the liquid. The surface tension decreases with increasing temperature. Table 4 shows different values of the surface tension of pure water in contact with air at various temperatures. The interfacial tension between water and another liquid that is immiscible with water is approximately equal to the difference between their surface tensions.

Gibbs' rule shows that the addition of a solute to a solvent leads to different behaviors, depending on the surface tension. If the solute at a low concentration has a weak surface tension it will be concentrated at the surface of the solvent and lower the surface tension of the solution. On the contrary, large amounts of a solute of high surface tension will concentrate away from the surface and will not increase the surface tension of the solution. This phenomenon is of great interest in the treatment of surface water and wastewater.

III. PHYSICAL PROPERTIES OF IMPURE WATER

The impurities of water come as solids, liquids, or gases, and they are dispersed in three states: particulate, colloidal, and dissolved (3). The distinction between these three states is based on size. Generally, the par-

Table 1   Relative Density of Pure Water at Atmospheric Pressure

Temp., °C	0	4	10	15	20	25	30	100
ρ	0,99987	1,00000	0,99973	0,99913	0,99823	0,99707	0,99567	0,95838

ticulate state is defined as any size larger than 0.1 μm, colloidal as sizes lower than 0.1 μm and larger than 1 nm. The dissolved state corresponds to sizes lower than 1 nm (4). Figure 3 gives the size ranges for various kinds of particulates.

A.   Particulates in Natural Waters

Mineral particulates include carbonates, oxides, and clays (Si–Al particles). Organic particulates are generally colloidal humic and fulvic matter. They are the product of the decay and leaching of organic debris and litter found in the water source. They frequency impart a color to the water.

B.   Inorganic, Biotic, and Organic Particles

Natural weathering of minerals produces a variety of particles. All such particles affect the quality of water. In colloidal forms they can affect the biological quality, because bacteria and viruses fall in this size range. Even if the mass of colloids is minor, their high specific surface area allows the adsorption of contaminants. Natural organic matter, viruses, metals, and other toxic substances are adsorbed mainly on the surfaces of the colloids. Bacteria and viruses may be adsorbed on mineral or organic materials, which may "shield" pathogens from disinfection during water treatment.

The presence of particulates is generally characterized by the presence of turbidity. Turbidity is a measure of the scattering of light by particulates. The scattering depends on the concentration, size, shape, wavelength of the incident beam, scattering angle, and optical properties of the particulates (5). The light-scattering behavior of particles whose diameter is the same as the wavelength of the incident light is complex. This relates to Mie theory, which predicts that light scattering as a function of particle size reaches a maximum when the diameter is roughly equal to the wavelength. Turbidity is sensitive to the optical index of the medium, which varies with the number of particles, and to the presence of colored organics. The effect of temperature on the optical index is small at the temperature of surface waters. The effect of dissolved compounds is to increase the optical index.

The turbidimeter design criteria stipulated in the *Standard Methods* (6) call for a tungsten-filament lamp and a spectral response for the detector of between 400 and 600 nm. Thus, turbidity measurements should be most sensitive to particles that are 400–600 nm in diameter.

C.   Particle Size Distributions

The turbidity measurement is a rough approximation of the size distribution. The methods for measuring size differ with the size range of the particulates. In the particle size range of roughly 1 μm to 1 mm, two methods exist: particle counting and light scattering. Figure 4 shows two graphs, of the number–size relationship and the volume–size relationship. Note that there are 10^5 particles of 1-μm size, whereas the maximum in volume is centered on 10 μm.

The number–size distribution is measured classically, by using particle counters and the electric sensing

Table 2   Viscosity of Water

Temp., °C	μ (poises)	Temp., °C	μ (poises)	Temp., °C	μ (poises)
0	0.01792	35	0.00723	70	0.00406
5	0.01519	40	0.00656	75	0.00380
10	0.01308	45	0.00599	80	0.00357
15	0.01140	50	0.00549	85	0.00336
20	0.01005	55	0.00506	90	0.00317
25	0.00894	60	0.00469	95	0.00299
30	0.00801	65	0.00436	100	0.00284

Source: Bingham and Jackson, 1917 (Ref. 3).

Table 3 Vapor Presure of Pure Water at Atmospheric Pressure

Temp., °C	0	10	15	20	25	30	40	60	100
Vapor pressure, atm	0.006	0.012	0.0168	0.0231	0.0313	0.0419	0.0728	0.1965	1.0
Latent heat, ΔH, kcal/mole	10.73	10.63	10.58	10.53	10.48	10.43	10.33	10.13	9.70
Free energy, $\Delta F°$, kcal/mole	2.77	2.48	2.34	2.20	2.054	1.91	1.63	1.08	0

zone method, or light blockage (7). Typically, the analyzed size range is from 1 μm to 1 mm.

The volume–size distributions are measured via light scattering, mainly photon correlation spectroscopy. This method is based on the time-variable aspects of light scattering that originate in the motion of suspended particles. This technique is used for particles greater than 2–3 nm and smaller than 1 μm. For particles whose size is close to the light wavelength, a rigorous application of Mie theory is required.

The photon correlation spectroscopy technique involves an analysis of the time-varying component of light scattering. The instantaneous intensity of light scattered by a suspension of colloidal material depends on the position and orientation of each scatterer. The position and orientation of small colloids fluctuate because of Brownian motion and produce fluctuations in the scattered intensity. These fluctuations may be correlated over small intervals of time, and the autocorrelation function $G(\tau)$ contains information on the dynamics of the scatterer:

$$G(\tau) = 1 + B \exp(-\Gamma \tau)$$

where τ is the lag time at which the autocorrelation is calculated, B is a function of the detection system, and Γ is the decay constant:

$$\Gamma = 2Dq^2$$

where D is the diffusion coefficient and q is the scattering vector:

$$q = \left(\frac{4\pi}{\lambda}\right) n \sin \frac{\Phi}{2}$$

where n is the refractive index of the medium and Φ is the scattering angle.

The diffusion coefficient is given as

$$D = \frac{kT}{f}$$

where k is Boltzmann's coefficient, T is the absolute temperature, and f is the friction factor.

For spherical particles, D can be calculated using the Stokes–Einstein equation:

$$D = \frac{kT}{3\pi \mu d_p}$$

The origin of D is the considerable velocity of translation of colloids. They are subject to diffusion (Brownian diffusion) and move in the liquid. They conform approximately to the equation of simple kinetic theory:

$$pV_m = \tfrac{1}{3}Nmu^2 = RT$$

where p is the osmotic pressure (dynes/cm^2), V_m is the volume (ml) occupied by 1 mole of the particles, N is Avogadro's number (6.06×10^{23}), m is the mass of each particle (g), u is the mean velocity of the particles, in cm/s, R is the gas constant, and T is the temperature, in K.

Colloids diffuse with a rate of diffusion limited by the liquid's drag on the particles. The rate of diffusion per unit area follows Fick's law (8):

$$c\frac{ds}{dt} = -D\frac{dc}{ds}$$

where $c(ds/dt)$ is the quantity of particles per second diffusing through 1 cm^2 of boundary in the direction of s, ds/dt is the diffusion velocity, $-dc/ds$ is the concentration gradient, and D is the diffusion coefficient.

For spherical particles that are large in comparison with the mean free path of the fluid, D can be calculated using the Stockes–Einstein equation:

Table 4 Surface Tension γ of Pure Water in Contact with Air

Temp., °C	0	10	20	30	40	50	60	70	80	100
γ, dyne/cm	75.6	74.22	72.75	71.18	69.56	67.91	66.18	64.4	62.6	58.9

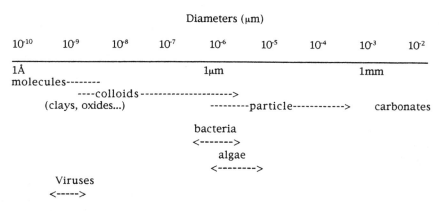

Fig. 3 Size ranges for various kinds of particles.

$$D = \frac{kT}{3\pi\mu d_p}$$

where μ is the fluid viscosity and d_p is the particle diameter.

D. Coagulation-Flocculation of Suspensions

In natural waters the particles and colloids make contact directly through their surfaces or the adsorbed species. These contacts provoke the formation of aggregates. Their behavior in water depends on the hydrodynamics, the forces developed between the particle surfaces, and so on.

The aggregates form particles of varying density or differing morphology. These parameters affect the settling of these aggregates. The porosity of such aggregates is an important factor, because it affects the volume of sediments.

The theory of Brownian aggregation has been developed by Smoluchowski. It describes the kinetics of the formation of aggregates. The rate at which two particles with masses m_i and m_j and concentrations n_i and n_j collide is given by $n_i n_j \beta_{ij}$, where β_{ij} is the coagulation kernel (9). New particles of mass $(m_i + m_j)$ are formed at a rate of $\alpha n_i n_j \beta_{ij}$, where α is the stickiness coefficient:

$$\frac{dn_i}{dt} = 0.5\alpha \sum \beta_j n_{i-j} n_{i-j} - \alpha N\iota \sum B_{\eta\varphi} nj(1 + d_{ij})$$

where d_{ij} is 1 if $i = j$ and 0 otherwise. The chemical effects are included in α and the physical ones in β_{ij}.

Particle contacts are due to:

Brownian motion
Shear (laminar and turbulent)

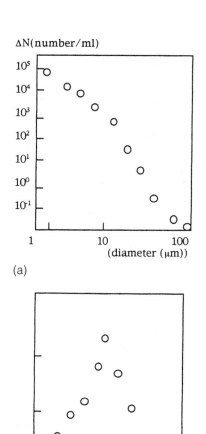

(a)

(b)

Fig. 4 Example of a size distribution of particles. (a) Number–size (μm) distribution; (b) volume–size (μm) distribution.

Differential sedimentation

The total collision kernel β_{ij} is calculated as the sum of the collision kernels describing each of these processes:

$$\beta_{ij} = \beta_{ds,ij} + \beta_{sh,ij} + \beta_{Br,ij}$$

where the subscripts ds, sh, and Br denote the terms for differential sedimentation, shear, and Brownian motion, respectively. For particles greater than 5 μm, the term $\beta_{Br,ij}$ can be neglected.

The coagulation kernels are calculated for solid spheres with various hydrodynamic models. The first simple hydrodynamic model corresponds to the use of fluid flow. This flow is known as rectilinear flow of differential sedimentation:

$$\beta_{ds,ij} = p(r_i + r_j)^2 (w_j - w_i)$$

where r_i and r_j are the radii of aggregates of particles of size i and j and w_j and w_i are the settling rates of the aggregates j and i, respectively.

For curvilinear flow:

$$\beta_{ds,ij} = 0.5\pi(r_i)^2 (w_j - w_i)$$

where $rj > ri$ (10).

For the rectilinear case of turbulent shear:

$$\beta_{sh,ij} = 1.3 \left(\frac{\epsilon}{\nu}\right)^{0.5} (r_i + r_j)^3$$

where ϵ is the energy dissipation rate and ν is the kinematic viscosity.

It is possible to derive a curvilinear version of the turbulent shear coagulation kernel:

$$\beta_{sh,ij} = \frac{9.8p^2}{(1 + 2p)^2 (\epsilon/\nu)^{0.5}(r_i + r_j)^3}$$

where $p = r_i/r_j$ and $r_j > r_i$.

The aggregates possess various properties, such as density, porosity, and morphology, that affect their removal by sedimentation or filtration. They are also important in water treatment in determining the dewatering and rheological characteristics of sludges, the permeation rate of water through membranes with cake deposits (11), and the development of head loss in granular media filters (12).

The study of floc morphology (13) shows that the surface is fractal in the case of bacteria aggregates. The mass distribution within a floc follows a power law (14) that corresponds to fractal theory (15). Those objects are self-similar. An important characteristic of self-similar fractal objects is that their morphology is invariant with an increase in magnification.

Self-similarity in fractals produces power law relationships between length scale and properties such as density, surface area, and connectivity. Many studies on natural or industrial flocs have shown that floc density is a function of floc size (13,16–20,22).

The relation between the mass density $\rho(r)$ and the radius of the aggregates or a sphere of radius r centered at some point in the flocs varies as

$$\rho(r) \sim r^{D-3}$$

where $1 \le D \le 3$.

The flocs come in a great variety of structures. That means that D takes different values. Very dense aggregates have a fractal dimension close to 3. An open-configuration aggregate is characterized by smaller fractal dimensions. The various physical mechanisms of aggregation lead to different D values. Simulations of such aggregations have been made by different authors (23,24).

Diffusion-limited aggregation (DLA) aggregate growth is produced by diffusive transport and irreversible adhesion. In the case of "particle–cluster aggregation," the fractal dimension is approximately 2.5 (25). In the case of collisions between clusters or aggregates, the "cluster–cluster aggregation" model, the fractal dimension is around 1.75.

If the trajectories are ballistic, the fractal dimension is larger. In the particle–cluster model, the fractal dimension approaches 3. In the cluster–cluster aggregation model, the fractal dimension increases from 1.75 to 1.95. If the probability of sticking decreases, the fractal dimension increases. Under these restrictive conditions and in a DLA model, the clustering of clusters leads to a fractal dimension between 1.9 and 2.1.

Diffusion transport tends to create aggregates with lower fractal dimensions and more fragile or porous flocs.

Breakup due to nondiffusive transport is at the origin of more compact flocs (26) with fractal dimension values between 2.1 and 2.5. Experimentally, some authors have shown that even in DLA clustering of clusters, some restructuring exists and the fractal dimension is larger than 1.75 (13,27). The fractal dimensions of aggregates formed in laminar and turbulent regimes have a fractal dimension larger than 2 (28).

The porosity of the flocs can be calculated from the mass distribution within a floc or from the number of units (29):

$$f = 1 - Az^{-g}$$

where f is the porosity, g is between 0.3 and 0.5, and

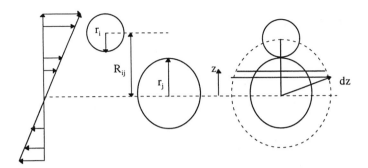

Fig. 5 Orthokinetic flocculation and differential settling.

IV. SEDIMENTATION

Sedimentation refers to the separation of particles from water by the force of gravity. The settling velocity is affected primarily by particle size, shape, and density and water viscosity. Stoke's law describes the terminal settling velocity of a sphere in a laminar-flow regime:

$$U_t = \frac{g(\rho_p - \rho)d_p^2}{18\mu}$$

where U_t is the settling velocity, g is the gravitational force constant, μ is the kinematic viscosity, ρ_p is the particle density, ρ is the liquid density, and d_p is the particle diameter.

The settling of fractal aggregates cannot be represented by Stoke's law.

Differential settling is at the origin of the flocculation in many natural suspensions if the particles or flocs settle at different velocities. Faster settling particles may collide with slower-settling particles. The aggregates will settle faster because of their increased mass, and they might experience further collisions and flocs. The sketch of orthokinetic flocculation in Fig. 5 is similar to a differential settling with a differential velocity $(v_i - v_j)$ in the direction of gravity (x-direction). Assuming that the particles settle according to Stoke's law and the $r_i = r_j = r_s$ (constant particle density), Smoluchowski's equation follows:

$$\frac{dN_{ij}}{dt} = \frac{N_i N_j \cdot 2\pi g(\rho_s - \rho)(r_i + r_j)^3(r_i - r_j)}{9\mu}$$

where N_{ij} is the number of particles of size ij, g is the gravity, ρ_s is the density of flocs, ρ is the density of the solvent, and μ is the kinematic viscosity of the solvent. This equation can be developed to allow the creation of k particles and their possible collisions with i particles and j particles and so on (30). It cannot be developed for $i = j$ or for an initially monodisperse suspension.

The settling velocities associated with settling flocculation may be relatively high. Marine snow (31) settles at 74–39 m/day. This value is in the range of industrial settling in conventional water treatment plants (32,33). Both systems correspond to the settling of flocs with fractal structures.

V. SOLUBLE IMPURITIES

Soluble impurities are elements dispersed in water as single molecules or as ions. Soluble polar or mineral compounds are held together in the solid state by ionic bonds. The acidic elements in the crystal have captured electrons from the metallic elements, resulting in negatively charged acidic radicals and positively charged metals. Ions of like charge repel each other, and those of unlike charge are attracted by coulombic force:

$$F = \frac{z^1 z^2 \epsilon^2}{Dr^2}$$

where F is the force in dynes, z^1 and z^2 are the valences

Table 5 Hydration Number for Some Ions in Water at Normal Pressure and Temperature Conditions

	H^+	Cs^+	K^+	Na^+	Li^+	Cl^-
Moles of H_2O per ion	1	4.7	5.4	8.5	14	4

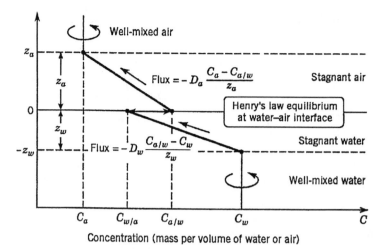

Fig. 6 Geometry and steady-state hypothesis for gas transfer across a water–gas interface. (From Ref. 34.)

of the two ions, ϵ is the charge of an electron, r is the distance in cm between the ions, and D the dielectric constant of the medium.

When a NaCl crystal is immersed in water, the high value of D reduces the bond energy of 120 kcal/mole in the crystal to 2.33 kcal/mole. Then the water molecules dissociate the crystal into ions.

Water molecules are attached to ions, leading to the hydration of ions. Table 5 gives values of the hydration number for some ions.

VI. GAS TRANSFER ACROSS A WATER–GAS INTERFACE

The rate of mass transfer of a substance across a water–gas boundary can be described as a diffusion film model. The two bulk phases are well mixed at a small distance from the interface. The flux through the film of thickness z is given by Fick's law:

$$F = -D \frac{dc}{dz}$$

where F is in mol-cm^{-3}, z is in cm, and D is in cm^2-s^{-1}. At a steady state,

$$F = F_a = F_w$$

that is, the flux in the air film (F_a) is equal to the flux in the water film (F_w). So, applying the preceding equation, we get

$$F = -\frac{D_a}{z_a}(c_a - c_{a/w}) = \frac{-D_w}{z_w}(c_{w/a} - c_w)$$

where a and w refer to air film and water film and $c_{a/w}$ and $c_{w/a}$ refer to the concentration in the air film at the air–water interface and the concentration in the water film at the water–air interface.

The ratio $c_{w/a}$ to $c_{a/w}$ is the Henry factor H (in the absence of reactions):

$$H = \frac{c_{w/a}}{c_{a/w}} = K_H RT$$

where K_H is the equilibrium constant. So

$$F = \frac{D_w}{z_w(c_w - c_{w/a})} = \frac{D_a}{z_a} \frac{c_{w/a}}{H - C_a}$$

Finally, the flux through the water film is

$$F = \frac{1}{V_w^{-1} + V_a^{-1}H}(c_w - c_a H)$$

where $V_w = D_w/z_w$ and $V_a = D_a/z_a$ are the transfer coefficients (cm-s^{-1}) or the mass transfer velocities through water and air, respectively. Figure 6 summarizes the geometry of the gas transfer.

REFERENCES

1. RA Horne. Marine Chemistry. New York: Wiley Interscience, 1969.
2. JD Bernal, RH Fowler. A theory of water and ionic solution. J Chem Phys 1:515–520, 1933.
3. TR Camp. Water and Its Impurities. New York: Reinhold, 1963.
4. J Buffle, W Stumm. General chemistry of aquatic systems. In: J Buffle, RR de Vitre, eds. Chemical and Bi-

ological Regulations of Aquatic Systems. Chelsea, MI: Lewis, 1994.

5. Kerker. The scattering of Light and Other Electromagnetic Radiation. New York: Academic Press, 1969.

6. American Public Health Association (APHA), American Water Works Association (AWWA), and Water Environment Foundation (WEF). Standard Methods for the Examination of Water and Waste Water. 18th ed. Washington, DC: American Public Health Association, 1992.

7. G Barth, ed. Modern Methods of Particle Size Analysis. vol 73. New York: Wiley, 1984.

8. HT Kruyt, HS van Klooster. Colloids. New York: Wiley, 1930.

9. CR O'Melia. Coagulation and Flocculation. In: WJ Weber Jr, ed. Processes for Water Quality Control. New York: Wiley Intersciences, 1972, pp 61–110.

10. AL Alldredge, P McGillivary. Deep-Sea Res 38:431–443, 1991.

11. S Veerapaneni, M Wiesner. Particle deposition on an infinitely permeable surface. Dependance of deposit morphology on particle size. J Colloid Interface Sci 162:110–122, 1994.

12. S Veerapaneni, M Wiesner. Hydrodynamics of fractal aggregates with radially varying permeability. J Colloid Interface Sci 177:45–57, 1996.

13. A Thill, A Veerapaneni, B Simon, MM Wiesner, JY Bottero. Fractal analysis of big aggregates by confocal microscope. J Colloid Interface Sci 204:357–362, 1998.

14. F Zartarian, C Mustin, G Villemin, Ait-Ettager, A Thill, JY Bottero. 3D Morphology of bacteria flocs. Langmuir 13:35–42, 1997.

15. BB Mandelbrot. Fractals: Form, Chance and Dimension. San Francisco: WH Freeman, 1977.

16. AL Lagvankar, RS Gemmel. A size-density relationship for flocs. J Am Wat Wks Ass 9:1040–1050, 1968.

17. N Tambo, Y Watanabe. Physical aspect of flocculation. Process—1. Fundamental treatise. Water Res 13:429–439, 1979.

18. DN Sutherland. A theoretical model of floc structure. J Colloid Interface Sci 25:373–380, 1967.

19. JY Bottero, M Axelos, D Tchoubar. Structure of silica–Al flocs by small angle x-ray scattering. Langmuir 117:47–53, 1991.

20. GA Jackson, SE Lochman. In: HP Van Leeuwen, J Buffle, eds. Environmental Particles. vol 2. Chelsea, MI: Lewis, 1993, pp 387–414.

21. A Thill, JY Bottero. Unpublished results, 1998.

22. K Wong, B Cabane, P Somasundaran. Highly ordered microstructure of flocculated aggregates. Colloids Surfaces 30:355–360, 1988.

23. P Meakin, R Jullien. The effects of restructuring on the geometry of clusters formed by diffusion-limited, ballistic, and reaction-limited cluster–cluster aggregation. J Phys Chem 89:246–250, 1988.

24. R Jullien, R Botet. Aggregation and Fractal Aggregates. World Scientific, 1987.

25. R Jullien, H Kolb. J Phys A 45:L977, 1985.

26. R Jullien, H Kolb. J Phys A 17:L639, 1984.

27. Y Adachi. Dynamic aspects of coagulation and flocculation. Advances Colloids Interface Sci 56:1–31, 1995.

28. B Lartiges, JY Bottero, L Michot. Structure of aggregated silica particles by Al polycations. Langmuir 7:1985–1989, 1996.

29. R Hogg, RC Klimpel, DT Ray. Structural aspects of floc formation and growth. Proceedings of the Engineering Foundation Conference on Flocculation, Sedimentation and Consolidation. Sea Island, GA, 1985.

30. LA Spielman. Hydrodynamic Aspects and Flocculation. In: KJ Ives, eds. The Scientific Basis of Flocculation. Alphen aan den Rijn, The Netherlands: Sijthoff and Noodhoff, 1978, p 63.

31. GA Jackson. Coagulation of marine algea. In CP Huang, CR O'Melia, JJ Morgan, eds. Advances in Chemistry Series 244. Washington DC, 1995, pp 203–217.

32. Memento Degremont. Tome 1. 1996.

33. Treatment Process Selection for Particle Removal, AWWARF Ed, Denver: 1997.

34. RP Schwarzenbach, PhM Gschwend, D Imboden. Environmental Organic Chemistry. New York: Wiley Interscience, 1993.

5

Chemical Parameters

Hilda Ledo de Medina

University of Zulia, Maracaibo, Zulia State, Venezuela

I. ACIDITY

A. Introduction

Common causes for acidic surface water are acid rainfall due to atmospheric carbon dioxide and other airborne pollutants, runoff from mining spoils, and decomposition of plant materials. Acidic groundwater can be caused by the same factors but is mostly controlled naturally by the equilibrium relationship with surrounding minerals.

Water quality can become acidified by bacteria, algae, chloroform (from chlorination), nitrates, and, not least, dangerous, heavy metals (cadmium and aluminum). These pollutants may come from factory discharges, agriculture, sewer pipes, water mains, and acid rain. Acid waters are of concern because of their corrosive characteristics and the expense involved in removing or controlling the corrosion-producing substances.

The corrosive factor in most waters is carbon dioxide, but in many industrial wastes it is mineral acidity. Carbon dioxide is a normal component of all natural water; it is an end product of both aerobic and anaerobic bacterial oxidation. It may enter surface waters by absorption from the atmosphere. Corrosion is a natural process involving chemical or electrical degradation of metals in contact with water. Acidic water with pH values in the range of 6–7 is more corrosive to the metals used in plumbing systems than is alkaline water. The rate of corrosion will vary depending on the acidity of the water, its electrical conductivity, oxygen concentration, and temperature.

Mineral acidity is present in many industrial wastes, particularly those of the metallurgical industry, and some comes from the production of synthetic organic materials. Acid mine drainage is one of the most severe environmental problems associated with mining. When water comes into contact with pyrite in coal and the rock surrounding it, chemical reactions take place that cause the water to gain acidity and to pick up in solution iron, manganese, and aluminum. Water that comes into contact with coal has a characteristic orange-red, yellow (sometimes white) color. The metals stay in solution beneath the earth due to the lack of oxygen. When water emerges from the mine, it reacts with the oxygen in the air or dissolved in the stream and deposits iron, manganese, and aluminum on rocks and the streambed. Each of the chemical characteristics of acid mine drainage (AMD) is toxic to fish and aquatic insects in moderate concentrations. At high concentrations all plant life is killed. Due to the oxidation of iron sulfides such as pyrite and pyrrhotite, acidity is produced that leads to a subsequent leaching of potentially toxic metal ions into the receiving waters (1).

B. Principle Behind the Method

Sample is titrated with standard 0.02 NaOH to pH 8.3, and results are reported as mg $CaCO_3$/L.

C.   Scope and Application

The method is applicable to drinking and surface waters, domestic and industrial wastes, and saline waters. The range is 5–500 mg/L as $CaCO_3$.

D.   Interferences

A. Dissolved gases contributing to acidity or alkalinity, such as CO_2, hydrogen sulfide, or ammonia, may be lost or gained during sampling, storage, or titration. Minimize such effects by titrating to the endpoint promptly after opening a sample container, avoiding vigorous shaking or mixing. Protect the sample from the atmosphere during titration, and let the sample become no warmer than it was at collection.

B. Suspended matter present in the sample, or precipitates formed during the potentiometric titration, may cause a sluggish electrode response. This may be offset by allowing a 15–20-s pause between additions of titrant or by slow, dropwise addition of the titrant as the endpoint pH is approached. Do not filter, dilute, concentrate, or alter sample.

C. Residual free available chlorine in the sample may bleach the indicator. Eliminate this source of interference by adding 1 drop of 0.1 M sodium thiosulfate ($Na_2S_2O_3$).

E.   Sampling Procedure and Storage

Collect samples in polyethylene or borosilicate glass bottles and store at a low temperature (4°C). Fill bottles completely and cap tightly. Because waste samples may be subject to microbial action and to loss or gain of CO_2 or other gases when exposed to air, analyze samples without delay, preferably within 1 d (2). If biological activity is suspected, analyze within 6 h. Avoid sample agitation and prolonged exposure to air.

F.   Apparatus

Electrometric titrator: Use any commercial pH meter or electrically operated titrator that employs a glass electrode and can be read to 0.05-pH units. Standardize and calibrate according to the manufacturer's instructions.

Magnetic stirrer and stirring bars.

Burettes, avoid borosilicate glass.

G.   Reagents

1.   Carbon Dioxide–Free Water

Prepare all stock and standard solutions and dilution water for the standardization procedure with deionized water that has been freshly boiled for 15 min and cooled to room temperature. The final pH of the water should be 6.0 or greater, and its conductivity should be less than 2 μmhos/cm.

2.   Potassium Hydrogen Phthalate Solution 0.01 N

Weigh 2.0423 $\pm$ 0.0001 g primary standard $KHC_8H_4O_4$, and dilute to 1000 ml with deionized water.

3.   Standard Sodium Hydroxide Titrant, 0.1 N

Dissolve 40.0 g of NaOH pellets in CO_2-free deionized water. Store in tightly stoppered pyrex bottle protected by a soda-lime tube.

4.   Standard Sodium Hydroxide Titrant, 0.02 N

Dilute 200 ml 0.1 N NaOH to 1000 ml and store in a polyolefin bottle protected from atmospheric CO_2 by a soda-lime tube or tight cap. Standardize by titrating 15.00 ml of $KHC_8H_4O_4$ solution 0.01 N, using a 50-ml burette. Titrate to the inflection point, which should be close to pH 8.3.

$$\text{Normality} = \frac{A \times B}{204.2 \times C}$$

where:

A = g $KHC_8H_4O_4$ weighed into 1-L flask,
B = ml $KHC_8H_4O_4$ solution taken for titration
C = ml NaOH solution used

Use the measured normality in further calculations, or adjust to 0.1000 N; 1 ml = 1.00 mg $CaCO_3$.

5.   Hydrogen Peroxide, H_2O_2, 30%

6.   Sodium Thiosulfate, 0.1 M

Dissolve 25 g $Na_2S_2O_3 \cdot 5H_2O$ and dilute to 1000 ml with distilled water.

H.   Procedure

1.   Sample Preparation

1. If sample is known or suspected to contain hydrolyzable metal ions or reduced forms of poly-

valent cations, it is necessary, to ensure oxidation of such substances, to pipette a suitable sample into titration flasks. Measure the pH. If the pH is above 4.0, add 5-ml increments of 0.02 N sulfuric acid (H_2SO_4) to reduce the pH to 4 or less. Remove electrodes. Add 5 drops 30% H_2O_2 and boil for 2–5 min. Cool to room temperature and titrate with standard alkali according to the next procedure ("Potentiometric Titration Curve").

2. If free residual chlorine is present, pipette a suitable sample into titration flasks. Add 0.05 ml (1 drop) 0.1 M $Na_2S_2O_3$ solution, or destroy with ultraviolet radiation, and titrate according to the next procedure ("Potentiometric Titration Curve").

2. Potentiometric Titration Curve

Pipette a sample aliquot into a titration vessel (100 ml). Immerse the electrodes and measure the sample pH. Add standard alkali in increments of 0.5 ml or less such that a change of less than 0.2 pH units occurs per increment. Continue adding titrant and measure pH until pH 9 is reached. Construct the titration curve by plotting observed pH values versus cumulative milliliters titrant added. A smooth curve showing one or more inflections should be obtained. A ragged or erratic curve may indicate that equilibrium was not reached between successive alkali additions. Determine acidity relative to a particular pH from the curve (3).

3. Automatic Potentiometric Titration to pH 8.3

Prepare sample and titration assembly as specified in the previous subsection. Titrate to a preselected endpoint pH (3.7 or 8.3) without recording intermediate pH values. As the endpoint is approached, make smaller additions of alkali and be sure that pH equilibrium is reached before making the next addition (3–5).

I. Calculation

Acidity, as mg $CaCO_3$/L

$$= \frac{[(A \times B) - (C \times D)] \times 50.000}{\text{ml sample}}$$

where:

A = ml NaOH titrant used
B = normality of NaOH
C = ml H_2SO_4 used
D = normality of H_2SO_4

II. ALKALINITY

A. Introduction

Alkalinity refers to the capability of water to neutralize acid. This is really an expression of buffering capacity. A buffer is a solution to which an acid can be added without changing the concentration of available H^+ ions (without changing the pH) appreciably. It essentially absorbs the excess H^+ ions and protects the water body from fluctuations in pH. So, generally, soft water is much more susceptible to fluctuations in pH from acid rains or acid contamination. The presence of calcium carbonate or other compounds, such as magnesium carbonate, contribute carbonate ions to the buffering system.

Alkalinity is significant in many uses and treatments of natural waters and wastewaters. Because the alkalinity of many surface waters is primarily a function of carbonate, bicarbonate, and hydroxide content, it is taken as an indication of the concentration of these constituents. The measured values also may include contributions from borates, phosphates, silicates, or other bases if these are present. Alkalinity measurements are used in the interpretation and control of water and wastewater treatment processes.

B. Principle Behind the Method

The phenolphthalein alkalinity is a measure of the alkalinity fraction, which is contributed by the hydroxide and half of the carbonate. The total alkalinity is the contribution due to all bicarbonates, carbonates, and hydroxides present in the sample.

The phenolphthalein alkalinity and the total alkalinity are determined by potentiometric titration of an unfiltered sample aliquot with a standard solution of strong acid to pH 8.3 and 4.5, respectively. They are expressed as mg/L $CaCO_3$. Phenolphthalein or metacresol purple may be used for alkalinity titration to pH 8.3. Bromcresol green, methyl orange, or a mixed bromcresol green–methyl red indicator may be used for pH 4.5.

C. Scope and Application

This potentiometric titration method is applicable to all types of waters in the range 0.5–500 mg/L alkalinity as $CaCO_3$. The upper range can be extended by dilution of the original sample.

D. Interferences

A. Suspended matter present in the sample, or precipitates formed during the potentiometric titration, may cause a sluggish electrode response. This may be offset by allowing a 15–20-s pause between additions of titrant or by slow, dropwise addition of titrant as the endpoint pH is approached. Do not filter, dilute, concentrate, or alter sample.

B. Analysis should be started as soon as possible to prevent the loss of any dissolved gases. Avoid shaking or mixing the sample.

E. Sampling Procedure and Storage

Collect samples in polyethylene or borosilicate glass bottles and store at a low temperature (6). The sample container should be tightly capped as soon as the sample has been collected, and the alkalinity should be determined as soon as possible after opening the container in the laboratory.

F. Apparatus

An automatic potentiometric titrator, including a sampler, potentiometer, automatic burette, and digital printer

pH meter

G. Reagents

1. Sodium Carbonate Solution, Approximately 0.02 N

Dry 2–4 g primary standard Na_2CO_3 at 250°C for 4 h and cool in a desiccator. Weigh 1.0600 ± 0.1000 g, dissolve in CO_2-free distilled water, and dilute to 1 L. Do not keep longer than 1 week.

2. Stock Sulfuric Acid (0.1 N)

Dilute 2.8 ml concentrated H_2SO_4 to 1 L with distilled water.

3. Standard Sulfuric Acid 0.02 N

Dilute 200.00 ml 0.1000 N standard acid to 1000 ml with distilled or deionized water. Standardize by potentiometric titration of 15.00 ml 0.05 N Na_2CO_3 to pH of about 5. Lift out electrodes, rinse into the same beaker, and boil gently for 3–5 min under a watchglass cover. Cool to room temperature, rinse cover glass into

beaker, and finish titrating to the pH inflection point. Calculate normality:

$$\text{Normality, N} = \frac{A \times B}{53.00 \times C}$$

where:

A = g Na_2CO_3 weighed into 1-L flask
B = ml Na_2CO_3 solution taken for titration
C = ml acid used

Use measured normality in calculations or adjust to 0.020 N; 1 ml 0.020 N solution = 1.00 mg $CaCO_3$.

4. Phenolphthalein Solution, Alcoholic, pH 8.3 Indicator

Dissolve 1 g of phenolphthalein in 50 ml of alcohol and add 50 ml of water.

5. Methyl Orange Indicator

Dissolve 1 g of methyl orange in 1 L of water. Filter, if necessary.

6. Sodium Thiosulfate, 0.1 N

Dissolve 2.5 g $Na_2S_2O_3 \cdot 5H_2O$ in distilled water, add 0.5 ml $CHCl_3$, and dilute to 100 ml.

H. Procedure

1. General Discussion

The results obtained from the phenolphthalein and total alkalinity determinations offer a means for stoichiometric classification of the three principal forms of alkalinity present in many waters. The classification ascribes the entire alkalinity to bicarbonate, carbonate, and hydroxide, and assumes the absence of other (weak) inorganic or organic acids, such as silicic, phosphoric, and boric acids.

The choice of pH 8.3 as the endpoint for the first step in the titration corresponds to the equivalence point for the conversion of carbonate ion to bicarbonate ion, which means the first half of the carbonate fraction and for the hydroxide fraction:

$$CO_3^{2-} + H^- \rightarrow HCO_3^-$$

for the first half of the carbonate fraction

$$OH^- + H^+ \rightarrow H_2O \quad \text{for the hydroxide fraction}$$

The use of a pH of about 4.5 for the endpoint for the second step of the titration (the titration from pH 8.3

to pH 4.5) corresponds approximately to the equivalence point for the conversion of bicarbonate ion to carbonic acid:

$$HCO_3^- + H^+ \rightarrow H_2CO_3$$

for the bicarbonate fraction originally present in the sample, and for the second half of the carbonate fraction if the original pH was greater than 8.3 (7)

The phenolphthalein will be suitable for the conversion of the sodium carbonate, while the methyl orange will change color only when the carbonate has been neutralized completely (the conversion from sodium carbonate to bicarbonate and then from bicarbonate to carbonic acid). On the other hand, if phenolphthalein indicator is employed, the change from pink to colorless will take place when the carbonate converts to bicarbonate. The volume of acid consumed in order to titrate the sodium carbonate until the final point with orange methyl indicator is double the volume necessary when phenolphthalein is employed (the equivalent number of hydrogen ions used in the process is also double).

2. Titration Methods

1. Indicator Color Change: For samples whose initial pH is above 8.3, the titration is made in two steps. In the first step the titration is conducted until the pH is lowered to pH 8.3, the point at which phenolphthalein indicator turns from pink to colorless. The second phase of the titration is conducted until the pH is lowered to about 4.5, corresponding to the methyl orange endpoint. When the pH of a sample is less than 8.3, a single titration is made to a pH of 4.5.

Prepare the sample (100–200 ml) and titration vessel. Measure the pH. If the pH of the sample is above 8.3, add phenolphthalein indicator. Add standard acid solution 0.02 N until a change to colorless is observed. Then add methyl orange indicator and continue the acid addition until the indicator turns from yellow to orange (8,9). If the pH of the sample is less than 8.3, a single titration is made to a pH of 4.5. Prepare and titrate an indicator blank.

Color indicators may be used for routine and control titrations in the absence of interfering color and turbidity and for preliminary titrations to select the sample size and the strength of the titrant.

2. Potentiometric titration curve: Pipette a sample aliquot into a titration vessel. Immerse the electrodes and measure the sample pH. Add standard acid solution in increments of 0.5 ml or less, such that a change of less than 0.2 pH units occurs per increment. Continue adding titrant and measuring pH until pH 4.5 or lower is reached. Construct the titration curve by plotting observed pH values versus cumulative milliliters titrant added. A smooth curve showing one or more inflections should be obtained. A ragged or erratic curve may indicate that equilibrium was not reached between successive acid additions. Determine alkalinity relative to a particular pH from the curve (9,10).

3. Potentiometric titration to preselected pH: Determine the appropriate endpoint pH (8.3 or 4.5). Prepare the sample and the titration assembly. Titrate to the endpoint pH without recording intermediate pH values and without undue delay (10).

I. Calculations

The different species (OH^-, CO_3^{2-}, HCO_3^-) can be easily determined using the volumes indicated in Table 1.

Phenolphthalein alkalinity, mg $CaCO_3$/L

$$= \frac{50.000 \times N \times V_1}{\text{ml sample}}$$

Total alkalinity, mg $CaCO_3$/L

$$= \frac{50.000 \times N \times V_2}{\text{ml sample}}$$

where:

N = normality of standard acid
V_1 = ml of acid to pH 8.3
V_2 = ml of acid to pH 4.5

J. Criteria

For protection of aquatic life, the buffering capacity should be at least 20 mg/L.

III. HARDNESS

A. Introduction

Hardness in water is caused principally by the presence of calcium and magnesium salts. *Temporary hardness*

Table 1 Alkalinity Volume Relationships

Ions present	Volume of acid at pH 8.3	Volume of acid at pH 4.5	Volume to determine		
			OH^-	CO_3^{2-}	HCO_3^-
OH^-	A	0	A	0	0
HCO_3^-	0	B	0	0	B
CO_3^{2-}	A	$B = A$	0	$2A$	0
$CO_3^{2-} + OH^-$	A	$B < A$	$A-B$	$2B$	0
$CO_3^{2-} + HCO_3^-$	A	$B > A$	0	$2A$	$B-A$

A = volume of acid at pH 8.3; B = volume of acid from pH 8.3 to pH 4.5 ($B = V_2 - V_1$).

is caused by the bicarbonates and can be destroyed by boiling. This form of hardness can be quantitatively determined by simple titration with a standard acid. *Permanent hardness* is caused principally by the sulfates and chlorides of calcium and magnesium, and is not affected by boiling. *Total hardness* is represented by the combined temporary and permanent hardness and is frequently determined by titration with ethylenediaminetetraacetic acid (EDTA).

Not only does the procedure determine the total hardness of water, but it also shows the relative amounts of calcium and magnesium present. They react with soap and produce a deposit called *soap curd* that remains on the skin and clothes and, because it is insoluble and sticky, cannot be removed by rinsing. Soap curd changes the pH of the skin and may cause infection and irritation. It also remains on the hair, making it dull and difficult to manage. A ring around the bathtub and spotting on glassware, chrome, and sinks are constant problems in the presence of hard water.

Cooking with hard water can also be difficult, producing scale on spots. Some vegetables cooked in hard water lose color and flavor.

Hard water may also shorten the life of plumbing and water heaters. When water containing calcium carbonate is heated, a hard scale is formed that can plug pipes and coat heating elements. Scale is also a poor heat conductor. With increased deposits on the unit, heat is not transmitted to the water fast enough, and overheating of the metal causes failure. Buildup of deposits will also reduce the efficiency of the heating unit, increasing the cost of fuel.

General hardness is commonly expressed in parts per million (ppm) of calcium carbonate ($CaCO_3$). Carbonate hardness is the measure of bicarbonate (HCO_3^-) and carbonate (CO_3^{2-}) ions in the water.

B. Principle Behind the Method

The complexing agent used for the titration is a salt of the tetraprotic acid H_4EDTA (ethylenediaminetetraacetate). When Ca^{2+} is treated with the disodium salt Na_2H_2EDTA, a very stable complex is formed:

$$Ca^{2+} + H_2EDTA^{2-} \rightarrow CaEDTA^{2-} + 2H^+ \quad \text{(a)}$$

Magnesium forms a similar complex, $MgEDTA^{2-}$, which is far less stable than the Ca^{2+} complex.

When a sample containing calcium and magnesium ions is titrated with a solution of H_2EDTA^{2-}, the calcium ions are first complexed, forming $CaEDTA^{2-}$. As more reagent is added and the calcium ions are all combined in the complex, the magnesium ions form $MgEDTA^{2-}$. The desired endpoint of the titration is the point at which all of the Ca^{2+} and Mg^{2+} ions have been complexed.

It has been found that the indicator eriochrome black T (H_3In) forms a colored complex with Mg^{2+} and that this complex is less stable than $MgEDTA^{2-}$. Consequently, reaction (b) occurs:

$$\underset{\text{Wine red}}{MgIn^-} + H_2EDTA^{2-} \rightarrow MgEDTA^{2-} + \underset{\text{Blue}}{HIn^{2-}} + H^+$$

$$\text{(b)}$$

A color change is observed when the last of the indicator ions is displaced from its magnesium complex. Since all calcium solutions do not necessarily contain magnesium ions, it is customary to prepare the H_2EDTA^{2-} solution with a small amount of magnesium ion. When added to a solution containing calcium ion and indicator, the magnesium complex first loses EDTA to calcium [Eq. (c)], and then the free Mg^{2+} reacts with the indicator [Eq. (d)]. After all the calcium ion is complexed, reaction (b) occurs and a color change is observed:

$$MgEDTA^{2-} + Ca^{2+} \rightarrow CaEDTA^{2-} + Mg^{2+} \quad (c)$$

$$Mg^{2+} + HIn^{2-} \rightarrow MgIn^- + H^+ \quad (d)$$

The MgEDTA titration must be carried out at quite an elevated pH (at a too low pH, the MgEDTA complex is not stable). This is achieved by the addition of ammonium chloride in ammonia solution and results in the formation of the soluble magnesium amine complex.

Water hardness is usually determined by titration methods. The titration is carried out in ammoniacal solution at pH 10 using eriochrome black T as the endpoint indicator. The endpoint provides a total concentration of calcium and magnesium ions in solution. If the titration is repeated at pH 13 (by the addition of sodium hydroxide and in the absence of ammonia), magnesium hydroxide is produced and will precipitate out of solution. The calcium hydroxide, however, is soluble and is titrated by the EDTA. The concentration of magnesium present can thus be calculated simply by difference.

C. Scope and Application

This method can be used for the determination of total hardness in all types of waters and wastewaters. The range is 5–250 mg/L as $CaCO_3$.

D. Interferences

Suspended or colloidal organic matter also may interfere with the endpoint. Eliminate this interference by evaporating the sample to dryness on a steam bath and heating in a muffle furnace at 550°C until the organic matter is completely oxidized. Dissolve the residue in 20 ml 1N hydrochloric acid (HCl), neutralize to pH 7 with 1N sodium hydroxide (NaOH), and make up to 50 ml with distilled water; cool to room temperature and continue according to the general procedure.

E. Sampling Procedure and Storage

Samples should be collected in polyethylene bottles.

F. Apparatus

Burette
Analytical balance

G. Reagents (11)

1. Buffer Solution

Dissolve 16.9 g ammonium chloride (NH_4Cl) in 143 ml concentrated ammonium hydroxide (NH_4OH). Add 1.25 g magnesium salt of EDTA and dilute to 250 ml with distilled water.

2. Complexing Agents

For most waters no complexing agent is needed. Occasionally water containing interfering ions requires adding an appropriate complexing agent to give a clear, sharp change in color at the endpoint. The following are satisfactory:

Inhibitor I: Adjust acid samples to pH 6 or higher with buffer or 0.1 N NaOH. Add 250 mg sodium cyanide (NaCN) in powder form. Add sufficient buffer to adjust to pH 10.0 ± 0.1. (**Caution**: NaCN is extremely poisonous. Take extra precautions in its use. Flush solutions containing this inhibitor down the drain with large quantities of water after ensuring that no acid is present to liberate volatile poisonous hydrogen cyanide.)

Inhibitor II: Dissolve 5.0 g sodium sulfide nonahydrate ($Na_2S \cdot 9H_2O$) or 3.7 g $Na_2S \cdot 5H_2O$ in 100 ml distilled water. Exclude air with a tightly fitting rubber stopper. This inhibitor deteriorates through air oxidation. It produces a sulfide precipitate that obscures the endpoint when appreciable concentrations of heavy metals are present.

MgCDTA (magnesium salt of 1,2-cyclohexanediaminetetraacetic acid): Add 250 mg per 100 ml of sample and dissolve completely before adding buffer solution. Use this complexing agent to avoid using toxic or odorous inhibitors when interfering substances are present in concentrations that affect the endpoint but will not contribute significantly to the hardness value. Commercial preparations incorporating a buffer and a complexing agent are available. Such mixtures must maintain pH 10.0 ± 0.1 during titration and give a clear, sharp endpoint when the sample is titrated.

3. Indicators

Many types of indicator solutions have been advocated and may be used if the analyst demonstrates that they yield accurate values. The prime difficulty with indicator solutions is deterioration with aging, giving indistinct endpoints. For example, alkaline solutions of eriochrome black T are sensitive to oxidants, and aqueous or alcoholic solutions are unstable. In general, use the least amount of indicator providing a sharp endpoint. It is the analyst's responsibility to determine individually the optimal indicator concentration.

Eriochrome black T [sodium salt of 1-(1-hydroxy-2-naphthylazo)-5-nitro-2-naphthol-4-sulfonic acid]: Dissolve 0.5 g dye in 100 g 2,2′,2″-nitrilotriethanol (also called triethanolamine) or 2-methoxymethanol (also called ethylene glycol monomethyl ether). Add 2 drops per 50 ml solution to be titrated. Adjust volume if necessary.

Calmagite [1-(1-hydroxy-4-methyl-2-phenylazo)-2-naphthol-4-sulfonic acid]: This is stable in aqueous solution and produces the same color change as eriochrome black T, with a sharper endpoint. Dissolve 0.10 g Calmagite in 100 ml distilled water. Use 1 ml per 50 ml solution to be titrated. Adjust volume if necessary.

The first two indicators can be used in dry powder form if care is taken to avoid excess indicator. Prepared dry mixtures of these indicators and an inert salt are available commercially.

If the endpoint color change of these indicators is not clear and sharp, it usually means that an appropriate complexing agent is required. If NaCN inhibitor does not sharpen the endpoint, the indicator probably is at fault.

4. Standard EDTA Titrant, 0.01 M

Weigh 3,723 g analytical reagent grade disodium ethylenediaminetetraacetate dihydrate, also called (ethylenedinitrilo)tetraacetic acid disodium salt (EDTA), dissolve in distilled water, and dilute to 1000 ml. Standardize against standard calcium solution as described in the next subsection.

Because the titrant extracts hardness-producing cations from soft-glass containers, store in polyethylene (preferable) or borosilicate glass bottles. Compensate for gradual deterioration by periodic restandardization and by using a suitable correction factor.

5. Standard Calcium Solution

Weigh 1.000 g anhydrous $CaCO_3$ powder (primary standard or special reagent low in heavy metals, alkalis, and magnesium) into a 500-ml Erlenmeyer flask. Place a funnel in the flask neck and add, a little at a time, 1

ml concentrated HCl for each 1 ml deionized water (1 + 1 HCl) until all $CaCO_3$ has dissolved. Add 200 ml distilled water and boil for a few minutes to expel CO_2. Cool, add a few drops of methyl red indicator, and adjust to the intermediate orange color by adding 3N NH_4OH or 1 + 1 HCl, as required. Transfer quantitatively and dilute to 1000 ml with distilled water; 1 ml = 1.00 mg $CaCO_3$.

6. Sodium Hydroxide, NaOH, 0.1 N

H. Procedure (11)

Titration of sample: Select a sample volume that requires less than 15 ml EDTA titrant, and complete titration within 5 min, measured from the time of buffer addition.

Dilute 25.0 ml sample to about 50 ml with distilled water in a porcelain casserole or other suitable vessel. Add 1–2 ml buffer solution. Usually 1 ml will be sufficient to give a pH of 10.0–10.1. The absence of a sharp endpoint color change in the titration usually means that an inhibitor must be added at this point or that the indicator has deteriorated.

Add 1–2 drops indicator solution or an appropriate amount of dry-powder indicator. Add standard EDTA titrant slowly, with continuous stirring, until the last reddish tinge disappears. Add the last few drops at 3- to 5-s intervals. At the endpoint the solution normally is blue. Daylight or a daylight fluorescent lamp is recommended highly because ordinary incandescent lights tend to produce a reddish tinge in the blue at the endpoint.

I. Calculation

Hardness (EDTA) as mg $CaCO_3$/L

$$= \frac{A \times B \times 1000}{\text{ml sample}}$$

where

A = ml titration for sample
B = mg $CaCO_3$ equivalent to 1.00 ml EDTA titrant

Hardness may be calculated based on 2.497 [Ca] + 4.118 [Mg] in mg/L, obtained from individual determinations of Ca and Mg by atomic absorption (AA), inductively coupled plasma (ICP), or other method (12).

IV. OXIDANT DEMAND AND THE CHLORINE IODOMETRIC METHOD (18)

There is a need to maintain the biological antiseptic quality of water. Addition of oxidants achieves this. Though an excess of oxidants is always effective, this is uneconomic and could produce damaging side effects on the associated larger animal life. Oxidants include oxygen, ozone, chlorine (or chlorine producing), bromine (or bromine producing), reverse osmosis, and ultraviolet radiation. The important thing to note is that the antiseptic effect is what should be measured and controlled.

Water is often disinfected before it enters a distribution system to ensure that dangerous microbes are killed. Chlorine, chloramines, or chlorine dioxide most often are used, because they are very effective disinfectants, and residual concentrations can be maintained to guard against biological contamination in the water distribution system.

Chlorination can effectively treat biological pathogens such as coliform bacteria and *Legionella*, though it is ineffective against hard-shelled cysts like those produced by *Cryptosporidium* and *Giardia lamblia*. *Cryptosporidium* is a protozoan that causes the gastrointestinal disease cryptosporidiosis. *Giardia* is very resistant to disinfection by chlorine and may cause gastrointestinal illnesses in humans. Chlorination of drinking water produces carcinogens, such as the trihalomethanes.

Ozone is a powerful disinfectant, but it is not effective in controlling biological contaminants in the distribution pipes. Disinfection by-products (DBPs) are contaminants that form when disinfectants react with organic matter that is in treated drinking water. Long-term exposure to some DBPs may increase the risk of cancer or other adverse health effects. The ozonation of natural source waters containing natural organic matter produces biodegradable by-products such as organic acids, aldehydes, and ketoacids. These organic by-products serve as a carbon source for bacteria, potentially causing regrowth problems in distribution systems. Removal of biodegradable dissolved organic carbon can also reduce the formation potential of chlorination DBPs such as trihalomethanes and haloacetic acids.

Ozone treatment oxidizes organic contaminants in much the same way that chlorine does. An ozone generator converts the oxygen found in air to O_3, or ozone. Ozone is effective for treating pathogens such as coliform bacteria and *Legionella*, but it is not effective against hard-shelled cysts like *Cryptosporidium* or *Giardia lamblia* without using high contact times and concentrations.

Ozone (O_3) is an allotropic, triatomic form of oxygen. It is more commonly known as "activated oxygen" and is becoming widely used in both air and water purification. Ozone's action is to react with organics to oxidize unpleasant odors and significantly to reduce bacteria, mold, mildew, and fungus. Ozone occurs when an electrical charge, such as corona discharge, molecularly disassociates a stable molecule (O_2) and splits it apart, leaving two unstable atoms (O_1) of oxygen. Seeking stability, these atoms attach to other oxygen molecules (O_1), creating ozone (O_3).

Ozonation has emerged as one of the most promising alternatives to chlorination. Ozonation, however, tends to oxidize bromide to bromate, which presents a potential problem, since bromide is naturally present in source waters. Bromate is a potential carcinogen, even at low μg/L levels. The following equations show the pathway by which bromide (Br^-) is oxidized by ozone to bromate (BrO_3^-):

$$Br^- + O_3 + H_2O \rightarrow HOBr + O_2 + OH^-$$

$$HOBr + H_2O \rightarrow H_3O^+ + OBr^-$$

$$OBr^- + 2O_3 \rightarrow BrO_3^- + 2O_2$$

$$HOBr + O_3 \rightarrow No\ reaction$$

Ozonation of surface waters that contain bromide result in the formation of bromate, which has been identified as a potential carcinogen.

Ultraviolet light destroys the genetic material of pathogens such as coliform bacteria and *Legionella*, which effectively neutralizes them by preventing them from reproducing. Ultraviolet light is not effective for the treatment of hard-shelled cysts like *Cryptosporidium* and *Giardia lamblia*.

Reverse osmosis is a process that works by forcing water under great pressure against a semipermeable membrane, where ion exclusion occurs. Reverse osmosis is effective for the reduction of a broad range of health and aesthetic contaminants, though it is typically not used for the reduction of biological pathogens.

Only one disinfection treatment is not effective to maintain the chemical and biological antiseptic quality of the water. The combination of two or more water disinfection methods seems to solve this problem. For example, with the combination of photocatalysis and ozonolysis, photocatalysis gives a constant decline in total organic carbon (TOC) and ozonolysis gives no buildup of high intermediate concentrations (13) and ozonolysis and chlorine dioxide (14). Another example is the application of ozonolysis and chlorine dioxide.

There is an oxyhalide analysis by ion chromatography that permits simultaneous iodate, chlorite, and bromate detection from a single injection of aqueous sample. The method utilizes anion chromatographic resolution of the oxyhalides iodate, chlorite, and bromate and their subsequent generation of the tribromide ion, via postcolumn derivation, which is detected at 267 nm in UV spectrophotometer (15).

One method, using ion chromatography, quantifies bromate to a low $\mu g/L$ level (the detection limit is 1 $\mu g/L$), in the presence of high (mg/L) levels of common anions such as chloride and sulfate (16); using the technique of preconcentration it is possible to detect trace quantities of bromate (17).

A. Introduction to the Chlorine Iodometric Method

Chlorine dioxide is a deep yellow, volatile, unpleasant-smelling gas that is toxic and under certain conditions may react explosively. It should be handled with care in a vented area. Chlorine, hypochlorous acid, and hypochlorite ion as free chlorine residuals, along with the chloramines, are called combined chlorine residuals. At a lower pH, the formation of HOCl is favored over OCl^-, which is more effective for disinfection. A greater concentration of combined chlorine residual than of free chlorine residual is required to accomplish a given kill in a specified time. It is important to know both the concentration and the kind of residual chlorine acting (19).

B. Principle Behind the Method

Potassium iodide reacts with chlorine and other oxidizers to form iodine, which is titrated with thiosulfate in the presence of a starch indicator (20):

$$Cl_2 + 2KI \rightarrow I_2 + 2KCl$$

$$I_2 + 2\ Na_2S_2O_3 \rightarrow 2NaI + Na_2S_4O_6$$

C. Scope and Application

The minimum detectable concentration is 20 μg ClO_2/L.

D. Interference

There is little interference in this method; temperature and strong light affect solution stability.

E. Sampling Procedure and Storage

Determine ClO_2 promptly after collecting the sample. Do not expose the sample to sunlight or strong artificial light and do not aerate to mix.

F. Apparatus

Gas generating system
Burette

G. Reagents

1. Stock Chlorine Dioxide Solution

Prepare a gas-generating and absorbing system. Connect an aspirator flask, 500-ml capacity, with rubber tubing to a source of compressed air. Let air bubble through a layer of 300 ml distilled water in flask and then pass through a glass tube ending within 5 mm of the bottom of the 1-L gas-generating bottle. Conduct evolved gas via glass tubing through a scrubber bottle containing saturated $NaClO_2$ solution or a tower packed with flaked $NaClO_2$, and finally, via glass tubing, into a 2-L borosilicate glass collecting bottle, where the gas is absorbed in 1500 ml distilled water. Provide an air outlet tube on the collecting bottle for the escape of air. For gas generation select a bottle constructed of strong borosilicate glass and having a mouth wide enough to permit insertion of three separate glass tubes: the first leading almost to the bottom for admitting air, the second reaching below the liquid surface for the gradual introduction of H_2SO_4, and the third near the top for the exit of evolved gas and air. Fit to the second tube a graduated cylindrical separatory funnel to contain H_2SO_4. Locate this system in a fume hood with an adequate shield.

Dissolve 10 g $NaClO_2$ in 750 ml distilled water; place this in the generating bottle. Carefully add 2 ml concentrated H_2SO_4 to 18 ml distilled water and mix. Transfer to a funnel. Connect the flask to the generating bottle, the generating bottle to the scrubber, and the latter to the collecting bottle. Pass a smooth current of air through the system, as evidenced by the bubbling rate in all bottles.

Introduce 5-ml increments of H_2SO_4 from the funnel into the generating bottle at 5-min intervals. Continue the air flow for 30 min after the last portion of acid has been added.

Store the yellow stock solution in a glass-stoppered dark-colored bottle in a dark refrigerator. The concentration of ClO_2 thus prepared varies between 250 and

600 mg/L, corresponding to approximately 500–1200 mg free chlorine/L.

2. Standard Chlorine Dioxide Solution

Use this solution for preparing temporary ClO_2 standards. Dilute the required volume of stock ClO_2 solution to the desired strength with chlorine-demand-free water. Standardize the solution by titrating with standard 0.01 N or 0.025 N $Na_2S_2O_3$ titrant in the presence of KI, acid, and starch indicator by following the procedure given next. A full or nearly full bottle of chlorine or ClO_2 solution retains its titer longer than a partially full one.

When repeated withdrawals reduce the volume to a critical level, standardize the diluted solution at the beginning, midway in the series of withdrawals, and at the end of the series. Shake contents thoroughly before drawing off the needed solution from the middle of the glass-stoppered dark-colored bottle. Prepare this solution frequently.

H. Procedure

Select the volume of sample, prepare it for titration, and titrate the sample and the blank. Let ClO_2 react in the dark with acid and KI for 5 min before starting titration.

I. Calculations

Express ClO_2 concentrations in terms of ClO_2 or as free chlorine content. Free chlorine is defined as the total oxidizing power of ClO_2 measured by titrating iodine released by ClO_2 from acidic solution of KI. Calculate the result in terms of chlorine itself.

For standardizing ClO_2 solution:

$$\text{mg } ClO_2/\text{ml} = \frac{(A \pm B) \times N \times 13,49}{\text{ml sample titrated}}$$

$$\text{mg } ClO_2 \text{ as } Cl_2/\text{ml} = \frac{(A \pm B) \times N \times 35,45}{\text{ml sample titrated}}$$

For determining ClO_2 temporary standards:

$$\text{mg } ClO_2/\text{L} = \frac{(A \pm B) \times N \times 13.490}{\text{ml sample}}$$

$$\text{mg } ClO_2 \text{ as } Cl_2/\text{L} = \frac{(A \pm B) \times N \times 35.453}{\text{ml sample}}$$

where:

A = ml titration for sample
B = ml titration for blank
N = normality of $Na_2S_2O_3$

J. Criteria

For the preoxidation and reduction of organic pollutants, required dosages are between 0.5 and 2 mg/L, with contact times usually as low as 15–30 min, depending on the water characteristics; in the case of postdisinfection, 0.2–0.4 mg/L of ClO_2 are generally used.

V. DISSOLVED OXYGEN

A. Introduction

Dissolved oxygen analysis measures the amount of gaseous oxygen (O_2) dissolved in an aqueous solution. Most of the dissolved oxygen in the water comes from the atmosphere. The amount of dissolved oxygen gas is highly dependent on temperature and atmospheric pressure. Solubility is greater for freshwater than for saltwater and greater for cold water than for warm water. The amount of oxygen that can dissolve in pure water (saturation point) is inversely proportional to the temperature of the water.

Dissolved oxygen is also supplied to the water from aquatic plants through photosynthesis. Dissolved oxygen levels rise from morning through the afternoon as a result of photosynthesis, reaching a peak in late afternoon.

The main factor contributing to changes in dissolved oxygen levels is the buildup of organic waste. Organic waste can enter rivers in many ways, such as in sewage, urban and agricultural runoff (fertilizers), and in the discharge of industrial sources. Fertilizers stimulate the growth of algae and other aquatic plants. In deeper waters photosynthesis is reduced due to poor light penetration and due to the fact that dead phytoplankton (algae) fall toward the bottom, where bacteria start to grow and consume the oxygen in the water.

Dissolved oxygen (DO) levels in natural waters and wastewaters are dependent on the physical, chemical, and biochemical activities prevailing in the water body. The analysis for DO is a key test in water pollution control activities and waste treatment process control.

B. Modified Winkler Method

1. Principle Behind the Method

The test is based on the addition of divalent manganese solution, followed by strong alkali, to the water sample in a glass-stoppered bottle. Any DO present in the sample rapidly oxidizes an equivalent amount of the dispersed divalent manganous hydroxide precipitate to hydroxides of higher oxidation states. In the presence of iodide ions and upon acidification, the oxidized manganese reverts to the divalent state, with the liberation of iodine equivalent to the original DO content in the sample. The iodine is then titrated with a standard solution of thiosulfate:

$$Mn^{2+} + 2OH^- \rightarrow \underset{\text{(white precipitate)}}{Mn(OH)_2(s)}$$

$$Mn(OH)_2(s) + O_2 \rightarrow \underset{\text{(brown precipitate)}}{2MnO(OH)_2(s)}$$

$$2MnO(OH)_2(s) + 4H^+ + 2I^- \rightarrow Mn^{2+} + I_2 + 3H_2O$$

$$I_2 + I^- \rightarrow I_3^-$$

$$I_3^- + 2S_2O_3^{2-} \rightarrow 3I^- + S_4O_6^{2-}$$

2. Scope and Application

This method is applicable to surface waters and most wastewaters. The detection limit is 0.1 mg/L DO.

3. Interferences

Ferrous iron greater than 1 mg/L.
Any reducing or oxidizing materials.
Nitrites. Nitrites are reduced by sodium azide (azide modification of the Winkler method) in acid solution as follows:

$$N_3^- \xrightarrow{H^+} NH_3 \xrightarrow{NO_2^-,\ H^+} N_2 + N_2O + H_2O$$

4. Sampling Procedure and Storage

1. Samples should be collected in 300-ml biochemical oxygen demand (BOD) bottles. Special precautions are required to avoid entrapment or dissolution of atmospheric oxygen. An APHA type of sampler is recommended for surface waters. Where other types of samplers are used, the BOD bottles should be filled by transferring the sample with the use of a length of flexible tubing that extends to the bottom of the BOD bottles.
2. Sample analysis should be completed either at the sampling site or within 4–8 h after preservation. Sample may be preserved by completing steps 1–5 of the Procedure (addition of the manganese sulfate solution, the alkaline iodide–azide solution, and the concentrated sulfuric acid).

5. Apparatus

BOD bottles: 300-ml capacity with tapered ground-glass pointed stoppers and flared mouths
Magnetic mixer and Teflon magnet

6. Reagents

a. Manganese Sulfate Solution. Dissolve 480 g $MnSO_4 \cdot 4H_2O$, 400 g $MnSO_4 \cdot 2H_2O$, or 364 g $MnSO_4 \cdot H_2O$ in distilled water, filter it, and dilute it to 1 L. The manganese sulfate solution should give no color with starch when added to an acidified solution of potassium iodide.

b. Alkali–Iodide–Azide Reagent. Dissolve 500 g sodium hydroxide, NaOH (or 700 g potassium hydroxide, KOH), and 135 g sodium iodide, NaI (or 150 g potassium iodide, KI) in distilled water and dilute to 1 L. To this solution add 10 g sodium azide, NaN_3, dissolved in 40 ml distilled water. Potassium and sodium salts may be used interchangeably. This reagent should give no color with starch solution when diluted and acidified.

c. Sulfuric Acid, concentrated.

d. Starch Solution. Make a suspension by adding 5 g soluble starch to a little distilled water in a mortar or beaker. Pour this into 1 L of boiling water, boil a few minutes, and let settle overnight. Pour off the clear supernate and preserve by adding 5 ml chloroform. Store in a refrigerator.

e. Stock Sodium Thiosulfate Solution (0.10 N). Dissolve 24.82 g $Na_2S_2O_3 \cdot 5H_2O$ in boiled and cooled distilled water and dilute to 1 L. Preserve by adding 5 ml chloroform.

f. Standard Sodium Thiosulfate Titrant (0.025 N). Dilute the appropriate volume of stock sodium thiosulfate (as determined in the standardization procedure of step h) to 1000 ml with distilled water. Add 5 ml chloroform as a preservative.

g. Standard Potassium Dichromate Solution (0.1 N). Dissolve 4.904 g $K_2Cr_2O_7$ dried at 103°C for 2 h to 1000 ml with distilled water.

h. Standardization of Sodium Thiosulfate. Dissolve approximately 2 g KI, free from iodate, in a 500-ml Erlenmeyer flask with 100–150 ml distilled water; add 10 ml 10% H_2SO_4 solution, followed by exactly 20.00 ml standard dichromate solution. Place in the dark for 5 min. Dilute to 300 ml with distilled water, and titrate the liberated iodine with the 0.10 N thiosulfate titrant, adding starch toward the end of the titration, when a pale straw color is reached. Calculate the volume of the 0.10 N stock sodium thiosulfate that must be diluted to 1 L to get exactly 0.025 N standard sodium thiosulfate solution:

$$\frac{\text{Vol. of stock Na}_2\text{S}_2\text{O}_3}{\text{solution required}} = \frac{\text{Vol. of titrant}}{20.0} \times 250$$

7. Procedure

1. To the sample collected in the 300-ml BOD bottle add 2 ml of the manganous sulfate solution followed by 2 ml of the alkaline iodide–azide solution below the surface of the liquid.
2. Stopper with care to exclude air bubbles, and mix by inverting the bottle at least 15 times.
3. When the precipitate settles, leaving a clear supernate above the manganese hydroxide floc, shake again.
4. After settling has produced at least 200 ml of clear supernate, carefully remove the stopper and immediately add 2.0 ml concentrated H_2SO_4 by allowing the acid to run down the neck of the bottle.
5. Restopper, and mix by gentle inversion until dissolution is complete and the iodine is uniformly distributed throughout the bottle.
6. Decant 203 ml into an Erlenmeyer flask.
7. Titrate with 0.025 N thiosulfate to a pale straw color.
8. Add 1–2 ml of starch solution and continue the titration to the first disappearance of the blue color.

8. Calculations

1.0 ml of 0.025 N sodium thiosulfate is equivalent to 1.0 ml DO when 203 ml of sample is titrated.

9. Criteria

The criteria for aquatic life require that the average dissolved oxygen remain above 5.0 mg/L. Oxygen levels that remain below 1–2 mg/L for a few hours can result in large fish kills.

C. Oxygen-Membrane Electrode Method

1. Principle Behind the Method

Oxygen-membrane electrodes have three main components: the membrane, the oxygen-sensing element, and the electrolyte solution. The oxygen probe contains a solution of potassium chloride (KCl), which will ab-

sorb oxygen. As more oxygen is diffused into the solution, more current will flow through the cell. Lower oxygen pressure (less diffusion) means less current.

A membrane-electrode probe is connected to a meter, and the probe is inserted into the water. Readings may be given in mg/L or percent saturation. A milligram per liter (mg/L) reading is the weight of oxygen in a liter of water. Percent saturation is the amount of oxygen present divided by the amount representing saturation, multiplied by 100. Thus, when the amount of DO measured is equal to the saturation value, saturation is 100%. The temperature of the water and the atmospheric pressure must be known in order to calculate ppm (parts per million) of dissolved oxygen.

The dissolved-oxygen probe is a galvanic measuring element that produces a millivolt output proportional to the oxygen present in the medium it is placed in. With the older style of electrodes, oxygen diffuses through the membrane onto the cathode ($O_2 + 4H^+ + 4e^- \rightarrow 2H_2O$), where it reacts chemically and combines with the anode ($2Pb + 2H_2O \rightarrow 2PbO + 4H^+ + 4e^-$). This chemical process develops an electric current, which flows through a built-in resistor. The resistor converts the current (microamps) into millivolts. The dissolved-oxygen electrode senses the oxygen concentration in water and aqueous solutions. A platinum cathode and a silver/silver chloride reference anode in KCl electrolyte are separated from the sample by a gas-permeable plastic membrane. A fixed voltage is applied to the platinum electrode. As oxygen diffuses through the membrane to the cathode, it is reduced:

$$\tfrac{1}{2}O_2 + H_2O + 2e^- \rightarrow 2OH^-$$

The oxidation taking place at the reference electrode (anode) is:

$$Ag^+ + Cl^- \rightarrow AgCl + e^-$$

Accordingly, a current will flow that is proportional to the rate of diffusion of oxygen and, in turn, to the concentration of dissolved oxygen in the sample. This current is converted to a proportional voltage that is amplified and read. The greater the oxygen partial pressure, the more oxygen diffuses through the membrane in a given time. This results in a current that is proportional to the oxygen in the sample.

2. Scope and Application

Applications for dissolved oxygen measurement include processes in which the amount of oxygen affects a reaction rate or process efficiency or indicates an environmental condition. Some important applications include wastewater treatment, wine production, bioreactions, and environmental water monitoring. The sensor generally has a range of 0–20 mg/L of dissolved oxygen.

3. Interferences

A. Some gases are known to interfere with DO readings. Check for significant concentrations of hydrogen sulfide, sulfur dioxide, halogens, neon, and nitrous and nitrite oxide.

B. Stirring: Consumption of oxygen by the probe can cause a lowering of the oxygen concentration at the boundary layer between the sample and the probe membrane. For this reason, sample stirring is recommended.

4. Sampling Procedure and Storage

See Sec. V.B.4, the procedure for the modified Winkler method.

5. Apparatus

Oxygen meter, consisting of an upper part with cathode, anode, and cable, and a cap with membrane

6. Reagents

Distilled air-saturated water
Electrolyte solution, KCl: provided by the manufacturers

7. Procedure

1. Fill the electrode and put in the membrane according to the instructions in the manufacturer's manual (22).

2. To calibrate the sensor, simply place the electrode in the zero-oxygen solution provided with the probe for the 0% calibration point. For the 100% calibration point, place the electrode in air (or, better yet, in water saturated with air, using a bottle that manufacturers provide). The electrode can be calibrated in any units you choose: % DO, mg/L, or ppm dissolved oxygen.

3. Place the electrode in the sample, avoiding air entrapment on the membrane surface. A correction must be made using salinity and temperature switches before measuring the dissolved oxygen in the samples. Correction charts are provided in most instruction manuals.

A method has been published that features automated analysis using oxygen sensors based on room-temperature phosphorescence (23).

8. Calculations

Instruments give the dissolved oxygen value (mg/L or percent saturation) by direct reading.

9. Criteria

See Sec. V.B.9 for the criteria for the modified Winkler method.

VI. FLUORIDE

A. Introduction

Fluoride is a naturally occurring compound. It is found in nearly all soils, plants, animals, and water supplies. Fluoride, in small quantities, is considered essential to the proper development of healthy teeth and bones.

A fluoride concentration of approximately 1.0 mg/L in drinking water effectively reduces dental caries without harmful effects on health. Fluoride may occur naturally in water, or it may be added in controlled amounts. Some fluorosis may occur when the fluoride level exceeds the recommended limits. In rare instances the naturally occurring fluoride concentration may approach 10 mg/L; such waters should be defluoridated.

Accurate determination of fluoride has increased in importance with the growth of the practice of fluoridation of water supplies as a public health measure. Maintenance of an optimal fluoride concentration is essential in maintaining the effectiveness and safety of the fluoridation procedure (24).

Fluorides are compounds containing the element fluorine. Some of the most common of these compounds include the following: sodium fluoride (NaF), sodium silicofluoride (Na_2SiF_6), and calcium fluoride (CaF_2). Fluorine is the most reactive nonmetallic element. It will form compounds with all elements except helium, neon, and argon. It will also form salts by combining with metals.

Fluoride ions may be present either naturally or artificially in drinking water and are absorbed to some degree in the bone structure of the body and in tooth enamel. Fluoride at extremely high levels can cause mottling (discoloration) of the teeth. Some fluoride compounds may also cause corrosion of piping and other water treatment equipment. Natural fluorides occur in rocks in some areas. Another source of fluorides in streams and reservoirs is released from sewage treatment plants, since most public water supplies add fluoride to drinking water to reduce dental decay.

Based on the NAS (National Academy of Sciences) review and other studies, there are no data available at this time to conclude that the fluoride drinking standards should be revised. There are studies of the biological effects of fluorides, especially on carcinogenic activity and on human fertility due to sodium fluoride (25–31). However, the U.S. National Toxicology Program (NTP) has conducted toxicity and carcinogenicity studies with sodium fluoride administered in the drinking water to rats and mice. Results showed that there was equivocal evidence of carcinogenic activity of sodium fluoride in male rats based on the occurrence of a small number of osteosarcomas in the treated animals (32).

B. Preliminary Treatment

For determining fluoride ion (F$^-$) in water, the electrode and colorimetric methods are subject to error due to interfering ions. It may be necessary to distill the sample before making the determination. When interfering ions are not present in excess of the tolerances for the method, the fluoride determination may be made directly without distillation (24). **Caution**: Regardless of the apparatus used, provide for a thorough mixing of sample and acid; heating a nonhomogenous acid–water mixture will result in bumping or possibly a violent explosion.

C. Sampling and Storage

Preferably use polyethylene bottles for collecting and storing samples for fluoride analysis. Glass bottles are satisfactory if they have previously not contained high-fluoride solutions. Always rinse bottles with a portion of the sample. Never use an excess of dechlorinating agent. Dechlorinate with sodium arsenite rather than sodium thiosulfate when using the sodium 2-(parasulfophenylazo)-1,8-dihydroxy-3,6-naphthalene disulfonate (SPADNS) method, because the latter may produce turbidity that causes erroneous readings.

D. Criteria

The Kentucky Water Quality Standards maximum for fluoride in streams is a concentration of 1 mg/L, or 1 part per million. Higher levels may be harmful to aquatic life. Fluoride concentration in water to be used for domestic water supply should not exceed 1.0 mg/L.

Due to the fact that people tend to drink more in warm weather, warmer-weather dosage levels are much lower than those for colder weather. The maximum contaminant level for fluoride is 4 mg/L. Dosage above this level have been known to cause cases of skeletal fluorosis (the weakening of bones due to the high levels of fluoride deposited in them). A maximum level of contaminant of 2 mg/L was set, to prevent against the occurrence of dental fluorosis (a graying of the teeth).

E. Ion-Selective Electrode Method

1. Principle Behind the Method

The fluoride-ion electrode is a solid-state sensor. In this electrode, the active membrane portion is a single crystal of LaF$_3$ doped with europium(II) to lower its electrical resistance and facilitate ionic charge transport. The LaF$_3$ crystal, sealed into the end of a rigid plastic tube, is in contact with the internal and external solutions. Typically, the internal solution is 0.1 M each in NaF and NaCl; the fluoride ion activity controls the potential of the inner surface of the LaF$_3$ membrane, and the chloride ion activity fixes the potential of the internal Ag/AgCl wire reference electrode. The electrochemical cell incorporating the LaF$_3$ membrane electrode is:

Ag│AgCl, Cl$^-$(0.1M), F$^-$(0.1M)│LaF$_3$ crystal│test

solution‖ reference electrode

It obeys a Nernst-type relation of the following form:

$$E = \text{constant} + \frac{RT}{F} \ln \frac{[\text{F}^-]_{\text{int}}}{[\text{F}^-]_{\text{ext}}}$$

which, because $[\text{F}^-]_{\text{int}}$ is constant, simplifies to

$$E = \text{constant} + 0.05916 \text{ pF} \quad \text{at } 25°\text{C}$$

The fluoride electrode can be used with a standard calomel reference electrode and almost any modern pH meter having an expanded millivolt scale. Calomel electrodes contain both metallic and dissolved mercury; therefore, dispose of them only in approved sites, or recycle them. For this reason, the Ag/AgCl reference electrode is preferred.

The fluoride electrode measures the ion activity of fluoride in solution rather than concentration. Fluoride ion activity depends on the solution total ionic strength and pH and on fluoride complexing species. Adding an appropriate buffer provides a nearly uniform ionic-strength background, adjusts pH, and breaks up complexes so that, in effect, the electrode measures concentration.

2. Scope and Application

This method is applicable to the determination of fluoride in all types of waters. The range is 10^{-1} M to 10^{-5} M (approximately 0.2–2000 mg/L) fluoride.

3. Interferences

Selective ion electrodes are subject to two types of interferences: method interference and electrode interference. Method interferences occur when some characteristic of the sample prevents the probe from sensing the ion of interest. For example, a fluoride electrode can detect only fluoride ion. In acid solution, however, fluoride forms complexes with the hydrogen ion and is thereby masked from the fluoride detector. Electrode interferences arise when the electrode responds to ions in the sample solution other than the ion being measured. The selectivity coefficient is an index of the ability of an electrode to measure a particular ion in the presence of another ion.

Fluoride forms complexes with several polyvalent cations, notably aluminum and iron. The extent to which complexation takes place depends on solution pH, relative levels of fluoride, and complexing species. However, CDTA (cyclohexylenediaminetetraacetic acid), a component of the buffer, preferentially will complex interfering cations and release free fluoride ions. Concentrations of aluminum, the most common interference, up to 3.0 mg/L can be complexed preferentially. In acid solution, F^- forms a poorly ionized $HF \cdot HF$ complex, but the buffer maintains a pH above 5 to minimize hydrogen fluoride complex formation. In alkaline solution, hydroxide ion also can interfere with electrode response to fluoride ion whenever the hydroxide ion concentration is greater than one-tenth the concentration of fluoride ion. At the pH maintained by the buffer, no hydroxide interference occurs.

4. Sampling Procedure and Storage

Samples should be collected in polyethylene bottles.

5. Apparatus

Expanded-scale or digital pH meter or ion-selective meter.

Sleeve-type reference electrode: Do not use fiber-tip reference electrodes, because they exhibit erratic behavior in very dilute solutions.

Fluoride electrode.

Magnetic stirrer, with TFE-coated stirring bar.

Timer.

6. Reagents

a. Stock Fluoride Solution. Dissolve 221.0 mg anhydrous sodium fluoride, NaF, in distilled water and dilute to 1000 ml; 1.00 ml = 100 μg F^-.

b. Standard Fluoride Solution. Dilute 100 ml stock fluoride solution to 1000 ml with distilled water; 1.00 ml = 100 μg F^-.

c. Fluoride Buffer. Place approximately 500 ml distilled water in a 1-L beaker and add 57 mL of glacial acetic acid, 58 g NaCl, and 4.0 g 1,2 cyclohexylenediaminetetraacetic acid (CDTA). Stir to dissolve. Place beaker in a cool water bath and slowly add 6N NaOH (about 125 ml) with stirring, until the pH is between 5.3 and 5.5. Transfer to a 1-L volumetric flask and add distilled water to the mark.

7. Procedure

a. Instrument Calibration. No major adjustment of any instrument normally is required to use electrodes in the range of 0.2–2.0 mg F^-/L. For those instruments with zero at center scale, adjust the calibration control so that the 1.0 mg F^-/L standard reads at the center zero (100 mV) when the meter is in the expanded-scale position. This cannot be done on some meters that lack a millivolt calibration control. To use a selective-ion meter, follow the manufacturer's instructions.

b. Preparation of Fluoride Standards. Prepare standards by serial dilution with deionized water of the 0.1 M or 1000 ppm standard, in the range of 10^{-1}–10^{-6} M (0.2–2000 mg/L).

c. Treatment of Standards and Sample. In 100-ml beakers or other convenient containers, add by volumetric pipette from to 25 ml standard or sample. Bring standards and sample to the same temperature, preferably room temperature. Add an equal volume of buffer. The total volume should be sufficient to immerse the electrodes and permit operation of the stirring bar.

d. Measurement with Electrode

1. Adjust the meter to measure millivolts.
2. Measure 100 ml of standard (or an adequate volume) into a 150-ml beaker. Add buffer solution. Stir thoroughly.
3. Rinse the electrodes with distilled water, blot them dry, and place them into the beaker. When a stable reading is displayed, record the millivolt value and corresponding standard concentration.
4. Repeat the previous procedure with each standard.
5. Using semilogarithmic graph paper, prepare a

calibration curve by plotting the millivolt values on the linear axis and the standard concentration values on the logarithmic axis.

6. Measure 100 ml of the sample into a 150-ml beaker. Add buffer solution. Stir thoroughly.

7. Rinse the electrodes with distilled water, blot them dry, and place them into the beaker. When a stable reading is displayed, record the millivolt value.

8. Using the calibration curve prepared in step 5, determine the unknown concentration.

8. Calculations

Plot potential measurement of fluoride standards against concentration on semilogarithmic graph paper. Plot milligrams of F^- per liter or M on the logarithmic axis (ordinate). Plot millivolts on the abscissa. The relationship should be linear in a range of 10^{-1} to 10^{-5} M. From the potential measurement for each sample, read the corresponding fluoride concentration from the standard curve. The slope must be approximately 59.0 ± 2.0 mV per decade change in fluoride concentration.

The detection limit (DL) is the smallest concentration of analyte that can be reliably discerned from background measurement noise. The detection limits with ion-selective electrode techniques are defined analogously to other signal-producing physicochemical techniques measuring ion concentration, at which the measured signal is exactly twice as large as the background noise:

(signal/noise) = 2 = DL

$(59/z) \log DL = (59/z) \log 2$

This is precisely the case when the deviation from the Nernst line is 18/z mV [$(59/z) \log 2 = 18/z$ at 25°C] (10). See Fig. 1.

F. Ion Chromatographic Method

1. Principle Behind the Method

Ion chromatography (IC) is a popular method for ion analysis because many anions can be determined quickly with high precision, simultaneously, and different chemical species of the same element (e.g., chlorite, chlorate, and chloride) can be separated. Anions in the test sample are separated by an ion chromatographic system containing a guard column, a separator column, and a suppressor device and are measured using a conductivity detector.

2. Scope and Application

This method is applicable to the determination of bromide, chloride, fluoride, nitrate-N, nitrite-N, orthophosphate, and sulfate in drinking water and wastewater in a wide range of concentrations.

3. Interferences

A. To elute fluoride together with other common inorganic anions such as chloride, nitrate, and sulfate within an acceptable time frame of less than 15 min, mixtures of sodium carbonate and sodium hydrogencarbonate are the most widely used eluents. However, under these chromatographic conditions fluoride elutes very close to the system void volume, making determination at concentration levels of less than 100 μg/L dif-

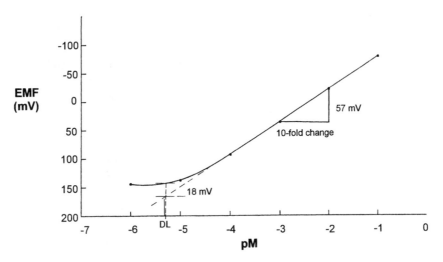

Fig. 1 Determination of the detection limit for a fluoride ion–selective electrode.

ficult if not impossible, owing to interference from the negative water dip. The water dip occurs when the injected water passes the conductivity cell, decreasing the background conductivity of the carbonic acid formed in the suppressor by exchanging the eluent cations with hydronium ions. Given the proximity of fluoride to the void dip, its absolute size needs to be minimized to allow maximum resolution of fluoride from the void dip.

There have been numerous attempts to circumvent this problem: by sample pretreatment, by changing the eluent conditions, or by tailoring the stationary-phase design. An easy way to compensate for the negative dip is to add carbonate to the sample, matching the carbonate concentration in the mobile phase, thus making the negative dip invisible. However, this approach does not work with real samples such as mineral waters. If the carbonate concentration in the sample is higher than the total carbonate concentration in the mobile phase, a positive signal that is almost indistinguishable from the fluoride peak is obtained within the void volume of the separator column. (see Sec. VI.F.7.b.)

B. Coelution of fluoride and chloride is observed in chromatograms of samples containing high concentrations of chloride. This chloride interference can be reduced by the precipitation of silver chloride with silver acetate or silver oxide or by passing the sample through a column containing a cation-exchange resin in the Ag^+ form (33), using an eluent containing the matrix anion (34) or using the "heart-cut" technique (35).

There is a method that uses a system consisting of a water-dip cutting column and two column-switching valves for avoiding water-dip interference (36). An anion-exchange column (ethylvinylbenzene-divinylbenzene substrate) was developed for the determination of inorganic anions (chloride, nitrite, bromide, nitrate, orthophosphate, sulfate), including fluoride and oxyhalides such as chlorite, chlorate, and bromate, under isocratic conditions (37).

4. Sampling Procedure and Storage

Collect samples in scrupulously clean sample bottles (60-ml high-density polyethylene bottles with polypropylene screw caps). Sample preservation and holding times for fluoride, bromide, and chloride are "none required" and 28 days; for nitrate-N, nitrite-N, and or-

thophosphate-P are cool to 4°C and 48 h; and for sulfate are cool to 4°C and 28 days, respectively (38).

5. Apparatus

Balance: analytical, capable of accurately weighing 0.0001 g

Ion chromatographic system: equipped with guard column, anion separator column (Dionex AS4A, Dionex Corp., Sunnyvale, CA, is suitable), anion-suppressor device, syringes, pumps to maintain flow rate of 2 ml/min, 50-μl sample loop, compressed air, and conductivity detector with ca. 1.25 μl internal volume; data system is recommended for measuring peak areas

0.22- or at least 0.45-μm-membrane filters

Note: (a) AS4A column must be used only in aqueous samples. An anion-separator column was developed that is solvent compatible (Dionex AS4SC, Dionex Corp., Sunnyvale, CA, is suitable), which can be used in waters with organic solvents. (b) A self-regenerating suppressor (SRS) for ion chromatography has been developed that improves detection limits by up to two orders of magnitude or more compared to nonsuppressed ion analysis techniques. This is accomplished by increasing the analyte conductivity signals of interest while decreasing eluent background noise and sample counter-ion interferences. The SRS continuously produces the ions required to neutralize the eluent by the electrolysis of water, making it absolutely maintenance free. The water supply is drawn from recycled eluent, eliminating the need for additional postcolumn reagents and maintenance time (Dionex SRS, Sunnyvale, CA).

6. Reagents

a. Reagent Water. Milli-Q-purified water, deionized and then filtered through a 0.2-μm membrane, must be used.

b. Eluent Solution: 1.7 mM Sodium Bicarbonate ($NaHCO_3$), 1.8 mM Sodium Carbonate (Na_2CO_3). Dissolve 0.2856 g $NaHCO_3$ and 0.3816 g Na_2CO_3 and dilute to 2 L in reagent water.

c. Regeneration Solution: 5.0 mM H_2SO_4. This solution is not necessary if the ion chromatograph is equipped with a self-regenerating suppressor (SRS), because deionized water is used as regenerant.

d. Stock Standard Solutions: 1000 mg/L (1 mg/mL). Stock standard may be purchased as certified solutions or prepared from ACS reagent-grade materials (dried at 105°C for 30 min). To prepare 1000 mg/

L stock standard solutions, dissolve and dilute to 1 L in reagent water the following: (a) bromide (Br^-): 1.2876 g sodium bromide (NaBr); (b) chloride (Cl^-): 1.6485 g sodium chloride (NaCl); (c) fluoride (F^-): 2.2124 g sodium fluoride (NaF); (d) nitrate ($NO_3^- -N$): 6.0679 g sodium nitrate (NaNO_3); (e) nitrite ($NO_2^- -N$): 4.9257 g sodium nitrite (NaNO_2); (f) phosphate ($PO_4^{3-}-P$): 4.3937 g potassium phosphate (KH_2PO_4); and (g) sulfate (SO_4^{2-}): 1.8141 g potassium sulfate (K_2SO_4). Stock standard solutions are stable for at least 1 month stored at 4°C. Prepare working standards fresh weekly, except prepare nitrite and phosphate working standards fresh daily (38). A mixed M stock standard solution must be prepared.

7. Procedure

a. Calibration (38). Use the eluent (1.7 mM sodium bicarbonate, 1.8 mM sodium carbonate) and regenerant (electrolyzed 18 MΩ/cm water) flow rates of 1.0 and 2.0 ml/min, respectively, and a conductivity sensitivity of 30 μS. Standards must be filtered through a 0.45-μm filter before injection.

Prepare calibration standards and a blank. Dilute the stock standards with reagent water in volumetric flasks, first individually to check the retention time of each one, and then prepare mix standard with all the anions together to save time and to check the resolution of peaks.

The linear ranges are 0.3–25 mg/L (all anions), except 2–100 mg/L for sulfate (38).

Inject 0.1–1.0 ml (determined by injection loop volume) of each calibration standard and plot the peak height or area responses against concentration.

b. Injection of Samples (38,39). Load and inject a fixed amount of well-mixed and filtered sample. Flush the injection loop thoroughly using each new sample. Use the same size loop for standards and samples. Record the resulting peak size in area, or peak-height units. An automated constant-volume injection system may also be used.

Dilute the sample with reagent water and reanalyze it if the response for peak exceeds the working range of the system.

The exact determination of fluoride is possible if the advantage of the simultaneous determination of other mineral acids is eliminated and if the chromatographic conditions are changed so that fluoride is separated from the carbonate-hydrogencarbonate traveling with the mobile phase. An increase in fluoride retention can be achieved by using an eluent of lower eluting power, such as sodium tetraborate. The suppressor column chemically converts the highly conductive species of the eluent (1.5 mM Na_2B_4O_7) into the significantly less conductive, weakly ionized species H_2B_4O_7, resulting in increased detection sensitivity for the analytes. Figure 2 shows that fluoride is well resolved from the system void volume and separated at the baseline from the other mineral acids and oxyhalides in about 5 min when sodium tetraborate was used as eluent, and using the same conditions as described in Sec. VI.F.7.a.

8. Calculations

Determine the peak-height or area counts of anions in the samples. Compare with the calibration curve to determine concentration.

VII. pH

A. Introduction

Practically every phase of water supply and wastewater treatment, e.g., acid–base neutralization, water softening, precipitation, coagulation, disinfection, and corrosion control, is pH dependent. *Buffer capacity* is the amount of strong acid base, usually expressed in moles per liter, needed to change the pH value of a 1-L sample by 1 unit.

pH as defined by Sorenson is $-\log [H^+]$; it is the intensity factor of acidity. Pure water is very slightly ionized and at equilibrium the ion product is

$$[H^+][OH^-] = K_w$$
$$= 1.01 \times 10^{-14} \quad \text{at } 25°C$$

and

$$[H^+] = [OH^-]$$
$$= 1.005 \times 10^{-7}$$

where

$[H^+]$ = activity of hydrogen ions, moles/L

$[OH^-]$ = activity of hydroxyl ions, moles/L

K_w = ion product of water.

The acidity or basic nature of a solution is expressed as the pH. The concentration of the hydrogen ion $[H^+]$ in a solution determines the pH. Mathematically this is expressed as

$$pH = -\log [H^+]$$

The pH value is the exponent to the base 10 of the hydrogen ion concentration. Each change in pH unit

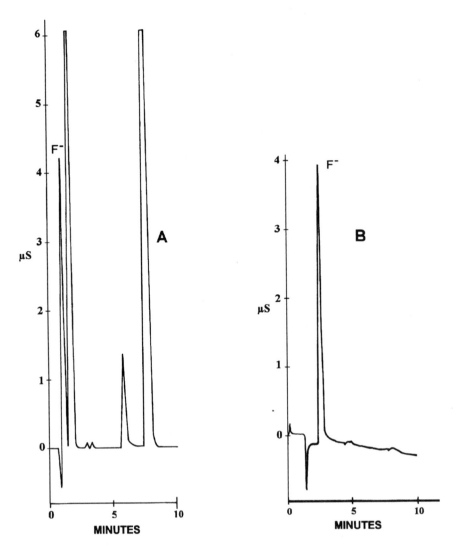

Fig. 2 Chromatograms showing the separation of fluoride. (A) Carbonate-bicarbonate as eluent; [F$^-$] = 1.026 mg/L; (B) sodium tetraborate as eluent; [F$^-$] = 1.000 mg/L.

represents a tenfold change in acidity. For example, a solution at pH 3 is ten times more acidic than one at pH 4.

Natural waters usually have pH values in the range of 4–9, and most are slightly basic because of the presence of bicarbonates and carbonates of the alkali and alkaline earth. Water contains both hydrogen (H$^+$) ions and hydroxyl (OH$^-$) ions. Each measured liquid or substance is given a pH value on a scale that ranges from 0 to 14. Pure deionized water contains equal numbers of H$^+$ and OH$^-$ ions and has a pH of 7. This means that it is neutral, neither acidic nor basic. If the water contains more H$^+$ ions than OH$^-$ ions, then the water is considered acidic and has a pH less than 7. If the

water contains more OH$^-$ ions than H$^+$ ions, the water is considered basic and has a pH of more than 7.

The pH scale is significantly changed by the increased amounts of nitrogen oxides (NO$_x$) and sulfur dioxides (SO$_2$) from automobiles and coal-fired power plant emissions. These emissions are converted to nitric acid and sulfuric acid in the atmosphere. If the waters are very acidic, it can cause heavy metals to be released into the water.

The pH of a water body results from the ratio of H$^+$ to OH$^-$. In natural waters this usually is dependent on the carbonic acid equilibrium. When carbon dioxide from the air enters freshwater, small amounts of carbonic acid are formed that then dissociate into hydrogen ions and bicarbonate ions, as shown in the following equations:

$$CO_2 + H_2O \rightarrow H_2CO_3 \text{ (carbonic acid)}$$

$$H_2CO_3 \rightarrow HCO_3^- + H^+$$

This increase in H^+ ions makes the water more acidic and lowers the pH. If CO_2 is removed (as in photosynthesis), the reverse takes place and pH rises.

B. Principle Behind the Method

Two parts are involved in a pH electrode system. The first is a pH sensor, which consists of a special pH-sensitive glass and whose voltage output is proportional to the pH. The second is a reference sensor, which provides a stable and constant reference point. Electrical contact is made with the solution using a saturated salt solution (usually KCl) that leaks slowly out of a porous junction. These two sensors are usually built into the same housing, hence the term "combination" pH electrode.

A pH meter is simply a device that measures the voltage from the electrodes and converts it to a pH reading on a display.

C. Scope and Application

This electrometric measurement is applicable to all waters and over the whole pH scale. The pH of most natural waters falls within the range 4–9.

D. Interferences

Factors that affect the output of pH electrodes include temperature changes, a blocked reference junction, and coatings on the pH glass. High sodium concentrations at a pH above 10 interfere with the measurement.

The glass electrode is relatively immune to almost all types of interfering material. The Nernstian slope increases with increasing temperature, and electrodes take time to achieve thermal equilibrium. This can cause a long-term drift in pH. Because chemical equilibrium affects pH, standard pH buffers have a specified pH at indicated temperatures.

E. Sampling Procedure and Storage

Plastic sample containers should be used and tightly sealed as soon as the sample has been collected. Keep the sample cool and analyze it as soon as possible.

F. Apparatus

pH meter.

Reference electrode consisting of a half-cell that provides a constant electrode potential. Commonly used are calomel and silver:silver chloride electrodes. Either is available with several types of liquid junctions. The calomel reference system is for use in applications where silver ions from the electrode are not permitted to flow into the sample or where there is a danger of ions in the sample contaminating the silver wire in the electrode. The calomel reference system consists of a silver wire immersed in a small quantity of Hg/HgCl2 (mercury in contact with mercurous chloride). This makes contact with the internal electrolyte via a fiber wick. This inner electrolyte is usually 1 M KCl (potassium chloride). The liquid junction of the reference electrode is critical, because at this point the electrode forms a salt bridge with the sample or buffer and a liquid junction potential is generated that in turn affects the potential produced by the reference electrode.

Glass electrode: The sensor electrode is a bulb of special glass containing a fixed concentration of HCl or a buffered chloride solution in contact with an internal reference electrode. Several types of glass electrodes are available. Combination electrodes incorporate the glass and reference electrodes into a single probe.

Beakers: Preferably use polyethylene or TFE (Teflon or equivalent) beakers.

Stirrer: Use either a magnetic, TFE-coated stirring bar or a mechanical stirrer with an inert plastic-coated impeller.

Flow chamber: Use for continuous-flow measurements for poorly buffered solutions.

G. Reagents

Standard pH buffer solutions of pH 7.0 and pH 10.0 (available commercially)

H. Procedure

1. Preparation of pH Meter and Electrodes for Measuring

1. Check that the calomel part of the electrode is filled with saturated potassium chloride and that no air bubbles are present. If necessary, remove the stopper and refill through the hole in the side

of the electrode. Always keep the electrode filled with saturated potassium chloride solution.

2. When a new, or dry, electrode is used, it must be soaked for at least 24 hours in buffer solution or according to the electrode manufacturer's manual.

3. If air bubbles are trapped in the inner liquid of the glass bulb, shake the electrode (as with a clinical thermometer) to remove the air pockets.

4. Test the batteries according to the manufacturer's instructions.

5. Occasionally it may be desirable to clean the glass electrode by rinsing it with distilled water prior to using again.

6. Closely follow the procedure provided by the manufacturer for setting up the instrument and preparing electrodes.

2. *Instrument Calibration*

1. For standardizing, select two different buffer solutions in the approximate pH range of the samples to be measured.

2. Set the temperature compensator to the temperature of the sample being measured. Buffers should also be at the same temperature as the sample.

3. Place a buffer solution into a beaker and, while gently stirring, place electrodes in the solution; wait until the meter needle stabilizes, and set the calibration control to the correct value of the buffer.

4. Repeat step 2 with the other buffer. The reading should be within 0.1 pH units.

5. Adjust the slope control.

6. Place the electrodes into a beaker containing the sample and take the reading.

I. Calculations

The pH meter reads directly in pH units.

J. Criteria

A pH range of 6.0–9.0 appears to provide protection for freshwater fish and bottom-dwelling invertebrates.

REFERENCES

1. S. Peiffer, C. Beierkuhnlein, A. Sandhage-Hofmann, M. Kaupenjohann, S. Bar. Impact of high aluminum loading on a small catchment area (Thuringia Slate Mining Area)—geochemical transformations and hydrological transport. Water, Air and Soil Pollution 94:401–416, 1997.

2. U.S. Environmental Protection Agency. Sample preservation. In: Methods for Chemical Analysis of Water and Wastes. EPA-600/4-79-020. Cincinnati, OH: U.S.E.P.A., 1983. pp xv–xvii.

3. American Public Health Association (APHA), American Water Works Association (AWWA), Water Environment Federation (WEF). Chap 2. Method 2310. In: Standard Methods for the Examination of Water and Wastewater. 19th ed. Washington, DC, 1995, pp 23–25.

4. Association of Official Analytical Chemists (AOAC). Waters and salt. Method 973.42. In: Official Methods of Analysis of AOAC International. 16th ed. Vol I. Gaithersburg, MD: AOAC International, 1997, p 2.

5. U.S. Environmental Protection Agency. Method 305.1. In: Compilation of EPA's Sampling and Analysis Methods. Boca Raton, Florida: CRC Lewis, 1996, p 19.

6. U.S. Environmental Protection Agency. Sample preservation. In: Methods for Chemical Analysis of Water and Wastes. EPA-600/4-79-020. Cincinnati, OH: U.S.E.P.A., 1983, p xvii.

7. R.K. Smith. Physical, biological and general chemical parameters. In: Handbook of Environmental Analysis. 3rd ed. New York: Genium, 1997, pp 199–200.

8. U.S. Environmental Protection Agency. Method 310.2. In: Compilation of EPA's. Sampling and Analysis Methods. City, State: CRC Lewis, 1996, pp 34–35.

9. American Public Health Association (APHA), American Water Works Association (AWWA), Water Environment Federation (WEF). Chap 2. Method 2320. In: Standard Methods for the Examination of Water and Wastewater. 19th ed. Washington, DC, 1995, pp 25–28.

10. Association of Official Analytical Chemists (AOAC). Waters, and salt. Method 973.43. In: Official Methods of Analysis of AOAC International. 16th ed. Vol I. Gaithersburg, MD: AOAC International, 1997, p 3.

11. American Public Health Association (APHA), American Water Works Association (AWWA), Water Environment Federation (WEF). Chap 2. Method 2340. In: Standard Methods for the Examination of Water and Wastewater. 19th ed. Washington, DC, 1995, pp 35–38.

12. R.K. Smith. Physical, biological and general chemical parameters. In: Handbook of Environmental Analysis. 3rd ed. New York: Genium, 1997, p 202.

13. T. Mueller, Z. Sun, M Kumar, K. Itoh, M. Murabayashi. The combination of photocatalysis and ozonolysis as a new approach for cleaning 2,4-dichlorophenoxyaceticacid polluted water. Chemosphere 36:2043–2055, 1998.

14. L. Liyanage, G. Finch, M. Belosevic. Sequential disinfection of *Cryptosporidium parvum* by ozone and chlorine dioxide. Ozone: Sci. Engin. 19:409–423, 1997.

15. H. Weinberg, H. Yamada. Post-ion-chromatography derivatization for the determination of oxyhalides at sub-PPB levels in drinking water. Anal. Chem. 70:1–6, 1998.

16. R. Joyce, H. Dhillon. Trace level determination of bromate in ozonated drinking water using ion chromatography. J Chromatogr. A 671:165–171, 1994.

17. H. Weinberg. Pre-concentration techniques for bromate analysis in ozonated waters. J. Chromatogr. 671:141–149, 1994.

18. American Public Health Association (APHA), American Water Works Association (AWWA), Water Environment Federation (WEF). Chap 4. Method 4500. In: Standard Methods for the Examination of Water and Wastewater. 19th ed. Washington, DC, 1995, pp 54–55.

19. C. Sawyer, P. McCarty. Residual chlorine and chlorine demand. In: Chemistry for Environmental Engineering. 3rd ed. Chap 9. New York: McGraw-Hill, 1978, pp 385–399.

20. R.K. Smith. Physical, biological and general chemical parameters. In: Handbook of Environmental Analysis. 3rd ed. New York: Genium, 1997, pp 223–230.

21. Methods Manual for Chemical Analysis of Water and Wastes. Alberta Environment. Environmental Protection Services Standards and Approvals Division Canada. NAQUADAT No 08101L, 1977, pp 1–4.

22. Laboratory Products Catalog and Electrochemistry Handbook. Beverly, MA: Orion Research, 1998.

23. F. Alava-Moreno, M. Valencia-González, A. Sanz-Medel, M. Díaz-García. Oxygen sensing based on the room temperature phosphorescence intensity quenching of some lead-8-hydroxyquinoline complexes. Analyst 122:807–810, 1997.

24. American Public Health Association (APHA), American Water Works Association (AWWA), Water Environment Federation (WEF). Chap 2. Method 4500C. In: Standard Methods for the Examination of Water and Wastewater. 19th ed. Washington, DC, 1995, pp 61–62.

25. J.A. Yiamouyiannis, B. Dean. Fluoridation and cancer: age dependence of cancer mortality related to artificial fluoridation. Fluoride 10:102–123, 1977.

26. J.C. Robins, J.L. Ambrus. Studies on osteoporosis IX. Effect of fluoride on steroid induced osteoporosis. Res. Communications Chem. Pathol. Pharmacol. 37:453–461, 1982.

27. N.J. Chinoy, M.V. Narayana. In vitro fluoride toxicity in human spermatozoa. Fluoride 27:231–232, 1994.

28. S.C. Freni. Exposure to high fluoride concentrations in drinking water is associated with decreased birth rates. J. Toxicol. Environ. Health 42:109–112, 1994. .

29. A.K. Susheela, P. Jethanandani. Circulating testosterone levels in skeletal fluorosis patients. J. Toxicol. Clin. Toxicol. 34:183–189, 1996.

30. A.K. Susheela, A. Kumar. A study of the effect of high concentrations of fluoride on the reproductive organs of male rabbits, using light and scanning electron microscopy. J. Reproduc. Fertil. 92:353–360, 1991.

31. J.R. Bucher, M.R. Hejtmancik, J.D. Toft, R.L. Persing, S.L. Eustis, J.K. Haseman. Results and conclusions of the National Toxicology Program's rodent carcinogenicity studies with sodium fluoride. Int. J. Cancer 48:733–737, 1991.

32. K. Cammann. Working with Ion-Selective Electrodes. Chemical Laboratory Practice. New York: Springer-Verlag, 1979, p 158.

33. R.G. Kelly, C.S. Brossia, K.R. Cooper, J. Krol. Analysis of disparate levels of anions of relevance to corrosion processes. J Chromatogr. 739:191–198, 1996.

34. M. Novic, B. Divjak, B. Pihlar, V. Hudnik. Influence of the sample matrix composition on the accuracy of the ion chromatographic determination of anions. J. Chromatogr. 739:35–42, 1996.

35. J.K. Killgore, S.R. Villaseñor. Systematic approach to generic matrix elimination via "heart-cut" column-switching techniques. J. Chromatogr. 739:43–48, 1996.

36. H. Kumagai, S. Tetsushi, K. Matsumoto, Y. Hanaoka. Determination of anions at the ng/L level by means of switching valves to eliminate the water-dip interference. J. Chromatogr. A 671:15–22, 1994.

37. J. Weiss, S. Reinhard, C. Pohl, C. Saini, L. Narayaran. Stationary phase for the determination of fluoride and other inorganic anions. J. Chromatogr. A 706:81–92, 1995.

38. Association of Official Analytical Chemist (AOAC). Waters, and salt. Vol. 1. Method 993.30. In: Official Methods of Analysis of AOAC International. 16th ed. Gaithersburg, MD: AOAC International, 1997, pp 28–30.

39. U.S. EPA. The Determination of Inorganic Anions in Water by Ion Chromatography. Method 300.0. Washington, DC: U.S. EPA, 1989.

40. American Public Health Association (APHA), American Water Works Association (AWWA), Water Environment Federation (WEF). Chap 4. Method 4500B. In: Standard Methods for the Examination of Water and Wastewater. 19th ed. Washington, DC, 1995, pp 65–69.

41. Methods Manual for Chemical Analysis of Water and Wastes. Alberta Environment. Environmental Protection Services Standards and Approvals Division Canada. NAQUADAT No 10301L, 1977, pp 1–2.

6

Organoleptical Properties of Water

Maria Anita Mendes and Marcos N. Eberlin
State University of Campinas—UNICAMP, Campinas, São Paulo, Brazil

I. INTRODUCTION

Water contaminated by organic compounds presents a major inconvenience to humans: undesirable organoleptical properties—odor, taste, and color. Consumers are sensitive to the slightest change in the organoleptical properties of water and use these esthetic parameters to judge water quality by association with the presence of possible hazardous contaminants (a reasonable concern), so the monitoring and control of the odor, taste, and color of water must be pursued as rigorously as for other chemical and biological water properties. Although high purity is demanded for most types of water, purity expectations and rules are even stricter for drinking water. Hence, continuous and accurate monitoring and control of drinking water organoleptical properties has been a great technical challenge and has called for the application of the most sensitive and precise analytical techniques (1).

Sources of water contamination by chemicals that affect the organoleptical properties of water are basically three: 1) water disinfection by chlorination that forms trihalomethanes (THMs); 2) domestic and industrial effluent discharge and agricultural and urban runoff that contaminates water with many synthetic organic compounds; and 3) natural sources such as the decomposition of vegetable matter by algae, actinomycetes, and other microorganisms that contribute mainly with humic acids and fulvic acids. Table 1 summarizes the acceptable levels set by the U.S. national primary drinking water contaminant standards, the most frequent sources, and the potential health effects of the chief organic water contaminants (2).

Groundwater is virtually free of organic contaminants; it contains total concentrations of organic carbon (TOC) less than 1 mg/L. Such TOC concentrations in drinking water are also low, normally below 10 mg/L, although a high TOC, which represents the presence of many kinds of organic compounds, including the natural humic and fulvic acids, does not necessarily indicate contamination by harmful organic compounds. Contaminated water, even with perceptible organoleptical properties, also displays very low concentrations of each contaminant; hence, a combination of very sensitive, sophisticated analytical techniques and sensory methods must be applied for reliable identification and quantitation of the organic compounds responsible for the organoleptical properties of water.

To correctly identify and remediate water organoleptical properties, several steps are often necessary: careful sampling and preservation of the water sample; efficient enrichment of the chemical contaminants; detection of individual chemicals by sensory and sensitive instrumental methods; determination of the origin of each contaminant; and estimation of the contribution of each contaminant to the total odor, taste, and color of water. Many thousands of organic compounds can be found as contaminants, so trace level detection and monitoring of water contaminants, especially on a routine basis, are not trivial tasks.

Table 1 U.S. Drinking Water Quality Standards, Frequent Sources, and Potential Health Effects of Common Water Contaminants

Contaminant	Maximum concentration (mg/L)	Potential health effects	Sources of drinking water contamination
Benzene	0.005	Cancer	Some foods, gas, drugs; pesticide, paint, and plastic industries
Carbon tetrachloride	0.005	Cancer	Solvents and their degradation products
p-Dichlorobenzene	0.075	Cancer	Room and water deodorants, "mothballs"
1,2-Dichloroethane	0.005	Cancer	Leaded gas, fumigants, paints
1,1-Dichoroethylene	0.007	Cancer, liver, and kidney effects	Plastics, dyes, perfumes, paints
Trichloroethylene	0.005	Cancer	Textiles, adhesives, and metal degreasers
1,1,1-Trichloroethane	0.2	Liver, nervous system effects	Adhesives, aerosols, textiles, paints, inks, metal degreasers
Vinyl chloride	0.002	Cancer	May leach from PVC pipe, formed by solvent breakdown
Acrylamide	TT[a]	Cancer, nervous system effects	Polymers used in sewage and wastewater treatment
Alachlor	0.002	Cancer	Runoff from use as herbicide on corn, soybeans, other crops
Aldicarb	—	Nervous system effects	Insecticide on cotton, potatoes, other crops; widely restricted
Aldicarb sulfone	—	Nervous system effects	Biodegradation of aldicarb
Aldicarb sulfoxide	—	Nervous system effects	Biodegradation of aldicarb
Atrazine	0.003	Mammary gland tumors	Runoff from use as herbicide on corn and noncropland
Carbofuran	0.04	Nervous, reproductive system effects	Soil fumigant on corn and cotton; restricted in some areas
Chlordane	0.002	Cancer	Leaching from soil treatment for termites
Chlorobenzene	0.1	Nervous system and liver effects	Waste solvent from metal degreasing processes
2,4-D	0.07	Liver and kidney damage	Runoff from use as herbicide on wheat, corn, rangelands, lawns
o-Dichlorobenzene	0.6	Liver, kidney, blood cell damage	Paints, engine cleaning compounds, dyes, chemical wastes
cis-1,2-Dichloroethylene	0.07	Liver, kidney, nervous and circulatory system effects	Waste industrial extraction solvents
trans-1,2-Dichloroethylene	0.1	Liver, kidney, nervous and circulatory system effects	Waste industrial extraction solvents
Dibromochloropropane	0.0002	Cancer	Soil fumigant on soybeans, cotton, pineapple, orchards

Table 1 Continued

Contaminant	Maximum concentration (mg/L)	Potential health effects	Sources of drinking water contamination
1,2-Dichloropropane	0.005	Liver, kidney effects; cancer	Soil fumigant, waste industrial solvents
Epichlorohydrin	TT[a]	Cancer	Water treatment chemicals, waste epoxy resins, coatings
Ethylbenzene	0.7	Liver, kidney, nervous system effects	Gasoline, insecticides, chemical manufacturing wastes
Ethylene dibromide	0.00005	Cancer	Leaded gas additives, leaching of soil fumigant
Heptachlor	0.0004	Cancer	Leaching insecticide for termites, very few crops
Heptachlor epoxide	0.0002	Cancer	Biodegradation of heptachlor
Lindane	0.0002	Liver, kidney, nervous system, immune system, circulatory system effects	Insecticides for cattle, lumber, gardens, restricted in 1983
Methoxychlor	0.04	Growth, liver, kidney, nervous system effects	Insecticides for fruits, vegetables, alfalfa, livestock, pets
Pentachlorophenol	0.001	Cancer, liver, kidney effects	Wood preservatives, herbicides, cooling tower wastes
PCBs	0.0005	Cancer	Coolant oils from electrical transformers, plasticizers
Styrene	0.1	Liver, nervous system damage	Plastics, rubber, resin, and drug industries; leachate from city landfills
Tetrachloroethylene	0.005	Cancer	Improper disposal of dry cleaning and other solvents
Toluene	1	Liver, kidney, nervous system, circulatory system effects	Gasoline additive, manufacturing, and solvent operations
Toxaphene	0.003	Cancer	Insecticide on cattle, cotton, soybeans; canceled in 1982
2,4,5-TP	0.05	Liver, kidney damage	Herbicide on crops; rights-of-way, golf courses; canceled in 1983
Xylenes (total)	10	Liver, kidney, nervous system effects	By-product of gasoline refining, paints, inks, detergents
Adiptae, [di(2-ethylhexyl)]	0.4	Decreased body weight	Synthetic rubber, food packaging, cosmetics
Dalapon	0.2	Liver, kidney effects	Herbicides on orchards, beans, coffee, lawns, roads, railways
Dichloromethane	0.005	Cancer	Paint stripper, metal degreaser, propellant, extractant
Dinoseb	0.007	Thyroid, reproductive organ damage	Runoff of herbicide from crop and noncrop applications
Diquat	0.02	Liver, kidney, eye effects	Runoff of herbicide on land and aquatic weeds
Dioxin	3×10^{-8}	Cancer	Chemical production by-product, impurity in herbicides

Table 1 Continued

Contaminant	Maximum concentration (mg/L)	Potential health effects	Sources of drinking water contamination
Endothall	0.1	Liver, kidney, gastrointestinal effects	Herbicide on crops and land aquatic weeds; rapidly degraded
Endrin	0.002	Liver, kidney, heart damage	Pesticides on insects, rodents, birds; restricted since 1980
Glyphosate	0.7	Liver, kidney damage	Herbicide on grasses, weeds, brush
Hexachlorobenzene	0.001	Cancer	Pesticide production waste by-product
Hexachlorocyclopentadiene	0.05	Kidney, stomach damage	Pesticide production intermediate
Oxamyl (vydate)	0.2	Kidney damage	Insecticide on apples, potatoes, tomatoes
PAHs [benzo(a)pyrene]	0.0002	Cancer	Coal tar coatings, burning organic matter, volcanoes, fossil fuels
Phathalate, [di(2-ethylhexyl)]	0.006	Cancer	PVC and other plastics
Picloran	0.5	Kidney, liver damage	Herbicide on broadleaf and woody plants
Simazine	0.004	Cancer	Herbicide on grass sod, some crops, aquatic algae
1,2,4-Trichlorobenzene	0.07	Liver, kidney damage	Herbicide production, dye carrier
1,1,2-Trichloroethane	0.005	Kidney, liver, nervous system damage	Solvent in rubber, other organic products; chemical production wastes
Bromodichloromethane	See TTHMs	Cancer; liver, kidney, reproductive effects	Drinking water chlorination by-product
Bromoform	See TTHMs	Cancer; nervous system, liver, kidney effects	Drinking water chlorination by-product
Chloral hydrate	Zero	Liver effects	Drinking water chlorination by-product
Chloroform	See TTHMs	Cancer; liver, kidney, reproductive effects	Drinking water chlorination by-product
Dibromochloromethane	See TTHMs	Nervous system, liver, kidney, reproductive effects	Drinking water chlorination by-product
Dichloroacetic acid	See HAA5	Cancer; reproductive, developmental effects	Drinking water chlorination by-product
Haloacetic acids (HAA5)	0.06	Cancer and other effects	Drinking water chlorination by-product
Trichloroacetic acid (P)	See HAA5	Liver, kidney, spleen, developmental effects	Drinking water chlorination by-product
Total trihalomethanes (TTHMs)	0.10	Cancer and other effects	Drinking water chlorination by-product

Source: Adapted from Ref. 2.
[a]TT: Treatment technique is required.

A. Odor and Taste

Odor and taste depend on contact of a stimulating substance with the appropriate human receptor cell. The stimuli are chemical in nature, and odor and taste are classified as chemical senses. Odor is a combination of olfactory (smell) and trigeminal (feeling) sensations detected by the nose while sniffing (3). Flavor is a combination of aromatic and basic tastes and feeling factors perceived while tasting. Tastes are classified primarily as sweet, sour, salty, and bitter; each of these classes of taste stimulates specific taste buds located on the front, back, and sides of the tongue. Thus, for a sample to be properly tasted, it must be dispersed over all the surfaces of the tongue (4).

Pure water produces no odor or taste sensations; hence, unlike most water-quality parameters, taste and odor are immediately perceived, and humans and other animals use adverse sensory response to avoid potentially toxic water (5–8). Odor and taste are major quality factors affecting the acceptability of drinking water, so extensive research efforts have been devoted to understanding the origins of, and ultimately to develop remediation techniques for, water odor and taste.

Odors and tastes of water can sometimes be related to the presence of specific micropollutants, such as phthalates (plasticizer or stagnant taste), aldehydes and ketones (sweet, solvent, or fishy taste), alkenes and fatty carboxylic acids (fatty or oily taste), and geosmin and methylisoborneol (earthy or musty taste) (9). But odors and tastes originating from biological or industrial sources often result from a complex mixture of volatile and other organic compounds; in some cases the inorganic contaminants most often associated with pipe leachate or corrosion products also confer a perceptible odor and taste on water. Because of the several orders of magnitude differences among the odor and taste threshold concentrations of organoleptical compounds, some chemical contaminants contribute much more than others to the global taste and odor of water. Often, the components imparting the characteristic odor or taste are present in the lowest concentrations.

B. Color

Pure water is colorless; hence, color in water also undoubtedly indicates some type of contamination, and consumers relate colored water to potentially toxic water. The most common contaminants imparting color to water are natural humic and fulvic acids, natural metallic ions such as iron and manganese, plankton, weeds, and industrial wastes. Owing to recent legisla-

tion and higher consumer expectations, more attention has been paid to water color, and excessive color is normally removed to make water suitable for general and industrial application and to eliminate potential health risks.

Color in water is usually expressed as °hazen, or mg/L Pt, and measured in a Nessleriser or comparator. Alternatively, absorbance measurement at 400 nm can be used. The term *color* is used to mean true color from which turbidity has been removed. The term *apparent color* includes the color due to solutes and that resulting from suspended matter. The apparent color is determined by measuring the sample as taken, whereas to measure the true color the sample is filtered first through a 0.45-μm filter. To remove the two forms of water color, different remediation methods are required.

II. WATER PURIFICATION

The quality of raw water that must be purified for drinking or other industrial and commercial applications varies widely, from almost pristine to highly polluted; further, many types and a broad range of pollutant concentrations are found. Owing to this wide spectrum of pollutants at variable concentrations, different water purification processes are normally applied (10,11). Figure 1 displays a pictorial representation of commonly used methods for water remediation.

A. Aeration

Aeration helps to remove dissolved gases and volatile organic compounds that may have a detectable odor, such as the foul-smelling H_2S. Water aeration also helps to reduce organoleptical contamination via oxidation of the most easily oxidized organic compounds. The remaining organic compounds can be removed by activated charcoal treatment, but this process is relatively expensive. Water treatment continues after aeration by capture of colloidal particles via coagulation with Fe or Al salts, sedimentation, dual-media (anthracite and sand) filtration, hardness removal, and disinfection by chlorine or ozone treatment.

B. Chlorination

Chlorination is performed mainly by dissolving molecular chlorine gas or calcium hypochlorite, $Ca(OCl)_2$ in the water; their equilibrium reactions with water (Eqs. 1 and 2) form hypochlorous acid (HOCl):

LIVERPOOL JOHN MOORES UNIVERSITY
LEARNING SERVICES

Aeration   **Setting and precipitation**   **Hardness removal**   **Disinfection**

Al or Fe salt to precipitate colloids   **Phosphate**

→ **Consumer**

Air   **Ca^{++} pptd as phosphate**   **Cl$_2$ or ozone**

○ **Water**
● **Suspended particles**

Fig. 1 Common stages of drinking water purification. (Adapted from Ref. 10.)

$$Cl_2(g) + H_2O \rightarrow HOCl(aq) + H^+ + Cl^- \quad (1)$$

$$OCl^-(aq) + H_2O \rightarrow HOCl(aq) + OH^- \quad (2)$$

HOCl readily permeates cell membranes, killing most microorganisms. One advantage of chlorination is that chlorine or calcium hypochlorite remain dissolved, so the water is protected from subsequent microorganism contamination. Most residual chlorine in water exists as chloramines (NH_2Cl, $NHCl_2$, and NCl_3), which are formed by its reaction with dissolved ammonia. Remediation by chlorination can, however, accentuate the problem of odor and taste when the water is contaminated with organic matter, particularly when phenols are present. Chlorine readily substitutes hydrogen atoms of phenol rings to yield toxic chlorinated phenols that often have an offensive odor and taste.

Chlorination of water also produces the undesirable thrihalomethanes (THMs, CHX_3) as by-products; hence, THM contamination is a major concern for monitoring drinking water quality. The main THM contaminant is chloroform, $CHCl_3$, which is produced when hypochlorous acid reacts with dissolved organic matter, that is, mainly humic acids (water-soluble, non-biodegradable components of decayed plant matter). Of particular importance are humic acids with 1,3-dihydroxibenzene rings, as in the elementary case shown here:

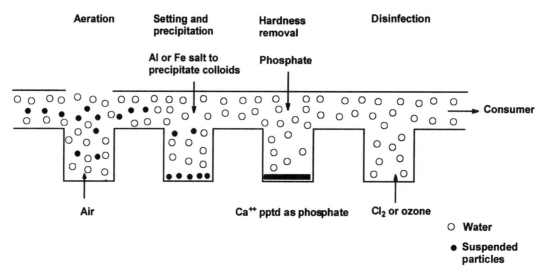

Chlorination occurs readily at the C-2 carbon atom. Ring opening between C-2 and C-3 forms an aliphatic chlorinated ketone, which is further trichlorinated at the terminal carbon by HOCl; then replacement of CCl_3 by OH occurs readily in water to yield chloroform.

Bromoform, $CHBr_3$, is similarly produced by the action on humic materials of hypobromous acid, HOBr, which is formed when bromide ions are present in water, displacing chlorine from HOCl (Eq. 3):

$$HOCl(aq) + Br^-(aq) \rightarrow HOBr(aq) + Cl^- \quad (3)$$

C. Ozonation

Ozonation, used primarily to rid drinking water of harmful bacteria and viruses, also has a dramatic effect on the organoleptical properties of water (12,13). Ozonation of the humic and fulvic acid matter and other naturally occurring compounds yields a wide spectrum of products of varying volatility. Musty and muddy odors caused by naturally occurring compounds such as geosmin and 2-methylisoborneol are at least partially removed through ozone oxidation. Ozonation, however, also results in the formation of new odor-causing compounds, some described as fruity or fragrant, although less pleasant compounds are also formed. Ozonation also often fails to eliminate phytoplakton odors, causing the musty fishy odors to be transformed into other odors.

III. ANALYSIS OF ORGANIC CONTAMINANTS

Classic schemes of organic analysis requiring milligram-sized quantities of the extracted organic mixture were used in the first approaches to the analysis of organic water contaminants (14). Extraction by carbon filter and solvent also required hundreds to thousands of gallons of water to be processed. Increasing concern with drinking water quality and the need to control its organoleptical properties has driven the development of a fast-growing number of preconcentration and instrument techniques that nowadays allow the efficient qualitative and quantitative analysis of trace chemical contaminants in very reduced amounts of water samples.

These sensory methods and sophisticated preconcentration and instrument methods are now in routine use detecting organics at low parts-per-trillion levels. For some analytes, even low parts-per-quadrillion detection limits have been reported. Recent developments in extraction and preconcentration methods, and of direct online and real-time systems, are leading the way to even greater analytical performance.

A. Sensory Methods

The analytical chemist must often rely on sensitive instrumentation to identify the chemicals responsible for the organoleptical properties of water. But in problems involving malodor identification, the nose is also a useful preliminary guide to the analytical effort (15).

1. Flavor Profile Analysis

Flavor profile analysis (FPA) is an empirically based, perceptual judgment of both the elements and the structure of the impressions of odor and taste (4,16). The FPA method, which was initially developed for use by the food industry, is becoming a major method for the study of sensory characteristics of water. It provides an overall sensory description of the water by using a taste-and-odor panel, which consists of several individuals, at least four, trained in sensory perception. The panel members work as a team to reach a composite judgment; flavor attributes are determined by tasting; odor attributes (aroma) are determined by sniffing.

The method often allows various flavor or odor attributes to be determined per sample and each attribute's strength to be measured. Initially, each panelist records his perceptions without discussion. Once each individual has made an independent assessment, the panel discusses its finding and reaches a consensus in developing a flavor profile for the sample. Since descriptors can be subjective in nature, and often various odors are present, the reproducibility and accuracy of results depend heavily on the training and experience of each panelist (17). Therefore, FPA analysis requires well-trained panelists and data interpreters.

Chemical analysis is often run in parallel with the FPA. When there are sufficient FPA and chemical analysis results for many samples on the same water source over a period of time, a statistical correlation can be developed (Fig. 2). The statistical correlation developed between FPA and chemical analysis is a presumptive that must be confirmed by testing the odor of the chemical identified, for instance, by sensory gas chromatography (GC) (17). The relationship between FPA and sensory GC helps to confirm the presumptive relationship developed between the chemical analysis and the FPA. A final confirmation of the chemicals that cause the characteristic odor are completed by panel testing of the chemicals identified by sensory GC (Fig. 3).

The FPA method requires reference standards reflecting the composition of real water samples for panel screening and training, to establish a common vocabulary to standardize the application of the method. A "flavor wheel," which has been adopted by many water utilities that use the FPA method, presents a common classification scheme of the main tastes and odors occurring in drinking water and the chemicals that are used to represent them (18). Four classes of tastes, eight classes of odors, and one class of "nosefeel-mouthfeel" are represented in the flavor wheel. Other FPA odor reference libraries, including odors commonly attributable to microbiological and industrial sources, have been developed for drinking water (19).

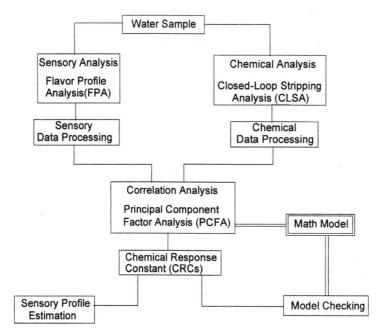

Fig. 2 A general mathematical approach to develop a correlation between sensory data and the corresponding chemical analysis of taste and odor. (Adapted from Ref. 7.)

2. Sensory Gas Chromatography

Studies on the origin of taste and odor in water are usually intended to connect the off-flavor to certain activities, processes, or organisms and ultimately to identify the compound causing the specific odor and taste. Sensory GC, which has been developed and used successfully by the perfume and food industries, is a powerful technique for odor identification. The sample extract is chromatographically separated, and chemicals are evaluated for odor response at an olfactory port of the individual GC peaks eluting from the GC column (20). A qualitative descriptor is recorded often by using the flavor wheel, and a quantitative level is assigned to each odors peaks. The sensory GC procedure makes a direct link from odors to instrumental response. Sensory GC analysis of individual chemicals can also account for the synergistic or antagonistic relationships among chemicals.

The GC effluent is transferred up to the glass cup where the olfactory port is located. The system contains a heated transfer section from the GC oven to a detection cone (Fig. 4). Air drawn from the gas chromatography oven is used to heat the transfer line to avoid condensation of the analytes that are being transferred. The glass cone is purged with humidified air to prevent a loss of olfactory sensitivity from the drying out of the nasal mucous membranes over long testing periods. Two aliquots of the sample are injected separately: the

first injection is instrumentally detected; 20 seconds later, the second injection is directed to the olfactory port for sensory detection. When the peak appears on the instrumentally detected chromatogram, the trained odor observer starts smelling the eluting compounds at the olfactory outlet.

B. Preconcentration Methods

Chemicals imparting organoleptical properties to water vary greatly in nature and number, and they are present in very low concentration, often in low part-per-trillion concentration; many of them have not been identified so far. Efficient trace organic analysis of water must therefore rely on a combination of very efficient extraction and enrichment methods for organic compounds in water with highly sensitive chemical and instrumental analysis, which are needed for establishing cause–effect relationships and for understanding the complex nature of the water organoleptical properties of odor, taste, and color (21).

1. Preliminary Sample Treatments

Analysis of organic material in water frequently involves three main steps: sample concentration to improve sensitivity of the analytical method; partioning of the complex organic mixtures into more homogeneous fractions for compound class analysis or as a

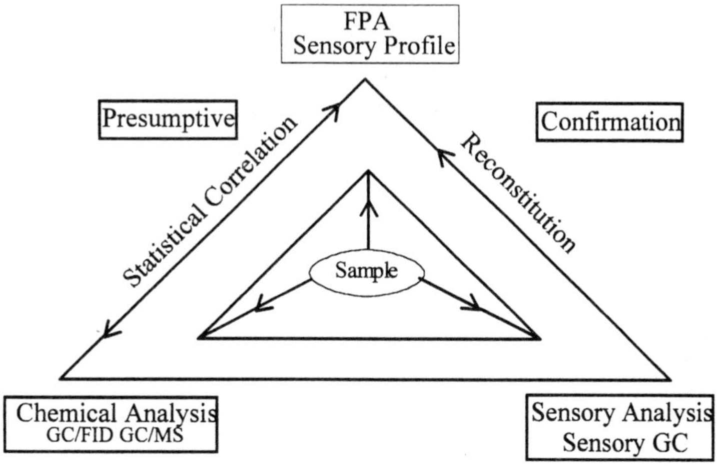

FPA
Sensory Profile

Presumptive

Statistical Correlation

Confirmation

Reconstitution

Sample

Chemical Analysis
GC/FID GC/MS

Sensory Analysis
Sensory GC

Fig. 3 Schematic diagram of the relationship between instrumental and sensory techniques. (Adapted from Ref. 17.)

preparatory step for more specific organic analysis; and isolation of the organic material of interest from the solvent and solute matrix. Table 2 presents examples of common ranges of concentrations for organic constituents in various types of water, which span more than five orders of magnitude (14). Certain organic constituents in waste waters are often sufficiently concentrated so that direct analysis is possible. But natural water and drinking water samples, whose total organic carbon (TOC) concentration is usually less than 10 mg/L and whose specific organic compound concentrations are in micrograms per liter (parts per billion) or even nanograms per liter (parts per trillion), often must be concentrated by two or three orders of magnitude before analysis. Sample preservation is also a major concern. Practical aspects of environmental sampling

must be observed so as to ensure the integrity of the sample from the time of collection until the time of analysis, and to minimize both loss and degradation of the analyte.

Therefore, correct execution of the concentration procedures are critical for the accuracy, precision, and sensitivity of the analytical process. Handling samples with such low contamination levels must be done with great care in order to avoid losses and contamination; when using solvents in the extraction procedure, they must be very pure. Sampling, storage, or enrichment must be performed carefully to avoid a loss of trace contaminants.

2. Headspace Methods

Static and dynamic headspace methods are powerful methods for the extraction of volatile organic compounds (VOCs) from water. Headspace methods rely on the establishment of equilibrium partitioning of an analyte between the solution and the gas phase and on the analysis of the vapor phase (22).

a. Static Headspace. The static headspace method is a rapid, simple, and easily automated method for extracting and concentrating VOCs from water (23). Commonly, a serum vial is partially filled with the sample and placed in a thermostated bath, and the closed airspace above the contaminated water is sampled and analyzed directly. Trace organics that favor the vapor phase are determined, often in the 10–100-ppb range, and most chemicals amenable to headspace concentration are also amenable to gas chromatography analysis.

The major advantage of static headspace over dynamic headspace methods, such as purge and trap (see

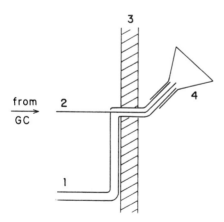

Fig. 4 A simple device used for chromatographic sniffing: 1—heated copper tubing; 2—capillary column; 3—GC oven wall; 4—sniffing funnel. (Adapted from Ref. 8.)

Table 2 Typical Total Organic Carbon, Dissolved Organic Carbon, and Particulate Organic Carbon Concentrations in Different Types of Water

Water type	Total organic carbon (mg/L)	Dissolved organic carbon (mg/L)	Particulate organic carbon (mg/L)
Groundwater	0.7	0.7	—
Seawater	1.1	1.0	0.1
Drinking water	2.0	—	—
Surface water (lakes)	7.7	7.0	0.7
Surface water (rivers)	8.0	5.0	3.0
Untreated domestic sewage	200	80	120
Ammoniacal liquors from coal and oil-shale conversion processes	10,000	—	—

Source: Adapted from Ref. 14.

later), is its simplicity. The method avoids the use of adsorbents and the carryover problems associated with the use of traps, and it is unaffected by the foaming problems caused by the presence of surfactants.

b. Dynamic Headspace. Detection limits in static headspace methods are restricted by both the equilibrium concentration of the organic chemical in the gas phase and the limited amount of headspace vapor that can be sampled and analyzed. In dynamic headspace methods, the most common of which are purge and trapping (P&T) and closed-loop stripping (CLS), improved detection limits are achieved by actively removing the analyte from the water via an inert gas stream bubbled through the sample. Purging is continued for a measured time to near quantitative removal of the organic contaminant from the water. The organics are therefore continuously trapped from the gas stream and subsequently released for analysis.

Purge and Trap: Typical detection limits from 1 to 100 ppb, with precision of 1–10%, are achieved with purge and trapping (P&T), a relatively simple, well-established technique for quantitative extraction of VOCs from water samples (Fig. 5). The method, first introduced by Bellar and Lichenberg (24), has become widely accepted, owing to little need for sample preparation, good reproducibility, and superior sensitivity to the static headspace technique. The VOCs are often extracted by placing the water sample in a purge vessel and by passing an inert gas (often helium or nitrogen) through the sample; and adsorbent traps the purged compounds, which are further thermally desorbed, removed to a cold trap via a heated, deactivated fused-silica column, and transferred for analysis by a rapid heating of the trap. Tenax, charcoal, and Carbopack-B/

Carbosive are the adsorbents most often used, owing to their thermal stabilities. However, Tenax thermal decomposition products (benzene and alkyl benzenes, acetophenone, benzaldehyde, benzoic acid, ethylene oxide, α-hydroxiacetophenone, and phenol) have been reported to interfere in P&T trace analysis (25). To ensure good reproducibility and high extraction efficiencies, the P&T extraction conditions must be closely controlled, and Environmental Protection Agency

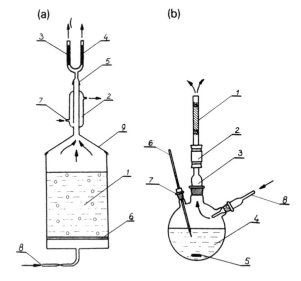

Fig. 5 Schematic diagrams of two purge and trap apparatuses for trapping volatiles on solid sorbents: (a) 1—sampler body; 2—condenser; 3,4—trap tubes; 5—tube holder; 6—glass frit; 7—reducer; 8—gas inlet; 9—reducer. (b) 1—trap tube; 2—PTFE union; 3—reducer; 4—sampler; 5—magnetic stirring bar; 6—thermometer; 7—thermometer adapter; 8—purged gas entrance. (Adapted from Ref. 22.)

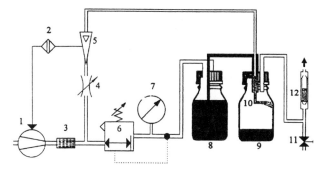

Fig. 6 Overview of pneumatic functions in the spray and trap apparatus: 1—pump; 2—coil sensor; 3—charcoal filter; 4—flow restriction; 5—flowmeter; 6—pressure regulator; 7—pressure gauge; 8—analyte reservoir; 9—recipient bottle; 10—spray nozzle; 11—settled-aerosol release; 12—Tenax trap. (Reproduced with permission from Ref. 26.)

(EPA) methods require strict adherence to the P&T extraction conditions. Several commercial P&T systems are available, and the technique is easily automated. A major limitation of P&T is that water-miscible VOCs are not efficiently extracted.

Spray and Trap: The P&T method uses gas bubbles for dynamic, nonequilibrium extraction of organic contaminants from water. An alternative, which increases the surface area of the liquid, is the spray and trap (S&T) method, by which organics are volatized by spraying (26). Figure 6 displays the main operating steps used in the S&T technique. The spray extraction is a concurrent atomization process performed in a simple spray nozzle aeration ejector. The dispersion is discharged and expanded through a venturi-type nozzle, and adjusted to produce optimum-size droplets nearly 100 μm in diameter. The extraction air is purified by activated charcoal.

Closed-Loop Stripping: Closed-loop stripping (CLS), shown schematically in Fig. 7a, is a common and simple dynamic headspace method for the ultratrace (low ppt) semiquantitative determination of organic compounds of medium volatility and molecular weight in water (27). The outstanding concentration factor of CLS, without the need of an evaporation step, makes the technique useful for the screening of a large number of contaminants.

As with static headspace and P&T techniques, CLS relies on the partitioning of analytes between the dissolved phase and the vapor phase. The organic substances are liberated from water and transferred to a very small amount of adsorbent (usually charcoal) in a hermetically closed circuit system, in which the carrier

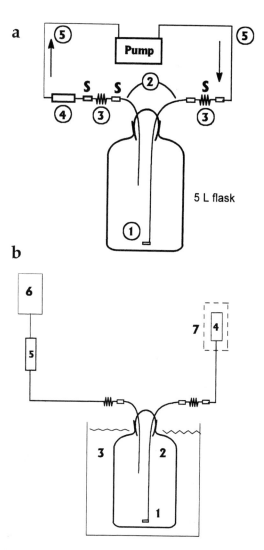

Fig. 7 Schematics of (a) a closed and (b) an open stripping system: (a) 1—coarse glass for gas inlet; 2—fused glass–metal connection; 3—coiled steel tubing; 4—filter holder; 5—stainless steel tubing, S-Swaglok fittings. (b) 1—coarse gas for inlet; 2—sample reservoir; 3—thermostatic water bath; 4—filter; 5—activated carbon filter for gas cleaning; 6—nitrogen gas cylinder; 7—oven. (Adapted from Ref. 27.)

may be an inert gas or water vapor. The organic compounds are then dissolved from the charcoal, separated by capillary gas–liquid chromatography, and identified, for instance, by either gas or liquid chromatography/mass spectrometry.

In CLS, the stripping and trapping are successfully achieved by using a closed circuit, in which substances that break through the filter are automatically recycled, therefore small absorbent filters can be used. The closed circuit also minimizes carrier gas contamination

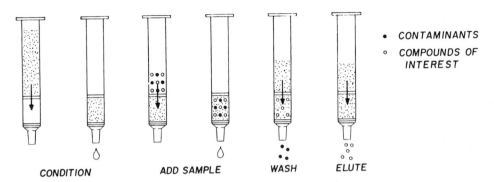

Fig. 8 Basic procedure of cartridge solid-phase extraction. [Adapted from *The Supelco Guide to Solid Phase Extraction*, 2nd ed., (Bellefonte, PA: Supelco Inc., 1988).]

problems. The first step in CLS extraction is to equilibrate the sample (usually 250–1000 ml) at the appropriate temperature (approximately 40°C). Gas is then purged through the sample to extract the analytes, which are collected on a small trap (1.5–5 mg). After an extraction period, the trap is removed and the VOCs are recovered with small amount (15 μl) of carbon disulfide or CH_2Cl_2. The small amounts of solvent required for extraction, the trap material, and the gas used

for extraction lead to low levels of contamination in CLS.

The CLS technique is limited, however, to the analysis of clean samples, such as drinking water. Samples with relatively high concentrations of VOCs overload the charcoal trap; CLS recoveries for highly volatile compounds are low, and analytes may coelute with the solvent during chromatographic analysis. Thermal desorption in CLS minimizes coelution problems of

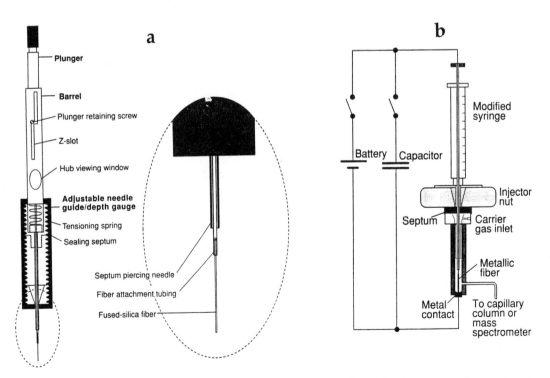

Fig. 9 (a) Typical device used in solid-phase microextraction. The fused-silica fiber is connected to, and protected by, a stainless steel tubing contained in a specially designed syringe. (b) Device for electrically heating an SPME fiber made from a metallic material. (Adapted from Ref. 30.)

Table 3 Main Techniques for Extraction/Concentration of Organic Contaminants from Water

Technique	Applicability	Detection lmit (MS)	Analysis time	Solvent use (ml)
Headspace	VOCs	ppb	30 min	0
P&T	VOCs, PVOCs	ppb	30 min	0
CLS	VOCs	ppt	2 h	0
LLE	SVOCs	ppt	1 h	200
SPE	SVOCs, POCs	ppt	30 min	50
SPME	VOCs, SVOCs	ppt	5 min	0
MIMS	VOCs	ppb	Real time	0
T-MIMS	VOCs, SVOCs	ppt	5–30 min	0

P&T: purge and trap; CLS: closed-loop stripping; LLE: liquid–liquid extraction; SPE: solid-phase extraction; SPME: solid-phase microextraction; MIMS: membrane introduction mass spectrometry; T-MIMS: trap MIMS; VOCs: volatile organic compounds; SVOCs: semivolatile organic compounds; PVOCs: polar volatile organic compounds.
Source: Adapted from Ref. 1.

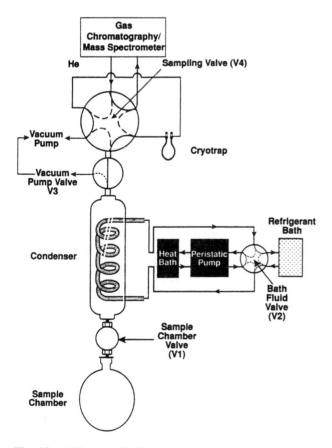

Fig. 10 A Vacuum distillation apparatus. (Reproduced with permission from Ref. 31.)

highly volatile compounds. Polar organic compounds are either poorly purged or not recovered at all.

With open systems, CLS purging may be undertaken at higher temperatures, thus increasing purging efficiencies. Purging experiments performed at temperatures ranging between 30 and 90°C have given comprehensives results concerning the effects of temperature on several groups of organic compounds, including phenols (28).

Figure 7b shows the straight-and-open system, a modification of closed-loop stripping that reduces the problem of contamination of the standard system. The open CLS system is also more flexible in regard to stripping temperature and purging gas flow rate, whereas it also allows the stripping technique to be utilized without sacrificing sensitivity or capacity. A condition for the success of the open system is the perfect performance of the filter. In closed CLS, compounds not adsorbed when passing the filter for the first time can be adsorbed later; in the open system, all compounds have to be adsorbed immediately.

3. Liquid–Liquid Extraction

For hydrophobic semi- and nonvolatile organic contaminants in water, liquid–liquid extraction is the most common method of extraction and concentration (29). The solvent must be water imiscible and have a high solvent/water partition coefficient; small solvent-to-water ratios are used to achieve maximum sensitivity. After extraction with solvent of the highest available purity, the organic phase is often dried, typically with anhydrous sodium sulfate, and the volume of the extract is reduced to improve sensitivity. Serial solvent extraction is required to obtain the best, near-quantitative recoveries of analytes. Salts such as NaCl are frequently added to the water to improve extraction efficiencies. Dichloromethane, CH_2Cl_2, is by far the most used solvent in liquid–liquid extraction, but its continued use has been discouraged because of its potential health hazards. However, a better or equivalent substitute has been hard to find (29).

Polar compounds that form hydrogen bonds with water, such as those with COOH and OH groups, are difficult to remove from water by liquid–liquid extraction. Acidification using pH near 2 are applied to force these compounds toward their less polar molecular forms. Interference problems with trace solvent contaminants or preservatives are always a concern in liquid–liquid extractions. The formation of emulsions is also a frequent complication in liquid–liquid extraction, which often calls for high-speed centrifugation.

4. Solid-Phase Extraction

Solid-phase extraction (SPE) is rapidly replacing liquid–liquid extraction in many analytical procedures. The advantages of SPE include simplicity, consumption of little solvent, selective extraction, ease of automation, and parts-per-trillion detection limits. In SPE, analytes are adsorbed from water onto a solid support and then extracted either by liquid extraction with a minimum volume of solvent or by thermal desorption. Many bonded-phase supports are used, such as XAD resins and silica-impregnated Teflon disks, which afford selective retention of either analytes or interferents.

The most popular SPE method employs commercially available cartridges of variable sizes filled with several phases bonded to a silica matrix. The silica-based sorbents provide controlled pore and particle size

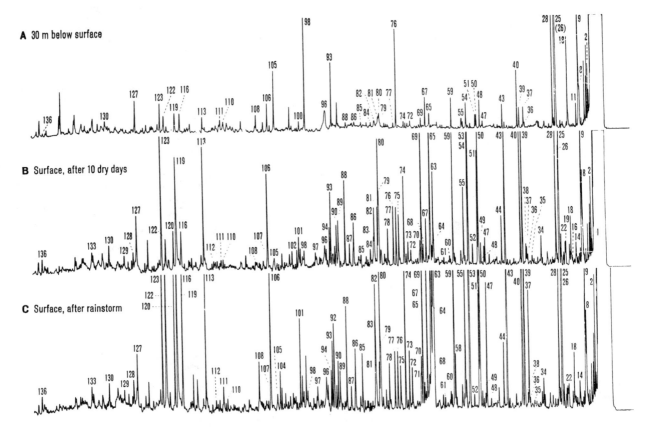

Fig. 11 Gas chromatograms from Lake Zürich water samples taken in July 1973 at the same location but at different depths or under different weather conditions. (Reproduced with permission from Ref. 33.)

and large surface area, whereas silica is both mechanically strong and rigid and chemically stable. Figure 8 summarizes the basic procedures used in SPE: The water sample is passed through the cartridge; the analytes are adsorbed by either London dispersion forces, hydrogen bonding, or electrostatic interactions; the cartridge is washed to remove interferents; and the analytes are finally desorbed with a small volume of solvent.

5. Solid-Phase Microextraction

Figure 9 shows a typical device used in solid-phase microextraction of VOCs from water (30). The method uses uncoated fibers or fibers coated with many types of chemicals, such as methyl silicon, poly(dimethylsiloxane), liquid crystal polyacrylate, and polyimide. The common procedure involves placing the fiber in the water sample; the sample is stirred, equilibrium is established in a few minutes, and the analyte is extracted into the fiber for a predetermined time. The fiber is transferred directly into the injection port of a gas chromatograph, where analytes are thermally released for analysis (Fig. 9b). With SPME, sub–ppb detection limits and good linearity and precision are routinely achieved.

Table 3 compares the overall performance of the most common techniques for the extraction/concentration of organic contaminants from water. The membrane introduction mass spectrometry (MIMS) and trap MIMS (T-MIMS) techniques, which allow the simultaneous performance of extraction, concentration, and analysis, will be discussed further in Sec. III.C.4.

6. Vacuum Distillation

Vacuum distillation has been developed as an alternative technique for determining VOCs in water and other environmental matrixes (31). The advantages of the technique compared with P&T includes the elimination of the adsorbent trap and the direct interface to the analytical instrument, such as a gas chromatograph.

Figure 10 shows a vacuum distillation apparatus. A sample chamber is connected to a condenser that is attached to a heated six-port sampling valve (V4). This valve is connected to a condenser, a vacuum pump, a cryotrap, and a gas chromatograph/mass spectrometer (GC/MS). The condenser coil temperature is controlled by a circulating system that comprises a cryogenic cooler with a cold reservoir and a temperature bath. The routing of the bath fluid, isopropanol, is controlled by valve V2. First, valve V4 is switched to the distil-

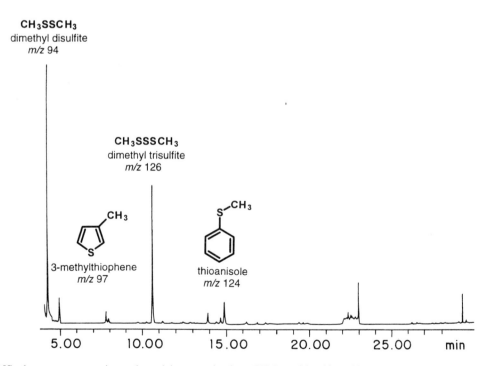

Fig. 12 GC/MS chromatogram using selected ion monitoring (SIM) to identify sulfur-containing compounds in water. The peaks for dimethyl disulfide, 3-methylthiophene, and thioanisole also match the authentic compounds for retention times. (Adapted from Ref. 35.)

Table 4 Water Contaminants Separated and Identified by GC
(See Fig. 11)

No. Substance	No. Substance
2. Heptane	75. C_4-Benzene
8. Octane	76. Dichlorobenzene
9. Benzene	77. Terpene $C_{10}H_{16}O$
11. Isononane	78. Dimethylethylbenzene
14. Carbon Tetrachloride	79. C_5-Benzene
18. Trichloroethylene	80. 1,2,4,5-Tretramethylbenzene
19. Nonane	81. C_5-Benzene
22. Isododecane	82. 1,2,3,5-Tetramethylbenzene
25. Tetrachoroethylene	83. Terpene $C_{10}H_{14}O$
26. Toluene	84. Tridecane
28. Dimethyl Disulphide	85. Terpene $C_{10}H_{16}O$
34. Isododecane	86. C_4-Benzene
35. Aminomethylpyridine	87. Terpene $C_{10}H_{16}O$
36. Decane	88. Camphor
37. Ethylbenzene	89. C_5-Benzene
38. 2-Methylpentanol-2	90. C_5-Benzene
39. 1,4-Dimethylbenzene	93. Internal standard (1-chlorodecane)
40. 1,3-Dimethylbenzene	94. C_5-Benzene
43. 1,2-Dimethylbenzene	97. Tetradecane
44. Isoundecane	98. Trichlorobenzene
47. *n*-Propylbenzene	101. Cyclocitral
48. Chlorobenzene	102. Caran-4-ol
49. Undecane	105. Naphthalene
50. 4-Ethyltoluene	106. Pentadecane
51. 3-Ethyltoluene	107. 1-Phenyl-2-thiapropane
52. Limonene	108. 1-Pentadecene
53. Cineol	110. 2-Methylnaphthalene
54. 1,3,5-Trimethylbenzene	111. Hexadecane
55. 2-Ethyltoluene	112. 1-Methylnaphthalene
59. 1,2,4-Ethylbenzene	113. Molecular weight 182
60. Isopropyltoluene	116. Dimethylnaphthalene
61. Isododecane	119. Heptadecene
63. Dimethyl trisulphide	120. Diphenyl
64. Propyltoluene	122. 1-Heptadecene
65. 1,2,3-Trimethylbenzene	123. Diphenyl Ether
67. Isobutylbenzene	127. Tri-*n*-butyl Phosphate
68. Dodecane	128. Octadecane
69. Dimethylethylbenzene	129. Acenaphthene
70. Methylpropylbenzene	130. tert-Butyl acetophenone
72. Dimethylethylbenzene	133. Nonadecane
73. Methylisopropylbenzene	136. Eicosane
74. Methylisopropylbenzene	

Source: Adapted from Ref. 33.

lation position, which allows the cryotrap and condenser to be evacuated. The vacuum distillation begins when valve V1 is opened, allowing sample vapors to pass over the condenser coil (chilled to below −5°C), resulting in vapor water condensation. The vapors not condensed are cryogenically collected in the cryotrap tube, chilled with liquid nitrogen. After 10 min of vacuum destilation, valve V4 is switched to the desorbing position, connecting the cryotrap to the GC/MS. The

condensate is then thermally desorbed and transferred using helium as the carrier gas.

C. Instrumental Analysis

1. Gas Chromatography

While many principles and techniques have been developed for the extraction and preconcentration of or-

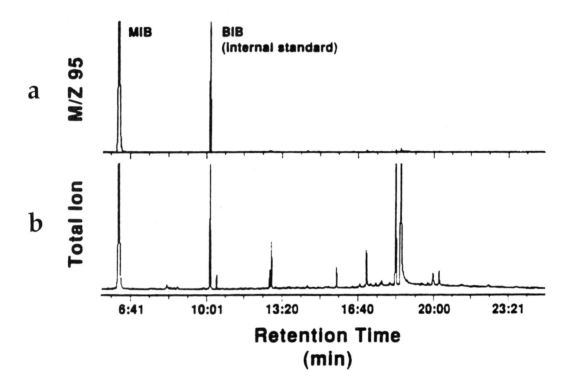

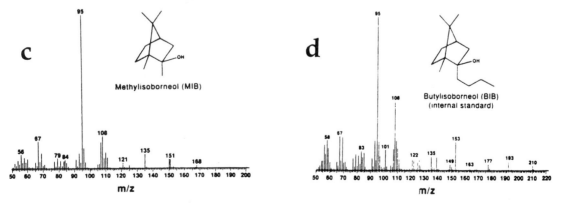

Fig. 13 (a) M/z 95 SIM and (b) total chromatograms of a contaminated water sample. Background subtracted mass spectrum of (c) methylisoborneol and (d) butylisoborneol. (Adapted from Ref. 36.)

ganic pollutants in water, by far the most used method for the subsequent analytical step is gas chromatography (GC) (32). Such wide application of GC results from its many advantages, such as reduced sample size, the capability of handling multiple components, ease of operation, and its high power to separate the complex organic mixtures that are often extracted from contaminated water. The organic compounds must, however, be capable of being volatilized without decomposition or chemical rearrangement.

Basically, GC analysis relies on packed or capillary columns that selectively retard compounds in a gaseous mixture as they are carried through by a gas flux. As each component exits the column, it is detected by techniques such as TCD (thermal conductivity detection), FID (flame ionization detection), ECD (electron capture detection), or (mass spectrometry). Reten-

tion times thus became a characteristic of each component and are used for their identification, whereas their response signals are used for quantitation.

Each GC analytical method for organic contaminants in water calls for specific extraction, cleanup, and, occasionally, derivatization procedures. Since analyte retention times can vary for different GC runs, depending, for instance, on the condition of the column, the temperature, the carrier-gas flow, and other organic constituents present, GC alone serves as a limited fingerprint method for the routine analysis of water samples. Standards are normally required to confirm compound identification by comparison of retention times, to determine detection limits, and for quantitation.

Figure 11 presents an illustrative example of a GC chromatogram from which a total of 132 chemical con-

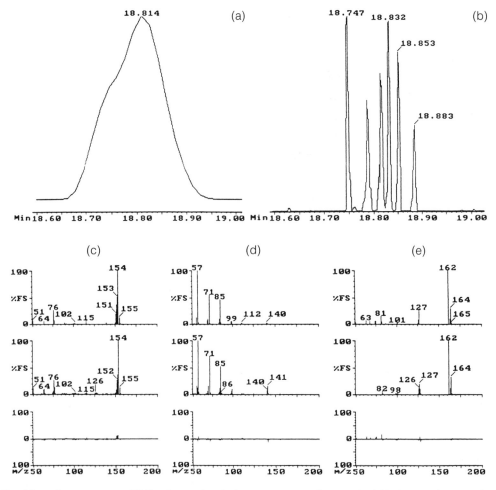

Fig. 14 (a) Total ion chromatogram (TIC) of a coeluted six-component mixture. (b) MS-deconvoluted TIC chromatogram. Library search results for deconvoluted spectra of (c) the third component, biphenyl; (d) the fourth component, *n*-tetradecane; and (e) the fifth component, 2-chloronaphathalene. (Adapted from Ref. 37.)

taminants extracted from water samples were separated, 86 of them (Table 4) identified by GC analysis (33).

2. *Gas Chromatography/Mass Spectrometry*

The outstanding separating power of gas chromatography (GC), together with the versatility, high sensitivity, and outstanding structural elucidation ability of mass spectrometry (MS), has created the most powerful instrument for the analysis of organic compound mixtures—the GC/MS (34). In water analysis, GC/MS is undoubtedly the "fingerprint" method: GC separates the contaminants and MS identifies the appearing peaks via the recording of their full mass spectra. The advantages of GC/MS include the capability to monitor specific compounds or classes of compounds, owing to the selectivity of the MS analysis. Diagnostic ions are selected for such monitoring. Figure 12 shows a GC/MS chromatogram in which selected ion monitoring (SIM) identifies several sulfur-containing compounds in water (35). Figure 13 shows a similar case, in which the total ion chromatogram identifies several chemicals, whereas SIM using the characteristic methylisoborneol fragment ion of m/z 95 specifically identifies the peak corresponding to this compound and an analogous internal standard, butylisoborneol (36).

The power of GC/MS analysis is also exemplified by the ability of MS data analysis to improve further the separation power of the chromatography column. Coeluted peaks can be deconvoluted by monitoring the variation of ion intensities in the superimposed mass spectra. Figure 14a and b shows an example of deconvolution by using mass chromatogram peak centroids of six components that were eluted within a 9 MS scan window. Figure 15c–e shows three of the respective deconvoluted mass spectra and their nearly perfect match with standard mass spectra.

3. *High-Performance Liquid Chromatography*

High-performance liquid chromatography (HPLC) is also a powerful technique used for the analysis of nonvolatile organic water contaminants, which are not amenable to GC/MS analysis. Although most attention has been paid to volatile and semivolatile organic contaminants, many organoleptical compounds are nonvolatile. When coupled with MS, HPLC has basically the same advantages and requires similar steps as GC/MS: extraction, concentration, fractionation, mass detection, and interpretation. The technique has been used mainly for the analysis of polynuclear aromatic hydrocarbons (PAHs), pesticides, and acrylamides.

4. *Membrane Introduction Mass Spectrometry*

a. Conventional Membrane Introduction Mass Spectrometry. Pre-extraction and concentration of the water sample are essential to improve the sensitivity of trace organic analysis. Ideally, these steps should be simple, rapid, and highly selective, use no solvent, and display performance independent of instrument design. Poor performance of the method on any of these parameters may compromise the accuracy, precision, and sensitivity of the analytical process. These difficulties are crucial, especially when analyzing volatile organic compounds (VOCs) in water.

Membrane introduction mass spectrometry (MIMS) (39–41) has emerged as a very efficient technique with

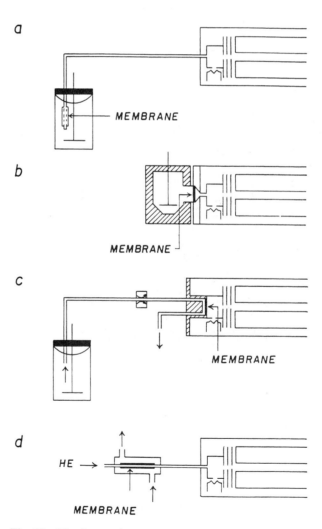

Fig. 15 The four main types of membrane inlets: (a) membrane inlet inserted directly into a reactor; (b) stirred sample cell; (c) flow-through inlet; (d) helium-purge inlet. (Reproduced with permission from Ref. 42.)

outstanding speed and trace-level detection capabilities for the direct analysis of VOCs in water. In MIMS, preliminary sample pre-extraction and preconcentration procedures are eliminated, and the sample is analyzed directly with online, real-time extraction/concentration. Many types of membrane inlets have been used (Fig. 15) (42), but the membrane probe is usually placed into the mass spectrometer in direct contact with the ionization source (Fig. 15d). Figure 16 shows a typical membrane probe (43) whereas, Figure 17 shows a helium-purge membrane system adapted to a commercially available GC/MS instrument (44).

In MIMS analysis, the VOCs are extracted and concentrated from the water sample via selective VOC migration through the membrane interface directly to the mass spectrometer. The most common membrane is made of silicone polymer, which shows the appropriate mechanical resistance to work as the interface between the liquid sample and the high-vacuum mass spectrometer. The hydrophobicity of the silicone membrane and its permeability to VOCs permit the extraction, concentration, and injection steps to be performed rapidly and simultaneously. The best performance for MIMS is for direct trace-water analysis of less polar and relatively light VOCs; detection limits of low or sub−parts-per-billion are normally achieved.

The MIMS technique provides a simple, direct, rapid, and sensitive method for online and real-time monitoring of VOCs in water. The analytes are not, however, subject to chromatographic separation before analysis, which compromises performance for more complex mixtures. Selective ion monitoring (SIM) or chemical ionization (CI) (which often generates mainly the protonated molecule for each contaminant) applied together with MS/MS for ion structural analysis can minimize such a limitation (43). Figure 18a presents an example of an application of CI-MIMS/MS. The three-dimensional spectra of an aqueous solution of benzene, toluene, and xylene (BTX) show, along the dashed line, mainly the respective protonated molecules of m/z 79 (benzene), m/z 93 (toluene), and m/z 107 (xylene), whereas the fragment ions produced by collision-induced dissociation of each mass-selected protonated molecule are displayed across the diagonal Q3 axis. Data extraction (Fig. 18b−d) permits easy visualization of each MS/MS spectrum, thus simplifying MS structural analysis. Similar CI-MIMS/MS experiments can be applied to more complex mixtures.

b. Trap Membrane Introduction Mass Spectrometry. To lower even further the detection limits of various VOCs in water samples, a series of interesting trapping strategies have recently been applied in conjunction with MIMS. By trapping the analyte ions, not the neutral analyte, in a highly sensitive ion-trap mass spectrometer fitted with a capillary membrane probe, parts-per-quadrillion (ppq) detection limits for a variety of VOCs have been attained (45).

Trapping the VOCs after membrane separation has also been shown to provide improved detection limits for MIMS. For instance, a membrane and trap gas chromatograph-mass spectrometer system has been shown to detect water soluble polar volatile organic com-

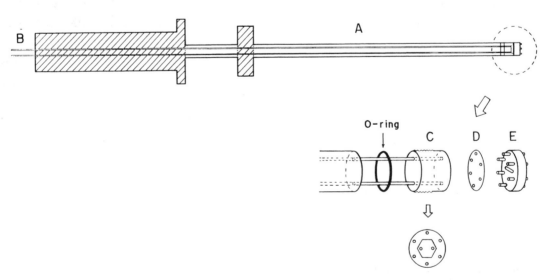

Fig. 16 Schematics of a MIMS probe: (A) ½″ tubing; (B) 1/24″ tubing; (C) probe head; (D) membrane; (E) membrane adaptor. Reproduced with permission from Ref. 43.

pounds in water at less than 100 ppb concentrations (46).

Although showing considerably improved detection limits, the use of trapping materials also has limitations. Background signals produced by thermal decomposition during the heating cycle or inefficient, slow, or inconstant release of the adsorbed chemicals compromise reproducibility and limit the sensitivity and applicability of these trapping techniques. A simple and efficient online preconcentration cryotrap-membrane introduction mass spectrometry (CT-MIMS) system that uses no trapping material has recently been reported (47). It allows the detection of VOCs in aqueous solution at a very low ppt level using a simple quadrupole MS analyzer. Trapping is performed via external liquid nitrogen cooling, followed by ballistic heating

that thermally releases the condensed VOCs into the mass spectrometer.

Figure 19 shows a diagram of the CT-MIMS system. The conventional MIMS probe is modified so that the membrane interface is placed about 15 cm away from the ion source. A U-shaped trap tube is then inserted between the membrane interface and the ion source. Cryotrapping is performed with liquid nitrogen for 15 min, followed by fast heating at a ratio of approximately $15°C\ s^{-1}$ that thermally releases the condensed VOCs into the ion source region of the quadrupole mass spectrometer. By applying 70 eV electron ionization and selective ion monitoring scan modes, very sharp and intense peaks are obtained. The performance of the CT-MIMS system was compared with that of conventional MIMS, and after reaching the best con-

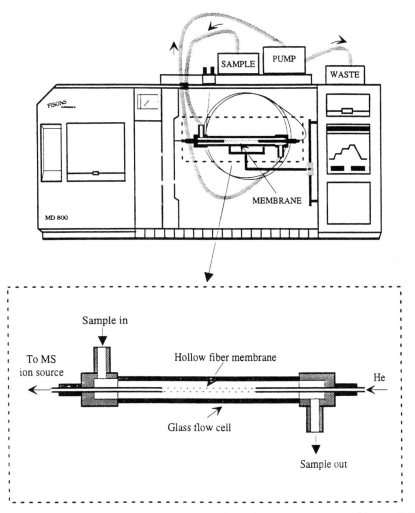

Fig. 17 Schematic of the system using a helium-purge trap-membrane inlet system mounted into a GC/MS instrument. (Reproduced with permission from Ref. 44.)

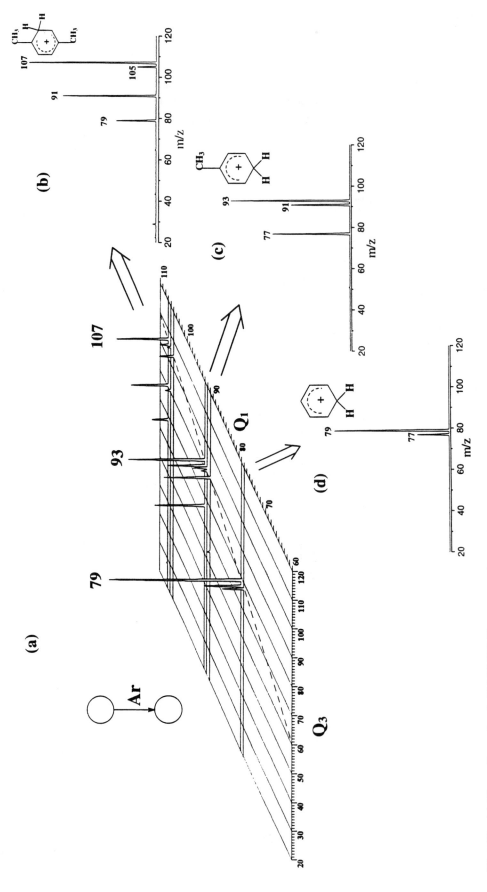

Fig. 18 (a) Three-dimensional CI-MIMS/MS spectrum for a BTX aqueous solution and extracted 2D MS/MS spectrum for each protonated molecule, that is, (b) protonated p-xylene of m/z 107, (c) protonated toluene of m/z 93, and (d) protonated benzene of m/z 79. (Adapted from Ref. 43.)

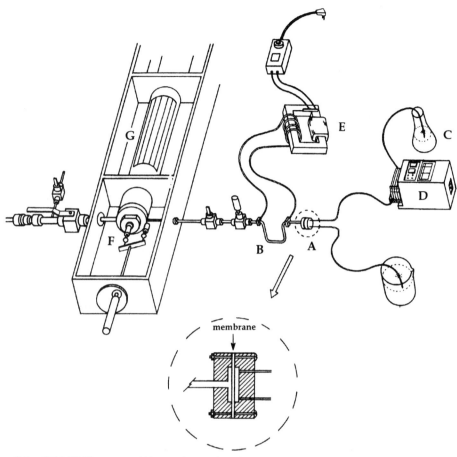

Fig. 19 Diagram of the CT-MIMS system: (A) membrane interface; (B) U-shaped trap tube; (C) water sample; (D) peristaltic pump; (E) heating system; (F) ion source; (G) mass spectrometer. (Reproduced with permission from Ref. 47.)

Table 5 CT-MIMS Gains and Detection Limits for a Series of VOCs

VOCs	Monitored ion (m/z)	MIMS signal (kcounts)	CT-MIMS signal (kcounts)	CT-MIMS gain	MIMS detection limit (ppb)	CT-MIMS detection limit (ppt)
Benzene	78	11.4	1099	96	1	10
Toluene	91	11.0	1139	103	1	10
Xylene	106	10.4	987	95	1	10
Chlorobenzene	112	14.5	1419	98	1	10
Benzaldehyde	106	1.8	170	95	15	150
Acetone	58	3.4	321	95	10	100
2-Butanone	43	7.9	836	106	5	50
Ethyl ether	59	10.9	1136	104	1	10
Tetrahydrofuran	72	0.4	35.4	96	50	500
Carbon tetrachloride	117	3.4	311	92	5	50
Chloroform	83	6.0	583	98	2	20
Dichloromethane	49	1.6	193	118	20	200
1,1,2,2-Tetrachloroethane	83	3.0	275	93	5	50
Chlorodibromomethane	127	5.3	579	110	2	20

Source: Adapted from Ref. 47.

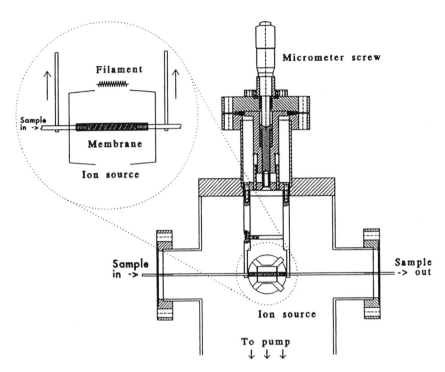

Fig. 20 Schematic of the trap and release MIMS system. (Reproduced with permission from Ref. 48.)

ditions for the trapping and heating cycles, an improvement factor in signal intensity of about a hundred was observed for a series of VOCs. Therefore, CT-MIMS decreases the conventional MIMS detection limits by an order of magnitude, and are typically on the order of 10–20 ppt (Table 5), with excellent linearity and reproducibility.

Semivolatile organic compounds (SVOCs), although not migrating very efficiently through the membrane, can also be analyzed with detection limits of low ppb using the trap and release membrane inlet MS system recently developed by Lauritsen (48–49). The SVOCs are preconcentrated on the membrane before they are thermally released into the ion source. Figure 20 shows a diagram of the T&R-MIMS system. A 10-mm-long tubular silicone membrane is mounted between two $^1/_{16}$-inch stainless steel tubes through which the sample flows. A micrometer screw is used to adjust precisely the position of the membrane relative to the ion source. The solution is pumped through the membrane inlet for 15 min, and then an air plug of typically 50 s is pumped through the system. Radiation from the ion source filament heats the membrane, causing the thermal release of the SVOCs. The signal rises, reaching a maximum after 20 s; after 40 s, the signal is almost vanished, and

the sample flow is restarted, rapidly cooling the membrane.

REFERENCES

1. C. J. Koester, R. E. Clement. Analysis of drinking water of traces organics. Critical Reviews in Analytical Chemistry 24:263, 1993.
2. Pontius, F. W. Future directions in water quality regulations. Journal of the American Water Work Association 89:40, 1997.
3. J. E. Amoore, J. W. Johnston Jr., M. Rubin. The Stereochemical Theory of Odor. Scientific American, February 1964.
4. S. W. Krasner, M. J. McGuire, V. B. Ferguson. Tastes and odors: the flavor profile method. Journal of the American Water Work Association 77:34, 1985.
5. A. E. Greenberg, L. S. Clesceri, A. D. Eaton. Standard Methods for the Examination of Water and Wastewater. 18th ed. Washington, DC: American Public Health Association, 1992.
6. A. Bruchet, C. Anselme, C. Jammes, J. Mallevialle. Development of a taste and odor expert system: present state, strengths, limitations and possible future evolution. Water Science and Technology 31:243, 1995.

7. A. K. Meng, L. Brenner, I. H. Suffet. Correlation of chemical and sensory data by principal components factor. Water Science and Technology 25:49, 1992.

8. R. Sävanhed, H. Borén, A. Grimvall. Stripping analysis and chromatography sniffing for the source identification of odorous compounds in drinking water. Journal of Chromatography Science 328:219, 1985.

9. S. Rigal. Odor and flavor in water: quantitative method for a new european standard. Water Science and Technology 31:237, 1995.

10. C. Baird. Environmental Chemistry. New York: W. H. Freeman, 1995.

11. E. E. Hargesheimer, S. B. Watson. Drinking water treatment options for taste and odor control. Water Research 30:1423, 1996.

12. C. Anselme, H. I. Suffet, J. Mallevialle. Effects of ozonation on tastes and odors. Journal of the American Water Work Association 80:45, 1988.

13. B. Thorell, H. Borén, A. Grimvall, A. Nyström, R. Sävenhed. Characterization and identification of odorous compounds in ozonated water. Water Science and Technology 25:139, 1992.

14. R. A. Minear, L. H. Keith. Water Analysis. New York: Academic Press, 1982.

15. G. Preti, T. S. Gittelman, P. B. Staudte, P. Luitweller. Letting the nose lead the way: malodorous components in drinking water. Analytical Chemistry 65:699A, 1993.

16. D. M. C. Rashash, A. M. Dietrich, R. C. Hoehn. FPA of selected odorous compounds. Journal of the American Water Work Association 89:131, 1997.

17. D. Khiari, L. Brenner, G. A. Burlingame, I. H. Suffet. Sensory gas chromatography for evaluation of taste and odor events in drinking water. Water Science and Technology 25:97, 1992.

18. D. Khiari, I. H. Suffet, S. E. Barret. The determination of compounds causing fishy/swampy odors in drinking water supplies. Water Science and Technology 31:105, 1995.

19. S. W. Krasner. The use of reference materials in sensory analysis. Water Science and Technology 31:265, 1995.

20. D. Khiari, S. E. Barret, I. H. Suffet. Sensory GC analysis of decaying vegetation and septic odors. Journal of the American Water Work Association 89:150, 1997.

21. B. Lundgren, H. Borén, A. Grimvall, R. Sävanhed. Isolation of off-flavor compounds in water by chromatography sniffing and preparative gas chromatography. Journal of Chromatography 482:23, 1989.

22. A. J. Núñhez, L. F. González, J. Janák. Pre-concentration of headspace volatiles for trace organic analysis by gas chromatography. Journal of Chromatography 300:127, 1984.

23. R. E. Kepner, H. Maarse, J. Strating. Gas chromatographic head space techniques for the quantitative determination of volatile components in multicomponent aqueous solutions. Analytical Chemistry 36:77, 1964.

24. T. A. Bellar, J. J. Lichtenberg. Determining volatile organics at the microgram per liter levels by gas chromatography. Journal of the American Water Work Association 66:739, 1974.

25. B. S. Middleditch, A. Zlatkis, R. D. Schwartz. Trace analysis of volatile polar organics: problems and prospects. Journal of Chromatographic Science 26:150, 1988.

26. G. Matz, P. Kesners. Spray and trap method for water analysis by thermal desorption gas chromatography/mass spectrometry in field applications. Analytical Chemistry 65:2366, 1993.

27. K. Grob. Organic substances in potable water and its precursors. Part I. Methods for their determination by gas–liquid chromatography. Journal of Chromatography 84:255, 1973.

28. H. Borén, A. Grimvall, J. Palmborg, R. Sävenhed, B. Wigilius. Optimization of the open stripping system for the analysis of trace organics in water. Journal of Chromatography 348:67, 1985.

29. J. P. Mieure. Determining volatile organics in water. Environmental Science & Technology 14:930, 1980.

30. Z. Zhang, M. J. Yang, J. Pawliszyn. Solid-phase microextraction. Analytical Chemistry 66:844A, 1994.

31. M. H. Hiatt, D. R. Youngman, J. R. Donnelly. Separation and isolation of volatile organic compounds using vacuum distillation with GC/MS determination. Analytical Chemistry 66:905, 1994.

32. Robert L. Grob (ed.). Modern Practice of Gas Chromatography. New York: Wiley, 1995.

33. K. Grob, G. Grob. Organic substances in potable water and its precursors. Part II. Applications in the area of Zürich. Journal of Chromatography 90:303, 1974.

34. F. W. Karasek. Basic Gas Chromatography–Mass Spectrometry: Principles and Techniques. New York: Elsevier, 1988.

35. B. G. Brownlee, S. L. Kenefick, G. A. MacInnis, S. E. Hrudey. Characterization of odorous compounds from bleached kraft pulp mill effluent. Water and Science Technology 31:35, 1995.

36. E. D. Conte, S. C. Conway, D. W. Miller, P. W. Perschbacher. Determination of methylisoborneol in channel catfish pond water by solid-phase extraction followed by gas chromatography–mass spectrometry. Water Research 30:2125, 1996.

37. B. N. Colby. Spectral deconvolution for overlapping GC/MS components. Journal of the American Society for Mass Spectrometry 3:558, 1992.

38. Mark A. Brown (ed.). Liquid Chromatography/Mass Spectrometry: Applications in Agricultural, Pharmaceutical, and Environmental Chemistry. ACS Symposium Series, No. 420. Washington, DC: American Chemical Society, 1990.

39. R. C. Johnson, R. G. Cooks, T. M. Allen, M. E. Cisper, P. H. Hemberger. Membrane introduction mass spectrometry: trends and appplications. Mass Spectrometry Reviews 19:00, 2000.

40. S. Bauer. Membrane introduction mass spectrometry: an old method that is gaining new interest through re-

cent technological advances. Trends in Analytical Chemistry 14:202, 1995.

41. F. R. Lauritsen, T. Kotiaho. Advances in membrane inlet mass spectrometry (MIMS). Reviews in Analytical Chemistry 15:237, 1996.

42. T. Kotiaho. On-site environmental and in situ process analysis by mass spectrometry. Journal of Mass Spectrometry 31:1, 1996.

43. M. A. Mendes, R. S. Pimpim, T. Kotiaho, J. S. Barone, M. N. Eberlin. The construction of a membrane probe and its application toward the analysis of volatile organic compounds in water via the MIMS and MIMS/MS techniques. Química Nova 19:480, 1996.

44. V. T. Virkki, R. A. Ketola, M. Ojala, T. Kotiaho, V. Komppa, A. Grove, S. Facchetti. On-site environmental analysis by membrane inlet mass spectrometry. Analytical Chemistry 67:1421, 1995.

45. M. Soni, S. Bauer, J. W. Amy, P. Wong, R. G. Cooks. Direct determination of organic compounds in water at part-per-quadrillion levels by membrane introduction mass spectrometry. Analytical Chemistry 67:1409, 1995.

46. J. A. Shoemaker, T. A. Bellar, J. W. Eichelberger, W. L. Budde. Determination of polar volatile organic compounds in water by membrane permeate and trap GC/MS. Journal of Chromatographic Science 31:279, 1993.

47. M. A. Mendes, R. S. Pimpim, T. Kotiaho, M. N. Eberlin. A cryotrap membrane introduction mass spectrometry system for analysis of volatile organic compounds in water at the low parts-per-trillion level. Analytical Chemistry 68:3502, 1996.

48. M. Leth, F. R. Lauritsen. A fully integrated trap-membrane inlet mass spectrometry system for the measurement of semivolatile organic compounds in aqueous solution. Rapid Communication in Mass Spectrometry 9:591, 1995.

49. F. R. Lauritsen, R. A. Ketola. Quantitative determination of semivolatile organic compounds in solution using trap-and-release membrane inlet mass spectrometry. Analytical Chemistry 69:4917, 1997.

7

Radioanalytical Methodology for Water Analysis

Jorge S. Alvarado

Argonne National Laboratory, Argonne, Illinois

I. INTRODUCTION

Many nations are facing the challenge of cleaning up and protecting our environment. The concern over radionuclides in the environment is based on the possible health hazard caused by the emitted ionizing radiation, especially in situations where radionuclides are taken up into the body by ingestion, inhalation, or absorption. The magnitude of the hazard depends on the distribution of the radionuclides within the body, the length of time they are retained in the body, and the type of energy of the emitted radiation. The initial reactions of radiation in the body will involve hydrogen-oxygen chemical species. Other interactions may produce cancer (bone cancer and leukemias), genetic effects, damage to the blood vessels, induction of cataracts of the eyes lenses, and infertility with exposure to large amounts of ionizing radiation.

Techniques and methods of chemical characterization and monitoring are essential in the execution of control programs to minimize health hazards. Research and development efforts can translate into new technologies for the detection of radionuclides and the improvement of environmental cleanup. These efforts are measured by (a) reductions in the unit cost (by doing fewer steps in the analysis), (b) reductions in the time required to provide the information to the user, or (c) improvements in the quality of the information provided. The desired characteristics of any new methods are the following: faster, to reduce turnaround times; cheaper, to reduce the cost of characterization; performs better, to achieve desired data quality objectives;

minimizes the generation of secondary mixed waste, through reduction of scale and elimination of steps; moves toward real-time analysis; and speeds site characterization.

The purpose of this review is to summarize the main aspects of the problems of radionuclides in the environment and to present new techniques for the determination of isotopes in water.

II. RADIONUCLIDES IN THE ENVIRONMENT

Radionuclides that can be found in the environment can be divided into two groups:

1. Naturally occurring radioisotopes that formed after the "big bang" in the evolution of the stars and that have such long half-lives that their and their daughters' products are still present, and those radionuclides that are being continuously produced by nuclear reactions between cosmic radiation and stable elements
2. Radionuclides released into the environment from man's activities

In natural sources all elements with an atomic number greater than 83 are radioactive, but most are present in very small concentrations. They belong to chains of successive decay, and all of the species in one such chain constitute a radioactive family series. Three of these families include all of the natural activities in this region of the periodic chart (1). The existence of branching decays in each of the three series should be noticed. These representations recognize only the most common decomposition elements, but new decompo-

sition branches have been discovered.

One family has ^{238}U as the parent substance and after a succession of transformations reaches ^{206}Pb as a stable product, as shown by the following:

$$^{238}U \rightarrow ^{234}Th \rightarrow ^{234}U \rightarrow ^{230}Th \rightarrow ^{226}Ra \rightarrow ^{222}Rn \rightarrow ^{218}Po \rightarrow ^{214}Pb$$
$$\downarrow \quad \downarrow$$
$$^{218}At \rightarrow ^{214}Bi \rightarrow ^{214}Po$$
$$\downarrow \quad \downarrow$$
$$^{210}Tl \rightarrow ^{210}Pb \rightarrow ^{210}Bi \rightarrow ^{210}Po$$
$$\downarrow \quad \downarrow \quad \downarrow$$
$$^{206}Hg \rightarrow ^{206}Tl \rightarrow ^{206}Pb$$

Thorium-232 is the parent substance in the thorium series, with ^{208}Pb as the stable end product, as shown

by the following:

$$^{232}Th \rightarrow ^{228}Ra \rightarrow ^{228}Ac \rightarrow ^{228}Th \rightarrow ^{224}Ra \rightarrow ^{220}Rn \rightarrow ^{216}Po \rightarrow ^{212}Pb \rightarrow ^{212}Bi$$
$$\downarrow \quad \downarrow$$
$$^{208}Tl \rightarrow ^{208}Pb$$

The third series is known as the actinium series and has ^{235}U as the parent element and ^{207}Pb as the stable end product. This series is as follows:

$$^{235}U \rightarrow ^{231}Th \rightarrow ^{231}Pa \rightarrow ^{227}Ac \rightarrow ^{227}Th$$
$$\downarrow \quad \downarrow$$
$$^{223}Fr \rightarrow ^{223}Ra$$
$$\downarrow \quad \downarrow$$
$$^{219}At \rightarrow ^{219}Rn$$
$$\downarrow \quad \downarrow$$
$$^{215}Bi \rightarrow ^{215}Po \rightarrow ^{215}At$$
$$\downarrow \quad \downarrow$$
$$^{211}Pb \rightarrow ^{211}Bi \rightarrow ^{211}Po$$
$$\downarrow \quad \downarrow$$
$$^{207}Tl \rightarrow ^{207}Pb$$

Other naturally occurring radioactive isotopes, such as ^{40}K, ^{87}Rb, ^{113}Cd, ^{115}In, ^{138}La, ^{144}Nd, ^{147}Sm, ^{148}Sm, ^{152}Gd, ^{176}Lu, ^{174}Hf, ^{187}Re, and ^{190}Pt, have been found. These elements occur in very low abundance or have such extremely long half-lives that their activity can be assumed constant during human evolution. The distribution of these elements changes. Potassium-40 has an abundance of 0.0118% in naturally occurring potassium. The average abundance of potassium is compa-

rable with that of uranium and is about one-quarter of that of thorium. Rubidium-87 is considerably more abundant by mass than ^{40}K but represents less risk because of its longer half-life and its lower β-radiation.

The artificial, or man-made, radioactive elements were discovered by I. Curie and F. Joliot in 1934 (1). Since then, many of these elements have been produced by nuclear explosions and nuclear reactors. The primary sources of radionuclides produced in the fission process and found in the environment are atmospheric testing of nuclear weapons and nuclear accidents, such as the well-known Chernobyl accident in April 1986 (2,3). After the Chernobyl disaster, the initial widespread radioactivity that was most noticeable was due to ^{131}I; however, its contribution became negligible after about two months because its half-life is only eight days. The remaining ^{137}Cs and ^{134}Cs were then the dominant radionuclides, with half-lives of 30 years and 2.1 years, respectively. Table 1 lists some of the cosmogenic, natural, and artificial radionuclides in the environment.

III. DETECTORS

In a radiation process, the first-order kinetics govern the decay of radioactive atomic nuclei. The common process includes the ejection of an alpha (^{4}He) or beta

Table 1 Some Cosmogenic, Natural, and Artificial Radionuclides in the Environment

Element	Half-life	Main radiation	Element	Half-life	Main radiation
^{3}H	12.4 years	β	^{99}Tc	2.14×10^5 years	β
^{7}Be	53 days	γ	^{222}Rn	3.8235 days	α
^{10}Be	2.7×10^6 years	β	^{220}Rn	55.6 sec.	α
^{14}C	5730 years	β	^{239}Pu	2.41×10^4 years	α
^{35}S	88 days	β	^{240}Pu	6.57×10^3 years	α
^{36}Cl	3.0×10^5 years	β	^{241}Pu	14.4 years	β
^{39}Ar	269 years	β	^{242}Pu	3.76×10^5 years	α
^{232}Th	1.4×10^{10} years	α	^{215}Po	1.78 ms	α
^{230}Th	8.0×10^4 years	α	^{40}K	1.3×10^9 years	β, γ
^{228}Th	1.9131 years	α	^{60}Co	5.27 years	β, γ
^{231}Th	25.52 hours	β	^{87}Rb	4.7×10^{10} years	β
^{238}U	4.5×10^9 years	α	^{89}Sr	50.5 days	β
^{235}U	7.0×10^8 years	α, γ	^{90}Sr	28.7 years	β
^{233}U	1.592×10^5 years	α	^{90}Y	64.1 hours	β
^{236}U	2.342×10^7 years	α	^{134}Cs	754.2 days	β, γ
^{226}Ra	1.6×10^3 years	α	^{242}Cm	162.8 days	α
^{228}Ra	5.76 years	β	^{241}Am	432 years	α
^{224}Ra	3.66 days	α	^{243}Am	7.37×10^3 years	α

(electron or positron) particle from the nucleus, though electron capture or spontaneous fission is also observed. Ejected alpha particles have an energy spectrum that is characteristic of the initial and final states of the decay process. In contrast, the beta particles possess a broad range of ejection energies. A neutrino is emitted to conserve energy during decay, and a gamma ray may be emitted after particle ejection if the product nucleus is in an excited state. In addition, the emitted gamma rays have an energy spectrum characteristic of the excited and ground states of the product nucleus. Routine radioanalytical measurements involve the detection of alpha, gamma, and beta emissions from the collection of decaying nuclei.

Many types of detectors are used for the detection of radioactive particles. The first and most simple method of detection was films, where the radioactivity was the general agent for blackening or fogging of photographic negatives (4). Since then, detectors have been developed in a wide variety of systems, including semiconductor detectors (scintillation counters and Cerenkov counters), track detectors (photographic film, cloud chambers, bubble chambers, spark chambers, and dielectric track detectors), neutron detectors (ionization chambers), and mass detectors (mass spectrometers). Although each method has important applications, the focus of this chapter will be on those techniques most useful for analytical applications.

A. Gamma Detectors

Many radionuclides decay with the emission of gamma rays (photons). These photons can be detected and identified by their characteristic energies by using either a germanium detector, a lithium-drifted germanium detector, or a scintillation detection system.

In the solid ionization chamber the energy difference between the valence band and the conduction band, called the bandgap (Eg), is small enough, allowing thermal excitation that produced conduction. Because the bandgap for germanium is 0.67 eV, it is necessary to use low temperatures (by using liquid nitrogen) to avoid excessive thermal noise. Despite this disadvantage, high-purity germanium detectors are the method of choice for detecting gamma photons because the spectra of the γ-emitters can be distinguished without using any radiochemical separation.

Lithium-drifted germanium detectors are available in many configurations and sizes. Prices for these devices are relative to their size. One of the disadvantages of these detectors is the necessity of low temperatures. Permanent damage may result if such a detector is warmed to room temperature.

Scintillation detectors are cheaper than the solid-phase detectors and have higher detection efficiencies than the germanium detectors. In these types of detectors, the gamma rays produce light in a suitable scin-

Table 2 Detection Limits for Long-Lived Radioisotopes by ICP-MS[a], in pg/L

Isotope	ETV-ICP-MS[b]	USN-ICP-MS[c]	PN-ICP-MS[d]
^{99}Tc	60	20	300
^{226}Ra	24	10	400
^{232}Th	80	800	6000
^{230}Th	56	—[e]	—
^{237}Np	—	20	800
^{236}U	36	—	—
^{238}U	200	500	7000
^{239}Pu	—	20	700

[a]ICP-MS = inductively coupled plasma–mass spectrometry.
[b]*Source*: Ref. 20. ETV = electrothermal vaporization.
[c]*Source*: Ref. 21. USN = ultrasonic nebulization.
[d]*Source*: Ref. 22. PN = pneumatic nebulization.
[e]Not reported.

tillator mounted on a photomultiplier tube, the light causes the ejection of photoelectrons from the scintillator onto the photosensitive electrode of the tube, and the output pulse from the multiplier is monitored. Various scintillators, each with particular advantages, have been developed, and many are still being used. The main disadvantage of the scintillation detector is its lower photo energy resolution; this lower resolution makes necessary radiochemical separation of samples that contain more than one gamma emitter.

The most widely used inorganic scintillator is the sodium iodide (Na(I)) scintillator with 0.1%–0.2% thallium. The height of the electrical pulse from the scintillation detector will be linearly proportional to the energy initially deposited in the crystal via the photoelectric Compton phenomena. The energies of the gamma rays coming from the radioactive source, as well as their intensities, can be measured. The background rates in this type of detector are high. Massive shielding can be effective in reducing the background effects caused by cosmic rays and by the gamma rays from surrounding material, and further background reduction can be achieved with anticoincidence arrangements. To minimize the background problem, some laboratories have used as shielding steel boxes made with material from old ships built before the nuclear era began.

With the advent of sodium iodide crystals and a single-channel analyzer, the analysis of gamma emitters was made easier. Some radiochemistry was still needed for the various radionuclides because multiple isotopes and gamma rays were present. The use of simultaneous equations made this analysis routine for the following isotopes: ^{131}I, barium, ^{140}La, and ^{137}Ce. Shorter-lived io-

Table 3 Commercial Resins and Their Application to Environmental Analysis

Resin	Applications
Sr resin	Sr, Pb
TRU resin	Th, U, Pu, Am, Cm, Fe(III)
UTEVA resin	U, Th, Np, Pu
TEVA resin	Th, Np, Pu, Tc, Am/Ln separations
Actinide resin	Group actinide separation, gross alpha measurements
RE resin	Rare-earth elements
Ln resin	Lanthanides, Pa, Ra
Pb resin	Pb
Nickel resin	Radioactive Ni measurements
Tritium column	Alternative to distillation for analysis of aqueous samples for tritium

Source: Ref. 41.

Table 4 Methods for Tc-99 Analysis

Method	Single sample time (h)	Average sample time (h)	Labor time (h/sample)	Secondary waste
^{99}Tc assay[a]	3	0.75	0.20	20 mL, mixed
HASL-300[b]	24	1.4	0.25	500 mL, mixed
Holm method[c]	216	26.4	3	97 mL, mixed
Rad-Disk[d]	2.7	0.50	0.20	Membrane

[a]*Source*: Ref. 45.
[b]*Source*: Ref. 46.
[c]*Source*: Ref. 47.
[d]*Source*: Ref. 48.

dine isotopes cause some interferences. According to the International Atomic Energy Agency (5), sample geometries must be selected and calibrated for the density of the samples of interest as a function of the gamma-ray energy. Calibration curves should be prepared from reliable and traceable sources.

B. Alpha Detectors

The short range of the alpha particles (25–30 μm) and the significant absorption by detector windows make the measurement of such particles more difficult. The use of samples with a thickness greater than 6 mg/cm^2 is a well-established technique but gives poor energy resolution. Thin windows, proportional counters, and pulse ionization chambers have also been used, but the methods of choice today are the surface-barrier and p-n junction detectors. These detectors usually require a vacuum and very thin electrodeposited samples for spectroscopy measurements. Surface-barrier detectors usually exhibit better energy resolution than the diffused p-n junction devices of similar size. The surface-barrier detectors are more sensitive to the ambient atmosphere, and they need an atmosphere free of chemical fumes and water vapor.

Surface-barrier and diffused p-n junction detectors are the best detectors available for low-energy and heavy, charged particles. Typical detector energy resolutions are on the order of 10–20 KeV with 100% detector efficiency. Practical limitations in the construction of this detector restrict the depletion depths to less than 2 mm. The cost of these detectors is low.

The U.S. Environmental Protection Agency (EPA) lists the regulations regarding limits for alpha emitters. Radium-226, ^{228}Ra, and gross alpha are analyzed by this method. Two other alpha emitters of importance are uranium and plutonium (6). The total uranium requires only separation of the uranium, followed by al-

pha counting. The isotopic uranium requires electroplating and alpha counting by using a solid-state detector. Currently, no "standard methods" exist for plutonium in drinking water. The current methods (7,8) usually require separation and addition of a tracer for the recovery, electroplating, and detection by alpha spectrometry to determine the isotope content and tracer recovery. Finally, radon is another important alpha emitter isotope. Two methods are used for the determination of radon in water (9,10). Radon in drinking water is found only in groundwater supplies.

C. Beta Detectors

The classical way to measure low-level beta-particle activity is with a Geiger–Müller gas-flow counter, proportional counters, and solid or liquid scintillation counters; although for low-energy-beta particles, liquid scintillation is the most frequently used method. The major disadvantage of these methods is the color quenching, impurities, and chemical quenching. The Geiger–Müller counter in conjunction with a guard detector is presented as a better option. This system gives very low background and a counting efficiency of approximately 40%. The major disadvantage is the inability to give energy resolution.

Solid samples suffer from self-absorption of the beta particles, and this problem must be dealt with by using thin samples and extrapolating to zero thickness of the maximum range of the beta particles in a particular sample material. Calibration sources should resemble the samples in thickness and absorption properties.

For beta particles with greater than 0.26 MeV of energy traveling through water, Cerenkov radiation is emitted. This can be counted in a liquid scintillation system and has the advantage of simple preparation. No other particles or photons will be detected.

The EPA's proposed drinking water standards for beta and photo emitters are limited to exposure of 4 mRem ede/yr. Levels of strontium and tritium are determined by this method. The strontium method (11) covers the measurement of total strontium, including ^{90}Sr and ^{89}Sr. Interferences from calcium and other radionuclides are removed by one or more precipitations of the strontium carrier as strontium nitrate. Barium and radium are removed as chromate. The ^{90}Y daughter of ^{90}Sr is removed by the hydroxide precipitation step, and the separated combined ^{89}Sr and ^{90}Sr are counted for beta-particle activity. Tritium is determined by liquid scintillation counting after distillation (12).

D. Mass Spectrometers

A quite common technique for the determination of many isotopes is mass spectrometry. Mass spectrometry is very sensitive and isotope specific. It is especially suitable for heavy elements such as the actinides, where isobaric disturbances are few. Three different types of mass spectrometers have been used for the determination of radionuclides in the environment. These types are the thermal ionization mass spectrometer (TIMS), the inductively coupled plasma–mass spectrometer (ICP-MS), and the accelerator mass spectrometer (AMS). The AMS is used mainly for the determination of geologic ages or the study of radionuclide production in the atmosphere. Thermal ionization mass spectrometry is a very sensitive technique with very low detection limits; however, TIMS instrumentation is expensive and requires very pure samples and extensive chemical separations, addition of isotope tracers, and operational finesse. In addition, sample preparation is more complicated and requires longer analysis times. The TIMS is not commonly used for the analysis of low-level samples.

Inductively coupled plasma–mass spectrometry is a very rapid technique for the determination of long-lives radionuclides. This technique is based on the ionization of elements in the plasma source. Typically, plasmas use radiofrequency and argon to reach excitation temperatures ranging from 4900 to 7000 K (13,14). The ions produced are introduced through an interface into a vacuum chamber and are analyzed by a quadrupole mass spectrometer. Other attempts are being made to use faster mass spectrometer detectors, such as time-of-flight mass spectrometers.

The ICP-MS offers advantages such as low detection limits (typically nanograms per liter), mass-selective detection, and multicomponent detection. In determining long-lived radionuclides, ICP-MS is surpassing other techniques, such as differential-pulse chromatography (15), radiochemical neutron activation (16), ion

Table 5 Radioanalytical Methods Employing Solvent Extraction

Analyte	Extraction condition (Ref.)
^{99}TcO$_4^-$	From dilute H$_2$SO$_4$ solutions into a 5% TnOA[a] in xylene mixture and back-extracted with NaOH[b] (55, 56)
^{210}Pb	As lead bromide from bone urine, feces, blood, air, and water with Aliquat-336[g] (57)
Actinides	From water after concentration by ferric hydroxide precipitation and group separation by bismuth phosphate precipitation; U extracted by TOPO,[c] Pu and Np extracted by TiOA[d] from strong HCl[e]; and Th separated from Am and Cm by extraction with TOPO (49)
Thorium	From aqueous samples after ion exchange with TTA,[f] TiOA, or Aliquat-336[g] (RP 570) (8)
Uranium	From waters with ethyl acetate and magnesium nitrate as salting agents (49); with URAEX™, followed by PEARLS spectrometry (58)

[a]TnOA: tri-n-octylamine.
[b]NaOH: Sodium hydroxide.
[c]TOPO: tri-n-octylphosphine oxide.
[d]TiOA: triisooctylamine.
[e]HCl: hydrochloric acid.
[f]TTA: 2-thenoyltrifluoroacetone.
[g]Aliquat-336: tricaprylyl-methylammonium chloride.

Table 6 Radioanalytical Methods Employing Extraction Chromatography

Analyte	Ligand	Method (Ref.)
Ni-59/63	Dimethylgloxime	Aqueous samples (8)
Sr-89/90	4,4′(5′)-bis(t-butyl-cyclohexano-18-crown-6 in octanol	Water; ASTM (59)
Sr-90	Octyl(phenyl)-N,N-diisobutyl-carbomoylmethylphosphine oxide [CMPO] in tributyl phosphate	Water (60)
Tc-99	Aliquat-336N	Water (61)
Pb-210	4,4′(5′)-bis(t-butyl-cyclohexano-18-crown-6 in octanol	Water (8)
Ra-228	Octyl(phenyl)-N,N-diisobutyl-carbamaymethylphosphine oxide [CMPO] in tributyl phosphate or diethyl-phosphoric acid [HDEHP] impregnated in Amberline XAD-7	Natural water (62)
Rare earths	Diamyl,amylphosphonate	Actinide-containing matrices (63)
Actinides	Octyl(phenyl)-N,N-diisobutyl-carbomoylmethylphosphine oxide [CMPO] in tributyl phosphate	Waters (64)
	Diamyl,amylphosphonate	Acidic media (65)
	Tri-n-octylphosphine oxide [TOPO] and di(2-ethylhexyl)phosphoric acid	Environmental (66)

chromatography (17), and classical photometry (18). The main advantage of ICP-MS is its ability to determine long-lived-radionuclides with low-intensity radiation, and alpha emitting radionuclides that required tedious radiochemical separations.

Despite the fact that ICP-MS itself was a major improvement, techniques of sample introduction are critical for the performance of the plasma source. Because of its simplicity and high reproducibility, pneumatic nebulization (PN) is the most common method of introducing aqueous samples; however, the efficiency is poor, with reported typical efficiencies for PN of 1% with ICP spectrometers (19). Ultrasonic nebulization has a more efficient production of droplets, ranging up to 30%, and better detection limits than PN. Electrothermal vaporization (ETV) has been introduced as an option because it has high analyte transport efficiency, producing higher sensitivities. In addition, ETV reduces polyatomic interferences, requires smaller samples (typically 25 μl), and has the ability to analyze organic liquids, strong acids, liquids high in solids, and slurries (20). Some of the detection limits reported for

the determination of long-lived radionuclides are listed in Table 2 (20–22).

E. Decay Counting and Inductively Coupled Plasma–Mass Spectrometry

In the early 1990s, as described by Crain and Alvarado (23), investigators started to report ICP-MS detection limits for elements, such as technetium-99, that were better than the detection limits reported for decay counting (24,25). However, Toole and coworkers (26) demonstrated that the differences in sensitivity between ICP-MS and decay counting were related to the radioactive half-life of the analytes. This relationship was modeled by Smith et al. (27). This model described the relation between sensitivity ratios obtained by using ETV-ICP-MS and decay counting as a function of radioisotope half-life. The sensitivity ratio becomes unity at half-lives of approximately 570 years. At half-lives higher than 570 years, ICP-MS showed better detection limits, and vice versa. However, sensitivity is not the

Table 7 Results for the Determination of Uranium-238 in Water

Sample	ETV-ICP-MS (μg/L)	α-Spectrometry (μg/L)
Tap water		
Chicago, IL	0.0015 $\pm$ 0.0004	<0.09
Lemont, IL	ND[a]	<0.09
River water		
Fox River, IL	0.006 $\pm$ 0.002	0.27 $\pm$ 0.12
Kankakee River, IL	0.0043 $\pm$ 0.0007	<0.06
Well water		
Lemont, IL	ND	<0.09
Borden, IN	0.012 $\pm$ 0.003	<0.06
Others		
Herrick Lake	0.003 $\pm$ 0.001	<0.09
Spring water	0.0015 $\pm$ 0.0004	<0.06

[a]ND = not detected.

only important area in which ICP-MS and radiation techniques differ. As noted previously, the energy bandpass of most of the alpha- and beta-radiation detectors is such that complex preparative procedures are sometimes required to ensure complete separation of the analytes from interferences. These procedures are not completely effective because it is possible for analyte isotopes to interfere with each other (28). Interferences are very uncommon in MS, but they are caused by the overlap of isobaric atomic ions or polyatomic ions with the analyte isotope. These mixtures of analytes and interferences can often be separated with less complex procedures; therefore, compared with radiation detection, mass spectrometry offers fewer preparation steps and greater isotopic selectivity.

IV. SEPARATION TECHNIQUES

As a principle, sample preparation should be kept to a minimum. For gamma emitters, carring out any preparation is often unnecessary. For alpha- and beta-emitting radionuclides, chemical separation is usually necessary. Alpha particles are monoenergetic, so radionuclides can be identified by their emission energies. But because the alpha ranges are very short ($\sim$50 μm), most of the particles are absorbed within the sample itself. To be able to detect them, alpha emitters must first be extracted from the sample. On the other hand, beta particles from a particular radionuclide are not monoenergetic but have a spectrum of energies up to a particular maximum. If the environmental sample contains more than one beta emitter, distinguishing

between the spectra can be very difficult. For the case of two beta emitters with similar energies, chemical separation will be necessary.

The radiometric determination of isotopes in the environment requires separation from large quantities of inactive matrix constituents and from a number of interfering radionuclides. A variety of methods have been described for effecting the necessary separations. Some of the procedures are based on precipitation (29,30), liquid–liquid extraction (31–34), ion exchange (35,36), and chromatography (37–39). All of these procedures, however, suffer from various limitations. Precipitation, for example, is tedious and must often be repeated several times to obtain adequate recoveries. Liquid–liquid extraction is too cumbersome for use with large numbers of samples and often requires the use of toxic solvents (e.g., dichloromethane for strontium determination). Ion-exchange procedure typically require careful pH control, because satisfactory separations (e.g., calcium) are achieved only within a narrow pH range. In addition, ion exchange is not suitable for samples containing a high concentration of acid. For the same reason, none of the chromatography methods have proved satisfactory.

In recent years, the most important advances in radiochemistry occurred with the creation of crown ethers to adsorb specifically the isotopes of interest. Horwitz et al. (40) showed a technique for the determination of strontium that can be adapted to produce a novel extraction chromatography resin by sorbing a solution of 4,4'(5')-bis(t-butylcyclohexano)-18-crown-6 (DtBuCH18C6) in octanol on an inert polymer substrate. The resultant material provides a simple and effective means of overcoming many of the limitations

Table 8 Results for the Determination of Thorium-232 in Water

Sample	ETV-ICP-MS (μg/L)	α-Spectrometry (μg/L)
Tap water		
Chicago, IL	0.17 ± 0.04	0.27 ± 0.05
Lemont, IL	0.27 ± 0.08	0.28 ± 0.06
River water		
Fox River, IL	0.8 ± 0.2	0.92 ± 0.09
Kankakee River, IL	1.4 ± 0.3	1.5 ± 0.1
Well water		
Lemont, IL	0.24 ± 0.07	0.30 ± 0.07
Borden, IN	0.07 ± 0.03	0.06 ± 0.02
Others		
Herrick Lake	0.19 ± 0.06	0.16 ± 0.05
Spring water	0.15 ± 0.05	0.13 ± 0.04

associated with other methods of isolating radionu-clides. Many other resins and substrates are being used today for the determination of other radionuclides in the environment. Table 3 (41) shows the commercial names for resins developed by using specific crown ethers or organic compounds in a solid substrate and the applications of these resins.

Recently, Smith et al. (42) demonstrated the use of solid-phase extraction disks for the determination of radiostrontium ($^{89/90}$Sr), technetium-99, and radium in surface water, groundwater, and drinking water. Solid-phase extraction disks (Empore™ technology) have proved to be highly effective for sample preparation in the analysis of organic compounds, wastewaters, and other aqueous samples (43,44). The isotopes of interest were easily isolated by pulling a sample aliquot through an appropriate Empore™ Rad-Disk with vac-uum. The disk is subsequently assayed for beta, gamma, or alpha activity. Radiometric interferences are minimal. The method is efficient, safe, reliable, and po-tentially deployable in the field. Sample preparation and counting source preparation steps may be con-densed into a single step, thereby reducing labor cost and eliminating many potential sources of laboratory

Table 9 Results for the Determination of Technetium-99 in Water

Sample	ETV-ICP-MS (ng/L)	Membrane/β-Counter[a] (ng/L)
Paducah-5920	1.4 ± 0.2	1.2 ± 0.1
Paducah-6275	26 ± 2	25 ± 3

[a]Low-background proportional counter.

error. Moreover, many of the hazardous chemicals as-sociated with traditional procedures are eliminated. Samples are easily batched, and a 1-L sample may be prepared with as little as 20 minutes of effort, and im-provement over traditional procedures, as illustrated in Table 4 (45–48).

V. APPLICATIONS TO EXTRACTION CHROMATOGRAPHY

Current techniques for the separation of low levels of actinide elements include precipitation, solvent extrac-tion, volatilization, and ion exchange. Some of the methods that are currently used are: the EPA sequential procedure for measuring radioactivity in drinking water (49), the Purex process (50), the Turex process (51), the TRU-Resin™ method of Horwitz et al. (52), and the PEARLS process (53). Some disadvantages in the current methods are poor recoveries, lack of complete separation of the elements, long turnaround times, poor resolution, and the generation of significant quantities of mixed waste. Boll et al. (54) have described a method that separates thorium, uranium, neptunium, plutonium, and americium into four spectroscopically distinct groups using the TRU-Resin™. TRU-Resin™ is an extraction chromatography resin composed of a solution of a bifunctional organophosphorus extractant, octyl(phenyl)-*N,N*-diisobutylcarbamoylme-thyl-phosphine oxide (CAPO), in tri-*n*-butyl phosphate (TBP) supported on an inert polymer substrate (am-berlite XAD-7). The sample is first loaded in the HCl solution with hydrogen peroxide. This allow the am-ericium and most matrix ions to pass through the col-

Table 10 Radioanalytical Methods Employing Inductively Coupled Plasma–Mass
Spectrometry

Analyte	Method	Ref.
^{99}Tc	Isotope dilution in aqueous samples, ICP-MS	72
	Effects of chemical form and memory effects in aqueous samples	73
^{99}Tc, ^{237}NP	Cyclohexanone solvent extraction and TTA–xylene solvent extraction	74
^{238}U	Flow injection and TRU™ in separation module, groundwater samples	75
Actinides	Isotopes in solution at low concentration (pg/L) by Quadrupole ICP-MS	76
Ta, U	Anodic and adsorptive stripping voltometry–ICP-MS	77
^{230}Th, ^{234}U ^{239}Pu, ^{240}Pu	Flow injection–ICP-MS with solid-phase extraction	78
Radium	ETV-ICP-MS with seawater as physical carrier	79
Uranium	Rainwater; high-resolution ICP-MS with ultrasonic nebulization (0.06 pg/L)	80
^{233}U, ^{239}Pu	Hydride interferences, ThH and UH	28

umn. The thorium is eluted by using dilute HCl, followed by the neptunium and plutonium, which are eluted together with oxalic acid in dilute HCl solution. Finally, the U is eluted with ammonium oxalate solution. A calcium oxalate coprecipitation is performed on the original load solution containing the americium ions, and the dissolved precipitate is then reloaded onto a TRU-Resin™ column in nitric acid with ascorbic acid. The procedure requires approximately 1.5 working days, reduces waste, and results in actinides recoveries of 80–100%. Table 5 (8,49,55–58) and Table 6 (8,59–66) show some radiochemical methods employing solvent extraction and extraction chromatography.

In two separate publications for the analysis of soil samples, Smith et al. (67) and Crain et al. (68), summarized two possible routes for the analysis of aqueous samples into chromatography extraction columns and detection by nonconventional radiometric techniques such as ICP-MS. In this procedure, TRU-Spec SPS™ columns were used for group separation of actinides, and TEVA-Spec™ columns were used to isolate the trivalent actinides from the lanthanide elements. A reduced solution (with ascorbic acid) was passed through a 1-ml TRU-Spec™ column equilibrated with 2 M nitric acid–0.5 M aluminum nitrate. The trivalent actinides, including americium, and the lanthanide elements were eluted from the column with 12 ml of 4 M HCl. Plutonium and thorium were removed with 30 ml of 0.1 M tetrahydrofuran-2,3,4,5 tetracarboxylic acid

(THFTCA). The trivalent actinides were separated by using TEVA-Spec™ resin. The lanthanide elements were removed by washing the column with 10 ml of 1 M NH$_4$SCN in 0.1 M formic acid. The trivalent actinides were eluted from the column with 15 ml of 2 M HCl. The THFTCA fraction containing plutonium (^{239,240}Pu) and thorium (^{230,232}Th) can be analyzed directly by ICP-MS. However, separation by using a Bio-Rad AG 1-X8 anion-exchange column was necessary prior to alpha spectrometry. The uranium ($^{233-236,238}$U) can be analyzed directly after the TRU-Spec™ column by ICP-MS.

VI. APPLICATIONS OF INDUCTIVELY COUPLED PLASMA–MASS SPECTROMETRY

Environmental and geochemical studies of radionuclides are closely related, because geochemical processes control the mobility of radioactive contaminants in the environment. However, environmental studies are generally focused on radiation protection and geochemical studies are focused on geochronology and chemical processes. Usually, the starting points of these studies are solid samples; but because of the nature of this chapter, we are going to start with water or aqueous solutions. Aqueous samples can have enough dissolved solid content to suppress the sensitivity of the mass spectrometer. In these cases, chemical separations may be used to remove the sample matrix, preconcentrate the analytes, and resolve any anticipated spectral interferences, thereby improving method detection limits.

If the chemical yield of the sample preparation procedure is less than unity (common in radiochemistry), then the accuracy and precision of the analytical procedure will be strongly correlated with the reproducibility of the yield. Yields may be determined in advance by multiple analysis or data may also be yield-corrected by using isotope dilutions or the method of standard additions. This approach is very convenient for the analysis as long as radioactivity may be added to the sample, but the chemical specification of the analyte and the calibration spike must be identical if the calibration techniques are to be effective. Traditional calibration techniques may also be used if the preparative yield is reproducible.

Preparative strategies have been a key factor in the development of new procedures for radionuclide determination by ICP-MS. Momoshima and coworkers (69) used ferric hydroxide coprecipitation to scavenge ^{99}Tc from seawater after technetium had been reduced to its

tetravalent oxidation state. Solvent extraction and ion exchange reduced interferences caused by spectral overlap and the sample matrix. Hollenbach et al. (70) determined ^{99}Tc, ^{230}Th, and ^{234}U in soil using a combination of flow injection and extraction chromatography–ICP-MS. High chemical yields were obtained, and detection limits were 20 ng/kg, 5 ng/kg, 3 ng/kg for ^{99}Tc, ^{230}Th, and ^{234}U, respectively. Electrothermal vaporization–ICP-MS was used by the author for the determination of ^{99}Tc, ^{236,238}U, ^{230,232}Th, and ^{226}Ra in tap water, river water, well water, and others. Detection limits in picograms were obtained for each element. Tables 7, 8, and 9 show some of the results obtained for the determination of levels of ^{238}U, ^{232}Th, and ^{99}Tc, respectively, by ETV-ICP-MS and these results compared with results obtained by using isotope-dilution alpha spectrometry, low-background proportional counters, and some of the new separations described in this chapter.

Shiraishi et al. (71) studied the distribution of uranium and thorium in freshwater samples collected in Ukraine, Russia, and Belarus. The analytes were detected directly by PN-ICP-MS, and the isotopes ratios

Table 11 Measurements by Empore Rad-Disks

Natural samples				
Sample	Volume (ml)	Activity (pCi/L)	Activity (pCi/L)	Accuracy (%)
		(Measured)	(Spiked)	
Well water[a] + ^{89}Sr	400	376 ± 7	387 ± 7	97
Tap water[a] + ^{89}Sr	1000	396 ± 8	387 ± 7	102
Mississippi River[b] + ^{90}Sr	500	9.9 ± 0.4	10.7 ± 0.5	93
Well water[a] + ^{99}Tc	8000	336 ± 16	347 ± 12	97
Deionized water[a] + ^{99}Tc	1000	532 ± 31	550 ± 31	97
Mono Lake water[c] + ^{99}Tc	1000	723 ± 21	723 ± 17	100
Performance evaluation samples				
Sample	Volume (ml)	^{90}Sr Activity (pCi/L)	^{90}Sr Activity (pCi/L)	Accuracy (%)
		(Measured)	(Reported)	
EMSL/LV water[d]	400	13.5 ± 0.5	15 ± 5	90
	400	11.4 ± 0.8	15 ± 5	76
EML water[e]	25	1,872 ± 32	1,854 ± 88	101
	25	1,924 ± 32	1,854 ± 88	104

[a]Argonne National Laboratory, Argonne, Illinois
[b]Mississippi River near Le Claire, Iowa,
[c]Mono Lake, California.
[d]U.S. Environmental Protection Agency program administered by Environmental Monitoring on Systems Laboratory—Las Vegas.
[e]U.S. Department of Energy program administered by Environmental Measurements Laboratory, New York, New York.

were determined by increasing the signal integration periods to obtain good counting statistics. The isotopic compositions found for uranium were consistent with fallout from the Chernobyl nuclear accident. Table 10 (28,72–80) shows some procedures for the determination of isotopes by ICP-MS.

VII. APPLICATION OF RAD-DISK TECHNOLOGY

A. Strontium Rad-Disk

The process for the Strontium Rad-Disk (AnaLig® Sr01) involved passing the sample, acidified with 2 M nitric acid, through a 47-mm disk positioned on a vacuum filter apparatus at a rate of 50 ml/min (42). For direct counting, the disk is dried with 20 ml of acetone and placed in a planchet for counting. Counting is done with a low proportional counting. Liquid scintillation and gamma spectroscopy are alternative counting techniques.

B. Radium Rad-Disk

The radium disk (AnaLig®Ra01) is the most simplistic method for the determination of ^{226}Ra and ^{228}Ra. Of the five radium isotopes, these two represent the most significant health hazard and concern for accurate detection. The sample is isolated by the same procedure described for the analysis of strontium. Once the sample is isolated, the investigator has a number of options for quantification. Radiation can be measured directly from the disk, but interpretation is difficult because of the multiple ingrowth paths occurring. Smith et al. (81) described two alternative analysis by using gamma or alpha spectrometry. Seely and Osterheim (82), in conjunction with Argonne National Laboratory scientists, developed a methodology for the simultaneous measurement of ^{226}Ra and ^{228}Ra. After the acidified sample is drawn through the disk, it is washed with 20 ml of nitric acid and dried. The dried disk is sealed in a 3.5-ml Mylar (polyethylene terephthalate) envelope and set aside for 21 days until equilibrium is reached between the radium isotope and their daughters. The envelope is then placed directly on a gamma counter. A multichannel analyzer is used to identify ^{214}Pb peaks, ^{226}Ra, ^{228}Ac, and ^{228}Ra. Ions at typical concentrations commonly found in environmental waters produced no interferences.

C. Technetium-99 Membrane

This method (48) describes the rapid isolation of Tc^{7+} and Tc^{4+} from aqueous samples by using an anion-exchange membrane disk and the subsequent membrane separation for measuring the ^{99}Tc beta activity. The method has been applied to water samples ranging in volume from 10 to 10,000 ml. The method requires minimal operator involvement and chemical manipulation, and produces virtually no chemical waste. A detection limit of 1.5 pCi/L was determined by using a low-background gas-flow proportional counter and a 1-L sample volume.

Table 11 shows some measurements and results by Empore™ Rad-Disks.

ACKNOWLEDGMENTS

This work was supported by the U.S. Department of Energy, Assistant Secretary for Environmental Management, under contract W-31-109-Eng-38.

REFERENCES

1. Friedlander G, Kennedy JW, Macias ES, Miller JM. Nuclear and Radiochemistry. 3rd ed. Wiley, New York, 1981.
2. International Atomic Energy Agency. Summary Report on the Post-Accident Review Meeting, Chernobyl Accident. Safety Series No. 75-INSAG-1, IAEA, Vienna, 1986.
3. United Nations Scientific Committee on the Effects of Atomic Radiations. Sources, Effects and Risk of Ionization Radiation. United Nations, New York, 1988.
4. Wang CH, Willis DL, Loveland WD. Radiotracer Methodology in the Biological, Environmental, and Physical Science, Prentice-Hall, Englewood Cliffs, NJ, 1975.
5. International Atomic Energy Agency. Measurement of Radionuclides in Food and the Environment: A Guide Book. IAEA Technical Series No. 295, IAEA, Vienna, 1989.
6. Greenberg AE, Clesceri LS, Eaton AD. Standard Methods for the Examination of Water and Wastewater. 18th ed. American Public Health Association, Washington, DC, 1992.
7. U.S. Dept. of Energy. Environmental Measurements Laboratory, ML-Procedure Manual. HASL-300. U.S. Department of Energy, Environmental Measurement Laboratory, New York, 1992.
8. Goheen SC, McCulloch M. DOE Methods for Evaluating Environmental and Waste Management Samples. NTIS, DOW-EM-0089. U.S. Department of Energy,

Office of Environmental Restoration and Waste Management, Washington, DC, 1993.

9. Lucas HF Jr. A Fast and Accurate Survey Technique for Both Radon-222 and Radium-226. In: The Natural Radiation Environment. JAS Adams, WM Lowder, eds. University of Chicago Press, Chicago, 1964.

10. Prichard HM, Gesell TF. Health Physics, 1977, 22, 577–581.

11. U.S. Environmental Protection Agency. Method 905.0. In: Prescribed Procedures for Measurement of Radioactivity in Drinking Water. Washington, DC, 1980, pp. 58–74.

12. U.S. Environmental Protection Agency. Method 907.0. In: Prescribed Procedures for Measurement of Radioactivity in Drinking Water. Washington, DC, 1980. pp. 75–82.

13. Blades MW, Caughlin BL. Spectrochimica Acta, 1985, 40B, 579.

14. Kalnicky DJ, Fassel VA, Kniseley RN. Applied Spectroscopy, 1977, 31, 137.

15. Daes SK, Kulkarni AV, Dhaneshwar RG. Analyst, 1993, 118, 1153.

16. Franek M, Krivan V. Analytical Chimica Acta, 1993, 274, 317.

17. Jackson PE, Carnevale J, Fuping H, Haddad PR. Journal of Chromatography A, 1994, 67, 181.

18. Rohr U, Meckel L, Ortner HM. Fresenius Journal of Analytical Chemistry, 1994, 348, 356.

19. Browner RF, Boorn AW. Analytical Chemistry, 1984, 56, 787A.

20. Alvarado JS, Erickson MD. Journal of Analytical Atomic Spectrometry, 1996, 11, 923.

21. Crain JS, Smith LL, Yaeger JS, Alvarado JS. Journal of Radioanalytical and Nuclear Chemistry, 1995, 194, 133.

22. Kim CK, Seki R, Morita S, Yamasaki S, Tsumura A, Takaka Y, Igarashi Y, Yamamoto M. Journal of Analytical Atomic Spectrometry, 1991, 6, 205.

23. Crain JS. Spectroscopy, 1996, 11, 30–39.

24. Brown RM, Long SE, Pickford CJ. Science of the Total Environment, 1998, 70, 265.

25. Igarashi Y, Kim CK, Takaku Y, Shiraishi K, Yamamoto M, Ikeda N. Analytical Science, 1990, 6, 157.

26. Toole J, Hursthouse AS, McDonald P, Sampson K, Baxter MS, Scott RD, McKay K. In: Plasma Source Mass Spectrometry. KE Jarvis, AL Gray, I Jarvis, J Williams, eds. Royal Society of Chemistry, Cambridge, England, 1990, p. 155.

27. Smith M, Wyse E, Koppenaal D. Journal of Radioanalytical and Nuclear Chemistry, 1992, 160, 341.

28. Crain JS, Alvarado JS. Journal of Analytical Atomic Spectrometry, 1994, 9, 1223–1227.

29. Weiss HV, Shipman WH. Analytical Chemistry, 1957, 29, 1764.

30. Fourie HO, Ghijsels JP. Health Physics, 1969, 17, 685.

31. Butler FE. Analytical Chemistry, 1963, 35, 2069.

32. Talvitie NA, Demint RJ. Analytical Chemistry, 1965, 37, 1605.

33. Veltar RJ. Nuclear Instrumental Methods, 1966, 42, 169.

34. Cahill DF, Lindsay GJ. Analytical Chemistry, 1966, 38, 639.

35. Noshkin VE, Mott NS. Talanta, 1967, 14, 45.

36. Porter CR, Kalin B, Carter MW, Vehnberg GL, Pepper EW. Environmental Science Technology, 1967, 1, 745.

37. Lada WA, Smulek W. Radiochemistry and Radioanalytical Letters, 1978, 34, 41.

38. Smulek W, Lada WA. Journal of Radioanalytical Chemistry, 1979, 50, 169.

39. Kremliakova NY, Novikov AP, Mysoedov BF. Journal of Radioanalytical and Nuclear Chemistry, 1990, 145, 231.

40. Horwitz EP, Chiarizia R, Dietz ML. Solvent Extraction and Ion Exchange, 1992, 10(2), 313.

41. Eichrom Industries, Inc. Web page, 1998, Innovations in Metals Separations. [URL http://www.eichrom.com (as of February 10, 1998)], Darien, IL.

42. Smith LL, Orlandini KA, Alvarado JS, Hoffmann KM, Seely DC, Shannon RT. Radiochimica Acta, 1996, 73, 165.

43. Hagen D, Markell C, Schmitt G. Analytical Chimica Acta, 1990, 236, 157.

44. Barcelo D, Durand G, Bouvot V, Nielen M. Environmental Science and Technology, 1993, 27(2), 271.

45. William R. Martin Marietta Utility Services, Inc., Paducah, KY, U.S.A., personal communication.

46. Environmental Measurement Laboratory. Technetium in Water and Vegetation. In: Environmental Measurement Laboratory Procedures Manual, HASL-300, 27th ed. vol. 1. U.S. Department of Energy, New York, 1992.

47. Holm E. Nuclear Instruments and Methods in Physics Research, 1984, 223, 204–207.

48. Orlandini KA, King JG, Erickson MD. Rapid Separation and Measurement of Technetium-99. Submitted to Radiochimica Acta, 1998.

49. U.S. Environment Protection Agency. Prescribed Procedures for Measurement of Radioactivity in Drinking Water. EPA 600-4-80-032. Environmental Monitoring and Support Laboratory, Cincinnati, 1980.

50. Freeman AJ, Keller C. Handbook on the Physics and Chemistry of the Actinides. vol. 6, Elsevier Science, New York, 1991.

51. Horwtiz EP, Kalina DG, Diamond H, Vandegrif GP, Schultz WW. Solvent Extraction and Ion Extraction, 1985, 3, 75.

52. Horwitz EP, Chiarizia R, Dietz ML, Diamond H, Nelson DM. Analytical Chimica Acta, 1993, 281, 361.

53. McDowell WJ, McDowell BL. Liquid Scintillation Alpha Spectrometry. CRC Press, Boca Raton, FL, 1994.

54. Boll RA, Schweitzer GK, Garber RW. Journal of Radioanalytical and Nuclear Chemistry, 1997, 220(2), 201.

55. Golchert NW, Sedlet J. Analytical Chemistry, 1969, 41(4), 669.

56. Chen Q, Dahlgaard H, Hansen HJM, Aarkrog A. Analytical Chimica Acta, 1990, 228, 163.

57. Morse RS, Welford GA. Health Physics, 1971, 21, 53.

58. Leyba JD, Vollmar JD, Fjeld RA, Devol TA, Brown DD, Cadieux JR. Journal of Radioanalytical and Nuclear Chemistry, 1995, 194(2), 337.

59. American Society of Testing Materials. Standard Test Method for Strontium-90 in Water. In: Annual Book of ASTM Standards, Philadelphia, D-5811-95, 1995.

60. Brines J. Radioanalytical and Nuclear Chemistry, Letters, 1996, 2112(2), 143.

61. Sullivan TM, Nelson DM, Thompson EG. Radioactivity and Radiochemistry, 1993, 4(2), 14.

62. Burnett WC, Cable PH, Moser R. Radioactivity and Radiochemistry, 1995, 6(3), 36.

63. Carney KP, Cummings DG. Journal of Radioanalytical and Nuclear Chemistry, 1995, 194, 41.

64. Berne A. Use of Eichrom's TRU Resin in the determination of Am, Pu, and U in Air Filters and Water Samples. Environmental Measurement Laboratory, EML-75. New York, 1995.

65. Horwitz EP, Chiarizia R, Dietz ML, Diamond H, Graczyk D. Analytical Chimica Acta, 1992, 266, 25.

66. Testa C, Desideri D, Meli MA, Roselli C. Journal of Radioanalytical and Nuclear Chemistry, Articles, 1995, 194, 141.

67. Smith LL, Crain JS, Yaeger JS, Horwitz EP, Diamond H, Chiarizia R. Journal of Radioanalytical and Nuclear Chemistry, Articles, 1995, 194, 151.

68. Crain JS, Smith LL, Yaeger JS, Alvarado JS. Journal of Radioanalytical and Nuclear Chemistry, Articles, 1995, 194, 133.

69. Momoshima N, Sayad M, Takashima Y. Radiochimica Acta, 1993, 63, 73.

70. Hollenbach M, Grohs J, Mamich S, Kroft M, Denoyer ER. Journal of Analytical Atomic Spectrometry, 1994, 9, 927.

71. Shiraishi K, Igarashi Y, Yamamoto M, Nakajima T, Los IP, Zelensky AV, Buzinny MZ. Journal of Radioanalytical and Nuclear Chemistry, 1994, 185, 157.

72. Beals DM. Determination of Technitium-99 in Aqueous Samples by Isotope Dilution Inductively Coupled Plasma–Mass Spectrometry. Westinhouse Savannah River Company, Aiken, SC.

73. Ritcher RC, Koirtyohann SR, Jurisson SS. Journal of Analytical Atomic Spectrometry, 1997, 12, 557.

74. Sumiya SH, Morita SH, Tobita K, Kurabayashi M. Journal of Radioanalytical and Nuclear Chemistry, Articles, 1994, 177(1), 149.

75. Aldstadt JH, Kuo JM, Smith LL, Erickson MD. Analytical Chimica Acta, 1996, 319, 135.

76. Liezers M, Tye CT, Mennie D, Koller D. In: Applications of Inductively Coupled Plasma–Mass Spectrometry. RW Morrow, JS Crain, eds. ASTM-STP1291, Philadelphia, 1995, p. 61.

77. Zhou F, Van Berkel GJ, Morton SJ, Duckworth DC, Adeniyi WK, Keller JM. In: Applications of Inductively Coupled Plasma—Mass Spectrometry. RW Morrow, JS Crain, eds. ASTM-STP1291, Philadelphia, 1995, p. 82.

78. Hollenbach M, Grohs J, Kroft M, Mamich S. In: Applications of Inductively Coupled Plasma–Mass Spectrometry. RW Morrow, JS Crain, eds. ASTM-STP1291, Philadelphia, 1995, p. 99.

79. McIntyre RC, Gregorie DC, Chakrabarti L. Journal of Analytical Atomic Spectrometry, 1997, 12, 547.

80. Tsumura A, Okamoto R, Takaku Y, Yamasaki S. Radioisotopes, 1995, 44, 85.

81. Smith LL, Alvarado JS, Markun FJ, Hoffmann KM, Seely DC, Shannon RT. Radioactivity and Radiochemistry, 1997, 8(1), 30.

82. Seely, DC; Osterheim JA. Radiochemical Analyses Using Empore™ Disk Technology. MARK IV Conference, Methods for Analytical Radiochemistry, Kona, Hawaii, April 1997.

8

Bacteriological Analysis

Chris W. Michiels and Els L. D. Moyson
Catholic University of Leuven, Leuven, Belgium

I. MICROORGANISMS AND WATER QUALITY

Natural waters may contain many different types of microorganisms in highly variable numbers. In fact, the presence of viable microorganisms is the rule rather than the exception, and a large variety of bacteria, yeasts, molds, algae, and protozoa are commonly found in fresh and marine surface waters and in waters from subterranean aquifers. Whether these microorganisms will affect the functional properties of this water for human use depends on the type of application of the water (e.g., irrigation, cooling of industrial machinery, swimming, drinking, pharmaceutical preparations for intravenous injection) and on the types and numbers of microorganisms. For instance, sulfur-oxidizing bacteria can produce sulfuric acid in water containing reduced sulfur compounds under aerobic conditions, and this is a well-known cause of quality deterioration that can occur in oil that is pumped up by water injection underground. It is also an important cause of the corrosion of steel pipes (1). Another important problem of microbiological origin in the industrial use of water is the formation of biofilms, which stems from the tendency of most waterborne microorganisms to attach to solid surfaces. Depending on the conditions, biofilms can develop into a millimeter-thick layer consisting of a high number of microorganisms entrapped in a slime matrix. The accumulation of this material can cause reduced water flow in pipes or reduced heat transfer in heat exchangers. In drinking water and recreational water, finally, microorganisms can cause sensory defects

(odor, color, taste) and, not the least, disease. A nonexhaustive overview of different problems with water quality and safety caused by microorganisms is given in Table 1.

II. SCOPE

The current chapter will give an overview of the state of the art in the analysis of the microbiological safety of drinking and recreational water. Drinking water includes untreated spring and well water, as well as treated tap water; recreational water includes both fresh and marine water. Problems related to industrial, agricultural, or other uses of water are not covered. Although many different types of waterborne microorganisms may cause disease, methods for the detection of viruses and protozoa will not be covered. This may seem odd, because these organisms do have a prominent place in waterborne disease. However, for reasons outlined next, the monitoring of microbiological water safety is generally based on the analysis of only a limited number of well-chosen parameters, instead of a screening for all possible pathogens.

III. MICROORGANISMS AND SAFETY OF DRINKING AND RECREATIONAL WATER

Clean water is the most essential need for human life. Throughout human history, however, contamination of

Table 1 Problems with Water Quality and Safety Caused by Microorganisms

Name of microorganism	Problems
Health-related problems in drinking and recreational water	
Salmonella, Shigella, enteropathogenic *E. coli, Campylobacter jejuni, Vibrio cholerae, Yersinia enterocolitica, Aeromonas hydrophilia*[a]	Gastrointestinal infection after ingestion
Leptospira, Pasteurella tularensis, Staphylococcus aureus, Pseudomonas aeruginosa,[a] *Aeromonas hydrophila*[a]	Infection of various tissues after contact with mucous membranes or wounds
Legionella pneumophila, Mycobacterium[a]	Infection of lungs after inhalation
Cyanobacteria (*Microcystis, Anabaena*), *Clostridium botulinum*	Formation of toxins that resist cooking
Sensory-quality-related problems in drinking water	
Cyanobacteria	Coloration due to release of pigments
Iron- and maganese-oxidizing bacteria and fungi	Rust-colored or black deposits
Actinomycetes, cyanobacteria	Taste and odor formation due to the production of geosmin or 2-methylisoborneol
Microalgae	Cloudiness
Problems related to industrial applications	
Iron-oxidizing bacteria (*Gallionella, Sphaerotilus*) and sulfate-reducing bacteria (*Desulfovibrio, Desulfotomaculum*)	Microbially induced corrosion
Sulfur-oxidizing bacteria (*Thiobacillus*)	H_2SO_4 formation
Biofilm-forming bacteria	Biofilm, fouling

[a]Opportunistic pathogens.

water with infectious agents has been a major cause of morbidity and mortality, and it continues to be. A most dramatic example is the devastating cholera epidemics that swept over Europe and North America during the 19th century. In the single year of 1849 the disease was estimated to have infected 440,000 and killed 110,000 people in the UK (2). Even today, contaminated water as a cause of cholera and other diarrheal diseases is estimated to kill about 2 million children and cause about 300 million episodes of illness each year (3). Not surprisingly, a vast majority of these waterborne diseases occur in nonindustrialized countries.

In the second half of the 19th century, microbiology evolved into a mature scientific discipline, and the microbial causes of cholera and other major infectious diseases of that time, as well as their routes of trans-

mission, were identified. Since then, a number of measures have been introduced in industrialized countries that have been successful in virtually eliminating the 19th century waterborne killer diseases. These measures include the improved separation of sewage effluents from drinking water resources, the nationwide development of water distribution networks, and the large-scale disinfection and monitoring of drinking water to eliminate pathogenic bacteria. As a result, most developed countries now enjoy a water supply that is safer than ever before. Nevertheless, it should be emphasized that this success can be maintained only by permanently keeping all the control measures in place at the same high standard. Any relaxation of these standards or malfunctioning of these measures will result in increased levels of pathogenic microorganisms and an increased incidence of disease.

The majority of waterborne diseases are caused by the ingestion of water, and result in gastrointestinal disorders. Contamination of water with the responsible pathogens occurs primarily through human or animal feces. A number of other pathogens causing infections of the respiratory tract, ears, eyes, or skin can also be transmitted by water, but since these are much less frequent, and for historical reasons as well, microbiological analysis of drinking or recreational water has always focused on pathogens of fecal origin. Table 2 gives an overview of the major diseases caused by waterborne bacteria.

IV. INDICATOR ORGANISMS

A. Rationale for Using Indicator Organisms

Drinking water and recreational water must be safe, i.e., not cause disease upon normal use. For drinking water it is generally agreed that any pathogenic microorganisms should be absent. This zero-tolerance principle differs from the no-toxic-effect-level approach generally used for chemical contaminants, for a number of reasons. First, absence cannot be measured for chemicals, because there is always a lower detection

Table 2   Overview of Major Waterborne Diseases

Name of microorganism	Symptoms	Most common source and route of infection
Salmonella	Gastroenteritis: diarrhea, abdominal cramps, vomiting, nausea	Oral ingestion of fecally contaminated water
Salmonella typhi	Fever, nausea, sepsis	
Shigella	Gastroenteritis: diarrhea, abdominal cramps, vomiting, nausea	Oral ingestion of fecally contaminated water
Enteropathogenic *E. coli*	Gastroenteritis: diarrhea, abdominal cramps, vomiting, nausea	Oral ingestion of fecally contaminated water
Campylobacter	Gastroenteritis: diarrhea, abdominal cramps, vomiting, nausea	Oral ingestion of fecally contaminated water
Vibrio cholerae	Gastroenteritis: diarrhea, abdominal cramps, vomiting, nausea, (cholera)	Oral ingestion of fecally contaminated water
Leptospira	Acute infections involving kidneys, liver, and central nervous system	Entering bloodstream through skin abrasions or mucous membranes
Pasteurella tularensis	Chills and fever, swollen lymph nodes	Entering bloodstream through skin abrasions or mucous membranes
Yersinia enterocolitica	Gastroenteritis: diarrhea, abdominal cramps, vomiting, nausea	Oral ingestion of fecally contaminated water
Legionella pneumophila	Legionnaires' disease	Inhalation in aerosol
Pseudomonas aeruginosa	Superficial or systemic infections; gastroenteritis	Contact with skin; ingestion in immunosuppressed patients
Aeromonas spp.	Superficial or systelic infections; gastroenteritis	Ingestion, contact with skin
Mycobacterium	Tuberculosis	Inhalation in aerosol, contact with skin

limit; cultural methods for bacterial detection, in contrast, allow in principle the detection of a single bacterium, and can therefore ensure absence, at least in the analyzed sample. Second, bacteria are living organisms that can multiply. Therefore, compliance with a nonzero tolerance level at the time of sampling would not guarantee compliance at the time of consumption. Third, the infective dose (the number of bacteria that need to be ingested to cause disease) can be as low as a few hundred cells for bacteria like *Salmonella* and *Shigella*, or even a single cell for protozoa like *Cryptosporidium*. Hence, safety requires absence for microbial pathogens.

If we accept this zero-tolerance principle, how then can the absence of pathogens be monitored? Although detection of all the possible pathogens in a water sample is today perhaps technically possible, it is not a realistic option, because the analyses are time consuming, can only be conducted in specialized laboratories, and would be prohibitively expensive. In addition, microbiological contamination often occurs in localized and sudden spikes, and it can be easily understood that sampling directly for pathogens would not be very successful in view of the very limited volume of water that can be sampled compared to the total volume of water distributed. For example, U.S. regulations prescribe the monitoring of tap water once per month and per 1000 people served. This means that only 100 samples are analyzed over a month and over the entire area of a city of 100,000 inhabitants. For all these reasons, the monitoring of microbiological water quality makes use of indicator organisms. The concept of using indicator organisms was already developed by late in the 19th century. At that time, the detection of specific pathogens in sewage water remained elusive, whereas other bacteria that are characteristic of the feces of both sick and healthy persons could be readily isolated. This was particularly true for a bacterium isolated from the feces of a cholera patient by Escherich in 1885 and named ''Bacterium coli'' and later renamed *Escherichia coli*. Subsequently, it was demonstrated that this *E. coli* is invariably and exclusively present in feces and that it survives better and is thus present in higher numbers than pathogenic bacteria in polluted water. Therefore, Schardinger in 1892 (4) proposed that the presence of *E. coli* could serve as an indication of fecal contamination and thus of the potential presence of fecal pathogens. A century later, though several improvements have been developed in the techniques for the detection of *E. coli*, and alternative indicator organisms for evaluating water quality have been proposed, the basic concept of using fecal indicator organisms is still

used. This concept can be summarized in a number of criteria that the ideal fecal indicator organism has to fulfill:

1. Always be present when the pathogens are present
2. Occur in high numbers in the feces of humans or animals that can be the source of intestinal pathogens
3. Survive equally well as or better than the pathogens in the aqueous environment, but without being able to grow, since the pathogens generally also will not grow in water
4. Be rapidly and easily detectable, also when high numbers of other microorganisms are present.

B. Commonly Used Indicator Organisms

Although *E. coli* was originally proposed as an indicator organism, methods in the first part of the 20th century did not allow the direct discrimination of *E. coli* from a number of related bacteria belonging to the genera *Klebsiella*, *Enterobacter*, and *Citrobacter* and even some species of *Aeromonas*. Collectively, this group of organisms was designated as the coliforms and was defined and isolated as aerobic or facultatively anaerobic, gram-negative, nonsporulating, rod-shaped bacteria that can ferment lactose at 35–37°C with the formation of gas. Many water-quality standards were therefore written in terms of the concentration of coliform bacteria, and it was believed that all coliforms were of fecal origin (5). Although today *E. coli* is the only coliform organism that is considered to be primarily of fecal origin, coliform testing is still widely in use for determining the hygienic quality of water as well as several food products.

Because total coliforms could be a misleading indicator of fecal pollution, the guidelines were rewritten in terms of fecal coliforms. Fecal coliforms are distinguished from other coliforms by their ability to ferment lactose with gas formation at elevated temperature (44.5–46°C). The correlation between this property of thermotolerance and fecal origin can be understood from an ecological point of view, since coliforms that normally inhabit the gastrointestinal tract of humans or warm-blooded animals are adapted to a higher temperature than environmental coliforms that live in the soil or on the vegetation. Although the use of fecal coliform counts as an indicator of water quality is a significant improvement, it must be emphasized that fecal coliforms are not exclusively *E. coli*, but also include a few other bacteria that are not necessarily of fecal origin, such as thermotolerant *Klebsiella* species, which are commonly found in the soil and on the vegetation (6). Therefore this group is better called *thermotolerant coliforms*.

Another group of bacteria that is taxonomically unrelated to the coliforms but that has been used as a fecal indicator in water since the beginning of the 19th century are the so-called fecal streptococci (7). The streptococci are a large family of gram-positive, nonsporulating, catalase-negative cocci currently including the genera *Streptococcus*, *Enterococcus*, and *Lactococcus*. The fecal streptococci (sometimes referred to as fecal enterococci or enterococci) are a subgroup consisting of those streptococci species that are exclusive inhabitants of the human or animal gastrointestinal (GI) tract. The most common representatives of this group are *E. faecalis*, *E. faecium*, *E. durans*, *E. hirae*, *S. bovis*, and *S. equinus*, but other species have been recently added (8). The fecal streptococci generally satisfy the criteria for fecal indicator organisms mentioned earlier. In comparison to *E. coli*, they are usually present in lower numbers in feces, but survive better in the aqueous environment, and are about twice as resistant to disinfection (8).

Over the last decades many studies have been performed to evaluate the correlation between the presence of indicator organisms (total and thermotolerant coliforms, *E. coli*, fecal streptococci) and the incidence of gastrointestinal illness symptoms such as vomiting, diarrhea, stomachache, and nausea of swimmers. These studies indicate that in marine waters, fecal streptococci are the best indicators (9,10). The poor performance of total coliforms, fecal coliforms, and *E. coli* for marine water was ascribed to their poor survival. For instance, it was demonstrated that coliform counts in seawater declined faster than those of viruses (11,12). In freshwater the occurrence of GI illnesses was not significantly correlated with fecal coliform concentrations but did correlate with both fecal streptococci and *E. coli* (5).

C. Other Possible Indicator Organisms

During the last two decades, we have been increasingly confronted with a number of emerging waterborne pathogens, which survive better in an aqueous environment and which are more resistant to water disinfectants than the traditional enteric bacterial pathogens and indicator organisms. These pathogens include enteroviruses and protozoa such as *Cryptosporidium* and *Giardia*. Accordingly, more resistant indicator organisms have been proposed. A first group that has been investigated in this respect are *E. coli*–specific viruses, the so-called coliphages. They are *E. coli*–specific, so their occurrence indicates that *E. coli* is or has been present. In particular, the F-specific coliphages are of interest, since they have similar resistance characteristics to the pathogenic enteroviruses (13). The monitoring of coliphages for water quality is, however, not yet routine needs to be supported by further validation. A second group of indicators that is even more resistant are the anaerobic sporeformers belonging to the genus *Clostridium*. Clostridia can be detected as sulfite-reducing anaerobes, but the value of this whole group as a fecal indicator is questionable, since many *Clostridium* species are of environmental origin (soil, putrifying organic matter). Therefore a better choice is *C. perfringens*, which is a common member of the human colon flora (14). However, the presence of *C. perfringens* should be interpreted with caution and not be taken as the only evidence of fecal pollution, because the spores of this organism can survive for years in the environment.

D. Heterotrophic Plate Count as a Measure of Water Quality

Although water can contain a variety of microorganisms of differing significance, the estimation of the total number of viable microorganisms can provide useful information for the assessment and surveillance of water quality. The majority of microorganisms found in water grow better in laboratory culture media at 22°C than at higher temperatures, because they are adapted to the normal conditions in soil and water as their natural environment. On the other hand, microorganisms that are isolated from water and that grow well at 37°C are more likely to come from other sources, including warm-blooded humans or animals. Therefore separate counts are usually made at two or more different temperatures. However, the heterotrophic plate count should not be used as a measure of the hygienic quality or safety of drinking water, since the majority of the organisms counted are ecologically unrelated to enteropathogens. It is useful mainly for monitoring the efficiency of water treatment processes, as an indication of the levels of accompanying flora that can interfere with coliform, thermotolerant coliform or *E. coli* tests, and to assess microbiological changes such as regrowth in water during storage or distribution (15).

E. Microbiological Standards for Drinking and Recreational Water

As an illustration of how the use of indicator bacteria for water-quality monitoring is applied, we will briefly compare some current legislation. In Europe, Council Directive 98/83/EC (16) sets minimal chemical and microbiological quality criteria for the quality of water for human consumption (Table 3a). The parameters for which maximum levels are given are *E. coli* and enterococci and, in the case of water in bottles or other packs, also *Pseudomonas aeruginosa* and the heterotrophic plate count at 22°C and 37°C. In addition, the monitoring of coliforms and, when it concerns water that is derived from or influenced by surface waters, of *Clostridium perfringens* is also mandatory, but only for control purposes, and there is no maximum value for these parameters. The directive also specifies minimum sampling frequencies in the volume of water distributed.

In the United States, criteria for drinking water quality and the monitoring requirements for drinking water quality are described in the National Primary Drinking Water Regulations (17). Microbiological monitoring is based mainly on presence/absence testing of coliforms, which means that coliforms do not have to be quantified. Depending on the number of people served, a water distribution system must analyze a certain number of 100-ml samples per month. No more than 5% coliform-positive samples should be found in a month. Further, any coliform-positive samples must be analyzed for fecal coliforms or *E. coli* and must be negative. Finally, if fecal-coliform-positive or *E. coli*–positive samples are found, a certain number of repeat samples have to be analyzed and have to be negative for coliforms and for fecal coliforms or *E. coli* for compliance.

The microbiological criteria for recreational water are normally less stringent than those for drinking wa-ter, but they may also vary from country to country. As an illustration, the criteria for swimming water in the EU (18), and for swimming pool water in Flanders (Belgium) (19) are given in Table 3b and 3c, respectively. Note that the relevant EU legislation dates from 1976 and prescribes the analysis of coliforms and fecal coliforms, not *E. coli*, as opposed to the more recent drinking water legislation (Table 3a).

V. METHODS OF ANALYSIS

A. General Techniques

The detection and enumeration of waterborne bacteria is generally performed by methods that rely on cultivation, either on solid media (such as the pour plate and spread plate techniques and the membrane filtration method) or by culture methods in liquid media (such as the presence-absence test and the multiple-tube test, known as the most probable number (MPN) method.

1. Pour Plate and Spread Plate Techniques

In the *pour plate technique*, a specified volume of the water (generally 1 ml) is added to a petri dish of 90–100-mm diameter. About 15 ml of agar containing growth medium previously melted and tempered to a temperature close to that of solidification is added to the test portion and mixed carefully. After solidification, the plates are incubated at the appropriate temperature. The colonies that develop within and on the surface of the medium within a specified time are counted (20).

A smaller volume of the water sample (0.1–0.5 ml) is spread over the surface of a solid agar medium in the *spread plate technique*. After incubation at the appropriate temperature and for the appropriate time, colonies developed on the medium are counted (20).

Table 3a Microbiological Criteria for Water for Human Consumption in the EU

	Parameter value		
Microbiological parameter	Unbottled	Bottled or packed	Recommended method
E. coli	0/100 ml	0/250 ml	ISO 9308-1
Enterococci	0/100 ml	0/250 ml	ISO 7899-2
Pseudomonas aeruginosa		0/250 ml	ISO 12780
Heterotrophic plate count 22°C		100/ml	ISO 6222
Heterotrophic plate count 37°C		20/ml	ISO 6222

Source: Ref. 16.

Table 3b   Microbiological Criteria for Water for Swimming in the EU

Microbiological parameter	Parameter value	
	Guide value	Imperative value
Total coliforms	500/100 ml	10,000/250 ml
Fecal coliforms	100/100 ml	2,000/250 ml
Fecal streptococci	100/100 ml	
Salmonella		0/L
Viruses		0 pfu/10 L

pfu: plaque-forming units.
Source: Ref. 18.

2. Most Probably Number (MPN) Method, or Multiple-Tube Test

The MPN method gives an estimate of the mean density of organisms in a sample, assuming a random dispersion. The precision of the estimate depends upon the number of tubes inoculated. Multiple test portions of the sample and/or dilutions of it are inoculated into tubes of liquid culture medium. It is assumed that, on incubation, each tube that received one or more organisms will show growth, resulting in a characteristic change in the medium (positive reaction, e.g., gas, acid). From the number and the distribution of tubes showing a positive reaction, the most probable number of organisms can be estimated from MPN tables (20).

Different inoculation systems can be used. In the usual "symmetrical" systems, the same number of replicate tubes is used for each decimal dilution. Usually three to five tubes are used in each series; but three replicate tubes of three successive dilutions are a minimum. The expected bacterial count will usually serve as a guide to selecting a suitable series of dilutions to yield three sets of results acceptable for calculation of the MPN value (20).

Table 3c   Microbiological Criteria for Water for Swimming Pools in Flanders (Belgium)

Microbiological parameter	Parameter value
Heterotrophic plate count 37°C	100/ml
Coagulase-positive staphylococci	0/100 ml
Pseudomonas aeruginosa	0/100 ml
Legionella pneumophila[a]	0/100 ml

[a]For whirlpools only, and one analysis yearly is sufficient.
Source: Ref. 19.

3. Presence–Absence (P/A) Test

A single test portion of the water sample is inoculated and incubated in a suitable volume of an appropriate liquid medium, and the presence of the organism is demonstrated by growth and/or specific changes in the medium. With this method the number or organisms is not quantified. But large numbers of samples can be analyzed in a short time, so it provides an ideal tool for comparative studies in routine sample analysis.

4. Membrane Filtration Method

A measured volume of the water sample (mostly 100 ml) is filtered through a sterile membrane, with a pore size that will retain the organisms to be enumerated. After filtration of the water sample, the membranes are placed face upwards on an agar medium or on a sterile absorbent pad saturated with culture broth. They can even be overlaid with a molten agar medium (20). The method is highly reproducible and highly sensitive, particularly when using membranes of a pore diameter of 0.45 μm, which allow rapid filtration of large water volumes, and it is therefore the most popular technique for the microbiological examination of water samples. Detection of some organisms requires filtration through a 0.22-μm membrane, which decreases test sensitivity because of the smaller volumes of water that can be filtered. In some cases, resuscitation of stressed organisms is required (see Sec. VI).

The most important advantage of the membrane filtration technique is the speed with which results are obtained. There is also considerable saving in labor, media, and the amount of glassware needed compared to other techniques. The membrane filtration method, however, is not recommended for water samples with a high turbidity or a high background flora.

5. Criteria for Choosing the Enumeration Technique

Most microorganisms in water can be enumerated by either of the four techniques. The choice of an appropriate technique will depend on the detection limit required, the nature of the water sample, and the cost of the technique.

When a low detection limit is required, the membrane filtration method is often preferred. If the analyzed water is clear, large volumes up to several liters can be filtered with this method, giving a detection limit, in principle, of a single organism in the whole sample.

The maximum volume that can be analyzed in the pour plate technique is 5 ml per plate. This detection limit can be improved by increasing the number of test portions. But since inoculation of more than five plates is usually considered impractical, the detection limit is therefore one bacterium in 25 ml for the pour plate technique. The maximum volume used on very dry spread plates is occasionally 1 ml, corresponding to a detection limit not higher than one organism in 5 ml for five replicate spread plates.

Suspended particles in water samples with high turbidity can cause problems by clogging of membranes in the membrane filtration method, and by being identified mistakenly for bacterial colonies in the plating methods. Therefore for very turbid water the MPN technique may sometimes be the only possible method, particularly if, as often is the case, the spread plate technique does not provide sufficient sensitivity.

The chemical composition of the water sample can in some cases interfere with detection. Techniques in which high volumes of the water sample are mixed with a relatively small amount of growth medium, such as the MPN method, will be more affected than techniques in which the water sample is added in a smaller proportion, such as the pour plate and spread plate techniques, or methods that separate the microorganisms from the sample, such as the membrane filtration method. Different types of interference can occur. The water sample can contain toxic soluble substances, inhibiting the growth of the target organisms. This can be suspected when different dilutions of the sample are analyzed and proportionally higher counts are obtained in the most diluted samples. This disadvantage can be avoided with the membrane filtration method, unless the toxic substances are retained on the membrane. Sometimes substances in the sample can adversely affect the characteristic reactions of the microorganisms sought without interfering with their growth. An ex-

ample is the presence of a fermentable sugar in the water sample that is not present in the detection medium and that can be fermented by nontarget microorganisms in the sample, resulting in a change in pH as would be expected to be produced by the target bacteria. Also, the initial pH of the water sample may interfere with observation of the results; e.g., the formation of acid by fecal streptococci in azide glucose broth is undetectable when the pH of the water sample is too low.

The physiological properties of the target organisms may dictate a preference for certain methods. Strictly aerobic organisms are preferentially cultured on spread plates or with the membrane filtration method, while for facultatively anaerobic microorganisms the pour plate technique may provide improved selective conditions and give better results. Deep and narrow tubes can be used with more anaerobic organisms. Many bacteria from an aquatic environment cannot withstand the thermal shock resulting from mixing a sample with melted agar at 44°C. Therefore the pour plate technique should in fact be avoided for their enumeration (15).

Finally, the nature and concentration of the accompanying microflora may interfere with successful detection of the target microorganisms due to competition. This occurs especially in liquid media, because on solid media the individual organisms form colonies that are separated from each other and therefore compete only to a limited extent, provided that there is no colony spreading or overcrowding by the accompanying microflora. Therefore, the use of solid media is generally preferable. However, when no satisfactory selective plating medium exists and the accompanying microflora are so numerous that it makes detection of typical colonies of the target organism impossible, the MPN is preferable above colony count procedures.

B. Determination of Heterotrophic Plate Count (HPC)

The heterotrophic plate count in water can be determined by the spread plate and pour plate techniques as well as by the membrane filtration method using a specified nutrient agar. The latter permits testing of large volumes of low-turbidity water and is recommended for waters with very low concentrations of microorganisms. The pour plate method, on the other hand, generally results in lower HPC counts, because many of the natural bacteria in the low-nutrient and cold aquatic environment are particularly sensitive to the heat shock they experience when they are mixed with 43–46°C melted agar.

An overview of the influence of various media and the incubation temperature and time has been given by Reasoner (15). Early plate counts were performed at two different temperatures, 20°C (room-temperature count) and 37°C (body-temperature count). The 37°C plate count was believed to give an indication of fecal pollution, because fast-growing microorganisms at this temperature were likely to be ecologically related to pathogens present in sewage. The 20°C plate count was used to enumerate slow-growing natural water bacteria. For practical reasons (saving in incubator space in laboratories performing both milk and water analyses) the temperature of the body-temperature count was later lowered to 35°C (15). Counts at 28°C are also sometimes used and are only slightly lower than those performed at 20°C, but give more rapid results.

The most commonly used media are probably plate count agar (21) and yeast extract agar (22) for pour plate and spread plate. There was an increasing need for lower detection limits, so Taylor and Geldreich (23) designed membrane heterotrophic plate count agar (m-HPC, formerly called m–SPC, agar), a nutrient-rich medium specifically for membrane filtration. This technique has been found to give results equivalent to the pour plate technique (24). Later on, to improve bacterial recoveries, low-nutrient agar media were introduced for bacterial enumeration of the heterotrophic plate count, such as R2A agar (Table 4), which can be used for the pour plate, spread plate, and membrane filtration methods (25). Although the total concentration of nutrients in this medium is lower, it contains a greater variety of nutrients and results in higher counts than the high-nutrient media (15). However, maximum counts on R2A medium are obtained only after incubation for up to 5–7 days at 20–28°C, because due to its lower nutrient concentration microorganisms grow slower. Another low-nutrient agar (NWRI agar) has more recently been developed with optimized nutrient composition, which is likely to produce higher colony counts than the three other media described (21). Although the low-nutrient media have proven superior for the recovery of heterotrophic bacteria, it can be useful to continue the use of high-nutrient plate count agar or yeast extract agar for some time to allow more detailed media comparisons or to extend the continuity of older data (21).

It can be concluded that the pour plate method and the use of a rich medium, an incubation temperature of 35–37°C, and a short incubation period result in reduced counts of bacteria present in potable water (15).

Table 4 Media and Culture Conditions for the Detection of Heterotrophic Flora

Ref.	Name of the medium	Incubation conditions
Pour plate and spread plate methods		
ISO 6222	Yeast extract agar (YEA)	37 ± 1°C for 24–48 h and 22 ± 1°C for 72 h
APHA	Plate count agar (PCA)	35°C for 48 h and 20–28°C for 5–7 days
	R2A	
	NWRI agar (HPCA)	
Membrane filtration method		
APHA	m-HPC	35°C for 48 h
	R2A	35°C for longer than 48 h
	NWRI agar (HPCA)	20°C for 7 days

Composition of the media:
YEA: 0.6% tryptone; 0.3% yeast extract; 1.2% agar; pH = 7.2.
PCA: 0.5% tryptone; 0.25% yeast extract; 0.1% glucose; 1.5% agar; pH = 7.0.
R2A: 0.05% yeast extract; 0.05% proteose peptone; 0.05% casamino acids; 0.05% glucose; 0.05% soluble starch; 0.03% dipotassium hydrogen phosphate; 0.005% magnesium sulfate heptahydrate; 0.3% sodium pyruvate; 1.5% agar; pH 7.2.
NRWI agar (HPCA): 0.3% peptone; 0.05% soluble casein; 0.02% dipotassium hydrogen phosphate; 0.005% magnesium sulfate; 0.0001% ferric chloride; 1.5% agar; pH = 7.2.
mHPC (membrane heterotrophic plate count): 2.0% tryptone; 2.5% gelatin; 1.0% glycerol; 1.5% agar; pH = 7.1.

Table 5 Selective Media for the Detection of Coliforms, Thermotolerant Coliforms, and *E. coli* Using the MPN Method

Name of the medium	Typical reactions	References
Presumptive isolation[a]		
Lactose broth	Gas formation in Durham tube	ISO 9308-2
MacConkey broth	Gas formation in Durham tube	ISO 9308-2
Improved formate lactose glutamate medium (= minerals modified glutamate medium)	Gas and acid formation when bromocresol purple is added	ISO 9308-2; 28
Lauryl tryptose lactose broth	Gas and acid formation when bromocresol purple is added	ISO 9308-2; APHA
Confirmation[b]		
Brilliant-green lactose (bile) broth	Gas formation in Durham tube	ISO 9308-2; APHA; 28
Lauryl tryptose lactose broth	Gas formation in Durham tube	28
EC medium[d]	Gas formation in Durham tube	ISO 9308-2
Lauryl tryptose mannitol broth with tryptophan (presumptive *E. coli*)	Gas formation in Durham tube and formation of a red ring after addition of Kovacs' reagent	ISO 9308-2; 28
Tryptone water (presumptive *E. coli*)	Formation of a red ring after addition of Kovacs' reagent	ISO 9308-2; 28
Optional tests[c]		
LES Endo agar	Dark red colonies with a golden-green metallic sheen	APHA
MacConkey agar	Red colonies sometimes surrounded by an opaque zone of precipitated bile	APHA; 28

[a]24–48 h at 35–37 ± 0.5°C.
[b]24–48 h at 35–37 ± 0.5°C (total coliforms); at 44–44.5 ± 0.25°C (thermotolerant coliforms, presumptive *E. coli*).
[c]18–24 h at 35–37 ± 0.5°C.
[d]Only for thermotolerant coliforms at 44–44.5°C.

The media and incubation conditions recommended for the determination of the heterotrophic plate count by the APHA and ISO are listed in Table 4.

C. Enumeration of Coliforms, Thermotolerant (Fecal) Coliforms, and *E. coli*

1. *Most Probable Number (MPN) Fermentation*

The MPN method proceeds in two steps. The first step is a presumptive isolation of the total group of coliforms, including the thermotolerant coliforms and *E. coli*. Positive tubes are then subjected to a confirmation step specific for coliforms, thermotolerant coliforms, or *E. coli*.

a. *Presumptive Isolation.* The number of presumptive coliforms is estimated by acid and/or gas formation from lactose in selective liquid isolation media after 24 and 48 h incubation at 35–37°C ± 0.5°C. Coliforms, by definition, are able to produce gas and acid from lactose. These characteristics can be demonstrated when an inverted vial (Durham tube) and a pH indicator are added to the broth. Various liquid media have been described that all contain lactose but that differ mainly in selectivity, i.e., the ability to inhibit the accompanying flora (Table 5).

According to ISO 9308-2, lactose broth (no inhibitory components), MacConkey broth (bile salts as selective component), improved formate lactose glutamate medium (no inhibitory components), or lauryl tryptose lactose broth (lauryl sulfate as selective component) can be used, together with a Durham tube (26).

Table 6 Selective Media for the Detection of Coliforms, Thermotolerant Coliforms, and *E. coli* Using the Membrane Filtration Method

Name of the medium	Typical colonies	References
Presumptive isolation[a]		
Lactose trifenyltetrazolium-chloride (TTC) agar with tergitol-7	Yellow, orange, or brick red colonies with a yellow halo in the medium under the membrane	ISO 9308-1
Lactose agar with tergitol-7	Yellow colonies with a yellow halo in the medium under the membrane	ISO 9308-1
Membrane enriched Teepol agar	Yellow colonies with a yellow halo in the medium under the membrane	ISO 9308-1
Membrane lauryl sulfate agar	Yellow colonies with a yellow halo in the medium under the membrane	ISO 9308-1; 28
Endo agar[c]	Dark red colonies with a golden-green metallic sheen	ISO 9308-1; APHA
LES Endo agar[c]	Dark red colonies with a golden-green metallic sheen	ISO 9308-1; APHA
Membrane fecal coliform (mFC) medium[d]	Blue colonies	ISO 9308-1; APHA
m-7h FC medium[e]	Yellow colonies (no confirmation needed)	APHA; 30
Confirmation[b]		
Lactose peptone water	Gas formation in Durham tube	ISO 9308-1; 28
Lactose broth + brilliant-green lactose (bile) broth	Gas formation in Durham tube	APHA
Lactose tryptose mannitol broth with tryptophan	Gas formation in Durham tube and formation of a red ring after addition of Kovacs' reagent	ISO 9308-1
Tryptone water (presumptive *E. coli*)	Formation of a red ring after addition of Kovacs' reagent	ISO 9308-1; 28

[a]Preincubation: 4 h at 30°C; 14–20 h at 35–37 ± 0.5°C (total coliforms) or at 44–44.5 ± 0.25°C (thermotolerant coliforms, presumptive *E. coli*).
[b]24–48 h at 35–37 ± 0.5°C (total coliforms) or at 44–44.5 ± 0.25°C (thermotolerant coliforms, presumptive *E. coli*).
[c]Only for total coliforms at 35–37°C.
[d]Only for thermotolerant coliforms at 44–44.5°C.
[e]7 h at 41.5°C.

The Durham tube can be omitted when bromocresol purple is added to demonstrate acid formation (21). Other sources recommend minerals modified glutamate medium for the isolation of coliforms, which is in fact the same as improved formate lactose glutamate medium (27,28). This medium has been found to be slightly superior to lauryl tryptose lactose broth by the American Public Health Association (APHA) (21), especially with chlorinated waters, where injured or stressed organisms might be present, and for samples with colony counts below 50 per 100 ml of water. The use of the MacConkey broth may suffer from variations in the inhibitory properties of different batches of bile salts and therefore have problems of reproducibility (28).

The tubes showing turbidity due to bacterial growth and gas formation, together with acid production if the medium contains a pH indicator, are scored as positive. This result should be considered only presumptive, because some spore-forming and other gram-positive bacteria are not sufficiently inhibited and can also produce gas from lactose.

b. Confirmatory Step. Presumptive results must be confirmed by inoculation in confirmatory (more selective) media and incubation at either 35–37°C ±

Table 7 Selective Components and pH Indicators Employed in Coliform Media for the Membrane Filtration Method

Name of the medium	Selective components	pH Indicator
Presumptive isolation		
Lactose trifenyltetrazolium-chloride (TTC) agar with tergitol	Tergitol-7 (sodium heptadecyl sulfate)	Bromothymol blue
Lactose agar with tergitol-7	Tergitol-7	Bromothymol blue
Membrane enriched Teepol agar	Teepol 610	Phenol red
Membrane lauryl sulfate agar	Sodium lauryl sulfate	Phenol red
Endo agar	Sodium lauryl sulfate; sodium desoxycholate	Basic fuchsin
LES Endo agar	Sodium lauryl sulfate; sodium desoxycholate	Basic fuchsin
Membrane fecal coliform (mFC) medium	Bile salts	Aniline blue
m-7h FC medium	Sodium lauryl sulfate; sodium desoxycholate	Phenol red and bromocresol purple
Confirmation		
Lactose peptone water	No inhibitory components	Phenol red (and acid fuchsin)
Lactose broth + brilliant-green lactose (bile) broth	Brilliant-green; bile salts	No pH indicator
Lactose tryptose mannitol broth with tryptophan	Sodium lauryl sulfate	No pH indicator
Tryptone water (presumptive *E. coli*)	No inhibitory components	No pH indicator

0.5°C during 48 h for total coliforms or at 44–44.5°C ± 0.25°C during 24 h for thermotolerant coliforms and for *E. coli*. Confirmatory media used are brilliant-green lactose (bile) broth and lauryl tryptose lactose broth for total (35–37°C) and thermotolerant (44–44.5°C) coliforms, and EC medium for thermotolerant coliforms. The selective components of those media (brilliant green, bile salts, or lauryl sulfate) inhibit the growth of most noncoliform organisms. Tryptone water and lauryl tryptose mannitol broth with tryptophan are used specifically for the confirmation of presumptive *E. coli* at 44–44.5°C ± 0.25°C.

Because brilliant-green lactose (bile) broth, lauryl tryptose mannitol broth with tryptophan, and EC medium do not contain pH indicators, fermentation of lactose is demonstrated only by gas production. The use

Table 8 Chromogenic and Fluorogenic β-D-Galactosidase Substrates Used for the Detection of Coliforms

Chromogenic-fluorogenic substrate	Properties of the reaction product	Refs.
Ortho-nitrophenyl-β-D-galactopyranoside (ONP-GAL)	Yellow, diffusible	32, 21
5-Bromo-4-chloro-3-indoxyl-β-D-galactopyranoside (X-GAL)	Blue, insoluble	33, 34
4-Methylumbelliferyl-β-D-galactopyranoside (MU-GAL)	Blue fluorescent, diffusible	35–37

Table 9 Chromogenic and Fluorogenic β-ᴅ-Glucuronidase Substrates Used for the Detection of *E. coli*

Chromogenic-fluorogenic substrate	Properties of the reaction product	Refs.
Para-nitrophenyl-β-ᴅ-glucuronide (PNP-GLU)	Yellow, diffusible	49, 39
5-Bromo-4-chloro-3-indoxyl-β-ᴅ-glucuronide (X-GLU), indoxyl-β-ᴅ-glucuronide (Y-GLU)	Blue, insoluble	50, 37
Phenolphtalein-β-ᴅ-glucuronide (PHE-GLU)	Red, diffusible	38, 47
8-Hydroxyquinoline-β-ᴅ-glucuronide (8HQ-GLU)	Black, insoluble	51
4-Methylumbelliferyl-β-ᴅ-glucuronide (MU-GLU)	Blue fluorescent, diffusible	52, 40, 53, 41, 42, 21

of lauryl tryptose mannitol broth with added tryptophan in principle allows both gas and indole production to be demonstrated in a single tube. But this medium may occasionally give false-negative results in the indole test; therefore, a negative indole reaction must be repeated in tryptone water (28).

c. *Optional Confirmatory Tests.* The APHA suggests a number of additional tests that can be performed to obtain further confirmation (21). This can involve streaking on LES Endo agar or MacConkey agar plates from each coliform-positive tube from the confirmatory phase. Isolated colonies grown on LES Endo agar during 18–24 h at 35–37°C ± 0.5°C are defined as typical (pink to dark red with a green metallic surface sheen), atypical (pink, red, white, or colorless colonies without a sheen), or negative (all others). Typical lactose-fermenting colonies developing on MacConkey agar are red and may be surrounded by an opaque zone of precipitated bile (21). Well-isolated typical and atypical coliform colonies are then reinoculated into lactose broth or lauryl tryptose broth, and gas production is examined after 24–48 h at 35–37°C ± 0.5°C. As a final confirmation, colonies grown on a nutrient agar slant should be gram-negative and oxidase-negative.

2. *Membrane Filtration*

a. *Presumptive Isolation.* After filtration of the sample, the 0.45-µm membrane filter is transferred to a selective medium for coliform organisms (Table 6). The selective compounds used in these media are diverse and include bile salts or synthetic detergents such as teepol 610 or sodium lauryl sulfate (Table 7). Despite their efficiency for the selective isolation of

coliform bacteria from feces, these compounds seem to reduce the recovery of stressed organisms commonly present in aquatic environments (29). For this reason, tergitol-7 has been introduced as alternative (see Sec. VI).

Sometimes a tetrazolium salt is added to the media that is reduced to a red, insoluble formazan by the bacteria to improve contrast of the bacterial colonies and facilitate counting.

Taking into account the operational definition of coliform organisms (production of acid and gas from lactose) and the fact that gas production cannot be demonstrated by the membrane filtration method, the proposed media should contain lactose and a pH indicator to demonstrate acid production (Table 7). ISO 9308-1 proposes seven media, some of which are also recommended by other organizations (Table 6). The same media can generally be used for the isolation of total coliforms at 35–37°C and thermotolerant coliforms at 44°C, except Endo and LES Endo media, which should not be used at 44°C. Membrane fecal coliform (mFC) is a medium designed exclusively for the examination of fecal coliforms at 44°C (Table 6).

A preincubation step for 4 h at 30°C is recommended to allow recovery of stressed organisms. The membranes are subsequently transferred to incubators at 35–37°C or at 44–44.5°C for 14–20 h (21,28). The APHA describes a rapid fecal coliform test (21), originally designed by Reasoner and his colleagues (30). It is a membrane filtration method, using a special medium (m-7h FC medium), containing lactose, mannitol, bromocresol purple, and phenol red as the most significant components. Incubation occurs at 41.5°C for 7 h.

At this temperature, lactose-fermenting organisms (thermotolerant coliforms) produce yellow colonies.

b. Confirmatory Step. The counts obtained at 35–37°C and at 44–44.5°C are only presumptive results for respectively, coliforms, and thermotolerant coliforms, because gas production is not detected. Therefore a representative number of colonies obtained at 35–37°C as well as at 44–44.5°C are subcultured in tubes of lactose peptone water (31) or lactose broth, followed by subculturing in brilliant-green lactose (bile) broth (21), and incubated at 35–37°C for 48 h (total coliforms) and at 44–44.5°C for 24 h (thermotolerant coliforms) and in tubes of tryptone water or lactose tryptose mannitol broth with tryptophan at 44–44.5°C for 24 h (presumptive *E. coli*). Gas production within this period confirms the presence of the total and thermotolerant coliform organisms. Indole formation in the tryptone water culture confirms the presence of presumptive *E. coli*.

c. Completed Test. While subculturing colonies from the membrane to tubes of the confirmatory media, it is also advisable to subculture on a nutrient agar slant for an oxidase test (28). All coliforms, including the thermotolerant species and *E. coli*, are oxidase-negative, while *Aeromonas* species, which are frequently isolated from water on coliform media, give a positive oxidase test.

3. Use of Chromogenic-Fluorogenic Substrates

The addition of chromogenic-fluorogenic substrates to selective isolation media offers new perspectives for direct detection and identification on a single medium without confirmation. These substrates can be applied in the MPN technique, in media used for membrane filtration, as well as in presence/absence tests. Cleavage by the target enzyme transforms the initially uncolored or nonfluorescent substrate into a colored or fluorescent reaction product. These carefully chosen substrates for specific target enzymes are expected to be more reliable than the detection of the end products of a metabolic pathway, of which the enzyme is a part. The chromogenic-fluorogenic substrates must of course not be toxic for the cell and must be metabolized in the same way as the normal substrates.

a. Enzyme Substrates for Detection of Coliforms. The target enzyme used for coliform detection is β-D-galactosidase, the enzyme breaking lactose into the monomers glucose and galactose. Detection of this enzyme will retrieve a number of coliform organisms that were ignored in the traditional methods based on gas and acid formation from lactose. For example, coliforms that fail to ferment lactose because of a defi-

ciency in the lactose transporting enzyme lactose permease, or that do not produce gas because of a deficient formate dehydrogenase, will be identified as β-D-galactosidase positive, and thus coliform bacteria.

An overview of the chromogenic and fluorogenic β-D-galactosidase substrates used for the detection of coliforms is presented in Table 8. Among the chromogenic substrates, X-GAL has the advantage of yielding an insoluble reaction product, which for colonies on solid media offers greater sensitivity and discrimination than a diffusible product. However, the chromogenic substrates are being gradually replaced by fluorogenic substrates, which have greater sensitivity and thus allow more rapid detection.

b. Enzyme Substrates for Detection of *E. coli*. The β-D-glucuronidase enzyme in *E. coli* was already detected by Buehler et al. in 1951 (38). Hansen and Yourassowsky (39) found that over 94% of *E. coli* strains and only a few strains of *Salmonella* and *Shigella*, but no other coliforms, express β-D-glucuronidase enzyme activity. This specificity was confirmed in more recent studies (40–42) and compares favorably with some of the traditional criteria for *E. coli* detection, such as indole and gas production, for which the fraction of negative strains is higher. On the other hand, several enterohemorrhagic *E. coli*, such as serotype 0157:H7, which are important pathogens, lack β-D-glucuronidase activity and thus escape detection (43,44). Furthermore it is clear that enzymatic detection does not obviate the need for selective media, for β-D-glucuronidase has also been demonstrated in some flavobacteria (45), staphylococci (46), streptococci (47), and clostridia (48). An overview of chromogenic and fluorogenic substrates for β-D-glucuronidase is presented in Table 9.

c. Chromogenic-Fluorogenic Media for Simultaneous Detection of Coliforms and *E. coli*. Various combinations of chromogenic-fluorogenic substrate systems for the simultaneous enumeration of coliforms and *E. coli* have been developed (54). Initially the chromogenic-fluorogenic substrates were used as a complementary step for the enumeration of total coliforms and *E. coli* in drinking water. Mates and Shaffer (52) described a preliminary membrane incubation on Endo or LES Endo agar for 24 h at 35–37°C, followed by a membrane transfer to nutrient agar supplemented with 4-methylumbelliferyl-β-D-glucuronide (MU-GLU) and incubation for an additional 4 h at the same temperature. In this way, from the typical coliform colonies grown on Endo or LES Endo agar, *E. coli* could be differentiated by their fluorescence after incubation on the MU-GLU medium.

Later, media were developed with enough selectivity for direct transfer on a single chromogenic-fluorogenic medium. Some media contained two active substrates for simultaneous detection of total coliforms and *E. coli*. Brenner et al. (37) developed a selective and specific medium, containing indoxyl-β-D-glucuronide (Y-GLU) and 4-methylumbelliferyl-β-D-galactopyranoside (MU-GAL). It is a simple and easy-to-use method for the detection of *E. coli* (Y-GLU) and coliforms (MU-GAL), together with noncoliform organisms, within 24 h at 35 $\pm$ 0.5°C. Colonies on this agar are inspected for blue color and/or fluorescence. Temperature-sensitive *E. coli* and anaerogenic strains, not recovered by the MPN method, can be detected using this single medium at 35°C in maximum 24 h, and the background counts are significantly lower than those of Endo agar.

Other media used only one chromogenic or fluorogenic substrate for detection, but in combination with traditional observation methods, such as acid formation from lactose visualized by a pH indicator. The m-LGA (membrane-lactose glucuronide agar) designed by Sartory and Howard (55) employs X-GLU and a pH indicator, resulting in yellow coliform colonies (lactose fermenting and glucuronidase negative) and green *E. coli* colonies (lactose fermenting and glucuronidase positive). Since methylumbelliferon is nonfluorescent at low pH (56), acid production cannot be detected in combination with MU substrates unless the colonies are briefly exposed to alkali before enumeration (57,58). Another disadvantage of methylumbelliferon is its rapid diffusion into agar media, necessitating immediate enumeration within 24 hours.

Finally, ONP-GAL and MU-GLU have also been incorporated in improved MPN procedures to allow the simultaneous detection of coliforms and *E. coli*, without further confirmation, in 18 hours. In comparison to the UK membrane reference methods, this procedure was demonstrated to be a suitable alternative (59).

4. Presence–Absence (P/A) Test

The P/A test, originally developed as complementary test to quantitative methods, can be used for routine purposes if absence of the target organisms is required (60). It is a simplified MPN technique: Only a single large sample is examined instead of a series of tubes with different volumes.

The recent success of chromogenic-fluorogenic substrates has allowed the development of several P/A tests that can detect a single coliform or *E. coli* in 100 ml of drinking water within a working day. These methods are based on instrumental rather than visual endpoint detection. The duration of the original 24-h test could be reduced by 2–6 h using spectrophotometry (61). Fluorometric detection even allowed recovery of one fecal coliform per 100 ml of water within 7 h (35,62). The detection time for total coliforms is substantially longer because of the limited β-galactosidase activity that is produced. Although these tests are unique in combining speed, simplicity, and sensitivity, a weak point remains that many β-galactosidase-positive organisms other than coliforms cause false-positive results, because these media allow rapid recovery and consequently are generally not highly selective.

D. Enumeration of Fecal Streptococci

1. Most Probable Number (MPN) Fermentation

a. Presumptive Isolation. The basic MPN protocol is used to inoculate the water sample in a series of test tubes containing azide glucose broth. This medium contains sodium azide, a respiratory chain inhibitor that inhibits strictly aerobic but also several facultatively anaerobic bacteria, including the Enterobacteriaceae. Most fecal streptococci are not affected by this compound, except some *S. bovis* and *S. equinus* strains (63). Fermentation of glucose by fecal streptococci acidifies the medium and changes the pH indicator, bromocresol purple, from purple to yellow. The incubation is at 35 or 37 $\pm$ 0.5°C for 44 $\pm$ 4 h (64,21,28).

b. Confirmatory Step. False-positive reactions caused by other gram-positive bacteria, not inhibited by azide, are eliminated in a more selective confirmatory medium (bile–esculin–azide agar (BEAA)). The bile salts in combination with azide and an incubation temperature of 44°C inhibit the growth of almost all bacteria, other than fecal streptococci. A further confirmation is based on hydrolysis of esculin into esculetin by streptococcal β-glucosidase. Fecal streptococci produce regular smooth colonies surrounded by a brownish-black halo due to precipitation of esculetin with ferric salts in the medium. The development of this color is usually evident within a few hours and will give rapid confirmation. Some *Bacillus* species give also a black coloration of the medium but form spreading irregular colonies unlike streptococci colonies. The APHA recommends the use of Pfizer selective enterococcus (PSE) agar, a slightly modified BEAA (21).

c. Completed Test. False-positive tests can be caused by some staphylococci, both in the presumptive media and in the confirmatory media. Therefore, the ISO procedure prescribes a catalase test on suspect colonies to distinguish streptococci (negative) from staphylococci (positive). Some selective media may interfere

Table 10 Media for the Detection of Fecal Streptococci Using the MPN Method

Name of the medium	Typical reactions	Refs.
Presumptive isolation[a]		
Azide glucose broth	Turbidity + acid formation (when bromocresol purple is added)	ISO 7899/1; APHA; 28
Confirmation[b]		
Bile–esculin–azide agar (BEAA)	Discrete colonies surrounded by a black halo	ISO 7899/1, 28
Pfizer selective enterococcus (PSE) agar	Discrete colonies surrounded by a black halo	APHA
Completed test[c]		
Nutrient agar	Negative catalase test after addition of hydrogen peroxide	ISO 7899/1
Brain-heart infusion broth + 6.5% NaCl	Growth in 6.5% NaCl and at 44°C	APHA

[a] 48 h at 35–37 ± 1°C.
[b] 48 h at 44 ± 0.5°C; 24 h at 35 ± 0.5°C for PSE.
[c] 24 h at 35–37°C.

with the catalase test, so suspect colonies are best subcultured on a nonselective nutrient agar. Catalase-positive colonies will cause effervescence when they are covered with a drop of 3% hydrogen peroxide solution. The APHA procedure on the other hand proposes testing for the ability of fecal streptococci to grow at 44°C and in 6.5% NaCl.

The different media and tests used for detection of fecal streptococci by the MPN method are listed in Table 10.

Table 11 Media for the Detection of Fecal Streptococci Using the Membrane Filtration Method

Name of the medium	Typical colonies	Refs.
Presumptive isolation[a]		
KF-streptococcus agar (Kenner)	Red or pink colonies	ISO 7899/2
Membrane enterococcus agar (Slanetz & Bartley)	Red or pink colonies	ISO 7899/2; APHA; 28
Oxolinic acid–esculin–azide agar (OEAA)	Discrete colonies surrounded by a black halo (no confirmation needed)	67
mE Agar + substrate test	Red or pink colonies with reddish-brown precipitate on the bottomside of the filter (no confirmation needed)	APHA
Confirmation[b]		
Bile–esculin–azide agar (BEAA)	Discrete colonies surrounded by a black halo	ISO 7899/2; 28
Kanamycin–esculin–azide agar (KEAA)	Discrete colonies surrounded by a black halo	28 (not recommended)
Completed test[c]		
Nutrient agar	Negative catalase test after addition of hydrogen peroxide	ISO 7899/2

[a] 44 h at 35–37 ± 1°C.
[b] 44 h at 44 ± 0.5°C.
[c] 24 h at 35–37°C.

2. Membrane Filtration

a. Presumptive Isolation. Several media have been proposed for the enumeration of fecal streptococci by the membrane filtration technique, and these are usually also based on the use of sodium azide to suppress the growth of the accompanying flora (Table 11). To enhance contrast, the media sometimes incorporate 2,3,5-triphenyltetrazolium chloride (TTC), which is reduced to a red formazan by fecal streptococci, resulting in red colonies. The most popular presumptive media are KF-streptococcus agar (Kenner) and in particular membrane enterococcus agar (Slanetz and Bartley) (65,21,28). On those media, colonies showing a red or pink color after 44 ± 4 h at 35–37°C ± 1°C are counted as presumptive fecal streptococci. Yoshpe-Purer (66) demonstrated that membrane enterococcus agar is more selective than KF-streptococcus agar since no gram-negative bacilli, such as *Vibrio alginolyticus*, commonly present in marine water, are isolated on the former medium.

OEAA is a selective medium developed by increasing the sodium azide concentration from 0.15 to 0.4 g liter^{-1} and replacing kanamycin by oxolinic acid in KEAA, which is commonly used as a confirmatory medium. This modification resulted in higher specificity, selectivity, and recovery efficiency of the medium, and allows OEAA medium to be used without further confirmation of typical colonies (67).

Confirmation can also be omitted with mE agar incubated for 48 h at 41°C ± 0.5°C, followed by a substrate test (21). The mE agar contains azide and actidione (for selectivity) and TTC and esculin, resulting in pink colonies with a black halo for enterococci. In the substrate test more esculin is added and also ferric citrate to demonstrate esculin hydrolysis. This test medium is used mainly to enumerate enterococci in recreational water.

b. Confirmatory Step. The confirmatory step is performed by subculturing a representative number of typical colonies onto bile–esculin–azide agar (BEAA) or kanamycin–esculin–azide agar (KEAA) at 44°C for 18–48 h. The appearance of typical colonies is as described for the confirmation in the MPN method. Fecal streptococci grow better at 35–37°C, but this temperature permits growth of other organisms. An elevated incubation temperature (44 ± 0.5°C) during the confirmatory stage enhances selectivity, but some fecal streptococci may be inhibited.

c. Completed Test. As with the MPN technique, the ISO procedure recommends a catalase test on suspect colonies to complete the confirmation.

3. Use of Chromogenic-Fluorogenic Substrates

Esculin is a chromogenic substrate that has been used for many years in media for the identification of streptococci. Besides esculin, a number of more advanced chromogenic-fluorogenic substrates have been designed for fecal streptococci and can be used with the MPN technique, with the membrane filtration method, as well as in presence/absence tests.

The target enzyme for the detection of fecal streptococci is β-D-glucosidase (esculinase). Additional selectivity is obtained by incubation at 44°C and selective components, as in the traditional media. Such specific membrane filtration media for fecal streptococci make a confirmatory step unnecessary (Table 12).

A semiautomated MPN method for the enumeration of fecal streptococci in bathing water was developed that reduced analysis times significantly compared to the membrane filtration method. The test uses MU-GLC to detect esculinase activity. The procedure involves preparing a mixture of the powdered growth medium with the water sample, pouring it out in a sterile plastic panel with 51 wells, and incubating for 24 h at 41.0 ± 0.5°C. Results are read under a 365-nm UV light. The procedure performed equally well as the APHA standard membrane filtration method (71), but it is much faster and simpler.

Table 12 Chromogenic and Fluorogenic β-D-Glucosidase Substrates Used for the Detection of Fecal Streptococci

Chromogenic-fluorogenic substrate for β-D-glucoside	Properties of the reaction product	Refs.
p-Nitrophenyl-β-D-glucopyranoside (PNP-GLC)	Yellow, diffusible	68
Indoxyl-β-D-glucoside (Y-GLC)	Blue, insoluble	69, 70
4-Methylumbelliferyl-β-D-glucoside (MU-GLC)	Blue fluorescent, diffusible	71

Table 13 Selective Media for the Detection of Sulfite-Reducing Clostridia, Including *Clostridium perfringens*, Using the MPN Method

Name of the medium	Typical reactions	Refs.
Differential reinforced clostridial medium (DRCM)	Blackening of the medium	ISO 6461/1
Lactose sulfite broth	Blackening of the medium	74

E. Enumeration of Sulfite-Reducing Clostridia and *Clostridium perfringens* Spores and Vegetative Cells

1. Most Probable Number (MPN) Fermentation

The detection and enumeration of sulfite-reducing clostridial spores, including those of *Clostridium perfringens*, occurs by inoculating the water sample after heating for 15 min at 75°C into differential reinforced clostridial medium (DRCM), containing cysteine (72) (Table 13). Incubation is performed anaerobically at 37 ± 1°C for 44 ± 4 h. Culture flasks in which blackening is observed, as a result of the reduction of sulfite and the precipitation of iron(II) sulfide, are scored as positive. Lactose sulfite broth, containing lactose, cysteine, sodium metabisulfite, and ferric ammonium citrate as the most significant components, is also employed (73). Since other anaerobic or facultatively anaerobic bacteria can also produce sulfide, and since both media lack selective ingredients to inhibit accompanying flora, heating of the samples is necessary, and

only spores, but no vegetative cells, can be counted. The media also lack specificity to distinguish *C. perfringens* from other sulfite-reducing clostridia.

Despite its inherent inaccuracy, the MPN method is useful when the number of viable cells in the sample is low, and this is generally the case for bacterial spores in water. Further, higher recoveries of clostridia have been obtained from liquid rather than from solid culture media, and the use of anaeroby-generating systems is not necessary when the headspace in the culture flasks is reduced to a very small volume.

2. Membrane Filtration and Use of Chromogenic or Fluorogenic Substrates

Several media have been employed for enumerating *Clostridium perfringens* spores and vegetative cells from water samples by the membrane filtration method (Table 14). Sartory (74) proposed egg-yolk-free tryptose sulfite cycloserine agar (TSC) for the enumeration of *Clostridium perfringens* in water, a medium also rec-

Table 14 Selective Media for the Detection of *Clostridium perfringens* Using the Membrane Filtration Method

Name of the medium or test used	Typical colonies	Refs.
Presumptive isolation[a]		
Membrane *Clostridium perfringens* medium (mCP)	Yellow colonies turning pink upon exposure to ammonia fumes	76, 77, 16
Tryptose sulfite cycloserine (TSC)	Black colonies due to sulfite reduction	ISO 6461/2, 74
Confirmation		
Tryptose sulfite agar (TS)	Black colonies due to sulfite reduction	78
Gram stain and morphology	Gram-positive bacilli	78
Motility medium	Nonmotile bacteria	78
Nitrate medium	Nitrate reduction	78
Gelatin medium	Gelatin liquefaction within 44 h	78
Lactose medium	Lactose fermentation	78

[a] 24–48 h at 44°C (37°C according to the International Standard).

ommended by the International Standard (75). Black colonies obtained after anaerobic incubation for 24–48 h at 44°C (37°C according to ISO) are considered to be presumptive *Clostridium perfringens*. This medium contains cycloserine to inhibit the majority of the accompanying flora, and can thus be used for counting both vegetative cells and spores of *C. perfringens* in unheated water samples.

Another medium used by several investigators, and that is now also prescribed by the European Guidelines (16), is mCP medium (membrane *Clostridium perfringens* medium) (76,77). This is a complex medium containing sucrose, bromocresol purple (pH indicator for acid formation from sucrose fermentation), indoxyl-β-D-glucoside (chromogenic substrate for β-D-glucosidase or cellobiase; *C. perfringens* is negative), and phenolphtalein diphosphate (for the detection of acid phosphatase). The addition of cycloserine and polymyxin B makes the medium inhibitory to nonclostridial accompanying flora, and thus allows analysis of both vegetative cells and spores. Further selectivity is provided by incubation under anaerobic conditions at 44°C for 21 ± 3 h. Yellow (cellobiase-negative) colonies becoming pink-red upon exposure to ammonia fumes for 30 seconds are considered to be presumptive *Clostridium perfringens*.

Evaluation of these two media demonstrated higher recoveries of spores and vegetative cells from water on TSC than on mCP (78). Although mCP was more selective and specific for laboratory-grown cultures of *Clostridium perfringens*, it was inferior to TSC for the analysis of river water or treated water containing many stressed or injured bacteria. With strongly polluted water containing high amounts of metabolizable substrates for the bacteria, comparable recoveries were obtained on the two media (78).

Color differentiation of presumptive colonies on mCP is sometimes difficult, so typical colonies (yellow colonies turning into pink upon exposure to ammonia fumes) as well as atypical colonies (green colonies or those that remained yellow upon exposure to ammonia fumes) are picked for confirmation. The confirmation as *Clostridium perfringens* is based on the following tests: sulfite reduction, gram-positive sporulating rods, nonmotile, reduction of nitrate, gelatin liquefaction, and lactose fermentation (78). Identical confirmation tests should be performed with presumptive black colonies from TSC isolation medium if identification as *C. perfringens* is desired, since this medium does not discriminate between different clostridial species. However, ISO 6461/2 does not prescribe any confirmation procedure.

It was already suggested in 1973 that the expression of acid phosphatase activity could be diagnostic for *Clostridium perfringens* amongst other clostridia (79). This principle is applied in the aforementioned mCP medium by the inclusion of phenolphtaleïne phosphate. Other, more recently developed chromogenic (5-bromo-4-chloro-3-indoxyl-phosphate) or fluorogenic (4-methylumbelliferyl-phosphate) substrates have also been proposed (80). However, at the pH of standard formulations for tryptose sulfite media, both acid and alkaline phosphatases are active, thus reducing the specificity of the assay. Lowering of the medium pH may restore this specificity but may inhibit some environmentally derived and stressed *Clostridium perfringens* strains.

F. Enumeration of *Staphylococcus aureus*

1. Most probable Number (MPN) Fermentation

Staphylococcus aureus can be enumerated by a modified most probable number procedure (21). Tubes of m-staphylococcus broth are inoculated with the water sample and incubated at 35 ± 1°C for 24 h. Turbid tubes are streaked on lipovitellin salt mannitol agar and incubated at 35 ± 1°C for 24–48 h (Table 15). Opaque (24 h) or yellow (48 h) zones around the colonies are positive evidence of lipovitellin-lipase activity (opaque) and mannitol fermentation (yellow), characteristic of *S. aureus*. Negative confirmation must be repeated from the original tube. Positive isolates should be further confirmed as catalase-positive, coagulase-positive, mannitol-fermenting, gram-positive cocci.

2. Membrane Filtration

The International Standard describes a method for the enumeration of *Staphylococcus aureus* in foods and feed (81). This method can be used for the analysis of swimming water when preceded by filtration of a 100-ml volume through a 0.45-μm membrane. The membrane is transferred to a Baird–Parker medium, containing potassium tellurite and egg yolk emulsion as characteristic components. If the presence of *Proteus* is suspected, sulphamezathine is added (81). Incubation is performed at 35–37 ± 1°C. The APHA (21) proposes the same medium and an incubation for 48 h at 35 ± 0.5°C (Table 16). After incubation, typical colonies are selected for confirmation. Typical colonies on Baird–Parker agar are black, shining, and convex, 1–1.5 mm and 1.5–2.5 mm in diameter after incubation for 24 h and 48 h, respectively, and surrounded by a clear halo. After 24 h of incubation, an opalescent ring immediately around the colonies may appear in the clear zone (81). The confirmation of the typical colonies is based on a positive coagulase reaction. Brain–heart infusion is inoculated with cells from a typical colony and incubated for 20–24 h at 35–37°C. Then 0.1 ml of each culture is added to 0.3 ml of rabbit plasma and incubated at 35–37°C. Clotting of the plasma is examined after 4 to 6 h (81).

It is recognized that some strains of *Staphylococcus aureus* give weakly positive coagulase reactions and may be confused with other bacteria. Therefore additional tests may be included, such as thermonuclease production, hemolysin production, and the production of acid from mannitol (81).

Klapes and Vesley (82) described a rapid thermonuclease test identifying colonies from *Staphylococcus aureus* among staphylococci isolated from swimming pool water by membrane filtration recovery on various selective and differential media. After incubation for 24–48 h at 37°C, the membrane filters are transferred face upwards to a petri dish containing plate count agar. The plates are then heated for 2 h at 60°C and subsequently overlaid with 6–8 ml of molten toluidine blue O-DNA-agar. After 3–4 h of incubation at 37°C, colonies with thermonuclease activity are surrounded by bright pink halos (82).

G. Presence–Absence Test for *Salmonella*

1. Standard Method

Because of its prominent role in food- and waterborne disease, *Salmonella* is the most sought after enteric pathogen in food and water. Accordingly, detection methods for *Salmonella* have been the subject of intensive research, and are rapidly changing. We will describe here the standard ISO method, along with some well-accepted modifications of this protocol, and give

a summary of the most important recent developments of more rapid techniques. This last is not meant to be exhaustive, and the reader is encouraged to consult the scientific literature for more specific information.

The method proposed by the International Standard (83) can be applied to all types of water except raw sewage. This method, as well as most other methods for *Salmonella* detection, consists of five consecutive steps: concentration, enrichment, isolation, presumptive detection, and confirmation.

a. Concentration. A low-turbidity water sample, typically 2 L, is filtered through a membrane filter (mostly 0.45 μm, sometimes 0.22 μm). For turbid water the membrane can be precoated with diatomaceous suspension as a filtration aid to prevent clogging (21).

b. Enrichment. After filtration the membrane filter is transferred to 50 ml of nonselective broth (buffered peptone water) for 16–18 h of incubation at 36 ± 2°C (83) (Table 17). Small water samples can be analyzed by direct addition to an equal volume of double-strength buffered peptone water. This nonselective enrichment step allows for the recovery and growth of injured salmonellae, along with other bacteria, but it is sometimes omitted, as is the case in the procedure described by the APHA (21).

c. Isolation. A volume of 0.1 ml of the pre-enrichment culture is transferred to 10 ml of selective liquid media, such as a modified Rappaport Vassiliadis medium or selenite cystine (83) (Table 17). The APHA proposes dulcitol selenite, tetrathionate, or selenite cystine broth (21). These media suppress the growth of accompanying flora, including coliform bacteria. As no single enrichment medium ensures optimum growth of all *Salmonella* serotypes, the use of two or more selective enrichment media in parallel is recommended. Incubation occurs typically for 18–24 h. But in order to detect slow-growing *Salmonella* species, it is recommended to prolong the incubation time for another 24

Table 15 Selective Media for the Enumeration of *Staphylococcus aureus* Using the MPN Method

Name of the medium	Typical reactions	Refs.
Presumptive isolation[a]		
m-Staphylococcus broth	Turbid medium due to growth	APHA
Confirmation[b]		
Lipovitellin salt mannitol agar	Colonies surrounded with opaque (after 24 h) or yellow (after 48 h) zones	APHA

[a] For 24 h at 35 ± 1°C.
[b] For 24–48 h at 35 ± 1°C.

Table 16 Selective Media for the Enumeration of *Staphylococcus aureus* Using the Membrane Filtration Method

Name of medium or test	Typical reactions	Refs.
Presumptive isolation[a]		
Baird–Parker medium	Black, shining, and convex colonies surrounded by a clear zone, which may be partially opaque	ISO 6888; APHA
Confirmation[b]		
Brain–heart infusion (BHI) followed by coagulase test in rabbit plasma	Clotting of the plasma after 4–6 h	ISO 6888; APHA

[a] For 24–48 h at 35–37 ± 1°C.
[b] For 20–24 h at 35–37 ± 1°C (BHI); for 4–6 h at 35–37 ± 1°C (rabbit plasma).

Table 17 Culture Media for the Detection of *Salmonella*

Name of the medium	Typical reactions	Refs.
Non-selective enrichment[a]		
Buffered peptone water	Turbid medium due to growth	ISO 6340
Isolation[b]		
Modified Rappaport Vassiliadis medium	Turbid medium due to growth	ISO 6340
Selenite cystine broth	Turbid medium due to growth	ISO 6340; APHA
Dulcitol selenite broth	Turbid medium due to growth	APHA
Tetrathionate broth	Turbid medium due to growth	APHA
Presumptive detection[c]		
Brilliant-green phenol red lactose agar	Red or slightly pink-white and opaque colonies with red surroundings	ISO 6340; APHA
Xylose lysine desoxycholate agar	Colorless to red colonies, usually with black center	ISO 6340
Bismuth sulfite agar	Black colonies surrounded by a metallic sheen	ISO 6340; APHA
Xylose lysine brilliant-green agar	Colorless to red colonies, usually with black center	APHA
Confirmation[d]		
Iron–two-sugar agar (Kligler)	Red slant and yellow butt (lactose-negative, glucose-positive reaction) with blackening of the agar (formation of hydrogen sulfide)	ISO 6340
Urea agar (Christensen)	No pink coloration (no urea degradation)	ISO 6340
L-Lysine decarboxylase medium (Falkow)	Purple color due to L-lysine decarboxylase activity	ISO 6340

[a] 16–18 h at 36 ± 2°C.
[b] 18–24 h at 42 ± 0.5°C.
[c] 18–24 h at 36 ± 2°C.
[d] 24 h at 36 ± 2°C.

h. Appropriate choice of the incubation temperature might further differentiate pathogens from the other bacteria. More *Salmonella* are obtained performing the selective enrichment and isolation at 41.5–42°C, but some species, e.g., *S. typhi*, will not grow at this elevated temperature.

   d.   Presumptive Detection.   Brilliant-green phenol red lactose agar (Edel & Kampelmacher), xylose lysine desoxycholate agar, and bismuth sulfite agar (Wilson & Blair) are selective solid media commonly used for the isolation and presumptive detection of *Salmonella* (83,21). Xylose lysine brilliant-green agar is recommended for *Salmonella* species from marine samples (21) (Table 17).

   The media are streaked with a loop from the selective enrichment cultures and incubated for 18–24 h at 36 ± 2°C. If no suspect *Salmonella* colonies are observed after 24 h, the plates are reincubated for another 24 h to allow for the development of slow-growing or partially inhibited organisms.

   e.   Confirmation.   Typical colonies of *Salmonella* are picked from the solid selective media and subjected to a set of biochemical tests, including lactose/glucose fermentation, hydrogen sulfide formation, urea degradation, and L-lysine decarboxylase activity (83) (Table 17). Definitive confirmation is finally obtained by reaction with O-specific antisera in a slide agglutination test (83).

   Detection of *Salmonella* in suspect potable water can also be performed as an extension of the routine total coliform analysis. In that case, membranes from Endo agar (see Table 6) showing sufficient coliforms and background growth are transferred into tetrathionate broth for selective enrichment (21). After enrichment, *Salmonella* is isolated on brilliant-green agar.

2.   *Overview of Alternative Methods for* Salmonella *Detection*

A growing number of rapid methods for detection of *Salmonella* can replace one or more of the slow steps in the standard procedure already described (84). However, because most of the rapid methods based on nucleic acid or immunological detection have limited sensitivity, a "slow" cultural-enrichment step remains necessary. Such a step provides the additional advantage of making the method detect only live, culturable organisms (85).

   A number of methods have been developed to speed up the nonselective- and selective-enrichment steps. These can be generally divided into improved media formulations and incubation conditions, which make it

possible to conduct both steps in 18–24 h, compared to 36–48 h using the standard procedure (86–88), and immunomagnetic enrichment methods, which use magnetic microbeads coated with antisalmonella antibodies to capture and concentrate bacteria, and which can replace the cultural selective enrichment (89). An important consideration is the ability of these methods to allow the recovery of injured cells, a process that inevitably takes time.

   A wide variety of rapid alternative methods have been developed to replace the standard presumptive detection and confirmation steps, and the majority of these methods owe their specificity to the use of either specific antisera or nucleic acid probes. Both groups of methods exist in many different formats, including for the immunological methods, latex agglutination, dipstick, enzyme-linked immunosorbent assay, immunofluorescence microscopy, flow cytometric detection, and immunoelectrochemical detection (84,21,90–92); and for the nucleic acid–based methods, hybridization assays, polymerase chain reaction, and other amplification methods (84,93,94,85,95).

VI.   RECOVERY OF STRESSED ORGANISMS

Pathogenic bacteria and indicator organisms of fecal origin whose primary habitat is the gut of warm-blooded animals are only transient members of aqueous ecosystems, and the aqueous environment is highly unfavorable for these organisms due to its generally low nutrient content and low temperature. Exposure to this stressful situation results in a gradual accumulation of physiological damage to the cells, which ultimately leads to cell death. This damage is reflected by an increased lag time before growth is resumed under favorable conditions and an increased sensitivity to other stresses that can be tolerated by uninjured cells. The important consequence for monitoring water quality is that at least some of the injured indicator bacteria are not able to recover and grow in the selective culture media that are used for their detection (96–98). This can lead to an underestimation of bacterial counts and to acceptance of a potentially hazardous condition (21). There is indeed evidence that pathogenic bacteria can recover from injury and cause disease when ingested (99).

   Stressed organisms can be present in water under most circumstances: in chlorinated drinking water, in wastewater effluents, in saline water, in polluted natural water, in relatively clean surface waters. They can be stressed due to disinfection chemicals, the presence of toxic pollutants such as heavy metals, extreme temperature and pH, or solar radiation. Therefore, analytical methods should allow for maximal recovery of injured cells. Residues of chlorine should be neutralized with an adequate dose of sodium thiosulfate at the time of sampling to prevent further damage to the bacteria. For the same reason, heavy metals should be neutralized by a chelating agent such as EDTA (21). Further recommendations to alleviate stress during sample handling are to limit sample transportation and handling time to 6 h if possible, to cool samples below 10°C during sample transportation, and to add some peptone to the sample dilution buffer (21).

If there is suspicion of severe injury, for instance, when different methods yield largely different results, it may be necessary to adopt one or more modifications to the procedure designed specifically to enhance recovery. For enumeration of total and fecal coliforms by the membrane filtration method, the following have been suggested:

1. Use of m-T7 agar, a mildly selective medium containing a low concentration of tergitol-7 and penicillin G as inhibitory ingredients, which allows improved recovery of coliforms (21). Cefsulodin is also a successful selective agent against *Aeromonas* spp. in mildly selective media (100).

2. Use of two-layer diffusion agar, consisting of a selective bottom layer covered with a nonselective top layer. When these plates are used within 1 h after pouring, the bacteria will have time to recover before diffusion of the inhibitory components from the bottom layer create the required selective conditions (21).

3. A nonselective enrichment in broth culture prior to the normal membrane filtration procedure. Incubation should not exceed 4 h, to avoid growth of bacteria after recovery (21,101).

4. Temperature acclimation. Recovery of fecal coliforms is enhanced when the first 4–6 h of incubation are performed at 35°C or even lower, before switching to the normal incubation temperature of 44.5°C. This technique can be applied together with one of the foregoing (21).

Since these modifications all result in relaxed selectivity towards the target organisms, it is important to pick a sufficient number of suspect colonies for subsequent confirmation.

Other approaches are based on the supplementation of isolation media with components that enhance recovery without reducing medium selectivity. Components that have been successfully used are catalase, pyruvate, and sterile but biochemically active microbial membrane preparations (102–105). These components are scavengers of oxygen or of reactive oxygen species, such as peroxide and superoxide, and thus alleviate oxidative stress to which the bacteria are exposed in the plating media. It has been demonstrated that peroxide and superoxide can be formed due to auto-oxidation and photochemical reactions during preparation and sterilization of culture media, and cause lower recoveries of stressed bacteria (106,103). Successful application of oxygen scavengers has been reported in selective media for coliforms (104,55), fecal streptococci (70), and *Clostridium perfringens* (107).

REFERENCES

1. W Lee, Z Lewandowski, WG Characklis. Microbial corrosion of mild steel in a biofilm system. In: GG Geesey, Z Lewandowski, HC Flemming, eds. Biofouling and Biocorrosion in Industrial Water Systems. Boca Raton, FL: Lewis, 1994, pp 205–212.

2. N Longmate. King Cholera: The Biography of a Disease. London: Hamish Hamilton, 1966.

3. TE Ford, RR Colwell. A global decline in microbiological safety of water: a call for action. A report from The American Academy of Microbiology, Washington, DC, 1996, pp, 1–40.

4. F Schardinger. Ueber das Volkommen Gährung erregender Spaltpilze im Trinkwasser und ihr Bedeutung für die Hygienische Beurtheilung desselben. (The occurrence of fermentative bacteria in, and their significance for the sanitary evaluation of drinking water.) Wien Klin Wschr 5:403–405, 421–423, 1892.

5. EA Laws. Aquatic Pollution: An Introductory Text. 2nd ed. New York: Wiley, 1993, pp 157–178.

6. EE Geldreich, BA Kenner, PW Kabler. Occurrence of coliforms, fecal coliforms, and streptococci on vegetation and insects. Appl Microbiol 12:63–69, 1964.

7. AC Houston. On the value of examination of water for streptococci and staphylococci with a view to detection of its recent contamination with animal organic matter. Supplement to the 29th Annual Report Local Government Board Containing Report of Medical Officer 1899–1900. London City Council, UK, 1900.

8. AF Godfree, D Kay, MD Wyer. Faecal streptococci as indicators of faecal contamination in water. J Appl Microbiol Symp Suppl 83:110S–119S, 1997.

9. VJ Cabelli, AP Dufour, LJ McCabe, MA Levin. A marine recreational water-quality criterion consistent with indicator concepts and risk analysis. J Water Pollution Control Federation 55:1306–1314, 1982.

10. D Kay, JM Fleisher, RL Salmon, F Jones, MD Wyer, AG Godfree, Z Zelenauch-Jacquotte, R Shore. Predicting the likelihood of gastroenteritis from sea bathing: results from a randomized exposure. Lancet 344: 905–909, 1994.

11. VJ Cabelli, AP Dufour, MA Levin, LJ McCabe, PW Haberman. Relationship of microbial indicators to health effects at marine bathing beaches. Amer J Pub Health 69:690–696, 1979.

12. AP Dufour. Bacterial indicators of recreational water quality. Can J Pub Health 75:49–56, 1984.

13. AH Havelaar. Bacteriophages as models of human enteric viruses in the environment. ASM News 59:614–619, 1993.

14. VJ Cabelli. Obligate anaerobic bacterial indicators. In: G Berg, ed. Indicators of Viruses in Water and Food. Ann Arbor, MI: Ann Arbor Science, 1978.

15. DJ Reasoner. Monitoring heterotrophic bacteria in potable water. In: GA McFeters, ed. Drinking Water Microbiology. New York: Brock/Springer Series in Contemporary Bioscience, 1990, pp 452–477.

16. Council Directive 98/83/EC. Official Journal of the European Communities L 330:32–54, 1998.

17. U.S. Code of Federal Regulations. Title 40, Protection of the environment, Part 141, National Primary Drinking Water Regulations. Washington, DC: U.S. Government Printing Office, 1998.

18. Council Directive 76/160/EC. Official Journal of the European Communities, 1976.

19. Vlarem II. Besluit van de Vlaamse regering houdende algemene en sectorale bepalingen inzake milieu-hygiëne van 1 juni 1995. Afdeling 5:32/9 Zwembaden Art. Nr 5/32/9.2.2 par. 4/1, Belgisch Staatsblad 31/7/95, 1995.

20. International standard (ISO 8199). Water quality—General guide to the enumeration of micro-organisms by culture. 1st ed., 1988.

21. AD Eaton, LS Clesceri, AE Greenberg, eds. Standard Methods for the Examination of Water and Wastewater. 19th ed. Washington, DC: American Public Health Association—American Water Works—Water Environment Federation, 1995.

22. International standard (ISO 6222). Water quality—Enumeration of viable micro-organisms—Colony count by inoculation in or on a nutrient agar culture medium. 1st ed., 1988.

23. RH Taylor, EE Geldreich. A new membrane filter procedure for bacterial counts in potable water and swimming pool samples. J Am Water Works Assoc 71:402–405, 1979.

24. CN Haas, MA Meyer, MS Paller. Analytical note: evaluation of the m-SPC method as a substitute for the standard plate count in water microbiology. J Am Water Works Assoc 74:322, 1982.

25. DJ Reasoner, EE Geldreich. A new medium for the enumeration and subculture of bacteria from potable water. Appl Environ Microbiol 49:1–7, 1985.

26. International standard (ISO 9308-2). Water quality—Detection and enumeration of coliform organisms, thermotolerant coliform organisms and presumptive *Escherichia coli*. Part 2: Multiple tube (most probable number) method. 1st ed., 1990.

27. Public Health Laboratory Service. A minerals modified glutamate medium for the enumeration of coliform organisms in water, by The Public Health Laboratory Service Standing Committee on the Bacteriological Examination of Water Supplies. J Hygiene 81:367–374, 1969.

28. Anon. The Microbiology of Water 1994: Part 1—Drinking water. Reports on Public Health and Medical Subjects No. 71. Methods for the Examination of Water and Associated Materials. London: HMSO, 1994.

29. GA McFeters, JS Kippin, MW LeChevallier. Injured coliforms in drinking water. Appl Environ Microbiol 51:1–5, 1986.

30. DJ Reasoner, JC Blannon, EE Geldreich. Rapid seven-hour fecal coliform test. Appl Environ Microbiol 38:229–236, 1979.

31. International standard (ISO 9308-1). Water quality—Detection and enumeration of coliform organisms, thermotolerant coliform organisms and presumptive *Escherichia coli*. Part 1: Membrane filtration method. 1st ed., 1990.

32. SC Edberg, MJ Allen, DB Smith (The National Collaborative Study). National field evaluation of a defined substrate method for the simultaneous enumeration of total coliforms and *Escherichia coli* from drinking water: comparison with the standard multiple tube fermentation method. Appl Environ Microbiol 54:1595–1601, 1988.

33. M Manafi, W Kneifel. A combined chromogenic-fluorogenic medium for the simultaneous detection of total coliforms and *E. coli* in water. Zbl Hyg 189:225–234, 1989.

34. AN Ley, S Barr, D Fredenburgh, M Taylor, N Walker. Use of 5-bromo-4-chloro-3-indoxyl-β-D-galactopyranoside for the isolation of β-D-galactosidase-positive bacteria from municipal water supplies. Can J Microbiol 39:821–825, 1993.

35. JD Berg, L Fiksdal. Rapid detection of total and fecal coliforms in water by enzymatic hydrolysis of 4-methylumbelliferone-β-D-galactoside. Appl Environ Microbiol 54:2118–2122, 1988.

36. G Cenci, G Caldini, F Sfodera, G Morozzi. Fluorogenic detection of atypical coliforms from water samples. Microbiologica 13:121–129, 1990.

37. KP Brenner, CC Rankin, YR Roybal, GN Stelma Jr, PV Scarpino, AP Dufour. New medium for the simultaneous detection of total coliforms and *Escherichia coli* in water. Appl Environ Microbiol 59:3534–3544, 1993.

38. HJ Buehler, PA Katzman, EA Doisy. Studies on β-glucuronidase from *E. coli*. Proc Soc Exp Biol Med 76:672–676, 1951.

39. W Hansen, E Yourassowsky. Detection of β-glucuronidase in lactose-fermenting members of the family Enterobacteriaceae and its presence in bacterial urine cultures. J Clin Microbiol 20: 1177–1179, 1984.

40. PA Hartman. The MUG (glucuronidase) test for *Escherichia coli* in food and water. In: A Turano, ed. Rapid Methods and Automation in Microbiology and Immunology. Brescia, Italy: Brixia Academic Press, 1989, pp 290–308.

41. M Manafi, W Kneifel, S Bascomb. Fluorogenic and chromogenic substrates used in bacterial diagnostics. Microbiol Rev 55;225–248, 1991.

42. EW Frampton, L Restaino. Methods for *Escherichia coli* identification in food, water, and clinical samples based on beta-glucuronidase detection. J Appl Bacteriol 74:223–233, 1993.

43. GW Chang, J Brill, R Lum. Proportion of β-D-glucuronidase-negative *Escherichia coli* in human fecal samples. Appl Environ Microbiol 55:335–339, 1989.

44. JS Thompson, DS Hodge, AA Borczyk. Rapid biochemical test to identify verocytotoxin-positive strains of *Escherichia coli* serotype O157. J Clin Microbiol 28:2165–2168, 1990.

45. JP Petzel, PA Hartman. A note on starch hydrolysis and β-glucuronidase activity among flavobacteria. J Appl Bacteriol 61:421–426, 1986.

46. LJ Moberg. Fluorogenic assay for rapid detection of *Escherichia coli* in food. Appl Environ Microbiol 50: 1383–1387, 1985.

47. TO Röd, RH Haug, T Midtvedt. β-Glucuronidase in the streptococci groups B and D. Acta Pathol Microbiol Scand Sect B 82:533–536, 1974.

48. Y Sakaguchi, K Murata. Studies on the β-glucuronidase production of clostridia. Zbl Bakteriol I Orig A254:118–122, 1983.

49. M Kilian, P Bülow. Rapid identification of Enterobacteriaceae. II Use of β-D-glucuronidase detecting agar medium (PGUA agar) for the identification of *E. coli* in primary cultures of urine samples. Acta Path Microbiol Scand Sect B 87:271–276, 1979.

50. EW Frampton, L Restaino, N Blaszko. Evaluation of the β-D-glucuronidase substrate 5-bromo-4-chloro-3-indolyl-β-D-glucuronide (X-GLUC) in a 24-h direct plating method for *Escherichia coli*. J Food Protect 51:402–404, 1988.

51. AL James, P Yeoman. Detection of specific bacterial enzymes by high contrast metal chelate formation. Part II. Specific detection of *Escherichia coli* on multipoint-inoculated plates using 8-hydroxyquinoline-beta-D-glucuronide. Zentralbl Bakteriol Mikrobiol Hyg A 267:316–321, 1988.

52. A Mates, M Shaffer. Membrane filtration differentiation of *E. coli* from coliforms in the examination of water. J Appl Bacteriol 67:343–346, 1989.

53. MJ Gauthier, VM Torregrossa, MC Babelona, R Cornax, JJ Borrego. An intercalibration study of the use

of 4-methylumbelliferyl-β-D-glucuronide for the specific enumeration of *Escherichia coli* in seawater and marine sediments. System Appl Microbiol 14:183–189, 1991.

54. K Venkateswaran, A Murakoshi, M Satake. Comparison of commercially available kits with standard methods for detection of coliforms and *Escherichia coli* in foods. Appl Environ Microbiol 62:2236–2243, 1996.

55. DP Sartory, L Howard. A medium detecting β-glucuronidase for the simultaneous filtration enumeration of *Escherichia coli* and coliforms from drinking water. Lett Appl Microbiol 15:273–276, 1992.

56. RH Goodwin, F Kavanagh. Fluorescence of coumarin derivatives as a function of pH. Arch Biochem Biophys 27:152–173, 1950.

57. JL Maddocks, MJ Greenan. A rapid method for identifying bacterial enzymes. J Clin Pathol 28:686–687, 1975.

58. TA Freier, PA Hartman. Improved membrane filtration media for enumeration of total coliforms and *Escherichia coli* from sewage and surface waters. Appl Environ Microbiol 53:1246–1250, 1987.

59. EJ Fricker, KS Illingworth, CR Fricker. Use of two formulations of Colilert and QuantiTray™ for assessment of the bacteriological quality of water. Water Res 31:2495–2499, 1997.

60. JA Clark. The presence-absence test for monitoring drinking water quality. In: GA McFeters, ed. Drinking Water Microbiology. New York: Brock/Springer Series in Contemporary Bioscience, 1990, pp 399–411.

61. JH Standridge, SM Kluender, M Bernhardt. Spectrophotometric enhancement of MMO-MUG (Colilert) endpoint determination. In: Proceedings of the Water Quality Technology Conference. American Water Works Association, Toronto, Canada, pp 157–162, 183–189, 1992.

62. SM Peterson, SC Apte, CM Davies, I Johns, P Fitch, G Swan, T Prus-Wisniowski, G Skyring. Rapid automated monitoring of coliforms in reticulated water. In: Proceedings of the Australian Water and Wastewater Association 16th Federal Convention, Australian Water and Wastewater Association, Sydney, 1995.

63. PA Hartman, RH Deibel, LM Sieverding. Enterococci. In: C Vanderzant and DF Splittstoesser, eds. Compendium of Methods for the Microbiological Examination of Foods. Washington, DC: American Public Health Association, 1992, pp 523–531.

64. International standard (ISO 7899/1). Water quality—Detection and enumeration of fecal streptococci. Part 1: Method by enrichment in a liquid medium. 1st ed., 1984.

65. International standard (ISO 7899/2). Water quality–Detection and enumeration of fecal streptococci. Part 2: Method by membrane filtration. 1st ed., 1984.

66. Y Yoshpe-Purer. Evaluation of media for monitoring fecal streptococci in seawater. Appl Environ Microbiol 55:2041–2045, 1989.

67. A Audicana, I Perales, JJ Borrego. Modification of Kanamycin-Esculin-Azide Agar to improve selectivity in the enumeration of fecal streptococci from water samples. Appl Environ Microbiol 61:4178–4183, 1995.

68. RW Trepeta, SC Edberg. Esculinase (β-glucosidase) for the rapid estimation of activity in bacteria utilizing a hydrolyzable substrate, p-nitrophenyl-β-D-glucopyranoside. Anton Leeuw Int J G Microbiol 53:273–277, 1987.

69. AP Dufour. A 24-hour membrane filter procedure for enumerating enterococci. Abstracts of the Annual Meeting of the American Society for Microbiology, 205, 1980.

70. DP Sartory, M Field, AM Pritchard. Towards 24 hour confirmed fecal streptococci enumeration from drinking water. In: R Morris, A Gammie, eds. Proceedings of the 2nd UK Symposium on Health-Related Water Microbiology. London: IAWQ, 1997, pp 207–213.

71. EJ Fricker, CR Fricker. Use of defined substrate technology and a novel procedure for estimating the numbers of enterococci in water. J Microbiol Meth 27:207–210, 1996.

72. International standard (ISO 6461/1). Water quality—Detection and enumeration of the spores of sulfite-reducing anaerobes (clostridia)—Part 1: Method by enrichment in a liquid medium. 1st ed., 1986.

73. H Beerens, CL Romond, C Lepage, J Criquelion. A liquid medium for the enumeration of Clostridium perfringens in food and feces. In: JEL Corry, D Roberts, FA Skinner, eds. Isolation and Identification Methods for Food Poisoning Organisms. Society for Applied Bacteriology, Technical series number 17, Academic Press, 1982, pp 137–149.

74. DP Sartory. Membrane filtration enumeration of faecal clostridia and Clostridium perfringens in water. Wat Res 20:1255–1260, 1986.

75. International standard (ISO 6461/2). Water quality—Detection and enumeration of the spores of sulfite-reducing anaerobes (clostridia)—Part 2: Method by membrane filtration. 1st ed., 1986.

76. JW Bisson, VJ Cabelli. Membrane filter enumeration method for Clostridium perfringens. Appl Environ Microbiol 37:55–66, 1979.

77. R Armon, P Payment. A modification of m-CP medium for enumerating Clostridium perfringens from water samples. Can J Microbiol 34:78–79, 1988.

78. DP Sartory, M Field, SM Curbishley, AM Pritchard. Evaluation of two media for the membrane filtration enumeration of Clostridium perfringens from water. Lett Appl Microbiol 27:323–327, 1998.

79. G Schallehn, H Brandis. Phosphatase-reagent for quick identification of Clostridium perfringens. Zentralbl Bakteriol Orig A 225:343–345, 1973.

80. J Watkins, J Xiangrong. Cultural methods of detection for micro-organisms: recent advances and successes. In: DW Sutcliffe, ed. The Microbiological Quality of Water. Ambleside, Freshwater Biological Association, 1997, pp 19–27.

81. International standard (ISO 6888). Microbiology—General guidance for enumeration of Staphylococcus aureus—Colony count technique. 1st ed., 1983.

82. NA Klapes, D Vesley. Rapid assay for in situ identification of coagulase-positive staphylococci recovered by membrane filtration from swimming pool water. Appl Environ Microbiol 52:589–590, 1986.

83. International standard (ISO 6340). Water quality—Detection of Salmonella species. 1st ed., 1995.

84. P Feng. Commercial assay systems for detecting food-borne Salmonella: a review. J Food Prot 55:927–934, 1992.

85. E Dupray, MP Caprais, A Derrien, P Fach. Salmonella DNA persistence in natural seawaters using PCR analysis. J Appl Microbiol 82:507–510, 1997.

86. S Pignato, AM Marino, MC Emanuele, V Iannotta, S Caracappa, G Giamanco. Evaluation of new culture media for rapid detection and isolation of salmonellae in foods. Appl Environ Microbiol 61:1996–1999, 1996.

87. C Wiberg, P Norberg. Comparison between a cultural procedure using Rappaport–Vassiliadis broth and motility enrichments on modified semisolid Rappaport–Vassiliadis medium for Salmonella detection from food and feed. Int J Food Microbiol 29:353–360, 1996.

88. D Blivet, G Salvat, F Humbert, P Colin. Development of a new culture medium for the rapid detection of Salmonella by indirect conductance measurements. J Appl Microbiol 84:399–403, 1998.

89. K Hanai, M Satake, H Nakanishi, K Venkateswaran. Comparison of commercially available kits with standard methods for detection of Salmonella strains in foods. Appl Environ Microbiol 63:775–778, 1997.

90. C Desmonts, J Minet, R Colwell, M Cormier. Fluorescent-antibody method useful for detecting viable but nonculturable Salmonella spp. in chlorinated wastewater. Appl Environ Microbiol 56:1448–1452, 1990.

91. RG McClelland, AC Pinder. Detection of low levels of specific Salmonella species by fluorescent antibodies and flow cytometry. J Appl Bacteriol 77:440–447, 1994.

92. JD Brewster, AG Gehring, RS Mazenko, LJ VanHouten, CJ Crawford. Immunoelectrochemical assays for bacteria: use of epifluorescence microscopy and rapid-scan electrochemical techniques in development of an assay for Salmonella. Anal Chem 68:4153–4159, 1996.

93. CC Somerville, IT Knight, WL Straube, RR Colwell. Simple, rapid method for direct isolation of nucleic acids from aquatic environments. Appl Environ Microbiol 55:548–554, 1989.

94. IT Knight, S Shults, CW Kaspar, RR Colwell. Direct detection of *Salmonella* spp. in estuaries by using a DNA probe. Appl Environ Microbiol 56:1059–1066, 1990.

95. RYC Kong, WF Dung, LLP Vrijmoed, RSS Wu. Co-detection of three species of water-borne bacteria by multiplex PCR. Marine Pollution Bull 31:317–324, 1995.

96. MW LeChevallier, GA McFeters. Recent advances in coliform methodology. Environ Health 47:5–9, 1984.

97. MW LeChevallier, GA McFeters. Enumerating injured coliforms in drinking water. J Am Water Works Assoc 77:81–87, 1985.

98. GA McFeters. Chapter 23: Enumeration, occurrence, and significance of injured indicator bacteria in drinking water. In: GA McFeters, ed. Drinking Water Microbiology. New York: Brock/Springer Series in Contemporary Bioscience, 1990, pp 478–492.

99. A Singh, R Yeager, GA McFeters. Assessment of in vivo revival, growth, and pathogenicity of *Escherichia coli* strains after copper- and chlorine-induced injury. Appl Environ Microbiol 52:832–837, 1986.

100. JL Alonso, I Amoros, MA Alsonso. Differential susceptibility of aeromonads and coliforms to cefsulodin. Appl Environ Microbiol 62:1885–1888, 1996.

101. S Massa, M Fanelli, MT Brienza, and M Sinigaglia. The bacterial flora in bottled natural mineral water sold in Italy. J Food Qual 21:175–185, 1998.

102. JP Calabrese, GK Bissonnette. Improved detection of acid mine water stressed coliform bacteria on media containing catalase and sodium pyruvate. Can J Microbiol 36:544–550, 1990.

103. JP Calabrese, GK Bissonnette. Improved membrane filtration method incorporating catalase and sodium pyruvate for detection of chlorine-stressed coliform bacteria. Appl Environ Microbiol 56:3558–3564, 1990.

104. DP Sartory. Improved recovery of chlorine-stressed coliforms with pyruvate supplemented media. Wat Sci Tech 31;255–258, 1995.

105. H Adler, G Spady. The use of microbial membranes to achieve anaerobiosis. J Rapid Methods Automation Microbiol 5:1–12, 1997.

106. RM Lee, PA Hartman. Optimal pyruvate concentration for the recovery of coliforms from food and water. J Food Protection 52:119–121, 1989.

107. AM Hood, A Tuck, CR Dane. A medium for the isolation, enumeration and rapid presumptive identification of injured *Clostridium perfringens* and *Bacillus cereus*. J Appl Microbiol 69:359–372, 1990.

9

Algal Analysis—Organisms and Toxins

Glen Shaw

National Research Centre for Environmental Toxicology (NRCET), The University of Queensland, Brisbane, Queensland, Australia

Maree Smith

Queensland Health Scientific Services, Brisbane, Queensland, Australia

I. INTRODUCTION

Cyanobacteria (blue-green algae) are a group of pro-karyotic organisms that occur worldwide in environments that include volcanic hot springs and under Antarctic ice. They can exist as solitary cells or in colonies, commonly as a filament known as a *trichome*. The trichome usually includes vegetative cells, heterocysts for nitrogen fixation, and akinetes, which act as spores for reproduction. Many species of cyanobacteria exist in many ecosystems and cause no problem. Indeed, some species, such as *Aphanizomenon flos-aquae*, are harvested and used for human food. A small group of genera, however, produce toxins that have caused both animal (1) and human poisonings (2,3). In addition, the microcystin group of toxins are believed to be responsible for increased levels of primary liver cancer in some areas of China (4).

The dangers of toxic cyanobacteria are enhanced by the fact that some species form surface scums that can be consumed by animals, thus producing high-level intakes of the toxins. Likewise, without proper water treatment to remove toxins, human populations can be at risk from the consumption of surface waters.

In addition to freshwater cyanobacteria, some marine cyanobacteria, including *Oscillatoria*, *Trichodesmium*, and *Lyngbya*, can cause deleterious human health effects. These genera occur worldwide and produce a range of different toxins with various toxico-logical effects, including dermatitis, neurotoxicity, and hepatotoxicity.

A. Toxic Cyanobacterial Genera

A number of genera are known to have produced toxins in various locations around the world. It is an interesting feature of these toxins that the same toxins can be produced by different cyanobacterial genera. Another enigma is that the same species of cyanobacteria can produce different toxins in different locations. For instance, the hepatotoxic alkaloid cylindrospermopsin has been found to be produced by *Cylindrospermopsis raciborskii* in many locations around the world (1), by *Umezakia natans* in Japan (5), and by *Aphanizomenon ovalisporum* in Israel (6) and in Australia (7). Likewise, the paralytic shellfish poisons (PSPs) are produced by marine dinoflagellates but are also produced in Australia by *Anabaena circinalis* (8). The cyclic peptide toxins microcystins are produced by *Microcystis*, *Anabaena*, *Nodularia*, *Nostoc*, and *Oscillatoria* genera (1). The main genera with potential toxins produced are listed in Table 1. Other genera and species, however, may produce toxins, and new toxic species are being discovered. Some confusion may exist over the identification of species based on morphological characteristics; genetic typing of blue-green algae may overcome these difficulties (7).

Table 1 Chief Potentially Toxic Cyanobacteria and Their Toxins

Cyanobacterium	Toxin
Anabaena	One or more of the following toxins
A. circinalis	occur in the various *Anabaena* species:
A. flos aquae	PSP toxins (saxitoxin and derivitives),
A. hassallii	anatoxin-a, anatoxin-a(s), microcystins
A. lemmermanni	
A. spiroides	
A. variabilis	
Aphanizomenon	
A. flos-aquae	PSPs
A ovalisporum	Cylindrospermopsin
Lyngbya	
L. majuscula	Lyngbya toxins
Microcystis	All can contain microcystins
M. aeruginosa	
M. viridis	
M. wesenbergii	
Nodularia	
N. spumigena	Nodularin
Oscillatoria	
O. nigro-viridis	Debromoaplysiatoxin, oscillatoxin
O. agardhii/rubescens	Anatoxin-a, homoanatoxin-a, microcystins
O. acutissima	Anatoxin-a, homoanatoxin-a, microcystins
O. formosa	Anatoxin-a, homoanatoxin-a, microcystins
Schizothrix	
S. calcicola	Debromoaplysiatoxin
Trichodesmium	
T. thiebautii	Neurotoxin similar to anatoxin-a
T. erythraeum	Unidentified neurotoxins and hepatotoxins

B. Factors Contributing to Toxin Production

A complication in the assessment of the potential effects of cyanobacterial blooms is the fact that blooms of potentially toxic organisms may not always be toxic. It is thus necessary to conduct bioassays to determine toxicity or to perform chemical analyses for specific toxins. Blooms can consist of toxic and/or nontoxic strains or of mixed cyanobacterial species (9).

Factors that determine whether a bloom is toxic are poorly understood. Some studies have shown that variations in environmental factors and genetic heterogeneity are important in determining toxicity (10). Research undertaken by the authors has shown that all blooms of *C. raciborskii* have produced cylindrospermopsin but that the amount of cylindrospermopsin produced does not correlate with cell counts. The highest concentrations of cylindrospermopsin in water coincide with the later stages of bloom development, suggesting that the toxin may accumulate in water over shorter

time periods. With microcystins, a rapid decline has been shown to occur in the later stages of the logarithmic growth phase, presumably due to rapid degradation (10). The reduction of nitrogen or inorganic carbon from culture has been shown to reduce microcystin production (11).

C. Types of Cyanobacterial Toxins

Cyanobacterial toxins are secondary metabolites (i.e., they are not used by the organism for its metabolism) and fall into four basic classes: neurotoxins, hepatotoxins, nonspecific toxins, and irritants (1). Chemically, most toxins are alkaloids, peptides, or lipopolysaccharides.

1. Neurotoxins

Neurotoxins are produced by species of the following genera: *Anabaena*, *Aphanizomenon*, *Nostoc*, *Oscillatoria*, and *Trichodesmium*. The chemical structures for the neurotoxins are discussed in the following subsections.

a. Anatoxin-a. This toxin is a tropene-related alkaloid, as shown in Fig. 1. It is synthesized in the cell from ornithine via putrescine, and the enzyme ornithine decarboxylase is involved in the biosynthesis (12).

b. Anatoxin-a(s). This toxin is produced by *Anabaena* species and is more toxic than anatoxin-a. It is structurally unrelated to anatoxin-a and is an *N*-hydroxyguanidine methyl phosphate ester, as shown in Fig. 1.

c. Saxitoxin and Neosaxitoxin. These two structurally related toxins are produced by *Aphanizomenon flos-aquae* and *Anabaena circinalis* as well as by some marine dinoflagellates. These compounds are the two main PSPs and chemically are tricyclic molecules with hydropurine rings, as shown in Fig. 1. It has recently been discovered that the freshwater filamentous cyanobacterium *Lyngbya wollei* produces a range of novel saxitoxin analogs.

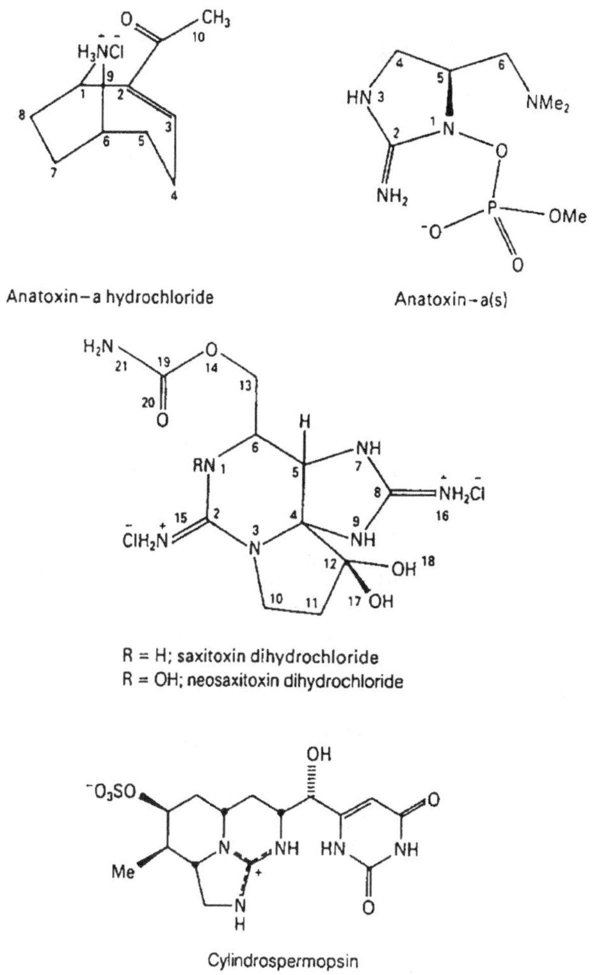

2. Hepatotoxins

A number of different hepatotoxins with varying chemical structures are produced by species and strains within the following genera: *Anabaena, Microcystis, Nodularia, Nostoc, Oscillatoria, Cylindrospermopsis, Umezakia, Aphanizomenon.* The following chemical types will be considered: cyclic peptides and alkaloid hepatotoxins.

a. Cyclic Peptides. The microcystins are a group of monocyclic heptapeptides that have been found in *Microcystis, Anabaena, Nodularia, Nostoc,* and *Oscillatoria.* A number of structural variants of microcystin have been determined (1). All microcystins contain L amino acids at two nonconserved positions on the molecule, while the D amino acids are highly conserved. The D amino acids are *N*-methyl-hydroanaline (Mdha) and a unique amino acid termed ADDA. The ADDA side chain is the structural component responsible for the toxicity of the microcystins. To date at least 60 microcystins have been characterized, with the most toxic being LR. Structures are shown in Fig. 2.

Nodularin is a monocyclic pentapeptide that is smaller than the microcystins and is produced by *Nodularia spumigena.* Nodularin also contains the amino acid ADDA, which is responsible for its toxicity. The structure of nodularin is shown in Fig. 2.

b. Alkaloid Hepatotoxins. The alkaloid cylindrospermopsin from *C. raciborskii, U. Natans,* and *A. ovalisporum* is chemically a cyclic guanidine group linked to hydroxymethyl uracil (5,13). The toxin is zwitterionic and as such is highly water soluble. A structural derivative, deoxycylindrospermopsin, that lacks the hydroxyl group on the methyl uracil has been found in *C. raciborskii* (14). The deoxy derivative has low toxicity relative to cylindrospermopsin. An isomer of cylindrospermopsin with a mass spectrometric fragmentation pattern identical to that of cylindrospermopsin has also been found by the NRCET research team. The toxicity of this isomer has not been established up to the present time. The structure of cylindrospermopsin is presented in Fig. 1.

3. Nonspecific Toxicants

Trichodesmium erythraeum from marine waters in Queensland, Australia has been shown to possess unidentified water-soluble toxins with neurotoxic effects

Anatoxin–a hydrochloride

Anatoxin–a(s)

R = H; saxitoxin dihydrochloride
R = OH; neosaxitoxin dihydrochloride

Cylindrospermopsin

Fig. 1 Structures of some common cyanobacterial toxins. (From Ref. 39.)

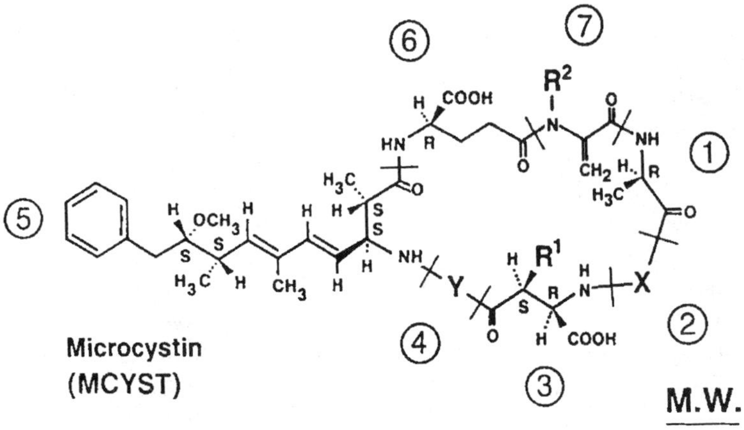

Microcystin (MCYST)

				M.W.
MCYST-LA:	X = Leu; R¹ = CH₃;	Y = Ala; R² = CH₃		909

MCYST-LA: X = Leu; R^1 = CH$_3$; Y = Ala; R^2 = CH$_3$ — 909
MCYST-YA: X = Tyr; R^1 = CH$_3$; Y = Ala; R^2 = CH$_3$ — 959
MCYST-LR: X = Leu; R^1 = CH$_3$; Y = Arg; R^2 = CH$_3$ — 994
desmethyl 3- MCYST-LR: X = Leu; R^1 = H; Y = Arg; R^2 = CH$_3$ — 980
MCYST-YM: X = Tyr; R^1 = CH$_3$; Y = Met; R^2 = CH$_3$ — 1035
MCYST-RR: X = Arg; R^1 = CH$_3$; Y = Arg; R^2 = CH$_3$ — 1037
desmethyl 3- MCYST-RR: X = Arg; R^1 = H; Y = Arg; R^2 = CH$_3$ — 1023
desmethyl 3,7- MCYST-RR: X = Arg; R^1 = H; Y = Arg; R^2 = H — 1009
MCYST-YR: X = Tyr; R^1 = CH$_3$; Y = Arg; R^2 = CH$_3$ — 1044

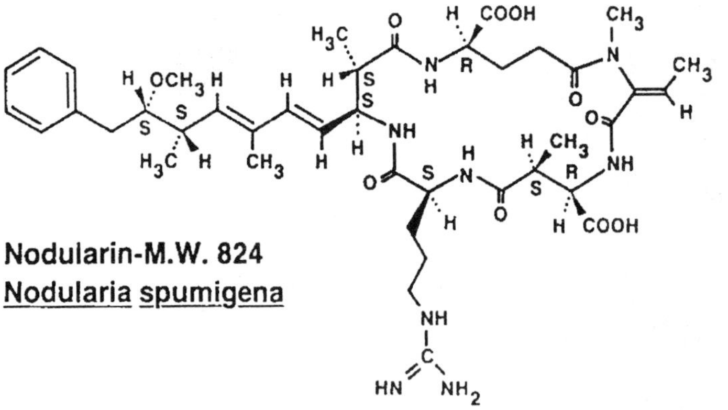

Nodularin-M.W. 824
Nodularia spumigena

Fig. 2 Structures of the cyclic peptide toxins, microcystins and nodularin. (From Ref. 39.)

and lipid-soluble toxins with hepatotoxic effects (15). The author has found that isolates of *T. erythraeum* associated with fish kills in Northern Queensland caused marked necrosis of subcapsular hepatocytes in livers of mice injected via the intraperitoneal (IP) route.

4. Irritants

Irritations, commonly caused swimmers itch, have been produced by various freshwater cyanobacteria and by the marine cyanobacteria *Trichodesmium erythraeum* and *Lyngbya majuscula*. The toxins in *T. erythraeum* that are responsible for irritation have not been identified. *Lyngbya majuscula* has been found to contain a series of substituted indole toxins collectively termed lyngbya toxins (16).

II. IDENTIFICATION AND ENUMERATION OF CYANOBACTERIA

A. Introduction

The routine biological monitoring of rivers, lakes, and dams for cyanobacteria is commonplace. These organisms are typically identified and enumerated under a microscope using a calibrated counting chamber and generally reported as cells per milliliter (17). Changes in the concentration of cells with location and time are monitored, enabling risk assessment to occur and contingency plans to be implemented (17,18). The aim of the following information is to provide the reader with an overview of the techniques commonly practiced in the identification and enumeration of cyanobacteria.

B. Sampling

One of the requirements for proficient enumeration of cyanobacteria is the development and implementation of reliable sampling techniques (19). These techniques will not be addressed in this chapter. However, due to their importance it is recommended that the reader investigate protocols relevant to this topic. Documents containing guidelines for sampling are too numerous to list in full, but Refs. 18 and 20 will provide the reader with useful information on these techniques.

C. Preservation of Samples

Samples that have been collected for cyanobacterial identification and enumeration should be preserved immediately at the sampling site by the addition of acidified Lugol's solution (19,20). Water quality in an un-

preserved sample can change within hours, and grazing zooplankton can reduce cyanobacterial numbers (21). These changes can consequently reduce the accuracy of the count.

Samples are preserved by the addition of 0.3 ml of acidified Lugol's solution to 100 ml of sample. The addition of this solution should give the sample a weak tea color. For long-term storage, 0.7 ml of Lugol's solution is added per 100-ml sample, with the addition of buffered formaldehyde to a minimum of 2.5% final concentration after 1 hour (22). If the sample has a high organic load, the iodine in the Lugol's solution may be readily consumed. If this occurs, additional solution is added to the sample until a weak tea color is achieved. The volume of additional solution is recorded. Samples preserved this way will keep for several years if stored in the dark and topped up regularly with the solution.

Acidified Lugol's solution can be made by dissolving 20 g potassium iodide (KI) and 10 g iodine crystals in 200 ml distilled water containing 20 ml glacial acetic acid (17,19). This solution must be stored in the dark in a glass bottle and will remain effective for at least 1 year (19,21).

The iodine in the solution will not only preserve the cyanobacterial cells in the sample, but will also increase their specific weight. This in turn will facilitate the sedimentation process that is commonly used in the concentration of cyanobacteria in water. The iodine present in solution will also stain these organisms. This of course presupposes familiarity with freshwater cyanobacteria, because loss of structural detail may occur (23). It is therefore recommended that a small portion of fresh sample be kept to aid in identification.

For the preservation of iodine, the preserved samples should be kept in the dark during transport and the usual care taken when handling chemicals (21,24).

D. Subsampling Preserved Samples

A subsample of the preserved sample is either transferred into a calibrated counting chamber or prepared for concentration. Subsampling errors can be minimized by inverting the sample thoroughly 20–30 times so that it is well mixed before subsampling. The breakup of filaments and colonies during mixing is often unavoidable (21). However, inverting the sample gently will reduce the degree of breakup.

E. Sample Concentration Techniques

Only very high concentrations of natural cyanobacteria can be enumerated directly. Therefore, the sample usually needs to be concentrated before enumeration and identification is performed (24). The three most common methods employed for sample concentration are sedimentation, centrifugation, and membrane filtration.

1. Sedimentation (Gravity)

Sedimentation is the benchmark phytoplankton concentration method. It is nonselective and nondestructive (22). A subsample of between 5 and 1000 ml of well-mixed preserved sample is taken and sedimented in a measuring cylinder of suitable size (21). The volume of sample concentrated varies inversely with the abundance of organisms present in the sample (22).

The measuring cylinder is filled with a subsample of preserved material without forming a vortex. It must be kept vibration free and moved carefully to avoid disturbance of settled matter (22). The sample is allowed to sediment for 2 hours for each 1 cm of water in the cylinder at 20°C. After sedimentation the top 90% of volume is carefully siphoned off without disturbing the sedimented cyanobacteria. The remainder of the sample is mixed gently and transferred into a suitable-sized sample vial with a secure lid ready for identification and enumeration (21).

Species of cyanobacteria containing gas vesicles may not settle fully despite the addition of acidified Lugol's iodine; therefore, the gas vesicles need to be collapsed before the sample is set up for sedimentation. One method of collapsing the gas vesicles involves exposing the sample to ultrasonication for 1 minute (21). After such treatment the sample is sedimented as previously described. It is suggested that the condition of cells and their relative abundance should be examined before and after such treatment to validate the procedure. An alternative to ultrasonication is siphoning the sample off from a point well below the meniscus. Cells floating on the surface will be pulled down into the concentrate and will be incorporated into the sample (21).

The measuring cylinder used to sediment cyanobacteria in water samples should have a height:diameter ratio not exceeding 5:1. If the cylinder is higher than five times its diameter, convection currents will occur. This may result in a considerable amount of cyanobacteria failing to settle, regardless of the settling time (21).

2. Centrifugation

Centrifugation is a rapid method for the concentration of cyanobacteria (24). Batch or continuous centrifugation can be applied. It is suggested that centrifugation

of batch samples be performed for 20 min at 1000 g. Although centrifugation expedites sedimentation, it may destroy fragile organisms (22). It is therefore suggested that if the sample is subjected to such methodology, the condition of cyanobacterial cells and their relative abundance before and after centrifugation be compared. Comparison counts with those obtained from preserved samples settled by sedimentation can also be performed (24).

3. Membrane Filtration

Membrane filtration is another alternative method used to concentrate samples. A measured volume of well-mixed preserved sample is poured into a funnel equipped with a membrane filter having a pore diameter of 0.45 μm. A vacuum of less than 50 kPa is applied to the filter unit until only a small amount of sample remains on the filter. The vacuum is gently broken and then reapplied at approximately 12 kPa to filter the remaining sample without allowing the filter to dry. When the content of detritus is high, the filter clogs quickly and silt may crush the organisms or obscure them from view (22). If this occurs, the sample can be analyzed unconcentrated or diluted so that cells are visible enough to identify and enumerate. This involves a compromise between diluting the subsample in order to see the cells and reducing cell concentrations below statistically acceptable levels (21).

Centrifugation and membrane filtration are not recommended as benchmark concentration methods. However, either method may be used if validated against the benchmark sedimentation method.

F. Microscope Requirements

A standard or an inverted compound microscope should be used for cyanobacterial enumeration and identification.

1. Standard Compound Microscope

Compound microscopes are equipped with a mechanical stage capable of moving a counting chamber or cell past the objective lens. Standard equipment comprises 10× or 12.5× oculars and 10×, 20×, 40×, and 100× objectives. With standard objectives, the Sedgwick–Rafter counting chamber limits magnification to 200–250× (22). However, with the addition of a long working distance objective, the Sedgwick–Rafter chamber can be used at 400× magnification. For the 20× and 40× magnifications, the use of phase contrast objectives is essential. The Lund cell can also be used with a standard compound microscope.

2. Inverted Compound Microscope

The inverted compound microscope is also used routinely for cyanobacterial enumeration and identification. This instrument is unique in that illumination comes from above and the objectives are below a movable stage (22).

The Utermöhl counting chamber is used with an inverted microscope. To use an inverted microscope with the Utermöhl chamber, three hairlines or threads are needed in the optical path of the microscope. Two hairlines are arranged in parallel, and the distance between them may be adjustable while the third one runs at right angle to the other two. A special part with adjustable lines can be obtained with most modern inverted microscopes. Otherwise, fixed hairlines can be inserted into one of the eyepieces. If hairlines or threads are not available, a Whipple grid can be used instead (21).

G. Microscope Calibration

Calibration of the microscope used for the identification and enumeration of cyanobacteria is essential. The procedure should be performed directly after microscope servicing. The usual equipment used for calibration is either a Whipple grid or a graticule placed in one of the eyepieces (oculars) of the microscope and a calibrated micrometer slide placed on the stage.

1. Microscope Calibration Using A Whipple Grid

The Whipple grid is comprised of an accurately ruled grid subdivided into 100 squares of equal area. One square near the center is subdivided into 25 smaller squares. The outer dimensions of the grid are such that with a $10\times$ objective and ocular, an area approximately 1 mm square on the microscope stage is covered. Because this area may differ from one microscope to another, the Whipple grid needs to be calibrated for each microscope. This is achieved by placing the ocular and stage micrometers in parallel and in part superimposed. The left edge of the Whipple grid is lined up with the zero mark on the stage micrometer scale. The width of the Whipple grid image is determined to the nearest 0.01 mm from the stage micrometer scale. Should the width of the image of the Whipple grid be exactly 1000 μm, the larger squares will be 100 μm on a side and each of the smaller squares 20 μm on one side (22).

2. Microscope Calibration Using a Graticule

The graticule has an accurately ruled micrometer etched into it and is divided into 100 equal divisions. Using a $10\times$ objective and ocular, the graticule micrometer and stage micrometer are positioned in parallel and in part superimposed. The zero mark on the graticule micrometer is matched with the zero mark on the stage micrometer. The width of one scale line on the graticule is determined from the stage micrometer scale, to the nearest 0.01 mm. If the width of the image of both scale lines is exactly equal, then one scale line on the graticule micrometer is equal to 10 μm. This procedure needs to be repeated at all magnifications used. If the width of one scale line on the graticule micrometer is equal to 2.5 scale lines on the stage micrometer at a $400\times$ magnification, then one scale line on the graticule micrometer is equal to 25 μm.

H. Choice of Counting Chamber

It is recommended that an Utermöhl chamber be used with an inverted microscope and that a Sedgwick–Rafter chamber or Lund cell be used with a standard compound microscope. The choice of counting chamber and microscope is optional and usually depends on the historical preference already implemented in the laboratory and the financial resources available if initial purchasing is required (21).

1. Sedgwick–Rafter (S–R) Chamber

This chamber is commonly used for cyanobacterial identification and enumeration. It is easily manipulated and provides reproducible data when used with a calibrated microscope. The S–R chamber consists of a 20-mm $\times$ 50-mm microscope slide, with a grid floor consisting of 1000 fields and a raised well capable of holding 1 ml (Fig. 3) (21).

Filling the chamber involves slowly dispensing a subsample of well-mixed preserved sample (original, concentrated, or diluted) into the corner of the chamber on which a cover slip is diagonally placed. The sample should fill the well of the chamber, under the cover slip, until the chamber is full. The water tension often rotates the cover slip to cover the entire S–R chamber. If necessary, the coverslip should be carefully maneuvered to cover the entire chamber, ensuring that no air bubbles are present in the corners of the chamber. If air bubbles are present, a small amount of sample is added to fill the area entirely. The chamber must not be overfilled, for this will make the field of depth greater than 1 mm and produce an invalid count (22).

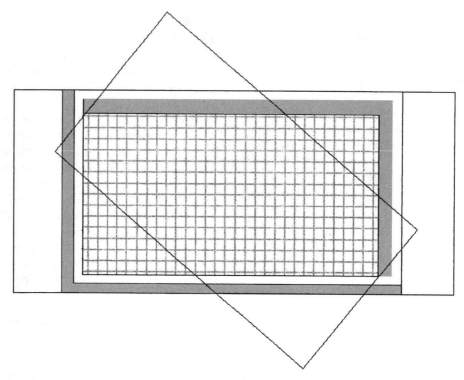

Fig. 3 Sedgwick–Rafter counting chamber.

It is not necessary to blot out excess sample from the chamber, because this may draw cyanobacteria toward the sides of the chamber. The sample should be discarded and the chamber cleaned and refilled. Large air spaces caused by evaporation may develop in the chamber during lengthy examinations. This can be prevented by placing a drop of distilled water on the edge of the coverslip before beginning the analysis. The sample should be allowed to sit for at least 15 minutes before examination so that all cyanobacteria settle to the bottom of the chamber (22). Glass S–R chambers are preferable to plastic ones because the latter are easily scratched (21).

a. Calibration of Sedgwick–Rafter Chamber. Each chamber needs to be calibrated before use. Before filling the S–R chamber with distilled water, the coverslip is positioned diagonally across the cell and placed on a calibrated four-place balance. The balance is tared and the chamber removed. The volume is determined by filling the chamber in the manner described in the preceding paragraph and reweighing the filled chamber. The procedure is repeated 10 times, making sure the chambers are completely dried between measurements. A record should be kept of the calibration measurements for each chamber. If the chamber volume differs from 1 ml by more than 5%,

a factor should be employed to correct the volume to 1 ml. This factor should be marked on the chamber and used when calculating the final cell-per-milliliter results (21).

2. Utermöhl Chamber

This counting chamber is a combined chamber, consisting of two sections. The first is bottom plate consisting of a round hole that is closed with a coverslip on one side, which forms the counting chamber; the second is a chamber cylinder (Fig. 4). These cylinders can vary in height and volume and are placed on the bottom plate, filled with preserved subsample, and removed after sedimentation (21,23).

A suitable volume of the original sample is placed in the assembled chamber and put aside for sedimentation. After a suitable sedimentation period, the cylinder can be removed, and enumeration and identification of the cyanobacteria collected on the bottom chamber plate is performed (23).

If the number of cyanobacteria in the original sample is known or found to be too small for accurate enumeration after direct sedimentation into the Utermöhl chamber, it may be necessary to carry out the concentration in two steps (23). A suitable volume can

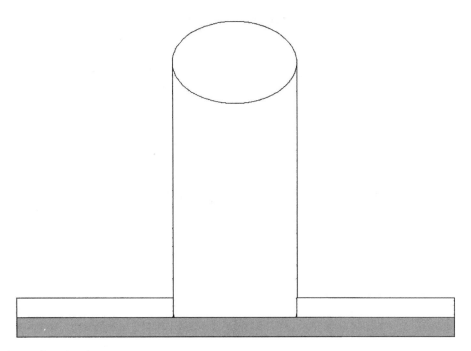

Fig. 4 Utermöhl counting chamber.

be concentrated via a larger cylinder, as described in Sec. II.E.1. After sedimentation the supernatant is removed and the remaining sample transferred into the Utermöhl chamber cylinder for a second sedimentation.

The commercially available Utermöhl chambers with a hole of 2.5-cm diameter situated in the bottom plate allows the direct sedimentation of volumes between 5 and 100 mL (19,21).

Usually, the whole bottom plate or counting chamber is examined, since the distribution of organisms is not always random. By moving the mechanical stage, the chamber bottom is traversed backwards and forwards along adjacent strips so that eventually the whole bottom is covered. For the abundant taxa present it might be sufficient to count every second or third strip. All cyanobacterial cells lying between the two parallel hairs, or within the edges of the Whipple grid if no hairs are present, are counted. Care needs to be taken not to count filaments that lie across both hairs twice (21,23). See Sec. II.I.4 for information on establishing a convention for tallying organisms lying across both hairs.

3. Lund Cell

The third type of chamber consists of an ordinary glass microscope slide with two long, thin pieces of brass or glass glued to the longer edges (Fig. 5). A rectangular coverslip is placed on top to form the counting cham-

ber. A subsample of the original, concentrated, or diluted sample is placed in the chamber by placing the tip of a pipette close to the open edge of the coverslip. The liquid is sucked under the coverslip by capillary action. The formation of air bubbles in the chamber should be avoided. The filled chamber should stand for 15 minutes, for the cyanobacteria to settle onto the surface of the slide, before identification and enumeration is performed (21).

The Lund cell must be calibrated before use. By placing the coverslip on the cell and placing both the cell and coverslip together on a four-place balance, the balance is tared and the chamber removed. The volume is determined by filling the cell in the manner described in Sec. II.H.1 and then reweighing the filled cell. The procedure is repeated 10 times. Make sure the cell is completely dried between measurements. A record should be kept of the calibration measurements for each cell, since it is used in the calculation of final counts.

A Whipple grid, located in one of the eyepieces, defines the field of view counted. A field conversion factor (FCF) needs to be calculated to relate the area counted to the total area of the chamber, for only part of the chamber is usually counted. By knowing the volume of the Lund cell and the area being examined, it is possible to relate the cyanobacterial count recorded per field of view to cells in the original sample. The

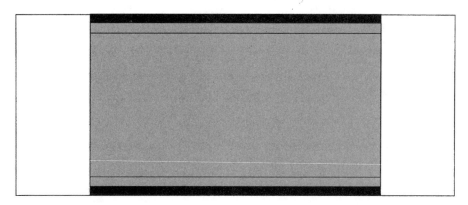

Fig. 5 Lund cell.

total number of fields is the total area of the chamber (mm^2) divided by the area of the Whipple grid (mm^2) (21).

I. Enumeration of Cyanobacteria

An experienced taxonomic phycologist and a good-quality microscope are required for accurate identification and enumeration of cyanobacteria. Enumeration is recommended for Lugol's preserved samples only. The basic procedure for identification and enumeration of cyanobacteria is to fill a calibrated counting chamber with a subsample of either the original, concentrated, or diluted sample; after a period of settling, the cyanobacterial cells are simultaneously identified and enumerated using phase microscopy. Cyanobacterial units may be filaments, such as *Cylindrospermopsis raciborskii*, or colonies, such as *Microcystis aeruginosa*, depending on the species' usual morphological form (21).

In low-density samples, counting stops when the counting chamber is completed. In high-density samples, counting continues until a fixed number of units have been identified (17).

The enumeration of cyanobacterial cells can be carried out in two stages. First, the bottom area of the counting chamber or cell is examined under high magnification to identify and enumerate the smaller cyanobacterial units, then a lower magnification is used to enumerate and identify the larger cyanobacterial units (19).

1. Number of Fields to Be Examined

The number of fields examined depends upon the cyanobacterial density of the sample and upon the level of accuracy required. The accuracy of the count varies indirectly with the square root of the number of units counted, multiplied by 100 to express it as a percentage (23). Thus a count of 100 units has an accuracy of ±20%. Lasett et al. (17) described the accuracy of a count as $\sqrt{2n} \times 100$, where n = number of units counted. Using this formula, a count of 23 units has an accuracy of ±20%. Both descriptions are acceptable; however, it is recommended that the analyst report the method of choice to facilitate adequate comparison of results from different studies. Determining the acceptable level of accuracy is a compromise between the identifiable health risk of an incorrect or low-precision count and the cost of obtaining results with increased precision and confidence (25).

A minimum of 30 fields must be counted to ensure that 90–95% of the cyanobacterial species present within the S–R chamber are detected (21,26).

2. Determination of Indistinguishable Cells Within a Trichome

The cells within each trichome should be counted (or estimated) where possible. Sometimes it is impossible to determine the individual cells within a trichome. This could be due to their small size and/or indistinguishable divisions. If this occurs, the trichome is counted as one unit. The average length of 30 cells is then determined at 1000× magnification, and the length of the trichome in the sample is measured during the analysis. Cell concentrations of the species in question can then be calculated from these two measurements (trichome length divided by the average cell length) (21). If the average cell length under 1000× magnification cannot be determined, then it is acceptable to use the average cell lengths referred to in reference documents.

3. Enumeration of Cells in Large Three-Dimensional Colonies

Under bloom conditions, some species form colonies so large that only an estimate of cells per colony can be made, unless the colonies are dispersed and individual cells are counted. *Microcystis aeruginosa* is an example of the cyanobacterial species that can produce such colonies. An estimated count should be acknowledged on both the worksheet and the report (21). More accurate counts of *Microcystis aeruginosa* can be achieved by using ultrasonic waves to disintegrate colonies. Colonies may not be totally broken up into single cells but may be sufficiently reduced to allow single cells to be counted (20). Generally, 1 minute at ~12 μm, 20 kHz, will disrupt the colonial structure without significant cell destruction.

The value of using ultrasonic waves should be assessed carefully. It may be preferable to leave *Microcystis aeruginosa* in the colonial form, because it may be difficult to distinguish from other organisms after colonial disruption (27).

4. Determination of Organisms Lying on Boundary Lines of Counting Chambers

A counting convention for tallying organisms lying on an outer boundary line of a counting chamber or cell needs to be established by the laboratory. For example, when counting within a field in an S–R chamber or a Whipple grid and Lund cell, the analyst designates the top and left boundaries as a no-count line and the bottom and right boundaries as count lines. The analyst would tally every cyanobacterial unit touching a count line from the inside or outside but ignore any touching a no-count line. In an Utermöhl chamber those organisms that lie across the upper horizontal lines are counted as lying within the hairs, but those across the lower are not (23).

5. Other Information Regarding Enumeration of Cyanobacteria

It is desirable to distinguish between live and dead cells, because dead cells should not be included in the count. However, this is difficult, particularly when no visible change in appearance occurs. Criteria for such borderline cases should be established and documented by the laboratory (23).

Computer programs for recording cell counts are available and can shorten counting time considerably (20).

The laboratory should participate in interlaboratory and intralaboratory collaborative trials on a regular basis. This process is implemented to evaluate the laboratory's skills in performing identification and enumeration of cyanobacteria in comparison to other laboratories and other algal taxonomists within the laboratory. The results from such trials should be documented.

J. Taxonomic Identification of Cyanobacteria

Due to the nature of environmental water samples, the morphology of cyanobacteria may change depending on environmental conditions and the phase of growth of the organism (28,29). This morphological variation may lead to difficulty and errors in cyanobacterial identification (29). Is is therefore recommended that each laboratory use standard bench references as well as in-house taxonomic reference documents. These documents should comprise photomicrographs of organisms observed in the laboratory from a variety of samples collected under various environmental conditions. These documents will not only aid the experienced analyst in the identification of cyanobacteria but will be a valuable training aid for new taxonomists.

The level of taxonomic identification required is dictated by the objectives of the biological monitoring program. Dominant cyanobacterial species and problem species, such as potentially toxic cyanobacteria, should be identified to species level. Identification of preserved cyanobacteria can be difficult in some circumstances. It is therefore suggested that some unpreserved or live sample be retained to aid identification.

K. Recording of Unidentified Cyanobacteria

A protocol for the effective recording of problematic identifications should be developed by the laboratory. It is suggested that a photomicrograph or drawing be made of the difficult organism. A written record should be kept containing information such as cell dimensions, color, and unusual features (21). It is recommended that the photomicrograph or drawing be labeled with a unique number for future reference.

Identification by an algal taxonomist can then be sought at a later stage, by sending either the photomicrograph, the drawing, or a subsample of the organism to another laboratory (21). Once the organism has been correctly identified, the photomicrograph or drawing can be labeled with the correct identification and incorporated into the laboratory's taxonomic reference documents for future use.

An effective data storage system must be able to accommodate taxonomic changes and eventual identification of unknown cyanobacteria (21).

L. Calculations

1. Sedgwick–Rafter Chamber

The final result, expressed as cells per milliliter, is calculated as

$$\text{Cells per ml} = \frac{\text{numbers of cells counted} \times 1000}{\text{sample concentration factor} \times \text{number of fields examined}}$$

or

$$\text{Cells per ml} = \frac{\text{numbers of cells counted} \times \text{sample dilution factor} \times 1000}{\text{number of fields examined}}$$

For example, 150 ml of original sample is concentrated to 15 ml (concentration factor of 10), and a 1-ml subsample is taken for enumeration. If 2000 cells of *Anabaena circinalis* were counted in 200 fields, then the concentration of this organism is calculated as

$$\text{Cells per ml} = \frac{2000 \times 1000}{10 \times 200} = 1000$$

2. Utermöhl Chamber

The cell-per-milliliter concentration for each cyanobacterium sighted is calculated as

$$\text{Cells per ml} = \frac{\text{numbers of cells counted}}{\text{sample concentration factor}} \times F$$

or

$$\text{Cells per ml} = \text{numbers of cells counted} \times F \times \text{sample dilution factor}$$

where F is the factor derived from the ratio of the chamber floor examined to the entire floor.

3. Lund Cell

The final result, expressed as cells per milliliter, is calculated as

$$\text{Cells per ml} = \frac{\text{numbers of cells counted}}{\text{sample concentration factor}} \times \text{FCF}$$

or

$$\text{Cells per ml} = \text{numbers of cells counted} \times \text{FCF} \times \text{sample dilution factor}$$

The field conversion factor (FCF) is a factor used to convert the area counted to the total area of the chamber (see Sec. II.H.3.a) (21).

M. Recording Results

It is suggested that a standard enumeration worksheet be used to record the results of enumeration and identification of cyanobacteria. The most common cyanobacterial taxa should be prelisted on the worksheet (21). The worksheet should also contain an area to record the total number of cells and units counted, number of fields or strips counted, the calculation/formulae used to calculate cell numbers, the concentration or dilution factor, the unique sample identifier, the analyst's name, and the date analysis was performed.

N. Reporting Results

It is recommended that the magnification used for the enumeration of the smallest identifiable type of cyanobacteria and the counting error be reported, in order for the results of different studies involving cyanobacterial enumeration to be adequately compared with one another (30).

O. Quality Assurance Techniques

In any sedimentation concentration method, the rate of sedimentation of the cyanobacteria present may vary. Some may settle after 24 hours and others may not settle at all. It is therefore recommended that the supernatant be checked regularly (annually) for cells that have not sedimented. This procedure will evaluate the adequacy of the procedure (24).

Errors incurred with subsampling should be investigated regularly by taking 10 successive subsamples from the same storage bottle and comparing the counts of five of the most dominant cyanobacterial species (19,21,23).

A regular check of microscope adjustment and measuring equipment is essential (31). It is suggested that microscopes are serviced at least every 12–24 months and the measuring eye micrometers calibrated with each microscope.

Accreditation authorities will provide further details regarding quality assurance requirements for the analysis of water samples for enumeration and identification of cyanobacteria.

P. Current Developments in Cyanobacterial Enumeration and Identification Techniques

Traditional techniques for the identification of cyanobacteria have relied primarily on morphological characteristics observed under the microscope. Morphology may change depending on environmental conditions and the organism's phase of growth (28,29). This morphological variation may lead to difficulty and errors in cyanobacterial identification (29).

1. Molecular Techniques

The foregoing limitations have prompted the development of molecular biological techniques and characterizations of cyanobacteria in an effort to circumvent some of these problems (28,32). Investigations to date have focused on examining ribosomal RNA (rRNA) repeat units, in particular, the small and large subunit rRNA genes. These repeat units have a mosaic of conserved and highly variable domains, which allow the design of oligonucleotide probes. Oligonucleotide probes are synthetic DNA molecules, which are generally 18–35 bases long. These probes can be labeled with radioactive, fluorescent, chemiluminescent, or colormetric molecules and are usually attached to the 5′ end, either directly or through a secondary labeling reaction. Assays that detect sequences within these rRNA molecules generally require an amplification step, such as the polymerase chain reaction (PCR) (32).

Randomly amplified polymorphic (RAPD) PCR techniques have been investigated to characterize cyanobacteria in pure culture (28,33). However, the requirement for axenic cultures restricts the applicability of the RAPD-PCR techniques (28). An alternative to RAPD-PCR is afforded by the PCR–restriction fragment length polymorphism (RFLP) analysis of the RNA repeat unit and the phycocyanin operon. These methods are more reliable and useful for the direct analysis of cyanobacteria isolated from complex natural populations (34). This technique requires only small quantities of impure DNA and can distinguish between the various genera and species of cyanobacteria by using the variability of the rRNA repeat unit. A number of cyanobacterial 16S rRNA genes and the internal transcribed spacer region between 16S and 23S RNA genes have been sequenced for taxonomic purposes

(28). References relevant to this topic include 28, 29, 32, 33, and 34.

2. Flow Cytometry

Another area currently under investigation in the field of cyanobacterial identification and enumeration is flow cytometry. At one or more wavelengths, the emission and excitation efficiency of individual cells is measured, together with the cells' light-scattering properties. The fluorescence signals resulting from excitation at these different wavelengths gives insight into the pigment composition of organisms present in the sample. These parameters enable taxonomic groups to be differentiated (35–37).

Many studies have indicated that environmental conditions such as light exposure and temperature may alter the flow cytometric appearance of phytoplankton cells. Therefore, the main challenge of flow cytometry will be to recognize cells in different physical, chemical, or nutritional environments. Future research will focus on this area (37).

Currently, most applications of flow cytometry make use of various physiological and morphological characteristics of the cell, such as pigment composition, cell size, and DNA content. These criteria are often not sufficient for the identification at the level of genus or species, unless a certain species dominates or can be easily distinguished because of obvious characteristics. It is therefore suggested that microscopic approaches be combined with flow cytometry during cyanobacterial analysis (38).

III. ANALYSIS FOR BLUE-GREEN ALGAL TOXINS

A. Background

A variety of techniques are available for the estimation of cyanobacterial toxins. These include animal bioassays, analytical methods, and biochemical effect measurements. The technique used often depends on the availability of suitable laboratory facilities, which range from a simple animal house for mammalian bioassays to sophisticated instrumentation such as HPLC-MS/MS. While sensitive instrumental assays, immunological techniques, and some biochemical methods such as protein phosphatase inhibition assays are suitable for toxins at levels in the low-microgram-per-liter range in water, the mammalian bioassays are usually suitable only for toxins in algal cellular material.

B. Sampling

As with all water analyses, cyanobacterial toxin determinations require adequate sampling procedures. In general, for toxin determination in water by instrumental or immunoassay methods, clean glass or plastic bottles are suitable for sampling water, with 1-L volumes being commonly sampled. Water samples should be transported on ice to the laboratory to prevent the possible decomposition of toxins. With samples for mammalian bioassays, it has been suggested by Carmichael (39) that about 0.5 g dry weight of cells is required. This equates to the collection of 1 L of light-to-moderate water bloom or 50–100 ml of a heavy surface scum. Chemical analysis for toxins, in comparison, can be accomplished using water samples with no obvious scum, since several of the analytical techniques can determine toxins at levels below 1 μg/L. Some analytical methods require preconcentration using solid-phase microcartridges; this can be regarded as a sampling procedure if performed in the field though it is usually undertaken in the laboratory.

C. Bioassays

1. Background

Animal bioassays are the oldest methods for the determination of toxins in cyanobacterial samples (40). Although developed primarily for use with dinoflagellate toxins in shellfish, they still have application to various types of cyanobacterial toxins, including the PSP toxins found in *Anabaena circinalis*, and also some of the hepatotoxins, such as the microcystins and cylindrospermopsin. The major advantage of a properly calibrated mammalian bioassay over chemical analyses and in vitro methods is that the toxicity result is more relevant to human health. This is due to the fact that the use of living mammals takes into account confounding factors such as metabolism and distribution of the toxins in the organism. Due, however, to the difficulty in some countries of gaining ethical approval for mammalian bioassays, the use of in vitro methods employing cell cultures is gaining popularity.

2. Mouse Bioassay for Paralytic Shellfish Poisons (PSPs)

The mouse bioassay was first applied to determine PSP toxicity in mussels from California by Sommer and Meyer (41). The PSP toxins are alkaloids that form three main groups: the carbamate toxins, decarbamoyl toxins, and *N*-sulfocarbamoyl toxins. These vary in

Table 2   Toxicities of PSP Toxins

Toxin	Toxicity range (MU/μmol)
GTXI	752–1975
GTXII	793–1150
GTXIII	1465–2234
GTXIV	602–1775
GTXV(B1)	125–354
GTXVI(B2)	175–180
dcGTXI	950
dcGTXII	380
dcGTXIII	380
dcGTXIV	950
Cl(epi-GTXVIII)	17–25
C2(GTXVIII)	180–430
C3	8
C4	57
STX	1656–2050
NeoSTX	1038–2200
dcSTX	1175–1220
dc-neoSTX(STXVII)	900

MU = mouse units.

chemical structure (Table 2), and toxicity (Fig. 6). Many countries require the monitoring of PSP toxins in mollusc flesh by the mouse bioassay (42). The mouse bioassay has been applied to blue-green-algal–derived PSPs in Australia, where the cyanobacterium *Anabaena circinalis* produces these toxins.

Sample preparation consists of the following:

1. Take 0.1 g of freeze-dried cellular material.
2. Resuspend in 10 mL of isotonic saline (4.5 g NaCl in 500 mL of pure water).
3. Sonicate 3–4 minutes.
4. Centrifuge and use supernatant for toxicity testing.

The most commonly used bioassay procedure for PSP toxins is that of the AOAC (43). This bioassay is regarded as being quantitative, and results are expressed as mouse units dependent on time of death after injection. Times of death of 1 minute or less score 100 mouse units, and death times of 15 minutes rate as 1.0 mouse units. In the AOAC procedure, the extract to be injected is pH adjusted to between 2.0 and 4.0. If direct comparison with AOAC tables is desired for cyanobacterial scum extracts, then the pH should be adjusted to this range before injection. Based on the AOAC procedure, the following protocol is recommended.

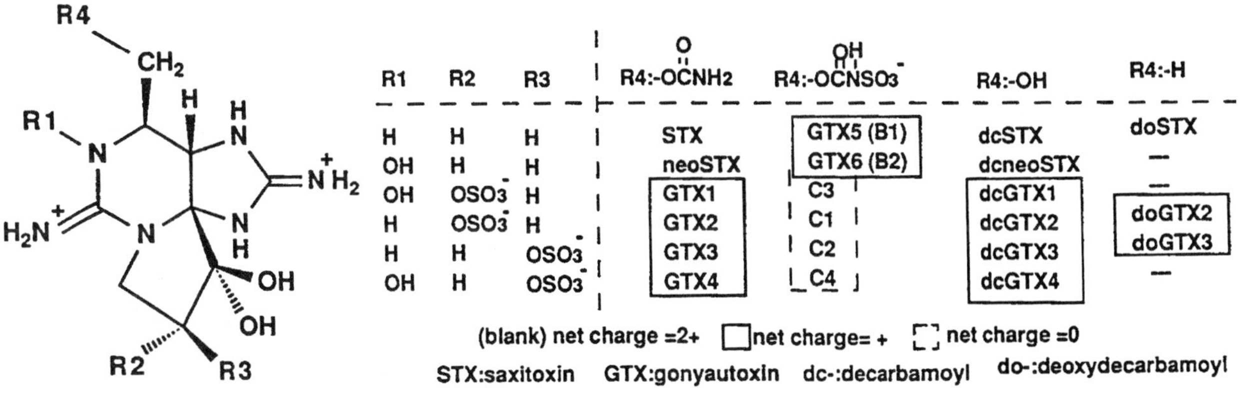

Fig. 6 Structures of PSP toxins. (From Ref. 59.)

Procedure. Intraperitoneally inject each test mouse (approximately 20 g each) with 1 ml of extract, note the time of dosing, and observe the mice for time of death, as indicated by the last gasping breath. It is preferable to dilute the extract so that the median time of death of several mice is between 5 and 7 minutes. The toxicity of the blue-green algal scum in mouse units is then calculated using "Sommer's Table," given in the AOAC Official Method (43). Mouse units can be converted to micrograms saxitoxin equivalents per gram of dried cyanobacterial scum. It is important to realize that hot acid extraction procedures used by some authors produce results that relate to the maximum potential toxicity of the scum, since this extraction protocol converts the low-potency *N*-sulfocarbamoyl toxins to the highly toxic carbamate analogs (40).

3. Mouse Bioassay for the Hepatotoxin Cylindrospermopsin

The hepatotoxin cylindrospermopsin (Cyn) has been shown to be produced by the following blue-green algae: *Cylindrospermopsis raciborski (13), Umezakia natans* (5), *Aphanizomenon ovalisporum* (7). It has been shown that the typical lesion in mice dosed via the intraperitoneal (IP) route (43,44) and the oral route (45) is lipid vacuolation followed by extensive hepatocyte necrosis at higher dose rates. The appearance of lipid vacuolation in the liver is determined via gross morphological observation and confirmed by histology. The 5-day LD_{50} is 0.2 mg/kg and the 24-h LD_{50} is 2 mg/kg. A mouse bioassay procedure has been developed for Cyn that can be used as a confirmatory test for the analytical determination of the toxin. The procedure is as follows.

1. The water sample is concentrated 100-fold via rotary evaporation.
2. Aliquots of the concentrate (1 ml) are dosed via IP injection in triplicate mice (approximately 20 g each).
3. One negative control is dosed for each sample. Control consists of tap water concentrated 100-fold with a dose of 1 ml via IP injection.
4. Any mice that appear lethargic and huddled should be euthanized and their livers examined for lipid accumulation.
5. A thin slice of liver lobe should be fixed in 10% buffered neutral formalin and allowed to stand for at least 24 h.
6. Paraffin wax blocks of the liver should be made and sectioned using standard histological techniques.
7. Slides should be prepared using haematoxylin-eosin staining.
8. Interpretation of histological slides should be performed by personnel trained in histopathology.

Note: Samples with high salt content may produce neurotoxic-like symptoms.

4. Mouse Bioassay for Microcystins

Mouse bioassays for microcystins have been developed, such as that described by Vezie et al. (46). The procedure is basically similar to that for cylindrospermopsin, but with the exception that an aqueous suspension of freeze-dried cells is used for IP injection. The onset of death with microcystins is faster than with cylindrospermopsin, and liver morphology is represented by swelling and a dark red color, corresponding to hemorrhage within the liver.

5. Bioassays Using Organisms Other Than Mammals

Due to ethical considerations, the use of mammals for bioassay purposes is not always acceptable. There are a number of studies that have proposed alternative organisms, in the main, invertebrates. In a number of instances, the brine shrimp *Artemia salina* has been used in a bioassay to determine the toxicity of blue-green algal blooms (46–48).

With this bioassay, dried extracts of lyophilized cyanobacterial cells are suspended in seawater and a series of dilutions made. Freshly hatched brine shrimp larvae are added to the samples and incubated under light for 24 and 48 h. Toxicity is expressed as the percentage of dead larvae minus the mortality in control samples.

Other organisms used for bioassays of cyanobacterial toxins include the desert locust, *Schistocera gregaria*, for saxitoxin and other PSPs (49), the plant *Sinapis alba L* for microcystins (50), and the commercial system (Microtox) employing the luminescent bacterium *Photobacterium phosphoreum* (Cohn) (46,51,52).

An evaluation of the mouse bioassay, Microtox, and brine shrimp assays for different cyanobacterial strains producing hepatotoxins (46) has shown that the *Artemia salina* larvae were sensitive to hepatotoxins and correlated well with the mouse bioassay. The Microtox responses, however, were not consistent with the mouse bioassay, and toxicities from other compounds were also observed with the Microtox assay.

6. Toxicity Testing Using Cell Cultures

Because of the increasing stringency of regulations concerned with the use of mammals for toxicity testing, the use of cell cultures for cyanobacterial toxins has been proposed (53). It is important that the cell type used be sensitive to the toxin being tested. Most cell culture methods have been developed for the hepatotoxic microcystins (53–55), although cell cultures have demonstrated the cytotoxicity of the hepatotoxic cylindrospermopsin (56).

The most commonly used cultures are primary rat hepatocytes, although liver slice cultures have also been used (54). It has been shown in a number of studies that only parenchymal liver cells respond to the toxins at concentrations analogous to doses effective in the whole animal (53). With the microcystins and nodularins, toxic effects on hepatocytes can be observed, with characteristically clustered blebs being distinguished (53). Recently, a biotest for microcystin hepatotoxins has utilized primary rat hepatocytes (55). This test compares the toxicity to hepatocytes with that in Chinese hampster ovary cells to distinguish between microcystin toxicity and effects from other harmful compounds that may be present in cyanobacterial extracts.

The use of analysis for biochemicals in conjunction with cultured cell lines for the determination of cyanobacterial hepatotoxins produces a sensitive assay that can also be quantitative for effect. In this system (54), leakage of a range of cytosolic enzymes is measured, and this is related to toxicity to the cell lines. The measurement of biochemical changes in cell lines has also been used by Runnegar et al. (56) to demonstrate the toxicity of the hepatotoxin cylindrospermopsin. It has thus been demonstrated that cell lines can be used as primary determinants of toxicity of blue-green algal toxins and also that the incorporation of biochemical measurement can be used to increase the sensitivity of the methods and to introduce a measurable quantity into toxicity estimations.

D. Chemical Assays for Algal Toxins

1. Paralytic Shellfish Poisons (PSPs)

The potential presence of a wide range of PSP toxins, as shown in Table 2, makes it difficult to develop an analytical method that can detect all of the toxins using a single technique and yet still be economically feasible. For the analysis of toxins in drinking water, the most commonly used technique for these hydrophilic compounds is that of high-performance liquid chromatography (HPLC).

There are a number of variations on HPLC methodology used in the literature, and many of the methods are based on the oxidation/fluorescence assay (42). The main difference between methods depends on whether the oxidation is performed before or after separation of components on the HPLC column. The precolumn oxidation method of Lawrence et al. (57,58) has been applied to 10 PSP toxins and uses oxidation at room temperature under mildly basic conditions with hydrogen peroxide or periodic acid. Reversed-phase columns are used that produce single peaks for most toxins, but all GTX toxins elute together and C-1/C-2 plus neosaxitoxin/B-2 are coeluting pairs. The detection limits of toxins range from 20 to 500 pg/injection.

One of the most applicable postcolumn methods for PSP toxins is that of Oshima (59). With this system, three different isocratic chromatographic conditions are used for the toxin groups categorized by their basicity. For the C-toxins, which were previously difficult to separate, tetrabutylammonium phosphate results in complete separation of four C-toxins. This method has shown a better correlation with the mouse bioassay for PSP toxins than other methods have (59).

An analytic method based on the postcolumn method of Oshima (59) is as follows.

Column: Reversed phase C8, 5 micron, dimensions 4.6 mm × 150 mm

Mobile-phase flow rate: 0.8ml/min

Mobile-phase composition:

C1–C4: 1 mM tetrabutyl ammonium phosphate adjusted to pH 5.8 with acetic acid

GTX1–GTX6, dcGTX2, and dcGTX3: 2 mM sodium 1-heptane sulfonate in 10 mM ammonium phosphate, pH 7.1

STX, neoSTX, and dcSTX: 2 mM sodium 1-heptane sulfonate in 30 mM ammonium phosphate, pH 7.1:acteonitrile = 10:5

Oxidizing reagent: 7 mM periodic acid in 50 mM potassium phosphate buffer, pH 9.0 at a flow rate of 0.4 ml/min

Reaction conditions: In 10 m of Teflon tubing (0.5-mm ID) at 65°C in a water bath

Acidifying reagent: 0.5 M acetic acid at a flow rate of 0.4 ml/min

Detection: Fluorescence at excitation of 330 nm and emission of 390 nm

2. Anatoxin-a and Anatoxin-a(s)

Anatoxin-a has been satisfactorily assayed by three analytical techniques: gas chromatography–mass spectrometry (GC-MS); gas chromatography with electron capture detection (GC-ECD); HPLC. With the GC-MS procedures, acetylation of the extracted compound is normally performed before the GC-MS step (60,61). The use of GC-ECD is regarded as more sensitive and generally incorporates an internal standard for more accurate quantitations (62,63). The GC-ECD method, however, requires a complicated cleanup step, and a derivitization of the toxin is still necessary. The use of pentafluorobenzyl bromide as the derivitization method with ECD permits a sensitivity as low as 2.5 pg (64). The HPLC methods involve the use of UV detection and reversed-phase columns (63,65). The use of photodiode array detection with the HPLC methods would no doubt add a degree of confirmation to the detection procedure for anatoxin-a.

Anatoxin-a(s) has proved difficult to analyze due to the lack of a suitable chromophore for HPLC analysis and lack of volatility for GC analysis. The use of HPLC-MS has been suggested as a possible procedure

(63). An analysis method involving the use of HPLC-MS/MS has been developed for the cyanobacterial toxin cylindrospermopsin (66). This procedure features high sensitivity and specificity and would no doubt be applicable to anatoxin-a(s) with suitable modifications.

3. Microcystins

a. Sample Extraction.　For the analysis of microcystins, water and cyanobacterial cellular material is normally used. With water extraction, an organic solvent is normally not necessary, but a cleanup procedure employing octadecyl reversed-phase solid-phase extraction (SPE) cartridges is often used (63,64). Preconcentration of the water is often performed by boiling, to reduce volume; alternatively, vacuum rotary evaporation can be used. With cellular material, the most effective extraction solvent is methanol (63,67). Methanol has the advantage of being a good solvent for both the highly water-soluble microcystins, such as microcystin LR, and the hydrophobic microcystins encountered in some strains of *Microcystis aeruginosa*. In samples dominated by *Microcystis* spp., a routine procedure has been established that uses 75% methanol in water for the extraction of lyophilized cellular material (68).

Two rapid procedures for microcystin extraction follow.

1. With filtered cellular material, freeze and thaw the sample to disrupt cells. Extract toxins in the filter device by passing aqueous methanol followed by methanol through the filter. Use a total of approximately 20 ml of solvent. The extract can be analyzed directly or cleaned up and concentrated by the use of SPE cartridges (69).
2. Extract membrane filters containing approximately 20 mg of freeze-dried cellular material in 2 ml microcentrifuge tubes with 1.5 ml of 75% methanol in water. Extraction is facilitated by sonication and shaking for 30 min followed by centrifugation. To achieve complete extraction of microcystins, the pellet is re-extracted twice (68).

b. Sample Cleanup and Concentration.　The purpose of the cleanup step is to remove coextracted impurities that may interfere with instrumental detection techniques. This usually results in improved detection limits. For microcystins and nodularin, ODS silica reversed-phase SPE cartridges are normally used. The SPE cartridges permit coextracted interfering compounds to pass through the column while the microcystins are retained and subsequently eluted with appropriate solvent (67,70). The procedure described next is based on that described by Harada et al. (63) and has been used with minor modifications in many studies.

1. The residue from the extraction of cellular material or the concentration of water by rotary evaporation or boiling is dissolved in 5 ml of methanol.
2. Prepare the SPE cartridge by conditioning with the passage of 10 ml of methanol followed by 10 ml of water. Note that the cartridges can be eluted with the assistance of a vacuum manifold. If this is the case, it is important to ensure that the cartridge is not permitted to go dry at any stage.
3. Pass filtered or concentrated filtered water samples through the cartridge. If a methanol extract is being cleaned up, dilute with water so that the methanol concentration is less than 10%.
4. When the sample has been loaded onto the cartridge, the cartridge is sequentially eluted with 10 ml each of 10% methanol, 20% methanol, and 30% methanol. The eluted fractions are discarded.
5. Elute the cartridge with 3 ml of 0.1% trifluoroacetic acid solution in methanol. This eluate is collected and dried under a stream of nitrogen. The sample can now be dissolved in a small quantity of methanol for instrumental analysis.

Note: Some investigators use 80% methanol in water for the final elution of the SPE cartridge to obtain the microcystins and nodularins (3).

c. Analytical Determinations. Instrumental methods most commonly used for microcystins are based on HPLC systems with either UV or, preferably, photodiode array (PDA) detection. The advantage of PDA detection is that some confidence of identification of microcystins is gained, although the identification of individual microcystins is not possible (63, 71). The use of HPLC-MS for detection has the advantage of the identification of individual microcystins (72). The use of various forms of mass spectrometry for microcystins is discussed in Kondo and Harada (73). A comprehensive account of choosing an analytical strategy for microcystins in given by Meriluto et al. (74).

The use of HPLC with PDA detection is, however, routinely practiced in many laboratories worldwide. An analytical method based on that described by Harada et al. (63) follows.

Apparatus

Gradient HPLC system with PDA detector

C_{18} reversed-phase column, such as 5-micron packing of size 4.6 $\times$ 250 mm
Column oven at 40°C

Procedure

1. Degas all solvents by filtration or other suitable procedure.
2. Establish a linear gradient system with a flow of 1 ml/min as follows:

Time (mins)	0	10	40	42	44	46	55
Eluant A, %	70	65	30	0	0	70	70
Eluant B, %	30	35	70	100	100	30	30

3. Monitor absorbance with the PDA detector from 200 to 300 nm.
4. Run blanks and calibration standards of microcystins of interest before, after, and during runs of samples.
5. Quantitation should be made using absorption at 238 nm, which is the maximum absorbance for most microcystins and nodularin.
6. Identification of microcystins is made via retention-time matching and PDA absorbance spectra. An example of the PDA spectrum of a sample together with microcystin-LR is given in Fig. 7.

4. Cylindrospermopsin

Cylindrospermopsin has only recently been structurally identified (13). The first analytical method for cylindrospermopsin in bloom samples was published in 1994 (5). This method employed HPLC with PDA detection and an ODS reversed-phase column. This method was, however, recognized by the authors as being deficient in terms of chromatographic efficiency. Recently a method employing HPLC-MS/MS has been published (66) that features rapid, sensitive analysis with confirmation. The limit of detection for cylindrospermopsin in water by this method is 0.2 μg/L, and the linearity of the assay is from 1 to 600 μg/L.

a. Sample Preparation. Water samples are prepared simply by freezing and thawing the water to lyse any cellular material. The sample is then mixed and centrifuged. Filtration of the sample through a 0.45-micron syringe filter is necessary before analysis. Cellular material is freeze-dried and extracted with methanol. The extract is diluted with water, to reduce methanol concentration to below 10%, and then centrifuged and filtered.

b. Preparation of Cylindrospermopsin Standard. The standard was prepared from lyophilized *Cylin-*

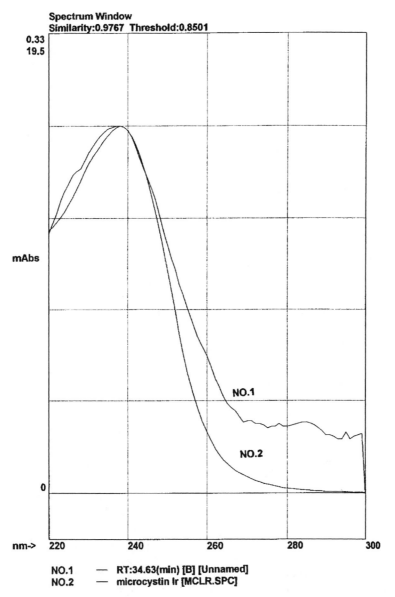

Fig. 7 Example of photodiode array spectrum of unknown sample (No. 1) and microcystin-LR (No. 2).

drospermopsis raciborskii. This material was extracted with methanol. The extract was then purified via SPE chromatography followed by preparative HPLC. The purity of the standard was confirmed by comparing the absorbance at 262 nm to literature values (13).

  c. Analysis

Column: C18 (250 × 4.6 mm, 5 micron)
Linear gradient: 60% methanol over 5 min with a final isocratic stage holding at 60% methanol for 1 min; flow 1.1 ml/min

Mobile phase: Buffered to 5 mM with ammonium acetate
Injection: 110 μl with postcolumn splitting to submit 20% of column effluent to the MS/MS interface
MS/MS: The transition from the M+H ion (416 m/z) to the 194 m/z fragment was monitored for quantitation using multiple-reactant-monitoring mode. Figure 8 shows the MS/MS spectrum of the two major product ions of cylindrospermop-

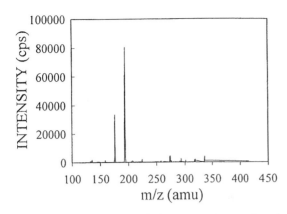

Fig. 8 MS/MS spectrum of the two major product ions from cylindrospermopsin.

sin. The ion at 194 m/z was monitored for quantitation.

E. Immunoassay Methods for Algal Toxins

The development of immunoassay methods for algal toxins provides a source of rapid and inexpensive methodology for the determination of some toxins in a variety of matrices, including water, liver, and cellular material. The most promising method that is currently available is the enzyme-linked immunosorbent assay (ELISA) (63). The use of ELISA methods, however, depends on the development of suitable antibodies for the toxin or toxins in the case of groups of toxins such as the microcystins.

1. Immunoassays for Microcystins

Most developmental work has been applied to the microcystins using both monoclonal (75) and polyclonal antibodies (76,77). With the availability of suitable antibodies, several methods have been published for the analysis of microcystins by ELISA techniques (76,78–81). Commercial kits for microcystin ELISA are available, and the evaluation of commercial kits and researcher-developed kits is currently being undertaken (82). Currently available commercial kits are listed next, although other companies are currently developing kits also:

Strategic Diagnostics Inc, Newark, Delaware
Waco Chemicals, USA Inc., Richmond, VA, or
    Waco Chemicals GmbH, Germany

The testing principles for direct competitive ELISA using polyclonal antibodies and indirect competitive

ELISA using monoclonal antibodies as defined by Carmichael et al. (3) are as given next.

a. Direct Competitive ELISA Using Polyclonal Antibody

1. Attach antimicrocystin antibody to a solid-phase device (high-binding-affinity microtiter plate).
2. Use microcystin-LR as the standard. Microcystin-LR competes with microcystin-LR enzyme conjugate (MC-LR-peroxidase or MC-LR–alkaline phosphatase) for the binding site of the antibody attached to the microtiter plate.
3. A color develops in each microtiter plate well in inverse proportion to the microcystin concentration. The color is read at 490 nm for the peroxidase conjugate and 405 nm for the alkaline phosphatase conjugate.

The competitive ELISA principle is shown diagrammatically in Fig. 9.

b. Indirect Competitive ELISA Using Monoclonal Antibodies

1. Attach microcystin-LR–bovine serum albumin to a microtiter plate.
2. Use microcystin-LR as the standard. The monoclonal antibody against microcystin-LR competes with microcystins in the sample for binding sites on the microtiter plates.
3. The color is developed with a second antibody, horse radish peroxidase conjugated goat to mouse IgG antibody. In this case the chromagen used is 3,3′,5,5′-tetramethyl benzidine (TMBZ). The intensity of the developed color is inversely proportional to the microcystin concentration.

This assay is shown diagrammatically in Fig. 10.

c. Procedure for Competitive ELISA with Polyclonal Antibody.   The procedure described next is based on that of Carmichael et al. (3).

1. Coating the plates. Measure 100 μl of dilute antibody to each well. (The total protein concentration, including antibody, is equivalent to an absorbance of 0.007 at 280 nm. Allow the plate to stand overnight at 4°C. The plate is then washed with 0.1% Tween 20 in 0.01 M phosphate-buffered saline (PBS). The plate should be washed four times and emptied after each wash.
2. Blocking the plates. Bovine serum albumin in 0.01 M PBS is added to each well at a volume of 150 μl. The plate is incubated at 37°C for 30 min. The plate is then washed four times with 0.1% Tween 20 in PBS.

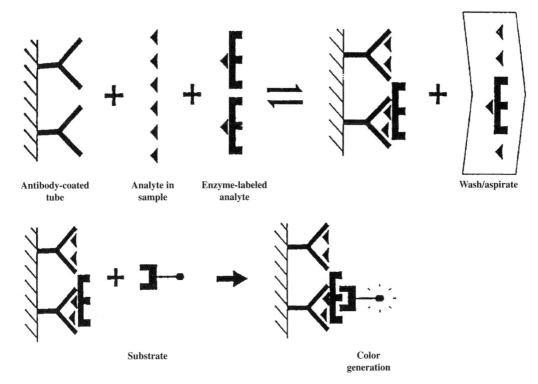

Fig. 9 Diagrammatic representation of competitive ELISA for microcystins using polyclonal antibodies and a colorimetric endpoint. (From Ref. 3.)

3. Colorimetric reaction. Add 50 μl of dilute micro-cystin–peroxidase conjugate and 50 μl of micro-cystin-LR standards at a range of concentrations from 1 to 20 μg/L. Unknown samples are added at different dilutions depending on the expected concentration range. Incubate the plate at 37°C for 1 h, and then wash six times with 0.1% Tween 20 in PBS. The final stage of the colorimetric reaction consists of the addition of 100 μL of enzyme substrate (30% hydrogen peroxide plus orthophenylene diamine) in buffer to each well to develop the color. The reaction is terminated by the addition of 100 μl of 1N HCl after 10 min.

4. Measurement of absorbance. The absorbance is measured at 490 nm using a microplate reader

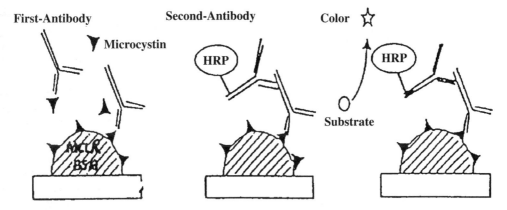

Fig. 10 Diagrammatic representation of the indirect competitive ELISA using monoclonal antibodies. HRP = horse radish peroxidase. (From Ref. 3.)

2. Immunoassays for PSP Toxins

With the PSP toxins, a number of ELISA methods have been developed but do not appear to have been applied to water and algal samples to the extent that the techniques have been adapted for microcystins. Polyclonal antibodies have been produced that cross-react with saxitoxin and neosaxitoxin (83), but many methods developed reliably detect saxitoxin but do not cross-react with the other PSP toxins (63,84). Most immunoassay systems for PSP toxins have been developed to replace the mouse bioassay for the routine monitoring of marine shellfish and are discussed in detail elsewhere (85).

F. Biochemical Effect Measurements Applied to Algal Toxins

The most studied class of cyanobacterial toxins, the microcystins, act by the inhibition of protein phosphatases 1 and 2A (86). The microcystins are potent and specific inhibitors of these protein phosphatases, and a number of radiometric and colorimetric assays have been developed that are applicable to microcystins (80,86–88). The sensitivities of the colorimetric assays correspond to limits of detection of approximately 10 ng/ml of microcystins. A good correlation between the protein phosphatase inhibition assay and HPLC methods for microcystins has been shown (86). Because nodularin also inhibits the protein phosphatases, methods estimating protein phosphatase inhibition are also applicable to nodularin. It should be noted that some methods in the literature assay the inhibition of protein phosphatases either 1 or 2A. The general principles of colorimetric protein phosphatase analysis as proposed by Carmichael et al. (3) are presented next.

1. Principles of Colorimetric Protein Phosphatase Analysis

Microcystin and nodularins covalently bind to the catalytic subunit of the protein phosphatases in an irreversible and competitive way.

Protein phosphatases will dephosphorylate p-nitrophenol phosphate (pNPP) to produce the yellow paranitrophenol, which absorbs at 405 nm. The reaction is carried out at alkaline pH (8.5).

The color production from the liberation of pNPP can be measured spectrometrically and quantitatively corresponds to protein phosphatase activity.

The concentration of microcystins or nodularin is related directly to the inhibition of protein phosphatase activity and thus inversely correlates with absorbance at 405 nm.

2. Procedure

The colorimetric procedure for the estimation of protein phosphatase activity applicable to microcystins and nodularin according to Carmichael et al. (3) is as follows.

1. Protein phosphatases 1 or 2A can be purchased from a commercial source, such as Gibco Life Technologies, Calbiochem, or Boehringer Mannheim.
2. The following is added to each well of a multiwell plate: 10 μl of unknown sample, control, or microcystin-LR standard (concentrations of microcystin 0.2, 1, 3, 10, 30 μg/L); 40 μl of diluted protein phosphatase (dilution specified by suppliers for inhibition studies).
3. Incubate at 37°C for 5 min.
4. Add 50 μl of substrate solution (40 mM pNPP in 50/50 1.5-mg/L bovine serum albumin in stock buffer/1.5 mM $MnCl_2$ in stock buffer). Stock buffer consists of: 40 mM tris HCl, 20 mM KCl, 30 mM $MgCl_2$, pH 8.6. Note: The substrate solution will start the enzymatic reaction.
5. Monitor the color production of each sample and standard at 405 nm at 5-min intervals for a total period of 40 min. The microplate reader should be set to generate the inhibition curve using the linear kinetic mode.
6. The microcystin or nodularin concentration is calculated in the samples using the standard curve generated in each run from the microcystin-LR standards.

Notes:

Extraction, concentration, and cleanup of samples before assay can be achieved by using the techniques involving SPE cartridges detailed in the instrumental section (Sec. III.D.3.b).

Run all controls, standards, and samples in triplicate. The coefficient of variation of each sample should not exceed 15%.

Interpolation of sample concentration is valid only when the sample concentration falls within the linear portion of the standard curve.

A total of 60 different microcystins have been identified to date, and most samples contain mixtures of microcystins. The protein phosphatase inhibition assay is calibrated on microcystin-LR and as such the results should be reported as microcystin-LR protein phosphatase inhibition equivalent units.

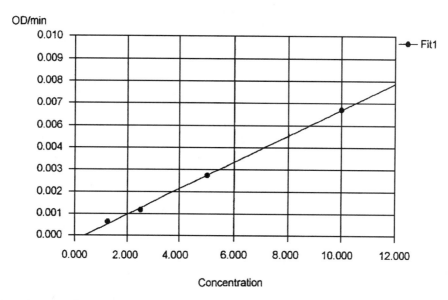

Fig. 11 Typical plot of the linear relationship between optical density and concentration for protein phosphatase 1 inhibition by microcystin-LR at concentrations ranging from 1 to 10 μg/L.

A typical graph of optical density versus concentration for protein phosphatase inhibition is presented in Fig. 11.

The colorimetric protein phosphatase inhibition assay provides a rapid measurement of the biochemical effect of microcystins and nodularin. It should be noted, however, that a comparison of the colorimetric method for inhibition of protein phosphatase 2A with the radiometric method showed that the colorimetric method has a shorter linear region; and the colorimetric procedure shows 80% inhibition of activity, while the radiometric method produces 100% inhibition (89).

methods involve the use of instrumental techniques such as HPLC with UV and photodiode array detectors and HPLC-MS/MS. These techniques are capable of analyzing the major toxins in water at limits of detection below proposed water-quality guidelines.

Rapid and sensitive analysis for various classes of toxins is being achieved via the use of immunoassay techniques such as ELISA. This technique is currently used mostly for microcystins and has not yet been developed for cylindrospermopsin. The use of biochemical effect assays such as the protein phosphatase inhibition assays provide an accurate estimation of the toxicity of water and are applicable to the microcystins and nodularin.

IV. CONCLUSIONS

Analysis for toxic blue-green algae may take many forms. In most applications, the first stage of evaluation of potential toxicity of water involves the identification and enumeration of the cyanobacteria present using microscopy. If potentially toxic species are present, bioassays can be used to determine the toxic potential of the water. Bioassays are, however, not rapid or sensitive.

In more recent years, highly precise and sensitive methods have been developed for most major cyanobacterial toxins, such as PSPs, microcystins and nodularins, anatoxins, and cylindrospermopsin. These

REFERENCES

1. W. Carmichael. *J. Appl. Bacteriol.*, *72*:445 (1992).
2. P. Hawkins, M. Runnegar, A. Jackson, I. Falconer. *Appl. Environ. Microbiol.*, *50*:1292 (1985).
3. W. Carmichael, J. An, G. Jones. *Rapid Screening Techniques for Microcystins and Nodularins*. CSIRO, Brisbane, 1998.
4. Y. Ueno, S. Nagata, T. Tsutsumi, A. Hasegawa, M. Watanabe, H. D. Park, G.-C. Chen, G. Chen, S. Z. Yu. *Carcinogenesis*, *17*:1317 (1996).

5. K. I. Harada, I. Ohtani, K. Iwamoto, M. Suzuki, M. Watanabe, M. Wantanabe, K. Terao. *Toxicon, 32*:73 (1994).

6. R. Banker, S. Carmeli, O. Hadas, B. Teltsch, R. Porat, A. Sukenik. *J. Phycol., 33*:613 (1997).

7. G. Shaw, A. Sukenik, A. Livne, R. Chiswell, M. Smith, A. Seawright, R. Norris, G. Eaglesham, M. Moore. *Environ. Toxicol., 14*:167 (1999).

8. A. Humpage, J. Rositano, A. Bretag, R. Brown, P. Baker, C. Nicholson, D. Steffensen. *Aust. J. Mar. Freshwater Res., 45*:761 (1994).

9. W. Carmichael, P. Gorham. In: *The Water Environment: Algal Toxins and Health* (W. Carmichael, ed.). Plenum Press, New York, pp. 161 (1981).

10. G. Codd, S. Bell, W. Brooks. *Water Sci. Tech., 21*:1 (1989).

11. G. Codd, G. Poon. In: *Proc. Phytochem. Soc. Europe, 1988* (J. Callon and L. Rodgers, Eds.), Oxford University Press, Oxford, 1998.

12. J. Gallon, K. Chit, E. Brown. *Phytochem., 29*:1107 (1990).

13. I. Ohtani, R. Moore, M. Runnegar. *J. Am. Chem. Soc., 114*:7941 (1992).

14. R. Norris, G. Eaglesham, G. Pierens, G. Shaw, M. Smith, R. Chiswell, A. Seawright, M. Moore. *Environ. Toxicol., 14*:163 (1999).

15. R. Endean, S. Monks, J. Griffith, L. Llewellyn. *Toxicon, 31*:1155 (1993).

16. R. Moore, P. Ragelis (ed.). *ACS Symposium Series.* American Chemical Society, Washington, DC, 1984.

17. G. M. Lasett, R. Malcolm, G. J. Jones. *Environmetrics, 8*:313 (1997).

18. G. A. Codd, I. Chorus, M. Burch. In: *Toxic Cyanobacteria in Water, A Guide to Their Public Health Consequences, Monitoring and Management* (I. Chorus and J. Bartram, eds.). E & FN Spon, London, pp. 313–328 (1999).

19. R. A. Vollenweider (ed.). *A manual on Methods for Measuring Primary Production in Aquatic Environments.* Blackwell Scientific Publication, Oxford, 1969.

20. L. Lawton, B. Marsalek, J. Padisak, I. Chorus. In: *Toxic Cyanobacteria in Water, A Guide to Their Public Health Consequences, Monitoring and Management* (I. Chorus and J. Bartram, eds.). E & FN Spon, London, pp. 346–367 (1999).

21. G. Hotzel, R. Croome. *A Phytoplankton Methods Manual for Australian Rivers.* Land and Water Resources and Development Corporation, Canberra, 1998.

22. American Public Health Association. *Standard Methods for the Examination of Water and Wastewater*, AMPH, Washington, DC, 1998.

23. J. W. G. Lund, C. Kipling, E. D. Le Cren. *Hydrobiologia, 11*:143 (1958).

24. SCOR Working Group 33. *A Review of Methods for Quantitative Phytoplankton Studies.* UNESCO, Paris, 1974.

25. G. Jones. *National Protocol for the Monitoring of Cyanobacteria and their Toxins in Surface Waters.* ARM-CANZ, Australia, 1977 (draft).

26. B. J. McAlice. *Limno. Oceanogr., 16*(1):19 (1971).

27. C. S. Reynolds, G. H. M. Jaworski. *Br. Phycol. J.* 13: 269 (1978).

28. W. Lu, E. H. Evans, S. M. McColl, V. A. Saunders. *FEMS Microbiol. Lett., 153*:141 (1997).

29. E. Murayama-kayano, S. Yoshimatsu, T. Kayano, T. Nishio, H. Ueda, T. Nagamune, *J. Fermentat. Bioeng.,* 85(3):343 (1998).

30. E. W. Wilde, W. R. Cody. *J. Freshwat. Ecol., 13*(1):79 (1998).

31. E. Willen. *Br. Phycol. J., 11*:265 (1976).

32. J. V. Tyrrell, P. R. Bergquist, D. J. Saul, L. Mackenzie, P. L. Bergquist. *N. Z. J. Mar. Freshwat. Res., 31*:51 (1997).

33. B. A. Neilan. *Am. Soc. Microbiol., 61*(6):2286 (1995).

34. B. A. Neilan. *Phycologia, 35*(6):147 (1996).

35. G. B. J. Dubelaar, H. W. Balfoort, H. W. Hofstraat. *Water Sci. Tech., 24*(10):285 (1991).

36. M. F. Wilkins, L. Boddy, C. W. Morris, R. Jonker. *CABIOS, 12*(1):9 (1996).

37. H. W. Balfoort, J. Snoek, J. R. M. Smits, L. W. Breedveld, J. W. Hofstraat, J. Ringelberg. *J. Plankton Res., 14*:575 (1992).

38. E. G. Vrieling, G. Vriezekolk, W. W. C. Gieskes, M. Veenhuis, W. Harder. *J Plankton, 18*(8):1503 (1996).

39. W. Carmichael, G. Hallegraeff, D. Anderson, A. Cembella, H. Enevoldsen (eds.). *Manual on Harmful Marine Algae.* UNESCO, Paris, Vol. 33, 1995.

40. M. Fernandez, A. Cembella, in *Manual on Harmful Marine Microalgae*, (G. Hallegraeff, D. Anderson, A. Cembella, H. Enevoldsen Eds.). UNESCO, Paris, Vol. 33, p. 213 (1995).

41. H. Sommer, K. Meyer. *Arch. Pathol., 24*:560 (1937).

42. L. Botana, M. Rodriguez-Vieytes, A. Alfonsa, C. Louazo. In: L. Nollet (ed.). *Handbook of Food Analysis.* Marcel Dekker, New York, Vol. 2, p. 1147, 1996.

43. *AOAC Official Methods of Analysis.* Assoc. Off. Anal. Chem., 1997, Vol. Suppl. March 1997.

44. K. Terao, S. Ohmori, K. Igarashi, I. Ohatani, M. Watanabe, K. Harada, E. Ito, M. Watanabe. *Toxicon, 32*: 833 (1994).

45. A. Seawright, C. Nolan, G. Shaw, R. Chiswell, R. Norris, M. Moore, M. Smith. *Environ. Toxicol., 14*:135 (1999).

46. C. Vezie, F. Benoufella, K. Sivonen, K. Bertru, A. Laplanche. *Phycologia, 35*:198 (1996).

47. J. Kivirnanta, K. Sivonen, S. Niemela, K. Huovienen. *Environ. Toxicol. Water Qual., 6*:423 (1991).

48. S. Hawser, J. O'Neil, M. Roman, G. Codd. *J. Appl. Phycol., 79* (1992).

49. J. McElhiney, L. Lawnton, C. Edwards, S. Gallacher. *Toxicon, 36*:417 (1998).

50. P. Kos, G. Gorzo, G. Suranyi, G. Borbely. *Anal. Biochem., 225*:49 (1995).

51. L. Lawton, D. Campbell, K. Beattie, G. Codd. *Lett. Appl. Microbiol.*, *11*:205 (1990).

52. K. Lahti, J. Ahtiainen, J. Rapala, K. Sivonen, S. Niemela. *Lett. Appl. Microbiol.*, *21*:109 (1995).

53. J. Eriksson, D. Toivola, M. Reinikainen, C. Rabergh, J. Meriluto. In: *Detection Methods for Cyanobacterial Toxins* (G. Codd, T. Jefferies, C. Keevil, E. Potter, eds.). Royal Society of Chemistry, Cambridge, p. 75, 1994.

54. R. Bhattacharya, P. Rao, A. Bhaskar, S. Pant, S. Dube. *Human Exptl. Toxicol.*, *15*:105 (1996).

55. R. Heinze. *Phycologia*, *35*:89 (1996).

56. M. Runnegar, S.-M. Kong, Y.-Z. Zhong, S. Lu. *Biochem. Pharmacol.*, *49*:219 (1995).

57. J. Lawrence, C. Menard. *J. Assoc. Off. Anal. Chem.*, *74*:1006 (1991).

58. J. Lawrence, C. Menard, C. Charbonneau, S. Hall. *J. Assoc. Off. Anal. Chem.*, *74*:404 (1991).

59. Y. Oshima, G. Hallegraeff, D. Anderson, A. Cembella, H. Enevoldsen. *Manual on Harmful Marine Algae.* UNESCO, Paris, 1995.

60. K. Sivonen, K. Himberg, R. Luukkainen, S. Niemela, G. Poon, G. Codd. *Tox. Assess*, *4*:339 (1989).

61. K. Himberg. *J. Chromatogr.*, *481*:358 (1989).

62. D. Stevens, R. Krieger. *Anal. Toxicol.*, *12*:126 (1988).

63. K. Harada, F. Kondo, L. Lawton. In: I. Chorus, J. Bartram (eds.). *Toxic Cyanobacteria in Water. A Guide to Their Public Health Consequences, Monitoring and Management.* World Health Organization, London, p. 369, 1999.

64. C. Bumke-Vogt, W. Mailahn, W. Rotard, I. Chorus. *Phycologia*, *36*:51 (1996).

65. K. Harada, Y. Kimura, K. Ogawa, M. Suzuki, A. Dahlem, V. Beasley, W. Carmichael. *Toxicon*, *27*:1289 (1989).

66. G. Eaglesham, R. Norris, G. Shaw, M. Smith, R. Chiswell, B. Davis, G. Neville, A. Seawright, M. Moore. *Environ. Toxicol.*, *14*:151 (1999).

67. L. Lawton, C. Edwards, G. Codd. *Analyst*, *119*:1525 (1994).

68. J. Fastner, I. Flieger, U. Neumann. *Water Res.*, *32*:3177 (1998).

69. H. Utkilen, N. Gjolme. In: G. Codd, T. Jefferies, C. Keevil, E. Potter (eds.). *Detection Methods for Cyanobacterial Toxins.* Special Publication No. 149. Royal Society of Chemistry, Cambridge, p. 168, 1994.

70. K. I. Harada, M. Watanabe, W. Carmichael, H. Fujiki (eds.). *Toxic Microcystis.* CRC Press, Boca Raton, FL, p. 103, 1996.

71. L. Lawton, K. Beattie, S. Hawser, D. Campbell, G. Codd. In G. Codd, T. Jefferies, C. Keevil, E. Potter (eds.). *Detection Methods for Cyanobacterial Toxins.* Special Publication No. 149. Royal Society of Chemistry, Cambridge p. 111, 1994.

72. D. Bourne, G. Jones, R. Blakely, A. Jones, A. Negri, P. Riddles. *Appl. Env. Microbiol.*, *62*:4086 (1996).

73. F. Kondo, K. Harada. *Journal of Mass Spectrom. Soc. Jpn.*, *44*:355 (1996).

74. J. Meriluto, A. Harmata-Brasken, J. Eriksson, D. Toivola, T. Lindholm. *Phycologia*, *35*:125 (1996).

75. R. Kfir, E. Johannsen, D. Botes. *Toxicon*, *24*:543 (1986).

76. F. Chu, X. Huang, R. Wei. *J. Assoc. Anal. Chem. 73*: 451 (1990).

77. S. Nagata, H. Soutome, T. Tsutsumi, A. Hasegawa, M. Sekijima, M. Sugamata, K. I. Harada, M. Suganuma, Y. Ueno, *Natural Toxins*, *3*:78 (1995).

78. Y. Ueno, S. Nagata, T. Tsutsumi, A. Hasegawa, M. Watanabe, H. D. Park, G.-C. Chen, G. Chen, S. Z. Yu. *Carcinogenesis*, *17*:1317 (1996).

79. S. Nagata, T. Tsutsumi, A. Hasegawaa, F. Yoshida, Y. Ueno. *Am. Org. Anal. Chem.*, *80*:408 (1997).

80. J. An, W. Carmichael. *Toxicon*, *32*:1495 (1994).

81. W. Brooks, G. Codd. *Environ. Technol. Lett.*, *9*:1343 (1988).

82. G. A. Codd: personal communication (1999).

83. J. Kralovec, M. Laycock, R. Richards, E. Usleber. *Toxicon*, *34*:1127 (1996).

84. S. Bell, G. Codd. *Iss. Environ. Sci. Technol.* 5:109 (1994).

85. A. Cembella, L. Milenkovic, G. Doucette, M. Fernandez. In: *Manual on Harmful Marine Microalgae* (Hallegraeff, D. Anderson, A. Cembella, eds.). UNESCO, Paris, IOC Manuals and Guides No. 33, p. 177 (1995).

86. C. Ward, A. Beattie, E. Lee, G. Codd. *FEMS Microbiol. Lett.*, *153*:465 (1997).

87. C. Holmes. *Toxicon*, *29*:469 (1991).

88. T. Lambert, M. Boland, C. Holmes, S. Hrudey. *Environ. Sci. Technol.*, *28*:753 (1994).

89. P. Lam: personal communication (1999).

10

Halogens

Geza Nagy
Janus Pannonius University, Pecs, Hungary

Livia Nagy
Research Group of Hungarian Academy of Sciences, Budapest, Hungary

I. INTRODUCTION: PHYSICAL AND CHEMICAL PROPERTIES

The halogens—fluorine, chlorine, bromine, iodine, and the astatine—form the 17 (or VIIa) column of the periodic table of elements. All astatine isotopes are radioactive. ^{210}At is the most stable of them, with a half-life of 8.3 hours. Therefore astatine does not appear in nature or in water samples. Its quantitative determination does not have practical importance. The halogens are reactive nonmetals with high electron negativities and electron affinities, forming diatomic molecules in the elemental stage. They can form a very large number of inorganic and organic compounds. Most of the halides can be classified into two categories. The fluorides and chlorides of many metallic elements, especially those belonging to the alkali metal and alkaline earth metal (except beryllium) families, are ionic compounds. Most of the halides of nonmetals such as sulfur and phosphorus are covalent compounds. Fluorine, being the most electronegative of all reactive elements, occurs with only 0 and −1 oxidation numbers. Chlorine, bromine, and iodine can have oxidation numbers of −1, 0, +1, +3, +5, +7 in ions or molecules. Table 1 summarizes some of the most important characteristics of the stable halogens.

In nature, because of their high reactivity, the halogens are always found combined with other elements.

Chlorine, bromine, and iodine occur most often as halides in seawater, in soil, and in minerals. Chloride is a major anionic component of the biomass. Fluorine occurs in sparingly soluble mineral deposits, such as fluorite and fluorspar (CaF_2), cryolite (Na_3AlF_6), and fluorapatite ($Ca_5(PO_4)_3F$). The most easily oxidized halogen element, iodine, is also found in iodates.

The halogens are toxic materials. Their toxicity together with their reactivity decreases from fluorine to iodine.

Except for astatine, the halogens are produced on industrial scale and used as reagent or oxidizing agents. Chlorine production is by far the largest. It is accomplished by electrochemical oxidation of aqueous sodium chloride solutions. Fluorine, however, cannot be obtained by electrochemical oxidation of aqueous solutions. Water decomposition would come at lower potential than the oxidation of fluoride. Even if an electrode with high overpotential could be found, the evolved fluorine would react immediately with the water content of the electrolysis cell. Therefore it is produced by electrolysis of KF dissolved in HF, as Henri Moissan worked it out in the end of the 19th century.

The chemical oxidation of bromide- or iodide-containing solutions such as seawater and some brines are used for bromine and iodine production. In this procedure chlorine is used as oxidizing agent.

Table 1 Characteristics of the Stable Halogens

Property	Fluorine	Chlorine	Bromine	Iodine
Melting point (°C)	−223	−102	−7	114
Boiling point (°C)	−187	−35	50	183
Appearance	Pale yellow gas	Yellow-green gas	Red-brown liquid	Dark violet vapor, dark metallic-looking solid
Ionization energy (kJ/mol)	1680	1251	1139	1003
Electronegativity	4.0	3.0	2.8	2.5
Standard reduction potential (V)	2.87	1.36	1.07	0.53
Bond energy (kJ/mol)	150.6	242.7	192.5	151.0

II. DETERMINATION OF HALOGENS AND THEIR DERIVATIVES IN WATER ANALYSIS

There are a few halogen-related inorganic species that are regularly analyzed in the practice of water analysis. Of the elemental halogens, fluorine is not stable in aqueous solutions. Therefore its analysis in water samples has no meaning.

Most often, chlorine is used as reagent or as oxidizing agent. Therefore it often appears in industrial effluents. Because it is very toxic to most microorganisms and its residues are not very toxic to men, chlorine is often used as disinfectant in drinking water treatments, in swimming pools, cooling waters, etc. It is a very strong oxidizing agent that reacts fast with reducing materials or with unsaturated organic molecules in the water. Therefore this "chemical chlorine demand" must be satisfied first to have chlorine excess for the disinfection. The detection or quantitative analysis of the excess or active chlorine is an everyday task in water treatment plants or other places using chlorine as disinfectant or as additive to control the growth of microorganisms.

Total active chlorine means the total amount of oxidative chlorine compounds that liberate iodine in slightly acidic media. *Free active chlorine* content is the sum of the amounts of hypochlorous acid (HOCl), hypochlorite ions (OCl⁻), and dissolved chlorine gas in the sample. The part of total active chlorine that is present in a form different than free active chlorine (e.g., chloramines) is called *bound active chlorine.*

Bromine is also a powerful industrial reagent. Since its residues are less irritating to the eyes, it is sometimes also used in swimming pools as disinfectant. Therefore bromine concentration measurements in water samples have practical importance.

Iodine has been used also for the disinfection of swimming pool water. The concentration range in which it shows strong bactericidal, virucidal, or amebicidal action is $5-50$ mg/cm³, leaving a $0.2-0.6$-mg/dm³ residual iodine concentration. Traces of iodine or iodide ions are found in raw waters. Iodine is an essential trace element for humans. The adult daily requirement of iodine or iodide is $80-150$ μg. The iodide content of drinking water is sometimes checked to decide on the amount of supplement needed. Iodide is usually supplied as a table salt additive in areas where it is needed. As a matter of fact, the determination of iodine or iodide content of water samples is not needed too often. However, the titration or back-titration of iodine content is an everyday trick in iodometric analytical procedures.

The halides are the most stable and abundant forms of the halogen elements. Many of the ionic halides are well soluble in water. Their concentration in water samples can be an important quality-determining parameter. The determination of halides in different kinds of water samples is an important analytical task. Many methods based on different principles and applicable to different concentration ranges have been worked out for the detection and quantitative determination of the concentration of the four halide ions in different water samples.

Fluorides appear in measurable concentration in wastewater of the aluminum, glass, and electronic industries as well as in some mineral waters. The fluoride content of the drinking water was found beneficial in inhibiting the tooth decay. Therefore it is recommended to add NaF up to a concentration of 1 mg fluoride/dm³. Above this, yellow spots show up in the teeth or other unwanted health problems can occur. Therefore, the checking of the fluoride content of the drinking water must be done with a high frequency.

The chloride ion concentration of natural water samples is usually high. In drinking water, 80–100-mg/dm^3 chloride concentration is ideal; however, 250–350-mg/dm^3 concentration can be accepted for potable water quality. High chloride content makes the taste unpleasant. Industrial and agricultural waste can increase the chloride concentration considerably. The holobity is an important characteristic of water used to characterize the dissolved inorganic content. The concentrations of four cationic and four anionic components are measured and used to describe this. Chloride is one of the four anions. However, in some water samples, e.g., boiler feed waters, chloride concentration is very low, making analysis difficult.

Bromide and iodide ions appear in mineral waters and in seawater samples in low concentration. Industrial effluents can contain higher concentrations of these ions.

Household bleach contains sodium hypochloride, so it is an active, decomposing waste constituent. In water treatment plants the more expensive chlorine dioxide is sometimes used as disinfectant to avoid any bad taste of the water. It is instable, so it is usually prepared in situ by reacting sodium chlorite with chlorine or hydrochloric acid. Thus it is of practical importance to include the analysis of these species in any water analysis. Small quantities of other oxohalides can be formed during water treatment as by-products if ozone is used for disinfection. The presence of oxohalides in drinking water can be a high health risk so the analysis of them is recommended. The tolerable oxohalide content in drinking water is at the ppb level.

A. Active Chlorine

Active chlorine detection or determination is usually carried out in waters to which chlorine or hypochlorite has previously been added, to check for any excess of this disinfecting agent. The analysis should be carried out immediately after sampling, at the spot. Strong illumination and agitation should be avoided after sampling and during analysis. If immediate analysis cannot be made, drinking water samples can be stored in completely full dark bottles in a refrigerator for 3 hours.

1. Detection

Mostly colorimetric methods are used for detection. They are based on the color change of an organic reagent upon oxidation.

a. *o*-Tolidine Test. *o*-Tolidine (3,3′-dimethyl-4,4′-diamino-difenyl) solution is colorless. After reaction with free chlorine it turns to yellow.

Procedure: 10 cm^3 of the sample is pipetted into a small test tube. To eliminate the interference of iron(III) ions, 2–3 drops of $\frac{1}{3}$ M phosphoric acid is added. After this, 2–3 drops of *o*-tolidine reagent is added. The color of the sample is compared to a reagent blank. Yellow indicates the presence of free chlorine. The detection limit is 0.05 mg/dm^3 chlorine. Strong oxidizing agents interfere, giving the same color change. The *o*-tolidine reagent is prepared by dissolving 2 g *o*-tolidine in 20 cm^3 1M hydrochloric acid and diluted to 1 dm^3 by distilled water.

b. Methyl Orange Test. Free chlorine reacting with the methyl orange decomposes it, making the solution colorless.

Procedure: 10 cm^3 of the sample is pipetted into a small test tube. 2–3 drops of 1M hydrochloric acid is added to acidify and then 0.1 cm^3 methyl orange indicator solution (0.01%) is pipetted to it. The color is compared with the color of a reagent blank prepared by substituting the sample with distilled water. A solution that is colorless or a lighter color in the sample tube indicates free chlorine. Strong oxidizing agents interfere, giving the same color change.

2. Quantitative Determination of Free Chlorine in Water Samples

Different methods worked out to measure free chlorine concentration mostly take advantage of its strong oxidizing character. A broad scale of volumetric and coulometric titrations with different endpoint detection, as well as voltammetric and colorimetric methods, have been worked out. In practice, water analysis often involves classical titrimetric procedures, such as titration with arsenous acid or an appropriate iodometric approach.

a. Volumetric Determination by Arsenous Acid Reagent. The chlorine in an aqueous solution oxidizes the arseneous acid in a quantitative reaction as follows:

$$Cl_2 + H_2O = HOCl + HCl$$

$$H_3AsO_3 + HOCl = H_3AsO_4 + HCl$$

So the free active chlorine content is titrated with arseneous acid reagent. Different endpoint detection techniques can be used here. The simplest is to prepare the titrant with a small amount of a color indicator dissolved in it. Methyl orange is a good choice for this. As long as the chlorine is in excess, it destroys the indicator. The endpoint is indicated by the orange color. It shows up more sharply if potassium bromide is added to the water sample. Then the free chlorine stoichiometrically liberates an equal quantity of bromine according to the reaction

$$Cl_2 + 2KBr = 2KCl + Br_2$$

and the bromine reacts faster with the methyl orange than does the chlorine.

Strong oxidizing agents and many organic molecules interfere with this method. It is suitable to determine $0.1–20$ mg/dm^3 chlorine.

Procedure: 100 cm^3 of the sample is introduced into the titration flask, a few potassium bromide crystals are added to it, and after dissolution the arsenous acid titrant is added to the intensively agitated solution as long as the color of the indicator remains stable.

If the organic content of the sample is high, then a certain volume of the titrant is added to the titration flask, together with potassium bromide crystals, and the sample is added from a burette. The disappearance of the color of the methyl orange indicates the endpoint.

The following formula is used to calculate the chlorine concentration:

$$Cl_2 \ [mg/dm^3] = \frac{a \times 0.1 \times 1000}{V}$$

where a is the volume of the titrant solution, in cm^3, and V is the volume of the water sample, in cm^3 (100 cm^3).

Reagents:

Hydrochloric acid: 10% (specific gravity = 1.05 g/cm^3)
Methyl orange solution, prepared by dissolving 0.1216 g methyl orange color indicator in 1 dm^3 distilled water
Arseneous acid, 0.25 M
  Preparation: 4.945 g arsenious oxide (As_2O_3) and a calculated amount (6.0 g) of sodium hydroxide are quantitatively introduced into a 1-dm^3 volumetric flask, dissolving it in $60–70$ cm^3 distilled water. A few drops of methyl orange indicator are added, and the solution is acidified with 10% HCl (the hydrochloric acid solution is added as long as the solution turns to red) and the volumetric flask is filled up with distilled water.
Arsenous acid titrant
  Preparation: 25.4 cm^3 of the 0.25 M arseneous acid solution and 30.3 cm^3 methyl orange

solution are pipetted into a 100-cm^3 volumetric flask, and the flask is filled up with 10% HCl; 1 cm^3 of this titrant measures 0.1 mg active chlorine.

b. *Iodometric Determination of the Active Chlorine.* Active chlorine gas liberates a stoichiometric amount of iodine from acidic iodide solutions:

$$Cl_2 + 2KI = 2KCl + I_2$$

The iodine can be titrated volumetrically with sodium tiosulfate reagent. In the classical procedures starch color indicator is used. However, amperometric, deadstop, bipotentiometric, and many other instrumental endpoint location techniques have been worked out and have been used in practice. Any component liberating or consuming iodine would interfere. To avoid interferences of nitrite ions, iron, or manganese, the determinations are usually carried out in dilute acetic acid (pH $3–4$). This method is very accurate in the case of high chlorine concentrations (>1 mg/dm^3). It measures the total active chlorine content. The lower limit of determination is 0.05 mg/dm^3 active chlorine. In the case of high organic content [chemical oxygen demand (COD) > 6 mg/dm^3], the iodine consumption can cause serious error, so applying the method is not recommended.

Procedure: A 1000-cm^3 sample volume is used if the expected active chlorine concentration is below 1 mg/1000 cm^3. In the range of $1–10$ mg/1000 cm^3, a 500-cm^3 sample volume is recommended. And for higher chlorine concentrations, the sample volume can be further decreased.

Five cm^3 glacial acetic acid and 1 g potassium iodide are added to the sample solution, measured into a right-size titration flask. To be able to see the color change clearly, a bright white background should be used under the flask. Using continuous agitation, sodium thiosulfate titrant is added gradually until the solution turns to light yellow. Then 5 cm^3 starch solution is added, and the dark blue solution is further titrated until the blue color disappears. If the sample is light yellow after the potassium iodide addition, then the starch indicator is added initially before the titration. It is highly recommended to titrate the same volume of distilled water as of blank solution and to use the obtained blank volume for correction.

The following equation is used to obtain the total active chlorine content of the water sample:

$$Cl_2 \ [mg/dm^3] = \frac{(a - b) \times f \times 354.5}{V}$$

where

a = volume of sodium thiosulfate titrant solution added in titrating the sample (cm^3)

b = volume of sodium thiosulfate titrant solution added in titrating the blank (cm^3)

V = volume of water sample (cm^3)

f = factor of the titrant

Reagents:

Glacial acetic acid

Potassium iodide crystalline solid

Sodium thiosulfate stock solution (0.1 M)

Preparation: 25.0 g Na$_2$S$_2$O$_3 \cdot$5H$_2$O is dissolved in freshly boiled, cold distilled water; 0.2 g sodium carbonate and 10 cm^3 *i*-buthyl or -amyl alcohol is added and filled up to 1000 cm^3. The solution is stored in a dark bottle.

Sodium thiosulfate titrant solution (0.01 M)

Preparation: 100 cm^3 of the stock solution is diluted to 1000 cm^3 with freshly boiled distilled water. Its concentration (factor, f) is checked titrating 20 cm^3 standard KH(IO$_3$)$_2$ solution after the addition of 80 cm^3 distilled water, 2 cm^3 1:1 sulfuric acid, 1 g potassium iodide, and, when the yellow color turns light, 2 cm^3 starch indicator solution. $f = 20/F$, where F is the volume of the titrant needed to titrate the standard (in cm^3).

KH(IO$_3$)$_2$ stock solution (8.33 mM)

Preparation: 3.247 g anhydrous KH(IO$_3$)$_2$ is dissolved in 1000 cm^3 distilled water. It can be stored for 6 months.

KH(IO$_3$)$_2$ standard solution (0.833 mM): 10 cm^3 KH(IO$_3$)$_2$ stock solution is diluted to 100 cm^3 with distilled water starch indicator solution:

Preparation: 5 g starch (from potato) is mixed well with 20 cm^3 distilled water; 1000 cm^3 boiling water containing 1 g dissolved salicylic acid is added to it. The clear part of the solution is used as indicator.

c. Diethyl-*p*-phenylene-diamine (DPD)–Based Photometric Method. This method can be used for the determination of both the free and the total active chlorine content in water samples. The active chlorine reacting with DPD produces a red-colored compound. The absorbance at about 515 nm of this red component is in a well-defined dependence with the active chlorine concentration of the sample in the range of 0.03–4.0 mg/dm^3. In the absence of iodide ions, only the free active chlorine reacts with the DPD reagent; in the presence of iodide ions, the total amount of the active chlorine takes part in the reaction. The lower limit of

determination and the broad scale of applicability are great advantages of the method. The drawback is that photometric apparatus must be used in the field.

Bromine, iodine, and bromoamines interfere, giving the same-colored product with DPD. Oxidizing agents such as ozone, chlorine dioxide, permanganate, iodate, chromate, and MnO$_2$ interfere if their concentration exceeds 0.03 mg/dm^3. The method cannot be used if ozone or chlorine dioxide had been used for treating the sample source water. The MnO$_2$ interference can be eliminated if a special blank solution is prepared and used for compensation. To prepare this blank, 5 cm^3 buffer solution, one crystal of potassium bromide, and 0.5 cm^3 sodium arsenite solution (500 mg NaAsO$_2$ dissolved in 100 cm^3 distilled water) are added to 100 cm^3 of the sample. The active chlorine content reacts with the sodium arsenite. Then 5 cm^3 of the DPD reagent is added to the solution, and it is used as reagent blank. In the case of colored samples, 5 cm^3 buffer solution is added to 100 cm^3 sample solution, and it is used as blank. Ethylenediaminetetraacetic acid (EDTA) can decrease the interference of the heavy-metal ions; the positive error caused by the monochlor-amine in free active chlorine determinations can be eliminated by adding thioacetamide.

Procedure for Determination of Free Active Chlorine: Five cm^3 buffer solution, 5 cm^3 DPD reagent, and 100 cm^3 of the water sample are introduced into an Erlenmeyer flask and homogenized with a short, intense shock. Immediately after this, if the presence of bound active chlorine cannot be excluded, 0.5 cm^3 thioacetamide solution (2.5 g/dm^3) is added and the solution homogenized. The absorbance is measured immediately, at the absorbance maximum of around 515 nm. The accurate determination of the wavelength of the absorbance maximum should be made with the spectrophotometer appropriate to the conditions employed. The blank solution is placed into the reference cuvette, or its absorbance value is used for correction. If the free active chlorine concentration is higher than 4 mg/dm^3, then the analysis has to be made with diluted sample. Above a 10 mg/dm^3 sample concentration, the application of iodometric titration is recommended.

Procedure for Determination of Total Active Chlorine Content: Five cm^3 buffer solution, 5 cm^3 DPD reagent, about 1 g potassium iodide, and 100 cm^3 of the water sample are introduced into an Erlenmeyer flask and homogenized. The absorbance is measured after a 2-min waiting time, as described earlier. If the total active chlorine concentration exceeds 4 mg/dm^3, then the sample has to be diluted before analysis.

Above a 10-mg/dm^3 sample concentration, the application of the iodometric method is recommended.

The active chlorine measurements are evaluated using an absorbance–concentration calibration curve prepared with potassium permanganate calibrating solutions. In order to prepare the calibration curves, solutions corresponding to active chlorine concentrations of 0, 0.1, 0.2, 0.5, 1.0, 2.0, 3.0, and 4.0 mg/dm^3 are prepared by diluting 0, 1.0, 2.0, 5.0, 10.0, 20.0, 30.0, and 40.0 cm^3 potassium permanganate calibrating solution, respectively, to 100 cm^3 with distilled water. Five cm^3 buffer solution and 5 cm^3 of the DPD reagent are added to each of these solutions (100-cm^3 volume) and the photometric measurements are carried out. The absorbance is plotted against the corresponding chlorine concentration value. The free or total active chlorine concentration is obtained from the calibration curve. The bound active chlorine is the difference between these two.

Reagents:

Buffer solution, pH = 6.5
    Preparation: 24 g anhydrous disodium-hydrogen-phosphate (Na$_2$HPO$_4$) and 46 g anhydrous potassium-dihydrogen-phosphate (KH$_2$PO$_4$) are dissolved in distilled water; 100 cm^3 0.02 M EDTA solution and 0.020 g mercury(II) chloride (HgCl$_2$) are added to it and the volume filled up to 1000 cm^3.
DPD reagent
    Preparation: 1.1 g anhydrous *N,N*-diethyl-*p*-phenylene-diamine-sulfate (DPD sulfate) is dissolved in the mixture of 250 cm^3 distilled water, 2 cm^3 concentrated sulfuric acid, and 25 cm^3 0.02 M EDTA solution and filled up to 1000 cm^3. If in a dark bottle, it can be stored for a month or as long as it is colorless.
Potassium permanganate calibrating solution
    Preparation: 10.0 cm^3 of a stock solution of 0.891 g potassium permanganate dissolved in distilled water (kept in a dark bottle) is diluted freshly to 1000 cm^3 with distilled water; 1 cm^3 of this calibrating solution corresponds to 10 μg active chlorine.

d. Diethyl-*p*-phenylene-diamine (DPD)–Based Volumetric Method. This method can be used for the determination of both free and total active chlorine content in water samples. As mentioned before, active chlorine reacting with DPD produces a red-colored compound. Ferrous ammonium sulfate (Fe(NH4)$_2$(SO$_4$)$_2$) reacts with this red compound in a quantitative reaction

resulting in a colorless product. So the red compound can be titrated with ferrous ammonium sulfate reagent solution using visual endpoint detection. In the absence of iodide ions, free active chlorine can be measured; in the presence of iodide ions, total active chlorine content can be measured in this way. The lower limit of measurement is 0.5 mg/dm^3. The titration can be carried out easily at the sampling site.

The interferences caused by the presence of oxidizing agents were discussed before. The titration can be carried out in samples with low MnO$_2$ or CrO$_4^{2-}$ concentrations.

Procedure: 100 cm^3 sample solution, 5 cm^3 DPD reagent solution, and 10 cm^3 buffer solution are mixed together in a titration flask. Immediately following this, the pinkish solution is titrated with ferrous ammonium sulfate reagent solution until the color disappears, to measure the free active chlorine content (reagent volume A cm^3). At the equivalence, 1 g crystalline potassium iodide is added. If the pinkish color returns, the titration is continued after a 2-min waiting time. The total reagent volume (B cm^3) reflects the total free active chlorine content of the sample.

In case of small bound active chlorine content ($B - A < 5$ cm^3), volume A correctly reflects the free chlorine. In the other case, the free active chlorine titration must be repeated with thioacetamide addition. Then 5 cm^3 DPD reagent solution and 10 cm^3 buffer solution are mixed together in a titration flask, 100 cm^3 sample solution is added to them; immediately after, a certain volume of the thioacetamide solution is added to the mixture. The volume needed to eliminate the interference of chloramines depends on the value of $B - A$, as shown in Table 2. The homogenized pinkish solution is titrated with ferrous ammonium sulfate reagent solution until the color disappears; 1 cm^3 of the titrant added corresponds to 1 mg/1000 cm^3 active chlorine.

Reagents: Most of the reagents needed are as described earlier.

Table 2 Volume of Thioacetamide Needed to Eliminate Chloramine Interference

$B - A$ (cm^3)	Volume of thioacetamide solution (cm^3)
<10	0.5
10–20	1.0
20–30	1.5
30–40	2.0

Ferrous ammonium sulfate (Fe(NH4)$_2$(SO$_4$)$_2$)

Preparation: 0.553 g Fe(NH4)$_2$(SO$_4$)$_2 \cdot$ 6H$_2$O is dissolves in freshly boiled, cold distilled water; 1 cm^3 1:1 sulfuric acid is added and filled up to 500 cm^3. The solution must be made fresh every day; 1 cm^3 of it measures 0.1 mg active chlorine.

e. *o-Tolidine–Based Photometric Method.* As described earlier, active chlorine reacting with colorless *o*-tolidin (3,3′-dimethyl-4,4′-diamino-difenyl) forms a yellow product. The absorbance of this product at the absorbance maximum (about 435 nm) is in a well-defined dependence with the total active chlorine concentration in the range of 0.01–2.0 mg/dm^3. In acidic media containing manganese(II) ions, the reaction is fast and straightforward. The lower limit of determination with this method is very small; the need for a spectrophotometer (which is not always available at the sample source) is a drawback.

Oxidizing components such as chlorine dioxide, ozone, bromine, iodine, Fe(III), Cr(VI), Mn(IV), MnO$_4^-$, and NO$_2^-$ interfere, as do yellow-colored dissolved components or turbidity. In the presence of Fe(III), Mn(IV), and NO$_2^-$, the use of a reagent blank is suggested for compensation in the photometric measurements. It is prepared by mixing 5 cm^3 sodium arsenite (5 g/dm^3), 5 cm^3 manganese(II) sulfate solutions, 100 cm^3 water sample, and finally 5 cm^3 *o*-tolidine reagent. The method cannot be used if ozone or chlorine dioxide had been used for treating the water.

Procedure: Five cm^3 manganese(II) sulfate and 100 cm^3 water sample is mixed together in an Erlenmeyer flask; 5.0 cm^3 *o*-tolidine reagent is added to it and homogenized. After a 5-min waiting time, but within 15 min, the absorbance is measured at about 435 nm. The wavelength of the absorbance maximum needs to be determined in the actual conditions.

If the total active chlorine content is higher than 2 mg/dm^3, then the analysis has to be repeated with diluted sample. A calibration curve is used for the evaluation. In order to prepare the calibration curves, solutions corresponding to active chlorine concentrations of 0, 0.05, 0.1, 0.2, 0.5, 1.0, 1.5, and 2.0 mg/dm^3 are prepared by diluting 0, 0.5, 1.0, 2.0, 5.0, 10.0, 15.0, 20.0 cm^3 potassium permanganate calibrating solution, respectively, to 100 cm^3 with distilled water. These calibrating solutions are handled as the water samples, and by plotting the absorbance values against the corresponding active chlorine concentration, the calibration curve is prepared. The total active chlorine concentration is taken from the calibration curve.

Reagents:

Manganese(II) sulfate solution

Preparation: 3.1 g MnSO$_4 \cdot$ H$_2$O is dissolved in about 200 cm^3 distilled water containing 3 cm^3 concentrated sulfuric acid and filled up to 1000 cm^3. The reagent can be stored without decomposing in a well-closed bottle.

Ortho-tolidine (3,3′-dimethil-4,4′-diamino-difenyl) solution

Preparation: 1.35 g *o*-tolidine-hydrochloride is dissolved in 500 cm^3 distilled water, and a mixture of 350 cm^3 distilled water and 150 cm^3 concentrated hydrochloric acid is added to it. The solution can be stored for 6 months in the dark.

Potassium permanganate calibrating solution

Preparation: as given before.

In other varieties of the photometric methods, the calibrating solutions are prepared from sodium hypochlorite. In this case a concentrated solution is first prepared, and its active chlorine concentration is determined with iodometric titration.

Residual Chlorine Measurements. A flow injection analysis (FIA) method with spectrophotometric detection has been described (1) for the routine analysis of the residual chlorine content of tap water samples. With it a measuring frequency of two analyses/min could be used. In the procedure, as little as 30 μl of sample is injected into a carrier stream of H$_2$O flowing at 4.2 ml/min in a flow injection system and treated with 4.5 μM Rhodamine 6G(I) containing 0.32 M HCl in a mixing coil before measuring the color fading of I at 524 nm. The calibration graph is linear in the range of 0.05–0.8 mg/dm^3 chlorine. With little interference, recoveries were 91–92% with a relative deviation of standards (RDS) of 0.8–1.3%.

Beltz and coworkers (2) built a fiber-optic-based residual chlorine monitor. The "smart sensor" consisted of a computer-controlled deuterium light source and an optical flow through an Al-coated capillary detection cell and a differential absorption UV spectrometer. The equipment utilized improved on UV performance optical fiber. At pH 9, chlorine and hypochlorous acid were detected as OCl$^-$. The detection limit was found to be 0.2 mg/dm^3 for dissolved chlorine.

Different colorimetric test kits are available for the estimation of active chlorine. Bosch and coworkers (3) compared the performance of different colorimetric reagents for the determination of residual chlorine in water samples. The 3,3′,5,5′-tetramethylbenzidine reagent produced the best results in the pH range of 1–2. In this case the absorbance was measured at 450 nm and the detection limit was 2 ng/cm^3.

Pantaler and coworkers (4) proposed a reactive indicator paper for semiquantitative active chlorine determination in potable water. The indicator paper was made by consecutive treatment of ordinary filter paper with EDTA and Michlers thioketone solutions. To determine active chlorine, a drop of water sample was added onto a strip of indicator paper, and the color produced after a few seconds was compared with a color scale. With the stripes, 0.1–3 mg/dm^3 active chlorine in drinking water samples could be estimated.

g. Chemical Chlorine Demand, or Chlorine Binding Capacity. Microorganisms and other components in water consume some chlorine gas. This property has to be measured to know how much disinfectant is needed for water treatment. The chlorine-binding capacity is expressed as mg/dm^3.

In the determination, increasing amounts of chlorine water are added to the water sample; 10 minutes later, potassium iodine is added. According to the following equation, a stoichiometric amount of iodine is liberated from the sample in the case of chlorine excess:

$$Cl_2 + 2I^- = I_2 + 2Cl^-$$

The liberated iodine can be titrated volumetrically with sodium thiosulfate reagent.

Procedure: 500-cm^3 water samples are introduced into dark glass containers and an increasing amount of chlorine water is added to them, in order to achieve a chlorine excess of 0.2–0.3 mg. Then 1 g potassium iodide is added to each flask, mixed, and the mixture stored for 10 min in the dark. The solution with the appropriate iodine content is transferred to a titration flask and titrated in the presence of starch indicator with 0.01 M sodium thiosulfate titrant solution. To be able to see the color change, a bright white background should be used under the titration flask.

The following formula is used to calculate the chemical chlorine demand, or the chlorine binding capacity:

$$Cl_2(mg/dm^3) = (a - b \times 0.355) \times 2$$

where

a = chlorine amount added to the sample (mg)
b = volume of 0.01 M sodium thiosulfate titrant added to titrate the sample (cm^3)

Reagents:

Sodium thiosulfate titrant solution (0.1M)
   Preparation: as described earlier.
Starch indicator solution

Preparation: as described earlier.
Potassium iodide, crystalline solid
Chlorine water (0.5 mg/cm^3)
   Preparation: Chlorine gas is bubbled through distilled water in a slow gas stream as long as a chlorine concentration of more than 0.5 mg/cm^3 is achieved. The actual chlorine concentration is determined via iodometric titration. For this, 50 cm^3 of chlorine water is measured into the titration flask, and 1 g potassium iodide is added, mixed well, and kept in the dark for 10 min to complete the reaction. The solution is titrated in the presence of starch indicator with 0.1 M sodium thiosulfate titrant solution. (1 cm^3 of 0.1 M sodium thiosulfate is equivalent to 3.55 mg of chlorine.) The concentration of the chlorine water is adjusted to achieve the active chlorine content of 0.5 mg/cm^3 with distilled water.

The chlorine excess (free chlorine) can also be determined by the previously described *o*-tolidine–based photometric method.

h. Free Chlorine Determination. Linear potential sweep voltammetry with a wax-impregnated carbon electrode could be used for the determination of free chlorine at the ng/dm^3 level. The pH change does not affect the results, since during determination the equilibrium between free and bound chlorine is not disturbed and the sum of the $NClO - ClO^-$ is measured (5).

Saunier and Regnier (6) describe a continuously operating amperometric apparatus for the measurement of free chlorine, hypochlorous acid, and combined chlorine in water samples. The apparatus contains two amperometric measuring units. One measures the total chlorine in untreated water; the other measures the combined chlorine content in water to which NO_3^- has been added as reducing agent. The free chlorine can be calculated from the difference.

Constant-current potentiometry seems to have the advantage of error-free operation, as compared to the conventional amperometric endpoint detection in the case of chlorine determinations in water. Barbolani and coworkers (7) employed 1-μA DC current between two identical platinum electrodes and measured the potential difference between them to detect the endpoint of the titrations. Chlorine was titrated with phenylarsine oxide at pH 7; chlorine and chlorine dioxide were titrated analogously in the presence of iodide ions; and all three components were titrated at pH 2 in the presence of iodide. The method was used for water samples (taken from a water purification plant) containing both chlorine and chlorine dioxide.

B. Chlorine Dioxide (ClO$_2$)

Chlorine dioxide is a greenish yellow gas with an irritating odor. It can form an explosive mixture with air. It is highly soluble in water, and a concentrated solution is stable in a closed container. Chlorine dioxide is used as a disinfectant in water treatment processes as a substitute for chlorine. Its advantage over chlorine is that it does not react with ammonia and does not produce trihalomethanes or chloramines. The recommended maximum dosage in drinking water production ranges between 0.3 and 1 mg/dm^3.

1. Chlorine Dioxide Determination

Water samples taken for chlorine dioxide analysis must be analyzed immediately after sampling. Most of the methods worked out for free chlorine measurements can be used for chlorine dioxide analysis if other oxidizing agents are not present.

a. Photometric Determination of Chlorine Dioxide with *o*-Tolidine Reagent.   Chlorine dioxide reacts with *o*-tolidine at pH 1.9. A yellow-colored product is obtained whose absorbance is measured at 420 nm. The free chlorine content interferes with the method. It can be eliminated by reacting it with malonic acid. Chlorites react in the same way as chlorine dioxide. Their interference cannot be eliminated with malonic acid.

Procedure: A 100-cm^3 sample containing no more than 1 mg/dm^3 chlorine dioxide is treated with 2 cm^3 malonic acid solution (1 g/dm^3) for 3 min. Then 1 cm^3 *o*-tolidine solution is added and the mixture homogenized. Approximately 3 min of reaction time is allowed, and the absorbance is measured at 420 nm. The measured value is reduced with the reagent blank absorbance value and used for evaluation, comparing it with the calibration curve prepared with standard solutions.

Reagents:

o-Tolidine solution (0.1%)
   Preparation: 1 g *o*-tolidine is mixed with 5 cm^3 HCl (HCl:water 1:4); 200 cm^3 water is added. After dissolution, 500 cm^3 HCl solution is added in a 1000-cm^3 volumetric flask, and it is filled up to the mark. The solution is stored in a dark container.

Chlorine dioxide stock solution
   Preparation: 5 g sodium chlorite is dissolved in 400 cm^3 water in a three-necked flask supplied with gas inlet and outlet tubes and a separatory funnel. Adding sulfuric acid:water (1:9) dropwise through the funnel produces chlorine dioxide gas. It is purged out with an air stream and bubbled through distilled water (700–800 cm^3). Between the reaction vessel and the absorber, a gas-washing bottle containing solid sodium chlorite is employed for eliminating the free chlorine. The absorbing solution is filled up to 1000 cm^3. It contains 0.2–0.4 mg ClO$_2$ with 1–2% free chlorine impurity per dm^3.

Chlorine dioxide standard solution
   Preparation: The ClO$_2$ concentration of the stock solution is determined via iodometric titration and diluted to 0.01 mg/1000 cm^3. Prepare freshly.

b. Simultaneous Determination of Chlorine Dioxide, Chlorine, and Chlorite Content of Water Samples via Iodometric Titrations.   First the sample is titrated iodometrically with thiosulfate in neutral media. The reagent consumption in this case corresponds to the sum of one-fifth of the ClO$_2$ and the Cl$_2$ content. A second titration follows in acidic media, in which the remaining four-fifths of the ClO$_2$ and the ClO$_2^-$ ions consume the thiosulfate. The ClO$_2$ and the Cl$_2$ are purged from another portion of the water sample buffered to pH 7. It is acidified and titrated following the conventional iodometric protocol. In this case the thiosulfate reagent measures the ClO$_2^-$ ion content of the sample separately.

Procedure: Into a titration vessel, 75 cm^3 distilled water, 5 cm^3 phosphate buffer, 2 g crystalling KI, and a known volume of the sample solution containing about 2–100 mg ClO$_2$ are introduced. After mixing, it is titrated with 0.05 M sodium thiosulfate reagent. *a* is the reagent volume consumed, in cm^3.

Then 50 cm^3 0.5 M sulfuric acid is added to acidify the titrated sample. This is homogenized and kept for 10 min of reaction time in a dark place. It is titrated again with 0.05 M sodium thiosulfate reagent. *b* is the reagent volume consumed in this time, in cm^3.

Then 100 cm^3 sample solution is introduced into a gas-washing vessel (impinger) and 5 cm^3 buffer solution is added to it. Observing the necessary health safety measures, an air stream is bubbled through the solution for 20–30 min to purge the ClO$_2$ and Cl$_2$ content of the sample. After this, 50 cm^3 0.5 M sulfuric acid is added to acidify, and it is homogenized and titrated with 0.05 M sodium thiosulfate reagent. c is the reagent volume consumed, in cm^3.

Chlorine dioxide concentration (mg/dm^3)

$$= \frac{\left(b - c\,\dfrac{V_1}{V_2}\right) \cdot k \cdot 674.5}{V_1}$$

Free chlorine concentration (mg/dm^3)

$$= \frac{\left(4a - b + c\,\dfrac{V_1}{V_2}\right) \cdot k \cdot 443}{V_1}$$

Chlorite ion concentration (mg/dm^3)

$$= \frac{c \cdot k \cdot 843}{V_1}$$

where V_1 and V_2 represent the reagent volumes used in the first two and the third titrations, respectively, and k is the factor of the titrant.

Reagents:

Phosphate buffer (pH 7)
  Preparation: 72.4 g $Na_2HPO_4 \cdot 2H_2O$ and 32.4 g KH_2PO_4 are dissolved in distilled water, filled up to 1 dm^3, and pH adjusted to 7.0
Sodium thiosulfate titrant, 0.05 M
  Preparation: 12.4 g $Na_2S_2O_3 \cdot 5H_2O$ is dissolved in 900 cm^3 water, 0.2 g Na_2CO_3 is dissolved in it, and filled up to 1000 cm^3.

The disinfecting action of chlorine dioxide was compared with that of liquid chlorine by Junli et al. (8). Testing with six different viruses, chlorine dioxide proved to be a much better agent for virus inactivation than chlorine within the pH range of 3.0–7.0. In case of algae and animal plankton, chlorine dioxide was better than or equal to chlorine as a disinfecting agent. Chlorine reacts with the proteins of the capsomers and destroys their semipermeability. It makes them disappear and reacts with the internal RNA. Finally, the viruses are decomposed. The superstrong virus-killing action of chlorine dioxide, however, is the result of its adsorption and penetration into the protein of capsomers and reaction with the internal RNA.

Smart and Freese (9) analyzed the chlorine dioxide content of different water samples with a rotating voltammetric electrode. In their measurements the electrode potential was kept at +0.5 V vs. an Ag/AgCl electrode, and a rotation rate of 400 rpm was employed. Using the amperometric current, an analytical signal

lower limit of determination of less than 1 mg/dm^3 could be achieved.

Aieta and coworkers (10) worked out electrometric titrations for the sequential determination of chlorine dioxide, chlorine, chlorite, and chlorate. Phenylarsine oxide or sodium thiosulfate titrants and potentiometric or amperometric endpoint detection are used in their method.

Elleouet and Madec (11) describe a differential pulse polarography (DPP) method for the determination of chlorine dioxide in drinking water. In the procedure, 0.5 cm^3 0.5 M phosphate buffer and 0.1 cm^3 indigo-carmine were added to 50 cm^3 tap water sample and a 25 cm^3 portion of the solution was introduced into the voltammetric cell. Adsorptive accumulation was performed on the surface of a hanging mercury-drop electrode at −0.1 V vs. Ag/AgCl for 1 min with intensive stirring before the DPP scan, which was done with 4 mV/s in the negative direction (pulse amplitude 20 mV). The detection limit with this method is 1 ppb; the range of concentration measurements is 10–300 nM.

Under normal water treatment conditions, different interfering species, such as metal ions or chlorinated organic compounds, can occur in the water samples. To eliminate these, Vatanabe et al. (12) separate the chlorine dioxide content of the samples from the matrix with a purge-trap technique using 600 cm^3/min N_2 purging gas stream for 15 min at 25°C. The separated sample is injected into an FIA system. Carried by a streaming buffer solution it merges with 0.8 mM 4-amino antipyrine reagent solution. After passing the sample through a heated reaction coil, the absorbance is detected at 503 nm.

Wang and Yuan (13) use a leucomethylene blue reagent for the determination of chlorine dioxide disinfectant residues in drinking water and wastewater samples. In their method the sample is mixed with 3 cm^3 reagent solution (20 mg/dm^3), filled up to 20 cm^3, and extracted with 1,2-dichloroethane at pH 1.3. After a reaction time of 10 min the absorbance is measured at 658 nm. The interfering chlorine and hypochlorite ions can be masked by adding oxalic acid.

C. Chloride

Chloride is one of the major ionic components of water samples of different origin. Under the usual conditions, the solubility of chloride ions is high and the chloride concentration of sample solutions is unaffected by biological or chemical processes, pH changes, or light radiation. According to EPA regulations the samples collected for chloride content measurements can be stored in closed containers for 28 days without need for any preservation measures.

The chloride ions are not toxic to human. The average daily intake of chloride ions is about 6 g, but daily intake as high as 12 g is not considered abnormal (WHO 1984).

Freshwater sources, groundwater, and surface water reservoirs and streams usually contain less than 10 mg/dm^3 chloride. Higher analysis results indicate contamination by industrial effluents (water softening, paper works, oil wells, galvanic industries), by sewage from hog farms or other agricultural or communal facilities, or by snow- or ice-melting road treatments.

A chloride content in drinking water of 500 mg/dm^3 causes an unpleasant salty taste. However, sensitive persons can notice it at 300 mg/dm^3 in water and at 40 mg/dm^3 in coffee. WHO (1963) listed 200 mg/dm^3 as an acceptable maximum and 600 mg/dm^3 as the maximum allowable chloride concentration for tap water. The European Community (1980) suggests 25 mg/dm^3 chloride concentration for drinking water and 200 mg/dm^3 as limiting value. When a value higher than this limit is confirmed, health authorities and the consuming public should be notified.

The WHO (1984) maximal value guideline for chloride concentration in drinking water is 250 mg/dm^3. For water used for irrigation, the maximal value is 100 mg/dm^3. Industrial applications (e.g., boiler feeding) often require water with much less chloride content. However, seawater, mixed water samples, mineral waters, and sewage contain chlorides in much higher concentration.

1. Detection

Ambient water samples usually contain chloride ions. Their presence can be detected through the silver chloride precipitate obtained with silver nitrate reagent in acidic solution.

Procedure: A 10-cm^3 sample is acidified slightly with chloride-free nitric acid in a test tube. Then 3–4 drops of 5% silver nitrate reagent is added. The formation of white precipitate or turbidity that disappears after the addition of ammonium hydroxide and reappears after acidification indicates the presence of chloride ions. Bulky white precipitate means a concentration higher than 3000 mg/dm^3; white turbidity shows a concentration higher than 700 mg/dm^3; and a slight opalization indicates a concentration of less than 100 mg/dm^3. The detection limit is 1 mg/dm^3.

2. Determination

Good reviews have been published about the very extensive literature dealing with the quantitative analysis of chloride ion content (14,15). A few well-established classical and instrumental methods are used frequently in water analysis. Some of these are also described in college textbooks on analytical chemistry.

Some of the analytical procedures are based on the formation of sparingly soluble silver chloride precipitate. Gravimetric methods (16,17) based on the addition of an excess amount of silver nitrate reagent under controlled conditions and weighing the dried precipitate can provide high accuracy; however, this approach is tedious and time consuming. It is much more efficient to use the argentimetric titration procedure for chloride analysis. In this case, an increasing amount of silver nitrate reagent is added to the sample volumetrically using a burette or generated coulometrically from a positively polarized silver metal electrode, and the stoichiometric endpoint is detected in an appropriate way. Several argentimetric methods with different endpoint detections have been worked out (18). In water analysis, most often a variation of the direct titrimetric method introduced by Mohr in 1856 is used.

a. Argentimetric Titration with Mohr Indication. The chloride ions are titrated with silver nitrate reagent in a neutral or slightly basic medium. The endpoint of the titration is indicated with potassium chromate, which, with the slight excess of the silver ions, forms a reddish brown precipitate (Ag_2CrO_4). The solubility of the silver chloride is quite high, so the lower limit of determination is about 2 mg/dm^3. In case of samples with a lower chloride concentration, a known amount of chloride is added to the samples before titration and the results are corrected with the blank value. Sample solutions in the concentration range of 0.5–300 mg/dm^3 can be analyzed in this way. The color and turbidity of the samples disturb the endpoint observation. To eliminate disturbance, an aluminum hydroxide suspension can be added and subsequent filtration can be used. Sulfites and sulfides also interfere. They can be eliminated by boiling the acidified samples. Usually, 0.5 cm^3 concentrated nitric acid is added to a 250-cm^3 sample volume, and it is boiled for 10 minutes. High concentrations of ferric (10 mg/dm^3) and phosphate (25 mg/dm^3) ions or organic matter (higher than 100 mg/dm^3 permanganate consumption) interfere.

Sample dilution often provides a good way to decrease the negative effect in these cases. If the concentration of the organic matter in the water sample exceeds 300 mg/dm^3, then the addition of 1–2 g sodium carbonate, drying, and ashing at 500°C is recommended to eliminate the interference. Then the chloride content of the filtered extract is analyzed.

If they are present, the Mohr titration measures bromide, iodide, and cyanide ions along with chloride. However, different, more or less complicated separation procedures are available for separate determination.

Procedure: If the sample concentration is higher than 2 mg/dm^3, then 1–2 drops of phenolphthalein indicator are added to 100 cm^3 sample solution, and the pH of the solution is adjusted to the color-change range of this indicator with 0.05 M sulfuric acid or 0.1 M sodium hydroxide. Then 1 cm^3 potassium chromate (10%) is added and the sample titrated with silver nitrate solution. At the endpoint, the original lemon yellow color changes to reddish brown. In order to get a blank value, 100 cm^3 distilled water is titrated in the same way.

The sample concentration is calculated with the following simple formula:

$$Cl = \frac{(a - b) \times 1000}{V}$$

where a and b are the reagent volumes added to the sample and to the blank, respectively, and V is the sample volume. If the reagent volume difference $a - b$ is larger than 30–40 cm^3, then the titration is repeated with diluted sample solution. If it is less than 0.2 cm^3, then the analysis is repeated using the slightly different procedure given next.

Procedure Used in the 0.5–2.0-mg/cm^3 Range: The pH of 100 cm^3 sample solution is adjusted to the range of phenolphthalein color change, as just described. Then 10 cm^3 sodium chloride and 1 cm^3 potassium chromate (10%) solutions are added, and the titration is performed with diluted silver nitrate reagent (1 cm^3 corresponds to 0.2 mg chloride). Then 100 cm^3 distilled water is titrated in the same way to obtain a blank reagent volume. The result is given by the formula Cl = 2 $(a - b)$, in mg/dm^3, where a and b are the reagent volumes added to the sample and to the blank, respectively, and 2 is a factor needed to obtain the results in mg/dm^3 for a sample volume of 100 cm^3.

Reagents:

Silver nitrate titrant

   Preparation: 4.792 g dry silver nitrate is dissolved in distilled water and filled up to 1000 cm^3. The concentration is checked by titrating 5 cm^3 sodium chloride; 1 cm^3 cor-

responds to 1 mg chloride ions. The solution is kept in a closed dark container.

Diluted silver nitrate titrant

   Preparation: 200 cm^3 of the silver nitrate titrant is diluted to 1000 cm^3 with distilled water.

Sodium chloride solution

   Preparation: 1.649 g dry sodium chloride is dissolved and filled up to 1000 cm^3; 1 cm^3 contains 1 mg chloride ions.

The other, very popular argentimetric titration, the *Volhard method*, uses a different approach. An excess of the silver-chloride-forming silver nitrate reagent is added to the sample and after the reaction

$$Cl^- + Ag^+ \rightarrow AgCl$$

the excess of silver nitrate is back-titrated with potassium thiocyanate standard solution, giving a very slightly soluble silver thiocyanate precipitate:

$$Ag^+ + SCN^- \rightarrow AgSCN$$

The thiocyanate reagent excess in the titration vessel is detected by adding Fe^{3+} ions and observing the intense color of the $FeSCN^{2+}$ ions formed after the endpoint. To prevent the hydrolysis of the iron(III) salt the medium is kept acidic during the determination. Unfortunately, the excess of the thiocyanate reagent can react with the silver chloride precipitate, forming the less soluble silver thiocyanate. This can result in endpoint fading or reagent overconsumption. Obviously, filtration of the silver chloride or other tricks, e.g., addition of an immiscible solvent to form a protective film around the precipitate, can prevent this. The lower limit of determination and the accuracy of the Volhard method are better than those of the Mohr titration. Concerning the sensitivity toward some interferences, the Volhard titration also has certain advantages. These advantages, however, are sometimes offset by the more complicated procedure and the need for two reagents. In water analysis, therefore, a direct titration such as the Mohr method is often preferred.

Adsorption indicators can be used for endpoint detection in argentimetric titrations. The mechanism of the adsorption indicators can be understood with the help of the following brief description. When chloride ions are titrated with silver nitrate reagent, the silver chloride adsorbs the chloride ions on its surface, since it is the available ion. In reagent excess, however, the other ion, the positive silver ion, is adsorbed. Indicator dyes of ionic character, such as fluorescein and dichlorofluorescein, can adsorb on the precipitate. Before the endpoint the precipitate surface is negative because of the adsorbed chloride ions; however, it gets positive in reagent excess. Therefore at the endpoint of the titration the fluorescein adsorbs as anion on the positively charged silver chloride surface, forming a red-silver fluoresceinate. The sensitivity of the endpoint detection with adsorption indicators can be affected by light. High chloride or indifferent salt concentration can initiate flocculation. Further on, the lower concentration limit of argentimetric chloride determinations with adsorption indicators is relatively high (about 0.8 mg/100 cm^3). In actual practice, water adsorption indicators are seldom used, despite the extensive literature dealing with them.

Different *electrochemical* methods have been worked out to indicate the endpoint of argentimetric titrations. They can be used for the analysis of chloride as well as other halides. Their application is especially advantageous in mechanized or automated titrators.

If classic (zero-current) potentiometry is used, then the potential difference between an indicator electrode and a reference is followed during titration. Silver metal electrode, silver/silver chloride electrode, or an ion-selective chloride electrode can be used as indicator. The reference can be a calomel or a silver/silver chloride electrode. To avoid contamination by the electrolyte of the internal filling solution of the reference electrode, the two half-cells are separated. A current bridge with potassium nitrate electrolyte is used to keep them in contact electrolitically. Usually the measured cell voltage is plotted against the added reagent amount, and the inflection point of the curve is taken as the endpoint. Iodide and bromide, forming a less soluble precipitate with the silver ions, can be titrated in the same way. Being present, they interfere with both the detection and the titrating reaction. Automatic titrators controlling the reagent addition can stop it when the cell voltage achieves a preset value corresponding to the stoichiometric equivalence.

The so-called *zero-point titration* also gained application in water analysis (19). In this method the potential difference between two identical indicator electrodes (e.g., two silver wire electrodes) is followed. One of them is in a half-cell containing solution corresponding to the equivalence point of the titration, and the other is dipped into the sample-containing titration vessel. The silver nitrate reagent (0.01 M) is added as long as potential difference exists. With this zero-point potentiometry a determination limit as low as 1.3 mg/dm^3 can be achieved.

The *constant-current potentiometry* is another endpoint detection technique that can be used in the titrations of very dilute chloride samples (20–22). In this case two silver metal or chloride-coated silver indicator electrodes are dipped into the intensively stirred titration vessel. A well-stabilized, constant low-intensity current is forced between the two electrodes, and the potential difference between them is measured. The titration endpoint is very sharp in this case, even at a very low concentration.

In case of argentimetric titrations, the addition of *coulometric* reagent is very advantageous, especially if the sample concentration is very low. Silver ions can easily be generated with close to 100% current efficiency from a positively polarized silver metal electrode. A current density of a few milliamps per centimeter is adjusted, and a background electrolyte in which the generated silver ions stay soluble is used. Most often an acidic background electrolyte is selected, since in basic media silver oxide is forming, which, as a coating, can passivate the silver electrode surface.

For the titration of the chloride content of water samples, the background electrolyte suggested by Cotlove and coworkers (23) can be used advantageously. It contains acetic acid, nitric acid, and gelatin. The background electrolyte is prepared by mixing a base solution (1 dm^3 of it contains 102 cm^3 glacial acetic acid and 10.3 cm^3 concentrated nitric acid dissolved in distilled water) with a gelatin solution (0.6 g gelatin, 0.01 g thymol, and 0.01 g thymol blue dissolved in 100 cm^3 distilled water in the volume ratio of 10:0.25). The gelatin serves as a protective colloid that, surrounding the precipitate, avoids the reduction of the generated silver ions at the cathode surface and also reduces the so-called adsorption error (24). Application of water–organic solvent mixtures, such as methanol:water 1:1 containing perchloric acid (e.g., 0.5 M), can reduce the solubility of the silver chloride and this can be advantageous in cases of low chloride concentration.

Usually an indifferent platinum cathode is used in argentometric titrations, along with electrolytically generated silver ion reagent. It is often placed in a separated half-cell.

Oxidizing agents or photodecomposition of the silver chloride, however, can bring some uncertainty into the absolute character of the coulometric chloride measurements. Usually the coulometry is combined with instrumental endpoint detection. Since reagents generating current can be controlled more easily than volumetric reagent addition devices, many different, more or less automatic titrators have been worked out and investigated, starting from the early days of instrumental analysis (25).

b. Determination of Chloride with Mercurimetric Titration. Chloride ions form a very strong, not dissociating complex with mercury(II) ions. The excess of mercury ions after the equivalence point can be detected with an appropriate indicator. Diphenylcarbazone is a good indicator that forms a blue-violet Hg(II)-diphenylcarbazone complex. So the chloride can be titrated with mercury(II) nitrate or mercury(II) perchlorate reagent. According to Cheng (26), the analysis is easy and accurate in an 80% alcoholic solution at pH 3.5. Bromide, iodide, sulfide, phosphate, chromate, and iron(III) interfere seriously with the mercurimetric titration. Fluoride, cyanide, and sulfate also should be eliminated before titration. Sulfate can be precipitated with barium nitrate, while with the addition of thorium nitrate both phosphate and fluoride can be precipitated and filtered off. The disturbing effect of sulfide and cyanide can be eliminated by reacting with hydrogen peroxide or potassium permanganate. The mercurimetric titration works well with sample solutions containing more than 10 mg/dm^3 chloride. In the lower concentration range the accuracy decreases, but with a 1–2% error 15 μg of chloride could be determined in a 20-cm^3 sample volume. The mercurimetric titration of chloride has been an effective, officially recommended method of water analysis (27,28).

Procedure: Two drops of bromophenol blue indicator solution (0.1% in ethanol) is added to the aliquot of the sample solution containing 1–1.5 mg chloride. If the sample is basic (the indicator is blue), then it is titrated with 0.1 M HNO$_3$ until a yellow color appears. If it is acidic (yellow), then 0.1 M sodium hydroxide is added until color change. Then 0.5 cm^3 0.1 M HNO$_3$, 100 cm^3 95% ethanol (pH should be about 3.6), and 0.5 cm^3 diphenylcarbazone indicator solution (0.1% in ethanol) are added to the titration vessel, and it is titrated with mercury(II) nitrate with intense agitation. Equivalence is indicated by the appearance of the violet color. The indicator blank should be determined and taken into consideration.

Reagents:

Mercury(II) nitrate titrant (0.005 M)
  Preparation: 3.4 g Hg(NO$_3$)$_2$·H$_2$O is dissolved in about 600 cm^3 nitric acid (0.01 M). It is kept for two days in a closed container after being filtered and filled up to 2 dm^3. The solution needs to be standardized. For this, a mixture of 5 cm^3 0.01 M sodium chloride and 10 cm^3 distilled water is titrated as given previously.
c. Spectrophotometric Chloride Determinations. For the determination of trace amounts of chlo-

ride, different spectrophotometric methods have been worked out. These are used in the analysis of highly purified waters, boiler feed waters, or other water samples.

Determination with Mercury(II) Thiocyanate Reagent (29): Here, mercury(II) thiocyanate reagent is added to the sample solution. The chloride ions form a sparingly dissociating mercuric chloride complex and liberate a stoichiometrically equivalent amount of thiocyanate ions. The thiocyanate is reacted with iron(III) ions, giving an intense red ferric thiocyanate complex according to the following equations:

$$2Cl^- + Hg(SCN)_2 \rightarrow HgCl_2 + 2SCN^-$$
$$SCN^- + Fe^{3+} \rightarrow Fe(SCN)^{2+}$$

The absorbance is measured at 460 nm. The presence of bromide, iodide, cyanide, thiosulfate, sulfide, thiocyanate, and nitrite ions and the original color of the sample interfere with the determinations. The method can be used in the range of 0.01–10 mg/dm^3.

Procedure: All the glassware to be used in the measurements needs to be washed thoroughly first with diluted nitric acid and subsequently with distilled water. It is good practice to use the same glassware in one measurement. The laboratory atmosphere should be free of hydrochloric acid fumes.

A 25-cm^3 water sample, 5 cm^3 iron(III) ammonium sulfate, and 2 cm^3 mercury(II) thiocyanate are mixed together in a closed Erlenmayer flask. After 25 minutes the absorbance is measured at 460 nm in 5–10-cm cuvettes against blank prepared with distilled water. A calibration curve is prepared for the evaluation by diluting 0.5-, 1.0-, 2,0-, 3.0-, 5.0-, 10.0-, 15.0-, 20.0-, and 25.0-cm^3 aliquots of freshly prepared 0.001-mg/cm^3 chloride solution up to 25 cm^3, and measuring their absorbance value, as described previously. These solutions contain 0.02, 0.04, 0.08, 0.12, 0.2, 0.4, 0.6, 0.8, and 1.0 mg/dm^3 chloride, respectively.

Reagents:

Iron(III) ammonium sulfate
  Preparation: 5.0 g iron(II) ammonium sulfate (Mohr salt) [Fe(NH$_4$)$_2$(SO$_4$)$_2$·6H$_2$O] is dissolved in 20 cm^3 distilled water; 38 cm^3 concentrated nitric acid is added, and it is boiled to oxidize the iron until the nitrous gases disappear. It is then filled up to 100 cm^3.
Mercury(II) thiocyanate solution
  Preparation: 0.3 g Hg(SCN)$_2$ is dissolved in methanol. The solution needs to be kept in

a dark bottle. It can be used for 1 month. The freshly prepared solution should be stored for one day before use.

Sodium chloride solution

Preparation: 0.165 g dry sodium chloride is dissolved in distilled water and filled up to 1 dm^3; 10 cm^3 of this solution is diluted to 1 dm^3 to obtain a solution that contains 0.001 mg/cm^3 chloride.

Determination with diphenylcarbazone: The chloride ion content of the sample is reacted with the excess of mercuric nitrate reagent. After formation of the mercuric chloride complex, the concentration of the free mercuric ions is detected spectrophotometrically with diphenylcarbazone reagent. The diphenylcarbazone reagent forms a violet-colored complex with the mercuric ions, with a 560-nm absorbance maximum. So the chloride content of the sample decreases the absorbance of the solution.

Copper, iron(III), bromide, iodide, thiocyanate, acetate, and oxalate interfere. However, in high-purity, demineralized water samples, these are not usually present. With this method chloride measurements can be made in the 0.05–0.06-mg/dm^3 range without any preconcentrating step.

Procedure: Ten cm^3 water sample, 2 cm^3 mercury(II) nitrate solution, 10 cm^3 buffer solution, and 1 cm^3 diphenyl-carbazone solution is mixed together. The absorbance of this solution is measured at 560 nm in a 5-cm cuvette against a reagent blank. The reagent blank is prepared by substituting the sample with distilled water (10 cm^3) in the mixture given previously. The readings should be taken after the absorbance stays constant for 2 minutes.

If the chloride concentration of the sample is higher than 0.6 mg/dm^3, then the analysis must be repeated with diluted sample.

The measurements are evaluated on the basis of a calibration curve prepared with a freshly diluted stock solution of 1-μg/cm^3 chloride concentration. An aliquot (0.5, 1.0, 2.0, 3.0, 4.0, 5.0, and 6.0 cm^3) of this solution is brought up to 10-cm^3 volume with distilled water to prepare calibrating solutions with 0.05, 0.1, 0.2, 0.3, 0.4, 0.5, and 0.6 mg/dm^3 chloride concentration, respectively.

Reagents:

Mercury(II) nitrate solution

Preparation: 0.25 g mercury metal is dissolved in a slight excess of nitric acid and filled up to 1 dm^3 with distilled water; 10 cm^3 of this solution is diluted to 100 cm^3. The solution contains 25 μg/cm^3 mercury(II) ions.

Diphenylcarbazone solution (0.05% in methanol solvent)

Buffer solution, borax buffer pH = 3

Preparation: Mixture of 1.2 cm^3 solution *A* (19.1 g/1000 cm^3 Na$_2$B$_4$O$_7 \cdot$ 10H$_2$O) and 98.8 cm^3 solution *B* (5.9 g/1000 cm^3 succinic acid).

d. Other Photometric Methods. By substituting the mercuric nitrate reagent with mercuric chloranilate, no separate reagent is needed for the measurement of the excess of mercuric ions. When the mercuric chloranilate reacts with the chloride ions of the sample, besides the formation of mercuric(II) chloride complex, stoichiometrically equal amounts of the reddish purple acid chloranilate ions are liberated. So by measuring the absorbance at 530 nm, the chloride content of the sample can be measured (30,31). The cation interference can be avoided by passing the sample through a cation-exchange column. The mercuric chloranilate can be administered in solid form, and after the reaction its excess can be eliminated by filtering.

In other, less frequently used methods, the chloride content of the sample is oxidized quantitatively to chlorine and the chlorine concentration is determined spectrophotometrically. Potassium permanganate oxidant and *o*-tolidine (32) or methyl orange (33) chlorine-measuring reagent can be used in these procedures.

The AgCl precipitate formed after the addition of silver nitrate reagent can be detected with nefelometric or turbidimetric techniques. In this way the chloride concentration can be measured in water samples using appropriate calibration curves (34–36). To be able to measure in the low concentration range, the measurements are carried out in water–organic solvent mixtures such as methanol–water (37).

Indirect methods using atomic absorption measuring techniques have been also worked out. These seldom-used procedures precipitate the chloride with silver reagent and detect the silver ions in the dissolved precipitate or the excess of them in the filtered reaction media (38,39). A little more complicated version forms a phenyl mercury(II) chloride complex, separates it by extraction with chloroform, and detects the mercury by AAS. This allows one to analyze very dilute (0.015 ppm) chloride samples (40).

e. Direct Potentiometric Chloride Measuring Methods. As mentioned earlier, potentiometry at zero current is often used for endpoint detection in argentimetric titrations. In the absence of interfering species, direct potentiometry can be used for chloride determinations. In this case a silver/silver chloride electrode of the second kind, or, more often, an ion-selective chloride-indicating electrode is used. The active measuring membrane of the ion-selective chloride electrode is made of silver chloride, a mixture of silver chloride and silver sulfide, or a mixture of mercurous chloride and mercuric sulfide (41). In these determinations the potential difference between the indicating electrode and a reference is measured. The internal filling solution of reference electrodes such as calomel and silver/silver chloride electrodes contain chloride ions in a high concentration. Therefore care must be taken to avoid the contamination of the sample solutions through the current bridge of the reference electrode. The use of a chloride-free reference electrode or a reference electrode with a double junction is recommended. Often the reference half-cell is separated from the sample-holding one and a chloride-free current bridge is used between them.

Most often the ion-selective chloride electrode (e.g., Orion, type 94-17-96-17) is calibrated with standard solutions. A calibration graph is prepared by plotting the cell voltage against the negative logarithm of the chloride concentration of the standards. The cell voltage is measured in the case of the sample solution, and the sample concentration is determined from this value by using the calibration data or the curve. To be able to get concentration data, the ionic strengths in the calibrating and sample solutions are kept equal and constant. It can be made in the most convenient way by adding a given volume of a concentrated indifferent electrolyte to a given volume of the sample or standard solutions. A 1:1 dilution with 2 M potassium nitrate is often used for the analysis of the chloride concentration of natural water samples. For high-precision analysis the temperature in the measurement cell must be controlled.

Boiler feed waters contain chloride ions in the 0.1–10-mg/dm^3 range. Torrance suggests using pH 4.7 acetate buffer background electrolyte in this range (42).

In the linear range of the electrode function, a standard addition or sample subtraction method can also be used successfully. An automatic monitoring method, e.g., was worked out by Nagy et al. (43) using coulometric sample subtraction for monitoring the chloride content of tap water.

Bromide, iodide, cyanide, and, especially, sulfide ions strongly interfere with the potentiometric determination of chloride ions. Fortunately these ions are seldom present in disturbing concentration in certain water samples. The solid-state ion-selective elec-

trodes are not sensitive to oxidizing agents; however, their function is influenced by strong reducing agents.

The chloride ion content of sample solutions can be deposited on the surface of hanging mercury-drop, mercury-film, or silver electrodes by anodic polarization. This allows one to work out highly sensitive cathodic stripping voltammetric methods for the analysis of halides. In everyday water analysis, however, these methods are not used.

f. Spectrophotometric Mercury(II)sulfocyanide–Based Chloride Measurement. The flow injection version of the spectrophotometric mercury(II)sulfocyanide-based chloride-measuring method was investigated by Zhao (44). Using 0.15-cm^3-volume tap water samples, 3.5 cm^3/min carrier and 1.7 cm^3/min reagent flow rates, and 448-nm detection, the dynamic range of calibration was found between 1 and 40 μg/cm^3.

g. Turbidimetric Chloride Measurement by FIA. A turbidimetric method was described by Pidade and coworkers (45) for the serial analysis of the chloride content of natural water samples. A flow injection apparatus was used for the measurements.

h. Other Chloride Measurements. Recently, Hong and Zhou (46) worked out an automatic analytical technique based on mercury(II) thiocyanate and iron(III) nitrate reagent. The fully automated apparatus uses the continuous-flow injection principle: a 20-μl sample is injected into the water carrier stream, which merges with the reagent stream containing mercury(II) thiocyanate and iron(III) nitrate polyoxyethylene glycol dodecyl ether in aqueous methanol. After it passes through the reactor section, the absorbance is measured at 480 nm.

Aleksandrova and Kletenik (47) proposed a voltammetric method for the determination of chloride concentration in different water samples. They use a renewable silver working electrode. Both direct anodic voltammetry and cathodic stripping voltammetry could be successfully used in 1 M sulfuric acid. In case of the cathodic stripping method, 3 min per electrolysis step was employed at −0.07 V to obtain a silver chloride film at the electrode surface.

D. Fluoride

Most types of water contain fluoride ions. The fluoride concentration of the different water samples ranges from traces to 10 mg/dm^3. Since health authorities (48) recommend drinking water fluoride levels of no more than 1 mg/dm^3, the concentration range of potable water samples is mostly 0.5–1.5 mg/dm^3. The wastewater from glass factories and brine waters contain very high fluoride concentrations.

The water sample collected for fluoride measurements requires no preservation. It can be stored in a special glass or plastic container for about 28 days.

1. Detection

a. Zirconium–Alizarin Red S Method. The complex between zirconium and alizarin red S (sodium alizarin sulphonate) gives a red-brown color in acid solution if alizarin red S is in excess and a violet color if the zirconium is in excess. The complex is depolarized by fluoride ions. Phosphate, arsenate, sulfate, thiosulfate, and oxalate as well as organic hydroxy acids interfere with this reaction.

2. Quantitative Determination of Fluoride Content in Water Samples

Before the introduction of the ion-selective fluoride electrode, measuring fluoride concentration was a quite difficult task. Fewer methods are available for this than for the analysis of other halides.

a. Spectrophotometric Determination of Fluoride Using Solochrome Cyanine R. This method is based on the bleaching action of the fluoride ion content of the sample. The color of the red Zr-Solochrome Cyanine R (Aldrich Cat. No. 23,406-0, Sigma Prod. No. E2502) (49) complex fades as $ZrOF_2$ is formed in the medium. As a matter of fact, no simple stoichiometric relationship exists between the fluoride and the zirconium complex and the dye. Therefore in order to obtain reliable results the reaction conditions need to be controlled very carefully. The absorbance of the reaction media is measured at 540 nm. The fluoride concentration is evaluated using an absorbance–fluoride concentration calibration curve prepared with standard solutions. The method can be used for samples containing 0–2.5 μg fluoride.

Procedure: The following solutions are pipetted into a 25-cm^3 volumetric flask in the following order: water sample or standard containing up to 2.5 μg fluoride, water to bring up the volume of the solution to 20 cm^3, and 2.00 cm^3 of acidic zirconium reagent. After mixing 2 cm^3 of solochrome cyanine R, reagent is added, homogenized, and filled up to the 25 cm^3 mark. A 5-min reaction time is allowed, and the absorbance is measured within 2 min at 540 nm against a water reference in a 2-cm cuvette.

b. Electrometric Methods

Fluoride Determination with Ion-Selective Electrode: If a selective electrode operating in the con-

centration range of the sample solutions is available for the analysis of an ionic species, no method can compete in simplicity with direct potentiometry. Frant and Ross (50) worked out quite a well-functioning fluoride-selective electrode in 1966. The active measuring membrane of this electrode is made of europium-doped LaF_3 crystals incorporated in a hollow electrode body. Inside the body are the internal filling solution and the internal reference electrode. To carry out potentiometric measurements at zero current, the internal reference electrode of the fluoride ion–selective electrode is connected to the high-impedance input of a special mV meter, often called pX meter or pH meter. The electrode is dipped into the measuring cell containing the sample or a standard solution. A constant-potential reference electrode connected to the other input of the meter is brought into contact with the electrolyte of the measurement cell. The potential difference between these two, i.e., the electromotive force (EMF) of the cell, is the analytical signal. The dependence of the electrode potential of a well-functioning ion-selective electrode on the sample concentration follows the Nernst equation. That means that in the dynamic range, a linear relationship exists between the negative logarithm of ionic activity and the electrode potential:

$$E = E° - \frac{RT}{F} \ln a_F \quad \text{that is, for 25°C}$$

$$E = E° - 0.05916 \log a_F$$

where:

E = electrode potential
$E°$ = normal electrode potential
R = universal gas constant
T = temperature
F = Faraday constant
a_F = activity of the fluoride ion in the solution

Fluoride ion–selective electrodes are made and commercialized by several companies. They are everyday tools of water analysis laboratories.

To be able to measure concentrations, the ionic strength must be kept constant in the calibrating and in the sample solutions. In this way the activity coefficient will also be constant. Therefore in direct potentiometry, a high concentration of an inert background electrolyte is added to the samples and standards before measurements for ionic strength adjustment. This background electrolyte can beneficially influence the conditions of the measurements. It can adjust the pH to the optimal value, can mask the interferences by complex formation, or, if needed, can avoid sample oxidation. For potentiometric fluoride measurements a special background electrolyte, the total ionic strength adjusting buffer (TISAB), is recommended (51).

Hydroxide ions strongly interfere with the function of the LaF_3-based fluoride ion–selective electrode. Therefore the pH of the sample and the calibrating solution need to be kept low. In case of low fluoride concentrations, the pH should be lower than 8.0. On the other hand, at low pH values, fluoride activity is affected by the undissociated HF and HF_2^- formations.

Fluoride samples of high concentration can be titrated potentiometrically (52) with lanthanum nitrate or thorium nitrate reagent. The potentiometric standard addition technique with NaF standard solution (53) was also found applicable; however, direct potentiometry using calibration curves is relied on most often in water analysis.

Direct Potentiometric Method: The sample and standard solutions are introduced into the potentiometric measuring cell mixed with background electrolyte. The fluoride ion–selective electrode and appropriate reference electrode (saturated calomel electrode or silver/silver chloride electrode) are dipped in and the electromotive force measured. An EMF–log (fluoride ion concentration) calibration curve is used for evaluation.

The hydroxide ions interfere with the measurements, so the pH of the standards and the samples needed to be adjusted to the necessary value. For high precision, the temperature of the measurement cell has to be controlled and kept constant.

Procedure: Calibrating standard solutions are prepared as shown in Table 3. The sample solution is diluted with TISAB 1:1. The diluted sample and the calibrating standards are introduced into the carefully washed measurement cell. The ISE and the reference electrode are dipped into the solution and the cell voltage is followed. The solution is stirred. When a stable cell voltage is achieved, its value is taken and used for preparation of the calibration curve or for evaluation of the sample concentration. Often, more or less automatic apparatus are employed in practice with automatic solution intake and automatic evaluation. Calibration also has to be done with standard solutions.

The proper operation of the electrode is checked three times doing the calibration: first with standard solutions following each other in decreasing order of concentration, then in increasing order, and finally in decreasing order again. Between measurements, the cell and the electrodes must be washed very carefully. The three readings taken in the same solution during the three calibrations must not differ by more than a few tenths of a millivolt. If they do, the electrode has to be renewed or replaced, or the washing needs to be done more intensively.

The calibration curve is prepared by plotting the cell voltage against the negative logarithm of the fluoride concentration of the standards (pF). The plot is a straight line between pF 5 and 2. It deviates from linearity between the pF values of 6 and 5. The sample concentration is determined from the calibration curve, taking the dilution factor into consideration.

Reagents:

Sodium fluoride solution 0.1 M
   Preparation: 4.2 g NaF, previously dried for 2 hours at 105°C, is dissolved in distilled water and filled up in a volumetric flask to 1000 cm^3.
TISAB (total ionic strength adjustment buffer)

Table 3 Suggested Standard Solutions for Calibrating Fluoride Ion-Selective Electrodes

Serial no. of standard solution	Volume added (cm^3)	TISAB (cm^3)	Water (cm^3)	Fluoride concentration (M)	Fluoride concentration (mg/dm^3)
I	From 10–1 M NaF 5.00	25.0	20.0	10^{-2}	190
II	From soln. I 5.00	22.5	22.5	10^{-3}	19
III	From soln. II 5.00	22.5	22.5	10^{-4}	1.9
IV	From soln. III 10.00	45.0	45.0	10^{-5}	0.19
V	From soln IV 25.00	12.5	12.5	5×10^{-6}	0.095
VI	From soln. IV 10.00	20.0	20.0	2×10^{-6}	0.038
VII	From soln. IV 5.00	22.5	22.5	10^{-6}	0.019

TISAB: Total ionic strength adjusting buffer.

Preparation: into a 1000-cm³ volumetric flask, 1.0 mole sodium chloride, 0.25 mole acetic acid, 0.75 mole sodium acetate, and 1 mmole trisodium citrate are introduced, and it is filled up to the mark. The pH of the solution is 5.0–5.5, with ionic strength of 1.75 M.

Potentiometric Cl⁻ and F⁻ Measurement: As has been demonstrated (54), chloride and fluoride ion concentration of potable water samples can be determined simultaneously with chloride and fluoride ion–selective electrodes built in a special flow-through apparatus. Using the sequential injection principle, cyclohexane-1,2-diamine-*NNN′N′*-tetraacetic acid, total ionic strength adjusting buffer, and an Ag/AgCl reference electrode, water samples of 0.2-cm³ volume could be analyzed in the range of 20–500 μg/cm³ (chloride) and 0.5–200 μg/cm³ (fluoride) concentration.

Sarma and Rao recently studied (55) the potentiometric fluoride measuring method by analyzing well waters with a commercial analyzer. In their experiments the detection limit of fluoride ions was 1 μM.

Spectrophotometric F⁻ Determination: A spectrophotometric method was worked out by Liao and co-workers (56) for the determination of trace-level fluoride concentrations in water samples. In this procedure the samples are mixed with a reagent mixture (alizarin-3-methylimino-*NN*-diacetic acid/sodium acetate/12.5% acetic acid buffer of pH 4.1/1 mM lanthanum nitrate (1:1:1:1)). After a reaction time, the colored complex is extracted with 5% *NN*-dimethylaniline solution in 3-methylbutan-1-ol, and the absorbance is detected in a special, long capillary at 580 nm.

E. Bromide

Drinking water usually contains less than 1 mg/dm³ bromide. The human taste threshold for bromide ions in water ranges from 0.17 to 0.23 mg/dm³. Seawater and some well waters contain more than 2 mg/dm³ bromide. Swimming pool water is sometimes disinfected with 2 mg/dm³ bromine, resulting in increased bromide concentrations. Industrial effluents also contribute to the bromide content of sewage waters.

For bromide analysis the samples (100 cm³) can be taken in glass or plastic containers. The water samples can be stored in closed containers for 28 days without special sample preservation measures.

The bromide content of water samples can be separated by oxidizing to bromine with a strong oxidizing agent and distilling it out. Bromide content also can be distilled out in the form of cyanogen bromide, which can be collected in sodium hydroxide. Preconcentration can be achieved in this way (57).

1. Determination of Bromide Content of Water

The different argentimetric titrations with color indicators or electrometric endpoint indications, using volumetric or coulometric reagent addition (discussed with chloride measurements), can be used for the quantitative analysis of bromide ions too. However, the bromide concentration in most water samples is too small for these methods. With these methods the separate determination of the different halides is also difficult. Bromide ions can be determined with mercurimetric titrations, also discussed in the section on chloride determinations. Nitroprusside, diphenylcarbazide, or diphenylcarbazone indicators serve well in this case. When doing the titration in a water–ethanol solvent mixture of 80% ethanol, as suggested in Ref. 26, the sensitivity is quite high. However, this method too is not suitable for mixed halides.

Spectrophotometric and some of the electrometric methods have the bromide-measuring range needed for water analysis.

2. Spectrophotometric Method Based on Rosaniline Reagent

Bromide ions are oxidized in slightly acidic media with hypochlorite to bromate. The excess of hypochlorite is taken away by reacting it with sodium formate, and bromide ions are added in excess. The reaction between the bromate and the bromide results in bromine (amplification). The bromine is reacted with the rosaniline reagent (Basic Fuchsin). The product of this reaction is dissolved in butanol, and the absorbance is measured at 573 nm.

The method can be used in the range of 0.1–2.0 mg/dm³ bromide concentration. Manganous ions and, naturally, bromates interfere with the determination.

Procedure: Into a 50-cm³ water sample containing 0.005–0.1 mg bromide are added 10 cm³ buffer solution and (dropwise) 5 cm³ hypochlorite reagent solution. The mixture is boiled for 10 minutes. Then 2.5 cm³ of sodium formate solution is added and the sample boiled further for 5 minutes. When it is cooled down it is transferred quantitatively into a 100-cm³ volumetric flask, 15 cm³ rosaniline solution is added, and the mixture is homogenized. Three minutes later, 25 cm³ t-butyl alcohol–water solvent mixture (specific density 0.8 g/cm³) is added and the flask is filled up to the mark. The absorbance is measured. Reagent blank is made in the same way, substituting the sample with distilled water, and the absorbance difference between the sample and the blank is used for evaluation. A calibration curve is prepared, plotting the absorbance measured in standard solutions against their concentration (0.0–2.0 mg/dm^3).

Reagents:

Phosphate buffer pH = 6.3
   Preparation: 18.2 g NaH$_2$PO$_4$·2H$_2$O and 3.6 g K$_2$HPO$_4$ are dissolved in 100 cm^3 water, and the pH is adjusted.
Sodium formate solution: 50 g/cm^3
Basic potassium hypochlorite solution: about 1.1 M for KOCl, and 0.08 M for KOH
Acidic rosaniline–potassium bromide reagent
   Preparation:
      Component *a*—0.05 g rosaline is dissolved in 250 cm^3 1 M sulfuric acid and stored in a refrigerator.
      Component *b*—0.6 g KBr is dissolved in 250 cm^3 distilled water and stored in a refrigerator.

Ten cm^3 component *a*, 10 cm^3 component *b*, and 80 cm^3 7.5 M sulfuric acid are mixed. Prepare fresh before use.

This method can be used for the determination of bromate also, in which case no oxidation is needed.

Another spectrophotometric method is based on the reaction of bromide ions with triphenylmethane dyes in the presence of chloramine B, chloramine T, or NaClO. The color of the dyes fades as consequence of the reaction. Brilliant green, acid violet (18), or crystal violet (16) serve well as dyes. The lower limit of determination is quite good. It is about 0.01 mg/dm^3. However, as can be seen in the iodide measurements section, iodide gives the same reaction.

3. Electrometric Bromide-Measuring Method

The application of direct potentiometry with silver bromide precipitate–based ion-selective electrodes for bromide measurements in water samples have been investigated (58). Cyanide, sulfide, and iodide ions represent the major interferences. A 20 times higher concentration of chloride can also cause a positive error. Therefore the applicability of direct potentiometry to the analysis of bromide concentration in water samples is limited.

Bromate ions can be reduced on the dropping-mercury electrode, so they can be determined with the polarographic technique (59). This can be utilized in the analysis of bromide concentration. The polarographic bromide determinations use the previously described oxidation to bromate by hypochlorite reaction. The solution is treated with formic acid to eliminate the excess of the oxidizing agent. The polarographic analysis is done in neutral lanthanum chloride background solution. The polarographic wave is in the electrode potential range of −0.8 to −1.6 V vs. mercury pool. Voltammetric determinations of bromide ions can also be carried out through oxidizing them on the surface of a pyrolytic graphite electrode (60).

The bromide content of water samples ranging between 2 and 20 mg/dm^3 can be measured with iodometric titration. In this procedure, the samples are pretreated with calcium oxide to eliminate the interfering iron, manganese, and organic matter. After this, one part of the sample is oxidized with bromine water and titrated iodometrically with phenylarsine oxide or sodium thiosulfate reagent. In this process the iodide is oxidized to iodate and determined as described in the iodide section. The other part of the pretreated sample is oxidized with calcium hypochlorite. The iodide is oxidized to iodate and the bromide to bromate in this way. The excess of the oxidizing agent is removed by reacting it with sodium formate. The sample is acidified, potassium iodate is added, and the formed iodine is titrated with phenylarsine oxide or sodium thiosulfate reagent. The sum of bromide and iodide content is obtained in this way. The bromide concentration of the sample can be calculated by comparing the results of the two titrations. A very important advantage of this method is that the measurement can be made in a large excess of chloride. Another advantage is the amplification. As is obvious from the stoichiometry, six thiosulfate ions measure one bromide ion.

Highly sensitive detection techniques such as MS or preconcentration steps help when the bromide concentration of the water sample is too low. Sub-ppb levels of bromate formed during the ozone treatment of bromide-containing drinking water can be determined, according to Charles et al. (61), by electrospray ion chromatography–tandem mass spectrometry. For the determination, 10 cm^3 drinking water sample was pretreated by passing it through sulfate-, chloride-, and bicarbonate-removing cartridges. Then 5 cm^3 of the eluate was analyzed on an IonPac AG9-SC column with aqueous 90% methanol/27.5 mg/dm^3 ammonium sulfate mobile phase. Negative ion electrospray MS-MS detection yielded an 0.1 μg/dm^3 lower limit.

4. Bromide Determination with Preconcentration Step

Low levels of bromide ions could be determined in freshwater samples by the method of Lundstrom and coworkers (62). In the first step, the bromide ion content of the slightly acidified sample is preconcentrated on an anion-exchange resin. Then it is eluted with 2 M $NaClO_4$ and the bromide content is oxidized to BrO_3^- with persulfate, which is determined spectrophotometrically. The detection limit of this method is 1.5 nM and the lower limit of determination is 5 nM.

The trace level bromide content of natural water samples was preconcentrated by coprecipitation as AgBr with AgCl in the work of Denis and Masschelin (63). The precipitate was oxidized to $AgBrO_3$ with NaClO at pH 7. After separation it was determined via differential pulse polarography (DPP) in 1 M $MgCl_2$ solution using a 50-mV pulse amplitude and $-0.2-1.8$ V sweep range. A 2-ng/cm^3 detection limit could be achieved with this method.

F. Iodide

Iodide ions can be determined quite well with argentimetric titrations. The Volhard method, electrometric endpoint indications, and adsorption indicators work well. The Mohr endpoint indication, however, does not give good results because of the adsorption of the chromate on the silver iodide precipitate. The presence of chloride and bromide ions disturbs the argentimeric iodide determination.

Several methods are based on the redox character of the iodide ions. They can be oxidized to iodine and titrated with sodium thiosulfate or phenylarsin oxide titrant. The oxidation can be made in acidic medium (64). The excess of the oxidizing agent can be taken away by adding urea to the solution. Bromide ions interfere with this method.

One of the best methods worked out for the determination of small amounts of iodide is based on oxidation to iodate by bromine water and reacting the iodate with iodide to form iodine, as shown in the following equations:

$$I^- + 3Br_2 + 3H_2O \rightarrow IO_3^- + 6Br^- + 6H^+$$

$$IO_3^- + 5I^- + 6H^+ \rightarrow 3I_2 + 3H_2O$$

The six-fold amplification that can be seen from the equations is highly beneficial if a small amount of iodide is to be determined.

The excess of bromine is destroyed by adding formic acid and boiling the solution. The iodine produced can be titrated. The bromine and chloride ions do not interfere with this method. Iron, manganese, and organic matter can interfere; however, pretreatment of the samples with calcium oxide removes these.

Procedure: For pretreatment, a visible excess of CaO is added to 400 cm^3 water sample, and intense agitation is employed for 5 minutes. The solution is filtered and the first 75 cm^3 is discarded. The pH of the collected solution is adjusted to 7.0 with sulfuric acid (H_2SO_4:water 1:4, v:v). Then 100 cm^3 is transferred to a 250-cm^3 iodine flask, 15 cm^3 of sodium acetate solution (275 g sodium acetate trihydrate in 1 dm^3 solution), 5 cm^3 acetic acid solution (glacial acetic acid: water 1:8, v:v), and 40 cm^3 bromine water are added. The solution is mixed and 5 min of reaction time is allowed. Then 2 cm^3 sodium formate solution (50 g/ 100 cm^3) is added. After a reaction time of 5 min, the bromine fumes are removed by purging with a nitrogen stream. About 1 g of potassium iodide and 10 cm^3 sulfuric acid (H_2SO_4:water 1:4, v:v) are added. A 5-min reaction time is allowed, keeping the sample in a dark place. The sample is titrated with sodium thiosulfate standard solution using starch indicator or electrometric endpoint location.

Reagents:

Bromine water
  Preparation: 0.2 g bromine is added to 500 cm^3 distilled water and stirred until dissolved.
Sodium thiosulfate, stock solution
  Preparation: 186.15 g $Na_2S_2O_3 \cdot 5H_2O$ is dissolved in water and diluted to 1000 cm^3; 5 cm^3 chloroform is added for preservation.
Sodium thiosulfate standard titrant
  Preparation: 50 cm^3 stock solution is diluted to 1000 cm^3; 5 cm^3 chloroform is added and standardized with potassium biiodate.
Sodium thiosulfate working standard
  Preparation: 100 cm^3 sodium thiosulfate standard titrant is diluted to 500 cm^3. Prepare fresh daily.

1. Bromide and Iodide Determination

Trace level determination of bromide and iodide ion content of mineral water samples can be measured with a gas chromatography (GC) method, as shown by Kirchner and coworkers (65). In their procedure, the ionic content of the samples is preconcentrated via evaporation. The bromide and iodide are derivatized with ethylene oxide in sulfuric acid to 2-bromo- and iodoethanol, respectively. The derivatives are extracted with cyclohexane/ethyl acetate (7:3), and 1 μl of the organic phase is analyzed by GC on a DB wax-coated, 30-m-long column operated at 100°C using an ECD detector and He carrier gas. Iodite can be analyzed in the same way after reducing to iodide with sodium nitrite.

2. *Iodine Measurements*

Titration of iodine with thiosulfate or phenylarsin oxide titrant is an everyday task in iodometric analysis. The iodine content of water samples, however, is much lower than the detection limit of this titration, even with amperometric endpoint location. Some of the highly sensitive electrometric inverse methods have been successfully used.

Scholz and coworkers (66) determined iodine in seawater with a differential pulse polarographic (DPP) method. In their work, 50-cm^3 seawater samples were extracted with 5 cm^3 benzene. To the extracts 12 cm^3 volumes of ethanolic 0.1 M–KOH/ethanol/ethanolic 1 M–acetic acid (5:14:5) were added. The solutions were transferred to a polarographic cell containing a hanging mercury-drop electrode, and after 10-s deposition time at 0 V electrode potential, DPP scans were performed with 50-mV modulation amplitude and 5-mV/s scan rate. With this method an iodine concentration as small as 1 nM could be detected.

Silver disc microelectrode and differential pulse voltammetry can be used successfully for trace level iodine determination in tap water samples. According to Fang and coworkers (67), the mixture of 4 cm^3 acetate buffer (pH 5), 1 cm^3 1 M EDTA solution, and 5 cm^3 water sample is separated, after the silver working electrode is preconditioned by cycling between +0.1 and 0.6 V. A 1-min electrolysis time follows, and the differential stripping pulse voltammetry step is performed with a 50-mV pulse amplitude. The peak at −0.32 V is evaluated. A calibration graph, which is linear in the range of 50 nM–1.2 μM, is used for the evaluation.

III. APPLICATION OF ION CHROMATOGRAPHY

Chromatography is considered one of the most powerful analytical techniques today. It is recognized that a high number of inorganic ions—among them the halides—can be separated and determined by ion chromatographic methods, and the ion chromatographs are getting more advanced and wider spread in analytical laboratories, so they are gaining application in the prac-

Table 4 RSD Values of IC Measurements of Different Halide Anions at Low Concentrations

Component	Range of concentration (ppm)	RSD (%)
Fluoride	0.010–1.000	4.7
Chloride	0.015–1.500	4.6
Bromide	0.025–2.500	3.3

RSD: relative standard deviation.

tice of water analysis. Methods have already been worked out for the simultaneous determination of different anion species in a wide variety of water samples. F^-, Cl^-, Br^-, I^-, ClO_2^-, ClO_3^-, BrO_3^-, and IO_3^- seem to be the anions most often analyzed in this way.

In the last decade, ion chromatographic methods have been proposed as standard analytical procedures for different water samples. The development of this separation technique has been quite rapid. Theoretical models also (68) support this advantage. Ion chromatography keeps up with the newest water treatment technology and health requirements (69). Therefore its importance in water analysis is expected to grow in the future.

The ion chromatographic methods used in water analysis usually employ electric conductivity detectors with some kind of suppressor (70). Its ease of use, simplicity to maintain, and broad dynamic concentration range make it popular. Solute-specific amperometric detectors have also been successfully used. The electric conductivity detectors are universal ion sensors. Less frequently, specific electrometric detectors are also selected. Among them, solute-specific amperometric detectors are getting popular. Colorimetric and UV spectrometric methods are also in use.

Table 5 Measuring Range of the ASTM Method for Different Halide Ions

Component	Range of concentration (mg/L)
Fluoride	0.26–8.49
Chloride	0.78–26.0
Bromide	0.63–21.0

Columns for suppressed-mode anion chromatography for the analysis of F^-, Cl^-, and Br^- ions) are available in the market. They utilize pellicular packings, e.g., substituted styrene-divinylbenzene copolymer coated with unique quaternary amine functional groups. In catalogs, trade named columns can be found, like IonPac AS4A from Dionex, STAR-ION A300 from Phenomenex, and Sarasep AN300* from Serasep. These columns meet all requirements for EPA Method 300.

Advantages are that the eluents usually employed in halide analyses are inexpensive: sodium carbonate, bicarbonate, and hydroxide or a mixture of them. Phenomenex (71) suggests a 3.6 mM Na_2CO_3 eluent for analyses of halogen anions (F^-, Cl^-, Br^-, BrO_3^-, ClO_3^-, and I^-) on a STAR-ION A300 column. 1.7 mM Na_2CO_3 and 1.8 mM $NaHCO_3$ buffer is the eluent when Dionex AS4A and AS9 columns are used for analyses of F^-, Cl^-, Br^- and ClO_2^-, BrO_3^-, Cl^-, ClO_3^- anions, respectively, in drinking water, reagent water, and

Table 6 Application of IC to the Halogen Analysis of Water Samples

Ion	Column	Eluent	Detector	Sample	Ref.
Br^-	Dionex AS3	3.0 mM $NaHCO_3^-$, 2.0 mM Na_2CO_3	Amp.	Groundwater	80
Br^-	Two HPIC AS4 columns in series	2.8 mM $NaHCO_3$, 2.3 mM NA_2CO_3	UV	Seawater	81
Br^-, Cl^-, (NO_3^-, SO_4^{2-}, HCO_3^-, NO_2^-, $S_2O_3^{2-}$, $H_2PO_4^-$, $HCOO^-$)	Vydac 302 IC	4.0 mM sodium hydrogen phthalate	Cond.	Drinking water	82
Cl^- (NO_3^-, SO_4^{2-})	TSK gel IC-anion PW	0.4 mM trimellitate	UV-VIS	River water	83
Cl^- (NO_3^-, SO_4^{2-})	Vydac 302 IC	4.0 mM phthalic acid (pH 5)	Cond.	Wastewater	84
F^-	Shodex IC I-52	2.5 mM phthalic acid pH 4.0	Cond.	Seawater	85
F^-, Cl^-, Br^- (NO_3^-, SO_4^{2-})	Dionex anion exchange	2.5 mM Na_2CO_3, 3.0 mM $NaHCO_3$	Cond.	Seawater	86
I^-	TSK gel IC anion PW	0.1 mM NaCl, 5 mM sodium phosphate pH 6.7	Amp.	Seawater	87
Cl^-, (NO_3^-, SO_4^{2-})	Waters IC-Pak-C	10 mM phenyl-ethylamine pH 5.5	Cond.	Surface water	88
F^-, Cl^-, Br^- (NO_3^-, SO_4^{2-})	Dionex anion exchange	2.0 mM Na_2CO_3, 3.0 mM $NaHCO_3$	Cond.	Pore water	89
Br^-, I^-, ($S_2O_3^{2-}$, SO_3^{2-}, and SCN^-)	Spherisorb SAX	20 mM $NaNO_3$, 10 mM NaH_2BO_3 pH 7	Amp.	Water	90
Br^-, BrO_3^-, IO_3^-, and I^-	Excelpak ICS-A13 $\times$ 2	5.0 mM Na_2CO_3, 1.0 mM $NaHCO_3$	ICP-MS	Raw water, ozonized water	91
Br^-, BrO_3^-, IO_3^-, and I^-	Dionex IonPac AG10	100 mM NaOH	ICP-MS	Drinking water	92
I^-	Dionex IonPac AS11	Methansulphinic acid, 5.84 g/L NaCl, 4,4'-bis(dimethylamino) diphenylmethane in methanol	UV/VIS postcolumn reaction	Seawater	79

Amp: amperometry.
Cond: conductivity.

wastewater (72). The order of elution is dependent on the column used. The retention time of the species is determined by the eluent as well.

Before analysis, natural and technological water samples usually need very simple pretreatment. Cooling down or filtration is required in some cases (79). A commercial syringe filter of 0.2-μm pore size used for direct seawater sample injection saves the sample filtering step.

The sample volume or sample injection loop used in ion chromatography for water analysis is in the range of 1–50 μl. In extreme cases, a 1-ml sample volume is possible as well. If the sample concentration is too high, then sample dilution or a smaller injection loop is generally recommended, as is a preconcentration step for very low sample concentrations.

It is important to mention that for the analysis of a low-halogen-anionic-concentration, high-purity water, so-called reagent water (73) is needed to dilute the sample, to prepare standard solution and eluent. Different kinds of water purification methods, like distillation, ion exchange, membrane filtration, reverse osmosis, electrodialysis, and their combination, can be successfully used for the production of reagent-grade water.

Standard methods are recommended for the determination of trace levels (μg/L) of fluoride, chloride, and bromide at the same time with other anions in high-purity water when online analysis is required (74,75).

Table 4 illustrates the accuracy of IC measurements for lower concentrations. Table 4 summarizes the results obtained in a recent survey (76). In the quoted experiments, Dionex 4SA4 and AG4 columns, isocratic separation with 1.7 mM Na_2CO_3–1.8 mM $NaHCO_3$ eluent, and suppressed conductometric detection were used. For injection, a 10-μl sample loop was applied.

The ASTM standard method for halides and the other major anionic component analyses in drinking water and wastewater, issued in 1997, involves a chemically suppressed IC method (77). The measuring concentration ranges of this method for different ions are listed in Table 5.

Water samples taken from sea or waste effluents contain a high salt concentration. The analysis of anionic components forming strong acids, such as chloride and bromide, can be accurately done in these samples (78) with suppressed ion chromatography. The generally difficult trace level iodide determination is not an easy task in IC either, especially in samples with a high background salt concentration. For the deter-

mination of a ppm level of iodide in saline water, Branda et al. (79) worked out the so-called on-column matrix elimination technique. Applying a 150-μl sample volume, they found a relative standard deviation of 3.2% for 5-ppm iodide. In their method the eluent solution contains the same background electrolyte as the sample.

There are numerous reports on the use of IC of common anions such as fluoride, chloride, bromide, iodide, chlorite, and bromite in a wide range of water samples, such as wastewater, drinking water, and seawater. A wide selection of IC separation and detection methods for halides are summarized in Table 6.

ABBREVIATIONS

AAS	atomic absorption spectroscopy
COD	chemical oxygen demand
DC	direct current
DPD	diethyl-p-phenylene-diamine
DPP	differential pulse polarography
ECD	electron capture detector
EDTA	ethylene-diamine-tetra-acetate
EMF	electromototive force
EPA	Environmental Protection Agency
FIA	flow injection analysis
GC	gas chromatography
IC	ion chromatography
ISE	ion-selective electrode
MS-MS	mass spectroscopy–mass spectroscopy
RNA	ribonucleic acid
RSD	relative standard deviation
TISAB	total ionic strength adjusting buffer
UV	ultraviolet
WHO	World Health Organization

REFERENCES

1. J. Y. Gao, A. H. Xian, Lihua Jianyan, Fene Huaxue. 33: 220–222, 1997.
2. M. Beltz, W. J. O. Boyle, K. F. Klein, K. T. V. Grattan. Sens. Actuators B 39:380–385, 1997.
3. S. F. Bosch, M. G. Bosch, P. Rodriguez. J. AOAC Int. 80:1117–1121, 1997.
4. R. P. Pantaler, L. A. Egorova, L. I. Avramenko, A. B. Blank. Zh Anal. Khim. 51:521–524, 1996.
5. H. C. Hu. J. Am. Water Works Assoc. 73:150–153, 1981.
6. B. Saunier, Y. Regnier. Eau. Ind. 56:43–49, 1981.

7. E. Barbolani, G. Piccardi, F. Pantani. Anal. Chim. Acta 132:223–228, 1981.

8. H. Junli, L. Wang, N. Ren, X. L. Liu, R. F. Sun, G. Yang. Wat. Res. 31:455–460, 1997.

9. R. B. Smart, J. W. Freese. J. Am. Water Works Assoc. 74:530–531, 1982.

10. E. M. Aieta, P. V. Roberts, M. Hernandez. J. Am. Water Works Assoc. 76:64–70, 1984.

11. C. L. Elleouet. Madec. Analusis 24:199–203, 1996.

12. T. Vatanabe, T. Ishii, Y. Yoshimura, H. Nakazawa. Anal. Chim Acta 341:257–262, 1997.

13. G. Z. Wang, L. Yuan. Anal. Lett. 30:1415–1421, 1997.

14. J. W. Williams. Handbook of Anion Determination. Butterworths, London, 1979.

15. H. N. Armstrong, H. H. Gill, R. F. Rolf. In: Treative on Analytical Chemistry (Ed. I. M. Kolthoff, P. J. Elving). Part II. Vol. 7. Wiley Interscience, New York, 1961, pp. 116–121.

16. I. M. Kolthoff, E. B. Sandell, E. J. Meehan, S. Bruckenstein. Quantitative Chemical Analysis. 4th ed. Macmillan, London, 1969.

17. K. Little. Talanta 18:927, 1971.

18. M. N. Desai, K. C. Desai, M. H. Gandhi. Lab. Prac. 18:939–1063, 1969.

19. British Standard 2690: Part 6 1968.

20. E. Bishop. Mikrochim. Acta 619, 1956.

21. E. Bishop. Analyst 83:212–221, 1958.

22. E. Bishop, R.G. Dhaneshwar. Analyst 87:845–852, 1962.

23. E. Cotlove, H. V. Trantham, R. L. Bowman. J. Lab. Clin. Med. 51:461–469, 1958.

24. H. A. Laitinen, J. M. Kolthoff. J. Phys. Chem. 45 (1941) 1079.

25. J. J. Lingane. Analyt. Chem. 26:622–629, 1954; A. Cedergren, G. Johansson. Talanta 18:917–923, 1971.

26. F. W. Cheng. Microchem. J. 3:537–543, 1959.

27. APHA. Standard Methods for the Examination of Water, Sewage and Industrial Waters. The American Public Health Association and the American Waterworks Association. 13th ed. American Public Health Association, New York, 1971.

28. Society for Analytical Chemistry. Official, Standardized and Recommended Methods of Analysis. Society for Analytical Chemistry, London, 1973, pp. 351–359.

29. ASTM. Annual Book of ASTM Standards. Part 23. Philadelphia, Am. Soc. for Test and Mat., 1971.

30. J. E. Barney, R. J. Bertolacini. Analyt. Chem. 29:1187–1195, 1957.

31. R. J. Bertolacini, J. E. Barney. Analyt. Chem. 30:202–211, 1958.

32. E. Scheubeck, O. Ernst. Z. Analyt. Chem. 249:370–378, 1970.

33. M. Taras. Analyt. Chem. 19:799–803, 1947.

34. D. F. Boltz, W. J. Holland. Colorimetric Determination of Non-metals (Ed. D. F. Boltz). Wiley Interscience, New York, 1958.

35. J. Raminez-Munoz. Analytica chim. Acta 74:309–317, 1975.

36. J. Chwastowska, Z. Marczenko, U. Stolarczyk. Chemia. Analit. 8:517–523, 1963.

37. A. B. Lamb, P. W. Carleton, W. B. Meldrum. J. Am. Chem. Soc. 42:251–259, 1920.

38. W. Reichel, L. Acs. Analyt. Chem. 41:1886–1893, 1969.

39. H. Fujinuma, K. Kasama, K. Takeuchi, S. Hirano. Japan Analyst 97:1487–1494, 1970.

40. R. Belcher, A. Nadjafi, J. A. Rodroguez-Vazquez, W. I. Stephen. Analyst 97:993–999, 1972.

41. J. F. Lechner, I. Sekerka. J. Electroanal. Chem. 57:317–325, 1974.

42. K. Torrance. Analyst 99:203–211, 1974.

43. G. Nagy, K. Toth, Zs. Feher, J. Kunovics. Anal. Chim. Acta 319:49–58, 1996.

44. Z. Y. Zhao. Fenxi Ceshi Xuebao. 16:32–34, 1997.

45. S. R. Piedade, C. C. Oliveira, E. A. G. Zagatto. Quim. Anal. (Barcelona) 16:233–237, 1997.

46. L. C. Hong, X. G. Zhou, Lihua Jianyan. Huaxue Fence. 33:11–13, 1997.

47. T. P. Aleksandrova, Y. B. Kletenik. Zavod. Lab. 63:7–10, 1997.

48. WHO Geneva 1982.

49. F. J. Green. The Sigma-Aldrich Handbook of Stains, Dyes and Indicators. Aldrich Chemical Company, Milwaukee, Wisconsin, 1991, p. 209.

50. M. S. Frant, J. W. Ross. Science 154:1553–1561, 1966.

51. M. S. Frant, J. W. Ross. Analyt. Chem. 40:1169–1174, 1968.

52. S. S. M. Hassan. Michrochim. Acta 889–895, 1974.

53. P. A. Evans, G. J. Moody, J. D. R. Thomas. Lab. Pract. 20:644–679, 1971.

54. J. Alpizar, A. Crespi, A. Cladera, R. Forteza, V. Cerda. Electroanalysis 8:1051–1054, 1996.

55. D. R. R. Sarma, S. L. N. Rao. Bull. Environ. Contam. Toxicol. 58:241–247, 1997.

56. Y. M. Liao, J. F. Wang, M. Z. Feng, W. Wang, Q. S. He, G. Y. Wu. Fenxi Huaxue 25:201–204, 1997.

57. J. D. Winefordner, Tin Maung. Analyt. Chem. 35:382–389, 1963.

58. D. C. White. Microchim. Acta 449–453, 1961.

59. A. Steyermark, R. A. Lalancette, E. M. Contreras. J. Ass. Off. Analyt. Chem. 55:680–687, 1972.

60. S. I. Gusev, E. V. Sokolova, I. A. Kozhevnikova. Zh. Analit. Khim. 17:499–503, 1962.

61. L. Charles, D. Pepin, B. Casetta. Anal. Chem. 68:2554–2558, 1996.

62. U. Lundstrom, A. Olin, F. Nydahl. Talanta 31:45–48, 1984.

63. M. Denis, W. J. Masschelein. Annuluses 9:429–432, 1981.

64. C. A. Abeledo, I. M. Kolthoff. J. Am. Chem. Soc. 53: 2893–2899, 1931.

65. S. Kirchner, A. Stelz, E. Muskat. Lebensm.-Forsch. 203:311–315, 1996.

66. A. Moller, M. Lovric, F. Scholz. Int. J. Environ. Anal. Chem. 63:99–106, 1996.

67. B. Fang, S. P. Li, H. Q. Fang, H. Y. Chen. Fenxi Huaxue 25:59–62, 1997.

68. P. Hajós, O. Horvath, V. Denke. Anal. Chem. 67:434–440, 1995.

69. WHO. WHO Guidelines for Drinking-water Quality. 2nd ed. vol. 1. Recommendations, World Health Organization, Geneva, 1993.

70. U.S. EPA. The Determination of Inorganic Anions in Water by Ion Chromatography, Method 300.0. EPA, Washington, DC, 1989.

71. Phenomenex for Chromatography Catalog. 1998.

72. U.S. EPA. The Determination of Inorganic Anions in Water by Ion Chromatography, Method 300.0. EPA, Washington, DC, 1991.

73. ASTM Standard 1997. D1193-91 Standard Specification for Reagent Water.

74. ASTM Standard 1997. D5542-92 Standard Test Methods for Trace Anions in High-Purity Water by Ion Chromatography.

75. ASTM Standard 1997. D5996-96 Standard Test Method for Measuring Anionic Contaminants in High-Purity Water by On-Line Ion Chromatography.

76. L. Nagy. Separation of inorganic anions in highly diluted solution by ion chromatographic method. Proceeding of Chemist Conference in Eger, Hungary, p. 83, 1996.

77. ASTM Standard 1997. D4327/97 Standard Test Method for Anions in Water by Chemically suppressed Ion Chromatography.

78. R. P. Singh, N. M. Abbas, S. A. Smesko. J. Chromatogr. A 733:73–91, 1996.

79. A. C. M. Brandao, W. W. Buchberger, E. C. V. Butler, P. A. Fagan, P. R. Haddad. J. Chromatogr. A 706:271–293, 1995.

80. G. S. Pyen, D. E. Erdman. Anal. Chim. Acta 149:47–55, 1987.

81. E. L. Johnson, K. K. Haak. Anion Analisis by Ion Chromatography. Chapter 6. In: Liquid Chromatography in Environmental Analysis (ed. J. F. Lawrence). Chap. 6. Humana Press, Clifton, NJ, 1984.

82. S. Dogan, W. Haerdi. Chimia 35:339–349, 1981.

83. S. Motomizu, I. Sawatani, T. Hironaka, M. Oshima, K. Toei. Bunseki Agaku 36:77–89, 1987.

84. J. A. Hern, G. K. Rutherford, G. W. van Loon. Talanta 30:677–6675, 1983.

85. T. Okutani, M. Tanaka. Bunseki Kagaku 36:169–175, 1987.

86. H. Itoh, Y. Shinbori. Bunseki Kagaku 29:239–246, 1980.

87. H. Itoh, H. Sunahara. Bunseki Kagaku 37:292–299, 1988.

88. C. Erkelens, A. H. Biliet, L. De Galan, E. W. B. De Leer. J. Chromatogr. 404:67–79, 1987.

89. G. S. Pyen, M. F. Fishman. In: Ion Chromatographic Analysis of Environmental Pollutants (ed. D. Mulik and E. Sawicki). Vol. 2. Ann Arbor Sci. Publ. Ann Arbor, MI, 1979, p. 235.

90. A. Liu, L. Xu, T. Li, S. Dong, E. Wang. J. Chromatogr. A 699:39–52, 1995.

91. M. Yamanaka, T. Sakai, H. Kumagai, Y. Inoue. J. Chromatogr. 789:259–269, 1997.

92. J. T. Creed, M. L. Magnuson, J. D. Pfaff, C. Brockhoff. J. Chromatogr. 753 (1996) 261.

11

Sulfate, Sulfite, and Sulfide

Shreekant V. Karmarkar
Zellweger Analytics, Milwaukee, Wisconsin

M. A. Tabatabai
Iowa State University, Ames, Iowa

I. INTRODUCTION

Among the various forms of sulfur in the biosphere, sulfate has received the most attention because of its mobility, availability to plants and microorganisms, precipitation reactions, and reactivity with positively charged surfaces in the soil environment. Sulfate occupies the center of the sulfur cycle, with a significant portion of the cycle being in the biosphere. Therefore, sulfate is widely distributed in nature, and its concentration in natural waters may range from a few to several thousand mg L^{-1}. Because of pyrite oxidation, mine drainage effluents may contain very high sulfate concentrations. Because Na and Mg sulfates exert a cathartic action, the recommended sulfate concentration in potable supplies is limited to 250 mg L^{-1} (1).

Although sulfite is not very stable and does not occur in natural waters, it may occur in certain industrial wastes and polluted waters. It is most commonly found in boiler and boiler feed waters to which Na_2SO_3 has been added to reduce dissolved oxygen to a minimum and prevent corrosion. Other materials that may contain sulfite are those treated with SO_2 as a preservative.

Sulfides occur in many well waters and some surface waters as a result of sulfur reduction of organic matter by bacterial action under anaerobic conditions. Various concentrations may also be found in waters receiving sewage or wastes from tanneries, paper mills, oil refineries, chemical plants, and gas manufacturing works

(1). Many water samples contain trace concentration of sulfides. It is important to use zinc acetate in a procedure to fix sulfide as zinc sulfide and zinc hydroxide and to preserve the sample. This treatment preserves the sample for up to 24 h.

This chapter will summarize various analytical methods for sulfate, sulfite, and sulfide. Further details on the individual methods are reported in the references cited.

II. SULFATE

The difficulties associated with the determination of microgram quantities of sulfate in water samples have long been recognized (2,3). Many of the methods used for the determination of the sulfate ion are either time-consuming or require considerable analytical skill. Several methods are available for the determination of sulfate. These include, but are not limited to, gravimetric, turbidimetric, nephelometric, titrimetric, colorimetric, and ion chromatographic methods. The method selected depends on the type of sample, the sulfate concentration, and the accuracy and precision desired.

A. Gravimetric Method with Ignition of Residue

The principle involved is the precipitation of sulfate as $BaSO_4 \cdot 2H_2O$ in an HCl medium by the addition of

$BaCl_2$. The precipitation is carried out near boiling temperature. The precipitate is filtered, washed with water until free of chloride, ignited or dried, and weighed as $BaSO_4$. This method is subject to many positive and negative errors, depending on the chemical composition of the water sample. The analyst should be familiar with the more common interferences. When mineral concentration is low, such as in potable water, the errors are of minor importance. The interferences that lead to positive errors are suspended matter, Si, $BaCl_2$ precipitate, NO_3^-, and SO_3^{2-}. Those that lead to negative errors are alkali metals and heavy metals such as Cr and Fe.

B. Turbidimetric Method

Determination of sulfate by turbidimetric methods appears to be the simplest approach for the analysis of water samples. These methods are usually rapid and sensitive and normally do not require special skill in handling a large number of analyses at one time. Most turbidimetric methods, however, have a serious shortcoming in that they do not give reproducible results; the formation of reproducible $BaSO_4$ suspensions under uniform precipitating conditions is difficult. To overcome some of the difficulties associated with turbidimetric methods, Tabatabai (4) developed a simple method for determination of sulfate in water samples, including rainfall and snowmelt. The method is based on the quantitative precipitation of sulfate as small, uniform-sized crystals of $CaSO_4$ in a suspension medium containing HCl and gelatin, and on measuring the turbidity formed in the water sample. The results obtained by this method are comparable to those obtained with the methylene blue reduction method of Johnson and Nishita (5). In this method, an aliquot of water sample (filtered through a 0.2-μm Metricel GA-8 membrane filter, Gelman Instrument Co., Ann Arbor, Mich) containing 5–100 μg of sulfate-S (2–20 ml of water) in a dry 50-ml Erlenmeyer flask is treated with 2.0 ml of 0.5 M HCl after adjusting the volume to 20 ml. The solution is mixed, by swirling the flask, and treated with $BaCl_2$–gelatin reagent. After 30 min, the flask is swirled again, and the resulting turbid mixture moved into a Klett–Summerson colorimeter cell (2-cm light path), where the turbidity is measured with a Klett–Summerson photoelectric colorimeter fitted with a blue (no. 42) filter. Sulfate-S of the analyzed solution is calculated from a calibration graph prepared with standard solutions containing 0, 20, 40, 60, 80, and 100 μg of sulfate-S per 20 ml. The turbidimetric method described has the advantages of being simple, rapid, and

subject to almost no interference from other ions commonly found in water samples. Tests indicated that the recovery of 40 and 80 μg of sulfate-S in 20 ml was quantitative (99.7–100.2%) when the solution analyzed was brought to 100 mg L^{-1} with respect to the following cations and anions: Ca^{2+}, Mg^{2+}, K^+, Na^+, NH_4^+, NO_3^-, NO_2^-, Cl^-, PO_4^{3-}, and HCO_3^-.

An automated method based on flow injection analysis (FIA) has been reported (6). In this FIA method, deionized water is used as a carrier, and it injects 520 μl of sample into a flowing stream of reagents consisting of barium chloride solution containing gelatin and polyvinyl alcohol (PVA) and hydrochloric acid. The barium sulfate precipitate is suspended as a colloid with gelatin and PVA, and it scatters light at 430 nm to produce a signal proportional to sulfate concentration. The method has a range of 1–100 mg SO_4^{2-}/L, and 45 samples can be analyzed in an hour.

C. Colorimetric Methods

Several colorimetric methods are available for the determination of sulfate (2). These include reaction with barium chloranilate in alcohol solution and spectrophotometric measurement of the highly colored chloranilate ion formed. The method is reported to be rapid and relatively free from interferences by other anions, such as phosphate, oxalate, bicarbonate, chloride, and nitrate. Cations interfere, but those can be removed by cation-exchange resin (7). The Technicon Auto-Analyzer has been adapted for the automatic determination of sulfate in natural waters by barium chloranilate; 15 determinations of water samples containing from 5 to 400 mg sulfate L^{-1} can be accomplished per hour (8).

Another colorimetric method is that involving barium chromate (9). In this method, an excess of barium chromate is added to precipitate barium sulfate and the excess Ba is then precipitated with dilute NH_4OH. After removing both precipitates, the free chromate, which is equivalent to the sulfate removed, is determined colorimetrically at 436 nm.

Another procedure is available for automatic colorimetric determination of sulfate in water by methylthymol blue. This method has been used by Lazrus et al. (10) for the determination of sulfate in rainwater. The interfering cations must be removed with cation-exchange resin to prevent complexation with the indicator. The system must be cleaned with EDTA to remove deposits of barium sulfate. This method is sensitive in a range from 0.5 to 50 mg sulfate L^{-1}. This method is too sensitive for the determination of sulfate in natural water containing 5–400 mg of sulfate L^{-1} (8).

Lastly, a colorimetric method based on FIA is also available. In this method, at pH 13.0 barium forms a blue color complex with methylthymol blue (MTB). This gives a dark blue baseline. The sample is injected into a low, but known, concentration of sulfate. The sulfate from the sample then reacts with the ethanolic barium–MTB solution and displaces the MTB from the barium to give barium sulfate and uncomplexed MTB. Uncomplexed MTB has a gray color. The pH is raised and the gray color of uncomplexed MTB is measured at 460 nm (11). The method has a range of 2–100 mg SO_4^{2-}/L, and 60 samples can be analyzed in an hour.

D. Ion Chromatography

Ion chromatography (IC) falls under the broad category of liquid chromatography. Typical components of an IC system include an optional autosampler, a high-pressure pump, an injection valve with a sample loop of suitable size (typically 10–250 μl), a guard column, an analytical column, an optional suppression or derivatization system, a flow-through detector and a data system ranging in complexity from a chart recorder to a computerized data system. A suitable mobile phase, called the *eluent*, is constantly flowing through the columns and the detector. Typically, all of the components in contact with the eluent and sample are made from inert components such as PEEK (polyetheretherketone). Following suitable sample preparation, usually filtering through a 0.45-μm membrane filter and diluting as required, the sample is introduced to the IC via the injection valve. The valve injects a fixed volume of sample onto the guard and analytical columns. The ions of interest are then separated by means of differing affinities for the column packing material as the ions are swept along by the flowing eluent.

Several IC methods, based on ion-exchange separation, are available for the determination of sulfate in waters. Among the seven commonly determined anions by IC (Br^-, Cl^-, F^-, NO_3^-, NO_2^-, PO_4^{3-}, and SO_4^{2-}), the ion-exchange material has the highest affinity toward SO_4^{2-}, and hence it elutes last. These IC methods essentially fall into two categories (12,13), suppressed and nonsuppressed.

In *suppressed* conductometric detection, the eluent exiting the analytical column is first fed to a chemical suppressor. Suppressed IC with conductivity detection was introduced in 1975 (14). The eluents commonly used are NaOH, NaHCO₃, and NA₂CO₃ at various proportions. The suppressor, in H^+-form ion exchanger,

lowers the background conductance of the eluent and at the same time converts the anions into their highly conducting acid forms. Several IC systems employing different forms of suppression devices are commercially available (12,13).

In the second type, called *nonsuppressed* conductometric detection, the analytical column effluent flows directly to a conductivity detector. Nonsuppressed IC was introduced in 1979 by Gjerde et al. (15). The typical eluents used in nonsuppressed IC of anions are phthalic acid and *p*-hydroxybenzoic acid for the determination of anions. The equivalent conductance values of chloride, sulfate, and other common anions are appreciably greater than that of the eluent anion, and hence a positive peak is detected as the anions are carried through the detector.

The nonsuppressed IC methods, however, did not gain as much widespread acceptance as did suppressed IC, especially in environmental analysis, for several reasons. The first was that the regulatory methods, such as USEPA method 300.0, are based on suppressed IC. The second reason was that the signal/noise ratio is much greater with suppressed IC than with nonsuppressed IC. Lastly, the modern suppression devices developed since 1981, a couple of years after the introduction of nonsuppressed IC, eliminated the drawbacks of the original packed-bed suppressor (13).

The regulatory IC methods for the analysis of waters include those published by ASTM (American Society for Testing and Materials), Standard Methods, and USEPA (13). The typical run times for suppressed and nonsuppressed IC methods are in the range of 8–16 minutes. Using a suppressed IC method with a 200-μl sample loop, sulfate can be quantified in the range of 0.1–100 mg SO_4^{2-}/L with a method detection limit of 0.016 mg SO_4^{2-}/L. In various waters, only Cl^-, NO_3^-, PO_4^{3-}, and SO_4^{2-} are present in appreciable concentrations (16). When determining only these four anions, the IC method for all seven anions is unnecessarily slow because of a throughput of six to eight samples per hour. To improve the throughput for analyses of such water samples, a rapid IC method, with a run time of 3 minutes, has been developed for the determination of Cl^-, NO_3^-, PO_4^{3-}, and SO_4^{2-} (13).

E. Other Methods

A number of highly specialized methods are available for the determination of sulfate in water samples or can be adapted for water samples. These include: (i) amperometry, (ii) atomic absorption spectrophometry, (iii) conductimetry, (iv) electrometry, (v) flame photometry,

(vi) high-frequency titration (oscillometry), (vii) infrared spectroscopy, (viii) polarography, (ix) radiometry, (x) thermogravimetry, and (xi) methylene blue method after reduction to H_2S (2).

III. SULFITE

A. Titration Method

An acidified water sample containing sulfite is titrated with a standard KI–KIO$_3$ solution. Free I_2 is released when the sulfite has been oxidized, resulting in the formation of a blue color in the presence of starch indicator. This method has a detection limit of 2 mg SO_3^{2-} L^{-1} (1).

B. Ion Chromatography

International Standard ISO 10304-3 (17) recommends both suppressed and nonsuppressed IC techniques for the determination of sulfite, along with chromate, iodide, thiocyanate, and thiosulfate, in waters. The working range, based on the conductivity detection is 0.1–50 mg SO_3^{2-}/L.

IV. SULFIDE

A. Methylene Blue Method

The zinc sulfide formed from the reaction of sulfide with zinc acetate–sodium acetate is treated with p-aminodimethylanaline–sulfuric acid solution and ferric ammonium sulfate–sulfuric acid solution to produce methylene blue. Details of the procedures and the reactions involved are described by Johnson and Nishita (5) and Tabatabai (18). A simplified and less technical procedure for very low concentrations of sulfide is also available (1).

B. Ion Chromatography

Sulfide is an anion of a weak acid, H_2S, with its pK values of 7.24 and 14.9. Suppressed IC cannot be used for the determination of sulfide, since the suppressed eluent, with pH of 4–5, lacks an ionic, and hence a conducting, form of sulfide. The HMSO analytical method is, therefore, based on amperometric detection. The IC method followed by amperometric detection, as documented in HMSO analytical methods, has a range of 0.005–1 mg S^{2-}/L (19). Sulfide could also be determined using ion-exclusion separation followed by

postcolumn derivatization with excess iodine and inverse detection at 350 nm (20). Using this method, sulfide can be determined in the range of 0.1–16 mg S^{2-}/L. This method could also be used for the determination of L-ascorbic acid, sulfite, sulfide and thiosulfate.

REFERENCES

1. American Public Health Association. 1969. Standard methods for examination of water and wastewater. American Public Health Assn., New York, pp. 287–293.
2. JD Beaton, GR Burns, J Platou. 1968. Determination of sulphur in soils and plant material. Tech. Bull. No. 14. Sulphur Institute, Washington, DC.
3. FH Rainwater, LL Thatcher. 1960. Methods for collection and analysis of water samples. U.S. Geol. Surv. Water-Suppl. ppa. 1454. U.S. Government Printing Office, Washington, DC, pp. 279–285.
4. MA Tabatabai. 1974. Determination of sulphate in water samples. Sulphur Institute Journal 10, No. 2, Summer 1974.
5. CM Johnson, H Nishita. 1952. Microestimation of sulfur in plant materials, soils, and irrigation water. Anal. Chem. 24:736–742.
6. D Diamond. 1997. Determination of sulfate by flow injection analysis, Trubidimetric method. QuikChem method 10-116-10-1-E, Lachat Instruments, 6645 West Mill Road, Milwaukee, WI 53218.
7. RJ Bertolacini, JE Barney II. 1957. Colorimetric determination of sulphate with barium chloranilate. Anal. Chem. 29:281–283.
8. ME Gales, WH Talyor, JE Longbottom. 1968. Determination of sulphate by automatic colorimetric analysis. Analyst 93:97–100.
9. K Nemeth. 1963. Photometric determination of sulphate in soil extracts. Z. PflErnahr. Dung. 103:193–196.
10. AL Lazrus, KC Hill, JP Lodge. 1965. A new colorimetric microdetermination of sulphate ion. Technicon Symposium 1965. Automation in Analytical Chemistry. Mediad, New York.
11. K Switala. 1996. Determination of sulfate by flow injection analysis colorimetry. QuikChem method 10-116-10-2-A, Lachat Instruments, 6645 West Mill Road, Milwaukee, WI 53218.
12. MA Tabatabai, WT Frankenberger, Jr. 1996. Liquid chromatography. In: DL Sparks et al. (eds). Methods of Soil Analysis. Part 3. Chemical Methods. Soil Sci. Soc. Am., Madison, WI, pp. 225–245.
13. SV Karmarkar. 1998. Ion chromatography. In: R. A.

Meyers (ed.). Encyclopedia of Environmental Analysis and Remediation. Wiley, New York, NY, pp. 2391–2404.

14. H Small, TS Stevens, WC Bauman. 1975. Novel ion exchange chromatographic method using conductimetric detection. Anal. Chem. 47:1801–1809.

15. DT Gjerde, G Schmuckler, JS Fritz. 1979. Anion chromatography with low-conductivity eluents. II. J. Chromatogr. 187:35–45.

16. MA Tabatabai, WA Dick. 1983. Simultaneous determination of nitrate, chloride, sulfate, and phosphate in natural waters by ion chromatography. Soil Sci. Soc. Am. J. 12:209–213.

17. International Standard Organization. 1997. Water Quality—Determination of dissolved anions by liquid chromatography of ions. Part 3: Determination of chromate,

iodide, sulfite, thiocyanate, and thiosulfate. ISO 10304-3. International Organization for Standardization, Case Postale 56, CH-1211, Geneve 20, Switzerland.

18. MA Tabatabai. 1996. Sulfur. In: DL Sparks et al. (eds). Methods of Soil Analysis. Part 3. Chemical Methods. Soil Sci. Soc. Am., Madison, WI, pp 921–960.

19. HMSO Publications. 1990. Special inorganic anions. In: The Determination of Anions and Cations, Transition metals, Other Complex Ions and Organic Acids and Bases in Waters by Ion Chromatography. HMSO Publications Center, PO Box 276, London, SW8 5DT, UK.

20. Y Miura, T Maruyama, T Koh. 1995. Ion chromatographic determination of L-ascorbic acid, sulfite, sulfide, and thiosulfate, using a cation-exchanger of low crosslinking. Anal. Sci. 11:617–621.

12

Analysis of Nitrates and Nitrites

Maria Teresa Oms, Amalia Cerdà, and Victor Cerdà
University of Balearic Islands, Palma de Mallorca, Spain

I. INTRODUCTION

Nitrogen is an essential nutrient for plants and animals. It exists in the environment in many forms as a part of the nitrogen cycle, with nitrate (NO_3^-) and nitrite (NO_2^-) being the most abundant ones in water systems (Fig. 1). Nitrate and nitrite are colorless and odorless ions. They are highly soluble in water and are retained weakly in soil, so they may be transported through the soil profile after rainfall or irrigation and reach underground water. In fact, nitrate is the most common contaminant in groundwater. However, its concentration depends on the land cover and soil type, groundwater level, land use, etc., with the highest concentrations being associated with carbonate-type soils, shallow groundwater, and urban and agricultural use of land. Nitrite is an intermediate oxidation state of nitrogen, occurring either in the reduction process of nitrate or in the conversion of ammonia to nitrate. Nitrite concentrations in water systems are generally below 0.5 mg/L nitrite-*N* because it oxidizes readily to nitrate. High concentrations of nitrite are usually linked to microbial activity, but they may also indicate polluted water in estuaries or near an outfall (1). When present, nitrate does not volatilize, and it is likely to remain in water until consumed by plants or other organisms. Because of its stability and solubility, many studies use nitrate as an indicator of potential water pollution, for an elevated level of nitrate in water systems may also indicate the presence of other contaminants, microbial pathogens, or pesticides. Nitrate concentrations are found in a very wide range in natural water and wastewater samples. While nitrate concentration in deep seawater and natural unpolluted groundwater is usually below 0.05 mg/L NO_3^--N, in shallow groundwater and surface streams the concentrations range from less than 0.1 mg/L NO_3^--N to 20 mg/L NO_3^--N depending on soil type, land use practices and well depth. Raw wastewaters usually have low levels of nitrate, while nitrate concentrations as high as 30 mg/l NO_3^--N are usually found in wastewater discharges and wastewater effluent plants.

Although nitrate and nitrite are naturally occurring forms of nitrogen, elevated levels in water systems usually result from human activities (2). Natural sources of nitrate and nitrite in the environment include gaseous nitrogen fixation due to the activity of microorganisms such as bacteria and blue-green algae, soil degradation, geological deposits, and, finally, decomposition of plant and animal residues in which organic nitrogen and ammonia are converted to nitrate and nitrite ions.

Anthropogenic sources of nitrate and nitrite include intensive use of chemical nitrogenous fertilizers, improper disposal of plant and animal waste, municipal and industrial wastewater discharge, sewage disposal systems, landfills, etc. (2). Agriculture is the largest contributor to the pollution of groundwater and surface water (2,3) because of the use of fertilizers and also because of the high concentration of livestock in small areas, which produces a great amount of manure. In order to prevent this type of pollution and to protect human health, certain regulations have been established (4). Atmospheric deposition of nitrogen-containing compounds has also been suggested to play an impor-

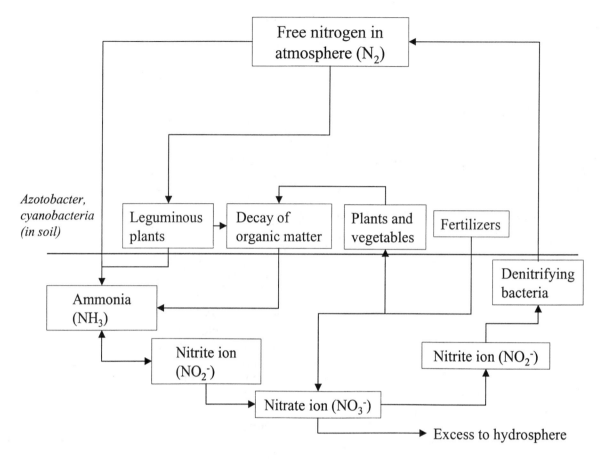

Fig. 1 Nitrogen cycle.

tant role in the nitrogen inputs to inland waters and marine systems (5–10).

High concentrations of nitrate and nitrite may have significant effects on human health as well as on the environment. In aquatic environments, the excess of nitrates may lead to eutrophication of the ecosystem due to overenrichment with nutrients. The phenomenon is of primary concern in lakes and estuarine and coastal systems, where nitrate is usually the limiting nutrient. Thus, in the presence of high nitrate levels, aquatic plants and algae are overproduced, eventually causing a number of negative impacts on the environment, such as dissolved oxygen depletion, which can lead to the death of fish and other aquatic species; sunlight blocking and the consequent elimination or reduction of photosynthesis by submerged aquatic vegetation; settling of dead algae and macrophytes to the bottom of the water system, stimulating bacteria proliferation, changes in color, odors, etc. (11). This situation is less dramatic in freshwater streams, where the limiting factor is not generally nitrate but light penetration, type

of substrate, flow velocity, and phosphate concentration. The main concern regarding surface water and groundwater containing high nitrate concentrations is derived from their use for human consumption and the potential health effects.

Ingested nitrate is the major source of nitrate in the body. The most important health effect associated with nitrate ingestion occurs when nitrate is reduced to nitrite in the digestive system. The nitrite may oxidize the iron in the hemoglobin molecule of the red blood cells to form methemoglobin, a molecule that lacks the capacity to bind and transport oxygen to the tissues. The resulting condition is known as methemoglobinemia or "blue baby syndrome" (12,13). Infants under six months of age, pregnant women, and, in general, those individuals with reduced gastric acidity or with a hereditary lack of methemoglobin reductase are at greater risk, because in most adults the enzyme system is able to convert methemoglobin back to oxyhemoglobin and hence the methemoglobin content remains low despite relatively high amounts of nitrate-nitrite intake.

Another concern about nitrate is its link to certain types of gastrointestinal cancer. It has been suggested that nitrate represents a potential risk due to the formation of *N*-nitroso compounds such as nitrosamine, which are strongly carcinogenic in animals. However, the available data are inconclusive or even contradictory (14).

Nitrate pollution is a significant problem worldwide, and recent reports from national and international organizations show that nitrate concentrations in water are increasing steadily (15–17). Due to concern about its potential hazards, most national and international organizations have set legal limits to nitrate content in drinking water and wastewater discharges (18–22) (see Table 1). Despite this, the water pollution caused by nitrates is still a serious problem. For instance, in Western Europe, where 65% of the public water supply is derived from groundwater, it is estimated that in more than 85% of groundwater sources the concentration of nitrate exceeds the EU guideline (23). The same concern applies to European rivers and lake waters, where a steady increase in nitrate concentration has been observed in the last years (24).

The literature on methods and techniques for nitrate and nitrite determination and their application to the analysis of water samples is probably one of the most extensive and increased during the 1990s. In a recent literature survey of *Analytical Abstracts* done by the authors, about 5000 papers on nitrate and nitrite analysis in water samples were found within the last 10 years. After their classification according to the detection technique applied, it was found that the methods and applications based on spectrophotometric measurements were the most frequently described, either for direct determinations or for indirect determinations after reaction with appropriate reagents. They have also been extensively and successfully coupled to chromatographic separation techniques (see later) and used for the development of automated procedures based on flow injection analysis and sequential injection analysis.

The second most frequently used group of methods were those involving electrochemical detection, especially those based on the construction and use of ion-selective electrodes (around 50% in the literature survey), followed by amperometry, voltammetry, and polarography. While potentiometry using ion-selective electrodes is the most common option for direct methods, there has been an increasing interest in the use of amperometric detection in ion chromatography, especially when high sensitivity and selectivity are required. Development of electrochemical techniques has

been also encouraged by the development of miniaturized systems and microsensors for online and in situ measurements (see later).

II. SAMPLE HANDLING AND PRESERVATION

Samples for the determination of nitrate and nitrite can be collected using glass or polyethylene bottles. Samples must be preserved by refrigeration at 4°C, and analysis should be made as soon as possible after collection, especially in the case of nitrite analysis to avoid its oxidation to nitrate (26). When samples cannot be immediately analyzed, they should be quickly frozen to $-20/-30$°C (1). If the sum of nitrite plus nitrate (often referred to as oxidized nitrogen) is to be analyzed and the samples must be stored for more than 24 hours, then they must be preserved in acidic medium (2 ml of concentrated sulphuric acid per liter) and refrigerated (27). In acidic medium, nitrite cannot be analyzed because of its fast conversion to nitrate. Addition of the appropriate reagents immediately after collection has been suggested as an alternative to avoid oxidation of nitrite (1). Addition of mercury chloride and chloroform after sample filtration on alumina has been reported to be a suitable method to preserve the nitrate and nitrite content of the samples for months (28), although this procedure is not suitable if reduction using copper-cadmium is to be used. In addition, the use of mercury salts is not recommended, for environmental reasons.

In order to avoid interference from suspended particles, it is usually necessary to filter the samples prior to analysis. This is particularly important in spectrophotometric measurements but also in those methods involving flow injection analysis, chromatographic separation, or cadmium column reduction, because suspended matter may restrict and even block sample flow by clogging the tubing and columns. Although water samples are fairly clean and it might appear that no filtration is required, in the case of chromatographic separations failure to perform the filtration step leads to a decrease in column lifetime.

Filtration can be made manually using a standard membrane filter of 0.45-μm pore size. Online filtration of large particles has also been included within the manifold for continuous analysis by inserting a column filled with XAD resins before sample injection (29), although when extremely low concentrations are to be analyzed, disposable filter units are recommended. The use of cleanup columns, especially for the determina-

Table 1 National and International Guidelines on Nitrate Content

	Guide level (in N)	Maximum level (in N)	Water use	Regulatory organization	Ref.
NO_3^-	5.6 mg/L	11.3 mg/L	Surface water for production of drinking water	European Union (EU)	18
NO_2^-	—	—			
NO_3^-	5.6 mg/L	11.3 mg/L	Drinking water	EU	19
NO_2^-	—	<0.1 mg/L			
NO_3^-	7.9 mg/L	10 mg/L	Surface water	WHO	21
NO_3^-	10 mg/L	10 mg/L	Drinking water	U.S. EPA	22
NO_2^-	1 mg/L	1 mg/L			
$NO_3^- + NO_2^-$	10 mg/L	10 mg/L			
Total Nitrogen[a]	—	10 mg/L[b] 15 mg/L[b]	Wastewater discharges	EU	25

[a]Total nitrogen = Kjeldahl nitrogen + nitrate + nitrite.
[b]15 mg/L for organic charge between 10 and 100,000 eq-h and 10 mg/L for organic charge higher than 100,000 eq-h.

tion of nitrate and nitrite by ion chromatography, has been found to be effective to remove the interference of humic substances (30). Dialysis techniques in which the selected ions are extracted and transferred through a membrane have been applied to the cleanup of samples, especially for ion chromatography, but the resulting sample dilution has limited its widespread use (31). The incorporation of a dialysis membrane in continuous flow analysis is a simple but effective option to minimize the interference of colloids and organic matter (32). Large concentrations of oil and/or grease in samples require pre-extraction using an organic solvent (33).

III. DIRECT DETECTION METHODS

A. Direct Determination by UV Spectroscopy

The direct measurement of nitrate absorbance at 220 nm has become a standard method of analysis (26,34) that does not require any additional reagent, and it is suitable for unpolluted samples with a low content of organic matter. The turbidity of samples due to suspended particles and colloidal matter, as well as organic compounds absorbing at the same analytical wave-length, may be serious interference, therefore samples should be filtered prior to analysis. Most saline ions do not absorb in the UV region, and only Br^-, SCN^-, I^-, CO_3^{2-}, Fe^{2+}, Fe^{3+}, and Cr(VI) may be an important interference in certain sample matrices. Many procedures have been described to eliminate or reduce these interferences, such as ion exchange for metals and retention on Amberlite type resins for organics (35), and precipitation and/or flocculation and filtration through activated carbon (36).

Spectral corrections to one or several wavelengths, as described by several authors (26,34,37), are frequently not enough, especially when complex samples with many interfering unknown species (i.e., wastewater) are to be analyzed.

The similarity between the spectra of nitrate and nitrite in the ultraviolet region prevents the separate determination of both ions by direct measurements in the region between 200 and 230 nm, when both ions coexist in the sample, especially if the concentration of one of them is much higher than that of the other. In that case, nitrate can be quantified alone after the elimination of nitrite interference by the addition of sulfamic acid (38). The spectral differences observed at longer wavelengths (300–360 nm) are the basis of a method developed by Wetters and Uglum in which both ions are simultaneously analyzed (39), with no need to apply separation techniques. It is based on the neglectable absorbance of the nitrate ion at 355 nm, whereas at 302 nm it displays a characteristic absorption band. The absorption ratio of nitrite ion at those wavelengths is constant, and is equal to 2.5, and this correction factor is applied to the nitrate determination at 302 nm.

The development of new spectrometers capable of recording the sample spectrum within a specific wavelength range allowed some authors new methods to eliminate the matrix interference based on substracting the matrix spectrum obtained after reducing the nitrate ions to ammonium with hydrazine hydrate (36), cupperized zinc (40), or Niquel–Raney alloy (41). A further step in direct UV measurement of nitrate has been the introduction of mathematical algorithms for spectral data treatment. Deconvolution methods have been proposed by Thomas et al. (42,43) and Karlsson et al. (44), who have applied chemometrics to the analysis of nitrates in wastewater. The main disadvantage is that the acquisition of accurate calibration spectra within the UV-visible region is complex and time-consuming and depends on the operational conditions of each plant. On the other side, the multicomponent analysis allows the application of the method to matrices with an elevated content in organic matter, such as wastewater, without any preliminary sample treatment.

B. Online Direct UV Measurement for Nitrate Determination

Online direct UV nitrate determination has been described by several authors based on the application of flow injection analysis arrangements in order to eliminate the potential interfering species. Slanina et al. (45) have proposed a method for analyzing nitrates in rainwater in which the sample is injected into a water carrier, merged with perchloric acid, and filtered online using activated carbon before reaching the UV detector. The sampling frequency is 18–35 samples per hour without interference of organic matter or metallic cations like Fe(III) or Cr(VI).

Thompson and Blankley (32) proposed a segmented flow–based system incorporating a dialysis unit to eliminate the interference of suspended particles, macromolecules, and colloids. As a result of the low effectiveness mass transfer through the dialysis membrane (<5%), the detection limit is increased to 0.091 mg/L NO_3^--N but the linear interval is extended to 30 mg/L NO_3^--N. Interference from nitrite is minimized by adding sulfamic acid to the acceptor channel of the dialysis cell and hydroxylamine sulphate to avoid the interfer-

ence of Cr(VI) and Fe(III). The method allows the analysis of 30 samples per hour.

The development of diode array detectors allows the scanning, almost instantaneously, of an entire spectrum in real time (even in flowing samples), leading to the development of general-purpose, low-cost, low-maintenance instruments but also to dedicated instruments for specific applications. This allows one to obtain a substantial amount of information and, together with the extremely fast development of computers and computational techniques, has led to a new approach in direct UV measurements for nitrate determination in water samples, based on the use of diode array detectors to obtain the UV spectrum of the sample, which is then processed by applying pattern recognition algorithms for the determination of nitrite and nitrate in complex matrices without any pretreatment. An extreme application of this method using up to 1024 wavelengths has been described for the determination of up to 15 nutrients, including nitrate, in hydroponics nutrient solutions (46,47). Very recently, commercial instruments have been designed based on this approach for online analysis of nitrate in drinking water (48) and also for the on-line analysis of more complex samples, such as wastewater, without any pretreatment. The possibility of rapid screening of spectra in real time has opened new possibilities for the utilization of direct UV measurements for monitoring purposes, e.g., in wastewater facilities (49,50).

The most recent developments in direct UV measurement for nitrate and nitrite determination in water samples are in the field of miniaturization: sensors based on fiber optics (51,52) for direct measurements or in combination with diode array detectors and further spectral data treatment by deconvolution techniques for multicomponent analysis (53).

C. Direct Electrochemical Methods

1. Potentiometric Methods

Potentiometric techniques applied to nitrate determination have been highly developed during the last 20 years, as reflected in the review by Liteanu et al. (54), in which a great variety of methods based on the use of ion-selective electrodes are described and applied to the analysis of water, plant extracts, air, and soil samples. One of the main advantages of those techniques is the possibility of the selective analysis of nitrate in complex matrices with little or no sample pretreatment, although ionic strength and pH adjustment are required. The main disadvantages is their relatively low selectivity, short lifetime, and poor stability, which have limited the application in routine water analysis (55–57). These negative aspects have been associated with gradual losses of sensor solution, favored by the high porosity of the supporting membrane initially used in commercial electrodes. Still, the use of ion-selective electrodes gives satisfactory results for drinking waters and it has become a standard method of analysis (26). Many applications to nitrate determination in drinking waters, natural waters, and residual waters have been described using porous membrane electrodes (55,58,59), allowing nitrate determination in the range 0.14–1400 mg/L NO_3^--N. Nitrite interferes with the nitrate determination and it can be masked by adding sulfamic acid to the sample solution if present (26). The interferences from chloride and bicarbonate are important and may be minimized by adding a buffer solution containing silver sulphate and maintaining a constant pH of 3, respectively.

In the 1970s a new concept in the construction of ion-selective electrodes arose, based on the development of new sensor groups by the immobilization of the ion exchanger in a polymeric matrix of PVC (60,61). The resulting nonporous-membrane ion-selective electrode had a similar working concentration range but better mechanical stability, longer lifetime, and better general performance than the porous-membrane commercial electrodes. Since then, further developments in nitrate nonporous-membrane electrodes have been based mainly either on improving the properties of the ionic groups of the membrane constituents (62,63) or on the research into new polymeric materials (64,65) to achieve higher selectivity and longer lifetimes.

Also in the '70s, solid-state electrodes were developed based on the replacement of the internal reference solution with a conductive solid element, initially a platinum wire connected directly to the electrode porous membrane (66). Following this idea a new generation of all-solid-state electrodes has been developed based on the use of either a conductive epoxy resin or a nonconductive epoxy resin with graphite dust, which is at the same time the support of a nonporous sensor membrane and the internal electrical contact (67). These electrodes have a longer lifetime than the traditional ones but also more pronounced potential drifts.

The use of automated continuous systems for the analysis of nitrate with selective electrodes has become very popular, because they offer many advantages compared to the manual batch methods, such as shorter analysis time, since it is not necessary to reach the equilibrium; greater selectivity and the minimization of interferences by kinetic discrimination; and better control of the operational parameters (speed of agitation, position of the electrode in the solution).

First attempts were done by housing a conventional liquid membrane nitrate-selective electrode into specially designed flow-through cells (68,69). Since then many continuous methods using a nitrate-selective electrode have been described, most of them based on flow injection analysis (FIA) (70–72). The development of all-solid-state electrodes and ISFETs (ion-selective field-effect transistors) electrodes has greatly facilitated the construction of robust and inexpensive electrodes and the design of sensors with special geometries—tubular or sandwich—suitable for their integration in flow systems (73–75), although stability problems have not been totally overcome yet.

In general, direct methods based on nitrate-selective electrodes offer moderate selectivity, ruled by the selectivity coefficients, and respond not only to nitrate but also to chloride, nitrite, bicarbonate, carboxylate, sulphate, and phosphate ions, among others. Most of them are commonly found in water samples and many efforts have been dedicated to the elimination of these interferences. Basically two possibilities have been applied: correction through mathematical calculations, as proposed by Langmuir and Jacobson (76), and sample pretreatment in order to eliminate organic matter, nitrite, chloride, and bicarbonate as mentioned earlier.

2. Voltammetric and Amperometric Methods

The electrochemical reactions of nitrate and nitrite on an electrode surface have led to a wide group of analytical methods for the determination of these ions. Some of them are based on the oxidation of the nitrite at moderate anodic potentials on platinum or glassy carbon electrodes for the amperometric determination of this ion (77–79). At the selected potential, the nitrate ion is not electrochemically active, but it can be analyzed after reduction to nitrite.

Chemically modified electrodes (e.g., carbon paste electrode chemically modified with an ion exchanger) for the determination of nitrite have been proposed to overcome the low sensitivity and passivating of platinum and glassy carbon electrodes (80,81).

Many methods are also described based on the electrochemical reduction of nitrate and nitrite. The reduction potential of both ions are very close, so their diffusion currents overlap. When standard electrodes such as glassy carbon, platinum, gold, and mercury, are used, the reduction step is very slow and requires extremely negative potentials, which limits its application to specific analytical problems. The presence of certain polyvalent cations in the electrolyte supports and the Cr(III)–glycine complex, activates the reduction processes, thus causing differences in the diffusion currents of nitrate and nitrite and allowing the analytical determination of these ions (78,82–84).

Some authors have proposed voltammetric methods using solid-state electrodes chemically modified to catalyze the reduction step (79,85,92), thus requiring lower potential and allowing higher selectivity.

While potentiometry is preferred for electrochemical analysis of nitrate, amperometry is the preferred technique for continuous methods for the determination of nitrite. Most continuous methods are based on the use of chemically modified electrodes that allow nitrate oxidation at moderate potential, with a wide linear range from 0.14 to 140 mg/L nitrate nitrogen (81,85–90). Amperometric sensors have been developed (52,91) for the FIA determination of nitrite and applied to the determination of nitrate as nitrite after reduction in a cadmium column (see later), with a detection limit of 0.05 mg/L NO_3^--N and a wide linear range from 1.0 to 190 mg/L NO_3^--N. The amperometric determination of nitrate has also been proposed to be achieved at a low potential by previous conversion of nitrate into a nitro derivative (92).

IV. INDIRECT DETECTION METHODS

Nitrate and nitrite ions are more frequently analyzed by indirect methods, in which a derivatization reaction is achieved to obtain other species with measurable optical or electrochemical properties.

A. Photometric Detection of Nitrite

Methods based on the reaction of nitrite with organic compounds require smooth conditions and are usually chosen for the determination of this ion because of their high selectivity and sensitivity (93). The most widely used is that based on diazotation and coupling reactions to form a highly colored azo dye. The nitrites present in the sample (or resulting from the reduction of nitrates) react with sulfanilamide to form a diazo compound that is then coupled with α-naftil-ethylenediamine hydrochloride in acidic medium to form the azo dye (reaction of Shinn). The absorbance of the azo dye is proportional to the nitrite content.

Most FIA methods for photometric measurement of nitrite are based on this reaction. Noteworthy among them are those based on the merging-zones approach

developed by Zagatto et al. (94) involving intermittent flows and merging sample and reagent flows; the kinetic method described by Koupparis et al. based on stopped flow analysis (95); and that by Ariza et al. (96) using a normal FIA configuration for determinations in the range 0.15–3 mg/L NO_2^--N and reversed FIA for nitrite concentrations in the range 0.03–0.6 mg/L NO_2^--N.

Some authors have also described continuous methods for nitrite in which the coupling reaction is eliminated and the diazo compound is monitored directly by reaction with an aromatic amine (97,98).

Recently the photometric determination of nitrite in seawater by reaction with proflavin has been described (99) for nitrite concentrations in the range 0.02–1.2 mg/L NO_2^--N.

Finally Motomizu et al. (100) have developed a continuous FIA method in which nitrite reacts directly with 3-amino-naphtalenedisulphonic acid at 90°C in hydrochloric acid to form a diazonium salt that later is transformed in alkaline medium into azoic acid, a fluorescent compound. The method is very sensitive and is applied to the analysis of nitrate in river and lake water and in seawater samples, with a detection limit of 1.4 ng/L NO_3^--N.

B. Reduction of Nitrate to Nitrite

Reactions of nitrate with aromatic organic compounds are the origin of analytical methods for indirect nitrate determination, but because of the strong experimental conditions required they are being replaced nowadays by simpler, less polluting, and more precise methods.

Most indirect methods for nitrate determination are based largely on the reduction of nitrate to nitrite and subsequent determination of nitrite. Reduction can be accomplished by adding either a homogenous or a heterogeneous reducing agent but also through photochemical or enzymatic reactions. This approach allows joint determination of nitrate plus nitrite (oxidized nitrogen) but also separate determination of both ions by carrying out the procedure first with and then without the reduction step. The reductions and type of detection in each case are specified in the Table 2.

1. Heterogeneous Reduction of Nitrate

The most frequently used reducing agent in solid state is zinc, cadmium alloys, or granulated copper-cadmium, added either directly to the sample or by passing the sample through a column containing the reducing agent.

Table 2 Methods for the Indirect Determination of Nitrate and Nitrite

Reduction step	Water type	Reducing agent	Linear range	Operation mode	Ref.
Nitrate to nitrite	Seawater	Zn	0.02–0.2 mg/L NO_3^-	Batch, photometric detection	102
Nitrate to nitrite	Natural waters	Zn	0–35 mg/L NO_3^-	Batch, photometric detection	101
Nitrate to nitrite	Seawater, river water, and tap water; industrial effluents	Cd-Zn	0.1–2.2 mg/L NO_3^-	Batch, photometric detection	109
Nitrate to nitrite	River water	Cupperized cadmium	Not specified	Batch, photometric detection	106
Nitrate to nitrite	Seawater	Cupperized cadmium	0.0002–0.3 mg/L NO_3^-	Batch, photometric detection	107
Nitrate to nitrite	Natural water	Cupperized cadmium	0–3 mg/L NO_3^-	Segmented flow, photometric	110
Nitrate to nitrite	Seawater	Cupperized cadmium	0.2–2 mg/L NO_3^-	Segmented flow, photometric	118
Nitrate to nitrite	River water, seawater	Cupperized cadmium	0–12.4 mg/L NO_3^-	FIA, fluorescence detection	100
Nitrite Nitrate to nitrite	Natural water	Cupperized cadmium	0.05–0.5 mg/L NO_2^- 0.06–0.6 mg/L NO_3^-	FIA stopped-flow, fluorescence detection	170
Nitrate to nitrite	Natural water	Cupperized cadmium	0.06–62 mg/L NO_3^-	FIA, voltammetry	91
Nitrite Nitrate to nitrite	River, lake water	Cupperized cadmium	0.08–6.6 mg/L NO_2^- 0.1–13 mg/L NO_3^-	FIA stopped-flow, photometry	95
Nitrite Nitrate to nitrite	Environmental waters	Cupperized cadmium	0.01–2.2 mg/L NO_2^- 0.1–3.5 mg/L NO_3^-	FIA, photometry	98
Nitrite Nitrate to nitrite	Natural water	Cupperized cadmium	0–1.6 mg/L NO_2^- 0–22 mg/L NO_3^-	FIA merging-zones, photometry	120
Nitrite Nitrate to nitrite	Wastewater Coastal marine water	Cupperized cadmium	0.06–4 mg/L NO_2^- 0.075–10 mg/L NO_3^-	FIA, photometry	99
Nitrate to nitrite	River, lake, well, rainwater	Cupperized cadmium	0–1.3 mg/L NO_3^-	FIA, photometry	121
Nitrate to nitrite	Estuarine, coastal water	Cupperized cadmium	0.06–4.5 mg/l NO_3^-	FIA, photometry	123
Nitrate to nitrite	Tap, rain, and river water	Cupperized cadmium	LOD: 0.04 mg/L NO_3^-	FIA, photometry, and coulometry	113
Nitrate to nitrite	River water	Cupperized cadmium	0–53 mg/L NO_3^-	FIA, photometry	124
Nitrite Nitrate to nitrite	Wastewater	Cupperized cadmium	0.05–5 mg/L NO_2^- 0.25–50 mg/L NO_3^-	FIA, photometry	111, 112
Nitrate to nitrite	Seawater, natural waters	Hydrazine	0.09–2.6 mg/L NO_3^-	Batch, photometric detection	128
Nitrate to nitrite	Seawater	Hydrazine	4.4–88 mg/L NO_3^-	Batch, photometric detection	129
Nitrate to nitrite	Seawater, estuarine water	Hydrazine	0–1.3 mg/L NO_3^-	Batch, photometric detection	130
Nitrate to nitrite	Groundwater, surface, tap water	Hydrazine	0–5 mg/L NO_3^-	Batch, photometric detection	131
Nitrate to nitrite	Natural water	Hydrazine	0.3–7 mg/L NO_3^-	Batch, photometric detection	132
Nitrate to nitrite	Natural water	Hydrazine	1–10 mg/L NO_3^-	FIA, photometric detection	133
Nitrate to nitrite	River water	UV reduction	0.02–0.7 mg/L NO_3^-	Batch, photometric detection	137

Table 2 Continued

Reduction step	Water type	Reducing agent	Linear range	Operation mode	Ref.
Nitrate to nitrite	Riverwater, seawater	UV reduction	0.003–0.6 mg/L NO_3^-	FIA, photometric detection	136
Nitrate to nitrite	River, tap water	Nitrate reductase	0.05–7.5 mg/L NO_3^-	Batch, fluorimetric detection	149
Nitrate to nitrite	River, tap water	Nitrate reductase	0.05–7.5 mg/L NO_3^-	FIA, fluorimetric detection	138
Nitrate to nitrite	Industrial effluents	Nitrate reductase	0.017–7 mg/L NO_3^-	FIA, photometric detection	139
Nitrate, nitrite to ammonium	Potable water	Ti(III)	0–20 mg/L NO_3^-	Batch, gas-phase spectrometry	148
Nitrate, nitrite to ammonium	River water	Titanium chloride	0.03–12.4 mg/L NO_3^-	FIA, fluorimetry	148
Nitrate, nitrite to ammonium	Potable water	Nitrate and nitrite reductase	3–620 mg/L NO_3^-	Batch, electrochemical detection	149
Nitrate, nitrite to ammonium	Tap, mineral water	Zn	0.9–6.6 mg/L NO_3^-	FIA, conductivity	143
Nitrate to NOCl	Hydroponic fluids	HCl/H_2SO_4 conc.	5.6–300 mg/L NO_3^-	FIA, amperometry	153

Metallic zinc has been used either alone or combined with a manganese salt, for the heterogeneous nitrate reduction to nitrite (101,102). The method requires a strict control of the temperature and reaction time to avoid reduction of nitrogen to lower oxidation states. Quantitative reductions are not obtained, with 85–90% being the best yields reported (103) in alkaline medium, whereas only 10% reduction is achieved in saline matrices (104).

The use of cadmium as reducing agent was proposed by Pötzl and Reiter in 1960 and other authors incorporated it to their methods later in granular form, dust (105,106), or in metal filings (107), achieving reduction yields higher than 90% in most cases. Results improved when metallic impurities were introduced on the reducing surface (108), particularly when a copper film was created on the cadmium surface (106,107), although other pairs, such as Cd-Zn (109), Cd-Hg (103), and Cd-Ag (108,110), have also been used. The effectiveness of the reduction process depends not only on the type of reductor but also on its grain size and amount and the compaction degree in the column, since these parameters define the active surface and the time of contact between the reducer and the sample (111). To avoid such drawbacks some authors have substituted the reducing column by a reducing wire (e.g., cadmium wire treated with copper or silver) (110), although then the reduction is not quantitative.

Some disadvantages of copper cadmium reduction is the degradation of the reducing surface caused by the gradual loss of copper, the precipitation of other species present in water samples on the surface, the pressure drop due to the compaction of particles (when using a column containing the reducing agent), and the adsorption of certain components, mainly organic matter (112). To avoid the precipitation of formed Cd^{2+}, ammonium chloride or EDTA may be added to the sample, although, even then, the degradation of the reducing agent persists and the column must periodically be regenerated to replace the metallic surface layer until it is finally substituted (because every regeneration shortens the life time of the reducer) (112).

Another possibility is to form the metallic surface in situ by electrochemical cadmium deposition on an inert surface of pyrolitic graphite (113). The method is very sensitive and is free of the interference of saline components, mainly chloride, which allows its use for seawater analysis. Nevertheless it undergoes interference from sulphide and phosphate (114–116), the latter generally being present in most surface and underground waters.

The cadmium reduction method has been adapted for continuous analysis (either segmented-flow or flow injection analysis modes), and there are many references in which the method is applied to the analysis of nitrate in all kinds of water samples (117). Basically, a column filled with the reducing agent (e.g., copper-cadmium granules) is inserted into the flow injection manifold to reduce the nitrates to nitrites. The resulting nitrite ion plus those originally present in the sample are, in general, colorimetrically monitored using the reaction of Shinn or similar. This approach has been extensively applied to the determination of low concentrations of nitrite and nitrate in seawater (118,119), wastewater plant effluents (111), and natural waters (95,120–122). The determination of both ions separately is achieved by first carrying out the procedure with the reduction step for the analysis of nitrate plus nitrite and then repeating the procedure without the reduction step for the analysis of nitrite alone. When the method is intended to be used for coastal and estuarine waters, variations in salinity must be taken into account for proper corrections (123). Automatic analyzers based on the cadmium reduction approach have also been developed for nitrate monitoring purposes in river water (124,125). Also, continuous-flow analyzers that can operate underwater either with peristaltic pumps or powered by osmotic pumps have been described for in situ nitrate determinations in the deep ocean (126,127).

2. Homogenous Reduction to Nitrite

The reduction of nitrate in alkaline medium by hydrazine sulphate and catalyzed by copper ion was first described in 1955 by Mullin and Riley (128) for the determination of nitrate in seawater. Since then several methods based on that principle have been developed and adapted to different types of water. Its main limitation is probably the strong dependency of conversion rates on the temperature, which may lead to poor reproducibilities (128,129). In addition this method is more prone to interference from cations, mainly Mg^{2+}, Ca^{2+}, and Fe^{3+}, than those using cadmium. The interference of these ions is more important in saline matrices, because they may precipitate as hydroxides in the alkaline medium required. Interfering cations may be eliminated prior to the analysis by using ion-exchange resins or by precipitation in alkaline medium and filtration (130), thus allowing the use of higher working temperatures and shortening the time of analysis. The addition of complexing agents is rarely used, for it inhibits the catalytic effect of copper. This phenomenon also occurs when natural chelates such as humic or fulvic acids are present in the sample matrix and at high concentrations they are an important interference. The interference from humic and fulvic acids has been described to diminish when an excess of zinc is added to the sample, although the results are not better than those obtained just by increasing the catalyst concentrations. In some cases it is possible to add a complexing agent to remove a specific interfering species (e.g., magnesium) without affecting the cata-

lyzed reaction (130,131), allowing determinations of nitrate in seawater with a detection limit of 0.003 mg/L NO_3^--N.

In most of the methods described, the reduction with hydrazine is followed by the photometric determination of the nitrite by the reaction of Shinn. Sawicki and Scaringelli (132) compared the performance of several reagents as diazotation and coupling agents, obtaining the best results when ANSA (amino-naphthyl sulfonic acid) is the coupling reagent. The method has been automated using FIA techniques (133).

3. Photochemical Reduction

Irradiation of the sample with ultraviolet produces the reduction of nitrate to nitrite by photolysis of this anion (134). The photochemical reactions involve one or more excited states and may evolve through several routes, through which dissociation and intramolecular reordering of nitrate takes place to form nitrite, oxygen, pernitrite, and several unstable intermediate radicals. The predominance of one reaction over another and the quantum yields of photolysis depend on pH, on the initial nitrate concentration, on the intensity of light, and on the radiation wavelength. The photolysis by sunlight (135) of the nitrate occurring naturally in aquatic environments may be produced in the laboratory by using a mercury discharge lamp for the conversion of nitrate to nitrite (112,136,137). The authors report a 93% reduction efficiency for nitrate after sample irradiation for 8 minutes at alkaline pH. The method also allows the determination of total nitrogen if organic nitrogen, ammonium, and nitrite are previously photo-oxidized in the presence of persulfate to nitrate.

4. Enzymatic Reduction

The reduction of nitrate to nitrite using the enzyme nitrate reductase has also been described (138,139). Nitrate is determined either through photometric determination of the resulting nitrite with the Shinn reaction (139) or indirectly through the decrease of NADH natural fluorescence when converted to NAD^+ during the enzymatic reduction (138). The methods are highly selective and allow the determination of low nitrate concentrations in the low-ppb range. The main difficulties with enzyme reductions are due to the manipulation of the enzyme, which is very sensitive to changes in temperature and pH. Hg(II) and Cu(II) reduce the activity of the enzyme and are the main interferences of this type of system.

C. Reduction of Nitrate and Nitrite to Ammonium

1. Heterogeneous Reduction

This is generally accomplished using the Devarda alloy for the reduction step (Cu-Zn-Al) and has become a standard method of analysis. The resulting ammonium generated after the reduction process can be determined either by distillation and further neutralization with standardized acid solution (140), by nesslerization (141), or potentiometrically using a gas-sensing electrode or an ammonium-selective electrode (142). The method is slow and tedious and in many cases leads to erratic results if ammonium is present in the sample matrix (141). Other potential drawbacks are the incomplete reduction of nitrate and the interference from organic nitrogen compounds, which may release ammonium during the alkaline distillation (141). The mentioned drawbacks are very important when analyzing wastewater samples, and in that case another reducing agent should be used. More recently an FIA method has been described based on the reduction of nitrate and nitrite to ammonium using a column filled with metallic zinc (143). The resulting ammonium is determined using a gas-diffusion cell coupled to a conductance-flow cell. The method allows the determination of up to 60 samples per hour and has been applied to the determination of nitrate in tap and mineral water.

2. Homogeneous Reduction

Cr(II) and Ti(II) have been proposed as reducing agents for nitrate and nitrite, either in acidic media or at alkaline pH (144–146) and have been reported to achieve reduction rates of up to 99%. The calculation of nitrate plus nitrite may be achieved by analyzing the resulting ammonium, as described in the Devarda method (145) (see preceding section), by adding an excess of Cr(II) and back-titration of the excess with standardized iron solution and by using specific chromogenic reagents or through potentiometric measurements (146). In acidic medium, nitrate is also reduced by Ti(III) to ammonia, which is either quantified in the gas phase through its molecular absorption at 200 nm (147) or separated using a PTFE membrane prior to its determination by fluorimetry (148).

3. Enzymatic Reduction

Nitrate and nitrite may be reduced to ammonium by the combined action of nitrate reductase and nitrite reductase. The reduction is reported to be quantitative when the two enzymes are immobilized in a single solid matrix. The resulting ammonium is potentiometrically monitored using either an ammonium-selective electrode (149) or a nonporous membrane in which the sensor group is nonactine absorbed in a silicone rubber matrix (150). The latter electrode also responds to potassium, which interferes with the determination.

D. Formation of Nitrosil Chloride

The indirect determination of nitrate and nitrite can be also based on the formation of nitrosil chloride, which is the quantified by direct measurement in the UV-wavelength range (151); by volatilization of this compound and its detection in the gas phase (152); by amperometry (153) and through derivatization reactions with organic reagents to form a colored compound colorimetrically detected (154–156).

The nitrite also is reduced under similar conditions, although it can be analyzed separately by using lower concentrations of sulfuric acid to prevent the reduction of nitrate. If nitrate is intended to be analyzed alone, then the nitrite can be eliminated by first reducing it to nitrosil chloride, which is removed from the solution using nitrogen as carrier (155). The interference of sulfide is also eliminated by this procedure.

E. Reduction to Hydroxylamine

Nitrate and nitrite are reduced to hydroxylamine with an amalgamated zinc reducing agent under acidic pH, later reoxidized with Fe(III). In the presence of ferrozine, a colored complex, Fe(II)-ferrozine, is formed and detected photometrically at 562 nm. The potential interference of the iron is removed by passing the sample through an ion-exchange resin. The method allows determination at levels as low as 0.2 μg/L N (157).

F. Reduction to Nitrogen Oxide

The conversion of nitrate and nitrite to nitrogen monoxide (NO) has been used for the determination of these ions with electrochemical or chemiluminiscent detection. Since the determination of NO occurs in the gas phase using an inert gas as carrier (158–160), the methods are free of interference from sample coloration, suspended particles, and dissolved, nonvolatile electroactive species. NO may be amperometrically detected by oxidation in a platinum electrode (161) or by differential pulse polarography (162).

Trojanek and Bruckenstein (163) have reported detection limits at the subnanogram level using flow injection analysis by means of an electrode fabricated by depositing a porous gold layer on one side of the membrane. Nitrite is first reduced in the carrier stream to NO, which is transported from the flowing solution through the membrane to the golden face, where it is electrolized.

The sensitivity and selectivity of the NO chemiluminiscence detection allows the determination of nitrite at the parts-per-billion level (159). In batch mode, a detection limit of 0.05 μg/L NO_2-N using a 20-ml sample is achieved. Trace analysis of nitrite in small sample volume by flow injection analysis has been accomplished by conversion of nitrite to NO, which is transported to the gas phase through a semipermeable membrane and subsequently monitored with a chemiluminescence detector (164). And 60–180 samples can be analyzed per hour, the detection limit for a 100 μL sample being 0.04 μg/L of nitrite ion.

Whereas the reduction of nitrite to nitrogen monoxide (NO) is obtained under smooth conditions with iodide, ascorbic acid, or hydroxyquinone, the reduction of nitrate to NO requires stronger conditions (158), and some methods have been proposed in which nitrate is first reduced to nitrite by hydrazine and the resulting nitrite is then reduced to NO by iodide. V(III) is a more effective reducing agent and can also be used under smoother conditions (159,165). With this reagent and under temperature control it is possible to achieve either the selective reduction of nitrite, at room temperature, or the joint determination of nitrate plus nitrite, at 80–95°C. Hydrazine and ascorbic acid have also been used for the reduction of nitrite and nitrate (166).

G. Kinetic Methods

Kinetic methods in which nitrite catalyzes the formation of a chromophore or a luminiscent product have been described for nitrite determination. Montes and Laserna (167) have developed a method for the kinetic determination of nitrite in drinking waters based on the bromation of pyridine-2-aldehyde-2-pyridylhydrazone in hydrochloric acid. Using fluorimetric detection, concentrations as low as 0.0046 mg/L NO_2^- can be analyzed, and the same method based on recording the disappearance of the organic reagent spectrophotometrically (168) has a detection limit of 0.021 mg/L NO_2^-.

The nitrite ion also catalyzes the oxidation of a number of organic compounds, such as thionine (169), fenosafranine (170), and pyrogallol red (171). The nitrite is analyzed by recording the oxidation rate of the organic compound either photometrically (thionine and pyrogallol red) or fluorimetrically (fenosafranine), the latter being the more sensitive and allowing determination in the range 0.9–13.8 μg/ml N. All of them are free from the interference of nitrate and ammonium. Certain oxidants, like chlorate, bromate, and iodate, may interfere, but they are not common in water samples.

The inhibitor effect of nitrite on the photochemical reaction between iodine and EDTA has been applied to the development of a kinetic method for the determination of nitrite with amperometric detection (172).

V. CHROMATOGRAPHIC METHODS FOR THE DETERMINATION OF NITRATE AND NITRITE

Nitrate and nitrite can readily be separated from other inorganic anions coexisting in the sample by using ion chromatography (IC). The separation techniques for nitrate and nitrite used in ion chromatography are based largely on ion exchange, ion interaction, or ion exclusion, although separations based on reversed phase or chelating stationary phases have been also described.

In addition, each of these separation techniques can be combined with a number of different detection methods, i.e., conductivity, electrochemical detection, potentiometry, spectrophotometric detection, and post-column derivatization detection. In this section, the methods are grouped first according to the separation mode and then by the detection method. However, this division is quite arbitrary, because generally the detection mode employed is the key factor ruling the type of eluent and column to be used. Finally a brief overview of sample preconcentration and pretreatment in order to avoid matrix interferences and enhance sensitivity is given. Table 3 summarizes the chromatographic methods of analysis of nitrate and nitrite.

A. Ion-Exchange Chromatography

The most frequently used anion-exchange materials are formed by quaternary amine functional groups as strong-base and less substituted amines for weak-base exchangers. The first materials developed for nonsuppressed IC had the functional groups chemically bonded to a silica backbone, although the introduction of functionalized polymers and copolymers such as polystyrene, methacrylate, and polystyrene-divinylbenzene either as silica particle coating or as a backbone for the ion-exchange resin readily increased the potential of the IC separations (31). A number of workers have reported the use of reversed-phase C_{18} columns or

Table 3 Chromatographic Methods of Analysis of Nitrate and Nitrite

Parameter	Water type	Eluent	Detection limit	Detection method	Ref.
Nitrite + nitrate	Industrial water	HCO_3^-/CO_3^{2-}	0.25 mg/L NO_2^- 0.1 mg/L NO_3^-	Conductivity	183
Nitrate	Estuarine water	HCO_3^-/CO_3^{2-}	<5 mg/L NO_3^-	Conductivity	209
Nitrite + nitrate	River water	HCO_3^-/CO_3^{2-}	<3 mg/L	Conductivity	211
Nitrite + nitrate	Natural water	KHP[a]	0.1 mg/L	Conductivity	180
Nitrite + nitrate	Brines	Gluconate/borate	0.1 mg/L NO_2^- 1 mg/L NO_3^-	Conductivity	216
Nitrite + nitrate	Atmospheric water	KHP/tetrabuthyl ammonium	0.1 mg/L NO_2^- 0.2 mg/L NO_3^-	Conductivity	205
Nitrate	Tap water	Tiron	<0.07 mg/L NO_3^-	Conductivity/UV	190
Nitrite + nitrate	Seawater	HCO_3^-/CO_3^{2-}	0.3 mg/L NO_2^- 0.04 mg/L NO_3^-	Conductivity/UV	184
Nitrite	Seawater	Sulphuric acid	0.2 mg/L NO_2^-	Conductivity/UV	207
Nitrite + nitrate	Ground-, rain water	Phosphate/ MeOH	0.01 mg/L NO_2^- 0.01 mg/L NO_3^-	Amperometry	141
Nitrite	Seawater	NaCl	0.0014 mg/L NO_2^-	Amperometry/UV	136
Nitrite + nitrate	Seawater	NaCl/phosphate	0.002 mg/L NO_2^- 0.008 mg/L NO_3^-	Amperometry/UV	174
Nitrite + nitrate	Natural water	Nonylammonium phosphate	0.002 mg/L NO_2^- 0.0015 mg/L NO_3^-	UV	204
Nitrite	Seawater, brines	NaCl	0.008 mg/L NO_2^-	UV	209
Nitrite + nitrate	Saline water	NaCl + Na_2SO_4	sub mg/L range	UV	212
Nitrite + nitrate	River water	Pyromellitate/ MeOH	0.02 mg/L NO_2^- 0.03 mg/L NO_3^-	UV	132
Nitrite + nitrate	Domestic water	CTACl[b]	0.02 mg/L NO_2^- 0.015 mg/L NO_3^-	UV	197
Nitrite + nitrate	Domestic water	$NaClO_4$	0.0002 mg/L NO_2^- 0.001 mg/L NO_3^-	UV	222
Nitrite + nitrate	River, lake water	KHP[a]	0.006 mg/L NO_2^- 0.008 mg/L NO_3^-	Indirect photometry	176
Nitrate	River, lake water	CTACl[b]	0.02 mg/L NO_3^-	Indirect photometry	173
Nitrite	Natural water	Sulfuric acid	0.0001 mg/L NO_2^-	Electrochemical	206
Nitrite + nitrate	City water, seawater	Succinic acid/ KHP	0.0064 mg/L NO_2^- 0.0091 mg/L NO_3^-	Fluorescence with postcolumn derivatization	142

[a]KHP is potassium hydrogen phthalate.
[b]CTACl is hexadecyltrimethylammonium chloride.

polystyrene-divinylbenzene resins for the separation of nitrate and nitrite. The packing material is coated either permanently or dynamically with hydrophobic molecules containing ionic functional groups (e.g., cetylpyridinium chloride and cetylpyridinium salicilate), which act as fixed ion-exchange sites of the packing material. Several quaternary ammonium (173,174) or phosphonium (175) salts are used with this purpose. The advantage of such columns is that they can be prepared with a great variety of resins, functional groups, capacities, etc., which has led to their widespread use.

Nonsuppressed ion chromatography (often referred to as single-column ion chromatography), in which the eluent is not chemically modified before entering the detector, has been described for the analysis of nitrate and nitrite (176,177,187,210). Eluents of low equivalent conductance are typically used for the separation, such as phthalate (176,180,210), benzoate, sulfobenzoate, and citrate (178), in either the salt or the acidic form. Also the use of gluconate-borate (179) has been suggested for the separation of nitrite from other weak-acidic anions, namely, bicarbonate and hydrogenphosphate. Dogan and Hoerdi (180) reported limits of detection of 0.1–0.2 mg/L for nitrate and nitrite using conductivity detection. These values were improved 10–50 times by preconcentrating the sample in an ion-exchange precolumn. Okada and Kuwamoto (181) have used potassium hydroxide as mobile phase for indirect conductivity detection of nitrate and nitrite, together with other common ions. They improved the sensitivity (limit of detection—LOD nitrate and nitrite 0.015 mg/L both) using phthalate and direct conductivity detection.

Suppressed ion chromatography, in which a device called a suppressor is inserted to modify the characteristics of the eluent and the solute, thus improving the detection, was first proposed with conductivity detection by Small et al. (182), and it is widely used and described elsewhere. It has become a standard method for the analysis of chloride, nitrite, phosphate, bromide, nitrate, and sulphate in water and wastewater and a proposed method for the analysis of inorganic anions in waters by the Environmental Protection Agency (EPA Proposed Method B1011). These ions are separated within 20 minutes using carbonate-bicarbonate as eluent. Detection limits in the range 0.1–0.01 mg/L are typically achieved. The use of "fast-run" analytical columns reduces the chromatographic time considerably, and separation of these ions may be achieved within 8 minutes. Mosko (183) carried out the determination using both types of analytical columns and concluded that "fast-run" columns could not be used for samples containing nitrite or bromide due to the poor resolution of adjacent peaks.

After conductivity, spectrophotometric detection, either in direct or indirect mode, is the most common detection method. Direct UV measurements at low wavelengths (200–230 nm) predominate in the literature as the mode of choice for the detection of nitrate and nitrite, owing to the strong absorption of both ions in the UV region. Carrozzino and Righini (184) monitored nitrate in seawater using UV detection at 210 nm in series with the conductivity detection, without interference of chloride. Thomsen and Cox (185) replace aromatic compounds of high UV absorbance used as eluents when indirect spectrophotometric detection is utilized (see later) by alkanesulphonates (methane-, 1-butane-, 1-hexane-, and 1-octanesulphonate) with lower absorption, for the separation of nitrate and nitrite on silica-based anion-exchange and polystyrene-divinylbenzene columns. Both ions are determined by direct UV absorption at 210 nm. The authors report a major influence of the chain length in the elution efficiency when using polystyrene-divinylbenzene ion exchange column, whereas with the silica-based column the hy-

drophobicity of the eluent was of minor importance, and the retention of nitrate and nitrite was affected mostly by the concentration of the eluent and the capacity of the column. Using 40 mM methanesulphonate as eluent with a flow rate of 2 ml/min, both ions were separated in less than 6 minutes. Increasing the flow rate from 2 to 4 ml/min, the nitrate and nitrite peaks were still completely separated in a 1-min assay. Nitrite and nitrate mixtures were baseline resolved even when these ions were in ratios 1:1000. High concentrations of chloride shifted the nitrite peak toward nitrate, and for concentrations above 1 mg/l of chloride resulted in poor resolution of the nitrate/nitrite pair.

Indirect spectrophotometric detection of nitrate and nitrite has been applied using aromatic acid salts with strong UV absorption as eluents. Among them, phthalate (186,189), sulfobenzoate (187), hydroxybenzoate, p-toluenesulfonate, and pyromellitate (188) have been employed with UV detection within the wavelength range 260–300 nm. The detection limits reported when using phthalate are 0.2 and 0.3 mg/L for nitrate and nitrite, respectively, comparable to those obtained with ion-suppressed conductivity (176,189). Ohta et al. (188) used a methanol solution of 1,2,4,5-benzenetetracarboxilate at pH 6.25 as eluent, the detection limit being then 20 μg/L for nitrite and 30 μg/L for nitrate. Elution has also been performed with solutions of tiron acid (1,2-dihydroxybenzene-3,5-disulfonic acid) or its sodium salt (190). Since this species have greater elution strength than phthalate, they provide higher sensitivity and shorter retention times, with lower eluent concentration.

Amperometry is a selective detection principle that is also used coupled to IC separations and applied in direct mode. Ito et al. (174) used a polymethacrylate-based anion-exchange column of low capacity and alternatively a cetyltrimethylammonium-coated octadecylsilane (ODS) column with higher exchange capacity. The eluent was a phosphate-buffered (pH 5.8) sodium chloride solution. Nitrite was monitored using both UV (225 nm) and amperometric detection, while nitrate was detected spectrophotometrically (since nitrate is electroinactive at the stated potential). The IC system with the low-capacity column could not be applied to nitrite and nitrate analysis at μg/L levels in seawater samples because of poor resolution for resulting peaks. The higher-capacity coated ODS column improved the separation between anions and allowed the determination in seawater by direct injection of the sample. The overlapping between the chloride and nitrite peaks is avoided using a potential at which chloride does not undergo oxidation (174,191,192). According to Pastore et al. (191), the tolerance toward chloride is thus three to five times higher compared to the use of the same eluent and photometric detection, whereas the detection limit for nitrite is four times lower, being 0.002 mg/L nitrite ion with amperometric detection and 0.004 mg/L with UV detection.

Amperometry has also been coupled to chromatographic systems for the determination of nitrate, although in that case a derivatization reaction is required (193). Alawi (194) proposed a method based on the separation of the o-nitrophenol derivative obtained by reaction between an excess of phenol and the nitrate ion, with the separation step performed on a reversed-phase column using a methanol/phosphate buffer (pH 5.4) and amperometric detection in the reduction mode (-0.47 V). The method is free from interferences and provides high sensitivities for the determination of nitrate and nitrite in the low μg/L level with an LOD of 5 ng/ml, although it requires several pretreatment steps apart from the nitration reaction, such as extraction of the o-nitrophenol with dichloromethane and evaporation to dryness.

Fluorescence detection offers both selectivity and sensitivity and is applied to the determination of nitrite and nitrate in indirect mode. An HPLC system with postcolumn oxidation of Ce(IV) to Ce(III) by nitrite and fluorescence detection of the resulting Ce(III) is used for the quantification of this anion at the low μg/L range (195). The method offers higher sensitivity and selectivity due to the specificity of the measurements. Nitrate can also be detected using this procedure by online incorporation of a reduction column, to convert nitrate into nitrite before adding the cerium solution. The eluent was borate-buffered succinic acid, because it provided better sensitivity among those evaluated (1.3 μg/L for nitrite and 3.2 μg/L for nitrate).

B. Ion-Interaction Chromatography

There are several reports dealing with the separation of nitrate and nitrite based on ion-interaction chromatography. Most of the separations are carried out on C_{18} reversed-phase materials or on neutral polystyrene-divinylbenzene. Quaternary ammonium salts with hydrophobic long-chain aliphatic or aromatic groups are normally used as ion-interaction reagents, dissolved in acetonitrile–water or methanol–water mixtures (196). Iskandarani and Pietrzyk (218) used a nonpolar polystyrene-devinylbenzene stationary phase with an ace-

tonitrile eluent containing tetrapentylammonium as ion-interaction reagent. Nitrate and nitrite were detected by both conductivity and UV detectors connected in series. An HPLC system with a Spherisorb ODS column and phosphate-buffered eluent containing micellar hexadecyltrimethylammonium have been reported by Mullins and Kirkbright (197). Another separation method (198) based on the same ammonium salt but combined with a cyano column (Polygosil-60-D-10 CN) has been described for the separation of nitrate and nitrite and spectrophotometric detection (205 nm). The detection limits of this method are 4 and 3 μg/L of nitrite and nitrate ions, respectively, which are lower than those obtained by Kok et al. (199) using reversed-phase and tetramethylammonium phosphate as interaction reagent with direct UV detection (214 nm). This latter author reported a detection limit of 0.1 mg/L for both nitrate and nitrite. Ion-interaction chromatography using a methanol solution of cetyltrimethylammonium and citrate buffer was proposed by Wheals (200) using four detection modes: UV absorbance, electrochemical, refractive index, and conductivity detection.

Other authors used long-chain primary aliphatic amines instead of the ammonium quaternary salts. Typical counterions are perchlorate, salicylate, and, most often, phosphate. Octylammonium orthophosphate was preferentially used by Skelly (201) and Gennaro et al. (202) in ion-interaction chromatographic methods with direct UV detection. Gennaro et al. (203) used octylammonium salicylate as the interaction reagent for the separation of chloride, nitrite, and nitrate and conductometric detection. Under such conditions the method was not suitable for the analysis of nitrite in seawater due to overlapping with the tail of the chloride peak. Baseline resolution, however, is obtained using octylammonium orthophosphate and spectrophotometric detection at 230 nm, even when analyzing solutions containing chloride to nitrite ratios of up to 10^6. Detection limit is 0.005 mg/L of nitrite in the presence of 0.6 M sodium chloride. Poboz et al. (204) used nonylammonium phosphate (pH 6.5) and UV detection at 205 nm to detect nitrate and nitrite ions selectively with detection limits of 0.0015 and 0.002 mg/L, respectively. Chloride and sulphate in concentrations up to 100 mg/L did not affect results. Separation of chloride, nitrite, nitrate, and sulphate on a reversed-phase Partisil 10 ODS-3 stationary phase in combination with an eluent consisting of a mixture of tetrabutylammonium iodide and potassium hydrogenphthalate has been reported (205). However, the limits of detection are 0.1–0.2 mg/L higher than those usually achieved with ion-interaction chromatographic methods.

C. Ion-Exclusion Chromatography

Based with ion-exclusion chromatography and amperometric detection, a chromatographic method for the selective determination of nitrite at $\mu g/L$ levels has been developed (206). Nitrite is separated from other weak acids (phosphate, carbonate, fluoride), which are eluted according to their pK values, as well as from anions of strong acids, which are not retained on the column (chloride, nitrate, sulphate, bromide), in less than 8 minutes and detected with a platinum working electrode ($+1V$ vs. Ag/AgCl).

Rokushika and Yamamoto (207) used polyvinyl alcohol (PVA) gel-based reversed-phase columns with dilute sulphuric acid as the eluent, although the mechanism of separation is attributed by the authors to some kind of interactions between HONO molecules and the alcoholic OH groups on the surface of the PVA beads, rather than to ion exclusion. At pH 2.3, nitrite and carbonate were well resolved from the matrix ions (e.g., bromate, bromide, and nitrate), which eluted near the void volume of the column. The species of interest were monitored with both conductivity and UV detectors connected in series, and a linear range of 0.2–100 mg/ml was attained at the chosen wavelength (210 nm). Chloride did not disturb the nitrite peak, even when its concentration was 30 mg/ml. However, when 10 mg/ml of nitrate or sulphate are present in the sample matrix, shifts of the retention time of nitrite peak and overlapping with the tail of a large matrix ion peak are observed. Sensitivity losses were observed due to decreased absorbance of nitrous acid in relation to nitrite ion. When increasing the pH by passing the eluted solution through a cation-exchange hollow fiber, nitrous acid is again converted to the ionized form and the sensitivity is enhanced.

D. Sample Handling in Ion Chromatography

The determination of low levels of nitrate and nitrite in the presence of large amounts of other inorganic anions, such as chloride, sulphate, bromide, phosphate, and bicarbonate, has been a subject of concern of numerous papers (208,215,184,209,210,187,211,212,185, 177,186,213). Dilution of the sample is sometimes applied, depending on the concentration of the ions of interest, but very often removal of the interference by suitable methods is needed. Peak overlapping between ions with close retention times (as are the chloride-nitrite and bromide-nitrate pairs) made the quantification of both ions difficult or even impossible.

The peak shape and the column efficiency varied with the ion concentration in the matrix. The injection of saline samples may cause shifts in the retention time of nitrate and nitrite as a result of a competition between the eluent and the matrix anion for the active "sites" of the stationary phase. This phenomenon was observed when a sample with high chloride content was injected into a suppressed chromatographic system with a carbonate/bicarbonate eluent. Owing to its higher affinity for the stationary phase, this eluent replaced completely the chloride retained, and the sample plug in the resulting chromatogram is followed by a modified eluent, which is very poor in carbonate/bicarbonate but enriched with chloride. The reason for the increased retention times of nitrate and nitrite is the weaker elution strength of the modified eluent. In the case of nitrite this effect is accompanied by loss of sensitivity due to ion exclusion in the suppressor column. Ion exclusion is a source of chromatographic interference in suppressed ion chromatography, as described in the Standard Procedure (1) and also in the papers of Mosko (183) and Koch (214), and accurate results are attained only by frequent recalibration, especially if nitrite is a minor constituent in a sample with a high saline content as increases the amount of hydrogen ions accumulated in the suppressor surface.

Novic et al. (215) analyzed samples containing 2 mg/L nitrite and 5 mg/L nitrate using a suppressed system with UV and conductivity detection connected in series in chloride matrices, with concentrations of chloride ranging from 0 to 10 g/L. Retention times of nitrite and nitrate increased when increasing the chloride concentration. Coelution of both ions occurred at chloride concentrations higher than 15 g/L. No change in the nitrate peak area was observed in the entire chloride concentration range, while decreases in nitrite peak signal were attributed to protonation in the suppressor column. Nitrite response was improved by the addition of sodium hydroxide for the neutralization of the eluent. The opposite trend was observed in a sulphate matrix. Nitrite and nitrate retention times were shortened proportionately to the increase in the concentration of sulphate in the sample. Moreover, the nitrite response remained constant and unaffected by the sulphate concentration.

Chloride and bromide interference may be removed by means of a silver-based cation-exchange resin directly added to the sample or packed in a column and incorporated online prior to the chromatographic system (209). Jackson and Jones (216) reported the use of a sulphonated Dupont Nafion fiber with silver *p*-toluenesulphonated counterion in nonsuppressed chromatographic systems with conductivity detection. Cations from the sample are exchanged for silver cations, which precipitate halide ions. The cleanup device could remove more than 99% chloride and bromide in brine samples. This sample pretreatment approach combined with conventional nonsuppressed ion chromatography using conductivity detection gives a detection limit for nitrite of approximately 0.1 mg/L in brine samples containing up to 5000 mg/L chloride.

Dähllof et al. (217) and Naish (186) overcome the interference of chloride in the determination of low concentrations of nitrate in seawater by using two pre-columns and an additional valve to elute chloride before the injection of the interference-free sample in the analytical column. The determination was performed using carbonate-bicarbonate gradient elution, a suppressed ion chromatography system, and conductivity detection. The pH and ionic strength of both eluents were optimized by factorial experimental design. With the chromatographic system a limit of detection of 0.5 μM nitrate (217) was attained. When the phosphate concentration reaches 30 μM, it might be necessary to dilute the sample.

The use of an eluent of the same composition as the major interference has been successfully used to avoid chloride removal (211,215). Pastore et al. (191) and Rokushika et al. (211) used chloride salt solutions, either potassium or sodium salts, for the elution of nitrate and nitrite and to overcome the interference of chloride in samples containing high levels of this ion. The methods proposed by these authors were applied to the analysis of seawater by direct injection of the sample, with no sample pretreatment. Pastore et al. (191) found the combination of glassy carbon amperometric detection and elution with NaCl to be well suited for the determination of nitrite in the presence of a large excess of chloride, providing a limiting chloride-to-nitrite ratio of about 715,000 and an LOD for nitrite of 0.0014 mg/L. Other mobile phases also evaluated in their work were: carbonate-bicarbonate with UV detection, borate-carbonate mixtures with conductivity detection, and sodium chloride using either spectrophotometric (210 nm), conductimetric, or amperometric detection. They allowed chloride-to-nitrate ratios of 5,000, 1,000 or 200,000 (by adding 500 mg/L of chloride to 0.1 mg/L of nitrite a complete resolution was not achieved but data deconvolution allowed a relation of 10,000) and detection limits for nitrite of 0.01, 0.05, and 0.005 mg/L, respectively. Higher detection limits were obtained with carbonate/bicarbonate and conductometric detection with a chloride-to-nitrite ratio of 200.

Similarly, phosphate-buffered chloride and sulphate eluents have been used to solve the problem of chloride and sulphate interference, respectively (212). The detection limits obtained using a sulphate as eluent are slightly higher than those reported when using phthalate. However, when using chloride as eluent they are between 0.03 mg/L without chloride and 0.06 with 20,000 mg/L chloride for nitrite and between 0.08 without chloride and 0.12 with 20,000 mg/L of chloride for nitrate, close to the values obtained with phthalate. Virtually no changes were observed up to 20,000 mg/L. For samples containing more than 20,000 mg/L chloride, the nitrite peak broadens slightly. There was a moderate increase in the detection limits with increased chloride content in the sample. In all cases the detection method used (UV detection at 210 nm) (211,212) or amperometric detection (191) can discriminate against the matrix ion. The method is restricted to the separation and detection of UV-absorbing anions, but would be equally applicable to other selective detection methods, such as amperometry.

Mutual interference between nitrite and nitrate can also be expected. Iskandarani et al. (218) proposed an ion-interaction chromatographic method with spectrophotometric detection that allows determination of nitrate and nitrite in ratios of nitrite/nitrate 1/200 and 300/1 and good separation of both ions in less than 15 minutes using a strong elution power. As ratios increase above these limits, peak overlap also increases. Better resolution of both ions can also be obtained for more disadvantageous nitrite/nitrate ratios using reduced eluting power, i.e., by decreasing the concentration of acetonitrile in the eluent while increasing the retention times.

Organic species may interfere in the chromatographic separation and detection of nitrate and nitrite, either due to coelution and interference in the UV detection or because of their adsorption on the IC column, and may deactivate the column (e.g., humic acids). To remove these substances selectively prior to the chromatographic separation, Marko-Varga et al. (30) suggest the use of a short cleanup column packed with a bonded-phase amine material. The pretreated samples are then chromatographed in a nonsuppressed system with hydrogenphthalate as eluent and with both conductimetric and spectrophotometric detection. The system allows 500 injections of 500-μl samples containing 50 ml/L of humic acids before saturation. Also an octadecylsilica cartridge loaded with cetyltrimethyl-ammonium p-hydroxybenzoate is found to be effective in removing the model organic compounds without reducing the precision of the analysis for common inorganic anions (219).

Online sample preconcentration allows detection limits far below the 0.1-mg/L level and is simple to apply and adequate for automation. By introducing an anion-exchange resin in the sample loop of the injection valve (180,220), a large sample volume may be loaded before injection in the chromatographic system, the increase in sensitivity being a function of the volume of sample loaded. A linear range for nitrate of 0.002–10 mg/L is attained using the proposed method. Another way to increase the sensitivity is the injection of large sample volumes (up to 2 ml), and this approach has been reported using indirect UV absorbance detection (221) and phthalate buffer. Detection limits obtained using 1-ml injection volume are 0.0083 mg/L of nitrite and 0.0115 mg/L of nitrate. Also, Okada used a high-capacity ion-exchange column to concentrate nitrite from a large sample volume (10–50 ml) and could detect 0.0001 mg/L nitrite by UV absorbance at 210 nm. Eek and Ferrer (222) use a high-capacity ion-exchange column, sodium perchlorate, and UV and refractive index detection for the isolation and quantification of nitrate and nitrite. Ultraviolet detection allowed limits of detection of 0.0002 mg/L, lower than those obtained using refractive index. The injection is of 100 μl. The use of sodium perchlorate as the eluent together with a high-capacity anion-exchange column and UV detector permits, without preconcentration, a 100-fold increase in the sensitivity for nitrite ion compared (LOD 0.0002 mg/L) with the existing methods based on single-column ion chromatography.

VI. RECENT DEVELOPMENTS

Recently, capillary electrophoresis (CE) has been developed and applied to the analysis of nitrate and nitrite in water samples (223–227). The methods, based on the mobility of the ions when an electrical field is applied to the sample in a capillary, are fast and very sensitive, and detection is usually either direct or indirect spectrophotometry. Typical analysis times are less than 5 minutes, with detection limits at the part-per-trillion range. Capillary electrophoresis has been applied to a variety of water samples, ranging from milligrams to micrograms per liter, including wastewater (228), tap and river water (229), and drinking water (230), although its sensitivity allows the monitoring of even trace levels of anions in high-purity water and steam (226,231).

Sequential injection analysis (SIA) was first introduced by Ruzicka et al. (232–234) and has opened new possibilities in the field of laboratory automation. Although there are close similarities between flow injection analysis and sequential injection analysis, the sequential injection approach offers distinct advantages, such as stability and robustness, versatility (the same manifold may be applied to different chemical determinations) and multiple detection methods, and decreased reagent and sample consumption. The method has been applied to the determination of nitrates and nitrites in water samples using several chemical approaches, such as homogeneous reduction with hydrazine sulphate (235) and heterogeneous reduction with copper-cadmium column (236) using the sequential injection sandwich technique. The methods allow determinations of nitrite and nitrate in the $\mu g/L$ range with unattended and automatic performance during extended periods, which makes them an ideal tool for routine analysis and monitoring of nitrates and nitrites in environmental samples.

REFERENCES

1. K. Grasshoff. In: Methods of Seawater Analysis. 2nd ed. (K. Grasshoff, M. Eberhardt, K. Kremling, eds.). Verlag Chemie, Weinheimer, Germany (1983).
2. L.J. Puckett. Environ. Sci. & Technol. 29:408–414 (1995).
3. European Topic Centre on Inland Waters. Annual Summary Report, Document number P08/95-2, Editor, T.J. Lack, p. 22 (1996).
4. Council Directive 91/676/CEE. Official Journal of the European Communities, L 375, 31–12-1991.
5. P. Brimblecombe. Nature 298:460–462 (1982).
6. S. Cornell, A. Rendell, T. Jickells, Nature 376:243–246 (1995).
7. N.J.P. Owens, J.N. Galloway, R.A. Duce. Nature 357:397–399 (1992).
8. H.W. Paerl, J. Rudek, M.A. Mallin. Marine Biology 107:247–254 (1990).
9. H.W. Paerl, J. Rudek, M.A. Malin. Marine Biol. 107:247–254 (1990).
10. H.W. Paerl, M.L. Fogel, P.W. Bates. Trends in Microbial Ecology, 459–464 (1993).
11. M. Allaby. In: Dictionary of the environment. 3rd ed. New York University Press, New York, p 423 (1989).
12. E.H.W.J. Burden. Analyst 86(1024):429–433 (1961).
13. C.J. Johnson, B.C. Kross. Am. J. Ind. Med. 18:449–456 (1990).
14. D. Forman, S. Al-Dabbaghand, R. Doll. Nature 313:620–625 (1985).
15. D.K. Mueller, D.R. Helsel. National Water-Quality Assessment Program. U.S. Geological Survey, open-file circular 1136.

16. C. Macilwain. Nature 377:4 (1995).
17. D. Stanners, P. Bourdeau, eds. Europe's Environment: The Dobris Assessment: An Overview. European Environment Agency, Copenhaguen, Denmark (1994).
18. Council Directive 79/869/EEC. Official Journal of the European Communities, L 271 (1979).
19. Council Directive 78/659/EEC, Official Journal of the European Communities, L 222 (1978).
20. Council Directive 80/778/EEC, Official Journal of the European Communities, L 229 (1980).
21. WHO. European Standards for Drinking Water. 2nd. ed. World Health Organization. Copenhaguen (1970).
22. U.S. EPA. Drinking Water Regulations and Health Advisories. U.S. Environmental Protection Agency: Office of water, open file EPA-822-B96-002, October (1996).
23. European Topic Centre on Inland Waters, Annual Summary Report, Document number P08/95-2, Editor, T.J. Lack, p. 9 (1996).
24. Environment in the European Union 1995. Report for the review of the fifth Environment Action Program. European Environment Agency, Copenhaguen, 1996.
25. Council Directive 91/271/CE. Official Journal of the European Communities, L 135 (1991).
26. APHA. Standard Methods for the Examination of Water and Wastewater. 18th ed. American Public Health Association (APHA), American Water Works Association (AWWA), Water Environment Federation publication (WPCF). APHA, Washington, DC (1992).
27. U.S. EPA. Methods for Chemical Analysis of Water and Waste. U.S. Environmental Protection Agency, Report EPA-600/4-79-020, USEPA. Cincinnati, (1983).
28. J. Luo, R. Sun. Lihua Jianyan, Huaxe Fenxi 25:127 (1989).
29. A. Cerdà, M.T. Oms, V. Cerdà, R. Forteza. Anal. Methods Inst. 2:330–336 (1995).
30. G. Marko-Varga, I. Csiky, J. Ake. Anal. Chem. 56: 2066–2069 (1984).
31. P.R. Haddad, P.E. Jackson. In: Ion Chromatography. Principles and Applications. Journal of Chromatography Library. Vol 36, chapter 14. pp 409–435, Elsevier, Amsterdam, 1990.
32. K.C. Thompson, M. Blankley. Analyst 109:1053–1056 (1984).
33. U.S. EPA. Methods for Chemical Analysis of Water and Waste. U.S. Environmental Protection Agency, method 353.2, USEPA, Cincinnati, (1983).
34. R.C. Hoather, R.F. Rackman. Analyst 84:548–551 (1959).
35. L. Brown, E.G. Bellinger. Water Res. 12:223–229 (1978).
36. P.O. Rennie, A.M. Summer, F.B. Basketter. Analyst 104:837–845 (1979).
37. E. Goldman, R. Jacobs. J. Am. Water Works Assoc. 53:187–191 (1961).

38. D.L. Mites, C. Espejo. Analyst 102:104–109 (1977).
39. J.H. Wetters, K.L. Uglum. Anal. Chem. 42:335–340 (1970).
40. P. Morries. Proc. Soc. Water Treat. Exam. 20:132–137 (1971).
41. J. Mertens, D.L. Massart. Bull. Soc. Chim. Belge 80: 151–158 (1971).
42. O. Thomas, S. Gallot, N. Mazas. Fresenius J. Anal. Chem. 338:234–237 (1990).
43. O. Thomas, S. Gallot, N. Mazas. Fresenius J. Anal. Chem. 338:238–240 (1990).
44. M. Karlsson, B. Karlberg, R.J.O. Olsson. Anal. Chim. Acta 312:107–113 (1995).
45. J. Slanina, F. Bakker, T. Bruyn-Hes, J.J. Möls. Anal. Chim. Acta 113:331–342 (1980).
46. K. Schlager. Absorption and Emission Spectrometry for On-line Chemical Analysis of Nutrient Solutions. Phase II, Final Report. NASA Contract NAS10-11796 (1993).
47. B.J. Beemster, S.J. Kahle. Applied Spectrometry Associates, Inc., Technical publication ASA#63 (1995).
48. B.J. Beemster, S.J. Kahle. Applied Spectrometry Associates, Inc., Technical publication ASA#28, Presented at AWWA '93 Annual Conference and Exposition, San Antonio, Texas, June 6–10, 1993.
49. O. Thomas, F. Theraulaz, V. Cerda, D. Constant, P. Quevauviller. Trends Anal. Chem. 16:419–424 (1997).
50. S.J. Kahle, B.J. Beemster. Applied Spectrometry Associates, Inc., Technical publication ASA#62 (1995).
51. L.A. Saari. Trends Anal. Chem. 6:85 (1987).
52. M.A. Stanley, J. Maxwell, M. Forrestal, A.P. Doherty, B.D. MacCraith, D. Diamond, J.G. Vos. Anal. Chim. Acta 299:81–90 (1994).
53. B.J. Beemster, S.J. Kahle. Proc. Water Qual. Technol. Conf. 1991. Part 1:727–740 (1992).
54. C. Liteanu, E. Stafaniga, E. Hopirtean. Rev. Anal. Chem. 5:159–184 (1981).
55. S.S. Potterton, W.D. Shults. Anal. Lett. 1:11–22 (1967).
56. N. Raikos, K. Fytianos, C. Samara, V. Samanidou. Fresenius J. Anal. Chem. 331:495–498 (1988).
57. S.P. Pande. J. Indian Water Works Assoc. Jan–March: 119–123 (1989).
58. L. Brown, E.G. Bellinger. Water Res. 12:223–229 (1978).
59. H. Shechter, N. Gruener. J. Am. Water Works Assoc. 68:543–546 (1976).
60. J.E.W. Davies, G.J. Moody, J.D.R. Thomas. Analyst 97:87–94 (1972).
61. A. Craggs, G.J. Moody, J.D.R. Thomas. J. Chem. Edu. 51:541–542 (1974).
62. H.P.J. Nielsen, E.H. Hansen. Anal. Chim. Acta 85:1–16 (1976).
63. M.G. Mitrakas, C.A. Alexiades, V.Z. Keramidas. Analyst 116:361–367 (1991).

64. L. Ebdon, J. Braven, N.C. Frampton. Analyst 115: 189–193 (1990).

65. L. Ebdon, J. Braven, N.C. Frampton. Analyst 116: 1005–1010 (1991).

66. D. Hulanicki, R. Lewandowsky, M. Maj. Anal. Chim. Acta 69:409–414 (1974).

67. A.A.S.C. Machado. Analyst 119:2263–2274 (1994).

68. P.J. Milham. Analyst 95:758–759 (1970).

69. A. Hulanicki, M. Zurawska. Z. Analit. Khim. 32: 7767–7774 (1977).

70. E.H. Hansen, A.K. Ghose, J. Ruzicka. Analyst 102: 705–713 (1977).

71. J. Ruzicka, E.H. Hansen, E.A. Zagatto. Anal. Chim. Acta 88:1–16 (1977).

72. E.H. Hansen, A.K. Ghose, J. Ruzicka. Analyst 102: 705–713 (1977).

73. J. Alonso, J. Bartroli, S. Jun, J.L.F.C. Lima, M. Conceiçao, B.S.M. Montenegro. Analyst 118:1527–1532 (1993).

74. S. Alegret, J. Alonso, J. Bartroli, J.M. Paulis, J.L.F.C. Lima, A.A.S.C. Machado. Anal. Chim. Acta 1964: 147–152 (1984).

75. M.M.J. Antonisse, R.J.W. Lugtenberg, R.J.M. Egberink, J.F.J. Engbersen, D.N. Reinhoudt. Anal. Chim. Acta 332:123–129 (1996).

76. D. Langmuir, R.L. Jacobson. Env. Sci. Technol. 4: 834–838 (1970).

77. R.J. Davenport, D.C. Johnson. Anal. Chem. 45:1979–1980 (1973).

78. G.L. Lundquist, G. Washinger, J.A. Cox. Anal. Chem. 47:319–322 (1975).

79. M.E. Bodini, D.T. Sawyer. Anal. Chem. 49:485–489 (1977).

80. K. Kalcher. Talanta 33:489–494 (1986).

81. T.J. O'Shea, D. Leech, M.R. Smyth, J.G. Vos. Talanta 39:443–447 (1992).

82. S.W. Boese, V.S. Archer, J.W. O'Laughlin. Anal. Chem. 49:479–484 (1977).

83. H. Emmi, K. Hasebe, K. Ohzki, T. Kambara. Talanta 31:319–323 (1984).

84. K. Marsukova. Anal. Chim. Acta 221:131–138 (1989).

85. A.P. Doherty, R.J. Forster, M.R. Smyth, J.G. Vos. Anal. Chim. Acta 255:42–52 (1991).

86. A.Y. Chamsi, A.G. Fogg. Analyst 113:1723–1727 (1988).

87. J.A. Cox, K.R. Kulkarni. Analyst 111:1219–1220 (1986).

88. M.M. Malone, A.P. Doherty, M.R. Smyth, J.G. Vos. Analyst 117:1259–1263 (1992).

89. A.P. Doherty, M.A. Stanley, D. Leech, J. Vos. Anal. Chim. Acta 319:111–120 (1996).

90. M. Noufi, Ch. Yarnitzky, M. Ariel. Anal. Chim. Acta 234:475–478 (1990).

91. A.G. Fogg, A.Y. Chamsi, M.A. Abdalla. Analyst 108: 464–470 (1983).

92. A.G. Fogg, S.P. Scullion, T.E. Edmonds, B.J. Birch. Analyst 116:573–579 (1991).

93. E. Sawicki, T.W. Stanley, J. Pfaff, A. D'Amico. Talanta 10:641–655 (1963).

94. E.A.G. Zagatto, A.O. Jacintho, J. Mortatti, H. Bergamin. Anal. Chim. Acta 20:399–403 (1980).

95. M.A. Koupparis, K.M. Walczak, H.V. Malmstadt. Anal. Chim. Acta 142:119–127 (1989).

96. A.C. Ariza, P. Linares, M.D. Luque de Castro, M. Valcarcel. J. Autom. Chem. 14:181–183 (1992).

97. M.F. Mousavi, A. Jabbari, S. Nouroozi. Talanta 45: 1247–1253 (1998).

98. M.J. Ahmed, C.D. Stalikas, S.M. Tzouwara-Karyanni, M.I. Karayannis. Talanta 43:1009–1018 (1996).

99. R. Segarra-Guerrero, C. Gomez-Bendito, J. Martinez-Calatayud. Talanta 43:239–246 (1996).

100. S. Motomizu, H. Mikasa, K. Toei. Anal. Chim. Acta 193:343–347 (1987).

101. D.L. Heanes. Analyst 100:316–321 (1975).

102. K. Matsunaga, M. Nishimura. Anal. Chim. Acta 45: 350–353 (1965).

103. A.W. Morris, J.P. Riley. Anal. Chim. Acta 29:272–279 (1963).

104. E. Foyn. Rep. Norw. Fishery Mar. Invest. 9:3–7 (1951).

105. R.S. Lambert, R.J. Dubois. Anal. Chem. 43:955–957.

106. M. Okada, H. Miyata, K. Toei. Analyst 104:1195–1197 (1979).

107. E.D. Wood, F.A.J. Armstrong, F.A. Richards. J. Mar. Biol. Ass. U.K. 47:23–31 (1967).

108. F. Nydahl. Talanta 23:349–357 (1976).

109. T. Zhou, Y. Xie. Intern. J. Environ. Anal. Chem. 15: 213–219 (1983).

110. R.B. Willis. Anal. Chem. 52:1376–1377 (1980).

111. D. Gabriel, J. Baeza, F. Valero, J. Lafuente. Anal. Chim. Acta 359:173–183 (1998).

112. A. Cerdà, M.T. Oms, V. Cerdà, R. Forteza. Anal. Meth. Instr. 2:330–336 (1995).

113. R. Nakata, M. Terashita, A. Nitta, K. Ishikawa. Analyst 115:425–430 (1990).

114. J. Vandenabeele, K. Verhaegen, S.Y. Avnimelech, O. Van Cleemput, W. Vertraete. Environ. Technol. 11: 1137–1142 (1990).

115. N.K. Cortas, N.W. Wakid. Clin Chem. 36:1440–1443(1990).

116. J.I. Skicko, A. Tawfik. Analyst 113:297–300 (1988).

117. A. Cerdà, M.T. Oms, V. Cerdà, R. Forteza. Anal. Meth. Instr. 2:330–336 (1995).

118. C. Oudot, Y. Montel. Marine Chem. 24:239–252 (1988).

119. K.S. Johnson, R.L. Petty. Limnol. Oceanogr. 28:1260–1266 (1983).

120. M.F. Giné, H. Bergamin, E.A.G. Zagatto, B.F. Reis. Anal. Chim. Acta 114:191–197 (1980).

121. S. Nakashima, M. Yagi, M. Zenki, A. Takahashi, K. Toei. Fresenius J. Anal. Chem. 319:506–509 (1984).

122. J. Maimó, A. Cladera, F. Mas, R. Forteza, J.M. Estela, and V. Cerdà. Int. J. Environ. Anal. Chem. 35:161–167 (1989).

123. T. McCormack, A.R.J. David, P. Worsfold, R. Howland. Anal. Proc. Inc. Anal. Comm. 31:81–83 (1994).

124. J.R. Clinch, P.J. Worsfold, H. Casey. Anal. Chim. Acta 200:523–531 (1987).

125. H. Casey, R.T. Clarke, S.M. Smith, J.R. Clinch, P.J. Worsfold. Anal. Chim. Acta 227:379–385 (1989).

126. K.S. Johnson, C.M. Sakamoto-Arnold, C.L. Beehler. Deep-sea Res. 36:1407–1413 (1989).

127. H.W. Jannasch, K.S. Johnson, C.M. Sakamoto. Anal. Chem. 66:3352–3361 (1994).

128. J.B. Mullin, J.P. Riley. Anal. Chim. Acta 12:464–480 (1955).

129. C.F. Bower, T. Holm-Hanses. Aquaculture 21:281–286 (1980).

130. A.J. Kempers, G. Van der Velde. Int. J. Environ. Anal. Chem. 47:1–6 (1992).

131. A.J. Kempers, A.G. Luft. Analyst 113:1117–1120 (1988).

132. C.R. Sawicki, F.P. Scaringelli. Microchem. J. 16:657–672 (1971).

133. B.C. Madsen. Anal. Chim. Acta 124:437–441 (1981).

134. G. Mark, H. Korth, H. Schuschmann, C. Von Sonntag. J. Photochem. Photobiol. Part A 101:89–113 (1996).

135. O. Zafiriou, M.B. True. Marine Chem. 8:33–42 (1979).

136. K. Takeda, K. Fijuwara. Anal. Chim. Acta 276:25–32 (1993).

137. Y. Zhang, L. Wu. Analyst 111:767–769 (1986).

138. C. Kiang, S.S. Kuan, G.G. Guilbault. Anal. Chem. 50:1323–1325 (1978).

139. D.R. Senn, P.W. Carr, L.N. Klatt. Anal. Chem. 48:954–958 (1976).

140. J.M. Brenmer, D.R. Keeney. Anal. Chim. Acta 32:485–495 (1965).

141. W.H. Evans, G.J. Stevens. Wat. Pollut. Control 71:98–104 (1972).

142. F.W. Allerton. Analyst 72:349–351 (1947).

143. L. Cardoso de Faria, C. Pasquini. Anal. Chim. Acta 245:183–190 (1991).

144. G. Pruden, S.J. Kalembasa, D.S. Jenkinson. J. Sci. Food Agric. 36:71–73 (1985).

145. J.J. Lingane, R.L. Pacsok. Anal. Chem. 21:622–625 (1949).

146. M.J. Hepher, R.H. Alexander, J. Dixon. J. Sci. Food Agric. 49:379–383 (1989).

147. M.S. Cresser. Analyst 102:99–103 (1977).

148. T. Aoki, S. Uemura, M. Munemori. Environ. Sci. Technol. 20:515–517 (1986).

149. C. Kiang, S.H. Kuan, G. Guilbault. Anal. Chem. 50:1319–1322 (1978).

150. W.R. Hussein, G.G. Guilbault. Anal. Chim. Acta 76:183–192 (1975).

151. F.A.J. Armstrong. Anal. Chem. 35:1292–1294 (1963).

152. A. Syty, R.A. Simmons. Anal. Chim. Acta 120:163–170 (1980).

153. A.G. Fogg, S.P. Scullion, T.E. Edmonds. Analyst 115:599–604 (1990).

154. M. Nakamura. Analyst 106:483–487 (1981).

155. N. Velghe, A. Claeys. Analyst 108:1018–1022 (1983).

156. N.A. Fakhri, S.A. Rahim, W.A. Bashir. Int. J. Environ. Anal. Chem. 16:131–138 (1983).

157. S.J. Bajic, B. Jaselskis. Talanta 32:115–118 (1985).

158. T. Maeda, K. Aoki, M. Munemori. Anal. Chem. 52:307–311 (1980).

159. R.D. Cox. Anal. Chem. 52:332–335 (1980).

160. K. Yoshizumi, K. Aoki, T. Matsukoa, S. Asakura. Anal. Chem. 57:737–740 (1985).

161. D.D. Nygaard. Anal. Chim. Acta 130:391–394 (1981).

162. W. Holak, J.J. Specchio. Anal. Chem. 64:1313–1315 (1992).

163. A. Trojanek, S. Bruckenstein. Anal. Chem. 58:866–869 (1986).

164. A.J. Dunham, R.M. Barkley, R.E. Sievers. Anal. Chem. 67:220–224 (1995).

165. R.S. Braman, S.A. Hendrix. Anal. Chem. 61:2715–2718 (1989).

166. Y. Kanda, M. Taira. Analyst 117:883–887 (1992).

167. R. Montes, J.J. Laserna. Anal. Sci. 7:467–471 (1991).

168. R. Montes, J.J. Laserna. Talanta 34:1021–1026 (1987).

169. M. Jiang, F. Jiang, J. Duan, X. Tang, Z. Zhao. Anal. Chim. Acta 234:403–407 (1990).

170. T. Perez-Ruiz, C. Martinez-Lozano, V. Tomas. Anal. Chim. Acta 265:103–110 (1992).

171. A.A. Ensafi, M. Samimifar. Talanta 40:1375–1378 (1993).

172. C. Sanchez-Pedreño, M.T. Sierra, M.I. Sierra, A. Sanz. Analyst 112:837–840 (1987).

173. K. Ito. J. Chromatogr. A 764:346–349 (1997).

174. K. Ito, Y. Ariyoshi, F. Tanabiki, H. Sunahara. Anal. Chem. 63:273–276 (1991).

175. Y. Michigami, K. Fujii, K. Ueda. J. Chromatogr. A 664:117–122 (1994).

176. N. Chauret, J. Hubert. J. Chromatogr. 469:329–338 (1989).

177. P.R. Haddad, P.E. Jackson. J. Chromatogr. 346:139–148 (1985).

178. D.T. Gjerde, J.S. Fritz. Anal. Chem. 53:2324–2327 (1981).

179. G. Schmuckler, A.L. Jagoe, J.E. Girard, P.E. Buell. J. Chromatogr. 356:413–419 (1986).

180. S. Dogan, W. Haerdi. Chimia 35:339–342 (1981).

181. T. Okada, T. Kuwamoto. Anal. Chem. 55:1001–1004 (1983).

182. H. Small, T.S. Stevens, W.C. Bauman. Anal. Chem. 47:1801–1809 (1975).

183. J.A. Mosko. Anal. Chem. 56:629–633 (1984).

184. S. Carrozzino, F. Righini. J. Chromatogr. A 706:277–280 (1995).

185. J.K. Thomsen, R.P. Cox. J. Chromatogr. 521:53–61 (1990).

186. P.J. Naish. Analyst 109:809–812 (1984).

187. H. Small, T.E. Miller Jr. Anal. Chem. 54:462–469 (1982).

188. K. Ohta, K. Tanaka, J. Fritz. J. Chromatogr. A 731: 176–186 (1996).
189. R.A. Cochrane, D.E. Hillman. J. Chromatogr. 241: 392–394 (1982).
190. H. Sato. Anal. Chim. Acta 206:281–288 (1988).
191. P. Pastore, I. Lavagnini, A. Boaretto, F. Magno. J. Chromatogr. 475:331–341 (1989).
192. K. Ito, Y. Ariyoshi. J. Chromatogr. 598:237–241 (1992).
193. G.A. Sherwood, D.C. Johnson. Anal. Chim. Acta 129: 101 (1981).
194. M.A. Alawi. Fresenius Z. Anal. Chem. 317:372–375 (1984).
195. S.H. Lee, L.R. Field. Anal. Chem. 56:2647–2653 (1984).
196. R.M. Cassidy, S. Elchuk. Anal. Chem. 54:1558–1563 (1982).
197. F.G.P. Mullins, G.F. Kirkbright. Analyst 109:1217–1221 (1984).
198. J.P. de Kleijn. Analyst 107:223–225 (1982).
199. S.H. Kok, K.A. Buckle, M. Wootton. J. Chromatogr. 260:370–374 (1983).
200. B.B. Wheals. J. Chromatogr. 262:61–76 (1983).
201. N.E. Skelly. Anal. Chem. 54:712–715 (1982).
202. M.C. Gennaro, P.L. Bertolo, A. Cordero. Anal. Chim. Acta 239:203–209 (1990).
203. M.C. Gennaro, P.L. Bertolo. Ann. Chim. (Rome) 80: 13 (1990).
204. E. Pobozy, B. Sweryda-Krawiec, M. Trojanowicz. J. Chromatogr. 633:305–310 (1993).
205. Q. Xianren and W. Baeyens. J. Chromatogr. 514:362–370 (1990).
206. H. Kim, Y. Kim. Anal. Chem. 61:1485–1489 (1989).
207. S. Rokushika, F.M. Yamamoto, K. Kihara. J. Chromatogr. 630:195–200 (1993).
208. R.P. Singh, N.M. Abbas, S.A. Smesko. J. Chromatogr. A 733:73–91 (1996).
209. R.D. Wilken, H.H. Kock. Fresenius Z. Anal. Chem. 320:477–479 (1985).
210. D.R. Jenke, G.K. Pagenkopf. Anal. Chem. 56:85–88 (1984).
211. S. Rokushika, K. Kihara, P.F. Subosa, W. Leng. J. Chromatogr. 514:355–361 (1990).
212. Marheni, P.R. Haddad, A.R. McTaggart. J. Chromatogr. 546:221–228 (1991).
213. N. Gros, B. Gorenc. J. Chromatogr. A 770:119–124 (1997).
214. W.F. Koch. Anal. Chem. 51:1571–1573 (1979).
215. M. Novic, B. Divjak, B. Pihlar, V. Hudnik. J. Chromatogr. A 739:35–42 (1996).
216. P.E. Jackson, W.R. Jones. J. Chromatogr. 538:497–503 (1991).
217. I. Dallhöf, O. Svensson, C. Torstensson. J. Chromatogr. A 771:163–168 (1997).
218. Z. Iskandarani, D.J. Pietrzyk. Anal. Chem. 54:2427–2431 (1982).
219. O. Zerbinati. J. Chromatogr. A 706:137–140 (1995).
220. R.A. Wetzel, C.L. Anderson, H. Scheleicer, G.D. Crook. Anal. Chem. 51:1532–1535 (1979).
221. A.L. Heckenberg, P.R. Haddad. J. Chromatogr. 299: 301–305 (1984).
222. L. Eek, N. Ferrer. J. Chromatogr. 322:491–497 (1985).
223. V. Pacáková, K. Stulik. J. Chromatogr. A 789:169–180 (1997).
224. W.R. Jones, P. Jandik. J. Chromatogr. A 546:445 (1991).
225. P. Jandik, W.R. Jones, A. Westonand, P.R. Brown. LC-GC 9:634 (1991).
226. G. Bondoux, P. Jandik, W.R. Jones. J. Chromatogr. A 602:79–88 (1992).
227. J.P. Romano, J. Krol. J. Chromatogr. A 640:403–412 (1993).
228. S.A. Oehrle, R.D. Blanchard, C.L. Stumpf, D.L. Wulfeck. Paper Presented at the Sixth International Symposium on High Performance Capillary Electrophoresis, January 31–February 3, 1994, San Diego.
229. F. Guan, H. Wu, Y. Luo. J. Chromatogr. A 719:427–433 (1996).
230. S.A. Oehrle. J. Chromatogr. A 733:101–104 (1996).
231. G. Bondoux, T. Jones. LC-GC 13:144–148 (1995).
232. J. Ruzicka, G.D. Marshall, G.D. Christian. Anal. Chem. 62:1861–1866 (1990).
233. T. Gubeli, G.D. Christian, J. Ruzicka. Anal. Chem. 63: 1680 (1991).
234. J. Ruzicka, T. Gubeli. Anal. Chem. 63:2407–2413 (1991).
235. M.T. Oms, A. Cerdà, V. Cerdà. Anal. Chim. Acta 315: 321–330 (1995).
236. A. Cerdà, M.T. Oms, R. Forteza, V. Cerdà. Anal. Chim. Acta 371:63–71 (1998).

13

Ammonia

Stuart W. Gibb

Environmental Research Institute, Thurso, Caithness, Scotland, United Kingdom

I. INTRODUCTION

Ammonia is a key parameter in water and wastewater measurement. The aims of this chapter are: (a) to provide readers with a knowledge of the basic chemistry of ammonia, its occurrence, and its significance in waters and wastewaters, and (b) to provide practical guidelines and protocols to facilitate successful determination of ammonia in a wide range of waters and wastewaters.

A. Historical Context

The perception that nitrogen is an important and dynamic ecological factor has roots going back over 200 years. Soon after its discovery in 1772, by Rutherford (or Cavendish or Scheele), Berthelot identified "azote" as a component in animal tissue and excreta. Boussingault proposed its role as a major controlling factor in plant nutrition and productivity, with Hillreigel and Wilfarth providing the first evidence of nitrogen fixation in 1880 (1).

Great changes have occurred in the nitrogen cycle over the last 100 years as a result of population expansion, increasing industrialization, and changing land use. For example, ammonia is industrially produced in great quantities by the Haber process (U.S. production in 1986, 12.7 megatons) and extensively in the manufacture of fertilizers, plastics, foams, and explosives. Consequently, environmental concentrations of ammonia have also been subject to great change. Ammonia is today recognized as an important and dynamic component of the nitrogen cycle and thus a key parameter in the assessment of water and wastewater quality (Fig. 1).

B. General Aqueous-Phase Chemistry of Ammonia

Nitrogen is one of around 30 bioessential elements incorporated into biogeochemical cycles. This involves the transfer of essential nutrients from living organisms to the physical environment and back to the organisms in a cyclical pathway. The biogeochemical cycle of nitrogen comprises both natural and anthropogenic components and is subject to great complexity due to the diversity of compounds and transformations involved (Fig. 1).

Nitrogen is found in a variety of inorganic forms in natural waters and wastes, ranging from the thermodynamically most stable (in the aerobic environment), nitrate (NO_3^-), with an oxidation state of $+V$, to reduced compounds such as ammonia ($-III$) (Fig. 2). Together, the inorganic species nitrite NO_2^-, NO_3^-, and NH_3 are referred to as *dissolved inorganic nitrogen* (DIN).

Dissolved organic nitrogenous compounds (DON) include humic substances, amino acids, amines, and nucleic acids. The oxidation state of nitrogen in all organic compounds is also $-III$.

In general, biological production and transformation dictate the concentrations of organic and inorganic nitrogen in the aquatic environment (3). Bacteria, plants,

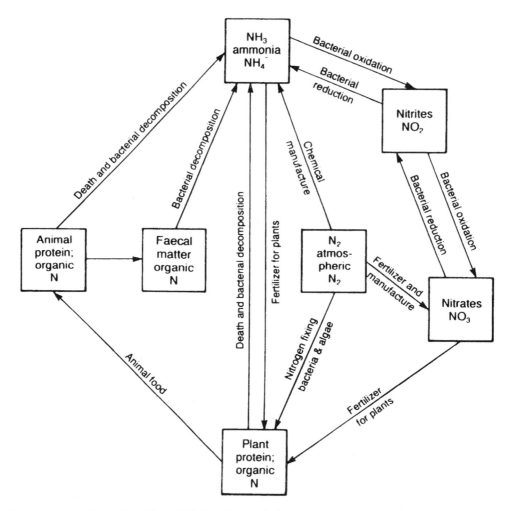

Fig. 1 The nitrogen cycle. (Reproduced from Ref. 2 with permission.)

and animals all play complementary roles in the turnover of ammonia: Bacteria dominate the regenerative processes, converting organic compounds to inorganic species; ammonia may serve as a primary or secondary source of nitrogen for plant life; and, in general, animals contribute to the cycle only through excretion (Figs. 1 and 2).

At room temperature, ammonia is a colorless gas with pungent odor detectable at levels above 50 ppm (4) (Table 1). It is highly water soluble, undergoing extensive hydrogen bonding to form a basic solution according to the following equilibrium:

$$NH_{3(g)} + H_2O \Leftrightarrow NH_{4(s)}^+ + OH^-$$

where $pK_a = 9.245$, ionic strength $I = 0$ mol dm^{-3}, $T = 20°C$.

In water, dissolved ammonia exists in equilibrium between free ammonia gas ($NH_{3(s)}$) and the solvated ammonium cation ($NH_{4(s)}^+$):

$$NH_{4(s)}^+ + H_2O \Leftrightarrow NH_{3(s)} + H_3O^+$$

where H_3O^+ is the hydronium ion.

The $NH_{3(s)}$–$NH_{4(s)}^+$ equilibrium is of profound importance to the water chemistry and biogeochemical cycling of ammonia. Over the pH range typical of natural waters, NH_4^+ is the dominant species (Fig. 3). However, it is the NH_3 that participates in phase-exchange processes such as volatilization, hydrophobic partitioning, and bioassimilation. The $NH_{3(s)}$–$NH_{4(s)}^+$ equilibrium is extremely dependant on pH and to a lesser extent on salinity (ionic strength, I) and temperature, T.

1. Effect of pH

High pH favors the formation of free gaseous $NH_{3(s)}$ (Fig. 3). The concentration of $NH_{3(s)}$ may be calculated

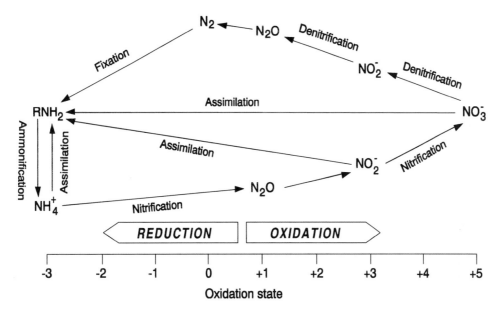

Fig. 2 Schematic representation of the aquatic nitrogen cycle showing the principle transformation processes between the various forms of nitrogen.

from the total dissolved concentration, $(NH_{3(s)}-NH_{4(s)}^+)$, according to the following equation (8):

$$[NH_{3(s)}] = \frac{[(NH_3 + NH_4^+)_{(s)}] \cdot [OH^-]}{K_b + [OH^-]}$$

where $[OH^-]$ is the hydroxide ion concentration, calculated from the measured pH using the expression $[OH^-] = 10^{-(14-pH)}$. The base dissociation constant, K_b, is calculated from the thermodynamic stability constant, K_a, using the expression $K_b = 10^{-(14-pKa)}$. Since the pK_a value reported in Table 1 refers to conditions of T = 20°C and I = 0 mol dm^{-3}, it must be conditioned before application of the foregoing equation to most waters and wastes.

2. Effects of Salinity and Temperature

The Debye–Huckel, Davis, and Guntelberg equations, which define the influence of I on pK_a, are only applicable in dilute solutions [0 < 0.1 mol dm^{-3} (9)] and are unsuitable for application to most natural waters and wastewaters. However, Khoo et al. (10) give an empirical formula for the salinity dependence of pK_a

Table 1 General Physicochemical Properties of Ammonia

Parameter	Value	Ref./notes
Formula weight, NH$_3$	17.03	5
Boiling point (°C)	−33.4	5
Melting point (°C)	−77.7	5
Solubility (g/L): cold water	899	5
hot water	74	
alcohol	132	
Solubility coefficient (298 K)	0.0420	6
Thermodynamic stability constant, pK_a	9.245 I = 0 mol dm^{-3}, T = 20°C	7
Henry's law constant (293 K)	1.8×10^{-3}, I = 0 mol dm^{-3}	8
Density (760 mm Hg)	0.7710 g/L	5
DOT ID no.	UN1005	
CAS no.	7664-4-7	

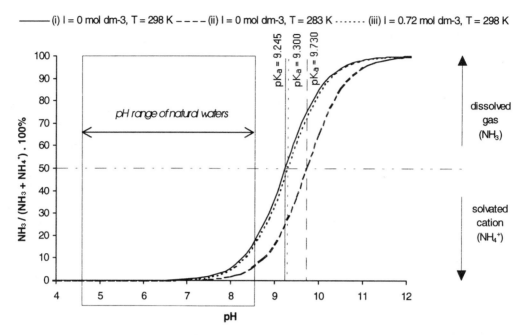

Fig. 3 Variation of the speciation of $NH_{3(s)}$–$NH_{4(s)}^+$ with ionic strength and temperature using pH as the master variable. (i) I = 0 mol dm^{-3}, T = 298 K (pK$_a$ = 9.245) (Ref. 6); (ii) I = 0 mol dm^{-3}, T = 283 K (pK$_a$ = 9.245); (iii) I = 0.72 mol dm^{-3}, T = 298 K (pK$_a$ = 9.245).

for the $NH_{3(s)}$–$NH_{4(s)}^+$ equilibrium between salinities of 0 and 45:

$$pK_a = pKw_a + (0.1552 - 0.003142T)I$$

where pKw$_a$ is the value of pK$_a$ in pure water (I = 0 mol dm^{-3}, T = 20°C) and I is obtained from the measured salinity (S, ‰) using (11):

$$I = 0.00147 + 0.01988S + 2.08357 \times 10^{-5} \cdot S^2$$

The speciation of ammonia has important implications for the assessment of analytical data, since some techniques respond only to $NH_{3(s)}$, others only to $NH_{4(s)}^+$, and some to both forms. It should be recognized that when using techniques that give a response specific to $NH_{3(s)}$ or $NH_{4(s)}^+$, it is imperative that pHs well removed from 9.245 are used during measurement, i.e., removed from the pK$_a$ value (Fig. 3).

C. Notation and Units of Measurement

To allow clear differentiation between the different forms of ammonia, "NH_3" will be used to refer to the dissolved free gas, and "NH_4^+" the solvated cation. Where differentiation between the two species is unnecessary, or reference is being made to the sum NH_3 + NH_4^+, the term "NH_x" will be used to indicate total ammoniacal nitrogen.

Although a number of different recognized units may be used to express aqueous concentrations of NH_3, NH_4^+, and NH_x, the units of mg/L or μg/L will be used predominantly in this text. The factors presented in Table 2 should allow straightforward conversion between these different units.

II. OCCURRENCE AND SIGNIFICANCE OF AMMONIA IN WATER AND WASTEWATERS

A. Sources of Ammonia

Though NH_x is produced both naturally and anthropogenically (Fig. 1), natural sources account for the greatest proportion of the global aqueous NH_x budget (12). The principle sources of NH_x in the aquatic environment may be classified as follows (4):

Biogenic: Bacteria play an important role in the nitrogen cycle, producing NH_x via the decomposition of proteins, amino acids, and other organic nitrogenous compounds in the water column and benthic environment (1,3). In addition to oxidizing NH_3 to nitrite (NO_2^-) and NO_3^-, bacteria are capable of fixing atmospheric N_2 and of reducing

Table 2 Factors for Conversion Between NH$_x$ Concentration Units

	Mass units			Molar units	
	mg/L NH$_3$–N	mg/L NH$_3$–NH$_3$	mg/L NH$_3$–NH$_4^+$	Micromolar (μM or μmol/L)	Nanomolar (nM or nmol/L)
1 mg/L NH$_3$–N =	1	1.216	1.288	71.4	7.14 $\times$ 10^4
1 mg/L NH$_3$–NH$_3$ =	0.823	1	1.059	58.7	5.57 $\times$ 10^4
1 mg/L NH$_3$–NH$_4^+$ =	0.777	0.944	1	55.4	5.54 $\times$ 10^4
1 μM NH$_3$ =	0.014	0.017	0.018	1	1000
1 nM NH$_3$ =	1.4 $\times$ 10^{-5}	1.7 $\times$ 10^{-5}	1.8 $\times$ 10^{-5}	0.001	1

Source: Based on Ref. 2.

NO$_2^-$ and NO$_3$ to NH$_x$ and even free N$_2$. Animal and human excrement also contribute (4) (Fig. 1).

Agricultural: Eighty percent of all man-made NH$_x$ is used in fertilizer (4). A third of this is applied directly as pure NH$_x$, with the remainder being applied as NH$_4^+$-containing fertilizer. Waste from livestock housing, courts, yards, feedlots, and grazing land also makes a significant contribution.

Industrial: NH$_x$ is used in a diverse array of industrial process, including:

    Coke production from coal
    Synthesis of nitric acid, synthetic fibers, plastics, and explosives
    Metallurgic and oil-refining operations
    Ceramics production
    Strip mining
    Production of refrigeration equipment
    Production of household cleaners
    Food processing

Residential and urban: NH$_x$ is introduced into air and waste waters in the urban and residential environments upon usage and disposal of a range of NH$_x$-containing cleaning and domestic products.

Atmospheric deposition: NH$_x$ in the atmosphere may derive from volatilization and combustion processes (e.g., waste disposal, internal combusion, and fuel combustion). Atmospheric deposition of NH$_x$ (via rain, particulates, aerosols, and gas exchange) contributes a significant proportion of total nitrogenous inputs to natural water systems. The significance of these inputs increases with the remoteness of the receiving water from anthropogenic influence.

There have been a number of assessments of the global nitrogen cycle and the significance of NH$_x$, but estimates are over wide ranges (12). Though the aquatic nitrogen cycle and the turnover and concentrations of NH$_x$ may once have been in steady state, anthropogenic activities are likely to have caused a significant perturbation of any natural balance (12). Anthropogenic nitrogen fixation now exceeds natural fixation; and as a result of fertilizer runoff, sewage dumping, and soil erosion, the anthropogenic inputs of nitrogen have overwhelmed natural inputs in many aquatic systems.

B. Why Measure Ammonia?

NH$_x$ is ubiquitous in waters and wastewater. Its concentrations span many orders of magnitude, from oligotrophic oceanic waters to heavily polluted wastewaters (Fig. 4).

1. Ammonia as a Micronutrient

Plants assimilate nitrogen (mainly NH$_x$, NO$_3^-$, and NO$_2^-$) to construct their cellular amino acids and proteins. In the assimilation process the inorganic nitrogen species is first transported into the cell via an active transport system involving ATP hydrolysis. The nitrogen is then transformed into metabolites, such as proteins, by a series of anabolic reaction (12):

$$\text{NO}_3^- + 2\text{H}^+ + 2e^- \rightarrow \text{NO}_2^- + \text{H}_2\text{O} \qquad (1)$$

$$2\text{NO}_2^- + 4\text{H}^+ + 4e^- \rightarrow \text{N}_2\text{O}_2^{2-} + 2\text{H}_2\text{O} \qquad (2)$$

$$\text{N}_2\text{O}_2^{2-} + 6\text{H}^+ + 4e^- \rightarrow 2\text{NH}_2\text{OH} \qquad (3)$$

$$\text{NH}_2\text{OH} + 2\text{H}^+ + 2e^- \rightarrow \text{NH}_3 + \text{H}_2\text{O} \qquad (4)$$

$$\alpha\text{-ketoglutaric acid} + \text{NH}_3 + 2\text{NADPH}$$
$$\rightarrow \text{glutamic acid} + 2\text{NADP} + \text{H}_2\text{O} \qquad (5)$$

$$\text{pyruvic acid} + \text{glutamic acid}$$
$$\rightarrow \text{alanine} + \alpha\text{-ketoglutaric acid} \qquad (6)$$

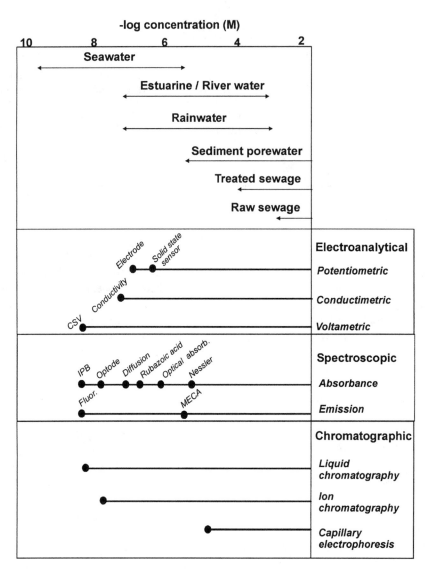

Ranges based on data from Abbreviations: IPB = Indophenol Blue; CSV = cathodic stripping voltammetry; MECA = molecular emission cavity analysis; GC = gas chromatography; LC = liquid chromatography; HPLC = high performance liquid chromatography; IC = ion chromatography

Methods: Electrode (13,14), Solid state (15), Conductivity (16), CSV (17), IPB (18), Optode (19), Misc. (20), Rubazoic Acid (21), Nessler (22), Fluorimetric (23), HPLC (24), IC (25), Capillary Electrophoresis (26)

Fig. 4 Concentration ranges of NH_x in selected waters and wastewaters, and lowest reported detection limits of techniques for the determination of NH_x in aqueous media.

If NO_3 is assimilated, the sequence commences with reaction (1). If NH_x is assimilated, anabolism is initiated by reaction (5). When NO_3^- or NO_2^- is assimilated, the nitrogen must first be reduced to the $-III$ oxidation state. Once in reduced form, e.g., NH_x, the nitrogen reacts with a carboxylic acid [e.g., α-ketoglutaric acid, reaction (5)] to produce an amino acid [e.g., glutamic acid, reaction (6)]. The 20 or so other amino acids re-

quired as protein building blocks are formed by transamination of glutamic acid.

Since NH_x is in the correct oxidation state, less energy needs to be expended in its assimilation and conversion to organic matter than for NO_3^- or NO_2^-. To take advantage of this saving in energy, many plants and phytoplankton have evolved transport mechanisms that favor uptake of NH_x over NO_3^- or NO_2^-.

2. Eutrophication

Eutrophication, the excessive growth of biomass as a consequence of high nutrient inputs, leads to the perturbation of the ecological balance of a body of water. The decay of biomass may lead to oxygen depletion or hypoxia, resulting in problems such as increased fish mortality. The role of NH_x as a preferred micronutrient thus makes it a key water-quality parameter in studies of eutrophication and eutrophication potential.

3. Rainwater Acidity

NH_x may be introduced into the atmosphere via combusion and volatilization processes. In the atmosphere NH_x can be incorporated into cloudwater or rainwater and can also react with H_2SO_4 aerosol particles to form NH_4^+-containing aerosol. These particles can act as cloud condensation nuclei, thereby incorporating the NH_4^+ into cloudwater. Due to the basic character of NH_x, these phenomena may be highly influential in the regulation of the acid–base chemistry of the atmosphere.

4. Environmental Effects

The NH_4^+ ion is nontoxic and is not of concern to organisms. However, un-ionized NH_3 is capable of crossing cell membranes, particularly at high pH values. In excess it is harmful to aquatic life and may accumulate in the organism and cause alteration in metabolism or increases in body pH (4). For example, fish may suffer a loss of equilibrium, hyperexcitability, increased respiratory activity and oxygen uptake, and increased heart rate. At extreme NH_3 levels, coma and death may occur. Experiments have shown that the lethal concentration for a variety of fish species ranges from 0.2 mg/L (trout) to 2 mg/L (carp) (27).

However, even levels of NH_x falling within the "acceptable" range may adversely affect aquatic life, including reduction in hatching success, reduction in growth rate and morphological development, and injury to gill tissue, liver, and kidneys (27). Studies have shown that exposure to NH_3 concentrations as low as 0.002 mg/L for 6 weeks causes hyperplasia of the gill lining in salmon fingerlings and leads to bacterial gill disease (4).

5. Human Health Effects

Despite being an essential micronutrient, NH_x may accumulate in the human body, as in other organisms, and cause deleterious secondary effects, such as alteration of the metabolism and increases in body pH (4).

At elevated levels NH_x is an irritant affecting eyes, nose, throat, and lungs. Ingestion may result in corrosion of the mouth lining, esophagus, and stomach (4). NH_x is thus an essential parameter in the assessment of potable water quality.

6. Chlorination Schemes

Chlorine is a widely used form of water disinfection, and its use is attributed to the virtual eradication of water-borne diseases. During treatment, chlorine reacts with the water to produce the disinfection agent hypochlorous acid, HOCl. At the pHs typical of potable waters, NH_x can react with the HOCl according to the following (2):

$$NH_3 + HOCl$$
$$\Leftrightarrow NH_2Cl \text{ (monochloramine)} + H_2O$$
$$NH_2Cl + HOCl$$
$$\Leftrightarrow NHCl_2 \text{ (dichloramine)} + H_2O$$
$$NHCl_2 + HOCl$$
$$\Leftrightarrow NHCl_3 \text{ (nitrogen trichloride)} + H_2O$$

These are competitive reactions, dependent on pH, initial concentrations, temperature, and time. Monochloramine is a less effective disinfectant than HOCl but is useful as a residual disinfectant after the main disinfection is complete. It is also capable of preventing the production of carcinogenic trihalomethanes in the presence of organics. Dichloramine and nitrogen trichloride impart objectionable taste and odor to potable waters and are hence undesirable (2). Thus, NH_x measurements are also used as part of chlorination schemes.

III. ANALYSIS OF AMMONIA

A. Overview

Although NH_x is one of the most commonly determined analytes, its precise and accurate determination should be considered nontrivial. The challenges posed in the determination of NH_x in aqueous media include the extreme range of concentrations at which it occurs (Fig. 4) and overcoming the polar and highly water-soluble nature of the analyte itself. The importance of overcoming these challenges and achieving reliable measurements of NH_x is reflected in the sheer number and diversity of techniques proposed in the literature. These techniques, classified according to the principle divisions of analytical chemistry techniques, are summarized in Table 3. The ranges of concentration to

Table 3 Classification of Techniques Proposed for Determination of NH_x in Aqueous Media

Classification	Type	Technique
Gravimetric: precipitation and volatilization techniques	None	None
Titrimetric: analyte determined from amount of reagent required to react with analyte completely	Volumetric	Acid–base titration
Electroanalytical: involve the measurement of electrical properties, such as potential, current, resistance, and quantity of electricity	Conductimetric	Conductimetry
	Potentiometric	Electrode
	Voltametric	Cathodic stripping voltammetry (CSV)
Spectroscopic: techniques based on measurement of the interaction between electromagnetic radiation analyte atoms, molecules, or ions or on production of such radiation by analyte (current usage extends this classification to include methods such as acoustic, mass, and electron spectroscopy)	Absorbance	Indophenol blue (IPB)
		Rubazoic acid
		Nesslerization
		Optodes
		Miscellaneous
	Emission	Chemiluminescence
		Fluorescence
		Molecular emission (MECA)
Chromatographic: separation of a mixture of components by their passage through or over a material that interacts differentially with them	Column	Gas chromatography (GC)
		Liquid chromatography (LC)
		Capillary electrophoresis (CE)[a]
	Planar	Thin-layer chromatography (TLC)
		Paper chromatography (PC)

[a]Although CE is not strictly a form of chromatography it is most conveniently and appropriately classified in this way.
Source: Based on the principle divisions of analytical techniques suggested in Ref. 28.

which each of these techniques is applicable is shown in Fig. 4.

B. Criteria for Selecting Analytical Method

In selecting a technique for the determination of NH_x, many different criteria can be used. A number of key criteria are collated in Table 4. In practice it is unlikely that any method will fulfill all criteria; thus, prioritization is most probable.

C. Selected Methods for the Analysis of Ammonia

Methods for the determination of NH_x in waters and wastewaters have been subdivided as follows:

Recommended methods: These are well-established methods that have been in active use for a number of years and are, with one exception, recommended by organizations such as the U.S. Environmental Protection Agency (USEPA) and the American Public Health Association (APHA) (Tables 5 and 6, Methods A and D). The remain-

ing method is a published technique suitable for the determination of trace NH_x (Tables 5 and 6 Method E). These methods are discussed in Sec. V.

Reviewed methods: In addition to the foregoing methods, there are other techniques. These are likely to be of more specialized interest, but may allow the use of existing laboratory apparatus and instrumentation. These methods are discussed in Section VI.

The methods presented should allow the successful analysis of most waters and wastewaters. All methods cited are fully referenced and credited. Readers seeking further information should consult the original document from which the method has been extracted.

In presenting several methods, the reader should have a choice, and is likely to be able to utilize existing laboratory equipment. Manual, automated and instrumental methods are presented, the intention being that most laboratories be able to achieve successful analysis of NH_x. However, the analyst should be aware that, in the spirit of scientific progress, the most suitable methods are likely to be modified, im-

Table 4 General Criteria for Selecting a Technique for Determination of NH_x in Aqueous Media

Qualitative aspects	Quantitative aspects
Minimal interference and matrix effects	Sensitivity—accurate quantification over required concentration range
Discrete/automated/stand-alone/ or continuous analysis	Limits of detection
Reliability	Linearity of response
Flexibility and scope for adaptation, e.g., use of apparatus for alternative analysis	Precision
	Low blanks
Stability and robustness (especially for field application)	Sample throughput
Simplicity and ease of use	Startup and running costs
Health and safety considerations	
Method recognition (government, EPA, etc.)	

LIVERPOOL
JOHN MOORES UNIVERSITY
AVRIL ROBARTS LRC
TITHEBARN STREET
LIVERPOOL L2 2ER
TEL. 0151 231 4022

proved, or superseded by techniques providing improved sensitivity, higher throughput, reduced susceptibility to interference, and the like.

IV. PRELIMINARY CONSIDERATIONS

The following section deals with protocols and practicalities that should be addressed before undertaking analysis of NH_x. These are relevant to most analytical techniques and include: the preparation of apparatus, the preparation of NH_x-free water, and the collection, preservation, preparation, and preliminary distillation of samples.

A. Preparation of Apparatus

It is important that all apparatus used in the analysis and coming into contact with samples and standards are free of NH_x. All apparatus should be thoroughly soaked clean using the following solutions (33):

Dissolve 100 g of potassium hydroxide in 100 ml of water. Cool the solution and add 900 ml of methylated spirits. Store in a polythene bottle and label TOXIC.

Apparatus should then be rinsed with copious amounts of NH_x-free water (next subsection) and thereafter reserved for NH_x determinations in a clean and dry storage environment.

B. Preparation of Ammonia-Free Water

In the preparation of reagents, blanks, and standards, NH_x-free water is required. Although in practice it is not possible to remove all NH_x from water, the minimization of its presence is essential to achieve optimum results. NH_x-free water may be prepared from laboratory-quality distilled water by ion exchange of distillation.

Ion exchange: NH_x-free water is prepared by passing distilled water through an ion-exchange column containing a mixed-bed resin consisting of a strongly acidic cation-exchange resin and a strongly basic anion-exchange resin. Resins should be selected that remove both NH_x, and species that might interfere with the determination of NH_x. Some anion-exchange resins release NH_x. If this occurs, use only strongly acidic cation-exchange resin. NH_x-free water should always be tested prior to use for the possibility of a high blank. Ion-exchange resins should be regenerated according to manufacturer's recommendations.

Distillation: The NH_x content of distilled water can be minimized through the addition of 0.1 ml concentrated H_2SO_4 per liter of distilled water and then redistilling

NH_x-free water should be prepared daily to avoid contamination through storage. If storage is necessary, gas-tight containers constructed from glass, low-density polythene (LDPE), or Teflon (PTFE) should be used. The addition of 5–10 g of a strongly acidic cation-exchange resin per liter of water may additionally reduce residual NH_x. If resin is added, it should be routinely replaced and regular blanks analyses performed on the container contents.

Table 5 Scope and Applicability of Methods Recommended for the Analysis of NH$_x$ in Aqueous Media

	Method					
	A: Distillation + titration (29)	**B.1:** Manual phenate (30,31)	**B.2:** Automated phenate (32)	**C:** Nesslerization (29)	**D:** Electrode (32)	**E:** Gas diffusion/ fluorescence (23)
Detection type	Titrimetric	Spectroscopic absorbance (colorimetric)	Spectroscopic absorbance (colorimetric)	Spectroscopic absorbance (colorimetric)	Electroanalytical/ potentiometric	Spectroscopic emission
LOD (mg/L)	5	0.01	0.01	0.01	0.03	0.00003
Range (mg/L)	>5	0.01–2	0.01–2	0.02–5	0.03–1400	0.00003–0.03
	Higher concentrations by dilution	Higher concentrations by dilution	Higher concentrations by dilution	Higher concentrations by dilution	Higher concentrations by dilution	Higher concentrations by dilution
Throughput (samples/h, approx.)	2–3		Up to 60	30 samples in 3 h	15–20	Up to 30
Interferences	Volatile alkaline compounds (hydrazine, amines)	Calcium, magnesium, turbidity, color, glycine, hydrazine, amines	Calcium, magnesium, turbidity, primary amines	Calcium, magnesium, turbidity, primary amines	Amines	Primary amines
APPLICATION						
Oligotrophic seawater/lake water						●
Coastal seawater		○	●	●	○	●
Estuarine/river water		●	●	●	●	○
Lake water	○	○	○	○	●	●
Sediment pore water	○	●	●	●	●	
Drinking water	○	●	●	●	○	○
Wastewater/ sewage	●	●	●	●	●	

● = applicable under most circumstances; ○ = applicable under limited circumstances.

Table 6 Cross-Referenced Methods Approved by Recognized Analytical and Environmental Organizations

Organization	Method					
	A: Distillation + titration	**B.1:** Manual phenate	**B.2:** Automated phenate	**C:** Nesslerization	**D:** Electrode	**E:** Gas diffusion/ fluorescence (23)
U.K Standing Committee of Analysts [SCA (33)]	SCA; NH_3 in waters; method B	SCA; NH_3 in waters; method E	SCA; NH_3 in waters; method F		SCA; NH_3 in waters; method C	
U.S. Environmental Protection Agency [USEPA (32)]	EPA 350.2		**EPA 350.1**		**EPA 350.3**	
American Society for Testing and Materials [ASTM (34)]				D1426-89; test method A	D1426-89; test method B	
Analytical Methods Committee of the Royal Society of Chemistry [AMC (35)]		W03; salicylate chemistry; see SCA	W74			
American Public Health Association, American Water Works Association; and Water Environment Federation [APHA (29)]	**APHA-4500-NH_3E**	APHA-4500-NH_3D	APHA-4500-NH_3H	**APHA-4500-NH_3C**	APHA-4500-NH_3F, APHA-4500-NH_3G	

Methods in bold are presented in this text.

In order to avoid matrix effects when analyzing sea-water, low-NH_x seawater should be used as a blank and in the preparation of standards. This should be collected from regions distant from sources of pollution, ideally from remote oceanic regions, preferably from depths below 300 m. If this is not possible, artificial seawater should be used, prepared using NH_x-free water; see previous discussion.

C. Sample Collection, Preservation, and Preparation

1. Sample Collection

Water samples should be collected in gas-tight bottles constructed from glass, LDPE, or PTFE. The volume of sample will depend on a number of factors, including the analytical technique used, the number of replicate analyses to be performed, and the need for standard additions. Sample bottles should be filled to the exclusion of headspace to minimize changes in concentration occurring from atmospheric contamination or volatilization.

2. Sample Preservation

Changes in the concentration of NH_x in stored samples can occur through a number of processes, including volatilization, algal or bacterial activity, and photoreactivity. Therefore analysis should be performed as soon as possible after sample collection. However, if analyses cannot be performed within 1–2 h of sampling, a number of preservation options are available allowing storage of samples up to 2 weeks:

> Cold storage at 4°C.
> Freezing at −15°C.
> Addition of 2 ml of phenol solution per 50 ml of sample. Prepare phenol solution by dissolving 20 g of analytical-grade phenol in 20 ml of 95% v/v ethyl alcohol (36).
> Addition of enough concentrated H_2SO_4 per liter to achieve pH 1.5–2, and refrigeration at 4°C (32). The pH must then be restored by the addition of concentrated sodium hydroxide or potassium prior to analysis.

Note:

1. Volumetric changes arising from preservation must be accounted for in the calculation of the final sample concentration.
2. Before sample preservation is undertaken, it should be ensured that the steps and additions involved are compatible with the analytical methodology to be used (see individual methods).
3. In the determination of trace NH_x, e.g., in oligotrophic lake water or seawater, preservation is not a realistic option and only immediate analysis is suitable.

3. Suspended Material and Filtration

Samples containing suspended material should be allowed to settle prior to analysis or preferably filtered through a glass-fiber or membrane filter prior to analysis. However, it should be recognized that NH_x may be absorbed or released during filtration, depending on the filter material in use. This should be checked prior to full sample analysis (29,32,33).

4. Dechlorination

Several analytical techniques require that free chlorine be removed from samples before analysis. Effective dechlorination may be achieved by the addition of a crystal of sodium thiosulphate to the sample at the time of collection.

D. Preliminary Distillation

Distillation may serve both to extend the viable operational range of the method in use and to reduce the effect of interferences. Distillation is often necessary in the analysis of polluted waters and wastes. It may be used in conjunction with most analytical techniques but is an essential prerequisite if titrimetry is the chosen method. Although several variations of distillation are to be found in the literature, a well-recognized method suitable for most waters and wastes has been chosen and is reproduced here (29).

1. Outline [Source: APHA Method 4500-NH_3B (29)]

The sample is buffered at pH 9.5 with a borate buffer to decrease hydrolysis of cyanates and organic nitrogen compounds. It is distilled into a solution of boric acid when nesslerization or titration is to be used or into H_2SO_4 when the phenate method is used. The NH_x in the distillate can be determined colorimetrically (nesslerization or indophenol blue), titrimetrically (with standard H_2SO_4 and a mixed indicator), or via a pH meter. The choice between the colorimetric and the acidimetric methods depends on the concentration of NH_x.

2. Apparatus

Distillation apparatus: Arrange a borosilicate glass flask of 800–2000-ml capacity attached to a vertical condenser so that the outlet tip is submerged below the surface of the receiving acidic solution. Use an all–borosilicate glass apparatus or one with condensing units constructed of block tin or aluminum tubes.

pH meter.

3. Reagents

Ammonia-free water: See Sec. IV.B, earlier. Use to prepare all reagents.

Borate buffer solution: Add 88 ml 0.1 N NaOH solution to 500 ml approximately 0.025 M sodium tetraborate $(Na_2B_4O_7)(9.5$ g $Na_2B_4O_7 \cdot 10H_2O$/liter) and dilute to 1 liter.

Sodium hydroxide, 6 N.

Dechlorinating agent: Use 1 ml of either of the following reagents to remove 1 mg/L residual chlorine in 500 ml sample:

1. *Sodium thiosulphate:* Dissolve 3.5 g $Na_2S_2O_3 \cdot 5H_2O$ in NH_x-free water and dilute to 1 liter. Prepare fresh weekly.
2. *Sodium sulphite:* Dissolve 0.9 g Na_2SO_3 in water and dilute to 1 liter. Prepare fresh daily.

Neutralization reagent: Prepare with NH_x-free water.

1. Sodium hydroxide, NaOH, 1 N
2. Sulphuric acid, H_2SO_4, 1 N

Absorbent solution, plain boric acid: Dissolve 20 g H_3BO_4 in water and dilute to 1 liter with NH_x-free water.

Indicating boric acid solution: Dissolve 20 g H_3BO_4 in NH_x-free distilled water, add 10 ml of mixed indicator solution, and dilute to 1 liter. Prepare monthly.

Mixed indicator solution: Dissolve 200 mg methyl red indicator in 100 ml 95% ethyl or propyl alcohol. Dissolve 100 mg methylene blue in 50 ml 95% ethyl or isopropyl alcohol. Combine solutions. Prepare monthly.

Sulphuric acid, 0.04 N: Dilute 1.0 ml concentrated H_2SO_4 to 1 liter in NH_x-free water.

4. Procedure

Preparation of equipment: Add 500 ml of NH_x-free water and 20 ml borate buffer to a distillation flask and adjust pH to 9.5 with 6 N NaOH solution. Add a few glass beads or boiling chips and use this mixture to steam out the distillation apparatus until the distillate shows no trace of NH_x.

Sample preparation: Use 500 ml dechlorinated sample or a portion diluted to 500 ml with NH_x-free water. When NH_x–N concentration is less than 100 μg/L, use a sample volume of 1000 ml. Remove residual chlorine by adding, at the time of collection, dechlorinating agent equivalent to the chlorine residual (see earlier). If necessary, neutralize to approximately pH 7 with dilute acid or base, using a pH meter.

Add 25 ml borate buffer solution and adjust to pH 9.5 with 6 N NaOH using a pH meter.

Distillation: To minimize contamination, leave the distillation apparatus assembled after steaming out and until just before starting sample distillation. Disconnect steaming out flask and immediately transfer sample flask to distillation apparatus. Distill at a rate of 6–10 ml/min, with the tip of the delivery tube below the surface of the receiving solution. Collect distillate in a 500-ml Erlenmeyer flask containing 50 ml plain boric acid solution for nesslerization method. Use 50 ml indicating boric acid solution for titrimetric method. Distill NH_3 into 50 ml 0.04 N H_2SO_4 for the phenate method and for the ion-selective electrode method. Collect at least 200 ml distillate. Lower distillation receiver so that the end of the delivery tube is free of contact with the liquid, and continue distillation during the last minute or two to cleanse condenser and delivery tube. Dilute to 500 ml with NH_x-free-water.

When the phenate method is used for determining NH_3–N, neutralize distillate with 1 N NaOH solution.

NH_x determination: Determine by nesslerization, phenate, titrimetric, or ion-selective method.

V. RECOMMENDED METHODS FOR THE DETERMINATION OF AMMONIA

A. Titrimetric Method

1. Outline [Source: APHA Method 4500-NH₃ E (29), Tables 5 and 6]

The titrimetric determination of NH_x is used only on samples that have been carried through preliminary distillation. Table 7 is useful in selecting the sample volume for distillation and titration.

Table 7 Sample Volumes Required for Titrimetric Analysis of NH$_x$

Ammonia nitrogen in sample (mg/L)	Sample volume (ml)
5–10	250
10–20	100
20–50	50.0
50–100	25.0

Table 8 Precision and Bias Data for Titrimetric Determination of NH$_x$

	Distillation plus titration	
NH$_x$ concentration (μg/L)	Relative standard deviation (%)	Relative error (%)
200	69.8	20.0
800	28.6	5.0
1500	21.6	2.6

Source: Ref. 29.

2. Apparatus

Distillation apparatus: Arrange a borosilicate glass flask of 800–2000-ml capacity attached to a vertical condenser so that the outlet tip is submerged below the surface of the receiving acidic solution. Use an all–borosilicate glass apparatus or one with condensing units constructed of block tin or aluminium tubes.

pH meter.

3. Reagents

Use NH$_x$-free water in making all reagents and dilutions.

Mixed indicator solution: Dissolve 200 mg methyl red indicator in 100 ml 95% ethyl or isopropyl alcohol. Dissolve 100 mg methylene blue in 50 ml ethyl or isopropyl alcohol. Combine solutions. Prepare monthly.

Indicating boric acid solution: Dissolve 20 g H$_3$BO$_3$ in NH$_x$-free water, add 10 ml mixed indicator solution, and dilute to 1 liter. Prepare monthly.

Sodium carbonate solution, approximately 0.05 N: Dry 3–5 g primary standard Na$_2$CO$_3$ at 250°C for 4 h and cool in a desiccator. Weigh out 2.5 ± 0.2 g (to the nearest milligram), transfer to a 1-L volumetric flask, fill flask to the mark with NH$_x$-free water, and dissolve and mix reagent. Do not keep longer than 1 week.

Standard sulphuric acid titrant, 0.02 N: Dilute 200.00 ml 0.1000 N standard acid to 100 ml with distilled or deionized water. Standardize against 15.00 ml 0.05 N Na$_2$CO$_3$ with about 60 ml in a beaker by titrating to a pH of about 5. Lift out electrodes, rinse into the same beaker, and boil gently for 3–5 min under a watchglass cover. Cool to room temperature, rinse cover glass into beaker, and finish titrating to the pH inflection point. Calculate normality:

$$N = \frac{A \times B}{(53.00 \times C)}$$

where A = g Na$_2$CO$_3$ weighed into 1-L flask, B = ml Na$_2$CO$_3$ solution taken for titration, and C = ml acid used.

4. Procedure

Proceed as for Preliminary Distillation method (Sec. IV.D) using indicating boric acid solution as absorbent for the distillate.

Titrate NH$_x$ in distillate with standard 0.02 N H$_2$SO$_4$ until indicator turns pale lavender.

Carry a blank through all steps of the procedure and apply the necessary correction to the results.

5. Calculation

$$\text{mg NH}_3\text{–N/L} = \frac{(A - B) \times 280}{\text{ml sample}}$$

where A = volume of H$_2$SO$_4$ titrated for sample (ml) and B = volume of H$_2$SO$_4$ titrated for blank (ml).

6. Precision and Bias

See Table 8 for precision and bias data for titrimetric determination of NH$_x$ [APHA (29)].

B. Indophenol Blue Colorimetry (Phenate and Salicylate Chemistries)

Berthelot first reported the determination of NH$_x$ by the indophenol blue (IPB) reaction in 1859 (37). Although there are many variations of the method, in general terms the technique involves reaction of NH$_x$ and a phenolic agent, which, under suitable oxidizing conditions, results in the formation of blue indophenol dyes. Such dyes are highly conjugated and strongly absorb light between 630 and 720 nm. Generation of the blue indophenol color requires an alkaline pH, since the reaction is slow, the kinetics of the process are normally accelerated by use of a catalyst (38).

Searle (38) has reviewed almost 400 variations and applications of the IPB method. The proliferation of interest in the technique stems largely from the fact that, though being a sensitive and selective reaction, the mechanism by which it proceeds is both complex and poorly understood.

Phenolic Agents. Although several different phenolic reagents have been employed in the reaction, only two are widely used in practice: phenol itself (e.g., see Refs. 18 and 30) and sodium salicylate (e.g., see Refs. 39–41). The use of sodium salicylate as the phenolic reagent is advantageous since both phenol itself and *o*-chlorophenol, a by-product of the reaction, are highly toxic; e.g., phenol is toxic through the skin, which it blisters, and is highly toxic by ingestion (2).

pH and Precipitation. The main advantage of the phenate chemistry is that it requires a pH of only 10.5 compared to that of 12.6 by the salicylate chemistry. In many aqueous media, the higher pH promotes the precipitation of alkaline earth metals (mainly Ca^{2+} and Mg^{2+}), causing turbidity and interference of spectrophotometric measurements. Thus, the phenate chemistry is more suitable for general usage, including application to saline waters and many wastewaters.

Precipitation, as well as causing spectrophotometric interference, can cause blockages in automated systems. However, the effects of precipitation can be minimized in a number of different ways:

Allowing the precipitate to settle and decanting the supernatant (42)

Distillation or diffusion (40) (see also section IV.D)

Treating with phenol at pH 3.4 and extracting the reaction intermediate (quinone monochloroimide) with hexanol generating the complex in the hexanol by addition of sodium hydroxide (18,43).

Addition of a chelating agent, such as citrate or EDTA, before alkalization of the sample (an approach that has been most widely adopted in the determination of NH_x in waters and wastewaters).

Oxidizing Agent. Although sodium hypochlorite is traditionally employed as the oxidizing agent, several authors report inconsistency and instability of this reagent (38). Alternatives, including Chloramine T (43) and dichloro-iso-cyanurate (DIC) (40,44), have been reported. Although DIC is now increasingly used in lieu of hypochlorite, it has been suggested that use of

DIC allows interference from amino compounds. However, Riley and Skirrow (45) report this adaptation to be subject to negligible interferences from amino compounds at their normal concentrations in waters.

Catalysts. Although several catalysts have been proposed, including manganese II (46) and potassium ferrocyanide (47), it is catalysts derived from nitroprusside that now find almost exclusive usage (41,44,48). Use of nitroprusside, also improves sensitivity and the stability of the indophenol complex (38).

Other Reaction Variables. Since IPB chemistries are pH sensitive, buffer solutions are commonly used. Optimum pH is dependent on the combination of reagents, catalysts, and concentrations involved, and many optimized pH values have been suggested (range = 9–13). Reaction temperature, light conditions, and time and order of reagent addition also need to be carefully controlled to achieve optimal sensitivity (38).

Interferences. Although the phenate colorimetry generally exhibits a high specificity for NH_x, a range of potential interferents exist (38):

Metals. The interference of dissolved metals can exert a positive or negative influence on the reaction, depending on the metal in question and its oxidation state. Metals appear to constitute a significant problem only when present at high concentrations with respect to NH_x and other nitrogen compounds. For many metals the interference effect can be minimized through the use of chelation agents such as citrate or EDTA (see earlier).

Nonmetallic elements. Sulphur, selenium, and the halogens (except Cl) are reported as interferents (38). However, due to the relatively unfavorable thermodynamics of their interfering side reactions and their typical concentrations, in general they are unlikely to constitute a significant source of uncertainty.

Nitrogenous species. Amines, amides, amino acids, nitrate, proteins, and urea all constitute potential interferents, e.g., high concentrations of NO_3^- have been found to suppress color development (49), and it is well recognized that low-molecular-weight amines are capable of exerting a positive interference on NH_x quantification. Although the degree of interference is dependent on the individual method variation used (38), in general nitrogenous compounds do not constitute a significant source of uncertainty at their characteristic concentrations.

For a more extensive review of all aspects of IPB colorimetry, including its conditions, mechanisms, interferences, and applications, the reader should consult the vast literature reserve available, possibly using Searle's 1984 review (38) as a convenient starting point.

Since IPB colorimetry is relatively straightforward, it is well suited to automation. This factor has, in recent years at least, had a major role in establishing phenate colorimetry as the method of choice in the general determinations of NH_x. Today, autoanalyzer techniques have extensive usage in the temporal and spatial mapping of NH_x in environmental studies. Manual and automated methodologies are presented next. Both use phenate chemistries since these are more widely applicable. However, should readers favor the safer salicylate chemistry, equivalent procedures are presented by Russell (2) and the SCA (33).

1. Manual Phenate Colorimetry

a. Outline (Source: Refs. 30 and 31; Table 5). This is a manual phenate colorimetric technique in which alkaline phenol and hypochlorite react with NH_x to form indophenol blue dye. The blue color formed is intensified with sodium nitroprusside and is proportional to the NH_x concentration. Note that in this method, molarity has been used as the unit of concentration.

b. Apparatus and Equipment.

Spectrophotometer (650 nm), 10-cm cell
Selection of Erlenmeyer flasks and miscellaneous glassware (all washed and thoroughly rinsed in NH_x-free water)

c. Reagents and Standards.

Phenol solution: Dissolve 10 g of analytical-grade phenol in 100 ml of 95% v/v ethyl alcohol.
Sodium nitroprusside solution (0.5%): Dissolve 1 g of sodium nitroprusside $(Na_2[FE(CN)_5NO] \cdot 2H_2O)$ in 200 ml of NH_x-free water. Store in an amber bottle for up to 1 month.
Alkaline citrate reagent: Dissolve 100 g of trisodium citrate and 5 g of sodium hydroxide (NaOH) in 500 ml of NH_x-free water.
Sodium hypochlorite solution: Use a solution of commercially available hypochlorite (e.g., "Clorox"). Solution should be at least 1.5 N. This solution decomposes slowly and requires regular checking as follows: Dissolve 12.5 g of sodium thiosulphate $(Na_2S_2O_3 \cdot 5H_2O)$ in 500 ml of water.

Add ~2 g of potassium iodate crystals (KI) to 50 ml of NH_x-free water and add 1.0 ml of the sodium hypochlorite solution by pipette. Add 5–10 drops of concentrated hydrochloric acid (HCl) and titrate the liberated iodine with the sodium thiosulphate solution until all the yellow coloration has disappeared. When <12 ml of thiosulphate is used in the titration, the solution should be discarded.
Oxidizing solution: Mix 100 ml of the sodium citrate solution with 25 ml of the sodium hypochlorite solution. Prepare fresh mixture each day.

d. Procedure.

Add 50 ml of water sample to an Erlenmeyer flask. Add to the water, in sequence, and swirling after each addition, 2 ml phenol solution, 2 ml of sodium nitroprusside solution, and 5 ml of oxidizing solution.
Stand flasks at room temperature (20–27°C) for 60 min to allow the color to develop. During this period the flask should be stoppered or sealed with aluminium foil. The color is stable for 24 hours after the reaction procedure. It may be noted that under high light intensity, overdevelopment of the blue color may occur. This may be avoided by using amber flasks or by wrapping flasks in aluminium foil.
Record absorbance at 640 nm in a spectrophotometer using a 10-cm cell.

e. Blanks. Use method of analysis to determine the absorbance in the "blank" matrix, e.g., NH_x-free water or low-nutrient seawater.

f. Calibration.

Preparation of NH_x standard solution: Dissolve 0.100 g of analytical grade ammonium sulphate $((NH_4)_2SO_4)$ in ~500 ml of NH_x-free water in a 100-ml volumetric flask, add 1 ml of chloroform, and fill to 100 ml.
Pipette 1.0 ml of this solution into a 500-ml flask, and fill to the mark with NH_x-free water or seawater. The resulting solution is 3 μM in NH_x.
Measure out 50-ml aliquots of 3 μM NH_x solution into each of five Erlenmeyer flasks and analyze as described in the Procedure just described. Repeat for 5 × 50-ml aliquots of the "blank" (diluent solution). Read all absorbances after 60 min.

g. Calculation.

Calculate the calibration factor $F = 3.0/(\varepsilon_s - \varepsilon_b)$, where ε_s is the mean absorbance of the five replicate 3 μM standards and ε_b is the mean absorbance of the five replicate blanks.

Subtract the absorbance recorded for the blank from that of the sample and calculate the concentration of NH_x using the expression $NH_x(\mu M) = F \cdot \varepsilon$, where ε is the corrected absorbance and F is the calibration factor.

h. Precision and Accuracy. For both freshwater and seawater: standard deviation $\pm 4.8\%$; relative error $+5\%$.

2. Automated Phenate Colorimetry

a. Outline (Source: EPA method 350.1 (Ref. 32); Tables 5 and 6).   Alkaline phenol and hypochlorite react with ammonia to form an indophenol blue dye, the intensity of which is proportional to the NH_x concentration. The blue color formed is intensified with sodium nitroprusside. Approximately 20–60 samples per hour can be analyzed.

Ca^{2+} and Mg^{2+} may be present in concentrations sufficient to cause precipitation problems during analysis. A 5% EDTA solution is used to prevent precipitation from river water and industrial waste. For seawater, a sodium potassium tartrate solution is used. Sample turbidity and color may interfere with this method. Turbidity must be removed by filtration prior to analysis. Sample color that absorbs in the photometric range used will also interfere. See also Sec. V.B.

b. Sample Handling and Preservation. Preservation by the addition of 2 ml concentrated H_2SO_4 per liter and refrigeration at 4°C (see also Sec. IV.C.).

c. Apparatus.

Technicon Autoanalyser Unit AAII consisting of:

1. Sampler
2. Analytical cartridge (AAII)
3. Proportioning pump
4. Heating bath with double-delay coil
5. Colorimeter equipped with 15-mm tubular flow cell and 630–660-nm filters.

Recorder
Digital printer for AAII (optional)

d. Reagents.

NH_x-free water: See Sec. IV.B. Use in preparation of all solutions.

Sulphuric acid 5 N: Air scrubber solution. Carefully add 139 ml of concentrated sulphuric acid to approximately 500 ml of NH_x-free distilled water and dilute to 1 liter with NH_x-free water.

Sodium phenolate: Using a 1-L Erlenmeyer flask, dissolve 83 g phenol in 500 ml of distilled water. In small increments, cautiously add, with agitation, 32 g of NaOH. Periodically cool flask under water faucet. When cool, dilute to 1 L with NH_x-free water.

Sodium hypochlorite solution: Dilute 250 ml of a bleach solution containing 5.25% NaOCl (such as "Clorox") to 500 ml with NH_x-free water. Available chlorine level should approximate 2–3%. Since "Clorox" is a propriety product, its formulation is subject to change. The analyst must be alert to detecting any variation in this product significant to its use in this procedure. Due to the instability of this product, prolonged storage should be avoided.

Disodium ethylenediamine-tetraacetate (EDTA) (5%): Dissolve 50 g of EDTA (disodium salt) and approximately six pellets of NaOH in 1 L of NH_x-free water. *Note:* On saltwater samples where EDTA solution does not prevent precipitation of cations, sodium tartrate solution may be used. It is prepared as follows: *Sodium tartrate solution (10% $NaKC_4H_4O_6 \cdot H_2O$):* To 900 ml of distilled water add 100 g sodium potassium tartrate. Add two pellets of NaOH and a few boiling chips, and boil gently for 45 minutes. Cover, cool, and dilute to 1 L with NH_x-free water. Adjust pH to 5.2 $\pm$ .05 with H_2SO_4. After allowing to settle overnight in a cool place, filter to remove precipitate. Add 0.5 ml Brij-34 solution (Technicon Corporation) and store in stoppered bottle.

Sodium nitroprusside (0.05%): Dissolve 0.5 g of sodium nitroprusside in 1 L of NH_x-free water.

Stock solution: Dissolve 3.819 g of anhydrous ammonium chloride NH_4Cl dried at 105°C in NH_x-free water and dilute to 1000 ml; 1.0 ml = 1.0 mg NH_x–N.

Standard solution A: Dilute 10.0 ml of stock solution (earlier) to 1000 ml with NH_x-free water; 1.0 ml = 0.01 mg NH_x–N.

Standard solution B: Dilute 10.0 ml of standard solution A (earlier) to 1000 ml with NH_x-free water; 1.0 ml = 0.0001 mg NH_3–N.

Using standard solutions A and B, prepare the standards listed in Table 9 in 100-ml volumetric flasks (prepare fresh daily). *Note:* When saline water samples are analyzed, substitute ocean wa-

Table 9 Preparation of Standards for Use in Automated Phenate Colorimetry [EPA method 350.1 (32)]

NH_x-N (mg/L)	ml Standard Solution/100 ml
	Solution A
0.01	1.0
0.02	2.0
0.05	5.0
0.10	10.0
	Solution B
0.20	2.0
0.50	5.0
0.80	8.0
1.00	10.0
1.50	15.0
2.00	20.0

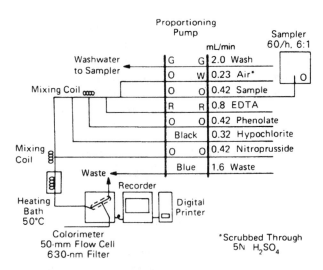

Fig. 5 Manifold for determination of NH_x using method EPA 350.1 (Ref. 32).

ter (SOW, Table 10) should be used for preparing the standards shown in Table 9. If SOW is used, subtract its blank background response from the standards before preparing the standard curve.

e. Procedure.

Since the intensity of the color used to quantify the concentration is pH dependent, the acid concentration of the wash water and the standard NH_x solutions should approximate that of the samples. For example, if the samples have been preserved with 2 ml concentrated H_2SO_4/liter, the wash water and standards should also contain 2 ml concentrated H_2SO_4/liter.

For a working range of 0.01–1.0 mg NH_x-N/L, set up the manifold as shown in Fig. 5. Higher concentrations may be accommodated by sample dilution.

Allow both colorimeter and recorder to warm up for 30 minutes. Obtain a stable baseline with all reagents, feeding NH_x-free water through the sample line.

Table 10 Preparation of Substitute Ocean Water (SOW)

NaCl	24.53 g/L	NaHCO$_3$	0.20 g/L
MgCl$_2$	5.20 g/L	KBr	0.10 g/L
Na$_2$SO$_4$	4.09 g/L	H$_3$BO$_3$	0.03 g/L
CaCl$_2$	1.16 g/L	SrCl$_2$	0.03 g/L
KCl	0.70 g/L	NaF	0.003 g/L

For the AAII, use a 60/h 6:1 cam with a common wash.

Arrange NH_x standards in the sampler in order of decreasing concentration of nitrogen. Complete loading of sampler tray with unknown samples.

Switch sample line from distilled water to sampler and begin analysis.

f. Calculations. Prepare appropriate standard curve derived from processing NH_x standards through manifold. Calculate concentration of samples by comparing sample peak heights with standard curve.

g. Precision and Accuracy.

In a single laboratory (EMSL), using surface water samples at concentrations of 1.41, 0.77, 0.59, and 0.43 mg NH_x-N/L, the standard deviation was ±0.005.

In a single laboratory (EMSL), using surface water samples at concentrations of 0.16 and 1.44 mg NH_x-N/L, recoveries were 107% and 99%, respectively.

Bibliography. Refs. 50–54.

C. Nesslerization

Nesslerization is the oldest of the colorimetric techniques and at one time was also the most widely applied. The technique involves addition of nessler reagent, an alkaline mixture of potassium iodide and mercury (I) chloride to the sample and photometric measurement of the developed chromgen at 425 nm (22,55). The reaction is fast and requires no temperature control (22).

Precision and reliability are reduced when the nessler reagent yields a turbid colloidal product through reaction with several inorganic ions (e.g., magnesium, iron, manganese, and sulphide (56)), causing nonlinearity of the calibration and unreproducible results (48,57). Thus distillation or pretreatment of samples is normally required.

1. Outline [Source: APHA Method 4500-NH₃C (Ref. 29); Tables 5 and 6)

Treatment before direct nesslerization with zinc sulphate and alkali precipitates calcium, iron magnesium, and sulphide, which form turbidity when treated with nessler reagent. The floc also removes suspended matter and sometimes colored matter. Addtion of EDTA or Rochelle salt solution inhibits precipitation of residual Ca^{2+} and Mg^{2+} in the presence of the alkaline nessler reagent. However, use of EDTA demands an extra amount of nessler reagent to ensure sufficient nessler reagent excess for reaction with the NH_x.

The graduated yellow/brown colors produced by the nessler–NH_x reaction absorb strongly over a wide wavelength range. The yellow color characteristic of low NH_x–N concentration (0.4–0.5 mg/L) can be measured with acceptable sensitivity in the wavelength region 400–425 nm when a 1-cm light path is available. A light path of 5 cm extends measurements into the nitrogen concentration range of 5–60 μg/L. The reddish brown hues typical of NH_x–N levels approaching 19 mg/L may be measured in the wavelength region of 450–500 nm. A judicious selection of light path and wavelength thus permits the photometric determination of NH_x over a considerable range.

Departures from Beer's law may be evident when photometers equipped with broadband color filters are used. For this reason, prepare the calibration curve under conditions identical with those adopted for the samples.

Direct nesslerization should be applied to domestic wastewaters only when errors of 1–2 mg/L are acceptable. Use this method only after it has been established that it yields results comparable to those after distillation. Check the validity of direct nesslerization measurements periodically.

2. Apparatus

Colorimetric equipment: One of the following is required.

1. Spectrophotometer, for use at 400–500 nm and providing a light path of 1 cm or longer.
2. Filter photometer, providing a light path of 1 cm or longer and equipped with a violet filter having a maximum transmittance at 400–425 cm. A blue filter can be used for higher NH_x–N concentrations.

pH meter, equipped with a high pH electrode.

3. Reagents

Use NH_x-free water for preparing all reagents, rinsing, and making dilutions. All the reagents listed in Sec. IV.B. (Preliminary Distillation) except the borate buffer and absorbant solution are required, plus the following:

Zinc sulphate solution: Dissolve 100 g $ZnSO_4 \cdot 7H_2O$ and dilute to 1 L with NH_x-free water.

Stabilizer reagent: Use either EDTA or Rochelle salt to prevent Ca^{2+} or Mg^{2+} precipitation in undistilled samples after addition of alkaline nessler reagent.

1. *EDTA:* Dissolve 50 g disodium ethylenediamine tetraacetate dihydrate in 60 ml water containing 10 g NaOH. If necessary, apply gentle heat to complete dissolution. Cool to room temperature and dilute to 100 ml.
2. *Rochelle salt solution:* Dissolve 50 g potassium sodium tartrate tetrahydrate, $KNaC_4H_4O_6 \cdot 4H_2O$, in 100 ml water. Remove NH_x usually present in the salt by boiling of 30 ml of solution. After cooling dilute to 100 ml.

Nessler reagent: Dissolve 100 g HgI_2 and 70 g KI in a small quantity of NH_x-free water and add this mixture slowly, with stirring, to a cool solution of 160 g NaOH dissolved in 500 ml NH_x-free water. Dilute to 1 L. Store in rubber-stoppered borosilicate glassware and out of sunlight to maintain reagent stability for up to one year under normal laboratory conditions. Check reagent to make sure that it yields a characteristic color with 0.1 mg NH_x–N/L within 10 min after addition and does not produce a precipitate with small amounts of ammonia within 2 h. (*Caution:* TOXIC—take care to avoid ingestion.)

Stock NH_x solution: Dissolve 3.819 g anhydrous NH_4Cl, dried at 100°C, in water and dilute to 1000 ml; 1.00 ml = 1.00 mg N = 1.22 mg NH_x.

Standard NH_x solution: Dilute 10.00 ml stock ammonium solution to 1000 ml with water; 1.00 ml = 10.00 μg N = 12.2 μg NH_x.

4. Procedure

Treatment of undistilled samples: If necessary, remove residual chlorine from the freshly collected sample by adding an equivalent amount of dechlorinating agent. (Do not store chlorinated samples.) Add 1 ml $ZnSO_4$ solution to 100 ml sample and mix thoroughly. Add 0.4–0.5 ml 6 N NaOH solution to obtain a pH of 10.5, as determined with a pH meter and a high-pH glass electrode, and mix gently. Let the treated sample stand for a few minutes, whereupon a heavy flocculent precipitate should form, leaving a clear and colorless supernate. Clarify by centrifuging or filtering. Pretest any filter paper used to be sure no NH_x is present as a contaminant. Do this by running water through the filter and testing the filtrate. (*Caution:* Samples containing more than about 10 mg NH_x–N/L may lose NH_3 during this treatment of undistilled samples because of the high pH. Dilute such samples to the sensitive range for nesslerization before pretreatment.)

Color development.

1. *Undistilled samples:* Use 50.0 ml sample or a portion diluted to 50 ml with water. If the undistilled portion contains sufficient concentrations of Ca^{2+}, Mg^{2+}, or other ions that produce turbidity or precipitate with nessler reagent, add 1 drop (0.05 ml) EDTA reagent of 1–2 drops (0.05–0.1 ml) Rochelle salt solution. Mix well. Add 2.0 ml nessler reagent if EDTA reagent is being used or 1.0 ml nessler reagent if Rochelle salt is being used.
2. *Distilled samples:* Neutralize the boric acid used for absorbing the NH_x distillate by adding either 2 ml nessler reagent, an excess that raises the pH to the desired high level, or, alternatively, neutralizing the boric acid with NaOH before adding 1 ml nessler reagent.
3. Mix samples thoroughly. Keep such conditions as temperature and reaction time the same in the blanks, samples, and standards. Let reaction proceed for at least 10 min after adding nessler reagent. Measure color in sample and standards. In NH_x–N is very low, use a 30-min contact time for sample, blank, and standards. Measure color photometrically as directed next.

Photometric measurement: Measure absorbance or transmittance with a spectrophotometer or filter photometer. When using a spectrophotometer,

read samples at 400–425 nm for 1-cm light path and at 450–500 nm for 5-cm light path. Prepare calibration curve at the same temperature and reaction time used for samples. Measure absorbance or transmittance readings against a reagent blank, and run parallel checks frequently against standards in the nitrogen range of the samples. Redetermine complete calibration curve for each new batch of nessler reagent.

For distilled samples, prepare standard curve under the same conditions as the samples. Distill reagent blank and appropriate standards, each diluted to 500 ml, in the same manner as the samples. Dilute 200 ml distillate plus 50 ml boric acid absorbent to 500 ml with water and take a 50-ml portion for nesslerization.

5. Calculation

Deduct amount of NH_x–N in water used for diluting original sample before calculating concentration. Deduct also reagent blank for volume of borate buffer and 6 N NaOH solutions used with sample. Calculate total NH_x–N by the following equation:

mg NH_3–N/L (51 ml final volume)

$$= \frac{A}{\text{ml sample}} \times \frac{B}{C}$$

where A = μg NH_x–N (51 final volume), B = total volume distillate collected, ml, including acid absorbent, and C = volume distillate taken for nesslerization, ml. The ratio B ÷ C applies only to distilled samples; ignore in direct nesslerization.

6. Precision and Bias

See Table 11 for precision and bias data for the determination of NH_x by nesslerization [APHA (29)].

D. Ion-Selective Electrode

Electrodes, in the form of gas-sensing probes, have over the last 20 years received considerable attention as simple, convenient devices for the determination of NH_x in a diverse range of aqueous media.

Russell (2) gives full consideration to the theory of operation of the gas-sensing electrode. In summary, electrodes are essentially galvanic cells consisting of a tube containing a reference electrode, an ion-specific electrode, and an electrolyte solution (Fig. 6). A thin, hydrophobic, gas-permeable membrane (typically PTFE) attached to the active end of the electrode acts

Table 11 Precision and Bias Data for the Determination of NH$_x$ by Nesslerization

NH$_x$ concentration (μg/L)	Direct nesslerization		Distillation plus nesslerization	
	Relative standard deviation (%)	Relative error (%)	Relative standard deviation (%)	Relative error (%)
200	22.0–38.1	0–8.3	15.7–46.3	2.0–10.0
800	11.2–16.1	0–0.3	16.3–21.2	3.1–8.7
1500	5.3–11.6	0.6–1.2	7.5–18.0	3.6–4.6

Source: Ref. 29.

as a barrier between the internal electrolyte (ammonium chloride) (58). Samples are treated to high pH, resulting in the deprotonation of NH$_4^+$ to NH$_3$, which diffuses through the membrane into the cell, where it alters the pH of the internal solution. The change in

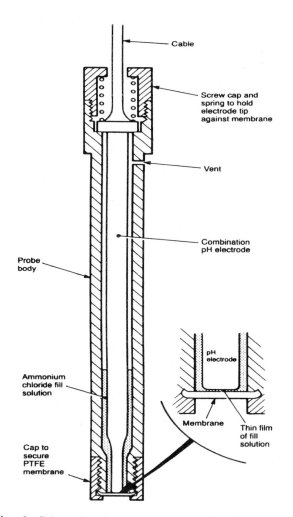

Fig. 6 Schematic of ammmonia electrode. (Reproduced from Ref. 2 with permission.)

pH is sensed by a pH electrode, giving rise to a potentiometric signal. The potential of the cell is related to the concentration of the NH$_3$. Since the unit is a complete electrochemical cell, it is more properly referred to as a "probe" than as an electrode (28).

The diffusion of NH$_3$ is driven by the difference in partial pressures across the membrane. Once the partial pressure of NH$_3$ in the internal solution is in equilibrium with that of the sample, the electrode voltage is stable. The probe is thus subject to an equilibration time, which is dependent upon the concentration of NH$_x$ in the sample.

Gas-sensing probes are resistant to salt effects and thus are suitable for application to seawaters and wastewaters. Since the electrode responds only to gases that are capable of diffusion across the membrane at high pH conditions, and affecting the internal pH, it exhibits relatively high selectivity. However, the potential interference of volatile amines is well documented. For example, Lopez and Rechnitz (59) report that relative selectivity coefficients of amines are considerably greater than unity and that the response times for these species are about the same as for NH$_3$. Nevertheless, since volatile amines are typically present at levels 10–1000 times lower than that of NH$_x$ in most waters and wastes, they are unlikely to interfere significantly (60).

The probes typically exhibit Nernstian response from 0.03 to 1400 mg NH$_x$–N/L. However, lower concentrations may be determined through the application of a standard addition electrode method (29). Alternatively, incorporation of a diffusion cell of "gas dialysis concentrator" into the apparatus will result in a significant improvement of the sensitivity, with a concurrent improvement in selectivity (58,60–62).

1. Outline (Source: EPA method 350.3 (Ref. 32); Tables 5 and 6)

Interferences. Volatile amines may act as positive interference. Since mercury interferes by forming a strong complex with NH_x, samples cannot be preserved with mercuric chloride. Color and turbidity have no effect on the measurements; thus, distillation may not be necessary.

Sample Handling and Preservation. Samples may be preserved with 2 ml of concentrated H_2SO_4 per liter and stored at 4°C. See also Sec. IV.C.

2. Apparatus

Electrometer (pH meter) with expanded millivolt scale or a specific ion meter
Ammonia-selective electrode, such as Orion Model 95-10 or EIL Model 8002-2
Magnetic stirrer, thermally insulated, and Teflon-coated stirring bar

3. Reagents

NH_x-free water: See Sec. IV.B. Use in preparation of all solutions.
Sodium hydroxide, 10 N: Dissolve 400 g of sodium hydroxide in 800 ml of NH_x-free water. Cool and dilute to 1 L with NH_x-free water.
Ammonium chloride stock solution: Dissolve 3.819 g of anhydrous ammonium chloride NH_4Cl, dried at 105°C in NH_x-free water, and dilute to 1000 ml; 1.0 ml = 1.0 mg NH_x–N.
Ammonium chloride standard solution: 1.0 ml = 0.01 mg NH_x–N. Dilute 10.0 ml of the stock solution to 1 L with NH_x-free water in a volumetric flask.

Note: When analyzing saline waters, standards must be made up in synthetic ocean water (SOW; see section V.B.2.)

4. Procedure

Preparation of standards: Prepare a series of standard solutions covering the concentration range of the samples by diluting either the stock or the standard solutions of ammonium chloride.
Calibration of the electrometer: Place 100 ml of each standard solution in clean 150-ml beakers. Immerse electrode into standard of lowest concentration and add 1 ml of 10 N sodium hydroxide solution while mixing. Keep electrode in the solution until a stable reading is obtained. *Note:* The pH of the solution after the addition of NaOH *must* be above 11. *Caution:* Sodium hydroxide must not be added prior to electrode immersion, for NH_3 may be lost from a basic solution.

Repeat this procedure with the remaining standards, going from lowest to highest concentration. Using semilogarithmic graph paper, plot the concentration of NH_x in mg NH_x–N/L on the log axis vs. the electrode potential developed in the standard on the linear axis.
Calibration of a specific ion meter. Follow the directions of the manufacturer for the operation of the instrument.
Sample measurement: Follow the earlier procedure for 100 ml of sample in 150-ml beakers. Record the stabilized potential of each unknown sample and convert the potential reading to the NH_x concentration using the standard curve. If a specific ion meter is used, read the NH_x level directly in mg NH_x–N/L.

5. Precision and Accuracy

In a single laboratory (EMSL), using surface water samples at concentrations of 1.00, 0.77, 0.19, and 0.13 mg NH_x–N/L, standard deviations were ±0.038, ±0.017, ±0.007, and ±0.003, respectively.
In a single laboratory (EMSL), using surface water samples at concentrations of 0.19 and 0.13 mg NH_x–N/L, recoveries were 96% and 91%, respectively.

6. Bibliography

Refs. 63–65.

E. Gas Diffusion Fluorescence

This is an automated fluorescence technique for the determination of trace concentrations of NH_x in natural waters (*source:* Ref. 23, Tables 5 and 6). The technique involves conversion of all NH_x in sample to NH_3 and subsequent diffusion of NH_3 across a gas-permeable membrane into a stream of *o*-phthaldialdehyde reagent. The reaction adduct formed is determined fluorimetrically.

With a sample throughput of 60 samples per hour, this method permits the fine-scale examination of temporal and spatial variations in NH_x concentrations. It thus constitutes a most attractive technique for the measurement of aqueous NH_x at concentrations below the detection limits of most other techniques (Table 5, Fig. 4).

1. Apparatus

Multichannel peristaltic pump (e.g., Ismatec 16-channel pump) and calibrated pump tubing

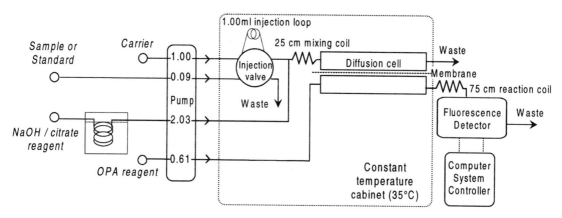

Fig. 7 Schematic of system for fluorescent determination of NH$_x$ in natural waters. (Based on Ref. 23.)

Fluorescence detector

Pneumatically actuated injection Teflon rotary valve (e.g., Rheodyne)

Diffusion cell: 30.1 cm AAI block diffuser (A-87-03, Technicon) fitted with nonlaminated Goretex (or similar) hydrophobic, gas-permeable membrane (0.45-μm pore size)

Gas-permeable microporous PTFE tubing (2.0-mm ID, 2.0-μm pore size)

Control unit (e.g., Zenith Data Systems Z151-PC) coupled to integrator or PC-based integration software

Constant-temperature cabinet capable of housing preceding apparatus

2. Reagents and Standards

o-phthaldialdehyde (OPA) solution: Dissolve 15.46 g of borate buffer (H$_3$BO$_3$) in 750 ml of NH$_x$-free water and adjust pH to 9.5 using 10 M NaOH. Add 100 mg *o*-phthaldialdehyde (OPA) and 500 μl of 2-mercaptoethanol (ME) to 2.0 ml of methanol and add mixture to buffer solution. Deoxygenate solution by bubbling with oxygen-free N$_2$ at a flow rate of 200 ml min^{-1} for 20 minutes. Store in a capped amber glass bottle for 24 hours to allow decay of background fluorescence. The reagent is stable for 72 hours after preparation.

Alkali-citrate reagent: Dissolve 200 g of trisodium citrate and 18.0 g of sodium hydroxide (NaOH) in 450 ml of NH$_x$-free water. Purge with oxygen-free N$_2$ (as for OPA solution), cap, and store at room temperature.

Carrier solution: Acidify 4000 ml of NH$_x$-free water with 8 ml of concentrated sulphuric acid. Purge

with oxygen-free N$_2$ (as for OPA solution), cap, and store at room temperature.

3. Procedure

Configure system according to Fig. 7 as follows.

Using a four-channel peristaltic pump equipped with calibrated pump tubing generate the following flow rates: carrier solution, 1.0 ml min^{-1}; alkaline-citrate reagent, 0.09 ml min^{-1}; OPA reagent, 0.61 ml min^{-1}; sample/standard: 2.03 ml min^{-1}. (In the original method this was achieved using an Ismatec 16-channel pump, operated at 25% of full speed with pump tubing rated at 0.60, 0.05, 0.32, and 1.20 ml min^{-1}, respectively.)

Flow lines in the system are completed using 0.8-mm ID PTFE tubing and T-fittings.

Residual NH$_x$ is reduced in the alkali-citrate reagent by passing through a coil of microporous PTFE tubing (1 m, 2.0-mm ID, 2.0-μm pore size) immersed in 10% H$_2$SO$_4$.

A rotary injection valve (PTFE internal construction) with pneumatic actuator and 1.0-ml sample loop is used to inject samples and standards into the carrier stream.

Carrier stream (with injected sample or standard) is merged with the alkaline citrate reagent and passed through a 25-cm mixing coil and then into one-half of the diffusion cell.

In the diffusion cell, under the high-pH conditions of the carrier stream, NH$_4^+$ from the sample is converted into gaseous NH$_3$, which may diffuse across the membrane into the OPA acceptor stream.

The OPA reagent is passed through the second half of the diffusion cell (where it receives NH_3), through a 75-cm PTFE reaction coil, and to a fluorescence detector (e.g., Hitachi F-1050; excitation 335 nm, emission 470 nm, time constant 3 s, sensitivity 100).

Valve operation and data collection are controlled by a Zenith Data System Z151-PC or similar unit coupled to a suitable integration package (e.g., E-Lab).

The complete system (excluding the detector) is kept in a constant-temperature cabinet).

Blanks. Replicate injections of NH_x-free water or low-nutrient seawater are used to establish the system blank. This is subtracted from the sample peak area before calculation of the NH_x concentration.

Calibration. Standards in the concentration range of interest are prepared from dilution of a 1 mM ammonium chloride (NH_4Cl) solution with either NH_x free water or filtered low-nutrient seawater. Integrated peak areas from replicate injections of standards are used to prepare a calibration plot and calculate a response factor. It should be noted that response factors for seawater are ~ 1.7 times greater than those for freshwater.

Precision and Accuracy. Relative standard deviation at 250 nM = 1.8%.

VI. REVIEW OF OTHER METHODS FOR THE DETERMINATION OF AMMONIA

The techniques discussed in the following section have been organized according to the categories presented in Table 2.

A. Electroanalytical Techniques

1. *Conductimetric Techniques*

The conductimetric detector is a device demonstrating high sensitivity together with simplicity, reliability, and operational stability. It is thus well suited to the determination of NH_4^+ in both laboratory and fieldwork. However, due to its lack of selectivity, it cannot be applied directly to the analysis of NH_4^+ in natural and wastewaters. The use of silicon hollow fibers or gas-permeable membranes, across which NH_3 can diffuse from high-pH sample stream into a stream of deionized water that is fed to the conductivity cell has been reported (67,68). Using such systems in an automated configuration it is possible to achieve a throughput of

60 samples per hour and a detection limit of 0.004 mg/L NH_x.

2. *Voltammetric Techniques*

Voltammetry comprises a group of electroanalytical methods in which information about the analyte is derived from measurement of current, as a function of applied potential, under conditions that encourage polarization of an indicator or working electrode (28). Cathodic stripping voltammetry (CSV) has been applied to the trace determination of NH_4^+ (17). The method involves reaction of NH_4^+ with formaldehyde at low pH to form an imino compound that is adsorbed onto a hanging-mercury-drop electrode. Through differential pulse scanning, the adsorption of the imino compound may be quantified and related to the concentration of NH_x in the sample. The CSV response was reported to be linear in the range 0.14–42 μg/L NH_x, and the method applied successfully to both fresh waters and saline waters. However, the method is reported to suffer from potential competitive adsorption of electroactive and surface-active compounds, particularly in the analysis of seawater.

B. Spectroscopic Techniques

1. *Absorbance Techniques*

a. Indophenol Blue (IPB). With a detection limit of ~ 0.01 mg/L NH_x, IPB is often too insensitive for many oceanographic and limnologic applications. However, using solvent extraction and back-extraction of the indophenol complex and its intermediates it is possible to achieve a limit of detection of ~ 0.0005 mg/L NH_x under field conditions (18). However, due to the additional extraction steps the method is significantly more complex and involved than conventional phenate and salicylate colorimetries and is unsuitable for automation.

b. Rubazoic Acid Technique. The rubazoic acid technique was first reported by Kruse and Mellon in 1953 (56). It is based on the acidic (pH 3.5–3.7) reaction of NH_x with Chloramine T in the presence of pyridine, pyrazolone, and bis-pyrazolone to form a purple-colored product. This species is extracted into carbon tetrachloride as yellow rubazoic acid, the absorbance of which is proportional to the concentration of NH_x in the sample. The process is sensitive to ~ 0.03 mg/L NH_x. However, it is time consuming and also involves the use of the toxic pyridine species.

Prochazkova (69) discovered that the use of pyridine was not essential for the reaction and reported an adapted version involving the reaction of NH_x with Chloramine T at pH 6.5. The solution was then buffered at pH 10 using sodium carbonate, and bis-pyrazolone and pyrazolone were added. The resultant rubazoic acid was then acidified and extracted with tri-chloroethylene for photometric measurement (450 mm). Although Kruse and Mellon (56) reported several interferents, Prochazkova (69) reported only Fe^{2+} to represent a problem, and even this could be eliminated through pretreatment with potassium permanganate. A field intercomparison of automated versions of Pro-chazkova's rubazoic method and an adaptation of So-lorzano's phenate colorimetry (30) (0.001–0.03 mg/L NH_x) demonstrated good agreement between tech-niques ($r = 0.96$) (21).

c. Miscellaneous Techniques. Van Son et al. (66) proposed a colorimetric flow injection method for the assay of aqueous NH_x. Under elevated-pH conditions, NH_3 diffuses across a gas-permeable membrane, caus-ing an absorbance change of a bromothymol blue in-dicator. A limit of ~0.01 mg/L NH_x and sample throughput of 100 per hour are reported. Willason and Johnson (20) report a similar automated technique but with phenol red as the indicator. The inclusion of citrate as a chelation agent allows the high-pH conditions re-quired to be achieved without precipitation of alkaline earth metal hydroxides, even in saline waters. Sixty analyses per hour are reported possible, with a signif-icantly improved detection limit of 0.7 μg/L NH_x.

Canale-Guierrez et al. (17) report a further flow analysis technique incorporating an immobilized en-zyme reactor system with spectrofluorimetric detection. Sample is mixed with a carrier stream consisting of 2-oxoglutarate and nicotinamide adenine dinucleotide (NADH) with which NH_x reacts in an immobilized glu-tamate dehydrogenase enzyme reactor. The consump-tion of NADH, which is a measure of the NH_x, con-centration is monitored at 340 nm. A detection limit of 0.07 mg/L NH_x is reported.

2. Emission Techniques

a. Chemiluminescence. Chemiluminescence de-tection exhibits exquisite selectivity for nitrogenous species and demonstrates a high mass sensitivity, per-mitting detection of 1–3 ng per sample. Due to the gas-phase reaction mechanism involved, chemilumines-cence detection is well suited to the quantification of gaseous species and has been used in atmospheric stud-ies, but it has not been applied to the determination of aqueous NH_x.

b. Spectrofluorimetry. Spectrofluorimetric deter-mination of derivatized NH_x offers a highly sensitive alternative to the bulk of colorimetric methodologies. Typically the reactions involved used the ternary re-action of NH_x with o-phthalaldehyde (OPA) under al-kaline-buffered conditions (pH 9–10) in the presence of a reducing agent to yield fluorescent isondole (71). 2-Mercaptoethanol is most commonly utilized as the reducing agent (72,73), although sulphite has also been used with reports of increased sensitivity and selectiv-ity (74,75). Excitation wavelengths range from 340–410 nm, depending on the specific conditions em-ployed, with emission measured between 425 and 486 nm.

The mechanistic pathways involved are also open primary amines and amino acids. These together with many inorganic species (e.g., sodium nitrate, calcium chloride) may constitute interferents (71,73). In order to utilize the sensitivity of the ammonium-OPA-mer-captoethanol (MCE) reaction while minimizing the ef-fects of interference, several authors have incorporated diffusion cells into their systems (23,71,75). In such systems NH_3 is formed under high-pH conditions and diffused across a gas-permeable membrane into a flow-ing stream of OPA to produce the fluoescent adducts. This methodology allows, with the exception of pri-mary amines, the exclusion of interferents (23,71). For-tunately, the concentration of primary amines relative to that of NH_x, together with their significantly higher pK values, should ensure that under the analytical con-ditions employed their contribution to the response is marginal.

c. Molecular Emission Cavity Analysis (MECA). This technique is applicable to the mea-surement of NH_3 or indeed of any other nitrogenous species that may be converted to NH_3 (76). The anal-ysis of aqueous media is achieved by conversion of NH_4^+ to NH_3 under high-pH conditions. At elevated temperature, the NH_3 is transported by a carrier stream (typically oxygen) into the MECA cavity. Here the white band emission intensity is measured as a function of time at 500 nm using an MECA spectrophotometer. The technique is rapid (~30 seconds per analysis), and concentrations to ~0.03 mg/L NH_x have been reported in a variety of sample matrices, including tap and spring waters and snow (77).

3. Optodes

Fiber-optic sensors or optodes represent one of the emergent tools available to the analytical chemist to-day. They are, by their nature, nonintrusive, nondestructive devices, which, by virtue of their operating mechanisms, exhibit good selectivity. Typically, sensors are constructed by trapping a thin layer of a colorimetric (78,79) or fluorescent (19,80,81) indicator solution between a gas-permeable membrane and a fiber-optic bundle. NH_3 from the samples diffuses across the membrane, enters the internal solution, and reacts with the protonated form of the indicator dye:

$$NH_3 + HIn \rightarrow NH_4^+ + In^-$$

The nonprotonated form of the indicator is detected by absorbance or fluoescence through a fiber-optic probe linked to a photomultiplier tube. Once a dynamic equilibrium is established between the NH_3 partial pressures on both sides of the membrane, the signal will stabilize.

Colorimetric optodes have employed dyes such as bromothymol blue (79) and *p*-nitrophenol (78) and exhibit detection limits of $0.01-0.07$ mg/L NH_x. However, the inherent sensitivity of fluorometric techniques employing dyes such as $2',7'$-dichlorofluorescein and 5-carboxy-$2'$-$7'$-dichlorofluorescein exhibit greatly improved detection limits, e.g., Rhines and Arnold (80) report a detection limit of ~1.3 μg/L NH_x, and Kar and Arnold (19) report an optode responsive over the range $0.1-40$ μg/L NH_x. Although the technique was developed for the measurement of extracellular NH_x, this concentration range correlates closely with that expected of many natural waters.

C. Chromatographic Techniques

Chromatography (from the Greek *khroma*, meaning "color," and *graphia*, meaning "writing") may be universally defined as the separation of a mixture of components by their passage through or over a material that interacts differentially with them. It is generally acknowledged that the foundation of chromatography was laid in 1906 when the Russian biologist Mikhail Tswett reported separating green plant pigments using a powdered chalk medium (82). However, it was not until 1941, when Martin and Synge (83) introduced the concept of partition column chromatography and established that the separation process was dependent on solute partition between the separating medium and the solvent flowing through it, that the technique began to gain momentum.

Gas chromatography (GC), was first reported in 1952 by James and Martin (84). The subsequent development of sensitive and versatile detectors, such as the thermal conductivity detector (TCD) and flame ionization detector (FID), allowed the technique to grow into a major analytical tool in the 1950s. The introduc-

tion of capillary GC by Golay in 1951 made more efficient analyses of more complex mixtures possible and gave extra impetus to the growth of the technique (85).

The development of liquid chromatography (LC), being inhibited by the lack of sensitive and versatile liquid-phase detectors, was slow by comparison. However, in the late 1960s the concurrent development of high-sensitivity detectors (e.g., photometric) and new polymeric stationary phases allowed high-performance LC (HPLC) (operating under high pressure with short columns) to emerge as a powerful analytical tool.

Ion chromatography effectively came into being as a distinct type of LC in 1975 with the innovative work of Small, Stevens, and Bauman, of the Dow Chemical Company (86). Since this time IC has flourished to become established as a highly efficient method for the determination of ionic species in solution.

1. Gas Chromatography (GC)

The published analysis of NH_x by GC spans the history of the technique, from James' and Martin's groundbreaking work in 1952 (84) to the present day. Conceptually, GC is well suited to the determination of gasphase NH_3. Typically the GC analysis of NH_3 has been performed using packed columns (rather than capillaries). A wide range of column support materials has been evaluated, including diatomite, graphitized carbon, and porous aromatic polymers. However, much of the work carried out has been focused upon overcoming the problems associated with the analysis of such a polar compound as NH_3, including:

Adsorptive effects: evident in peak asymmetry and decomposition artefacts
"Ghosting" phenomena
Reduced detector response (quenching)
Column bleed

Such effects are most critical in studies of low levels of NH_x, e.g., those typical of environmental samples. Attempts to overcome these problems have included saturation of adsorption sites by stationary-phase additions, removal of active sites by acid and base washing, and silanization and coating with an inert material. In addition, a range of tail reducers have been employed, including KOH and tetrahydroxyethylethyldiamine (THEED) (60).

Despite the considerable problems reported by some workers, others appear to have used identical systems with appreciable success. This apparently paradoxical situation is difficult to understand fully and therefore to draw clear conclusions from. Nevertheless it is possible to select three columns that have had the greatest reported success in the analysis of NH_3 and other volatile bases in aqueous media (60):

Chromosorb 103 (87,88)
Carbopack B + Carbowax 20M + KOH (89)
Chromosorb W + Amine 220 + KOH (87)

A wide choice of detectors is also applicable to the quantification of NH_3. The flame ionization detector (FID) (90), through the most rugged and versatile detector, lacks the sensitivity and selectivity of the nitrogen phosphorus–selective detector (NPD) (89) or the chemiluminescence detector (CLD) (88). Both of these detectors have excellent linear response characteristics and high mass sensitivity factors, making only small samples necessary (<10 μl). However, these detectors are often susceptible to moisture, e.g., certain designs of NPD suffer from response quenching upon injection of certain solvents, including water.

Since a disparity exists between environmental levels of NH_x and the detection limits capable of GC techniques, a preconcentration or derivatization method is normally required. The analysis of aqueous NH_x by GC is thus fundamentally inhibited by the nature of the analyte and the medium.

2. Liquid Chromatography I: High-Performance Liquid Chromatography (HPLC)

A selection of HPLC techniques applicable to the determination of NH_x in aqueous sample media is presented in Table 12. In general, reverse-phase HPLC using octyl (C-8) or octadecyl siloxane (C-18) packed columns is used for the resolution of NH_4^+. It has been estimated that over 75% of all HPLC studies are now performed on such materials (28).

Typical mobile phases include water, low-molecular-weight alcohols (methanol, ethanol), and acetonitrile. Both isocratic (constant eluent composition throughout chromatographic separation) and gradient elutions (stepwise or continual change of eluent composition) have been applied (Table 12).

NH_4^+ does not absorb energy in the ultraviolet or visible region of the electromagnetic spectrum, nor is it a fluorescent compound. Thus it is largely "invisible" to conventional HPLC absorbance and emission detectors. It must therefore be chemically tagged through derivatization before it can be detected and quantified. Generally precolumn, rather than postcolumn, derivatization has been favored (Table 12).

Derivatization reactions should meet the following general basic criteria:

Compound specificity and selectivity.
Kinetic and thermodynamic efficiency: Derivatization should be rapid, resulting in the formation of species thermodynamically stable under the analytic conditions.
Formation of derivatives that demonstrate high detector response factors: Require optimum molar detector response output with the minimum of interference.

The most commonly employed fluorogenic derivatizing reagent for the derivatization of NH_x is *o*-phthalaldehyde (OPA) (92,96,98,99) (Table 12), which reacts with NH_4^+ in the presence of certain thiols (typically 2-mercaptoethanol). The reaction is buffered using either borate (e.g., Ref. 98) or phosphate (e.g., Ref. 92). However, borate is favored, since fluorescence quenching of the OPA adduct occurs with phosphate. Since the OPA derivatization mechanism is viable for all primary amines. It is also suitable for coanalysis of amino acid and other primary amine assays in natural waters (Table 12).

Other agents for forming fluorescent derivatives with NH_4^+ include:

Halogenonitrobenzofurazans (95).
Ferrocene tagging agents, e.g., ferrocenecarboxylic acid chloride and ferrocenesulfonyl chloride (100).

Chromophoric agents utilized for derivatization of NH_x include:

m-Tolouoyl chloride (MTA): Reacts with aliphatic and aromatic, primary and secondary amines to form *m*-toluamides (detected in the range 230–254 nm), e.g., Chen and Farquharson (91) used MTA to assay in wastewaters.
Phenyl isocyanate (PHI): Reacts with amino compounds to form strongly UV-absorbing *N*,*N'*-disubstitute ureas, e.g., Ref. 94.
Dabsyl chloride: Slow reaction with primary and secondary amines may be exploited by adding the derivatizing reagents to the mobile phase prior to separation and detecting the reaction products once resolved (93).

Such derivatization procedures exhibit excellent limits of detection, often at the nano- and picogram levels (Table 12).

The ability to resolve NH_x chromatographically means HPLC can be used simultaneously to analyze other species of interest in the same run, e.g., amino acids (92,96,99,100) and alkyl amines, allyl amines,

Table 12 Selected HPLC Techniques for the Determination of NH$_x$ in Aqueous Sample Media

Authors (year)	Limit of detection (LOD) or lowest reported measurement (LR) where given (ng on column, unless otherwise stated)	Derivatization	Column (packing particle size) (mode: RPLC or NPLC)	Detector mobile phase (elution: isocratic/gradient)	Notes
Chen, Farquharson (1979) (91)	10 (LR)	Precolumn derivatization with *m*-toluoyl chloride	μBondpak C$_{18}$ (10 μm) (RPLC)	PM detection Acetonitrile/water (isocratic)	Used in wastewater analysis
Lindroth, Mopper (1979) (92)		Precolumn derivatization with *o*-phthaldialdehyde/ 2-mercaptoethanol	Nucleosil RP-18 (5 μm) (RPLC)	Fluorescence detection Methanol/citrate/borate/ phosphate (gradient and isocratic)	Only suitable for primary amino species
Lin, Lai (1980) (93)	2 (LR)	Precolumn derivatization with dabsyl chloride	μBondpak C$_{18}$ (unknown μm) (RPLC)	SPM detection 1. Ethanol/acetonitrile/water 2. Acetonitrile/water 3. Methanol/water 4. Ethanol/water (isocratic)	Employed in the analysis of amines in fish
Bjorkquist (1981) (94)	1 (LOD)	Precolumn derivatization with phenylisocyanate	Sperisorb 10 ODS (5 μm) (RPLC)	PM detection Acetonitrile/water (gradient)	Used for ammonia, primary and secondary amines; tested on fertilizers
Nishikawa, Kuwata (1984) (95)		Precolumn derivatization with 7-chloro-4 nitro-2,1,3-benzoxadiazole	LiChrosorb RP-18 (10 μm) (RPLC)	Fluorescence detection Methanol/acetonitrile/water (isocratic)	Primary and secondary amines only; used in air analysis
Mopper, Lindroth (1987) (96)		Precolumn derivatization with *o*-phthaldialdehyde/ 2-mercaptoethanol	C18 porous silica (RPLC)	Fluorescence detection Methanol (isocratic)	Based on Ref. 92
Simon, Lemacon (1987) (97)		Precolumn derivatization with *m*-toluoyl chlorides	1. Silica Nucleosil C$_{18}$ (RPLC) 2. Silica Spherosil XOA600 (C$_{18}$) + ODS (RPLC) 3. Silica Sperosil XOA 600 (C$_{18}$) (NPLC) 4. Nucleosil CN (5 μm) (NPLC)	PM detection 1 & 2: Acetonitrile/water 3 & 4: Isoctane/methylene chloride/2-propanol (gradient and isocratic)	Primary and secondary amines only; used in air analysis
Delmas, Frikta, Linley (1990) (98)	100 nM (LR)	Precolumn derivatization with *o*-phthaldialdehyde/ 2-mercaptoethanol	C$_{18}$ porous silica (5 μm) (RPLC)	Fluorescence detection Methanol (isocratic)	Based on Ref. 92; DFAA analysis
Gorzelska, Galloway (1990) (99)	5 nM (LR)	Precolumn derivatization with *o*-phthaldialdehyde/ 2-mercaptoethanol	Rainin Microsorb C$_{18}$ (5 μm) (RPLC)	Fluorescence detection Sodium acetate/methanol/ tetrahydrofurane (gradient)	Adapted from Ref. 92 for analysis of ammonia, primary amines, and amino acids

HPLC: high-performance liquid chromatography: RPLC: reverse-phase HPLC; PM: photometric; NPLC: normal-phase HPLC; SPM: spectrophotometric.

aromatic amines, di-, tri-, and polyamines, and heterocyclic amines and azines (91,93–95,99).

3. Liquid Chromatography I: Ion Chromatography (IC)

Ion chromatography, as a distinct type of chromatography, refers to any modern and efficient method of separating and determining ions. There are three distinguishable modes of ion chromatography: ion-exchange chromatography (IC); ion-exclusion chromatography (IEC); and ion-pair chromatography, also known as mobile-phase ion chromatography or ion-interaction chromatography (MPIC). However, IC is the most commonly used type of ion chromatography. It is well suited to the determination of the NH_4^+ cation in aqueous media.

The essential principle of this technique is an ion exchange between the mobile phase (eluent) and the exchange groups covalently bonded to the stationary phase. A selection of IC techniques applicable to the determination of NH_x in aqueous sample media is presented in Table 13.

In IC, NH_4^+ is separated from other cations, including alkali metals and amines, using cation-exchange columns and strongly acidic eluents (e.g., HCl, HNO_3). Thus, the polar, water-soluble characteristics, which can inhibit the analysis of NH_x by GC, and to a lesser extent by HPLC, are effectively exploited in IC.

Two stationary-phase materials (sorbents) are commonly used in the IC of NH_4^+ (Table 13):

Surface sulphonated styrene–divinylbenzene (SS-DVB) copolymeric resins: The most commonly used stationary phases in the determination of NH_4^+ (Table 13). SS-DVB columns have low ion-exchange capacities and are stable from pH 1 to pH 14 and in the presence of many solvents and are thus highly durable (85).

Silica-based ion cation-exchange columns: Exhibit many of the advantageous characteristics of polymeric columns, but, in general, are less rugged and their exchange capacity tends to decrease with usage. They are also stable only over a more limited pH range (2–8) and more susceptible to hydration effects than their polymeric counterparts (85,118).

Conductimetric detection is most commonly used in the quantification of NH_4^+ by IC (Table 13), providing sensitive although nonselective detection. Generally conductimetric detection is carried out with the aid of a suppressor. Suppressors are incorporated into the IC

system intermediate between the separation column and the detector and may take the form of ion-exchange columns, hollow-fiber membranes, or, more recently, plate membranes. Such techniques are termed *suppressed conductivity "dual-column" IC.*

Suppression is intended to produce the maximum difference between the conductimetric signals arising from eluent and analyte ions. For example, in the case of HCl eluents, selective exchange of Cl^- counterions for OH^- ions (from an anion-exchange column or through transfer across a membrane) is used to convert the acidic eluent into water. This results in reduced background conductivity against which analytes, such as NH_4^+, may be quantified, i.e., increased analyte sensitivity.

As well as the workhorse conductivity detector, photometric, spectrophotometric, and fluorescence detectors have all been used in the detection of NH_4^+. However, NH_4^+ neither fluoresces nor absorbs strongly, and thus derivatization is required for its detection, as is the case for its analysis by HPLC, e.g., Gardner and St. John (115) used postcolumn derivatization with *o*-phthalaldehyde in the presence of mercaptoethanol with fluorescence detection to study NH_x in natural water samples.

A further alternative to suppressed conductivity detection is to employ light-absorbing eluents (e.g., photometric or fluorimetric) and matched detectors. The appearance of nonactive analyte ions such as NH_4^+ is thus signaled by dips or troughs in the baseline signal, because "transparent" ions displace light-absorbing ions of the eluent. This technique is known as *single-column IC* (SCIC) with indirect detection, e.g., Sherman and Danielson (114) used the fluorescent CeIII ion in the mobile phase and quantified NH_4^+ through reductions in the measured fluorescent signal.

Ion chromatography also allows the introduction of far greater sample volumes (up to 1 ml) than GC (1–10 μM) and, to a certain extent, than HPLC. This is advantageous in the determination of the low analyte concentrations characteristic of natural waters. However, in the determination of NH_4^+, high concentrations of Na^+ (which elutes just ahead of NH_4^+) and K^+ (which elutes just after NH_4^+) can interfere with resolution and quantification. This is not problematic in the analysis of low-ionic-strength media such as rainwater. However, in seawater, where concentrations of Na^+ exceed those of NH_4^+ by several orders of magnitude, it is not possible to reliably determine NH_4^+ due to "swamping" effects. Thus the analysis of NH_4^+ in more complex aqueous media, such as seawater and wastes, requires the use of derivatization or a selective sam-

Table 13 Selected IC Techniques for the Determination of NH$_x$ in Aqueous Sample Media

Authors (year)	Limit of detection (LOD) or lowest reported measurement (LR) where given (ng on column, unless otherwise stated)	Separation column	Mobile phase	Pre-/post column reaction	Suppression mode	Detector	Notes (matrix)
Small, Stevens, Bauman (1975) (86)	100 (LR)	SS-DVB cation exchange	0.001 N HCl (isocratic)	None	Dual-column–anion-exchange column (OH-form)	Suppressed conductivity	First modern IC paper to use pellicular resins and suppressed conductivity (aqueous)
Bouyoucos (1977) (101)	250 (LR)	SS-DVB cation exchange	0.01 N HCl (isocratic)	None	Dual-column–anion-exchange column (OH-form) + anion-exchange column (Cl-form)	Suppressed conductivity	Problem of nonlinear response overcome by conversion chloride salts using postsuppressor anion-exchange column (aqueous)
Gardner (1978) (102)	0.68 (LOD)	"Durrum DC-4A" cation-exchange resin	Sodium chloride/boric acid/sodium hydroxide (isocratic)	Postcolumn reaction with OPA + 2-ME	Single column	Suppressed conductivity	Method termed microfluorometric determination (aqueous)
Mulik, Estes, Sawicki (1978) (103)	50 (LR)	SS-DVB cation exchange	HNO$_3$ (isocratic)	None	Dual-column–anion-exchange column (OH-form)	Suppressed conductivity	Applied to determination of ammonium particulates (gaseous)
Zweidinger et al. (1978) (104)	53 μM (LR)	SS-DVB cation exchange	0.0025 N HNO$_3$ (isocratic)	None	Dual-column–anion-exchange column (OH-form)	Conductivity	Vehicle exhaust emissions
Fritz et al. (1980) (105)	12 μM (LR)	SS-DVB cation exchange	0.0015 M HNO$_3$ (isocratic)	None	Single-column–cation exchange	Indirect conductivity	Applied to the analysis of tap water (aqueous)

Reference	Detection limit	Column packing	Eluent (mobile phase)	Sample prep	Column configuration	Detection	Comments
Buechele, Reutter (1982) (106)	50 (LR)	SS-DVB cation exchange	0.01 M HCl + methanol/water additions (isocratic)	None	Dual-column—anion-exchange column	Suppressed conductivity	Investigates effect of methanol and water mobile-phase additions; 6 μl conductivity cell volume (aqueous)
Small, Miller (1982) (107)	3400 (LR)	"Dowex 50" resin	0.01 M copper sulphate (isocratic)	None	Single column	Indirect photometric	First paper on indirect IC detection (aqueous)
Bouyoucos, Melcher (1983) (108)	170 (LR)	SS-DVB cation exchange	0.015 M HCl (isocratic)	None	Dual-column—anion-exchange column (OH-form)	Conductivity (direct suppressed)	Applied to atmospheric analysis (gaseous and aqueous)
Bag (1985) (109)	29 μM (LR)	2 Silica Gel, 1 SS-DVB cation exchange	0.01 M HNO$_3$ (isocratic)	None	Dual-column—anion-exchange column (OH-form)	Suppressed conductivity	Qualitative optimization study; best results with multiple separation columns; 4-h suppressor lifetime (aqueous)
Iskandarani, Miller (1985) (110)	69 (LR)	SS-DVB cation exchange + PMA silica gel anion exchange	Sulpobenzoic acid/copper hydroxide (isocratic)	None	Single column	Indirect photometric	Simultaneous coanalysis of cations and anions using two separation columns (aqueous)
Johnson (1986) (111)	5001 (LR)	SS-DVB cation exchange	0.005 M HCl (isocratic)	None	Dual-column—micromembrane	Suppressed conductivity	Dionex textbook application (aqueous)
Sithole, Guy (1980) (112)	1000 (LOD)	"Dowex 50" "Aminex A-9" "Aminex A-8"	0.01 M copper sulphate (isocratic)	None	Single column	Indirect photometric	Applied to soil analysis (aqueous/soil extracts)
McAleese (1987) (113)	4.0 cond. det. + 29 photo. det. (LOD)	SS-DVB cation exchange	0.0075 M benzyl tri-methylammonium chloride (isocratic)—photo. det. + 0.01 M HCl (isocratic)—cond. det.	None	Dual-column—fiber	Indirect photometric + suppressed conductivity	Comparison of two contrasting IC techniques employing two different modes of detection (aqueous)
Sherman, Danielson (1985) (114)	0.08 photo. + 0.09 Fluor (LOD)	SS-DVB cation exchange	0.01 mM cerium sulphate (isocratic)	None	Single column	Indirect photometric + indirect fluorescence	Employs fluorescence of CeII; aqueous sample media and urine

Table 13 Continued

Authors (year)	Limit of detection (LOD) or lowest reported measurement (LR) where given (ng on column, unless otherwise stated)	Separation column	Mobile phase	Pre-/post column reaction	Suppression mode	Detector	Notes (matrix)
Gardner, St. John (1991) (115)	0.85 (LOD)	SS-DVB cation exchange	Sodium chloride/boric acid/sodium hydroxide (isocratic)	Postcolumn reaction with OPA/2-MA	Single column	Direct fluorescence	Aqueous
Williams et al. (1992) (25)	0.06 μM (LOD)	SS-DVB cation exchange	Not given	None	Dual-column–micromembrane	Suppressed conductivity	Used as part of an intercomparison of measurements for atmospheric ammonia (atmospheric)
Gibb et al. (1995) (116,117)	0.04 μM (LOD)	SS-DVB cation exchange	40 mM HCl or methane sulphonic acid	None	Dual-column–micromembrane	Suppressed conductivity	Used coupled to a flow injection system as part of an automated system (aqueous)

LR: lowest reported value; LOD: limit of detection; cond. det.: conductivity detection; photo. det.: photometric detection; SS-DVB: surface-sulphonated styrene–divinylbenzene copolymeric resin; PMA: polymethylacrylate; OPA: o-phthalaldehyde; 2-ME: 2-mercaptoethanol.

pling, extraction, or isolation procedure (e.g., Refs. 116 and 117).

4. Planar Chromatography

Planar chromatographic methods include thin-layer chromatography (TLC) and paper chromatography (PC). Currently most planar chromatography techniques are based upon the thin-layer approach, which is faster, has greater resolving powers, and is more sensitive (28). It has been used in the analysis of NH_x and primary and secondary amines as their DANS-amide derivatives with spectrophotometric detection. However, Budd (120) used paper chromatography to determine NH_x and amines extracted by a microdiffusion technique. Components were separated by a phenol/water descending solvent and revealed by spraying with ninhydrin in acetone. The resultant color intensity was used as a semiquantitative estimation of concentration.

The analysis of NH_x by planar chromatography is, like its analysis by HPLC, largely dependent upon derivative formation. Planar chromatography is further restricted by the lack of suitable multidimensional detectors.

5. Capillary Zone Electrophoresis (CZE)

Strictly speaking, CZE is not a chromatographic technique, since analytes do not move by dynamic equilibrium along a mobile stream. The method is instead based on the migration of charged particles, colloids, or ions through a buffer solution contained within a capillary (50–100 cm in length) under the influence of an electric field. The technique is characterized by excellent mass sensitivity, low sample consumption, and high resolution. Because of combined electro-osmotic flow and electrophoretic separation, all species normally travel in the same direction, allowing the detection of positive, neutral, and negatively charged species at one point on the capillary (121). Gross and Yeung (122) used an 18-μm-ID column to achieve elution of ammonia-C_4 amines in under 5 minutes at concentrations of 0.1–1.4 mg/L NH_x (with limits of detection being established in the 0.1–0.5-fmol range). The limitation to the technique, however, is the very low volume of sample that can be introduced onto the capillary. This in turn limits the minimum absolute concentration that can be detected.

ACKNOWLEDGMENTS

Thanks to DG Cummings for proofreading this manuscript.

REFERENCES

1. EJ Carpenter, CP Capone. Nitrogen in the Marine Environment. London, Academic Press, 1983.
2. S Russell. Ammonia: WRc Instrument Handbooks. WRc, Swindon, UK, 1994.
3. JP Riley, R Chester. Introduction to Marine Chemistry. Academic Press, London, 1971.
4. National Research Council (NRC), Subcommittee on Ammonia. University Park Press, Baltimore, 1979.
5. CRC Handbook of Chemistry and Physics. 67th ed. CRC Press, Boca Raton, FL, 1993.
6. GM Hidy. Aerosols: An Industrial and Environmental Science. Academic Press, New York, 1984.
7. RG Bates, GD Pinching. Acidic dissociation constant of the ammonium Ion at 0 to 50°C, and the base strength of ammonia. J Res Nat Bur Stds 4:419–430, 1949.
8. A Van Neste, RA Duce, C Lee. Methylamines in the marine atmosphere. Geophys Res Lett 14(7):711–714, 1987.
9. W Stumm, JJ Morgan. Aquatic Chemistry. 2d ed. Wiley Interscience, New York, 1981.
10. KH Khoo, CH Culberson, RG Bates. Thermodynamics of the dissociation of ammonium ion in seawater from 5 to 40°C. J Soln Chem 6(4):281–290, 1977.
11. J Lyman, RH Fleming. Composition of seawater. J Mar Resources 3:134–136, 1940.
12. SM Libes. An Introduction to Marin Biogeochemistry. Wiley, New York, 1992.
13. C Garside, G Hill, S Murray. Determination of submicromolar concentrations of ammonia in natural waters by a standard addition method using a gas-sensing electrode. Limnol Oceanogr 23(5):1073–1076, 1978.
14. H Hara, A Motoike, S Okazaki. Continuous flow determination of low concentrations of ammonium ions using a gas dialysis concentrator and a gas electrode detector system. Analyst 113:113–115, 1988.
15. F Winquist, A Spetz, I Lundstrom, B Danielsson. Determination of ammonia in air and aqueous samples with a gas-sensitive semiconductor capacitor. Anal Chim Acta 164:127–138, 1984.
16. POJ Hall, RC Aller. Rapid small volume, flow injection analysis for CO_2 and NH_4^+ in marine and freshwaters. Limnol Oceanogr 37(5):1113–1119, 1992.
17. AM Harabin, CMG van den Berg. Determination of ammonia in seawater using catalytic cathodic stripping voltammetry. Anal Chem 65(23):3411–3416, 1993.
18. MA Brzezinski. Colorimetric determination of nanomolar concentrations of ammonia in seawater using solvent extraction. Mar Chem 20:277–288, 1987.
19. S Kar, MA Arnold. Fiber-optic ammonia sensor for measuring synaptic glutamate and extracellular ammonia. Anal Chem 64(20):2438–2443, 1992.
20. SW Willason, KS Johnson. A rapid sensitive technique for the determination of ammonia in seawater. Mar Biol 91:285–290, 1986.

21. G Slawyk, JJ MacIsaac. Comparison of two automated ammonium methods in a region of coastal upwelling. Deep-Sea Res 19:521–524, 1972.

22. MM Santos Filha, BF dos Reis, H Bergamin, N Baccan. Flow-injection determination of low levels of ammonium ions in natural waters employing preconcentration with a cation-exchange resin. Anal Chim Acta 261:339–343, 1992.

23. R Jones. An improved fluorescence method for the determination of nanomolar concentrations of ammonium in natural waters. Limno Oceanogr 36(4):814–819, 1991.

24. K Gorzelska, JN Galloway, K Watterson, WC Keene. Water-soluble primary amine compounds in rural continental precipitation. Atmos Env 26A:1005–1018, 1992.

25. EJ Williams, ST Sandholm, JD Bradshaw, JS Schendel, AO Langford, PK Quinn, PJ LeBel, SA Vay, PD Roberts, RB Norton, RB Watkins, MP Buhr, DD Parrish, JG Calvert, FC Fehsenfeld. An intercomparison of five ammonia measurement techniques. J Geophys Res 97(D11):11591–11611, 1992.

26. I Gross, ES Yeung. Indirect fluorometric detection of cations in capillary zone electrophoresis. Anal Chem 62:427–431, 1990.

27. U.S. Environmental Protection Agency. Quality Criteria for Water. EPA Publication 440/5-86-001. U.S. Gov. Prin. Office, Washington, DC, 1987.

28. DA Skoog, DM West, FJ Holler. Fundamentals of Analytical Chemistry. 6th ed. Saunders College, Orlando, FL, 1992.

29. APHA. Standard Methods for the Examination of Water and Wastewater. 18th ed. American Public Health Association, American Water Works Association and Water Environment Federation, Washington, DC, 1992.

30. L Solorzano. Determination of ammonia in natural waters by the phenolhypochlorite method. Limnol Oceanogr 14:799–801, 1969.

31. K Grasshoff, M Ehrhardt, K Kremling, eds. Methods of Seawater Analysis. Verlag Chemie, Weinheim, 1976. JDH Strickland, TH Parsons. A practical handbook of seawater analysis. Bull Fish Res Bd Can 167:1–31, 1968.

32. U.S. Environmental Protection Agency. Methods for Chemical Analysis of Water and Wastes. Washington, DC, 1983.

33. Standing Committee of Analysts. Methods for the Examination of Waters and Associated Materials—Ammonia in Water. Her Majesty's Stationery Office, London, 1981.

34. ASTM Manual on Industrial Water and Industrial Waste Water. 2nd ed. Washington, DC, 1966, p. 418.

35. Analytical Methods Committee of the Royal Society of Chemistry. Official and Standardized Methods of Analysis. 3rd ed. Royal Society of Chemistry, Cambridge, 1994.

36. D Degobbis. Limnol Oceanogr 146–150, 1973.

37. MPE Berthelot. Rep Chim Appl 1:284, 1859.

38. PL Searle. The Berthelot or indophenol reaction and its use in the analytical chemistry of nitrogen. Analyst 109:549–568, 1984.

39. H Verdouw, CJA Van Echteld, EMJ Dekkers. Ammonia determination based on Indophenol formation with sodium salicylate. Water Res 12:399–402, 1978.

40. MD Krom, S Grayer, A Davidson. An automated method for the determination of ammonia for use in mariculture. Aquaculture 44:153–160, 1985.

41. S McCleod. Micro-distillation unit for the use in continuous flow analyzers. Its construction and use in determination of ammonia and nitrate in soils. Anal Chim Acta 266:107–112, 1992.

42. RT Emmet. Direct spectrophotometric analysis of ammonia in natural water by the phenol-hypochlorite reaction. Nav Ship Res Develop Cent Rep 2570, 1968.

43. BS Newell. The determination of ammonia in seawater. J Mar Biol Ass UK 47:271–280, 1968.

44. G Catalano. An improved method for the determination of ammonia in seawater. Mar Chem 20:289–295, 1987.

45. JP Riley, G Skirrow. Chemical Oceanography. 2nd ed. Vol. 3. Academic Press, London, 1975.

46. JP Riley. The spectrophotometric determination of ammonia in natural water with particular reference to sea water. Anal Chem Acta 9:575–589, 1953.

47. MI Liddicoat, S Tibbitts, EI Butler. The determination of ammonia in seawater. Limnol Oceanogr 20:131–132, 1974.

48. D Scheiner. Determination of ammonia and kjeldahl nitrogen by indophenol method. Water Res 10:31–36, 1976.

49. BL Hampson. Relationship between total ammonia and free ammonia in terrestrial and ocean waters. J Cons Int Explo Mer 37(2):117–122, 1977.

50. A Hiller, D Van Slyke. Determination of ammonia in blood. J Biol Chem 102:499, 1933.

51. B O'Connor, R Dobbs, B Villiers, R Dean. Laboratory distillation of municipal waste effluents. JWPCF 39 R:25, 1967.

52. J Fiore, JE O'Brien. Ammonia determination by automatic analysis. Wastes Engineering 33:352, 1962.

53. RL Booth, LB Lobring. Evaluation of the Auto-Analyzer II: a progress report. In: Advances in Automated Analysis: 1972 Technicon International Congress, v. 8, pp. 7–10, Mediad Inc., Tarrytown, NY, 1973.

54. Standard Methods for the Examination of Water and Wastewater. 14th ed. p. 616, Method 604, 1975.

55. DD Siemar. Use of solid boric acid as an ammonia absorbent in the determination of nitrogen. Analyst 111(9):1013–1016, 1986.

56. JM Kruse, MG Mellon. Colorimetric determination of ammonia and cyanate. Anal Chem 25(8):1188–1190, 1953.

57. JA Tetlow, AL Wilson. An absorptiometric method for determining ammonia in boiler feed-water. Analyst 89: 453–461, 1964.

58. ME Meyerhoff. Polymer membrane electrode based potentiometric ammonia gas sensor. Anal Chem 52(9): 1532–1534, 1980.

59. ME Lopez, GA Rechnitz. Selectivity of the potentiometric ammonia gas sensing electrode. Anal Chem 54(12):2085–2089, 1982.

60. SW Gibb. PhD Thesis, University of East Anglia, 1994.

61. HL Lee, ME Meyerhoff. Comparison of tubular polymeric pH and ammonium ion electrodes as detectors in the automated determination of ammonia. Analyst 110:371–376, 1985.

62. DM Prantis, ME Meyerhoff. Continuous monitoring of ambient ammonia with a membrane-electrode-based detector. Anal Chem 59:2345–2350, 1987.

63. RL Booth, RF Thomas. Selective electrode determination of ammonia in water and wastes. Env Sci Tech 7:523–526, 1973.

64. WL Banwart, JM Bremner, MA Tabatabai. Determination of ammonium in soil extracts and water samples by an ammonium electrode. Comm Soil Sci Plant 3:449, 1952.

65. D Midgley, K Torrance. The determination of ammonia in condensed steam and boiler feed-water with a potentiometric ammonia probe. Analyst 97:626–633, 1972.

66. M Van Son, RC Schothorst, G and Den Boef. Determination of total ammoniacal nitrogen in water by flow injection analysis and a gas diffusion membrane. Anal Chim Acta 153:273–275, 1983.

67. RM Carlson. Automated separation and conductimetric determination of ammonia and dissolved carbon dioxide. Anal Chem 50(11):1528–1531, 1978.

68. LC De Faria, C Pasquini. Flow-injection determination of inorganic forms of nitrogen by gas diffusion and conductimetry. Anal Chim Acta 245:183–190, 1991.

69. L Prochazkova. Spectrophotometric determination of ammonia as rubazoic acid with bispyrazolone reagent. Anal Chem 36(4):865–871, 1964.

70. L Canale-Guitierrez, A Maquieira, R Puchades. Enzymatic determination of ammonia in food by flow injection. Analyst 115:1243–1246, 1990.

71. T Aoki, S Uemura, S Munemori, M Makoto. Continuous flow fluorometric determination of ammonia in water. Anal Chem 55(9):1620–1622, 1983.

72. M Roth. Fluorescence reaction for ammonia. Anal Chem 43(7):880–882, 1971.

73. SS Goyal, DW Rains, RC Huffaker. Determination of ammonium ion by fluorimetry or spectrophotometry after on-line derivatization with o-phthalaldehyde. Anal Chem 60(2):175–179, 1988.

74. Z Genfa, PK Dasgupta. Fluorometric measurement of aqueous ammonium ion in a flow injection system. Anal Chem 6(5):408–412, 1989.

75. MT Jeppesen, EH Hansen. Flow-injection fluorometric assay of nitrogen-containing substrates by on-line enzymatic generation of ammonia. Anal Chim Acta 245:89–99, 1991.

76. R Belcher, SL Bogdanski, AC Calokerinos, A Thownshend. Determination of ammoniacal nitrogen in fertilizers by molecular emission cavity analysis. Analyst 102:220–221, 1977.

77. IMA Shakir, SY Atto, NA Jawkal. Determination of ammonia and ammonium ions in air and in artesian wells in arbil governate using thermal chemiluminescence by molecular emission cavity analysis (MECA). J Univ Kuwait (Sci) 15(2):269–279, 1988.

78. MA Arnold, TJ Ostler. Fiberoptic ammonia gas sensing probe. Anal Chem 6(58):1137–1140, 1986.

79. GF Kirkbright, R Narayanaswamy, NA Welti. Fiberoptic pH probe based on the use of an immobilized colorimetric indicator. Analyst 109:1025–1028, 1984.

80. TD Rhines, MA Arnold. Simplex optimization of a fiber-optic ammonia sensor based on multiple indicators. Anal Chem 60(1):76–81, 1992.

81. OS Wolfbeis, HE Posch. Fiberoptic fluorescing for ammonia. Anal Chim Acta 185:321–327, 1986.

82. M Tswett. Adsorptionsanalyse und chromatographische methode. anwendung auf die chemie des chlorophylls. Ber Deut Botan Ges 24:384–393, 1906.

83. AJP Martin, RLM Synge. A new form of chromatogram employing two liquid phases. 1. A theory of chromatography; 2. Application to the microdetermination of the higher monoamino-acids in proteins. Biochemical J 35:1358, 1941.

84. AT James, AJP Martin. Gas-liquid partition chromatography. A technique for the analysis of volatile materials. Int Cong on Anal Chem 77(8):915–931; Analyst 77:915–931, 1952.

85. Shpigun, YA Zolotov. Ion Chromatography in Water Analysis. Ellis Horwood, Chichester, U.K., 1988.

86. H Small, TS Stevens, WC Bauman. Novel ion-exchange chromatographic method using conductometric determination. Anal Chem 47(11):1801–1809, 1975.

87. AR Mosier, CE Andre, FG Viets Jr. Identification of aliphatic amines volatilized from cattle feedyard. Env Sci Tech 7(7):642–644, 1973.

88. N Kashihira, K Makino, K Kirita, Y Watanabe. Chemiluminescent nitrogen detector—gas chromatography and its application to measurement of atmospheric ammonia and amines. J Chrom 239:617–624, 1982.

89. X-H Yang, C Lee, MI Scranton. Determination of nanomolar concentrations of individual dissolved low molecular weight amines and organic acids in seawater. Anal Chem 65:572–576, 1993.

90. E Glob, S Sorensen. Determination of dissolved and exchangeable trimethylamine pools in sediments. J Microbiol Meth 6:347–355, 1987.

91. ECM Chen, RA Farquharson. Analysis of trace quantities of ammonia and amines in aqueous solutions by reverse phase high performance liquid chromatography using *m*-toluoyl derivatives. J Chrom 178:358–363, 1979.

92. P Lindroth, K Mopper. High performance liquid chromatographic determination of subpicomole amounts of amino acids by precolumn fluorescence derivatization with *o*-phthaldialdehyde. Anal Chem 51(11):1667–1674, 1979.

93. J-K Lin, CC Lai. High performance liquid chromatographic determination of naturally occurring primary and secondary amines with dabsyl chloride. Anal Chem 52:630–635, 1980.

94. B Bjorkqvist. Separation and determination of aliphatic and aromatic amines by high performance liquid chromatography with ultraviolet detection. J Chrom 204:109–114, 1981.

95. Y Nishikawa, K Kuwata. Liquid chromatographic determination of low molecular weight aliphatic amines in air via derivatization with 7-chloro-4-nitro2,1,3-benzoxadiazole. Anal Chem 56:1790–1793, 1984.

96. K Mopper, P Lindroth. Diel and depth variations in dissolved free amino acids and ammonium in the Baltic Sea determined by shipboard HPLC analysis. Limnol Oceanogr 27(2):336–347, 1982.

97. P Simon, C Lemacon. Determination of aliphatic primary and secondary amines and polyamines in air by high-performance liquid chromatography. Anal Chem 59:480–484, 1987.

98. D Delmas, MG Frikha, EAS Linley. Dissolved primary amine measurement by flow injection analysis with *o*-phthalaldehyde: comparison with high performance liquid chromatography. Mar Chem 29:145–154, 1990.

99. H Gorzelska, JN Galloway. Amine nitrogen in the atmospheric environment over the North Atlantic Ocean. Global Biogeochemical Cycles 4 3:309–333, 1990.

100. RL Cox, TW Schnieder, MD Koppang. Ferrocene tagging of amines, amino acids and peptides for liquid chromatography with electrochemical detection. Anal Chim Acta 262:145–159, 1992.

101. SA Bouyoucos. Determination of ammonia and methylamine in aqueous solutions by ion chromatography. Anal Chem 49(3):401–403, 1977.

102. WS Gardner. Microfluorometric method to measure ammonium in natural waters. Limnol Oceanogr 23(5):1069–1072, 1978.

103. JD Mulik, E Estes, E Sawicki. Ion chromatographic analysis of ammonium ion in ambient aerosols. In: E Sawicki, JD Mulik, E Wittgenstein, eds. Ion Chromatographic Analysis of Environmental Pollutants. Ann Arbor Science, Ann Arbor, MI, 1978.

104. RB Zweidinger, J Tejada, JE Sigsby Jr, RL Bradow. Application of ion chromatography analysis of ammonia and alkylamines in automobile exhausts. In: E Sawicki, JD Mulik, E Wittgenstein, eds. Ion Chromatographic Analysis of Environmental Pollutants. Ann Arbor Science, Ann Arbor, MI, 1978.

105. JS Fritz, DT Gjerde, RS Becker. Cation chromatography with a conductivity detector. Anal Chem 52:1519–1522, 1980.

106. RC Buechele, DJ Reutter. Effect of methanol in the mobile phase on the chromatographic determination of some monovalent cations. J Chromatog 240:502–507, 1982.

107. H Small, TE Miller Jr. Indirect photometric chromatography. Anal Chem 54:462–469, 1982.

108. SA Bouyoucos, RG Melcher. Collection and ion chromatographic determination of ammonia and methylamines in air. Am Ind Hyg Assoc J 44(2):119–122, 1983.

109. SP Bag. Determination of trace organic dicarboxylic acids and amines by ion chromatography. Talanta 32(8b):779–784, 1985.

110. Z Iskandarani, TH Miller Jr. Simultaneous independent analysis of anions and cations using indirect photometric chromatography. Anal Chem 57:1591–1594, 1985.

111. EL Johnson, ed. Handbook of Ion Chromatography. Dionex Corp., Sunnyvale, CA, 1986.

112. BB Sithole, RD Guy. Determination of alkylamines by indirect photometric chromatography. Analyst 111:395–397, 1980.

113. DL McAleese. Indirect photometric chromatography of cations and amines on a polymer-based column. Anal Chem 59:541–543, 1987.

114. JH Sherman, ND Danielson. Indirect cationic chromatography with fluorescence detection. Anal Chem 59:1483–1485, 1985.

115. WS Gardner, PA St. John. High-performance liquid chromatographic method to determine ammonium ion and primary amines in seawater. Anal Chem 63:537–540, 1991.

116. SW Gibb, RFC Mantoura, PS Liss. Analysis of ammonia and methylamines in natural waters by flow injection gas diffusion coupled to ion chromatography. Anal Chim Acta 316:291–304, 1995.

117. SW Gibb, JW Wood, RFC Mantoura. Automation of flow injection gas diffusion-ion chromatography for the nanomolar determination of methylamines and ammonia in seawater and atmospheric samples. J Auto Chem 17(6):205–212, 1995.

118. JS Fritz. Ion chromatography. Anal Chem 59(4):335A–344A, 1987.

119. EH Gruger. Chromatographic analyses of volatile amines in marine fish. J Agr Food Chem 20(4):781–785, 1972.

120. JA Budd. Catabolism of trimethylamine by a marine bacterium, *Pseudomonas* NCMB 1154. Mar Biol 4:257–266, 1969.

121. X Huang, T-KJ Pang, MN Gordon, R Zare. On-column conductivity detector for capillary zone electrophoresis. Anal Chem 59:2747–2749, 1987.

122. I Gross, ES Yeung. Indirect fluorometric detection of cations in capillary zone electrophoresis. Anal Chem 62:427–431, 1990.

14

Determination of Organic Nitrogen

Amalia Cerdà, Maria Teresa Oms, and Victor Cerdà
University of Balearic Islands, Palma de Mallorca, Spain

I. INTRODUCTION

Most knowledge of the quantities, sources, and distribution of nitrogen in water systems is focused on dissolved inorganic nitrogen (DIN) species, such as nitrite, nitrate, and ammonium, for they are well-known pollutants causing adverse health and environmental effects, whereas organic nitrogen compounds are frequently ignored. However, organic nitrogen is a ubiquitous component found in all locations, including urban, rural, and remote sites, and also in all types of water matrices, including snow and rainwater, lakes and estuaries, seawater, and rivers. Several studies have shown that the contribution of dissolved organic nitrogen (DON) may represent more than half of total nitrogen loading (1–4). The amount of DON in water samples depends largely on the type of sample and on the location, although concentrations are usually in the low-ppm range or below.

Organic nitrogen is defined as organically bound nitrogen in the trinegative oxidation state. It includes a wide variety of organic molecules containing nitrogen, such as urea and other amines, amino acids, proteins, and other nitrogen-containing macromolecules (1). The nature and importance of organic forms of nitrogen in water systems is still poorly understood, in fact, the term *organic nitrogen* is generally applied to the nitrogen fraction remaining after nitrate, nitrite, and ammonium have been substracted.

Although organic nitrogen was once considered to be biologically unavailable, several studies have demonstrated that its impact on estuarine and lake eutroph-

ication, massive algal bloom, extended bacterial production, and poor water quality may be more important than previously realized (5–7). Further research is clearly needed to assess the quantity and characterization of DON inputs and their potential influence on aquatic ecosystems, for there is a wide range of potential sources of organic nitrogen. Studies based on the use of ^{15}N tracer procedures have been conducted to determine the relative contribution of different sources (8). Potential sources of organic nitrogen to the environment can be divided into point sources, such as wastewater effluents, and nonpoint sources, such as atmospheric deposition, agriculture, combustion, and runoff from forest, urban areas, and the like (2,4,6,9). Dissolved organic nitrogen may also be produced by phytoplankton and bacteria (2,5,7,10); for instance, in oceanic, coastal, and estuarine environments, an average of 25–41% of dissolved inorganic nitrogen (ammonia and nitrate) taken by phytoplankton is estimated to be released as dissolved organic nitrogen (7). Typical organic nitrogen concentrations may vary from less than 1 mg/L in lakes and marine environment to more than 20 mg/L in raw wastewater.

The procedures described in the literature for the determination of specific organic nitrogen compounds, such as urea, aliphatic and aromatic amines, amino acids, and nitrophenols, usually involve either enzyme reactions (11–13) or complex chromatographic separation (14–16). In contrast, the determination of the so-called organic nitrogen fraction for monitoring purposes or routine analysis is frequently achieved by using much simpler methods based on a preliminary digestion step to convert organic species into an inorganic form of nitrogen. The resulting inorganic nitrogen, most often in the form of nitrate or ammonium, is then determined with any of the procedures described elsewhere.

II. SAMPLE COLLECTION AND PRESERVATION

Concentrations of dissolved organic nitrogen are influenced by biological activity and the balance between release and consumption, and therefore accurate processes in sampling and conservation must be taken into account to ensure the integrity of the samples for later analysis. In addition, the expected concentrations are usually very low, and special precautions should be taken to avoid foreign contamination of the samples. Contamination sources such as organic leaching from the sampling bottles, rubber seals, and natural and general pollutants around the sampling point can be avoided by using appropriate devices and clean handling practices.

Sampling bottles are usually made of glass, PVC, or polyethylene, although aluminum or stainless steel is used when large sample volumes at sea depth are required. Contamination from sampling bottles is avoided by using careful cleaning procedures consisting of washing first with a diluted detergent, rinsing with distilled water, washing again with diluted hydrochloric acid, and finally rinsing extensively with distilled water (17). This procedure should be applied to any part in contact with the sample.

Unless samples are analyzed online and on-site, it is necessary to store the collected samples for later analysis. Preservation techniques must then be used with the aim of delaying the chemical and biological transformations once the sample is removed from its original environment. Immediate sample freezing is one of the simplest methods for storing the sample and ensuring its integrity for later analysis (18). Potential errors due to flocculation during freezing can be reduced by intensive mixing before analysis.

Filtration is one of the most often-used pretreatments of samples, and it can be done either before or after sample freezing. The purpose of filtration is usually, on one hand, to separate particulate materials such as humic acids and, on the other, to eliminate or reduce biological activity (bacteria and phytoplankton) that could interfere and alter the balance between organic and inorganic forms of nitrogen. As with sampling devices, a careful pretreatment of filter and filtration equipment is required, because equipment can be a source of sample contamination, especially when pressure or vacuum is necessary (17).

After filtration, it is essential that samples be frozen or at least refrigerated. Acidification of samples may also help to eliminate bacterial activity; in this case HCl is the most appropriate acid. However, acidification may sometimes be undesirable for organic nitrogen analysis because it may induce a higher sorption of atmospheric ammonia and volatile amines.

III. ANALYTICAL METHODS FOR ORGANIC NITROGEN DETERMINATION

The first step for the determination of organic nitrogen is the digestion of the sample in order to convert organic compounds into an inorganic form of nitrogen, mainly nitrate, nitrite, or ammonium, although transformation to other forms of nitrogen, such as NO_x or N_2, have also been described. Once the sample is mineralized, the total nitrogen can be determined; from it, the organic nitrogen fraction is quantified by substraction. The preliminary digestion, also referred to as *mineralization*, is the most tedious and time-consuming step and also the greatest source of errors in the whole analytical procedure for organic nitrogen.

The main procedures for organic nitrogen determination are depicted in Fig. 1. They can be grouped into four categories according to the mineralization procedures:

Catalyzed acid digestion (Kjeldahl method), in which organic forms of nitrogen are converted to ammonia
Photochemical oxidation
High-temperature combustion
Alkaline persulfate oxidation, in which the resulting inorganic nitrogen form is nitrate

Selection of the most adequate technique will depend on the characteristics of the sample to be analyzed (sample matrix, volume), on the operational conditions (pH, concentration of the oxidizing reagent and/or catalyst), as well as on the compatibility between the digestion process and the analytical method for the determination of the inorganic species.

Developments in organic nitrogen determination are focused mainly on the automation of digestion procedures and on continuous digestion and analysis using

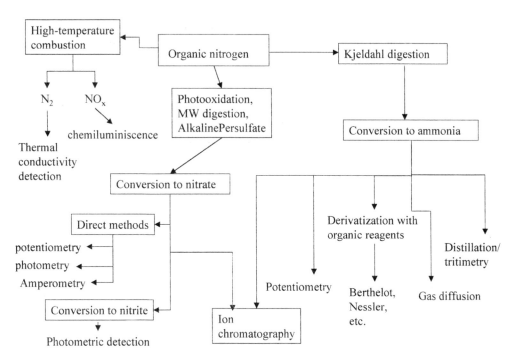

Fig. 1 Procedures for organic nitrogen determination.

flow injection analysis (FIA) techniques, with the aim of overcoming the main drawback of those procedures, which is the time needed for complete determination. In those systems, continuous digestion takes place in helicoidal reactors while sample is flowing, and, usually, aluminum blocks, thermostatic water, or oil baths are required (19).

Quite recently, microwave- (MW-) assisted digestion has been developed, and it has quickly been incorporated into research and routine laboratory procedures due to the relatively short time required for the digestion and also to the low sample and reagent volumes needed as compared to traditional digestion methods (20–27). The development of microwave ovens especially designed for such applications has also favored the application of microwave-assisted digestion to a great variety of environmental matrices (27) and to on-line digestion procedures using flow injection analysis (FIA) (28,29). All these aspects will be considered in the following sections.

A. Kjeldahl Method: Conversion to Ammonia

Kjeldahl digestion is clearly the most popular and frequently used method for organic nitrogen determination, and it is a standard AWWA-APHA procedure (30).

In brief, sample digestion is achieved by reaction of organic compounds with hot, concentrated sulphuric acid (temperature 360–380°C). Potassium sulphate is added to increase the boiling point, and copper (31,32), mercury (33–35), titanium (36), or selenium (31–33,37,38) salts are added as catalysts. The highest efficiency is obtained using mercury as a catalyst (39), followed by selenium, although because of their high toxicity and the difficulties in the later elimination of residues, they are being replaced by copper salts because its efficiency is generally good for the determination of organic nitrogen in most environmental samples. Under described digestion conditions, the organic nitrogen compounds are converted into ammonium sulphate within 1–2 hours. The resulting ammonium is then converted to ammonia in alkaline medium and, together with the ammonium initially present in the sample, separated by distillation and collected in a solution of diluted boric acid. Quantification is usually achieved by titration with either diluted hydrochloric or sulphuric acid. The sum of organic nitrogen and ammonia is usually referred as *total Kjeldahl nitrogen* (TKN). Organic nitrogen is determined after subtraction of the initial ammonium content in the sample. The method can be applied to samples containing either high or low concentrations of organic nitrogen and requires a relatively large sample volume for low concentrations. Although the method has been largely proved and is applicable to the analysis of drinking waters, saline waters, and urban and industrial wastewater, it is tedious and time-consuming, toxic metals are used as catalysts, and very aggressive reagents are needed.

Depending on the sample volume and the expected concentrations, either a macro- or a micro-Kjeldahl approach can be selected. The term *macro-Kjeldahl* is applied when the sample volume is in the range of 25 ml (for 50–100 mg/L N) to 500 ml (1 mg/L N or below). The term *micro-Kjeldahl* is applied for sample volumes in the range 5 ml (for 40–400 mg/L N) to 50 ml (4–40 mg/L N). In fact, the only difference is the dimensions of the laboratory glass material involved (micro or macro scale).

The Kjeldahl method was first developed for the determination of nitrogen in biological samples, and certain aspects must be taken into account when applying it to other matrices. For example, the method is well adapted for the determination of primary amines and amino acids in proteins; however, certain organic nitrogen compounds, such as nitroderivatives, semicarbazones, and some refractory tertiary amines, are not quantitatively converted into ammonium, thus causing incomplete recoveries. Although TKN is sometimes referred to as total nitrogen, the method does not convert either nitrate or nitrite to ammonium; therefore, if the total nitrogen (TN) content of the sample has to be determined, an additional reduction of these anions to ammonium is necessary (40–49).

Several procedures have been described for overcoming the main drawbacks of this method, namely, the use of toxic reagents and the long time required for the determination. Among them the use of hydrogen peroxide allows a much shorter digestion period and even the complete elimination of metallic catalysts (40,50,51). Increased precision of the method is also described, probably by its avoiding ammonium-metal complex formation.

The combined use of a H_2SO_4–H_2O_2 and microwave radiation has been applied to the determination of Kjeldahl nitrogen in food (52–54) and soil (55). However, its application to water samples is still scarce (21).

The ammonium resulting from Kjeldahl digestion can also be determined using other well-known methods, such as those based on the Berthelot and Nessler reactions or potentiometry using ammonium-selective electrodes. The purpose of this chapter is not ammonium determination, so the following is only a brief description of these methods.

The Berthelot method is based on the reaction between ammonium in alkaline medium, hypochloride, and phenol, which results in the formation of indo-phenol blue, which can be spectrophotometrically detected. The reaction mechanisms and experimental conditions established by different authors were extensively reviewed by Searle (56) in 1984.

The Nessler method (57) is based on the reaction between ammonia and tetraiodomercuriate (II), the latter being formed on site from mercury (II) iodide in alkaline potassium iodide medium. As a result, a colloidal dispersion is formed, and it is detected by absorbance measurements at 400 and 425 nm.

There are few references describing the use of the FIA approach for the determination of Kjeldahl nitrogen with continuous digestion. Schmitt et al. (58) describe a microwave-assisted method for the determination of urea, and they apply it to the determination of ammonium and urea in natural water. In this method, the resulting ammonium is converted to ammonia and photometrically analyzed using gas-diffusion techniques.

Several segmented flow systems have also been described in which the digestion is conducted in a thermostated helicoidal reactor at 300°C (33). Metallic catalysts are replaced by a mixture of H_2SO_4–H_2O_2, and the resulting ammonium is analyzed online using the Berthelot method. The digestion time is only 5 minutes, and recoveries between 70% and 100% are achieved for the nitrogen compounds tested (ammonium sulphate, nicotinic acid, nicotinamide, urea, uric acid, alanine, tryptophan, and creatinine). The method has been applied satisfactorily to the analysis of total N in natural waters (channel and lake) within a range of 0.01–0.5 mg/L N.

Some of the most common problems in FIA Kjeldahl-digestion procedures are, on one hand, the clogging of tubing by the deposit of salts inside the digester and, on the other, the lower recoveries caused by the substitution of the catalyst salts (33). In order to overcome these problems in the routine analysis of wastewater, some authors describe a semiautomatic system, with the digestion step carried out batchwise in block digesters while the determination of resulting ammonium is achieved using FIA techniques based on the use of ammonia gas–sensing electrodes (59), ammonium-selective electrodes (41), the Nessler reaction (40,60), or methods based on the reaction of Berthelot (34,61–63). Several devices have been proposed for the automatic distillation of ammonium before FIA analysis, with the aim of reducing potential interferences (48,64,65). Another approach has been the use of gas-diffusion techniques coupled to photometric (40), potentiometric (41), and conductimetric (31,66) detection.

B. Photo-Oxidation

In this group of methods, decomposition of organic matter takes place under ultraviolet radiation in the presence of small amounts of oxidant. Under UV radiation, organic nitrogen and ammonium are converted into nitrite and nitrate, which can easily be quantified by molecular spectrophotometry, e.g., by first reducing all nitrogen forms to nitrite and then analyzing it using the reaction of Griess (see the chapter on nitrate and nitrite determination).

As described in a review on photochemical methods (74), the photo-oxidation reaction is complex. It involves radical species, and reaction paths depend on several factors, such as pH, and they may be enhanced if certain photosensitizers are present. Depending on the pH, they can result in the formation of many intermediate species, such as hydrogen peroxides, singlet oxygen, hydroxyl radicals, or hydrated electrons. The studies reveal the formation of a great number of intermediate molecules, potentially responsible for the oxidative processes, when a sample is exposed to UV radiation. This could explain the discrepancies in the results obtained under different operational conditions.

Anyway, it is generally accepted that the hydroxyl radicals are the most important species in the oxidation of organic matter. They can be generated directly from hydrogen peroxide under radiation with short UV; for this reason, H_2O_2 and persulfate are the most frequently used oxidants for UV mineralization, either for the quantification of total nitrogen, phosphorus, and carbon or for the determination of metal ions. In those cases, the UV radiation acts as a catalyst for the oxidation reaction. For instance, when exposed to the action of UV light, hydrogen peroxide decomposes, forming hydroxyl radicals, which initiate radical chain reactions involving organic substances in the sample.

The method was first proposed in 1966 by Armstrong et al. (68) for the determination of carbon, nitrogen, and organic phosphorus and was based on the oxidation of the sample by hydrogen peroxide under ultraviolet radiation and subsequent reduction to nitrite using amalgamated cadmium filings. A high-pressure mercury lamp (1200-W mercury arc tube), was used to irradiate a 100-ml sample volume introduced in stoppered fused-silica tubes clustered around the UV lamp. Hydrogen peroxide is added to ensure an excess of oxygen and to facilitate the oxidation of organic nitrogen compounds and ammonium to nitrate and nitrite. Mineralization is achieved after 3 hours for up to 12 samples simultaneously. Quantitative oxidation of the following compounds is reported: ammonium chloride,

2,2′-bipyridine, casein, thiourea, adenine, guanidine, and even pyridine, this last one considered by diverse authors as highly refractory.

Under the experimental conditions, nitrate is the predominant form of nitrogen (70%), although, by prolonged irradiation, the oxidizing reagent may be fully decomposed and photochemical reduction of nitrate to nitrite may occur (see the chapter on the determination of nitrate and nitrite). The method has been applied to the determination of organic nitrogen in seawater within the range of 0.002–0.015 mg/L N.

In 1968 Armstrong and Tibbits (67) published a modification of this method consisting of the substitution of the high-pressure lamp by a medium-power mercury lamp (380 W). Quantitative mineralization (>98%) of organic compounds in saline water is achieved after longer irradiation time (12 hours) using 100 ml of sample. Among the nitrogen compounds evaluated (ammonium chloride, formamide, oxamic acid, glycine, pyridine, etc.), urea is the most refractory, and its complete oxidation is only achieved after 24 hours of irradiation. The method has been applied to the determination of organic nitrogen in several points of the western English Channel (including midchannel and coastal waters). The proposed method allows determination within the range of 2–10 μg/L N, with the average concentration in that location being 5.1 μg/L N.

Although mineralization based on UV radiation was proposed as early as the 1960s, it is not widely used for the routine analysis of water samples, in spite of its important advantages, such as smooth conditions, no need for strong heating and corrosive reagents, and, last but not least, the absence of significant amounts of toxic residues. At the moment there is an increasing tendency toward incorporating this digestion mode into automatic analyzers, although it must be noted that this method can only be applied to the determination of dissolved organic nitrogen, because the particulate fraction requires more drastic conditions for oxidation.

The method initially proposed by Armstrong et al. (68) has been adapted by Henriksen (69) to the determination of nitrogen and phosphorus in natural waters. The required irradiation time as well as other experimental conditions are similar to those already described. At a working pH range of 6.5–9 and using a 900-W high-pressure lamp, quantitative digestion is achieved within 4 hours as long as the nitrogen concentration does not exceed 0.5 mg/L. This method is semiautomatic, for the determination of nitrate formed after the photochemical oxidation is carried out by reduction with cupperized cadmium in a segmented-flow system. The sampling frequency is 24 samples a day, lower than the typical sampling frequency using continuous methods of analysis, because the sample mineralization is carried out in a batch mode and is the bottleneck of the analytical procedure.

Gustafsson (70) describes a similar system, in which the sample digestion takes place in two steps: the first one under sulphuric acid conditions for 1 hour and a second one with hydrogen peroxide at alkaline pH (8–9) for 3 hours more. Quantitative digestion of nitrogen compounds is reported. The method is applied to fresh and brackish waters containing 0.02–0.03 mg/L N, and the results are comparable to those obtained using the Kjeldahl method. Gustaffson also studied the effect of organic matter and also of certain inorganic ions (chloride and bromide) on the photo-oxidation yields. The conclusions of Gustaffson, in agreement with previous studies by Semenov et al. (71), showed that after 5 hours of irradiation (three at pH 9 and two at pH 4), complete digestion of alanine, tryptophan, adenine, thymine, uracil, DNA, cytosine, diethylamine, and urea is accomplished.

Manny et al. (72) describe a method for the determination of total dissolved nitrogen in lake waters at acidic pH adjusted with boric acid. The irradiation time for quantitative sample digestion is 1.5–3 hours, depending not only on the structure of the nitrogen compound to be analyzed but also on the composition of the sample matrix, on the content in oxidizable organic matter, and on the concentration of inorganic salts, which may slow down the decomposition process (73). Under the digestion conditions established by these authors, the reaction product contains a mixture of nitrate, nitrite, and ammonium.

Several authors have used potassium persulfate as oxidizing agent instead of hydrogen peroxide, because the ultraviolet radiation at 254 nm catalyzes the decomposition of this compound in sulphate and other reactive species, such as oxygen and hydroxyl radicals, which in turn oxidizes the organic compounds (74). For instance, Zhang and Wu (75) use persulphate for the determination of total nitrogen by photo-oxidation of the organic nitrogen compounds, ammonium, and nitrite to nitrate.

Online sample UV oxidation has been described using FIA systems (76–78) and segmented flow (79–84) by integrating them in nitrate analyzers in which the formed nitrate is determined either by direct measurement in ultraviolet (76) or indirectly by reduction to nitrite using copperized cadmium (77,79,81,82) or hydrazinium sulphate (83,84) or reduction to ammonium (80). In all cases, reactors for online photo-oxidation are usually coils made of quartz or, even better, from

Teflon, which is UV-transparent, less fragile and easier to manipulate than quartz. These systems reduce the time of digestion to a few minutes, and some authors (79,80) have designed experimental schemes for digestion taking place in two steps, as proposed by Gustafsson. The scope of application of these methods is very wide, from natural waters to polluted aquatic systems and wastewater.

The incompatibility between the conditions required for digestion and those required for the analysis of the resulting nitrate and/or nitrite must be carefully considered. The indirect determination of nitrate by reduction to nitrite is one of the most common approaches, but that achieving reduction with copperized cadmium is especially troublesome; several authors have noticed that the cadmium column is irreversibly damaged when in contact with the oxidant (77,78).

Inserting a cadmium wire before the copperized cadmium column has been proposed by several authors (82,85,86) in order to extend the lifetime of the reducing column. Also, the addition of metabisulphite in the FIA system has been proposed by McKelvie et al. (77) to eliminate the excess of persulfate. It is also possible to avoid this problem by using another reduction procedure, such as the homogeneous reduction using hydrazine sulphate described by Mullin and Riley (87).

Most of the described systems do not allow differentiation between inorganic and organic nitrogen forms; thus, the resulting value is the dissolved total nitrogen content in the sample. Some methods allowing separate determination of organic and inorganic forms of nitrogen provide more information on sample speciation and composition. In the FIA system developed by Cerdà et al. (78), nitrate formed after the UV oxidation of the sample is reduced to nitrite by hydrazine sulphate, which in turn is analyzed using the reaction of Griess modified by Shinn. The system is developed as a multiparametric analyzer and, through the combination of several valves, allows the selection of the proper channel for the determination of nitrite alone, nitrite plus nitrate, and total nitrogen (the sum of nitrate, nitrite, ammonia, and organic nitrogen). The efficiency and reproducibility of the proposed methods have been tested using several types of model nitrogen compounds, including ammonia, urea, EDTA, barbituric acid, amino acids, and aromatic compounds such as nitrophenol and carboxylic acids. Efficiency is good up to 10 mg/L N for all compounds, but it clearly falls at higher concentrations, except for EDTA, aspartic acid, and barbituric acid. Reproducibility is 1.5–3% for the tested compounds.

Interference due to organic carbon competing for the

same oxidant as organic nitrogen compounds was also studied. Standard solutions of urea and ammonium (10 mg/L N) were spiked with increasing concentrations of glucose (10, 100, 200, and 500 mg L^{-1} C). No interference was observed at a ratio of 1:20 N/C, but it was a negative interference at higher organic carbon content and must be taken into account and corrected, e.g., by increasing the concentration of the oxidant agent.

This multiparametric FIA system has been applied to the determination of nitrogen species, with the exception of ammonia, in wastewater without previous treatment. Comparative studies showed a good correlation between the results obtained with the proposed method and those obtained using the Kjeldahl method (ammonia plus organic nitrogen) and standard methods for nitrate and nitrite, the reproducibility being superior for the FIA multiparametric system. Sample frequency is 30 samples per hour, which represents a much shorter analysis time than that required by the conventional Kjeldahl method.

The segmented-flow system proposed by Oleksy-Frenzel and Jekel (84) is based on UV oxidation of the sample after separation of the organic fraction using gel permeation. The system has been designed to allow the sequential determination of the dissolved nitrogen and carbon fractions as well as the dissolved organic halide (DOX) content of the samples and is applied to the characterization of wastewaters from chemical industries as well as to evaluate the performance of the treatment plant.

C. High-Temperature Combustion (HTC)

The determination of organic nitrogen is achieved by oxidative pyrolysis either at high temperature (900–1100°C) or at a slightly lower one (650–900°C), with the latter case being usually catalyzed by platinum. During pyrolysis, all nitrogen forms, but molecular nitrogen gas, are transformed into NO, which is quantified by chemiluminiscence and from which the total amount in the original sample is deduced (88,89). Organic nitrogen is obtained from the difference between total nitrogen content and inorganic nitrogen content (nitrate, nitrite, and ammonium). As in previous methods, the effectiveness of the pyrolysis is closely tied to the experimental conditions, the chemical form of the nitrogen, and the sample matrix. The method is extremely attractive because it allows the joint determination of all the nitrogen forms, but N_2, using automatic equipment with a high sampling frequency (2–6 minutes per sample) and high sensitivity.

High-temperature combustion at 950–1000°C ensures the quantitative conversion of all nitrogen compounds, without matrix effects and without the use of strong acids or toxic chemicals. Automated commercial systems based on combustion and chemiluminiscence detection have been developed and specially designed for unattended performance, with the analysis time being less than 5 minutes for any type of sample. The conversion rate to NO is variable, depending upon the instrument conditions and sample constituents, although this dependence is less critical than with the photo-oxidation methods (see earlier), due probably to the stronger conditions required (temperature and/or catalysts).

Clifford and McGaughey (89) have studied the oxidation rates of several nitrogen model compounds and have reported quantitative conversions to NO in the concentration range of 0–10 mg/L N when using platinum as catalyst, even in the presence of up to 350 mg/L of dissolved organic carbon, with good reproducibility, with a detection limit of 0.1 mg/L N. Conversion of 100% was achieved for most of the model compounds (ammonium chloride, dimethyl glyoxime, sodium nitrate, sodium nitrite, alanine, ammonium hydroxide, thiourea, glycine, and pyridine), while recoveries for sulfamic acid, EDTA, and sodium cyanide were close to 90%, with those being the most recalcitrant compounds. The typical analysis time is 2 minutes.

The analytical range can be increased up to 40 mg/L N, with no interference of organic carbon, by increasing the amount of oxygen in the carrier gas. Comparison between Kjeldahl digestion and high temperature showed excellent agreement for wastewater.

Jones and Daughton (90) have studied the pyrolysis process on 56 nitrogen compounds, including those refractory to the Kjeldahl digestion method, and have reported 90–110% conversion for most of them. When applied to industrial effluents, the values obtained using the HTC method were 10% higher than those obtained using the Kjeldahl method, probably because certain refractory compounds can be analyzed only by using the HTC approach. The HTC method is also adequate for low nitrogen concentrations [0.07–0.6 mg total dissolved nitrogen (TDN)], such as those found in seawater. Walsh (88) compares the performance of this method with photo-oxidation. Apparently there are no big differences among the results obtained; however, 18–22 hours of irradiation time is required for quantitative mineralization with UV light.

Compared to other mineralization methods, high-temperature combustion is more effective, although the equipment needed for this method is more sophisticated than that for Kjeldahl, UV oxidation, or alkaline persulfate digestion. The HTC is usually applied to heavily polluted waters, like industrial effluents, where refractory organic compounds are expected.

Sample speciation has been proposed by Daughton et al. (91) by combining chromatographic separation and gas-diffusion methods with high-temperature combustion. First, separation of polar and nonpolar nitrogen compounds is achieved by reversed-phase chromatography on C_{18} columns. Polar compounds are mainly inorganic salts and are eluted from the column while less polar ones, such as aromatic amines and N-heterocycles, are retained. Methanol eluates can then be analyzed for total nitrogen. The result of this operation underestimates organic nitrogen, since only the nonpolar organic nitrogen is in fact quantified. Then the separation of volatile and nonvolatile compounds is accomplished through gas-diffusion membranes. This yields the content of "nonvolatile organic nitrogen" in the sample. Neither the first approach nor the second strictly represents the organic nitrogen in wastewater samples. However, the quantification of nonvolatile organic nitrogen compounds in the polar fraction, together with the nonpolar nitrogen, provides a higher estimate of the organic nitrogen in the original sample, even if certain organic substances that are simultaneously polar and volatile, such as aliphatic amines, are excluded. The method, according to the authors, correlates well with the Kjeldahl method.

Shi et al. (92) propose the separation of the organic components of the sample in a chromatographic system using supercritical fluid conditions (CO_2 as the mobile phase) and introduction of the eluate in the pyrolysis furnace. The nitrogen compounds are transformed into NO and detected by chemiluminiscence. The analytical range is 3–850 mg/L N, with the detection limit being 1.4 mg/L.

Determination of organic nitrogen after reduction to N_2 has also been proposed. After oxidative pyrolysis at high temperature (850–950°C) in a quartz tube, the resulting nitrogen oxides are reduced to molecular nitrogen by passing them through a furnace at 650°C containing metallic copper (reduction is catalyzed by this metal) (93). Detection is by thermal conductivity. Following this procedure, Pietrogrande et al. (94) have studied extensively the experimental conditions for oxidation and their effect on nitrogen compounds of different chemical structures and have found that some of them, namely, benzilic alcohol and molecules containing an imidazol or pirrol group, are resistant to the combustion.

D. Alkaline Persulfate Oxidation

In 1969, Koroleff (95) developed a new method for the determination of total nitrogen in natural water based on the oxidation of nitrogen compounds to nitrate using potassium persulfate in strongly alkaline medium (NaOH, pH 12.5–13.2). In the proposed method, known as alkaline persulfate digestion and also as the Koroleff method, mineralization is achieved under high pressure and temperature conditions (2 bar, 120°C). Although it is simpler and shorter than the Kjeldahl method, it still requires 30–60 min of autoclaving time. The Koroleff method has been successfully applied to the analysis of natural water (96–98), including seawater with nitrogen concentrations below 1 mg/L (20–750 μg/L N) (99).

Under the described experimental conditions, persulfate decomposes according to the following reaction:

$$K_2S_2O_8 + H_2O \rightarrow 2KHSO_4 + \tfrac{1}{2}O_2$$

With released oxygen providing the optimal conditions for the quantitative oxidation of nitrogen compounds in the sample, nitrate is obtained as the only reaction product. A number of methods can then be utilized for nitrate determination (see the chapter on nitrite and nitrate), such as heterogeneous reduction with copperized cadmium, generally after cooling the sample and adjusting it to the proper pH.

The method provides high recovery and good reproducibility for a wide variety of nitrogen-containing compounds, such as urea, EDTA, inorganic ammonium salts, nicotinic acid, and a wide range of amino acids and proteins, including refractory compounds not mineralized using the Kjeldahl method, like pyridine nitro and nitrous compounds. However, as with other digestion methods, persulfate alkaline digestion fails when applied to compounds containing nitrogen-to-nitrogen bonds (hydrazine, antipyrine, benzotriazole, methyl orange, etc) or HN=C groups (guanidine, creatinine, etc.), which yield low recoveries or even prevent the oxidation to nitrate completely (97,103).

Dahl has coordinated an intercomparison study between different laboratories to verify the performance of the Koroleff method regarding precision and reproducibility (100). For this purpose, synthetic water samples containing 0.76 mg/L N of ammonium and glycine were prepared to test the performance of the inorganic and organic nitrogen compounds, respectively.

Oxidation with alkaline persulfate can also be used for phosphorous compounds (this is an official method for the determination of total phosphorus). However, determination of total phosphorus (TP) is carried out under acidic conditions; therefore, both determinations (total nitrogen and total phosphorous) must be done separately. Several authors have developed manual or semiautomatic methods for the simultaneous determination of TN and TP within a single digestion step by taking advantage of the pH change happening during digestion (101–104). Indeed, initial alkaline conditions (pH 12.6–12.8) become acidic (pH 2.0–2.1) due to the formation of potassium hydrogenosulfate as reaction by-product.

The scope of application of the modified method is wide and includes the analysis of TN and TP in river (103) and lake water (104), with detection limits of 0.01 mg/L for both TN and TP. In addition, Langer and Hendrix (102) and Hosomi and Sudo (104) showed that the method can be applied to the determination of particulate organic nitrogen and organic phosphorous in freshwaters, which is composed mainly of algae, bacteria, organic detritus, and suspended sediments. Reference materials used by these authors were plant leaves, algae, and sediments.

When the autoclave is replaced by a microwave furnace, quantitative oxidation of refractory compounds can be achieved in less than 45 minutes. This microwave radiation is focused directly on the sample, and losses due to absorption of energy on the container walls are avoided. Recently, Johnes and Heathwaite, modifying previous similar procedures, have developed a method allowing the simultaneous determination of total nitrogen and total phosphorous in natural water samples (river, lake, and groundwater) using potassium persulfate as oxidant agent and microwave radiation as heating source. The mineralization step is carried out in a Teflon vessel, to withstand high pressure and high temperature, and allows the effective determination of nitrogen compounds at a concentration range between 2 and 50 mg/L, with a detection limit of 0.05 mg/L N. The system is semiautomatic, since after the manual digestion the mineralized sample is introduced in a segmented-flow system for the determination of the resulting nitrate using homogeneous reduction with hydrazine sulphate.

Cerdà et al. (105) have developed an FIA system for the determination of total nitrogen in which sample digestion takes place online, under focused microwave radiation. The sample and the reagent lines are merged, and flow is directed into a flow-through Teflon digestion cell placed into the microwave oven. There the mixture is irradiated, and mineralization takes place while flowing through the digester. The resulting nitrate

is reduced to nitrite using hydrazinium sulphate, and nitrite is finally analyzed using the reaction of Shinn–Griess. The linear interval is 0.3–20 mg/L N, the detection limit is 0.2 mg/L, and the relative standard deviation of the method is 3% for a 5-mg/L N sample. The complete analysis (digestion process plus the later determination of formed nitrate as nitrite) is achieved in less than 2 minutes, thus allowing a sample frequency of 45 samples per hour.

High mineralization rates within less than 5 minutes have also been reported with FIA and microflow systems and using selenium (106) and platinum (19,107) as catalysts. Platinum wires are introduced into a heated capillary reactor. Aoyagi (19) applied it to the determination of TN and TP in wastewater, with a sampling rate of 15 samples per hour and with the nitrate being detected directly at 220 nm. The detection limit is reported to be 5 μg/L N, and the relative standard deviation is 0.85%. The method described by Aoki et al. (106) is based on the reduction of formed nitrate to ammonia by titanium (III) chloride. Ammonia is then separated using a tubular microporous Teflon membrane and quantified by fluorescence using orthophthaldialdehyde (OPA) and 2-mercaptoethanol. The detection limit for this method is 9 μg/L N, and the sampling frequency is six samples per hour.

REFERENCES

1. K. Mopper, R. G. Zika. Nature 325:246–249 (1987).
2. S. Cornell, A. Rendell, T. Jickells. Nature 376:243–246 (1995).
3. D. A. Bronk, M. A. Sanderson, D. J. Koopmans. Paper Presented at the Santa Fe 99, Aquatic Science Meeting of the American Society of Limnology and Oceanography, Santa Fe, NM, February 1–5, 1999.
4. A. Knap, T. Jickells, A. Pszenny, J. Galloway. Nature 320:158–160 (1986).
5. K. D. Hammer. Zentralbl. Hyg. Umweltmed. 194:321–341 (1993).
6. S. Cornell, T. Jickells, C. Thornton. Atmos. Environ. 32:1903–1910 (1998).
7. D. A. Bronk, P. A. Glibert, B. B. Ward. Science 265:1843–1846 (1994).
8. C. Kendall. Tracing nitrogen sources and cycling in catchments. In: Isotope Tracers in Catchment Hydrology. C. Kendall, J. J. McDonnell (eds.). Elsevier Science, Amsterdam, Chap 16, pp 519–576 (1998).
9. L. J. Puckett. Environ. Sci. Technol. 29:408A–414A (1995).

10. S. P. Pavlou, G. E. Friederich, J. J. Macisaac. Anal. Biochem. 6:16–24 (1974).

11. H. Hara, T. Kitagawa, Y. Okabe. Analyst 118:1317–1320 (1993).

12. D. S. Rogers, K. H. Pool. Anal. Lett. 6:801–804 (1973).

13. H. Mana, U. Spohn. Anal. Chim. Acta. 325:93–104 (1996).

14. M. E. L. R. de Queiroz, K. Wuchner, R. Grob, J. Mathieu, Analysis 20:12–18 (1992).

15. S. Lartiges, P. Garrigues. Analysis 21:157–165 (1993).

16. M. Ahel, K. M. Evans, T. W. Fileman, R. F. C. Mantoura. Anal. Chim. Acta 268:195–200 (1992).

17. J. H. Sharp, E. T. Peltzer, M. J. Alperin, G. Cauwet, J. W. Farrington, B. Fry, D. M. Karl, J. H. Martin, A. Spitzy, S. Tugrul, C. A. Carlsson. Mar. Chem. 41:37–49 (1993).

18. J. E. Dore, T. Houlihan, D. V. Hebel, G. Tien, L. Tupas, D. M. Karl. Mar. Chem. 53:173–185 (1996).

19. M. Aoyagi, Y. Yasumasa, A. Nishida. Anal. Sci 5:235–236 (1989).

20. A. Abu-Samra, M. J. Steven, S. R. Koirtyohann. Anal. Chem. 47:1475–1477 (1975).

21. L. B. Jassie, H. M. Kingston, eds. Introduction to Microwave Sample Preparation: Theory and Practice. American Chemical Society, Washington, DC, 1988.

22. D. Didenot. Spectra 2000 146:44–50 (1990).

23. H. Matusiewicz, R. E. Sturgeon. Prog. Analyt. Spectrosc. 12:21–39 (1989).

24. A. Sinquin, T. Görner, E. Dellacherie. Analysis 21:1–10 (1993).

25. C. Demesmay, M. Olle. Spectra Analyse 175:27–30 (1993).

26. H. M. Kuss. Fresenius J. Anal. Chem. 343:788 (1992).

27. A. Zlotorzynski. Crit. Rev. Anal. Chem. 25:43–76 (1995).

28. M. De la Guardia, A. Salvador, J. L. Burguera, and M. Burguera. J. Flow Injection Anal. 5:121–127 (1988).

29. M. Burguera, J. L. Burguera. Anal. Chim. Acta 366:63 (1998).

30. Standard Methods for the Examination of Water and Wastewater. 18th ed. American Public Health Association (APHA), American Water Works Association (AWWA), Water Environment Federation Publication (WPCF). APHA, Washington, DC (1992).

31. J. J. Rodrigues, C. Pasquini. Analyst 116:841–845 (1991).

32. W. E. Baethgen, M. M. Alley. Commun. Soil Sci. Plant Anal. 20:961–969 (1989).

33. J. Davidson, J. Mathieson, A. W. Boyne. Analyst 95:181–193 (1970).

34. W. D. Basson. Fresenius Z. Anal. Chem. 311:23 (1982).

35. L. R. McKenzie, P. N. W. Young. Analyst 100:620–628 (1975).

36. D. Lerique. Analysis 20:M21–22 (1992).

37. D. Scheiner. Water Res. 10:31 (1976).

38. P. C. Williams. Analyst 89:276–281 (1964).

39. P. L. Kirk. Anal. Chem. 22:354–358 (1950).

40. J. H. Ginkel, J. Sinnaeve. Analyst 105:1199 (1980).

41. J. L. F. C. Lima, A. O. S. S. Rangel, M. R. S. Souto. Fresenius J. Anal. Chem. 358:657 (1997).

42. G. Pruden, S. J. Kalembasa, D. S. Jenkinson. J. Sci. Food Agric. 36:71–73 (1985).

43. A. Devarda. Chem. Ztg. 16:1952 (1892).

44. R. M. Carlson, R. I. Cabrera, J. L. Paul, J. Quick, R. Y. Evans. Commun. Soil Sci. Plant Anal. 21:1519–1529 (1990).

45. L. C. de Faria, C. Pasquini. Anal. Chim. Acta 245:183 (1991).

46. M. J. Hepher, R. H. Alexander, J. Dixon. J. Sci. Food Agric. 49:379–381 (1989).

47. M. S. Cresser. Analyst 102:99–103 (1977).

48. J. Keay, P. M. A. Menage. Analyst 95:379–382 (1970).

49. T. Aoki, S. Uemura, M. Munemori. Environ. Sci. Technol. 20:515 (1986).

50. O. Elkei. Anal. Chim. Acta 86:63–68 (1976).

51. K. H. Nicholls. Anal. Chim. Acta 76:208–212 (1975).

52. A. Bermond, C. J. Ducauze. Analysis 19:64–66 (1991).

53. C. L. Suard, M. H. Feinberg, J. Ireland-Ripert, R. M. Mourel. Analysis 21:287–291 (1993).

54. M. H. Feinberg, J. Ireland-Ripert, R. M. Mourel. Anal. Chim. Acta 272:83–90 (1993).

55. X. T. He, R. L. Mulvane, W. L. Banwart. Soil Sci. Soc. Am. J. 54:1625–1629 (1990).

56. P. L. Searle. Analyst 109:549–568 (1984).

57. J. Nessler. Chem. Centr. 27:529 (1856).

58. A. Schmitt, L. Buttle, R. Uglow, K. Williams, S. Haswell. Anal. Chim. Acta 284:249–255 (1993).

59. R. J. Stevens. Water Res. 10:171–175 (1976).

60. E. A. G. Zagatto, B. F. Reis, H. Bergamin, F. J. Krug. Anal. Chim. Acta 109:45 (1979).

61. J. W. B. Stewart, J. Ruzicka. Anal. Chim. Acta 82:137–144 (1976).

62. J. W. B. Stewart, J. Ruzicka, H. Bergamin, E. A. Zagatto. Anal. Chim. Acta 81:371 (1976).

63. J. A. Bietz. Anal. Chem. 46:1617–1618 (1974).

64. S. McLeod. Anal. Chim. Acta 266:107–112 (1992).

65. S. McLeod. Anal. Chim. Acta 266:113–117 (1992).

66. X. L. Su, L. H. Nie, S. Z. Yao. Talanta 44:2121–2128 (1997).

67. F. A. J. Armstrong, S. Tibbitts. J. Mar. Biol. Ass. U.K. 48:143–152 (1968).

68. F. A. J. Armstrong, P. M. Williams, J. D. H. Strickland. Nature 211:481–483 (1966).

69. A. Henriksen. Analyst 95:601–608 (1970).

70. L. Gustafsson. Talanta 31:979–986 (1984).

71. A. D. Semenov, V. G. Soier, V. A. Bryzgalo, L. S. Kosmenko. Zh. Analit. Khim. 31:2030 (1976).

72. B. A. Manny, M. C. Miller, R. G. Wetzel. Limnol. Oceanogr. 16:71–85 (1971).

73. L. J. Heidt, J. B. Mann, H. R. Schneider. J. Am. Chem. Soc. 70:3011–3015 (1948).

74. J. Golimowski, K. Golimowska. Anal. Chim. Acta 325:111–133 (1996).

75. Y. Zhang, L. Wu. Analyst 111:767–769 (1986).

76. S. Hinkamp, G. Schwedt. Z. Wasser-Abwasser-Forsch. 24:60–65 (1991).

77. I. D. McKelvie, B. T. Hart, M. Mitri, I. C. Hamilton, A. D. Stuart. Anal. Chim. Acta 293:155–162 (1994).

78. A. Cerdà, M. T. Oms, R. Forteza, and V. Cerdà. Analyst 121:13–17 (1996).

79. B. K. Afghan, P. D. Goulden, J. F. Ryan. Adv. Autom. Anal. 2:291–297 (1970).

80. J. H. Lowry, K. H. Mancy. Water Res. 12:471–475 (1978).

81. H. Kroon. Anal. Chim. Acta 276:287–293 (1993).

82. V. J. G. Houba, I. Novozamsky, J. Uittenbogaard, J. J. van der Lee. Landwirtsh. Forsch. 40:295–302 (1987).

83. P. Kutscha-Lissberg, F. Prillinger. Plant Soil 64:63–66 (1982).

84. J. Oleksy-Frenzel, M. Jekel. Anal. Chim. Acta 319: 165–175 (1996).

85. J. F. van Staden, A. E. Joubert, H. R. van Vliet. Fresenius Z. Anal. Chem. 325:150–152 (1986).

86. M. J. Ahmed, C. D. Stalikas, S. M. Tzouwara-Karayanni, M. I. Karayannis. Talanta, 43:1009–1018 (1996).

87. J. B. Mullin, J. P. Riley. Anal. Chim. Acta 12:464–480 (1955).

88. T. W. Walsh. Mar. Chem. 26:295–311 (1989).

89. D. A. Clifford, L. M. McGaughey. Anal. Chem. 54: 1345–1350 (1982).

90. B. M. Jones, C. G. Daughton. Anal. Chem. 57:2320–2325 (1985).

91. C. G. Daughton, B. M. Jones, R. H. Sakaji. Anal. Chem. 57:2326–2333 (1985).

92. H. Shi, J. T. B. Strode III, L. T. Taylor, E. M. Fujinari. J. Chromatogr. 734:303–310 (1996).

93. R. W. Jenkins, C. H. Chi, V. J. Linnenh. Anal. Chem. 38:1257–1258 (1966).

94. A. Pietrogrande, M. Zancato, A. Guerrato, G. C. Porretta. Analysis 21:353–357 (1993).

95. F. Koroleff. Determination of total nitrogen in natural waters by means of persulphate oxidation. International Congress for Exploration of the Sea (ICES), 1969, Poster C:8.

96. L. Solórzano, J. H. Sharp. Limnol. Oceanogr. 25:751–754 (1980).

97. F. Nydahl. Water Res. 12:1123–1130 (1978).

98. M. M. Smart, F. A. Reid, J. R. Jones. Water Res. 15: 919–921 (1981).

99. C. F. d'Elia, P. A. Steudler, N. Corwin. Limnol. Oceanogr. 22:760–764 (1977).

100. I. Dahl, Vatten. 30:180–186 (1974).

101. J. C. Valderrama. Mar. Chem. 10:109–122 (1981).

102. C. L. Langer, P. F. Hendrix. Water Res. 16:1451–1454 (1982).

103. J. Ebina, T. Tsutsui, T. Shirai. Water Res. 17:1721–1726 (1983).

104. M. Hosomi, R. Sudo. Int. J. Environmental Studies 27: 267–275 (1986).

105. A. Cerdà, M. T. Oms, R. Forteza, V. Cerdà. Anal. Chim. Acta 351:273–279 (1997).

106. T. Aoki, S. Uemera, M. Munemori. Bunseki Kagaku 35:32 (1986).

107. M. Goto, S. Murofushi, D. Ishii. Bunseki Kagaku 37: 47–51 (1988).

15

Phosphates

Ian D. McKelvie

Monash University, Clayton, Victoria, Australia

I. PHYSICAL AND CHEMICAL PROPERTIES

Although phosphorus is the eleventh most abundant element in the earth's crust, where it forms approximately 1120 mg/kg, it is geochemically classed as a trace element (1,2). In the lithosphere, it occurs as phosphates, and these may be leached by weathering processes into the hydrosphere. Phosphorus may then be precipitated as insoluble metal phosphates that are incorporated into sediments and cycled on a geological time scale (millions of years), or it can participate in the rapid terrestrial and aquatic biological phosphorus cycles.

In aquatic systems, phosphorus occurs in a wide variety of inorganic and organic forms (Fig. 1) (3). While these may exist in either the dissolved, colloidal, or particulate form, the predominant species is orthophosphate in either the mono- or diprotonated form (HPO_4^{2-}, $H_2PO_4^-$). The dissolved component is operationally defined by filtration, and for this reason the term *filterable* is used in preference to either *dissolved* or *soluble*, both of which are used extensively and interchangeably in the literature.

There may also be significant amounts of organic or condensed phosphates present. *Filterable condensed phosphates* (FCP) are comprised of inorganic polyphosphates, metaphosphates, and branched ring structures. The *filterable organic phosphorus* (FOP) fraction consists of nucleic acids, phospholipids, inositol phosphates, phosphoamides, phosphoproteins, sugar phosphates, aminophosphonic acids, phosphorus-containing pesticides, and organic condensed phosphates (4–7).

Phosphorus in aquatic systems may originate from natural sources such as the mineralization of algae and the dissolution of phosphate minerals, from anthropogenic point source discharges of sewage and industrial effluents, and from diffuse inputs from grazing and agricultural land. Environmental interest in phosphorus stems from its critical role in the process of eutrophication. In many aquatic systems, phosphorus may be a limiting nutrient for the growth of algae. Given that phosphorus may exist in a variety of dissolved and particulate forms, there has been considerable emphasis in the analytical and ecological literature on the determination of the amount of bioavailable phosphorus (BAP).

The analysis of phosphorus in waters has historically been based on the photometric measurement of 12-phosphomolybdate or the phosphomolybdenum blue species produced when phosphomolybdate is reduced. Phosphorus species that are determined in this manner are referred to as *reactive*, and much of the nomenclature of phosphorus speciation derives from this origin.

Figure 2 shows the commonly analyzed fractions of phosphorus that can be determined for the various sample pretreatment procedures available. Of these fractions, total phosphorus (TP) and filterable reactive phosphorus (FRP) are perhaps the most commonly measured. Total phosphorus is frequently used to measure discharge compliance for wastewaters, and it represents the maximum *potentially* bioavailable phosphorus discharged; filterable reactive phosphorus, comprising mostly orthophosphate, provides an indi-

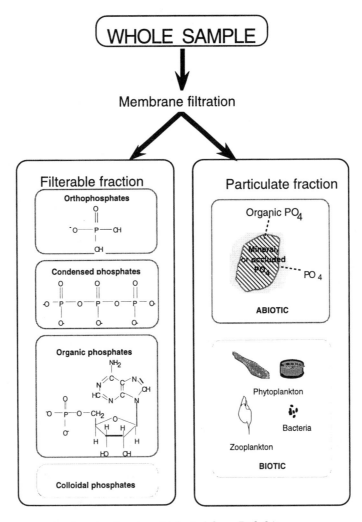

Fig. 1   Forms of phosphorus occurring in natural waters. (Adapted from Ref. 3.)

cation of the amount of most readily bioavailable phosphorus.

A number of reviews have appeared in the environmental analytical literature that focus on the analysis of phosphorus in the aquatic environment (3–6,8,9).

II.   SAMPLE PRESERVATION, STORAGE, AND PRETREATMENT

A.   Preservation and Storage

The storage regime used for water samples is dictated by the forms of phosphorus to be determined. Some commonly used approaches are listed in Table 1.

For samples used for the determination of the total phosphorus concentration, acidification and/or deep

freezing are recommended; for dissolved components, filtration at the time of sampling is recommended. The use of an antimicrobial agent such as mercuric chloride or chloroform, which has been common practice in the past, is no longer favored because of difficulties associated with the disposal of these toxic materials.

There is clear evidence that the various forms of phosphorus can change rapidly between the time of sampling and analysis. Lambert et al. (13) showed that very rapid decreases in both filterable reactive phosphorus (FRP) and total filterable and total reactive phosphorus fractions (TFP, TRP) occurred within 2 hours when samples were refrigerated prior to filtration and analysis. This is supported by the work of Haygarth et al. (14), who studied the effects of different storage regimes and container types on the stability of soil water samples analyzed for MRP (molybdate-reactive

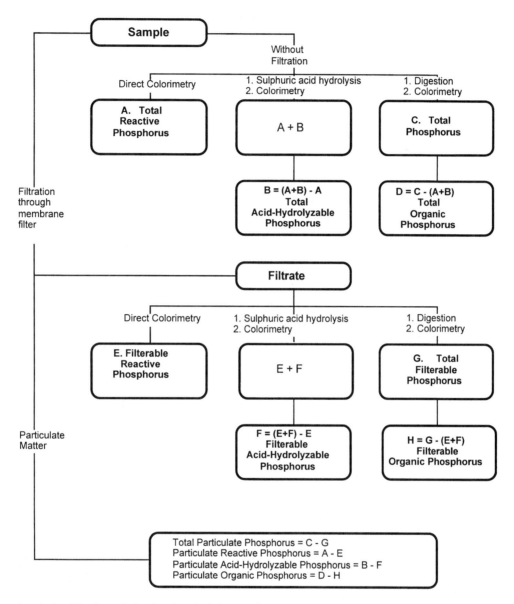

Fig. 2 Operational classification of physicochemical forms of phosphorus in water, based on filtration, digestion method, and molybdate reactivity. (Adapted from Ref. 10.)

phosphorus = total reactive phosphorus, TRP) over a period of 32 days. Extended storage of unfiltered samples at 4°C pending analysis is therefore *not* recommended.

If samples require extended storage prior to analysis, freezing is the preferred method. On the basis of long-term stability studies of samples stored frozen, Clementson and Wayte (15) recommended that samples for FRP analysis should be analyzed within 4 months, but even this was dependent on the size and nature of the storage container. Haygarth et al. also found that the lowest losses of P species occurred in those samples stored in larger-volume containers (>100 ml). Phosphate losses were less in PTFE sample containers than in polyethylene or polystyrene containers. While APHA-AWWA-WEF Standard Methods (10) recommend the use of glass vessels for the sampling and storage of phosphates, other workers have found significant adsorption of phosphate to glass and have recommended surface treatment to avoid these losses (16).

Table 1 Recommended Preservation and Storage Conditions for Phosphorus Samples

	Treatment	Ref.
Total phosphorus (TP)	Acidify with 1 ml conc. HCl/L	10
	or	
	Freeze at $\leq -10°C$	10
	or	
	Preserve with 40 mg HgCl$_2$/L	10
Total filterable phosphorus (TFP)	Filter immediately after collection, freeze	10
Filterable reactive phosphorus (FRP)	Filter immediately after collection, freeze	10
Filterable acid-hydrolyzable phosphorus (or filterable condensed phosphorus, FCP)	Preserve with HgCl$_2$	11
	or	
	Filter immediately after sampling, store at $\leq 4°C$, analyze within 24 h	12

B. Filtration

1. Membrane Filtration

Water samples may be separated into "dissolved" or "soluble" and "particulate" fractions, which are arbitrarily defined by the pore size of the filter used. Membrane filters of 0.45- and 0.2-μm nominal pore size are commonly used for this purpose, as well as glass fiber filters (GF/F $\approx$ 0.7 μm and GF/C $\approx$ 1.2 μm) (5). However, the filtrate obtained using these membranes or filters may contain significant amounts of colloidal phosphates as well as the truly dissolved phosphates. While there is widespread acceptance of 0.45-μm membranes for filtration, to the extent that this pore size is prescribed in some standards for sampling and analysis of phosphates, e.g., Australian Standards (17), there is a strong argument that 0.2 μm should be used in preference, because of the ability of 0.2 μm membranes to exclude bacteria, which may comprise a sizable fraction of the biotic particulate phosphate.

There are a number of potential problems inherent in the filtration process, including: transmission of larger particle sizes than nominal pore diameter (18), rupturing of cells during filtration (which is minimized by limiting the applied pressure), retention of small aggregates because of Van der Waal's forces, destabilization and aggregation of colloidal material, and progressive diminution of pore size because of filter cake formation and clogging (5). This last effect may be minimized by the use of a cross-flow filtration, in which the surface of the membrane is actively cleaned by the turbulent flow of sample tangential to the surface of the membrane (19).

Because the nature of the filtrate is highly dependent on the nature of the membrane used, on the conditions under which the filtration is performed, and on the nature of the sample itself, the use of the terms *soluble* and *dissolved* reactive phosphorus is misleading. A less ambiguous nomenclature would be the term *filterable reactive phosphorus (membrane pore size, in μm)*, e.g., FRP (0.2), where a 0.2-μm filter was used.

2. Ultrafiltration

Tangential flow filtration systems, either in stirred-cell configuration, or continuously pumped cross-flow configuration, permits the rapid filtration of larger volumes of water more quickly, with minimization of pore clogging, and overcomes a number of the potential filtration problems described earlier. Broberg and Persson (5) have proposed the potential use of ultrafilters, such as ca. 500 Da, as a means of assessing the truly dissolved P fraction, i.e., excluding the colloidal fraction. However, this process is very time-consuming, and for this reason it is doubtful that it will be widely adopted as a practical method for the measurement of dissolved phosphorus.

C. Preconcentration

In oceanic and pristine freshwater systems, phosphate concentrations of less than 0.1 μg P/L may be encountered, and some means of sample preconcentration may be employed to enable detection. Preconcentration techniques involving anion-exchange resins for use in low-ionic-strength waters have been described. Camarero used batch extraction of sample through anion exchanger Sep-Paks (Millipore) under vacuum to achieve preconcentration factors of over 30-fold, leading to detection limits of ca. 10 ng/L (20). Freeman et al. described a rapid, automated preconcentration system that employed a small anion-exchange column (Bio-Rad AG 1-X8) in the injection valve of a flow injection system; elution of phosphate preconcentration from 2.9 ml of sample resulted in a detection limit of ca. 3 nM (0.1 μg P/L) and resolved from interfering silica (21). Such methods cannot be applied to high-ionic-strength waters; preconcentration of orthophosphate by *magnesium hydroxide–induced coprecipitation* (MAGIC) has been proposed as a suitable technique for use in marine waters (22). Precipitation using lanthanum nitrate was used by Stevens and Stewart to preconcentrate dissolved phosphorus species from 100 L of water to a final volume of 100 ml (23). However, a lengthy (3–4-day) filtration of the precipitated lanthanum phosphate was required to achieve this thousand-fold preconcentration, and removal of iron and other cations by the use of a cation-exchange resin was also necessary.

Reverse osmosis has been used as a concentration technique for dissolved organic phosphorus in freshwaters (24). Prefiltered water (100 L) was concentrated to 2.5 L prior to analysis of the >300-Da material that was retained. An intermediate cation-exchange step was required to prevent precipitation of Ca and Mg salts.

Other methods, involving preconcentration of phosphate as phosphomolybdate or phosphomolybdenum blue, are described later.

D. Digestion

The determination of total phosphorus, total filterable phosphorus, and condensed phosphorus necessitates predigestion of the water sample prior to detection of orthophosphate (see the analytical schema in Fig. 2). Complete conversion of particulate and filterable components requires conditions that are conducive to the dissolution of phosphate mineral phases, hydrolysis of phosphate esters, and oxidation of organic phosphorus species. Numerous methods have been proposed, but whichever procedure is selected for the determination of TP or TFP, the digestion efficiency should be assessed by using a range of appropriate organic or condensed phosphorus model compounds and standard reference materials. A range of suitable model compounds

for this purpose has been suggested by Kérouel and Aminot (25).

1. Thermal Digestion Methods

a. Wet Chemical Digestion.   Wet chemical digestion involving peroxydisulphate alone (26) or acidified peroxydisulphate (27) is perhaps the most widespread method for determining TP. However, other, more rigorous digestion procedures developed for sediment digestion may be necessary, because incomplete digestion has been reported using peroxydisulphate (28). Nitric-sulphuric acid, or nitric-sulphuric-perchloric acid digestion may prove necessary if preliminary digestion efficiency testing reveals the peroxydisulphate digestion procedures to be inadequate (10). Digestion may be performed at ambient pressure, or at elevated pressure and temperature using a pressure cooker or autoclave. Lambert and Maher (29) compared autoclave peroxydisulphate and nitric-sulphuric acid digestion methods for determination of TP in waters with turbidities of up to 200 units (NTU). In the most turbid waters, recovery of TP at >100 μg P/L was incomplete using the peroxydisulphate method, and they recommend dilution to ca. 100 μg P/L to overcome this problem.

b. High-Temperature Combustion and Fusion.   As alternatives to the wet chemical methods just described, high-temperature combustion with magnesium sulphate followed by acid leaching (30) or high-temperature fusion with magnesium nitrate have been proposed (31). The latter method has been shown to decompose phosphonates, which are quite refractory (25).

c. Microwave Digestion.   A number of workers have reported the use of microwave heating, which was performed in both batch (32) and online flow injection modes (33–35). Williams et al. (34) overcame the observed low digestion efficiency of condensed phosphate by the addition of a hydrolytic enzyme.

2. Ultraviolet Photo-Oxidation

Ultraviolet photo-oxidation may be employed to mineralize organic phosphorus to phosphate prior to detection, and this has been the subject of a comprehensive review by Golimowski and Golimowska (36). Such UV photo-oxidation may be performed either in batch mode, using a high-wattage UV source and a quartz reactor vessel (37,38), or in a continuous-flow mode, using either a quartz or a Teflon photoreactor (39). Batch UV radiation systems usually involve the use of high-wattage UV lamps (ca. 1000 W) and extended irradiation times. Under these conditions, condensed phosphates are hydrolyzed, but this is almost certainly an artifact of the elevated temperature and gradual acidification of the sample as peroxydisulphate degrades to form sulphuric acid. Solórzano and Strickland (40) have noted that UV photo-oxidation alone is insufficient to convert condensed phosphates to orthophosphate, and they have suggested that the use of UV photo-oxidation provides a basis for discrimination between the organic and condensed phosphorus fractions.

Photo-oxidation of organic phosphorus may be performed by UV irradiation of the untreated sample, but it is more common for hydrogen peroxide, potassium peroxydisulphate, ozone, or other oxidizing agents that enhance the oxidation process to be added.

When H_2O_2 is exposed to UV light, it forms hydroxyl radicals:

$$H_2O_2 + h\nu \rightarrow 2 \; OH^{\cdot}$$

The hydroxyl radical is among the strongest oxidizing agents found in aqueous systems (41), and these initiate radical chain reactions with organic substances present, resulting in mineralization of the sample.

Photo-oxidation using peroxydisulphate also produces hydroxyl radicals and oxygen, by the following route:

$$S_2O_8^{2-} \rightarrow 2SO_4^{\cdot}$$

$$SO_4^{\cdot} + H_2O \rightarrow HSO_4^- + OH^{\cdot}$$

$$S_2O_8^{2-} + OH^{\cdot} \rightarrow HSO_4^- + SO_4^{\cdot} + \tfrac{1}{2}O_2$$

$$SO_4^{\cdot} + OH^{\cdot} \rightarrow HSO_4^- + \tfrac{1}{2}O_2$$

Many organic compounds can be converted to carbon dioxide using long-wavelength UV (black-light lamp) and TiO_2 as a catalyst. Excitation of an electron from the valence band (v) into the conduction band (c) creates an electron–hole pair, which may then react with, e.g., oxygen adsorbed to the TiO_2 surface to form radicals such as $O_2^{\cdot-}$ and $OH^{\cdot}$. This approach has been applied to the determination of dissolved organic carbon, but it has also been shown to mineralize organic phosphates (42,43).

3. Combined Thermal Hydrolysis and Photo-Oxidation Digestion

In order to determine the *total* phosphorus concentration in water, the digestion process must involve both oxidative and hydrolytic processes in order to hydrolyze P—O—P linkages (e.g., polyphosphates) and to oxidize phosphoesters and C—P compounds to inorganic phosphate. For example, in an online TP digestion system that involved both thermal digestion and UV photo-oxidation (44), it was found necessary to use a mixture of perchloric acid and peroxydisulphate to form Caro's acid (H_2SO_5) in order to obtain high recoveries of both organic and condensed phosphorus. Caro's acid produces both sulphuric acid and hydrogen peroxide on decomposition (45):

$$H_2SO_5 + H_2O \rightarrow H_2SO_4 + H_2O_2$$

It was observed that the ability to digest phosphorus decreased with time (days) as the hydrogen peroxide formed underwent decomposition, but also that the efficiency of thermal condensed phosphate digestion did not show any decrease with time because of the stability of the sulphuric acid.

III. MEASURES OF BIOAVAILABLE PHOSPHORUS

A. Algal Bioassay

The analysis of phosphorus species in natural waters is driven largely by a need to assess the likely potential for eutrophication, and a number of phosphorus analysis parameters have been employed as estimators of bioavailable phosphorus (BAP). Traditionally, BAP has been determined by the use of algal bioassays (46). However, these are slow (7–21 days), labor intensive, susceptible to large statistical variability, and relatively insensitive (see Fig. 3). For this reason, numerous workers have sought to replace algal bioassay by some chemical parameter that will provide a rapid and more convenient means of determining the bioavailable phosphorus concentration.

B. Total Phosphorus and Filterable Reactive Phosphorus

Because of the time and labor involved in algal assays, FRP (0.45 or 0.2 μm) has often been used as a de facto measure of *ready* BAP and TP, which includes the particulate, condensed, and organic phosphorus components, as a measure of *potential* BAP. However, Shalders et al. found, in a comparison study, that there was little organic or condensed P that was readily hydrolyzable and that could be utilized by *Anabaena circinalis*, which had not already been hydrolyzed and included in the FRP measurement (47). Other workers have found, however, that some of the filterable P is quite refractory and hence is not bioavailable. For example, an average 22% of filterable *unreactive* phosphorus (= TFP − FRP; see Fig. 2) and 5% of total particulate phosphorus were not bioavailable to *Selenastrum capricornutum* during bioassays conducted over 2–4 weeks. On this basis, TP alone was not considered an adequate criterion for use in eutrophication control measures of domestic sewage discharge (48).

C. Iron Oxide Adsorption Methods

An alternative approach to bioassay and chemical extraction techniques has been the use of iron-strip adsorption techniques used in soil science to measure plant available phosphorus (49). This approach involves the equilibration of dissolved and particulate-bound phosphorus with an Fe oxide–impregnated paper strip over a period of hours (Fig. 4), followed by acid leaching and spectrophotometric analysis. When applied to the determination of exchangeable P in waters (50,51), these tests have been shown to be well correlated with BAP determined by algal bioassay (52). However, this technique is not without ambiguity. Coadsorbed organic phosphorus may also be hydrolyzed at the Fe oxide surface (53,54) or conceivably during the acid leaching and colorimetry stages of the process (55), leading to an overestimation of the

amount of BAP. Recent research by Dils and Heathwaite also suggests that the technique may be confounded when applied to BAP determination in turbid waters because of adherence of particulate material to the paper strips during the equilibration phase (56).

An alternative approach to iron-impregnated strips is the use of diffusive gradients in thin films (DGT). Phosphate diffuses through a thin polyacrylamide gel and is bound in a second gel layer containing Fe oxide (57). After in situ deployment, the gels are sectioned, leached, and analyzed for molybdate-reactive phosphorus. Using the measured gel concentration and a knowledge of the diffusion coefficient of phosphate in the gel, the bulk solution phosphate concentration can be calculated from Fick's first law of diffusion. This enables a time-integrated measurement of phosphate concentration and could also be used to determine sediment–water fluxes of phosphate.

D. Enzymatic Methods

Algae and bacteria are known to exude exocellular alkaline phosphatase, which is thought to facilitate utilization of otherwise-unavailable dissolved and particulate organic phosphorus species. This phenomenon

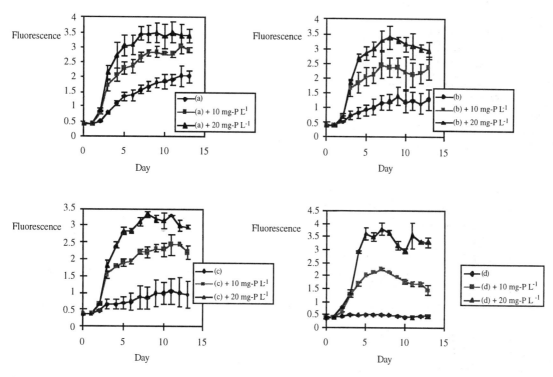

Fig. 3 The growth of *Selenastrum capricornutum* culture in the standard algal assay bottle test with (a) adenosine 5′-triphosphate, (b) sodium tripolyphosphate, (c) DL—glycerol phosphate, or (d) phytic acid as the only phosphorus source.

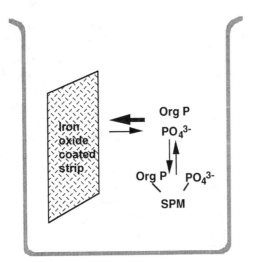

Fig. 4 Diagrammatic representation of the iron-strip method for the measurement of suspended particulate matter (SPM) and desorbable and free orthophosphate phosphate. Other forms of P, e.g., organic phosphorus (Org P), may also be desorbed from SPM or directly adsorbed from the water column.

has been used as the basis for another group of analytical procedures, aimed at determining bioavailable phosphorus in both the dissolved fractions. Alkaline phosphatase has been used in both soluble (58,59) and immobilized forms to hydrolyze filterable organic phosphorus in natural waters, and the resultant orthophosphate has been detected along with the FRP in a flow injection system (60). Use of the soluble enzyme technique was largely discarded because of product inhibition of the alkaline phosphatase by orthophosphate already present in the water. However, this is not problematic when the enzyme is used in the immobilized form in a flow injection configuration, and the technique has been successfully applied to the determination of alkaline phosphatase hydrolyzable phosphorus (APHP) in a range of natural waters and wastewaters (61). A surprising outcome of this work was the apparent paucity of APHP in the dissolved fraction alone, even when there was appreciable TFP. It is thought that this was due to the rapid hydrolysis of dissolved alkaline phosphatase substrates under normal conditions. Use has also been made of the enzyme phytase to hydrolyze organic phosphates in waters (62,63). However, it has also been observed that there is an appreciable amount of enzyme-available material in sediments, which is potentially bioavailable if released from the sediments because of change in redox, salinity, microbial, bioturbation, or other conditions.

IV. ANALYSIS METHODS: HISTORICAL, RECENT, FUTURE

A. Classical Methods of Analysis

Gravimetric methods of analysis were classically used for analysis of phosphate. Phosphate was directly precipitated as magnesium pyrophosphate, magnesium ammonium phosphate hexahydrate, ammonium phosphomolybdate, or nitratopentammine cobalt phosphomolybdate (64). Alternatively, phosphate could be determined volumetrically by titration of ammonium phosphomolybdate with sodium hydroxide. However, the poor sensitivity of these methods precludes their use for all but the most phosphate-enriched waters or wastewaters.

B. Colorimetry/Spectrophotometry

1.   Direct Photometry, Based on Formation of Phosphomolybdate or Phosphomolybdenum Blue

The determination of phosphate is most commonly based on the formation of the heteropoly acid, 12-molybdophosphoric acid (12-MPA) under acidic conditions. In nitric acid, the reaction is thought to be:

$$H_3PO_4 + 6\ Mo(VI) \rightleftharpoons (12\text{-}MPA) + 9H^+$$

while in sulphuric acid, the stoichiometry is:

$$H_3PO_4 + 5\ Mo(VI) \rightleftharpoons (12\text{-}MPA) + 7H^+$$

The absorbance of the 12-MPA dimer may then be measured, or it may be reduced to form the highly colored phosphomolybdenum blue (PMB), using a variety of reductants. The optimum wavelength and sensitivity are a function of both molybdate and acid concentrations (65,66). Because molybdenum blue may also form through the direct reduction of Mo(VI) if the pH is 0.7 or less, even in the absence of phosphorus, the acid normality is kept in the range of 0.3–0.5 N (65). The selectivity of this reaction for phosphate is also highly pH dependent, and again the reaction acidity must be strictly controlled.

a.   Detection as Unreduced Vanadomolybdophosphoric Acid.   In the presence of ammonium metavanadate and acidic ammonium molybdate, phosphate forms vanadomolybdophosphoric acid. The absorbance of this yellow-colored product is commonly measured at 470 nm. The sensitivity is generally less than that for reduced PMB methods, but it is quite tolerant of interfering ions, and is suitable for monitoring wastewaters and contaminated waters. A detection limit of

200 μg P/L (1-cm cell) has been reported for this method (10).

  b. Detection as Reduced 12-MPA (= Phosphomolybdenum Blue). Better sensitivity can be achieved if 12-MPA is reduced to form phosphomolybdenum blue. A wide variety of reductants have been employed for this purpose, including: tin + copper sulphate + HCl (66,67), Sn(II) chloride (68), Sn(II) chloride and hydrazine sulphate (69), 1-amino-2-naphthol-4-sulphonic acid (70), ascorbic acid, and potassium antimonyl tartrate (71).

While the $SnCl_2$ reduction method gives very sensitive results, it is susceptible to a salt interference and has only a short-lived colored product. For this reason, the ascorbic acid method is preferred to the Sn(II) method; however, longer color development time is necessary when ascorbic acid alone is used. The Murphy and Riley (71) method introduced the use of antimonyl tartrate, which catalyzes the reduction step, suppresses the interference from silicate, and avoids problems of chloride interference. Consequently this method is used as the basis for many batch and automated techniques in current use (cf. Table 2).

  c. Detection as 12-MPA Ion-Association Complexes. Phosphomolybdate forms strong ion-association complexes with basic dyes at low pH. For example, the sensitivity of a method based on spectrophotometric determination of the 12-MPA–malachite green complex (77,94) was approximately 30 times that of a reduced phosphomolybdate determination (95). Other dyes used for this purpose include Saffranin, brilliant green, fuchsine red, methylene blue, methyl violet, rhodamine B (9), and crystal violet (96). Surfactants such as polyvinyl alcohol are frequently used to avoid precipitation of the ion-association complex.

  d. Solvent Extraction of Phosphomolybdenum Blue or Phosphomolybdate Ion-Association Complexes. Enhanced sensitivity may be achieved through solvent extraction of phosphomolybdenum blue or phosphomolybdate ion-association complexes prior to spectrophotometric measurement. For example, extraction of PMB with iso-butanol enabled a detection limit of 0.2 μg P/L to be achieved (97). Motomizu et al. (98) extracted the phosphomolybdate–malachite green ion pair into a mixture of toluene and 2-methylpentane-2-one to obtain a detection limit of 0.1 μg P/L. Other solvents used include iso-butanol + benzene (99) and ether (100). Solid-phase extraction of PMB has also been reported (101).

  e. Solvent Extraction of Phosphomolybdate. Improved sensitivity may also be achieved by solvent extraction of phosphomolybdate without the formation of PMB. Sugawara and Kanamori (102) used n-butanol/chloroform to extract phosphomolybdate. Molybdenum was then determined spectrophotometrically with thiocyanate after decomposition of phosphomolybdate, to give a detection limit of 0.08 μg/LP.

  f. Interferences in Photometric Techniques Based on Formation of Phosphomolybdate and Phosphomolybdenum Blue. While methods based on the formation of phosphomolybdenum blue or its ion-association complexes are the most commonly used methods for phosphorus determination, they may be susceptible to interference from a number of sources.

Sjösten and Blomqvist (103) have reported that the rate of formation of phosphoantimonyl blue was reduced by decreasing temperature and decreasing phosphate concentrations. At low temperatures ($<5°C$) and concentrations (5 μg P/L), reaction times of ca. 50 minutes were required to reach complete color development. The authors note that these effects may cause significant nonlinearity in the calibration of automated instruments (FIA, segmented-flow analysis) at low concentrations, or underestimation in samples that have not been allowed to reach ambient temperature prior to analysis.

Interferences Due to Hydrolysis of Labile Phosphorus Species. It has been shown that both the acid conditions used and the presence of molybdate (104,105) can enhance hydrolysis of dissolved organic and condensed phosphates to give an overestimate of orthophosphate. Similarly, colloidal phosphates in the filterable fraction may be molybdate reactive, which again will lead to an overestimation of the orthophosphate concentration (106). Rigler (107) observed that this overestimation of the true orthophosphate concentration may be as much as 10–100 times the true concentration of orthophosphate. In attempts to avoid these hydrolytic effects, a "6-second extraction method" was developed in which phosphomolybdate formed was rapidly removed from the acidic environment (108) or excess molybdate was complexed with a citrate–arsenite reagent (109).

Interferences in the Formation of Molybdenum Blue Species. Silicate, arsenate, and germanate also form heteropoly acids, which on reduction yield molybdenum blue species with similar absorption maxima (110). This positive interference in the determination of phosphate is particularly pronounced for silicate because of its relatively high concentration in many waters. However, the formation of silicomolybdate may be suppressed by the addition of tartaric or oxalic acid to the molybdate reagent (111). [If the organic acid is added *after* the formation of the heteropoly acid, the phosphomolybdate is destroyed; this is used as the basis for the determination of silicate in the presence of phosphates (111).] Kinetic discrimination between phosphate and silicate, arsenate, and germanate is also possible because of the faster rate of formation of phosphomolybdate. Thus, the widely adopted Murphy and

Riley method employs a reagent mixture of acidic molybdate and antimonyl tartrate (71) at concentrations known to enhance the kinetics of phosphomolybdate and suppress the formation of silicomolybdate.

Fluoride concentrations of >100 mg/L were also shown to inhibit the formation of phosphomolybdate (112), but this effect was shown to be lessened at higher silicate concentrations.

Table 2 Examples of Methods for Determination of Phosphorus Species, with Indicative Detection Limits

Technique/method	Species detected	Typical detection limit		Comments	Ref.
		$\mu g\ L^{-1}$ P	μM		
Molecular spectroscopic techniques					
Visible photometry	MRP	150	4.8	10-mm cell	10
Phosphomolybdenum blue–batch method		10	0.32	100-mm cell	
Visible photometry	MRP	10	0.32		72
Phosphomolybdenus blue–FIA method	MRP	0.4	0.013	Lower detection limit possible; detection limit defined by extent of preconcentration used	21
Visible photometry Phosphomolybdenum blue–FIA ion-exchange preconcentration technique					
Visible photometry Phosphomolybdenum blue–FIA reagent-injection technique	MRP	12	0.39	In situ monitoring system, LED photodiode detector	73
Visible photometry Phosphomolybdenum blue–FIA reagent-injection technique	MRP	2.5	0.08	Shipboard monitoring system, LED photodiode detector	74
Visible photometry Phosphomolybdenum blue–segmented continuous-flow system	MRP	0.4	0.013	50-mm-path-length detection cell	75
Visible photometry Phosphomolybdate–malachite green ion-pair FIA method	MRP	10	0.32		76
Visible spectrophotometry Phosphomolybdate–malachite green ion-pair FIA solvent extraction method	MRP	0.1	0.003		77
Long-path-length capillary spectrophotometry	MRP	0.03	0.001	Off-line color development	78
Thermal lens spectroscopy	MRP	0.005	1.6×10^{-4}		79
Fluorescence quenching of phosphomolybdate	MRP	2	0.065	Quenching of Rhodamine 6G by phosphomolybdate; fluorescence-FIA method	80
Atomic spectroscopic techniques					
Inductively coupled plasma–atomic emission spectrometry	Total phosphorus	200	6.5	For most sensitive emission line	81
Inductively coupled plasma–atomic emission spectrometry-F1	Total phosphorus + MRP	200	6.5	With 200-μl injection in FIA mode	82
		70	2.3	Continuous-aspiration mode	

Table 2 Continued

Technique/method	Species detected	Typical detection limit		Comments	Ref.
		$\mu g\ L^{-1}$ P	μM		
Inductively coupled plasma–mass spectrometry	Orthophosphate	8	0.26	Liquid chromatographic separation of model phosphate compounds with ICPMS detection	83
Electrochemical techniques					
Potentiometry-FIA	Orthophosphate + tripolyphosphate	310	10	Indirect detection using Pb(II) electrode—better selectivity for SO_4^{2-}	84
Enzyme electrode	Orthophosphate	775	25	Biosensor based on glucose 6′phosphate inhibition of hydrolysis by potato acid phosphatase; high selectivity for F^-	85
Enzyme electrode-FIA	Orthophosphate	3	0.1	Amperometric detection of H_2O_2 produced by interaction of phosphate with co-immobilized nucloside phosphorylase and xanthine oxidase	86
Voltammetry-FIA	MRP	20	0.65	Amperometric detection of phospho-molybdate species	87
Voltammetry	MRP	9	0.29	Differential pulse polarographic detection of catalytic reduction of perchlorate or nitrate by solvent extracted phosphomolybdate	88
Separation techniques					
High-performance liquid chromatography	$H_2PO_4^-$	750	24	HPLC determination of hypoxanthine produced by nucleoside phosphorylase catalyzed reaction of phosphate with inosine	89
Ion chromatography	$H_2PO_4^-$	14.7	0.47	Unsuppressed IC - Indirect UV detection, 1 mL injections	90
Ion chromatography	$H_2PO_4^-$	2	0.06	Suppressed IC, conductivity detection, concentrator column	91
Capillary electrophoresis	$H_2PO_4^-$	0.6	0.02	Preconcentration by isotachophoresis, with conductimetric detection	92
Capillary electrophoresis	$H_2PO_4^-$	0.3	0.01	Electromigrative preconcentration, UV detection	93

Source: Modified from Ref. 3.

Negative interferences in the tin(II) chloride reduction method may also be caused by the presence of higher concentrations of iron(III), aluminium, calcium, and chloride (113). The Fe, Al, and Ca interferences are presumably due to competitive complexation of the phosphate, while that for chloride is probably due to inhibition of the phosphomolybdate reduction. The chloride interference in this method is particularly problematic, especially for the determination of phosphate in marine and estuarine waters, and for this reason the ascorbic acid reduction method of Murphy and Riley (71) is often favored.

2. Indirect Photometry

A method for the indirect determination of phosphate has been reported (100). This is based on the absorbance measurement of displaced chloroanilate at 530 nm following the formation of lanthanum phosphate:

$$La(C_6Cl_2O_4)_2 + PO_4^{3-} \rightarrow LaPO_4^{3-} + C_6Cl_2O_4$$

The sensitivity of this technique is poor, however, with a reported detection range of 3–100 mg P/L, making it suitable only for wastewater analysis.

C. Atomic Spectroscopic Methods

An indirect atomic absorption spectrometry (AAS) method based on the measurement of molybdenum after solvent extraction of phosphomolybdate has been reported (114). A more recent variation of this method involved flotation of the malachite green–phosphomolybdate ion pair at an aqueous–diethyl ether interface (115). After dissolution in methanol, the molybdenum was determined using flame AAS (nitrous oxide flame) at 313.26 nm. The method was successfully applied to the measurement of seawater containing ca. 40 μg P/L.

Inductively coupled plasma atomic emission spectrometry (ICP-AES) has also been applied to the analysis of phosphorus species. Manzoori et al. (82) demonstrated a flow injection system that enabled the colorimetric determination of total reactive phosphorus (using unreduced phosphovanadomolybdate, $\lambda_{max} = 470$ nm) prior to aspiration into an ICP-AES system where TP was measured. A detection limit of ca. 200 μg P/L was achieved for the TP measurement using the 177.49-nm phosphorus line, and the method was applied to the analysis of wastewaters. Interference from background argon emission lines tends to limit the sensitivity and thus to limit the application of the ICP-AES technique to the analysis of wastewaters. How-

ever, the increasing availability of high-resolution inductively coupled plasma mass spectrometry (ICP-MS) systems is likely to result in wider use of this technique for TP determination.

D. Chromatographic Methods

1. High-Performance Liquid and Ion Chromatography

a. Orthophosphate. Phosphate is commonly determined by ion chromatography, both in suppressed and unsuppressed modes. The sensitivity of the conductivity detection techniques commonly used is generally inadequate for direct application to the analysis of pristine waters (cf. Table 2), and some form of preconcentration is required (91).

b. Condensed Phosphates. Anion-exchange chromatography and ion-exchange chromatography have been used extensively for the separation and quantitation of condensed phosphates. Because phosphate is a poor UV chromophore, common practice has been to use large anion-exchange columns and to collect fractions for subsequent acid hydrolysis and detection as MRP (116). However, the use of HPLC/ion chromatography with postseparation hydrolysis and detection via an FIA (117,118) has the advantages of both speed of analysis and sensitivity. Typical separations obtained from a hyphenated system of this type are shown in Fig. 5.

c. Organic Phosphates. Interest in characterizing organic phosphorus present in natural waters has prompted the development of ion chromatographic separation systems for compounds such as inositol phosphates. Online UV photo-oxidation has been utilized for oxidation and subsequent detection of these organic phosphate species (12,119,120).

2. Gel Filtration/Size Exclusion Chromatography

Separation using gel filtration gained popularity during the 1970s and 1980s as a means of separating high- and low-molecular-weight phosphorus fractions and as a means of estimating "true" orthophosphate concentrations (23,121–125). Most efforts involved the use of large separation columns, long elution times, and fraction collection and off-line digestion/digestion to measure total or reactive phosphorus, and as such they were unsuitable for routine monitoring applications. However, an online postcolumn flow injection detection system for detection of organic P species has also been described (126).

(a)

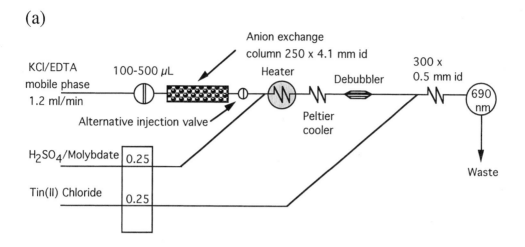

(b)

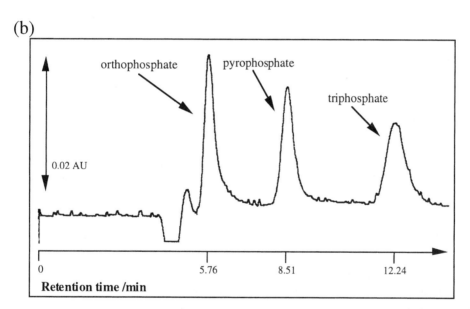

Fig. 5 (a) Ion-exchange chromatography FIA system for the separation of polyphosphates. (b) Optimized separation of ortho-phosphate (50 μg P/L), diphosphate (50 μg P/L), and triphosphate (50 μg P/L) with an injection volume of 500 μL. (Adapted from Ref. 118.)

Potential difficulties associated with the use of gel filtration include early elution due to anion exclusion (minimized by the use of eluent with ionic strength over 0.01), late elution due to hydrophobic interactions, and specific adsorption.

3. Capillary Electrophoresis

Capillary electrophoresis (CE) techniques have been applied to orthophosphate analysis. While generally offering much faster separations of anions in waters, CE with conventional UV detection suffers from a lack of

sensitivity. However, recent advances in on-capillary preconcentration using isotachophoresis (92,93) have enabled sub-μg/L detection limits to be achieved in high-ionic-strength matrices, and this approach is a promising one for water analysis.

E. Automated Methods

Automated methods are clearly favored where there are large numbers of samples to be analyzed, when the unit processes, such as digestion and separation, are slow, and where the analytical measurements must be made online and unattended. While this approach is still widely used, there has been a tendency to use flow injection analyzers for the same applications.

1. Segmented-Flow Analysis

Continuous-flow techniques have been widely used for automated phosphorus analysis of waters since the introduction of the segmented continuous-flow analysis systems in the 1950s. Segmented-flow manifolds described include those that are suitable for the determination of phosphorus in water in the presence of high silica (127), for highly sensitive detection of FRP in the presence of mercuric chloride preservative (75), and for the determination of total filterable phosphorus (128).

2. Flow Injection Analysis

a. Filterable Reactive Phosphorus. The analysis of FRP by FIA using the Murphy and Riley ascorbic acid reduction chemistry has been reported. Better sensitivity may be achieved using $SnCl_2$ because the reduction reaction kinetics are faster (21,129), but this may result in susceptibility to chloride interference.

A recent advance in the flow analysis of FRP is the use of a capillary-flow technique that involves the use of electro-osmotic flow for the delivery of reagents in a miniaturized flow injection manifold etched into a glass plate (130). This approach, while still in its infancy, holds considerable promise as the basis for miniature automated water monitoring devices.

b. Total Phosphorus and Total Filterable Phosphorus. The automation of digestion processes is highly desirable, and a number of flow injection digestion techniques suitable for the detection of organic and total phosphorus have been described. Online total filterable phosphorus measurement systems that use strong acids and oxidants and thermal- (131–134) or microwave-assisted digestion (33–35) have been shown to be effective. Other methods involving the use of photooxidation (39) have been described, and a combined UV–thermal system for the determination of TP has been demonstrated and employed for online monitoring (44). Because the latter systems involve photo-oxidation in the presence of high concentrations of peroxydisulphate, oxygen bubbles are randomly generated and must be removed online by the use of either a membrane degasser or a hydrophobic hollow-fiber membrane.

3. Automated Batch Analyzers

A novel approach to the determination of TP has been described by Dong and Dasgupta (135). An automated microbatch analyzer that comprised a sealed digestion vessel containing a fiber-optic light-emitting diode detector system and a number of reagent addition and waste lines was employed for high-temperature digestion of wastewaters. Total phosphorus measurements took approximately 9 minutes per sample and gave results comparable to results obtained using the ASTM block or autoclave digestion techniques.

V. DETECTION METHODS: HISTORICAL, RECENT, FUTURE

A. Photometric Detection

The chemical basis for photometric phosphomolybdate and phosphomolybdenum blue methods is well established. For the analysis of most natural waters, the use of 4-cm or even 10-cm cells is necessary to achieve adequate sensitivity when batch techniques and conventional filter photometers or spectrophotometer are employed (10) (see Table 2). However, increasing use is being made of solid-state detectors, especially in low-cost flow injection systems (21,73,136). These involve the use of a light-emitting diode as the red light source (usually with a broad-band emission peak centered at about 650 nm) and a photo diode as the light detector, in lieu of the more conventional photomultiplier. In some cases the light path is across the FIA flow tubing, and the optical path length through the sample may be as little as 1 mm. But despite this, analytical sensitivity is high, because of the inherently high reproducibility of the sample and reagent mixing and transport processes, and the ability continually to reference the light output from the light-emitting diode (137).

In an attempt to achieve very high-sensitivity phosphate analysis, some researchers have investigated the use of long-path capillary detectors (78,138,139) or even thermal lensing techniques (79,140). However, while these detection systems gave considerably improved sensitivity (see Table 2) compared with conventional batch or flow methods, their application to routine phosphate analysis is probably limited at this stage.

The use of gel-phase photometry has also been investigated as a means of improving sensitivity. In both cases this involved the adsorption of PMB or phosphomolybdate–malachite green onto Sephadex gel

(141,142). The approach was rapid (20 samples per hour) and enabled detection of phosphorus at <1 μg/L.

B. Photoluminescent Detection

A number of authors have described the detection of orthophosphate based on measurement of the quenching of rhodamine fluorescence by phosphomolybdate (80,143,144). Detection limits of ca. 0.1 μg P/L have been reported (Table 2), and while this approach offers little enhancement in selectivity, it is potentially more sensitive than spectrophotometry.

An indirect method for the detection of phosphate that involves the demasking of the Al–morin complex by PO_4^{3-} has also been described (100); this approach has been utilized in the detection of phosphorus oxyacids separated by ion chromatography (145).

The use of chemiluminescence for the determination of orthophosphate species with greater sensitivity and selectivity is an area that warrants further investigation. The immobilized enzyme method of Kawasaki et al. (146), which involves the production of peroxide and subsequent determination with luminol, may be a suitable approach.

C. Electrochemical Detection

Potentiometric methods for the detection of phosphate include both direct methods using phosphate-ion-selective electrodes (147,148) and indirect methods using lead (84,149), calcium (for phosphate measurement in detergents) (150), cadmium (151), silver (152), or cobalt (153) electrodes. These methods, which have detection limits in the range of 30–300 μg P/L are generally too insensitive for all but wastewater analysis.

While potentiometric enzymatic electrodes for the detection of phosphate have been developed (85,154), no application has been made of these to water analysis because of the relatively poor sensitivity (cf. Table 2).

Detection of phosphomolybdate by voltammetric techniques based on the determination of the phosphomolybdate moiety have been described by Fogg and others (87,155,156). While the approach is convenient and rapid (Table 2), it is less sensitive than photometric detection of PMB and is similarly nonselective for orthophosphate.

A sensitive enzyme electrode for the detection of phosphate has been described by D'Urso and Coulet. It is based on the amperometric detection of hydrogen peroxide produced by membrane-coimmobilized nucleoside phosphorylase and xanthine oxidase (86). This

electrode is quite sensitive for phosphate (Table 2) and has promise as a fast, selective, and portable means of determining phosphate in water.

D. Enzymatic Detection

The use of methods not involving phosphomolybdate has been proposed as an approach to the determination of true inorganic phosphate concentration, because organic or condensed P species in the sample are not subjected to hydrolytic conditions. Pettersson (157) used the inhibition by orthophosphate of alkaline phosphatase hydrolysis of substrates such as 4-methylumbelliferyl phosphate as the basis for determination of phosphate. As little as 0.1 μg/L P could be detected in natural waters by measuring the decrease in fluorescence of the product of enzymatic hydrolysis. Interference was not caused by trace metals or oxyacids such as arsenate or silicate, but did not occur when phosphomonoesters and other organic phosphates were present.

Stevens (158) also applied an enzymatic technique to the determination of orthophosphate in the range of 5–200 μg L^{-1} P in natural waters. The basis for this determination was the reaction of orthophosphate with glyceraldehyde-3-phosphate to produce 1,3-diglycerophosphate in the presence of glyceraldehyde-3-phosphate dehydrogenase and oxidized nicotinamide-adenine dinucleotide. Reduced nicotinamide-adenine dinucleotide (NADH) formed by this process reacts with a tetrazolium salt in the presence of meldola blue to produce formazin, which was detected spectrophotometrically and used as a measure of orthophosphate. Both of these enzymatic orthophosphate methods gave consistently lower results than phosphomolybdate-based FRP when applied to natural waters, which was ascribed to acidic molybdate hydrolysis of labile P species in the latter technique.

E. Development of Portable and In Situ Analysis Systems

The ability to perform on-site analysis at high frequency is highly desirable because it obviates problems associated with the loss of sample integrity due to hydrolysis, microbial action, or adsorption during transport and storage. The deployment of portable or unattended analysis systems would permit discharge monitoring to be performed with greater frequency and reliability than is possible with hand sampling and off-line laboratory analysis. Process control of wastewater treatment plants could be enhanced by the improved process monitoring that these systems could provide. In response to this perceived need, a number of researchers have developed in situ or remote monitoring systems suitable for water and wastewater analysis of phosphate (73,159–161). While most of these have only ever been demonstrated as research prototypes, some are now being marketed commercially, including a submersible flow injection analysis (FIA) system developed at the University of Plymouth (now produced by Chelsea Instruments, Surrey, UK) and a flow injection system for total phosphorus designed by Monash University (44) and produced by Greenspan Technology (Warwick, Queensland, Australia).

VI. APPLICATIONS IN WATER ANALYSIS

Much of the interest in determining phosphorus in waters stems from its crucial role in the eutrophication process or in monitoring wastewaters that may contribute to this process. Table 3 shows indicative P concentration ranges for waters of various trophic classifications. What is obvious is that the analytical techniques used to determine P in even eutrophic waters must be quite sensitive. In pristine waters, very low concentrations are observed, e.g., 1 μg P/L or less of FRP (163), and it is generally only in polluted waters and wastewaters that concentrations in the mg/L range are found. The nature and origin of the sample therefore dictates the techniques that can be used. Table 2 shows typical detection limits for some of the techniques described, and this may be used as an approximate guide for selection of a method of analysis appropriate to the sample concentration.

A. Potable Water

Because phosphorus, as phosphates, are not deleterious to human health, guidelines, e.g., WHO (164), do not

typically list criteria for acceptable P concentrations for drinking water quality. While the presence of high phosphate concentrations in drinking water may be indicative of sewage contamination, more appropriate methods of detection, e.g., *E. coli*, are usually employed. However, the presence of high phosphorus concentrations in reservoir waters may lead to the occurrence of nuisance blooms of algae, which can release toxins and cause taste and odor problems, resulting in the need for expensive water treatment. Consequently, the measurement of phosphorus concentrations is usually performed on waters in the catchments or reservoir waters themselves as part of an overall nutrient management strategy. Both TP and FRP are commonly measured, and sensitive batch and automated photometric methods are most frequently used. The use of portable test kits designed for wastewater monitoring for monitoring potable waters is not recommended because of the potentially large errors that may occur.

B. Wastewaters

Total phosphorus, as a measure of the potentially bioavailable P, is most frequently used to monitor the compliance of wastewater discharges with license agreements, whereas FRP is more commonly used to measure the efficiency of P removal in biological and chemical processes. The ability to perform frequent or even online determination of these parameters provides the potential for improved process control. A number of flow injection (160, 161) and continuous-flow (e.g., Dr. Lange, Bran & Luebbe, Tytronics) systems for measurement of FRP or TRP have been developed for this purpose.

C. Brackish and Estuarine Waters

Analysis of brackish and estuarine waters, by virtue of their widely varying salinity, can be problematic, especially if tin(II) chloride reduction of phosphomolybdate is employed. Preparation of standards in a matrix of the same salinity as the sample matrix, sample salinity adjustment, or standard addition may be necessary to compensate for salinity errors. Conventional flow injection manifolds with sample injection for the determination of reactive phosphorus in estuarine waters are limited by the Schlieren, or refractive index (RI) effect, which can also cause major errors in quantitation. A simple flow injection analysis (FIA) manifold that obviates this RI error in reactive phosphorus measurement has been described (165). It involves the injection of acidic phosphomolybdate reagent into a

Table 3 Total Phosphorus Concentrations and Levels of Lake Productivity

General level of lake productivity	Total phosphorus (μg/L)
Ultraoligotrophic	<5
Oligo–mesotrophic	5–10
Meso–eutrophic	10–30
Eutrophic	30–100
Hypereutrophic	>100

Source: After Vollenweider, modified by Wetzel (Ref. 162).

carrier stream of sodium chloride of similar refractive index, which is then merged with sample (the salinity of which may vary widely from sample to sample) and tin chloride reductant. This approach also has the advantage of automatically compensating for any sample background color. Reactive phosphorus was measured in samples with salinities ranging from 0 to 34% using calibration standards prepared in deionized water, with a detection limit of 6 μg P/L. Salinity interference was suppressed by the use of a high-chloride carrier; an improved method based on ascorbic acid reduction has also been reported (166). Other techniques for refractive index correction have been reported that involve the use of dual-wavelength detection with the application of a correction algorithm (167–169) or the use of large injection volumes (170).

D. Seawater

The high salinity of seawater may give rise to a number of interferences, either in sample pretreatment or in the detection steps. For example, in the determination of TP in estuarine and marine waters, there is the potential problem of chlorine formation during digestion with peroxydisulphate:

$$K_2S_2O_8 + 2Cl^- \rightarrow 2K^+ + Cl_{2(g)} + 2SO_4^{2-}$$

This is not problematic if the sample is digested in an open vessel, where the chlorine is boiled off. If digestion is performed in a closed vessel in a microwave oven or autoclave, the chlorine is trapped and subsequently interferes in the detection process involving the ascorbic acid reduction step. This problem is readily avoided by introducing sodium sulphite into the reaction vessel (171).

Similarly, the determination of TP in samples containing large amounts of salt may be complicated by salt precipitation as the sample is evaporated during digestion. Under these circumstances, separate digestions should be performed to determine the particulate phosphorus and total filterable phosphorus and the TP determined as the sum of these (cf. Fig. 2) (10).

Enhanced sensitivity in the determination of reactive P in seawater has been achieved by extracting phosphomolybdenum blue from volumes of up to 1000 L onto a synthetic acrylic cation-exchange medium (Acrilan) using a 2-L volume of the resin fibers. Filterable unreactive P could be adsorbed to Fe(III) hydroxide–coated acrylic fibers in a similar manner (172). While the extraction efficiency of this technique was over 95%, it appears not to have been exploited to gain maximum sensitivity.

A flow injection system for the analysis of reactive phosphate in seawater was introduced by Johnson and Petty (173). This involved the concept of reverse or reagent injection FIA, which they showed to be inherently more sensitive than the conventional sample injection flow injection approach. A major advantage of the system was that it could be used for the underway analysis.

VII. CONCLUSIONS

While most water analysis for phosphate is laboratory based, it is predicted that the emergence of robust, sensitive, and commercially available portable and online instruments for the analysis of phosphate and TP will replace a major part of this analytical load. Such a move is likely to be enhanced by the development of sensitive phosphate-selective enzyme electrodes using amperometric detection, which would provide a viable and selective alternative to PMB spectrophotometry. Further advances towards miniaturized flow systems are also expected.

However, in the foreseeable future most small-to-medium-sized laboratories will continue to use either batch or automated PMB-based spectrophotometry (FIA, segmented continuous-flow analysis) techniques, with the emergence of ICP-MS as a possible alternative in larger laboratories.

The development of P-specific or higher-sensitivity detection systems for capillary electrophoresis and liquid chromatography is seen as essential if further developments in the speciation of aquatic phosphorus using these approaches is to occur.

ABBREVIATIONS

AAS	atomic absorption spectrometry
BAP	bioavailable phosphorus
Da	daltons
DGT	diffusive gradients in thin films
FCP	filterable condensed phosphorus
FIA	flow injection analysis
FOP	filterable organic phosphorus
FRP	filterable reactive phosphorus
HMWP	high-molecular-weight phosphorus
ICP-AES	inductively coupled plasma atomic emission spectrometry

ICP-MS	inductively coupled plasma mass spectrometry
MPA	molybdophosphoric acid
MRP	molybdate-reactive phosphorus
N	normality
NTU	nephelometric turbidity units
Org P	organic phosphorus
PMB	phosphomolybdenum blue
RI	refractive index
TFP	total filterable phosphorus
TP	total phosphorus
TRP	total reactive phosphorus

REFERENCES

1. VE McKelvey. Abundance and distribution of phosphorus in the Lithosphere. In: EJ Griffith, A Beeton, JM Spencer, DT Mitchell, eds. Environmental Phosphorus Handbook. New York: Wiley-Interscience, 1973, p 718.
2. NN Greenwood, A. Earnshaw. Chemistry of the Elements. Oxford: Pergamon Press, 1984.
3. ID McKelvie, D Peat, PJ Worsfold. Techniques for the speciation and quantification of phosphorus in natural waters. Anal Proc 32:437–445, 1995.
4. DE Armstrong. Analysis of phosphorus compounds in natural waters. In: M Halmann, ed. Analytical Chemistry of Phosphorus Compounds. New York: Wiley-Interscience, 1972, pp 744–769.
5. O Broberg, G Persson. Particulate and dissolved phosphorus forms in freshwater: composition and analysis. Hydrobiologia 170:61–90, 1988.
6. K Robards, ID McKelvie, RL Benson, PJ Worsfold, N Blundell, H Casey. Determination of carbon, phosphorus, nitrogen and silicon species in waters. Anal Chim Acta 287:143–190, 1994.
7. W Stumm, JJ Morgan. Aquatic Chemistry. 3rd ed. New York: Wiley, 1996.
8. RA Kimerle, W Rorie. Low-level phosphorus detection methods. In: EJ Griffith, AM Beeton, JM Spencer, DT Mitchell, eds. Environmentsl Phosphorus Handbook. New York: Wiley, 1973, pp 367–379.
9. O Broberg, K Pettersson. Analytical determination of orthophosphate in water. Hydrobiologia 170:45–59, 1988.
10. APHA-AWWA-WEF. Standard Methods for the Examination of Water and Wastewater. 18th ed. Washington, DC: American Public Health Association, 1992.
11. AC Rossin, JN Lester. An evaluation of some of the methods available for the preservation of condensed phosphates in samples of waste water. Environ Technol Lett 1:9–16, 1980.
12. DJ Halliwell. Speciation and analysis of condensed and inositol phosphates in wastewaters. PhD dissertation, Monash University, Melbourne, Australia, 1998.
13. D Lambert, W Maher, I Hogg. Changes in phosphorus fractions during storage of lake water. Water Res 26:645–648, 1992.
14. PM Haygarth, CD Ashby, SC Jarvis. Short-term changes in the molybdate reactive phosphorus of stored soil waters. J Environ Qual 24:1133–1140, 1995.
15. LA Clementson, SE Wayte. The effect of frozen storage on open-ocean seawater samples on the concentration of dissolved phosphorus and nitrate. Water Res 26:1171–1176, 1992.
16. W Hassenteufel, R Jagitsch, FF Koczy. Impregnation of glass surface against sorption of phosphate traces. Limnol Oceanogr 8:152–156, 1963.
17. Standards Australia, Standards New Zealand (ed). Australian/New Zealand Standard®. Water quality–Sampling. Part 1: Guidance on the design of sampling programs, sampling techniques and the preservation and handling of samples. Standards Australia and Standards New Zealand, 1998.
18. JG Stockner, ME Klut, WP Cochlan. Leaky filters: a warning to aquatic ecologists. Can J Fish Aquat Sci 47:16–23, 1990.
19. S Vigneswaran, R Ben Ain (eds). Water, Wastewater and Sludge Filtration. Boca Raton, FL: CRC Press, 1989.
20. L Camarero. Assay of reactive phosphorus at nanomolar levels in non-saline waters. Limnol Oceanogr 39:707–711, 1994.
21. PR Freeman, ID McKelvie, BT Hart, TJ Cardwell. A flow injection analysis method for determination of low levels of phosphorus in natural waters. Anal Chim Acta 234:409–416, 1990.
22. DM Karl, G Tien. MAGIC: A sensitive and precise method for measuring dissolved phosphorus in aquatic environments. Limnol Oceanogr 37:105–116, 1992.
23. RJ Stevens, BM Stewart. Concentration, fractionation and characterization of soluble organic phosphorus in river water entering Loch Neagh. Water Res 16:1507–1519, 1982.
24. BM Stewart, C Jordan, DT Burns. Reverse osmosis as a concentration technique for soluble organic phosphorus in fresh water. Anal Chim Acta 244:267–274, 1991.
25. R Kérouel, A Aminot. Model compounds for the determination of oroganic and total phosphorus dissolved in natural waters. Anal Chim Acta 318:385–390, 1996.
26. DW Menzel, N Corwin. The measurement of total phosphorus in seawater on the liberation of organically bound fractions by persulfate oxidation. Limnol Oceanogr 10:280–282, 1965.

27. ME Gales, EC Julian, RC Kroner. Method for quantitative determination of total phosphorus in water. J Am Wat Wks Ass 58:1363–1368, 1966.

28. D Zeigler, M Readnour. Comparison in water. Trans Miss Acad Sci 9:144–149, 1975.

29. D Lambert, W Maher. An evaluation of the efficiency of the alkaline persulphate digestion method for the determination of total phosphorus in turbid waters. Water Res 29:7–9, 1995.

30. L Solórzano, JH Sharp. Determination of total dissolved phosphorus and particulate phosphorus in natural waters. Limnol Oceanogr 25:754–758, 1980.

31. AD Cembella, NJ Antia, FJR Taylor. The determination of total phosphorus in seawater by nitrate oxidation of the organic component. Water Res 20:1197–1199, 1986.

32. L Woo, W Maher. Determination of phosphorus in turbid waters using alkaline peroxodisulphate digestion. Anal Chim Acta 315:123–135, 1995.

33. S Hinkamp, G Schwedt. Determination of total phosphorus in waters with amperometric detection by coupling of flow-injection analysis with continuous microwave oven digestion. Anal Chim Acta 236:345–350, 1990.

34. KE Williams, SJ Haswell, DA Barclay, G Preston. Determination of total phosphate in waste waters by on-line microwave digestion incorporating colorimetric detection. Analyst 118:245–248, 1993.

35. RL Benson, ID McKelvie, BT Hart, IC Hamilton. Determination of total phosphorus in waters and wastewaters by on-line microwave-induced digestion and flow injection analysis. Anal Chim Acta 291:233–241, 1994.

36. J Golimowski, K Golimowska. UV-photooxidation as pretreatment step in inorganic analysis of environmental samples. Anal Chim Acta 325:111–133, 1996.

37. FAJ Armstrong, PN Williams, JDH Strickland. Photooxidation of organic matter in seawater by ultraviolet radiation, analytical and other applications. Nature 211:481, 1966.

38. A Henriksen. Determination of total nitrogen, phosphorus and iron in freshwater by photo-oxidation with ultraviolet radiation. Analyst 95:601–608, 1970.

39. ID McKelvie, BT Hart, TJ Cardwell, RW Cattrall. Spectrophotometric determination of dissolved organic phosphorus in natural waters using in-line photo-oxidation and flow injection. Analyst 114:1459–1463, 1989.

40. L Solórzano, JD Strickland. Polyphosphate in seawater. Limnol Oceanogr 13:515–518, 1968.

41. J Hoigné, H Bader. Ozone and hydroxyl radical-initiated oxidations of organic and organometallic trace impurities in water. In: FE Brinckman, JM Bellama, eds. Organometallics and Organometalloids: Occurrence and Fate in the Environment. Washington, DC: American Chemical Society, 1978, pp 292–313.

42. RW Matthews, M Abdullah, GK-C Low. Photocatalytic oxidation for total organic carbon analysis. Anal Chim Acta 233:171–179, 1990.

43. GK-C Low, R Matthews. Flow-injection determination of organic contaminants in water using an ultraviolet-mediated titanium dioxide film reactor. Anal Chim Acta 231:13–20, 1990.

44. RL Benson, ID McKelvie, BT Hart, YB Truong, IC Hamilton. Determination of total phosphorus in waters and wastewaters by on-line UV/thermal induced digestion and flow injection analysis. Anal Chim Acta 326:29–39, 1996.

45. I Kolthoff, IK Miller. The chemistry of persulphate. I. The kinetics and mechanism of the decomposition of the persulphate ion in aqueous medium. JACS 73:3055–3059, 1951.

46. WE Miller, JC Greene, T Shiroyama. The *Selenastrum capricornutum Printz* algal assay bottle test—experimental design, application and data interpretation protocol.

47. R Shalders, ID McKelvie, BT Hart. The Measurement of Bioavailable Nutrients. ed. CRC for Freshwater Ecology, 1997, pp XX.

48. P Ekholm, K Krogerus. Bioavailability of phosphorus in purified municipal wastewater. Water Res 32:343–351, 1998.

49. SEATM Van der Zee, LGJ Fokkink, WH van Reimsdjik. A new technique for assessment of reversibly adsorbed phosphate. Soil Sci Soc Amer J 51:599–604, 1987.

50. AN Sharpley. An innovative approach to estimate bioavailable phosphorus in agricultural runoff using iron oxide–impregnated paper. J Environ Qual 22:597–601, 1993.

51. AN Sharpley. Estimating phosphorus in agricultural runoff available to several algae using iron-oxide paper strips. J Environ Qual 22:678–680, 1993.

52. RL Oliver, BT Hart, GB Douglas, R Beckett. Phosphorus speciation in the Murray and Darling Rivers. Water 20:23–26, 1993.

53. DS Baldwin, JK Beattie, LM Coleman, DR Jones. Phosphate ester hydrolysis facilitated by mineral phases. Environ Sci Technol 29:1706–1709, 1995.

54. DS Baldwin, JK Beattie, DR Jones. Hydrolysis of organic phosphorus compound by iron oxide–impregnated filter papers. Water Res 30:1123–1126, 1996.

55. JS Robinson, AN Sharpley. Organic phosphorus effects on sink characteristics of iron oxide–impregnated filter paper. Soil Sci Soc Am J 58:758–761, 1994.

56. RM Dils, AL Heathwaite. Development of iron oxide–impregnated paper strip technique for the determination of bioavailable phosphorus in runoff. Water Res 32:1429–1436, 1998.

57. H Zhang, W Davison, R Gadi, T Kobayashi. In-situ measurement of dissolved phosphorus in natural wa-

ters using DGT. Anal Chim Acta 370:29–38, Submitted.

58. JD Strickland, L Solórzano. Determination of monoesterase hydrolyzable phosphorus and phosphomonoesterase activity in seawater. In: J Barnes, ed. Some Contemporary Studies in Marine Science. London: Allen and Unwin, 1966, pp 665–674.

59. RZ Chróst, W Siuda, D Albrecht, J Overbeck. A method for determining enzymatically hydrolyzable phosphate in natural waters. Limnol Oceanogr 31: 662–667, 1986.

60. Y Shan, ID McKelvie, BT Hart. Characterization of immobilized Escherichia coli alkaline phosphatase reactors in flow injection analysis. Anal Chem 65:3053–3060, 1993.

61. Y Shan, ID McKelvie, BT Hart. Determination of alkaline phosphatase hydrolyzable-phosphorus in natural water systems by enzymatic flow injection. Limnol Oceanogr 39:1993–2000, 1994.

62. SE Herbes, HE Allen, KH Mancy. Enzymatic characterization of soluble organic phosphorus in lake water. Science 187:432–434, 1975.

63. ID McKelvie, BT Hart, TJ Cardwell, RW Cattrall. Use of immobilized phytase and flow injection for the determination of dissolved phosphorus species in natural waters. Anal Chim Acta 316:277–289, 1995.

64. TP Whaley, LW Ferrara. Gravimetric analysis of phosphorus compounds. In: EJ Griffith, AM Beeton, JM Spencer, DT Mitchell, eds. Environmental Phosphorus Handbook. New York: Wiley, 1973, pp 313–326.

65. SR Crouch, HV Malmstadt. A mechanistic investigation of molybdenum blue for determination of phosphate. Anal Chem 39:1084–1089, 1967.

66. F Osmond. Sur une réaction pouvant servir au dosage colorimétrique du phosphore dans les fontes, les aciers, etc. Bull Soc Chem Paris 47:745–748, 1887.

67. G Denigés. Reaction de coloration extremement sensible des phosphates et arseniates. Ses applications. C.r. Acad 171:802–804, 1920.

68. C Juday, EA Birge, GI Kemmerer, RJ Robinson. Phosphorus content of lake waters of northeastern Wisconsin. Trans Wis Acad 23:233–248, 1928.

69. HL Golterman. Studies on the cycle of elements in fresh water. Acta Bot Neerlandica 9:1–58, 1960.

70. APH Association. Standard Methods for the examination of water and wastewater including bottom sediments and sludges. 12th ed. New York: American Public Health Association, 1965, pp 769.

71. J Murphy, JP Riley. A modified single solution method for the determination of phosphorus in natural waters. Anal Chim Acta 27:31–36, 1962.

72. JJ Pauer, HR van Vliet, JF van Staden. Determination of phosphate at low concentrations in surface waters by flow injection analysis. Water SA 14:125–130, 1988.

73. PJ Worsfold, JR Clinch, H Casey. Spectrophotometric field monitor for water quality parameters. The determination of phosphate. Anal Chim Acta 197:43–50, 1987.

74. KS Johnson, RL Petty, J Thomsen. Flow injection analysis for seawater micronutrients. In: A Zirino, ed. Mapping Strategies in Chemical Oceanography. Washington, DC: ACS, 1985, pp 7–30.

75. MT Downes. An automated determination of low reactive phosphorus concentrations in natural waters in the presence of arsenic, silicon and mercuric chloride. Water Res 12:743–745, 1978.

76. S Motomizu, T Wakimoto, K Toei. Determination of trace amounts of phosphate in river water by flow injection analysis. Talanta 30:333–338, 1983.

77. S Motomizu, M Oshima. Spectrophotometric determination of phosphorus as orthophosphate based on solvent extraction of the ion associate of molybdophosphate with malachite green using flow injection. Analyst 112:295–300, 1987.

78. FI Ormaza-González, PJ Statham. Determination of dissolved inorganic phosphorus in natural waters at nanomolar concentrations using a long capillary cell detector. Anal Chim Acta 244:63–70, 1991.

79. K Fujiwara, W Lei, F Shimokoshi, K Fuwa, T Kobayashi. Determination of phosphorus at the parts per trillion level by using a laser-induced thermal lensing colorimetry. Anal Chem 54:2026–2029, 1982.

80. W Fusheng, W Zhongxiang, T Enjiang. The determination of trace amounts of phosphate in natural water by flow injection fluorimetry. Analyt Lett 22:3081–3090, 1989.

81. A Varma. Handbook of Inductively Coupled Plasma Atomic Emission Spectroscopy. Boca Raton, FL: CRC Press, 1991.

82. JL Manzoori, A Miyazaki, H Tao. Rapid differential flow injection of phosphorus compounds in wastewater by sequential spectrophotometry and inductively coupled plasma atomic emission spectrometry using a vacuum ultraviolet emission line. Analyst 115:1055–1058, 1990.

83. S-J Jiang, RS Houk. Inductively coupled plasma mass spectrometric detection for phosphorus and sulphur compounds separated by liquid chromatography. Spectrochimica Acta 43B:405–411, 1988.

84. JF Coetzee, CW Gardner. Determination of sulphate, ortho-phosphate, and triphosphate ions by flow injection analysis with the lead ion selective electrode as detector. Anal Chem 58:608–611, 1986.

85. F Schubert, R Renneberg, FW Scheller, L Kirstein. Plant tissue hybrid electrode for determination of phosphate and fluoride. Anal Chem 56:1677–1682, 1984.

86. EM D'Urso, PR Coulet. Phosphate sensitive electrode: a potential sensor for environmental control. Anal Chim Acta 239:1–5, 1990.

87. AG Fogg, NK Bsebsu. Flow injection voltammetric determination of phosphate: direct injection of phos-

phate into molybdate reagent. Analyst 107:566–570, 1982.

88. SC Hight, F Bet-Pera, B Jaselskis. Differential pulse polarographic determination of orthophosphate in aqueous media. Talanta 29:721–724, 1982.

89. DK Morgan, ND Danielson. Enzymatic determination of phosphate in conjunction with high-performance liquid chromatography. J Chromatogr 262:265–276, 1983.

90. AL Heckenberg, PR Haddad. Determination of inorganic anions at parts per billion levels using single column ion chromatography without sample preconcentration. Journal of Chromatogr 299:301–305, 1984.

91. RA Wetzel, CL Anderson, H Schleicher, GD Crook. Determination of trace levels of ions by ion chromatography with concentrator columns. Anal Chem 51:1532–1535, 1979.

92. D Kaniansky, I Zelensky, A Hybenova, FI Onuska. Determination of chloride, nitrate, sulphate, nitrite, fluoride, and phosphate by on line coupled capillary isotachophoresis–capillary zone electrophoresis with conductivity detection. Anal Chem 66:4258–4264, 1994.

93. G Bondoux, P Jandik, WR Jones. New approach to the analysis of low levels of anions in water. J Chromatogr 602:79–88, 1992.

94. S Motomizu, T Wakimoto, K Toei. Spectrophotometric determination of phosphate in river waters with molybdate and malachite green. Analyst 108:361–367, 1983.

95. CH Fiske, Y Subbarow. The colorimetric determination of phosphorus. J Biol Chem 44:375, 1925.

96. DT Burns, D Chimpalee, N Chimpalee, S Ittipornkul. Flow-injection spectrophotometric determination of phosphate using crystal violet. Anal Chim Acta 254:197–200, 1991.

97. K Stephens. Determination of low phosphate concentrations in lake and marine waters. Limnol Oceanogr 8:361–362, 1963.

98. S Motomizu, T Wakimoto, K Toei. Solvent extraction-spectrophotometric determination of phosphate with molybdate and malachite green in river and seawater. Talanta 31:235–240, 1984.

99. JB Martin, DM Doty. Determination of inorganic phosphate. Anal Chem 21:965–967, 1949.

100. DF Boltz. Spectrophotometric, spectrofluorimetric and atomic absorption spectrometric methods for the determination of anions in water. In: S Ahuja, EM Cohen, TJ Kneip, JL Lambert, D Sweig, eds. Chemical Analysis of the Environment and Other Modern Techniques. New York: Plenum Press, 1973, pp 201–228.

101. N Lacy, GD Christian, J Ruzicka. Enhancement of flow injection optosensing by sorbent extraction and reaction rate measurement. Anal Chem 62:1482–1490, 1990.

102. K Sugawara, S Kanamori. Spectrophotometric determination of submicromolar quantities of orthophosphate in natural waters. Bull Chem Soc Japan 34:526–531, 1961.

103. A Sjösten, S Blomqvist. Influence of phosphate concentration and reaction temperature when using the molybdenum blue method for determination of phosphate in water. Water Res 31:1818–1823, 1997

104. H Weil-Malherbe, RH Green. The catalytic effect of molybdate on the hydrolysis of organic phosphate bonds. Biochem J 49:286–292, 1951.

105. SJ Tarapchak. Soluble reactive phosphorus measurements in lake water: evidence for molybdate-enhanced hydrolysis. J Environ Qual 12:105–108, 1983.

106. MP Stainton. Errors in molybdenum blue methods determining orthophosphate in freshwaters. Can J Fish Aquat Sci 37:472–478, 1980.

107. FH Rigler. Further observations inconsistent with the hypothesis that molybdenum blue method measures orthophosphate in lake water. Limnol Oceanogr 13:7–13, 1968.

108. W Chamberlain, J Shapiro. On the biological significance of phosphate analysis: comparison of standard and new methods with a bioassay. Limnol Oceanogr 14:921–927, 1969.

109. WA Dick, MA Tabatabai. Determination of orthophosphate in aqueous solutions containing labile organic and inorganic phosphorus compounds. J Environ Qual 6:82–85, 1977.

110. RA Chalmers, AG Sinclair. Analytical applications of β-heteropoly acids. Part I. Determination of arsenic, germanium and silicon. Anal Chim Acta 33:384–390, 1965.

111. RA Chalmers, AC Sinclair. Analytical applications of β-heteropoly acids. Part II. The influence of complexing agents on selective formation. Anal Chim Acta 34:412–418, 1966.

112. S Blomqvist, K Hjellström, A Sjösten. Interference from arsenate, fluoride and silicate when determining phosphate in water by the phosphoantimonylmolybdenum blue method. Int J Environ Anal Chem 54:1993.

113. RL Benson, YB Truong, ID McKelvie, BT Hart. Monitoring of dissolved reactive phosphorus in wastewaters by flow injection analysis. Part 1. Method development and validation. Water Res 30:1959–1964, 1996.

114. WS Zaugg, RJ Knox. Indirect determination of inorganic phosphate by atomic absorption spectrophotometric determination of molybdenum. Anal Chem 38:1759–1760, 1966.

115. T Nasu, M Kant. Determination of phosphate, arsenate and arsenite in natural water by flotation-spectrophotometry and extraction–indirect atomic absorption spectrometry using malachite green as an ion-pair reagent. Analyst 113:1683–1686, 1988.

116. D Jolley, W Maher, P Cullen. Rapid method for separating and quantifying orthophosphate and polyphos-

phates: application to sewage samples. Water Res 32: 711–716, 1998.

117. N Yoza, H Hirano, Y Baba, S Ohshi. Characterization of enzymatic hydrolysis of inorganic polyphosphates by flow injection analysis and high-performance liquid chromatography. J Chromatogr 325:385–393, 1985.

118. DJ Halliwell, ID McKelvie, BT Hart, R Dunhill. Separation and detection of condensed phosphates by IC-FIA. Analyst 121:1089–1093, 1996.

119. CM Clarkin, RA Minear, S Kim, JW Elwood. An HPLC postcolumn reaction system for phosphorus-specific detection in the complete separation inositol phosphate congeners in aqueous samples. Environ Sci Technol 26:199–204, 1992.

120. PR Haddad, PE Jackson. Ion Chromatography—Principles and Applications. Amsterdam: Elsevier Science Publishers, 1990.

121. MT Downes, HW Pearl. Separation of two dissolved reactive phosphorus fractions in lake water. J Fish Res Bd Can 35:1636–1639, 1978.

122. C Hino. Characterization of orthophosphate released from dissolved phosphorus by gel filtration and several hydrolytic enzymes. Hydrobiologia 174:49–55, 1989.

123. DRS Lean. Movement of phosphorus between its biologically important forms in lake water. J Fish Res Bd Can 30:1525–1536, 1973.

124. RA Minear. Characterization of naturally occurring dissolved organophosphorus compounds. Environ Sci Technol 6:431–437, 1972.

125. DS Baldwin. Reactive "organic" phosphorus revisited. Water Res 32:2265–2270, 1998.

126. ID McKelvie, BT Hart, TJ Cardwell, RW Cattrall. Speciation of dissolved phosphorus in environmental samples by gel filtration and flow-injection analysis. Talanta 40:1981–1993, 1993.

127. FR Campbell, RF Thomas. Automated method for determining and removing silica interference in determination of soluble phosphorus in lake and stream waters. Environ Sci Technol 4:602–604, 1970.

128. MD Ron Vaz, AC Edwards, CA Shand, MS Cresser. Determination of dissolved organic phosphorus in soil solutions by an improved automated photo-oxidation procedure. Talanta 39:1487–1497, 1992.

129. TAHM Janse, PFA Van der Wiel, G Kateman. Experimental optimization procedures for the determination of phosphate by flow injection analysis. Anal Chim Acta 155:89–102, 1983.

130. RNC Daykin, SJ Haswell. Development of a micro flow injection manifold for the determination of orthophosphate. Anal Chim Acta 313:155–159, 1995.

131. T Korenaga, K Okada. Automated system for total phosphorus in waste waters by flow injection analysis [in Japanese]. Bunseki Kagaku 33:683–686, 1984.

132. M Aoyagi, Y Yasumasa, A Nishida. Rapid spectrophotometric determination of total phosphorus in indus-

trial wastewaters by flow injection analysis including a capillary reactor. Anal Chim Acta 214:229–237, 1988.

133. M Aoyagi, Y Yasumasa, A Nishida. Simultaneous determination of total phosphorus and total nitrogen by using a flow injection system. Anal Sci 5:235–236, 1989.

134. M Aoyagi, Y Yasumasa, A Nishida. Digestion method for FIA of phorphorus and nitrogen in hydrogen peroxide solution. Bunseki Kagaku 39:131–133, 1990.

135. S Dong, PK Dasgupta. Automated determination of total phosphorus in aqueous samples. Talanta 38:133–137, 1990.

136. D Betteridge, EL Dagless, B Fields, NF Graves. A highly sensitive flow-through phototransducer for unsegmented continuous-flow analysis demonstrating high-speed spectrophotometry at the ppb level and a new method of refractometric determinations. Analyst 103:897–908, 1978.

137. PK Dasgupta, HS Bellamy, H Liu, JL Lopez, EL Loree, K Morris, K Petersen, KA Mir. Light emitting diode based flow-through optical absorption detectors. Talanta 40:53–74, 1993.

138. W Lei, K Fujiwara, K Fuwa. Determination of phosphorus in natural waters by long-capillary-cell absorption spectrometry. Anal Chem 55:951–955, 1983.

139. K Fuwa, W Lei, K Fujiwara. Colorimetry with a total-reflection long capillary cell. Anal Chem 56:1640–1644, 1984.

140. K Nakanishi, T Imasaka, N Ishibashi. Thermal lens spectrophotometry of phosphorus using a near-infrared semiconductor laser. Anal Chem 57:1219–1223, 1985.

141. K Yoshimura, M Ishi, T Tarutani. Microdetermination of phosphate in water by gel-phase colorimetry with molybdenum blue. Anal Chem 58:591–594, 1986.

142. K Yoshimura, S Nawata, G Kura. Gel-phase absorptiometry of phosphate with molybdate and malachite green and its application to flow analysis. Analyst 115:843–848, 1990.

143. S Motomizu, M Oshima, N Katsumura. Fluorimetric determination of phosphate in sea water by flow injection analysis. Anal Sci Technol 8:843–848, 1995.

144. M Kan, T Nasu, M Taga. Fluorophotometric determination of phosphate as an ion pair of molybdophosphate with Rhodamine 6G. Analyt Sci 7:87–91, 1991.

145. SE Meek, DJ Pietrzyk. Liquid chromatographic separation of phophorus oxo acids and other anions with post column indirect fluorescence detection by aluminium-Morin. Anal Chem 60:1397–1400, 1988.

146. H Kawasaki, K Sato, J Ogawa, Y Hasegawa, H Yuki. Determination of inorganiz phosphate by flow injection method with immobilized enzymes and chemiluminescence detection. Anal Biochem 182:366–370, 1989.

147. SA Glazier, MA Arnold. Phosphate-selective polymer membrane electrode. Anal Chem 60:2540–2542, 1988.

148. J Liu, Y Masuda, E Sekido, S-I Wakida, K Hiiro. Phosphate ion–sensitive coated-wire/field effect transistor electrode based on cobalt phthalocyanide with polyvinyl(chloride) as the membrane matrix. Anal Chim Acta 224:145–151, 1989.

149. H Hara, S Kusu. Continuous-flow determination of phosphate using a lead ion–selection electrode. Anal Chim Acta 261:411–417, 1992.

150. PW Alexander, J Koopetngarm. Flow-injection determination of phosphate species in detergents with a calcium ion–selective electrode. Anal Chim Acta 197:353–359, 1987.

151. DE Davey, DE Mulcahy, GR O'Connell. Flow-injection determination of phosphate with a cadmium ion-selective electrode. Talanta 37:683–687, 1990.

152. JF van Staden. Behavior of silver orthophosphate as the electroactive sensor of a coated, tubular solid-state phosphate-selective electrode in flow injection analysis. S Afr J Chem 46:14–19, 1993.

153. Z Chen, R De Marco, PW Alexander. Flow-injection potentiometric detection of phosphates using a metallic cobalt wire ion-selective electrode. Anal Commun 34:93–95, 1997.

154. T Katsu, T Kayamoto. Potentiometric determination of inorganic phosphate using a salicylate-sensitive membrane electrode and an alkaline phosphatase enzyme. Anal Chim Acta 265:1–4, 1992.

155. AG Fogg, GC Cripps, BJ Birch. Static and flow injection voltammetric determination of total phosphate and soluble silicate in commercial washing powders at a glassy carbon electrode. Analyst 108:1485–1489, 1983.

156. AG Fogg, NK Bsebsu. Differential-pulse voltammetric determination of phosphates as molybdovanadophosphate at a glassy carbon electrode and assessment of eluents for the flow-injection voltammetry. Analyst 106:1288–1295, 1981.

157. K Pettersson. Enzymatic determination of orthophosphate in natural waters. Int Rev Ges Hydrobiol 64:585–607, 1979.

158. RJ Stevens. Evaluation of an enzymatic method for orthophosphate determination in freshwaters. Water Res 13:763–770, 1979.

159. S Motomizu, M Oshima, L Ma. On-site analysis for phosphorus and nitrogen in environmental water samples by flow-injection spectophotometric method. Anal Sciences 13:401–404, 1997.

160. RL Benson, YB Truong, ID McKelvie, BT Hart, G Bryant, W Hilkmann. Monitoring of dissolved reactive phosphorus in wastewaters by flow injection analysis. Part 2. On-line monitoring system. Water Res 30:1965–1971, 1996.

161. KM Pedersen, M Kümmel, H Søeberg. Monitoring and control of biological removal of phosphorus and nitrogen by flow injection analyzers in a municipal pilot-scale wastewater treatment plant. Anal Chim Acta 238:191–199, 1990.

162. RG Wetzel. Limnology. Philadelphia: Saunders College Publishing, 1983.

163. BT Hart, PR Freeman, ID McKelvie, S Pearse, DG Ross. Phosphorus spiralling in Myrtle Creek, Victoria, Australia. Verh Int Verein Limnol 24:2065–2070, 1991.

164. Guidelines for Drinking Water Quality. 2nd ed. Geneva: World Health Organization, 1993.

165. ID McKelvie, DMW Peat, GP Matthews, PJ Worsfold. Elimination of the Schlieren effect in the determination of reactive phosphorus in estuarine waters by flow-injection analysis. Anal Chim Acta 351:265–271, 1997.

166. S Auflitsch, DMW Peat, PJ Worsfold, ID McKelvie. Determination of dissolved reactive phosphorus in estuarine waters using a reversed flow injection manifold. Analyst 122:1477–1480, 1997.

167. EAG Zagatto, MAZ Arruda, AO Jacintho, IL Mattos. Compensation of the Schlieren effect in flow-injection analysis by using dual-wavelength spectrophotometry. Anal Chim Acta 234:153–160, 1990.

168. H Liu, PK Dasgupta. Dual-wavelength photometry with light emitting diodes. Compensation of refractive index and turbidity effects in flow-injection analysis. Anal Chim Acta 289:347–353, 1994.

169. A Daniel, D Birot, M Lehaitre, J Poncin. Characterization and reduction of interferences in flow-injection analysis for the in situ determination of nitrate and nitrite in sea water. Anal Chim Acta 308:413–424, 1995.

170. T Yamane, M Saito. Simple approach for elimination of blank peak effects in flow injection analysis of samples containing trace analyte and an excess of another solute. Talanta 39:215–219, 1992.

171. I Dellien, LA Johansson. A comment on the Swedish Standard Method for the Determination of Total Phosphorus Concentration in Water. Vatten 37:349–353, 1981.

172. T Lee, E Barg, D Lal. Techniques for extraction of dissolved inorganic and organic phosphorus from large volumes of sea water. Anal Chim Acta 260:113–121, 1992.

173. KS Johnson, RL Petty. Determination of phosphate in seawater by flow injection analysis with injection of reagent. Anal Chem 54:1185–1187, 1982.

16

Main Parameters and Assays Involved with the Organic Pollution of Water

Angel Cuesta, Antonio Canals, and José Luis Todolí
University of Alicante, Alicante, Spain

I. INTRODUCTION

The estimation of the organic contamination in a water sample is a complex and delicate problem that involves several determination assays, because the organic matter is present under diverse chemical compounds and degradation states. The global organic matter balance cannot be obtained by considering a single method, but it must be done by a comparison of the results obtained by different methods (1,2). An additional difficulty lies in the fact that, in general terms, no single parameter can be used to quantify the organic matter content.

In principle, the carbonated matter is used as a nutrient by aerobic germs, and it is oxidized to carbon dioxide and water, while species such as nitrite and nitrate are used as food by, for example, nitrobacteria. In an oxygen-deficient environment, such as sewer or stale water, bacteria take oxygen not only from nitrates and nitrites, but also from sulfates, with sulfur hydrogen as residual product. These oxidation phenomena that take place in nature are very difficult to reproduce on a laboratory scale. However, some tests (e.g., biochemical oxygen demand, BOD) allow a biological appreciation of the phenomena, although there are some inherent problems, which will be discussed later.

Several chemical methods have been developed in order to get a more complete and reproducible oxidation of organic matter. Some of them are based on the use of chemical reagents and a methodology that avoids the ambiguity of biological methods. In this way, chemical oxygen demand (COD) has become one of the obtained parameters. Nevertheless, the degradation (i.e., extent and velocity) of organic substances by means of biological methods can be different from that produced by chemical methods. Therefore, the results obtained with both sorts of methods may be difficult to compare. In particular, the extent of oxidation reached when using a strong oxidizing agent (e.g., potassium dichromate) is more complete for many organic compounds than is biological oxidation, although in some cases it is not fully accomplished. As a result, the COD values obtained by this method are so high that, under biological conditions, the complete oxidation of organic matter takes a long time, and it is not always reached.

Another way to evaluate organic matter content is to measure the carbon present in a sample. Total organic carbon (TOC) is, in this case, the employed parameter. The rapid evolution of relatively complex techniques introduced in the last years has promoted the development of these methods. These techniques show, as the most relevant advantage, applicability to almost every category of organic products, even to the most resistant oxidizing compounds. Besides, the results are obtained quickly and the determinations can be automated easily.

There are other parameters for estimating water contamination. Besides those already mentioned (i.e., BOD, COD, and TOC) total organic halide (TOX) is a very useful quantity. The combination of all these parameters would give complete information for the characterization of the organic matter present in a sample.

II. BIOCHEMICAL OXYGEN DEMAND (BOD)

Biochemical oxygen demand (BOD) is the amount of oxygen (measured in mg/L) required for the oxidation of the organic matter by biological action under specific standard test conditions (3).

In the test for BOD determination, the oxygen required to degrade the organic compounds of a water sample by biological means is measured. The test is frequently used to evaluate the efficiency of organic matter removal after a given wastewater treatment process. Due to its biological character, both the application of the method and the interpretation of the results are often difficult. In addition, its reproducibility is sometimes unsuitable. Other problems that can emerge are that BOD changes with time, up to 25 days. These facts led to the development of some BOD variations.

Carbonaceous BOD (CBOD) is one of these parameters. In this case, the effect of the nitrifying bacteria, which can also consume some more dissolved oxygen, is avoided by means of chemical inhibition (i.e., by the addition of 2-chloro-6-(trichloro methyl) pyridine).

Nevertheless, among the BOD variations, BOD_5 is the most widely used parameter. It is defined as the BOD obtained after an incubation period of 5 days. In some cases, the water is seeded with a given mass of microorganisms that will depend on their initial amount in the water sample.

Two different kinds of methodologies have been developed for BOD_5 determination: the dilution method (classic method) and the instrumental methods. In both cases, the measurement of the concentration of dissolved oxygen (DO) is of crucial importance. The initial DO drops with time due to the oxidation of the organic matter by the microorganisms present either in natural or seeded water.

A. Dilution Method

In the classic method, also called the *dilution method*, the DO is determined via a Winkler titration. Thus, manganese sulfate is added to the sample. After the redissolution of the precipitate, the solution is titrated with thiosulfate until the color of solution changes from dark blue to clear, employing starch indicator. The volume of titrant employed corresponds to the DO value.

The classic BOD assay is rather long and requires multiple steps. First, a given sample volume is placed inside a volumetric flask and made up to a fixed total volume with distilled water. The role of dilution is to ensure that there is a mass of oxygen sufficient to avoid any decline in bacterial activity. The flask can be shaken in order to ensure that the water is saturated with oxygen. To keep an appropriate (optimum) medium for development of the microorganisms, the pH of the sample should be between 6 and 8. A blank (i.e., dilution water) is also prepared, and the same treatment as for the sample is applied. When necessary, both sample and blank must be seeded with a volume of water with microorganisms. A fraction of the total solution content is stored in a covered flask, avoiding the presence of air bubbles. The sample should be kept away from all light and at 20°C. Then DO is measured in both diluted sample and blank, at the beginning and 5 days after the preparation. Finally, the BOD_5 value can be obtained by applying the following equation:

$$BOD_5 = \frac{(D_0 - D_5) - (B_0 - B_5)f}{P} \quad (1)$$

where D_0 (mg/L) is the DO of the diluted sample after preparation, D_5 (mg/L) is the DO of the diluted sample after 5 days of incubation at 20°C, B_0 (mg/L) is the DO of the dilution water before incubation, B_5 (mg/L) is the DO of the dilution water after 5 days of incubation at 20°C, P is the decimal volumetric fraction of sample used (note that the sample is diluted), and f is the ratio of seed in the diluted sample to seed in the dilution water [i.e., f = (% seed in diluted sample)/(% seed in dilution water)].

If the sample and the dilution water are not seeded, Eq. (1) can be simplified to:

$$BOD_5 = \frac{D_0 - D_5}{P} \quad (2)$$

By applying this method, a suitable detection limit for environmental purposes should be 1 mg/L. One factor that must be taken into account is that the presence of chlorine in the sample produces interferences in the determination of the DO value. In this case, the addition of sodium sulfite leads to a reduction in the amount of this species.

B. Instrumental Methods

Due to the multiple drawbacks that the dilution method presents, an effort has been made to develop alternative methods in order to overcome or at least alleviate them. These processes are devoted to the reduction of the total analysis time and to the improvement of the reproducibility. Moreover, the correlation between the obtained parameter and BOD_5 should be as good as possible. For this purpose, different strategies have been proposed, the most popular among them being respirometry and biosensors.

1. Respirometric Methods

These were the first employed for the rapid determination of BOD. The respirometric methods are based on the addition of microorganisms to the water in the same flasks employed in the dilution method (i.e., one for the diluted sample water and another for the blank). The oxygen decrease produced by the oxidation of the water organic matter by the bacteria is continuously measured by means of a galvanic cell oxygen probe. Unlike the dilution method, the solution is aerated. From this information a value that can be related to BOD_5 is obtained (4).

Application of this method involves the measurement of the rate of decrease of the oxygen content versus time. First, a graph of oxygen concentration against time is plotted (Fig. 1). Obviously, oxygen concentration decreases with time. After that, the rate of decreased oxygen concentration at a given time is obtained by measuring the slope of the line at that point. Then the rate of oxygen consumption is plotted against time, and the area under the curve gives an indication of the total amount of oxygen employed in oxidation of the organic matter (Fig. 2). By applying this method, an analysis can be performed in less than 2 hours.

2. Biosensors

As before, the rate of oxygen consumption is measured, but the sample solution is not continuously aerated. Microorganisms are adsorbed on the electrode surface, supported by different kinds of materials. The electrode is then introduced into the water sample and DO is measured. Many works on this topic have been published (Table 1), with encouraging results. Unfortunately, BOD_5 cannot yet be substituted by the parameter supplied by these electrodes (i.e., BODS).

3. Other Methods

In addition to the previously mentioned methodologies for determining BOD, some authors have suggested alternative systems. This section summarizes the most relevant ones found in the literature.

a. Headspace BOD. Logan et al. (12,13) employed the headspace technique combined with gas chromatography (GC) to obtain a parameter that can be correlated with BOD_5 (i.e., headspace biochemical oxygen demand, HBOD). The sample is placed inside a container similar to those used in classic BOD determination, and the remaining space is filled with oxygen. Under these conditions, a high concentration of oxygen is reached, with subsequent growth in the velocity of the oxidation of the organic matter. Next, the air in the headspace is sampled with a syringe and then introduced into the gas chromatograph. Finally, the oxygen concentration is determined. The water sample is left for 3 days under the same conditions as those fixed for BOD determination, and after this period of time

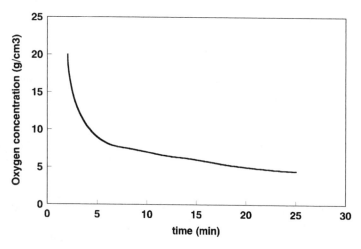

Fig. 1 Variation of oxygen concentration with time (generic curve).

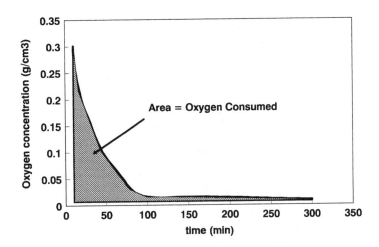

Fig. 2 Method for estimating consumed oxygen.

the oxygen concentration in the headspace is again measured. The correlation obtained by Logan et al. is not good, but they give an estimating value of the rate of disappearance of organic matter (12,13).

b. Direct Measurement of Absorbance. In order to consider a BOD determination method ideal, it should meet several characteristics: (a) it must be easy to handle; (b) the sample pretreatment must be avoided; and (c) the organic matter content must be measured instantaneously. A method close to the ideal one should be the direct measurement of any water physical property that changes as a function of the level of organic compounds. Although some good results have been obtained, they are applicable only to very particular cases (i.e., a region, river, lake, etc.)

Brookman (14) measures water absorbance at 280 nm. The absorbance values are further correlated with BOD. The values of the square of the regression coefficient, R^2, corresponding to the correlation line are contained within the 0.74–0.76 range. By adopting this procedure, the complexity of the BOD determination is minimized. Furthermore, no extra biological media (i.e., bacteria) are required. One severe drawback of this method is the interferences produced by the solid particles present in the sample water, leading to significant light diffraction and, thus, to irreproducible BOD values.

Reynolds and Ahmad (15) correlate the fluorescent signal at 340 nm emitted by a wastewater sample irradiated at 280 nm with BOD. In this case R^2 takes values up to 0.89. The improvement of the method is due to the fact that Reynolds and Ahmad use the Raman emission at 309 nm as an internal standard. With

this procedure, many of the errors produced by the absorbance methods are mitigated.

III. CHEMICAL OXYGEN DEMAND (COD)

Chemical oxygen demand (COD) is the amount of oxygen needed to reduce the organic matter present in a water sample by chemical methods (3). This parameter is of great importance in monitoring water quality, and it is widely employed in analytical laboratories. The advantages of this assessment are that the COD determination takes less time and is simpler and more reproducible than that for BOD. Besides, BOD can be estimated from COD, although the mathematical relationship can vary greatly from one sample to another. This lack of correlation is due to the fact that some bacteria produce more complete oxidation of organic matter than might any chemical oxidizing agent. Thus, for instance, *Acetobacter* is able to oxidize acetate more quickly and efficiently than it does potassium dichromate (i.e., the common oxidant used in COD determination). Meanwhile, for other organic compounds (e.g., substituted aromatic hydrocarbons) chemical oxidation is more complete than biological oxidaton, since there are no bacteria capable of oxidizing them.

In the COD assay, a known excess of strong oxidant is added to the sample and, by means of an indirect determination, the mass (or concentration) of oxidant that has not been reduced by reaction with the organic matter is derived.

Table 1 Significant Data on Various Biosensors for BOD Tests

Ref.	Measuring time	pH	Electrode life	Yeast	$BOD_5/BODS^a$	Detection	Linear range[b] (mg O_2/L)
5	18 min	7.0	17 days	*Trichosporon Cutaneum*	0.2–7.5	Amperometric	0–60
6	30 s	6.8	48 days	*Trichosporon Cutaneum*	0.2–6	Amperometric	0–100
7	4 min stabilization; measure 15–30 s	7.2	2 days	*Bacillus subtilis* and *B. licheniformis 7B*	0.25–8	Amperometric	0–80
8	25 min stabilization; measure 10 min	6.8	1–2 weeks	*Trichosporon Cutaneum*	0.72–1.66	Fluorescent	0–110
9	15 min	7.0	2–8 days	*Trichosporon Cutaneum*	0.1–0.2	Amperometric	0–150
10	7–20 min	7.0	3 days	*Trichosporon Cutaneum*	0.6–2.7	Ultraviolet Absorption	0–18
11	20–30 min	7.2	Several months	*Bacillus subtilis*	1.1–2.9	Amperometric	0–70

[a]BODS/BOD: Ratio of BOD_5 to BOD supplied by the sensor.
[b]BOD range for which there is a linear relationship between the measured magnitude and the BOD value.

LIVERPOOL JOHN MOORES UNIVERSITY
LEARNING SERVICES

Various methods have appeared in the literature dealing with the determination of COD in water samples. A brief description of those most commonly used is presented next.

A. Classic (Opened Reflux) Method

In the first method described for COD determination, the sample is placed in a refluxing flask (16). Then the oxidizing agent is set in contact with the water sample. Nowadays, potassium dichromate is the most accepted oxidant, although some others (e.g., permanganate, Ce(IV), persulfate) have also been used. Oxidation of the organic matter could be now performed. Nevertheless, there are several considerations that must be taken into account in order to improve the speed and accuracy of the method. On this subject, it has been observed that the use of a catalyst is advisable to increase the reaction rate. The addition of silver sulfate has been shown to reduce significantly the time required to complete organic matter oxidation. As regards the accuracy of the method, it must be borne in mind that the presence in the water sample of inorganic species that might be oxidized by dichromate (i.e., mainly chloride) could give rise to higher COD values than the real value. The addition of $HgSO_4$ has proven to be a good way to eliminate the chloride interference, since mercury generates very stable complexes with this anion.

Finally, in order to enhance the oxidizing capability of dichromate, a given volume of sulfuric acid is also added. Then follows a further heating step for a period that normally reaches 2 hours. After cooling the mixture and washing down the condenser with distilled water, the dichromate that has not been reduced is titrated with ferrous ammonium sulfate, employing ferroine sulfate as indicator. The endpoint of the titration is detected by a change in the indicator color from blue-green to reddish brown. In general, the COD value is obtained by applying the following relationship:

$$COD \ (mg \ O_2/L) = \frac{(a - b)cf'8000}{ml_{sample}} \qquad (3)$$

where a (ml) is the volume of titrating solution used with the blank, b (ml) is the volume of titrating solution used with the sample, c (eq/L) is the normality of the titrating solution, f' is the titrating solution correction factor, 8000 is the equivalent weight of O_2 expressed in mg O_2/eq, and ml_{sample} is the volume of the water sample analyzed.

B. Semimicro (Closed Reflux) Method

The principle of this method is the same as for the opened reflux (classic) method previously described. In this case, various culture tubes sealed with PTFE caps are used. Sample and the four reagents mentioned earlier are placed inside the tubes. Blanks (i.e., reagents with distilled water) are also prepared. All the tubes are mounted on a heating block or placed inside an oven at a 150°C for 2 hours. After this period of time, the excess of dichromate that has not been reduced is determined against a ferrous ammonium sulfate standard solution, using ferroine as indicator.

Many variations of the semimicro method have been suggested. Most involve modifying the detection step. In this way, titration can be substituted by the spectrophotometric determination of chromium(VI) (17–19). This is the so-called *colorimetric closed reflux* method. When the organic matter digestion step has been completed, the suspended solids are left to settle before the absorbance is read. Therefore, the interferences caused by sample turbidity due to the presence of inorganic particles are attenuated. Once the solid particles are removed, the absorbance signal is measured at either 445 nm (17) or 600 nm (19), and finally the concentration of nonreduced dichromate or chromium(III) that appears by reduction, respectively, is obtained. Absorbances of the sample, bank, and standards are also determined. Sample COD is obtained by interpolation from the calibration curve. Previous studies show that, with this method, the linear dynamic range (i.e., the COD range for which there is a linear correlation between the COD and the difference between the blank and sample absorbances) reaches maximum COD values of 900 mg O_2/L (18). Hence, when the colorimetric method is selected, samples whose COD values are higher than this value should be diluted.

The closed reflux method is more efficient in the oxidation of volatile organic compounds than is the opened reflux method. The reason for this is that the oxidant is in contact with these compounds for a longer time (16,17). Besides, the closed method is cheaper, because only 2 ml of sample are required, since 5 ml is the total volume of the mixture. This means that the amounts of reagents and sample are reduced by a factor of over 20 compared to the opened reflux method. In addition, this fact makes the method less contaminating and allows for the simultaneous digestion of a great number of samples (e.g., by employing an oven, up to 40–50 samples can be digested in 2 hours).

The precision of this method depends on various factors, such as the chloride content. Hence, the relative standard deviation (RSD) values range from 5.6% in water without chlorides to 4.8% in water with 100 mg/L of chloride. The precision of the colorimetric method is slightly poorer than that for the chromium titration–based procedure. Nevertheless, in other investigations it has been indicated that this method affords precision values up to seven times better than the opened reflux method (19). The limits of detection (LOD) are about 3 mg O_2/L and 5 mg O_2/L for the closed and opened reflux methods, respectively.

C. Other Discontinuous Methods

Many modifications of the two methods just described have been reported in the literature. The main goals of these are: (a) to reduce the amount of reagents employed; (b) to increase the sample throughput; and (c) to increase the efficiency of the oxidation step. Table 2 presents the most relevant discontinuous methods for COD determination.

D. Methods Based on Flow Injection Analysis

Over the past 30 years, the methods based on flow injection analysis (FIA) have experienced enormous development in analytical laboratories (33). The COD has also been one of the applications of this methodology. In these cases, the digestion and detection steps are carried out online. Hence, the analyte concentration is continuously determined from a liquid stream. Small volumes of sample and reagents are added at strategic points of the system.

Several FIA manifolds have been developed for COD determination. The most significant characteristics of these systems are summarized next.

The first attempt to develop an FIA system for COD determination was performed in 1980 by Korenaga (34). In this method, the heating step consists of a thermostated bath of water, oil, or polyethylenglycol in which a 20–50-m Teflon® capillary is immersed. The reagents are mixed into the flow by connections, while a given volume of the sample is inserted in the flowing carrier through injection valves. Figure 3 shows the scheme of an FIA system similar to that employed by Korenaga. Due to the high length of the capillary, the pressure needed for the mixture to flow through the system is high and an appropriate pump is required.

When the sample leaves the bath, the absorbances are obtained spectrophotometrically (34–37). For this purpose a flow cell is adapted at the end of the line (Fig. 3). The absorbance is measured at wavelengths

that depend on the oxidant employed. To date, this methodology has been applied using potassium permanganate (34,35), cerium(IV) (37), and dichromate (36,38) as oxidizing agents. A recent application of the FIA methodology to COD determination describes the electrochemical generation of the oxidant [i.e., Co(III)] (39). Following the FIA method, a calibration curve is obtained from the absorbances of the standard solutions. Note that, when the oxidizing agent is monitored, the absorbance obtained for the blank is always higher than that obtained for the water sample, since a fraction of the oxidant is spent in the reaction with the organic matter.

Recently, some FIA systems based on microwave heating of the sample have been proposed for COD determination (40–42). In these methods, the digestion step is accelerated with respect to the method of Korenaga and coworkers. Besides, the methods based on spectrophotometric detection have some problems (Sec. III.B). These were recently overcome by the use of a flame atomic absorption spectrometer (FAAS) as the detector (41). Since this is a nonspecific detector for Cr(VI), an anionic-exchange resin was placed before the FAAS instrument. The chromium(VI) that was not reduced in the digestion step is retained in the anionic resin. Then it is eluted with the appropriate solution (e.g., 10 mol/L nitric acid) and the FAAS signal is recorded as a peak. Figure 4 shows an outline of this system. Moreover, with this setup, the linear dynamic range lies between 50 and 10,000 mg O_2/L. The LOD are about 6 mg O_2/L, and up to 50 samples per hour can be analyzed. In a more recent study, the resin was changed by a selective Cr(VI) organic solvent extraction step (42).

In summary, the proposed FIA-based methods for the determination of COD have several important advantages as compared with the conventional methods discussed in Secs. III.A & III.B: (a) higher sample throughput, (b) enhanced response times, producing shorter startup and shutdown times, (c) simpler methodology, (d) higher precision, and (e) higher dynamic range.

IV. TOTAL ORGANIC CARBON (TOC)

Total organic carbon (TOC) is defined as the amount of carbon covalently bonded in organic compounds in a water sample (3). The TOC is a more suitable and direct expression of total organics than either BOD or COD, but it does not provide the same kind of information. If a reproducible empirical relationship is es-

Table 2 Discontinuous Methods Employed for COD Determination

Refs.	Change	Modification and benefits/drawbacks
17	Reagents	$KCr(SO_4)_2 \cdot 12H_2O$ as a chloride interference suppressor
20	Reagents	Potassium permanganate is used as oxidant; results are poorer than those obtained with dichromate.
21, 22	Reagents	Cerium(IV) sulfate is used as oxidant; oxidation is less complete than with dichromate.
23	Reagents	Mn(II) is used as catalyst instead of Ag(I); process cost is reduced.
24	Detection	Colorimetric determination of chromium(VI) with diphenylcarbazide; sensitivity is improved.
25	Titration mode/detection	Cr(VI) titration is carried out by potentiometric means.
26	Titration mode/detection	Cr(VI) titration is carried out by amperometric means.
27	Reagents and titration mode	Chemiluminescence emitted by bacteria is decreased due to the presence of toxic products.
28–30	Reagents and titration	Absorbance of water is correlated with COD.
31	Instrumentation	A discontinuous microanalyzer is used.
32	Reagents and titration	Turbidity is correlated with COD.

tablished between TOC values and either COD or BOD, the TOC can be used to estimate the respective BOD or COD values. Typical TOC values range from 1 μg C/L to 50 mg C/L. To determine the content of organically bonded carbon, the organic molecules must be broken down to single carbon units and converted into a simple molecular form that can be quantitatively measured (2,43–62). The instruments employed to determine TOC could be classified as online and off-line. The first category has several advantages over the off-line methods, among them (a) simplicity of the method and (b) avoidance of the errors induced by the dramatic change of TOC with time, since the online instruments are able to take measures very quickly.

The TOC is included within the total carbon (TC) and has many fractions that can be analyzed separately. Table 3 summarizes the definitions of the different TC portions. Figure 5, in turn, gives an overview of them and their distribution.

In order to determine TOC, IC must be eliminated from TC. Several methods have been proposed to this end (62,63). Eliminating or compensating for IC makes the TOC method more complicated, since an extra acidifying-gas purging step is required (44,50). The interference of IC can be eliminated by acidifying samples to pH 2 or less in order to convert all the fractions included in this category (see Table 3) to CO_2, which is more easily removed from the water sample. None-

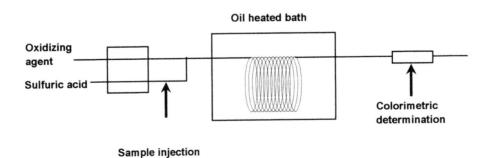

Fig. 3 Scheme of an FIA manifold for COD determination similar to that employed by Korenaga (Ref. 34).

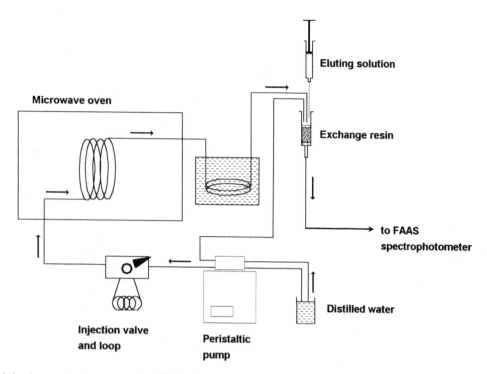

Fig. 4 Flow injection analysis system for COD determination assisted by microwave radiation, followed by an anionic-exchange resin and a flame atomic absorption spectrometry system.

theless, when a gas stream is passed to purge CO_2, volatile compounds can be dragged as well. In this case the measure corresponds in fact to the organic carbon that cannot be purged (i.e., NPOC). Hence, the POC values must be determined in order to know the true TOC value (Table 3). However, the POC value is usually very low and so can be neglected, assigning the NPOC value as the TOC value (Fig. 5). In addition, if the solid fraction is not significant, then the dissolved organic carbon (i.e., DOC) value is similar to the TOC value. Finally, volatile organic carbon (VOC) and non-volatile organic carbon (NVOC) are other parameters included with the POC and NPOC.

A. High Combustion Temperature (HCT) Methods

The most widely used method to accomplish the oxidation of carbon-containing species to CO_2 and H_2O is the catalytic (i.e., Pt, CuO) oxidation in gas phase at temperatures ranging from 680 to 950°C. First, the sample is injected in a gas flow with a syringe, thus requiring very low sample volumes (i.e., from 10 to 2000 μl). Second, the carbon dioxide generated is determined by means of a nondispersive infrared analyzer (43–49). Finally, the milligrams of carbon per liter are

obtained from a calibration curve. In some instances, oxidation of the organic matter is assisted by a wet oxidation (50). Hence, an oxidizing agent is added to the sample. Figure 6 shows the outline of a simplified system employed for TOC determination. An ultrapure gas stream (e.g., oxygen (50), helium (51)) is used to drive the CO_2 toward the detector.

The inorganic carbon can be either eliminated, to avoid interferences or measured. For IC determination, the sample can be injected into a separate reaction chamber packed with phosphoric acid–coated quartz beads, where all the IC is converted to CO_2, which is then measured. Under these conditions organic carbon is not oxidized and only IC is measured, as before. The TOC value can be obtained by substracting IC from TC.

The POC is a very interesting parameter in order to survey whether there are some synthetic organics. Note that these compounds are only slightly soluble in water and that by passing a gas stream they can be eliminated. The POC is determined by sparging the sample at ambient temperature. The purgeable components are further trapped and thermally desorbed and driven to the high-temperature zone where oxidation to CO_2 is produced. The determination of NPOC involves the injection of the sample, its acidification, and sparging.

Table 3 Different Total Carbon Fractions

Carbon fraction	Abbreviation	Definition
Total carbon	TC	—
Inorganic carbon	IC	Carbonate, bicarbonate, and dissolved CO_2
Total organic carbon	TOC	Carbon atoms covalently bonded in organic molecules
Purgeable organic carbon	POC	Referred to as VOC (volatile organic carbon): the fraction of TOC removed from an aqueous solution by gas sparging under specified conditions
Nonpurgeable organic carbon	NPOC	Fraction of TOC not removed by gas sparging
Dissolved organic carbon	DOC	Fraction of TOC not retained in a 0.45-μm-pore-diameter filter
Nondissolved organic carbon	NDOC	Fraction of TOC retained in a 0.45-μm-pore-diameter filter

Some commercially available TOC analyzers have appeared (64,65). These systems can be used to determine any TC fraction by selecting the appropriate automatic program. The reported significant figures are: dynamic range: 0.2–50,000 mg C/L; time for analysis: 2–3 min; precision: 1–2% (RSD).

As regards the detection system, some other choices have been investigated in order to get better sensitivities, such as an ion chromatograph placed after the CO_2 scrubbing on a KOH solution (50) or the conversion of CO_2 to methane and its further detection via a hydrogen–flame ionization detector (51). By using these sensitive techniques, the LODs are reduced from 2 mg C/L (44) to 2 μg C/L (50). By modifying the

sample volume, the dynamic linear range goes from 10 μg C/L to 5 mg C/L (44).

B. Other Methods

The main problem encountered in determining TOC by the method just described (Sec. IV.A) is that the system is very expensive. It is difficult to maintain the pyrolisis tube at 950°C, and this increases the cost. For this reason, various alternative methods have been described (2), among them oxidation of the organic matter by ultraviolet radiation (also applied to DOC determination) (46,52–56) and oxidation of the organic matter by persulfate (57).

In the most common method, the solution is irradiated with near-ultraviolet radiation (200–400 nm) to decompose organic matter by means of a radical formation mechanism. Then the generated CO_2 is trans-

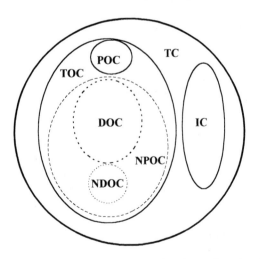

Fig. 5 Distribution of the different fractions of total carbon.

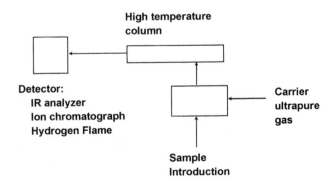

Fig. 6 Simplified scheme of a TOC determination system.

ported toward the detector with a carrier gas. In order to eliminate some ionic compounds that can interfere with the measurement, a membrane is placed before the detector. The detection is carried out either by the measurement of conductivity via a sensor or by a non-dispersive infrared analyzer. In this online system the sample analysis takes around 6 minutes. Other systems based on the same principle have also been described. In this case the oxidation and detection are produced in the same chamber. In this "batch" apparatus the sample is trapped and analyzed for 3–30 minutes. With this latter system some ionic species other than H^+ and HCO_3^- can interfere with the conductivity readings. Species such as TiO_2 (52,58) and persulfate (51,58–60) have been used as catalysts present as a diluted suspension in water. The TOC is obtained from the difference between the conductivities for the irradiated and nonirradiated samples.

The chloride is a source of interferences in TOC determination, since it scavenges the free radicals that are the principal agents of oxidation. In addition, the oxidation of chloride to chlorine can produce detector failure. Therefore, these methods are not advisable to determine TOC in seawater. The photodecomposition method is not suitable for refractory nonpurgeable organic compounds.

Dissolved organic carbon (DOC) is a parameter of great interest to the determination of the effectiveness of wastewater treatment procedures. The radiative methods employed in DOC determination (49,52–54,57,59) are basically the same as for TOC determination. The FIA technique has also been applied to determine DOC (2,55). In this case the sample and persulfate are mixed online and driven to a UV photoreactor. The hydroxyl radicals generated by persulfate are strong oxidants, and the DOC is transformed to CO_2 that, under the rather high pH, is present as bicarbonate and carbonate. The method is followed by the acidification of the stream and the separation of the CO_2. Finally, the decrease in the absorbance at 552 nm is measured. By applying this system, the LOD has been 0.09–0.22 mg C/L, with a dynamic range from 0.1 to 2 mg C/L (55). Precision values as good as 3% (RSD from 16 replicates) are reported. With this system, up to 45 samples/h can be analyzed.

Sharp (59) has concluded that the HCT methods are better than the chemical or radiation methods, since they are easy to handle and exhibit better reproducibility. In addition, the thermal combustion method can be applied to oxidize both volatile and nonvolatile organic compounds, whereas the wet oxidation methods are applicable only to nonvolatile species. This can par-

tially account for the discrepancies in results found between the various methods employed (2). On this subject, a solution has been developed consisting of the use of a diffusion technique (60). The radiation-based methods, in turn, do not need an oxidant, but only trace amounts of a catalyst.

V. TOTAL ORGANIC HALIDE (TOX)

Water treatment with chlorine leads to the formation of trihalomethane polychlorinated biphenils (PCBs) and other volatile and nonvolatile organic halides (1,66,67). Due to their high toxicity, it is very important to control the concentration of these compounds. Total organic halide (TOX) is the parameter used to estimate the total amount of organic halide present in water (3). The TOX value is a complex function of several parameters (i.e., pH, temperature, amount of organic matter and halogen, etc.) (68). As for carbon, there is also a parameter called dissolved organic halide (DOX), which reflects the amount of organic halogenated matter dissolved in a water sample. As with some parameters just discussed, TOX does not give any information about the structure or bonds between halogen and carbon.

The TOX is useful for screening a large number of samples before performing a specific analysis. In addition it is used for: (a) extensive field surveying of the degree of pollution by certain classes of synthetic organic compounds in natural waters; (b) mapping the extent of organohalide contamination in groundwater; (c) monitoring the breakthrough of some synthetic organic compounds in water treatment processes; and (d) estimating the level of formation of chlorinated organic by-products after treatment with chlorine. There is always a possibility of overestimating TOX concentration, because the inorganic halide contribution should be always considered when interpreting results.

The method most often employed has four steps and is called the adsorption-pyrolisis-titrimetric method (67,69,70).

A. Adsorption-Pyrolisis-Titrimetric Method

The four processes involved are: (a) all the halogenated compounds are adsorbed on activated carbon; (b) adsorbed inorganic halogens are displaced by means of nitrate; (c) the activated carbon with only the halogenated organic compounds adsorbed is pyrolized, and the halogens bonded to carbon are transformed to their corresponding halides (X^-), whereas the carbon yields

Table 4 Some "Official" Methods for Determination of Organic Pollution Parameters

Parameter/matrix	Normal method reference
BOD in wastewater	USEPA Method 405.1
COD	AE Greenburg, LS Clesceri, AD Eaton. Standard Methods for the Examination of Water and Wastewater. 18th ed. American Public Health Association, 1992.
TOC in raw water and groundwater	EPA Methods 9060 and 415.1; ASTM Methods D2479, D4779, and D4839; 1997 European Standards ISO/CEN prEN 1484
TOX, EOX, and POX in wastewater	EPA Method 600/4-82-057, 1982; LS Clesceri, AE Greenburg, TR Rhodes. Standard Methods for the Examination of Water and Waastewater. 17th ed. American Public Health Association, 1989.
TOX in groundwater	USEPA Method 450.1
AOX	October 1996 European Standards ISO/CEN prEN 1485. Water Quality—Determination of Adsorbable Organically Bonded Halogens (AOX)

CO_2; and (d) X^- is driven to a microcoulometric titration cell, where it is quantified by measuring the current produced by silver ion precipitation of the halides.

In this method the pH of the sample must be adjusted to a value of 2 with nitric acid. If there are suspended solids, the sample must be filtered. Then the treated solution is forced to pass through a carbon-active microcolumn. To drag the inorganic halide adsorbed, 5 ml of potassium nitrate are passed through the microcolumn filled with the activated carbon. Finally, the microcolumn is pyrolized and the mass of HX is determined. The mass of HX is equivalent to the milligrams of X in the injected sample. The same treatment is applied to both blanks and samples.

The determination of adsorbable organic halide (AOX) (71–73) is also very useful for samples with high levels of solids, since the columns employed in the TOX test can be blocked. The AOX assay includes a procedure in which carbon is shaken with the sample in a flask. Then it is filtered. Both filter and carbon are combusted. The method is sensitive to inorganic chlorine, bromine, and iodine, but it is not sensitive to fluorine compounds. For that reason, solutions involving the addition of sulfite and nitrite have been proposed. In general, this method cannot be applied for chloride concentrations greater than 500 mg/L, although the concentrations of halogenated compounds in water are usually smaller than 100 μg/L (67,69). The AOX includes some specific compounds (e.g., AOCl, AOBr, AOI). In a recent work (74) these species have been determined by coupling the AOX analyzer with an ionic chromatograph. Besides TOX and AOX, extractable organic halide (EOX) is also used; it is the fraction that can be removed from the sample by solvent extraction (3).

As in TOC determination, when samples are initially purged with inert gas, nonpurgeable organic halide (NPOX) is determined. The purgeable fraction is named purgeable organic halide (POX). The TOX value should be POX + EOX. Nevertheless, with the methods employed in the literature it has been found that the TOX values are much greater than the sum of POX and EOX. This indicates that a large quantity of organic halides is present in the sample that is carbon adsorbable but not solvent extractable. In conclusion, the EOX test is not suitable for the determination of some polar organic halides. Also, to determine these parameters some instruments are available [64].

B. Direct Measurement of Absorbance

This method is based on the fact that natural organic matter exhibits an aromatic character that produces a strong radiation absorption in the UV spectrum region. When these compounds react with halogens, a sharp decrease in absorbance is observed. Therefore, TOX can be correlated with this absorbance drop, obtaining very good results (75). In these studies the wavelength can be measured at either 272 or 254 nm.

Finally, to conclude the current chapter, Table 4 references some of the multiple methods established by "official" organisms to determine the parameters presented up to this point. A more complete list of reference methods can be drawn from Ref. 76.

ABBREVIATIONS

a	volume of titrating solution employed with the blank
AOX	adsorbable organic halide

B	volume of titrating solution employed with the sample
B_0	dissolved oxygen in the dilution water before incubation
B_5	dissolved oxygen in the dilution water after 5 days of incubation
BOD	biochemical oxygen demand
BOD_5	biochemical oxygen demand after an incubation period of 5 days
BODS	BOD supplied by sensors
C	Normality of the titrating solution
CBOD	carbonaceous BOD
COD	chemical oxygen demand
D_0	dissolved oxygen in the diluted sample after preparation
D_5	dissolved oxygen in the diluted sample after 5 days of incubation
DO	dissolved oxygen
DOC	dissolved organic carbon
DOX	dissolved organic halide
EOX	extractable organic halide
f	ratio of the % seed in the diluted water to the % seed in the dilution water
f'	titrating solution correction factor
FAAS	flame atomic absorption spectrometry
FIA	flow injection analysis
GC	gas chromatography
HBOD	headspace biochemical oxygen demand
HCT	high combustion temperature
IC	inorganic carbon
LOD	limit of detection
ml_{sample}	volume of the water sample analyzed
NDOC	nondissolved organic carbon
NPOC	nonpurgeable organic carbon
NPOX	nonpurgeable organic halide
NVOC	nonvolatile organic carbon
P	decimal volumetric fraction of sample used
PCB	polychlorinated biphenil
POC	purgeable organic carbon
POX	purgeable organic halide
R	regression coefficient
TC	total carbon
TOC	total organic carbon
TOX	total organic halide
VOC	volatile organic carbon

ACKNOWLEDGMENTS

The authors wish to express their appreciation to the DGICyT (Spain) for the financial support of this work (Project No. PB95-0693). Also A. Cuesta gratefully thanks the Consellería de Educación y Ciencia (Generalitat Valenciana) for the grant.

REFERENCES

1. RE Clement, ML Langhorst, GA Eiceman. Environmental analysis. Anal. Chem. 63:270R–292R, 1991.
2. K Robards, ID McKelvie, RL Benson, PJ Worsfold, NJ Blundell, H Casey. Determination of carbon, phosphorous, nitrogen and silicon species in waters. Anal. Chim. Acta 287:147–190, 1994.
3. WQA Glossary of Terms. Water Quality Association web site. http://www.wqa.org. October 1999.
4. J Suschka, E Ferreira. Activated sludge respirometric measurements. Water Res. 20:137–144, 1986.
5. M Hikuma, H Suzuki, T Yasuda, I Karube, S Suzuki. Amperometric estimation of BOD by using living immobilized yeasts. European J. Appl. Microbiol. Biotechnol. 8:289–297, 1979.
6. K Riedel, KP Lange, HJ Stein, M Kühn, P Ott, F Sheller. A microbial sensor for BOD. Water Res. 24:883–887, 1990.
7. TC Tan, F Li, KG Neoh. Measurement of BOD by initial rate response of a microbial sensor. Sens. Act. B 10:137–142, 1993.
8. C Preininger, I Klimant, OS Wolfbeis. Optical fiber sensor for biochemical oxygen demand. Anal. Chem. 66:1841–1846, 1994.
9. E Praet, V Reuter, T Gaillard, JL Vasel. Bioreactors and biomembranes for biochemical oxygen demand estimation. Trends Anal. Chem. 14:371–378, 1995.
10. Z Yang, H Suzuki, S Sasaki, I Karube. Disposable sensor for biochemical oxygen demand. Appl. Microbiol. Biotechnol. 46:10–14, 1996.
11. Z Qian, TC Tan. Response characteristics of a dead-cell BOD sensor. Water Res. 32:801–807, 1998.
12. BE Logan, GA Wagenseller. The HBOD test: a new method for determining biochemical oxygen demand. Water Environ. Res. 65:862–868, 1993.
13. BE Logan, R Patnaik. A gas chromatographic–based headspace biochemical oxygen demand test. Water Environ. Res. 69:206–214, 1997.
14. SKE Brookman. Estimation of Biochemical Oxygen Demand in slurry and effluents using ultra-violet spectrophotometry. Water Res. 31:372–374, 1997.
15. DM Reynolds, R Ahmad. Rapid and direct determination of wastewater BOD values using a fluorescence technique. Water Res. 31:2012–2018, 1997.
16. FJ Baumann. Dichromate reflux chemical oxygen demand. Anal. Chem. 46:1336–1338, 1974.
17. J Hejzlar, J Kopácek. Determination of low chemical oxygen demand values in water by the dichromate semi-micro method. Analyst 115:1463–1467, 1990.

18. DR Himebaugh, MJ Smith. Semi-micro tube method for chemical oxygen demand. Anal. Chem. 51:1085–1087, 1979.

19. AM Jirka, MJ Carter. Micro semi-automated analysis of surface and wastewater for COD. Anal. Chem. 47:1397–1402, 1975.

20. L Karlgren, G Ekedahl. Intercalibration of methods for chemical analysis of water: permanganate methods for determining Chemical Oxygen Demand. Vatten 1:32–43, 1971.

21. J Simal, MA Lage, I Iglesias. Determinación de la demanda Química de oxígeno (DQO) y demanda inmediata de oxígeno (DIO) de un agua, mediante sales de cerio(IV). Anales de Bromatol. 37:125–142, 1985.

22. HL Golterman, AG Wisselo. Ceriometry, a combined method for chemical oxygen demand and dissolved oxygen (with a discussion on the precision of the Winkler technique). Hydrobiologia 77:37–42, 1981.

23. P Selvapathy, JS Jogabeth. A new catalyst for COD determination. Ind. J. Environ. Health 33:96–102, 1991.

24. J Brunini, DA Nalin, DF Angelis, EA Silva-Filho, MJ Correia. Determinação colorimétrica com difenilcarbazida da demanda química de oxigênio em águas e efluentes. Eclética Química 16:1–8, 1991.

25. S Edwards, M Allen. Rapid and precise method for the determination of the chemical oxygen demand of factory effluent samples. Analyst 109:671–672 (1984).

26. M Novic, B Pihlar. M Dular. Use of flow injection analysis based on iodometry for automation of dissolved oxygen (Winkler method) and chemical oxygen demand (dichromate method) determinations. Fresenius Z. Anal. Chem. 332:750–755, 1988.

27. C Billings, M Lane, A Watson, TP Whitehead, G Thorpe. A rapid and simple chemiluminescent assay for water quality monitoring. Analusis 22:27–30, 1994.

28. S Suryani, F Theraulaz, O Thomas. Deterministic resolution of molecular absorption spectra of aqueous solutions: environmental applications. Trends Anal. Chem. 14:457–463, 1995.

29. RJ García-Villanoca, JA Gómez, R Sangrador. La correlación DQO. Absorbancia de las aguas: una cómoda opción donde se ensaye su viabilidad. Anal. Bormatol. XLIV-1:59–63, 1992.

30. E Naffrechoux, C Fachinger, J Suptil. Diode-array ultraviolet detector for continuous monitoring of water quality. Anal. Chim. Acta 270:187–193, 1992.

31. PK Dasgupta, K Petersen. Kinetic approach to the measurement of chemical oxygen demand with an automated micro batch analyzer. Anal. Chem. 62:395–402, 1990.

32. NS Abuazid, MH Al-Malack, AH El-Mubarak. Alternative method for determination of the COD for colloidal polymeric wastewater. Bull. Environ. Contam. Toxicol. 59:626–633, 1997.

33. J Ruzicka, EH Hansen. Flow injection analyses, Part I: A new concept of fast continuous flow analysis. Anal. Chim. Acta 78:145–157, 1975.

34. T Korenaga. Flow injection analysis using potassium permanganate: an approach for measuring chemical oxygen demand in organic wastes and waters. Anal. Lett. 13:1001–1011, 1980.

35. T Korenaga, H Ikatsu. Continuous flow injection analysis of aqueous environmental samples for chemical oxygen demand. Analyst 106:653–662, 1981.

36. T Korenaga, H Ikatsu. The determination of chemical oxygen demand in wastewaters with dichromate by flow injection analysis. Anal. Chim. Acta 141:301–309, 1982.

37. T Korenaga, X Zhou, O Kimiko, T Moriwake, S Shinoda. Determination of chemical oxygen demand by a flow injection method using cerium(IV) sulphate as oxidizing agent. Anal. Chim. Acta 272:237–244, 1993.

38. JMH Appleton, JF Tyson, RP Mounce. The rapid determination of chemical oxygen demand in wastewaters and effluents by flow injection analysis. Anal. Chim. Acta 179:269–278, 1986.

39. H Tanaka. Application of pretreatment methods using electrolysis-generated redox reagent for flow analysis. Bunseki Kagaku 47:79–91, 1998.

40. ML Balconi, M Borgarello, R Ferraroli, F Realini. Chemical oxygen demand determination in well and river waters by flow-injection analysis using a microwave oven during the oxidation step. Anal. Chim. Acta 261:295–299, 1992.

41. A Cuesta, JL Todolí, A Canals. Flow injection method for the rapid determination of chemical oxygen demand based on microwave digestion and chromium speciation in flame atomic absorption spectrometry. Spectrochim. Acta Part B. 51:1791–1800, 1996.

42. A Cuesta, JL Todolí, J Mora, A Canals. Rapid determination of chemical oxygen demand by a semiautomated method based on microwave sample digestion, chromium(VI) organic solvent extraction and flame atomic absorption spectrometry. Anal. Chim. Acta 372:359–370, 1998.

43. HAC Montgomery, NS Thom. The determination of low concentrations of organic carbon in water. Analyst 87:689–697, 1962.

44. CE Van Hall, J Safranko, VA Stenger. Rapid combustion method for the determination of organic substances in aqueous solutions. Anal. Chem. 35:315–319, 1963.

45. AEJ Miller, RFC Mantoura, MR Preston. Shipboard investigation of DOC in the northeast Atlantic using platinum-based catalysts in a Shimadzu TOC-500 HTCO analyzer. Mar. Chem. 41:215–221, 1993.

46. R Benner, JI Hedges. A test of the accuracy of freshwater DOC measurements by high-temperature catalytic oxidation and UV-promoted persulfate oxidation. Mar. Chem. 41:161–165, 1993.

47. R Benner, M Strom. A critical evaluation of the analytical blank associated with DOC measurements by high-temperature catalytic oxidation. Mar. Chem. 41:153–160, 1993.

48. T Fukushima, A Imai, K Matsushige, M Aizaki, A Otsuki. Freshwater DOC measurements by high-temperature combustion: comparison of differential (DTC-DIC) and DIC purging methods. Water Res. 30:2717–2722, 1996.

49. JB Christensen, DL Jensen, C Gron, Z Filip, TH Christensen. Characterization of the dissolved organic carbon in landfill leachate-polluted groundwater. Water Res. 32:125–135, 1998.

50. YS Fung, Z Wu, KL Dao. Determination of total organic carbon in water by thermal combustion–ion chromatography. Anal. Chem. 68:2186–2190, 1996.

51. RA Dobbs, RH Wise, RB Dean. Measurement of organic carbon in water using the hydrogen-flame ionization detector. Anal. Chem. 39:1255–1258, 1967.

52. MI Abdullah, E Eek. Automatic photocatalytic method for the determination of dissolved organic carbon (DOC) in natural waters. Water Res. 30:1813–1822, 1996.

53. PD Goulden, P Brooksbank. Automated determination of dissolved organic carbon in lake water. Anal. Chem. 47:1943–1946, 1975.

54. RA Van Steenderen, JS Lin. Determination of dissolved organic carbon in water. Anal. Chem. 53:2157–2158, 1981.

55. RT Edwards, ID McKelvie, PC Ferret, BT Hart, JB Bapat, K Koshy. Sensitive flow-injection technique for the determination of dissolved organic carbon in natural wastewaters. Anal. Chim. Acta 261:287–294, 1992.

56. W Stefanska, S Rubel. Potentiometric and conductimetric determination of TOC in water samples. Chem. Anal. 41:1015–1020, 1996.

57. JH McKenna, PH Doering. Measurement of dissolved organic carbon by wet chemical oxidation with persulfate: influence of chloride concentration and reagent volume. Mar. Chem. 48:109–114, 1995.

58. GCK Low, RW Matthews. Flow-injection determination of organic contaminants in water using an ultraviolet-mediated titanium dioxide film reactor. Anal. Chim. Acta 231:13–20, 1990.

59. JH Sharp. Marine dissolved organic carbon: are the older values correct? Mar. Chem. 56:265–277, 1997.

60. CE Van Hall, D Barth, VA Stenger. Elimination of carbonates from aqueous solutions prior to organic carbon determinations. Anal. Chem. 37:769–771, 1965.

61. I Gacs, K Payer. Determination of total organic carbon in water in the form of carbamate by means of a simple denuder/electrolytic conductivity flow cell system. Anal. Chim. Acta 220:1–11, 1989.

62. SA Huber, FH Frimmel. Flow injection analysis of organic and inorganic carbon in the low-ppb range. Anal. Chem. 63:2122–2130, 1991.

63. SW Kubala, DC Tilotta, MA Busch, KW Busch. Determination of total inorganic carbon in aqueous samples with a flame infrared emission detector. Anal. Chem. 61:1841–1846, 1989.

64. Dohrmann Web Site. *http://www.tekmar.com/products*, October 1998.

65. Shimadzu Web Site. *http://www.shimadzu.de/products*, October 1998.

66. Z Huixian, Y Sheng, X Xu, X Ouyong. Formation of POX and NPOX with chlorination of fulvinc acid in water: empirical models. Water Res. 31:1536–1541, 1997.

67. L Hureiki, JP Croué and B Legube. Chlorination studies of free and combined amino acids. Water Res. 28:2521–2531, 1994.

68. GL Amy, PA Chadik, ZK Chowdhury. Developing models for predicting trihalometane formationpotential and kinetics. J. Am. Water Works Assn. 79:89–97, 1987.

69. DL Berger. Screening procedure for total organic halogen. Anal. Chem. 56:2271–2272, 1984.

70. G Sansebastiano, V Rebizzi, E Bellelli, F Sciarrone, S Reverberi, S Saglia, A Braccaioli, P Camerlengo. Total organic halide (TOX) measurement in ground waters treated with hypochlorite and with chlorine dioxide. Initial results of an experiment carried out in Emilia-Romagna (Italy). Water Res. 29:1207–1209, 1995.

71. DW Francis, PA Turner, JT Wearing. AOX reduction of kraft bleach plant effluent by chemical pretreatment-plo scale trials. Water Res. 31:2397–2404, 1997.

72. Behavior of DOC and TOX using advanced treated wastewater for groundwater recharge. Water Res. 32:3125–3133, 1998.

73. F Archibald, L Roy-Arcand, M Methot. Time, sunlight and the fate of biotrated kraft mill organochlorines (AOX) in nature. Water Res. 31:85–94, 1997.

74. J Oleksy-Frenzel, S Wischnack, M Jekel. Determination of organic group parameters AOCl, AOBr and AOI in municipal wastewater. Vom Wasser 85:59–67, 1995.

75. GV Korshin, L Chi-Wang, MM Benjamin. The decrease of UV absorbance as an indicator of TOX formation. Water Res. 31:946–949, 1997.

76. State of the Art Web Site. *http://www.stateoftheart.it*, October 1998.

17

Organic Acid Analysis

Sigrid Peldszus

University of Waterloo, Waterloo, Ontario, Canada

I. INTRODUCTION

A. General Background

The common characteristic of organic acids is a carboxylate function on a hydrocarbon structure. This structure can vary considerably from aliphatic to aromatic, saturated to unsaturated, and straight chain to branched. Di- and tricarboxylic acids as well as substituted organic acids (e.g., substitution with hydroxyl or keto groups) play important roles in the metabolism of living organisms. However, the focus in the analysis of organic acids in water is mainly on short-chain organic acids. The most prevalent ones are listed, together with pKa values and structures, in Table 1.

A wide range of analytical methods are available for the determination of organic acids in various matrices. These methods range from traditional distillation and titration to gas chromatography (GC) methods, which usually include extraction and derivatization steps, to the use of ion chromatography or capillary electrophoresis. This chapter presents an overview of the currently available literature. The reader is encouraged to consult the cited references for further detail.

B. Need for Analysis

Organic acids can be found in very different water matrices, ranging from wastewater to drinking water to ultrapure water used in industry. With each water type, the concentrations of these acids vary from a few micrograms per liter to several hundred milligrams per liter. These two factors, water matrix and organic acid concentration, largely determine which analytical methods are used. Before discussing these methods in detail, the motivation behind organic acid analysis is briefly described.

1. Drinking Water

Short-chain organic acids (e.g., formic, acetic, oxalic, glyoxylic, pyruvic, and ketomalonic acids) can be found during drinking water treatment as the result of the utilization of ozone (1–9). As a strong oxidant, ozone is applied mainly for disinfection purposes but also for the destruction of taste and odor compounds such as geosmin or 2-methylisoborneol (MIB) during drinking water treatment (10). Other applications of ozone include color removal and pretreatment. During treatment, organic matter, which is present in the raw water, is oxidized by ozone, leading to the formation of ozonation by-products. As a consequence, aldehydes and organic acids are formed in μg/L concentrations, with organic acids being present in significantly higher concentrations. Aldehydes are normally completely removed during filtration, whereas organic acids are often only partially removed (6). Smaller concentrations of these acids are sometimes passed on into the finished drinking water, where they can serve as nutrients for microorganisms. This might lead to bacterial regrowth in the drinking water distribution system.

Only in the recent past has ozonation gained more in interest, at least in North America. It is used more and more in drinking water treatment to substitute chlorine at least partially, thus lowering chlorinated disinfection by-products. Hence, interest developed to de-

Table 1 Structures and pKa Values of Organic Acids

	Formula	Structure	pKa 1	pKa 2	MW
Monocarboxylic acids					
Formic acid	CH_2O_2	HCOOH	3.75^a	—	46.02
Acetic acid	$C_2H_4O_2$	CH_3–COOH	4.75^a	—	60.05
Propionic acid	$C_3H_6O_2$	CH_3–CH_2–COOH	4.87^a	—	74.08
Butyric acid	$C_4H_8O_2$	CH_3–$(CH_2)_2$–COOH	4.82^a	—	88.10
Valeric acid	$C_5H_{10}O_2$	CH_3–$(CH_2)_3$–COOH	4.86^a	—	102.13
Hydroxyacids					
Glycolic acid[1]	$C_2H_4O_3$	H_2COH–COOH	3.83^b	—	76.05
Lactic acid[2]	$C_3H_6O_3$	CH_3–HCOH–COOH	3.89^a	—	90.08
Ketoacids					
Glyoxylic acid[3]	$C_2H_2O_3$	CHO–COOH	3.34^c	—	74.04
Pyruvic acid[4]	$C_3H_4O_3$	CH_3–CO–COOH			88.06
α-Ketobutyric acid	$C_4H_6O_3$	CH_3–CH_2–CO–COOH	2.49^c	—	102.08
Dicarboxylic acids					
Oxalic acid[5]	$C_2H_2O_4$	HOOC–COOH	1.23^b	4.19^b	90.04
Malonic acid[6]	$C_3H_4O_4$	HOOC–CH_2–COOH	2.83^b	5.69^b	104.06
Oxalacetic acid[7]	$C_4H_4O_4$	HOOC–CO–CH_2–COOH	2.22^a	3.89^a	132.07

[a]Ref. 127.
[b]Ref. 128.
[c]Ref. 129.
[1]Hydroxyacetic acid.
[2]2-Hydroxypropanoic acid.
[3]Oxoacetic acid or formylformic acid.
[4]2-Oxopropanoic acid = α-ketopropionic acid = acetylformic acid = pyroracemic acid.
[5]Ethanedioic acid.
[6]Propenedioic acid = methanedicarboxylic acid.
[7]Oxobutanedioic acid = oxosuccinic acid = ketosuccinic acid.

termine the identity and quantity of by-products generated during ozonation, even though they are not yet regulated.

The challenge in analyzing these organic acids in drinking water lies in the fact that the organic acids are present in μg/L concentrations. Also, the analysis may be complicated by the presence of other organic compounds and inorganic anions at mg/L concentrations. Different approaches have been taken in developing suitable methods in order to achieve the required sensitivity (Table 2).

2. Wastewater

Volatile organic acids or volatile fatty acids (VFA) are defined as water-soluble fatty acids that can be distilled at atmospheric pressure (11). They consist of aliphatic, monocarboxylic acids with chain length from C1 to C7. The VFA are formed during wastewater treatment when using anaerobic digesters (12,13). Here, organic carbon is degraded into methane in two consecutive steps by two different groups of bacteria under anaerobic conditions. First, acidogenic bacteria convert organic car-

bon into VFA, mainly acetic acid. This is followed by VFA consumption through methanogenic bacteria, producing methane. These two processes have to be in balance to ensure smooth operation. Methanogenic bacteria have a relatively small population in these reactors. In addition, they are more susceptible to a pH change than are acidogenic bacteria. Hence, if unbalanced conditions are experienced, such as a rapid pH change, methanogenic bacteria are inhibited first, whereas acidogenic bacteria keep growing. This leads to a significant increase in VFA production, ultimately lowering the pH even further and thus endangering the whole treatment process. Usually changes in VFA concentration can be detected long before pH changes, thus making them an indicator for disturbances in the biological degradation process. Therefore, monitoring the concentration of VFA in anaerobic digesters is viewed as a crucial control parameter and is typically measured on a daily basis (12–14).

Wastewater possesses a complex matrix, with high concentrations of inorganic and organic compounds. The VFA are present in high mg/L concentrations, thus simplifying analysis, in the sense that only moderate detection limits and accuracy are required. The VFA have been analyzed regularly on a routine basis, with the consequence that a range of methods exists, from traditional distillation and titration procedures to direct aqueous injections into gas chromatographs equipped with special columns.

3. Atmospheric Precipitation

a. Rain.   More recently, organic acids, especially acetate, formate, and oxalate, have been measured in environmental samples such as rain, snow, ice, and fog in μmol/L concentrations (15–24). Various sources are being discussed as contributors to the presence of organic acids in rain. Direct emission from anthropogenic sources include vehicle exhausts (25,26) and combustion of coal, wood, or animal waste (27,28). Organic acids emitted through these sources into the air may then be washed out of the atmosphere by wet precipitation. Although little is known about natural biogenic sources, they are suspected to contribute significantly to the overall organic acid concentration in rain (29). Finally, organic precursors (e.g., hydrocarbons emitted from vehicles or other sources of emission) are oxidized in the atmosphere by photochemical reactions (radical chain reactions) (30). Detailed mechanisms, which involve complex interactions between radicals in the gaseous phase, the aqueous phase, and adsorption on particles, are still under discussion. One theory is that aldehydes, which presumably act as organic acid precursors, are oxidized by OH radicals and hydrogen peroxide to organic acids (23,31,32). It is reported that on average organic acids account for about a third of the total free acidity in North American rain (17,23). Consequently, an interest is developing to determine the extent to which organic acids are contributing to acidic rain and the damages this may cause. Current research continues to catalog organic acid precipitation patterns further and to elucidate the origin and formation of organic acids in rain (33). However, organic acids are not regulated as air pollutants.

Beside organic acids, rainwater matrix consists of inorganic anions (with sulfate, chloride, and nitrate being the major ones), cations, and other organic compounds. The overall matrix composition of rainwater is considerably less complex, and the concentrations of these rainwater constituents are about three orders of magnitude lower than in wastewater. As a result, efforts are currently being made successfully to determine inorganic anions and organic acids within one method rather than focusing on organic acids exclusively.

b. Fog, Mist, Snow, and Ice.   As expected, organic acids are found in fog, snow, and ice core samples (18,19,21,22,24,25,30,33,35,36). The formation and occurrence of these acids and the motivation for their analysis are similar to the ones described for rainwater. The study of ice core samples has the added advantage of putting current results into a historical context (22).

Fog, snow, and ice core samples require matrix-specific sample collection and preparation, which will not be discussed here. Nevertheless, these matrices are very similar to rainwater in their overall composition and the concentrations of their constituents, once phase transformation from the solid state (ice, snow) (21,22,24,25,36) or aerosol (mist, fog) (18,19,30,33–35) into water has taken place. Thus, the actual organic acid measurements are similar to the ones developed for rainwater, which is why these methods are included in this chapter.

4. Other Waters

a. Seawater.   The main focus in the analysis of seawater used to be on inorganic compounds. To get a more complete picture of this environment, organic components have recently been analyzed as well. However, only rarely is this matrix monitored for organic acids. Seawater contains high concentrations of inorganic ions plus very low concentrations of organic acids, thus turning organic acid analysis into a challenge. Only very few methods exist that attempt to analyze this difficult matrix for organic acids (37–42).

b. Groundwater.   Organic acids in groundwater have been measured in connection with organic contamination, such as hydrocarbon spills. Biological degradation of these contaminants under anaerobic conditions leads to the formation of organic acids as intermediates, with concentrations of up to 1 mmol/L (43–46). However, organic acid concentrations can vary significantly, depending on factors such as proximity to the original contaminant, availability of oxygen, availability of nutrients, type of microbial population, and presence of microbial inhibitors.

Moderate amounts of inorganic ions, the original contaminants, and further intermediates may be found in this type of groundwater in addition to organic acids. Several methods have been developed for this kind of analysis (43,47,48).

c. Landfill Leachates.   During the anaerobic degradation of organic waste in a landfill, organic acids are formed in high concentrations in the acidogenic phase (49). These intermediates, predominantly VFA, are

Table 2 Overview of Organic Acid Methods, Sorted by Matrix

Sample preparation	Instrumentation	Acids determined	Detection limits	Comments	Refs.
Drinking water					
Aqueous PFBHA oximation, extraction, CH_2N_2 methylation	GC/ECD	Gyoxylic, pyruvic, ketomalonic acids	na		1–3
Aqueous PFBHA oximation, extraction, CH_2N_2 methylation	GC/ECD	Gyoxylic acid	na	Simultaneous detection of several aldehydes	60
Aqueous PFBHA oximation, extraction, BF_3/methanol methylation, MTBSTFA sylilation	GC/MS (EI, CI)	Gyoxylic, pyruvic, ketomalonic, 4-oxobutenoic acids and others	na	Simultaneous detection of aldehydes and hydroxyl substituted compounds; focused on identification of compounds	4, 66
Direct injection	AIC/cond	Formic, acetic, oxalic acids	2–3 μg/L; MRL 15 μg/L	Removal of $HgCl_2$ with H^+ cartridge in line	5
Direct injection	AIC/cond	Formic, acetic, butyric, β-hydroxybutyric, glycolic, pyruvice, α-ketobutyric, oxalic acids	1–9 μg/L	Oxalate determination in matrices with high ionic strength requires switching technique	7–9
Filtration; sulfate, chloride and phosphate removal	AIC/cond?	Formic, acetic, oxalic, glycolic acids	20–40 μg/L; with concentrator 0.1–0.5 μg/L	No testing on "real" drinking water samples described	66
Wastewater					
Anion-exchange extraction/CH_2N_2 methylation	GC/ECD	Acetic acid and other VFA	na		65
Centrifugation and/or filtration, acidification	GC/FID	C2–C5 monocarboxylic acids	na	Direct aqueous injection	61, 62, 64, 123–126
Centrifugation, acidification	HPLC/UV	Formic acid	na	Complements GC/FID method for VFA	123, 124
Rain					
Concentration; derivatization to p-bromophenacyl esters; cleanup	GC/FID	C1–C7 aliphatic monocarboxylic acids	~0.1 μmol/L	Time-consuming sample preparation procedure; but confirmation with GC/MS possible	18
Concentration; derivatization to p-bromophenacyl esters; cleanup	GC/FID	Formic, acetic acid	~0.1 μmol/L	Time-consuming sample preparation procedure; but confirmation with GC/MS possible	23
Concentration by evaporation at pH8–9, derivatization with BF_3/butanol; removal of excess butanol	GC/FID or GC/MS	More than 20 C2–C10 dicarboxylic acids	na, but <0.8 μM	Rain, fog and mist, very time-consuming sample preparation	19
Concentration by evaporation; derivatization with BF_3/butanol	GC/FID or GC/MS	C2–C10 α-oxacids, including glyoxylic (C2) and pyruvic acid (α-carboxylic acid)	0.05 μg/L	Simultaneous determination of the aldehydes glyoxal and methylglyoxal	34

Sample preparation	Technique	Acids analyzed	Detection limit	Comments	Ref.
Concentration; derivatization with α,p-bromoacetophenone	GC/FID or HPLC	C1–C5 monocarboxylic acids	~20–50 μg/L	Used for rain, sewage, and soil pore water	71
Direct injection	AIC/cond	2- and 3-Hydroxybutyric, lactic, acetic, glycolic, propionic, formic, butyric, pyruvic, valeric, oxalic acids	0.5–1 μmol/L	Simultaneous detection of inorganic anions (F^-, Cl^-, NO_2^-, CO_3^{2-}, PO_4^{3-}, Br^-, SO_3^{2-}, SO_4^{2-})	48
Direct injection	AIC/cond	Formic, acetic, oxalic acids	20–100 μg/L	Simultaneous detection of inorganic anions (F^-, Cl^-, NO_2^-, NO_3^-, Br^-, SO_3^{2-}, SO_4^{2-}, PO_4^{3-})	24
Direct injection	AIC/cond	Formic, acetic, pyruvic acids	0.2 μmol/L	Method optimized for formic and acetic acids only	58
Direct injection	AIC/cond	Formic, acetic, pyruvic acids	0.02–0.1 μmol/L	Mist samples, method info sparse	59
Direct injection	IEC	Formic, acetic acids	na, but <0.6 mg/L	Describes need for sample preservation with $CHCl_3$	17
Addition of HCl to sample; direct injection	IEC	Formic, acetic; citric, lactic, glycolic, propionic, butyric, and valeric acids	na, but <0.6 mg/L	Focuses on the analysis of the more abundant acids, formate and acetate; sample preservation with $CHCl_3$	16
Direct injection or sample stacking	CZE/indirect UV	Formic, acetic, oxalic, malonic acid; capability to determine another 10, mainly dicarboxylic acids	Sample stacking: 30–10 nmol/L	Simultaneous detection of inorganic anions (Cl^-, NO_3^-, SO_4^{2-}, NO_2^-, Br^-, PO_4^{3-})	56
Direct injection	(a) micro HPLC; (b) CZE	(a) and (b) Formic, acetic acids	(a) 13–20 μg/L (b) 180–450 μg/L	Simultaneous detection of Cl^-, NO_3^- and SO_4^{2-}; determination in single raindrops	57
Derivatization with OPD	HPLC/UV	Oxalic, pyruvic, and glyoxylic acids	na; but <2 μM	Rain, fog and mist; no preconcentration necessary	38
Other environmental samples					
Filtration; direct injection	IEC/cond	Acetic, propionic, i-butyric, n-butyric acids	0.5 μmol/L	Groundwater	43
Outer part of core removed, sample melted	AIC/cond	Formic, acetic, oxalic, glycolic acids	0.2–0.6 ng/g	Ice core; simultaneous detection of inorganic anions (F^-, Cl^-, NO_2^-, NO_3^-, SO_4^{2-})	21
Anion-exchange preconcentrator column	IEC/UV	Formic, acetic, propionic, butyric acid	5.6–9.4 μg/L with 10-mL inj. vol.	Antarctic ice, for example	36
Centrifugation, concentrated at pH 9, acidification	GC/FID	Acetic, propionic, butyric acids	μM to high nM concentrations	Supernatant from wetland sludge; direct aqueous injection	63
Centrifugation, acidification	GC/FID	C2–C7 monocarboxylic acids	2–5 μmol/L	Sediment pore water, direct aqueous injection	47
Concentrated using static diffusion	GC/FID	Acetic, propionic, 1- and 2-butyric, 1- and 2- valeric, pyruvic, acrylic, benzoic acids	Down to 10 nmol/L dep. on conc. factor; acetate ~750 nmol/L	Seawater	37

Table 2 Continued

Sample preparation	Instrumentation	Acids determined	Detection limits	Comments	Refs.
Other matrices					
None	IC/cond	Formic, acetic, oxalic acids	Below 1 μg/L	Ultrapure water, use online preconcentration, simultaneous determination of inorganic anions	55
Centrifugation, filtration, removal of huminlike substances	(a) IEC/UV (b) IEC/cond	Formic to n-valeric acid, pyruvic, glyoxylic, glycolic, lactic, glyceric, succinic acids	na; analyzed from 50 to 50,000 μmol/L	Landfill leachates; developed and compared two IEC methods for organic acids	50
Acidification, distillation	HPLC/UV	Formic, acetic, propionic, n-butyric, and iso-butyric acids	na; analyzed high mg/L concentrations	Landfill leachates; less sensitive than direct GC/FID but capable to analyze formic acid	51
Acidification, distillation	GC/FID	C2–C4 monocarboxylic acids	na; analyzed mg/L concentrations	Landfill leachates; direct aqeous injection	53
Acidification, extraction with diethylether	GC/FID	C2–C7 monocarboxylic acids	na; analyzed mg/L concentrations	Low-level radioactive waste leachates	52
Vacuum distillation; acidification	GC/FID	Acetic, propionic, iso-butyric and n-butyric, methylvaleric (IS), isovaleric, valeric, heptanoic acids	0.05 mmol/L	Biological samples; direct aqeous injection; compared different GC columns and conditions	95
Dilution with oxalic acid (1:10)	GC/FID	C2–C7 monocarboxylic acids; lactic acid	na; analyzed mg/L concentrations	Silage juice; direct aqeous injection; compared different GC columns and conditions	90

Abbreviations used in table: see the Abbreviations list at the end of the chapter.

found to contribute significantly to the dissolved organic matter in landfill leachates. Therefore, measurement of organic acids in leachates can serve as an indicator of the degree of degradation inside the landfill. Another reason for monitoring organic acid concentrations is the possibility of heavy-metal mobilization through acidification. This is especially of concern in some older landfills, which have either no liner or only an incomplete seal towards the groundwater, thus posing a potential hazard to the environment.

Landfill leachates have the character of a high-strength wastewater, with extreme pH and high BOD and COD values, thus making it a complex matrix difficult to analyze. Several methods with very different approaches have been utilized for organic acid analysis in landfill leachates (50–53).

d. *Ultrapure Water Used in Industry.* Industries such as the electronic industry and power-generating plants require ultrapure water. Even very low concentrations (sub-μg/L concentrations) of impurities can cause significant damage either to the product itself or to the equipment utilized. Hence, methods were developed to monitor water quality, thus quantifying ppb concentrations of inorganic ions and/or organic acids in a very "clean" matrix (54,55).

II. ANALYSIS

A large number of methods are available to determine organic acids in various water matrices. Which method to choose for a particular application is to a large extent dependent on sample characteristics, such as the sample matrix and the expected concentration. The sample matrix affects the amount of cleanup and sample preparation necessary, while the analyte concentration dictates the sensitivity requirement of the overall method. The sensitivity of a method can be influenced by sample preparation and the instrumentation utilized. The purpose for which the data are generated determines the accuracy required and the time frame in which data have to be analyzed. The overall goal is to achieve correct results within a set time frame with a defined accuracy. Combinations of these factors result in specific overall methods. Table 2 gives an overview of methods available for organic acid analysis in water, organized by matrix and instrumentation.

A. Sample Preservation

Due thought has to be given to sample storage and preservation, considering that organic acids are biode-

gradable. The approach taken depends on the analytical methods used, the timing of the analysis, and the matrix investigated. A summary of currently used preservation techniques is given in Table 3.

Wastewater samples are usually not preserved, since for process control they are measured immediately after the sample is taken. If wastewater samples have to be stored, they may be filtered and then stored at 4°C, although it is best if they are frozen (50). Most methods investigating rainwater or drinking water samples call for immediate preservation after sampling. It was observed that the organic acid content, especially for formic and acetic acids, dropped dramatically when rainwater or drinking water samples were not preserved (5,7,8,17). Hence, it is strongly recommended to use appropriate preservation techniques.

The prevalent preservatives are chloroform and mercuric chloride. Freezing samples has also been applied successfully to rainwater (56,57). Chloroform tends to be used for samples that will be analyzed by liquid chromatography (7–9,17,58,59), mainly ion-exchange or ion-exclusion chromatography, whereas mercuric chloride is used predominantly for samples analyzed by gas chromatography (GC) (18,19,34,35). Drinking water samples that undergo direct aqueous derivatization with PFBHA (0-(2,3,4,5,6-pentafluorobenzyl)-hydroxylamine) have no preservative added. The reagent PFBHA may be added to sample vials prior to sampling. Then, once the sample is taken, the reaction sets in, instantaneously converting ketoacids and withdrawing them from immediate biological degradation (1–3,60). All preserved samples are stored at 4°C, sometimes in brown glass bottles.

B. Sample Preparation

Sample preparation is very matrix dependent. It may range from a simple filtration to a sequence of steps that can include filtration or centrifugation, extraction or concentration, and finally derivatization. Whenever possible, less time-consuming methods with a minimum of sample preparation should be preferred.

1. Filtration/Centrifugation

Methods involving direct injections, such as IC or GC, often require the filtration of turbid samples, usually through filters with a standard pore size of 0.45 μm (43,48). For some samples, such as ultraclean water or drinking water, this step may be omitted (5,7–9,55). Samples with high solids content such as wastewater, are often centrifuged before the supernatant is processed further (47,61–64).

Table 3 Sample Preservation and Preparation

Preservation	Preparation	Instrumentation	Comments	Refs.
Drinking water				
Processed within 4 h of collection	Aqueous derivatization, extraction, methylation	GC/ECD	Drinking water	1–3
Processed immediately	Aqueous derivatization, extraction, methylation	GC/ECD		60
Addition of 10 μL 1% $HgCl_2$ to 10 mL sample = 10 mg/L $HgCl_2$ in sample	None, direct injection	IC	Removal of Hg^{2+} through H^+-cartridge on line before sample enters IC	5
40 μL neat $CHCl_3$ to 40 mL sample	None, direct injection	IC	Oxalate determination in matrix with high ionic strength requires switching technique	7–9
Atmospheric precipitation				
Addition of 10–20 mg/L $HgCl_2$ (cleaned by extraction with CH_2Cl_2); stored in brown glass at 4°C	Concentration through evaporation to dryness at pH9, derivatization	GC/FID; GC/MS	Rain, time consuming sample preparation	18, 19, 34, 35
Storage at 4°C	Turbid samples, filtered with prewashed filters (0.45 μm; 2 h 180–200°C)	IC	Rain, fog; measured samples immediately	48
Storage at 4°C	None, direct injection	IC	Rain; did not freeze samples to prevent changes in redox equilibria especially NO_2/NO_3	24
Preservation with $CHCl_3$, storage at 4°C	None, direct injection	IC	(a) rain; (b) mist	(a) 58 (b) 59
Addition of 0.5 mL $CHCl_3$ for 250 mL sample, stored at 4°C in glass containers	None, direct injection	IEC	Rain; unpreserved sample: no organic acids after 62 days storage, preserved sample: reproducible result	17
Addition $CHCl_3$ to sample, stored at 4°C in glass containers	Addition of HCl, direct injection	IEC	Rain	16
Frozen	None, direct injection	(a) micro HPLC (b) CZE	Single rain drops, simultaneous detection of Cl^-, NO_3^- and SO_4^{2-}	57
Frozen	None, direct injection or sample stacking	CZE	Single raindrops; simultaneous detection of inorganic anions	56
Kept frozen	Outer part of core removed, sample melted	IC	Ice core samples; simultaneous detection of inorganic anions	21
Other Matrices				
Storage at 4°C	Pore water and microcosm samples filtered; otherwise direct injection	IEC	Groundwater samples; samples from microcosm study	43
Sample centrifuged, supernatant frozen	Acidification, 1 μL formic acid (50%) to 100 μL sample	GC/FID	Sediment pore water	47
Filtration 0.22 μm filter, acidified, storage at 4°C	Concentrated using static diffusion	GC/FID	Sea water; time consuming due to long sample preparation time	37
Storage at 4°C after centrifugation and acidification	Centrifugation and acidification	GC/FID	Wastewater or sludge supenatant	64
Frozen at −18°C in PE screw-top tubes	Centrifugation, filtration 0.1-μm cellulose nitrate filters; removal of strongly absorbing huminlike substances by PVP filter cartridges	(a) IEC/UV (b) IEC/cond	Landfill leachates	50

Abbreviations used in table: see Abbreviations list at the end of the chapter.

2. Extraction/Concentration

The concentrations of organic acids in sample matrices such as rainwater and drinking water are relatively low, in which case a concentration step is usually required for their determination. Different approaches to concentrate samples have been taken. They depend on the sample matrix, the specific organic acids to be analyzed for, and the instrumentation to be used for their measurement.

a. Gas Chromatography. Traditionally many gas chromatography (GC) preparation methods consist of an extraction step, which transfers the analytes into an organic phase, and a derivatization step, which makes them suitable for GC measurement. Exceptions are direct aqueous injections of water samples onto specialty GC columns (see Sec. II.C.1).

Liquid–liquid extraction of short-chain organic acids, ketoacids, or dicarboxylic acids often result in low, unreproducible extraction yields due to the hydrophilic character of the analytes. Thus, this type of extraction is usually unsuitable for the analysis of organic acids (63). Successful liquid–liquid extraction of these acids, such as the following one, are the exception (52). Low-level radioactive waste leachate is extracted with diethylether at pH 2 and then dried with anhydrous magnesium sulfate prior to the direct aqueous GC injection. Monocarboxylic acids (C2–C7) are quantified at mg/L concentrations with a precision of 5%. However, when this type of extraction is applied to wetland pore water samples, which contain organic acid in low μmol/L concentrations, contradictory results are achieved (63).

Alternatives to liquid–liquid extraction include evaporation at high pH, solid-phase extraction on anion-exchange resins, or changing the properties of the organic acids through aqueous derivatization (see Sec. II.B.3.a.), thus making them accessible to liquid–liquid extraction.

In various GC methods (18,19,23,34) organic acids are concentrated by first adjusting the sample pH to approximately 9, ensuring that the organic acids are dissociated. The sample is then concentrated to about 2 ml using a rotary evaporator under vacuum and then blown off to dryness with nitrogen. The dry extract is redissolved in solvent and derivatized. When determining volatile monocarboxylic acids in rain, good recoveries of 73–107% were reported (18). Nevertheless, it has to be pointed out that under these conditions, decomposition of higher molecular weight organic compounds might occur, possibly leading to an increase in short chain organic acids. In addition, utilizing the rotary evaporator to concentrate water is very time consuming and consequently a disadvantage, as well.

Solid-phase extraction on anion-exchange resins is very rarely used to concentrate organic acids (65). Recoveries for the organic acids may show great variability, which is probably caused by incomplete removal from the column, especially at low concentrations. Another factor to consider is the competition of the inorganic and carboxylate anions for the active sites on the resin. Problems may arise when the inorganic anion concentrations are higher than the organic acid concentrations. This may lead to a breakthrough of the carboxylate ions, which show a lesser affinity to the anion-exchange resin, resulting in irreproducible losses.

A unique approach to concentrate nanomolar concentrations of organic acids in seawater employs static or dynamic diffusion using membranes (37). Although very time consuming, this procedure ensures the removal of the majority of salts that might interfere with subsequent analysis.

b. Ion Chromatography. Most methods using anion-exchange chromatography allow for the direct injection of samples (5,7–9,24,28,58,59). In cases where the organic acid concentrations are very low, samples can be concentrated by using a concentrator column, usually an anion-exchange resin, online. This approach works well for samples with low inorganic anion concentrations (21,55). Problems arise when inorganic anions are present in abundance (e.g., drinking water). The inorganic anions compete with the carboxylate anions for the anion-exchange sites in the resin. Inorganic anions possess a greater affinity to these sites, so carboxylate anions are eluted first when all the sites are occupied. The following breakthrough of carboxylate anions leads to irreproducible results. However, if the predominantly interfering anions, chloride and sulfate, are removed, preconcentration can be applied successfully to the analysis of organic acids (66).

An alternative for increasing method sensitivity is the injection of larger sample volumes (see Sec. II.C.2.a) (7–9). If the column capacity is sufficiently high, volumes of up to 1 ml may be injected without significant peak broadening. The so called "relaunch" effect ensures that the analytes are collected as a relatively small sample band at the start of the column, a prerequisite for achieving sharp peaks (67–69).

3. Derivatization for Gas Chromatography, Liquid Chromatography, and Other Techniques

An abundance of derivatization techniques are available for organic acids (70). In general, a derivatization

Table 4 Derivatizations Utilized in GC or HPLC Applications

Sample preparation	Reagent	Organic acid type	Reaction products	Reaction conditions	Further processing	Instrumentation	Comments	Refs.
Dried residue, no water present	14% BF$_3$/butanol	Nonvolatile ω-oxacids and α-keto-carboxylic acid	Butylesters	Dried concentrated organic acids + 14% BF$_3$/butanol, 30 min @ 100°C	Extract with hexane, wash hexane with water	GC/FID, GC/MS	Simultaneous determination of the aldehydes glyoxal and methylglyoxal	34
Dried residue, no water present	14% BF$_3$/butanol	C2–C10 dicarboxylic acids	Butylesters	Dried concentrated organic acids + 14% BF$_3$/butanol	Extract with hexane, remove excess butanol with TFAA	GC/FID or GC/MS	Rain, fog, and mist: very time-consuming sample preparation	19
None	(a) PFBHA (b) CH$_2$N$_2$	Ketoacids	Oximes with methylester function	(a) 20 ml sample + 1 ml PFBHA (6 mg/ml), 105 min @ 45°C, acidify, extraction with 4 ml MTBE (b) + 0.25 ml CH$_2$N$_2$, 15 min @ 4°C + 15 min @ RT	Quenching of CH$_2$N$_2$ with silica gel	GC/ECD	Drinking water	1–3
None	(a) PFBHA (b) CH$_2$N$_2$	Ketoacids	Oximes with methylester function	(a) 20 ml sample + 2 ml PFBHA (1 mg/ml), 2 h @ RT, acidify, extraction with 4 ml diethylether (b) + CH$_2$N$_2$ + anhydrous Na$_2$SO$_4$	None	GC/ECD, GC/MS	Drinking water	60
None	(a) PFBHA (b) BF$_3$/methanol (c) MTBSTFA	Aldehydes, ketones, ketoacids, and hydroxyl-substituted carbonyl compounds	Oximes with methylester function and/or *tert*-butyldimethylsilyl groups	(a) 10 ml sample at pH 7 PFBHA; RT 24 h, extraction with hexane or MTBE (b) Aliquot of extract (a) evaporated to dryness + 250 μl 14% BF$_3$/MeOH, 1 h @ 70°C, extraction with hexane (c) 200 μl extract (b) + 100 μl MTBSTFA, 30 min @ 60°C	None	GC/MS	Ozonated drinking water and others, focused on identification and gives mass spectra	4

Sample preparation	Reagent	Analyte	Derivative	Procedure	Cleanup	Detection	Application	Ref.
None	(a) PFBHA (b) CH₂N₂ (c) MTBSTFA	Aldehydes, ketones, ketoacids, and hydroxyl-substituted carbonyl compounds	Oximes with methylester function and/or *tert*-butyldimethylsilyl groups	(a) 250 ml sample at pH 4 + PFBHA, 3 h @ 45°C, RP18 solid-phase extraction (b) 1 ml of extract (a) + 250 μl CH₂N₂, 30 min @ 4°C (c) 500 μl extract (b) + 100 μl MTBSTFA, 30 min @ 60°C	None	GC/MS	Ozonated drinking water; focus on identification	66
Cation-exchange procedure to replace cations with K⁺	α,*p*-bromoacetophenone; catalyst: dicyclohexyl-18-crown-16	Volatile, monocarboxylic acids	*p*-bromo-phenacylesters	Organic acids in acetonitrile + reagent + catalyst, 2 h @ 80°C in ultrasonic bath	Removal of excess reagent through SiO₂ column	GC/FID, GC/MS, HPLC	Rain; time-consuming; confirmation with GC/MS possible	18, 71
None	OPD	I. α-keto-acids II. oxalic acid	(I) quinoxilinol (II) hydroxyquinoxilinol	0.5 ml sample + 0.25 ml OPD (20 mg/ml in 6N HCl); 3 h at 110°C	Cool, adjust pH to 7 ± 1 with NaOH and H₃PO₄; freeze until measurement	HPLC	Rain, fog, and mist	35

Abbreviations used in table: see Abbreviations list at the end of the chapter.

should be specific to the compounds of interest, result in one end product, and achieve reproducible, preferably high yields. The excess reagent should not interfere with the determination of the derivative, and ideally the procedure should be easy and quick to perform. Derivatizations are applied to analytes to make them suitable for a chosen instrumentation and/or to increase the sensitivity of the overall method. A summary of the derivatizations applied to various organic acids in different types of samples is presented in Table 4.

a. Derivatizations for Gas Chromatography. Many short-chain carboxylic acids are thermostable and sufficiently volatile, thus fulfilling key requirements for GC measurements. Nevertheless, their high polarity makes it difficult to achieve satisfactory chromatograms with standard capillary columns. Only when using specialty columns is it possible to analyze these acids directly (see Sec. II.C.1.). Hence, carboxylic acids are often derivatized to their less polar, corresponding esters, which can be measured without difficulty on standard GC columns.

Almost all of the following reactions have to be performed in nonaqueous solutions. Thus, prior to their derivatization, the organic acids have to be transferred into suitable solvents by either a concentration step or an extraction procedure (see Sec. II.B.2.a.).

Esterification with alcohols using the Lewis catalyst boron trifluoride is a well-established standard derivatization method (70). Short-chain carboxylic acids are derivatized with butanol to their corresponding butylesters and then extracted into hexane. Compared to methylesters, the butylesters are separated more easily from the solvent peak due to their higher boiling point. Dicarboxylic acids as well as α-ketoacids were quantified with this procedure in rainwater samples (19,34). In the case of the dicarboxylic acids, an additional step for the removal of excess butanol was introduced to prevent interference with the determination of the analytes (19).

Methylation with diazomethane is rarely applied as the only derivatization step to short-chain carboxylic acids (65). It is usually preferred to generate esters with longer carbon chains, which have higher boiling points. Once the diazomethane has been generated, it is added to the extract, which can then be injected in the GC without any further sample preparation, thus making this reagent rather convenient. Heating of the sample mixture is not required, since the reaction takes place at low temperatures. The major drawbacks of diazomethane are its hazardous and explosive nature and the necessity to generate it before it can be applied.

A less common derivatization is the reaction of organic acids with α,p-bromoacetophenone in presence of dicyclohexyl-18-crown-16 as a catalyst to their p-bromophenacyl esters (18,23,71). The samples have to be conditioned by a cation-exchange procedure prior to the reaction, and excess reagent must be removed by passing the sample through a SiO_2 column after the reaction. Although this makes the procedure rather time consuming, it may be of advantage since p-bromophenacyl esters can be measured by GC as well as by HPLC.

A procedure specifically developed for ketoacids in drinking water utilizes a rather uncommon aqueous derivatization as a first step (1–4,60,66). The complete method involves two consecutive derivatizations. First, the ketofunction of the analyte is derivatized with PFBHA (O-(2,3,4,5,6-pentafluorobenzyl)hydroxylamine) to the corresponding oxime, which is much less polar than the original ketoacids and may be extracted with polar solvents such as MTBE (methyl-t-butylether) from the water sample. The extract is dried with sodium sulfate, and the carboxylate function is then methylated with either diazomethane or boron trifluoride/methanol. The extracts are measured with GC/ECD (electron capture detector), which detects halogenated derivatives with a high sensitivity.

Efforts have been made to use combinations of different extraction and derivatization steps for the identification of unknown substituted organic acids (4,66). These methods also start with the aqueous oximation of carbonyl functions with PFBHA. This is followed either by liquid–liquid extraction or by solid-phase extraction. By measuring this extract, carbonyl compounds such as aldehydes and ketones can be identified. Further derivatization of the same extract with a methylation reagent (here: boron trifluoride/methanol) can lead to the identification of ketoacids. When another derivatization to the same extract using syliation with MTBSTFA (n-($tert$-butyldimethylsilyl)-N-methyl-fluoracetamide) is applied, the hydroxyl functions are marked. All of these derivatized extracts can be measured by GC/MS (mass spectrometer) thus giving valuable information about the structure of the compounds through their mass spectra. In spectra acquired by electron impact ionization (EI) it was found that fragments m/z 181 are characteristic of the marked carbonyl function, whereas m/z 59 represents the methylated carboxylate groups and m/z 75 the sylilated hydroxyl functions. Measurements by GC/MS using chemical ionization (CI) usually result in a very dominant M^+ ion, from which the molecular weight of the unidentified compound can be deduced. Samples such as ozon-

Table 5 Overview of GC Methods, Including Sample Preparation, Derivatization, and Detailed GC Conditions

Extraction/preparation	Derivatization: reagent—product	Injection	Column	Temperature program	Detector	Carrier gas	Compounds	Detection limits	Matrix	Refs.
2. Extraction with MTBE	1. Aqueous oximation with PFBHA 3. Methylation with CH₂N₂	na	SPB5, 30 m, 0.32-mm ID, 0.25-μm df	na	ECD	na	Ketoacids	na	Drinking water	1—3
2. Extraction with diethylether	1. Aqueous oximation with PFBHA 3. Methylation with CH₂N₂	na	(a) CPSil 5, 30 m, 0.32-mm ID, df? (b) OV1701, 25 m 0.25-mm ID, df?	(a) 50°C (2 min); 3°C/min to 300°C (4 min); (b) 50°C (2 min); 4°C/min to 300°C (5 min)	(a) ECD (b) MS	na	Glyoxylic acid	na	Ozonated fulvic acid solutions	60
2. Extraction with MTBE 4. Extraction with hexane	1. Aqueous oximation with PFBHA 3. Methylation with BF₃/MeOH 5. Silylation with MTBSTFA	na	DB5, 30 m, 0.25-μm df, ID?	50°C (2 min); 5–10°C/min to 250°C (2 min)	MS (EI; CI with CH₄)	He?	C1–C5 ketoacids and others	na	Ozonated drinking water	4
2. RP18 solid-phase extraction	1. Aqueous oximation with PFBHA 3. Methylation with CH₂N₂ 4. Silylation with MTBSTFA	180°C, 1 μl, splitless for 30 s	DB5 30 m, 0.25-mm ID, 0.25-μm df	50°C (1 min), 4°C/min to 220°C, 50°C/min to 250°C	MS (EI, CI); m/z 50–650	He 1.5 mL/min	Ketoacids and others	na	Ozonated drinking water	66
Concentrated	α,p-bromoaceto-phenone/bromophenacyl esters	190°C, 1 μl; on column	Packed glass column; 2 m, 3-mm ID	170°C (10 min); 5°C/min to 210°C (5 min)	FID: 220°C	N₂, 30 mL/min	C1–C5 monocarboxylic acids	~20–50 μg/L	Rain, sewage, soil pore water	71
Concentrated at pH 8.5–9	α,p-bromoacetophenone/bromophenacyl esters	200°C, 1 μl; splitless	DB5, 30 m 0.25-mm ID, df?	40°C (6 min) to 30°C/min to 160°C to 8°C/min to 290°C	(a) FID: 300°C (b) MS	na	C1–C7 aliphatic acids	~0.1 μmol/L	Rain, fog	18, 23
Concentrated by evaporation under vacuum	BF₃/butanol-butylesters	300°C	(a) HP5, 25 m, 0.32-mm ID, 0.52-μm df (b) DB5, 30 m, 0.25-mm ID, 0.25-μm df	50°C (2 min); 30°C/min to 120°C; 8°C/min to 300°C (15 min)	(a) FID (b) MS: EI, m/z 35–550	na	C2–C10 ω-oxoacids, pyruvic acid	0.05 μg/L	Rain, snow, aerosols	34
Concentrated by evaporation under vacuum at pH 8–9	14% BF₃/butanol-butylesters	250°C, 1 μl, splitless	DB5, 30 m, 0.25-mm ID, df?	(a) 40°C (6 min); 6°C/min to 295°C (b) 35°C (6 min); 4°C/min to 280°C; 2°C/min to 310°C	(a) FID (b) MS	na	More than 20 C2–C10 dicarboxylic acids	na, but <0.8 μmol/L	rain, fog, and mist	19
Acidification to pH 2 with H₂SO₄	None	Direct aqueous injection; 250°C	Stabilwax-DA, 15 m, 0.53-μm ID, 0.5-μm df	105°C (2 min), 20°C/min to 145°C	FID 275°C	H₂, 10 ml/min	C2–C6 monocarboxylic acids = VFA	na; quantification in mg/L conc.	wastewater, bench scale experiments	123, 124

Table 5 Continued

Extraction/preparation	Derivatization: reagent—product	Injection	Temperature program	Column	Detector	Carrier gas	Compounds	Detection limits	Matrix	Refs.
Filtration 0.45 μm; acidification to pH 3 with H₃PO₄	None	Direct aqueous injection; 200°C	80°C (5 min), 10°C/min to 130°C (4 min)	HP-FFAP 10 m, 0.53-mm ID	FID 250°C	He	C2–C7 monocarboxylic acids = VFA	na; quantification in mg/L conc.	Wastewater, UASB reactor effluent	61
Centrifugation; supernatant filtered through 0.45-μm filters	None	Direct aqueous injection; 250°C	120°C isothermal	HP-Innowax, 15 m, 0.25-mm ID, 0.15 μm df	FID 300°C	na	C2–C5 monocarboxylic acids = VFA	na; quantification in mg/L conc.	Wastewater, bench scale	126
Centrifugation; acidification: 1 ml supernatant + 100 μl H₃PO₄	None	Direct aqueous injection; 150°C	120°C isothermal	0.3% Carbowax 20 M/0.1% H₃PO₄ on Supelco Carbopack	FID 200°C	He, 20 ml/min	C2–C4 monocarboxylic acids = VFA	na; quantification in mg/L conc.	Wastewater from thermophilic anaerobic digester	64
na	None	Direct aqueous injection	na	Glass column 2 m, 3-mm ID, B-DA/4% Carbowax 20 M on 80/120 mesh Carbopack	FID	He, 20 ml/min	C2–C4 monocarboxylic acids = VFA	na; quantification in mg/L conc.	Wastewater, bench scale	125
Centrifugation; supernatant acidified with 3% formic acid	None	Direct aqueous injection; 200°C	130°C isothermal	Glass column 2 m, 4-mm ID, 10% Fluorad FC 431 on 100–120 mesh Supelcoport	FID 280°C	N₂ saturated with formic acid; 50 ml/min	C2–C4 monocarboxylic acids = VFA	na; quantification in mg/L conc.	Wastewater, benchscale, UASB reactor	62
Acidification, distillation	None	Aqueous injection; 215°C	95° C (3 min), 2°C/min to 113°C, to 190°C (2 min)	Glass column: 152 mm, 2.0-mm ID; 10% SP-1200-1% H₃PO₄ on 80–100 mesh Chromosorb WAW	FID 250°C	N₂ 9.3 ml/min	C2–C4 monocarboxylic acids	na; analyzed mg/L conc.	Landfill leachates	53
Acidification, extraction with diethylether	None	250°C, 10 μl	70°C, 10°C/min to 130°C, 5°C/min to 180°C (1 min)	Stainless steel column GP 10% SP-1200/1% H₃PO₄ on 80–100 mesh Chromosorb WAW	FID, 250°C	He, 30 mL/min	C2–C7 monocarboxylic acids	na; analyzed mg/L conc.	Low-level radioactive waste leachates	52

Sample preparation	Derivatization	Injection	Column	Temperature program	Detector	Carrier gas	Analytes	Detection	Sample	Ref
Centrifugation, supernatant pH raised to 11, dried at 95°C, redissolved in 3 N H_3PO_4	None	Direct aqueous injection; 1 μl, 180°C	Precolumn, deactivated?, 1 m, 0.53 μm; Nukol, 15 m, 0.53 μm i.d., df?	120° isothermal	FID 180°C	N_2	C2–C4 monocarboxylic acids = VFA	na: quantification in μM to high nM	Wetland sediment pore water	63
Centrifugation, storage frozen supernatant, acidified	None	Direct aqueous injection; splitless, 0.5–3 μl (a) 155°C (b) 225°C	FFAP-CBwax (HP): (a) 10 m, 0.53-μm ID, 1 μm df (b) 25 m, 0.32-mm ID, 0.33-μm df	(a) 70°C (1.1 min), 10°C/min to 105°C (0.25 min) (b) 80°C (1.1 min), 15°C/min to 105°C (1 min), 10°C/min to 140°C (1 min)	FID (a) 200°C (b) 260°C	He (a) 20 mL/min (b) 4 mL/min	C2–C7 monocarboxylic acids = VFA	2–5 μM	Sediment pore water, (a) better than (b)	47
Concentration	None—sample adjusted to pH 2	Direct aqueous injection 200°C, 1 μL	FFAP, 30 m, 0.53-μm ID: df?	120°C (8 min); 20°C/min to 200°C	FID, 220°C	He, 10 mL/min	C2–C5 monocarboxylic acids, pyruvic, acrylic, benzoic acids	Down to 10 nM; except acetate only 750 nM	Seawater	37
Vacuum distillation; acidification with formic acid	None	Direct aqueous injection; (a) 1 μL; 220°C; split 1:30 (b) 1 μL; 200°C; splitless (c) 5 μL	(a) HP Supelcowax 10; 30 m, 0.32-mm ID, 0.25-μm df (b) DB-Wax 15; 15 m, 0.53-mm ID, 1.0-μm df (c) Column packed with Chromosorb 101	(a) 100°C, 10°C/min to 230°C (b) 100°C, 20°C/min to 215°C (c) 150°C (1 min), 5°C/min to 225°C (2 min)	FID (a) 260°C (b) 210°C (c) na	He (a) 20 cm/s (b) 30 cm/s (c) He saturated with formic acid	C2–C7 monocarboxylic acids = VFA	0.05 mM	Biological samples, (b) better than (a)	95
Filtration, dilution with oxalic acid (1:10)	None	Direct aqueous injection; 0.5 μL; on column	(a) DB-Wax, 15 m, 0.53-μm ID, 1-μm df (b) FFAP 10 m, 0.53-μm ID; 1-μm df	(b) 85°C; 6°C/min to 180°C (5 min)	FID, 200°C	He, 31 cm/s	C2–C7 monocarboxylic acids = VFA, lactic acid	na: analyzed mg/L conc.	Silage juice, (b) better than (a)	90

Abbreviations used in table: see Abbreviations list at the end of the chapter.

ated drinking water, ozonated paper pulp, and oxidized isoprene have been investigated with this methodology (4,66).

b. High-Performance Liquid Chromatography. High-performance liquid chromatography (HPLC) methods are seldom applied to the analysis of short-chain organic acids. Due to the poor UV absorbance and the nonfluorescent character of these compounds, only high concentrations can be measured directly by HPLC, in combination with UV, diode array, or fluorescence detector. To enhance the method sensitivity, organic acids may be derivatized in a pre- or postcolumn reaction. Although an abundance of derivatization methods for various compounds are available, especially for physiologically important acids in biological fluids (72–74), only very few have been applied to short-chain organic acids in a water matrix (Table 4).

As mentioned earlier, under GC derivatization, p-bromophenacyl esters of monocarboxylic acids may be measured by HPLC with UV detection (18,23,71). o-Phenylenediamine dihydrochloride (OPD) has been used to derivatize α-ketoacids and to oxalate to their corresponding quinoxilinols and hydroxy-quinoxilinol (35). The reagent is added directly to the aqueous sample and then heated. Following neutralization of the cooled mixture, the derivatized sample is injected into the HPLC. The derivatives formed absorb UV light and are fluorescent, allowing for either detection method. The direct reaction in aqueous medium makes this method simple to carry out and still sensitive enough to measure oxalic, pyruvic, and glyoxylic acids in rain, even when using UV detection (35).

C. Separation Techniques

Separation of the analytes is accomplished by chromatographic techniques, with GC being historically the first one to be established. The later-developing ion chromatography has gained in importance, compared to GC, for the analysis of organic acid in aqueous samples. Hence, ion chromatography is discussed in more detail than GC. Capillary electrophoresis is another, newer separation technique gaining interest and is therefore included in the discussion.

1. Gas Chromatography

Gas chromatography is a well-established technique with a wide range of applications. An abundance of literature is available describing its theoretical background and giving practical advice (75–78). Key re-

quirements for compounds to be analyzed by GC are volatility and thermostability.

In general, organic acids are quite stable at high temperatures, and many of the short-chain organic acids are volatile as well. Consideration has to be given to the very polar character of the organic acids, which can cause substantial problems in their chromatography. To circumvent this problem, organic acids are often derivatized to less polar compounds, usually esters, which are easy to chromatograph. The derivatization of organic acids for GC determinations is discussed in detail in Sec. II.B.3.a. Gas chromatographic conditions used for these derivatized compounds are listed in Table 5.

Alternatives to derivatization are specialty GC columns that allow for the direct aqueous injection of organic acid samples without time-consuming sample preparation, as long as appropriate GC conditions are chosen and the system is maintained regularly. This methodology is usually applied to monocarboxylic acids with carbon chain length from C2 to C8. However, longer-chain acids may be determined as well according to product information supplied by the manufacturer. Attempts to determine dicarboxylic or aromatic acids with these columns resulted in poor chromatography (79).

Usually a flame ionization detector (FID) is used in these applications. Its limited sensitivity leads to relatively high detection limits, so only samples with high concentrations of organic acids, such as wastewater samples, can be determined by direct aqueous injection. As reported by many authors, formic acid (47,80–84) cannot be detected with this method. Only when replacing the FID with a thermal conductivity detector (TCD) is it possible to determine formic acid by GC (85).

Packed columns and capillary fused-silica columns with high polarity are available to accommodate separation of organic acids. Specialty capillary columns usually contain a chemically bound film of polyethylenglycol (brand names: DB-Wax, HP Wax, Stabilwax, Supelcowax 10) or acid-modified polyethyleneglycol (brand names: DB-FFAP, HP-FFAP, Stabilwax-DA, Nukol). These type of columns are more sensitive to heat and oxygen than others with a less polar film. Maximum operation temperatures are approximately 280°C for Carbowax-type columns and approximately 200°C for FFAP-type columns. Packed columns contain packings coated with polar films, which are similar in character to the ones in capillary columns.

The aqueous samples are usually acidified prior to injection. Organic (47,90) as well as inorganic acids are used. By acidifying samples, the organic acids are in their protonated form, which increases their volatility, reduces adsorption effects, and results in better peak shapes. Some authors report that acidification of samples leads to the deterioration of films in capillary columns (79,85), others did not observe any adverse effects from overacidification (91). In general it can be expected that bonded phases are much more rugged than nonbound liquid phases.

A major problem in the direct analysis of organic acids are "memory effects," or "ghosting." During a GC run, organic acids are adsorbed onto the column or parts of the instrument and are released more or less at random during a later GC run. This leads to irreproducible retention times and unreliable quantification. Some measures are available to control or reduce this problem. Packed GC columns should be produced from glass (86), and the packing should be treated with phosphoric acid to reduce possible adsorption (84,87–89). It has been suggested that saturation of the carrier gas with formic acid is an alternative to acid treatment of the packing (92–94). This alternative is not recommended, since formic acid is known to pose an explosion hazard under the conditions used (95). For capillary columns, recommendations to prevent "ghosting" are contradictory. Some authors report it to be sufficient to end a GC run with heating the column up to 180–200°C, thus purging any adsorbed organic acids or impurities (91). Others recommend conditioning the column every morning by injecting formic acid (1%) ten times, thus saturating the film in the column and reducing possible irreversible adsorption (47). Elsewhere it is reported that none of this is necessary (90). These differences may be due to factors such as sample matrix, pretreatment of the sample, column quality, and especially GC conditions chosen. Therefore, no general recommendations can be made. All these factors have to be taken into account when evaluating the options available to arrive at the best conditions for a specific application. Some examples of GC conditions used by others are given in Table 5.

Regular maintenance of the GC system is crucial for achieving reliable results with this methodology. Most methods include a sample filtration step to reduce the particulate matter introduced into the GC system. Nevertheless, regular cleaning of the injection port insert, if using a split injector, is necessary (47,79) when using capillary GC. When column performance deteriorates, cutting off one loop of the front end of a capillary column often restores the chromatographic performance. A retention gap should be considered as an alternative. The retention gap is a deactivated piece of fused-silica tubing that acts as a precolumn. It is easily connected to the analytical column by using a press-fit glass connector. By shortening the retention gap, or by replacing it altogether, chromatographic performance of the system can be improved while extending the life of the analytical column (91). The deactivation in the precolumn should be of medium-to-high polarity to accommodate the high polarity of the solvent injected, such as, in this instance, water. When using packed columns, exchanging the column packing is necessary once column performance starts to deteriorate.

2. Liquid Chromatography

In LC, analytes are separated due to their affinities toward the solid phase and the liquid phase. Different techniques have been developed based on the type of solid phase utilized. High-performance liquid chromatography (HPLC) is used as a synonym for reversed-phase and normal-phase liquid chromatography, whereas ion chromatography includes ion-exchange chromatography (IC) and ion-exclusion chromatography (IEC). Organic acids have been determined with all of these techniques; however, IC and IEC remain the predominant methodologies. Various books about ion chromatography have been written in the last decade giving theoretical and practical advice (96–98).

a. Ion-Exchange Chromatography. In ion-exchange chromatography (IC), separation is based on the partitioning of the analyte, an ion, between the mobile phase and the ion-exchange groups bound to the stationary phase. Ions with hydrophobic characteristics involve adsorption processes as a secondary separation mechanism. Depending on their pKa values, organic acids deprotonate readily at a higher pH to their corresponding carboxylate anions, thus making them accessible to anion-exchange chromatography. An overview of IC methods used for organic acid analysis is given in Table 6.

The most common anion-exchange columns consist of polystyrene/divinylbenzene resins on which anion-exchange-site-carrying latex particles are bound. Quaternary ammonium groups substituted with alkyl or alkanol groups form these anion-exchange sites. Columns are characterized through particle diameter, amount of crosslinking, capacity, and properties of their exchange sites (96,97).

The predominant detection used in anion-exchange chromatography is suppressed conductivity, where the eluent is transformed to a lesser conductive species through protonation just before entering the conductivity cell. This is accomplished by passing the eluent

Table 6 Overview of IC Methods, Including Sample Preparation and Detailed IC Conditions

Injection	Column	Eluent	Detector	Compounds	Detection limits	Matrix	Comments	Ref.
20 μl	Anion trap column ATC-1; AG 11; AS 11	NaOH gradient; 0.1–10 mM	Autospr. cond. in external water mode, cleaned daily with 0.5 N H_2SO_4	Formic, acetic, oxalic acids	2–3 μg/l; MRL 15 μg/L	Drinking water	Removal of Hg^{2+} through H^+ cartridge essential; daily maintenance of suppressor imperative	5
760 μl	Anion trap column ATC-1; AG 10; AS 10	NaOH gradient; 7–125 mM	Autospr cond. in external water mode (flow 1.8 ml/min)	Formic, acetic, butyric, β-hydroxybutyric, glycolic, pyruvic, α-ketobutyric, oxalic acids	1–9 μg/L	Drinking water	Oxalate determination in matrices with high ionic strength requires switching technique	7–9
25 μl, loop 5–10 ml concentrated	AG 11 concentrator; AG 11; AS 11	NaOH gradient 4.5–200 mM	Spr. cond.?	Formic, acetic, glycolic, oxalic acids	20–40 μg/L; concentrator 0.1–0.5 μg/L	Ozonated model water	No testing on "real" drinking water samples described	66
na	AG 4?, AS4A	$Na_2B_4O_4$ 1.5 mM; isocratic? flow?	Chem. spr. cond. using 5.0 mM H_2SO_4	Formic, acetice (pyruvic) acids	0.2 μM	Rain	Focus on formic and acetic acids	58
10, 25, or 50 μl	Anion trap column ATC-1 (a) AG 11; AS 11 (b) AG 10; AS 10	(a) Borate gradient; 1–45 mM (b) Borate 7 mM isocratic	Chem. spr. cond.; 25 mM H_2SO_4; or self-generating spr. cond.	2- and 3-Hydroxybutyric, lactic, acetic, glycolic, propionic, formic, butyric, pyruvic, valeric, oxalic acids	0.5–1 μM	Fog, lake, sediment pore water, rain	Simultaneous detection of inorganic anions (F^-, Cl^-, NO_2^-, CO_3^{2-}, PO_4^{3-}, SO_3^{2-}, SO_4^{2-})	48
na	AS 4	$NaHCO_3$ 0.4 mM; isocratic? flow?	Chem. spr. cond.; 25 mM H_2SO_4	Formic, acetice, pyruvic acids	0.02–0.1 μM	Mist	Detection limits for 60-min sampling interval: 5–20 pptv	59
100 μl	AG 9; AS 9	Na_2CO_3/$NaHCO_3$ gradient	Chem. spr. cond.; 12.5 mM H_2SO_4	Formic, acetic, oxalic acids	20–100 μmg/L	Rain, snow	Simultaneous detection of inorganic anions (F^-, Cl^-, NO_2^-, NO_3^-, Br^-, SO_3^{2-}, SO_4^{2-}, and PO_4^{3-})	24
5 ml	Anion trap column ATC-1; TAC1 concentrator (a) AG 5; AS 5 (b) CH 10, Pax 500	(a) NaOH gradient; 0.5–30.5 mM; flow 1.8 ml/min (b) NaOH gradient, 1–29 mM; flow 1 ml/min	Chem. spr. cond.; 0.025 M H_2SO_4; 1.6 ml/min	Formic, acetic, oxalic, glycolic acids	0.2–0.6 ng/g	Ice cores	Simultaneous detection of inorganic anions (F^-, Cl^-, NO_2^-, NO_3^-, SO_4^{2-})	21
5 or 10 ml	AC 10 concentrator; AS 10	NaOH 0.085 mM isocratic; flow 1 ml/min	Chem. spr. cond.; 25 mM H_2SO_4 (flow 10 ml/min)	Formic, acetic, oxalic acids	Below 1 μg/L	Ultrapure water	Online preconcentration, simultaneous determination of inorganic anions	55

Abbreviations used in table: see Abbreviations list at the end of the chapter.

along a membrane that is permeable only for protons. Countercurrent to the eluent, on the other side of the membrane, flows an acid, usually sulfuric, which provides the necessary protons for the transformation of the eluent. This process reduces the background conductivity of the eluent significantly and leads to a substantial gain in sensitivity. Direct and indirect UV detection are rarely used with IC due to their inferior sensitivity. Amperometric detection has recently been gaining some interest (96–98).

Eluent choice is another important parameter. Eluents have to show an affinity to both the sample ion and the stationary phase, and at the same time they have to be suitable for the suppression process. Borate, bicarbonate, carbonate, and sodium hydroxide eluents are most commonly used for organic acid and other analyses.

The IC of organic acids competes directly with IEC. In IC, inorganic and organic anions can be determined simultaneously in one run, whereas in IEC inorganic anions elute with the system peak in the beginning of the chromatogram and only organic acids can be quantified. Depending on the sample matrix this can be an advantage or a disadvantage for IC. In samples with low ionic strength, inorganic anions and organic acids may be quantified within one run using IC. However, in samples with high ionic strength, inorganic anions can interfere significantly with the determination of organic acids (96–98).

The specifics of organic acid methods depend to a large extent on the organic acid concentration expected and on the characteristics of the matrix. Hence, methods will be discussed in connection with the sample matrix.

Rainwater samples contain moderate concentrations of inorganic anions and organic acids, usually in the micromolar range. Sample preparation for these samples is minimal and consists of filtration only if visible particles are present. Samples are injected directly into the sample loop of the anion-exchange chromatograph. These chromatographic systems are usually equipped with an anion trap column, a guard column, an analytical anion-exchange column, and suppressed conductivity detection. The anion trap column ensures that the eluent is free of inorganic impurities, whereas the guard column protects the analytical column. Separation of anions is accomplished through their interaction with the eluent and the analytical column, and sensitive detection is ensured by suppressed conductivity. In order to elute all anions in a reasonable time frame, gradient elution is preferably employed. The direct injection of rainwater samples is only possible because the sensitivity of the overall method is sufficient, detection limits are at low μg/L concentrations, and inorganic anion concentrations present will not interfere with the chromatography of the organic acids. All in all, these methods are sensitive, quick, and easy to handle (24,48,58).

Other matrices, such as sediment pore water, may also be analyzed with these type of methods (48). Sample collection and preparation of fog, mist, or even air samples, which uses a filter trap with subsequent aqueous extraction, result in aqueous solution that may be analyzed by the preceding methods as well (48,59).

Extremely low concentrations of inorganic anions and organic acids can be found in sample matrices such as ice cores, water used in the power production industry, and process water in the electronics industry. In order to determine organic acids in these type of samples, a concentration step has to be introduced, since direct IC injection methods are not sensitive enough. Concentrator columns are powerful tools for increasing method sensitivity, if handled correctly within their limitation (21,55).

Concentrator columns used in combination with IC are usually anion-exchange resins themselves. When loading a water sample on these resins, inorganic and carboxylate anions are strongly retained due to the minimal elution power of water on this resin. Transferring the trapped anions to the analytical column is usually performed with sodium hydroxide, which removes the trapped inorganic and carboxylate anions from the concentrator while ensuring their separation during the following chromatographic process.

The simplest way to use concentrator columns is to install them in place of the injection loop. Loading with sample should be done countercurrent to subsequent sample transfer onto the analytical column. Thus, the initial sample band entering the column is kept narrow, a prerequisite for acceptable peak shapes in the final chromatogram (97).

To achieve reproducible measurements with this technique, the volume introduced into the concentrator column has to be exactly the same in each injection and the analytes of interest have to be retained quantitatively on the concentrator column. The use of an autosampler ensures reproducible injection, if care is taken that transfer lines are flushed in between injections to avoid "carryover" from one injection to the next. Moreover, the external pump of the autosampler must be strong enough to overcome the increased backpressure caused by the injection into the concentrator column. To ensure complete and quantitative trapping of the analytes, the capacity of the concentrator column has to be appropriate. This means the capacity has to be high enough to ensure trapping of all the anions, but not too high to hinder the removal of the trapped anions from the concentrator column onto the analytical column. Hence, capacities of concentrator columns are usually moderate to high. The sample matrix discussed here is of low ionic strength. Breakthrough of carboxylate anions is not very likely, although it should be considered possible—at least in the method development phase and later, if significant changes in the sample matrix are experienced. High sensitivities can be achieved with detection limits of 1 μg/L and below by using this methodology (21,55).

Drinking water samples differ from the previously discussed matrices in that their organic acid concentrations are low (μg/L levels), whereas inorganic anions appear in much higher concentrations (mg/L levels). The challenge is to separate the carboxylate anions from the dominating inorganic anions. Different approaches have been taken, coming close to the current limits of anion-exchange chromatography.

To achieve the appropriate sensitivities, the use of concentrator columns was attempted. Direct loading of drinking water samples can lead to breakthrough of the more weakly retained carboxylate anions (7). If, or when, breakthrough occurs depends on the capacity of the concentrator, the sample volume loaded, and the actual concentration of the strongly retained inorganic anions. Irreproducible results can be expected with this direct approach.

If inorganic anions are removed prior to loading on a concentrator column, retention of carboxylate anions can be secured. This can be accomplished by filtering a sample through a cartridge containing silver cations in sequence with a cartridge containing barium cations, thereby removing the majority of chloride, phosphate, and sulfate, respectively. These commercially available cartridges fit on a syringe, and the sample is simply pressed through. A large sample volume, e.g., 5 or 10 ml, is then injected onto the concentrator column and further analyzed by IC (66). Detection limits for acetic, formic, and oxalic acids for this method were given with 0.1–0.5 μg/L. However, these detection limits were determined in deionized water, thus avoiding the critical salt removal step, which is not discussed in great detail. Further information regarding the removal of chloride and sulfate can be found in descriptions of bromate methods, where this kind of sample pretreatment is required prior to measurement (99–101).

As an alternative to concentrator columns, direct sample injections were successfully utilized for organic acid analysis in drinking water. However, these methods require regular system maintenance to ensure prime performance of the IC system.

A method analyzing for formic, acetic, and oxalic acids starts off with the removal of mercury cations, which were initially added for preservation purposes (5). Failing to do so results in strong disturbances of the chromatographic process, making quantification of organic acids impossible. Mercury removal is facilitated by placing an H^+ cation-exchange cartridge, which has to be replaced daily, in line between autosampler and injection valve. The detailed configuration of the IC system is described in Table 6. A low-capacity organic acid–specific anion-exchange column is utilized as analytical column (AS 11, Dionex). Although the injection volume is only 50 μl, method detection limits of 2–3 μg/L and consequently minimum reporting levels of 15 μg/L could be achieved. It is emphasized that regular, daily maintenance of the suppressor was crucial for the reliable performance of this method. This method is used on a routine basis to monitor organic acids in a very large drinking water facility in North America.

Determination of seven organic acids can also be accomplished by another drinking water method using direct injection (7–9). To increase method sensitivity, a large sample volume (760 μl) is injected. Injections of up to 1 ml are possible without any adverse effect on the chromatography due to the so called "relaunch," or zone compression, effect (97). Taking advantage of this effect, injections of large sample volumes have been used in the past for the analysis of inorganic anions (67–69). To accommodate the relatively large amount of anions entering the analytical system, an analytical column with medium capacity is utilized, thus achieving good chromatography (AS10, Dionex). When using a low-capacity column as analytical column (AS11, Dionex) in preliminary experiments, only unsatisfactory chromatograms were obtained. Sample pretreatment other than occasional filtration is not necessary, although samples are preserved with chloroform. Only small volumes of neat chloroform are added (e.g., 50 μl into 50 ml sample), thus having no negative impact on this solvent-compatible column. Method detection limits ranged from 1 to 5 μg/L (7–9).

Initially this method did not include oxalic acid, an important ozonation by-product (7). However, subsequent work extended this method to include oxalate (8,9). Inorganic anions, specifically sulfate, bromide, and phosphate, elute close to oxalic acid. At moderate concentrations of these inorganic anions, the previously introduced method had only to be optimized. It is emphasized that regular and proper maintenance of the eluent purification trap (ATC-1) and the guard column are key factors to method optimization. The resulting, optimized method has been routinely applied to a wide range of different drinking water samples from bench, pilot, and full-scale water treatment facilities.

However, in specific cases when large concentrations of sulfate, bromide, and/or phosphate are present, oxalic acid coelutes. To resolve this problem for these rather rare cases, a two-phase approach has been taken involving a column switching procedure (8,9). In the first phase, the sample is injected and processed as usual. At the time when oxalic acid has been known to elute, the effluent is redirected onto a concentrator column. In the second phase, the trapped anions are eluted from the concentrator column onto the same analytical column. Oxalic acid is then separated from the prior interfering inorganic anions by a different gradient developed specifically for this purpose. All organic acids are quantified with the chromatogram of the first phase, except for oxalic acid, which is quantified with the chromatogram from the second phase. Careful adjustment of the switching window is a key requirement for quantitative recovery. This method is best suited for samples with a stable matrix; otherwise, readjustment of the collection window might be necessary. Switching methods similar to the one described have been applied in the past for the determination of inorganic anions in mineral acids (130).

In summary, the analysis of organic acids in drinking water by IC is pushing the current limits of this technique. The IC system performance has to be excellent so that the required low detection limits are achieved. Although time consuming, this can be ensured by regular, daily maintenance of the system. Nonetheless, sample preparation is relatively short, making these organic acid methods relatively fast.

Inherent to all of the described IC methods are coelution and contamination issues. In fact they are not restricted to IC but can be encountered in almost all of the other techniques applied to organic acid analysis. Acetic, propionic, and pyruvic acids are known to coelute with other short-chain organic acids (7,8,48). As a consequence, the identity of organic acids should be confirmed with a second independent analytical method, whenever possible. Unfortunately, this coelution issue is not always given due consideration.

Contamination with organic acids during sampling or sample preparation is of importance, especially at lower analyte concentrations (i.e., μg/L). Chemicals or glassware used during sample preparation may contain organic acids (37,79). Hence, efforts must be made to minimize these contaminations while monitoring background concentrations of contaminants. Organic acids, especially acetate and to a lesser extent formate, have also been found to be present on skin (8). Wearing gloves reduces the risk of contamination through this source considerably.

Overall, little or no sample preparation is required for IC methods used for the analysis of organic acids in water. These IC methods are quick, reliable, and suitable for different types of water samples (Table 6). In the recent past, they have gained in popularity over IEC and GC.

b. Ion-Exclusion Chromatography. Separation in IEC is a complex process involving Donnan exclusion, steric exclusion, and adsorption/partitioning processes. The eluent, usually water spiked with an acid, is passed through a cation-exchange column, commonly a sulfonated polymer. The water will hydrate the sulfonated polymer surface, thus forming a negatively charged membrane, the Donnan membrane. Strong acids, which are completely dissociated into their anions, are repelled by the negatively charged Donnan membrane, thus eluting with the void volume. Large molecules will not be able to penetrate into the polymer due to steric hindrance. Only relatively small, neutral molecules, such as weak organic acids in their undissociated form, can penetrate the membrane and, hence, undergo partitioning between the polymer and the eluent. This separation mechanism makes ion exclusion a very suitable method for the analysis of weak organic acids, especially in complex matrices. This manifests itself in the traditionally widespread use of IEC in, for example, food analysis (96,97).

In brief, anions of strong acids such as sulfate and chloride elute with the void volume, whereas weak organic acids are separated. The elution order of these acids is quite predictable. The higher the pKa value (i.e., the lower the acid strength) and the higher the molecular weight, the later the acids in a homologue series elute. Monocarboxylic acids elute before dicarboxylic acids, saturated acids before unsaturated acids, and branched acids before their straight-chain isomers. Due to strong interactions with the resin, aromatic acids always display long retention times (96,97).

The IEC system is a typical liquid chromatography system consisting of injection port, pump, eluent supply, analytical column, and detector. Columns are usually comprised of completely sulfonated polystyrene-divinylbenzene polymers with different degrees of crosslinking and a high capacity. More recently, silica gel columns have been investigated for their use in IEC, with promising results (102). This material is inert toward solvents, thus allowing for the addition of high amounts of eluent modifiers, which, in the past, was accompanied by problems such as swelling when using the old type of polymer. However, more work is required before being applicable to routine analysis.

Ultraviolet detection as well as conductivity detection are used in IEC. It is possible to suppress background conductivity, although a somewhat different mechanism than in anion-exchange suppression is employed, using tetrabutylammonium hydroxide as the regenerant. Low background conductivities are achieved with aqueous solutions of aliphatic sulfonic acids and perfluorobutyric acid as eluents (96).

Initially, only water was used as eluent, which led to severely tailing peaks caused by partially dissociated acids. When decreasing the pH of the eluent, the dissociation equilibrium of weak acids is shifted to the nonionized form and narrow peaks are obtained. Mineral acids (H_2SO_4, HCl) are preferably used as eluents in connection with UV detection, whereas strong organic acids, especially aliphatic sulfonic acids such as methanesulfonic and octanesulfonic acid, are used as eluents in combination with suppressed conductivity detection (97). More recent investigations show that it is possible to use nonsuppressed conductivity, if employing eluents with very low background conductivity, such as polyvinylalcohol/water (103), butanol (104), and sucrose/methanol (105). Although these eluents are promising, they have not yet been applied to routine analysis.

Organic solvents, especially acetonitrile, are used as eluent modifiers to reduce retention times of more hydrophobic analytes, such as aromatic organic acids. The eluent modifiers compete with the analytes, thus reducing interaction between the polymer and the more hydrophobic analytes. Gradient elution is usually not applied in IEC, since it has been found that concentration gradients give only very little benefit (106). However, gradients with increasing modifier amounts achieved better and faster separation of the more hydrophobic organic acids than isocratic conditions (106).

With IEC being a very suitable method for weak organic acids, it is surprising that it has been used so little in water analysis. Organic acids determined in water samples include not only short-chain, aliphatic, monocarboxylic acids, but also hydroxy- and ketoacids as well as di- and tricarboxylic acids. Problems were encountered with pyruvic acid and especially oxalic acid. Both acids have relatively low pKa values, leading to very close elution to the system peak, if not coelution. Another factor to be considered in IEC is that weak inorganic acids (e.g., carbonate, borate, phosphate) elute between weak organic acids. If the inorganic acid concentration is significantly larger than the concentration of the organic acids, they may interfere

with the quantification of organic acids. An overview of IEC methods applied to organic acid analysis in water is given in Table 7.

Atmospheric precipitation samples have been analyzed quite successfully with IEC. These methods focus mainly on short-chain, aliphatic, monocarboxylic acids (formic to butyric acid), which can be determined down to μmol/L concentrations (16,17,36,43). When combining an anion concentrator with an ion-exclusion analytical column, even lower detection limits of 7–10 μg/L were achieved (36). The difficulty in combining these two techniques lies in the choice of eluent. The eluent must be able to remove the anions from the concentrator columns and at the same time be suitable for IEC. Methanesulfonic acid at pH 9 removes the organic acids, which are dissociated at this pH, from the concentrator column. Changing the pH of methanesulfonic acid to 2.7 allows for the ion-exclusion chromatography of the organic acids, which are then predominantly in their undissociated form. Nevertheless, it has to be kept in mind that use of anion-exchange concentrator columns is restricted largely to samples with relatively low ionic strength (see Sec. II.C.2.a).

Only rarely is IEC applied to wastewater and landfill leachates. Ion-exclusion chromatography used for this purpose includes a system with sulfuric acid eluent/UV detection and a system with perfluorobutyric acid eluent/suppressed conductivity detection (50). Both systems accomplish the separation of VFA, including formic acid, in this rather complex matrix. At the same time, it is possible to determine other acids, such as hydroxy- and ketoacids as well as di- and tricarboxylic acids. However, problems are encountered in the determination of pyruvic and, especially, oxalic acids. These peaks are being masked by the large system peak containing inorganic anions of strong acids.

It has been proposed to couple IC with IEC so that organic and inorganic anions are quantified within one run. This has been done in matrices such as coffee, biological materials, and brine (107,108), but it has not found widespread use in the routine analysis of water. As demonstrated by a method for ultrapure water, a combined IC/IEC configuration may be complicated to operate and is prone to contamination problems (54). Newer methods involving IC with or without a concentrator column are capable of determining inorganic and organic anions in one run, at least for ultrapure water and precipitation samples (23,24,48,55).

In general, IEC methods for organic acid determination are not as common as IC methods. Still, they are very suitable for the determination of a wide range of different types of organic acids, even in complex

Table 7 Overview of IEC Methods, Including Sample Preparation and Detailed IEC Conditions

Sample preparation	Injection	Column	Eluent	Detector	Compounds	Detection limits	Matrix	Comments	Ref.
Addition of 0.408 ml 0.1 N HCl to 20 ml sample	500 μl	Separator: ICE 30580; suppressor: ICE 30960 (25-mm length)	0.0020 N HCl; pump rate 10%	Conductivity, sample passed through water bath prior to detection	Formic, acetic, citric, lactic, glycolic, propionic, butyric, and valeric acids	na; significantly below 0.6 mg/L	Rain	Sample preservation with CHCl$_3$	16
Filtration of turbid samples	na	ICE-AS 1, Dionex	1 mM HCl isocratic with 0.8 ml/m	Conductivity	Acetic, propionic, i-butyric, and n-butyric acids	0.5 μM	Groundwater	—	43
Direct sample injection	10 ml; anion-exchange concentrator	HPX-87H, 300 mm, 7.8-mm ID, Biorad	Methanesulfonic acid (MSE) at pH 2.7	UV at 200 nm	Formic, acetic, propionic, and butyric acids	5.6–9.4 μg/L	Antartic ice	Concentrator column conditioned with MSE at pH 9	36
Centrifugation, filtration, removal of huminlike substances	(a) 20 μl (b) 25 μl	(a) Polyspher OA-HY Merck (b) HPICE AS-6; Dionex	(a) H$_2$SO$_4$; 0.5 ml/min; 5 mM at 45°C and 50 mM at 10°C (b) PFBA (perfluorobutyric acid); 1.0 ml/min; 0.4 mM at 60°C and 1.6 mM at 10°C	(a) UV at 210 nm (b) Spr. cond.; TBAOH regenerant (5 mM)	Formic to n-valeric acid, pyruvic, glyoxylic, glycolic, lactic, glyceric, and succinic acids	na; analyzed from 50 to 50,000 μM	Landfill leachates	Sample stored frozen	50

Abbreviations used in table: see Abbreviations list at the end of the chapter.

matrices. The methodology is restricted to weak organic acids, and problems are encountered when organic acids with low pKa values have to be determined.

c. High-Performance Liquid Chromatography. High-performance liquid chromatography, which is usually performed on reversed-phase columns, is well suited for the determination of hydrophobic compounds. Its separation mechanism is based on partitioning of the analyte between reversed-phase particles and the eluent, usually solvent buffer mixtures, followed by UV or fluorescence detection. These characteristics make it unlikely that direct HPLC methods will be used for the analysis of hydrophilic, short-chain organic acids, which show poor UV absorbance and do not fluoresce.

However, it is possible to separate and quantify VFA, including formic acid, on a Spherisorb 5 ODS column using a methanol/water gradient at pH 4, provided VFA are present at mg/L concentrations (51) (Table 8). Analysis is preceded by a distillation of the sample—in this instance, a landfill leachate—to remove interfering matrix (51,53). Results achieved by HPLC compared well to those achieved by GC/FID, with HPLC having the added advantage of being able to determine formic acid, which is not possible with direct GC/FID (51).

Physical properties of organic acids may be changed, through derivatization, to more hydrophobic compounds that absorb UV light or show fluorescence. Derivatization (see Sec. II.B.3.b.) with α,p-bromoacetophenone (18,23,71) or OPD (35) leads to these kind of compounds, which are then measured by HPLC. More details about HPLC methods can be found in Table 8.

d. Capillary Electrophoresis. Capillary electrophoresis (CE) is a relatively new, fast-developing technique that has generated considerable interest over the last few years. Separation is based on differences in the electrophoretic mobility (depending on m/z ratios) of the analytes, thus making its selectivity completely different from other chromatographic methods. Migration times are very short; hence, analyses are completed within minutes. Separation efficiencies are very high and almost comparable with capillary GC. Other advantages include minimum solvent consumption, small sample volumes, and a simple system configuration, which makes this technique quite economical. The main disadvantages include relatively high detection limits and the lack of routine applications, although both are changing rapidly. The focus in this section is on capillary zone electrophoresis (CZE), which is the electrophoresis option best suited to the analysis of

smaller ions. Excellent books are available describing CE and its possibilities (109–111).

A CZE system consists of an open fused-silica capillary (ID 25–75 μm; no film or deactivation) filled with electrolyte, usually a buffer. The ends of the capillary are immersed into reservoirs that contain electrodes, with a detector placed at the cathode end of the capillary. High voltage, up to 30 kV, is applied to the system so that the electrolyte starts flowing toward the cathode, generating an electro-osmotic flow (EOF). The apparent or net electrophoretic mobility with which an analyte moves toward the cathode in a particular system is constituted of the overall EOF and the individual electrophoretic mobility of the analyte. Cations are attracted by the cathode, thus flowing faster than the EOF, whereas anions are attracted to the anode, which is opposite to the EOF direction. However, the EOF is usually great enough to move the anions forward to the cathode, which opens up the possibility of determining cations and anions within the same run (109–111).

Injection and detection are crucial for successful identification and quantification of analytes when using CZE. Injection volumes are in the nanoliter range to avoid system overloading, since the total volume of the capillary is in the microliter range. Direct injection techniques have been developed to ensure efficient and reproducible injection. Techniques employed are electrokinetic injection (= electromigration injection) and hydrodynamic injection by pressure, vacuum, or gravity (= hydrostatic injection). The most widely applied detection is direct and indirect UV detection, although fluorescence, amperometric, and conductivity detection are utilized as well.

Separation and selectivity of a CZE system can be influenced by altering parameters such as applied voltage, system pH, type of buffer employed, and addition of electro-osmotic modifiers. All these parameters influence the net electrophoretic mobilities of the analytes. When investigating anions, the EOF is often reversed toward the anode instead of the cathode. This can be achieved through the addition of electro-osmotic modifiers, usually cationic surfactants (e.g., tetradecyltrimethylammonium bromide = TTAB).

Although CZE is an attractive technique, only a few applications have been found for its use in the analysis of organic acids in water. It has been applied to the analysis of single raindrops for formate, acetate, and oxalate acids as well as a whole range of other di- and tricarboxylic acids (56,57). The extremely low detection limits of 10–30 nmol/L can be achieved through sample stacking and indirect UV detection. Injection via sample stacking allows for concentration of the an-

Table 8 Overview of HPLC Methods, Including Sample Preparation, Derivatization, and Detailed HPLC Conditions

Sample preparation	Derivatization: reagent—product	Injection (µl)	Column	Eluents	Flow	Detector	Compounds	Detection limits	Matrix	Comments	Ref.
Concentrated	α,p-Bromoacetophenone—bromophenacyl esters	5	Analytical: RP 18, 10 µm, 25 cm, 3.2-mm ID; guard: Pherisorb RP-18, 30 µm, 4 cm, 1-mm ID	Methanol:water, 50:50 (v/v)	1.30 ml/min	UV, 254 nm	Cl–5 monocarboxylic acids	~0.05 mg/L	Rain, sewage, and soilpore water	Can also be measured by GC (18,23)	71
No	OPD—quinoxilinol, hydroxy—quinoxilinol	10–50	C18 column (Alltech)	A: 0.02 M NaH$_2$PO$_4$ B: acetonitrile 97:3 (A:B) to 25:75 (A:B) in 37 min	1.0 ml/min	UV, 320 nm	Oxalic, pyruvic, glyoxylic acids	na; but <2 µM	Rain, fog, and mist	Fluorimetric detection possible	35
Acidification and distillation	None	20	Spherisorb5 ODS; 250 mm, 4.6-mm ID	Methanol/water (3:97) adjusted to pH 4 with H$_2$SO$_4$; isocratic	Gradient 1.0–2.0 ml/min	UV, 210 nm	Formic, acetic, propionic, n-butyric, and isobutyric acids	na: analyzed high mg/L conc.	Landfill leachates	Less sensitive than direct GC/FID; but formic acid analysis possible	51

Abbreviations used in table: see Abbreviations list at the end of the chapter.

alytes at the start of the capillary before migration starts. The sensitivity of this method can be even further enhanced by indirect UV detection with aminobenzoate as the background electrolyte. Further details can be found in Table 9.

Considering that CZE has found extensive application in the analysis of organic acids in matrices such as biological fluids and food (112–118), it is anticipated that in the near future CZE will increasingly be used in water analysis as well.

D. Other Methods

Organic acid analysis using traditional methodology has been and still is applied in the routine operation of wastewater facilities using anaerobic digesters. Volatile fatty acids are determined by column chromatography or distillation followed by titration (11).

For column chromatography, the sample is filtered, acidified to pH 1, and then adsorbed onto a silicic acid column. The organic acids are eluted from the column with n-butanol in chloroform and then titrated with methanolic sodium hydroxide against a phenolphthalein indicator. All short-chain organic acids are reported together as mg acetic acid/L (11,12,119,120).

This method covers organic acid concentrations from 200 to 5000 mg/L, which is adequate for operation control purposes of anaerobic digesters. If concentrations are higher, a second elution step has to be applied in order to recover the organic acids quantitatively. The VFA are determined to almost 100%, in addition to other organic acids, e.g., pyruvic, lactic, and oxalic acids, to mention a few. However, for the operation of anaerobic digesters it is more important to monitor changes in VFA concentration than the total concentrations. Determination of additional, less prominent organic acids is also not necessary.

Sample preparation for the distillation method consists of filtration and acidification. The distillation itself is performed under standardized conditions and is followed by a titration with sodium hydroxide (11,12,121,122). Naturally, only volatile organic acids are retrieved. Their recovery varies with the carbon chain length of the organic acid and is dependent on the specific distillation conditions used. Hence, it is crucial to follow the method description closely in order to achieve reproducible results. To reflect the true concentrations, a recovery factor should be determined and then applied to the sample results. Results are reported in mg acetic acid/L, and concentrations measured with this method are in the high mg/L range. Although this method is more of an empirical method,

it can be, and is, used for monitoring VFA concentration in anaerobic digesters.

Although both methods, the column chromatography and distillation methods, are labor intensive, they are suitable alternatives to methods involving direct aqueous GC injections. These more traditional methods are still in use for the routine monitoring of VFA in anaerobic digesters in wastewater treatment plants.

III. APPLICATIONS

A wide range of analytical methods is available to determine organic acids in water (Table 2). The question of which analytical method to choose for a certain type of matrix is briefly discussed in the following sections.

A. Drinking Water

An interest in the quantification of organic acids in drinking water has only recently developed, since more research has been done on ozonation as part of the water treatment process. It has been found that organic acids are among the ozonation by-products formed during this process (1–3,6,60). To date, relatively few methods exist for this type of analysis (Table 2).

Organic acids in drinking water are usually present at $\mu g/L$ concentrations, whereas inorganic anions have mg/L concentrations. Different approaches to the analysis of these ozonation by-products have been taken, resulting in the development of fundamentally different methods involving either GC or IC.

When using GC methods, liquid–liquid extractions have traditionally been applied to transfer the analyte of interest into an organic phase, which is then injected into the GC. The highly polar and hydrophilic character of most short-chain organic acids makes this approach inadequate, resulting in low extraction yields. Instead, aqueous derivatization has been used to transform organic acids into less hydrophilic compounds. These derivatives can then be extracted with sufficient yield into an organic solvent and after further sample preparation steps (see Sec. II.B.3.a.) can be measured by GC. This method achieves low and adequate detection limits for drinking water. Unfortunately, this GC method is restricted to the analysis of ketoacids due to the reaction mechanism employed in the aqueous derivatization. Simple, aliphatic mono- and dicarboxylic acids cannot be determined with this method (1–3,60).

As an alternative, ion chromatography may be used. The major difficulty encountered here is caused by high inorganic anion concentrations, which are approxi-

Table 9 Overview of CZE Methods

Sample preparation	Injection system	Capillary	Electrolyte	Conditions	Detector	Compounds	Detection limits	Matrix	Ref.
Frozen sample	Hydrodynamic injection (sample stacking) 30 s, vacuum, 1.5 psi	Untreated fused silica, 75-μm ID, effective capillary length 50 cm	p-Aminobenzoic acid 3 mM + Na⁻ p-aminobenzoate 4.5 mM; cation modifier: Ba(OH)₂ 0.76 mM; electro–osmotic flow modifier: tetradecyltrimethylammonium hydroxide (TTAH)	Separation voltage: −30 kV; T = 25°C	Indirect UV detection at 264 nm	Formic, acetic, oxalic, malonic acids; capability to determine another 10 acids, mainly dicarboxylic	Sample stacking: 30–10 nmol/L	Single raindrops	56
Frozen sample	Hydrostatic injection: 10 cm for 30 s	Untreated fused silica, 75-μm ID; effective capillary length: 55 cm	K_2CrO_4 5 mM; tetradecyltrimethylammonium bromide 0.2 mM	22 kV	Indirect UV detection at 276 nm	Formic, acetic acids	180 μg/L and 450 μg/L	Single raindrops	57

Abbreviations used in table see Abbreviations list at the end of the chapter.

mately three orders of magnitude higher than organic acid concentrations. Inorganic anions may cause significant interference with the determination of the organic acids.

Initially, IEC seems to be the method of choice. Anions of strong acids, e.g., chloride and sulfate, elute in the front of the chromatogram within the void volume, and weak acids, which includes most organic acids, are separated mainly according to their pKa values. However, no application of IEC for the determination of organic acids in drinking water has been found in the current literature. The main reason may be that oxalate, which is one of the major ozonation by-products, elutes very close to the void volume. In the past it has been found to be difficult and often impossible to achieve satisfactory separation of oxalate when high concentrations of chloride or sulfate were present (50). Another fact to keep in mind is that weak acids such as carbonate and phosphate elute not with the void volume, but rather between the other weak organic acids.

Anion-exchange chromatography has been employed mainly for the detection of formic, acetic, and oxalic acids in drinking water (5,7–9,66), although other acids have been determined as well (7–9). Sample pretreatment is kept to a minimum; however, one of these methods requires the removal of chloride, sulfate, and phosphate prior to injection, which is accomplished by pushing the samples through cartridges filled with silver or barium cations (66). Removal of mercuric cations, added to the sample for preservation purposes, proves to be necessary in another method, and is accomplished in a time-saving manner by using a H^+-cation exchanger in line (5). Nevertheless, a third method requires no sample pretreatment other than filtration if samples are turbid (7–9). The aqueous samples are then injected directly into IC systems, where the organic acids are separated. All of these relatively fast methods achieve detection limits at very low $\mu g/L$ concentrations, thus being sensitive enough to detect organic acids in drinking water.

Both GC and IC are currently in use in research groups and water facilities dealing with the application of ozone during water treatment. When comparing the available GC methods (1–3,60) to the IC methods (5,7–9,66) it becomes apparent that the IC methods are less time consuming, due to their significantly shorter sample preparation procedures. However, current GC methods quantify ketoacids, whereas IC methods focus on formic, acetic, and oxalic acids, including a few other acids in one of these methods. Hence, GC and IC methods can be seen as complementary to each other. When using both methods, a more complete picture of type and quantity of organic acids formed during ozonation may be achieved.

B. Wastewater

Organic acids in wastewater are measured in the form of volatile fatty acids on a regular basis. Changes in the normally observed background concentration and the composition of VFA are parameters crucial to the day-to-day operation of wastewater treatment plants. The VFA have been measured for a long time; consequently a range of established methods is available (12,13).

The wastewater matrix is highly complex, with high concentrations of organic and inorganic constituents. Individual organic acids are present in high mg/L concentrations. Methods used for VFA analysis include traditional wet chemistry, direct aqueous injection into CG/FID, and other methods.

The traditional methods (see Sec II.D), which include procedures such as distillation and titration, are labor intensive and time consuming; however, their low instrumentation costs are of advantage (11,119–122). These methods have been employed for a long time and operators have gained considerable experience. Thus, these methods are probably still performed directly on site in wastewater treatment plants.

The predominant technique used for VFA analysis is direct aqueous injection into GC/FID (14,61–64,123–126). Little sample pretreatment is necessary and consists of centrifugation and/or filtration followed by acidification. The samples are then injected directly into a GC/FID designated solely for this analysis. Many variations of this method exist using packed columns or capillary columns. Usually, aliphatic monocarboxylic acids with carbon chain length from C2 to usually C5 are measured. Formic acid cannot be determined by GC/FID, although alternative methods using GC/TCD (83) or HPLC (51) are available. Overall method sensitivity is not very high but has proven to be sufficient for the determination of these rather high concentrations of organic acids. These GC/FID methods have the added advantage over traditional methods of being able to monitor specific acids. Valuable information can be gained in this way about the status of the wastewater treatment process, thus helping to achieve its optimum operation (14). In summary, direct aqueous injection into GC/FID delivers immediate results with appropriate accuracy and sensitivity, if care is taken to use appropriate GC operating conditions and to perform system maintenance on a regular basis.

An alternative for the determination of organic acids in wastewater are methods that are applied to landfill leachates. These methods are most likely suitable for wastewater, since both matrices show similarities in possessing high concentrations of inorganic and organic compounds.

C. Atmospheric Precipitation

Starting in the late 1970s, rain samples were analyzed for organic acids (15). Subsequently, a wide range of methods became available that use mainly GC and ion chromatographic techniques (Table 2).

The GC methods tend to be more time consuming, due to the necessary sample preparation steps, and their detection limits for organic acids in rain samples are often somewhat lower than those for other techniques used (Table 2). In order to identify organic acids, sample extracts are often measured by GC/MS. This advantage will be of less importance once LC/MS combinations are more readily available.

Ion-exchange chromatography and IEC are the preferred techniques for large sample numbers (16,17,24,48,58,59). They usually allow for direct sample injection without further sample preparation other than filtration of turbid samples.

To streamline analytical procedures further, successful attempts have been made to determine inorganic anions and organic acids simultaneously. This approach was to a large extent possible due to the fact that inorganic anions and organic acids in rain are generally present in the same concentration range (low to sub-ppm). Methods published using anion-exchange chromatography are capable of measuring fluoride, chloride, nitrite, nitrate, sulfate, sulfite, and phosphate, besides formate, acetate, and oxalate acids (24). In the case of Amman et al. (49) it even includes additional acids, such as lactic, glycolic, propionic, butyric, pyruvic, and valeric. Methods utilizing capillary electrophoresis show a similar potential and achieve comparable detection limits (56,57). Due to the minimized sample preparation and the simultaneous analysis of organic acids and inorganic anions, these approaches may be used to analyze large numbers of rain samples with minimum effort and within a short time frame.

Fog, snow, and ice core samples require matrix-specific sample collection and preparation, which is not discussed in this chapter. The actual organic acid measurements are virtually identical to the ones developed for rainwater, which is why they are included in the tables (23,36,48,59).

D. Other Applications

Other matrices analyzed for organic acids include seawater, groundwater, landfill leachates, and ultrapure water.

Only recently have attempts been made to analyze organic acids in seawater. Very high concentrations of inorganic ions and low concentrations of organic acid make the determination of these organic acids a challenge. Membrane dissociation is one of the more unconventional approaches used to concentrate and separate organic acids from inorganic ions (37), whereas other methods rely on using GC after vacuum distillation and/or derivatization (38–42).

Groundwater and soil pore water usually contain moderate amounts of inorganic ions and varying concentrations of organic acids. A special case gaining more interest lately is ground- and soil pore water samples from sites contaminated with organic compounds. During anaerobic, biological degradation of these organic contaminants, organic acids can be formed as intermediates or endproducts. Techniques applied predominantly to analyze for organic acids are IC and to a lesser extent IEC (43,47).

Landfill leachates have a very high concentration of inorganic and, especially, organic compounds, thus making them a somewhat similar matrix to wastewater. Measurements of organic acids in this type of sample are not an established routine procedure. However, a collection of very different methods are utilized to determine predominantly VFA as well as other acids.

Most of the landfill leachate methods involve a more time-consuming sample preparation step using either distillation (51,53) or extraction (52), which is then followed by GC/FID or HPLC measurement. It has been shown that it is also possible to use IEC with a somewhat shorter sample preparation involving centrifugation, filtration, and removal of humiclike substances through polyvinylpyrrolidine (PVP) cartridges (50). Which method to use is dependent mainly on which organic acids have to be monitored and what kind of instrumentation is available. Due to the similarity in their matrix, all of these methods are likely to be applicable to wastewater samples.

Certain industries, e.g., the power generating industry and the electronics industry, have to use ultrapure water for their cooling and production processes and therefore require frequent monitoring of their water quality. Although inorganic ions are the main compounds of concern, in some cases organic acids are determined as well. A combination of IEC and IC has been applied in the past (54); however, this specific method is quite complicated and requires additional instrumentation. Newer methods utilize an anion-exchange concentrator column in line with an IC system (55).

IV. SUMMARY AND OUTLOOK

Organic acid analysis in aqueous matrices is accomplished by a variety of different techniques. The GC methods requiring derivatization are in general more time consuming than the IC or IEC methods, which often allow for direct sample injection. Hence, IEC and, especially, IC are gaining in popularity over GC methods. An exception is the measurement of high concentrations of VFA by direct aqueous injection into the GC. This method is well established in the wastewater field. In contrast, CE is rarely used on a routine basis for organic acid analysis in water, probably due to its lack of sensitivity and lack of established applications. However, it is expected that CE will be utilized considerably more in the near future due to recent advances in this field.

If suitable methods for a certain water matrix do not exist, methods developed for other matrices, such as food and biological fluids, may at least give a good lead on how to approach this problem. Conditions for these analyses can serve as a starting point for water analysis and after further refinement may result in an appropriate method.

ABBREVIATIONS

AIC	anion-exchange chromatography
autospr.	autosuppressed
BOD	biochemical oxygen demand
CE	capillary electrophoresis
chem.	chemical
CI	chemical ionization
COD	chemical oxygen demand
conc.	concentration
cond.	conductivity (detector)
CZE	capillary zone electrophoresis
df	film thickness
ECD	electron capture detector
EI	electron impact ionization
EOF	electro-osmotic flow
FID	flame ionization detector
GC	gas chromatography
HPLC	high-performance liquid chromatography
IC	ion-exchange chromatography

IEC	ion-exclusion chromatography
IS	internal standard
LC	liquid chromatography
MRL	minimum reporting level
MS	mass spectrometer
MSE	methanesulfonic acid
MTBE	methyl-t-butylether
MTBSTFA	n-($tert$-butyldimethylsilyl)-N-methylfluoracetamide
na	not available
OPD	o-phenylenediamine dihydrochloride
PE	polyethylene
PFBHA	O-(2,3,4,5,6-pentafluorobenzyl)-hydroxylamine
PVP	polyvinylpyrrolidine
spr.	suppressed
TBAOH	tetrabutylammonium hydroxide
TCD	thermal conductivity detector
TFAA	trifluoroacetic acid anhydride
TTAB	tetradecyltrimethylammonium bromide
UASB	upflow anaerobic sludge blanket (Reactor)
UV	ultraviolet (detector)
VFA	volatile fatty acids

REFERENCES

1. Y. Xie, D.A. Reckhow, Identification and quantification of ozonation by-products: ketoacids in drinking water. Paper 5. Proceedings of IOA Pan American Committee Pasadena Conference: Ozonation for Drinking Water Treatment, Pasadena, (A, 1992).
2. Y. Xie, D.A. Reckhow. Proceedings of the American Water Works Association Annual Conference, 1992, pp 251–265.
3. Y. Xie, D.A. Reckhow. Ozone Sci. Engin. 14:269 (1992).
4. R.M. Le Lacheur, L.B. Sonnenberg, P.C. Singer, R.F. Christman, M.J. Charles. Environ. Sci. Technol. 27: 2745 (1993).
5. C.-Y. Kuo, H.-C. Wang, S.W. Krasner, M.K. Davis. ACS Symposium Series, 649:350 (1996).
6. G.A. Gagnon, S.D.J. Booth, S. Peldszus, D. Mutti, F. Smith, P.M. Huck. J. AWWA 89:88 (1997).
7. S. Peldszus, S.A. Andrews, P.M. Huck. J Chromatogr. A 723:27 (1996).
8. S. Peldszus, P.M. Huck, S.A. Andrews. Determination of carboxylic acids in drinking water at low μg/L concentrations: method development and application. Paper P1h. Proceedings of AWWA Water Quality Technology Conference (WQTC), Boston, 1996.
9. S. Peldszus, P.M. Huck, S.A. Andrews. J. Chromatogr. A 793:198 (1998).

10. J.M. Montgomery. Water Treatment: Principle and Design. New York: Wiley, 1985.

11. Standard Methods for the Examination of Water and Wastewater. Method 5560 Volatile Organic Acids. 5th ed. APHA, AWWA, WEF, pp 5–50, 1995.

12. C.N. Sawyer, P.L. McCarty, G.F. Parkin. Chemistry for Environmental Engineering. 4th ed. New York: McGraw-Hill, 1994.

13. Metcalf & Eddy Inc. Wastewater Engineering—Treatment/Disposal/Reuse. 3rd ed. New York: McGraw-Hill, 1991.

14. B.K. Ahring, M. Sandberg, I. Angelidaki. Appl. Microbiol. Biotechnol. 43:559 (1995).

15. J.N. Galloway, G.E. Likens, E.S. Edgerton. Science 194:722 (1976).

16. W.C. Keene, J.N. Galloway, J.D. Holden. J. Geophys. Res. 88:5122 (1983).

17. W.C. Keene, J.N. Galloway. Atmos. Environ. 18:2491 (1984).

18. K. Kawamura, I.R. Kaplan. Anal. Chem. 56:1616 (1984).

19. K. Kawamura, S. Steinberg, I.R. Kaplan. Int. J. Environ. Anal. Chem. 19:175 (1985).

20. W.C. Keene, B.W. Mosher, D.J. Jacob, J.W. Munger, R.W. Talbot, R.S. Artz, J.R. Maben, B.C. Daube, J.N. Galloway. J. Geophys. Res. 100:9345 (1995).

21. M. Legrand, M. De Angelis, F. Maupetit. J. Chromatogr. 640:251 (1993).

22. M. Legrand, M. De Angelis. J. Geophys. Res. 100:1445 (1995).

23. H. Sakugawa, I.R. Kaplan, L.S. Shepard. Atmos. Environ. 27B:203 (1993).

24. P. Hoffmann, V.K. Karandashev, T. Sinner, H.M. Ortner. Fresenius J. Anal. Chem. 357:1142 (1997).

25. K. Kawamura, I.R. Kaplan. Atmos. Environ. 20:115 (1986).

26. K. Kawamura, I.R. Kaplan. Environ. Sci. Technol. 21:105 (1987).

27. R.W. Talbot, K.M. Becher, R.C. Harris, W.R. Cofer. J. Geophys. Res. 93:1638 (1988).

28. G. Helas, H. Bingemer, M.O. Andreae. J. Geophys. Res. 97:6187 (1992).

29. W.C. Keene, J.N. Galloway. J. Geophys. Res. 92:14466 (1987).

30. K. Kawamura, K. Ikushima. Environ. Sci. Technol. 27:2227 (1993).

31. T.E. Graedel, K.I. Goldberg. J. Geophys. Res. 88:10865 (1983).

32. T.E. Graedel, M.L. Mandlich, C.J. Weschler. J. Geophys. Res. 91:5205 (1986).

33. C.G. Nolte, P.A. Solomon, T.Fall, L.G. Salmon, G.R. Cass. Env. Sci. Tech. 31:2547 (1997).

34. K. Kawamura. Anal. Chem. 65:3505 (1993).

35. S. Steinberg, K. Kawamura, I.R. Kaplan. Int. J. Environ. Anal. Chem. 19:251 (1985).

36. P.R. Haddad, P.E. Jackson. J. Chromatogr. 447:155 (1988).

37. X.-H. Yang, C. Lee, M.I. Scranton, Anal. Chem. 65:572 (1993).

38. R. Kondo, H. Kitada, A. Kawai, Y. Hata. Nippon Suisan Gakkaishi 56:519 (1990).

39. A. Van Neste, R.A. Duce, C. Lee, Geophys. Res. Lett. 14:711 (1987).

40. A. Vairavarmurthy, K. Mopper. Anal. Chim. Acta 237:215 (1990).

41. D. Christensen, T.H. Blackburn. Mar. Biol. 71:193 (1982).

42. A. Michelson, M.E. Jacobson, M.I. Scranton, J.E. Mackin. Limnol. Oceanogr. 34:747 (1989).

43. F.J. Sansone. Geochim. Cosmochim. Acta 50:99 (1986).

44. P.M. Bradley, F.H. Chapelle, D.A. Vroblesky. Geomicrobiol. J. 11:85 (1993).

45. P.B. McMahon, D.A. Vroblesky, P.M. Bradley, F.H. Chapell, C.D. Gullett. Ground Water 33:207 (1995).

46. I.M. Cozzarelli, R.P. Eganhouse, M.J. Baedecker. Environ. Geol. Water Sci. 16:135 (1990).

47. I.M. Cozzarelli, J.S. Herman, M.J. Baedecker. Environ. Sci. Technol. 29:458 (1995).

48. C.A. Hordijk, I. Burgers, G.J.A. Phylipsen, T.E. Cappenberg. J. Chromatogr. 511:317 (1990).

49. A. Amman, T.B. Rüttimann. J. Chromatogr. A 706:259 (1995).

50. E.A. McBean, F.A. Rovers, G.F. Farquhar. Solid Waste Landfill Engineering and Design. Englewood Cliffs, NJ: Prentice Hall, 1995.

51. K. Fischer, A. Chodura, J. Kotalik, D. Bienik, A. Kettrup. J. Chromatogr. 770:229 (1997).

52. J.F. Jen, C.W. Lin, C.J. Lin. J. Chromatogr. 629:394 (1993).

53. G. Manni, F. Caron. J. Chromatogr. A 690:237 (1995).

54. C.-T. Yan, J.-F. Jen. Anal. Chim. Acta 259:259 (1992).

55. W.R. Jones, P. Jandik, M.T. Schwartz. J. Chromatogr. 473:171 (1989).

56. M. Toofan, J.R. Stillian, C.A. Pohl, P.E. Jackson. J. Chromatogr. A 761:163 (1997).

57. A. Röder, K. Bächmannn. J. Chromatogr. A 689:305 (1995).

58. K. Bächmannn, I. Haag, T. Prokop, A. Röder, P. Wagner. J. Chromatogr. 643:181 (1993).

59. J.A. Morales, H.L. de Medina, M.G. de Nava, H. Velasquez, M. Santana. J. Chromatogr. A 671:193 (1994).

60. R.W. Talbot, B.W. Mosher, B.G. Heikes, D.J. Jacob, J.W. Munger, B.C. Daube, W.C. Keene, J.R. Maben, R.S. Artz. J. Geophys. Res. 100:9335 (1995).

61. F. Xiong, J.-P. Croue and B. Legube, Env. Sci. Tech., 25:1059 (1992).

62. H.H.P. Fang, H.K. Chui. J. Environ. Eng. 119:103 (1993).

63. A. Viser, I. Beeksma, F. van der Zee, A.J.M. Stams, G. Lettinga. Appl. Microbiol. Biotechnol. 41:549 (1993).

64. P. Westerman. FEMS Microbiol. Ecol. 13:295 (1994).

65. A. Chu, D.S. Mavinic, H.G. Kelly, W.D. Ramey. Water Res. 28:1513 (1994).

66. J.J. Richard, C.D. Chriswell, J.S. Fritz. J. Chromatogr. 199:143 (1980).

67. H.S. Weinberg, W.H. Glaze. Water Res. 31:1555 (1997).

68. J.P. Ivey, D.M. Davies. Anal. Chim. Acta 194:281 (1987).

69. D.M. Davies, J.P. Ivey. Anal. Chim. Acta 194:275 (1987).

70. T. Okada, T. Kuwamoto. J. Chromatogr. 350:317 (1985).

71. K. Blau, J. Halket. Handbook of Derivatives for Chromatography. 2nd ed. Chichester: Wiley, 1993.

72. M.J. Barcelona. H.M. Liljestrand, J.J. Morgan. Anal. Chem. 52:321 (1980).

73. H. Naganuma, Y. Kawahara. J. Chromatogr. 478:149 (1989).

74. S. Allenmark, M. Chelminska-Bertilsson, R.A. Thompson. Anal. Biochem. 185:279 (1990).

75. W. Jennings. Analytical Gas Chromatography. 2nd ed. San Diego: Academic Press, 1997.

76. K. Grob. Classical Split and Splitless Injection in Capillary Gas Chromatography. 2nd ed. Heidelberg: Hüthig Verlag, 1988.

77. K. Grob. On-Column Injection in Capillary Gas Chromatography. Basic Technique, Retention Gaps, Solvent Effects, Heidelberg: Hüthig Verlag, 1987.

78. D. Rood. A Practical Guide to the Care, Maintenance, and Troubleshooting of Capillary Gas Chromatographic Systems, Heidelberg: Hüthig Verlag, 1991.

79. J. Biehoffer, C. Ferguson. J. Chromatogr. Sci. 32:102 (1994).

80. F. Pacholec, D.R. Eaton, D.T. Rossi. Anal. Chem. 58:2581 (1986).

81. J.C. Dupreez, P.M. Lategan. J. Chromatogr. 124:63 (1978).

82. D.M. Ottenstein, D.A. Bartley. J. Chromatogr. Sci. 9:673 (1971).

83. B.A. Schaefer. J. Chromatogr. Sci. 13:86 (1975).

84. W.R. White, J.A. Leenheer. J. Chromatogr. Sci. 13:386 (1975).

85. B.J. Allen, M.H. Spence, J.S. Lewis. J. Chromatogr. Sci. 25:313 (1987).

86. C. Remesy, C. Demigne. Biochem. J. 141:85 (1974).

87. A. Di Corcia, R. Samperi. Anal. Chem. 46:140 (1974).

88. V. Mahadevan, L. Stenross. Anal. Chem. 39:1652 (1969).

89. D.P. Collin, P.G. McCormick, M.G. Schmitt. Clin. Chem. 20:1235 (1974).

90. D.V. McCalley. J. High Res. Chrom. 12:465 (1989).

91. S.E. Fleming, H. Trailer, B. Koellreuter. Lipids 22:195 (1987).

92. G. Gray, A.C. Olson. J. Agric. Food Chem. 32:192 (1985).

93. J.B. Zijlstra, J. Beukema, B.G. Wolthers, B.M. Byrne, A. Groen, J. Dankert. Clin. Chim. Acta 78:243 (1977).

94. R.G. Ackman. J. Chromatogr. Sci. 10:506 (1972).

95. F.J. Duisterwinkel, B.G. Wolthers, W. van der Slik, J. Dankert. Clinica Chimica Acta 156:207 (1986).

96. J. Weiss. Ion Chromatography. 2nd ed. Weinheim, Germany: VCH, 1994.

97. P.R. Haddad, P.E. Jackson. Ion Chromatography: Principles and Applications. Journal of Chromatography Library. Vol. 46. Amsterdam: Elsevier, 1990.

98. D.T. Gjerde, J.S. Fritz. Ion Chromatography. Heidelberg: Hüthig Verlag, 1987.

99. Application Note 101: Trace Level Determination of Bromate in Ozonated Drinking Water Using Ion Chromatography. Dionex Corporation, Sunnyvale, CA, 1995.

100. C.-Y. Kuo, S.W. Krasner, G.A. Stalker, H.S. Weinberg. Analysis of inorganic disinfection by-products in ozonated drinking water by ion chromatography. Proceedings of AWWA: Water Quality and Technology Conference (WQTC), San Diego, 1990, pp 503–525.

101. B.K. Koudjonou, M.C. Müller, E. Costentin, P. Racaud, H. Van der Jagt, J.S. Vilaro, J. Hutchinson. Ozone Sci. Eng. 17:561 (1995).

102. K. Ohta, K. Tanaka, P.R. Haddad. J. Chromatogr. A 739:359 (1996).

103. K. Tanaka, K. Ohta, J.S. Fritz. J. Chromatogr. A 770:211 (1997).

104. J. Morris, J.S. Fritz. Anal. Chem. 66:2390 (1994).

105. K. Tanaka, K. Ohta, J.S. Fritz, Y.-S. Lee, A.-B. Shim. J. Chromatogr. A 706:385 (1995).

106. R. Widiastuti, P.R. Haddad. J. Chromatogr. 602:43 (1992).

107. W. Rich, E. Johnson, L. Lois, P. Kabra, B. Stafford, L. Marton. Clin. Chem. 26:1492 (1980).

108. M. Pimminger, H. Puxbaum, I. Kossina, M. Weber. Fres. Z. Anal. Chem. 320:445 (1985).

109. D.R. Baker. Capillary Electrophoresis. New York: Wiley, 1995.

110. S.F.Y. Li. Capillary Electrophoresis: Principles, Practice and Applications. Journal of Chromatography Library. Vol. 52. Amsterdam: Elsevier, 1992.

111. J.P. Landers, ed. Handbook of Capillary Electrophoresis. Boca Raton, FL: CRC Press, 1994.

112. M. Arellano, J. Andrianary, F. Dedieu, F. Couderc, Ph. Puig. J. Chromatogr. A 765:321 (1997).

113. W. Buchberger, C.W. Klampfl, F. Eibensteiner, K. Buchgraber. J. Chromatogr. A 766:197 (1997).

114. P.R. Haddad, A.H. Harakuwe, W. Buchberger. J. Chromatogr. A 706:571 (1995).

115. C.W. Klampfl, W. Buchberger. Trends Anal. Chem. 16:221 (1997).

116. P.J. Oeffner. Electrophoresis 16:46 (1995).

117. T. Soga, G.A. Ross. J. Chromatogr. A 767:223 (1997).

118. C.H. Wu, Y.S. Lo, Y.-H. Lee, T.-I. Lin. J. Chromatogr. A 716:291 (1995).

119. A.F. Westerhold. J. Water Pollut. Control Fed. 35:1431 (1963).

120. W.H.J. Hattingh, F.V. Hayward. Int. J. Air Water Pollut. 8:411 (1964).

121. W.H. Olmstead, C.W. Duden, W.M. Whitaker, R.F. Parker. J. Biol. Chem. 85:115 (1929-1930).

122. H. Heukelkian, A.J. Kapovsky. Sewage Works J. 21:974 (1949).

123. A.M. Eilersen, M. Henze, L. Kloft. Water Res. 28:1329 (1994).

124. A.M. Eilersen, M. Henze, L. Kloft. Water Res. 29:1259 (1995).

125. I.-C. Kong, J.S. Hubbard, W.J. Jones. Appl. Microbiol. Biotechnol. 42:396 (1994).

126. O. Yeniguen, K. Kizilguen, G. Yilmazer. Environ. Technol. 17:1269 (1996).

127. K. Ohta. J. Chromatogr. A 739:359 (1996).

128. R.C. Weast, ed. CRC Handbook of Chemistry and Physics. 57th ed. Cleveland, OH: CRC Press, 1976–1977.

129. S. Budavari, ed. Merck Index. 11th ed. Rathway, NJ: Merck, 1989.

130. C. Umile, J.F.K. Huber. Talanta 41:110 (1994).

18

Determination of Phenolic Compounds in Water

Magnus Knutsson and Jan Åke Jönsson
Lund University, Lund, Sweden

I. INTRODUCTION

Phenol and substituted phenolic compounds are toxic to humans and aquatic organisms, thus becoming a cause for serious concern in the aquatic environment as they enter the food chain as water pollutants (1). The impact that the different phenolic compounds have in our aquatic environment is readily apparent even at low concentrations. The chlorophenols, for instance, affect the taste and odor of water at concentrations down to 1 μg/L.

The chemical nature of phenolic compounds found in water varies a lot, from polar compounds like phenol itself, to very unpolar compounds such as pentachlorophenol. This chemical diversity makes their determination in water very challenging for the analytical chemist.

Determination of phenolic compounds in water matrices is a subject that has received much attention lately (2,3). This chapter seeks to give an overview of the major different analytical techniques that can be fruitfully applied in this field. The text will focus on various chromatographic approaches for the determination of phenolic compounds in water, since these are the most commonly used and permit determination of individual compounds. Chromatographic separation of phenols is normally performed with either gas or liquid chromatography, and the use of these two techniques is discussed in some detail.

To enable the detection of target phenols present at low concentrations in complex water samples it is normally necessary to include some preparation of the sample prior to chromatographic separation. This serves to separate target compounds from matrix constituents, preferably yielding some degree of analyte enrichment, along with providing a consistency in the background composition of analyzed samples. Therefore some attention is given to the many different sample preparation techniques for extracting phenolic compounds from water samples.

II. CLASSIFICATION AND CHEMICAL CHARACTERIZATION

There are several different groups of phenolic compounds that can be found in water. Besides phenol itself, the following groups are of main interest: alkyl-, chloro-, hydroxy-, and nitrophenols.

Most of the substituted phenols are used or formed in different industrial processes. The determination of chlorophenols in water has been studied extensively, and they are most likely the group of phenols responsible for the largest impact on our aquatic environment. There are 21 chlorophenolic compounds altogether, with widely varying chemical behavior and properties, going from the monosubstituted chlorophenols to pentachlorophenol. Pentachlorophenol was previously used as a wood preservative and is now considered the highest-priority pollutant within the group. Other chlorophenols that are analyzed in industrial wastewater come from the bleaching process in the pulp industry and, because of their high toxicity, are of great interest.

Moreover, chlorinated phenols are major hydrolysis and photolysis products of the chlorinated phenoxy acid herbicides (4).

Of the different phenolic compounds just mentioned, 11 are currently listed by the U.S. Environmental Protection Agency (EPA) as priority pollutants (see Fig. 1) (5). Obviously a large number of the analytical methods found in the literature addressing the determination of phenolic compounds in water focus on these 11 priority pollutants.

European Community directive 75/440/EEC states that the maximum levels of phenolic compounds in surface water for drinking purposes should lie within the range 1–10 μg/L, depending on the required treatment (6).

III. EPA METHODS AND OTHER OFFICIAL METHODS

Most official methods for the determination of the 11 priority pollutant phenols in water are based on liquid–liquid extraction followed by separation with gas chromatography (GC). No official methods based on liquid chromatography have been found within the U.S. EPA, ASTM, ISO, or "Standard Methods" (7) compilations of accepted analytical methods. The EPA method 604 involves a serial extraction of an acidified sample with dichloromethane (8). An alternative description is found as method 6420 in "Standard Methods" (7). The extract is dried, and the solvent is exchanged to 2-propanol. The phenols are then determined by GC using a packed column and flame ionization detection (GC-FID). The method also provides a derivatization procedure with pentafluorobenzyl bromide and column chromatographic cleanup followed by GC determination using electron capture detection (GC-ECD). This lowered the method detection limit (MDL) for some of the compounds. The MDL values are in the range 0.14 μg/L (phenol) to 16 μg/L (4-methyl-4,6-dichlorophenol) for the different compounds and the two GC procedures. An equivalent EPA method is 8040A (9).

Alternatively, EPA methods 625 (8), 6410 (7), and 8250A (9) for extractable bases/neutrals and acids can also be used for the determination of phenols in water samples. These methods are also based a serial extraction with dichloromethane, first at pH >11 and then at pH <2. After drying the extract, the phenols are determined by GC using a packed column and mass spectrometry detection (GC-MS). The MDL values are approximately two times larger than for method 604.

Alternative GC-MS methods using capillary columns are 1625C (8) and 8270B (9), also applicable to soil and sludges. In these method descriptions, no MDL values are given. Method 1653 (10) provides conditions for acetylation of the phenols before extraction with hexane and GC-MS determination. For the latter method, detection limits are in the range of 0.15 μg/L (2,4-dichlorophenol) to 0.71 μg/L (2,4,6-trichlorophenol). Only a few of the compounds listed in Fig. 1 are covered here, but this technique provides MDL values using GC-MS in the same order of magnitude as with GC-ECD. The ISO methods 8165-1:1992 and 8165-2 (11) are generally equivalent to the aforementioned EPA methods.

These methods for the determination of phenols in water samples are regarded by many analysts as very time consuming and labor intensive, with many extraction and solvents exchange steps. Also, the use of hazardous chlorinated solvents is regarded as a limitation of these methods, since dichloromethane will be or already is forbidden for use in many countries.

The use of liquid–solid extraction, typically using hydrophobic extraction disks, bypasses the use of large volumes of solvents. For phenols, the EPA official methods describe this extraction technique only for pentachlorophenol (PCP), which together with other organic compounds can be measured by GC-ECD (method 515.2) or GC-MS (method 525.2) (12). For these methods, the MDL for PCP are given as 0.16 μg/L (GC-ECD) and 0.72-1.0 μg/L (GC-MS; different instruments).

A method involving the direct injection of water samples into a GC column is described in ASTM standard D2580 (13). This method obviously bypasses the use of solvents for extraction, but the lowest concentration for which this method can be used is 1 mg/L, which is considerably larger compared with the extraction methods already mentioned.

The total content of phenols in natural waters and wastewaters can be determined by using the 4-aminoantipyrine (4-AAP) colorimetric procedure. There are many descriptions of essentially the same method: EPA methods 420 (14) and 9065 (9), Standard Methods 5530 (7), ASTM D1783-91 (13), ISO 6439:1990 (11). Different procedures, involving, e.g., chloroform extraction and distillation, in some cases automated using flow injection analysis (FIA), are described in these standards. Another reagent, MBTH (3-methyl-2-benzothiazolinone hydrazone), is also used for the same purpose in EPA method 9067 (9). The 4-aminoantipyrene reacts with phenol and ortho- and meta-substituted phenols and, under proper pH conditions, also with phe-

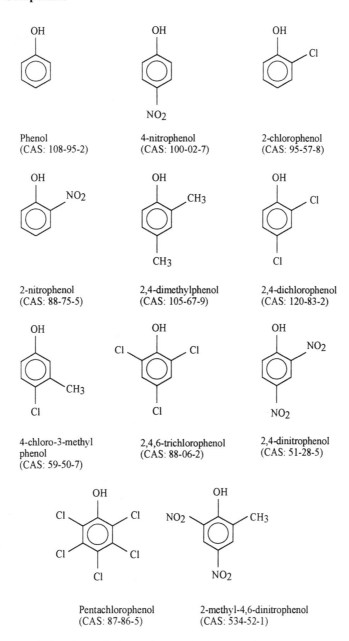

Fig. 1 Eleven phenolic compounds listed as priority pollutants by the U.S. EPA, along with their chemical name, structure, and CAS number.

nols with an alkyl, aryl, nitro, benzyl, nitroso, or aldehyde group. The methods can be used in the low μg/ L range. However, they cannot differentiate between different substituted phenols and thus give a total phenol content (or "phenol index"), provided that the phenols present react with the reagent. Due to this low specificity, detailed descriptions of these methods are beyond the scope of this chapter.

IV. LIQUID CHROMATOGRAPHIC DETERMINATION OF PHENOLIC COMPOUNDS IN WATER

Liquid chromatography (LC) is often used for the determination of phenolic compounds. Especially for the determination of phenols in various water samples, liquid chromatography is the choice over gas chromatography (GC), since it is more suitable for aqueous samples. Furthermore, since no derivatization of the phenols is required and the online connection between solid-phase extraction and the LC column is fairly straightforward, this approach appears very suitable. However, the conventional ultraviolet (UV) detector is much less sensitive than most of the GC detectors. This has promoted the search for more sensitive LC detection devices as well as improvements and alternative methods in sample preparation. Several detectors, more or less sensitive toward phenol compounds, are used today in the liquid chromatographic determinations of phenolic compounds in water: UV, diode array (DAD), electrochemical, fluorescence, and mass spectrometric (MS) detection. Their advantages and disadvantages, together with some applications, are summarized later.

Separation of phenols with liquid chromatography is normally performed with reversed-phase liquid chromatography (RPLC). The mobile phase consists of a mixture of a polar organic solvent (methanol or acetonitrile) and an aqueous buffer, and in most cases different types of silica C_{18} or C_8 columns are used as analytical columns.

The separation and retention of 29 phenolic and related compounds on different RPLC columns has been investigated by Marko-Varga and Barceló (15). The columns studied were LiChrospher 100, PLRP-S, Vydac, and Hypercarb. Also, the effects of various acetonitrile/buffer mixtures and the pH of the mobile phase on the retention and separation of the phenolic compounds on the different columns were evaluated. For this application it was found that the silica C_{18} column (LiChrospher 100) gave the best separation, probably due to a mixed retention mechanism.

A. Liquid Chromatography with Ultraviolet and Diode Array Detection

Although ultraviolet detection is regarded inferior when it comes to sensitivity, it is frequently used for phenolic determination (16–19), often together with a solid-phase extraction (SPE) sample preparation step (see Sec. VIII.B) in order to increase the overall system performance. If UV detection is used, phenols are normally monitored at a wavelength of around 280 nm.

Diode array detection (DAD), where a large span of wavelengths are monitored at the same time, has also been frequently used for the determination of phenols in water in combination with an SPE sample work-up step (20–25). Moving from UV detection at a single wavelength to DAD, a small sacrifice in sensitivity is

made up for by a much better peak identification, with available spectral libraries to confirm analyte presence.

Liquid chromatography with DAD was used together with online SPE for the determination of phenolic compounds in the River Meuse (25). Sample volumes of 10 ml gave detection limits of below 0.1 μg/ L for phenol and m-cresol in surface water.

B. Liquid Chromatography with Electrochemical Detection

Electrochemical detection of phenolic compounds is regarded as a more sensitive detection technique than UV, and it has frequently been used together with various sample preparation steps for the determination of phenolic compounds at the ng/L level (26–31). Several different modes of electrochemical detection have been used, with amperometric detection (26,27,30,32–36) being the one most frequently employed. Coulometric detection (31,37) has also been used (for a summary of different electrochemical detection of phenols in water, see Table 1).

The coulometric detector converts 100% of the analyte, since the oxidation of phenols occurs in the high-porosity electrode, whereas an amperometric detector normally converts only about 10% at the electrode surface (3).

Amperometric detection is used in conjunction with LC separation with a glassy carbon working electrode at an oxidizing potential around +1000 mV vs. Ag/ AgCl reference electrode. However, this type of electrochemical detection exhibits the problem of phenols fouling the electrode. The problem can be partly solved by cleaning the electrode, using two additional pulses, one oxidizing and one reducing, between each measurement pulse (pulsed amperometric detection).

Liquid chromatography with different modes of amperometric detection has frequently been used for the determination of phenols in water, with or without a preconcentration step. Without preconcentration, amperometric detection at +1150 mV vs. Ag/AgCl was tested with six different RPLC columns (33). Of the studied columns, the Spherisorb C_8 gave the best chromatographic behavior of the 11 tested phenols.

Pulsed amperometric detection using a glassy carbon electrode (+1200 mV vs. Ag/AgCl) in combination with online SPE with C_{18} material has been used for the determination of the 11 priority pollutant phenols at sub-μg/L levels (26). Another application where pulsed amperometric detection has proved successful is for phenols in seawater (27). After passing 1000 ml of seawater through a polymeric SPE material and de-

Table 1 Electrochemical Detection Procedures Used with LC Separation for Determination of Phenolic Compounds in Different Water Matrices

Electrochemical mode	Working electrode	Working potential	Sample preparation	Type of water sample	Ref.
Pulsed amperometric	Glassy carbon	+1250 mV vs. Ag/AgCl	SPE with polymeric sorbent	Seawater	27
Amperometric	Glassy carbon	+1150 mV vs. Ag/AgCl	None	Environmental water	33
Amperometric	Glassy carbon	+1200 mV vs. Ag/AgCl	None	River water	34
Amperometric	Glassy carbon	+1000 mV vs. Ag/AgCl	SPE with different sorbents	River water	28
Amperometric (dual electrode)	Glassy carbon	Various	None	Wastewater	32
Coulometric (dual electrode)	Glassy carbon	Various	SPE with polymeric sorbent	Groundwater	31
Amperometric	Glassy carbon	+1000 mV vs. Ag/AgCl	SPE with C_{18} sorbent	Drinking and river water	29
Coulometric (multielectrode)	Porous graphite	Various	SPE with S_{18} sorbent	Tap and mineral water	37
Pulsed amperometric	Glassy carbon	+1200 mV vs. Ag/AgCl	SPE with C_{18} sorbent	Tap water	26
Amperometric	Glassy carbon	+1100 mV Ag/AgCl	SPE with C_{18} sorbent	River water and wastewater	30
Amperometric	Glassy carbon	+900 mV vs. Ag/AgCl	Both LLE and SPE	Wastewater	36
Coulometric	Not stated	+750 mV vs. Pd	SPE with C_{18} and polymeric sorbents	Seawater	38
Amperometric	Glassy carbon	+1000 mV vs. Ag/AgCl	Supported liquid membrane	River water	129
Amperometric	Glassy carbon	+600 and +900 mV vs. Ag/AgCl	SPE with polymeric sorbent	Industrial wastewater	69

tecting at +1.25 V vs. Ag/AgCl, the phenols could be quantified at ng/L levels.

The use of multielectrode electrochemical detection in combination with solid-phase extraction using C_{18} material and LC separation was described for the identification of 27 phenolic compounds in water samples (37). The multielectrode consisted of four coulometric array cells, each containing four electrochemical detector cells. These employed porous graphite working sensors with palladium as reference and counter electrodes, and were arranged in series after the analytical column. Tap water and mineral water were analyzed; the authors reported very low detection limits for the phenols.

Dual coulometric detection was used with online solid-phase extraction with LiChrolut EN (31) for the determination of polar priority phenols at ng/L levels.

The first electrode was intended for sample cleanup (normally set at a low potential), and the detection of the phenols was made on the second electrode.

For the determination of phenols in seawater after enrichment using SPE cartridges and disks, the LC detection was performed using a large-surface-area coulometric electrode at +750 mV vs. the Pd reference electrode (38).

Furthermore, biosensors have been used for the electrochemical detection of phenols in combination with flow injection analysis (FIA) and LC separation (39,40). The biosensors are normally working at a much lower potential, and they are also very analyte specific, since several enzymatic steps may be involved in the detection. Normally the enzyme is immobilized onto solid graphite electrodes or in carbon paste electrodes.

C. Liquid Chromatography with Fluorescence Detection

Liquid chromatography in conjunction with fluorescence detection has been used to improve the sensitivity and selectivity for the determination of phenolic compounds in water. All the different techniques for the determination of phenolic compounds in water that use LC and fluorescence detection are summarized in Table 2.

Precolumn dansylation with dansyl chloride in combination with postcolumn photolysis has been described in a couple of papers (41,42). In one of the papers, peroxyoxalate chemiluminescence detection was also used, yielding detection limits as low as 0.01–0.1 μg/L for several phenols in surface water (41). Either way, the phenolic anions are extracted as an ion pair with tetrabutylammonium into an organic phase containing dansyl chloride. Precolumn derivatization with 2-(9-anthrylethyl) chloroformate has been described for the determination of phenols in industrial wastewater (43).

Postcolumn reactions with 4-aminoantipyrine and potassium ferricyanide has been used in combination with SPE sample preparation for phenols in wastewater (44). The reagent 4-aminoantipyrine was employed in a similar fashion for the fluorescent derivatization of 22 monohydric phenols (45). Another postcolumn reaction involved the coupling of diazotized sulfanilic acid with the phenols to form highly colored azo dyes (46). However, this setup merely showed a minor improvement (16-fold) compared to conventional UV detection. Postcolumn reaction with N-methylbenzothiazole-2-hydrazone and $Ce(NH_4)_2(SO_4)_3$ and detection at 500 nm is yet another approach that was used to de-termine 30 hydroxyaromatic compounds in wastewater (47).

D. Liquid Chromatography with Mass Spectrometric Detection

The combination of LC separation and mass spectrometry (MS) is described in several papers, and the number of applications where LC-MS is used is rapidly increasing with the availability of less expensive benchtop instruments. The superiority of the mass spectrometer compared to other LC detectors is undisputed, for it offers unsurpassed selectivity and also, to some degree, structure identification, thus being a powerful tool for the characterization of complex water samples. Several ionization techniques, such as atmospheric pressure chemical ionization (APCI), electrospray/ion spray (ESP/ISP), and thermospray (TSP) (48), have been employed in the MS determination of phenols at low concentration.

A comparison between positive- and negative-ion modes in TSP LC-MS with a quadrupole instrument showed that the negative-ion mode gave better sensitivity for the chlorophenols than did the positive mode (49). The APCI and ISP techniques in the negative-ion mode were used for the identification of 19 priority phenols (50). Some of these phenols (phenol, 4-methylphenol, and 2,4-dimethylphenol) could be detected only with ISP-MS. Following preconcentration of 50–100 ml river water with SPE, detection limits for the different phenols from 0.1–5 μg/L and 0.1–25 ng/L were found using full-scan and time-scheduled single-ion monitoring modes, respectively (for chromatogram, see Fig. 2).

Table 2 Fluorescence Detection with LC Separation for Determination of Phenolic Compounds in Different Water Matrices

Derivatization reagent	Mode	Type of water sample	Ref.
N-Methylbenzothiazole-2-hydrazone and $Ce(NH_4)_2(SO_4)_3$	Postcolumn	Wastewater	47
2-(9-Anthrylethyl) chloroformate	Precolumn	Wastewater	43
Dansyl chloride	Precolumn	River water	42
4-Aminoantipyrine		Wastewater	45
4-Aminoantipyrine and potassium ferricyanide	Postcolumn	Wastewater	44
Diazotized sulfanilic acid	Postcolumn	River water	46
Dansyl chloride	Precolumn	River water	41

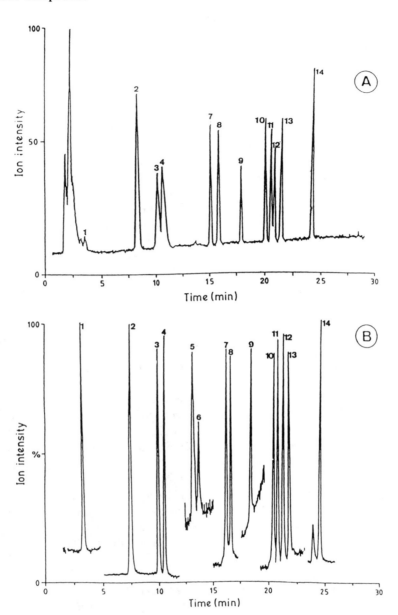

Fig. 2 Online SPE using OSP-2 followed by LC/APCI-MS of 50 ml of river water sample: (A) spiked at the 5-μg/L level under full-scan mode; (B) spiked at 0.06 μg/L under SIM conditions, using negative-ion mode of operation of (1) catechol, (2) 4-nitrophenol, (3) 2,4-dinitrophenol, (4) 2-nitrophenol, (5) 4-chlorophenol, (6) 2-chlorophenol, (7) 2,6-dinitro-4-methylphenol, (8) 4,6-dinitro-2-methylphenol, (9) 2,4-dichlorophenol, (10) 2,3,4-trichlorophenol, (11) 2,4,6-trichlorophenol, (12) 2,4,5-tri-chlorophenol, (13) 2,3,5-trichlorophenol, (14) pentachlorophenol. For other details, see Ref. 50. (From Ref. 50 with permission from the authors and the publisher.)

Both APCI-MS and ESP-MS were used in the negative-ion mode for the determination of chloro- and nitrophenols in tap water and seawater (51). After extraction of 250 ml seawater with polymeric SPE disks, detection limits in the low μg/L range were found for most of the phenols using LC-APCI-MS. Pentachlorophenol and 2,4-dinitrophenol were determined together with acidic pesticides in river water and drinking water using ESP-MS in combination with SPE using graphitized carbon packing material (52).

V. GAS CHROMATOGRAPHIC DETERMINATION OF PHENOLIC COMPOUNDS IN WATER

As described in Sec. III, all of the official methods for the determination of phenolic compounds in water are based on gas chromatography (GC). The GC methods are normally more sensitive than the LC methods, but because of the high polarity and low vapor pressure of the phenols, a derivatization step is normally necessary prior to the final GC analysis. GC separation of underivatized phenols using capillary columns with conventional phases is difficult, for phenols (in particular, nitrophenols) exhibit severe tailing. Highly deactivated capillary columns have been used for the direct separation of phenols (53,54), but in most cases the phenols are derivatized in order to improve their chromatographic performance. Several different derivatization agents have been used, e.g., pentafluorobenzyl bromide (55,56), pentafluorobenzyl chloride (57), acetic anhydride (58,59), and heptafluorobutyric anhydride (60).

Several different detectors have been used in combination with GC for the determination of phenols, e.g., the flame ionization detector (FID) (58,61), the electron capture detector (ECD) (55,57,59,60,62,63), and the mass spectrometer (MS) detector (56,64–66).

In a series of papers, Lee and coworkers described the use of pentafluorobenzyl bromide as derivatization agent for the determination of 22 phenols in water samples (55,56). Prior to derivatization, the phenols were extracted from the water sample into dichloromethane. In the first paper, six different columns were tested, and the OV-101 fused-silica capillary column with Carbowax-deactivated surface was found to give the most efficient separation (55). Detection was carried out using both the ECD and MS. A similar approach, using derivatization with pentafluorobenzyl chloride and ECD, was described for the analysis of monochlorinated and brominated phenols in aqueous samples (57).

The use of acetic anhydride for the acetylation of phenols has been described in several papers (58,59). Determination of chlorophenols in freshwater, wastewater, and seawater using acetylation and electron capture detection was reported by Abrahamsson and Xie (59). They compared two derivatization procedures, pentafluorobenzoylation vs. acetylation, and concluded that the acetylated derivatives gave better separation on the capillary column. Derivatization using heptafluorobutyryl in combination with GC-ECD (60) involves an extraction of the acidified sample into benzene prior to derivatization. Another method describes the conversion of eight phenols (phenol, cresols, and xylenols)

into corresponding bromophenols after reaction with bromine followed by an analysis using GC-ECD (63).

Gas chromatography with Fourier transform infrared spectroscopy (FTIR) has been used for determination of chlorophenols in drinking water (67). Before the GC-FTIR analysis the phenols were acetylated with acetic anhydride followed by off-line SPE using graphitized carbon cartridge. Gas chromatography with microwave-induced plasma atomic emission spectroscopy was used in combination with two different off-line SPE procedures (68). Derivatization with 3,5-bis(trifluoromethyl)benzyldimethylphenylammonium fluoride in combination with MS detection in negative chemical ion mode has been used for the determination of chlorophenols in industrial wastewater (66).

As can be seen earlier, solid-phase extraction sample preparation is a commonly integrated part of the overall system setup in GC analysis. The technique is treated in more detail in upcoming Sec. VIII.3.2.

Many of the papers already cited describing GC determination of phenols are fairly old (from the end of the 1970s and the beginning of the 1980s). Puig and Barceló remarked that there has been a general trend to change the overall procedure, viz., the use of liquid–liquid extraction and separation by GC is being replaced by solid-phase extraction and LC procedures (3). This seems to be a general trend, not a change in just phenol analysis. One of the reasons for this is that the derivatization step is regarded as very tedious and time consuming. On the other hand, the sensitivity and separation power of GC is still unsurpassed by even the latest developments in LC.

VI. ALTERNATIVE SEPARATION TECHNIQUES

Other separation techniques, such as capillary zone electrophoresis (CZE) and supercritical fluid chromatography (SFC), have been shown to perform well for the separation of phenols.

Several papers describing the use of CE for the separation of phenolic compounds in water samples have been published lately (19,69–72). The majority of these employ UV detection, but ESP-MS in the negative-ion mode (73) and indirect fluorescence detection (74) have also been used. In one study, a comparison between LC and CZE was performed to assess their suitability for the determination of the 11 priority pollutants in water (19). The authors claim that CZE gave a shorter analysis time and smaller matrix effects.

However, it was not possible to achieve the desired detection limits without a preconcentration on solid-phase material.

Micellar electrokinetic chromatography (MEKC) has been used by several authors for the separation of phenolic compounds (75) and in some cases for the determination in water (76). Off-line SPE using polymeric sorbents and MEKC with electrochemical detection were used for the determination of chlorinated phenols in a river at a low μg/L level (76). Generally, one problem associated with miniaturized techniques such as CE when combined with UV detection is the limitation to small injection volumes. Therefore, efficient enrichment steps in the sample preparation is necessary.

Supercritical fluid chromatography (SFC) has also been used for the separation of phenols (77–79). The supercritical fluid normally used is carbon dioxide with some modifier, e.g., methanol or chlorodifluoromethane (Freon 22) (78). Berger and Deye tested binary and tertiary supercritical mixtures, among them methanol/carbon dioxide mixtures containing very polar additives (79). Ong and coworkers used chlorodifluoromethane as the supercritical fluid (77). In most studies, UV detection was used and did not measure up to the required sensitivity. To bypass this problem, an online system with solid-phase extraction connected to the SFC instrumentation was designed. Some of these systems are presented in upcoming Sec. VIII.B.

VII. NONCHROMATOGRAPHIC TECHNIQUES

For direct measurements of o-nitrophenol, a selective optical chemical sensor has been developed by Wang and coworkers (80). Determination of o-nitrophenol in tap water was presented, but the sensitivity of the sensor is poor compared to the earlier-presented chromatographic systems.

An immunoassay kit for the measurement of pentachlorophenol has been developed with a limit of detection around 60 ng/L (81). The sample matrix had little influence on the immunoassay, but 2,4,5,6-tetrachlorophenol and 2,3,4,6 tetrachlorophenol show some cross-reactivity. The methodology can be used as an initial screening of phenols, and it normally does not require any sample preparation. The immunoassay methodology has also been applied for the determination of 4-nitrophenol and substituted 4-nitrophenols (82).

Phenol-specific immunoassay has also been used as a detection system in liquid chromatographic separation systems (83). The connection between LC and immunoassay detection was regarded by the authors to be more labor demanding than conventional LC-UV, but the payoff in selectivity and sensitivity is claimed to be immense.

Spectrophotometry utilizing the reagents 4-aminoantipyridine and 3-methyl-2-benzothiazolinone hydrazone is the classical technique for the nonspecific determination of phenolic compounds. It is the basis for several official methods and is discussed earlier (see Sec. III).

VIII. SAMPLE PREPARATION PROCEDURES

The fact that different substituted phenols even at very low concentrations affect our aquatic environment demands selective and sensitive determination systems. The first step in such a system is efficient sample preparation. Over the last two decades much focus has been given to sample preparation in chromatographic analysis, since this step is regarded as critical, error prone, and normally time consuming (84).

There are two main objectives with sample preparation step:

1. Cleanup of the sample to avoid deterioration of the chromatographic system (column, detectors, etc.) and degradation of the analytes
2. Concentration of the analytes, which normally is necessary before introduction to the final chromatographic instrument

Some of the commonly used sample preparation techniques for phenolic compounds are briefly described next, and references to specific applications to phenolic compounds are given.

A. Liquid–Liquid Extraction

Liquid–liquid extraction (LLE) is the classical sample preparation technique. Reviews of LLE are found in general papers on sample preparation (85,86), and it is still frequently used by the environmental analyst. This is mainly due to the fact that LLE is used in many of the official methods (see Sec. III). The technique is based on partitioning of the analyte between an aqueous and an organic phase, contained inside a bottle or a separatory funnel. The analyte is extracted from the aqueous phase to the water-immiscible organic phase, and, after extraction, the two phases are allowed to separate. If necessary the organic phase is dried with a suitable drying agent. Before introduction into the analytical instrument, the organic extract can be concentrated by a volume reduction. Also, solvent change is often made after evaporation to dryness.

The selectivity in LLE can be controlled by changing the organic solvent, by using ion-pairing or derivatization reagents, and by adjustment of pH in the aqueous phase.

For polar analytes (e.g., phenol and monosubstituted phenols) polar solvents such as ethyl acetate and methyl chloride are favored, whereas for unpolar analytes (e.g., higher substituted phenols) more unpolar solvents such as hexane and toluene are used.

However, conventional LLE is often regarded as having some severe drawbacks:

It is laborious and time consuming.
It uses large quantities of organic, often hazardous, solvents.
It is difficult to automate.

1. Liquid–Liquid Extraction in Combination with Liquid Chromatography

Liquid–liquid extraction with dichloromethane has been used before LC with UV and fluorescence detection (87). Dinitrophenols are detected with UV absorption followed by oxidation with cerium(IV) in an open tubular reactor, allowing fluorescence measurement of cerium(III).

2,4,5-Trichlorophenols and 4-nitrophenol, both degradation products of pesticides, have, together with some pesticides, been extracted with an online continuous-flow extraction in combination with LC with UV and MS detection (48). The enrichment factors were lower than those obtained with online precolumn systems using solid adsorbents, but the LLE suffered less from memory effects.

The combination of LLE and normal-phase LC with UV fluorescence has been used for phenol and cresols in rain (88). The rain samples were adjusted to pH 2 with sulfuric acid continuously extracted with dichloromethane, giving detection limits at the μg/L level.

2. Liquid–Liquid Extraction in Combination with Gas Chromatography

The combination of liquid–liquid extraction and gas chromatographic separation has frequently been applied to phenolic compounds in water samples. As already discussed, many of the official methods for

phenol determination in water are based on this combination.

In combination with derivatization with pentafluorobenzyl bromide and GC determination, LLE has been used for the determination of phenolic compounds in water (55,56). A similar approach for phenol and monochlorinated/monobrominated phenols in complex aqueous samples has been described by Booth and Lester (57).

Continuous liquid–liquid extraction has been connected online to gas chromatography for the determination of phenols in aqueous samples (89). The technique has been used in two different modes. The first one involves simultaneous extraction and derivatization. Acetate esters of phenol, cresols, and chlorophenols were formed by continuous extraction into n-hexane containing acetic anhydride. Another approach was first to derivatize the phenols with acetic anhydride. The esters were extracted with a pentane/diethyl ether mixture before capillary GC determination with both FID and ECD detection (90).

Continuous liquid–liquid extraction with ethyl acetate in combination with direct GC separation using highly deactivated capillary columns has been used for the determination of nitrophenols in groundwater (54). When GC-MS was used for identification of the nitrophenols and with a nitrogen-phosphorus detection, limits of detection in the μg/L range were found.

B. Solid-Phase Extraction

The use of solid-phase extraction (SPE) as a sample preparation procedure has developed very rapidly over the last 20 years, and it has been thoroughly described by several authors (91–96). The technique is based on sorption of the analyte onto a sorbent packed in a small column. As the water sample passes the SPE column, the analytes are trapped on the sorbent and subsequently desorbed and eluted with a small volume of solvent. Ultimately, the result is total matrix elimination and analyte enrichment.

One problem with trying to cover a wide range of phenolic compounds using SPE is finding a suitable sorbent. The breakthrough volumes of the different substituted phenols varies considerably, depending on the sorbents used and the properties, especially the polarity, of the phenols extracted.

Several sorbent materials have found use in the extraction of organic pollutants, including phenolic compounds, in environmental water samples. These were summarized by Hennion (see Table 3) (94). As can be seen from the table, the choice of sorbent for the ex-

Table 3 Different Sorbents Used for Sample Preparation of Organic Compounds in Environmental Water Samples, Including Their Mechanism, Type of Analytes Used, and Example Applications

Sorbent	Separation mechanism	Elution solvent	Nature of analyte	Environmental applications
Octadecyl- or octyl-bonded silicas	Reversed phase	Organic solvent	Nonpolar and weakly polar	PCBs, PAHs, PCBs, organophosphorus and organochlorine pesticides, alkylbenzenes, **polychlorophenols**, phthalate esters, polychloroanilines, apolar herbicides, fatty acids, aminoazobenzene, aminoanthraquinone
Porous styrene-divinylbenzene copolymer	Reversed phase	Organic solvent	Nonpolar and medium polar	**Phenol, monchlorinated phenols**, aniline, chloroaniline, moderately polar herbicides (phenoxy acids, triazines, phenylureas)
Graphitized carbon	Reversed phase	Organic solvent	Nonpolar and relatively polar	Alcohols, **nitrophenol**, relatively polar herbicides
Silica- and polymer-based ion exchangers	Ion exchange	Water (pH adjusted)	Cationic and anionic organics	**Phenol**, nitrolotriacetic acid, phenoxy acids, phenylenediamines, aniline and polar derivatives, sulphonic acids, phthalic acids, **aminophenols**
Metal-loaded sorbents	Ligand exchange	Complexing aqueous solution	Metal complexation property	Aniline derivatives, amino acids, 2-mercaptobenzimidazole, carboxylic acids, buturon

Source: Ref. 94 with permission from the authors and the publisher.

traction depends mainly on the polarity of the analyte. For unpolar analytes (e.g., polychlorinated phenols) silica C_{18} packing material is normally used, whereas for more polar analytes (e.g., phenol, monochlorinated phenols, and nitrophenols) either polymeric packing material (PRP-1 or PLRP-S) or graphitized carbon material are recommended.

The SPE technique can be used off-line as well as online. For the extraction of phenolic compounds from water samples using SPE, liquid chromatography is the preferred separation method, although applications utilizing GC, CE, and SFC are also found. For the off-line procedure, disposable cartridges are used, if the detection limits need to be lowered, larger amounts of sorbent can be used. Off-line extraction normally requires a volume reduction of the eluting solvent before injection onto the LC-system, and during that process losses of volatile phenols can occur. In the online approach, the sorbent is packed in small stainless steel precolumns. The application of this technique to envi-

ronmental water samples is the topic of several review papers (91–93).

One of the main problems when analyzing natural water matrices is the influence from humic substances, which can be partly overcome by acidification of the sample (97).

The stability of the phenols on polymeric sorbents during storage has been investigated (98). It was found that the stability of the phenols depended on the water matrix, storage temperature, and physical chemical properties. After storage at $-20°C$ for 2 months and at $4°C$ for 2 weeks, complete recovery of the phenols was observed.

An alternative way to achieve the same end is to use solid-phase extraction disks instead of a cartridge or a precolumn (99,100). A membrane or disk of polytetrafluoroethylene (PTFE) fibrils is impregnated with small particles or adsorbing materials, such as C_{18} silica or polymeric sorbents. Such SPE disks with different polymeric sorbents were used for the off-line extraction of 16 phenols, including the priority pollutants, from different water matrices (99). This procedure using a disk size of 47 × 0.5 mm, allowed extremely rapid flow rates of about 200 ml/min. A comparison between extraction of phenols using C_{18} and polymeric SPE disks has been made (100). The authors state that the polymeric disks yielded improved results for the investigated phenols, especially for the polar phenols, as would be expected.

The SPE disks have also been used in precolumns for online extraction (101). Taking this approach, the disks have to be cut in small pieces and placed into the precolumn. Online extraction with a stainless steel precolumn (diameter 4.6 mm) (packed in this fashion was used for phenolic determinations using LC with an electrochemical detector. Large water volumes (250 ml) could be extracted without analyte losses and resulted in detection limits between 0.01 and 0.1 μg/L in tap water and between 0.1 and 1.0 μg/L in river water.

A miniaturized form of SPE, solid-phase microextraction (SPME) (102,103), was introduced by Pawliszyn and coworkers and has been described as a solvent-free sample preparation technique. The technique has been used in combination with GC for polar phenols in water samples (104). The polar phenols are adsorbed to a polyacrylate material coated on the fused-silica fiber. Following adsorption, the fiber is transferred to a modified injector of a gas chromatograph, where the phenols are thermally desorbed, separated, and detected with both an FID and an MS detector. Acetylation of the phenols in the water and adsorption of the acetates improved the system performance.

A fiber coated with polyacrylate was also tried in combination with GC-MS for phenols from strongly contaminated wastewater (105). The SPME approach was compared with LLE using dichloromethane, revealing that LLE generally gave higher extraction values. Alternative coatings, e.g., a bonded sol-gel layer of poly(dimethyl siloxane) (PDMS), again in combination with GC separation, have found applicability to the determination of alkylphenols in water (106).

1. Solid-Phase Extraction in Combination with Liquid Chromatography

As previously discussed, many different types of packing materials can be used for the extraction of phenolic compounds from water samples, depending on the polarity of the phenols of interest. The most frequently found methodology in the contemporary literature for determination of phenols in water samples is SPE in combination with LC. A compilation of published work in this field is given later. As already discussed, the polarity of the phenolic compounds does govern the choice of sorbent material, but a general trend of using polymeric sorbents is apparent. As already seen, these materials have proved most successful for this particular class of compounds.

Puig and Barceló have tested several different sorbents for online SPE-LC. In their first paper, eight different sorbents were tested, indicating that PLRP-S (styrene-divinylbenzene copolymer) is the most suitable sorbent for the majority of phenolic compounds, although problems are encountered for the most polar molecules (100). In the second paper, three different polymeric packing materials (PLRP-S, LiChrolut EN, and Isolute ENV) and one porous graphitic sorbent (PGC) were interrelated (107). Of these four sorbents, LiChrolut EN and Isolute ENV gave the highest breakthrough volumes for the most polar phenols (catechol and phenol). The authors then conclude that LiChrolut EN will be the sorbent of choice if the whole range of phenols is to be monitored, and only in the case of nitro- or highly chlorinated phenols is PLRP-S to be preferred. However, a serious drawback in using these polymeric resins for SPE purposes is the more pronounced band-broadening of the eluting analytes due to a wide injection profile on C18 analytical columns (97).

Masqué and coworkers have compared three different sorbents for the extraction of pesticides and phenols (phenol, 4-nitrophenol, and 2,4-dinitrophenol) from natural and tap water (108). Carbopack B (a graphitized carbon black), Bond Elut PPL (a functionalized polymeric resin), and HYSphere-1 (a more highly cross-linked polymeric resin) were the three investigated sorbents. The breakthrough volume for phenol, the most polar compound, was highest with the HYSphere-1 material, hence promising better extraction recovery. With this sorbent, detection limits in the low μg/ml range were realized after extracting 100 ml of tap water.

A number of publications covering similar topics as the examples illustrated earlier is found in the literature (18,97,109). The combination of two precolumns has been described, one containing PLRP-S and the other ENVI-Chrom-P material in an online LC precolumn-based column switching system together with diode array detection (24). With this system phenolic compounds could be detected at low to sub-μg/L levels after preconcentration of 50 ml surface water samples.

The use of ion pairing of polar phenolic compounds before extraction with different polymeric packings has been described in a series of papers (21,23). By using ion pairing with tetrabutyl bromide and thus extracting the phenols as ion pairs, the retention in the precolumn is enhanced, thereby increasing the breakthrough volumes of the compounds (for chromatogram, see Fig. 3).

For the analysis of several microcontaminants from the River Meuse, online SPE-LC-DAD was used (25). It was found that the interferences caused by matrix components when using a single precolumn (PLRP-S) revealed the necessity of incorporating two precolumns coupled in series. The highly polar phenols not retained

by the first column were trapped on the second, while the potentially interfering humic substances were all efficiently retained on the first precolumn.

A number of different types of graphitized carbon are efficient sorbents for very polar phenols, e.g., phenol, 4-nitrophenol, which are poorly held by conventional silica C_{18} packing materials (110–115). High flow rates (70 ml/min) were used for the off-line extraction of 11 phenols from large volumes (0.5–2 L) of river and drinking water samples using graphitized carbon black (110). The phenols were eluted in a back-

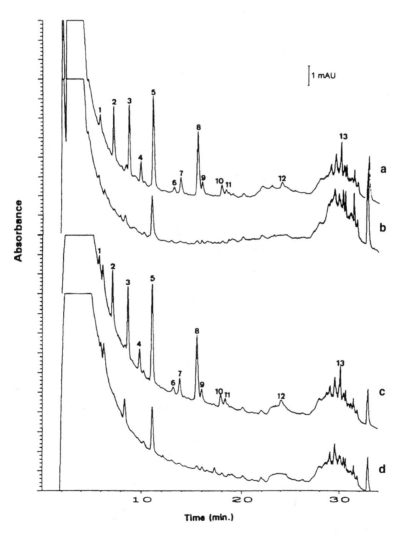

Fig. 3 Chromatograms obtained after preconcentration of 5 ml of real sample with a 2-mm-ID PLRP-S precolumn: (a) tap water spiked with 2 μg/L of each compound studied; (b) tap water; (c) Ebro River water spiked with 2 μg/L of each compound studied; (d) Ebro River water. Peak designation: (1) phenol, (2) 4-nitrophenol, (3) 2,4-dinitrophenol, (4) 2-chlorophenol, (5) 2-nitrophenol, (6) 2,6-dimethylphenol, (7) 2,4-dimethylphenol, (8) 2-methyl-4,6-dinitrophenol, (9) 4-chloro-3-methylphenol, (10) 2,4-dichlorophenol, (11) 2,4,6-trimethylphenol, (12) 2,4,6-trichlorophenol, (13) pentachlorophenol. (From Ref. 23 with permission from the authors and the publisher.)

flush mode with acidic dichloromethane/methanol, which is then dried before injection onto the LC system. The stability of the extracted phenols on this type of packing material was investigated; it was found that the cartridges could be stored for some days without any loss of the phenols. Porous graphitic carbon (PGC) has been used for aminophenols from drinking and river water samples (111). The phenolic compounds are retained to a higher extent on this type of a sorbent than on silica C_{18}, requiring more volume of eluting solvent, thus yielding augmented band broadening on the separation column. The authors circumvented this by using a PGC analytical column. A tandem extraction combining graphizited carbon black with a strong anion exchanger has been described by DiCorcia and co-workers (116).

Solid-phase extraction with C_{18} material has rarely been used for the extraction of phenolic compounds, mostly due to the problem of efficiently trapping a wide range of phenols. However, this is not to say never, for indeed some work involving SPE on C_{18} cartridges and LC-EC determination has been published (38).

2. Solid-Phase Extraction in Combination with Gas Chromatography

Large-volume sample preparation with SPE has been used in combination with sensitive GC determinations. Both online (61,62,117–119) and off-line extraction procedures have been evaluated for a wealth of phenolic compounds from water samples.

To be able to combine the aqueous SPE technique online with the nonaqueous GC separation, a drying step is necessary. Generally the SPE column is dried with a gentle stream of N_2 prior to analyte desorption with a GC-compatible solvent. The online SPE-GC with electron capture detection proved successful for chlorophenols in water samples (62). The SPE procedure involved drying the XAD-2 sorbent with N_2 and eluting the phenols with ethyl acetate. The analytical performance of this method was described as being proportional to the number of chlorines substituted; i.e., it was most sensitive for pentachlorophenol, having a detection limit of about 2 ng/L. In another approach the investigated phenols were first acetylated with acetic anhydride before enrichment on the SPE material, connected online to GC-FID analysis (61). The precolumn was again dried with N_2, and ethyl acetate was used to elute the analytes into a retention gap under partially concurrent solvent evaporation conditions (for chromatogram, see Fig. 4).

Another setup used extraction onto PLRP-S and GC-MS as final identification, where detection limits for phenol, 2-methylphenol, and 2-chlorophenol at the ng/L level with selected ion monitoring were achieved (118). A procedure with direct acetylation, extraction onto PLRP-S, drying, elution with ethyl acetate into a retention gap, and GC-MS established an analytical method for 26 phenols in river water (119).

The combination of off-line SPE to GC, hyphenated with various detection techniques for the elucidation of chlorophenols in drinking water, has recently been reviewed (120). The use of direct acetylation of phenols in water followed by off-line extraction of the acetylated phenols has been the topic of several papers (64,121–124). Extraction of 1000 ml of river water onto C_{18} disks combined with GC-MS analysis gave detection limits at low μg/L levels (64) (for chromatogram, see Fig. 5).

A similar approach using graphitized carbon cartridges and GC with a microwave-induced plasma atomic emission detector was used for the analysis of tap water (122). Analysis with GC-MS-MS after acetylation and off-line SPE on graphitized carbon of 1000 ml drinking water resulted in very low detection limits for the investigated chlorophenols (123). A polystyrene resin (XAD-4) was used for extraction and combined with off-line GC-MS determination of wastewater (121). Derivatization of more than 50 substituted phenols with N-(t-butyldimethylsilyl)-N-methyl-trifluoro-acetamide, thus forming their t-butyldimethylsilyl derivatives after off-line SPE using a polymeric sorbent, has successfully been joined together with GC-MS analysis (65). Two different off-line approaches were compared for the extraction of 16 chlorophenols from water with GC and microwave-induced plasma atomic emission detection (68). In one case acetylation of the chlorophenols was conducted prior to sorption onto the graphitized carbon SPE, whereas in the second case the phenols were acetylated after the elution from a PLRP-S sorbent. With the method using PLRP-S, larger volumes could be preconcentrated in a shorter time, and this was thus found to be the superior approach. Extraction of chlorophenols from industrial effluence was carried out on a polymeric sorbent (LiChrolut EN) with derivatization using 3,5-bis(trifluoromethyl)benzyl-dimethylphenylammonium (66). Detection of the benzyl derivatives via MS (using negative chemical ion mode) gave detection limits at low μg/L levels.

An off-line approach with SPE using C_{18} material and direct GC separation (without derivatization) on highly deactivated capillary columns has been pre-

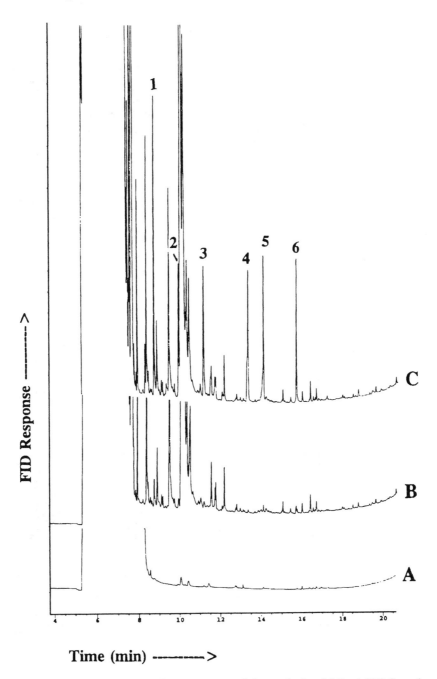

Fig. 4 Derivatization SPE-(on-column)-GC-FID chromatograms of the analysis of 2.2 ml HPLC-grade water. Injection temperature was 74°C (ethyl acetate). (A) SPE blank, (B) derivatization blank, (C) water sample spiked with 1 $\mu g/L$ of six chlorophenols. Peak designation: (1) phenol acetate; (2) 2-chlorophenol acetate; (3) 2,6-dichlorophenol; (4) 2,3,4-trichlorophenol acetate; (5) 2,3,5,6-tetrachlorophenol acetate; (6) pentachlorophenol acetate. (From Ref. 61 with permission from the authors and the publisher.)

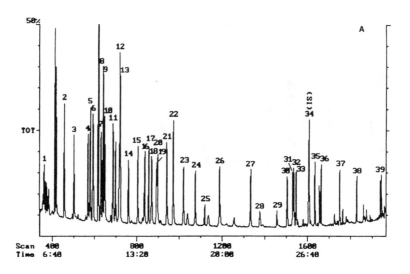

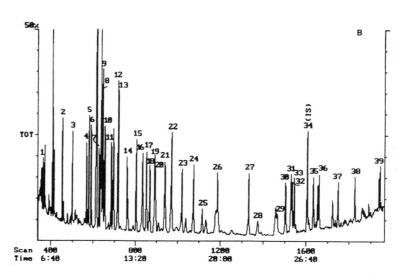

Fig. 5 Total ion current chromatograms of extracts obtained after direct acetylation and C₁₈ disk extraction of 1-L samples of (A) tap water and (B) river water, both spiked with 0.5 μg/L of each of the phenolic compounds. For peak numbering, see Ref. 64. (From Ref. 64 with permission from the authors and the publisher.)

sented for the determination of 27 phenols from aqueous samples (53).

3. Solid-Phase Extraction in Combination with Other Chromatographic Techniques

As mentioned previously, SFC has been used for the separation of phenolic compounds. Online connection of SPE to SFC for the determination of phenols in water samples has been described (125,126). However, to be able to connect the precolumn to the SFC equip-

ment, an additional drying step must be incorporated. To increase the breakthrough volume on the precolumn, tetrabutylammonium bromide was selected for ion pairing with the phenols. Benefiting from the separation characteristics of SFC, the matrix interferences were much less than with RPLC determination (125). The 11 priority pollutant phenols were determined in different water samples with SPE-SFC and diode array detection using ion pairing and polymeric sorbents (PLRP-S and LiChrolut EN) (126).

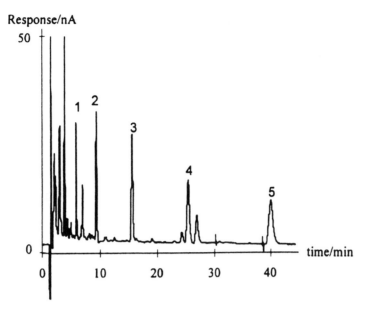

Fig. 6 Typical chromatogram after determination of phenols by SLM-LC-EC. Chromatogram obtained after 30 min. SLM enrichment of spiked Kävlinge River water: (1) 4-chlorophenol (0.05 μg/L); (2) 2,5-dichlorophenol (0.1 μg/L); (3) 2,4,5-trichlorophenol (0.1 μg/L); (4) 2,3,5,6-tetrachlorophenol (0.1 μg/L); (5) pentachlorophenol (0.1 μg/L). (From Ref. 129 with permission from the authors and the publisher.)

C. Other Sample Preparation Techniques for Determination of Phenolic Compounds in Water

The supported liquid membrane (SLM) extraction technique can serve as alternative sample preparation whenever dealing with difficult matrices and dirty samples, this being the case with many water samples (127,128). The SLM extraction utilizes a porous hydrophobic membrane impregnated with a water-immiscible organic solvent. The membrane is placed between two blocks in which sample channels are formed on both sides of the membrane; the donor and the acceptor. The analytes are extracted from the aqueous donor phase into the membrane and then back-extracted to the second aqueous phase, the acceptor. The process is normally driven by differences in pH between the two aqueous phases. By pumping the water sample in the donor and keeping the acceptor stagnant, an enrichment of the analytes in the acceptor is achieved.

The technique has been used in combination with LC using electrochemical detection for the determination of phenolic compounds with a large variety in polarity, viz., phenol, 4-chlorophenol, 2,5-dichlorophenol, 2,4,5-trichlorophenol, 2,3,5,6-tetrachlorophenol, and pentachlorophenol, in natural water samples (129). The

membrane provides very efficient cleanup, and detection limits below 0.1 μg/L for the phenols were obtained (for chromatogram, see Fig. 6). The technique has also been used for the extraction of five nitrophenols (2-nitrophenol, 3-nitrophenol, 4-nitrophenol, 2,3-dinitrophenol, and 2,4-dinitrophenol) (130).

Supercritical extraction (SFE) has been used mostly in environmental analysis (131) for the extraction of unpolar organic pollutants from solid samples, e.g., sediments. But the technique has also been used together with SPE disks (silica C18, polymeric, and ion-exchanger disks) for the extraction of phenols from water samples followed by GC-MS (132). After adsorbing the phenols onto the SPE disk, they are eluted with a supercritical fluid (carbon dioxide).

Another way of getting the adsorbed phenolic compounds out of silica C_{18} SPE disks was given by Chee and coworkers (133). They used closed-vessel microwave extraction prior to final LC-UV analysis.

IX. CONCLUSIONS

We hope that we have been able to give you an overview of the many possibilities for determining phenolic compounds in various water matrices. The area of phenolic analysis in water has been thoroughly investigated in the past, as is readily seen in this chapter, and the field will surely be a focus of future research as well, especially with the environmental concerns of today.

ACKNOWLEDGMENTS

Dr. György Marko-Varga is acknowledged for his help with literature search and fruitful discussions. Mr. Eddie Thordarson is gratefully acknowledged for revision of the language and other fruitful comments.

REFERENCES

1. PA Realini. J. Chromatogr. Sci. 19:124, 1981.
2. G Marko-Varga. In: D. Barceló, ed. Environmental Analysis: Techniques, Applications and Quality Assurance. Elsevier: Amsterdam, 1993, p 225.
3. D Puig, D Barceló. Trends Anal. Chem. 15:362, 1996.
4. GJ Sirons, ASY Chau, AE Smith. In: ASY Chau, BK Afghan, eds. Analysis of pesticides in water. Vol. II. CRC Press: Boca Raton, FL, 1982, chap 3.
5. 40 CFR, Part 423 App A. Effluent Guidelines and Standard (1995), Federal Register, U.S. GPO, Washington, DC.
6. Official Journal of the E.C. No. L 194/26, 1975.
7. AD Eaton, LS Clesceri, AE Greenberg eds. Standard Methods for the Examination of Water and Wastewater. 19th ed. American Public Health Association, Washington, DC, 1995.
8. 40 CFR, Part 136, Methods of Organic Chemical Analysis of Municipal and Industrial Wastewater (Revised as of July 1, 1995), Federal Register, U.S. GPO, Washington, DC.
9. EPA 530/SW-846, Test Methods for Evaluating Solid Waste: Physical/Chemical Methods 3rd ed. 4 vols. November 1986, U.S. GPO, Washington, DC (Final Update I, July 1992, Final Update II, September 1994, IIA August 1993, Final Update IIB and Proposed Update III, January 1995).
10. EPA 821/R-93-017, Analytical Methods for the Determination of Pollutants in Pulp and Paper Industry Wastewater. October 1993. National Technical Information Service, Springfield, VA.
11. International Standards Organization, Geneva, Switzerland.
12. EPA/600/R-95/131, Methods for the Determination of Organic Compounds in Drinking Water—Supplement III, August 1995. National Technical Information Service, Springfield, VA.
13. Annual Book of ASTM standards, Section 11. American Society for Testing and Materials, Philadelphia, 1997.
14. EPA/600/4-79-020, Methods for Chemical Analysis of Water and Wastes, Revised 1993. National Technical Information Service, Springfield, VA.
15. G Marko-Varga, D Barceló. Chromatographia 34:146, 1992.
16. N Masqué, E Pocurull, RM Marcé, F Borrull. Chromatographia 47:176, 1998.
17. E Pocurull, RM Marcé, F Borrull. J. Chromatogr. A 738:1, 1996.
18. N Masqué, M Galià, RM Marcé, F Borrull. Analyst 122:425, 1997.
19. I Rodriguez, MI Turnes, MH Bollain, MC Mejuto, RJ Cela. J. Chromatogr. A 778:279, 1997.
20. C Aguilar, F Borrull, RM Marcé. Chromatographia 43:592, 1996.
21. E Pocurull, M Calull, RM Marcé, F Borrull. J. Chromatogr. A 719:105, 1996.
22. MWF Nielen, UATh Brinkman, RW Frei. Anal. Chem. 57:806, 1985.
23. E Pocurull, RM Marcé, F Borrull. Chromatographia 41:521, 1995.
24. ER Brouwer, UATh Brinkman. J. Chromatogr. A 678:223, 1994.
25. AC Hogenboom, I Jagt, JJ Vreuls, UATh Brinkman. Analyst 122:1371, 1997.
26. DA Baldwin, JK Debowski. Chromatographia 26:186, 1988.
27. N Cardellicchio, S Cavalli, V Piangerelli, S Giandomenico, P Ragano, Fresenius J. Anal. Chem. 358:749, 1997.
28. P Trippel, W Maasfield, A Kettrup. Int. J. Environ. Anal. Chem. 23:97, 1985.
29. J Ruana, I Urbe, F Borrull. J. Chromatogr. A 655:217, 1993.
30. J Lehotay, M Baloghova, S Hatrik. J. Liq. Chromatogr. 16:999, 1993.
31. D Puig, D Barceló. J. Chromatogr. A 778:313, 1997.
32. A Hagen, J Mattusch, G Werner. Fresenius J. Anal. Chem. 339:26, 1991.
33. B Paterson, CE Cowie, PE Jackson. J. Chromatogr. A 731:95, 1996.
34. AG Huegsen, R Schuster. LC-GC Int. 4:40, 1991.
35. ECV Butler, G Dal Pont. J. Chromatogr. 609:113, 1992.
36. RE Shoup, GS Mayer. Anal. Chem. 54:1164, 1982.
37. G Achill, GP Cellerino, G Melzi d'Eril, S Bird. J. Chromatogr. A 697:357, 1995.
38. MT Galceran, O Jáuregui. Anal. Chim. Acta 304:75, 1995.
39. F Ortega, E Dominguez, E Burestedt, J Emnéus, L Gorton, G Marko-Varga. J. Chromatogr. A 675:65, 1994.
40. J Wang, F Lu, SA Kane, Y-K Choi, MR Smyth, K Rogers. Electroanalysis 9:1102, 1997.

41. PJM Kwakman, DA Kamminga, UATh Brinkman, GJ De Jong. J. Chromatogr. 553:345, 1991.

42. C de Ruiter, JF Bohle, GJ de Jong, UATh Brinkman, RW Frei. Anal. Chem. 60:666, 1988.

43. WJ Landzettel, KJ Hargis, JB Caboot, KL Adkins, TG Strein, H Veening, H-D Becker. J. Chromatogr. A 718: 45, 1995.

44. FP Bigley, RL Grob. J. Chromatogr. 350:407, 1985.

45. G Blo, F Dondi, A Betti, C Bighi. J. Chromatogr. 257: 69, 1983.

46. SK Ratanathanawongs, SR Crouch. Anal. Chim. Acta 192:277, 1987.

47. O Fiehn, M Jekel. J. Chromatogr. A 769:189, 1997.

48. A Farran, JL Cortina, J de Pablo, D Barceló. Anal. Chim. Acta 234:119, 1990.

49. D Barceló. Chromatographia 25:295, 1988.

50. D Puig, I Silgoner, M Grasserbauer, D Barceló. Anal. Chem. 69:2756, 1997.

51. O Jáuregui, E Moyano, MT Galceran. J. Chromatogr. A 787:79, 1997.

52. C Crescenzi, A Di Corcia, M Marchetti, R Samperi. Anal. Chem. 67:1967, 1995.

53. P Mussmann, K Levsen, W Radeck. Fresenius J. Anal. Chem. 348:654, 1994.

54. L Wennrich, J Efer, W Engewald. Chromatographia 41:361, 1995.

55. H-B Lee, ASY Chau. J. Assoc. Off. Anal. Chem. 66: 1029, 1983.

56. H-B Lee, L-D Weng, ASY Chau. J. Assoc. Off. Anal. Chem. 67:1086, 1984.

57. RA Booth, JN Lester. J. Chromatogr. Sci. 32:259, 1994.

58. RT Coutts, EE Hargesheimer, FM Pasutto. J. Chromatogr. 179:291, 1979.

59. K Abrahamsson, TM Xie. J. Chromatogr. 279:199, 1983.

60. LL Lamparski, TJ Nestrick. J. Chromatogr. 156:143, 1979.

61. AJH Louter, PA Jones, JD Jorritsma, JJ Vreuls, UATh Brinkman. J. High Res. Chromatogr. 20:363, 1997.

62. MA Crespin, E Ballesteros, M Gallego, M Valcárcel. Chromatographia 43:633, 1996.

63. Y Hoshika, G Muto. J. Chromatogr. 179:105, 1979.

64. ML Bao, F Pantini, K Barberi, D Burrini, O Griffini. Chromatographia 42:227, 1996.

65. T Heberer, H-J Stan. Anal. Chim. Acta 341:21, 1997.

66. J Cheung, RJ Wells. J. Chromatogr. A 771:203, 1997.

67. I Rodriguez, MH Bollaín, CM García, R Cela. J. Chromatogr. A 733:405, 1996.

68. I Rodriguez, MC Mejuto, MH Bollain, R Cela. J. Chromatogr. A 786:285, 1997.

69. D Martinez, E Pocurull, RM Marcé, F Borrull, M Calull. Chromatographia 43:619, 1996.

70. WE Rae, CA Lucy. J. Assoc. Off. Anal. Chem. 80: 1308, 1997.

71. D Martinez, E Pocurull, RM Marcé, F Borrull, M Calull. J. Chromatogr. A 734:367, 1996.

72. MI Turnes, MC Mejuto, R Cela. J. Chromatogr. A 733: 395, 1996.

73. C-Y Tsai, G-R Her. J. Chromatogr. A 743:315, 1996.

74. Y-C Chao, C-W Whang. J. Chromatogr. A 663:229, 1994.

75. CP Ong, CL Ng, NC Chong, HK Lee, SFY Li. J. Chromatogr. 516:263, 1990.

76. M van Bruijnsvoort, SK Sanghi, H Poppe, WTh Kok. J. Chromatogr. A 757:203, 1997.

77. CP Ong, HK Lee, SFY Li. Anal. Chem. 62:389, 1990.

78. CP Ong, HK Lee, SFY Li. J. Chromatogr. Sci. 30:19, 1992.

79. TA Berger, JF Deye. J. Chromatogr. Sci. 29:4, 1991.

80. Y Wang, K-M Wang, G Shen, R-Q Yu. Talanta 44: 319, 1997.

81. CS Hottenstein, SW Jourdan, MC Hayes, FM Rubio, DP Herzog, TS Lawruk. Environ. Sci. Technol. 29: 2754, 1995.

82. QX Li, MS Zhao, SJ Gee, MJ Kurth, JN Seiber, BD Hammock. J. Agric. Food Chem. 39:1685, 1991.

83. PM Krämer, QX Li, BD Hammock. J. Assoc. Off. Anal. Chem. 77:1275, 1994.

84. RE Majors. LC-GC Int. 4:10, 1991.

85. FI Onuska. J. High Res. Chromatogr. 12:4, 1989.

86. CJ Koester, RE Clement. Crit. Rev. Anal. Chem. 24: 263, 1993.

87. G Lamprecht, JFK Huber. J. Chromatogr. A 667:47, 1994.

88. J Czuczwa, C Leuenberger, J Tremp, W Giger, M Ahel. J. Chromatogr. 403:233, 1987.

89. E Ballesteros, M Gallego, M Valcarcel. J. Chromatogr. 518:59, 1990.

90. I Harrison, RU Leader, JJW Higgo, JC Tjell. J Chromatogr. A 688:181, 1994.

91. D Barceló, M-C Hennion. Anal. Chim. Acta 318:1, 1995.

92. RW Frei, MWF Nielen, UATh Brinkman. Int. J. Environ. Anal. Chem. 25:3, 1986.

93. M-C Hennion, V Coquart. J. Chromatogr. 642:211, 1993.

94. M-C Hennion. Trends Anal. Chem. 10:317, 1991.

95. M-C Hennion, V Pichon. Environ. Sci. Technol. 28: 576A, 1994.

96. MWF Nielen, RW Frei, UATh Brinkman. In: RW Frei, K Zech, eds. Selective Sample Handling and Detection in High Performance Liquid Chromatography. Elsevier: Amsterdam, 1988, pp 5.

97. I Liska, ER Brouwer, H Lingeman, UATh Brinkman. Chromatographia 37:13, 1993.

98. M Castillo, D Puig, D Barceló. J. Chromatogr. A 778: 301, 1997.

99. L Schmidt, JJ Sun, JS Fritz, DF Hagen, CG Markell, EE Wisted. J. Chromatogr. 641:57, 1993.

100. D Puig, D Barceló. Chromatographia 40:435, 1995.

101. O Jáuregui, MT Galceran. Anal. Chim. Acta 340:191, 1997.
102. J Pawliszyn. Trends Anal. Chem. 14:113, 1995.
103. AA Boyd-Boland, M Chai, YZ Luo, MJ Yang, J Pawliszyn, T Gorecki. Environ. Sci. Technol. 28:13, 1994.
104. KD Buchholz, J Pawliszyn. Anal. Chem. 66:160, 1994.
105. M Möder, S Schrader, U Franck, P Popp. Fresenius J. Anal. Chem. 357:326, 1997.
106. SL Chong, D Wang, JD Hayes, BW Wilhite, A Malik. Anal. Chem. 69:3889, 1997.
107. D Puig, D Barceló. J. Chromatogr. A 733:371, 1996.
108. N Masqué, RM Marcé, F Borrull. J. Chromatogr. A 793:257, 1998.
109. N Masqué, M Galià, RM Marcé, F Borrull. J. Chromatogr. A 771:55, 1997.
110. A Di Corcia, S Marchese, R Samperi, G Cecchini, L Cirilli. J. Assoc. Off. Anal. Chem. 77:446, 1994.
111. S Guenu, M-C Hennion. J. Chromatogr. A 665:243, 1994.
112. C Borra, A Di Corcia, M Marchetti, R Samperi. Anal. Chem. 58:2048, 1986.
113. A Di Corcia, A Bellioni, MD Madbouly, S Marchese. J. Chromatogr. A 733:383, 1996.
114. V Coquart, J-C Hennion. J. Chromatogr. 600:195, 1992.
115. A Di Corcia, S Marchese, R Samperi. J. Chromatogr. 642:163, 1993.
116. A Di Corcia, S Marchese, R Samperi. J. Chromatogr. 642:175, 1993.
117. MA Crespin, E Ballesteros, M Gallego, M Valcárcel. J. Chromatogr. A 757:165, 1997.
118. AJH Louter, S Ramalho, RJJ Vreuls, D Jahr, UATh Brinkman. J. Microcol. Sep. 8:469, 1996.
119. D Jahr. Chromatographia 47:49, 1998.
120. I Rodriguez, R Cela. Trends Anal. Chem. 16:103, 1997.
121. TM Pissolatto, P Schossler, AM Geller, EB Caramão, AF Martins. J. High Res. Chromatogr. 19:577, 1996.
122. I Rodriguez, MI Turnes, MC Mejuto, R Cela. J. Chromatogr. A 721:297, 1996.
123. I Turnes, I Rodriguez, CM Garcia, R Cela. J. Chromatogr. A 743:283, 1996.
124. V Janda, H van Langenhove. J. Chromatogr. 472:327, 1989.
125. E Pocurull, RM Marcé, F Borrull, JL Bernal, L Toribio, ML Serna. J. Chromatogr. A 755:67, 1996.
126. JL Bernal, MJ Nozal, L Toribio, ML Serna, F Borrull, RM Marcé, E Pocurull. Chromatographia 46:295, 1997.
127. JÅ Jönsson, L Mathiasson. Trends Anal. Chem. 18: 318, 1999.
128. M Knutsson, G Nilvé, L Mathiasson, JÅ Jönsson. J. Chromatogr. A 754:197, 1996.
129. M Knutsson, L Mathiasson, JÅ Jönsson. Chromatographia 42:165, 1996.
130. L Chimuha, MM Nindi, MEM El Noor, H Franh, C Velasco. J High Res Chromatogr 22:417, 1999.
131. V Janda, KD Bartle, AA Clifford. J. Chromatogr. 642: 283, 1993.
132. PH Tang, JS Ho. J. High Res. Chromatogr. 17:509, 1994.
133. K-K Chee, M-K Wong, H-K Lee. Mikrochim. Acta 126:97, 1997.

19

Cyanides

Kazunori Ikebukuro, Hideaki Nakamura, and Isao Karube
University of Tokyo, Tokyo, Japan

I. INTRODUCTION

Cyanide is well known as a deadly poison inhibiting respiration. Nevertheless, it is widely utilized in industrial applications, especially for electroplating, metal cleaning, pharmaceuticals, plastics, and coal coking. Occasional accidents have taken place when industrial plants discharge cyanide into environmental water. Plants dealing with cyanogenic glycosides also produce cyanide in nature (1). In some cases, such accidents have caused serious problems to the ecosystem in the water area, but the consequences can also affect humans through drinking water. Acute cyanide poisoning in humans can lead to convulsions, vomiting, coma, and death. The lethal dose is in the range of 0.5–3.5 mg/kg body weight (2). Therefore, to regulate the discharge of cyanide into the environment, the World Health Organization established the standard of 50 μg/L of hydrogen cyanide (HCN) for drinking water. Additionally, the Water Pollution Control Law in Japan stipulates 1000 μg/L of cyanide (38.5 μM) as the maximum concentration of cyanide allowed in wastewater. For this reason, analytical methods for cyanide detection have been required and examined in various studies (3–9). In particular, monitoring cyanide in environmental water and wastewater is the most efficient way to determine contamination rapidly.

This chapter discusses the methods used for the determination of cyanide, including standard methods, other conventional methods, and new methods using biosensors.

II. STANDARD METHODS

Several standard methods for the determination of cyanide compounds in environmental water and wastewater have been established employing titrimetric, colorimetric (spectrophotometric), or cyanide-selective electrode methods (3–4).

A. Pretreatment

Pretreatments for determining cyanide are classified into several methods. In the standard methods of the American Public Health Association (APHA), these pretreatments are divided into four methods: distillation (pH 2.0) for total cyanide (4500-CN$^-$ C); chlorination of cyanide after distillation (4500-CN$^-$ G); chlorination of cyanide without distillation (4500-CN$^-$ H); and dissociation of cyanide with weak acid (pH 4.5–6.0) (4500-CN$^-$ I) (3).

Distillation for total cyanide (4500-CN$^-$ C) is carried out by the acidification and boiling of the sample solution and the absorption of HCN gas into an alkaline acceptor solution using a distillation apparatus (3). The sample solution alkalinized by sodium hydrate (NaOH) solution is then acidified to pH 2.0 by the addition of sulfamic acid (NH_2SO_3H) and sulfuric acid (H_2SO_4) after the distillation apparatus is set up. Gaseous HCN volatilized by boiling the sample solution (for at least 1 hour) is then collected by absorption in an alkaline absorber solution (NaOH) to form free CN ions.

After distillation, cyanogen chloride (CNCl) is produced by the reaction of chloramine-T and several forms of cyanide (Fig. 1a) (4500-CN⁻ G). This can be expressed as a cyanide amenable to chlorination by the difference in cyanide concentration between the chlorinated and untreated samples. The method for cyanide amenable to chlorination without distillation (4500-CN⁻ H) is used as a shortcut method. This method, which also provides an estimation of the cyanide ion content by the chlorination of cyanide without distillation, is useful for natural waters and heat-treating effluents. The weak-acid dissociable cyanides procedure (4500-CN⁻ I) can also determine the cyanide amenable

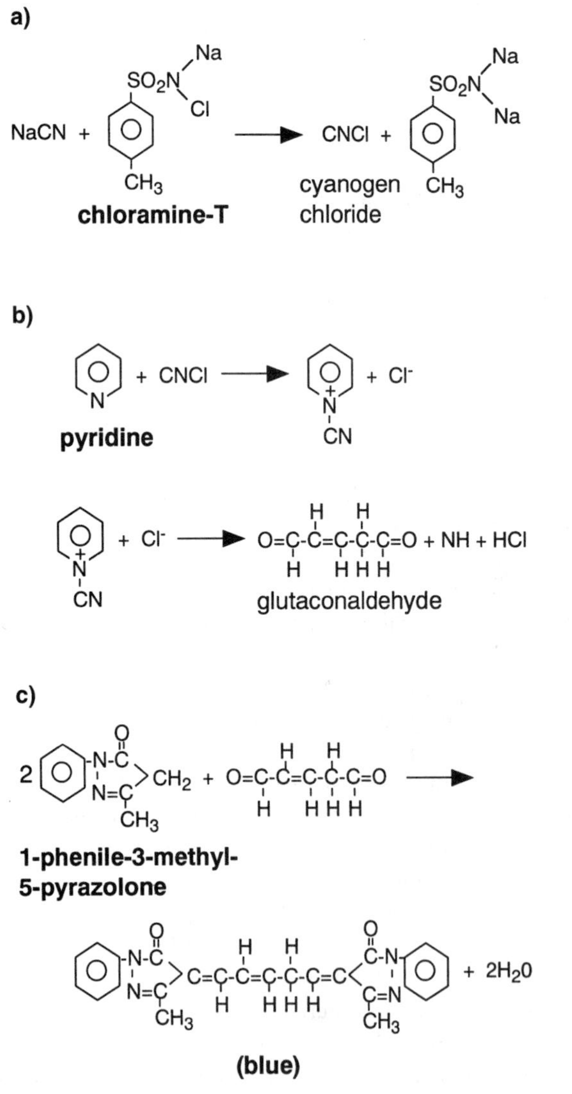

to chlorination by freeing HCN from the dissociated cyanide.

On the other hand, in the JIS method approved by the Japanese Standard Association, the pretreatments are classified into three methods: distillation under a condition of pH 2.0 or pH 5.5 or dissociation under a condition of pH 5.0 (4).

B. Measurements of Cyanide

1. Titrimetric Methods

The titrimetric method is used as the standard method of APHA (4500-CN⁻ D) (3). When the cyanide concentration is above 1000 μg/L (38.5 μM), the titrimetric method is used. Cyanide ion in a sample solution is titrated with a silver nitrate (AgNO$_3$) solution to form a soluble cyanide complex, Ag[Ag(CN)$_2$], by the Liebig reaction (Fig. 2a) (10). After a small excess of Ag ion is added, the excess of Ag ion is detected by a silver-sensitive indicator, p-dimethylaminobenzalrhodanine. Then the color of the solution turns from yellow to red (Fig. 2b). A detection limit of Ag ion by this indicator is about 0.1 mg/L (0.93 μM) with the blank test, and the coefficient of variation (CV) for cyanide determination is 4–40%.

2. Spectrophotometric Methods

Cyanide can be determined mainly by using a König reaction by the spectrophotometry method. The König reaction (Fig. 1b) (11–12), in which glutaconic aldehyde is obtained quantitatively when CNCl reacts with pyridine, is used in the pyridine–barbituric acid

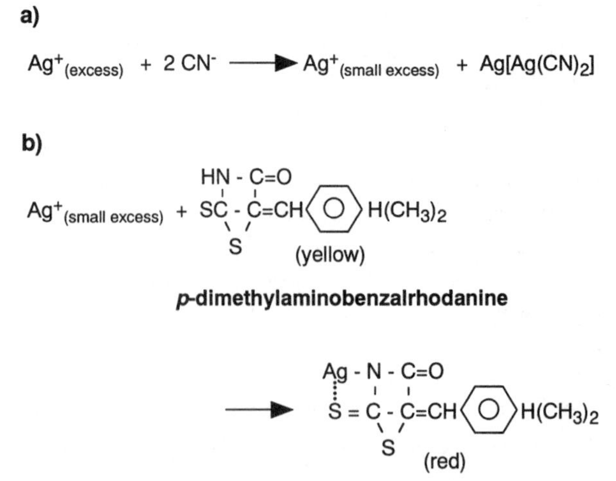

Fig. 1 Reaction scheme of the titrimetric method. (a) chlorinating reaction of cyanide, (b) König reaction, (c) color reaction.

Fig. 2 Reaction scheme of the colorimetric method. (a) Liebig reaction, (b) color reaction.

method (4500-CN⁻ E) (3), the pyridine–pyrazolone method (4), or the pyridine–carboxylic acid–pyrazolone method. When using the pyridine–barbituric acid reagent, the standard method of APHA, cyanide concentration is measured as follows: NaOH solution, acetate buffer solution, and chloramine-T solution are added to the sample solution containing cyanide, and the mixed solution stands for exactly 2 min. The pyridine–barbituric acid reagent is added to the sample solution. After being thoroughly mixed, the solution is allowed to stand for exactly 8 min. The cyanide concentration is measured by determining absorbance against distilled water at 578 nm using spectrophotometry. The detection range in this method is between 5 and 200 μg/L (0.19–7.7 μM) cyanide ion, and the CV is 3–11%.

In the pyridine–pyrazolone method (Fig. 1c), the JIS method (4), a sample solution containing cyanide is acidified by sodium dihydrogen phosphate. Then a sodium–potassium phosphate buffer solution, a chloramine-T solution, and a pyridine–pyrazolone reagent are added to the cyanide solution. After 30 minutes, absorbance of 620 nm is measured by spectrophotometry. The detection range in this method is between 10 and 180 μg/L (0.38–6.9 μM) cyanide ion, and the CV is 2–10%.

Interferences to these spectrophotometric methods are eliminated or reduced to a minimum in the case of treatment by distillation.

3. *Electrochemical Method*

A cyanide-selective electrode is used as another standard method for determining cyanide (3–4,13–16). Cyanide ion can be determined using a cyanide-ion-selective electrode in combination with a double-injection reference electrode and a pH meter, or an ion meter (4500-CN⁻ F) (3). A membrane type made on the basis of AgI is generally used as the cyanide-ion-selective electrode. The cyanide-ion-selective electrode and the double-injection reference electrode are immersed in an alkaline sample solution containing cyanide. The sample solution is well mixed in a beaker with a magnetic stirrer at 25°C. After reaching and maintaining equilibrium for at least 5 min, the ion meter records the value of the potential as the cyanide-ion concentration. The detection range in this method is between 0.05 and 10 mg/L (1.9–380 μM) cyanide ion, and the CV is 1–8%.

In the JIS method, an indicator electrode and a reference electrode are soaked in the alkaline sample solution (pH 12–13) containing cyanide. Then the poten-

tial between the two electrodes is measured as the concentration of cyanide ion. The detection range in this method is between 0.1 and 100 mg/L (3.8 μM–3.8 mM) cyanide ion, and the CV is 5–20% (4).

These methods tend to be affected by sulfide ion.

III. OTHER METHODS

In recent years, many methods for determining cyanide have been developed using spectrophotometric, fluorometric, atomic adsorption spectrometric, electrochemical, chromatographic, and flow injection analysis (FIA) methods (5–9).

A. Spectrophotometric Methods

The spectrophotometric methods are used most extensively for the sensitive determination of cyanide in water analysis. Techniques using the formation or destruction of colored complexes or metal complexes in the presence of cyanide are employed in this method. The König reactions by derivatives of pyridine described earlier are the most basic method, and many studies have been performed using pyridine–pyrazolone (abs (absorbance wave length): 620 nm), pyridine–barbituric acid (abs: 578 nm), or pyridine–carboxylic acid–pyrazolone (abs: 638 nm) (17–20. As for the other techniques of spectrophotometric methods, Ag complexes with porphin (abs: 423–431 nm) or 3-hydroxybenzaldehyde azine with $K_2S_2O_8$ catalyzed with Cu ion (abs: 465 nm) are used for determining cyanide ion by the destruction of the colored complexes (21–22).

Generally speaking, spectrophotometric methods require several steps; therefore, a combination of FIA and the spectrophotometric method was examined in order to shorten the measurement time. Those studies will be described later. This method also includes the problem of interference of existing substances such as NO_2^-, NO_3^-, SO_3^{2-}, or SCN^- in a sample.

B. Fluorometric Methods

Fluorometric methods to determine cyanide ion generally consist in replacing a fluorescent ligand from a cyanide complex, as in the case of the spectrophotometric methods. These methods employ the formation of fluorescent substances in the presence of cyanide using *p*-benzoquinone (23), pyridoxal (24), pyridine–barbituric acid (25), naphthalene-2,3-dialdehyde (NDA), or *o*-phthalic aldehyde (OPA) (26–28. In these fluorophores, NDA or OPA is usually used for fluorometric methods as a selective and sensitive method.

Cyanide ion reacts with NDA and glycine and then forms fluorescent substances (ex: 420 nm, em: 490 nm). The detection limit of cyanide ion in this method is improved between 0.2 and 20 μg/L (7.7 nM–0.77 μM) by using high-performance liquid chromatography (HPLC) (27). In the same way, cyanide ion is determined by reaction with OPA (ex: 330 nm, em: 400 nm), and the detection range of cyanide ion is between 0.5 and 2000 μg/L (19 nM–77 μM) (28).

These fluorometric methods are superior to the spectrophotometric methods in their detection limit, but their selectivity to cyanide ion might be insufficient. Therefore, some kinds of pretreatments are required to solve this problem.

C. Atomic Adsorption Spectrometry (AAS) Methods

The AAS methods determine cyanide indirectly and are based on the detection of the cyanide complex with metal.

When using silver, cyanide is dissociated by UV irradiation from various forms and converted to silver cyanide by passing through a silver filter. Silver cyanide can be determined as a cyanide ion to several micrograms per liter (29). When using copper, cyanide is converted to the cyanide complex of copper with 2-benzoylpyridinethiosemicarbazone. After extraction with isoamil acetate, cyanide ion is determined up to 4.8 μg/L (0.18 μM) (30).

Thus, the AAS methods can determine cyanide with high sensitivity. However, these methods also require expensive equipment and complicated steps.

D. Electrochemical Methods

Many electrochemical methods for determining cyanide ion have been developed using ion-selective electrodes (ISEs), potentiometry, amperometry, polarography, and coulometry (7–9). In particular, potentiometry is the most widely used one, with the aid of ion-selective electrodes.

An ISE is covered with a membrane that does not contain cyanide and is made of AgI or Ag_2S (31). This electrode can determine 18 μg/L (0.69 μM) cyanide ion, including cyanide complexes of several metals, for example, Zn, Cu, and Cd. However, S^{2-}, SCN^-, and metal ion such as Ni^{2+}, CO^{2+}, and Fe^{3+} interfere with the determination of cyanide. The other ISEs—for example, graphite electrodes coated with Ag_2S, AgI, or

Ag_2S/AgI (32)—are also interfered with by several substances, such as sulfide, ferrocyanide, and ferricyanide. Therefore, this method requires additional devices to prevent these negative effects.

A gas-permeable Teflon membrane is used for the selective determination of cyanide ion using an ISE coated with an Ag_2S membrane and an Ag/AgCl reference electrode (33). This system can determine the concentration of cyanide ion in a linear range between 0.025 and 10 mg/L (0.96–385 μM).

This method for determining cyanide by using ISEs still requires preventing the effects of interfering substances such as chlorine and hydrogen sulfide on the electrode.

E. Chromatographic Methods

Chromatographic methods used for the determination of cyanide are gas chromatography (GC), high-performance liquid chromatography (HPLC), and ion chromatography (IC). These methods employ electron-capture detectors, electrochemical detectors, UV/VIS detectors, fluorescence detectors, conductivity detectors, or amperometric detectors (7–9). The IC method is the most extensively used of the three (34).

Ion chromatography can determine all cyanide species, free and complexes from metal, by separation. Moreover, it obviates the need for distillation to convert cyanide complexes from metal to HCN. The determination of cyanide by IC is performed using ion-exchange chromatography, ion-pair chromatography, reverse-phase chromatography, or size exclusion chromatography.

Ion chromatography has been used mainly to separate modified cyanide ion by a combination of the pretreatment methods for the modification of cyanide and the detection methods described earlier. For example, a sensitive technique for determining cyanide by IC is based on converting CN^- to SCN^- under a reaction with polysulfide. In this method, the interference of Fe^{2+} and Cu^{2+} is observed. The detection limits of cyanide ion are 10 μg/L by employing a UV detector and 100 μg/L by employing a conductmetric detector (35).

Chromatographic methods are selective and sensitive.

F. Flow Injection Analysis (FIA) Method

The FIA method, which was established by J. Ruzicka and E. H. Hanssen in 1975 (36), is a rather simple analytical method with high reproducibility. This method offers the additional advantages of not only controlling the measuring conditions precisely but also measuring continuously. A buffer solution for the measurement is delivered continuously into the system. A sample solution containing target molecules is injected into a flow tube, where it reacts with some chemicals. Then products that can be detected by a method are produced. When these products reach the detector, the target molecules in the sample are measured. Therefore, this method is used in many methods for determining cyanide.

In the spectrophotometric method, the principle of molecular diffusion of HCN through a Teflon membrane is most frequently used to improve selectivity. Cyanide ion was determined by using a pyridine–barbituric reagent with a detection limit of 20 μg/L (0.77 μM) (37) and by using the formation of a cyanide complex with Ni (38) with a detection limit of 50 μg/L (1.9 μM) (39).

In the fluorometric methods, cyanide ion was determined using an NDA (naphthalene-2,3-dialdehyde) reagent with a detection limit of 1.5 μ/L (73 nM) (40).

In the electrochemical methods, cyanide ion was amperometrically determined using an ID (indicator electrode) made of Ag with a detection limit of 2.6 μg/L (0.1 μM) (41). Cyanide ion was also determined potentiometrically using an ISE with a detection limit of 26 μg/L (1.0 μM) (42).

The methods described earlier have been improved in reproducibility and rapidity by combination with the FIA method.

IV. BIOSENSORS

A. Introduction

The first biosensor was made by using an oxygen electrode covered with a membrane-immobilized enzyme (43). Showing extraordinary selectivity, this new, powerful method was the starting point for revolutionary developments with applications to chemical, clinical, food, and environmental analysis (44–47). The biosensor consists of a combination of a molecular recognition element and a transducer. The molecular recognition elements can generally be classified into four groups: proteins, organelles, cells, and tissues. As a transducer, an electrode, semiconductor, photon counter, sound detector, or piezoelectric device can generally be used. For example, for an enzyme biosensor, a purified enzyme is immobilized as a reusable element. This type of biosensor is highly selective by substrate specificity; however, its disadvantages are its instability and decreasing sensor response for the de-

termination of analytes. On the other hand, the responses of microbial sensors are more stable than those of enzyme sensors, even though the microbial sensor has inferior selectivity than the enzyme sensor because the sensor response is obtained by total metabolic function in intact cells. In the case of a biochemical oxygen demand (BOD) sensor using yeast cells (*Clostridium butyricum*) and an oxygen electrode, the features are utilized to determine the oxygen consumed as a result of the ingestion of nutrients, especially organic compounds (48). Because each biosensor has specific characteristics, it is extremely important to combine several types of them in order to determine the effects of cyanide.

Biosensors are suitable for application to environmental monitoring because they have a rapid response, high selectivity, and a pollution-free procedure and are relatively inexpensive and convenient to use. On the other hand, these methods rely on chemical and physical procedures that can often be slow and complex and require the use of expensive equipment and environmental loading reagents, and, in addition, cannot estimate the effects of toxicity to living bodies. Therefore, the damages as the total effects to the environment surrounding the polluted area can not be recognized using the chemical methods. An integrated biosensor would eliminate all these problems.

For this purpose, several types of cyanide biosensors have been developed for determining cyanide and estimating toxicity using stable biomaterials and several techniques. Table 1 (I–VII) shows the profiles of cyanide biosensors developed for practical application (49–55).

B. Biosensors Using Inhibition of Microbial Respiration

Cyanide acts as a nonspecific enzyme inhibitor and exerts powerful toxic effects by inhibiting cytochrome oxidase in the respiratory chain. Therefore, the toxicity of cyanide to aerobic cells can be estimated by measuring levels of respiration. When cyanide binds to cytochrome oxidase, the cell respiration is stopped. As a result, consumption of oxygen by respiratory cells decreases. Several cyanide biosensors have been developed following this principle (49–51).

1. Batch-and-Membrane Type of Sensor

The first cyanide sensor made by employing an oxygen electrode entrapped in yeast with an oxygen-permissi-

Table 1 Profiles of Cyanide Biosensors

Sensor type	Biomaterial[a]	Immobilizing material	Principle	Detector	Measuring condition	Linear detection range (CN ion) (Detec. limit)	Response time	Interfering substance	Stability	Ref.
I: Batch and memb.	Yeast	Porous membrane	Respiratory inhibition	Oxygen electrode	Tris buf. (150 mg/L glucose; (G), pH 8.0, 30°C	8.0–4,000 µg/L, r = 0.997	—	Not examined	Not examined	49
II: Flow and reactor	Yeast	Controlled pore glass	Respiratory inhibition	Oxygen electrode	Tris buf. (150 mg/L G, pH 8.0, 4.5 ml/min, 30°C	0–400 µg/L (4.0 µg/L)	3 min	Not examined	16 days	50
III: Flow and reactor	Yeast	Chitopearl HP-5020 beads	Respiratory inhibition	Oxygen electrode (two)	River water (150 mg/L G), 4.5 ml/min ~20°C	0–400 µg/L in tris buf. pH 8.0	7 min	Simazine (not affected to NaCl, Fe, Mn, Zn, BOD)	9 days	51
IV: Batch and memb.	Bacteria	Porous membrane	Activating respiration	Oxygen electrode	River water, 30°C	40–400 µg/L in phos. buf. pH 8.0, RSD 8% (n = 5)	2 min	Simazine (not affected to NaCl, Cr, Cd, Pb, Fe, Mn, Zn, LAS)	2 weeks[b]	52
V: Batch and memb.	Bacteria	Gas-permeable membrane	Activating respiration	Oxygen electrode	River water adjusted to pH 2, 30°C	40–400 µg/L, r = 0.995 (n = 4, with JIS method)	—	Acetate, ethanol (not affected to glucose, glutamate, Cu, Cd, Pb, Fe, Zn)	1 month	53
VI: Flow and reactor	Bacteria	Alginate gel	Activating respiration	Oxygen electrode	River water, 2.5 ml/min, 25°C	20–400 µg/L in tris buf. (1% CaCl2), pH 9.0	5 min	BOD and Cl ion	30 days	54
VII: FIA and reactor	Enzymes	Reacti-gel	Chemiluminescence	Photomultiplier	Phos. buf. pH 8.0, 2.30 ml/min, ~20°C	3.1–100 µg/L (0.31 µg/L), RSD 5% (n = 6)	2 min	Not affected to NaCl, Fe, Mn, Zn, LAS, BOD	150 times	55

[a]Yeast: *Saccharomyces cerevisiae* IFO 0337; bacteria: *Pseudomonas fluorescens* NCIMB 11764; enzymes: rhodanese (E.C. 2.8.1.1) and sulfite oxidase (E.C. 1.3.2.1).
[b]Half of initial response.

ble membrane was confirmed as a cyanide sensor (Table 1—I) (49).

The yeast (*Saccharomyces cerevisiae* IFO 0337) was selected from six kinds of aerobic microorganisms, *Micrococcus luteus* IFO 3342, *E. coli* IFO 14249, *Pseudomonas aeruginosa* IFO 1095, *Trichosporon cutaneum* IFO 10466, and *Bacillus subtilis* IAM 1069, considered to be the most sensitive microorganisms to cyanide (Table 2). Yeast was cultivated in a YM (yeast mold) medium consisting of 10 g/L glucose, 5 g/L peptone, 3 g/L yeast extract, and 3 g/L malt extract at pH 5.8, harvested by centrifugation, and then washed in a tris-HCl buffer. The yeast was entrapped between two porous cellulose nitrate membranes (0.45 μm) as previously described (56) (Fig. 3a). The biomembrane incorporating the yeast was placed on a Teflon membrane cover of a Clark-type oxygen electrode (Type U-1, ABLE Co., Tokyo) and covered with nylon mesh as a protective layer. The yeast in the biomembrane can take up oxygen, glucose, and other nutritious substances through the porous membrane, being exposed to cyanide as well when it exists in the solution.

Figure 4a shows the principle of the microorganism sensor using respiratory inhibition for determining cyanide. As a sensor output, the current was high when the dissolved oxygen in the solution was not taken up by a starved microorganism in the biomembrane. After adding a nutritional element such as glucose to the solution, the current decreased by the activated microorganism's consumption of oxygen. Therefore, glucose was necessary to maintain the respiratory level of immobilized microorganisms as minimum nutrition. When the current stabilized (C_0), cyanide was added to the solution, then the output current of the sensor in-

Table 2 Relative Activities of Respiration and Responses of Several Microorganisms to cyanide

Microorganism	Activity of respiration (%)	Response to cyanide (%)
S. cerevisiae	100	100
T. cutaneum	60	100
E. coli	70	50
B. subtills	5	—
M. luteus	5	—
P. aeruginosa	5	—

The several microorganisms were immobilized on the oxygen electrode, and their respirational activity was determined by adding 100 mg/L glucose, followed by addition of 400 μg/L (15 μM) cyanide ion in 0.01 M tris-HCl buffer (pH 8.0, 30°C).

creased (C). An absolute value of these differences was measured as a sensor response to cyanide (ΔC).

The system of the sensor is shown in Fig. 5a. The sensor was placed in a thermostatic circulating water jacket containing a buffer solution (30°C). The buffer was magnetically stirred during the measurement. The output current from the oxygen electrode was measured with an electronic recorder. When the output current became stable, the sample solution containing cyanide was added to the buffer solution. Then the current change was recorded as a sensor response (ΔC).

Consequently, this sensor was able to determine cyanide ion with a linear range between 8.0 and 4000 μg/L (0.3–150 μM), which indicates that this cyanide biosensor employing a flow system and a reactor has possible applications for the monitoring of cyanide.

2. Flow-and-Reactor Type of Sensor

The second sensor employed a flow system (Fig. 5b, Table 1—II) (50). The yeast was immobilized on CPG (controlled pore glass) beads acting as support; then the beads were packed into a column acting as a reactor. The CPG beads (20 g) with immobilized yeast [6.9 × 10^8 colony-forming units (c.f.u.)] were prepared by cultivating yeast in a YM (yeast and malt extract) medium and adding CPG beads. The system employed two electrodes, and the reactor was put between them. A buffer solution containing 150 mg/L glucose was saturated with oxygen and carried by using a peristaltic pump.

When the cyanide solution was passed through the reactor, the respiration of yeast immobilized on CPG was inhibited, and the amount of oxygen consumed by yeast decreased. As a result, the difference of current output between preelectrode and postelectrode was reduced and considered as a response to cyanide.

This system was able to detect cyanide ion with a linear range between 0 and 400 μg/L (0—15 μM) at a flow rate of 4.5 ml/min at 25°C. These results indicate the possible use of this sensor in the construction of a flow-sensor system that can be applied to continuous monitoring for protecting the discharge of cyanide from wastewater.

3. Flow-and-Reactor Type of Sensor for Real Samples

To ascertain applicability, a third sensor was developed for determining cyanide in river water using an improved previous sensor system (Fig. 5b, Table 1—III) (51).

At the beginning of this study, immobilization methods were examined using several supports: CPG, algin-

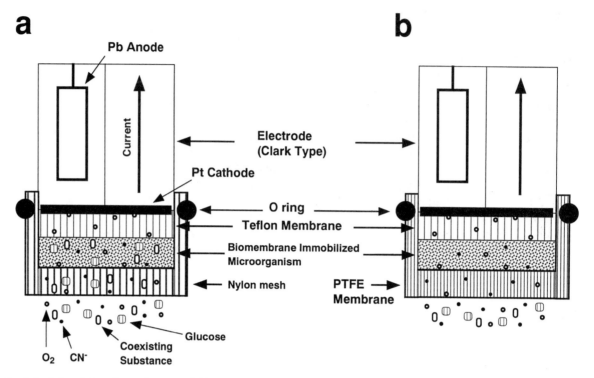

Fig. 3 Principle of the electrode type of biosensors immobilized microorganism. The biomembrane is made of the porous membrane covered with (a) nylon mesh or (b) gas-permeable membrane.

ate gel, and Chitopearl® (HP50-20; pore size 120 μm, sphere diameter 0.4–1.0 mm, Fuji spinning Co., Shizuoka, Japan). Then Chitopearl beads were chosen as the most sensitive support to cyanide. The reactor used had a 22-mm diameter, a cross-sectional area of 381 mm^2, and a length of 90 mm. Temperature had little effect on the sensor responses between 7°C and 29°C. Under optimized conditions consisting of a flow rate of 4.5 ml/min, a 0.01 M tris-HCl solution (pH 8.0) containing glucose at 150 mg/L, and at ambient temperature (approximately 20°C), the sensor responded to cyanide ion at a linear range from 0 to 400 μg/L (0–15 μM) (Fig. 6). The sensor was sensitive enough to detect cyanide contamination from industrial plants in river water. The maximal permissible concentration of cyanide in wastewater in Japan is 1000 μg/L (38.5 μM) (Water Pollution Control Law).

Next, in order to determine the application of the sensor to river water, distilled water (pH 5.7) was used instead of a phosphate buffer, and the results were examined. The change did not affect the sensor response to cyanide.

In this condition, the interference of other compounds was investigated using Na^+, Fe^{2+}, Mn^{2+}, Zn^{2+}, and chloride ion as a counterion. The concentration of

sodium chloride is extremely dependent on the sampling location, fluctuating between around 10 mg/L at a mountainside to 10 g/L near the sea. However, in the range of the concentration, sodium chloride did not affect the sensor response. The maximum concentration of Fe^{2+}, Mn^{2+}, and Zn^{2+} in the Tama River did not have a significant effect on the sensor response. Simazine could be determined with this sensor as well, the sensor response showing a linear relationship with its concentration. Toxic compounds such as herbicides can affect the sensor response by respiratory inhibition of *S. cerevisiae*. The sensor was able to maintain the response for 9 days but lost it after that (Fig. 7). A sensor with such stability can be used for monitoring river water.

Six river water samples were used for cyanide detection, all of them showing a linear response between 0 and 400 μg/L (0–15 μM) cyanide ion. The responses to 400 μg/L CN ion, Cl^- concentrations, BOD values, and pH of river water samples are shown in Table 3. Similar responses to cyanide were obtained in spite of differences in BOD values, Cl^- concentrations, and the pH of river water samples. This cyanide sensor thus proved its suitability for monitoring river water.

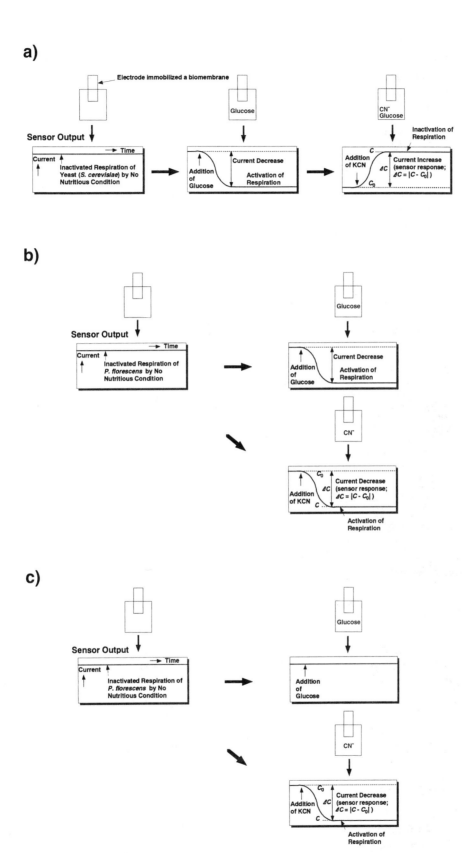

Fig. 4 Principle of detecting cyanide by the batch and microorganism membrane biosensors. The sensor detects cyanide (a) by the inhibition of yeast respiration, or (b) by the activation of the respiration of the cyanide-degrading bacteria using the porous membrane, or (c) by using the gas-permeable membrane.

a) The Batch and Membrane Type Microorganism Sensor

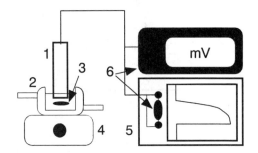

b) The Flow and Microorganism Reactor Type Sensor (two electrode system)

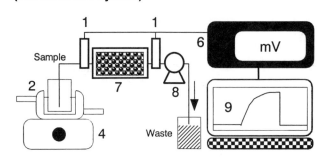

c) The Flow and Microorganism Reactor Type Sensor (one electrode system)

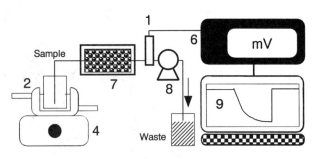

d) The Chemiluminescent FIA and Enzyme Type Sensor

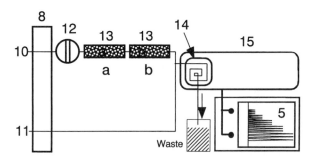

Fig. 5 Schematic diagram of cyanide biosensors: (1) Oxygen electrode, (2) thermostat, (3) microorganism membrane, (4) magnetic stirrer, (5) recorder, (6) digital multimeter or 10-kΩ resistor, (7) microorganism reactor, (8) peristaltic pump, (9) computer, (10) sample carrier solution, (11) chemiluminescence reaction mixture, (12) sample injector, (13) enzyme reactor of (a) rhodanase and (b) sulfite oxidase, (14) flow cell, (15) chemiluminescence detector.

Biosensors using inhibition of the microbial respiration are also affected by other toxic compounds, such as pesticides and herbicides. Therefore, this kind of sensor can be used for estimating total toxicity around a polluted water area. However, it is also important to increase the selectivity of the sensor as a cyanide detection system. Subsequently, biosensors using cyanide-degrading bacteria for the selective determination of cyanide were developed.

C. Biosensors Using Microbial Degradation of Cyanide

Many cases of microbial degradation of cyanide have been reported. For example, Harris and Knowles described the isolation of many kinds of cyanide utilizing *Pseudomonas* from soil (57–58). One of them, *P. fluorescens* NCIMB 11764, biodegrades cyanide aerobi-

cally as a sole nitrogen source. Oxidative cyanide degradation by bacteria is shown in Fig. 8. Cyanide oxidase produces cyanate from cyanide consuming oxygen. Then, the cyanate is hydrolyzed by cyanase, and ammonia and carbon dioxide are produced. Using this mechanism for microbially degrading of cyanide, several biosensors were developed (52–54).

1. Batch-and-Membrane Type of Sensor

The first microbial sensor using this bacterial degradation of cyanide was developed (Table 1—IV) (52). Cyanide can be measured using an oxygen electrode to determine the decrease of dissolved oxygen by degrading cyanide of *P. fluorescens* NCIMB 11764 (Fig. 4b). The system of this sensor employs an electrode, a batch system, and a digital multimeter instead of a 10-kΩ resistor (Fig. 5a). *Pseudomonas fluorescens* was immobilized between two membranes. Subsequently, this

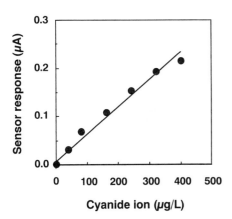

Fig. 6 Calibration curve for cyanide obtained by the flow-and-reactor type of yeast sensor. As a yeast immobilization support, Chitopearl HP50-20 was used. 10-mM Tris-HCl buffer (pH 8.0) containing 150 mg/L glucose was used as a carrier solution. The measurement was performed at a flow rate of 4.5 ml/min and at ambient temperature (∼20°C).

was placed on a PTFE membrane and fixed in place using 200-mesh nylon and an O-ring (same as Fig. 3a).

The *P. fluorescens* was incubated in a liquid medium by shaking at 30°C for 24 hours, and the harvested cells (40 mg), which were grown up to the stationary phase, were entrapped between two membranes.

The electrode immobilizing the intact cells was inserted into 30 ml of a 50 mM phosphate buffer (pH 8.0) saturated with oxygen by continuous stirring with a magnetic bar, and KCN concentration was measured at 30°C.

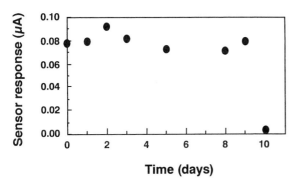

Fig. 7 Stability of the flow-and-reactor type of yeast sensor. As a yeast immobilization support Chitopearl HP50-20 was ussed. Distilled water containing 150 mg/L glucose was used. The measurement was performed at a flow rate of 4.5 ml/min and at ambient temperature (∼20°C). The reactor was stored in 150-mg/L glucose solution at the ambient temperature.

A typical curve of response to different concentrations of cyanide ion (40–400 μg/L; 1.5–15 μM) was obtained by the cyanide sensor. When the output current of the sensor stabilized (Fig. 4b, C_0), cyanide was added to the solution. Then the current was increased (C). The absolute value of these differences was measured as a sensor response to cyanide (ΔC). In the sensor responses, a linear relationship was obtained between the current decrease and cyanide ion concentration for 40–400 μg/L (1.5–15 μM). The reproducibility of this sensor was within ±8% for samples of 400 μg/L (15 μM) cyanide ion solution ($n = 5$).

The optimal conditions of temperature, pH of the buffer solution, and amount of intact cells were investigated. These optimal values were determined to be 30°C, pH 8.0 of the buffer solution, and 40 mg (wet weight) of intact cells.

The selectivity of this sensor to other toxic substances, which were linear alkylate sulfonate (LAS), Cd^{2+}, Cr^{3+}, Pb^{2+}, and simazine (up to 1 mg/L), were investigated in a 50 mM phosphate buffer (pH 8.0, 30°C) as well. The sensor response did not have an effect, except for simazine, which just gave a slight and negligible response opposite in sign to 400 μg/L (15 μM) cyanide ion. Consequently, it can be concluded that the sensor showed good selectivity to cyanide.

Subsequently, the effect of NaCl or heavy-metal ion on the sensor response was also investigated as coexisting substances in river water. The concentration of each heavy-metal ion used in this experiment was much higher than that in normal river water. However, neither NaCl of 30 mg/L nor each heavy-metal ion of 1 mg/L had an effect on the sensor response to 400 μg/L (15-μM) cyanide ion (pH 8.0, 30°C). These results showed that the microorganisms used have a resistance to metal ions similar to the characteristics of many bacteria (59).

After this sensor was optimized and its characteristics investigated, the multiple effects of numerous substances on the sensor response were investigated using water from the Watarase River. The sensor responses to cyanide ion concentrations between 80 and 400 μg/L (3.0–15 μM) were determined.

2. Batch-and-Membrane Type of Sensor Using a Gas-Permeable Membrane

The membrane-type sensor just described (52) was affected by the concentration of nutrients such as glucose and glutamate. Therefore, to improve selectivity, a cyanide-selective sensor using a gas-permeable mem-

Table 3 Sensor response of Flow-and-Reactor Type of Yeast Sensor to Cyanide in River Water Samples

	Sensor response (μA)	Chloride ion (mg/L)	BOD (mg/L)	pH
Edo River (Sekiyado Bridge)	0.086	—	1.50	7.20
Edo River (Bridge of Tozai Line)	0.083	12,100	10.10	7.00
Watarase River (confluence with Akiyama River)	0.077	6.9	1.02	7.41
Watarase River (water gate of Yaba River)	0.088	7.6	3.61	7.31
Watarase River (Namai Bridge)	0.086	8.6	2.10	7.20
Watarase River (Mikuni Bridge)	0.077	11.2	1.68	7.40

Cyanide ion was added to the river water samples to give a final concentration of 400 μg/L, and glucose was also added to give a final concentration of 150 mg/L. The measurement was performed at a flow rate of 4.5 ml/min and at ambient temperature (~20°C).

brane [PTFE (polytetrafluoroethylene), Nihon Milli-pore Ltd., Japan)] with a batch system was developed (Table 1—V) (53). The basic construction of this sensor was the same as that of the previous sensor (Fig. 5a).

Cyanide in water forms as molecular acid hydrogen cyanide (HCN) and/or as free cyanide ion (CN⁻) (60). Above pH 4, cyanides are easily converted into HCN. Liquid HCN can be volatile from water in the gas phase.

This gas-phase biosensor is based on the cyanide-degrading mechanism of *P. fluorescens* NCIMB 11764. The principle of this sensor is to measure gaseous cyanide, which is generated by acidification of the sample solution and passed through a gas-permeable membrane, by determining the oxygen consumed by the bacteria-degrading cyanide. By using the gas-permeable membrane, *P. fluorescens* immobilized in the biomembrane was exposed to only volatilized gases such as oxygen and HCN (Fig. 3b). Therefore, an ideal detection of cyanide was expected (Fig. 4c). In this system, the concentration of nutrients did not affect the sensor response, and only oxygen and HCN as volatilized gas were taken up by *P. fluorescens* in the biomembrane. The sensor response was obtained as an absolute value of the differences (ΔC; as a sensor response) of output current between the baseline current (C_0) and the response current (C).

After confirmation of the sensor response to cyanide, this sensor system was optimized. The optimum conditions were 30°C of determining temperature, pH 2 of phosphoric acid buffer, and 110 mg wet weight of immobilized microorganisms.

At the optimized conditions, the selectivity of the sensor, the effect of heavy-metal ion on the sensor response, and the lifetime were investigated.

The previous sensor without a gas-permeable membrane was affected by glucose and glutamate more than by cyanide ion. On the other hand, the gaseous sensor was not affected by nutrients that were not permeable to the membrane. Subsequently, the effects of heavy-metal ion, which is contained in most river waters as Cu^{2+}, Zn^{2+}, Pb^{2+}, Cd^{2+}, and Fe^{2+}, were investigated and compared with the JIS method (pyridine method) (4). The results are shown in Table 4. Sensor responses to 400 μg/L (15 μM) cyanide ion were not affected by heavy-metal ion due to the decomposing of weakly complexed metal cyanides at pH 2, while the JIS method was affected by Cu^{2+} and Fe^{2+}. The stability of the sensor response to 400 μg/L (15 μM) cyanide ion was investigated. The response decreased within 4 days to 73% of the initial response. However, the response could be maintained at the same level from the 4th to the 30th day. The lifetime of this sensor was satisfactory for practical use.

A comparison of the JIS method (pyridine method) (4) with the gas-phase biosensor was carried out using several samples of river water at a cyanide ion concentration between 40 and 400 μg/L (1.5–15 μM). Consequently, a good correlation was obtained by the

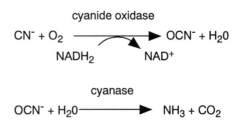

Fig. 8 Reaction scheme of cyanide degradation by *P. fluorescens* NCIMB 11764.

Table 4 Effects of Heavy-Metal Ions on Sensor Response and the JIS Method

Heavy-Metal ions	Sensor response (%)	JIS method (%)
Control	100	100
Fe	100	83
Cu	100	11
Cd	96	94
Pb	96	99
Zn	96	90

All divalent cations were used as sulfate compound at a final concentration of 1 mg/L. Cyanide ion was added to the sample solution to give a final concentration of 400 μg/L. The measurement was performed at 30°C.

two methods (correlation coefficient of 0.995, $n = 4$) (Fig. 9).

In conclusion, the sensitivity, selectivity, and stability of the gas-phase biosensor were considered adequate for practical use, suggesting the usefulness of this sensor. Therefore, cyanide biosensors employing a reactor with a flow system have been implemented for the continuous monitoring of cyanide in river water.

3. Flow-and-Reactor Type of Sensor for Real Samples

A flow-type cyanide sensor using an immobilized *P. fluorescens* column was developed. For the environ-

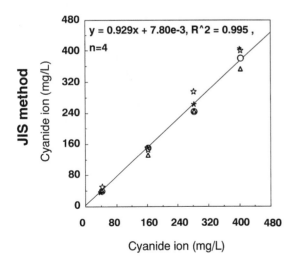

Fig. 9 Comparison of the gas-phase biosensor (pH 2, 30°C) with the JIS method.

mental monitoring of water using a microbial sensor, a flow-and-reactor type of sensor is the most suitable method (Table 1—VI) (54).

The sensor system is shown in Fig. 5c. Since the total oxygen consumption by bacteria immobilized onto beads was larger than that immobilized in the membrane, a reactor-type sensor was chosen. The *P. fluorescens* (2.5 g) was entrapped into calcium alginate gel beads, and then the beads were packed into a glass column (7-mm ID × 150-mm length). After being filled up with beads with immobilized bacteria, the reactor was positioned in a thermostatic chamber. The buffer solution used as a carrier, 50 mM tris-NaOH containing 1% $CaCl_2$, flowed through the reactor. After steadying the sensor output, the cyanide solution was pumped into the reactor using a peristaltic pump (MINIPLUS 3, Gilson, France). Cyanide dissolved in the sample solution was degraded by *P. fluorescens* immobilized in the reactor, and then consumed oxygen was detected by a Clark-type oxygen electrode as a decrease in current.

After optimization of the sensor system, a condition for determining cyanide was obtained by adding 400 μg/L (15 μM) cyanide ion at 25°C, 50-mM tris-NaOH buffer of pH 9.0, and a flow rate of 2.5 ml/min. Under these conditions, a calibration curve for cyanide ion (Fig. 10) was obtained at a linear response between 20 and 400 μg/L (0.75–15 μM) and a response time of 5 min. The result obtained had a wider range than that of previous cyanide-degrading sensors. Subsequently, sensor stability was examined under these optimum conditions, with a stable response to 400 μg/L (15-μM) cyanide ion being obtained at least 30 days after the first few days (Fig. 11). These results show that the range, sensitivity, and stability of this sensor were sufficient for practical use.

Using several river waters, sensor responses to cyanide were compared with the buffer solution under optimized sensor conditions. The samples were taken from the Watarase River, the Ayase River, and the Naka River. Table 5 shows the results when sensor response was obtained by adding 400 μg/L (15 μM) cyanide ion into each river. The sensor responses showed the same values except for the case of the Ayase River. Therefore, the sensor response might be influenced by a high BOD level and chloride ion concentration.

In summary, this sensor demonstrated its possible application for continuous detection of cyanide and on-line monitoring in river water, in the same manner as the flow-and-reactor type of yeast sensor (51). However, this sensor's selectivity needs to be improved. Therefore, the development of a gas-phase biosensor

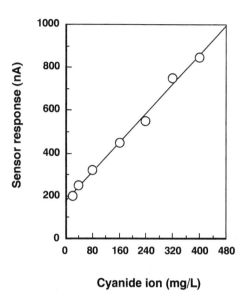

Fig. 10 Calibration curve for cyanide ion obtained by the flow-and-reactor type of cyanide-degrading microorganism sensor. As a microorganism immobilization support Chitopearl HP50-20 was used. 50-mM Tris-NaOH buffer (pH 9.0) was used as a carrier solution. The measurement was performed at a flow rate of 2.5 ml/min and at 25°C.

employing a flow system is required during further studies to avoid the effect of nutrients in river water.

D. Biosensors Using Enzyme-Decomposing Cyanide

Several bacterial sensors for determining cyanide have been developed. These sensors, which were superior to many methods, determine cyanide in a simple, rapid, and low-cost manner and will provide an on-site monitoring system in areas such as rivers. However, for monitoring cyanide, it is also important for the sensor to be more sensitive and selective to cyanide in river water when other toxic substances are drained out from other sources. Therefore, many enzyme sensors have been developed to improve selectivity using rhodanese with an electrical system (61) and an enzyme thermistor system (62) and injectase with an enzyme thermistor system (62). Unfortunately, these sensors lack sensitivity for cyanide detection.

In view of this, a highly sensitive bienzyme sensor system combined with a luminol chemiluminescence reaction using an FIA system was developed and applied to river water analysis (Table 1—VII) (55).

Rhodanese (from bovine liver, E.C. 2.8.1.1) and sulfate oxidase (from chicken liver, E.C. 1.3.2.1) were chosen for determining cyanide. The sulfite was gen-

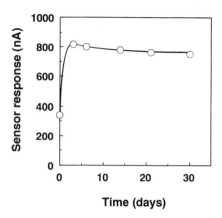

Fig. 11 Stability of the flow-and-reactor type of cyanide-degrading microorganism sensor. As a microorganism immobilization support, alginate gel was used. 50-mM Tris-NaOH buffer (pH 9.0) was used as a carrier solution. 400-μg/L cyanide ion was added to the sample solution. The measurement was performed at a flow rate of 2.5 ml/min and at 25°C.

erated by the rhodanese catalyzing the reaction of cyanide, and sodium sulfate was converted into sulfate and hydrogen peroxide by sulfate oxidase (Fig. 12). Then hydrogen peroxide was reacted with luminol catalyzed by peroxidase (POD, from *Arthromyces ramosus*, E.C. 1.11.1.7).

The chemiluminescence reaction is known as a highly sensitive method in the luminol chemiluminescence reaction for determining hydrogen peroxide (63). The detection limit of hydrogen peroxide is approximately 10 nM (64). The FIA system employed in this sensor, which was used for the simple and rapid determination of cyanide, can be applied to continuous monitoring with high reproducibility. In this system, subsequently generated chemiluminescence was detected.

Table 5 Comparison of Sensor Response with Buffer Solution and River Water Samples

	Sensor response (%)	Chloride ion (mg/L)	BOD (mg/L)	pH
Buffer	100	—	—	7.0
Watarase River	100	17	0.81	7.2
Naka River	97	18.5	2.50	6.9
Ayase River	40	32.8	4.39	6.9

Cyanide ion was added to the river water samples to give a final concentration of 400 μg/L. The measurement was performed at a flow rate of 2.5 ml/min and at 25°C.

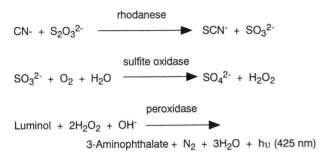

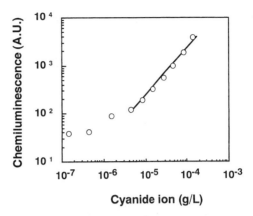

Fig. 12 Reaction scheme for bienzyme reaction and luminol chemiluminescence reaction.

Fig. 13 Calibration curve for cyanide obtained by the FIA chemiluminescence enzyme sensor. As a enzyme immobilization support the reacti-gel was used. As a carrier buffer for the cyanide reagent a 10 mM phosphate buffer (pH 8.0) was used, and as a chemiluminescence reagent a 0.8 M carbonate buffer containing 7 μM luminol and 5 mg/L POD was used. The measurement was performed at a flow rate of 2.30 ml/min and at ambient temperature ($\sim$20°C).

The sensor system shown in Fig. 5d consists of two columns with packed rhodanese or sulfate oxidase immobilized on beads, a mixing joint for the chemiluminescence reaction, and a photomultiplier tube (PMT, Model 825-CL, JASCO, Japan). A peristaltic pump (Minipuls 3, Gilson, France) delivered each reagent separately to each pathway at the same flow rate.

The rhodanese and sulfite oxidase were immobilized by covalent bonding using the carbodi-imidazole groups contained in reacti-gel (sphere diameter: 45–165 μm, Rockford, USA), and then these were stored at 4°C in 3.2 M ammonium sulfate before use. Teflon tubing was used as a column to pack the rhodanese or sulfite oxidase immobilized on beads.

The carrier buffer used as the cyanide reagent was a 10 mM phosphate buffer (pH 8.0), and the chemiluminescence reagent was a 0.8 M carbonate buffer containing 7 μM luminol and 5 mg/L POD. As another substrate for the rhodanese catalyzing reaction, $Na_2S_2O_3$ was added to the sample solution to give a final concentration of 1 mM, and the sample solution (20 μl) was injected from a sample injector. The chemiluminescence, which occurred when the cyanide reagent and the chemiluminescence reagent were mixed, was immediately carried to a flow cell located directly in front of the PMT. The chemiluminescence detected by the PMT was then recorded by a recorder.

To obtain the optimized conditions for cyanide detection, the concentration of the $Na_2S_2O_3$, the flow rate, and the pH of the rhodanese and sulfite oxidase reaction reagent were investigated. Results indicated that residual $Na_2S_2O_3$ in the sample solution after rhodanese reaction was influenced by a subsequent chemiluminescence reaction. Therefore, 1 mM $Na_2S_2O_3$ in the sample solution used as a concentration did not affect the chemiluminescence reaction. The peak flow rate and the pH of the enzyme reagent were obtained at 2.30 ml/min and pH 8.0, respectively.

Under the optimized conditions for this sensor, a calibration curve (Fig. 13) was obtained at a linear range of cyanide ion concentration between 3.1 and 100 μg/L (120 nM to 3.8 μM) with a relative standard deviation for 10 μg/L (380 nM) lower than 5% ($n = 6$). The detection limit was 0.31 μg/L (12 nM). This is sufficient sensitivity for determining cyanide in natural river water. These experiments were carried out at ambient temperature. Additionally, the stability of the immobilized rhodanese and sulfite oxidase column was examined, and this column can be used 150 times.

To examine selectivity, the effects of 5 mg/L atrazine, simazine, and LAS on the sensor response were investigated. Such a high concentration of these compounds had not been detected in the Tama River in 1990 (65). However, the sensor showed no response to them, though these compounds can affect a microbial sensor (49,52). Additionally, NaCl, Fe^{2+}, Mn^{2+}, Zn^{2+}, and BOD were investigated at above the highest concentrations determined in the Tama River in 1990. However, the sensor did not show any response to them either. Therefore, this sensor can be used for river water analysis for cyanide detection without any interference from these compounds.

Next, the cyanide sensor was applied to river water analysis using samples from the Edo River and the Watarase River (each 3 points). The JIS method (pyridine method) (4) did not detect cyanide in any of the river water samples. Therefore, 3.1 μg/L (120-nM) cyanide ion was added to each river sample, and then the measurements were performed using cyanide. Table 6 shows that results with similar responses were obtained in spite of the quite different contents of each river sample.

Consequently, this biosensor, which showed high selectivity and high sensitivity to cyanide ion, can be used in a river water monitoring system when cyanide ion exists at above 2.6 μg/L (100 nM). Additionally, this sensor can detect only free cyanide (CN^-). This is also important for determining the real concentration of cyanide ion, a substance affecting living things in an environment polluted by cyanide.

E.   Other Biosensors and Biosensing Systems for Determining Cyanide

Recently, new enzyme sensors for determining cyanide have been reported. An ISFET-based peroxidase biosensor (66) and an amperometric biosensor using peroxidase (67–68) used inhibition of peroxidase reaction by cyanide. Unfortunately, these sensors lack selectivity for that reason. A sensor based on methemoglobin incorporated into bilayer lipid membranes has been developed for cyanide detection (69). However, it had a matrix effect. Therefore, even though it had the highest sensitivity reported to date, the sensor was prohibited for use on real samples due to nonselective interaction with the membrane by coexisting substances in a water sample. On the other hand, new warning systems for monitoring the quality of water have been developed using information conveyed by the continuous electric organ discharges of the tropical fish *Apteronotus albifrons* (70–71). The system detected 14 μg/L (0.54 μM)

cyanide ion. However, this is a large-scale sensing system, and the conditions of the fish might be affected by the climate and the water quality. In addition, the response to cyanide is somewhat slow, taking approximately half an hour. When cyanide contamination occurs, a rapid detection is not possible with this response time. Especially in the case of Japan, where most rivers are short and many houses are crowded along the river, quick responses to cyanide contamination are required.

F.   Development of Integrated Cyanide-Monitoring Systems for Practical Utilization

Three types of cyanide biosensors were developed to respond to the requests of the Ministry of Construction. Biosensor techniques are applicable methods for onsite cyanide monitoring in rivers, for they offer a simple and rapid measuring method and a stable, compact, and low-cost sensing system.

The development of integrated cyanide-monitoring systems for practical utilization has been planned for several years. It is financially supported by the River Department, the Kanto Regional Construction Bureau, and the Ministry of Construction with the help of the Association of Electrical Engineering. The first step for the application of the cyanide-monitoring system consisted in examining microbial sensors using the inhibition of yeast respiration employing a reactor-and-flow system. The sensor system was fitted compactly into a buoy (1-m diameter) that would float on the water surface. After determining its applicability to river water, the system was moored as an experiment in a branch of the Edo River. However, the electrical power sup-

Table 6   Sensor Response to Cyanide in River Water Samples

	Sensor response (%)	Chloride ion (mg/L)	BOD (mg/L)	pH
Control	100	—	—	8.0
Edo River (Sekiyado Bridge)	99	—	1.50	7.2
Edo River (Unga Bridge)	107	57.9	8.52	7.0
Edo River (Shin-Katsushika Bridge)	94	23.1	3.05	7.4
Watarase River (confluence with Akiyama River)	91	16.9	1.02	7.4
Watarase River (water gate of Yaba River)	115	17.6	3.61	7.3
Watarase River (Namai Bridge)	92	8.6	2.10	7.2

Cyanide ion was added to the river water samples to give a final concentration of 3.1 μg/L. As a carrier buffer for the cyanide reagent a 10 mM phosphate buffer (pH 8.0) was used, and as a chemiluminescence reagent a 0.8 M carbonate buffer containing 7 μM luminol and 5 mg/L POD was used. The measurement was performed at a flow rate of 2.30 ml/min and at ambient temperature ($\sim$20°C).

plied by solar batteries set on a cap was not sufficient and had to be supplied from land by a cord. At present, this monitoring system is halfway into its development, and the current results will be reported in the near future.

The final goal is to realize integrated cyanide-monitoring systems combining three types of biosensors utilizing inhibition of yeast respiration, degradation of cyanide by bacteria (*P. fluorescens* NCIMB 11764) using a gas-permeable membrane, and bienzyme reaction for cyanide detection. This integrated system can compensate for the disadvantages of each biosensor. For example, yeast respiration is affected by other toxic compounds, such as pesticides and herbicides, the gas-permeable membrane might be affected by other volatile substances, and the bienzyme chemiluminescence biosensor might be affected by several reducing species existing in river water. However, the effects of these interferences can be avoided by integrating the characteristics of each biosensor in the system. Therefore, when the integrated monitoring system is realized, it will not only detect cyanide contamination in river water with high sensitivity by means of a bienzyme sensor but also estimate the effect of toxicity by monitoring yeast respiration. Then a yeast respirational sensor will be provided for estimating toxicity, and a bacterial degradation sensor will be provided as a long-lived sensing system and a supplement of the bienzyme sensor for cyanide detection.

Additionally, there is a project to decompose cyanide immediately in the case of cyanide contamination. The cyanide-degrading bacterium *Pseudomonas stutzeri* AK61 has been isolated from wastewater at a metal-plating plant and the cyanide-degrading enzyme (cyanidase) purified (72). By distributing these biomaterials, damage from such accidents is expected to be kept to a minimum.

In conclusion, this integrated monitoring system is expected as a new method to estimate the total effects of cyanide on the environment.

REFERENCES

1. MX Fuller. Cyanide and environment. Proc Conf, Tucson, AZ, Dec 11–14, 1984; Fort Collins, CO, pp 19–46, 1985.
2. GD Muir. Hazards in the Chemical Laboratory. Chemical Society, London, 1977.
3. American Public Health Association. Standard Methods for the Examination of Water and Waste Water. 18th ed. Washington, DC: APHA, AWWA, WPCF, 1992, pp 4-18–4-31.
4. Japanese Standard Association. Testing Methods for Industrial Waste Water. JIS K-0102, Tokyo, 1993, pp 127–135.
5. P MacCarthy, RW Klusman, SW Cowling, JA Rice. Water analysis. Anal Chem 63:301R–342R, 1991.
6. P MacCarthy, RW Klusman, SW Cowling, JA Rice. Water analysis. Anal Chem 65:244R–292R, 1993.
7. NK Kutseva, AN Kashin. Determination of cyanide in the environment. Industrial Laboratory 61:1–11, 1995.
8. M Nonomura. Analytical methods of cyanide compounds in environment. Bunseki 11:917–923, 1994.
9. M Nonomura. Progress of cyanide determination procedure in water. Kogyo Yosui 401:12–25, 1992.
10. J Liedig. Process for determining the amount of hydrocyanic acid in medical prussic acid, bitter almond water, and laurel water. Quarterly Chem Soc London 4:219–221, 1852.
11. W König. Über eine Klasse neuer Farbstoffe. J Prakt Chem 69:105–137, 1904.
12. W König. Verfahren zur Darstellung neuer stickstoffhaltiger Farbstoffe. Z Angew Chem 115, 1905.
13. MS Frant, JW Ross, Jr, JH Riseman. Electrode indicator technique for measuring low levels of cyanide. Anal Chem 44:2227–2230, 1972.
14. H Clysters, F Adams. Potentiometric determination with the silver sulfide membrane electrode, Part I. Determination of cyanide. Anal Chim Acta 83:27–58, 1976.
15. M Hofton. Continuous determination of free cyanide in effluents using silver ion selective electrode. Environ Sci Technol 10:277280, 1976.
16. T Koshimizu, K Takamatsu, Y Sekiguchi, Y Saiki, S Fukui, S Kanno. Studies on the practical application of an ion electrode methods (I): Fundamental analytical conditions in the determination of cyanide. Anzen Kougaku 12:179–186, 1973.
17. J Epstein. Estimation of microquantities of cyanide. Anal Chem 19:272–274, 1947.
18. Y Hirose, J Morimoto, T Okitsu, N Maeda, S Kanno. Problems of analytical method of cyanide ion in tap water and its improvement. Eisei Kagaku 34:65–69, 1988.
19. NP Kelada. Automated direct measurements of total cyanide species and thiocyanate, and their distribution in wastewater and sludge. J Water Pollut Contr Fed 61:350–356, 1989.
20. E Nakamura, M Kokubo, H Namiki. Determination of cyanide in formaldehyde cyanohydrin. Bunseki Kagaku 41:T131–T134, 1992.
21. H Ishii, K Kohata. Indirect spectrophotometric determination of trace cyanide with cationic porphyrins. Talanta 38:511–514, 1991.
22. A Velasco, M Valcarcel. Kinetic spectrophotometric determination of nanogram amounts of cyanide. Talanta 38:303–308, 1991.

23. A Ganjeloo, GE Isom, RL Morgan, JL Way. Fluorometric determination of cyanide in biological fluids with *p*-benzoquinone. Toxicol Appl Pharmacol 55:103–107, 1980.

24. S Takanashi, Z Tamura. Fluorometric determination of cyanide by the reaction with pyridoxal. Chem Pharm Bull 18:1633–1635, 1970.

25. P Lundquist, H Rosling, B Sorbo, L Tibbling. Cyanide concentrations in blood after cigarette smoking, as determined by a sensitive fluorometric method. Clin Chem 33:1228–1230, 1987.

26. A Sano, M Takezawa, S Takitani. Fluorometric determination of cyanide with 2,3-naphthalenedialdehyde and taurine. Talanta 34:743–744, 1987.

27. K Gamoh, H Sawamoto. Determination of cyanide ion by high performance liquid chromatography with fluorometric determination. Anal Sci 4:665–666, 1988.

28. K Gamoh, S Imamichi. Postcolumn liquid chromatographic method for the determination of cyanide with fluorometric detection. Anal Chim Acta 251:255–259, 1991.

29. JJ Rosentreter, RK Skogerboe. A method development for the routine analytical monitoring of aqueous cyanide species. Water Sci Tech 26:255–262, 1992.

30. S Chattaraj, AK Das. Indirect determination of free cyanide in industrial waste effluent by atomic absorption spectrometry. Analyst 116:739–741, 1991.

31. R Rubio, J Sanz, G Rauret. Determination of cyanide using a microdiffusion technique and potentiometric measurement. Analyst 112:1705–1708, 1987.

32. P Riyazuddin, M Kamalutheen. Inexpensive cyanide selective electrodes for application in routine monitoring and analysis. Bull Electrochem 4:417–419, 1988.

33. Y Asano, S Ito. Development of potentiometric continuous monitoring system for cyanide ion in aqueous solution utilizing hydrogen cyanide gas sensor. Bunseki Kagaku 39:693–698, 1990.

34. EO Otu, JJ Byerley, CW Robinson. Ion chromatography of cyanide and metal cyanide complexes (Review). Int J Environ Anal Chem 63:81–90, 1996.

35. H Satake, H Segawa, S Ikeda. Nonsuppressed ion chromatography of cyanide ion using the reaction with polysulfide. J Chem Soc Jpn, Chem Ind Chem 9:1587–1590, 1988.

36. J Ruzicka, EH Hanssen. A new concept of continuous flow analysis. Anal Chim Acta 78:145–153, 1975.

37. A Tanaka, K Mashiba, T Deguchi. Simultaneous determination of cyanide and thiocyanate by the pyridine/barbituric acid method after diffusion through a microporous membrane. Anal Chim Acta 214:259–269, 1988.

38. AT Haj-Hussein. Flow injection analysis, cyanide determination, tetracyanonikelaate(II) complex ion, ultraviolet-spectrophotometry. Anal Lett 21:1285–1296, 1988.

39. H Sulistyarti, TJ Cardwell, SD Kolev. Determination of cyanide as tetracyanonickelate(II) by flow injection and

40. D Narinesingh, S Saroop, TT Ngo. A flow injection method for the quantitation of cyanide: its application to the quantitation of bound cyanide in cassava using an immobilized linamarase bioreactor. Anal Chim Acta 354:189–196, 1997.

41. SD Nikolic, EB Milosavljevic, JL Hendrix, JH Nelon. Flow injection amperometric determination of cyanide on a modified silver electrode. Analyst 117:47–50, 1992.

42. O Elsholz, W Frenzel, CY Liu, J Möller. Evaluation of experimental conditions on the response of fluoride and cyanide selective electrodes in flow injection potentiometry. Fresenius' Z Anal Chem 338:159–162, 1990.

43. LC Clark, C Lyons. Electrode system for continuous monitoring in cardiovascular surgery. Ann NY Acad Sci 102:15–22, 1962.

44. F Scheller, F Schubert, D Pfeiffer, R Hitsche, I Dransfeld, R Renneberg, U Wollenberger, K Riedel, M Pavlova, M Kühn, HG Müller, PM Tan, W Hoffman, W Moritz. Research and development of biosensors (Review). Analyst 114:653–662, 1989.

45. AEG Cass, ed. Biosensors. A Practical Approach. Oxford: IRL Press, 1990, pp 155–170.

46. CY Chen, I Karube. Biosensors and flow injection analysis: current opinion in biotechnology. Anal Biotechnol 3:31–39, 1992.

47. I Karube. Biosensors. In: PH Sydenham, R Thorn, eds. Handbook of Measurement Science. Wiley, 1992, pp 1721–1756.

48. I Karube, T Matsunaga, S Mistuda, S Suzuki. Microbial electrode BOD sensors. Biotechnology and Bioengineering. Wiley, 1977, pp 1535–1547.

49. K Nakanishi, K Ikebukuro, I Karube. Determination of cyanide using a microbial sensor. Appl Biochem Biotechnol 60:97–106, 1996.

50. K Ikebukuro, M Honda, K Nakanishi, Y Nomura, Y Masuda, K Yokoyama, Y Yamauchi, I Karube. Flowtype cyanide sensor using an immobilized microorganism. Electroanal 8:876–879, 1996.

51. K Ikebukuro, A Miyata, SJ Cho, Y Nomura, SM Chang, Y Yamauchi, Y Hasebe, S Uchiyama, I Karube. Microbial cyanide sensor for monitoring river water. J Biotechnol 48:73–80, 1996.

52. JI Lee, I Karube. A novel microbial sensor for the determination of cyanide. Anal Chim Acta 313:69–74, 1995.

53. JI Lee, I Karube. Development of a biosensor for gaseous cyanide in solution. Biosens Bioelectron 11:1147–1154, 1996.

54. JI Lee, I Karube. Reactor type sensor for cyanide using an immobilized microorganism. Electroanalysis 8:1117–1120, 1996.

55. K Ikebukuro, M Shimomura, N Onuma, A Watanabe, Y Nomura, K Nakanishi, Y Arikawa, I Karube. A novel biosensor system for cyanide based on a chemiluminescence reaction. Anal Chim Acta 329:111–116, 1996.

56. I Karube, M Suzuki. In: AEG Cass, ed. Biosensor. IRL, Oxford, 1990, pp. 155–170.

57. RE Harris, CJ Knowles. The conversion of cyanide to ammonia by extracts of a strain of *Pseudomonas fluorescens* that utilizes cyanide as a source of nitrogen for growth. FEMS Microbiol Lett 20:337–341, 1983.

58. RE Harris, CJ Knowles. Isolation and growth of a *Pseudomonas* species that utilizes cyanide as a source of nitrogen. J Gen Microbiol 129:1005–1011, 1983.

59. JT Trevors, KM Oddie, BH Belliveau. Metal resistance in bacteria. FEMS Microbiol 32:39–54, 1985.

60. C Pohlandt, EA Jones, AF Lee. A critical evaluation of methods applicable to the determination of cyanides. J S Afr Inst Min Metall 83:11–19, 1983.

61. CA Groom, JHT Luong. A flow through analysis biosensor system for cyanide. J Biotechnol 21:161–172, 1991.

62. B Mattiasson, K Mosbach. Application of cyanide-metabolizing enzymes to environmental control: enzyme thermistor assay of cyanide using immobilized rhodanese and injectase. Biotechnol Bioeng 19:1643–1651, 1977.

63. K Akimoto, Y Shinmen, M Sumida, S Asam, T Amachi, H Yoshizumi, Y Saeki, S Shimizu, H Yamada. Luminol chemiluminescence reaction catalyzed by a microbial peroxidase. Anal Biochem 189:182–185, 1990.

64. K Hayashi, S Sasaki, K Ikebukuro, I Karube. Highly sensitive chemiluminescence flow injection analysis system using microbial peroxidase and a photodiode detector. Anal Chim Acta 329:127–134, 1996.

65. River Bureau. The Ministry of Construction. Japan. Annual Report on River Water Quality in Japan (Suishitu Nenkan), 1990.

66. V Volotovsky, N Kim. Cyanide determination by an ISFET-based peroxidase biosensor. Biosens Bioelectron 13:1029–1033, 1998.

67. J Zhao, RW Henkens, AL Crumbliss. Mediator-free amperometric determination of toxic substances based on their inhibition of immobilized horseradish peroxidase. Biotechnol Prog 12:703–708, 1996.

68. TM Park, EI Iwuoha, MR Smyth. Development of a sol-gel enzyme inhibition-based amperometric biosensor for cyanide. Electroanal 9:1120–1123, 1997.

69. CG Siontorou, DP Nikolelis. Cyanide ion minisensor based on methemoglobin incorporated in metal supported self-assembled bilayer lipid membranes and modified with platelet-activating factor. Anal Chim Acta 355:227–234, 1997.

70. M Thomas, A Florion, D Chretien, D Terver. Real-time biomonitoring of water contamination by cyanide based on analysis of the continuous electric signal emitted by a tropical fish: *Apteronotus albifrons*. Water Res 30:3083–3091, 1996.

71. M Thomas, D Chretien, A Florion, D Terver. Real-time detection of potassium cyanide pollution in surface water using electric organ discharges wave emitted by the tropical fish *Apteronotus albifrons*. Environ Technol 17:561–574, 1996.

72. A Watanabe, K Yano, K Ikebukuro, I Karube. Cyanide hydrolysis in a cyanide-degrading bacterium, *Pseudomonas stutzeri* AK61, by cyanidase. Microbiol 144:1677–1682, 1998.

20

Humic Substances in Water

Juhani Peuravuori and Kalevi Pihlaja
University of Turku, Turku, Finland

I. CLASSIFICATION OF DISSOLVED ORGANIC MATTER

Organic water quality studies have become popular during recent decades with the rise of environmental concerns. Regular studies and observations are made on specific organic compounds, such as pesticides, herbicides, trihalomethanes, and other chlorinated compounds, and on industrial chemicals that may be harmful to natural fauna and flora and to human health and cause damage to the natural ecology. In practice these compounds have been considered equal to organic water quality. In fact, the amount of the foregoing pollutants represents at most less than 1–2% of the organic constituents in water. On the other hand, it has been gradually accepted that the majority of organic constituents in water represent natural organic matter (NOM) that is essential in all aspects of water ecology and quality, e.g., as buffers for the acid–base effect of freshwaters and as a solid support for pollutant compounds and causing interference with analyses (1–3). These natural organic solutes (in particular, humic solutes) are also important precursors to harmful side effects (formation of mutagens) from the chlorination of drinking water.

In general the dissolved organic matter (DOM) in natural water is classified (4) roughly into two groups: (a) nonhumic solutes, consisting of compounds belonging to the well-known classes of organic substances, such as amino acids, hydrocarbons, carbohydrates, fats, waxes, resins, and low-molecular-weight acids, and (b) very complicated heterogeneous humic solutes. These two groups are not completely distinguished from one another, neither physically nor chemically, because some natural nonhumic solutes, such as carbohydrates, can be an integral part of the structural composition of humic solutes. Apparently, there is no strict chemical boundary between humic and nonhumic solutes; rather, a natural continuity seems to prevail. The mean ages of isolated humic solutes vary widely (4–7) from ca. 30 to 2500 years, depending on the source (aquatic or terrestrial) of the sample.

The amount and quality of the NOM in water varies widely with climate, geographical zone, and a number of other environmental factors. However, humic-matter fractions isolated from similar sampling sites show a given uniformity regardless of their areal location. Nevertheless, humic solutes occurring in surface waters, pore waters, and groundwaters and in seawaters have their own special properties. Thus, the use of environmental descriptors or source indicators for isolated humic-matter fractions is strongly recommended; e.g., in the case of seawater we should talk about marine aquatic humic solutes.

The carbon cycling in aquatic ecosystems is extremely complicated, and the origin of aquatic humic solutes can be dated back to many complex interacting sources, e.g., Steinberg and Münster in Ref. 6. The original main source may partially be both allochthonous (produced outside the system) and autochthonous (produced within the system, e.g., via enzyme-mediated oxidations). The formation of aquatic humic solutes occurs as the result of several processes in the aquatic environment, and it is a dynamic process with no unidirectional vector. Nevertheless, the final outcome seems not to be randomly regulated, even though some seasonal variation occurs. On the basis of fact finding, the similarities of different humic-matter fractions are more pronounced than their differences. Thus it is well founded to classify aquatic humic solutes into their own chemically unique main category, which will split further into specific subgroups according to a given water sampled and the isolation–fractionation method adopted.

The nomenclature and definition of aquatic humic substances is not simple, and they have features of scientific hair-splitting. Originally, in soil chemistry humic substances were classified (6) as "a general category of naturally occurring, biogenic heterogeneous organic substances that can generally be characterized as being yellow to black in color, of high molecular weight, and refractory." Humic substances are most frequently extracted from the solid NOM with aqueous bases (7). The basic solutions of humic substances so obtained are further partitioned into humic acids (HA) and fulvic acids (FA) based on their solubilities in aqueous acids and bases. Thus, *humic acids (HA)* are "the fraction of humic substances that is not soluble in water under acid conditions (below pH 2), but becomes soluble at greater pH," and *fulvic acids (FA)* are "the fraction of humic substances that is soluble under all pH conditions" (6). Accordingly, the third fraction obtained from the solid NOM contains *humins* ("the fraction of humic substances that is not soluble in water at any pH value"). The definition of humic substances in soil is not chemically unambiguous, and many alterations may take place during the isolation and purification procedures. There is also no way to remove coadsorbed organic impurities (carbohydrates, proteinaceous compounds, etc.) from the true humic substances. Despite the problems connected with the classification and definition of terrestrial humic substances, the same terminology has been adopted for aquatic DOM. It must be emphasized that all classifications and definitions of humic substances—regardless of the nature of the sample (dissolved or solid matter)—are only operational, based on the procedures used for their isolation.

In water chemistry the term *aquatic humus* is very popular. However, this term is as indefinite as *humic solutes*, and it remains open what is dealt with—DOM, humic substances in full without partition, or something else. In addition to NOM and DOM, the literature of water chemistry contains numerous other terms and abbreviations for expressing the amount and type of organic matter in water. The term *particulate organic carbon (POC)* is used for the carbon that will be retained on the 0.45-μm-membrane filter. The carbon concentration of the filtered water (DOC, dissolved organic carbon) and the carbon concentration of the unfiltered water (TOC, total organic carbon) may be determined rather accurately by carbon analyzers. The terms *total organic matter (TOM), DOM,* and *particulate organic matter (POM)* are analogous to TOC, DOC, and POC. The former terms refer to the entire organic matter and include the contributions of elements such as oxygen, hydrogen, and nitrogen in addition to carbon. That means that the values of TOM, DOM, and POM are about two times higher than those obtained when only the carbon content is determined (the TOC, DOC, and POC parameters).

Humic solutes are present in water as a true solution (ionic or molecular), in a colloidal state, or as suspended particles. The filtered water phase has the main impact on the chemistry and biology of water. Thus, dissolved humic solutes, in molecular and colloidal states, form the material that has received the most interest. The DOM contains not only humic solutes but also some organic, nonhumic solutes. Since humic solutes contain a large number of diverse chemical functionalities, there is no unambiguous analytical method that reveals the actual concentration of the special humic solutes (humic substances) in the water sample.

Several textbooks (1–17) discuss isolation, fractionation, chemical characterization, etc., of natural organic matter obtained from different environments as well as thousands of conference reports and scientific papers written for making oneself familiar with the natural organic matter. One should keep in mind that water chemistry is a developing area that is continuously searching for new and more reliable methods to explain better the nature of aquatic dissolved matter. The common characteristic of organic solutes classified as aquatic humic substances, regardless of the isolation method adopted, is that they are somewhat acidic. Therefore it is often more illustrative to simplify a given approach to discuss organic acids in the aquatic environment instead of or parallel to aquatic humic substances. It is essential to keep in mind that aquatic humic substances are never equal to 100% of the DOM and do not account alone for every feature connected with the aquatic environment. Furthermore, no ideal system is available for the isolation of aquatic humic substances that would satisfy every scientist.

II. ANALYTICAL METHODS FOR THE DETERMINATION OF AQUATIC HUMIC SUBSTANCES

The major difficulty in aquatic humus chemistry is how to separate the humic substances selectively from other organic and inorganic solutes as they occur in their original state in the aquatic environment. Because of the dilute solutions of natural aquatic organic matter, they must be concentrated for further studies. Figure 1 gives an overall view of the traditional isolation and fractionation procedures for aquatic humic substances and offers a conceptual summary of their nomenclatures.

A. Pretreatment

Filtration of the TOM in the water sample into the POM and DOM fractions is extremely critical in the study of aquatic samples. Filtration through a 0.45-μm filter has been arbitrarily adopted as the standard procedure for separating dissolved and particulate components (18). However, the inadequacy of this pore size for removal of colloidal species has been strongly criticized. The use of 0.45-μm filters (usually depth-type filters) is evidently a compromise between flow rate

and rejection of clay minerals. Aquatic water samples have also been filtered successfully in humus chemistry (19,20) with the aid of polycarbonate filter cartridges (screen-type filter; very large capacity of the filtrate without clogging of the filter) with 0.2-μm cutoffs. All filters vary in uniformity of pore size, chemical composition, and flow characteristics (21). Some filters are better suited to the study of aquatic humic substances than others, and the use of an unsuitable filter produces very inaccurate results, e.g., Aiken in Ref. 6.

B. Concentration

All organic and inorganic constituents from the aquatic sample can be concentrated by vacuum distillation, freeze-drying, or freeze-concentration. These methods are very slow and, therefore, expensive and laborious. Furthermore, humic substances must be separated, in one way or other, from the bulk of organic and inorganic matter. Some of these methods (especially freeze-drying) have also been used in conjunction with other special concentration methods as the final step in the isolation of humic substances.

Reverse osmosis (a membrane process) is widely used in water purification systems, producing ultrapure water. Reverse osmosis concentrates all solutes except

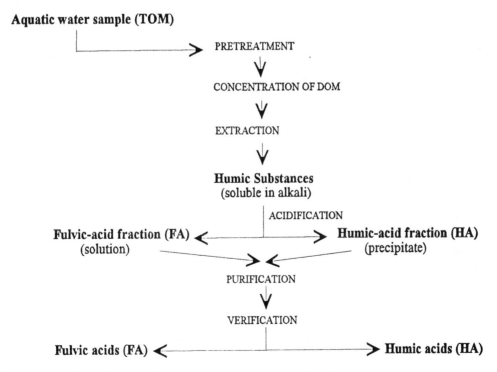

Fig. 1 Flowchart showing the isolation and nomenclature of aquatic humic substances.

certain organic compounds, such as phenols. More than 90% of the NOM present in a water sample can be retained via reverse osmosis (22). Large volumes of water (several hundred liters) can be processed without difficulty. However, reverse osmosis is an expensive and fairly equipment-intensive method; in addition, low-molecular-weight organic compounds and inorganic salts need to be separated, one way or other, from the bulk of the NOM to obtain humic substances.

Ultrafiltration concentrates dissolved matter via a membrane according to molecular size. In theory, aquatic humic solutes can be roughly separated with ultrafiltration from low-molecular-weight organic and inorganic solutes (23). The most commonly applied method (24–26) for separating humic solutes of larger molecular masses from those of smaller is to pump the water sample vertically through the membrane (batch operation under pressure). To avoid the fouling of membranes and membrane–solute interactions associated with the traditional batch process under pressure (25,27), the tangential-flow process (continuous operation) was applied (28–31) in aquatic humus chemistry. In addition to the fact that tangential ultrafiltration will optimize (23,28) a number of the problems connected with the conventional "dead-ended" filtration under pressure, large volumes of water (several hundred liters) can easily be processed for obtaining gram quantities of the solid NOM with different molecular masses, as was also the case with reverse osmosis. On the other hand, the equipment and membranes required for the powerful tangential-flow processing are fairly expensive. Furthermore, the separated humic-matter fractions with different molecular masses must be refined to obtain "pure" humic substances.

Various types of coprecipitation have been tried, during the period of humus chemistry, for concentrating and isolating humic substances using, e.g., iron, manganese, aluminium, and lead salts and calcium carbonate. Liquid extraction has also been used with some success to isolate humic substances from water. Column sorption techniques, such as alumina, nylon and polyamide powder, charcoal, and especially both weakly basic anion-exchange resins and nonionic macroporous sorbents, have proved to be quite effective. For a more detailed description about different concentration and isolation techniques, including their advantages and certain disadvantages, see Ref. 4, Aiken in Ref. 6, and Ref. 10. An illustrative description of aquatic humic substances based on different concentration and isolation methods has been compiled by Malcolm in Ref. 6. The history of research on aquatic humic substances has been divided into four periods: (i) prior to 1950, (ii) the awakening period (1950–1964), (iii) the Sephadex period (1964–1973), and (iv) the resurgence period (1973–present).

The noteworthy advantage of the column chromatographic methods is that they simultaneously concentrate and fractionate specific organic solutes (labeled as humic substances) from most other organic and inorganic constituents. Sorption of organic solutes onto a wide variety of different sorbents takes place via different mechanisms at given chemical conditions. Thus, organic matters retained on two different sorbents do not necessarily have the same chemical and physical properties, although they possess a somewhat acidic character. It must be emphasized again that the classification of the DOM into humic and nonhumic fractions is operational, based nowadays practically on adsorption chromatography and ion exchange. Besides, the chemical conditions prevailing in each isolation and fractionation procedure determine the degree of separation of humic from nonhumic substances. Thus, each isolation scheme with different modifications gives its own operational definition to the extract to be studied as humic substances. It is impossible to distinguish isolation procedures sharply from fractionation procedures, because most isolation procedures, such as adsorption chromatography, also partly fractionate aquatic humic substances.

At present the most powerful sorbents are apparently nonionic macroporous resins (such as the Amberlite XAD resins or analogous) and some weakly basic anion-exchange resins. The most popular isolation procedure is the application of nonionic macroporous sorbents (XAD series) at preadjusted acidity, and their use has become almost standard for isolating aquatic humic substances. However, the specific isolation of aquatic humic solutes from the water sample by nonionic sorbing solids at the preadjusted acidity has some risks. Based on the results of numerous isolations by sorbing solids at preadjusted acidities, it has been suspected (Frimmel in Ref. 11) that the striking similarity of many samples from different origins might be due to the method of isolation. Second, it has been argued (Shuman in Ref. 11) that the utilization of sorbing solids, such as XAD sorbents, alone may lead to serious errors in modeling the DOM, especially for acidity and metal complexation.

C. Extraction

Organic matter concentrated onto different sorbents is eluted with dilute bases (mostly 0.1 molar aqueous sodium hydroxide). This alkali-soluble organic fraction is currently defined as aquatic humic substances.

D. Acidification

The alkali-soluble extract (humic substances) is commonly acidified to pH ≈ 1 with concentrated hydrochloric acid and allowed to precipitate. Cooling the samples speeds up this fractionation procedure. Centrifugation is needed to separate the precipitated humic-acid-HA-fraction from the soluble fulvic-acid-FA-fraction.

E. Purification

This step usually refers to the removal of inorganic elements and some organic "impurities" (such as loose lipids, polysaccharides, amino acids, and fatty acids) by physical methods. Only poorly bound compounds are removed from humic extracts. Thus, "pure" humic substances or their HA and FA fractions contain variable amounts of known organic compounds as structural subunits.

F. Verification

Usually the simplest step here is to carry out some basic determinations, such as elemental composition, acidity, and some UV-visible spectra. More complicated and comprehensive analyses are required to investigate the chemical character of humic substances in detail.

III. ISOLATION BY CHROMATOGRAPHIC METHODS

A. Nonionic Macroporous Sorbents

The sorption onto the surface of a nonionic macroporous sorbent is based on hydrophobic properties of the organic matter studied. Nonionic macroporous copolymers classify organic solutes in a water sample at preadjusted acidity (ca. pH 2) into different artificial hydrophobic and hydrophilic fractions according to their ability to adsorb onto the nonionic macroporous sorbent in question. This operational classification of aquatic humic substances based on certain hydrophobic–hydrophilic interactions between organic solutes and the sorbing solid under the preadjusted conditions is in principle clear and thoroughly discussed elsewhere, e.g., Ref. 4 and Aiken in Ref. 6. It is highly notable that this hydrophobic–hydrophilic distinction is somewhat artificial, because to be dissolved, the organic matter must in reality be quite hydrophilic anyhow. In the case of an aquatic organic solute the hydrophobic sorption onto the surface of the solid sorbing particle is a function of the acidity (pH) of the solution. The operational classification of aquatic humic substances based on certain hydrophobic–hydrophilic interactions between organic solutes and solid sorbent is visualized diagrammatically in Fig. 2.

For organic solutes without any ionizable functional groups (noncharged), the acidity of the solution has no significant effect on the adsorption. To adsorb aquatic organic acids onto the sorbing solid, the pH of the solution should be (4) two pH units below their lowest

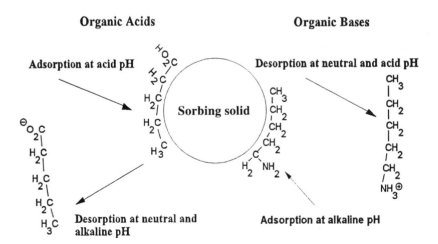

Fig. 2 Effect of acidity (pH) on hydrophobic–hydrophilic interactions of organic acids and bases with a particular sorbing solid.

acidity constant (pK$_a$). In contrast, for efficient desorption, the pH of the solution should be two pH units greater than the highest pK$_a$. It has been estimated (4) that pK$_a$ values of organic acids in natural waters range from ca. 1.2 to 13, the lowest averaging at ca. 4.2. At acidic conditions, for example, carboxylic groups are nonionic (i.e., –COOH). Therefore, the acidic organic solutes can with the aid of their certain hydrophobic carbon skeletons coat the surface of the sorbing solid (adsorption). At acidic conditions, organic bases cannot coat the surface of the sorbing solid, since the neutral forms of the basic functional groups (e.g., –NH$_2$) are now protonated (e.g., –NH$_3^+$). Acidic functional groups of the organic solutes retained onto the sorbing solid ionize in alkaline conditions (e.g., COO$^-$) and accordingly organic acids begin to elute into the solution (desorption).

It has been stated (4) that carboxyl- and hydroxyl-rich organic matter with the mole ratios of acidic functional groups, e.g., COOH/C of less than ca. 1:3, has a low affinity to sorbing solids (this organic matter is relatively too hydrophilic) and will pass through the column. It has also been postulated that the organic matter retained onto the sorbing solid with the mole ratio COOH/C of more than ca. 1:12 will not be eluted from the sorbing solid by the base (this organic matter is relatively too hydrophobic). It is possible to elute this organic fraction from the sorbing solid with organic solvents (e.g., methanol). It is essential to note that organic matter (macromolecular organic acids) of the DOM retained at a preadjusted pH of 2 onto the given sorbing solid and eluted with the base at pH 13 is, worldwide, adopted as aquatic humic substances. Figure 3 summarizes the analytical procedure that par-

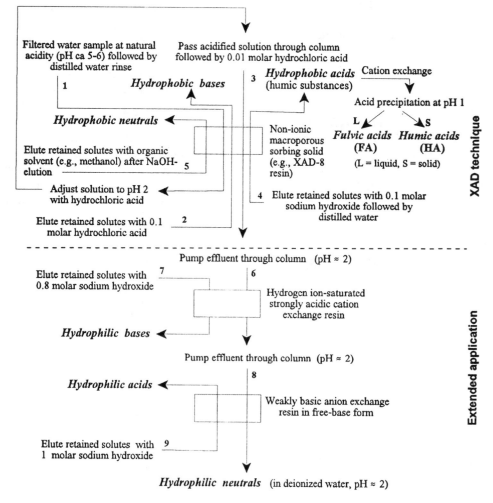

Fig. 3 Analytical procedure for classification of the DOM into different artificial hydrophobic and hydrophilic groups.

titions the DOM at preadjusted acidity into artificial hydrophobic and hydrophilic fractions.

Briefly, the strategy of the so-called XAD technique developed to isolate aquatic humic substances by nonionic sorbing solids is to (32–35; Aiken in Ref. 6): (i) acidify the filtered (mostly 0.2–0.45 μm) water sample to ca. pH 2 with concentrated hydrochloric acid (HCl); (ii) pass it through a column of the XAD sorbent; (iii) elute the retained organic matter with 0.1 M NaOH; (iv) acidify it with a strongly acidic cation-exchange resin; and (v) freeze-dry the isolated products. The organic matter eluted by base at pH 13 from the sorbing solid is called humic substances (so-called hydrophobic acids). It is also possible to elute the sorbing solid by pH-gradient desorption. For instance, the fraction eluted at pH 8 contains more free carboxylic groups than that eluted at pH 13 (phenolic fraction). Nonionic macroporous and cation-exchange resins separate, to some extent, certain loosely (physically) bound compounds, e.g., carbohydrates and amino acids and their derivatives from humic substances. Organic matter eluted from the XAD sorbent with the base at pH 13 (humic substances) is usually further fractionated with the aid of concentrated HCl at strongly acidic conditions into so-called humic acids (HA) and fulvic (FA) acids.

The exact mechanism of the XAD isolation is not fully known, and the chemical-physical properties of the different kinds of nonionic sorbents available have a very powerful influence on, e.g., the kinetics of sorption and irreversible sorption. The molecular size ranges of the retained humic solutes of interest are also strongly dependent on the chemical-physical nature of the nonionic sorbents in question. It has been discovered (29,31) by means of ultrafiltration that isolation–fractionation of aquatic humic substances with the XAD technique (more accurately XAD-8 sorbent) appears, under the same conditions, to occur with nearly the same mechanism, independent of the molecular size or of the degree of polydispersity of the solutes, though the adsorption apparently is more effective when the solution is less polydispersive. This phenomenon supports the opinion that isolated humic substances must play a role as certain definite entities in the DOM and that they cannot be mere accidental products of the isolation processes.

The regulated pH range (within the values of 1–13) of the water sample used for the XAD technique is very wide. This is a major disadvantage for teaching the method. Furthermore, the adsorption forces involved with these nonionic sorbents are primarily of the very weak and unspecific van der Waals type. Many organic acids present in the DOM, such as volatile fatty acids and hydroxy acids as well complex acids containing many hydroxylic and carboxylic groups, do not absorb fully onto nonionic sorbing solids, even if the acidity of the solution is lowered to pH 2. In Fig. 3, those organic acids (so-called hydrophilic acids) that are not retained onto the sorbing solid have been isolated by a weakly basic anion-exchange resin. Altogether, it is possible to obtain six different hydrophobic–hydrophilic fractions from the original DOM. It is notable that many natural waters do not contain so-called hydrophobic basic solutes, and after prefiltration the water initially can be acidified to pH 2 (Fig. 3, step 3) and passed through the three-column system without sample recycle through the sorbing solid column at natural acidity. However, with certain waters, such as wastewaters, it is worthwhile first to separate weak phenolic acids (fictional hydrophobic bases) from strong carboxylic acids. Hydrophilic basic solutes retained onto the strongly acidic cation-exchange resin at pH 2, most likely amphoteric proteinaceous and amino constituents of the DOM (33), are not so important in freshwater ecosystems (32,36–40). Only the neutral, so-called hydrophilic organic species, such as simpler sugars, two- and three-carbon alcohols, and ketones, will come through all three columns connected in sequence (Fig. 3: nonionic sorbing solid → cation exchanger → anion exchanger).

Nonionic polymeric adsorbents, such as the Amberlite XAD series, do not carry any ion-exchange groups. These sorbing solids are hard, insoluble spheres of porous polymers of high surface area, and are available in a variety of polarities and surface characteristics (41). Some typical properties of Amberlite polymeric adsorbents are shown in Table 1: Acrylic-ester sorbents are more hydrophilic, wet more easily, and adsorb more water than styrene-divinylbenzene sorbents. It has been stated (32,40; Aiken in Ref. 6) that hydrophobic styrene-divinylbenzene sorbents were found more difficult to elute than hydrophilic acrylic-ester sorbents arising from hydrophobic interactions and possibly from the occurrence of strong π–π bonding between the adsorbed organic solutes and the aromatic matrix of the sorbing solid. For nearly 20 years, XAD-2, XAD-4, and XAD-8 have been the most applied sorbing solids for the isolation of aquatic humic substances from different environments. XAD-2 has been widely used for the isolation of humic substances from seawater. XAD-2 has also been used for the isolation of humic substances from freshwater, surface water, and groundwater. XAD-4 has been applied for the isolation of humic substances from tap water. It can roughly be said that the molec-

Table 1 Typical Properties of Amberlite Polymeric Adsorbents

Sorbing solid	Chemical nature	Porosity (vol. %)	True wet density (g/cc)	Area (m²/g)	Average pore diameter (Å)	Skeletal density (g/cc)	Nominal mesh sizes
			Nonpolar				
XAD-1	Polystyrene	37	1.02	100	200	1.07	20–50
XAD-2	Polystyrene	42	1.02	330	90	1.07	20–50
XAD-4	Polystyrene	51	1.02	750	50	1.08	20–50
			Intermediate polarity				
XAD-7	Acrylic ester	55	1.05	450	80	1.24	20–50
XAD-8	Acrylic ester	52	1.09	140	250	1.23	20–50
			Polar				
XAD-9	Sulfoxide	45	1.14	250	80	1.26	20–50
XAD-11	Amide	41	1.07	170	210	1.18	16–50
XAD-12	Very polar	45	1.06	25	1300	1.17	20–50

ular weight of the aquatic humic solutes retained on the sorbent increases in the following order: XAD-4 → XAD-2 → XAD-8. It has been commented (42) that XAD-8 can be substituted by Supelite DAX-8 (Sigma Chemical Co., Poole, Dorset, England), and that the technical specifications of DAX-8 differ very little from those of XAD-8.

It is possible to increase the assortment of humic-matter fractions (humic substances) with the aid of different XAD-sorbent columns connected in sequence. The first column may be XAD-8/DAX-8 sorbent that is specific for larger-molecular-weight humic solutes, and the second may be XAD-4 sorbent that will retain smaller-molecular-weight humic solutes. This kind of isolation procedure has recently been carried out (43,44) using columns of XAD-8 and XAD-4 in sequence at the acidity of pH 2. The organic matter retained at pH 2 onto the XAD-8 sorbent was labeled, as usual, humic substances (representing artificial hydrophobic acids). Conversely, the organic matter retained (at pH 2) from the effluent of the previous XAD-8 column onto the XAD-4 sorbent was labeled, in this instance, "XAD-4 acids" and classified into the category of artificial hydrophilic acids.

However, this kind of adoption of new definitions/terms is an example of the difficulty in defining and classifying aquatic humic substances. The adsorption of organic matter onto the XAD-4 sorbent takes place by exactly the same mechanism as onto the XAD-8 sorbent, caused mainly by hydrophobic–hydrophilic interactions between the resin and solutes at the given acidity (i.e., organic solutes that are hydrophobic enough relative to the sorbing solid will adsorb).

It is generally agreed upon in the humic literature (33) that the DOM retained onto nonionic sorbing solids at a given acidity by single or combined sorbent systems, independent of their chemical-physical nature, is defined and classified as humic substances or hydrophobic acids. Analogously, the effluent of different nonionic sorbing solids is generally called the hydrophilic fraction (acids, bases, and neutrals). Although the categorization of the DOM into humic and nonhumic substances based on hydrophobic–hydrophilic interactions between a given sorbing solid and the DOM is fairly arbitrary, and there is no guarantee that the DOM retained, e.g., onto the XAD-8 sorbent represents solely aquatic humic substances, it is very important to append the acknowledged classification to avoid unnecessary confusion. The most critical factors in defining aquatic humic substances are concentration, fractionation, and purification. Therefore, the question "Does a distinct humic molecule exist?" might never be answered in an exact chemical sense.

B. Ion-Exchange Chromatography

Anion-exchange resins have also been applied to the isolation of aquatic humic substances (macromolecular organic acids). There are two types of anion-exchange resins: strongly basic and weakly basic ones. The resin matrix in both groups is usually crosslinked polystyrene or polyacrylic acid, as shown in Table 2. The basic nature of the resin depends chiefly on the active functional groups and also on their location. For instance, a nuclear amine group will confer a less basic character than a similar group in a side chain. The weakly basic

Table 2 Nature of Anion-Exchange Resins

Type of resin matrix	Functional group(s)	Characteristic
Crosslinked polystyrene or polyacrylic acid	RNR_3^+ OH^-	Strongly basic
Crosslinked polystyrene or polyacrylic acid	RNH_2, $RNHR_1$, RNR_1R_2	Weakly basic
Phenolic	ROH, RNH_2, $RNHR_1$, RNR_1R_2	Weakly basic

R = resin matrix. Materials based on different kinds of cellulose matrix are also available, consisting of functional groups bound to microgranular or fibrous cellulose. $-NH_2$ = primary amine. $-NHR_1$ = secondary amine. $-NR_1R_2$ = tertiary amine. R_1, R_2, and R_3 = substituents attached to amine groups.

exchangers contain amine (NH_2) or its mono- or disubstituted derivatives as functional groups, while the quaternary ammonium group ($-NR_3^+OH^-$) produces a strongly basic resin, comparable in strength to the caustic alkalis. Anion-exchange resins have high capacities for aquatic organic acids. High quantities (80–90% of the DOM) of organic acids have been obtained (37) from freshwater samples by anion-exchange columns, though the application of strongly basic anion-exchange resins bearing quaternary ammonium groups is not recommended (Aiken in Ref. 6) because of the many serious disadvantages, such as irreversible sorption, among other things. Interactions of organic solutes with polystyrene resin matrix (as is also the case with certain nonionic sorbing solids) can also cause elution problems.

Because aquatic humic substances (HS) contain numerous functional groups (the most important, especially from the isolation point of view, are free carboxylic acid and phenolic hydroxyl groups), a chromatography based on the types and amounts of these functional groups is a logical choice. An interesting resin for this purpose is a weakly basic anion exchanger with secondary or tertiary amine groups. The weakly basic anion-exchange resins are active at pH value ca. 6 or below. At this acidity the amine groups of the resin contain positive charges and the resin operates as an anion exchanger. At pH ca. 7 or above, the amine groups lose their positive charges and the resin turns into its neutral form. Thus, the charge and ion exchange ability of weakly basic resins is a function of the acidity. Weakly basic resins can be used only for organic (e.g., HS—COO^-) or inorganic anions ($-X^-$) in neutral or acidic solutions, having little or no exchange capacity under alkaline conditions.

Another important fact is that the functional amine groups of weakly basic resins are highly selective for phenolic functional groups of the solute. There exists a strong interaction between the neutral form of the amine groups and the phenolic hydroxyl groups of the solute. Weakly basic resins retain phenolic functional groups of the solute at pH 7–8. At these pH values, amine groups of the resin are not charged (neutral form) and phenolic parts of the HS solute will attach to the resin: RR_1HN—HO–Ar–HS-COO^-, (Ar = aromatic moiety in HS). Organic solutes also contain carboxylic acids (HS–COO^-H^+) as ionic functional groups. These acidic groups do not interact with amine groups of the resin at the pH range ca. 7–8, or if they do, only slightly (4). Elution of the organic matter retained onto weakly basic resin, at pH 7–8, can be carried out as a pH gradient. Elution at pH 8 will give a fraction relatively high in carboxylic acid and low in phenolic groups. Elution at pH 13 generates a fraction that is high in phenolic and low in carboxylic acid groups: RR_1HN + ^-O–AR–HS-COO^-.

It is possible to isolate almost all organic acids from fresh water without any pH adjustment of the original water sample using weakly basic anion exchangers. The most popular product for this purpose is the DEAE-cellulose. DEAE-cellulose (diethylaminoethyl-cellulose: $-OC_2H_4N(C_2H_5)_2$) is a weakly basic anion exchanger with tertiary amine functional groups bound to a hydrophilic matrix. The quantities of ionized organic solutes isolated with the DEAE procedure have generally been relatively great in freshwater: ca. 80% to nearly 100% of the DOM (32,40,45–49). It has been stated (46) that the optimum recovery of organic acids of the DOM occurs within the acidity of pH 4–6. At more acidic conditions, the recovery will be strongly decreased because the protonated organic acids are no longer anionic and hence are not retained by the DEAE-cellulose. Accordingly, at basic conditions the efficiency also decreases because of neutralization of the amino moieties on the DEAE-cellulose. The retention of organic solutes to the DEAE-cellulose is probably based on charge rather than hydrophobic interactions. However, it is possible that the mechanism of sorption of organic anions onto the DEAE-cellulose also includes hydrogen bonding through phenolic hydroxyl groups of organic solutes. Inorganic anions, such as chloride and bicarbonate, are not significantly concentrated by this method, and thus the DEAE-cellulose serves both to concentrate and to separate organic acids from most other anions, cations, and neutral species in natural freshwaters.

The DEAE procedure efficiently concentrates and isolates macromolecular organic acids from natural freshwaters. On the other hand, it has been discovered (50) by studying marine aquatic humic substances that even the relatively low salinity of brackish water will decrease the retention of acidic solutes, apparently of those with lower molecular sizes, onto the DEAE cellulose. Many chemical and physical characteristics support the assumption (32,40) that the structural composition of macromolecular organic acids (humic solutes) isolated by the DEAE-cellulose at natural acidity is likely a certain average combination of the four different acidic fractions isolated at preadjusted acidity by the XAD technique [i.e., so-called (i) "hydrophobic" fulvic acids (FA) and (ii) humic acids (HA), (iii) "hydrophobic neutral" solutes (also slightly acidic character), and (iv) "hydrophilic" acids]. Besides, the quantity (ca. 80% of the DOC) of organic acids retained onto the DEAE-cellulose was practically equivalent to the total amount of the various organic acids obtained by the XAD technique in connection with different adsorbents. This points out that the DEAE procedure also isolates, in addition to natural complex macromolecular acids (classical humic substances), the main part of the colorless low-molecular-weight specific organic acids from the freshwater sample. Thus, it is highly probable, that the DEAE procedure isolates more representative macromolecular organic acids as "real" humic substances than the XAD technique does. It is possible to fractionate the acidic organic matter obtained by the DEAE procedure, e.g., with the aid of the XAD technique at acidic conditions (pH 2 and 1) into so-called HA- and FA-type fractions (49). However, the exact chemical meaning of such secondary fractionation at some arbitrarily chosen acidity is unclear.

Cation-exchange resins remove inorganic elements from the organic extracts and saturate the organic acids with hydrogen atoms. This is the main use of cation exchangers. Cation-exchange resins also remove trace metals and other cations associated with organic solutes. Any amines or amino acids associated with organic solutes are retained onto cation-exchange resins. These are the usual applications of cation-exchange resins for isolation–fractionation procedures of the DOM in the context of humic substances.

IV. STRUCTURAL CHARACTERIZATION OF HUMIC SUBSTANCES

A. Functional Groups of the DOC

Isolation–fractionation of aquatic humic substances is based chiefly on different functional groups of dissolved organic compounds. The chemical and physical properties of organic solutes are closely related to the major functional groups. Functional groups are generally classified (4) according to their acidic, basic, and neutral properties. This refers to the ability of the functional group to donate or accept a proton in water. Neutral functional groups neither donate nor accept a proton. Table 3 shows different kinds of acidic, basic, and neutral functional groups and the approximate types of organic compounds containing these functional groups.

In addition to chemically active functional groups of various natures, the macromolecular structure of humic solutes comprises aromatic and heterocyclic building blocks that are randomly condensed or linked by aliphatic, oxygen, nitrogen, or sulphur bridges. Macromolecular humic solutes also contain hydrophilic as well as hydrophobic sites. Surface activity is a very important property of aquatic humic substances promoting interactions, especially with hydrophobic organic pollutants. Different binding forces and the types of mechanisms (including ionic, hydrogen, and covalent bonding, charge-transfer or electron donor–acceptor mechanisms, Van der Waals forces, ligand exchange, and hydrophobic bonding or partitioning) in the adsorption processes of several pesticides onto aquatic humic substances operate mostly simultaneously. The nature of interactions between organic pollutant chemicals and dissolved humic substances has been thoroughly discussed elsewhere (51; Senesi in Ref. 16).

B. Quantities of Organic Solutes Obtained by the XAD Technique

Aquatic humic substances form only part of the DOC in water and are never equivalent to the total DOC. In uncolored freshwater streams, aquatic fulvic acids (FA) and humic acids (HA) obtained by nonionic sorbing solids (like XAD resins) commonly account for approximately 40 and 45%, respectively, of the DOC content of 3–6 mg C/L (e.g., 35), and the ratio of FA to HA is commonly 9:1. In organically colored waters, common to Nordic countries, Northern Russia, and Canada, humic substances as the percentage fraction of the DOC increase with increasing DOC concentrations, and may even account for 60–80% of the DOC. In

Table 3 Important Functional Groups of Dissolved Organic Carbon

Functional group	Structure	Where found
	Acidic groups	
Carboxylic acid	(Ar-)R-CO$_2$H	90% of all dissolved organic carbon
Phenolic OH	Ar-OH	Aquatic humic substances, phenols
Enolic hydrogen	(Ar-)R-CH=CH-OH	Aquatic humic substances
Quinone	Ar=O	Aquatic humic substances, quinones
	Basic groups	
Amine	(Ar-)R-CH$_2$-NH$_2$	Amino acids
Amide	(Ar-)R-C=O(-NH-R)	Peptides
Imines	CH$_2$=NH	Humic substances (unstable, forming polymeric derivatives)
	Neutral groups	
Alcoholic OH	(Ar-)R-CH$_2$-OH	Aquatic humic substances, sugars
Ether	(Ar-)R-CH$_2$-O-CH$_2$-R	Aquatic humic substances
Ketone	(Ar-)R-C=O(-R)	Aquatic humic substances, volatiles, keto-acids
Aldehyde	(Ar-)R-C=O(-H)	Sugars
Ester, lactone	(Ar-)R-C=O(-OR)	Aquatic humic substances, hydroxy acids, tannins
Cyclic imides	(R-)O=C-NH-C=O(-R)	Aquatic humic substances

R is aliphatic backbone and Ar is aromatic ring. In addition to certain amino acids sulphur (S) may also occur as mercapto/thiol compounds (-SH) and sulfonic acid derivatives (Ar-)R-SO$_3$H. It is notable that the backbone may be R or Ar but mostly it is their combination.

highly colored waters, the HA fraction consists of a much higher fraction of the DOC (e.g., 20) and the ratio of FA to HA will decrease to 4:1 or less.

Figure 4a shows the average distributions of organic constituents, based on classical splitting of the DOC into different artificial hydrophobic–hydrophilic groups, of approximately 100 surface water samples from the United States (35). Figure 4b shows the corresponding representative distribution obtained (32,52) for Finnish highly colored brown-water lakes (DOC ca. 19 mg C/L). Figure 4c shows, for comparison, the hydrophobic–hydrophilic DOC splitting obtained for approximately 25 drinkable groundwater samples (DOC ca. 1 mg C/L or less) collected from aquifers in the United States (35). It is notable that none of these three examples contain so-called hydrophobic basic solutes. Furthermore, the quantities of the average hydrophobic–hydrophilic distributions of the DOC obtained for river and lake waters, originating from geographically different environments, are practically identical. On the other hand, the hydrophobic–hydrophilic distribution as part of the total DOC obtained for groundwater was completely different, especially in the case of so-called hydrophobic and hydrophilic acids and hydrophilic neutrals, from that of surface water samples, forming its own environmentally special category.

The DOC concentration of deep uncontaminated ocean waters has generally been reported (35) to be nearly 1 mg C/L. The concentration of marine aquatic humic substances, which are primarily of the FA type, is 10–20% of the DOC. However, the precise splitting of the marine DOC into humic solutes (substances) or more specifically into different artificial hydrophobic–hydrophilic groups is not so straightforward as in the case of terrestrial waters, being caused by, among other things, the salinity of the water and the specific nature of the DOM (35,50,53; Harvey and Boran in Ref. 6). The quantity of the low-molecular-weight hydrophilic acid fraction is especially strongly method dependent, remaining most frequently underestimated.

C. Characterization of Aquatic Humic Substances

Humic substances are heterogeneous compounds forming extremely complicated mixtures of hundreds or even thousands of organic constituents without exact chemical structure. Thus, all analytical approaches applied to the investigation of the chemical character of humic substances are at best only semiquantitative and contain more or less severe pitfalls, e.g., Frimmel in Ref. 11. Nearly every technique available to the analytical chemist has been utilized to explain the complex properties and behavior of humic substances. Some of the more widely applied methods are listed in Table 4 and described in detail in Refs. 6 and 9. Although many

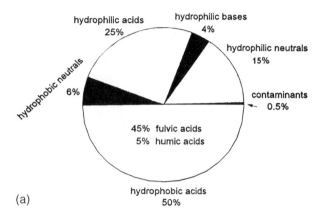

(a)

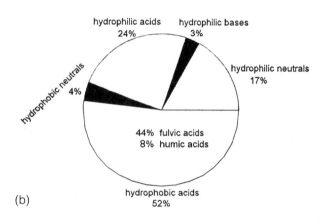

(b)

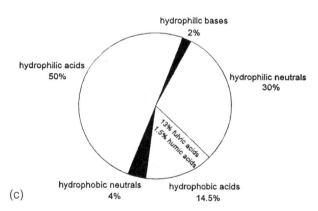

(c)

Fig. 4 (a) Representative distribution of surface water DOC in rivers of the United States. (b) Representative distribution of lake water DOC in highly colored Finnish lakes. (c) Representative distribution of groundwater DOC in aquifers of the United States.

analyses carried out on aquatic humic substances are typical of fundamental structural chemistry, the compilation demonstrates that modern humic-substances research not only is directed to unraveling their ill-defined chemical structure, but is broadly exploring the interconnected chemical, biological, and physical processes that maintain the ecological equilibrium. This equilibrium is closely related, especially in the case of freshwater systems, to the soil chemistry and agricultural production, as stated in Ref. 54.

Carboxylic acid groups are without doubt the most important functional groups in natural organic matter, because they contribute chiefly to the solubility (aqueous) and acidity of an organic constituent. Analysis of functional groups such as carboxyl, phenol, and methoxyl, in addition to, e.g., elemental analysis and nominal molecular weight determinations, increases the understanding of the chemical character of humic substances. Table 5 shows average representative values of some informative variables determined for FA- and HA-type substances (hydrophobic acids) isolated–fractionated with the XAD technique from stream water (Malcolm in Ref. 6) and lake water (20,30,32). A close similarity exists between the basic characteristics determined for humic solutes isolated from river and lake waters in geographically different environments.

Nondegradative approaches applied to humic substances are based on the basic idea that all analyses are performed without any chemical alteration. Spectroscopic techniques such as NMR (nuclear magnetic resonance spectroscopy) of proton (liquid state) and carbon-13 (liquid or solid state) will give very helpful information about the chemistry of humic substances. From NMR spectra it is possible to estimate carbon and hydrogen (nonexchangeable) distributions of humic substances. Figures 5a and b show the average representative ^{13}C-NMR [solid state; CPMAS, cross-polarization magic angle spinning) and ^{1}H-NMR (liquid state, in deuterium oxide (D_2O)] spectra for humic acids (HA) and fulvic acids (FA) isolated–fractionated with the XAD technique from some organically colored Finnish lakes (20). One should not forget that due to the very heterogenous macromolecular nature of various humic substances, the nuclei of NMR spectra can experience a wide variety of chemical environments, producing a wide variety of chemical shifts. Therefore, it is difficult to define exactly specific resonance regions for humic substances, since different types of carbon atoms and protons can resonate within a wide shift range, producing great amounts of overlaps.

In ^{13}C-NMR spectra, the region 0–50 ppm (ppm = chemical shift, δ, as parts per million downfield from

Table 4 Methods Applied Most Widely to the Analysis and Characterization of Humic Substances

Molecular weight determination	Functional group characterization	Binding studies
Viscosity	Fourier transform infrared spectroscopy	Cation exchange
Vapor pressure osmometry	^{13}C and ^{1}H (solid and liquid state) nuclear	Fluorescence
Ultracentrifugation	magnetic resonance	UV-visible spectroscopy
Gel filtration, high-performance liquid	Electron spin resonance	Dialysis
size-exclusion chromatography	Pyrolysis–gas chromatography	Potentiometric titration
Laser light scattering	Pyrolysis–mass spectrometry	
Field desorption mass spectrometry	Pyrolysis–Fourier transform infrared spectroscopy	
	pH titration	
	Electrophoretic techniques, capillary zone	
	electrophoresis	

the internal tetramethylsilane reference resonance) consists primarily of aliphatic (methyl, methylene, and methine carbons) carbon resonances (31,40,53,55–59; Malcolm in Ref. 9). The specific region 50–60 ppm consists, e.g., of resonances of the methoxyl carbons (–OCH$_3$). The region 60–90 ppm is believed to be due primarily to C–O, because the nitrogen content of

aquatic humic-matter fractions is invariably low. Although carbohydrate-type compounds resonate within this region, other C–O resonances, such as those of ethers, also fall in this region. The region 90–110 ppm consists of resonances of the dioxygenated carbons (e.g., anomeric carbons of polysaccharides). The full range 60–110 ppm has generally been supposed to rep-

Table 5 Characterization of Stream and Lake Hydrophobic Acids

Fulvic acids (FA)					
	Stream	Lake		Stream	Lake
1. Elemental analysis (in percent on a moisture- and ash-free basis)					
Carbon =	54.56	54.34	Nitrogen =	0.87	0.74
Hydrogen =	4.97	3.97	Sulphur =	0.74	0.60
Oxygen =	38.24	40.35	Phosphorus =	0.62	
Ash-% =	0.86	1.9			
2. Acidic functional groups (exchangeable hydrogens in meq/g)					
COOH (titration) =	6.4	5.2	Phenolic OH (titration) =	1.6	1.5
3. Molecular weight =	650–950	660–1100			

Humic acids (HA)					
	Stream	Lake		Stream	Lake
1. Elemental analysis (in percent on a moisture- and ash-free basis)					
Carbon =	55.94	55.89	Nitrogen =	1.27	1.50
Hydrogen =	4.13	3.91	Sulphur =	0.93	0.88
Oxygen =	37.48	37.82	Phosphorus =	0.25	
Ash-% =	1.13	6.1			
2. Acidic functional groups (exchangeable hydrogens in meq/g)					
COOH (titration) =	4.7	3.9	Phenolic OH (titration) =	1.9	2.0
3. Molecular weight =	2000–3000	1100–2000			

A "good" aquatic humic substance sample should have less than 2% ash (Frimmel in ref. 11). The content of oxygen is taken as the difference from 100%. Molecular weight represents the number-average value of distribution, from a few hundred up to several hundred thousand measured by vapor pressure osmometry.

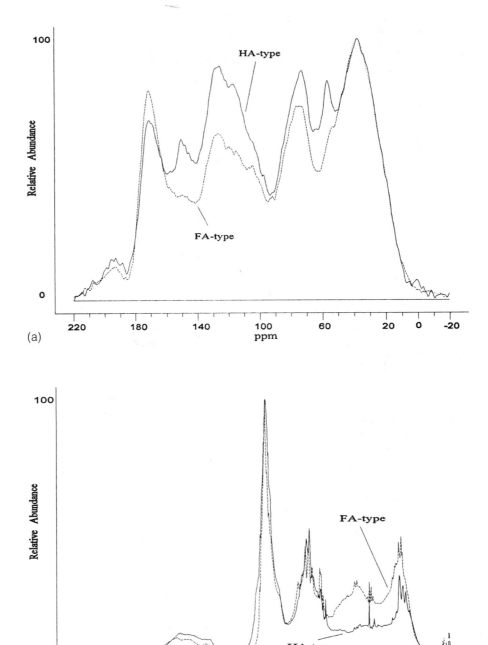

Fig. 5 (a) Representative ^{13}C-NMR (solid) spectra for lake humic acids (HA) and fulvic acids (FA). (b) Representative ^{1}H-NMR (liquid) spectra for lake humic acids (HA) and fulvic acids (FA).

resent chiefly carbohydrate structures. However, it is quite certain that the majority of the carbons in the 60–110-ppm region does not consist merely of "normal" polysaccharide material. It has been assumed (60) that only a minor portion of the ca. 60–92-ppm region is

due to carbohydrates, and that, e.g., aliphatic ethers and oximes make up the major part. On the whole, the region ca. 50–110 ppm corresponds to carbons bonded to an electronegative atom (O, N, Cl, etc.). Furthermore, accurate integration of the "anomeric peak"

(90–110 ppm) is complicated. It has been argued (60) that 50% of the intensity in the ca. 95–110-ppm region should be assigned to anomeric carbons and the rest to aromatic carbons. The aromatic and unsaturated carbons resonate within the range ca. 110–160 ppm, which can be subdivided into two specific categories. The first region, 110–140 ppm, is assigned to the resonances of unsubstituted and alkyl-substituted aromatic carbons: (a) protonated aromatic carbons (aryl H) resonate within the range ca. 110–120 ppm; (b) aromatic carbons ortho to the oxygen-substituted aromatic carbons resonate within the range ca. 118–122 ppm; (c) unsubstituted and alkyl-substituted aromatic carbons resonate within the range ca. 120–140 ppm. The second region, 140–160 ppm, in turn consists of the resonances of the aromatic carbons substituted by oxygen and nitrogen (e.g., phenols, aromatic ethers, or amines). The region 160–190 ppm largely represents resonances due to carboxyl carbons. The lowest field region, 190–220 ppm, is characteristic of the carbonyl carbons (aldehyde and ketone carbons).

In ^{1}H-NMR spectra, the small resonance at ca. 8.3 ppm (δ ppm; relative to tetramethylsilane) occurring in all spectra arises (56) possibly from the proton of the formate ion ($H-COO^-$), a decomposition product of humic materials dissolved in deuterated sodium hydroxide in deuterium oxide ($NaOD-D_2O$). The sharp absorptions sometimes found around 0.9, 1.2, 1.9, and 3.5 ppm were probably due to the presence of some external impurities. The region ca. 0.2–1.5 ppm (31,40,53,56,61–64) consists chiefly of methyl and methylene protons of carbons bound directly to other carbons. The region ca. 1.5–3 ppm is assigned to methylene and methine protons α to aromatic rings, or carboxyl and carbonyl groups. The full region, ca. 0.2–3 ppm, represents aliphatic protons. The region ca. 3–4 ppm consists chiefly of protons attached to carbon atoms bound to oxygen. This region is attributed (64) to methoxyl and carbohydrate protons, the most likely alternatives (61) for this region being carbohydrates, ether linkages, amino acids, and peptides. The broad resonance in the range ca. 4–5.5 ppm, caused by the proton of the partly deuterated water molecule (HDO) present in the solvent system (D_2O) as an impurity ($H_2O + -OD = HDO + -OH$) or by reactions of humic substances (HS) with sodium deuteroxide (HS + NaOD = NaS + HDO, NaS is the sodium salt of HS), is removed from the spectra of Figure 5b. The region ca. 5.5–8 ppm is assigned to the protons attached to olefinic and aromatic carbons. It has been stated (62) that most olefinic hydrogens of humic substances resonate within the range 5–7 ppm. The aliphatic region, ca. 0.2–3 ppm, can be subdivided (64) into three categories (the aliphatic protons γ, β, and α to aromatic carbon rings), even though it is difficult to integrate these ranges separately. Furthermore, in the literature (64), the zone 4.0–5.5 ppm is attributed to lactone protons β to aromatic carbon rings (solvent deuterium dimethylsulfoxide, $DMSOd_6-100\%$ d). This region also contains other chemical shifts of protons on attached carbons, e.g., to two heteroatoms (primarily oxygen, cf. region 3–4 ppm above) in some special structural constituents.

Table 6 shows carbon and hydrogen (nonexchangeable) distributions for the stream and lake FA- and HA-type substances characterized in Table 5. Integration of the peak areas for assessing the relative contents (%) of different types of carbon atoms in ^{13}C-NMR (solid) and hydrogen atoms in ^{1}H-NMR (liquid) spectra is based on the assumption of straight baselines in areas of spectral intensity. A similar resemblance also exists between the different kinds of carbon distributions estimated for humic materials of river and lake waters originating from geographically different environments, as noted in Table 5.

Tables 5 and 6 indicate certain differences between average basic properties of HA- and FA-type humic fractions. It is noteworthy that the differences in elemental compositions (carbon, hydrogen, oxygen, and nitrogen) are fairly minor. The FA is slightly more polar than the HA, and the carbon content is accordingly a little lower. It can also be concluded that the HA is formed from larger constituents than the FA and is also more heterogeneous. These two practically physical features differentiate most clearly the HA and the FA. The carbon distribution indicates that the contents of aliphatics hydrocarbons and carbohydrates were somewhat greater in the FA than in the HA. Aliphatic hydrocarbons of the FA consisted of more terminal methyl groups and methylene chains than those of the HA. Carbohydrates of the HA consisted of more nitrogen-containing derivatives than those of the FA. The HA was more aromatic than the FA; however, the degree of substitution of aromatic rings was greater for the FA than for the HA. The overall relative content of lignin-related units (both phenols and methoxyphenols) was greater in the HA than in the FA, and more than approximately half of the aromatic part of the HA consisted of lignin-related units. In sum, the differences in chemical characteristics for the HA and FA are unquestionable but relatively small.

There is a lot of other nondegradative methods applied to the study of humic substances, e.g., electron spin resonance (ESR) spectroscopy for the study of free

Table 6 Carbon and Proton Distributions of Stream and Lake Hydrophobic Acids

Fulvic acids (FA)					
	Stream	Lake		Stream	Lake
1. Carbon distribution by solid-state ^{13}C-NMR (in percent)					
0–50 ppm =	36	35	110–140 ppm =	12	13
50–60 ppm =	8	5	140–160 ppm =	5	5
60–90 ppm =	16	19	160–190 ppm =	16	12
90–110 ppm =	3	8	190–220 ppm =	4	3
2. Hydrogen (non-exchangeable) distribution by liquid-state ^{1}H-NMR (in percent)					
0.2–1.5 ppm =		31	3–4 ppm =		27
1.5–3 ppm =		29	5.5–8 ppm =		13

Humic acids (HA)					
	Stream	Lake		Stream	Lake
1. Carbon distribution by solid-state ^{13}C-NMR (in percent)					
0–50 ppm =	23	26	110–140 ppm =	21	20
50–60 ppm =	8	7	140–160 ppm =	9	9
60–90 ppm =	12	18	160–190 ppm =	16	9
90–110 ppm =	4	9	190–220 ppm =	7	2
2. Hydrogen (nonexchangeable) distribution by liquid-state ^{1}H-NMR (in percent)					
0.2–1.5 ppm =		23	3–4 ppm =		36
1.5–3 ppm =		22	5.5–8 ppm =		19

radicals, infrared (IR) spectroscopy, possibilities of Raman spectroscopy, absorption of radiation in the ultraviolet and visible regions, fluorescence, applications of x-ray photoelectron spectroscopy. These techniques yield above all additional information that may be valuable in determining structural aspects of humic substances. Infrared spectroscopy is widely adopted for the study of oxygen-containing functional groups. For a more thorough discussion about different nondegradative analytical techniques in the study of humic substances, see Ref. 9.

Degradative methods are applied to aquatic humic substances to give structural information about the carbon distribution by trying to simplify the complex humic substances to specific individual compounds. Figure 6 shows a general layout of analytical procedures for the degradative approach in the study of humic substances.

Chemical degradation methods have been applied widely to the study of humic substances. The main chemical degradation methods are oxidation, hydrolysis, and reduction. Several oxidation techniques have been applied to humic substances, e.g., alkaline permanganate (KMnO$_4$), alkaline copper(II) oxide (CuO–NaOH), alkaline nitrobenzene, chlorine including aqueous hypochlorite, peracetic acid, nitric acid, and hydrogen peroxide oxidations. With the exception of, e.g., peracetic and nitric acids, most oxidations are carried out under alkaline conditions, which may result in both the potential alteration of the original material and the formation of undesirable by-products. For that reason, alkaline oxidations are carried out under nitrogen atmosphere.

Hydrolysis has been shown to be effective in removing protein and carbohydrate constituents associated with humic substances. Alkaline treatment hydrolyzes mainly ester and ether linkages within humic substances. Acids in addition to bases catalyze the hydrolysis of esters, amides, and substituted amides. Boiling water has been observed to extract polysaccharides, polypeptides, and small quantities of relatively simple phenolic acids and aldehydes. Reductive methods saturate olefinic bonds (C=C), resulting in changes in oxygen-containing functional groups (reduction of COOH) and in cleavage of some ether linkages. Reductions with zinc dust distillation and zinc dust fusion and sodium amalgam have been the most useful. Other approaches (such as sodium in liquid ammonia, red phosphorus, and hydriodic acid and reduction by hydrogen in high temperatures and pressures) have been

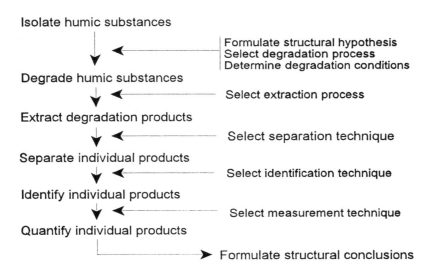

Fig. 6 Analytical procedure for the degradative approach.

also attempted. In addition to the classical reagents and conditions for chemical degradations of humic substances, where emphases are placed on oxidative and reductive processes, attention has been also focused on the high inputs of energy needed to cleave nonhydrolyzable pieces of humic substances. Degradations with sodium sulphide solutions and with phenol using *p*-toluenesulphonic acid as a catalyst have been applied to the study of humic substances for a better understanding of the degradation mechanisms.

In principle, chemical degradation methods are practicable for the study of humic substances. However, several problems arise. With mild procedures, yields of identifiable compounds are extremely low, usually only a few percent or even less from the starting material. In contrast, with too drastic methods the material is broken down into such small fragments (e.g., CO_2, H_2O, monobasic acids like formic, acetic, and propanoic acids and dibasic acids like oxalic, malonic, and succinic acids) that have lost almost any resemblance to the original structure. It is quite usual that high yields of degradation products have been reported, but the results are no longer realistic. The ideal chemical degradation method should be mild but at the same time yield high amounts of actual degradation products of moderate molecular complexities.

The most essential problem with all degradation methods is sorting out which degradation products are really artifacts. In this context an *artifact* can be defined as an identified degradation product whose formation pathway is incorrectly constructed and leads to false structural conclusions. Many theories of artifact formation have been presented; it may be said that al-

most every degradation product is a possible candidate for artifacts. It is impossible to demonstrate the absence of artifact formation during the chemical degradation of complex natural constituents like humic substances. However, there appears to be little scientific evidence for the critical statements presented for artifact formation. For instance, one example about a possible artifact is the formation of *m*-dihydroxy aromatic compounds by many different degradation processes of various humic substances, e.g., Norwood in Ref. 10. The appearance of *m*-dihydroxy aromatics as structural units in aquatic humic substances is potential but not definite. This is important to water chemists, since *m*-dihydroxy aromatic structures react sensitively with aqueous chlorine to produce chloroform (a suspected carcinogen) in drinking water.

The separation, identification, and quantification of hundreds of chemical degradation products have been carried out with combined gas chromatography (GC) and mass spectrometry (MS). Because many degradation products are acids, methylation-silylation increases their volatility and decreases polarity to help the GC-MS identification of the compounds. Sometimes carboxyl and hydroxyl groups of the original humic substances are premethylated before chemical degradation. Premethylation has been shown to increase product yields by, e.g., the oxidative degradation of various humic substances. However, the most significant disadvantage of premethylation is the frequent occurrence of questionable side reactions, so some degradation products are not indicative of the original humic substances. Separation, identification, and quantification of degradation products have been carried out in some cases with the aid of high-performance liquid chromatography (HPLC). However, only some water-soluble degradation products (acid derivatives) can be analyzed with this method (also, resolution is poorer than with GC).

Through different chemical degradation methods, a great number of definite low-molecular-weight organic compounds (mono-, tri-, and tetracarboxylic acids, including their hydroxy derivatives, almost all possible benzene carboxylic acids, phenolic acids, including their methoxy derivatives, different kinds of aldehydes and aliphatic hydrocarbons, etc.) have been obtained, and many of these compounds may be authentic structural units in humic substances. It is not reasonable to reproduce a humic molecule or even a basic structure of humic substances merely based on this scattered mosaic of degradation products (structural units). For more detailed discussion of chemical degradation of humic substances see Refs. 4, 7, and 9 and Norwood in Ref. 10.

Thermal degradation (pyrolysis, Py) is a powerful tool in studying structural units of humic substances. There are two main pyrolysis techniques: (a) controlling the decomposition kinetics by programming the temperature rise to the final temperature relatively slowly, and (b) carrying out the thermal decomposition "instantaneously" at constant temperature (cf. flash pyrolysis). In the latter technique the final temperature is reached within a few milliseconds with the aid of, e.g., inductively heated wire filaments (the so-called Curie-point method) or with resistively heated wire filaments. Thermal degradation in air leads predominantly to oxidation. Because of the variety of reactions possible, oxidative-thermal degradations are difficult to analyze. Thermal degradation in a vacuum and in an inert atmosphere are similar in that the thermal stability of the "polymer" is being measured in both cases. In a vacuum, however, the primary decomposition fragments are instantaneously removed from the reaction zone. In an inert atmosphere, the primary radicals can attack other "polymer" molecules or fragments of molecules, giving rise to induced decomposition reactions. Therefore, decomposition products generated in a vacuum and in inert atmosphere are very likely different. The final pyrolysis temperature chosen is normally in the range 500–700°C. By applying enough thermal energy to break chemical bonds, the molecule is fragmented in a way consistent with the relative strengths of the bonds involved. Higher temperatures give degradation products of too low a molecular weight to be indicative of the original material. Low-temperature pyrolysis at 300–400°C is used in special

cases to activate low-energy reactions, such as some decarboxylation, dehydration, or decomposition of certain amines.

The principal aim of thermal degradation is to maximize the quantity of molecular ions of the pyrolysis products and to make them escape quickly from the reaction zone to minimize secondary thermal fragmentation. Traditionally, thermal degradation is carried out with a pyroprobe connected at the injection port of a gas chromatography (thermal degradation in an inert atmosphere) using a selected technique (time programmed or flash pyrolysis, with special applications). Gas chromatographically (GC) separated individual pyrolyzates are identified with mass spectrometry (MS) or Fourier transform infrared spectrometry (FTIR) connected at the GC (labeled Py-GC-MS and Py-GC-FTIR, respectively). It is noteworthy that volatilization tends to decrease with increasing polarity (oxygen content) due to intermolecular forces, and thus polar degradation products are weakly represented in pyrolyzates and also poorly eluted from the GC column. To overcome this problem, various derivatization techniques using tetramethylammonium hydroxide (TMAH) have been developed, e.g., Ref. 65, del Rio and Hatcher in Ref. 17, and Saiz-Jimenez in Ref. 54, to enable hydroxy and carboxyl groups to be detected as methyl ethers and methyl esters, respectively. It is possible to expand greatly the number of degradation products to be analyzed by connecting the pyroprobe directly to the mass spectrometer (thermal degradation in a vacuum). In this case the thermal degradation occurs in the ion source of the MS (inside the analyzer, labeled Py-MS).

By the Py-MS method using inductively or resistively heated wire filaments and even low-energy (12–17 eV) electron ionization (EI), the content of some easily fragmented molecules can be low or absent (decomposition to CO_2), and pyrolytically formed molecular fragments higher than m/z (mass to charge ratio) of 350 are seldom observed. This is due to the extensive thermal fragmentation under flash pyrolysis and to the mass spectrometric fragmentation of the pyrolytically formed fragments even with low-energy electron ionization. For obtaining larger thermal fragments by the Py-MS method from humic substances (up to m/z 500 and even 3000), more specialized methods have been adopted for thermal degradation and low-energy ionization of pyrolytically formed fragments. For this purpose, field ionization (FI) and field desorption (FD) have been intensively applied (66,67). With the aid of the high-mass fragments obtained by these methods researchers have tried to gain insight into the composition and structure of complex humic substances. On the other hand, no ionization technique, EI, CI (chemical ionization), or FI, is ideal, and certain advantages and disadvantages will accrue with each of these ionization techniques (cf. Ref. 68).

The mechanism of thermal fragmentation of humic substances is not known. It has been postulated that pyrolysis reactions of different structural organic subunits (e.g., polysaccharides, polypeptides and proteins, lipids, lignins, polycarboxylic acids, aromatics, and phenolic compounds) of humic substances take place, chiefly as they would occur in isolated organic compounds. In other words, heteroatom scission at $-O-$ followed by elimination of OH as H_2O takes place in the case of, e.g., polysaccharides and lignins, elimination and chain scission are the dominant processes for lipids, and carboxylic acids decarboxylate readily. Based on this hypothesis, certain fragment ions from the enormous chaos of those obtained by the Py-MS method have been selected as "fingerprints" to reflect different types of single organic molecules formed during pyrolysis (acetic acid, furfurals, levoglucosan, furans, pyrroles, methoxyphenols, etc.) that will speak further for a certain structural composition (polysaccharides, polypeptides and proteins, lipids, lignins, aromatics, etc.) in humic substances, e.g., Ref. 69 and Bracewell, Haider, Larter, and Schulten in Ref. 9.

The Py-MS method has been widely applied to uncomplicated computerized data analysis. The fragment ions (m/z values) and their intensities have been mathematically manipulated into multidimensional space. The method will assemble samples (HA- or FA-type substances) obtained from different sources into their own clusters according to their uniformities–dissimilarities. The multifactor analysis of the data obtained by the Py-MS method is most often based on the amount and composition of carbohydrate, phenol, hydrocarbon, aromatic, and amino acid moieties characteristic of humic substances from different sources. However, it is much more complicated to conclude a structural composition of humic substances from the data of the Py-MS method; cf. Schulten in Ref. 1 and Refs. 67 and 70–76. Ideally, each fragment ion in the Py-MS spectrum represents the molecular ion of at least one pyrolysis product. However, it is not so straightforward, even when low EI energies (12–17 eV) in flash pyrolysis–MS are used, to suppress the mass spectrometric fragmentation of pyrolytically formed fragments. Furthermore, a given fragment ion obtained from the MS does not necessarily belong exclusively to one class of molecules, since ions of different structures may have the same mass-to-charge ratio and can contribute to the same mass peak in the mass spectrum. Thermal degradation and identification

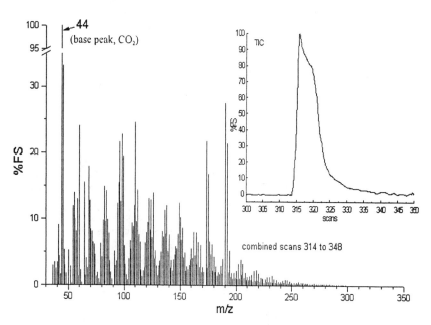

Fig. 7 Py-MS spectra of a lake fulvic acid (FA). The final spectrum is obtained by summing spectra over the range indicated on the inset profile of total ion current (TIC). % FS = percent of full scale. *Conditions*: vacuum flash pyrolysis (resistively heated wire filament) at 510°C (1 s), EI at 14 eV, mass range 35–350, scan time 0.16 s, interscan time 0.01 s, 2000 amu/s, ion source 180°C, temperature rise 10°C/ms, sample in methanol 5 mg/ml, sample size 60 μg.

of the degradation products with the MS are equipment-sensitive tasks (e.g., Ref. 77). This is, among other things, one reason that theoretical speculation is so possible about the different source of the fragment ions obtained by the Py-MS method. Figure 7 shows a Py-MS spectrum obtained (20) by quadrupole mass spectrometry from a lake FA fraction isolated–fractionated with the XAD technique.

The Py-MS method is very sensitive. It is, e.g., possible to observe certain minor constituents that may be contaminants. For example, a fragment ion of m/z 149 may reflect phthalates, which according to public opinion are the most probable contaminants introduced during the isolation of humic substances. On the other hand, it has been stated (78) that phthalates are also natural constituents of some plants and microorganisms. Furthermore, it has been demonstrated (79) that m/z values of 215 and 217 may reflect the presence of some atrazine (herbicide). For more detailed fundamental information about thermal degradation and the identification of the degradation products by MS concerning different kinds of biomaterials see Refs. 68 and 80.

REFERENCES

1. N Senesi, TM Miano, eds. Humic Substances in the Global Environment and Implications on Human Health. Amsterdam: Elsevier, 1994.
2. IH Suffet, P MacCarthy, eds. Aquatic Humic Substances—Influence on Fate and Treatment of Pollutants. Washington DC: American Chemical Society, 1989.
3. GG Choudhry. Humic Substances—Structural, Photophysical, Photochemical and Free Radical Aspects and Interactions with Environmental Chemicals. Vol. 7. In: Current Topics in Environmental and Toxicological Chemistry. New York: Gordon & Breach, 1984.
4. EM Thurman. Organic Geochemistry of Natural Waters. Dordrecht: Martinus Nijhoff/W Junk, 1985.
5. RF Christman, ET Gjessing, eds. Aquatic and Terrestrial Humic Materials. Ann Arbor, MI: Ann Arbor Science, 1983.
6. GR Aiken, DM McKnight, RL Wershaw, P MacCarthy, eds. Humic Substances in Soil, Sediment and Water—Geochemistry, Isolation, and Characterization. New York: Wiley, 1985.
7. FJ Stevenson. Humus Chemistry—Genesis, Composition, Reactions. New York: Wiley, 1982.
8. ET Gjessing. Physical and Chemical Characteristics of Aquatic Humus. Ann Arbor, MI: Ann Arbor Science, 1976.
9. MHB Hayes, P Mac Carthy, RL Malcolm, RS Swift, eds. Humic Substances II—In Search of Structure. New York: Wiley, 1989.
10. FH Frimmel, RF Christman, eds. Humic Substances and Their Role in the Environment. New York: Wiley, 1988.
11. EM Perdue, ET Gjessing, eds. Organic Acids in Aquatic Ecosystems. New York: Wiley, 1990.
12. CE Clapp, MHB Hayes, N Senesi, SM Griffith, eds. Humic Substances and Organic Matter in Soil and Water Environments—Characterization, Transformations and Interactions. Birmingham, STATE: International Humic Substances Society, 1996.
13. RA Baker, ed. Organic Substances and Sediments in Water. Chelsea, STATE: Lewis, 1991.
14. MHB Hayes, WS Wilson, eds. Humic Substances, Peats and Sludges—Health and Environmental Aspects. Cambridge: Royal Society of Chemistry, 1997.
15. C Matthess, FH Frimmel, P Hirsch, HD Schulz, E Usdowski, eds. Progress in Hydrogeochemistry—Organics, Carbonate Systems, Silicate Systems, Microbiology, Models. Berlin: Springer-Verlag, 1992.
16. AJ Beck, KC Jones, MHB Hayes, U Mingelgrin, eds. Organic Substances in Soil and Water—Natural Constituents and Their Influences on Contaminant Behavior. Cambridge: Royal Society of Chemistry, 1992.
17. JS Gaffney, NA Marley, SB Clark, eds. Humic and Fulvic Acids—Isolation, Structure, and Environmental Role. Washington, DC: American Chemical Society, 1996.
18. LG Danielsson. On the use of filters for distinguishing between dissolved and particulate fractions in natural waters. Water Res 16:179–182, 1982.
19. H De Haan, RI Jones, K Salonen, K. Does ionic strength affect the configuration of aquatic humic substances? Freshwater Biol 17:453–459, 1987.
20. J Peuravuori. Isolation, fractionation and characterization of aquatic humic substances—does a distinct humic molecule exist? Finnish Humus News 4(1):1–99, 1992.
21. DPH Laxen, IM Chandler. Comparison of filtration techniques for size distribution in freshwaters. Anal Chem 54:1350–1355, 1982.
22. SM Serkiz, EM Perdue. Isolation of dissolved organic matter from the Suwannee River using reverse osmosis. Water Res 24:911–916, 1990.
23. AS Michaels. Ultrafiltration. In: ES Perry, ed. Progress in Separation and Purification. New York: Wiley, 1968, pp 297–334.
24. ET Gjessing. Ultrafiltration of aquatic humus. Environ Sci Technol 4:437–438, 1970.
25. J Buffle, P Deladoey, W Haerdi. The use of ultrafiltration for the separation and fractionation of organic ligands in fresh waters. Anal Chim Acta 101:339–357, 1978.
26. PJ Shaw, RI Jones, H De Haan. Separation of molecular size classes of aquatic humic substances using ultrafiltration and dialysis. Environ Technol 15:765–774, 1994.

27. GR Aiken, RL Malcolm. Molecular weight of aquatic fulvic acid by vapor pressure osmometry. Geochim Cosmochim Acta 51:2177–2184, 1987.

28. MT Ganzerli Valentini, L Maggi, R Stella, G Ciceri. Metal-humic and fulvic acid interactions in fresh water ultrafiltration fractions. Chem Ecol 1:279–291, 1983.

29. J Peuravuori, K Pihlaja. Isolation and characterization of natural organic matter from lake water—Comparison of isolation with solid adsorption and tangential membrane filtration. Environ Int 23:441–451, 1997.

30. J Peuravuori, K Pihlaja. Molecular size distribution and spectroscopic properties of aquatic humic substances. Anal Chim Acta 337:133–149, 1997.

31. J Peuravuori, K Pihlaja. Multi-method characterization of lake aquatic humic matter isolated with sorbing solid and tangential membrane filtration. Anal Chim Acta 364:203–221, 1998.

32. J Peuravuori, K Pihlaja, N Välimäki. Isolation and characterization of natural organic matter from lake water —two different adsorption chromatographic methods. Environ Int 23:453–464, 1997.

33. JA Leenheer. Comprehensive approach to preparative isolation and fractionation of dissolved organic carbon from waters and wastewaters. Environ Sci Technol 15: 578–587, 1981.

34. EM Thurman, RL Malcolm. Preparative isolation of aquatic humic substances. Environ Sci Technol 15:463– 466, 1981.

35. RL Malcolm. Factors to be considered in the isolation and characterization of aquatic humic substances. In: B Allard, H Borén, A Grimvall, eds. Humic Substances in the Aquatic and Terrestrial Environment. Lecture Notes in Earth Sciences. Berlin: Springer-Verlag, 1991, 33, pp 9–36.

36. D McKnight, EM Thurman, RL Wershaw. Biogeochemistry of aquatic humic substances in Thoreau's bog, Concord, Massachusetts. Ecology 66:1339–1352, 1985.

37. MB David, GF Vance. Chemical character and origin of organic acids in streams and seepage lakes of central Maine. Biogeochemistry 12:17–41, 1991.

38. MB David, GF Vance, P Kortelainen. Organic acidity in Maine (U.S.A.) lakes and in HUMEX Lake Skjervatjern (Norway). Finnish Humus News 3(3):189–194, 1991.

39. GF Vance, MB David. Chemical characteristics and acidity of soluble organic substances from a northern hardwood forest floor, central Maine, USA. Geochim Cosmochim Acta 55:3611–3625, 1991.

40. J Peuravuori, K Pihlaja. Multi-method characterization of lake aquatic humic matter isolated with two different sorbing solids. Anal Chim Acta 363:235–247, 1998.

41. BDH. Ion Exchange Resins. 6th ed. Poole, England: BDH Chemicals, 1981.

42. JJ Farnworth. Comparisons of the sorption from solution of a humic acid by Supelite DAX-8 and by XAD-

43. RL Malcolm, P MacCarthy. Quantitative evaluation of XAD-8 and XAD-4 resins used in tandem for removing organic solutes from water. Environ Int 18:597–607, 1992.

44. P Kortelainen, MB David, T Roila, I Mäkinen. Acid–base characteristics of organic carbon in the Humex lake Skjervatjern. Environ Int 18:621–629, 1992.

45. RF Packham. Studies of organic color in natural water. Proc Soc Water Treatment Exam 13:316–334, 1964.

46. CJ Miles, JR Tuschall, PL Brezonik. Isolation of aquatic humus with diethylaminoethylcellulose. Anal Chem 55:410–411, 1983.

47. N Paxéus. Studies on aquatic humic substances. PhD dissertation, Göteborg, Sweden: Chalmers University of Technology and University of Göteborg, Sweden, 1985.

48. C Pettersson, I Arsenie, J Ephraim, H Borén, B Allard. Properties of fulvic acids from deep groundwater. Sci Tot Environ 81/82:287–296, 1989.

49. C Pettersson, J Ephraim, B Allard. On the composition and properties of humic substances isolated from deep groundwater and surface waters. Org Geochem 21: 443–451, 1994.

50. C Pettersson, L Rahm. Changes in molecular weight of humic substances in the Gulf of Bothnia. Environ Int 22:551–558, 1996.

51. N Senesi. Binding mechanism of pesticides to soil humic substances. Sci Tot Environ 123/124:63–76, 1992.

52. P Kortelainen. Contribution of organic acids to the acidity of Finnish Lakes. PhD dissertation, Helsinki: University of Helsinki, 1993.

53. RL Malcolm. The uniqueness of humic substances in each of soil, stream and marine environments. Anal Chim Acta 232:19–30, 1990.

54. A Piccolo, ed. Humic Substances in Terrestrial Ecosystems. Amsterdam: Elsevier Science, 1996.

55. RH Newman, KR Tate. Use of alkaline soil extracts for ^{13}C NMR. Characterization of humic substances. J Soil Sci 35:47–54, 1984.

56. AH Gillam, MA Wilson. Pyrolysis-GC-MS and NMR studies of dissolved seawater humic substances and isolates of a marine diatom. Org Geochem 8:15–25, 1985.

57. AM Vassalo, MA Wilson, PJ Collin, JM Oades, AG Waters, RL Malcolm. Structural analysis of geochemical samples by solid-state nuclear magnetic resonance spectrometry—role of paramagnetic material. Anal Chem 59:558–562, 1987.

58. RL Wershaw, DJ Pinckney, EC Llaguno, V Vicente-Beckett. NMR characterization of humic acid fractions from different Philippine soils and sediments. Anal Chim Acta 232:31–42, 1990.

59. P Conte, A Piccolo, B van Lagen, P Buurman, PA de Jager. Quantitative differences in evaluating soil humic substances by liquid- and solid-state ^{13}C-NMR spectroscopy. Geoderma 80:339–352, 1997.

8 resins. IHSS Newsletter (Georgia Institute of Technology) 13:8–9, 1995.

60. RL Malcolm. ^{13}C-NMR spectra and contact time experiment for Skjervatjern fulvic and humic acids. Environ Int 18:609–620, 1992.

61. PG Hatcher. ^{1}H and ^{13}C NMR of marine humic acids. Org Geochem 2:77–85, 1980.

62. RC Averett, JA Leenheer, DM McKnight, KA Thorn, eds. Humic Substances in the Suwannee River, Georgia —Interactions, Properties, and Proposed Structures. Denver: U.S. Geological Survey, Open-File Report 87-557, 1989, 377 p.

63. GG Choudhry, GRP Webster. Soil organic matter chemistry. Part 1—characterization of several humic preparations by proton and carbon-13 nuclear magnetic resonance spectroscopy. Toxic Envir Chem 23:227–242, 1989.

64. K Yonebayashi, T Hattori. Chemical and biological studies on environmental humic acids. Soil Sci Nutr 35: 383–392, 1989.

65. C Saiz-Jimenez, B Hermonsin, JJ Ortega-Calvo. Pyrolysis/methylation: a method for structural elucidating of the chemical nature of aquatic humic substances. Water Res 27:1693–1696, 1993.

66. N Simmleit, H-R Schulten. Analytical pyrolysis and environmental research. J Anal Appl Pyrolysis 15:3–28, 1989.

67. H-R Schulten, M Schnitzer. A contribution to solving the puzzle of the chemical structure of humic substances—pyrolysis-field ionization mass spectrometry. Sci Tot Environ 117/118:27–39, 1992.

68. HLC Meuzelaar, J Haverkamp, FD Hileman. Pyrolysis Mass Spectrometry of Recent and Fossil Biomaterials. Amsterdam: Elsevier Science, 1982.

69. J Peuravuori, K Pihlaja. Pyrolysis electron impact mass spectrometry in studying aquatic humic substances. Anal Chim Acta 350:241–247, 1997.

70. H-R Schulten, B Plage, M Schnitzer. A chemical structure for humic substances. Naturwissenschaften 78: 311–312, 1991.

71. H-R Schulten, M Schnitzer. A state of the art structural concept for humic substances. Naturwissenschaften 80: 29–30, 1993.

72. C Saiz-Jimenez, JJ Ortega-Calvo, B Hermosin. Conventional pyrolysis—a biased technique for providing structural information on humic substances? Naturwissenschaften 81:28–29, 1994.

73. H-R Schulten, M Schnitzer. Three-dimensional models for humic acids and soil organic matter. Naturwissenschaften 82:487–498, 1995.

74. H-R Schulten. The three-dimensional structure of humic substances and soil organic matter studied by computational analytical chemistry. Fresenius J Anal Chem 351:62–73, 1995.

75. H-R Schulten. The three-dimensional structure of soil organo-mineral complexes studied by analytical pyrolysis. J Anal Appl Pyrolysis 32:111–126, 1995.

76. H-R Schulten, P Leinweber. Characterization of humic and soil particles by pyrolysis and computer modeling. J Anal Appl Pyrolysis 38:1–53, 1996.

77. C Saiz-Jimenez. Analytical pyrolysis of humic substances—pitfalls, limitations, and possible solutions. Environ Sci Technol 28:1773–1780, 1994.

78. F Gadel, A Bruchet. Application of pyrolysis–gas chromatography–mass spectrometry to the characterization of humic substances resulting from decay of aquatic plants in sediments and waters. Water Res 21:1195–1206, 1987.

79. H-R Schulten. Pyrolysis and soft ionization mass spectrometry of aquatic/terrestrial humic substances and soils. J Anal Appl Pyrolysis 12:149–186, 1987.

80. WJ Irwin. Analytical Pyrolysis, A Comprehensive Guide. New York: Marcel Dekker, 1982.

21

Major Metals

Payman Hashemi

Lorestan University, Khoramabad, Iran

I. INTRODUCTION

Sodium, potassium, magnesium, and calcium are usually considered major metal ions in natural waters because their concentrations are considerably higher than those of other cations. Manganese, iron, zinc, and copper are also abundant elements with relatively high concentrations in waters, compared to trace metals. Hence, in most cases they can also be considered major metals. Their concentrations, however, might be quite low (at the $\mu g/l$ level or even less) in certain unpolluted natural waters. Distinct care may be necessary in this case to avoid contamination problems during sample handling, pretreatment, and analysis because these elements can be found almost everywhere. In the higher concentration ranges, however, ordinary precautions suffice. The selection of a method for final analysis as well as the procedure used for sample handling and pretreatment, hence, will be dependent on a knowledge of the concentration ranges and nature of the analytes and the analytical methods available.

In this chapter some properties of the major metals, methods of sample pretreatment, and different analytical techniques for their determination in natural waters will be discussed.

II. PROPERTIES AND IMPORTANCE

Accurate determination of metal ions in waters is important from the environmental point of view. The major metals are usually essential nutrients, and a deficiency of them can produce disease for humans, animals, and plants. Some of them have toxic effects and must not exceed certain levels in water.

Sodium and potassium are abundant elements, with concentrations ranging from <1 mg/L to >500 mg/L in aquatic systems. They are both important nutrients. In agriculture and human pathology, the ratio of sodium to total cations is of importance, because it can affect soil permeability (1).

Magnesium and calcium are necessary in animal and human nutrition. Calcium is an essential component of bones, teeth, shells, and plant structure. Ca and Mg concentrations in waters can vary from <1 mg/L to several hundred milligrams per liter, depending on the source and treatment of the water. They play a major role in water hardness; e.g., they precipitate soap. Calcium carbonate might be useful at low levels in water because it will produce a protective coating on pipes and distribution lines. At high concentrations, however, it can clog pipes or produce harmful scales in boilers (1). Magnesium is also a major cause of boiler scale when hard water is used in heating systems.

Manganese is an essential element for plants and, hence, is included in certain fertilizers. It exists in a number of oxidation states. In groundwater, due to the lack of oxygen, it exists almost entirely in divalent form. In surface water, the quadrivalent and trivalent states exist, respectively, in the suspension and soluble complex forms (1). Manganese cations, e.g., Mn^{2+}, are more toxic than the anionic form, e.g. MnO_4^-, and Mn(II) is more toxic than Mn(III) (2). High levels of manganese give an unplesant taste to water and produce deposits on bread during baking.

Iron exists in the oxidation states of 2+ and 3+ in water. A substantial amount of iron is in suspended or colloidal form. The solubility of Fe(III) increases with complexation with organic compounds. Hence, a major part of iron is in the form of dissolved complexes (2). In samples of oxygenated surface waters, concentrations of dissolved iron seldom exceed 1 mg/L. A higher level of iron gives a bittersweet, astringent taste to water. It also causes difficulties in distribution systems by supporting the growth of iron bacteria (1,2).

Zinc is an important essential element in both human and animal growth. It exists with only the valence of 2+ in natural waters. An important source of zinc in natural waters is the runoff of the Zn^{2+} from agricultural fields (1). It also enters water from the deterioration of galvanized iron pipes. The presence of zinc in drinking water in concentrations up to 40 mg/L appears to have no health consequences. A concentration higher than 5 mg Zn/L, however, gives a bitter, astringent taste to water, and it might precipitate as $Zn(OH)_2$ or $ZnCO_3$ in alkaline water to produce a milky turbidity (1).

Copper is essential to humans. The adult daily requirement has been estimated at 2 mg (1). Copper occurs in three oxidation states in water, 1+, 2+, and 3+; among them only the 2+ form is stable. In freshwater the predominant form of copper is the complex form with colloidal organic matter. Copper is not considered a cumulative poison, as is lead or mercury, and most of it is excreted by the body and little is retained (2). Problems of a technical or esthetic nature may occur for copper concentrations exceeding 0.2 mg/L; health effects, especially for bottle-fed children, appear for concentrations above 2 mg/L (3). Figure 1 compares the concentrations of particulate, colloidal, and dissolved Fe, Cu, and Zn in a river water sample (4).

III. SAMPLE PREPARATION

A. Sampling and Storage

Sampling is the first step, and probably of the most important one, in the analysis of a water sample. Any error at this stage renders the whole analysis useless. It is, however, not possible to present a general method. Different sampling methods might be used depending on the concentration ranges, sample composition, and the purpose of the analysis. The essential aspects of accurate sample collection for different water types have been outlined, for instance, by Nürnberg and Mart

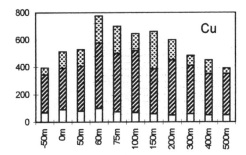
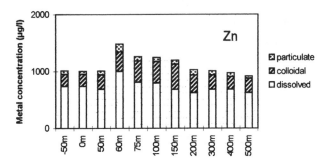
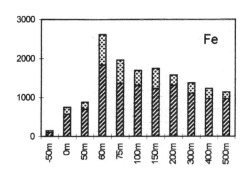

Fig. 1 Concentrations of particulate, colloidal, and dissolved Fe, Cu, and Zn in a polluted section of the Tambo River, Australia. (From Ref. 4.)

(5). The collected sample must be representative of the real water composition. The vessles used for sample collection must be cleaned carefully and must not be made of a material that leaches the elements of interest. In general, the shorter the time between the collection of a sample and its analysis, the more reliable will be the analytical results. Suspended matter, turbidity, the method used for their elimination, and chemical changes are important factors to be considered.

Despite the relatively high concentrations of the major metals, difficulties in storage have been observed by several workers. Sodium and potassium might leach from some glass containers. Hence, borosilicate glass or polyethylene bottles must be used for sampling and storage. It is also recommended to acidify the sample with nitric acid to pH 2 in order to avoid adsorption on vessel walls (1). For calcium and magnesium storage also, the customary precautions are sufficient. Any change in the pH–alkalinity–carbon dioxide balance might result in the precipitation of calcium carbonate. Hence, the sample should be acidified, and any precipitate of calcium carbonate must be redissolved before analysis.

Manganese, iron, zinc, and copper are more subject to losses due to adsorption and/or precipitation as compared to alkali and alkaline earth metals. Hence, it is recommended to analyze them as soon as possible after sample collection. If the purpose is the measurement of total concentrations, the sample should be acidified with HCl or HNO_3. Acid digestion or other digestion methods might be necessary before analysis. This is especially true for iron, which is mostly in particulate and colloidal forms (see Fig. 1). For the determination of the dissolved part, the sample is filtered through a 0.45-μm membrane filter in the field before acidification. Sample containers have to be cleaned carefully with acid and rinsed with distilled water.

Dissolved manganese might be oxidized to a higher oxidation state and precipitate. Hence, it is important to determine it very soon after collection. Acidification with HNO_3 can be used when total Mn is going to be determined (1). The adsorption of easily hydrolizable cations (e.g., Fe^{3+}, Cu^{2+}, and Zn^{2+}) on oxide/hydroxide and silicate surfaces are generally very pH dependent. Over a critical range, changing 1 or 2 pH units may increase the adsorption from about 0 to 100% (6) as

exemplified in Fig. 2. The pH of adsorption generally increases with decreasing hydrolysis constants.

B. Filtration and Digestion

In waters with relatively high particulate and colloidal content, filtration is necessary to prevent sorption or desorption of metal ions, especially during long-term storage. Filtration, as well as refrigeration at about 4°C, will also reduce bacterial sorption or interference in analysis. Filtration, however, is also a potential source of contamination and loss due to adsorption processes. Hence, it is recommended to precondition the filter with 50 ml deionized water and then at least 100 ml sample before filtration (1). The dissolved part of water is usually defined as the part that passes through a 0.45-μm filter. This, however, does not seem to be a strictly justified filter size, considering that most natural colloids exist predominantly in the size range 100–300 nm (7). For the determination of suspended metals, the filter is digested by acid or the filter is used directly for analysis.

When the total concentration of a metal in an aquatic system is of interest, usually the sample has to be digested without preliminary filtration. Acid digestion is performed with nitric acid or a combination of nitric and perchloric, hydrochloric, or sulfuric acid for complete digestion (1). The acidified sample will be evaporated to the lowest volume possible before precipitation. The heating and addition of acid will be continued until a clear solution is obtained. When perchloric acid is used, caution must be taken to avoid explosion. The organic matter must be oxidized first with nitric acid before the addition of $HClO_4$. An alternative method to acid digestion is dry-ashing. In this method the sample will be evaporated in a platinum or high silica glass crucible and then made into ashes in a muffle furnace at 500°. If only sodium has to be determined, a temperature of up to 600°C can be used. The ash is then dissolved in HNO_3 and warm water, filtered, and diluted to volume (1).

Digestion might also be performed after filtration of a water sample. This is usually the case when the total dissolved concentration of a metal is determined either directly, by, say, a voltametric method, or after preconcentration on a solid-phase adsorbent, by, say, an atomic spectrometric technique. In such cases a milder digestion condition is usually required in order to release the metal ion from organic complexes. Ultraviolet digestion with a mercury lamp has frequently been used for this purpose (5,8,9). The sample is usually filled in a quartz tube and acidified to pH 1.5–2.0 in

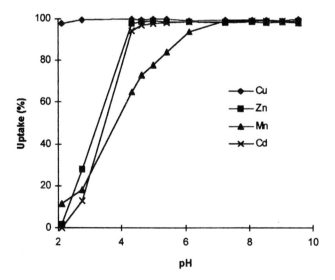

Fig. 2 Batch mode adsorption of metal ions on IDA-Novarose adsorbent as a function of pH. (From Ref. 10.)

order to avoid adsorption or precipitation of metal oxide and hydroxides. A small amount of hydrogen peroxide is often added to the sample to aid the photo-oxidation of the organic material (8). Recently, Hashemi et al. (10) showed that online buffering of primarily acidified samples just before accumulation on a preconcentration column can be used as an alternative to UV digestion. This method, which is much simpler and faster than the UV digestion technique, was tested for some natural water samples and promising results were obtained. The method, however, should be practiced with caution, because certain metal complexes may not be dissociated in the acidic pH used or may be reformed quickly after buffering the sample.

C. Preconcentration

Concentrations of Na, K, Mg, and Ca are usually high enough to be measured directly by a moderately sensitive method. On the other hand, this is not usually the case for the major transition metals, i.e., Mn, Fe, Zn, and Cu. The concentrations of these elements in open oceans and other unpolluted natural waters may be at the $\mu g/L$ or even ng/L levels. Determinations by a moderately sensitive method at such levels usually require a preconcentration step. In addition, even the most sensitive techniques are subject to interferences from matrix ions. A selective preconcentration technique can largely reduce or even completely eliminate such interferences.

A number of different methods have been used for the preconcentration of Mn, Fe, Zn, and Cu in aquatic systems. Coprecipitation, solvent extraction, and chelation on solid-phase sorbents are among the most commonly used preconcentration methods. These are described briefly next.

1. Coprecipitation

In this technique, a metal oxide, or alternatively an organic coprecipitant agent, is added to the water sample at a certain (usually an alkaline) pH. The analyte ions will then coprecipitate with the agent. The precipitate isolated from a large volume of original water sample is then filtered and dissolved in a small volume of acid to provide a concentrate for analysis. The analysis might also be performed directly on the filter by, for instance, an energy dispersive x-ray fluorescence spectrometer (XRFS) (11). Chakravatry and Van Grieken (12) used this method when they coprecipitated Mn, Cu, Zn, Ni, and Pb with ferric hydroxide in natural waters. After an hour of equilibrium time, the solution

was filtered on a membrane filter and analyzed with an energy-dispersive XRFS instrument. The recovery is largely pH dependent, and quantitative recoveries of Zn and Cu may be obtained at a pH above 7–8 and 9, respectively. Manganese is recovered only 60%, even at a pH in excess of 10. Coprecipitation of Cu in potable waters with zirconium hydroxide (13) and Cu and Mn in seawater with indium hydroxide (14) are other examples of preconcentration by coprecipitation.

Organic complexants also have been used as coprecipitation agents. An insoluble complex is formed in this method that might be left to age prior to a careful filtration under controlled conditions. The precipitate is then either dissolved in an organic solvent (15) or dried prior to a final analysis by electrothermal atomic absorption spectrometry (ETAAS) (15–17), neutron activation (18), etc. Cobalt pyrolidine dithiocarbamate (15), 8-hydroxyquinoline (8-HQ) (16), and 1-(2-pyridylazo)-2-naphthol (17,18) are examples of ligands that have been used for coprecipitation of Cu, Zn, and Mn in natural waters.

2. Solvent Extraction

Solvent extraction is another commonly used method of preconcentration. In this method a suitable complexing agent, dissolved in a small volume of a water-immiscible organic solvent, is thoroughly mixed with the aquatic sample. The organic solvent is then separated and analyzed, either directly or after back-extraction of the metal ions with an aqueous acid.

A variety of different complexing agents and organic solvents have been used in solvent extraction preconcentration of Mn, Fe, Zn, and Cu prior to their determination by different methods (19). Sodium diethyl-dithiocarbamate is a commonly used ligand in this technique prior to atomic absorption spectrometric determination of the transition elements (20–22). Organic solvents such as carbon tetrachloride (20), methylisobutyl ketone (21), and isoamylalcohol (22) have been used for dissolving this complexing agent. The analysis has been made either directly on the solvent or after back-extraction with nitric acid. The kinetics of the back-extraction is, however, generally slow, and the efficiency of the method for copper and iron is poor (23). Table 1 summarizes examples of the reagents used in the solvent extraction technique and their application to the determination of the major metals in natural waters.

3. Chelation by Solid-Phase Sorbents

Numerous articles, with a rapid increase in their number in recent years, describe the synthesis, properties,

Table 1 Examples of Complexing Agents and Organic Solvents Used for Solvent Extraction Preconcentration of Metal Ions in Natural Waters

Complexing agent	Organic solvent	Analytes	Water sample	Analytical finish[a]	Ref.
Ammonium pyrrolidinedithio-carbamate	Methylisobutyl keton	Mn, Fe, Cu	Potable	AAS	24
	2,6-dimethyl-4-heptane	Fe, Zn, Cu	Natural	AAS	25
	Methyl isobutyl ketone	Zn, Cu	River	AAS	26
	Xylene	Mn, Fe, Zn, Cu	River	ICP-AES	27
Sodium diethyldithio-carbamate	Methyl isobutyl ketone	Cu	Natural	AAS	21
	Isoamyl alcohol	Mn, Fe, Zn, Cu	River	AAS	22
	Carbon tetrachloride	Fe, Cu	Natural	AAS	20
1-Pyrrolidine-dithiocarbamate	Methyl isobutyl keton	Cu	Natural	AAS	28
Dithizone	Ethyl propionate	Fe(II), Zn	Natural	AAS	29
Diantipyryl-methane	Chloroform or dichloromethane	Zn, Cu	Mine	AES	30
Thenoyltrifluoro-acetone	Methyl isobutyl ketone	Mn(II)	Natural	AAS	31
8-Hydroxy-quinoline	Chloroform	Mn	Sea	AAS	32
	Chloroform	Mn, Fe, Zn, Cu	Sea	XRFS	33

[a] Abbreviations: AAS, atomic absorption spectrometry; ICP, inductively coupled plasma; AES, atomic emission spectrometry; XRFS, x-ray fluorescence.

and application of solid-phase chelating sorbents for the preconcentration of metal ions in waters. A number of books and review articles are also available concerning this subject (34–40). Table 2 lists examples of applications of different metal sorbents to the preconcentration of the major transition metals in water samples.

The usefulness of a metal adsorbent in an analytical application depends on the characteristics of the sor-

bent. Distribution coefficients and complex stability, selectivity, rate of adsorption and desorption, loading capacity, and acid–base behavior are some important properties to be considered. The conditional stability constants of the functional groups are often highly pH dependent, and the sorbent is useful only in a narrow pH region. The chelating sorbents usually have good selectivity toward transition elements. Some of them, however, are charged in the pH used and therefore col-

Table 2 Examples of Support Materials and Functional Groups Used in Chelating Adsorbents for Preconcentration of Metal Ions in Natural Waters

Support material (or sorbent)	Functional group	Analytes	Water sample	Analytical finish	Ref.
CC-1 Metpac	IDA	Mn, Zn, Cu	Reference	ICP-MS	41
		Mn, Zn, Cu	Sea	Spectrophotometer	42
Muromac A-1	IDA	Fe(III)	Standard	ICP-AES	43
		Mn, Fe(III), Zn, Cu	Reference	ICP-AES	44
Chelex-100	IDA	Zn, Cu	River	ICP-AES	45
		Mn, Fe(III), Zn, Cu	Sea	AAS	46
IDA-Novarose	IDA	Mn, Fe(III), Zn, Cu	Natural	ICP-AES	8,10
Controlled pore glass (CPG)	8-HQ	Mn, Zn, Cu	Reference	ICP-MS	47
Silica	8-HQ	Mn, Fe, Zn, Cu	Sea	ETAAS	48
PEI-Novarose	Polyethylene-imine	Cu	Tap	FAAS	49
Organic polymer	Tetraaza-macrocycles	Mn	Sea	ETAAS	50
Melamine-formaldehyde	Methylol melamine	Fe(II), Fe(III)	Natural	Spectrophotometer	51

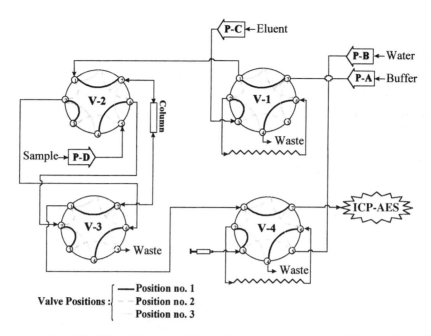

Fig. 3 Schematic diagram of an FIA-ICP-AES system. V-1 to V-4, seven port valves; P-A and P-B, cylindrical displacement pumps; P-C, peristaltic pump; P-D, reciprocating pump. The valves are set for reconditioning of the column. (From Ref. 52.)

lect matrix ions as well. In such cases, the weakly bound ions are usually replaced by a buffer before elution of the analytes (52).

Chelating adsorbents have been used in both the batch and the column modes. Column techniques have the advantage that the procedure can be automated and used in an online mode. In an automated preconcentration system, the adsorbent is packed in a small column, typically 4-mm ID $\times$ 50 mm, which forms a part of a low-pressure flow system. The preconcentration, enrichment, intermediate washing, elution, and regeneration steps, which compose the working cycle of the column, are automated and computer controlled. In an online system the outlet of the column is connected directly, for instance, to the nebulizer of a flame atomic absorption spectrometry (FAAS) or inductively coupled plasma (ICP) instrument. A transient signal will be obtained in this case. Larger sensitivity enhancement factors, less risk of contamination, less consumption of reagents, and faster and less laborious analysis are achieved when such online systems are used.

Among the preconcentration systems discussed earlier, chelating adsorbents, also known as chelating ion exchangers, seem to be superior in many respects. The coprecipitation and solvent extraction methods are usually laborious and time consuming, with a large risk of contamination due to sample handling. Chelating sorbents, on the other hand, can be packed in columns and easily used in automated online flow systems. This will reduce the sample handling and total analysis time compared to the batch mode coprecipitation and solvent extraction techniques. Figure 3 depicts a typical FIA system used for the preconcentration of metals in water (8).

IV. METHODS OF ANALYSIS

A number of different analytical techniques are available for the determination of metals. Since the concentrations of the major metals are relatively high in water, classical techniques were mostly used for their determination in early experiments. Such techniques, however, have been replaced chiefly by many faster, more sensitive, and sophisticated instrumental techniques in recent years. In this section the most important analytical techniques used for the detection and quantitation of the major metals are discussed. Examples of analytical applications of each method to the analysis of different aquatic systems will be given, as well. Many of these applications have been summarized in tables that can help the reader find a certain application of an analytical method to the determination of an specific metal ion.

A. Classical Methods

Classical methods of analysis have been used for the determination of the major metals for several decades. Even today when it is necessary to obtain maximum accuracy and precision in the determination of the higher-concentration constituents, classical techniques are often the methods of choice. They are, however, not usually employed in the modern water quality laboratory, because these methods are slow and tedious and often require a skilled chemist to analyze a large number of samples routinely. Among the classical techniques, the volumetric methods appear to have been the most widely used in water analysis that will be discussed here.

Volumetric methods are titrimetric procedures that involve a chemical reaction between a known concentration of a reagent (titrant) and an accurately known volume of sample. The reaction types used in this method include acid–base neutralization, oxidation–reduction, precipitation, and complexation. The volume of the titrant, added incrementally, that is required to reach the equivalence point is used to calculate the original concentration of the analyte constituent. An appropriate indicator is used to determine the equivalence point. In comparison to the gravimetric procedures (53), volumetric techniques are rapid, convenient, and readily automated.

Modern laboratories often employ automated titrators, which greatly improve convenience, efficiency, and even reliability of the determinations. The highest sensitivity and precision is usually obtained by using spectrophotometric, potentiometric, or amperometric methods to determine the endpoint rather than visual techniques. Such methods will also facilitate the automation of volumetric procedures (54).

The volumetric method has been employed in the determination of water hardness for several years. Water hardness is defined as the sum of the concentrations of calcium and magnesium, both expressed as calcium carbonate, in milligrams per liter (1). It is ordinarily determined by an EDTA (ethylenediamine tetraacetic acid) titration after the sample has been buffered to pH 10. Since the stability of Mg-EDTA is less than that of all other common divalent cations, it will not be titrated until enough EDTA has been added to titrate all the other cations in the water sample. If a small amount of magnesium ion indicator, such as Calmagite or Erichrom Black T, is added to a water containing Mg^{2+} and Ca^{2+} at a pH of 10.0 ± 0.1, the solution becomes wine red. At the equivalence point when the whole amount of Ca^{2+} and Mg^{2+} are complexed with EDTA,

the solution will turn from red wine to dark blue (1,55). This method was applied to seawater analysis by Pate and Robinson (56). To eliminate subjective errors in the determination of the endpoint, photometric endpoint detection may be used (57).

In order to determine calcium and magnesium individually, calcium may be evaluated by EDTA titration using a suitable indicator, e.g., Patton and Reeder's or Calcon (55). At a sufficiently high pH, 12–13, magnesium will largely be precipitated as the hydroxide and only Ca^{2+} will consume the EDTA (1,58). Recently, Rathore and coworkers (59) reported a new indicator for EDTA titration of Ca^{2+} and total Ca^{2+} and Mg^{2+} in water. Using 2-[(4-phenylthioacetic acid-azo]-1,8-dihydroxynaphthalene-3,6-disulphonic acid as indicator, they claim that no interferences have been observed by other metals and no masking agent is required. Calcium is titrated at pH 12–13, and for tiration of total calcium and magnesium the pH is adjusted to 10. The method has been applied to the determination of Ca and Mn concentrations in spiked tap water and diluted water samples with satisfactory results.

Calcium can also be determined by titration with EGTA [ethyleneglycol-bis(2-aminoethyl)-*N*, *N*, *N'*, *N'*-tetraacetic acid] using zincon (1-carboxyl-2'-hydroxy-5'-sulphoformacylbenzene) as indicator. Since the stability constant of Ca-EGTA is 10^6 times larger than Mg-EDTA, magnesium does not interfere with the reagent. Zincon here gives rise to an indirect endpoint with calcium. A photometric titration of seawater by EGTA/zincon reagents were performed by Kulkin and Cox (60). The sample pH was adjusted to 9.5, and the titration was performed at a wavelength of 500 nm. Other reagents, e.g., GHA [glyoxal-bis(2-hydroxyanil)], also have been used as indicator for calcium titration (57).

Methods for complexometric titration of magnesium in water samples have been well established as well (61–63). Greenhalgh et al. (64) developed a cation-exchange scheme using Amberlite CG 120 for separation and photometric titration of magnesium in seawater. After elution of Na and K, Mg is titrated directly in the ammonium chloride eluate using EDTA as titrant and Eriochrome Black T as indicator. The ion-exchange method is highly reproducible but is rather time consuming. Karolev (61) recently proposed a method based on a substitution reaction between Mg-EDTA and 8-hydroxyquinoline and titration of the liberated EDTA with calcium. The reaction is performed in homogeneous 50–60% dioxan medium at pH 10, in which magnesium forms a soluble and stable 8-quinolinol complex.

B. Spectrophotometric Method

Spectrophotometric methods are based on the formation of colored compounds with appropriate and usually specific reagents. In classical colorimetric techniques, the concentration of the colored complex in solution is estimated by visual comparison of the color intensity with that of several standard solutions of known concentration. In spectrophotometric methods, which have replaced almost entirely the visual techniques, a spectrophotometric instrument is used. In spectrophotometers, radiant energy of a very narrow wavelength range, in the visible or ultraviolet region, is selected from a source and passed through the sample solution, which is contained in a quartz cell. The amount of radiation adsorbed at a certain wavelength is proportional to the concentration of the light-absorbing chemical in the sample.

Spectrophotometry is a simple and selective technique, with a moderate sensitivity and relatively inexpensive instrumentation. These advantages in addition to its flexibility due to the many reagents available make it an appropriate method for the determination of metal ions in aqueous samples. Spectrophotometric methods are among the most widely used techniques for the determination of the major transition elements (see Table 3). They have, however, not been applied so often to the determination of the alkali and alkaline earth metals, apparently because of the lack of suitable chromogenic reagents available (82).

The use of an ion-association complex formed between a sodium or potassium crown ether complex and an anionic dye has been practiced by some researchers (94–100) for spectrophotometric determination of the alkali metals. Cromogenic macrocyclic reagents such as (2-hydroxy-3,5-dinitrophenyl)oxymethyl-15-crown-5 (98), 12-crown-4 (picrate as counterion, 1,2-dichloroethane as solvent) (99), and cryptand(2.1.1) (100) have been used for the spectrophotometric determination of sodium. Motomizu and coworkers (82) found that crown ether complexes of potassium are 10–100 times less soluble in water than the alkali metal ions themselves when they form ion-association complexes with tetraphenylborate (TPB) as a counterion. They proposed a flow injection analysis (FIA) method for the spectrophotometric determination of potassium on the basis of the precipitation reaction. Samples were injected into a carrier stream, which merged with a reagent stream containing TPB, 18-crown-6, EDTA, and saturated potassium ion. The calibration graph was linear in the range $0-10^{-4}$ MK^+. The procedure was successfully applied to river and tap waters.

Magnesium forms a colored complex with the dye Brilliant Yellow (101). The samples and standards are measured at a wavelength of 540 nm some 15–20 minutes after the addition of the reagent. Concentrations of Mg down to 0.5 mg/L can be determined by this technique. Interferences from varying concentrations of Ca and Al may be avoided by the addition of relatively large quantities of these elements. Calcium forms a colored complex with glyoxal bis-(2-hydroxyanil) di-(o-hydroxyphenylimino)ethane (101). The absorption of the buffered sample at pH 12.6 is measured at least 25 minutes after the addition of the agent at a wavelength of 520 nm. Interference of Cu can be suppressed by adding sodium diethyldithiocarbamate before the addition of the reagent. Low mg/L concentrations of Ca can be determined by this method. Recently, van Staden and Taljaard (72) proposed a sequential injection analysis (SIA) technique with spectrophotometric detector for the determination of calcium in water. The method is based on the fast complexation reaction between cresolphthalein and Ca. A frequency of 43 samples per hour was reported for the automated method, with a relative standard deviation of <1.4%. The calibration graph is linear up to 20 mg/L, and the detection limit is 0.05 mg/L.

An FIA method using pyridylazo resorcinol (PAR) as the reagent and multivariate calibration with diode array multiwavelength data evaluation with the partial least-squares method has recently been proposed by Hernandez et al. (102) for the simultaneous determination of calcium and magnesium. With this technique they could avoid some steps of the classical determinations (e.g., separation, preconcentration). The method was successfully applied to some natural samples and dialysis liquids, and a precision better than 5% was obtained.

A number of reagents are available for the spectrophotometric determination of manganese, iron, zinc, and copper. Table 3 lists examples of the reagents used in this method and their applications to water analysis. Persulfate method has been used for Mn determination for many years (103,104). In this method, dissolved Mn is oxidized to permanganate by potassium persulfate ($K_2S_2O_8$) in the presence of silver nitrate. Chloride interference can be prevented by adding mercuric sulfate. The persulfate method determines only dissolved Mn^{2+}. For the determination of total Mn it is necessary to dissolve manganese oxide (MnO_2) by adding hydrogen peroxide and phosphoric acid. The addition of $NH_4OH \cdot HCl$ is also sometimes recommended to reduce higher oxidation states and to dissolve, completely, manganese as Mn^{2+} (105).

Table 3 Examples of Reagents Used in Spectrophotometric Determination of Major Transition Elements in Water

Spectrophotometric reagent	Analyte(s)	Water sample	Conditions	Detection limit (μg/L)	Ref.
Tetraphenylborate and 18-crown-6	K	River, tap	liquid–liquid extraction	NR	82
Methylthymol blue (pH 11)	Ca, Mg	Mineral	Multivariate partial least-squares regression	NR	85
Cresolphthalein	Ca	Natural	SIA	50	72
Ferrozine	Fe(II)	Tap, river	FIA–anion-exchange column	4.3	65
SCN⁻	Fe(II), Fe(III)	Natural	FIA–optical sensor	2	66
Thiocyanate and a surfactant	Fe(III)	Rain	FIA	8	67
Ferrozine	Fe(II)	Brackish, sea, fresh	Solid-phase extraction	NR	69
Ferrozine	Fe(II), Fe(III)	Natural, waste	Solid-phase extraction	0.12	71
KSCN	Fe(III)	Fresh, sea	SIA-Chelex-100 column	20	73
1,10-Phenantroline	Fe(II)	Mineral	Standard addition	5×10^3	79
Ferrozine	Fe(II)	Natural	FIA–stopped flow	4×10^4	75
2-Hydroxy nicotionic acid	Fe(III)	—			76
3-(2-Pyridyl)-5,6-bis(4-phenylsulphonic acid)-1,2,4-triazine	Fe(II)	Sea	Solid-phase adsorption	88	83
Ferrozine	Fe(II), Fe(III)	Sea	Solid-phase adsorption	10–20	84
4,7-Diphenyl-1,10-phenantroline disulfonate	Fe(II)	Purified	Ion-exchange phase absoptimetry	10	86
2,2',2''Tripyridine	Fe(II)	Natural	Ethylenediamine	20	87
2-(5-bromo-2-pyridylazo)-5-(N-propyl-N-sulfopropyl-amino) aniline	Fe(II), Cu	Natural	FIA, double-beam spectrophotometry	50 (for Fe)	88
Ferrozine	Fe(II)	Sea	C18 cartridge preconcentration	30	89
Di-2-pyridyl ketone benzoylhydrazone	Fe(II), Fe(III)	Cloud	Chloroform water extraction	200	90
1-Amino-4-hydroxyantraquinone	Fe(III)			$<10^3$	91
Lead dioxide reactor	Mn	Natural	FIA–solid-phase oxidation	6×10^2	81
6-Methoxy-3-methyl-2[4-(N-methyl-anilino)phenylazo]-	Zn	Natural	Solvent extraction	2	92
benzothiazolinium chloride p-PAN	Zn	Natural		NR	68
N-(2,5-Dimethylphenyl)-p-toluimidoylphenyl hydrazine	Cu	River, pond		(μg/L level)	74
Potassium Pr xanthate	Cu	Waste	Solvent extraction	~10^3	70
3-(4-Phenyl-2-pyridinyl)-5-phenyl-1,2,4-triazine and picrate	Cu	Natural	Solvent extraction	2.3	77
3-Hydroxypicolinic acid	Cu	Natural	Masking by fluoride	NR	78
Dicupral	Cu	Tap		30	93
N-p-Nitro-(2-mercapto)-propionanilide	Cu	Natural	Adsorption on microcrystaline naphtalene	NR	80

NR: not reported.

An alternative method of Mn determination is the oxidation of Mn^{2+} by KIO_4 at pH 4.1–4.2 (acetate buffer) and then the addition of Leucomalachite green solution (0.08%). The produced permanganate will oxidize Leucomalachite, and its absorbance will be measured at 620 nm. A detection limit of 0.1 μg/L has been reported for the method (105). Manganese in alkaline medium gives a reddish-brown product with formadoxime, $H_2C{=}NOH$ (106). The complex is formed at a pH of about 10.5 in the presence of citrate. Total manganese is determined after heating the sample to 80°C and oxidation of the sample with persulphate.

Recently, online oxidation, spectrophotometric determination of manganese has become feasible via FIA systems (107). van Staden and Kluever (81) modified an FIA procedure for the determination of Mn^{2+}. In this method, Mn^{2+} reacts with a solid phase containing PbO_2 embedded in silica gel beads to produce the permanganate ion, which is determined spectrophotometrically. The reaction is pH dependent, and the MnO_4^- ions are measured at 526 nm. A detection limit of 0.56 mg/ L and a relative standard deviation of better than 1.8% were obtained.

Determination of iron has commonly been practiced by spectrophotometric techniques. Many of the reagents, however, form colored complexes only with dissolved Fe(II). The determination of total iron in this case is feasible by acid digestion of organic complexes and reduction of Fe(III) to Fe(II) by boiling the sample with $NH_2OH \cdot HCl$ or ascorbic acid. 1,10-Phenanthroline is a commonly used reagent of this type (1,101,108–110). It forms an orange-red complex with Fe^{2+}, which can be determined at 510 nm. The main interferences are strong oxidizing agents such as cyanide and nitrite and strong complexing agents such as phosphate and humic substances. Interferences from relatively high concentrations of chromium, zinc, cobalt, copper, and nickel also have been observed (108). Boiling with acid, adding excess hydroxylamine, and liquid–liquid extraction are among the methods used for reducing such interferences (1,108). Bathophenanthroline also forms a colored complex with Fe(II) and has been used to distinguish between Fe(II) and Fe(III) (1,111). The complex is extracted into hexanol and determined colorimetrically. Fe(III) is calculated from the total iron (after digestion) and Fe(II) concentrations by difference. Another commonly used method utilizes 2,4,6-tripyridyl,1,3,5-triazine (TPTZ) (112). TPTZ forms a violet complex with Fe(II) at pH 3.5–5.8, and ascorbic acid is used to reduce Fe(III) to Fe(II). Detection limits as low as 1 μg/L may be obtained.

Ferrozine [3-(2-pyridyl)-5,6-diphenyl-1,2,4-triazine-p,p′-disulfonic acid, monosodium salt monohydrate] as used originally by Stookey (113) and by Carter (114) is probably the most common chelating agent used for the spectrophotometric determination of Fe(II) in natural waters (108,115). King et al. (89) used a C_{18} Seppak cartridge loaded with ferrozine to determine Fe(II) in seawater. The retained ferrozine complex on the column was eluted by methanol and measured at 562 nm. The detection limit of Fe(II) was 0.6 nM in seawater. Copper(I) interferences were minimized by the addition of neocuproine to the column effluent. Blain and Treguer (84) used the same principle in an automated, shipboard determination of both Fe(II) and Fe(III) concentrations in seawater. In order to minimize contamination, filtration, acidification, reduction by ascorbic acid, preconcentration, and detection steps were all automated and included in the FI manifold. The limits of detection by this procedure were 0.1 nM for Fe(II) and 0.3 nM for Fe(III).

Zincon (2-carboxy-2′-hydroxy-5′-sulfoformazylbenzene) has been used for more than five decades for the photometric and spectrophotometric determination of zinc (116–118). Zincon reacts at pH 9 with ions to form a blue complex (116). Interferences due to heavy metals that also form complexes with zincon are removed by masking with cyanide (117). Cyclohexanone (1) or chloral hydrate (5) is added to free zinc selectively from its cyanide complex so that it can be complexed with zincon. Interferences from manganese is avoided by reduction to Mn(II) with ascorbic acid. Interferences by copper, when present in a significant amount, may be removed by ion exchange (116).

Ditizone is another reagent that provides an extremely sensitive technique for the spectrophotometric determination of zinc. Many other metals, however, also react with ditizone and produce colored complexes. To overcome the interferences, the pH is adjusted to 4.0–5.5 and sufficient sodium thiosulfate is added. It is important to use identical conditions for standard and sample, because zinc also reacts slowly with the sodium thiosulfate (1). An alternative procedure is to extract the zinc–ditizone complex with carbon tetrachloride. The extraction is performed from a nearly neutral solution containing bis(2-hydroxyethyl) dithiocarbomyl ion and cyanide ion to avoid complexation of other metals by ditizone (1,105). Detection limits down to 1 μg/L are obtained.

Many other reagents have been used for the determination of zinc. Kish and Zimomyra (92) recommended 6-methoxy-3-methyl-2[4-N-methyl-anilino) phenylazo]-benzothiazolinium chloride, which forms a blue complex with zinc in the presence of thiocyanate ion. The complex is extracted with a solution of tributylphosphate in benzene and measured at 640 nm. Concentrations of zinc in the range of 2 μg–3 mg/L can be determined with minimum interferences. Recently, Basyoni Salem (68) determined zinc in water samples and pharmaceutical preparations through complexation with 4-(2-pyridylazo)-1-naphthol (p-PAN). Complexation of zinc with p-PAN gives a dark red color at pH 2.5, which can be determined spectrophotometrically at 550 nm.

A number of reagents are available for the spectrophotometric determination of copper. Dicupral (tetraethylthiuramidisulfide) (119) forms a yellow-brown complex with Cu(II) in hydrochloric acid medium. The complex is soluble in aqueous ethanol or can be extracted into chloroform. The reaction is quite selective for copper(II), but Hg(II), Se(IV), and Ag(I) interfere seriously. High levels of NO_2^-, CN^-, and NO_3^- also interfere, but they can be removed by evaporation with concentrated HCl. A detection limit of about 30 μg/L copper is obtained. Smith and McCurdy (120) introduced Neocuproin (2,9-dimethyl-1,10-phenanthroline) in 1952. Neocuproine forms a yellow copper(I) complex in neutral or slightly acidic medium (pH 3–9) that can be extracted by organic solvents, e.g., chloroform–methanol mixture. Cu(II) must be reduced to Cu(I) with hydroxylammonium chloride or ascorbic acid. The reaction is highly selective for copper in the presence of citrate (119).

Bathocuproine disulfate (1,10-dimethyl-4,7-diphenyl1,10-phenanthroline disulfonate) (119) is another derivative of 1,10-phenanthroline for the spectrophotometric determination of copper. It forms a water-soluble red complex with Cu(I) in the pH range 3.5–11. The recommended pH range, however, is 4.3–4.5. Large amounts of Co, Cr(III), Ag, Cd, Hg, Sn, and Sb might interfere, but they are unlikely to be present in such levels in most water samples.

Recently, Kang et al. (88) developed an FIA method for the simultaneous spectrophotometric determination of Cu(II) and Fe(II). The FIA system is equipped with a double-beam spectrophotometric detector with two flow cells. 2-(5-Bromo-2-pyridylazo)-5-(N-propyl-N-sulfopropylamino) aniline was used as the reagent that forms water-soluble chelates with Cu(II) and Fe(II) in an acetate-buffered medium at pH 4.5. In one reaction coil, Fe(II) is oxidized to Fe(III) by KIO_4, so the Fe(II) chelate is not formed and only the colored complexes with Cu(II) are monitored at 578 nm. In a similar procedure in a second reaction coil, sodium ascorbate reduces Cu(II) and Fe(III) to Cu(I) and Fe(II). Thus, only

the Fe(II) complex is measured at 558 nm. The reported sample throughput is 30/h, with a precision of 1.2%.

Catalytic Spectrophotometric Technique. The ability of traces of the transition elements to catalyze reactions of other species present at much higher concentrations has been used as the basis of some very sensitive spectrophotometric techniques in recent years. Kolotyrkina et al. (121) proposed an FIA method based on the catalytic effect of Mn(II) on the oxidation of N,N-diethylaniline by potassium periodate in a neutral medium. An in-valve ion-exchange microcolumn was coupled directly to the spectrophotometer for the preconcentration of manganese in seawater. This shipboard FIA method allowed the determination of manganese in the concentration range of 10 ng/L to 20 μg/L, depending on the column enrichment time.

Hirayama and Unohara (122) studied the Fe(III) catalytic effect on the oxidation at N,N'-dimethyl-p-phenylene diamine (DmPD) by H_2O_2 and its application to the determination of trace Fe(III) in aqueous solutions. The concentration of the product is directly proportional to the time and Fe^{3+} concentration. A detection limit of 4 nmol/L was obtained for Fe(III). Batch catalytic procedures, however, are laborious and require strict control of reaction time and other conditions for accurate and reproducible determinations.

Kolotyrkina and coworkers (123) developed two FIA spectrophotometric procedures based on catalytic indicator reactions for determination of Fe(III) in seawater. A comparative study of the oxidation of p-phenylenediamine (PD) and two of its derivatives, i.e., DmPD and N,N'-diethyl-p-phenylenediamine (DePD), by hydrogen peroxide was also performed by them. In a simple flow system, the lowest detection limit of iron was obtained for DmPD, which was 0.08 μg/L. Using the second FIA system equipped with a preconcentration column packed with fibrous cellulose–diethyltriaminetetraacetic acid reduced the detection limit to 0.03 μg/L. It is claimed that the methods are virtually free of interferences. The only interference was that of copper, which could be masked by adding triethylenetetramine.

A catalytic spectrophotometric method for the determination of low μg/L concentrations of copper is based on the catalytic effect of Cu(II) on the oxidative coupling of 3-methyl-2-benzothiazolinone hydrazone with N-ethyl-N-(2-hydroxy-3-sulfopropyl)-3,5-dimethoxyaniline (124). The oxidation reaction is performed in the presence of hydrogen peroxide and pyridine, as an effective activator, to produce an intense color at $\lambda_{max} = 525$ nm. In this method, copper(II) can be determined at the 0.002–0.1-μg/L level with a precision of about 3%. The method was applied to the determination of copper in tap water and biological material, and reasonable results were obtained. A related group (125) designed a similar procedure using an FIA manifold. The method is based on the catalytic effect of Cu(II) on the oxidative coupling of N-phenyl-p-phenylene diamine with m-phenylenediamine in the presence of H_2O_2. Pyridine and NH_3 were used as activators to increase the sensitivity, and a nonionic surfactant stabilized the dye formed. A sampling rate of 30 h^{-1} was obtained. Fe(III) interference was suppressed by masking with citric acid. In general, automated FIA systems are superior and even necessary when catalytic methods are used, because they increase the reproducibility of the results and reduce the analysis time considerably. Table 4 presents examples of the catalytic reactions and their application to water analysis.

C. Flame Atomic Absorption Spectrometry (FAAS) Techniques

Among the spectrometric methods of metal analysis, FAAS is particularly suited to water analysis. It is a relatively inexpensive method, with a moderate sensitivity that is sufficiently high for the determination of the major metals in most aquatic systems. It may handle liquid samples directly with little or no chemical pretreatment. Most atomic absorption instruments are also equipped for operation in an emission mode. Sodium, potassium, and other alkali metals are typically determined by flame photometry or flame atomic emission spectrometry (FAES), because of their relatively low excitation and the simplicity of the emission techniques. For many metals difficult to determine by flame emission, atomic absorption exhibits superior sensitivity. The technique is relatively free from spectral interferences, and because of its versatility and simplicity of operation it has become the most extensively used method for the determination of metals in water.

Both absorption and emission spectrometric techniques are particularly suited to the determination of the major metals in water. Concentrations of the major transition elements, i.e., Mn, Fe, Zn, and Cu, in certain unpolluted natural waters are too low to be determined directly by these methods. A simple preconcentration technique might be included in this case to increase the sensitivity of the determinations (see Sec. III.C). Table 5 presents the limits of detection of the selected metal ions by the FAAS technique and the optimum concentration ranges. The limits of detection reported by manufacturers for their instruments, however, are in the best conditions valid for pure aqueous solutions. Field samples may contain high concentrations of matrix ions, which produce a high background. The background absorbance values are usually within the range that can be handled using a normal deuterium lamp background corrector.

In the air/acetylene flame, Na and K cause an enhancement of the absorption of both Ca and Mg. Anions such as SO_4^{2-}, PO_4^{3-}, SiO_2^-, and Al^{3+} can suppress the dissociation of the metal ions. This effect might be largely reduced in a hotter nitrous oxide acetylene flame, but ionization of Ca and Mg is more serious in this case. This latter effect is corrected by adding 2–5 g/L K^+, 5 g/L Sr^{2+}, or 10 g/L La^{3+}.

Measurement of Na, K, Mg, and Ca by the FAES or FAAS techniques usually require dilution of the water sample because their natural concentrations exceed the linear range of the methods. Recently, Lima and coworkers (135) equipped an FIA system with a dialysis unit for the automatic dilution of Ca and Mg by FAAS and Na and K by flame photometry in wastewater. Using this technique, no sample preparation was required. The results obtained were in agreement with standard methods within 2%, and the sampling rate was 140–180 samples/h.

Olsen et al. (136) used a simple flow injection system (FIAstar unit) to inject samples of seawater into a flame AAS instrument, allowing the determination of Cu, Zn, Cd, and Pb at the mg/L level at a rate of 180–250 samples/h. In order to enhance the sensitivity of the determination, an automated preconcentration system was developed by adding a small column packed with Chelex-100 to the FIA manifold. In this way metal ions in the low μg/L levels in seawater were determined. Chelex-100, however, undergoes drastic volume changes (up to 100%) in different media, which can make voids or too tight packings in the column (136). Hirata et al. (44) used a highly crosslinked macroporous version of this ligand, called Muromac A-1, which undergoes little or no volume changes. Preconcentration factors of 90–120 and detection limits in the range of 0.14–2.1 μg/L were obtained for Mn(II), Fe(III), Zn, Cu, Cd, Cr, and Pb.

A suitable ligand may be added to an aquatic sample in order to complex metal ions before their accumulation on a hydrophobic adsorbent. Fang et al. (137) proposed a continuous-flow preconcentration technique for the determination of some transition elements, including copper in tap water and seawater. A small column of 100-μL capacity was packed by C_{18} and used to collect diethylammonium diethylthiocarbamate com-

Table 4 Examples of Catalytic Reactions Used in Spectrophotometric Determination of Major Transition Elements in Water

Catalytic reaction	Analyte(s)	Water sample	System	Detection limit (μg/L)	Ref.
Oxidation of Na pyrogallol-5-sulfonate by H_2O_2	Fe(III)	Tap	FIA	2	126
Oxidation of MeOH by H_2O_2	Fe(III)	Natural, industrial	FIA	2×10^4	127
Oxidation of p-amino N,N-dimethyl-aniline dihydrochloride by H_2O_2	Fe(III)	Tap		0.05	129
Oxidation of chloropromazine by H_2O_2	Fe(total)	Ground, tap			132
Oxidation of p-phenetidine by periodate	Fe(III)	Sea	FIA	0.05	123
Oxidation of N,N-dialkyl-p-phenylene diamines by H_2O_2	Fe(III)	Sea	FIA	0.03	123
Oxidation of N,N-dimethyl-p-phenylenediamine by H_2O_2	Fe(III)	Standard		0.06	122
Oxidation of hydroxamine by dissolved oxygen	Fe(III)	Natural	FIA	2	134
Oxidation of chloropromazine by H_2O_2	Fe(II), Fe(III)	Tap, fresh		<5	133
Reaction of 2,3-dihydroxynaph-thalene and ethylenediamine in presence of H_2O_2	Mn	Potable		<0.01	131
Oxidation of N,N-diethylaniline by potassium periodate	Mn	Sea	FIA preconcentration	0.001	121
Oxidative coupling of N-phenyl-p-phenylene-diamine with m-phenylene-diamine	Cu	Natural	FIA	0.1	128
Oxidation of hydroquinone by H_2O_2	Cu	Boiler	FIA	NR	130
Oxidative coupling of 3-methyl-2-benzothiazolinone hydrogen with N-ethyl-N-(2-hydroxy-3-sulfopropyl)-3,5-dimethoxyaniline	Cu	Tap		~0.002	124

NR: not reported.

Table 5 FAAS Conditions and Detection Limits for Selected Metals Using Air/Acetylene Flame

Element	Wavelength (nm)	Detection limit (mg/L)	Optimum concentration range (mg/L)
Ca	422.7	0.003	0.2–20
Cu	342.7	0.01	0.2–10
Fe	248.3	0.02	0.3–10
K	766.5	0.005	0.1–2
Mg	285.2	0.0005	0.02–2
Mn	279.5	0.01	0.1–10
Na	589.0	0.002	0.03–1
Zn	213.9	0.005	0.05–2

plexes of the analysis in the aqueous samples. The enriched samples on the column were eluted by EtOH or MeOH directly into the nebulizer of the spectrometer. Enrichment factors of 19–25 with a sampling rate of 120/h and a precision of 1.3% for Cu were obtained.

Hashemi and Olin (49) developed a simple on-site preconcentration and sampling method for the analysis of copper in tap water. The slightly conical outlet of a 50-ml syringe was used as a small column, which was packed with only 0.05 ml of a PEI-Novarose adsorbent with polyethyleneimine (PEI) functional groups. In this technique the syringe is filled with tap water, on-site, and the sample is passed through the column by means of the plunger. No sample pretreatment is required, and the enriched column is sent to a laboratory for the elution and FAAS analysis of copper. The method was tested by distributing the devices among 22 households; satisfactory results were obtained for copper concentrations ranging from 10 to 650 μg/L. Comparison of the results with a standard analytical method resulted in a recovery of 99 $\pm$ 10% ($n = 22$) for the determination of Cu in the tap water samples.

D. Electrothermal Atomic Absorption Spectrometry

Graphite furnace, or electrothermal atomic absorption spectrometry (ETAAS), is among the most sensitive techniques for the determination of metals. The principle of the method is similar to that of FAAS, but it differs in the method of atomization. A small volume (5–80 μl) of the sample is injected into a small graphite tube that is heated electrically by a temperature program consisting of drying, charring, atomization, cleanup, and cooling steps. An argon flow protects the tube from oxidization by air. Fast heating of the graph-

ite furnace in the atomization step produces a high density of analyte atoms in the light path, which produces an unusually high sensitivity for most of metals.

The sensitivity of ETAAS is about two orders of magnitude higher than that of the FAAS method for the determination of most of metals. The analytical capabilities of this technique, however, cannot be deduced simply from the quoted limits of detection, because the analytical signals are much more strongly dependent on the matrix composition than in FAAS. It is, hence, critical to use a deuterium lamp or Zeeman effect background correction. Careful selection of the temperature and time of pyrolisis in the charring step is important for reduction of the background from the matrix. Sturgeon et al. (138) reported that by careful choice of pyrolisis conditions, most of the seawater matrix could be removed without loss of manganese and that the determination was possible by the standard addition technique. According to Nakahara and Chakrabarti (139), NaCl may be volatilized slowly at 950°C.

Volatilization of matrix salts can be facilitated by the addition of matrix modifiers. Ammonium nitrate, for instance, forms NH_4Cl and $NaNO_3$ with NaCl, which sublime or decompose at 350 and 380°C, respectively (140–142). Using this modifier, the background adsorption may be reduced to manageable levels for manganese, copper, etc. (142). Since NH_4NO_3 attacks pyrolitic graphite coating (140) and may even reduce the peak height sensitivity (138), it has been more or less replaced by other modifiers, such as $NH_4H_2PO_4$ and $(NH_4)_2HPO_4$ (143). These modifiers not only help to eliminate chloride but also stabilize volatile analytes such as Zn, Pb, and Cd, so higher char temperatures can be used (144). Guevremont (145) used citric or ascorbic acid as the matrix modifier in the determination of Zn. This modifier lowers the atomization temperature of Zn so that it is evaporated before sodium chloride, thus freeing the analytical signal from the background. The use of pyrolytic graphite and a L'vov platform is also recommended, for they facilitate the removal of matrix ions and increase the reproducibility of the signals. Recently, Xiao-quan et al. (146) used ascorbic acid as a modifier for the direct determination of Mn in river water and seawater with a tungsten atomizer. The tungsten tube atomizer is heated transversely and rapidly. Thus, it is possible to achieve relatively isothermic conditions both spatially and temporally. This might be advantageous for the analysis of samples with matrices that cause nonspectral interferences. A detection limit of 1.2 μg/L was obtained by this method for a 10-μl seawater sample.

The nominal sensitivity of ETAAS is sufficient for the direct determination of Fe, Zn, Cu, and Mn in natural waters. Saline waters, however, produce a background that is too high to be compensated for using a deuterium lamp or even Zeeman effect background correction (147). These effects reduce the sensitivity and increase the detection limits of the determination of the metals; hence, some kind of separation would be desirable. Chelating adsorbents have frequently been used for this purpose. They not only remove most of the interfering matrix salts, but also increase the sensitivity of the determination by preconcentration of the analytes. The preconcentration and separation of trace metals by chelating resins using Chelex-100 was described in detail by Kingston et al. (148) and Sturgeon et al. (149). The pH of the sample was adjusted to 5.0–5.5 and the sample passed through a Chelex-100 column. The alkali and alkaline earth metals were eluted with two 5-ml aliquots of 2.5 M HNO_3 and measured by ETAAS.

Sung et al. (150) developed a flow injection accessory for ETAAS determination of Cu and Mo in seawater. The sample was passed through a Muromac A1 column. It was then eluted by 20% v/v HNO_3 into the graphite tube by modifying the autosampler.

Tetraaza macrocycles have been found to be especially suitable ligands for the column preconcentration of manganese. Blain et al. (50) used poly(vinylbenzyltetraaza-1,4,8,11-cyclotetradecane) immobilized on an organic polymer for the preconcentration of Mn(II) from seawater prior to its determination by ETAAS. The method was applied to Mn determination in seawater, and a precision of 8% was obtained.

The analysis of sodium and potassium by ETAAS has rarely been practiced because of their naturally high concentrations and the existence of cheaper and more suitable techniques, such as flame photometry

and FAES. Contamination, however, can be better controlled in furnaces than in flames, although using high atomization temperatures might release contamination in the graphite itself (151). The determination of calcium and magnesium by ETAAS is often limited by contamination from the environment, and some type of clean-room facility is essential.

E. Inductively Coupled Plasma Methods

Inductively coupled plasma (ICP) is a partially ionized gas (typically Ar) produced in a quartz-torch using a 1–2.5-kW radio frequency power supply. Temperatures measured in Ar ICPs range from 5000 to 8000 K, depending on the region of the plasma and on the conditions used. Samples are typically introduced into the center of the plasma as aerosols (152). Inductively coupled plasma–AES has proven to be an excellent tool for the analysis of metals in aqueous samples. Its sensitivity is an order of magnitude or more better than that of FAAS. It possesses a large dynamic range, high stability, good reproducibility, and low background. Water samples may often be analyzed directly with little or no pretreatment.

Because most elements exist predominantly as singly charged ions in the plasma, ICP can be effectively used as an ionization source for mass spectrometry. The ICP-MS technique has rapidly become one of the most sensitive, accurate, and reliable techniques for elemental analysis in water. The ICP-MS instruments became commercially available in the early 1980s. A two- or three-stage differentially pumped interface is used to extract ions from the atmospheric-pressure plasma into the low-pressure mass analyzer. The mass analyzer is usually a quadrupole system (108). The ICP-MS detection limits are often up to three orders of magnitude superior to those of ICP-AES, primarily because there is little or no source of background in ICP-MS. Detection limits obtained in both ICP-AES and ICP-MS are sample dependent and are degraded if background increases or spectral overlaps are present (153). Thus, detection limits determined from solutions containing only one analyte are often not attainable when samples with complex matrices are analyzed.

The large dynamic range of the ICP techniques and the potential for simultaneous analysis provide the possibility of analysis of both major and trace elements in a single analysis run. Hence, when trace elements are determined in water samples, determination of the major cations is also recommended, to check for matrix interferences. Injection of highly saline samples may clog the nebulizer or sampling cone of the interface in the ICP-MS method. In such cases, the samples may be diluted or injected as a plug by a flow of the carrier solution in an FIA system (154). Transient signals will be obtained from the samples in this case.

Enrichment of the sample on chelating adsorbents also has frequently been used for both matrix modification and preconcentration of the samples prior to their determination by ICP-AES (8,10,43,155) or ICP-MS (41,156–158) techniques. Vermeiren and coworkers (45) determined Cu, Zn, Pb, and Cd in river water by ICP-AES after enrichment of the samples on Chelex-100. By the use of thermospray nebulization and 30-fold enrichment, a detection limit of 0.03 μg/L was obtained for Cu and Zn. Matrix effects are, however, fairly serious with this nebulization technique, and a standard addition method has to be used. Hirata et al. (43) used a small column packed with Muromac A-1 in an FIA-ICP-AES system for preconcentration of some metal ions, including Fe(III). Signal enhancement factors of 34–113 were obtained, compared to a conventional continuously aspirated system. Since Fe is not stable in the natural pH of water, a pH of 3–4 was used for preconcentration of this element. Enrichment of Cu on an 8-HQ column in an FIA-ICP-AES system was reported by Lan and Yang (155). Using an ultrasonic nebulizer equipped with a desolvation system, a detection limit of 0.07 μg/L was obtaine for Cu. The technique was applied to the determination of concentration–depth profile determination of copper in seawater.

Hashemi and Olin (8) used an iminodiacetic acid (IDA) based kinetically fast-chelating adsorbent, IDA-Novarose, which allowed high enrichment flow rates (up to 100 ml/min) in an FIA-ICP-AES system. Signal enhancement factors up to 1300 were obtained, with a detection limit of 8 ng/L for Cu after enrichment of 400 ml sample in 10 min. The effect of the specific capacity of the adsorbent on the adsorption of copper and cadmium was also studied for IDA-Novarose (159). It was found that a low-capacity IDA-Novarose with a capacity of only 10 μmol/ml adsorbent can still accumulate copper, quantitatively. The use of a low-capacity adsorbent significantly reduces the concentrations of the matrix ions and can be regenerated more quickly than a high-capacity sorbent.

Considerable efforts has been also invested in the development of online matrix separation techniques for the ICP-MS instrument using chelating ion-exchange resins. This is because of the principle disadvantage of low tolerance to dissolved solids (<0.2% m/v) and polyatomic interferences of this technique (158). Bloxham et al. (41) used Metpac CC-1 (an IDA-based) ad-

sorbent in an FIA system before the determination of Mn, Cu, Zn, and Pb in seawater by ICP-MS. The results showed that the effects of interferences can be overcome, and the method was validated by using certified reference materials. Nelms and coworkers (158) used an IDA chelating adsorbent, immobilized onto a controlled-pore glass support, for the same purpose. The use of the rigid support material reduced the conditioning time, with no dimensional change when changing solution composition. The reagent, however, showed a very low recovery (<0.35%) for Mn under the compromised conditions used.

The precision and accuracy of the ICP-MS measurements can also be significantly improved by using the isotope dilution method. Since another isotope of the same element is used as the internal standard, accurate results can be obtained even in complicated matrices. When there is sample loss during the preparation or pretreatment processes also, high accuracy is expected to be obtained by this method. Hwang and Jiang (157) recently used a flow injection isotope dilution ICP-MS method for the determination of Zn in several water samples. The isotope ratio for each injection was calculated from the peak areas of the flow injection peaks. They also equipped the flow system with an SO_3^- quinoline-8-ol carboxymethyl-cellulose column and gained enrichment factors up to 17 for zinc determination in water samples.

F. Electrochemical Methods

Electrochemical methods have been used extensively in the analysis of major and minor elements in aquatic systems. In this section the application of the two most important electrochemical techniques, i.e., voltammetric and potentiometric methods, in the analysis of the major metals in water will be discussed.

1. Voltammetric Techniques

Voltammetry and polarography are methods for studying the composition of electrolytic solutions by plotting current–voltage curves. The voltage applied to a small polarizable electrode (relative to a reference electrode) is increased negatively over a span of 1 or 2 V, and the resulting change in current is noted. *Voltammetry* is a general name for the method; the term *polarography* is usually restricted to applications of the dropping-mercury electrode. The attributes of the methods and their applications to water analysis have been reviewed, for instance, by Nürnberg (160) and van den Berg (161).

Voltammetric methods have the advantages of sensitivity, relatively small sample volume required, and amenability toward real-time shipboard analysis (162). They appear to provide one of the best opportunities for experimental modeling of the bioavailability of transition elements and their complexes with organic and inorganic ligands (163). In order to increase the sensitivity of the method, a preconcentration step is usually used. This preconcentration is carried out in situ at the working electrode without additional contamination risks (5). The detection limits obtained will be dependent on the experimental conditions and sample matrix. Wang et al. (Ref. 61 from 108) report a detection limit of 0.2 nmol/L of Fe(III) with a 2-min preconcentration time onto a hanging mercury drop. Farias et al. (164), on the other hand, report a detection limit of 1.8 nmol/L of Fe(III) with a 5-min preconcentration time using the same electrode. With the voltammetric methods, the sample usually has to be buffered to a certain pH. This may limit the usefulness of the method, because in the process of pH adjustment the speciation of metals may change.

Differential pulse anodic stripping voltammetry (DPASV) using a hanging mercury drop (HMD) or a rotating glassy carbon thin mercury film (TMF) electrode has been effectively utilized for the determination of Cu, Fe, Zn, and Mn concentration and speciation in natural waters (162,165,166). The TMF electrode usually provides a much greater sensitivity than the mercury-drop electrode. This is because of the large surface-to-volume ratios obtained in the TMF electrodes (165). While the detection limit of Cu and Zn by DPASV is typically 0.1–0.05 μg/L for an HMD electrode, it is in the range of <0.001 to 1 μg/L when a TMF electrode is used (166). The mercury film is formed in situ (10–100-nm thickness) on a glassy carbon substrate, usually from an $Hg(NO_3)_2$ solution.

The solubility of Cu in Hg is very low. Hence, the determination of Cu with the TMF electrode requires special consideration because the plating time of the mercury film and the instrumental gain affect the anodic stripping response. Recently, Daniele et al. (165) reported an approach to the calibrationless determination of copper and lead by ASV at TMF microelectrodes. Microelectrodes have some advantages over the conventional-size electrodes. Nonplanar diffusion occurs, because of their small size, and a steady state is reached rapidly. This will provide ideal deposition conditions, with no need for stirring devices during the preconcentration step. The authors derived an equation to be employed for the calculation of the concentrations

in a method not demanding calibration. The sample, however, has to be acidified, and the method can be applied only to water samples containing <1 mM chloride.

Electrochemical determination of Mn(II) has been reported by both anodic (167–169) and cathodic (170,171) stripping voltametry. A cathodic stripping voltametry (CSV) technique was used by Tanaka et al. (172) for the determination of Mn(II) in an ammonium nitrate solution. No interferences were reported, and the pH range required (6–9) is appropriate for the analysis of natural waters. Roitz and Bruland (170) applied a similar method to the analysis of Mn(II) in seawater. This method involved a deposition step whereby dissolved Mn(II) is oxidized to Mn(IV) oxide on a glassy carbon electrode, then the measurement of the cathodic stripping current is performed as deposited Mn(IV) oxide is reduced back to Mn(II). The differential pulse gives a much larger peak height than does a linear scan. A detection limit of 6 nM may be obtained for the Mn cathodic peak at +0.23 V.

Zhiquiang et al. (173) utilized chemically modified carbon paste electrodes containing 1,10-phenantroline and Nafion as modifiers for preconcentrating Fe(II) from dilute solutions. Iron(II) is rapidly collected on the modified carbon paste electrode, and the resulting surfaces are characterized by cyclic and differential pulse voltammetry. The differential pulse peak current at +0.83 V (versus standard calomel electrode, SCE) was found to be more suitable for practical analysis. A preconcentration period of 5 min permitted convenient measurements at concentrations down to 10^{-8} mol/L with a relative standard deviation (RSD) of 4.0% ($n = 10$). No major interferences were observed when the analysis was carried out in the presence of an excess of various coexisting metal ions (108). Differential pulse polarographic determination of iron(II) based on the formation of a 5,5-dimethylcyclohexane-1,2,3-trione-1,2-dioxime-3-thiosemicabazone-iron(II) complex was reported by Diaz et al. (174), and the method was applied to the determination of iron(II) in acids, waters, wines, and fruit juices (108).

Safavi and coworkers (175) studied copper preconcentration on a modified carbon paste electrode with 1,2 - bismethyl(2 - aminocyclopentane - carbodithioate) ethane. They mixed the reagent with the graphite powder and nujol oil for construction of the electrode. The electrode was used in a flow system for reproducible preconcentration and subsequent determination of Cu(II) accumulated by differential pulse voltammetry. The electrode could be rapidly renewed by acid. The chemically modified electrodes (CMEs) represent sufficiently sensitive and selective voltammetric sensors, and it is assumed that they allow the determination of free metal ions as well as those gradually released from kinetically labile complexes by dissociation during the accumulation step (176).

A disadvantage of loaded electrodes by immobilization of complexing agents is the two-phase reaction zone and the adsorption or penetration of complexes or ligands from the solution into the modifying layer. A Nafion film, containing a complexing agent, may be used for coating the electrode for solving this problem. This method was used by Labuda et al. (177) for the determination of copper in river water. Concentrations in the range of μM to nM copper could be measured with an RSD of 1–6%. They used three types of reagents, i.e., derivatives of 4-acylpyrazolone (1-phenyl-methyl-4-octanoylpyrazol-5-on and 1-phenyl-4-strea-roylpyrazol-5-on), which react specifically with copper in a weak acidic medium, and oxine, as the electrode modifiers. The use of CMEs for the analysis of natural waters containing organic chelators, however, might be limited because of the competitive equilibrium between the natural and CME chelators.

The amperometric determination of magnesium after complexation by an electroactive ligand also has been reported (178,179). Downard et al. (179) determined magnesium in a flow injection system by formation of the magnesium-Eriochrome Black T (EBT) complex at pH 11.5 (5% 1,2-diamino-ethane in 0.1 M KCl) and amperometric measurement of the excess of EBT at +.175 V on a glassy carbon electrode. pH and concentrations of EBT and masking agent were chosen so as to minimize interference from calcium.

2. Potentiometric Techniques

In this section two common potentiometric techniques based on ion-selective electrodes and stripping methods will be discussed.

Ion-selective electrodes (ISEs) provide rapid and selective potentiometric techniques for the determination of the major cations in water samples. The ISEs fall into three categories: solid state, liquid ion exchangers, and neutral carriers. Certain glass electrodes, which belong to the first category of electrodes, have been commonly used for the determination of Na^+, K^+, or total concentration of univalent cations (180). Copper-selective electrodes are typically made from precipitates of insoluble copper salts, e.g., CuS (mixed with Ag_2S to increase conductivity) (181,182). Calcium-selective ISEs based on liquid ion exchangers are commercially available. A hydrophobic liquid membrane, such as

dioctylphenylphosphonate, is used as the sensing element, while an alkyl phosphate ester is used as the chelating agent (180). Maj-Kurawska (183) described an ISE based on the neutral carrier N,N''-octamethyl-enebis(N'-heptyl-N'-methyl malonamide) for the determination of water hardness. The sum of Ca^{2+} and Mg^{2+} concentrations can be determined directly with this electrode (divalent ISE), without the addition of any auxiliary complexing agent such as acetylacetone. With high levels of acetylacetone (0.1 M) the electrode can also be used for the selective calcium determinations. Sun et al. (184) developed a copper(II)-selective electrode based on a molecular deposition technique in which water-soluble copper phthalocyaninetetrasulfate was alternatively deposited with a bipolar pyridine salt ($P_yC_6BPC_6P_y$) on a 3-mercaptopropionic acid–modified Au electrode. A detection limit of 7.0×10^{-6} mol/L and a linear dynamic range between 1×10^{-5} and 0.1 mol/L were obtained for copper. The electrode response is fast, with low resistance and good reproducibility. Fe^{2+}, Fe^{3+}, and Co^{2+}, however, interfere in the determination of Cu^{2+}.

The ISEs have been also used in FIA systems for the automated determination procedures for metal ions (185–187). Van Staden and Wagener (187) used a homemade coated tubular solid-state copper(II)-selective electrode in a flow injection system for the copper determination. Interferences of matrix ions in the FIA system were found to be similar, but less severe, compared to those found in the batch analysis. A similar construction using a tubular membrane calcium-selective electrode was used by a Chinese group (185) for FIA potentiometric determination of Ca in high-purity waters. A detection limit of 2.8×10^{-8} mol/L was established for Ca, when a metal ionic buffer was used for the elimination of interferences.

Sun et al. (188) described a new method for the determination of Fe(III) based on both nonlinear regression calibration plots and parabolic interpolation using a fluoride ISE. This technique is based on the strong complexation of Fe(III) with fluoride. The appropriate pH is 3, and the suitable total concentration of fluoride in order to complex all the iron(III) is equal to three times the highest concentration of Fe(III) in the standard series. This method has been used successfully in the determination of iron in mineral waters.

Potentiometric stripping analysis (PSA) is another commonly used potentiometric technique in water analysis. The facts that in most cases this technique can be applied directly to the analysis of aqueous samples without previous treatment and that it is virtually free from interferences of dissolved oxygen make it highly attractive for elemental analysis. Both PSA and ASV techniques are based on the same principle; i.e., the analytes are first deposited on the electrode surface while the solution is stirred, and then stripped back to the solution in the measurement step (189). The ASV technique works on a film electrode, usually made of electrochemically deposited mercury or gold on a glassy carbon support. One drawback of ASV is that an adsorbed organic layer is likely to diminish the amount of electroplated metal and may cause nonlinearity in the stripping current–deposition time relationship (190). The use of PSA largely overcomes this drawback because an adsorbed organic layer on the electrode would exhibit similar resistance to mass transfer of the analyte during the electrolytic desorption step (191,192). The presence of organic dipoles in the sample may generate "tensammetric" peaks in pulsed ASV, which can be mistaken for metal stripping peaks (193), while such problems are not expected to occur with PSA. This technique also requires simpler equipment compared to ASV (163). The PSA technique can compete with nonelectroanalytical techniques in terms of price and possibility of automation, but its general use is still rather limited (189).

Jagner et al. (194) described an automated constant-current stripping potentiometric method for the determination of Fe(III) as its solochrome violet complex. A mercury film electrode was used here. The method can be utilized over a large concentration range, 0–500 μg/L, by an automated adoption of the experimental conditions to the concentrations found in the samples. The method was applied to the determination of iron(II) in tap water and rainwater samples. Hassan and Marzouk (195) described a ferroin membrane sensor for the potentiometric determination of Fe(III) and/or Fe(II) for both batch and FIA determination. The technique is based on the formation and monitoring of ferroin with a PVC membrane sensor containing ferroin-TPB as an electroactive component plasticized with 2-nitrophenyl phenyl ether. The ferroin sensor was successfully applied to the determination of iron in water, alloys, rocks, and pharmaceuticals.

The analytical signal in PSA, unlike that in ASV, is independent of the electrode size, so reduction of the electrode size for use in flow injection systems is feasible (196). Considerable enhancement of PSA sensitivity can be achieved in the FIA systems through flow reversal. Multiple passes of the sample plug over the working electrode gives much increased electroplating efficiency. Based on this principle Chow's group (163,196) developed a technique called oscillating flow injection stripping potentiometry and applied it to the

determination of copper in natural waters. They compared this method with other analytical techniques, e.g., ICP-AES, ASV, and batch PSA, and calculated that it is suitable for fast and inexpensive trace heavy-metal analysis in natural waters with unknown matrices. Gil and Ostapczuk (189) described a PSA method with a gold-film working electrode for simultaneous determination of four elements, including copper in river water and reference materials. Gold is a good alternative to mercury since it is a noble material that can either be used as a disc or fiber electrode or deposited from a solution of Au(III) to form a stable and reproducible film. Many elements are soluble in gold, thus enabling its wide use in electrochemical stripping analysis.

Portable PSA utensils also have been developed. A portable potentiometric stripping analyzer was described by Jagner et al. (197) for the determination of Cu and Pb in tap water. Tap water samples of 400 μl were added to 200 μl matrix-modifying solution consisting of hydrochloric acid, calcium chloride, mercury (II), ethanol, Triton X-100, and bismuth(II) as internal standard, contained in disposable 2-ml Eppendorf sample tubes. Copper in the range of 0.1–5 mg/L was determined in less than 1 min, with a precision of ca. 15%.

G. Ion Chromatography

Ion chromatography (IC) is usually characterized by selectivity, sensitivity, and the possibility of determining many cations in a single run. This technique has been most commonly used for the determination of anionic species, apparently due to the fact that other methods, e.g., atomic absorption and emission spectrometry, can successfully compete with IC in the determination of metal cations. A number of articles, however, describe the use of IC for the separation, speciation, and determination of metal ions as well. Early applications of ion chromatography were limited to the determination of alkali and alkaline earth metals. Nevertheless, today it is widely used for the separation and quantitation of the transition elements as well.

In the ion chromatographic techniques, metal ions are separated typically on a low-capacity cation-exchange resin. In single-column IC, the operation of low-capacity resins provides the possibility of using dilute solutions with a reasonably low conductivity as eluents. In suppressed IC, on the other hand, a second column is used in order to convert the ions of the eluent to molecular species of limited ionization, without affecting the analyte ions. This efficiently reduces the background conductivity before conductometric detection. In this method a high-capacity anion-exchange column in hydroxide (198), nitrate (199), sulfate (200), IO_3^- (201), or other anionic forms is usually used. One problem with such suppressor columns is the need to regenerate them periodically. Recently, fiber membrane suppressors have become available that work continuously.

Both single- and suppressed-column IC methods have been utilized for the separation and determination of the major cations in water. Singh (202) described a suppressed IC method with Dionex CS-1 and CS-2 separator columns and a Dionex cation micromembrane suppressor. Using a mixture of 2.5 mM ethylenediamine, 5 mM hydrochloric acid, and 1 mM Zn as eluent, Na, Sr, Mg, and Ca ions were separated in high-salinity subsurface water and seawater. Sometimes a third column is added in suppressed IC, between the suppressor and the detector. This column will replace the analyte ions with another ion in order to improve the detection sensitivity. Nordmeyer et al. (200) used a column packed with high-capacity H-form cation-exchange resin between the suppressor and the conductometric detector to replace Mg and Ca with proteins. Owing to the high equivalent conductance of the proton, conductometric detection is more sensitive with this method. Such techniques, however, make the regeneration of the suppressors more complicated.

The separation and determination of alkali metals in IC is usually straightforward. In suppressed IC a dilute solution of strong acids [(e.g., 3 mM HNO_3 (203) or 5 mM HCl (204)] usually is used. The cations are detected conductometrically in the form of their corresponding hydroxides (205). Single-column IC with spectrophotometric (206–208) and both direct (209) and indirect (210,211) conductometric detection have also been utilized in the analysis of Na^+, K^+, and other alkali metal ions in environmental waters. In single-column conductometric determination of Na^+ and K^+, a dilute solution of strong acids (210,211) or a lithium salt (209) may be used. Since the equivalent conductivities of Li^+ are less than those of Na^+ and K^+, the detection of these cations is based on positive peaks. Alkali metals don't adsorb in the UV, so they are detected by indirect method. Hayakawa and coworkers (206) used a dilute solution of Cu(II) sulfate as the eluent. The separation was carried out on a Zipax SCX column followed by detection at 220 nm. Figure 4 depicts the chromatogram of the alkali metal cations (207). Haddad and Foley (212) studied the properties of a number of aromatic bases to be used as eluents in single-column IC both with conductometric and indirect UV detectors. Table 6 presents the detection limits

obtained by different eluents for sodium and potassium. As seen in the table, nitric acid gives the best results for conductometric detection of sodium, but for potassium the detection limits are similar for nitric acid and the aromatic bases used.

Calcium and magnesium determination by different IC techniques have commonly been practiced. The doubly charged cations are adsorbed on cation-exchange resins much more strongly than are singly charged cations; hence, they require high-capacity eluents and low-capacity cation exchangers. The elution is usually carried out with doubly charged cations of some transition metals or organic bases, e.g., Pb salts (200,201), $Zn(NO_3)_2$ (198), ethylenediaminechloride (199), ethylenetartarate (199), n-phenylenediamine (213), dipiconilic acid (214), and 2,6 pyridinedicarboxilic acid (215). In order to reduce the retention time of cations and increase the separation efficiency, some type of ligands are added to the eluent (B9). A chromatogram of pond water is shown in Fig. 5 (201). Ca and Mg here are separated using a 1 mM $Pb(NO_3)_2$, pH 4 solution as eluent with a flow rate of 2.3 ml/min. Conductometric detection and a suppressor column in the IO_3^- form were used.

Ohta et al. (215) described a single-column IC for

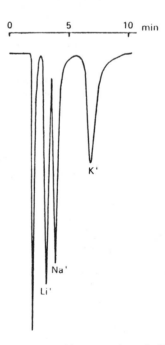

Fig. 4 Ion chromatographic separation of alkali metal cations on a Zipax SCX, 4.6 × 120-mm column. Indirect UV detection at 220 nm. Eluent: 0.25 mM copper sulfate. Flow rate 0.5 ml/min. (From Ref. 207.)

Table 6 Detection Limits of Na^+ and K^+ Determinations by IC with Different Eluents and Injection of 100 μl Sample

Eluent	Detection mode	Detection limits, $\mu g/L$	
		Na^+	K^+
2,6-Dimethylpyridine	Conductivity	2.1	0.6
	Indirect UV	0.8	0.9
2-Methylpyridine	Conductivity	1.1	0.6
	Indirect UV	1.2	3.0
2-Phenylethylamine	Conductivity	0.6	0.4
	Indirect UV	2.0	1.8
3-Methylbenzylamine	Conductivity	0.8	0.4
	Indirect UV	0.8	0.7
Benzylamine	Conductivity	0.4	0.4
	Indirect UV	1.3	2.3
Nitric acid	Conductivity	0.4	0.4

Source: Ref. 212.

UV-photometric determination of Ca and Mg and inorganic anions in river water samples. Pyromellitate–CH_3OH–water was used as the eluent with a SiO_2-based anion-exchange column. Anions were detected indirectly, while Mg^{2+} and Ca^{2+} were detected directly because complexes having strong UV absorption were formed between pyromellitate and the cations. Electrochemical detection of Ca and Mg was practiced, for instance, by Hadded et al. (216). They used a copper electrode, which is sensitive to ions and molecules that form strong complexes with Cu. A mixture of 1 mM tartarate and 1 mM diethylenetriamine (pH 4.6) was used as eluent here. This will produce a high background signal, which is reduced by Ca, Mg, etc. Indirect detection of the cations is, hence, achieved.

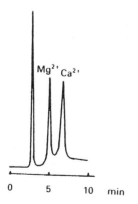

Fig. 5 Ion chromatographic separation of alkaline earth metals in pond water on a high-capacity IO_3^- form anion exchanger. Conductometric detection. Eluent: 1.0 mM $Pb(NO_3)_2$, pH 4.0. Flow rate 2.3 ml/min. (From Ref. 201.)

The separation and determination of Na, K, Ca, and Mg in a single run by IC also has been achieved by many research groups (217–219). A Japanese article (217) describes the separation and quantitation of these ions in ultrapure waters. Shim-Pack IC GC2 (4.6-mm ID × 10 mm) and Shim-Pack IC (4.6-mm ID × 125 mm) were used for preconcentration and separation, respectively. Up to 500 ml sample was injected, and detection limits <1 ng/ml were obtained. The eluent was a solution of 5 mM tartaric acid and 1 mM 2,6-pyridine carbonic acid.

The determination of transition elements, including Cu, Mn, Fe, and Zn, are more often performed by single-column (220–226) rather than suppressed IC (199,200). Spectrophotometric detectors have frequently been used for this purpose (220–222), usually with a postcolumn reaction, to convert the cations into intensely colored compounds, which are detected in the visible spectral region (205). Other types of detection methods, such as conductometric (199–200), electrochemical (223,224), coulometric (225), and indirect AAS (226), also have been used.

Jones et al. (220) determined Mg, Fe, Zn, Co, and Ni in the cooling fluids of a nuclear reactor. The cations were separated with tartaric acid on the column packed with Aminex A9 cation exchanger and then converted into colored compounds with postcolumn reaction by Eriochrome Black T and determined spectrophotometrically. Electrochemical (223,224) and coulometric (225) detection of the transition elements has also been performed using tartaric acid as the eluent.

The separation of a number of transition and rare-earth elements on a Nucleosil SA-10 column has been realized employing gradient elution (222). Fe, Cu, Zn, Ni, Co, Pb, and Mn were eluted with a gradient concentration from 0.04 to 0.12 M of a sodium lactate solution, followed by a postcolumn reaction with PAR-Zn-EDTA (pyridylaro resorcinol) and spectrophotometric detection at 510 nm. Rare-earth metals were eluted with a 0.01–0.04 M solution of 2-methylacetic acid and reacted with Arsenazo I before detection at 600 nm. Table 7 summarizes examples of applications of IC to water analysis.

H.   Other Chromatographic Methods

Chromatographic methods other than IC have also been used extensively for the separation and quantitation of the major metals in water. Their application, however, is limited mostly to the complex-forming metals. The chromatographic separation of metal ions started by utilizing paper chromatography in the 1930s. Thin-layer chromatography (TLC), whose applications were initially limited to organic compounds, was used extensively for metal ion separation. Numerous neutral metal complexes have been examined by TLC on a variety of adsorbents. Among the most widely used ligands are dithizone and substituted dithiocarbamates, which show good reactivities toward metal ions (228,232–234). A common application of TLC is its use as a quick method for predicting the behavior of metal complexes in high-performance liquid chromatography (HPLC) (235,236).

High-performance LC have been widely used for the separation of transition metals' complexes with different ligands. The complex-forming ligands are usually added to either the mobile or the stationary phase (237). One limitation of this technique for metal analysis has been the lack of suitable detection devices for the determination of metal ions. Matrix interferences and the low sensitivity of the conventional HPLC detectors have led to the development of online coupling of HPLC with AAS, AES, or mass spectrometry (238–240). With the selection of suitable complexing agents, however, conventional detectors such as spectrophotometric (241–245) and electrochemical (242) have also been successfully used as well.

The separation of the major transition metals by reversed-phase HPLC have become feasible after precolumn derivatization with ligands such as dithiocarbamic acids (246), 8-hydroxyquinol (247) *B*-diketones (248), and 4-(2-pyridylazo) resorsinol (249). Nagaoso et al. (243) separated trace levels of Fe(III) and Cu(II) on an ODS column after derivatization with cupferron. Amperometric detection was used for the simultaneous de-

termination of the analytes, and a better sensitivity was obtained with this method compared to the usual spectrophotometric detection. The method was applied to the determination of Fe(III) and Cu(II) in biological samples and natural waters. A related group (242) tried the separation of Fe(III), Mn(II), and Al(III) in natural waters on an ODS column after this complexation by 8-HQ. A 2:3 mixture of acetonitrile/20 mM acetate buffer containing 5 mM 8-HQ reagent was used as the eluent. Both spectrophotometric and amperometric detectors were used, but the sensitivity of the former was better than that of the latter in this case, while the latter was more specific. For the amperometric detection, a thin-layer flow cell and a glassy carbon working electrode was used. Analytical data obtained on river and seawater samples were in agreement with expected values.

An alternative HPLC method for the separation of metal ions that require no precolumn derivatization is ion-pair or ion-interaction chromatography. In this method a reversed-phase column is converted into a low-capacity ion exchanger by using an eluent containing an anionic surfactant agent. One can easily achieve the desired separation efficiency in this method by varying the concentration of either the organic modifier or the ligand. In a recent paper, Zappli et al. (241) described an application of this technique to the determination of μg/L levels of Fe(III), Cu, Zn, Mn, Pb, Ni, and Co in natural waters. The eluent used was a mixture of an alkanesulfonate (as the ion interaction reagent), tartaric acid, and acetonitrile (or methanol). Tartaric acid was also added to the samples before their filtration in order to improve the recoveries.

A gel filtration chromatographic method was described by Itoh and coworkers (250,251) for separation with sequential UV adsorption and ICP-MS detection of large organic complexes of metal ions in pond water. The sample was preconcentrated by ultrafiltration prior to the chromatographic separation. More than 40 trace metals were found in the organic molecules with molecular weights of approximately 300,000 and 10,000–50,000.

I.   Luminescence Techniques

Photoluminescence and chemiluminescence methods have been most commonly used for the determination of organic compounds, but they are also used in the field of inorganic speeds (252). Very few inorganic compounds in natural waters exhibit fluorescence or phosphorescence of sufficient intensity for analytical use. There are, however, many metal ions that can be

Table 7 Examples of Applications of Different Modes of IC to Water Analysis

Analytes	IC method	Eluent	Detection mode	Water sample	Ref.
Na, K, . . .	Suppressed	3 mM HNO$_3$	Conductivity	Geothermal	203
Na, K, . . .	Suppressed	5 mM HCl	Conductivity	Snow and ice	204
Na, K, . . .	Suppressed	3 mM HNO$_3$	Coulometric	Natural	229
Na, K, . . .	Single column	1.2 M HNO$_3$	Conductivity (indirect)	Tap	210
Na, K, . . .	Single column	2.5 mM LiCl	Conductivity	Standards	209
Na, K, . . .	Single column	0.25 mM CuSO$_4$	Conductivity (indirect)	Drinking	207
Na, K, . . .	Single column	5 mM HCl	Conductivity (indirect)	Underground	211
Mg, Ca	Single column	Methane sulfonic acid	Conductivity	Environmental	230
Mg, Ca	Single column	Dipiconilic acid + HNO$_3$	Conductivity	Environmental	214
Mg, Ca	Suppressed	Pb(II) and Ba(II) solutions	Conductivity	Natural	200
Mg, Ca	Suppressed	0.1 mM Pb(NO$_3$)$_2$ + HNO$_3$	Conductivity	Natural	201
Mg, Ca	Suppressed	Zn(NO$_3$)$_2$, HNO$_3$	Conductivity	Standard	198
Ca, . . .	Suppressed	*m*-phenylene-diamine + HNO$_3$	Coulometric	—	229
Mg, Ca, . . .	Single column	1 mM ethylenediamine, HNO$_3$ (pH 6)	Conductivity	Tap	210
Mg, Ca	Single column	2 mM ethylenediamine, 2 mM citrate (pH 5)	Conductivity	Drinking, mineral, river	231
Mg, Ca	Single column	2.5 mM Cu(SO$_4$)	Photometric	Drinking	207
Mg, Ca	Single column	1 mM tartarate + 1 mM diethylenetriamine	Electrochemical	—	216
Cu, ZN, Mn, . . .	Suppressed	Ethylenediamine, tartarate	Conductivity	—	199
Mg, Fe, Zn	Single column	Tartaric acid	Photometric	Nuclear reactor cooling	220
Mn, Fe, Zn, Cu	Single column	Sodium lactate	Photometric	Standard	222
Fe, Cu, . . .	Single column	Tartaric acid + citric acid	Electrochemical	Sea	223,224
Zn, Cu, . . .	Single column	Na tartarate + tartaric acid	Coulometric	—	225
Zn, Mn, . . .	Single column	Cu(II), 1 mM	AAS (direct)	—	226

induced to form fluorescent or phosphorescent complexes with suitable ligands. A representative list of fluorometric reagents is given in Table 8.

Photoluminescence procedures have been used in both the direct and indirect modes. In a direct procedure, a fluorescence chelate is formed with the analyte, and its emission is measured. In an indirect method, the influence of the analyte on the luminescence of other species (e.g., by quenching) is measured (252). The instrumentation in the photoluminescence technique is simple, and the detection limits are in the range of 0.1–10 μg/L (144). It is often important to remove interfering elements before analysis using extraction or ion-exchange techniques.

Quin-2 as a fluorimetric reagent has been shown (257) to provide good sensitivity and selectivity for cal-

cium (log k_{Cal} = 7.1) against magnesium (log k_{MgL} = 2.7). Using the reagent in a flow injection fluorimetric system showed few matrix effects (253). Highly concentrated sodium chloride solution (ca. 5 M) could be injected directly into the carrier with no significant change in the sensitivity. With the invention of the pulsed laser, there has been a considerable improvement in the limit of detection in fluorescence analysis. The pulsy nature of the source provides the possibility of improving the signal-to-noise ratio by rejecting scattered radiation and luminescence from impurities in the sample (258).

Chemiluminescence is based on the catalysis or inhibition of the alkaline oxidation of a luminescent reagent by (usually) metal ions (144). The instrumentation for chemiluminescence measurement is remarkably

Table 8 Some Fluorometric Reagents used for Complexation of the Major Metals

Metal	Fluorometric reagent	Ref.
Ca	2-[(2-Amino-5-methylphenoxy)methyl]-6-methoxy-8-aminoquinoline-N,N,N',N'-tetraacetic acid	253
	Calcein	252
Cu	2,2′-Bipyridine ketone azine	252
	Rose Bengal + 1,10-phenantroline	252
	2,9-Dimethyl-4,7-diphenyl-1,10-phenantroline	252
	p-Hydroxybenzylidene-5-(8-quinolylazo)-8-aminoquinoline	254
	7-(8-Hydroxy-3,6-disulfonaphethylazo)-8-hydroxyquinoline-5-sulfonic acid	255
Fe	Eosin + 1,10-phenantroline-sulfonic acid	252
Mg	8-Hydroxyquinoline-5-sulfonic acid	256
	Pontachrome blue-black R	252
Mn	8-Hydroxyquinoline-5-sulfonic acid	252
	2,3-Diketogulonate + peroxidase	252
Zn	2-(4-Methyl-2-pyridyl)-5(6)-phenylbenzinidazole	252

simple and may consist of only a suitable reaction vessel and a photomultiplier. The sample is usually mixed with reagent in an FIA systems. The reagent is usually a buffered solution of luminol containing hydrogen peroxide together with some additives that may enhance the luminescence or suppress interfering elements. Luminol is probably the most common reagent that has been used in this technique, for instance, for the determination of Cu(II), Fe(II), Mn(II), and Zn(II) (252). In general, chemiluminescence provides lower detection limits than do fluorometric techniques, but it tends to be more susceptible to interference and less precise than fluorescence or phosphorescence methods.

J. Other Methods

In addition to the methods already discussed, there are a number of other, probably less extensively used methods that have been applied to the analysis of the major metals in water. A brief discussion of some of them with examples of their applications to water analysis will be given here.

X-ray fluorescence spectrometry (XRFS) has been used for the determination of the major and minor metals in water since the 1970s. The method has a large dynamic range for metals, from low mg/L to 100%. Its detection limit in the direct analysis is in the low mg/L level in solution and between 0.01 and 0.1 μg in thin film. The detection limit of the method for heavy elements (down to about 0.1 mg/L) is lower than for light elements (up to about 50 mg/L) (144). Chelating ion exchangers in column packing or membrane forms have been used for the preconcentration of water samples prior to the XRFS determination of metal ions.

A membrane has the advantage that it can be mounted directly in the spectrometer for analysis. Enrichment factors in excess of 1000 have been achieved for the preconcentration of K, Ca, Mn, Cu, Zn, etc. using a chelex-100 membrane (259). Fe, Zn, Cu, Cr, and Pb were determined in drinking water after preconcentration on Ostsorb OXIN, a chelating ion exchanger (260). The method here was radionuclide XRFS with a ^{238}Pu source, a Si/Li semiconductor detector, and a multichannel analyzer. Solvent extraction also has been used for preconcentration prior to XRFS determinations. Holynska et al. (261) preconcentrated Cu, Hg, and Pb in drinking water by complexation using a mixture of carbamates followed by solvent extraction with methylisobutyl ketone. They used the total-reflection XRFS method for the final determination of the metals, and a minimum detection limit of 40 μg/L was obtained for Cu.

Neutron activation analysis (NAA) has different sensitivities for different metals. Its sensitivity for some metals is better than that of all the analytical techniques. Factors such as sample composition, neutron flux, irradiation time, decay time, counting time, and detector efficiency affect the final sensitivity of measurement. Detection limits of 0.2, 2, 5, and 0.25 μg/L have been reported, respectively, for the determination of Mn, Fe, Zn, and Cu in freshwater by this method (262). Different preconcentration methods have also been practiced in the NAA of metal ions. For instance, a coprecipitation method was recently used by Rao and Chatt (263) for the neutron activation determination of some elements, including Mn, Cu, and Zn. 1-(2-Thiazolylazo)-2-naphthol, pyrrolidine-dithiocarbamate, and N-nitrosophenylhydroxylamine were used as coprecipitation reagents. Quantitative recoveries were obtained for the elements with preconcentration factors on the order of 104. A precision of 5–10% and detection limits at the μg/L level were obtained. The method was applied to seawater and drinking water samples and biological material. Another common preconcentration method is the use of chelating adsorbents. An 8-HQ-loaded Amberlite XAD-2 column was used by a related group for preconcentration of Zn, Cu, Co, Hg, Cd, and V prior to their determination by NAA method (264). Quantitative recoveries were obtained for most of the metals at pH 6.0. Irradiation and neutron activation analysis was performed directly on the resin, without eluting the metals from the column. Detection limits of 0.01–3 μg/L were obtained.

Capillary zone electrophoresis (CZE) also has been practiced for the determination of metal ions, especially the major alkali and alkaline earth metals, in water (265–269). The determination of Na, K, Mg, and Ca ions in mixtures of seawater and formation water from oil wells by CZE was recently reported by Tangen et al. (265). A solution of 5.0 mM 4-methylbenzylamine, 6.5 mM hydroxyisobutyric acid, 6.2 mM 18-crow-6, and 25% (vol/vol) methanol was used as the carrier electrolyte. Using an indirect photometric detection technique, detection limits of 0.14, 0.15, 0.08, and 0.08 mg/L, respectively, for Na, K, Mg, and Ca were obtained. Determination of iron in rain, lake, and tap waters by CZE was performed after precolumn complexation with 1,10-phenatroline (268). Using 50 mM ammonium acetate–acetic acid at pH 5, as the best buffer system studied, for the separation of the Fe(II) complex resulted in a detection limit of 5×10^{-9} M iron with S/N = 5. The detection was performed with a photometric detector at 270 nm.

V. SUMMARY

A number of different methods are available for the determination of the major metals in water. Selection of a particular technique depends on the analytical purpose, the concentration range of the analytes, the concentration of matrix ions, the precision required, the amount of sample available, etc. Some techniques are also equally appropriate for an analytical purpose and, depending on the availability of the instrumentation required, the scientist will make the final decision for the method to be used.

When sophisticated instruments are not readily available and the concentration ranges of the analytes of interest are sufficiently high, a simple and low-cost spectrophotometric, electrochemical, flame photometric, or even classical volumetric method may be used. The FAAS instruments, equipped with the emission technique, are relatively inexpensive tools with moderate sensitivity that suffice, in most cases, for the analysis of the major metals in waters. A preconcentration procedure may be used before analysis of the major transition metals in order to decrease their detection limits for FAAS or another analytical method of analysis. Chelating resins, packed in small preconcentration columns, have the advantage that the enrichment analysis procedure can easily be automated using a flow injection manifold. The sample matrix also will be modified at the same time, which is particularly important when samples with high salinity are to be analyzed with, for instance, ICP-AES or ICP-MS techniques. Clogging of the nebulizer and/or chemical interferences may otherwise be encountered. The ICP techniques are 1–3 orders of magnitude more sensitive than, for instance, FAAS for the analysis of the major metals. They have the additional advantages of the possibility of simultaneous analysis, large dynamic ranges, and good reproducibility. The instrumentation and running of the techniques, however, are both expensive. Alternatively, an ETAAS machine may be used that possess comparative or better detection limits for most of the metals, but both the instrumentation and cost per analysis are cheaper. Ion chromatography also may be used for the separation and determination of the major alkali and alkaline earth metals as an alternative to flame emission. It has a moderate sensitivity, with the possibility of the determination of a number of cations in a single run. Other chromatographic methods are most suited for the speciation and quantitation of the transition metals, usually in complexed forms. While the methods just mentioned are most suited for the determination of the dissolved fraction of the metals, particulate forms of the metals may be determined by XRFS or NAA techniques. The analysis in this case may be performed directly on a filter, after passing a certain volume of the sample through it.

Regardless of the analytical techniques used, special care must be taken when Na, K, Mg, Ca, Mn, Fe, Zn, and Cu, the major metals, are being determined in unpolluted or high-purity waters. This is because of the abundance of these elements and their existence almost everywhere.

REFERENCES

1. A.E. Greenberg, R.R. Trussell, L.S. Clesceri, M.A.H. Franson. Standard Methods For the Examination of

Water and Wastewater. 16th ed. APHA, AWWA, WPCF, Washington, DC, 1985, part 300.

2. B.A. Eichenberger, K.Y. Chen. Origin and nature of selected inorganic constituents in natural waters. In: R.A. Minear, L.H. Keith, eds. Water Analysis. Vol. 1. Academic Press, New York, 1982, pp. 1–44.

3. National Food Administration. Livsmedelsverkets Kungörelse om dricksvatten. Uppsala, Sweden, 1993, pp. 62–63.

4. H. Borg. Trace elements in lakes. In: P.B. Salbu, P.E. Steinnes, eds. Trace Elements in Natural Waters. CRC Press, Boca Raton, FL, 1995, 177–202.

5. H.W. Nürnberg, L. Mart. Voltametric methods. In: T.S. West, H.W. Nurnberg, eds. Blackwell Scientific, London, 1988, pp. 123–149.

6. B. Allard. Groundwater. In: P.B. Salbu, P.E. Steinnes, eds. Trace Elements in Natural Waters. CRC Press, Boca Raton, FL, 1995, pp. 151–176.

7. A. Ledin. Colloidal carrier substances—properties and impact on trace metal distribution in natural waters. Ph.D. dissertation, Linköping University, Linköping, Sweden, 1993.

8. P. Hashemi, Å. Olin. Talanta 44:1037, 1997.

9. G. Mattsson. Investigation of cathodic stripping voltametric method for the determination of selenium in waters. PhD. dissertation, Uppsala University, Uppsala, Sweden, 1994.

10. P. Hashemi, B. Noresson, Å. Olin. Submitted to Talanta.

11. T.R. Crompton. The Analysis of Natural Waters. Vol. 2. Oxford University Press, New York, 1993, pp. 92–119.

12. R. Chakravarty, R. Van Grieken. Int J. Environ. Anal. Chem. 11:67, 1982.

13. S. Nakashima, M. Yagi. Anal. Lett. 17:1693, 1984.

14. M. Hiraide, T. Ito, M. Babo. Anal. Chem. 52:804, 1980.

15. I.A. Boyle, J.M. Edmond. Anal. Chim. Acta 91:189, 1977.

16. K. Akatsuka, I. Atsuya. Anal. Chim. Acta 202:223, 1987.

17. I. Atsuya, K. Itah. Fresenius Z. Anal. Chem. 329:750, 1988.

18. H. Bem, D.E. Ryan. Anal. Chim. Acta 166:189, 1984.

19. T.R. Crompton. The Analysis of Natural Waters. Vol. 1. Oxford University Press, New York, 1993, pp. 37–109.

20. D. Chakraborti, F. Adams, W. Van Mol, K.J. Irgolic. Anal. Chim. Acta 196:23, 1987.

21. J. Sourova, A. Kapkova. Vadni. Hospoarstvi Series B 30:133, 1980.

22. T.N. Tweeten. Anal. Chem 48:64, 1976.

23. B. Magnusson, S. Westerlund. Anal. Chim. Acta 131: 63, 1981.

24. K.S. Subramanian, J.C. Meranger. Int J. Environ. Anal. Chem. 7:25, 1979.

25. K.M. Bone, W.D. Hibert. Anal. Chim. Acta 107:219, 1979.

26. A. Tessier, P.G.C. Campbell, M. Bisson. Int. J. Environ. Anal. Chem. 7:41, 1979.

27. H. Tao, A. Miyazaki, K. Bansho, Y. Umezaki. Anal. Chim. Acta 156:159, 1984.

28. S. Bradshaw, A.J. Gascoigne, J.B. Headbridge, J.H. Moffett. Anal. Chim. Acta 197:323, 1987.

29. S.L. Sachdev, P.W. West. Environ. Sci. Tech. 4:749, 1970.

30. B.I. Petrov, A.P. Oshchepkova, U.P. Zhipovistev, B.B. Nemkovskii. Soviet J. Water Chem. Tech. 3:51, 1981.

31. K. Kato. Talanta 24:503, 1977.

32. G.P. Klinkhammer. Anal. Chem. 42:117, 1980.

33. B. Armitage, H. Zeitlin. Anal. Chim. Acta 53:47, 1971.

34. L. Ebdon, A.S. Fisher, S.J. Hill, P.J. Worsfold. J. Automatic Chemistry 13:281, 1991.

35. D.E. Leyden, W. Wegscheider. Anal. Chem. 53: 1059A, 1981.

36. G.C. Myasoedova, S.B. Savvin. CRC Crit. Rev. Anal. Chem. 17:1, 1986.

37. R.A. Nickson, S.J. Hill, P.J. Worsfold. Anal. Proc. Incl. Anal. Commun. 32:387, 1995.

38. M. Torre, M.L. Marina. Crit. Rev. Anal. Chem. 24: 327, 1994.

39. C. Kantipuly, S. Katragadda, A. Chow, H.D. Gesser. Talanta 37:491, 1990.

40. W. Wegscheider, G. Knapp. Crit. Rev. Anal. Chem. 11:79, 1981.

41. M.J. Bloxham, S.J. Hill, P.J. Worsfold. J. Anal. Atomie Spectrom. 9:935, 1994.

42. R. Caprioli, S. Torcini. J. Chromatogr. 640:365, 1990.

43. S. Hirata, Y. Umezaki, M. Ikeda. Anal. Chem. 58: 2602, 1986.

44. S. Hirata, K. Honda, T. Kumamaru. Anal. Chim. Acta 221:65, 1989.

45. K. Vermeiren, C. Vandecasteele, R. Dams. Analyst 115:17, 1990.

46. S. Pai, P. Whung, R. Lai. Anal. Chim. Acta 211:257, 1988.

47. S.M. Nelms, G.M. Greenway, R.C. Hutton, J. Anal. Atomic Spectrom. 10:929, 1995.

48. R.E. Sturgeon, S.S. Berman, S.N. Willie, J.A.H. Desaulniers. Anal. Chem. 53:2337, 1981.

49. P. Hashemi, Å. Olin. Int. J. Environ. Anal. Chem. 63: 37, 1996.

50. S. Blain, P. Appriou, H. Chaumell, H. Handel. Anal. Chim. Acta 232:331, 1990.

51. H. Filik, B.D. Ozturk, M. Dogutan, G. Gumus, R. Apak. Talanta 44:877, 1997.

52. P. Hashemi. Agarose based chelating adsorbents for preconcentration of trace and ultratrace heavy metals in water. Ph.D. dissertation, Uppsala University, Uppsala, Sweden, 1997.

53. L. Erdy. Gravimetric Analysis. Oxford, Pergamon, 1965.

54. Z. Feng, D. Dan (abstr). Lihua Jianyan, Huaxue Fence 32:343, 1996.
55. A.I. Vogel. Vogel's Textbook of Quantitative, Chemical Analysis. 5th ed. Longmans, London, 1989, pp. 228–238.
56. J.B. Pate, R.J. Robinson. J. Marine Res. 19:12, 1961.
57. K. Kremling. Determination of the major constituents. In: K. Grasshoff, M. Ehrhardt, K. Kremling, eds. Methods of Seawater Analysis. Verlag Chemie, Weinheim, Germany, 1983, pp. 247–267.
58. H. Diehi, J.L. Ellingboe. Anal. Chem. 28:882, 1956.
59. D.P.S. Rathore, P.K. Bhargava, M. Kumar, R.K. Tarla. Anal. Chim. Acta 281:173, 1993.
60. F. Kulkin, R.A. Cox. Deep-Sea Res. 13:789, 1966.
61. A. Karolev. Talanta 39:1575, 1992.
62. B.C. Sinha, S.K. Roy. Analyst 105:720, 1980.
63. B.C. Sinha, S.K. Roy. Analyst 107:965, 1982.
64. R. Greenhalgh, J.P. Riley, M. Tungudai. Anal. Chim. Acta 36:439, 1966.
65. P. Richter, M.I. Toral. Microchem. J. 53:413, 1996.
66. A.C. Lopes da Conceicao, M.T. Tena, M.M. Correia dos Santos, M.L. Simos Goncalves, M.D. Luque de Castro. Anal. Chim. Acta 343:191, 1997.
67. A.N. Tripathi, S. Chikhalikar, K.S. Patel. J. Autom. Chem. 19:45, 1997.
68. F. Basyoni Salem. Alexandria J. Pharm. Sci. 11:25, 1997.
69. M. Okumura, Y. Seike, K. Fujinaga, K. Hirao. Anal. Sci. 13:231, 1997.
70. T. Balaji, G.R.K. Naidu. Pollut. Res. 15:311, 1996.
71. M.L. Fernandez-de Cordova, A. Ruiz-Medina, A. Molina-Diaz. Fresenius' J. Anal. Chem. 357:44, 1997.
72. J.F. van Staden, R.E. Taljaard. Anal. Chim. Acta 323:75, 1996.
73. E. Rubi, R. Forteza, V. Cerda. Lab. Rob. Autom. 8:149, 1996.
74. K. Sharma, K.S. Patel. J. Indian Chem. Soc. 73:443, 1996.
75. P. Richter, M.I. Toral, P. Hernandez. Anal. Lett. 29:1013, 1996.
76. M.S. Saleh, K.A. Idriss, E.Y. Hashem. Bull. Fac. Sci., Assiut Univ. B 24:47, 1995.
77. M.I. Toral, P. Richter, B. Munoz. Environ. Monit. Assess. 38:1, 1995.
78. M.S. Saleh, K.A. Idriss, H.A. Azab, E.Y. Hashem. Bull. Fac. Sci. Assiut Univ. B 24:31, 1995.
79. C. Sarbu, V. Liteanu, R. Grecu. Rev. Roum. Chim. 40:829, 1995.
80. R.M.C. Sutton, S.J. Hill, P. Jones. J. Chromatogr. A 739:81, 1996.
81. J.F. van Staden, L.G. Kluever. Anal. Chim. Acta 350:15, 1997.
82. S. Motomizu, K. Yoshida, K. Toei. Anal. Chim. Acta 261:225, 1992.
83. L. Zhang, K. Terada. Anal. Chim. Acta 293:311, 1994.
84. S. Blain, P. Treguer. Anal. Chim. Acta 308:425, 1995.
85. F. Blasco, M.J. Medina-Hernandez, S. Sagrado, F.M. Fernandez. Analyst 122:639, 1997.
86. U. Hase, K. Yoshimura. Analyst 117:1501, 1992.
87. M.L. Moss, M.G. Mellon. Ind. Eng. Chem. Anal. Ed. 14:862, 1942.
88. S.W. Kang, T. Sakai, N. Ohno, K. Ida. Anal. Chim. Acta 261:197, 1992.
89. D.W. King, J. Lin, D.R. Kester. Anal. Chim. Acta 247:125, 1991.
90. Y. Erel, S.O. Pehkonen, M.R. Hoffmann. J. Geophys. Res. Part A 98:18423, 1993.
91. M.S. Abu-Bakr, H. Sedaira, E.Y. Hashem. Talanta 41:1669, 1994.
92. P.P. Kish, I.I. Zimomyra. Zavod. Lab. 35:541, 1969.
93. J. Michal, J. Zyka. Cem. Listy. 48:915, 1954.
94. K. Kina, K. Shiraishi, N. Ishibashi. Talanta 25:295, 1978.
95. T. Iwachido, M. Onoda, S. Motomizu. Anal. Sci. 2:493, 1986.
96. S. Motomizu, M. Onoda, M. Oshima. Analyst 113:743, 1988.
97. Z. Marczenko. Separation and Spectrophotometric Determination of Elements. Ellis Horwood, London, 1986, pp. 123–128.
98. H. Nakamura, H. Nishida, M. Takagi, K. Ueno. Anal. Chim. Acta 126:185, 1981.
99. G.E. Pacey, Y.P. Wu. Talanta 31:165, 1984.
100. M. Takagi, H. Nakamura, Y. Sanui, K. Ueno. Anal. Chim. Acta 126:185, 1981.
101. H.L. Golterman, R.S. Clymo, M.A.M. Ohnstad. Methods for Physical and Chemical Analysis of Fresh Water. 2d. ed. Blackwell Scientific, London, 1978, pp. 67–77.
102. O. Hernandez, F. Jimenez, A.I. Jimenez, JJ. Arias, J. Havel. Anal. Chim. Acta 320:177, 1996.
103. F. Nydahl. Anal. Chim. Acta 3:144, 1949.
104. E.B. Sandell. Colorimetric Determination of Traces of Metals. 3rd ed. Wiley Interscience, New York, 1959, Chap. 26.
105. H.L. Golterman, R.S. Clymo, M.A.M. Ohnstad. Methods for Physical and Chemical Analysis of Fresh Water. 2d. ed. Blackwell Scientific, London, 1978, pp. 121–138.
106. K. Koroleff. Determination of trace metals. In: K. Grasshoff, M. Ehrhardt, K. Kremling, eds. Methods of Seawater Analysis. Verlag Chemie, Weinheim, Germany, 1983, pp. 236–242.
107. R. Ruter, B. Neidhart. Mikrochim. Acta 18:271, 1984.
108. S. Pehkonen. Analyst 120:2655, 1995 (review).
109. W.B. Fortune, M.G. Mellon. Ind. Eng. Chem. Anal. Ed. 10:60, 1938.
110. R.P. Mehlig, R.H. Hulett. Ind. Eng. Chem. Anal. Ed. 14:869, 1942.
111. G.F. Lee, W. Stumm. J. Am. Water Works Ass. 52:1567, 1960.
112. P.F. Collins, H. Diehl, G.F. Smith. Anal. Chem. 31:1862, 1959.

113. L.L. Stookey. Anal. Chem. 42:119, 1970.
114. P. Carter. Anal. Biochem. 40:450, 1971.
115. G. Zhuang, Z. Yi, R.A. Duce, P.R. Brown. Nature, London 355:537, 1992.
116. R.M. Rush, J.H. Yoe. Anal. Chem. 26:1345, 1954.
117. D.G. Miller. J. Water Pollut. Control Fed. 51:2402, 1979.
118. J.A. Platte, V.M. Marcy. Anal. Chem. 31:1226, 1959.
119. G. Den Boef, A. Hulanicki, D.T. Burns. Spectrophotometric and fluorimetric methods. In: T.S. West, H.W. Nurnberg, eds. Blackwell Scientific, London, 1988, pp. 123–149.
120. G.F. Smith, W.H. McCurdy. Anal. Chem. 24:371, 1952.
121. I.Y. Kolotyrkina, L.K. Shpigun, Y.A. Zolotov, G.I. Tsysin. Analyst 116:707, 1991.
122. K. Hyrayama, N. Unohara. Anal. Chem. 60:2573, 1988.
123. I.Y. Kolotyrkina, L.K. Shpigun, Y.A. Zolotov, A. Malahoff. Analyst 120:201, 1995.
124. S. Ohno, N. Teshima, T. Watanabe, H. Itabashi, S. Nakano, T. Kawashima. Analyst 121:1515, 1996.
125. I.M. Fitsev, A.V. Zolotukhin, N.A. Ermolaeva, V.F. Toropova, A.R. Garifzyanov, G.K. Budnikov. J. Anal. Chem. 52:514, 1997.
126. I.M. Fitsev, A.V. Zolotukhin, N.A. Ermolaeva, V.F. Toropova, A.R. Garifzyanov, G.K. Budnicov. J. Anal. Chem. 52:514, 1997.
127. S.S. Mitic, G.Z. Miletic, M.V. Obradovic. Talanta 42:1273, 1995.
128. S. Nakano, K. Nakaso, K. Noguchi, T. Kawashima. Talanta 44:765, 1997.
129. D. Yu, S. Liu, Z. Liu, K. Liu. Chin. J. Geochem. 14:134, 1995.
130. S. Cao, J. Zhong, K. Hasebe, W. Hu. Anal. Chim. Acta 331:257, 1996.
131. K. Watanabe, K. Rokugawa, M. Itagaki (abstr). Bunseki Kagaku 44:933, 1995.
132. T. Tomiyasu, H. Sakamoto, N. Yonehara. Anal. Sci. 12:507, 1996.
133. T. Tomiyasu, H. Sakamoto, N. Yonehara. Anal. Sci. 10:761, 1994.
134. A. Cladera, E. Gomez, J.M. Stela, V. Cerda. Analyst 116:913, 1991.
135. J.L.F.C. Lima, C. Delerue-Matos, E.M. Carmo, V.F. Vaz. Rev. Port. Farm. 46:24, 1996.
136. S. Olsen, L.C.R. Pessenda, J. Ruzicka, E.H. Hansen. Analyst 108:905, 1983.
137. Z. Fang, T. Guo, B. Welz. Talanta 38:613, 1991.
138. R.E. Sturgeon, S.S. Berman, A. Desaulniers, D.S. Russell. Anal. Chem. 51:14, 1979.
139. T. Nakahara, C.L. Chakrabarti. Anal. Chim. Acta 104:99, 1979.
140. J.M. McArthur. Anal. Chim. Acta 93:77, 1977.
141. J.R. Montgomery, G.M. Peterson. Anal. Chim. Acta 117:397, 1980.
142. M. Hoenig, R. Wollast. Spectrochim. Acta, Part B 37:399, 1982.
143. R. Guevremont. Anal. Chem. 52:1574, 1980.
144. L.R.P. Butler, A. Strasheim. Atomic-, mass-, x-ray-spectrometric methods, electronparamagnetic and luminescence methods. In: T.S. West, H.W. Nurnberg, eds. Blackwell Scientific, London, 1988, pp. 63–121.
145. R. Guevremont. Anal. Chem. 53:911, 1981.
146. S. Xiao-quan, B. Radziuk, B. Welz, O. Vyskocilova. J. Anal. Atomic Spectrom. 8:409, 1993.
147. J.Y. Cabon, A.L. Bihan. J. Anal. Atomic Spectrom. 9:477, 1994.
148. H.M. Kingston, I.L. Barnes, T.J. Brady, T.C. Rains, M.A. Champ. Anal. Chem. 50:2064, 1978.
149. R.E. Sturgeon, S.S. Berman, A. Desaulniers, D.S. Russell. Talanta 27:85, 1980.
150. Y. Sung, Z. Liu, S. Huang. J. Anal. Atomic Spectrom. 12:841, 1997.
151. T.C. Rains. Atomic absorption spectrometry. In: R.A. Minear, L.H. Keith, eds. Water Analysis. Vol. II. Academic Press, New York, 1984, pp. 62–104.
152. J.W. Olesik. Anal. Chem. 63:12A, 1991.
153. J.P. Riley, R. Chester. Introduction to Marine Chemistry. Academic Press, New York, 1971, pp. 228–247.
154. J.A.C. Broekaert, F. Leis. Anal. Chim. Acta 109:73, 1979.
155. C. Lan, M. Yang. Anal. Chim. Acta 278:111, 1994.
156. A.G. Coedo, M.T. Dorado, I. Padilla. F.J. Alguacil. J. Anal. Atomic Spectrom. 11:1037, 1996.
157. T. Hwang, S. Jiang. Analyst 122:233, 1997.
158. S.M. Nelms, G.M. Greenway, D. Koller. J. Anal. Atomic Spectrom. 11:907, 1996.
159. B. Noresson, P. Hashemi, Å. Olin. Talanta 46:1051, 1998.
160. H.W. Nürnberg. Sci. Tot. Environ. 37:9, 1984.
161. C.M.G. van den Berg. The electroanalytical chemistry of seawater. In: J.R. Riley, E. Chester, eds. Chemical Oceanography. Vol. 9. Academic Press, London, 1988.
162. J.R. Donat, K.W. Bruland. Trace elements in oceans. In: P.B. Salbu, P.E. Steinnes, eds. Trace Elements in Natural Waters. CRC Press, Boca Raton, FL, 1995, pp. 177–202.
163. C.W.K. Chow, S.D. Kolev, D.E. Davey, D.E. Mulcahy. Anal. Chim. Acta 330:79, 1996.
164. P.A.M. Farias, A.K. Ohara, S.L.C. Ferreira Anal. Lett. 25:1929, 1992.
165. S. Daniele, C. Bragato, M.A. Baldo. Anal. Chim. Acta 346:145, 1997.
166. L. Mart, H.W. Nurnberg, P. Valenta. Z. Anal. Chem. 366:350, 1980.
167. G. Van Dijck, F. Verbeek. Anal. Chim. Acta 54:475, 1971.
168. G.W. Luther, D.B. Nuzzio, J. Wu. Anal. Chim. Acta 284:473, 1994.
169. R.J. Ohalloran. Anal. Chim. Acta 140:51, 1982.
170. J.S. Roitz, K.W. Bruland. Anal. Chim. Acta 344:175, 1997.

171. A. Trojanek, F. Opekar. Anal. Chim. Acta, 126:15, 1981.

172. T. Tanaka, T. Muramatsu, N. Takada, A. Mizuike. Bunseki Kagaku, 42:587, 1993.

173. G. Zhiquiang, L. Peibiao, W. Guangqing, Z. Zaofan. Anal. Chim. Acta 241:137, 1990.

174. M.E.V. Diaz, J.C.J. Sanchez, M.C. Mochon, A.G. Perez. Analyst 119:1571, 1994.

175. A. Safavi, M. Pakniat, N. Maleki. Anal. Chim. Acta 335:275, 1996.

176. J. Labuda. Selective Electrode Rev. 14:33, 1992.

177. J. Labuda, M. Vanickova, E. Uhleman, W. Mickler. Anal. Chim. Acta 284:517, 1994.

178. J. An, J. Zhou, X. Wen. Talanta 32:479, 1985.

179. A.J. Downard, J.B. Hart, H.K.J. Powell, S. Xu. Anal. Chim. Acta 269:41, 1992.

180. A.L. Bard, L.R. Faulkner. Electrochemical Methods. Wiley, New York, 1980, pp. 73–81.

181. R. Jasinski, J. Trachtenberg, D. Andrychuk. Anal. Chem. 46:364, 1974.

182. D.J. Crombie, G.J. Moody, J.D.R. Thomas. Talanta 21:1094, 1974.

183. M. Maj-Zurawska, M. Rouilly, W.E. Morf, W. Simon. Anal. Chim. Acta 218:47, 1989.

184. C. Sun, Y. Sun, X. Zhang, H. Xu, J. Shen. Anal. Chim. Acta 312:207, 1995.

185. T. Xu (abstr). Huaxue Chuanganqi 15:128, 1995.

186. P.W. Alexander, T. Dimitrakopoulos, D.B. Hibbert. Talanta 44:1397, 1997.

187. J.F. Van Staden, C.C.P. Wagener. Anal. Chim. Acta 197:217, 1987.

188. B.Y. Sun, Y.Z. Ye, H.W. Huang, Y. Bai. Talanta 40:891, 1993.

189. E.P. Gil, P. Ostapczuk. Anal. Chim. Acta 293:55, 1994.

190. F. Scholz, L. Nitschke, G. Henrion. Anal. Chim. Acta 199:167, 1987.

191. T.M. Florence. Analyst 111:489, 1986.

192. G.A. Bhat, R.A. Saar, R.B. Smart, J.H. Weber. Anal. Chem. 53:2275, 1981.

193. G.E. Batley, T.M. Florence. J. Electroanal. Chem. 72:121, 1976.

194. D. Jagner, L. Renman, S.H. Stefansdottir. Anal. Chim. Acta 281:305, 1993.

195. S.S.M. Hassan, S.A.M. Marzouk. Talanta 41:891, 1994.

196. S.D. Kolev, C.W.K. Chow, D.E. Davey, D.E. Mulcahy. Anal. Chim. Acta 309:293, 1995.

197. D. Jagner, e. Sahlin, B. Axelsson, R. Ratana-Ohpas. Anal. Chim. Acta 278:237, 1993.

198. J.W. Wimberley. Anal. Chem. 53:2138, 1981.

199. O.A. Shpigun, O.D. Choporova, Y.A. Zolotov. Anal. Chim. Acta 172:341, 1985.

200. F.R. Nordmeyer, L.D. Hansen, D.J. Eatough, D.K. Rollins, J.D. Lamb. Anal. Chem. 52:852, 1980.

201. J.D. Lamp, L.D. Hansene, G.G. Patch, F.R. Nordmeyer. Anal. Chem. 53:749, 1981.

202. R.P. Singh. Analyst 116:409, 1991.

203. R.P. Lash, C.J. Hill. Anal. Chim. Acta 108:405, 1979.

204. M. Legran, M. De Angelis, R.J. Delmas. Anal. Chim. Acta 156:181, 1984.

205. O. Shpigun, Y.A. Zolotov. Ion Chromatography in Water. Ellis Horwood, London, 1988, pp. 148–166.

206. K. Hayakawa, H. Hiraki, M. Miyazaki. Bunseki Kagaku, 32:504, 1983.

207. M. Miyazaki, K. Hayakawa, Seung-Gi-Choi. J. Chromatogr. 323:443, 1985.

208. R.C.L. Foley, P.R. Haddad. J. Chromatogr. 366:13, 1986.

209. R.L. Smith, D.J. Pietrzyk. Anal. Chem. 56:610, 1984.

210. J.S. Fritz, D.T. Gjerde, R.M. Becker. Anal. Chem. 52:1519, 1980.

211. R.P. Singh, N.M. Abbas. J. Chromatogr. A 733:93, 1996.

212. P.R. Haddad, R.C.F. Foley. J. Chromatogr. 61:1435, 1989.

213. J.W. Wimberley. Anal. Chem. 53, 1709, 1981.

214. K. Ohta, K. Tanaka (abstr). Mizu Shori Gijutsu 37:161, 1996.

215. K. Ohta, K. Tanaka, J.S. Fritz. J. Chromatogr. A 731:179, 1996.

216. P.R. Haddad, P.W. Alexander, M. Trojanowicz. J. Chromatogr. 294:397, 1984.

217. N. Hamada, T. Gotoh, T. Yagi (abstr). Kogyo Yosui 454:58, 1996.

218. V.D. Nguyen. Fresenius' J. Anal. Chem. 354:738, 1996.

219. D. Wu, Y. Yan, H. Han (abstr). Huanjing Huaxue 15:441, 1996.

220. P. Jones, K. Barron, L. Ebdon. Anal. Proc. 22:373, 1985.

221. D. Yan, G. Schwedt. Z. Anal. Chem. 320:252, 1985.

222. W. Wang, Y. Chen, M. Wu. Analyst 109:281, 1984.

223. H. Hojabri, A.G. Lavin, G.G. Wallace. Anal. Proc. 23:26, 1986.

224. H. Hojabri, A.G. Lavin, G.G. Wallace, J.M. Riviello. Anal. Chem. 59:54, 1986.

225. J.E. Girard. Anal. Chem. 51:836, 1979.

226. S. Maketon, E.S. Otterson, J.G. Tarter. J. Chromatogr. 368:395, 1986.

227. J.J. Byerley, J.M. Scharer, G.F. Atkinson. Analyst 112:41, 1987.

228. N. Singh, K. Rastogi, R. Kumar, T.N. Srivastava. Analyst 106:599, 1981.

229. K. Tanaka, Y. Ishihara, K. Nakajima. Bunseki Kagaku, 32:626, 1983.

230. S.A. Atwood, D.D. Wallwey. J. Chromatogr. A 739:265, 1996.

231. D. Yan, G. Schwedt. Z. Anal. Chem. 320:121, 1985.

232. A.R. Timerbaev, O.M. Petrukhin, Y.A. Zolotov (abstr). Zh. Anal. Khim. 37:1360, 1982.

233. G. Soundararajan, M. Subbaiyan. Indian J. Chem. Sec. A. 22:399, 1983.

234. A.L.J. Rao, S. Chopra. J. Inst. Chem. India 57:197, 1985.

235. P. Heizmann, K. Ballschmiter. J. Chromatogr. 137: 153, 1977.

236. M. Lohmuller, P. Heizman, K. Ballschmiter. J. Chromatogr. 137:165, 1977.

237. K. Robards, P. Starr, E. Patsalides. Analyst 116:1247, 1991 (review).

238. R. Kalani, S.P. Mathur. Chem. Environ. Res. 2:301, 1993.

239. L. Ebdon, S. Hill, R.W. Ward. Analyst 112:1, 1987.

240. P.C. Uden. Trends Anal. Chem. (Pers. Ed.) 6:238, 1987.

241. S. Zappoli, L. Morselli, F. Osti. J. Chromatogr. A 721: 269, 1996.

242. Y. Nagaosa, H. Kawabe, A.M. Bond. Anal. Chem. 63: 28, 1991.

243. Y. Nagaosa, T. Suenaga, A.M. Bond. Anal. Chim. Acta 235:279, 1990.

244. S. Comber. Analyst 118:505, 1993.

245. Y. Shijo, H. Sato, N. Uehara, S. Aratake. Analyst 121: 325, 1996.

246. R.M. Smith, A.M. Butt, A. Thakur. Analyst 110:35, 1982.

247. A. Berthod, M. Kolosky, J.L. Rocca, O. Vittori. Analysis 7:395, 1979.

248. R.C. Gurira, P.W. Carr. J. Chromatogr. 20:461, 1982.

249. M. Tabata, M. Tanaka. Anal. Lett. 13(A6):427, 1980.

250. A. Itoh, C. Kimata, H. Miwa, H. Sawatari, H. Haraguchi. Bull. Chem. Soc. Jpn. 69:3469, 1996.

251. H. Haraguchi, A. Itoh, C. Kimata. Anal. Sci. Technol. 8:405, 1995.

252. E.L. Wehry. Molecular luminescence methods. In: R.A. Minear, L.H. Keith. Water Analysis. Vol. II. Academic Press, New York, 1984, pp. 194–201.

253. H. Wada, H. Atsumi, G. Nakagawa. Anal. Chim. Acta 261:275, 1992.

254. Y. Yu, R. Nie, J. Huang, L. Su (abstr). Huaxue Xuebao 54:709, 1996.

255. H. Shen, Y. Tang (abstr). Yejin Fenxi 15:1, 1995.

256. S. Yu, Z. Liu, G. Wang, J. Du, Y. Yang (abstr). Fenxi Huaxue 24:433, 1996.

257. R.Y. Tsien. Biochemistry 19:2396, 1980.

258. K. Hiraki, K. Morashige, I. Nishikawa. Anal. Chim. Acta 97:121, 1978.

259. R.E. Van Grieken, C.M. Bresele, B.M. Van der Goe. Anal. Chem. 49:1326, 1977.

260. A. Bumbalova, A. Pikulikova, M. Komova, A. Muchova. J. Radioanal. Nucl. Chem. 164:357, 1992.

261. B. Holynska, B. Ostachowicz, C. Wegrzynek. Spectrochim. Acta Part B 51B:769, 1996.

262. E. Steinnes. Neutron activation analysis. In: T.S. West, H.W. Nurnberg, eds. Blackwell Scientific, London, 1988, pp. 123–149.

263. R.R. Rao, A. Chatt. J. Radioanal. Nucl. Chem. 168: 439, 1993.

264. R.R. Rao, D.G. Goski, A. Chatt. J. Radioanal. Nucl. Chem. 161:89, 1992.

265. A. Tangen, W. Lund, R. Buhl Frederiksen. J. Chromatogr. A 767:311, 1997.

266. K. Fukushi, K. Hiiro. Fresenius' J. Anal. Chem. 365: 150, 1996.

267. Y.H. Lee, T.I. Lin J. Chromatogr. A 675:227, 1994.

268. J. Xu, Y. Ma. J. Microcolumn Sept. 8:137, 1996.

269. P. Blatny, F. Kvasnicka, E. Kenndler. J. Chromatogr. A 757:297, 1997.

22

Heavy Metals

Elena González-Soto, Elia Alonso-Rodríguez, and Darío Prada Rodríguez
University of A Coruña, A Coruña, Spain

Nowadays the contamination of water resources concerns not just the scientific community. In recent years a general environmental awareness has developed that is related to the actual knowledge of and control of water pollution.

Metal ions in water can occur naturally from leaching ore deposits and from anthropogenic sources, which include mainly industrial effluents and solid waste disposal. Solid wastes may contain significant concentrations of metal ions that can be dissolved and incorporated into water systems.

The term *heavy metal* includes both essential and nonessential trace metals that may be toxic to living organisms, depending on their own properties, availability, and levels of concentration. Frequently, urban and industrial areas present high concentrations of heavy metals that can be dissolved, causing toxic effects to some microorganisms or that can lead to health effects when they are taken in through drinking water or accumulated through the food chain. The possible synergistic effects of some heavy metals that can considerably increase their toxic potential must be taken into account, along with the fact that the persistence of some heavy metals in the environment can lead to transformations to more toxic compounds.

Due to the high development of industrial activity in recent years, the levels of heavy metals in water systems have substantially increased over time. This makes it necessary to develop methods that allow one to detect and quantify the usually low but important levels of heavy metals in water.

This chapter deals with the crucial steps of these analytical procedures, including sample collection, storage, and preservation, sample pretreatment, and instrumental techniques for the determination of heavy metals in water.

I. SAMPLING

The first requirement for a water sample to be analyzed is that it be representative of the water system at the time of collection. Usually, a larger volume of water is collected to get a representative sample, and then it is homogenized for subsampling and subsequent analysis of the aliquots. When no volatile components are to be analyzed, it is recommended not to fill the bottle completely in order to be able to shake the water and homogenize the sample (1).

When the composition of the water to be studied remains unchanged over time, frequently a discrete sample is used, which is collected directly from the water to be analyzed and which represents the state of the sample at that moment. On the other hand, when the aim of the analysis is to know the average concentration of any component during a certain period, it is advisable to get a composite sample by mixing different water samples collected at different times.

When heavy metals are analyzed, extreme care should be taken, because contamination during sampling is an important and frequent source of error. The errors increase with decreasing concentrations of the elements to be analyzed, due to losses by adsorption at the surface of the storage vessels or contamination of the sample by the containers if they are inadequately washed.

The first requirement for any sampling equipment is that it not alter the chemical composition of the sample during contact, handling, or shipping. Sampling materials, sampling devices, cleaning of bottles, and sample preparation to avoid contamination are the subject of several studies (2–6) and also of Chapters 1 and 2 of this handbook.

It is important to select sampling bottles that will avoid either adsorption or desorption effects at their surface influenced mainly by the pH of the water, the presence of complexing agents, and the time the sample is in contact with the container. The bottles can be either glass or plastic. When heavy metals are analyzed, metal containers must be avoided, because they may alter the sample by the leaching of metals. Quartz or tetrafluorethylene (TFE) are the best materials for the containers, but they are expensive. So water samples for heavy-metal analysis are collected and stored in plastic bottles such as polyethylene and polyvinyl chloride (PVC). Vessels and filters must be washed with diluted hydrochloric acid and rinsed with distilled water to achieve the total removal of heavy metals from former samples. For mercury, polytetrafluoroethylene (PTFE) or glass bottles are the most suitable; plastic bottles are not recommended because reduction to metallic mercury and diffusion through the walls can take place (7); for methyl mercury, the bottles have to be protected against light to avoid degradation processes (8).

When chemical speciation is required, extreme care in sampling, handling, and storage is necessary because chemical reactions may happen. These reactions would alter the species present in the original sample due to changes in pH, redox potential, oxygen content, etc. Because of that, these kinds of samples must be stored in the dark at low temperature or even frozen and analyzed as soon as possible.

II. STORAGE AND PRESERVATION

As a rule, samples should be analyzed immediately after collection, because low concentrations of heavy metals decrease with time. Water content can change very quickly, mainly when it has a high content of organic and suspended material as, e.g., with waste water. When a quick analysis is not possible, samples should be stored away from any potentially contaminating source, such as concentrated solutions or contaminated atmospheres, because soot and airborne dust may contain heavy metals that will increase the true levels in the sample. Special care must be taken with metal contamination due to distilled water, filters, and containers (chemical reaction between the sample and the container, contamination from previous samples, etc.). Such care is especially important when dealing with samples containing microgram levels of heavy metals.

When a quick analysis is not possible, samples should be preserved by adding ultrapure HNO_3 to achieve a pH less than 2 to prevent precipitation of metal hydroxides or adsorption of metal ions on the walls of the container. After this, samples should be cooled down to 4°C, which will lower microbial activity. Refrigeration is an ideal preservation method because it does not affect the sample composition and does not interfere with any analytical method. Moreover, refrigeration helps retain in solution some elements, such as Hg, As, Se, Cd, and Zn, that may be lost due to volatilization when the temperature increases. If Hg is analyzed, because it may be easily lost, the addition of 2 ml of 10% $K_2Cr_2O_7$ per liter of sample is recommended to conserve it for a few days.

When different species of an element are to be determined (e.g., Fe(II), Fe(III), As(III), As(V)), samples should be analyzed as soon as they have been collected. When this is not possible, they should be quickly cooled down to 4°C, which was found to be effective for preserving some arsenic and tin species for several weeks (9). Prior to refrigeration, it is advisable to filter the samples, because particulates may alter the oxidation state of some species. As a general rule, samples must be maintained at their natural pH in order not to alter the chemical form of the metal. Acids must not be added, because in some cases they may give redox reactions and change the proportion of the species. For example, at moderate to high pH values Cr(VI) ions are stable, but at pH less than 4 Cr(VI) can autoreduce to Cr(III) (10). On the other hand, in inorganic arsenic speciation the addition of sulfuric acid is suggested to avoid changes in arsenate and arsenite species (11).

III. PRETREATMENT

Pretreatment consists of the steps necessary to separate interferences and concentrate the analyte to within the analytical technique sensitivity. Pretreatment depends to a great extent on the analytical technique to be used. Most of the modern techniques (HG-AAS, ETAAS, ICP-MS) have slight interferences and allow very low detection limits. Nevertheless, legislation concerning the contamination of freshwater, seawater, groundwater, etc. becomes more urgent every day. Therefore, even when using very sensitive techniques, it is sometimes necessary to make preconcentrations to allow the quantification of some metals that are found at very low levels.

Independent of the analytical technique to be employed, when metals are to be analyzed, there are three different fractions that will need different pretreatment. Samples for total-metal analysis should be acidified with nitric acid to a pH of less than 2 without previous filtration. Samples for dissolved-metal analyses should be filtered through a 0.45-μm-pore-size membrane filters before acidifying for removing the particulate matter. Samples for suspended-metal analyses must be filtered through 0.45-μm membrane filters without acidifying, and the substances retained on it analyzed.

It is now generally accepted that, besides the classification into dissolved and suspended metals, it is also important to consider the metals retained in the colloidal fraction, because they may control the distribution of an element between dissolved and particulate phases due to their potential adsorptive capacity.

Ultrasound (12), ultraviolet photolysis (13), and microwave digestion (14) have been used as pretreatment for the releasing of metals bonded to organic matter.

After the fraction of interest is separated, preconcentration or separation of the interfering ions prior to determination of the analyte are frequently necessary. The most frequently employed methods are based on the use of coprecipitation with different compounds (15–19), evaporative procedures (20,21), or the use of chelating agents that originate a complex that is extracted into a solvent (22–30). Ion-exchange resins or other charged sorbents in packed columns or cartridges are well suited for selective preconcentration because of the opposite charges of the metal ions. The analyte of interest is then eluted with a suitable solvent (31–34). Solid-phase extraction methods are also being widely used (35–43).

In recent years, biological organisms such as yeast, bacteria, and algae have been used to preconcentrate heavy metals, exploiting their capacity to adsorb different elements. Most of the methods developed use non-immobilized substrates, although microorganism immobilization procedures are being developed (44–46).

One significant improvement is the systems in which online pretreatment is coupled with detection of analytes, because they are simpler and less time consuming than other options and allow for improvement of analytical precision (47–55).

IV. ANALYTICAL METHODS

A. Cadmium

Reactions of Cd with reagents that generate color can be used for its determination. So highly spectrophotometric methods, with the elimination of interfering ions by adding a mixing masking agent, are described (56,57).

Flame atomic absorption spectrometry (FAAS) is used for the determination of Cd in water samples after preconcentration (58–60). Some authors have compared this technique with others, such as neutron activation analysis (NAA) (61) or fluorescence determination (62). A method for the generation of volatile Cd with sodium tetraethylborate and ICP-AES determination has also been described (63). The spectroscopic interferences in Cd determination encountered when a conventional pneumatic nebulizer is used are abbreviated with the application of vapor generation ICP-MS (64).

Electrothermal atomic absorption spectrometry (ETAAS) has found wide application for the determination of Cd in different water samples. The detection limits are excellent and, under favorable conditions, may reach ppb concentrations. In order to avoid the effects of the matrix, different modifiers were evaluated for Cd determination in natural water (65–68) and seawater (63,69).

Anodic stripping voltammetry (ASV) is the most widely applicable electrochemical technique for Cd determination, and extremely high sensitivity can be obtained (70). Although DPP is used for the measurement of Cd (71), when natural waters are analyzed one must take into account that polarography is limited to concentrations of the element about 10^{-7} M.

Highly sensitive ion-selective electrodes have been developed for the determination of Cd in industrial wastewater (72).

B. Lead

The determination of Pb in water is of considerable current interest in the environmental sciences, so the development of quick, simple, and cheap spectrophotometric methods is of great interest for analytical practice (73–75).

One relatively simple method that competes well with other flameless techniques, which require more expensive equipment and are usually much more sensitive to interferences, is FAAS with a previous solid-phase extraction step (37,76,77). A 10–100-fold improvement in sensitivity for Pb, as compared to conventional FAAS, is obtained by the direct determination of Pb in water using an atom-trapping technique (78,79).

Because Pb generates volatile hydrides, HG-AAS offers a sensitive method for its detection (80-82).

Cabrera et al. (83) recommend and use ETAAS, in conjunction with a stabilized temperature platform furnace (STPF), for the determination of Pb in potable, irrigation, and wastewaters because of its sensitivity, versatility, speed, and specificity. This technique can be used directly to quantify Pb levels, without the need for conventional separation and preconcentration steps. And ETAAS is also used for the determination of Pb in seawater after different preconcentration methods (84–86), achieving more feasibility when flow injection analysis (FIA) is used (87). In order to avoid the effects of the matrix, different modifiers have been evaluated (88,89).

The advances obtained by ICP-AES and ICP-MS detectors, coupled with HG techniques, explain their use for the determination of Pb (32,90,91).

Laser-excited AFS has a high sensitivity and multi-element detection capability, being used for the determination of Pb in freshwater (92) and seawater (93) without preconcentration.

Feldman et al. (94) present a rapid, accurate, and precise method for the determination of Pb based on isotope-dilution GC-MS detection of tetraethyl lead. The method requires a small volume of sample and uses commonly available equipment and materials.

Anodic stripping voltammetry is widely recognized as one of the most sensitive methods in electroanalytical chemistry for Pb determination (95,96). Direct determination of concentrations of Pb in freshwater samples has been also made by a sensitive CSV method (97).

Lead-210 is a naturally occurring radionuclide of the uranium-238 series, and its concentration in water can be measured directly by liquid scintillation counting. However, this direct method does not achieve suitable sensitivity, because previous complexation with EDTA is necessary (98,99).

C. Mercury

Selective and sensitive spectrophotometric methods for the determination of Hg in wastewater, based on the selective extraction of Hg(II) at neutral pH, are described (100,101).

Although FAAS has been used for the determination of Hg in freshwater and seawater samples after a previous preconcentration step (102), it is not often used because sensitivities are not very good due to the high volatility of this element.

The most widely used method for Hg determination in waters is CV-AAS, with or without a preconcentration step before the Hg generation (103–105), because it is a simple and sensitive method and is relatively free from interferences. Continuous-flow systems have also been used for Hg CV generation (106,107).

Preconcentration on coated graphite tubes has been successfully used for Hg CV generation and subsequent determination by ETAAS (108), and a number of matrix modifiers have been tested (109,110).

Atomic fluorescence spectrometry (AFS) has been proposed for Hg determination (111,112), usually after elemental Hg generation (113,114).

A comparison of different analytical techniques, such as atomic fluorescence, atomic absorption, and atomic emission, in the determination of Hg in water samples has been made (115,116).

Detection via ICP-AES is not widely used in the determination of Hg, because its sensitivity is not very good and it needs a previous concentration step (117). The huge advances obtained by ICP-MS detectors explains their increasing use in speciation studies of Hg, mainly coupled with CV generation (118), LC (119,120), and isotope dilution (121,122).

Although photoacoustic spectroscopy is a not very common technique, it has been applied to Hg(II) determination in water samples after a previous concentration into latex microparticles (123).

Gas chromatography (GC) coupled with FAAS (124) and reversed-phase liquid chromatography (125) is used for the speciation of inorganic and organic Hg.

Trace concentrations of Hg are determined in a flow system by constant current stripping chronopotentiometry in coulometric mode (126).

In recent years, the potential utility of biological substrates for metal speciation has been demonstrated. Aller et al. (127) use bacteria cells such as *Pseudomonas putida* and *Escherichia coli* for the selective retention of Hg(II) and Hg(I) ions. Speciation of methyl mercury and Hg(II) using baker's yeast biomass (*Saccharomyces cerevisiae*) is performed by continuous-flow Hg-CV-AAS (128).

D. Arsenic

A method for the spectrofluorimetric determination of As in water samples has recently been proposed (129).

Hydride generation (HG) is the most common method used for determining As, mainly coupled with atomic spectrometric detectors. It is a highly selective and sensitive technique with low detection limits, and is relatively free from interferences. Prereduction of As(V) to As(III) with KI–ascorbic acid (130,131), or L-cysteine (132) is usually required, because these inorganic As forms present different sensitivities. A method of nonflame AAS for total As concentration by the HG technique has recently been developed (133).

A flow injection nondispersive AAS device has been used for differential determination of As(III) and total As with L-cysteine as prereductant (134,135).

An usual approach to specify As compounds is the use of chromatographic separation techniques, often IC coupled with HG-AAS for organic and inorganic As species in tap water (136) and seawater (137), IEC-HG-AAS in geothermal waters (138), HPLC-HG-AAS in estuarine waters (139–141), and HPLC-HG-AFS in waters from drilled wells (142). Some organic As species (AsBe and AsCh) require a previous digestion, such as online photo-oxidation HPLC-HG-AAS (143), online thermo-oxidation HPLC-HG-AAS (144), or online microwave oxidation HPLC-HG-AAS (145).

When ETAAS is used for As determination, one must consider this technique as not interference free and possibly influenced by the molecular form in which As appears in the sample. Modifiers are added to the sample to prevent volatilization of As during calcination and to ensure a uniform decomposition and atomization. A comparison of different chemical modifiers for the direct determination of As in seawater by ETAAS has been reported (146). And FIA-HG is also used prior to the in situ preconcentration of As in a graphite furnace (147,148).

The robustness of ICP and the capacity for multielement analysis are attractive characteristics for interfacing these detectors with systems for generating hydrides. And ICP-AES is used as As detector (149), because the sensitivity is improved by preconcentrating in an anion-exchange resin (150). Different chromatographic methods are used to separate As species previous to HG-ICP-AES detection (151,152).

When HG is combined with MS interfaced with ICP detection, limits are much better than those obtained with ICP-AES. Recent advances in ICP-MS combined with the increased transport efficiency of As hydride species make HG-ICP-MS one of the most sensitive techniques for As determination (153–155). Direct determination of As in fresh and saline waters by electrothermal vaporization ICP-MS has also been successfully applied (156). Capillary electrophoresis has

recently been interfaced to ICP-MS detectors to speciate As in drinking waters (157).

Among the different electrochemical techniques available for As determination, amperometry and coulombmetry are not commonly used due to their high detection limits. The most used techniques are CSV (158) and ASV (159).

E. Chromium

Depending on its valence, Cr may be either beneficial [Cr(III)] or toxic [Cr(VI)]. Determination of total Cr does not give full information about the health hazard of a certain Cr pollution, and selective determination of both species is usually required. Sensitive spectrophotometric methods have been proposed for both Cr species determination (160–166). In order to avoid inaccurate results due to matrix interferences, Cr(VI) in drinking water, groundwater, and industrial wastewater effluents is removed from the matrix before derivatization and colorimetric determination (167).

Online preconcentration methods coupled with FAAS determination are rapid and suitable for routine determinations of the element (168–172). A simple method for the determination of Cr(III) and total Cr in natural waters, using a flow injection system comprising chelating ion-exchange and FAAS, has been described (173). Speciation of Cr is also achieved with an automated two-column ion-exchange system and detection by online FAAS (170). Cr(III), Cr(VI), and total Cr are separated by adsorption on a resin and determined by FAAS in lake waters (174).

One of the most common analytical techniques used for Cr determination in waters is ETAAS, although determination in seawater has numerous difficulties related to the salinity of the samples and the condition of the graphite tubes (175). Improvement in sensitivity and selectivity in Cr speciation by ETAAS can be attained when using a preconcentration step, such as complexation and extraction with suitable reagents (176–179), coprecipitation (180,181), or electrodeposition on a L'Vov platform (182). Frequently, they are laborious and slow methods, although they offer low detection limits and minimization of matrix interferences. The FIA-ETAAS coupling for the determination of Cr(III) and total Cr is described (183). A study of different chemical modifiers for the determination of Cr in rainwaters by ETAAS has been presented (184).

Chemiluminescence techniques have been used for the selective determination of Cr(III) and Cr(VI) (185,186). Flow injection analysis has been coupled with oxidation of luminol (187) to achieve a more automated method.

Thermal lens spectrometry is applied for quantification and routine determination of Cr(VI) in water (188).

Cr(III) and Cr(VI) are determined by NAA after a two-step coprecipitation process (189).

Nonchromatographic methods for Cr speciation are generally laborious and time consuming. The HPLC separation procedures appear to be simpler and more rapid, so HPLC coupled with ICP-AES detection (190) or HPLC with flame emission spectrometry (191) for Cr speciation have been used. A recent development has been the creation of methods coupling HPLC with ICP-MS, due to the excellent sensitivity and selectivity (192,193). The use of a direct injection nebulizer (DIN) has recently been proposed (194) instead of the conventional sample introduction system, to avoid sample dispersion in the spray chamber.

Isotope dilution–mass spectrometry (ID-MS) proves to be a reliable method for Cr speciation that combines high accuracy and precision with good sensitivity (195). The time consumed in the separation of the species is not much longer than that for ETAAS, which is normally much faster in total element analysis.

Adsorptive catalytic stripping (196,197) and CSV can be used to determine both Cr(III) and Cr(VI) in seawater (198,199) after preconcentration.

Gas chromatography (200) and reversed-phase ion-pair HPLC with UV detection are employed for Cr(III) and Cr(VI) determination after chelation with EDTA and separation on a chromatographic column (201).

Baraj et al. present an enhancement of the sensitivity of Cr(VI) determination by CZE using the stacking effect (202).

F. Nickel

A highly selective, sensitive procedure for flotation separation followed by spectrophotometric detection of Ni(II) in freshwater and seawater has been proposed (203).

Electrothermal AAS also allows Ni determination in a subnanogram range after extraction (204). Other analytical techniques used are chemiluminiscence (205) and liquid scintillation counting, which carries out the determination of the radioactive isotope ^{63}Ni (206). Ni has also been determined using different preadsorptive voltammetric techniques making use of a variety of organic complexing agents for trace and ultratrace metal determinations, which both maximize sensitivity and selectivity (207,208).

G. Cobalt

Four usual methods stand out for the analysis of Co in waters: AAS using the flame mode (35), AAS using the graphite furnace (209,210), laser-excited AFS with a graphite electrothermal atomizer (211), and adsorptive voltammetry (212,213).

Although analysis with chemiluminiscence detection shows the great advantages of low detection limit, simplicity of instrumentation and wide dynamic range (214), the chemiluminiscence method lacks selectivity for an objective metal ion. A significant improvement is achieved using flow injection analysis (215).

H. Selenium

There are many available methods to determine Se. In general, preconcentration methods must be applied, because Se levels in water are very low. Different reactions of Se with compounds that yield color can be used for its determination by spectrophotometry (216–220).

Fluorometric methods are inherently applied to concentration ranges lower than those obtained in spectrophotometric determinations, so they are one of the most sensitive analytical chemistry techniques that can be used and have been widely applied for the determination of Se in natural waters (221). Different atomic fluorescence spectrometers for the determination of the hydride-forming Se in water have also been developed (222). A simple, sensitive, and highly selective automatic spectrofluorometric method for the simultaneous determination of Se(IV) and Se(VI) by FIA has been developed (223).

Se determination in a subnanogram range by AAS can be performed using the graphite furnace and HG techniques, proving that both correlated satisfactorily. The best absolute detection limit can be observed in ETAAS, but HG-AAS is faster and cheaper. Different methods for determination of Se, with a previous preconcentration step by FI-HG-AAS, are reported (224–230). Se speciation is determined by adsorption onto copper oxide particles followed by HG-AAS and ion chromatography (231). Direct determination of Se by ETAAS (232–234) and HG-DCP-AES (235) are described. A comparison of FAAS, ETAAS, and ICP-MS in the determination of Se has been performed (236).

Single-wavelength-dispersive XRF is a nondestructive, rapid technique that has been applied in the determination of Se species in water for agricultural use (237).

The ICP-MS techniques present poor detection limits and low precision for Se. An improvement is achieved with different HG-ICP-MS systems (238,239) and also using ion-pairing reversed-phase or anion-exchange liquid chromatography followed by ICP-MS (240).

Ion chromatography can be used for the separation of selenate in the presence of other anions (241) in drinking water.

Gas chromatography has been proposed for the determination of Se in water (242). However, the sensitivity of the technique is not as good as that obtained with AAS. A method in which Se(IV) was determined by GC with the sensitive electron capture detector has been presented (243). Recently, a method for the determination of Se by GC-ICP-MS has been reported (244), offering the advantage of transferring the total analyte into the ICP-MS instrument without any loss by nebulization. A method for the selective determination of the volatile Se species, dimethyl selenide and dimethyl diselenide, in lake waters using a purge-and-trap injection system coupled to capillary GC-MIP-AES has been presented (245).

Cathodic stripping voltammetry (CSV) involves the cathodic stripping of an insoluble film, usually the Hg salt of the analyte deposited on the working electrode. Differential pulse CSV is used to determine Se in water directly (246–250) and for the determination of Se speciation in river and natural waters (251,252).

Capillary electrophoresis (CE) methods have the advantages of high speed and efficiency, and are used in the determination of seleno compounds in thermal waters (253).

I. Copper

The determination of Cu in a variety of water samples can be conventionally achieved by complexation with specific chelating agents followed by spectrophotometric measurement (254–261).

Different FAAS methods are proposed for Cu determination in water, most of them using preconcentration steps (262–264).

The ETAAS technique is widely used in the determination of Cu in water, having excellent detection limits (265). Determination of Cu in potable water by ETAAS after electrodeposition on a graphite disk electrode seems to be an alternative to the commonly used ASV for the preconcentration and determination of Cu (266).

Cu determination was also performed using flow injection coupled with microwave plasma torch AES (267).

Chemiluminescence methods, using different complexation agents, have been widely applied to the determination of Cu in different types of waters (268).

The determination of Cu by voltammetric methods (DPP or ASV) is widely used for analytical measurements (269–273). Although these methods are very valuable tools, they should not be used without critical evaluation (274). The construction and behavior of a chemically modified carbon press–formed electrode system appropriate for the chemical preconcentration and voltammetric quantitation of Cu has been described. The results were in agreement with those of FAAS (275). Farias et al. (276) describe different complexing ligands in order to investigate their use for the determination of Cu using repetitive cyclic and ASV. A remote electrode, operated in the potentiometric stripping mode, has operated continuously to obtain Cu distribution patterns (277). Recently, a new method for the determination of Cu in natural waters by batch and oscillaty flow injection stripping potentiometry (OFISP) has been proposed (278).

Potentiometry with a cupric ion-selective electrode is the only technique that allows the selective determination of the cupric ion activity without disturbing equilibria in the sample (279,280). De Marco (281) presents a comparison of the behavior of three types of Cu(II)-ion-selective electrodes (copper sulfide, copper selenide, and copper/silver sulfide) in seawaters.

J. Iron

Methods that involve the preconcentration of water samples followed by a spectrophotometric measurement using different reagents have been widely used in the determination of very low Fe concentrations in different types of water samples (282–292).

Rubi et al. (293) present a preconcentration and FAAS determination of Fe by sequential injection analysis.

A highly sensitive technique for the rapid determination of Fe(II) and total dissolved Fe in seawater has been developed. The technique employs FIA and chemiluminiscence detection, and lower detection limits are attainable with larger sample volumes (294). Obata (295) presents an analytical method for determining Fe(III) in oceanic and hydrothermal waters based on selective column extraction using a chelating resin and chemiluminiscence detection in a closed flowthrough system.

An isocratic, rapid, sensitive, and reproducible reversed-phase HPLC analysis using a UV/visible spectrophotometer detector has been developed to determine concentrations of Fe in rainwater and seawater after complexation with ferrozine (296). A similar method with column derivatization is presented (297).

Electrochemical methods to determine Fe in aqueous solution include DPV (298,299), catalytic-ASV (300), and the sensitive CSV with adsorptive deposition of complexes (301,302). This last method is more sensitive than the existing electroanalytical procedures, and it is shown that the dissolved Fe concentration in standard seawater can be successfully determined. Other techniques, such as cyclic voltammetry (303) and the adsorption of a Fe complex on a hanging Hg drop electrode (304), are used to determine total Fe.

The speciation of particulate forms of Fe in natural waters has received little attention, because the speciation techniques are labor intensive and expensive in both capital outlay and maintenance of instruments. Electron spectroscopy is potentially an ideal technique for the study of the oxidation states of solid Fe compounds retained by membrane filtration of dam water samples (305).

Chemical sensors of different sensitivities and selectivities are used for Fe(II) and Fe(III) determination (306–309).

The determination of Fe in aqueous samples can be readily accomplished by laser-induced breakdown spectroscopy (310,311) or by coupling capillary isotachophoresis and CZE with UV detection (312).

K. Manganese

Mn is arguably the most studied element in the marine environment. Work has been focused on colorimetric techniques suitable for use in an FIA system. The FIA reduces the contamination of the sample while increasing both the speed and reproducibility of analysis, and the colorimetric techniques are advantageous because of the widespread availability of spectrophotometric detectors (313–317).

Mn was analyzed by laser-enhanced ionization spectrometry using the 279.5-nm line (318) after a previous extraction.

Some investigations have been made on the determination of Mn by catalytic fluorometry (319). The method is a new way to determine Mn content with high sensitivity and selectivity. Generally, common ions do not interfere with the determination.

A highly sensitive and relatively inexpensive technique for the rapid determination of Mn employing FIA and chemiluminiscence detection has been developed (320).

Mn determination by FAAS has been performed using the flame (321) and the graphite furnace (322–324) techniques.

Zaw and Chiswell (305) present a study on the speciation of Mn in water using electron spectroscopy for chemical analysis, which is potentially an ideal technique for studying the oxidation states of Mn.

A rapid determination of Mn by batch stripping potentiometry in industrial wastewaters and seawater is described (325). The particular advantages of this procedure include a short preconcentration time, the ability to determine low concentrations of Mn without the requirement for deoxygenation, and the use of instrumentation with fast data-sampling rates. The consequence of these advantages is a shorter overall analysis time, which is significant for the routine monitoring of Mn. DP-CSV is also used for the determination of Mn(II) and Mn(VII) in seawater (326) and in coastal and estuarine waters (327).

L. Zinc

The automation of a spectrofluorometric method for the determination of Zn based on the formation of a fluorescent complex is described (328,329). The method is very selective and has been successfully applied to determine Zn(II) in tap water. Sensitive techniques for the determination of Zn in seawater (330) and drinking water (331), coupling FIA with fluorometric detection, have also been developed.

The determination of Zn(II) concentrations in waters by automated analytical methods has usually been carried out by preconcentration followed by FAAS (332), ETAAS (333,334), or ICP-AES (335). Nevertheless, these techniques frequently lack sensitivity when determining very low concentrations of this element in water, and they must be performed in shore-based ultraclean laboratories by highly trained personnel.

Analysis of total Zn in seawater by ASV is problematic because of interference by the hydrogen wave in acidified samples and the inability to detect organically complexed Zn at natural pH. Zn detection can be achieved by ASV at a rotating disk electrode (336) or by exchange with added EDTA and determination by DP-ASV (337).

Zn can be determined by FIA with ion-selective electrode detection after preconcentration on a cation exchanger (338).

M. Multiple Metals

Various methods used for the determination of multiple metals are shown in Table 1.

Table 1 Determination of Multiple Metals

Metals	Analytical techniques	Refs.
Hg, Pb	PT	339
Ni, Cu	SP	340
Co, Fe	SP	341
Zn, Hg	SP	342
Co, Mn	CL	343
Cd, Pb	FAAS, AFS, P, MS, SMVEA, GFAAS	344–350
Cu, Zn, Cd, Ni	FAAS	26
Cu, Zn, Cd, Pb, Fe	FIA-AAS	351
Cd, Cu, Pb	FAAS, ETAAS, ICP-MS, DP-ASV, CA, ASV	17, 51, 352–359
Zn, Pb, Cd, Ag	FAAS	345
Cu, Pb, Ni, Fe, Cr, Co	FAAS	25
Cr, Cu, Mn	AAS	360
Fe, Ni, Cr, Mn	FAAS	361
Cu, Cd, Mn, Co, Pb, Ni, Fe	FAAS	23
Cu, Co	FAAS	362
Cu, Pb, Fe, Zn	FAAS	370
As, Se	HG-AAS, ICP-MS, -ICP-AES, ETAAS, ICP-AAS, -MS, CE	363–370
Sb, As, Se	HG-AAS, ETV-ICP-MS, HG-ICP-MS, MIP-AES	371–375
Mn, Fe, Zn, Ni, Cr, Cd, Pb	ETAAS	376
Cd, Se, Sn	ETAAS	50
Cu, Cd	ETAAS	37, 378
Cu, Ni, Cd	ETAAS	379, 380
Ag, Cd, Pb, Sb	ETAAS	76
As, Cd, Pb, Cu	ETAAS	17, 381, 382
Ag, Ni, Co, Cr	ETAAS	383
Cd, Co, Cu, Fe, Ni, Pb, Zn	ETAAS	384
Mn, Ni, Zn	ETAAS	385
Cd, Cu, Pb, Zn	ETAAS, ASV, DPV	359, 386–391
Ni, Co, Cr, Cu, Pb, Zn, Mn, Cd	ICP-AES	392–394
Cd, Cu, Fe, Ni, Pb, Zn	ICP-AES	395, 396
Cd, Co, Cu, Fe, Mn, Ni, Pb, Zn, Cr	ICP-AES	393, 397
As, Sb, Bi, Hg	HG-ICP-AES, HG-ICP-MS	398, 399
Cr, Ni	ICP-AES	400
Cd, Cu, Fe, Mn, Ni, Zn	ASV, ETAAS, ICP-AES	401, 402
Co, Cu, Fe, Ni	ICP-AES	403
Co, Mo, V	ICP-AES	404
Cu, Cd, Co, Ni, Pb	ICP-MS	405, 406
Al, As, Cd, Cr, Cu, Mn, Ni, Pb, Zn	ICP-MS	407
Bi, Sb, Se, Te, Hg, As	HG-ICP-MS	408
Sb, As, Se	HG-ICP-MS	375
V, Ni, As	ICP-MS	409
Fe, Mn, Ni, Co	ICP-MS	410
Tl, Pb	Laser Excited-AFS	411
Cu, Co, Pb, Cd, Ni, Fe, Cr, Zn	XRF	27
Ti, Cr, As, Pb, Th	XRF	412
Cu, Pb, Fe, Hg, Cr	XRF	413
Cu, Hg, Pb	XRF	22
Pb, Tl, Ce	LIF	414
Te, Cd, Hg	ASV, CP	415
Cu, Sb, Bi, Pb	DP-ASV	416

Table 1 Continued

Metals	Analytical techniques	Refs.
Se, Hg, Cu, Pb	PS	417
Cu, Pb	CP, DP-ASV, ASV	418–420
Hg, As	ASV, NAA	421–423
Cu, Ag, Au, Hg	DPV	424
Ni, Co	PS, DPP	425–427
Co, Ni, Zn	V	426
Ni, Cu, Zn	V	428
Cu, Hg	P	429
Fe, Ti	PS	430
Co, Ni, Cu, Zn, Hg, Cd, Pb	IC	41, 431
Pb, Cd, Cu, Co, Zn, Ni	IC	432
Cd, Ni, Pb, Cu, Hg, Co, Bi	IC	431
Zn, Pb, Ni, Cu	IC	433
Cu, Ni, Zn, Mn	IC	434
Zn, Cu, Ni, Co, Mn	CIC	435
Ba, Cd, Pb, Cr	CS	436
Cu, Fe, Cr, Pb, Cd, Hg	MS	437
Pb, Sn	HPLC-ICP-MS	438
Be, Al, Cr	HPLC	439
Cu, Ni, Pd, V	HPLC	440
Cu, Ni, V	HPLC	441
V, Cr, Fe	HPLC	442
Co, Cu, Hg, V, Zn	NAA	34
Cr, Cs	γ-Ray spectrometry	443
Fe, Co, Ni	GPA-MP	444
Co, Cu, Mn, Ni, V	EV-ICP-MS	445
Zn, Cd, Pb, Cu, Ni, Sn	ASV	446
Fe, Zn, Cd, Mo	MPT-AES	447
Cd, Pb, Ni, Cu, Zn	ICP-MS	53
As, Sb, Se, Ge	HG-ICP-MS	448
Cu, Mo	ETAAS	54
Cu, Pb, Cd, Zn	ASV	449
As, Sb, Bi, Se, Te	FANES	450
Fe, Cu, Pb, Zn, Ni, Co	IIC	451
Ni, Co, Mn	FAAS	55
Mn, Mo	HPLC	452
Cd, Co, Cr, Mn, Ni, Pb	ETAAS	453
Cd, Co, Cu, Ni, Pb, U, Y	ICP-MS	454
Cd, Co, Cu, Mn, Ni, Pb, Zn	ICP-MS, GFAAS	455
Ni, Co, Cd, Pb, Cu	AV	406
Fe, Ni, Cu, Zn, Pb, Mn, Co, As	XRF	456
As, Bi, Hg, Sb, Se, Sn	ICP-AES, ETAAS	457
Ge, As, Se, Sn, Sb, Te, Bi	HG-ICP-MS	458
Co, Cu, Ni	ETAAS	459
Cu, Mn	ETAAS	460
Cu, Fe, Pb, Mn, Zn, Cd, Ni, Bi, Cr	FAAS	19

ABBREVIATIONS

AAS	atomic absorption spectrometry
AES	atomic emission spectrometry
AFS	atomic fluorescence spectrometry
ASV	anodic stripping voltammetry
AV	adsorptive voltammetry
CA	chronoamperometry
CE	capillary electrophoresis
CIC	chelation ion chromatography
CP	chronopotentiometry
CS	chemical sensor
CSV	cathodic stripping voltammetry
CV-AAS	cold vapor–atomic absorption spectrometry
CZE	capillary zone electrophoresis
DCP	direct current plasma
DP	differential pulse
DPP	differential pulse polarography
DPV	differential pulse voltammetry
ETAAS	electrothermal atomic absorption spectrometry
EV	electrothermal vaporization
FAAS	flame atomic absorption spectrometry
FANES	furnace atomic nonthermal excitation spectrometry
FIA	flow injection analysis
GC-MS	gas chromatography–mass specrometry
GPA-MP	gas-phase atomization in a microwave plasma
HG-AAS	hydride generation–atomic absorption spectrometry
HPLC	high-performance liquid chromatography
ICP-AES	inductively coupled plasma–atomic emission spectrometry
ICP-MS	inductively coupled plasma–mass spectrometry
ID-MS	isotope dilution–mass spectrometry
IEC	ion-exchange chromatography
IIC	ion-interaction chromatography
LIF	laser-induced fluorescence
MIP	microwave-induced plasma
MPT	microwave plasma torch
MS	microsensor
NAA	neutron activation analysis
P	polarography
PS	stripping potentiometry
PT	potentiometric titration
SMVEA	sequential metal vapor elution analysis
SP	spectrophotometry
UV	ultraviolet
V	voltammetry
XRF	x-ray fluorescence

REFERENCES

1. RN Reeve. In: JD Barnes, ed. Environmental Analysis. London: Wiley, 1994, pp 49–51.
2. L Mart, HW Nürnberg, P Valenta. Fresenius J Anal Chem 300:350–362, 1980.
3. L Mart. Talanta 29:1035–1040, 1982.
4. JR Moody, RM Lindstrom. Anal Chem 49:2264–2267, 1977.
5. G Capodaglio, G Toscano, P Cescon, G Scarponi, H Muntau. Ann Chim 84:329–345, 1994.
6. G Capodaglio, C Barbante, C Turetta, G Scarponi, P Cescon. Mikrochim Acta 123:129–136, 1996.
7. K May, K Reisinger, R Flucht, M Stoeppler. Vom Wasser 55:63–76, 1980.
8. R Ahmed, K May, M Stoeppler. Sci Total Environ 60:249–261, 1987.
9. GE Batley. In: GE Batley, ed. Trace Element Speciation Analytical Methods and Problems. Boca Raton, FL: CRC Press, 1990, pp 1–24.
10. K Vercoutere, R Cornelis. In: PH Quevauviller, AM Maier, B Griepink, eds. Quality Assurance for Environmental Analysis. Amsterdam: Elsevier Science, 1995, pp 195–212.
11. V Cheam, H Agemian. Analyst 105:737–743, 1980.
12. F Chmilenko, A Baklanov, V Chuiko, Khim Tekhnol Vody 12(11):1039–1042, 1990.
13. M Kolb, P Rach, J Schäfer, A Wild. Fresenius J Anal Chem 342(4–5):341–349, 1992.
14. S Sander, G Henze. GIT Fachz Lab 39(12):1120, 1995.
15. T Nakamura, H Oka, M Ishii, J Sato. Analyst 119(6):1397–1401, 1994.
16. M Hartmann, H Lass, D Puteanus. Colloq Atomspektrom Spureanal 5th, 1989, pp 703–709.
17. Z Zhuang, C Yang, X Wang, P Yang, B Huang. Fresenius J Anal Chem 355(3–4):277–280, 1996.
18. ZS Chen, M Hiraide, H Kawagushi. Mikrochim Acta 124(1–2):27–34, 1996.
19. L Elci, U Sahin, S Oztas. Talanta 44(6):1017–1023, 1997.
20. C Suzuki, J Yoshinaga, M Morita. Anal Sci 7(2):997–1000, 1991.
21. N Fudagawa, A Hioki, M Kubota, A Kawase. Bunseki Kagaku 41(3):39–44, 1992.
22. B Holynska, B Ostachowicz, C Wegrzynek. Spectrochim Acta 51B(7):769–773, 1996.
23. L Elci, M Soylak, M Dogan. Fresnius J Anal Chem 342(1–2):175–178, 1992.
24. U Naidu, G Naidu. Indian J Environ Prot 11(4):282–283, 1991.
25. R Saran, T Baul, P Srinivas, D Khathing. Anal Lett 25(8):1545–1557, 1992.
26. M Mallet, M Mehra. Orient J Chem 8(1):12–17, 1992.
27. A Granzhan, A Charykov. Vestn Leningr Univ 4:107–108, 1990.

28. GJ Batterham, DL Parry. Mar Chem 55:381–388, 1996.

29. R Tezcan, H Tezcan, Fresenius Environ Bull 5(3/4): 156–160, 1996.

30. MY Khuchawar, P Das. J Chem Soc Pak 18(1):6–8, 1996.

31. T Kiriyama. Bunseki Kagaku 42(4):223–228, 1993.

32. RA Remier, A Miyazaki. J Anal At Spectrom 7(8): 1238–1244, 1992.

33. MC Yebra, A Bermejo, P Bermejo. Mikrochim Acta 109(5–6):243–251, 1992.

34. R Rao, D Goski, A Chatt. J Radioanal Nucl Chem 161(1):89–99, 1992.

35. M Trojanowicz, K Pyrzynska. Anal Chim Acta 287(3): 247–252, 1994.

36. T Wang. Lihua Jianyan Huaxue Fence 29(5):273–274, 1993.

37. S Dadfarnia, I Green, CW McLeod. Anal Proc 31(2): 61–63, 1994.

38. NV Semenova, EI Morosanova, IV Pletnev, YA Zolotov. Zh Anal Khim 49(5):477–480, 1994.

39. ED Suttle, EW Wolff. Anal Chim Acta 258(2):229–236, 1992.

40. D Chambaz, P Edder, W Haerdi. J Chromatogr 541(1–2):443–452, 1991.

41. D Chambaz, W Haerdi. J Chromatrogr 600(2):203–210, 1992.

42. E Vassileva, I Proinova, K Hadjiivanov. Analyst 121(5):607–612, 1996.

43. E Vassileva, B Varimezova, K Hadjiivanov. Anal Chim Acta 336(1–3):141–150, 1996.

44. M Shengjun, JA Holcombe. Talanta 38(5):503–510, 1991.

45. CA Mahan, JA Holcombe. Spectrochim Acta 47B(13): 1483–1495, 1992.

46. A Maquieira, H Elmahadi, R Puchades. Analyst 121: 1633–1640, 1996.

47. Ph Quevauviller, KJM Kramer, T Vinhas. Mar Poll Bull 28(8):506–511, 1994.

48. Q Jin, H Zhang, F Liang. J Anal At Spectrom 10(10): 875–879, 1995.

49. JD Willis, KE Jarvis, JG Williams. Sci Total Environ 135(1–3):137–143, 1993.

50. M Sperling, X Yin, B Weiz. J Anal At Spectrom 6(4): 295–300, 1991.

51. R Ma, W Van Mol, F Adams. At Spectrosc 17(4):176–181, 1996.

52. DB Taylor, HM Kingston, DJ Nogay, D Koller, R Hutton. J Anal At Spectrom 11(3):187–191, 1996.

53. AP Packer, MF Giné, CES Miranda, BF Dos Reis. J Anal At Spectrom 12:563–566, 1997.

54. YH Sung, ZS Liv, SD Huang. J Anal At Spectrum 12: 841–847, 1997.

55. R Ma, F Adams. Anal Chim Acta 317(1–3):215–22, 1995.

56. Y Zhu, Ch Wang, J Chen, W Jiang, G Jin. Analyst 120(12):2853–2856, 1995.

57. B Vaidya, MD Porter, MD Utterback, RA Bartsch. Anal Chem 69(14):2688–2693, 1997.

58. D Tsalev, C Li, M Ivanova. Colloq Atmosp Spurenanal, 5th, 1989, pp 465–474.

59. C Hee, K Young. Bull Korean Chem Soc 17(4):338–342, 1996.

60. C García, JL Pérez, B Moreno, E Beato, S García. J Anal At Spectrom 11(1):37–41, 1996.

61. P Devi, G Naidu. J Radional Nucl Chem 146(4):267–272, 1990.

62. L Ebdon, P Goodall, SJ Hill, PB Stockwell, KC Thompson. J Anal At Spectrom 8(50:723–729, 1993.

63. H Chuang, S Huang. Spectrochim Acta 49B(3):283–288, 1994.

64. T Hwang, SJ Jiang. J Anal At Spectrom 12:579–584, 1997.

65. Y Ma, J Bai, J Wang, Z Li, L Zhu, Y Li, H Zheng, B Li. J Anal At Spectrom 7(2):425–432, 1992.

66. J Cabon, A Le Bihan. J Anal At Spectrom 7(2):383–388, 1992.

67. M Hiraide, T Usami, H Kawaguchi. Anal Sci 8(1):31–34, 1992.

68. G Zhang, J Li, D Fu, D Hao, P Xiang. Talanta 40(3): 409–413, 1993.

69. C Lan. Analyst 118(2):189–192, 1993.

70. PM Linnik, IV Iskra. Arch Hydrobiol 113(1–4):559–564, 1996.

71. A Patwardhan, A Joshi. Indian J Environ Prot 9(4): 297–300, 1989.

72. S Ito, Y Asano, H Wada. Talanta 44(40:697–704, 1997.

73. S Savvin, T Petrova, T Dzherayan, M Reikhshtat. Fresenius J Anal Chem 340(4):217–219, 1991.

74. E Rakhman 'ko, G Tsvirko, A Gulevich. Zh Anal Khim 46(8):1525–1529, 1991.

75. A Afkhami, A Safavi, A Massoumi. Anal Lett 24(9): 1643–1655, 1991.

76. A Naghmush, K Pyrzynska, M Trojanowicz. Talanta 42(6):851–860, 1995.

77. I Sekerka, J Lechner. Anal Chim Acta 254(1–2):99–107, 1991.

78. S Han, Y Li, Z De. J Anal At Spectrom 11(4):265–269, 1996.

79. HW Sun, LL Yang, DQ Zhang, WX Wang, JM Sun. Fresenius J Anal Chem 358(5):646–651, 1997.

80. M Chikuma, H Aoki. Microchem J 49(2–3):368–377, 1994.

81. H Chen, R Zhu, J Wu. J Environ Sci Health A29(5): 867–882, 1994.

82. G Samanta, D Chakraborti. Environ Technol 17(12): 1327–1337, 1996.

83. C Cabrera, M López, C Gallego, M Lorenzo, E Lillo. Sci Total Environ 159(1):17–21, 1995.

84. J Shiowatana, J Matousek. Talanta 38(4):375–383, 1991.

85. X Yan, Z Ni. J Anal At Spectrom 6(6):483–486, 1991.

86. V Granadillo, J Navarro, R Romero. J Anal At Spectrom 8(4):615–622, 1993.

87. M Sperling, X Yan, B Weltz. Spectrochim Acta 51B(14):1891–1908, 1996.

88. Y Liang, Y Xu. J Anal At Spectrom 12:855–858, 1997.

89. Y Xu, Y Liang. J Anal At Spectrom 12(4):471–474, 1997.

90. A Miyazaki, R Reimer. J Anal At Spectrom 8(3):449–452, 1993.

91. L Halicz, J Lam, J McLaren. Spectrochim Acta 49B(7):637–647, 1994.

92. V Cheam, J Lechner, I Sekerka, R Desrosiers, J Nriagu, G Lawson. Anal Chim Acta 269(1):129–136, 1992.

93. V Cheam, J Lechner, I Sekerka, R Desrosiers. J Anal At Spectrom 9(3):315–320, 1994.

94. B Feldman, H Mogadeddi, J Osterloh. J Chromatogr 594(1–2):275–282, 1992.

95. J Wang, J Lu, C Yarnitzky. Anal Chim Acta 280(1):61–67, 1993.

96. Q Wu, G Batley. Anal Chim Acta 309(1–3):95–101, 1995.

97. K Yokoi, A Yamaguchi, M Mizumachi, T Koide. Anal Chim Acta 316(3):363–369, 1995.

98. J Lebecka, S Chalupnik. Rare Nucl Processes Proc Europhys Conf Nucl Phys 14th, 1192:336–342, 1990.

99. D To. Anal Chem 65(19):2701–2703, 1993.

100. M Deb, N Nashine, RK Mishra. Ann Chim 86(7–8):381–391, 1996.

101. MTM Zaki, MA Esmaile. Anal Lett 30(8):1579–1590, 1997.

102. Z Kwokal, K May, M Branica. Sci Total Environ 154(1):63–69, 1994.

103. E Munaf, T Takeuchi, H Haraaguchi. Fresenius J Anal Chem 342(1–2):154–156, 1992.

104. E Munaf, T Takeuchi, D Ishili, H Haraguchi. Anal Sci 7(4):605–609, 1991.

105. ML Wang, GQ Gan, SH Qian, JS Jiang, YT Wan, YK Chan. Fresenius J Anal Chem 358(7–8):856–858, 1997.

106. C Hanna, J Tyson, S McIntosh. Anal Chem 65:653–656, 1993.

107. M García, R García, N García, A Sanz. Talanta 41:1833–1839, 1994.

108. P Bermejo, J Moreda, A Moreda, A Bermejo. J Anal At Spectrom 12(3):317–321, 1997.

109. A Le Bihan, J Cabon. Talanta 37(12):1119–1122, 1990.

110. P Bermejo, J Moreda, A Moreda, A Bermejo. Mikrochim Acta 124(1–2):111–122, 1996.

111. C Chan, R Sadana. Anal Chim Acta 282(1):109–115, 1993.

112. MJ Bloxham, SJ Hill, PJ Worsfold. J Anal At Spectrom 11(7):511–514, 1996.

113. W Jian, C McLeod. Talanta 39(11):1537–1542, 1992.

114. D Cossa, J Sanjuan, J Cloud, P Stockwell, W Corns. Water Air Soil Pollut 80(1–4):1279–1284, 1995.

115. M Okomura, K Fukushi, S Willie, R Sturgeon. Fresenius J Anal Chem 345(8–9):570–574, 1993.

116. R Swift, J Campbell. Spectroscopy 8(2):38–47, 1993.

117. G Aizpun, M Fernández, A Sanz. J Anal At Spectrom 8(8):1097–1102, 1993.

118. E Debrah, E Denoyer. J Anal At Spectrom 11(2):127–132, 1996.

119. M Bloxham, A Gachanjha, S Hill, P Worsfold. Anal At Spectrom 11(2):145–148, 1996.

120. C Wan, C Chen, S Jiang. J Anal At Spectrom 12:683–687, 1997.

121. R Smith. Anal Chem 65(18):2485–2488, 1993.

122. J Camuna, R Pereiro, J Sánchez, A Sanz. Spectrochim Acta 49B(5):475–484, 1994.

123. V VanderNoot, E Lai. Anal Chem 64(24):3187–3190, 1992.

124. H Emteborg, H Sinemus, B Radziuk, D Baxter, W French. Spectrochim Acta 51B(8):829–837, 1996.

125. Y Wang, C Whang. J Chromatogr 628(1):133–137, 1993.

126. E Beinrohr, M Cakrt, J Dzurov, P Kottas, E Kozakova. Fresenius J Anal Chem 356(3–4):253–258, 1996.

127. AJ Aller, JM Lumbreras, LC Robles, GM Fernández. Anal Proc 32(12):511–514, 1995.

128. Y Madrid, C Cabrera, T Pérez, C Cámara. Anal Chem 67:750–754, 1995.

129. A Pal, NR Jana, TK Sau, M Bandyopadhyay, T Pal. Anal Commun 33(9):315–317, 1996.

130. W Driehaus, M Jekel. Fresenius J Anal Chem 343(4):352–356, 1992.

131. I Brindle, H Alarabi, S Karshamn, X Le, S Zheng, H Chem. Analyst 117(3):407–411, 1992.

132. H Chen, I Brindle, X Le. Anal Chem 64(6):667–672, 1992.

133. J Janjic, L Conkic, J Kiurski, J Benak. Water Res 31(3):419–428, 1997.

134. X Yin, E Hoffmann, C Luedke. Fresenius J Anal Chem 355(3–4):324–326, 1996.

135. S Nielsen, EH Hansen. Anal Chim Acta 343(1–2):5–17, 1997.

136. E González, E Alonso, P López, S Muniategui, D Prada. Anal Lett 28(15):2699–2718, 1995.

137. E González, E Alonso, D Prada, E Fernández. Anal Lett 29(15):2701–2712, 1996.

138. T Yokoyama, Y Takahashi, T Tarutani. Chem Geol 103(1–4):103–111, 1993.

139. L Hunt, A Howard. Mar Poll Bull 28(1):33–38, 1994.

140. M Gómez, C Cámara, MA Palacios, A López. Fresenius J Anal Chem 357(7):844–849, 1997.

141. FH Ko, SL Shun, MH Yang. J Anal At Spectrom 12(5):589–585, 1997.

142. Z Mester, P Fodor. J Chromatogr A 756(1 + 2):292–299, 1996.

143. A Howard, L Hunt. Anal Chem 65:2995–2998, 1993.

144. MA López, MM Gómez, MA Palacios, C Cámara. Fresenius J Anal Chem 346:643–647, 1993.

145. MA López, MM Gómez, C Cámara, MA Palacios. J Anal At Spectrom 9:291–295, 1994.

146. P Bermejo, J Moreda, A Moreda, A Bermejo. Fresenius J Anal Chem 355(2):174–179, 1996.

147. Y An, S Willie, R Sturgeon. Spectrochim Acta 47B(12):1403–1410, 1992.

148. M Burguera, J Burguera. J Anal At Spectrom 8(2):229–233, 1993.

149. Y Feng, J Cao. Anal Chim Acta 293(1–2):211–218, 1994.

150. P Schramel, L Xu. Fresenius J Anal Chem 343(4):373–377, 1992.

151. Y Li, M Fernández, E González, A Sanz. J Anal At Spectrom 8(6):815–820, 1993.

152. R Rubio, A Padró, J Albertí, G Rauret. Anal Chim Acta 283(10:160–168, 1993.

153. S Branch, W Corns, L Ebdon, S Hill, P O'Neill. J Anal At Spectrom 6(2):155–158, 1991.

154. J Creed, M Magnuson, C Brockhoff, I Chamberlain, M Sivaganesan. J Anal At Spectrom 11(7):505–509, 1996.

155. M Huang, S Jiang, C Hwang. J Anal At Spectrom 10(1):31–35, 1995.

156. DC Greogire, ML Ballinas. Spectrochim Acta 52B(1):75–82, 1997.

157. ML Magnuson, JT Creed, CA Brockhoff, J Anal At Spectrom 12:689–695, 1997.

158. H Li, RB Smart. Anal Chim Acta 325(1–2):25–32, 1996.

159. SB Adeloju, TM Young. Anal Lett 30(1):147–161, 1997.

160. R Gao, Z Zhao, Q Zhou, D Yuan. Talanta 40(5):637–640, 1993.

161. R Escobar, Q Lin, A Guiram, F de la Rosa. Int J Environ Anal Chem 61(3):169–175, 1995.

162. M Lunar, S Rubio, D Pérez. Int J Environ Anal Chem 56(3):219–227, 1994.

163. G Piying, G Xuexin, Z Tianze. Anal Lett 29(4):651–659, 1996.

164. P Tarafder, S Singh, D Rathore. Fresenius J Anal Chem 354(1):124–125, 1996.

165. V Ososkov, B Kebbekus, D Chesbro. Anal Lett 29(10):1829–1850, 1996.

166. J Manzoori, M Sorouraddin, F Shemiran. Anal Lett 29(11):2007–2014, 1996.

167. K Edgell, J Longbottom, R Joyce. J AOAC Int 77(4):994–1004, 1994.

168. A Shah, S Devi. Anal Chim Acta 236(2):469–473, 1990.

169. J Posta, H Berndt, S Luo, G Schaldach. Anal Chem 65(19):2590–2595, 1993.

170. P Sule, J Ingle. Anal Chim Acta 326(1–3):85–93, 1996.

171. H Zou, S Xu, Z Fang. At Spectrom 17(3):112–118, 1996.

172. E Beinrohr, A Manova, J Dzurov. Fresenius J Anal Chem 355(5–6):528–531, 1996.

173. R Crespón, MC Yebra, P Bermejo. Anal Chim Acta 327(1):37–45, 1996.

174. B Demirata, J Tor, H Filik, M Afsar. Fresenius J Anal Chem 356(6):375–377, 1996.

175. S Apte, S Comber, M Gardner, A Gunn. J Anal At Spectrom 6(2):169–172, 1991.

176. E Beceiro, J Barciela, P Bermejo, A Bermejo. Fresenius J Anal Chem 344(7–8):301–305, 1992.

177. E Beceiro, P Bermejo, A Bermejo, J Barciela, C Barciela. J Anal At Spectrom 8(4):649–653, 1993.

178. M Gardner, J Ravenscroft. Fresenius J Anal Chem 354(5–6):602–605, 1996.

179. Z Li, Y Shi, P Gao, X Gu, T Zhov. Fresenius J Anal Chem 358(4):519–522, 1997.

180. K Lee, H Choi, Y Kim. Anal Sci Technol 3(3):419–425, 1990.

181. Y Kim, S Park, J Choi. Bull Korean Chem Soc 14(3):330–335, 1993.

182. J Vidal, J Sanz, J Castillo. Fresenius J Anal Chem 344(6):234–241, 1992.

183. M Sperling, X Yin, B Welz. Analyst 117(3):629–635, 1992.

184. K Thomaidis, E Piperaki, C Polydorov, C Efstathiv. J Anal At Spectrom 11(1):31–36, 1996.

185. B Gammelgaard, O Joens, B Nielsen. Analyst 117(3):637–640, 1992.

186. H Beere, P Jones. Anal Chim Acta 293(3):237–243, 1994.

187. R Escobar, Q Lin, A Guiraum, F de la Rosa. Analyst 118(6):643–647, 1993.

188. M Sikovec, M Franko, F Cruz, S Katz. Anal Chim Acta 330(2–3):245–250, 1996.

189. C Lan, C Tseng, M Yang, Z Alfassi. Analyst 116(1):35–38, 1991.

190. A Cox, C McLeod. Mikrochim Acta 109(1–4):161–164, 1992.

191. J Posta, A Gaspar, R Toth, L Ombodi. Microchem J 54(3):195–203, 1996.

192. F Byrdy, L Olson, N Vela, J Caruso. J Chromatogr A 712(2):311–320, 1995.

193. J Posta, A Alimonti, F Petrucci, S Caroli. Anal Chim Acta 325(3):185–193, 1996.

194. M Powell, D Boomer, D Wiederin. Anal Chem 67(14):2474–2478, 1995.

195. R Nusko, K Heumann. Anal Chim Acta 286(3):283–290, 1994.

196. J Wang, J Lu. Analyst 117(12):1913–1917, 1992.

197. Z Gao, KS Siow. Electroanalysis 8(6):602–606, 1996.

198. M Boussemart, C Van der Berg, M Ghaddaf. Anal Chim Acta 262(1):103–115, 1992.

199. M Boussemart, C Van der Berg. Analyst 119(6):1349–1353, 1994.

200. R Mugo, K Orians. Anal Chim Acta 271(1):1–9, 1992.

201. J Jen, Ou-Yang, C Chen, S Yang. Analyst 118(10):1281–1284, 1993.

202. B Baraj, M Martínez, A Sastre, M Aguilar. J High Resolution Chromatogr 18(10):675–678, 1995.

203. MA Kabil, SE Ghazy, MR Lasheen, MA Shallaby, NS Amar. Fresenius J Anal Chem 354(3):371–373, 1996.

204. Y Shijo, T Shimizu, T Tsunoda, S Tao, S Shiquan. Anal Chim Acta 242(2):209–213, 1991.

205. X Lu, M Lu, G Zhao. Toxicol Environ Chem 38(1–2):73–79, 1993.

206. J Lo, B Cheng, C Tseng, J Lee. Anal Chim Acta 281(2):429–433, 1993.

207. Z Zhang, Z Cheng, S Cheng, G Yang. Talanta 38(12):1487–1491, 1991.

208. P Farias, A Ohara, I Takase, S Ferreira, J Gold. Talanta 40(8):1167–1171, 1993.

209. M Sperling, X Yin, B Welz. J Anal At Spectrom 6(8):615–622, 1991.

210. M Hiraide, Z Chen, H Kawaguchi. Talanta 43(7):1131–1136, 1996.

211. A Yuzefovsky, R Lonardo, M Wang, R Michel. J Anal At Spectrom 9(11):1195–1202, 1994.

212. Z Zhang. Mikrochim Acta 1(1–2):89–95, 1991.

213. M Vega, B Dan, MG Constant. Anal Chem 69(5):874–881, 1997.

214. S Hirata, Y Hashimoto, M Aihara, G Mallika. Fresenius J Anal Chem 355(5–6):676–679, 1996.

215. V Sukhan, O Zaporozhets. Zh Anal Khim 46(12):2342–2348, 1991.

216. K Ramachandran, R Kaveeshwar, V Gupta. Talanta 40(6):781–784, 1993.

217. B Kasterka. Chem Anal 37(3):361–367, 1992.

218. S Lee, J Choi, H Choi, Y Kim. Korean Chem Soc 38(5):351–358, 1994.

219. AS Amin, MN Zareh. Anal Lett 29(12):2177–2189, 1996.

220. KN Ramachandran, GS Kumar. Talanta 43(10):1711–1714, 1996.

221. D Wang, G Alfthan, A Aro, L Kauppi, J Soveri. Trace Elem Health Dis Proc Jt Nord Trace Elem Soc/Union Pure Appl Chem Int Symp, 1990–91, pp 49–56.

222. W Corns, P Stockwell, L Ebdon, S Hill. J Anal Atom Spectrom 8(1):71–77, 1993.

223. MJ Ahmed, CD Stalikas, PG Veltsistas, SM Tzovwara, MI Karayannis. Analyst 122(3):221–226, 1997.

224. U Dernemark, J Pettersson, A Olin. Talanta 39(9):1089–1096, 1992.

225. M Cobo, M Palacios, C Cámara. Anal Chim Acta 283(1):386–392, 1993.

226. U Oernemark, A Olin. Talanta 41(1):67–74, 1994.

227. G Tao, E Hansen. Analyst 119(2):333–337, 1994.

228. J González, ML Fernández, JM Marchante, JE Sánchez, A Sanz. Spectrochim Acta 51B (14):1849–1857, 1996.

229. JL Burguera, P Carrero, M Burguera, C Rondon, MR Brunetto, M Gallignani. Spectrochim Acta 51B(14):1837–1847, 1996.

230. EM Flores, SR Mortari, AF Martins. J Anal At Spectrom 12(3):379–381, 1997.

231. K Reddy, Z Zhang, M Blaylock, G Vance. Environ Sci Technol 29(9):1754–1759, 1995.

232. L Tian, F Peng, X Wang. Anal Sci 2:1155–1158, 1991.

233. B Welz, G Bozsai, M Sperling, B Radziuk. J Anal At Spectrom 7(3):505–509, 1992.

234. YZ Liang, M Li, Z Rao. Fresenius J Anal Chem 357(1):112–116, 1997.

235. ID Brindle, E Lugowska. Spectrochim Acta 52B(2):163–176, 1997.

236. G Koelbl. Mar Chem 48(3–4):185–197, 1995.

237. MI Castilla, JV Llopis. Agrochimica 39(5–6):233–239, 1995.

238. E McCurdy, J Lange, P Haygarth. Sci Total Environ 135(1–3):131–136, 1993.

239. H Tao, J Lam, J McLaren. J Anal At Spectrom 8(8):1067–1073, 1993.

240. Y Cai, M Cabanas, J Fernández, M Abalos, J Bayona. Anal Chim Acta 314(3):183–192, 1995.

241. C Sarzanini, O Abollino, E Mentasti, V Porta. Chromatographia 30(5–6):293–297, 1990.

242. K Johansson, A Olin. J Chromatogr 598(1):105–114, 1992.

243. K Johansson, U Oernemark, A Olin. Anal Chim Acta 274(1):129–140, 1993.

244. S Gallus, K Heumann. J Anal At Spectrom 11(9):887–892, 1996.

245. MB de la Calle, M Ceulemans, C Witte, R Lobinski, F Adams. Mikrochim Acta 120(1–4):73–82, 1995.

246. F Seby, M Potin, A Castetbon. J Fr Hydrol 24(1):81–90, 1993.

247. G Mattsson, L Nyholm, A Olin, U Ornemark. Talanta 42(6):817–825, 1995.

248. M Potin, F Seby, M Astruc. Fresenius J Anal Chem 351(4–5):443–448, 1995.

249. L Campanella, T Ferri, B Pentronio. Analusis 24(2):35–38, 1996.

250. SB Adeloju, D Jagner, L Renman. Anal Chim Acta 338(3):199–207, 1997.

251. T Ferri, P Sangiorgio. Anal Chim Acta 321(2–3):185–193, 1996.

252. C Elleouet, F Quentel, C Madec. Water Res 30(4):909–914, 1996.

253. N Gilon, M Potin. J Chromatogr A 732(2):369–376, 1996.

254. M Kan, H Sakamoto, T Nasu, M Taga. Anal Sci 7(6):913–917, 1991.

255. K Yoshimura, S Matsuoka, Y Inokura, U Hase. Anal Chim Acta 268(2):225–233, 1992.

256. C Hill, K Street, W Philipp, S Tanner. Anal Lett 27(13):2589–2599, 1994.

257. F Theraulaz, O Thomas. Quim Anal 13(4):191–195, 1994.

258. K Sharma, SK Patel. J Indian Chem Soc 73(8):443–444, 1996.

259. M Toral, P Richter, B Muñoz. Environ Monit Assess 38(1):1–10, 1995.

260. S Nakano, K Nakaso, K Noguchi, T Kawashima. Talanta 44(5):765–770, 1997.

261. S Ohno, N Teshima, T Watanabe, H Itabashi, S Nakano, T Kawashima. Analyst 121(10):1515–1518, 1996.

262. M Comber, G Eales, P Nicholson, S Henn. J Autom Chem 14(1):5–8, 1992.

263. R Santelli, M Gallego, M Valcarcel. Talanta 41(5): 817–823, 1994.

264. P Hashemi, A Olin. Int J Environ Anal Chem 63(1): 37–46, 1996.

265. M Kan, T Nasu, M Taga. Anal Sci 2:1115–1116, 1991.

266. J Komarek, P Stavinoha, S Gomiscek, L Sommer. Talanta 43(8):1321–1326, 1996.

267. Y Madrid, M Wu, Q Jin, G Hieftje. Anal Chim Acta 277(1):1–8, 1993.

268. K Coale, K Johnson, P Stout, C Sakamoto. Anal Chim Acta 266(2):345–351, 1992.

269. F Quentel, C Elleouet, C Madec. Electroanalysis 6(8): 683–688, 1994.

270. E Carrera, A Momberg, M Toral, P Richter. Anal Lett 24(1):83–92, 1991.

271. A Economon, PR Fielden. Analyst 121(12):1903–1906, 1996.

272. A Safavi, M Pakniat, N Maleki. Anal Chim Acta 335(3):275–282, 1996.

273. JF Van Staden, M Matoetoe. Fresenius J Anal Chem 357(6):624–628, 1997.

274. M Goncalves, L Sigg. Electroanalysis 3(6):553–557, 1991.

275. G Zhang, C Fu. Talanta 38(12):1481–1485, 1991.

276. P Farias, S Ferreira, A Ohara, M Bastos, M Goulart. Talanta 39(10):1245–1253, 1992.

277. J Wang, N Foster, S Armalis, D Larson, A Zirino, K Olen. Anal Chim Acta 310(2):223–231, 1995.

278. C Chow, SD Kolev, DE Davey, DE Mulcahy. Anal Chim Acta 330(1):79–87, 1996.

279. B Hoyer. Talanta 38(1):115–118, 1991.

280. S Belli, A Zirino. Anal Chem 65:2583–2589, 1993.

281. R DeMarco. Anal Chem 66(19):3202–3207, 1994.

282. S Abe, M Endo. Anal Sci 7:1181–1184, 1991.

283. A Cladera, E Gómez, J Estela, V Cerdá. Analyst 116(9):913–917, 1991.

284. S Pehkonen, Y Erel, M Hoffmann. Environ Sci Technol 26(9):1731–1736, 1992.

285. U Hase, K Yoshimura. Analyst 117(9):1501–1506, 1992.

286. D King, J Lin, D Kester. Anal Chim Acta 247(1):125–132, 1991.

287. S Nakano, M Sakai, M Kurachi, T Kawashima. Microchem J 49(2–3):298–304, 1994.

288. L Zhang, K Terada. Anal Chim Acta 293(3):311–318, 1994.

289. I Suárez, S Pehkonen, M Hoffman. Environ Sci Technol 28(12):2080–2086, 1994.

290. S Blain, P Treguer. Anal Chim Acta 308(1–3):425–432, 1995.

291. X Gu, T Zhou. Anal Lett 29(3):463–476, 1996.

292. G Asgedom, BS Chandravanshi. Ann Chim 86(9–10): 485–494, 1996.

293. E Rubi, MS Jiménez, F De Mirabo, R Forteza, V Cerdá. Talanta 44(4):553–562, 1997.

294. V Elrod, K Johnson, K Coale. Anal Chem 63(9):893–898, 1991.

295. H Obata. Anal Chem 65:1524–1528, 1993.

296. Z Yi, G Zhuang, P Brown, R Duce. Anal Chem 64(22):2826–2830, 1992.

297. H Inoue, K Ito. Microchem J 49(2–3):249–255, 1994.

298. Z Gao, P Li, Z Zhao. Talanta 38(10):1177–1184, 1991.

299. Z Gao, P Li, G Wang, Z Zhao. Anal Chim Acta 24(1): 137–146, 1990.

300. Z Gao, KS Siow. Talanta 43(5):727–733, 1996.

301. K Yokoi, C Van den Berg. Electroanalysis 4(1):65–69, 1992.

302. C Van den Berg, M Nimmo, O Abollino, E Mentasti. Electroanalysis 3(6):477–484, 1991.

303. P Farias, A Kohara, S Ferreira. Anal Lett 25(10): 1929–1939, 1992.

304. L Wang, C Ma, X Zhang, J Wang. Anal Lett 27(6): 1165–1173, 1994.

305. M Zaw, B Chiswell. Talanta 42(1):27–40, 1995.

306. P Pulido, J Barrero, M Pérez, C Cámara. Quím Anal 12(1):48–52, 1993.

307. I. Kuselman, O. Lev. Talanta, 1993, 40(5):749–756.

308. J Barrero, M Moreno, M Pérez, C Cámara. Talanta 40(11):1619–1623, 1993.

309. AC López, MT Tena, MM Correia, ML Simoes, MD Luque. Anal Chim Acta 343(3):191–197, 1997.

310. Y Ito, O Ueki, S Nakamura. Anal Chim Acta 299(3): 401–405, 1995.

311. S Nakamura, Y Ito, K Sone, H Hiraga, K Kaneko. Anal Chem 68(17):2981–2986, 1996.

312. P Blatny, F Kvasnicka, E Kenndler. J Chromatogr A 757(1 + 2):297–302, 1997.

313. C Chin, K Johnson, K Coale. Mar Chem 37(1–2):65–82, 1992.

314. S Nakano, M Nozawa, M Yanagawa, T Kawashima. Anal Chim Acta 26(1–2):183–188, 1992.

315. J Resing, M Motti. Anal Chem 64(22):2682–2687, 1992.

316. J Mattusch, G Werner, H Mueller. Analyst 116(1):53–57, 1991.

317. L Mallini, A Shiller. Limnol Oceanogr 38(6):1290–1295, 1993.

318. A Miyazaki, H Tao. J Anal At Spectrom 6(2):173–177, 1991.

319. G Zhang, D Cheng, S Feng. Talanta 40(7):1041–1047, 1993.

320. T Chapin, K Johnson, K Coale. Anal Chim Acta 249(2):469–478, 1991.

321. M Billah, T Honjo, K Terada. Anal Sci 9(2):251–254, 1993.

322. X Shan, B Radziuk, B Welz, O Vyskocilova. J Anal At Spectrom 8(3):409–413, 1993.

323. C Lan, Z Alfassi. Analyst 119(5):1033–1035, 1994.

324. E Beinrohr, M Rapta, M Lee, P Tschoepel, G Toig. Mikrochim Acta 110(1–3):1–12, 1993.

325. G Scollary, G Chen, T Cardwell, V Vincente. Electroanalysis 7(4):386–389, 1995.

326. SB Khoo, MK Soh, Q Cai, MR Khan, SX Guo. Electroanalysis 9(1):45–51, 1997.

327. JS Roitz, KW Bruland. Anal Chim Acta 344(3):175–180, 1997.

328. P Fernández, C Pérez, A Gutiérrez, C Cámara. Fresenius J Anal Chem 342(7):597–600, 1992.

329. R Compano, S Hernández, L García. Anal Chim Acta 255(2):325–328, 1991.

330. J Nowicki, K Johnson, K Coale, V Elrod, S Lieberman. Anal Chem 66(17):2732–2738, 1994.

331. N Garran, F Sánchez, JM Cano. Fresenius J Anal Chem 355(1):88–91, 1996.

332. Y Dedkov, S Kel'ina, V Mashcehenko. Zh Anal Khim 46(5):853–857, 1991.

333. K Akatsuka, T Katoh, N Nobuyama, T Okanaka, M Okumura, S Hoshi. Anal Sci 12(2):109–113, 1996.

334. S Huang, K Shih. Spectrochim Acta 50B(8):837–846, 1995.

335. D Yuan, P Yang, X Wang, B Huang. Chin Chem Lett 1(3):237–238, 1990.

336. F Muller, K Kester. Mar Chem 33(1–2):71–90, 1991.

337. H Xue, L Sigg. Anal Chim Acta 284(3):505–515, 1993.

338. IA Gurev, LF Zyuzina, YI Rusyaeva. J Anal Chem 52(6):518–521, 1997.

339. S Pervez, G Pandey. J Inst Chem 65(3):89–90, 1993.

340. P Wang, S Shi, D Zhou. Microchem J 52(2):146–154, 1995.

341. M Toral, P Richter, L Silva, A Salinas. Microchem J 48(2):221–228, 1993.

342. X Peng, Q Mao, J Cheng. Fresenius J Anal Chem 348(10):644–647, 1994.

343. Q Lin, A Guiraum, R Escobar, F de la Rosa. Anal Chim Acta 283(1):379–385, 1993.

344. M Yebra, A Bermejo, P Bermejo. Analyst 116(10):1033–1035, 1991.

345. U Naidu, G Naidu. Indian J Environ Prot 11(4):282–283, 1991.

346. M Bolshov, V Koloshnikov, S Rudnev, C Boutron, U Gorkach, C Patterson. J Anal At Spectrom 7(2):99–104, 1992.

347. C Boutron, M Bol'shov, S Rudnev, F Harmann, B Hutch, N Barkov. J Phys 1:695–698, 1991.

348. N Tercier, J Buffle. Anal Chem 68(20):3670–3678, 1996.

349. K Ohta, H Taniguti, SI Itoh, T Mizuno. Mikrochim Acta 127(1–2):51–54, 1997.

350. M Colognesi, O Abollino, M Aceto, C Sarzanini, E Mentasti. Talanta 44(5):867–875, 1997.

351. A Maquieira, H Elmahadi, R Puchades. Anal Chem 66(21):3632–3638, 1994.

352. A López, E Blanco, A Sanz. Mikrochim Acta 112(1–4):19–29, 1993.

353. M Wang, A Yuzefovsky, R Michel. Microchem J 48(3):326–342, 1993.

354. P Lu, K Huang, S Jiang. Anal Chim Acta 284(1):181–188, 1993.

355. R Ma, W Van, F Adams. Anal Chim Acta 293(3):251–260, 1994.

356. V Cuculic, M Branica. Analyst 121(8):1127–1131, 1996.

357. S Kounaves, W Deng, P Hallock, G Kovacs, C Storment. Anal Chem 66(3):418–423, 1994.

358. G Williams, C D'Silva. Analyst 119(1):2337–2341, 1994.

359. S Adeloju, E Sahara, D Jagner. Anal Lett 29(2):283–302, 1996.

360. J Cabon, A Le Bihan. Spectrochim Acta 50B(13):1703–1716, 1995.

361. D Babu, P Naidu. Talanta 38(2):175–179, 1991.

362. K Murty, D Rama, G Naidu. Indian J Chem A 33A(2):180–182, 1994.

363. C Chan, R Sadana. Anal Chim Acta 270(1):231–238, 1992.

364. M Abdullah, Z Shiyu, K Mosgren. Mar Poll Bull 31(1–3):116–126, 1995.

365. M Veber, K Cujes, S Gomiscek. J Anal At Spectrom 9(3):285–290, 1994.

366. J Goossens, L Moens, R Dams. J Anal At Spectrom 8(6):921–926, 1993.

367. T Sawidis. Arch Environ Contam Toxicol 30(1):100–106, 1996.

368. H Narasaki, JY Cao. Anal Sci 12(4):623–627, 1996.

369. WW Ding, RE Sturgeon. Spectrochim Acta 51B(11):1325–1334, 1996.

370. H Narasaki, JY Cao. At Spectrosc 17(2):77–82, 1996.

371. G Cutter, L Cutter. Mar Chem 49:295–306, 1995.

372. B Fairman, T Catterick. J Anal At Spectrom 12:863–866, 1997.

373. J Bowman, B Fairman, T Catterick. J Anal At Spectrom 12:313–316, 1997.

374. E Bulska, E Beinrohr, P Tschopel, JAC Broekaert, G Tolg. Chem Anal 41(4):615–623, 1996.

375. C Haraldsson, M Poliak, P Dehman. J Anal At Spectrom 7(8):1183–1186, 1992.

376. M Hartmann, H Lass, D Puteanus. Colloq Atomspektrom Spurenanal 5th, 1989, pp 703–709.

377. Z Li, S Huang. Anal Chim Acta 267(1):31–37, 1992.

378. M Sperling, X Yin, B Welz. Fresenius J Anal Chem 343(9–10):754–755, 1992.

379. M Dai, J Martin, G Cauwet. Mar Chem 51:159–175, 1995.

380. P Yeats, S Westerlund, A Flegal. Mar Chem 49:283–293, 1995.

381. E Cimadevila, K Wrobel, A Sanz. J Anal At Spectrom 10(2):149–154, 1995.

382. Z Liu, S Huang. Spectrochim Acta 50B(2):197–203, 1995.

383. G Bozsai, M Melegh. Microchem J 51(1–2):39–45, 1995.

384. H Zhang, W Davison, G Grime. ASTM Spec Tech Publ STP 1293:170–181, 1995.

385. R Laslett, P Balls. Mar Chem 48:311–328, 1995.

386. A Kamenev, I Viter, E Gorshkova, G Gus'kov. Gig Sanit 11:93–94, 1990.

387. G Hall, J Valve. J Chem Geol 97(3–4):295–306, 1992.

388. P Buldini, D Ferri, D Nobili. Electroanalysis 3(6):559–566, 1991.

389. G Williams, C D'Silva. Analyst 119(11):2337–2341, 1994.

390. E Suttle, E Wolff. Anal Chim Acta 258(2):229–236, 1992.

391. J Galimowski, T Szczepanska. Fresenius J Anal Chem 354(5–6):735–737, 1996.

392. E Mentasti, V Porta, O Abollino, C Sarzanini. Ann Chim Anal 81(7–8):343–355, 1991.

393. G Panteleev, G Tsizin, A Formanovskii, N Starshinova, E Sdeyckh, N Juz'min, Y Zolotov. Zh Anal Khim 46(2):355–360, 1991.

394. M Goergen, V Murshak, P Rettger, I Murshak, D Edelman. At Spectrosc 13(1):11–18, 1992.

395. C Lan, M Yang. Anal Chim Acta 287(1–2):111–112, 1994.

396. Z Zhuang, X Wang, P Yang, C Yang, B Huang. J Anal At Spectrom 9(7):779–784, 1994.

397. H Taylor, J Garbarino, D Murphy, R Beckett. Anal Chem 64(18):2036–2041, 1992.

398. L Jia. Huanjing Kexue 14(1):82–86, 1993.

399. C Chen, S Jian. Spectrochim Acta 51B(14):1813–1821, 1996.

400. F Petrucci, A Alimonti, A Lasztity, Z Horvath, S Caroli. Can J Appl Spectrosc 39(5):113–117, 1994.

401. O Abollino, M Aceto, G Sacchero, C Sarzanini, E Mentasti. Anal Chim Acta 305(1–3):200–206, 1995.

402. V Porta, C Sarzanini E. Mentasti O. Abollino. Anal Chim Acta 258(2):237–244, 1992.

403. E Kitazume, N Sato. Anal Chem 65:2225–2228, 1993.

404. I Steffan, G Vuijicic. Mikrochim Acta 110(1–3):89–94, 1993.

405. H Yang. Anal Chim Acta 292(2):437–443, 1993.

406. EN Iliadou, ST Girousi, U Dietze, M Otto, AN Voulgaropoulos, CG Papadopoulos. Analyst 122(6):597–600, 1997.

407. M Leiterer, U Muench. Fresenius J Anal Chem 350(4–5):204–209, 1994.

408. A Stroh, U Voellkopf, Anal At Spectrom 8(1):35–40, 1993.

409. L Alves, L Allen, R Houk. Anal Chem 65(18):2468–2471, 1993.

410. R Rosenberg, R Ziliacus, P Manninen. J Anal At Spectrom 9(6):713–717, 1994.

411. V Cheam, J Lechner, R Desrosiers, J Azcue, F Rosa, A Mudroch. Fresenius J Anal Chem 355(3–4):336–339, 1996.

412. M Eltayeb, R Van Grieken. Anal Chim Acta 268(10:177–183, 1992.

413. C Vázquez. Ann Asoc Quim Argent 82(2):91–96, 1994.

414. J Kubitz, U Uebel, A Anders. Proc SPIE-Int Soc Opt Eng 2503:14–22, 1995.

415. A Kamenev, I Viter. Zh Anal Khim 48(7):1197–1204, 1993.

416. F Wang, S Li, S Liu, Y Zhang, Z Liu. Anal Lett 27(9):1779–1787, 1994.

417. E Gil, P Ostapczuk. Anal Chim Acta 293(1–2):55–65, 1994.

418. B Myasoedov, E Krivoshei, A Kamenev. Zh Anal Khim 48(7):1151–1157, 1993.

419. V Argent, J Southall, E D'Costa. Proc Ann Conf Am Water Works Assoc 43–54, 1994.

420. S Daniele, C Bragato, MA Baldo. Anal Chim Acta 346(2):145–156, 1997.

421. L Svintsova, A Kaplin, S Vartan'yan. Zh Anal Khim 46(5):896–903, 1991.

422. VS Nguyen, MH Mai, VH Nguyen. J Radioanal Nucl Chem 213(1):65–70, 1996.

423. EA Viltchinskaia, LL Zeigman, DM García, PF Santos. Electroanalysis 9(8):633–640, 1997.

424. A Almon. Anal Chim Acta 249(2):447–450, 1991.

425. C Nan, T Cardwell, V Vincente, I Hamilton, G Scolary. Electroanalysis 7(11):1068–1074, 1995.

426. J Pérez, J Hernández, J Herrera, C Collado, C Van den Berg. Electroanalysis 6(11–12):1069–1076, 1994.

427. P Sharma, S Kumbhat, C Rawat. Int J Environ Anal Chem 48(3–4):201–207, 1992.

428. E Achterberg, C Van den Berg. Mar Poll Bull 32(6):471–479, 1996.

429. C Wang, D Luo. Mikrochim Acta 111(4–6):257–266, 1993.

430. D Jagner, L Renman, S Stefansdottir. Anal Chim Acta 281(2):305–314, 1993.

431. S Ichinoki, M Yamazaki. J Chromatogr Sci 29(5):184–189, 1991.

432. N Ryan, J Glennon. Anal Proc 29(1):21–23, 1992.

433. B Paull, M Foulkes, P Jones. Analyst 119(5):937–941, 1994.

434. M Vasconcelos, C Gomes. J Chromatogr A 696(2):227–234, 1995.

435. R Caprioli, S Torcini. J Chromatogr 640(1–2):365–369, 1993.

436. D Edlund, D Friesen, W Miller, C Thornton, R Wedel, G Rayfield, J Lowell. Sens Actuators B10(3):185–190, 1993.

437. A Legin, E Bychkov, Y Vlasov. Sens Actuators B24(1–3):309–311, 1995.

438. S Hill, A Brown, C Rivas, S Sparkes, L Ebdon. Tech Instrum Anal Chem 17:411–434, 1995.

439. L Shoupu, Z Mingqiao, D Chuanyue. Talanta 41(2): 279–282, 1994.

440. M Khuhawar, A Soonro. Talanta 39(6):609–612, 1992.

441. Y Shijo, H Sato, N Uehara, S Aratake. Analyst 121(3): 325–328, 1996.

442. S Liu, M Zhao, C Deng. J Chromatogr 598(2):298–302, 1992.

443. G Riel in S King Ed. Proceedings of the Workshop on Monitoring Nuclear Contamination in Artic Seas. Washington, DC: Naval Research Laboratory, 1995, pp IV/18–IV/29.

444. V Rigin. Anal Chim Acta 283(2):895–901, 1993.

445. G Chapple, J Byrne. J Anal At Spectrom 11(8):549–553, 1996.

446. T Maxwill, WF Smyth. Electroanalysis 8(8–9):795–802, 1996.

447. J Quinhan, H Zhang, F Liang, J Qun. J Anal At Spectrom 10(10):875–879, 1995.

448. SJ Santosa, H Mokudai, S Tanaka. J Anal At Spectrom 12:409–415, 1997.

449. SB Adeloju, E Sahara, D Jagner. Anal Lett 29(2):283–302, 1996.

450. K Di Hrich, T Franz, R Wennrich. Spectrochim Acta 50B(13):1655–1667, 1995.

451. S Zappoli, L Morselli, F Osti. J Chromatogr A 721(2): 269–277, 1996.

452. Y Nagaosa, T Kobayashi. Int J Environ Anal Chem 61(3):231–237, 1995.

453. CG Bruhn, FE Ambiado, HJ Cid, R Woerner, J Tapia, R García. Quím Anal 15(2):191–200, 1996.

454. KE Jarvis, JG Williams, E Alcántara, JD Wills. J Anal At Spectrom 11(10):917–922, 1996.

455. A Bortoli, M Gerotto, M Marchiori, F Mariconti, M Palonta, A Troncon. Microchem J 54(4):402–411, 1996.

456. V Mazo, L Sbriz, M Alvarez. X-Ray Spectrom 26(2): 57–64, 1997.

457. S Arpadjan, L Vuchkova, E Kostadinova. Analyst 122(3):243–246, 1997.

458. LS Zhang, SM Combs. J Anal At Spectrom 11(11): 1043–1048, 1996.

459. K Cundeva, T Stafilov. Anal Lett 30(4):833–845, 1997.

460. K Cundeva, T Stafilov, S Atanasov. Analysis 24(9–10):371–374, 1996.

23

Trace Elements
Li, Be, B, Al, V, Cr, Co, Ni, Se, Sr, Ag, Sn, Sb, Ba, and Tl

Rosa Cidu

University of Cagliari, Cagliari, Italy

I. INTRODUCTION

The elements considered in this chapter, Li, Be, B, Al, V, Cr, Co, Ni, Se, Sr, Ag, Sn, Sb, Ba, and Tl, occur at the trace or ultratrace level in the crust, with the exception of aluminum, which is a major component. Alteration processes involving crustal material bring such elements into the aquatic environment, where they are usually present, including aluminum, at parts-per-billion (ppb $\cong$ μg/L) or parts-per-trillion (ppt $\cong$ ng/L) concentrations. However, much higher, undesirable concentrations may occur in some environments, and are caused by: the weathering of mineralized and metal-rich rocks; mining and industrial activities, which release trace elements at a rate greater than the natural weathering processes; and high-temperature waters, which favor the solubility of many trace elements.

Table 1 shows the estimated concentrations of trace elements in the aquatic environment, as compared to their abundance in the crust (1–6). These values can vary significantly, due either to natural processes or to the release from contaminated areas. Nowadays, aqueous inorganic contaminants represent an environmental hazard; industrialized countries in particular face serious contamination levels (7–9), and in undeveloped countries there is hardly any attempt at environmental protection. Although the effects of most trace elements in water on the biosphere and human health are not well known yet, many of the elements considered are believed dangerous or potentially harmful. Guidelines

for the protection of aquatic life and human health have been established by national and international organizations, such as the U.S. Environmental Protection Agency, the World Health Organization, and the Commission of the European Communities, and values of their recommended concentration in drinking water are shown in Table 2 (10–13). The guideline values have been changing through time, and more severe regulation has been established as knowledge of the effects of trace elements on the biosphere increases.

As can be observed from comparison of the data reported in Tables 1 and 2, the natural concentrations usually found in freshwater and seawater are at least one order of magnitude lower than recommended values. However, their occurrence at concentrations higher than guidelines does not seem exotic at all. For example, data from studies by the U.S. Department of Interior on effects induced by irrigation drainage showed that Se, B, and Mo are the trace elements most commonly found in surface water at concentrations exceeding chronic criteria for the protection of aquatic life (14); contaminant levels of Cr were observed in 12 out of 180 monitored waters from the Snake River Plain, Idaho (15); and recently cases of water contamination by antimony have been reported throughout Japan (16). Moreover, some elements could be harmful at much lower levels than previously believed. On the other hand, deficiency—like abundance—of some trace elements may cause perturbations in the aquatic system, which could lead to the selection of resistant

Table 1 Estimated Occurrence of Trace Elements in the Crust and Aquatic Environment

Element	Crust (mg/kg)	Seawater (μg/L)	Freshwater (μg/L)	Thermal water* [μg/L (range)]	Mine water* [mg/L (range)]
Lithium	42[a]	180[a]	12[a]	100–10,000	0.01–2
Beryllium	2[b]	0.0006[c]	<0.05	<0.1–30	0.01–0.2
Boron	20[d]	4,500[d]	15[d]	100–10,000	0.1–3
Aluminium	69,300[a]	0.5[a]	50[a]	20–10,000	0.1–10
Vanadium	97[a]	2.5[a]	1[a]	<0.1–5	0.01–1
Chromium	71[a]	0.3[a]	1[a]	<0.1–10	0.1–5
Cobalt	13[a]	0.05[a]	0.2[a]	<0.1–4	0.01–5
Nickel	49[a]	0.2[a]	0.5[a]	<0.1–40	0.1–5
Selenium	0.05[e]	0.09[c]	<0.05	<0.1–10	0.01–5
Strontium	278[a]	8,000[a]	60[a]	100–10,000	0.1–1
Silver	0.07[a]	0.04[a]	0.3[a]	<0.1–12	0.01–2
Tin	2[e]	0.81[c]	0.1[f]	<0.1–10	0.01–1
Antimony	0.9[a]	0.24[a]	1[a]	<0.1–200	0.01–1
Barium	445[a]	20[a]	10	10–1,000	0.01–0.3
Thallium	0.45[e]	<0.01[e]	<0.05	<1–100	0.01–0.1

*Data for mine and thermal waters from a variety of sources published in the past decade; ranges derived from about 800 water analyses from different areas of the world.
Sources: [a]Ref. 1; [b]Ref. 2; [c]Ref. 3; [d]Ref. 4; [e]Ref. 5; [f]Ref. 6.

species and hence to a degradation of biodiversity. Therefore, it appears important to know the "real" concentration of trace elements in the water.

The effective protection and wise use of water resources require the collection of accurate and precise data that can be used to characterize and assess water quality conditions, to understand processes that control the behavior, and cycling of chemical constituents. Understanding whether harmful trace elements in aqueous media are transported as conservative components or

are retarded by geochemical and biological processes will contribute to the design of effective strategies to remedy the compromised resources. The determination of aqueous elements at the trace and ultratrace level is useful in regional studies to distinguish the natural background concentration from anthropogenic inputs and also to highlight the significant variations in long-term analytical programs.

II. PHYSICAL AND CHEMICAL CHARACTERISTICS

The wide range of trace element concentrations observed in water depends on a number of factors. Regional lithology (13,17) and the occurrence of ore deposits influence the aqueous concentration of trace elements. The solubility of the solid phases and their rate of dissolution, the presence of gas phases and their transfer into water, the conditions of the environment (temperature, pH, redox potential), the distribution of species in the aqueous phase, and the interaction time between the aqueous solution and other phases are the prominent factors affecting the dissolved concentrations of elements (18–20). And the anthropogenic activities can significantly change the local concentration of trace elements in water.

The physical, chemical, and biological weathering of crustal material influences the aqueous transport of

Table 2 Proposed Standards for Trace Elements in Drinking Water

	USEPA (μg/L)	WHO (μg/L)	EC (μg/L)
Beryllium	1	1	—
Boron	2000	300	1000
Aluminium	50	200	50
Chromium	120	50	50
Nickel	100	—	50
Selenium	100	10	10
Silver	50	—	10
Tin	50	—	—
Antimony	5	5	—
Barium	2000	700	100
Thallium	2	1	—

Source: Refs. 10–13.

trace elements—as well as of major constituents. Physical properties may affect the behavior of each element during water–rock interaction processes. Some physical characteristics of the elements considered in this chapter that might affect their chemical form are shown in Table 3, together with their isotopic abundance (3,5).

Trace elements in water may exist in different forms: either as sorbates associated with colloids and with fine materials in suspension or as soluble species. Trace elements in the form of colloidal particles can travel long distances in subsurface environments (21). The presence of an air–water interface retards the colloid transport significantly; however, the retardation of element transport by colloids is highly dependent on the properties of the element and the colloidal surface (22). Adsorption, absorption, and coprecipitation of dissolved trace elements onto neoformed solid phases, colloidal particles, and very fine materials occur at the water–sediment interface (23), but these processes often represent a temporary sink, since sorbed elements may be back-released into the aqueous phase under favorable conditions (24).

The soluble species may persist longer in the aquatic environment, and therefore appear more important. Since the toxicological and biological importance of many elements depends on their form, *speciation* has been receiving an increasing attention in the scientific community. This term refers to element occurrence in various chemical forms, or species; these can be defined as specific compounds with peculiar physical-chemical properties, which influence the biological availability of elements. Several reviews on trace element speciation have been published recently (25–29). Indirect speciation of the total trace element concentration in water are derived by computer programs when the appropriate thermodynamic constants can be calculated, whereas direct speciation methods involve the separation of the species of interest from the other components; some of these will be considered further in the upcoming section on analytical methods.

Free ions and inorganic and organic complexes are the most common soluble forms. The transport of trace metals in solution is generally favored under acidic conditions as well as at high levels of organic matter in the environment (30,31). As a factor favoring migration, complex formation is an important process in the control of trace element mobility (32). Complex formation is affected mainly by the amount of dissolved solids, the relative concentration of suitable ligands present in solution, and the redox state of the environment. Complex formation is facilitated in waters with high ionic strength. The OH^-, Cl^-, SO_4^{2-} and

reduced sulphur species, CO_3^{2-}, HCO_3^-, and several organic compounds (e.g., humic and fulvic acids), represent the dominant ligands in complex formation. Redox-sensitive elements can have different chemical reactivities or physical characteristics depending on their oxidation state, hence, their concentration and distribution can be very different in oxic and anoxic environments (33). For example, in lake waters it has been observed that free divalent cations are dominant in the oxic surface water, while metal-sulphide complexes prevail in the anoxic bottom waters (34).

For each element, the most common species found in the aquatic environment are next considered in some detail, together with the main characteristics and parameters that may be important in the control of their mobility.

A. Lithium

Lithium is a mobile element, and in the aqueous environment it is present as Li^+, at concentrations usually in the range of 10–100 $\mu g/L$. Higher concentrations are frequently observed in thermal waters. The concentration of lithium in seawater reflects its conservative behavior. According to present information, there is no evidence of toxic effects from lithium intake; this element must instead be considered a trace element that in all probability is essential to humans, but it must be regarded as potentially phytotoxic (35).

B. Beryllium

In natural waters, Be occurs at the sub-$\mu g/L$ level (36), while enhanced concentrations have been observed in acid waters (37,38). The predominant aqueous species below pH 5.5 is Be^{2+}, and above this the hydroxide or poly-hydroxide species dominate. Beryllium oxides and hydroxides have low solubility and can act as a control on aqueous Be concentration (39).

C. Boron

The most important species in both freshwater and saltwater are boric acid $B(OH)_3$ and the borate anion $[B(OH)_4^-]$; their proportions are sensitive to variation in pH. The borate anion is important in saline waters (10) and dominates at alkaline conditions, whereas the reverse holds under acidic conditions (4). $H_2BO_3^-$ usually constitutes an appreciable proportion of total B in groundwater below 100°C, and it can be the dominant species in some environments (40). The oceans are an important reservoir for B, and here boron shows a con-

Table 3   Physical Characteristics of Trace Elements and Their Isotopic Abundance

Name	Symbol	Atomic number	Atomic weight	Electron configuration	Oxidation states	Ionic radius^a (Å)	1st Ionization potential (V)	Isotope mass	Isotope abundance (%)
Lithium	Li	3	6.94	$He2s^1$	+1	**0.76**	5.39	**7**	92.6
								6	7.4
Beryllium	Be	4	9.01	$He2s^2$	+2	**0.45**	9.32	**9**	**100**
Boron	B	5	10.81	$He2s^22p^1$	+3	**0.27**	8.30	**11**	**80.4**
								10	19.6
Aluminium	Al	13	26.98	$Ne3s^23p^1$	+3	**0.53**	5.98	**27**	**100**
Vanadium	V	23	50.94	$Ar4s^23d^3$	**+5**, +2, +3	**0.54**	6.74	**51**	**99.8**
								50	0.2
Chromium	Cr	24	52.00	$Ar5s^13d^5$	**+3**, +6	**0.62**	6.76	**52**	83.8
								50	4.3
								53	9.6
								54	2.4
Cobalt	Co	27	58.93	$Ar4s^23d^7$	**+2**	**0.65**	7.86	**59**	**100**
Nickel	Ni	28	58.71	$Ar4s^23d^8$	**+2**	**0.69**	7.63	**58**	**67.8**
								60	26.2
								61	1.2
								62	3.7
								64	1.1
Selenium	Se	34	78.96	$Ar4s^23d^{10}4p^4$	**+4**, −2, +3, +6	**0.50**	9.75	**80**	**49.8**
								74	0.9
								76	9.0
								77	7.6
								78	23.5
								82	9.2

Element	Symbol	Atomic number	Atomic weight	Electron configuration	Oxidation state	Ionic radius	Ionization	Mass number	Abundance (%)
Strontium	Sr	38	87.62	$Kr5s^2$	$+2$	1.18	5.69	**88**	**82.6**
								84	0.5
								86	9.9
								87	7.0
Silver (argentum)	Ag	47	107.87	$Kr5s^14d^{10}$	$+1, +2$	1.15	7.57	**107**	**51.8**
								109	48.2
Tin (stanum)	Sn	50	118.69	$Kr5s^24d^{10}5p^2$	$+2, -4, +4$	0.69	7.33	**120**	**32.8**
								112	1.0
								114	0.6
								115	0.3
								116	14.3
								117	7.6
								118	24.0
								119	8.6
								122	4.9
								124	5.9
Antimony (stibium)	Sb	51	121.75	$Kr5s^24d^{10}5p^3$	$+3, -3, +5$	0.76	8.64	**121**	**57.3**
								123	42.7
Barium	Ba	56	137.34	$Xe6s^2$	$+2$	1.35	5.81	**138**	**71.7**
								130	0.1
								132	0.1
								134	2.4
								135	6.6
								136	7.8
								137	11.3
Thallium	Tl	81	204.37	$Xe6s^24f^{14}5d^{10}6p^2$	$+1, +3$	1.50	6.11	**205**	**70.5**
								203	29.5

[a]Ionic radius in coordination six, referred to the bold, prevalent oxidation state.
Source: Refs. 3 and 5.

servative behavior, occurring in near-constant proportion to Cl. High boron concentrations are found in waters leaching sediments and soils derived from evaporitic and argillaceous sediments and in waste waters. Water-soluble boron compounds commonly used in detergent and cleaning products are discharged with domestic effluents into sewage treatment plants, where little or no boron is removed. Hence, the anthropogenic boron load is released almost entirely into the aquatic environment (41) and may represent a potential hazard (42).

D. Aluminum

The occurrence of Al in the aquatic environment is strongly influenced by pH, as can be observed in Fig. 1 (43), and very low concentrations occur at the pH range of most natural waters. At pH > 6.5, the species $Al(OH)_4^-$ dominates, while the $Al(OH)_2^+$, $Al(OH)^{2+}$, and Al^{3+} species, respectively, prevail as hydrolysis proceeds with increasing acidity. Therefore, Al speciation depends on pH and on the presence of ligands other than OH^- able to form stable complexes (e.g., AlF_n^{3-n}). Aluminum compounds are used extensively in water treatment and may persist in the finished water, either used as drinking water or returned into the aquatic system, posing a potential hazard in both cases. Acid soil conditions exacerbated by acid precipitation may lead to increased concentrations of Al in surface water (2).

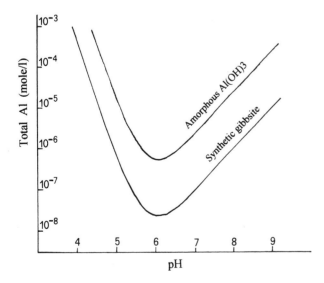

Fig. 1 Influence of pH on aqueous aluminum concentration. (From Ref. 43.)

E. Vanadium

Vanadium concentrations in natural waters range widely from several tens ppt to about 100 ppb (44). The pentavalent form V^{5+} is the most soluble in water (10). In aqueous solutions V forms 10 different oxides and hydroxides. In oxic waters the primary species is protonated vanadate $H_2VO_4^-$, while in anoxic waters the hydrolyzed trivalent form $V(OH)_2^+$ occurs (34).

F. Chromium

The main forms in freshwater are $Cr(OH)^{2+}$, $Cr(OH)_2^+$, and $Cr(OH)_4^-$; these complexes are generally stable and kinetically inert. The trivalent state Cr^{3+} is stable in a reducing acidic environment (45). However, trace amounts of Cr(III) have been found in surface seawater, likely due to photochemical reduction (46). Dissolved chromium may be also present as organic–Cr^{3+} complexes (47). The toxic species Cr^{6+} is water soluble, and under oxidizing conditions is present in solution as a component of complex anions, such as CrO_4^{2-}, $HCrO_4^-$, and $Cr_2O_7^{2-}$, depending on pH.

G. Cobalt

The dominant species in freshwater are Co^{2+}, $CoCO_3^0$, and $Co(OH)_3^0$; in a reducing environment, CoS occurs, while chloride complexes prevail in seawater (10). Adsorption of trace amounts of cobalt onto solid phases or colloids is a common process; the Co^{+3}-bearing complexes are less strongly adsorbed than are Co^{+2}-bearing complexes (48).

H. Nickel

The principal species present in water is Ni^{2+}, followed by $Ni(OH)_2$; under anoxic conditions the solid phase millerite (NiS) occurs (10,49). Nickel is easily mobilized during weathering, especially under acidic conditions. However, sorption processes and coprecipitation with iron and manganese oxides may occur in soils, while sulphide-rich reducing environments can limit the Ni concentration in seawater (5).

I. Selenium

Inorganic and organic compounds containing selenium in four possible oxidation states exist in waters. Selenite, SeO_3^{2-}, and selenate, SeO_4^{2-}, are the dominant dissolved inorganic species in freshwater and also in waters leaching mining wastes (50,51). Selenate is thermodynamically stable in the unsaturated zone, but Se^0 may also occur as metastable species whose solubility increases in the presence of thiols and inorganic sulfides (52). Selenate may be sorbed onto amorphous ferric hydroxides (10,39). Speciation and other aspects related to the environmental geochemistry of selenium have been reviewed (53). The bioavailability of Se is dependent on the dissolution of different inorganic and organic Se components in soils, and both excess and deficiency of selenium can cause adverse health effects.

J. Strontium

This element is always detected in water, usually at increasing concentration as salinity increases. Concentrations of strontium are strongly dependent on its availability in the rocks forming the aquifer or the drainage area (54). It occurs overall in the +2 valence state, and the free ion dominates in any environment, but the ion pair $SrSO_4^0$ is significantly present in sulphate-rich waters.

K. Silver

Silver is often not detected in natural water, its occurrence in the aquatic environment being confined to mining and geothermal (55,56) areas, and to anthropogenic inputs. Under reducing conditions, soluble Ag is not stable, and native silver and alcantite, Ag_2S, occur, respectively, at neutral to alkaline pH and acid pH. The main species in oxic environments is Ag^+, while the chloride complexes, such as $AgCl_2^-$ and $AgCl_3^{2-}$, are the dominant forms under a wide pH range in Cl-rich waters (49), and may transport silver in solution for a long distance (57).

L. Tin

The prevalent oxidation state in aquatic environments is Sn^{4+}, but under reducing conditions the Sn^{2+} form occurs. All forms of tin are readily sorbed to suspended solids and clay. Tin can be methylated in the environment to form organotin species, which accumulate in sediments. Organotin species are biologically active and may be highly toxic to aquatic life (10).

M. Antimony

A very large field in Eh–pH space is occupied by the ionic species $SbO_{3(aq)}^-$, predicting the relatively high mobility of antimony under oxidizing conditions, be it acidic or alkaline (58). Aqueous concentrations up to 200 μg/L Sb have been measured in geothermal (59) and mine (60) areas. Under strongly reducing and alkaline conditions, the most common antimony mineral, stibnite, Sb_2S_3, may dissolve and be converted into the ionic species $Sb_2S_{4(aq)}^{2-}$. Thioantimonite complexes may be present as colloidal particles in aqueous environments (61).

N. Barium

On the basis of its ionic potential, Ba^{2+} is considered a mobile ion. Its concentration in the hydrosphere is highly variable, depending mostly on the regional availability of barium sources (62). In the presence of sulphate ions, aqueous Ba concentration is controlled by its equilibrium with respect to the mineral barite. A major process of confining this element in solid phases at near-neutral pH is sorption; however, the adsorption rate decreases greatly with declining pH (10).

O. Thallium

Even though the large domain in the Eh–pH space for Tl^+ indicates that thallium is mobile under subsurface conditions (63), aqueous concentrations are found at the ppt level, with the exception of higher concentrations in some thermal and mine waters (17,64,65). The species containing Tl^0 and Tl^{3+} may exist in natural environments, respectively, under extremely reducing and oxidizing conditions (10).

III. SAMPLE PREPARATION

A. Sampling

Avoiding any disturbance of natural conditions is the first concern during water sampling. Unpumped open wells and stagnant waters are generally considered poor investigation tools and are usually disregarded. In contrast, excessive pumping may cause mixing of different waters, leading to misinterpretation of the data. In many cases, groundwater composition shows major variation with depth. Even on a small scale (12), variation may affect trace-element concentration as well as speciation. Water stratification due to differing salinity may occur in groundwater, lakes, and large rivers. In all the foregoing situations, depth-specific sampling is required to study the behavior of trace elements in detail (66).

The determination of trace elements in water requires the use of adequate samplers and storage containers. Materials to be avoided include metals or plastic-coated metals, rubber, and soft glass. Fluorinated ethylene polymers (e.g., PTFE, FEP, PFA), polypropylene, and high-density polyethylene are the most desirable materials, while polyvinyl chloride (PVC) and structural nylon should be avoided (67). Even the best plastic materials must be cleaned before use: Recommended procedures involve leaching for several days with dilute (e.g., first 10%, then 3%) acid solutions, copious rinsing with pure water, and, finally, copious rinsing with the water to be sampled prior to collection. In order to prevent potential contamination by sample handling during the analysis of major components, specific aliquots dedicated to the determination of trace elements are collected.

The 0.45-μm-pore-size filters are conventionally used to remove matter in suspension, and the elements determined in the filtered portion are frequently regarded as "in solution." Furthermore, the 0.45-μm filterable concentration of elements is generally thought to be a better approximation of the bioavailable fraction of water-borne metals than is the "total recoverable" concentration (68). Filters of other pore sizes, e.g., 0.2 μm (31,69,70), have also been used to distinguish "dissolved" components from suspended load, while the ultrafiltration (0.001-μm pore size) is required to remove metals present as colloids or very fine materials (21,30,31). Filtration is carried out in situ close to the sampling site, possibly under oxygen-free conditions (e.g., under N_2 pressure). In order to avoid contamination from the filter, the first filtered portion is discarded. The volume of filtered sample must not exceed the clogging limit of the filter; this volume may vary considerably in rivers and turbid waters, so a test of filter clogging should be performed for each study (71). Under unfavorable conditions that do not allow in situ operations, filtration should be carried out within a few hours after collection to minimize the loss of soluble compounds by sorption processes onto the suspended material and/or container walls.

B. Stabilization

Acid addition to give pH < 1 prevents the precipitation of dissolved components and/or sorption processes. Ultrapure reagents are necessary to minimize water contamination due to acidification. In order to monitor contamination, during each sampling campaign blank solutions should be prepared in the field using ultrapure water (e.g., 18 MΩ by the Millipore Milli-Q system) and following the same procedure as for water samples. Different acids at different concentrations, depending on the element to be determined and on the technique

to be used, are employed for sample stabilization. The addition of 1% HNO_3 is often used in the determination of most metals (% refers to the volume of concentrated acid with respect to the volume of water). The use of HNO_3 is preferred for analysis by ICP-MS because other acids may cause higher interference problems. The addition of 0.2% HCl is required for the analysis of Sb and Se by hydride generation. The addition of specific reagents may be required to stabilize the species of interest prior to their direct determination.

Water samples should be stored out of the light and kept refrigerated at 4°C until the analysis is to be carried out. The stability of trace elements in solution may vary significantly from one species to another, and storage time will be established accordingly. In any case it is better to carry out the trace element analyses as soon as possible after water collection.

Sometimes the total recoverable amount of elements needs to be quantified, especially in waters having high suspended load. When the total amount of trace elements in water has to be determined, unfiltered samples are collected and immediately thereafter acidified to 1% HNO_3. In the laboratory, 1 ml of concentrated HNO_3 and 0.5 ml of concentrated HCl are added to an aliquot of 100 ml; the sample is heated at 85°C until the volume has been reduced to approximately 20 ml; after cooling the sample is made up to a 50-ml volume, mixed, and allowed to stand before analysis (72). The digestion of unfiltered, acidified water samples in a microwave-closed system prior to analysis prevents losses of the total recoverable amount of trace elements (73).

IV. ANALYTICAL METHODS

A. Techniques

The most sensitive analytical techniques for the determination of trace elements in water are inductively coupled plasma mass spectrometry (ICP-MS), instrumental neutron activation analysis (INAA), graphite furnace (electrothermal) atomic absorption spectrometry (GFAAS or ETAAS), and inductively coupled plasma atomic (optical) emission spectrometry (ICP-AES or ICP-OES). In recent years, new, more sensitive generations of plasma source MS instruments have been developed using alternative plasma generation methods and high-resolution mass spectrometers (74). The principles and characteristics of these techniques have been extensively documented (e.g., Refs. 67, 75, and 76). The ICP-MS, ICP-AES, and INAA techniques have the advantage of multielement capability and linear response over several orders of magnitude, while

Table 4 Conditions for Determination of Trace Elements in Water by GFAAS[a] Using Pyrolytically Coated Graphite Tube with Platform and Zeeman Background Correction

Element	Al	V	Cr	Co	Ni
Wavelength (nm)	309.3	318.4	357.9	242.5	232.0
Slit (nm)	0.7	0.7	0.7	0.2	0.2
Lamp energy (mA)[b]	25	25	25	30	25
Injection volume (μl)	10	50	50	25	25
Drying (°C)[c]	80–130	80–130	80–130	80–130	80–130
Charring (°C)	1500	1200	1200	900	1000
Atomisation (°C)[d]	2500	2650	2400	2200	2300
Integration time (s)	5	5	5	4	5
Linearity (μg/L)	20[e]	10[f]	10[f]	20[e]	20[e]
Detection limit (μg/L)[g]	0.5	0.2	0.2	0.2	0.4

[a]Perkin Elmer Zeeman/3030 with autosampler.
[b]Hollow cathode lamp.
[c]Two steps.
[d]Gas (argon) stop; argon flow at drying and charring 300 ml/min.
[e]Calibration standards (μg/L): [1, 2], 5, 10, 20; peak area mode.
[f]Calibration standards (μg/L): 0.5, 1, 2.5, 5, 10; peak area mode.
[g]Calculated on the basis of 5 times the standard deviation of blank measurements made at regular intervals during the analytical cycle.

the small (usually 10–50-μl) sample volume required for analysis and the possibility of partial removal of the matrix components prior to the atomization step, are the main advantages of GFAAS. Flame atomic absorption spectrometry (AAS) continues to be important (77,78), especially for the determination of the alkali elements. Electrochemical methods have been used for single-element determination at the ppb or ppt level, for example: (a) stripping voltammetry for Sb (79), Sn (80), and the speciation of Al and Cr (46) and Se (81); (b) selective electrodes for B (82), Co and Ni in natural (83) and industrial (84) waters, and Tl (85). Notwithstanding the merits of other methods, ICP-MS, ICP-

AES, and GFAAS will be considered here in more detail, since these are nowadays the most suitable (86,87) and frequently used techniques for the quantification at the ppb and ppt levels of trace elements in water.

B. Instrument Operating Conditions

In order to achieve the best instrumental response, whatever the technique used, the operating conditions must be carefully established.

In the GFAAS technique, the intensity and stability of signals depend on the lamp alignment and stabilization, and on the sample introduction system. Owing to the small volume used, the uncorrected deposition of sample onto the graphite tube, or platform, causes poor precision. The conditions prior to the atomization step—in particular, drying and charring (pyrolysis) temperature, ramp and holding time for each step, and gas flow—are established according to the element to be determined and the matrix of sample to be analyzed. A few examples are shown in Tables 4 and 5 (88).

In ICP techniques, the intensity and stability of signals strongly depend on the plasma and aerosol conditions, so the radio frequency power, gas flow (auxiliary, coolant, and carrier), and sample uptake need to be optimized for each instrument (Table 6), and the best compromise between high sensitivity and good stability must be found.

A further concern is the data acquisition mode. High integration times usually allow an increase in sensitiv-

Table 5 Conditions for Determination of Trace Elements in Water by GFAAS[a] Using Pyrolytically Coated Graphite Tube with Platform and Zeeman Background Correction

Element	Ag	Sb
Wavelength (nm)	328.1	217.6
Injection volume (μl)	20	20
Drying (°C)[b]	110–130	110–130
Charring (°C)	800	1300
Atomisation (°C)[c]	1500	1900
Integration time (s)	5	5
Detection limit (μg/L)	0.07	1.3

Source: Ref. 88.
[a]Perkin Elmer SIMAA 6000 with autosampler.
[b]Two steps.
[c]Gas (argon) stop; argon flow during drying and charring 250 ml/min.

Table 6 Conditions for Multielement Determination of Aqueous Traces by ICP-AES[a]

Element	Li	B	Co	Sr	Ba
Wavelength (nm)	670.784	249.773	228.616	407.771	455.403
Background correction (nm)	+0.060	+0.028	+0.029	+0.050	+0.047
	−0.044	−0.027	−0.024	−0.048	−0.037
Forward power (W)	650	650	650	650	650
Coolant flow (L/min)	7.5	7.5	7.5	7.5	7.5
Auxiliary flow (L/min)	0.8	0.8	0.8	0.8	0.8
Nebulizer flow (L/min)	0.9	0.9	0.9	0.9	0.9
Sample uptake (ml/min)[b]	1	1	1	1	1
Integration time (s)	5	5	5	3	3
PMT[c] power	12	12	14	10	10
No. of replicates	3	3	3	3	3
Detection limit (μg/L)[d]	1	1	0.5	0.5	0.5

[a]ARL-FISONS 3520 with Meinhard nebulizer and Fassel minitorch.
[b]Gilson Minipuls peristaltic pump.
[c]Photomultiplier tube.
[d]Calculated on the basis of 5 times the standard deviation of blank measurements made at regular intervals during the analytical cycle.

ity. In GFAAS the peak height or peak area is chosen, depending on the matrix components that might influence the peak shape. The data acquisition mode is particularly important in ICP-MS. The faster the quadrupole scansion, the less the time spent at any given mass position and the lower the signal integrated for the measurement. The total number of ions counted at the spectral peak maximum depends on the scan speed, i.e., on the dwell time used. The ability to resolve small differences in concentration is obviously greater the longer the dwell time; in other words, the detection limit (see further on) is generally better for longer dwell times. A way of improving precision, detection limit, and operating range (linearity) with fast scanning (i.e., with low dwell time) is to increase the number of sweeps; by summing sweeps, more ions are counted. For a finite transient signal profile in ICP-MS, as in the case of flow injection or electrothermal vaporization, the results from many fast sweeps could be as good as from a few slow sweeps. If the signal summed from multiple sweeps is referred to as a *reading*, slower scan speeds will result in fewer readings, but each reading will consist of a larger number of ions counted. Faster scanning will result in more readings per transient signal profile, but each reading will consist of fewer ions counted. The former case would appear more desirable, since counting more ions per reading provides better precision, as long as a sufficient number of readings are taken to characterize the signal profile. A few examples of operating conditions for continuous sample uptake in ICP-MS are shown in Table 7, and for a discrete-volume introduction in Table 8.

During an analytical cycle, significant variation of the instrumental signal may occur, and this implies the deterioration of precision. When the signal changes as a linear function of time, an external drift correction can be applied. In practice, a standard solution is analyzed at regular intervals; the relative change in signal is recorded and used to correct the measured concentration of the samples. However, external drift correction does not always work well enough, especially for changes in the instrumental ICP-MS response over time. If this is the case, several elements (e.g., Sc, Rh, In, Pt) can be used as internal standards for drift correction. To be used as internal standard, an element must be free from interference and not present in the samples to be analyzed. The suitability of any element as a potential internal standard should be established in the matrix of interest, prior to its use for long-term drift correction (67). When the internal standard method is used for correcting instrumental drift, the addition of the internal standard solution is made using a volume exactly equal to the blank, standard, and sample solutions, each solution being of the same volume in order to minimize random fluctuations among solutions derived from differences in the internal standard concentration. The efficiency of internal standards also depends on the concentration of the element to be analyzed. The standard deviation of a signal measured by means of pulse counting can be used as a measure of instrumental noise. According to Poisson statistics, the standard deviation can be estimated as the square root of the signal. This means that precision should improve with increasing signal intensity. Since signal

Table 7 Conditions for Simultaneous Determination of Trace Elements in Water by ICP-MS[a]

Element	Li	Be	B	Al	V	Cr	Co	Sr	Ag	Ba	Tl
Mass (a.m.u.)	7	9	11	27	51	52	59	88	107	138	205
Forward power (W)	1100	1100	1100	1100	1100	1100	1100	1100	1100	1100	1100
Coolant flow (L/min)	15	15	15	15	15	15	15	15	15	15	15
Auxiliary flow (L/min)	0.9	0.9	0.9	0.9	0.9	0.9	0.9	0.9	0.9	0.9	0.9
Nebulizer flow (L/min)	0.8	0.8	0.8	0.8	0.8	0.8	0.8	0.8	0.8	0.8	0.8
Sample uptake (ml/min)[b]	1	1	1	1	1	1	1	1	1	1	1
Dwell time (ms)	100	75	120	150	150	150	75	50	200	50	200
No. of sweeps	9	9	9	9	9	9	9	9	9	9	9
No. of replicates	3	3	3	3	3	3	3	3	3	3	3
Detection limit (μg/L)[c]	1	0.2	1	0.5	1	1	0.1	0.1	0.05	0.05	0.02

[a]Perkin Elmer Elan 5000 with cross-flow nebulizer; normal resolution and peak hope scanning mode.

[b]Gilson Minipuls II peristaltic pump.

[c]Calculated on the basis of 5 times the standard deviation of blank measurements made at regular intervals during the analytical cycle.

variations at levels above 10,000 counts per second (cps) tend to be highly correlated over time and those below 10,000 cps are usually random, internal standardization should improve precision at high signal intensities, but is unlikely to improve precision at very low signal levels (89). In fact, at the lowest signal levels, precision could be degraded due to errors from measuring two isotopes instead of only one. Maximizing the amount of signal integrated and minimizing noise are especially important in the measurement of a transient signal.

C. Interference Effects

Interference in atomic and mass spectrometry can be classified under a number of general headings: physical effects that arise from factors affecting nebulizer and/or aerosol transport efficiency; spectral or isobaric overlaps that result from inadequate selectivity between spectral lines of analytes and those of matrix or background species, or between ions of a similar mass. Each signal is the complex one, and it is composed of the element-specific component as well as the background component derived from interference and matrix effects. The terms *interference effects* and *matrix effects* are often used interchangeably. A distinction has been made on the basis of how the concomitant interference and matrix affect the analytical working curve of concentration versus the analytical signal: an *interference effect* is any effect that changes the intercept value of the working curve, whereas a *matrix effect* changes the slope of the working curve. Combined matrix and interference effects are also encountered. The problem arising from the presence of interference is not so much its existence as that interference affects samples and

standards to differing degrees. Matrix effects in GFAAS can include chemical vaporization effects or other factors that affect atomization or ionization efficiency (90). A matrix effect specific to ICP-MS includes factors that affect ion transfer in the interface region of the instrument.

Methods for the correction or removal of interference are an important part of the development of analytical procedures. Indeed, much of the literature is devoted to overcoming interference problems. Simple approaches to the alleviation of interference involve the choice of optimal operating conditions, the selection of unaffected isotope peaks or elemental lines, the use of elemental equations based on the ratio of isotope abundance, blank subtraction, background correction, and dilution of water sample (91). Some of these approaches work efficiently, as in the case of the Zeeman background correction used in GFAAS if the appropriate ramp and isotherm time of charring are chosen to remove any organic matter prior to the atomization step; sample dilution may reduce matrix effects significantly, but this implies a deterioration of detection limits, whereas some of the previously mentioned approaches may be not appropriate in very complex matrices. Other approaches consist of alternative sample introduction systems that have been developed and used in an effort to eliminate or reduce interference by separating the analyte from the interfering matrix or by changing the form of the sample before it is introduced into ICP-MS or ICP-OES instruments. For example, the use of cryogenic desolvation (92) and ultrasonic nebulization in ICP-MS allows one to reduce potential interferences due to oxide formation as well as to increase the sensitivity with respect to cross-flow nebulization (75).

Table 8 Conditions for Determination of Se, Sb, and Sn in Water by Online Hydride Generation ICP-MS[a]

| | FIAS program (sample loop: 500 μl) | | | | Signal acquisition (peak hope transient) | |
Step	Time (s)	Pump 1 (rev/min)	Pump 2 (rev/min)	Valve position		
Presample (rinse)	10	100	0	1	Dwell time (ms)	20
1	3	100	0	1	Sweeps	10
2	3	100	0	1	Readings	15
3	10	0	120	2	Points across peak	3
4	3	75	0	1	Replicates	3
Postrun	30	75	0	1	Forward power (W)	1100
					Coolant flow (L/min)	15
					Auxiliary flow (L/min)	0.9
					Nebulizer flow (L/min)	1.0

[a]Perkin Elmer Elan 5000, online Perkin Elmer FIAS 200.

Accurate determinations can be made in the presence of both interference and matrix effects by using a standard addition procedure. This method consists of splitting the sample into at least three aliquots of the same volume, and adding to each aliquot known concentrations of the element to be analyzed. Spikes should be of similar concentration as expected in the sample. Each spiked aliquot is then analyzed using the chosen technique, and the concentration in the sample is derived as follows: A graph of the signal measured in each aliquot versus the concentration of the added element is constructed, and the line best fitting the spikes is found. The intercept of this line on the x-axis (a negative value) gives the concentration of the sample (Fig. 2). Provided that the volume changes introduced by the additions can be kept small, the standard addi-

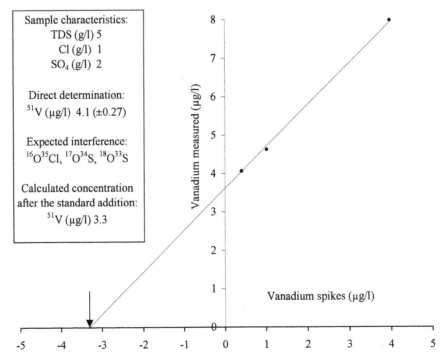

Fig. 2 Determination of vanadium in mine water by ICP-MS using the standard addition method (the size of points includes the standard deviation on three replicates).

tion approach ensures that both matrix and physical effects influence the samples and standards in a similar manner. The negative aspect of this procedure is that it is best suited to a relatively small number of elements (67) and is time consuming to perform. However, it can be used in a few, representative samples, especially when relevant interference is expected (e.g., high-salinity waters and wastewaters). Devices for automated standard additions that allow considerable saving of analytical time are also available (93).

D. Separation and Preconcentration Methods

Much research has concentrated on the removal of interference with chemical pretreatment or on separation methods carried out prior to the analysis being performed. Various types and combinations of separation and/or preconcentration methods have been described and used either for the removal of interference or for the direct quantification of species, as well as for the achievement of the lowest detection limits. Some criteria have been used by the author in making the following choice of methods used for the separation and preconcentration of trace elements in water. Briefly, methods using apparatus or materials not commercially available and unclear or ambiguous methodologies were generally excluded; routine methods have been considered if combined with sensitive techniques; and special attention was paid to methodologies validated by including data on precision, accuracy, and limits of detection.

Off-line separation and preconcentration of trace components involving absorption-elution, ion-exchange, or solvent-extraction batch procedures are effective in improving the selectivity and sensitivity of spectrometric methods. The success of this approach depends on a number of factors, the most important perhaps being the control of blank solutions for the achievement of the lowest detection limits. The separation and preconcentration of trace elements may be also carried out online with the analytical technique. Objectives for the online treatment of water samples may be either removal of interfering components or enhancement of sensitivity or both. Among online devices, the flow injection (FI) system has undergone rapid development. Higher efficiency and precision, lower consumption of sample and reagent, and reduced risks of contamination are the main advantages of FI online preconcentration as compared to conventional batch operations (94–96).

The separation of trace elements from high-salt and/or wastewaters significantly contributes to removal of matrix interference. The direct quantification of species can be achieved when each step involving separation and detection has been carried out, avoiding any species change during the analytical procedure. Among the separation techniques, ion chromatography is appropriated to characterize single samples for a wide range of ionic species, even at trace concentrations, and a review of applications in environmental research has recently been published (97). In regard to the elements considered in this chapter, the liquid chromatographic technique in particular has been used in the determination of dissolved Ni and Co (98) and in the separation of V (99), Cr (100,101), and Se species (102). However, the capability of ion chromatography to speciate Se is limited at high sulphate concentration by shifting and overlapping of the Se peaks (51).

Separation is often used as the first step in many preconcentration methods. Procedures based on selective precipitation (103), selective flotation (104), catalytic spectrophotometry (105), and electrophoresis (106,107) have been used for the separation and preconcentration of microamounts of V, Cr, Co, Ni, and Se ions in different waters.

Preconcentration by ion exchange has been used extensively for the determination of many metals at trace and ultratrace levels (108–110). In this procedure, the sample is poured through a column loaded with an ion-exchange material; or, alternatively, such material is added to the sample and successively recovered; then the elements of interest are eluted, sometimes back-extracted, and finally analyzed. Several materials have the ability selectively to adsorb the ions of the elements of interest. Ion-exchange columns have been employed for Al determination in freshwater (111) and seawater (112). Ion-exchange and chelating resins (e.g., Amberlite, Chelex 100, Dowex, Ostion) have been used for the preconcentration of trace amounts of Be (113), Al, Cr (114,115), Co, Ni (98), and Sn and V (95,116) in natural waters and wastewaters.

Solvent extraction has played an important role in the development of trace-element analysis, and has been used for the preconcentration of different metals. In this procedure the element of interest is first complexed or chelated, then extracted into an organic phase that can easily be separated from the aqueous phase, eventually back-extracted to enhance the concentration factor or to convert the species into a phase suitable for the technique used, and finally determined. Among the complexing/chelating agents, dithiocarbamate (DC), diethyl-dithiocarbamate (DDC), sodium-DDC,

ammonium pirrolydine-dithiocarbamate (APDC), alone
or in combination with organic agents, such as methyl-
isobutyl-ketone (MIBK), di-isobutyl-ketone (DIBK),
and trioctyl-methyl-ammonium chloride (TOMA.Cl)
have been used for extracting a variety of metals: V
(117), Cr, Co, Ni (34,118–120), Sn, Sb, Se (121–124),
and Tl (64).

Evaporation is a very simple and potentially accurate preconcentration method, but it is time consuming,
and low concentration factors are usually attained
(125). Matrix effects are enhanced; hence this method
is best suited for low-salinity waters. Evaporation for
the analysis of Be in natural waters by ICP-AES
achieving a detection limit of 0.05 μg/L has been used
(37). Another simple preconcentration method used in
GFAAS is multiple injection prior to the atomization
step; a detection limit of 0.05 μg/L has been achieved
using this method for the determination of Cr in seawater (126).

E. Implemented Techniques

By combining different devices with analytical techniques, the advantages of some may be used to overcome interference effects observed in others as well as
to enhance the sensitivity. Electrothermal vaporization
(ETV), hydride generation (HG), flow injection, and
ion chromatography have been used successfully as online devices with ICP-MS, ICP-AES, and AAS (Fig.
3). An important aspect of alternative sample introduction techniques, as compared to continuous nebulization, is that the signal of the analyte to be determined
varies as a function of time, due to the discrete volume
of sample introduced. The time available to carry out
an analytical measurement in a transient signal is restricted to the duration of the transient; therefore, appropriate care must be taken in the way the measurement is carried out.

The major advantages of ETV as a means of sample
introduction are the microvolume of sample required
together with the possibility of removing the organic
phase as well as some matrix components prior to analysis. This device is particularly useful when a small
quantity of water sample is available and in high-salinity waters and wastewaters having a complex matrix,
which may interfere in the determination of trace elements. Electrothermal vaporization as an online device
with ICP-MS allows one to improve the transport efficiency and sensitivity, and removes isobaric overlaps
due to oxide formation (127,128).

Some of the elements considered can be reduced to
volatile compounds (hydrides), separated from the ma-

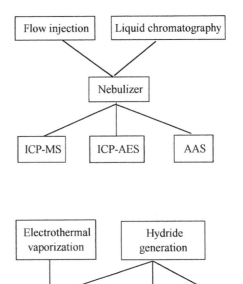

Fig. 3 Combination of techniques for the determination of
trace elements in water.

trix, and determined. Thiourea, and potassium iodide
have been used as reducing agents during sample preparation. The Sb, Se, and Sn hydrides can easily be generated from aqueous solutions by using the acid-boro-
hydride reaction (67). Hydride generation as an online
device with the ICP-MS (129–131), ICP-AES
(132,133), and AAS (134) techniques has been used
for the determination of trace concentrations of Sb, Se,
and Sn in different waters.

Flow injection systems used online with spectrometric techniques for the determination of trace elements have undergone rapid development in the past
decade (67,96,111,112,131,134,135).

The capability of chromatography in the separation
of dissolved components combined with the high sensitivity of ICP-MS makes a powerful tool for the direct
speciation of trace elements. Liquid chromatography
online with ICP-MS has been used for the speciation
of microamounts of Cr (100), V (99), and Se (136).
Gas chromatography after derivatization/sorption combined with ICP-MS has been used to determine ng/L
concentrations of organotin species (137).

F. Applications in Different Waters

It has been pointed out that the main problems in the
quantification of trace elements in water are: (a) their
occurrence at concentrations often not detectable by the

most sensitive techniques, and (b) the effects of the major components present in solution. The preconcentration of trace elements and their separation from the interfering components may overcome these difficulties. Therefore, an accurate determination of trace elements in water can be obtained using different procedures according to the main chemical composition of each water sample. The methods and techniques currently used for the determination of trace elements in different waters are summarized in Fig. 4, and a selection of separation and preconcentration methods is reported in Table 9.

Drinking waters are supplied by surface water and groundwater, which generally show low salinity. This results in low interference effects during trace element analysis, which allows their direct determination (13,70,75,77,88,138,139). Table 10 shows the range of concentration of some trace elements determined directly by GFAAS in 96 bottled waters from Italy (140). It can be observed that only a few samples had detectable concentrations of Be, Co, Se, and Ag. Therefore, the quantification of many trace elements, due to their extremely low concentrations in the low-salinity water, usually requires a preconcentration method prior to analysis (141).

Seawater has salinity in the range of 10–40 g/L, but sub-μg/L concentrations of Be, Al, Cr, Co, Ni, Se, Ag, Sn, Sb, and Tl are usually found. Sample dilution allows the determination of Li, B, Sr, and Ba in marine waters with an accuracy of up to 10% (Table 11). Dilution prior to cryogenic desolvation-ICP-MS overcomes the interference on the determination of V and Ni at the ng/L level (92), and preconcentration by the multiinjection method in GFAAS allows the determination of Cr in seawater (126). However, the high concentration of major components causes serious interference in the quantification of most trace elements in seawater, so separation and preconcentration methods are required to remove the matrix components prior to analysis (46,112,119,142,143).

Wastewaters derived by the leaching of mining wastes and tailings may have a high concentration of metals but low salinity. In this case many trace elements can be determined directly or after dilution by ICP-MS (38,144) and ICP-AES (23). Waters from municipal waste show high amounts of organic matter, and those from industrial activities have a variable range of metals and dissolved solids. These characteristics often result in a complex matrix, which may cause difficulties similar to those found in seawater, hampering the direct determination of trace elements. Hence, sample pretreatment is usually required prior to the quantification of trace elements in wastewaters (47,78,98,115,145).

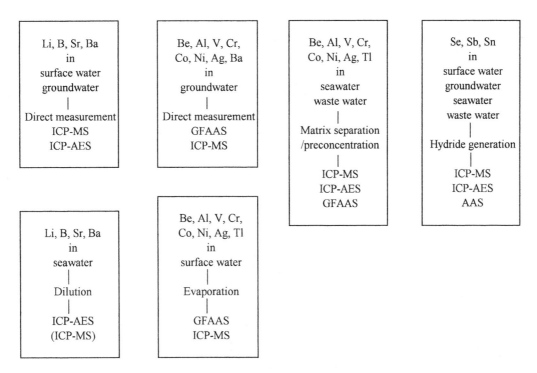

Fig. 4 Most common procedures for the determination of trace elements in different waters.

Table 9 Preconcentration and Separation Methods for Determination of Trace Elements in Different Waters

Element/species	Method[a]	Technique[b]	Detection limit (μg/L)	Application	Ref.
Be	Evaporation	ICP-MS	0.05	River and groundwater	37
V, Cr, Ni, Co	Coprecipitation with APDC	GFAAS	0.1–0.02	Lake water	34
Co, Ni	APDC/DDDC extraction	GFAAS	0.01, 0.05	River water	118
Co, Ni	DC/DIBK extraction	ICP-MS	0.002	Seawater	119
V, Ni, Sb, Co	DC/MIBK extraction	ET-ICP-MS	0.3–0.002	Seawater	122
Se, Sn, Sb	DC/MIBK extraction	GFAAS	0.06–01	Natural water	121
V, Co, Sn, Sb	TOMA.Cl extraction	ICP-AES	1–0.01	Natural water	124
Se	Anion-exchange resin	HG-AAS	0.2	Lake, river, and tap water	134
V species	Separation by chromatography	ICP-MS	0.03	Seawater	99
Cr species	Ion exchange (Ostion)	GFAAS	0.06	Groundwater	115
Cr species	Cation exchange (Chelex-100)	ICP-AES	3	Groundwater	47
Cr species	Cation exchange (Amberlite)	FI-AAS	1	Wastewater	96
Cr species	Ion exchange (Dowex 1 × 2)	AAS	70	Wastewater	115

[a]APDC: ammonium pirrolydinedithiocarbamate; DDDC: diethylammonium diethyl-dithiocarbamate; DC: dithiocarbamate; DIBK: di-isobutyl-ketone; MIBK: methyl-isobuthyl-ketone; TOMA.Cl: trioctylmethylammonium chloride.

[b]ICP-MS: inductively coupled plasma mass spectrometry; GFAAS: graphite furnace atomic absorption spectrometry; ET: electrothermal vaporization; ICP-AES: inductively coupled plasma atomic emission spectrometry; HG: hydride generation; FI: flow injection; AAS: atomic absorption spectrometry.

V. DATA QUALITY

A. Analytical Protocol

Good precision and accuracy can be obtained following procedures that are as clean as possible and protocols aimed at minimizing analytical errors. The accurate preparation of blank and standard solutions is the first step in obtaining a reliable calibration curve. The blank solution must be prepared using the same reagents and procedures as for samples prior to analysis. Since the blank response is the main drawback limiting instru-mental sensitivity, ultrapure reagents must be used to achieve the lowest detection limits.

Due to the low concentration used for calibration, standard solutions are prepared fresh prior to analysis. Appropriate dilutions of single-element standard solutions are used in GFAAS and AAS. Due to the low range of linear response in GFAAS and AAS, the use of several standards is required to achieve a good calibration curve (as an example, see Fig. 5), and measurements are made on the linear part of the curve. In ICP-MS and ICP-OES techniques, the use of commer-

Table 10 Trace Element Concentrations Determined by GFAAS in 96 Bottled Waters from Italy

Element	Number of samples above detection limit	Range (μg/L)	Mean (μg/L)	SD
Be	3	0.20–3.10	—	—
Al	96	1.00–38.9	7.15	4.6
V	29	0.50–21.0	1.8	0.7
Cr	96	0.10–11.9	0.45	0.25
Co	9	0.10–1.47	0.25	0.15
Ni	32	0.20–5.70	1.9	0.4
Se	9	0.50–1.95	1.1	0.6
Ag	11	0.05–16.5	8.65	3.25
Ba	96	2.4–397	32.9	18.7

Source: Ref. 140.

Table 11 Dissolved (<0.4 μm) Li, B, Sr, and Ba in Near-Shore Seawater Determined after Dilution

	LI (μg/L)	B (μg/L)	Sr (μg/L)	Ba (μg/L)
Sample 1[a]				
ICP-MS	194	5000	8500	7
ICP-AES	170	4700	8200	6
AAS	210		7200	
Sample 2[b]				
ICP-MS	142	4850	8100	22
ICP-AES	150	4200	7250	21
AAS	180		7000	

[a]Gulf of Cagliari (Italy), salinity 38.4 g/L.
[b]Funtanamare (W. Sardinia, Italy), salinity 33.8 g/L.

cial multistandard solutions is preferred, to prevent incompatibility among elements, which could arise from the mixing of different standard solutions. In ICP techniques, the range of linearity is very high, usually more than three orders of magnitude, and a very low number of standard solutions is required (e.g., the use of 0, 10, and 100-μg/L standard solutions is adequate for measurements in the 1–100-μg/L range).

After the best instrumental conditions have been assessed and a good calibration curve has been obtained, the quantitative analysis is carried out following an established analytical sequence. The analysis of the field blank solutions is performed prior to that of any other sample. When significant, the mean value of element concentration measured in the blanks must be subtracted from the element concentration measured in the samples. The quantification of major elements prior to trace element determination is very useful to evaluate their potential interference, and allows samples to be sorted according to increasing salinity, which helps to reduce memory effects between samples. When high concentrations of elements are randomly recorded, the potential memory effect needs to be checked by running a blank solution. If necessary, approximately 2-minute flush times of an acid washing solution in between samples usually reduce element counts to background level. The samples with high total dissolved solids (TDS) are diluted to approximately 1 g/L TDS. This procedure will help reduce either matrix effects or memory effects, as well as allow one to avoid partial occlusion at the nebulizer due to salt deposition.

The analysis of a standard reference solution having element concentrations close to those found in the sam-

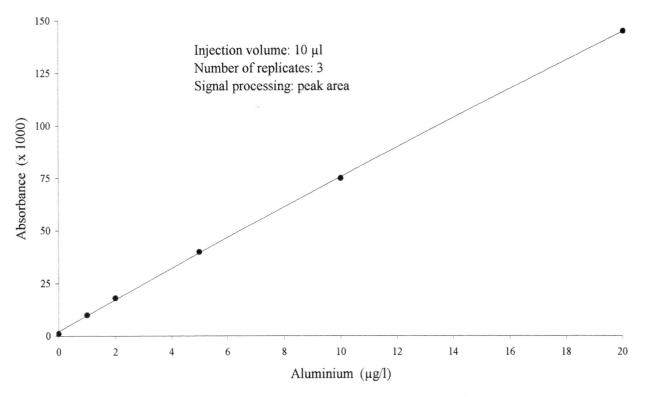

Injection volume: 10 μl
Number of replicates: 3
Signal processing: peak area

Fig. 5 Calibration curve in GFAAS (Perkin Elmer 3030) with Zeeman background correction.

ples, made at regular intervals, is useful in monitoring instrument performance during an analytical cycle; the derived results may be used for external drift correction. A recalibration procedure may be necessary when the standard response over time shows differences of more than 10% with respect to the nominal value.

B. Analytical Control

When the determination of element concentration is carried out at the trace and ultratrace levels, the lowest measurable concentration must be established. The lowest measurable concentration can be defined as the detection limit (DL) or the limit of quantitation (LOQ), both being calculated on the basis of the blank solution response over time. The routine DL (as well as the LOQ) of a specific analytical method is conventionally calculated as follows: A blank solution prepared following the procedure used for samples is analyzed at regular intervals within an analytical cycle and the 3σ (10σ for the LOQ) standard deviation value is calculated, considering all blank results. The use of the LOQ provides more realistic lowest concentrations, since the 10σ value takes into account almost all the background fluctuations. Measurements very close to the DL and LOQ are obviously less precise than those at high concentrations. The lowest detection limits can be achieved using only ultrapure reagents and procedures that are as clean as possible (146,147).

It is impossible to avoid analytical errors completely, despite all efforts minimize them, therefore errors must be estimated. Several approaches can be used to validate a set of data. Field controls include sample duplicates, sample splits, and field equipment blanks. Laboratory analytical controls consist of instrument calibration verification, sample holding times, blank analyses, matrix spike recoveries, and serial dilutions. Interlaboratory quality assurance (148) is a powerful but costly tool. Standard quality control requirements for evaluating inorganic analyses (149) and methods dedicated to the determination of trace elements in water (86,87,142) have been established by the U.S. Environmental Protection Agency, and many workers have been using these recommended procedures.

Precision both within a laboratory and across laboratories may be estimated using duplicate pairs (150) and carrying out duplicate analysis. Short-term precision is referred as to *repeatability*, long-term precision as to *reproducibility*. Accuracy for each element is more difficult to quantify, but may be assessed by the degree of convergence of results obtained using independent analytical methods and by the analysis of cer-

tified reference materials of a composition close to that of the samples being analyzed. Comparison between INAA and ICP-MS (151,152), ICP-AES and ICP-MS (91), and ICP-AES and GFAAS (153) have been used for estimating the accuracy of trace-element analysis.

The comparison of results derived from different analytical methods and techniques, and from the analysis of certified reference solutions carried out in parallel with water sample analysis, is the most common method for evaluating data accuracy. A few certified reference solutions are presently available for trace elements in waters; the most frequently used are listed next. The SRM1643d reference solution simulates the elemental composition of freshwater and contains the trace elements considered in this chapter, with the exception of tin, certified at concentrations in the 1–300-μg/L range. The SLRS-3 reference solution is the water from Ottawa River at Chenaux, Ontario, in which 18 trace metals (including Al, Sb, Ba, Be, Cr, Co, Ni, Sr, V) are certified, most being present at sub-μg/L concentrations. The CASS-3 reference solution is nearshore seawater collected close to Halifax, Nova Scotia; salinity is 30.2 parts per thousand (g/kg), and 13 elements are certified at μg/L or sub-μg/L concentrations. The NASS-4 reference solution is open-ocean seawater, with a salinity of 31.3 g/kg and with 13 elements certified at μg/L or sub-μg/L concentrations. SRM1643 is supplied by the U.S. National Institute of Standards & Technology, while SLRS, CASS, and NASS are supplied by the National Research Council of Canada.

REFERENCES

1. R. Chester. Marine Geochemistry. London: Unwin Hyman, 1990.
2. BJ Alloway, DC Ayres. Chemical Principles of Environmental Pollution. London: Blackie Academic & Professional, 1997.
3. G Ottonello. Principi di Geochimica. Bologna: Zanichelli, 1991.
4. WP Leeman, VB Sisson. Geochemistry of boron and its implications for crustal and mantle processes. In: ES Grew, LM Anovitz, eds. Boron Mineralogy Petrology and Geochemistry. Reviews in Mineralogy 33. Washington, DC: Mineralogical Society of America, 1993, pp 645–707.
5. KH Wedepohl, ed. Handbook of Geochemistry. Berlin: Springer-Verlag, 1969–1978.
6. A Giblin. Natural waters as sample media in drainage geochemistry. In: M Hale, JA Plant, eds. Handbook of Exploration Geochemistry. Vol. 6. GJS Govett, ed.

Drainage Geochemistry Amsterdam: Elsevier Science, 1994, pp 269–303.

7. JO Nriagu, JM Pacyna. Quantitative assessment of worldwide contamination of air, water and soil by trace metals. Nature 133:134–139, 1988.

8. I Thornton. Impact of mining on the environment; some local, regional and global issues. Appl Geochem 11:355–361, 1996.

9. E Helios Rybicka. Impact of mining and metallurgical industries on the environment in Poland. Appl Geochem 11:3–9, 1996.

10. JM Moore. Inorganic Contaminants of Surface Water. Research and Monitoring Priorities. Berlin: Springer-Verlag, 1990.

11. USEPA. National primary and secondary drinking water regulations; synthetic organic chemicals and inorganic chemicals. Federal Register 55(143), July 25 1990. ICP Information Newsletter 16:473–475, 1991.

12. CAI Appelo, D Postma. Geochemistry, Groundwater and Pollution. Rotterdam: Balkema, 1994.

13. D Banks, C Reimann, O Royset, H Skarphagen, OM Saether. Natural concentrations of major and trace elements in some Norwegian bedrock groundwaters. Appl Geochem 10:1–16, 1995.

14. RL Seiler. Synthesis of data from studies by the national irrigation water-quality program. Water Res Bull 32:1233–1245, 1996.

15. MJ Liszewski, LJ Mann. Concentrations of 23 trace elements in groundwater and surface water at and near the Idaho National Engineering Laboratory, Idaho, 1988–91. USGS Open-File Report 93-126, 1991.

16. Y Nakamura, T Tokunaga. Antimony in the aquatic environment in north Kyushu district of Japan. Water Sci Technol 34:133–136, 1996.

17. F Frau. Selected trace elements in groundwaters from the main hydrothermal areas of Sardinia (Italy) as a tool in reconstructing water–rock interaction. Miner Petrogr Acta 36:281–296, 1993.

18. JI Drever. The Geochemistry of Natural Waters. Englewood Cliffs, NJ: Prentice-Hall, 1982.

19. G Michard. Equilibres Chimiques dans Les Eaux Naturelles. Paris: Editions Publisud, 1989.

20. W Stumm, JJ Morgan. Aquatic chemistry. 3rd ed New York: Wiley, 1996.

21. BA Kimball, E Callender, EV Axtmann. Effects of colloids on metal transport in a river receiving acid mine drainage, upper Arkansas River, Colorado, USA. Appl Geochem 10:285–306, 1995.

22. HC Choi, MY Corapcioglu. Transport of a non-volatile contaminant in unsaturated porous media in the presence of colloids. J Contaminant Hydrol 25:299–324, 1997.

23. YTJ Kwong, CF Roots, P. Roach, W Kettley. Post-mine metal transport and attenuation in the Keno Hill mining district, central Yukon, Canada. Environm Geol 30:98–107, 1997.

24. JW Morse, T Arakaki. Adsorption and coprecipitation of divalent metals with mackinawite (FeS). Geochim Cosmochim Acta 57:3635–3640, 1993.

25. P Quevauviller. Atomic spectrometry hyphenated to chromatography for elemental speciation: performance assessment within the standards, measurements and testing programme (Community Bureau of Reference) of the European Union. J Anal Atom Spectr 11:1225–1231, 1996.

26. LA Ellis, DJ Roberts. Chromatographic and hyphenated methods for elemental speciation analysis in environmental media. J Chromatography 774:3–19, 1997.

27. R Koplik, E Curdova, O Mestek. Trace element speciation in water, soils, sediments and biological material. Chemicke Listy 91:38–47, 1997.

28. AK Das, R Chakraborty. Electrothermal atomic spectrometry in the study of metal ion speciation. Fresenius J Anal Chem 357:1–17, 1997.

29. P Smichowski, Y Madrid, C Camara. Analytical methods for antimony speciation in waters at trace and ultratrace levels. A review. Fresenius J Anal Chem 360:623–629, 1998.

30. A Itoh, H Haraguchi. Dissolved states of trace metal ions in natural water as elucidated by ultrafiltration size exclusion chromatography ICP-MS. Anal Sci 13S:393–396, 1997.

31. J Viers, B Dupré, M Polvé, J Schott, JL Dandurand, JJ Braun. Chemical weathering in the drainage basin of a tropical watershed (Nsimi-Zoetele site, Cameroon): comparison between organic-poor and organic-rich waters. Chem Geol 140:181–206, 1997.

32. DR Turner, M Whitfied, AG Dickson. The equilibrium speciation of dissolved components in fresh water and seawater at 25°C and 1 atm pressure. Geochim Cosmochim Acta 45:855–881, 1981.

33. E Viollier, D Jézéquel, G Michard, M Pèpe, G Sarazin, P Albéric. Geochemical study of a crater lake (Pavin Lake, France): trace element behavior in the monimolimnion. Chem Geol 125:61–72, 1995.

34. LS Balistrieri, JW Murray, B Paul. The geochemical cycling of trace elements in a biogenic meromictic lake. Geochim Cosmochim Acta 58:3993–4008, 1994.

35. U Schafer. Essentiality and toxicity of lithium. J Trace Microprobe Technology 15:341–349, 1997.

36. PN Bhat, KC Pillai. Beryllium in environmental air, water and soil. Water Air Soil Pollution 95:133–146, 1997.

37. WM Edmunds, JM Trafford. Beryllium in river baseflow, shallow groundwaters and major aquifers of the U.K. Appl Geochem SI 2:223–233, 1993.

38. R Caboi, R Cidu, L Fanfani, P Zuddas. Abandoned mine sites: implications for water quality. Proceedings of Fourth International Conference on Environmental Issues and Waste Management in Energy and Mineral Production, Cagliari, Italy, 1996, pp 797–805.

39. CW Fetter. Contaminant Hydrogeology. New York: Macmillan, 1993.

40. S Arnorsson, A Andresdottir. Processes controlling the distribution of boron and chlorine in natural waters in Iceland. Geochim Cosmochim Acta 59:4125–4146, 1995.

41. S Barth. Application of boron isotopes for tracing sources of anthropogenic contamination in groundwater. Water Res 32:685–690, 1998.

42. RO Nable, GS Banuelos, JG Paull. Boron toxicity. Plant Soil 193:181–198, 1997.

43. GS Plumlee. The environmental geology of mineral deposits. Proceedings of the Short Course on Environmental Geochemistry of Mineral Deposits, Denver, 1993, p. 41.

44. Y Sakai, K Ohshita, S Koshimizu, K Tomura. Geochemical study of trace vanadium in water by preconcentrational neutron activation analysis. J Radioanal Nuclear Chem 216:203–212, 1997.

45. G Godgul, KC Sakai. Chromium contamination from chromite mine. Environm Geol 25:251–257, 1995.

46. CMG Van der Berg, M Boussemart, K Yokoi, T Prartono, MLAM Campos. Speciation of aluminium, chromium and titanium in the NW Mediterranean. Marine Chem 45:267–282, 1994.

47. A Davis, JH Kempton, A Nicholson, B Yare. Groundwater transport of arsenic and chromium at a historical tannery, Woburn, Massachusetts, USA. Appl Geochem 9:569–582, 1994.

48. L Sigg, B Nowak, H Xue. Role of strong ligands for the infiltration of trace metals into groundwater. Miner Magazine 58A:838–839, 1994.

49. DG Brookins. Eh–pH diagrams for geochemistry. Berlin: Springer-Verlag, 1988.

50. DL Naftz, JA Rice. Geochemical processes controlling selenium in ground water after mining, Powder River Basin, Wyoming, U.S.A. Appl Geochem 4:565–575, 1989.

51. S Sharmasarkar, GF Vance, F Cassel-Sharmasarkar. Analysis of speciation of selenium ions in mine environments. Environm Geol 34:31–38, 1998.

52. Weres, AR Jaouni, L Tsao. The distribution, speciation and geochemical cycling of selenium in a sedimentary environment, Kesterson Reservoir, California, U.S.A. Appl Geochem 4:543–563, 1989.

53. WT Frankenberg, S Benson, eds. Selenium in the Environment. New York: Marcel Dekker, 1994.

54. J Zhang, K Takahashi, H Wushiki, S Yabuki, JM Xiong, A Masuda. Water geochemistry of the rivers around the Taklimakan Desert (NW China): crustal weathering and evaporation processes in arid land. Chem Geol 119:225–237, 1995.

55. AW Mann, JJ Fardy, JW Hedenquist. Major and trace element concentrations in some New Zealand geothermal waters. Proceedings of V International Volcanological Congress, Auckland, 1986, pp 61–66.

56. EN Pentcheva, L Van't dack, E Veldeman, H Hrstov, R Gijbels. Hydrogeochemical characteristics of geothermal systems in South Bulgaria. University of Antwerp UIA, Department of Chemistry, Antwerp, 1997.

57. R Cidu, L Fanfani. Influence of mine watering on groundwater quality at Monteponi (Sardinia, Italy). WRI-9 (Arehart and Hulston, eds.) Balkema, Rotterdam, 1998, pp 969–972.

58. BW Vink. Stability relations of antimony and arsenic compounds in the light of revised and extended Eh–pH diagrams. Chem Geol 130:21–30, 1996.

59. M Brondi, R Gragnani, M Prosperi. Impiego dell'ETA-AAS Zeeman nello studio di distribuzione e delle modalità di circolazione dell'antimonio in alcune sorgenti del Lazio e dei Campi Flegrei. In: C Minoia, S Caroli, eds. Applicazioni dell'ETA-AAS Zeeman nel Laboratorio Chimico e Tossicologico. Vol. 1. Acque, Alimenti, Ambiente. Padua, Italy: Edizioni Libreria Cortina, 1989, pp 149–173.

60. AR Zanzari, R Caboi, R Cidu, A Cristini, L Fanfani, P Zuddas. Hydrogeochemistry in the abandoned mining area of Tafone Graben (Italy). Proceedings of 8th International Symposium Water Rock Interaction, Vladivostok, 1995, pp 905–908.

61. RE Krupp. Solubility of stibnite in hydrogen sulfide solutions, speciation, and equilibrium constants, from 25 to 350°C. Geochim Cosmochim Acta 52:3005–3015, 1988.

62. WS Moore. High fluxes of radium and barium from the mouth of the Ganges-Brahmaputra River during low river discharge suggest a large groundwater source. Earth Plan Sci Letters 150:141–150, 1997.

63. BW Vink. The behaviour of thallium in the (sub)surface environment in terms of Eh and pH. Chem Geol 109:119–123, 1993.

64. M Dall'Aglio, M Fornaseri, M Brondi. New data on thallium in rocks and natural waters from Central and Southern Italy. Insights into applications. Miner Petrogr Acta, 37:103–112, 1994.

65. P Shand, WM Edmunds, J Ellis. The hydrogeochemistry of thallium in natural waters. Proc WRI-9 (Arehart and Hulston, eds.) Balkema, Rotterdam, 75–78, 1998.

66. G Michard, E Viollier, D Jézéquel, G Sarazin. Geochemical study of a crater lake: Pavin Lake, France. Identification, location and quantification of the chemical reactions in the lake. Chem Geol 115:103–115, 1994.

67. KE Jarvis, AL Gray, RS Houk, eds. Handbook of Inductively Coupled Plasma Mass Spectrometry. London: Blackie, 1994.

68. GEM Hall, GF Bonham-Carter, AJ Horowitz, K Lum, C Lemieux, B Quemerais, JR Garbarino. The effect of using different 0.45-μm filter membranes on "dissolved" element concentrations in natural waters. Appl Geochem 11:243–249, 1996.

69. B Dupré, J Gaillardet, D Rousseau, CJ Allègre. Major and trace elements of river-borne material: The Congo Basin. Geochim Cosmochim Acta 60:1301–1321, 1996.

70. KO Konhauser, MA Powell, WS Fyfe, FJ Longstaffe, S Tripathy. Trace element chemistry of major rivers in Orissa State, India. Environm Geol 29:132–141, 1997.

71. L Yan, RF Stallard, DA Crerar, RM Key. Experimental evidence on the behavior of metal-bearing colloids in low-salinity estuarine water. Chem Geol 100:163–174, 1992.

72. TD Martin, JT Creed, SE Long. Method 200.2 Sample preparation procedure for spectrochemical determination of total recoverable elements. ICP Information Newsletter 18:141–144, 1992.

73. T Paukert, Z Sirotek. A study of the microwave treatment of water samples from the Elbe River, Bohemia, Czech Republic. Chem Geol 107:133–144, 1993.

74. I Rodushkin, T Ruth. Determination of trace metals in estuarine and sea-water reference materials by high resolution inductively coupled plasma mass spectrometry. J Anal Atom Spectr 12:1181–1185, 1997.

75. KJ Stetzenbach, M Amano, DK Kreamer, VF Hodge. Testing the limits of ICP-MS: determination of trace elements in ground water at the part-per-trillion level. Ground Water 32:976–985, 1994.

76. C Minoia, S Caroli, eds. Applicazioni dell'ETA-AAS Zeeman nel Laboratorio Chimico e Tossicologico. Vol. 1. Acque, Alimenti, Ambiente. Padua, Italy: Edizioni Libreria Cortina, 1989.

77. CP Bosnak, ZA Grosser. The analysis of drinking water and bottled water by flame AA and GFAA. Atom Spectr 17:218–224, 1996.

78. JC Latino, ZA Grosser. Modern flame atomic absorption analysis of wastewater. Atom Spectr 17:215–217, 1996.

79. D Rurikova, M Pocuchova. Voltammetric determination of antimony in natural waters. Chem Papers Chemicke Zvesti 51:15–21, 1997.

80. J Bubnik. Voltammetric determination of small amounts of As, Sb and Sn in waters, leaches and materials with a complex matrix. Chemicke Listy 91:200–207, 1997.

81. C Elleouet, F Quentel, C Madec. Determination of inorganic and organic selenium species in natural waters by cathodic stripping voltammetry. Water Res 30:909–914, 1996.

82. J Wood, K Nicholson, RS Sokhi, JB Ellis, JD Burton GJL Leeks. Boron determination in water by ion-selective electrode. Proceedings of the Inland and Coastal Water Quality, Stevenage, UK, 1993, pp 237–243.

83. N Gassama, G Michard, C Beaucaire, G Sarazin. Behaviour of nickel and cobalt in natural waters of granitic areas: a first approach. Chem Geol 107:417–421, 1993.

84. M Pleniceanu, M Preda, N Muresan, C Spinu. New liquid-membrane electrodes used for potential determination of copper and nickel. Bull Soc Chim France 134:177–182, 1997.

85. I Svancara, P Ostapczuk, J Arunchalam, HE Emons, K Vytras. Determination of thallium in environmental samples using potentiometric stripping analysis-method development. Electroanalysis 9:36–31, 1997.

86. TD Martin, CA Brockhoff, JT Creed, SE Long. Method 200.7 Determination of metals and trace elements in water and wastes by inductively coupled plasma–atomic emission spectrometry. ICP Information Newsletter 18:147–169, 1992.

87. SE Long, TD Martin. Method 200.8 Determination of trace elements in waters and wastes by inductively coupled plasma–mass spectrometry. ICP Information Newsletter 16:460–473, 1991.

88. JC Latino, DC Sears, F Portala, IL Shuttler. The simultaneous determination of dissolved silver, cadmium, lead, and antimony in potable waters by ETAAS. Atom Spectr 16:121–126, 1995.

89. ER Denoyer. Optimization of transient signal measurements in ICP-MS. Atom Spectr 15:7–16, 1994.

90. G Holeczyova, M Matherny, N Pliesovska. Optimization of the atomic absorption spectrometric methods. 2. Matrix effects in electrothermal atomization of waters in W-tube. Chem Papers Chemicke Zvesti 51:84–90, 1997.

91. R Cidu. Comparison of ICP-MS and ICP-OES in the determination of trace elements in water. Atom Spectr 17:155–162, 1996.

92. LC Alves, LA Allen, RS Houk. Measurement of vanadium, nickel, and arsenic in seawater and urine reference materials by inductively coupled plasma mass spectrometry with cryogenic desolvation. Anal Chem 65:2468–2471, 1993.

93. M Selby. Approaches to interference-free elemental analysis with ICP-MS. Atom Spectr 15:27–35, 1994.

94. Z Fang. Flow injection on-line column preconcentration in atomic spectrometry. Spectrochim Acta Rev 14:235–259, 1991.

95. T Yamane, Y Osada, M Suzuki. Continuous flow system for the determination of trace vanadium in natural waters utilizing in-line preconcentration/separation coupled with catalytic photometric detection. Talanta 45:583–589, 1998.

96. MS Jimenez, L Martin, JM Mir, JR Castillo. Automatic chromium(III/VI) preconcentration and speciation by flow injection flame AAS. Atom Spectr 17:201–207, 1996.

97. C Woods, AP Rowland. Applications of anion chromatography in terrestrial environmental research review. J Chromatography 789:287–299, 1997.

98. N Cardellicchio, P Ragone, S Cavalli, J Riviello. Use of ion chromatography for the determination of transition metals in the control of sewage-treatment-plant

and related waters. J Chromatography 770:185–193, 1997.

99. CC Wann, SJ Jiang. Determination of vanadium species in water samples by liquid chromatography inductively coupled plasma mass spectrometry. Anal Chim Acta 357:211–218, 1997.

100. M Pantsarkallio, PKG Manninen. Simultaneous determination of toxic arsenic and chromium species in water samples by ion chromatography and inductively coupled plasma mass spectrometry. J Chromatography 779:139–146, 1997.

101. B Gammelgaard, YP Liao, O Jons. Improvement on simultaneous determination of chromium species in aqueous solution by ion chromatography and chemiluminescence detection. Anal Chim Acta 354:107–113, 1997.

102. JF Jen, YJ Yang, CH Cheng. Simultaneous speciation of aqueous selenium(IV) and selenium(VI) by high performance liquid chromatography with ultraviolet detection. J Chromatography 791:357–360, 1997.

103. N Belzile, HQ Chen, J Huang, YW Chen. Determination of trace metals in lake waters by x-ray fluorescence after a precipitation preconcentration. Can J Anal Sci Spectr 42:49–56, 1997.

104. MA Kabil, SE Ghazy, AA Elasmy. Simultaneous separation and microdetermination of cobalt(II), nickel(II) and copper(II). Fresenius J Anal Chem 357:401–404, 1997.

105. B Arikan, M Tuncay, R Apak. Sensitivity enhancement of the methylene blue catalytic-spectro-photometric method of selenium(IV) determination by CTAB. Anal Chim Acta 335:155–167, 1996.

106. JF Jen, MH Wu, TC Yang. Simultaneous determination of vanadium (IV) and vanadium(V) as EDTA complexes by capillary zone electrophoresis. Anal Chim Acta 339:251–257, 1997.

107. A Padarauskas, G Schwedt. Capillary electrophoresis in metal analysis investigations of multi-elemental separation of metal chelates with aminopolycarboxylic acids. J Chromatography 773:351–360, 1997.

108. D Beauchemin, JW McLaren, AP Mykytiuk, SS Berman. Determination of trace metals in an open ocean water reference material by inductively coupled plasma mass spectrometry. J Anal Atom Spectrom 3:305–308, 1988.

109. G Godgul, KC Sahu. Chromium contamination from chromite mine. Environm Geol 25:251–257, 1995.

110. JM Deely, DS Sheppard. Whangaehu River, New Zealand: geochemistry of a river discharging from an active crater lake. Appl Geochem 11:447–460, 1996.

111. J Quintela, M Gallego, M Valcarcel. Flow injection spectrophotometric method for the speciation of aluminium in river and tap water. Analyst 118:1199–1204, 1993.

112. JA Resing, CI Measures. Fluorometric determination of Al in seawater by flow injection analysis with in-line preconcentration. Anal Chem 66:4105–4111, 1994.

113. C Valencia, S Boudra, JM Bosque-Sendra. Determination of trace amounts of beryllium in water by solid-phase spectrometry. Analyst 118:1333–1336, 1993.

114. M Kuhn, C Niewohner, M IsenbeckSchroter, HD Schulz. Determination of major and minor constituents in anoxic thermal brines of deep sandstone aquifers in Northern Germany. Water Res 32:265–274, 1998.

115. D Oktavec, J Lehotay, E Hornaskova. Determination of Cr(III) and Cr(VI) in underground water and wastewater by flame and graphite furnace AAS. Atom Spectr 92-96, 1995.

116. XJ Chang, ZX Su, XY Luo. Synthesis and efficiency of an epoxy–urea chelating resin for preconcentrating and separating trace Bi, In, Sn, Zr, V and Ti from solution samples. Anal Lett 30:2611–2623, 1997.

117. A Adachi, K Ogawa, Y Tsushi, N Nagao, T Kobayashi. Determination of vanadium in environmental samples by atomic absorption spectrophotometry. Water Res 31:1247–1250, 1997.

118. J Zhang, WW Huang, JH Wang. Trace-metal chemistry of the Huanghe (Yellow River), China—examination of the data from in situ measurements and laboratory approach. Chem Geol 114:83–94, 1994.

119. GJ Batterham, NC Munksgaard, DL Parry. Determination of trace metals in seawater by inductively coupled plasma mass spectrometry after off-line dithiocarbamate solvent extraction. J Anal Atom Spectr 12:1277–1280, 1997.

120. P Kump, M Necemer, M Veber. Determination of trace elements in mineral water using total reflection x-ray fluorescence spectrometry after preconcentration with ammonium pyrrolidinedithiocarbamate. XR Spectr 26:232–236, 1997.

121. S Arpadjan, L Vuchkova, E Kostadinova. Sorption of arsenic, bismuth, mercury, antimony, selenium and tin on dithiocarbamate loaded polyurethane foam as a preconcentration method for their determination in water samples by simultaneous inductively coupled plasma atomic emission spectrometry and electrothermal atomic absorption spectrometry. Analyst 122:243–246, 1997.

122. SJ Santosa, S Tanaka, K Yamanaka. Sequential determination of trace metals in sea water by inductively coupled plasma mass spectrometry after electrothermal vaporization of their dithiocarbamate complexes in methyl isobutyl ketone. Environ Monitoring Assessment 44:515–528, 1997.

123. D Atanasova, V Stefanova, E Russeva. Preconcentration of trace elements on a support impregnated with sodium diethyldithiocarbamate prior to their determination by inductively coupled plasma atomic emission spectrometry. Talanta 45:857–864, 1998.

124. XY Zhang, K Satoh, A Satoh, K Sawada, T Suzuki. Simultaneous ion-pair solvent-extraction preconcen-

tration of 12 elements in natural water samples for the determination by inductively coupled plasma atomic emission spectrometry. Anal Sci 13:891–895, 1997.

125. A Lucaciu, L Staicu, S Spiridon, N Scintee, D Arizan. Multielement determination in some water samples by neutron activation method. J Radioanal Nuclear Chem 216:29–31, 1997.

126. P BermejoBarrera, J MoredaPineiro, A MoredaPineiro, A BermejoBarrera. Chromium determination in sea water by electrothermal atomic absorption spectrometry using Zeeman effect background correction and a multi-injection technique. Fresenius J Anal Chem 360: 208–212, 1998.

127. B Fairman, T Catterick. Simultaneous determination of arsenic, antimony and selenium in aqueous matrices by electrothermal vaporization inductively coupled plasma mass spectrometry. J Anal Atom Spectr 12: 863–866, 1997.

128. D Pozebon, VL Dressler, AJ Curtius. Determination of arsenic, selenium and lead by electrothermal vaporization inductively coupled plasma mass spectrometry using iridium-coated graphite tubes. J Anal Atom Spectr 13:7–11, 1998.

129. J Bowman, B Fairman, T Catterick. Development of a multi-element hydride generation inductively coupled plasma mass spectrometry procedure for the simultaneous determination of arsenic, antimony and selenium in waters. J Anal Atom Spectr 12:313–316, 1997.

130. T Tanizaki, K Baba, K Kadokami, R Shinohara. Simultaneous determination of arsenic, selenium and antimony by hydride generation ICP-MS. Bunseki Kagaku 46:849–855, 1997.

131. A Stroh, U Völlkopf. Optimization and use of flow injection vapour generation inductively coupled plasma mass spectrometry for the determination of arsenic, antimony and mercury in water and sea-water at ultratrace levels. J Anal Atom Spectr 8:35–40, 1993.

132. A Morrow, G Wiltshire, A Hursthouse. An improved method for the simultaneous determination of Sb, As, Bi, Ge, Se, and Te by hydride generation ICP-AES: application to environmental samples. Atom Spectr 18:23–28, 1997.

133. H Narasaki, JY Cao. Determination of arsenic and selenium in river water by hydride generation ICP-AES. Atom Spectr 17:77–82, 1996.

134. PE Carrero, JF Tyson. Determination of selenium by atomic absorption spectrometry with simultaneous retention of selenium(IV) and tetrahydroborate(III) on an anion-exchange resin followed by flow injection hydride generation from the solid phase. Analyst 122: 915–919, 1997.

135. S Caroli, A Alimonti, F Petrucci, Zs Horvath. On-line preconcentration and determination of trace elements by flow injection-inductively coupled plasma atomic emission spectrometry. Anal Chim Acta 248:241–249, 1991.

136. A Woller, H Garraud, J Boisson, AM Dorthe, P Fodor, OFX Donard. Simultaneous speciation of redox species of arsenic and selenium using an anion-exchange microbore column coupled with a micro-concentric nebulizer and an inductively coupled plasma mass spectrometer as detector. J Anal Atom Spectr 13:141–149, 1998.

137. L Moens, T DeSmaele, R Dams, P VandenBroeck, P Sandra. Sensitive, simultaneous determination of organomercury, -lead, and -tin compounds with headspace solid phase microextraction capillary gas chromatography combined with inductively coupled plasma mass spectrometry. Anal Chem 69:1604–1611, 1997.

138. R Ramesh, KS Kumar, S Eswaramoorthi, GR Purvaja. Migration and contamination of major and trace elements in groundwater of Madras City, India. Environm Geol 25:126–136, 1995.

139. KS Subramanian, R Poon, I Chu, JW Connor. Antimony in drinking water, red blood cells, and serum: development of analytical methodology using transversely heated graphite furnace atomization atomic absorption spectrometry. Arch Environm Contamination Toxicol 32:431–435, 1997.

140. C Minoia, L Vescovi, S Canedoli, A Ronchi, P Apostoli, L Pozzoli, E Sabbioni, L Manzo. Determinazione diretta di elementi in traccia in acque minerali mediante analisi in ETA-AAS Zeeman. In: C Minoia, S Caroli, eds. Applicazioni dell'ETA-AAS Zeeman nel Laboratorio Chimico e Tossicologico. Padua, Italy: Libreria Cortina, 1989, pp 219–237.

141. EM Cameron, GEM Hall, J Veizer, HR Krouse. Isotopic and elemental hydrogeochemistry of a major river system: Fraser River, British Columbia, Canada. Chem Geol 122:149–169, 1995.

142. SE Long, TD Martin. Method 200.10 Determination of trace elements in marine waters by on-line chelation preconcentration and inductively coupled plasma-mass spectrometry. ICP Information Newsletter 18: 171–178, 1992.

143. IB Gornushkin, BW Smith, JD Winefordner. Use of laser-excited atomic fluorescence spectrometry with a novel diffusive graphite tube electrothermal atomizer for the direct determination of silver in sea water and in solid reference materials. Spectrochim Acta 518: 1355–1370, 1996.

144. R Cidu, R Caboi, L Fanfani, F Frau. Acid drainage from sulfides hosting gold mineralization (Furtei, Sardinia). Environm Geol 30:231–237, 1997.

145. USEPA. Inductively coupled plasma–mass spectrometry Method 6020 CLP-M Version 7.0. ICP Information Newsletter 18:584–596, 1993.

146. G Benoit, KS Hunter, TF Rozan. Sources of trace metal contamination artifacts during collection, han-

dling, and analysis of freshwaters. Anal Chem 69:
1006–1011, 1997.

147. SL Tong, CY Ho, FY Pang. Monitoring of Ba, Mn,
Cu, and Ni during estuarine mixing. Anal Sci 13S:
373–378, 1997.

148. H Alkema, J Simser, L Hjelm. Interlaboratory quality
assurance studies: their use in certifying natural waters
for major constituents and trace elements. Fresenius J
Anal Chem 360:339–343, 1998.

149. USEPA. Region I—Laboratory data validation func-
tional guidelines for evaluating inorganic analyses.
U.S. Environmental Protection Agency, Washington,
DC, 1988.

150. GEM Hall. Capabilities of production-oriented laboratories in water analysis using ICP-AES and ICP-MS.
J Geochem Explor 49:89–121, 1993.

151. M Khalis, A Sekkaki, JL Irigaray, F Carrot, G Revel.
Environmental control of Boufekrane river water ba-
sins by neutron activation analysis and ICP/MS. Water
Res 31:2930–2934, 1997.

152. MA Veado, G Pinte, AH Oliveira, G Revel. Application of instrumental neutron activation analysis and
inductively coupled plasma mass spectrometry to studying the river pollution in the State of Minas Gerais. J
Radioanal Nuclear Chem 217:101–106, 1997.

153. S Recknagel, A Chrissafidou, D Alber, U Rosick, P
Bratter. Determination of beryllium in the primary
cooling water of the BER II research reactor by inductively coupled plasma optical emission spectrometry and Zeeman electrothermal atomic absorption
spectrometry. J Anal Atom Spectr 12:1021–1025,
1997.

24

Determination of Silicon and Silicates

Bieluonwu Augustus Uzoukwu
University of Port Harcourt, Port Harcourt, Nigeria

Leo M. L. Nollet
Hogeschool Gent, Ghent, Belgium

I. INTRODUCTION

Silicate compounds recently have been found very useful in many applications intended for domestic or industrial uses, and in many cases have been used either as partial contribution to the performance of a product, as additives in a wide range of applications, or to affect the overall performance of the product or as a source from which further important industrial products can be derived. These domestic and industrial applications of silicates have been attributed to the utility factors inherent in the properties of silicates that make such applications feasible. These properties include the non-toxicity and nonflammability of silicates, their high hydrophilic properties, their wide range of available compositions or proprietary modifications, their immense alkaline buffering capacity, and their adhesive and film-forming capabilities. Hence silicates have been found useful in the manufacture of detergents, soaps, and cleansing agents and in foundries. The production of detergents and soaps represents the major industrial use of silicates. This is because silicates enhance the effectiveness of surfactants by sequestering metal ions in solution, thus lowering the activities of these metal ions. Silicates act as a source of buffered alkalinity, thus enhancing the saponification process during soap production. Silicates also provide crisp and easily handled detergent granules due to their glassy nature upon dehydration.

Water samples may contain silicon as dissolved silicate species or as insoluble colloidal substances. Subsequently, a separation of the two species forms may be required during water analysis. The silicate content of a water sample can be determined through gravimetry or colorimetry. The gravimetric procedure will give the total silicon content of a water sample, while that of colorimetry will give the soluble silicate content after filtration.

In the 18th edition of *Standard Methods for the Examination of Water and Wastewater*, the methods on silica determination are 4500-Si D (molybdosilicate), 4500-Si E (heteropoly blue), and 4500-Si F (automated for molybdate-reactive silica) (26). Method 200.7 for silica is worked out in *Methods for the Determination of Metals in Environment Samples, Supplement 1* (27). Other methods are tabulated in Table 1.

II. DETERMINATION OF SILICA (SiO₂) CONTENT OF WATER BY COLORIMETRY USING MOLYBDENUM BLUE (1)

This method is based on the formation of a yellow silicomolybdate ion when silicate or silicic acid ions are added to an acid solution of molybdate ions. An intense molybdenum blue color is obtained when the silicomolybdate ion is reduced. The interference from

Table 1 Recent Methods on Silicon Determination

Analyte	Method	Detection limit	Refs.
Silicate in waters	Voltammetry with gold microdisc electrodes	3 μM	3
Silicon in tap water	Normal-phase HPLC	10 ng	4, 5
	Column: 8 μm		
	Diasorb 130 CN		
	Column (80 $\times$ 3-mm ID)		
	Mobile phase: tetrahydrofuran/Chloroform		
	(2:1) + 0.2% trioctylamine		
Silicon in water	ICP-emission spectrometry generation of silicon tetrafluoride	98 ng/ml	6
Silicate in water	RAM (random assembly of microdisks) electrodes	1 μM	7, 8
Silicon in water	Spectrophotometric assay		10
	HACH Silica Testing Kit		
Silicon in wastewater	Sequential injection analysis	0.9 ng/L	11
	Absorption at 400 nm		
Silicon in seawater	IEC + ICP-MS	$\pm$2.3 ng/L	12
	Column: Dionex		
	IonPac ICE-ASI (25 cm $\times$ 9-mm ID)		
	Mobile phase: H_2O (0.8 ml/min)		
Silicate ions in demineralized water	Potentiometry		13
	Fluoride-selective electrode		
Silicate in natural waters	Flow injection spectrophotometry with Malachite Green at 645 nm	14 ng/ml	15
Silicon in river and tap water	Flow-based analysis system/laser diode at 700 nm	0.0019 mg/L	16
Silicon in seawater	ICP-AES	0.3 μM	17
Silicon in water	Adsorption stripping voltammetry C electrode	1.2 μg/L	18, 19
Silicon	FIA + molybdenum blue method	7 mg/L	20
Silicon	FIA + molybdenum blue method		21
Silicon	FIA + molybdenum blue method	0.17 mg/L	22
Silicon	Ion-pair reaction of molybdosilicate with malachite green	0.15 μg/L	23
Silicon	Catalytic method	8.5 μg/L	24
Silicon in natural waters	Isotope dilution MS	ppb level	25

IEC: Ion-exchange chromatography; ICP-MS: inductively coupled plasma–mass spectrometry; ICP-AES: inductively coupled plasma–atom emission spectrometry; FIA: flow injection analysis.

phosphate ions can be prevented by control of the pH of the solution or by the addition of oxalic or tartaric acid as masking agent. The blue-colored solution obtained is used for determination of the SiO_2 content of water sample colorimetrically.

A. Reagents and Apparatus

1. Prepare a standard stock solution of silica (50 mg·L^{-1} SiO_2) by fusing 0.050 g of SiO_2 (dry silica powder) with 0.5 g Na_2CO_3 in a platinum crucible at 900°C until a clear melt is obtained. Allow the crucible and content to cool to room temperature before immersing the crucible and content in 250 ml of distilled water in a polyethylene beaker. If the melt did not dissolve com-

pletely, ensure that a solution is obtained by warming the beaker. Cool the solution and transfer it to a 1-liter polythene volumetric flask; mix and dilute it to mark with distilled water. Shake and store in a polythene container.

2. Prepare standard solutions of SiO_2 in the 0–5 mg·L^{-1} SiO_2 concentration range in 100-ml polythene volumetric flasks by pipetting appropriate volumes of standard stock solution to get the desired concentration of SiO_2 in the volumetric flasks, but don't top.

3. Prepare 11% ammonium molybdate reagent by dissolving 20 g of $(NH_4)_6Mo_7O_{24}\cdot4H_2O$ in some quantity of distilled water before diluting to 180 ml with distilled water. Filter the solution if necessary.

4. Prepare 30% H_2SO_4 by adding 12 ml of concentrated sulphuric acid carefully to 30 ml of distilled water in a container.

5. Prepare ammonium molybdate–sulphuric acid reagent by transferring the solution from step 3 and the preparation from step 4 into a 250-ml volumetric flask. Mix and dilute to mark with distilled water.

6. Prepare the reducing agent fresh before use as follows: Dissolve 2.4 g of $Na_2SO_3 \cdot 7H_2O$ and 0.2 g of 1-amino-2-naphthol-4-sulphonic acid in 50–80 ml of distilled water in a 100-ml volumetric flask before adding 14 g of $K_2S_2O_5$ gradually. Mix properly to ensure complete dissolution before diluting to mark with distilled water.

7. Prepare 28% tartaric acid solution by dissolving 28 g of tartaric acid in a small quantity of distilled water in a 100-ml flask before diluting to mark with distilled water.

8. A spectrophotometer or colorimeter.

B. Procedure

1. *To determine both soluble SiO_2 (as silicate or silicic acid) and insoluble SiO_2*: Transfer a known quantity v ml (ca. 20–40 ml) of water sample into a platinum crucible and reduce the volume to about a quarter of its size by evaporation. Add 15 ml of 25% NaOH to the reduced volume of sample in the crucible, mix properly before finally removing all the water from the mixture. Cover the crucible and fire its content at 600°C for 10 minutes over a Bunsen burner. Allow the melt to cool before immersing the crucible and its content in 50 ml of distilled water in a 250-ml polythene beaker. Wash the crucible into the beaker with a small quantity (20 ml) of distilled water. Transfer the solution and wash into a 100-ml polythene volumetric flask and neutralize with drops of concentrated HCl, but don't top yet.

2. *To determine only soluble SiO_2 (as silicate or silicic acid)*: In this case, no pretreatment is required. Filter if the sample is turbid. Transfer a known quantity v ml (ca. 40–80 ml) of clear water sample into a 100-ml polythene volumetric flask. Continue with water samples from step 1 or 2 as described in step 3 (next).

3. Treat both standards and the water sample from this point the same way. Neutralize the standard solutions in the 100-ml volumetric flasks with drops of concentrated HCl. The water sample from step 1 has already been neutralized. The

water sample from step 2 may not require neutralization unless the pH is over 7. Add to each of the flasks 2.5 ml of ammonium molybdate–sulphuric acid reagent; mix and leave for 10 minutes. Add to each of the flasks 2.5 ml of tartaric acid solution; mix and allow to stand for 5 minutes. Add to each of the flasks 2 ml of reducing agent; mix and top the flasks to the 100-ml mark with distilled water. Mix and leave the solutions for about 15 minutes to enable the intense blue color to develop. Obtain the absorbances of the solutions at 810 nm with a spectrophotometer or colorimeter against distilled water blank. Prepare a calibration curve by plotting the absorbances of the standards against their corresponding SiO_2 concentration. Compare the absorbance of the water sample with those of the calibration curve to determine the concentration c mg·L^{-1} of SiO_2 in the water sample:

$$\text{Concentration of } SiO_2 \atop (\text{mg·L}^{-1}) \text{ in water sample} = \frac{100 \text{ ml} \times c \text{ mg·L}^{-1}}{v \text{ ml}}$$

$$\text{Concentration of } SiO_2\text{-}Si \atop (\text{mg·L}^{-1}) \text{ in water sample} = \frac{46.7 \text{ ml} \times c \text{ mg·L}^{-1}}{v \text{ ml}}$$

The equation gives the concentration of both soluble and insoluble SiO_2 (mg·L^{-1}) in the water sample if the water sample was pretreated, as described in step 1, or the concentration of only soluble SiO_2 (mg·L^{-1}) in the water sample if the water sample was not pretreated, as described in step 2.

$$\text{Concentration of } SiO_3^{2-} \atop (\text{mg·L}^{-1}) \text{ in water sample} = \frac{126.6 \text{ ml} \times c \text{ mg·L}^{-1}}{v \text{ ml}}$$

III. DETERMINATION OF SILICA (SiO_2) CONTENT OF WATER BY GRAVIMETRIC METHOD

This method gives the total silica content of a water sample, and essentially it involves the removal of organic matter in the sample through ashing, followed by digestion with hydrochloric acid. Apart from removing the remaining remnants of organic matter in the sample, the acid will also remove metal ions that may be contained in the sample as metal silicates, leaving a residue of insoluble silica after a combined process of evaporation, baking, and ignition. This method, however, gives approximate results, so it is important that a large amount of water sample be used to increase the reliability of the result.

A. Reagents and Apparatus

1. Concentrated hydrochloric acid
2. Beaker and glassware washed thoroughly with a 1:1 mixture of HNO_3 and H_2SO_4
3. Steam bath

B. Procedure

Since the method requires a large volume v ml of water sample, about 400–1000 ml of sample is introduced into a beaker and evaporated to dryness on a steam bath. The beaker is fired at 600°C on a Bunsen burner until complete removal of organic matter in the sample is achieved. Collect the residue in the beaker by adding 10–20 ml of distilled water and an equal volume of concentrated hydrochloric acid. Cover the beaker with a washglass and simmer its content for about 10 minutes before removing the washglass. Evaporate the content of the beaker to dryness before baking for about 10–20 minutes. For a second time, collect the residue in the beaker with 10 ml of water and 10 ml of hydrochloric acid. Cover the beaker with a washglass and simmer for about 15 minutes. After cooling slightly, filter through ashless filter paper and wash the residue on filter paper with hot distilled water. Fold the paper around the residue and transfer it to a previously weighed crucible. Remove water from the paper by heating gently on a Bunsen burner before firing at 600°C on the burner or furnace to ignite the paper and residue until a white ash of silica residue is left. Weigh the residue as silica (SiO_2) to obtain the weight w g:

$$\text{Concentration of } SiO_2 \text{ (mg·L}^{-1}\text{) in water sample} = \frac{w \text{ g} \times 10^6}{v \text{ ml}}$$

$$\text{Concentration of } SiO_2\text{-Si (mg·L}^{-1}\text{) in water sample} = \frac{0.467 \times w \text{ g} \times 10^6}{v \text{ ml}}$$

IV. RECENT METHODS

Table 1 gives an overview of recent methods on the determination of silicon and silicate. As can be seen in the table, a variety of methods exist. Zhang and Ortner (9) reported the effect of thawing conditions on silicic acid in water. The best condition was in darkness at 4°C for several days.

REFERENCES

1. IR Morrison, AL Wilson. Analyst 88:100, 1963.
2. American Public Health Association. Standard Methods for the Examination of Water and Wastewater. Washington, DC: APHS, 1971.
3. NG Carpenter, AWE Hodgson, D Pletcher. Electroanalysis 9:1311–1317, 1997.
4. EM Basova, EN Dorokhova. J. Anal. Chim. 53:430–435, 1998.
5. EM Basova, E Kondik, E Yu, EN Dorokhova. J. Anal. Chem. 53:133–139, 1998.
6. A Lopez Molinero, L Martinez, A Villareal, JR Castillo. Talanta 45:1211–1217, 1998.
7. AWE Hodgson, D Pletcher. Electroanalysis 10:321–325, 1998.
8. AWE Hodgson, D Pletcher. Electroanalysis 17:1311, 1997.
9. J-Z Zhang, PB Ortner. Water Res. 32:2553–2555, 1998.
10. BR Manno, IK Abukhalaf, JE Manno. J. Anal. Toxicol. 21:503–505, 1997.
11. F Mas Torres, A Munoz, JM Estela, V Cerda. Analyst 122:1033–1038, 1997.
12. S Hioki, JWH Lam, JW Mclaren. Anal. Chem. 69:21–24, 1997.
13. VG Derkasova, VA Karelin. Zh. Anal. Khim. 51:1093–196, 1996.
14. Y Takaku, K Masuda, T Takahashi, T Shimamura. Anal. At. Spectrom. 9:1385–1387, 1994.
15. J Saurina, S Hernandez-Cassou. Analyst 120:2601–2604, 1995.
16. T Tsuboi, T Nakamura, K Yonezawa, A Matsukura, S Motomizu. J Flow Injection Anal. 12:85–90, 1995.
17. K Abe, Y Watanabe. J Oceanogr 48:283–292, 1992.
18. GV Prokhorova, EA Osipova, OI Gurentsova. Zh. Anal. Khim. 48:1621–1631, 1993.
19. OI Gurentsova, GV Prokhorova, EA Osopova, IY Torshin. Vestn Mosk Univ Ser Khim 33:466–471, 1992.
20. J Von Staden, J Paver, H Van Vliet, P Kempster. Water SA 16:205–210, 1990.
21. T Takenaka. Bunseki Kagaku 40:425–428, 1991.
22. F Mas, J Estela, V Cerda. Anal. Chim. Acta 239:151–155, 1990.
23. S Motomizu, M Oshima, K Araki. Analyst 155:1627–1630, 1990.
24. X Huang, M Ji, H Tan. Fenxi Huaxue 20:419–422, 1992.
25. M Rener, A Lamberty, P De Bievre. Analysis 20:229–234, 1992.
26. APHA Standard Methods for the Examination of Water and Wastewater. 18th ed. Washington, DC: APHA, AWWA, 1992.
27. USEPA Methods for the Determination of Metals in Environmental Samples Supplement I. EPA 600/R 94/111, 1994.

25

Analysis of Urea Herbicides in Water

Antonio Di Corcia

University "La Sapienza," Rome, Italy

I. PHYSICOCHEMICAL PROPERTIES

Herbicides derived from urea form a large group of chemical compounds widely used in agriculture to control weeds in cereal, vegetable, and fruit tree crops. On the basis of their chemical natures, use, and mode of action, substituted urea herbicides can be divided into two main groups: phenylureas and sulfonylureas.

Although phenylurea herbicides were introduced more than 40 years ago, they are still widely used. Phenylureas can be divided into three subgroups:

1. N-phenyl-$N'N'$-dialkylureas, such as chlorotoluron, isoproturon, and diuron
2. N-phenyl-N'-alkyl-N'-methoxyureas, such as linuron, monolinuron and metobromuron
3. Phenylureas containing a heterocyclic group, the major exponent being methabenzthiazuron

The chemical structures of the main representatives of the three phenylurea subgroups are depicted in Fig. 1; the common names, water solubility, half-life in soil, and leaching potential through the soil (when available) of the most widely used phenylureas are presented in Table 1.

Phenylurea herbicides (PUHs) are generally absorbed through the roots of plants and transported via the transpiration system. The mode of action of phenylurea herbicides seems to be due to the combined effects of the inhibition of photosynthesis and the irreversible injury of the plant photosynthesis system via inhibition of NADPH$_2$ (1).

Phenylureas reaching the environment are gradually decomposed over a short or longer period, the steps and the rate of decomposition depending on the stability of the molecule and on the medium. The active substance on the soil surface or reaching aquifers is chemically decomposed by UV radiation or components of the soil. PUHs absorbed by the plants or in the soil are biodegraded by stepwise demethylation or demethoxylation of the urea moiety followed by generation of aromatic amines. These species are the end products of microbial activity (1). Some of the phenylureas and their related end products are suspected to induce cancer (2,3).

Sulfonylureas are relatively new herbicides, introduced in the 1980s. Chlorsulfuron was the first sulfonylurea marketed in the United States, in 1982. Worldwide, 19 sulfonylureas had been commercialized by 1994, and five more are being developed. This rapid increase is due to their very high and specific herbicidal activity, which results in extremely low application rates of 10–40 g/ha. Furthermore, as compared to other herbicides, sulfonylureas are less toxic and degrade more rapidly. Chemical structures of some representative sulfonylureas are presented in Fig. 2. From a chemical point of view, these herbicides are labile and weakly acidic compounds. The common names, chemical formulas, water solubility, pKa, half-life in soil, and leaching potential through the soil (when available) of the most representative sulfonylureas are reported in Table 2.

Sulfonylureas are systemic herbicides absorbed by the foliage and roots. They act by inhibiting acetolac-

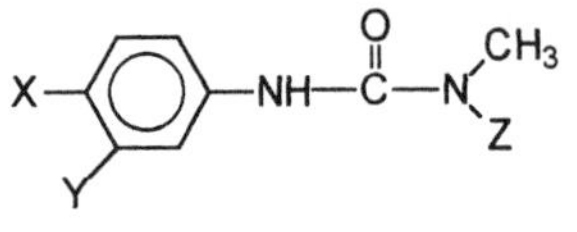

	X	Y	Z
Metoxuron	OCH_3	Cl	CH_3
Chlorotoluron	Cl	Cl	CH_3
Isoproturon	$i\text{-}C_3H_7$	H	CH_3
Diuron	Cl	Cl	CH_3
Neburon	Cl	Cl	C_4H_9
Linuron	Cl	Cl	OCH_3
Metobromuron	Br	H	OCH_3
Monilinuron	Cl	H	OCH_3

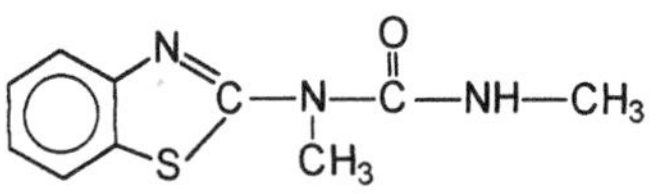

Methabenzthiazuron

Fig. 1 Structures of some selected phenylureas.

tate synthase, a key enzyme in the biosynthesis of branched chain aminoacids (4). This results in stopping cell division and plant growth. The most important degradation pathways of sulfonylureas are chemical hydrolysis and microbial degradation.

II. REGULATIONS

Published information shows that the consumption of pesticides for agricultural and industrial purposes is increasing. According to a report published by the United States Environmental Protection Agency (USEPA), a total of $5 \cdot 10^8$ kg of pesticides was used and dispersed into the environment in 1985. By various transport mechanisms, pesticides can reach and contaminate surface waters, groundwaters, and, ultimately, drinking waters. In the United States, 101 pesticides and 25 related degradation products are included in the list of priority pollutants that are to be monitored in water destined for human consumption (5,6). Four phenylurea herbicides—namely, diuron, fluometuron, linuron, and neburon—are present in this list. So far, no sulfonylurea herbicide has been included in this list. The selection of the different pesticides was based on the use of at least $7 \cdot 10^6$ kg in 1982, a water solubility larger than 30 mg/L, and hydrolysis half-life longer than 25 weeks. Pesticides and pesticide degradation products previously detected in groundwater, as well as pesticides regulated under the Safe Drinking Water Act,

were automatically included in the list of priority analytes.

In the 15 European countries making up the European Community, several priority lists of contaminants, which include many pesticides, have been published to protect the quality of drinking and surface waters. Linuron and monolinuron are included in the so-called "black list" of the 76/464/EEC Council Directive on pollution caused by certain dangerous substances discharged into the aquatic environment of the Community (7). In order to prevent the contamination of groundwater and drinking water in Western Europe, a priority list, which considers pesticides used over 50,000 kg per year and their capacity for probable or transient leaching, was recently published (8). Chlorotoluron, diuron, isoproturon, and methabenzthiazuron are present in this list.

The 80/779/EEC Directive on the Quality of Water Intended for Human Consumption states a maximum admissible concentration of 0.1 μg/L for individual pesticides and 0.5 μg/L for total pesticides, regardless of their toxicity (9).

Table 3 lists the acute oral toxicity of phenylurea and sulfonylurea herbicides for rats.

III. ANALYTICAL METHODS

A. Sampling

The volume of water sample to be collected depends on the amount of water needed to perform the analysis

Table 1 Physicochemical Properties of Phenylurea Herbicides

Common name	IUPAC name, M.F.[a]	Molecular weight	Water solubility (mg/L)	DT_{50}[b] (days)	K_{oc}[c] (ml/g)
Chlorbromuron	3-(4-Bromo-chlorophenyl)-1-methoxy-1-methylurea $C_9H_{10}BrClN_2O$	293.5	35 (20°C)	56–196	908
Chlorotoluron	3-(3-Chloro-p-tolyl)-1,1-dimethylurea $C_{10}H_{13}ClN_2O$	212.7	74 (25°C)	30–40	n.f.[d]
Diuron	3-(3,4-Dichlorophenyl)-1,1-dimethylurea $C_9H_{10}Cl_2N_2O$	233.1	36.4 (25°C)	90–180	400
Fluometuron	1,1-Dimethyl-3-(α,α,α-trifluoro-m-tolyl)urea $C_{10}H_{11}F_3N_2O$	232.2	110 (20°C)	10–100	31–117
Isoproturon	3-(4-Isopropylphenyl)-1,1-dimethylurea $C_{12}H_{18}N_2O$	206.3	65 (22°C)	6–28	n.f.
Linuron	3-(3,4-Dichlorophenyl)-1-methoxy-1-methylurea $C_9H_{10}Cl_2N_2O_2$	249.1	64 (25°C)	82–150	75–250
Methabenziathiazuron	1-(1,3-Benzothiazol-2-yl)-1,3-dimethylurea $C_{10}H_{11}N_3OS$	221.3	59 (20°C)	n.f.	n.f.
Metobromuron	3-(4-Bromophenyl)-1-methoxy-1-methylurea $C_9H_{11}BrN_2O_2$	259.1	330 (20°C)	30	184
Monolinuron	3-(4-Chlorophenyl)-1-methoxy-1-methylurea $C_9H_{11}ClN_2O_2$	214.6	735 (25°C)	45–60	250–500
Neburon	1-Butyl-3-(3,4-dichlorophenyl)-1-methylurea $C_{12}H_6Cl_2N_2O$	275.2	5 (25°C)	n.f.	n.f.

Source: Ref. 4.
[a] M.F. = molecular formula.
[b] DT_{50} = time for 50% loss.
[c] K_{oc} = distribution coefficient (K_d) between soil and water adjusted for the proportion of organic carbon in water. It is a measure of the relative affinities of the pesticide for water and soil surface. As such, it indicates the tendency of a certain pesticide to leach through the soil and reach groundwaters. Roughly, K_{oc} values higher than 100 indicate a low potential leaching.
[d] n.f. = not found.

at the required limit of quantification and, eventually, to duplicate the analysis.

As sampling containers, the best choice is amber glass bottles. When practicable, nonfragile and lighter containers, such as those made of plastic, can be a desirable alternative. However, it should be remembered that the latter type of containers, except for Teflon, can leach analytical interferences. Whatever the type of aqueous matrix and sample containers, it is good practice to triple-rinse the containers three times with the sample and then to collect it. Another good practice is that of sampling by using new containers, to avoid memory effects.

1. Well Water Sampling

When conducting a pesticide monitoring campaign, water in shallow wells located within an area of heavy pesticide use should not be sampled. The collection of a well water sample is usually performed after eliminating stagnant water. The collection of a homogeneous sample can be accomplished by measuring the stability of some parameters of interest (pH, conductivity). The stability of such parameters implies sample homogeneity. Usually, a representative sample is obtained after purging 3–10 well volumes. Other details of well water sampling can be found elsewhere (10).

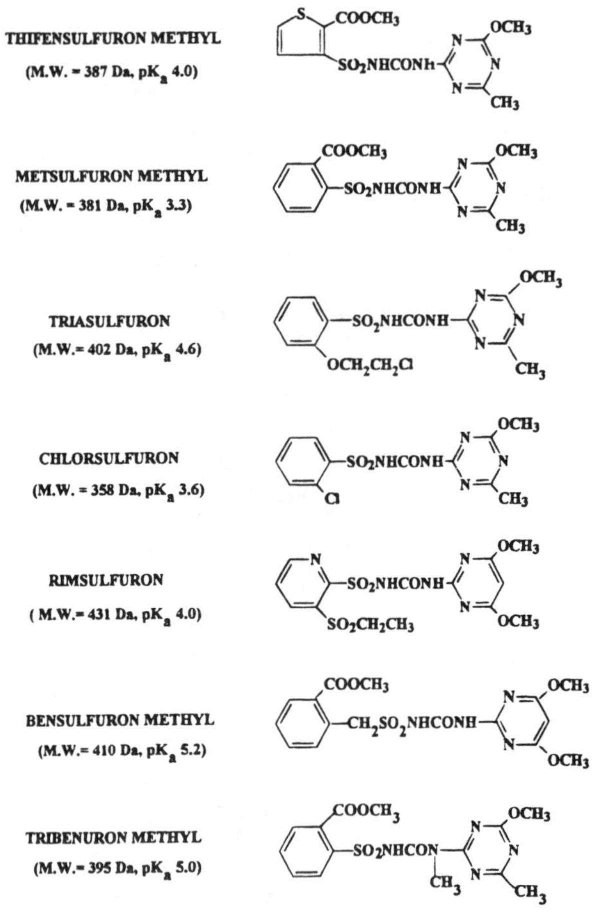

THIFENSULFURON METHYL
(M.W. = 387 Da, pK$_a$ 4.0)

METSULFURON METHYL
(M.W. = 381 Da, pK$_a$ 3.3)

TRIASULFURON
(M.W.= 402 Da, pK$_a$ 4.6)

CHLORSULFURON
(M.W. = 358 Da, pK$_a$ 3.6)

RIMSULFURON
(M.W.= 431 Da, pK$_a$ 4.0)

BENSULFURON METHYL
(M.W.= 410 Da, pK$_a$ 5.2)

TRIBENURON METHYL
(M.W.= 395 Da, pK$_a$ 5.0)

Fig. 2 Structures of some selected sulfonylureas.

2. Potable Water Sampling

Sampling from potable water is usually simplified by collecting water from an existing tap. Before sampling, water is flushed for about 10 min to eliminate sediments and gas pockets in the pipes. If any water treatment device exists, representative water sampling should be made before the treatment unit. These devices contain ion exchangers and active carbons able to strongly adsorb organic compounds.

3. Surface Water Sampling

The composition of stream water is both flow- and depth-dependent. Analyte concentration gradients are not present in shallow lakes, because of the action of wind, as well as in rapidly flowing shallow streams. When sampling water from deep water bodies and a single intake point is used, it should be located at about 60% of the stream depth, where complete mixing occurs. Samples from surface waters can be collected by

automatic sample devices. Depending on the device, samples can be collected at individual specified times or a composite sample can be accumulated over a specified time period (24 hours, usually). In some studies, manual collection could be made, making sure that the sampler entering the water approaches from downstream of the sample point. When representative depth-integrated water samples are to be collected, the methods of Nordin (11) and Meade (12) should be followed.

B. Sample Storage

Extensive environmental surveys require the analysis of a large number of samples. Once samples are collected, containers are shipped to a laboratory, where the rest of the analytical procedure is carried out. In order to avoid possible chemical and biochemical analyte alterations, field samples should be analyzed immediately after collection. Since it is impossible to do this for many environmental laboratories, serious problems of sample stability arise. A traditional way of preserving samples is that of placing them immediately after collection in insulated bags filled with ice, "blue" ice, or dry ice until arrival at the laboratory and then storing bottles in a refrigerator at 4°C. Hypochlorite in drinking water samples can continue to degrade pesticides by oxidation or chlorination reactions. To avoid this, thiosulfate should be added to samples. Transfer of a groundwater sample from an anaerobic ambient to an aerobic one may initiate biodegradation of some pesticides, which continues during transportation and storage. In this case, addition of biological inhibitors can prevent analyte biodegradation. During the National Pesticide Survey conducted by the EPA over a 2-year period, 1349 well water samples spiked with phenylureas were preserved by the addition of 10 mg/L HgCl$_2$ (13). Stability study results showed that no significant loss of fluometuron, diuron, linuron, and neburon occurred by storing HgCl$_2$-containing sample bottles at 4°C for 14 days.

C. Extraction

Before accomplishing aqueous sample extraction, one or more surrogates should be added to the sample. A surrogate analyte is defined by the EPA as "a pure analyte(s), which is extremely unlikely to be found in any sample aliquot in known amount(s) before extraction and is measured with the same procedure used to measure other sample components. The purpose of a surrogate analyte is to monitor method performance with each sample." Surrogates have an important role in as-

Table 2 Physicochemical Properties of Sulfonylurea Herbicides

Common name	IUPAC name, M.F.[a]	Molecular weight	pK_a	Water solubility (g/L, 25°C, pH 7)	DT_{50}[b] (days)	K_{oc}[c] (ml/g)
Bensulfuron-methyl	α(4,6-Dimethoxypyrimidin-2-ylcarbamoylsulfamoyl)-o-toluic acid $C_{16}H_{18}N_4O_7S$	410.4	5.2	0.12	28–140	n.f.[d]
Chlorsulfuron	1-(2-Chlorophenylsulfonyl)-3-(4-methoxy-6-methyl-1,3,5-triazin-2-yl)urea $C_{12}H_{12}ClN_5O_4S$	357.8	3.6	3.1	28–42	40
Metsulfuron-methyl	2-(4-Methoxy-6-methyl-1,3,5-triazin-2-ylcarbamoylsulfamoyl)benzoic acid $C_{14}H_{15}N_5O_6S$	381.4	3.3	2.8	7–35	35
Nicosulfuron	2-(4,6-Dimethoxypyridin-2-ylcarbamoyl-sulfamoyl)-N,N-dimethylnicotinamide $C_{15}H_{18}N_6O_6S$	410.4	4.6	12	26–67	n.f.
Primisulfuron-methyl	2-[4,6-Bis(difluoromethoxy)-2-pyrimidinyl]-amino]carbonyl]amino]sulfonyl]benzoic acid $C_{15}H_{12}F_4N_4O_7S$	468.3	3.5	0.24	4–29	n.f.
Rimsulfuron	1-(4,6-Dimethoxypyrimidin-2-yl)-3-(3-ethylsulfonyl-2-pyridylsulfonyl)urea $C_{14}H_{17}N_5O_7$	431.4	4.0	7.3	10–20	n.f.
Sulfometuron-methyl	2-(4,6-Dimethylpyrimidin-2-yl)benzoic acid $C_{15}H_{16}N_4O_5S$	364.4	5.2	0.24	ca 28	85
Thifensulfuron-methyl	3-(4-Methoxy-6-methyl-1,3,5-triazin-2-ylcarbamoylsulfamoyl)thiophen-2-carboxylic acid $C_{12}H_{13}N_5O_6S_2$	387.4	4.0	6.3	6–12	n.f.
Triasulfuron	1-[2-(2-Chloroethoxy)phenylsulfonyl]-3-(4-methoxy-6-methyl-1,3,5-2-yl)urea $C_{14}H_{16}ClN_5O_5S$	401.8	4.6	0.82	19	n.f.
Tribenuron-methyl	2-[4-Methoxy-6-methyl-1,3,5-triazin-2-yl-(methyl)carbamoylsulfamoyl]benzoic acid $C_{15}H_{17}N_5O_6S$	395.4	5.0	2.0	1–7	n.f.

Source: Ref. 4.
[a] M.F. = molecular formula.
[b] DT_{50} = time for 50% loss.
[c] K_{oc} = distribution coefficient (K_d) between soil and water adjusted for the proportion of organic carbon in water. It is a measure of the relative affinities of the pesticide for water and soil surface. As such, it indicates the tendency of a certain pesticide to leach through the soil and reach groundwaters. Roughly, K_{oc} values higher than 100 indicate a low potential leaching.
[d] n.f. = not found.

sessing the effectiveness of a sample preparation procedure. For drinking water analysis of phenyurea herbicides by GC techniques, a surrogate suggested by the EPA is 1,3-dimethyl-2-nitrobenzene.

Methods for the extraction of pesticides from water exploit the partitioning of analytes between the aqueous phase and a water-immiscible solvent (LLE) or an adsorbent material (SPE). The first method is still the most popular in many environmental laboratories. However, SPE is constantly increasing in popularity

because of its numerous advantages over LLE, which will be later illustrated.

1. Liquid–Liquid Extraction (LLE)

Among the solvents used for the extraction of phenylureas, dichloromethane is the most preferred, owing to its effectiveness of extracting compounds having a broad range of polarity. Solvent extraction is usually carried out in a separatory funnel, which is vigorously

Table 3 Toxicity Data and Tolerances in Drinking Water of Selected Phenylurea and Sulfonylurea Herbicides

Compound	Toxicity[a]		MAC[b] (ng/L) in drinking water	
	Rats (LD$_{50}$[c] mg/kg)	Rainbow trout (LC$_{50}$[d] (96 h), mg/L)	EC[e]	US
Phenylureas				
Chlorbromuron	>5,000	5.0	100	n.c.[f]
Chlorotoluron	>10,000	35		n.c.
Diuron	3,400	5.6		To be set
Fluometuron	>6,000	47		To be set
Isoproturon	2,420	37		n.c.
Linuron	4,000	3.2		To be set
Methabenziathiazuron	>2,500	16		n.c.
Metobromuron	2,623	36		n.c.
Metoxuron	3,200	19		n.c.
Monolinuron	2,215	56–75		n.c.
Neburon	>11,000	0.6–0.9		To be set
Sulfonylureas				
Bensulfuron-methyl	>5,000	>150		n.c.
Chlorsulfuron	>5,000	>250		n.c.
Metsulfuron-methyl	>5,000	>150		n.c.
Nicosulfuron	>5,000	>1,000		n.c.
Primisulfuron-methyl	>5,000	70		n.c.
Rimsulfuron	>5,000	>390		n.c.
Sulfometuron-methyl	>5,000	12.5		n.c.
Thifensulfuron-methyl	>5,000	>100		n.c.
Triasulfuron	>5,000	>100		n.c.
Tribenuron-methyl	>5,000	>1,000		n.c.

[a]*Source:* Ref. 4.
[b]MAC = maximum admissible concentration.
[c]LD$_{50}$ = dose required to kill 50% of the test organism.
[d]LC$_{50}$ = concentration required to kill 50% of the test organism.
[e]EC = European community.
[f]n.c. = not considered as a priority pollutant.

shaken to increase the contact area between the two liquids. This operation enhances the extraction rate and yield. The LLE technique has been involved in developing two official methods for analyzing phenylureas (14,15) and adopted for isolating sulfonylureas (16–18) from water samples. In al these methods, dichloromethane has been used as the extracting solvent. For efficiently extracting sulfonylureas, which are weakly acidic in nature, the pH of the water sample was adjusted in advance to 3.

Table 4 summarizes selected LLE-based sample preparation methods for urea herbicides.

The drawbacks of this technique are that it is labor intensive and time consuming. When performing trace analysis of pesticides, the extensive use of glassware may result in cumulative loss by adsorption on glass of hydrophobic pesticides. This technique requires the use of relatively large amounts of pesticide-grade solvents that are expensive as well as flammable and toxic. Even by using pesticide-grade solvents, the concentration step by a factor 1000 or more can introduce analyte interferences by residual solvent impurities. Vigorous shaking of solvent and water, especially surface water, may create serious problems of emulsions,

Table 4 Selected Liquid–Liquid Extraction Procedures for Extracting Urea Herbicides from Water Samples

Compound	Water volume	Solvent	Solvent exchange	Quantification technique	Recovery (%)	Ref.
Phenylureas	1 L	3 × 60 ml CH$_2$Cl$_2$	CH$_3$OH	LC-UV	89–94	15
Chlorsulfuron	1 L (pH 3)	3 × 70 ml CH$_2$Cl$_2$	Toluene	GC-ECD[a]	101–105	16
Sulfonylureas	0.5 L (pH 3)	100 ml CH$_2$Cl$_2$	Ethyl acetate	GC-ECD[a]	80–92	18

[a]ECD = electron capture detector.

owing to the presence in the sample of natural or synthetic surfactants. Emulsions can be eliminated only by additional time-consuming operations.

2. Solid-Phase Extraction (SPE)

Since the 1970s, as an alternative to LLE, the method of combined extraction and preconcentration of organic compounds in water by passing the sample through a short column of an adsorbing medium followed by desorption with a small quantity of an organic solvent has attracted the attention of many researchers. In the past decade, the availability of small-size particle (ca. 40 μm) adsorbents in inexpensive cartridges has largely contributed to the dramatic expansion of the SPE technique (Fig. 3). This technique appears especially appealing to researchers and analysts, and it is rapidly replacing LLE in official methods (19). Besides solving many problems associated with LLE, the SPE technique is particularly attractive because it lend itself to coupling with chromatographic systems for online applications.

Sample stability and storage space are problems that many environmental laboratories must address when collecting, storing, and analyzing water samples. With the large numbers of samples typical of environmental studies, the use of bulky glass bottles for sampling, transport, and storage also becomes a hindrance. One of the most impressive features of the SPE technique is that small adsorbent traps can be deployed in the field by using newly available submersible instrumentation. In this way, combined sampling, extraction, and preconcentration are done at the sampling site, thus eliminating most contamination and handling problems associated with sample collection. The small-volume trap could be sealed and shipped to the laboratory for elution and chromatographic analysis. Or it could be frozen in a small storage place, until analysis (20,21). Phenylurea herbicides extracted from a river water sample by means of an extraction cartridge filled with a sample of graphitized carbon black (GCB), namely, Carbograph 1, cartridge showed them to be stable on

this adsorbent for over 15 days of storage, even at ambient temperature (22).

a. Adsorbing Materials for Solid-Phase Extraction. Typical adsorbents for SPE are silica chemically modified with a C$_{18}$ alkyl chain, commonly referred to

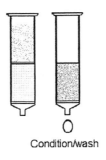

Condition/wash

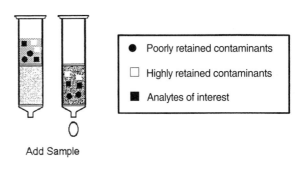

● Poorly retained contaminants
□ Highly retained contaminants
■ Analytes of interest

Add Sample

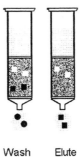

Wash   Elute

Fig. 3 Schematic of the off-line SPE process with cartridges.

as C-18, highly crosslinked styrene-divinylbenzene co-polymers (PS-DVB), commonly referred to as PRP-1, Envichrom P, or Lichrolut, and GCBs, commonly referred to as Carbopack, Envicarb, Carbograph 1, and Carbograph 4. All these materials are commercially available in medical-grade polypropylene housing and polyethylene frits. In spite of some limitations in extracting polar compounds from large water volumes, C-18 is still the most commonly used material, and it has been considered for introduction into official methods by European and American environmental agencies.

Recently, there has been a certain interest in developing and employing selective adsorbents based on analyte–antibody interactions achieved by immunosorbents. In the immunosorbent, the antibody is immobilized onto a silica support and used as an affinity ligand to extract selectively the target analyte from complex matrices. Taking advantage of the cross-reactivity of polyclonal antibodies, selective extraction can be achieved for a group of analytes having similar structures, such as a class of pesticides. Solid-phase extraction by immunosorbents containing either anti-isoproturon or antichlorotoluron antibodies succeeded in efficiently extracting many, but not all, of the phenylureas from spiked samples of river waters (23–25). The production of antibodies is, however, laborious, time consuming, and expensive. Furthermore, antisera obtained by different researchers may have different affinity and selectivity, making it difficult to standardize and implement such methods at present.

Very recently, Baker's yeast cells (*Saccharomices cerevisae*) were successfully immobilized onto silica gel and used in selective online trace enrichment of selected pesticides, including linuron, in various types of natural waters (26). This technique relies upon the fact that microorganisms are able to absorb pesticides from water in the environment. Cell membranes contain many classes of lipids and lipoproteins that contribute to pesticide absorption. Since the diffusion rate across membranes is inversely proportional to molecular size, low-molecular-weight compounds, such as pesticides, can be extracted from water and isolated by naturally occurring high-molecular-weight substances, such as humic acid, that are abundantly present in environmental waters.

b. Off-Line Solid-Phase Extraction with Cartridges. Off-line SPE of analytes from water samples with cartridges is commonly accomplished by attaching the cartridge to the outlet of a separatory funnel containing the sample. More simply, water can be transferred from the sample bottle to the cartridge by attaching it to Teflon tubing put in the water sample.

Water is then forced to pass through the cartridge by vacuum created by a water pump. Before pumping water, the cartridge is first washed with the eluent phase, to eliminate possible contaminants, and then with distilled water. After the water sample is passed through, the cartridge is washed with a little distilled water. Following this passage, water contained in the cartridge is in part eliminated by decreasing the pressure in the extraction apparatus. Analyte desorption is accomplished by flowing slowly 4–8 ml of a suitable solvent or a solvent mixture through the adsorbent bed and collecting the eluate in a vial. This is placed in a water bath at 25–50°C (depending on the analyte volatility), and, via a gentle stream of nitrogen, the extract is concentrated down to dryness, or to 50–100 μl if analytes are rather volatile.

Depending upon the type of adsorbent and the final destination of the extract, various solvents or solvent mixtures are used to re-extract pesticides from adsorbent cartridges. With both C-18 and PS-DVB materials, methanol or acetonitrile is the eluent of choice, when analyzing via liquid chromatography (LC) instrumentation. With C-18 cartridges and gas chromatography (GC) instrumentation, ethyl acetate is preferred, although methylene chloride was chosen in EPA Method 525 for eluting nonpolar and medium polar compounds (19). With GCB cartridges, a CH_2Cl_2/CH_3OH (80:20, v/v) mixture offers quantitative desorption of neutral pesticides having a broad range of polarity. For eluting acidic analytes, which are not desorbed by the foregoing solution, CH_2Cl_2/CH_3OH (80:20, v/v) acidified with trifluoroacetic acid (TFA), 10 mmol/L, can be used. When analyzing weakly acidic pesticides, such as sulfonylureas, in humic acid–rich aqueous environmental samples, more selective elution can be obtained by replacing TFA with a weakly acidic agent, such as acetic acid (27). By doing so, a large fraction of humic acids coextracted with weakly acidic pesticides are not re-extracted from the GCB cartridge and do not interfere with the analysis (Fig. 4).

Solid-phase extraction cartridges filled with C-18 material followed by GC (28) or LC-MS (29,30) analysis have been adopted for the off-line extraction of some phenylurea herbicides from drinking water. Carbograph 1 cartridges proved to be a valuable material for quantitatively extracting phenylureas from large volumes of tap water (22,31). For analyzing phenylurea herbicides, two methods involving the use of Carbograph 4 cartridges have recently been developed (32,33). With this material, quantitative extraction of several phenylurea herbicides for 4 L drinking water, 2 L groundwater, and 0.5 L river water has been

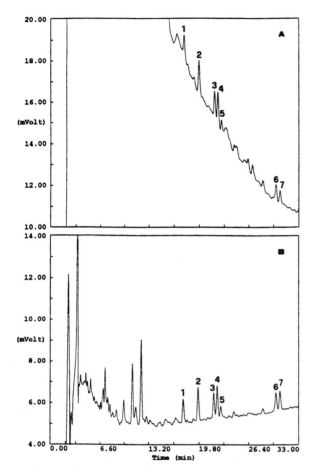

Fig. 4 LC-UV chromatograms obtained by extracting seven sulfonylureas added at the level of 2 μg/L to 0.2 L of a simulated surface water sample and reextracting with a CH₂Cl₂/CH₃OH mixture (80:20, v/v) acidified with (A) 10 mmol/L trifluoroacetic acid and (B) 10 mmol/L acetic acid. *Column*: Alltima 250 × 4.6-mm ID containing 5-μm C-18. *Elution*: CH₃CN/H₂O (both containing 3 mmol/L trifluoroacetic acid) linear gradient elution from 32:68 to 62:38 (v/v) in 40 min. *Peak numbering*: 1, thifensulfuron methyl; 2, metsulfuron methyl; 3, triasulfuron methyl; 4, chlorsulfuron; 5, rimsulfuron; 6, bensulfuron methyl; 7, tribenuron methyl. (From Ref. 27.)

achieved. With respect to a C-18 cartridge, Carbographs had a far better extraction efficiency for polar phenylureas (22). Carbograph 1 cartridges have been involved in two multiresidue methods (MRMs) for analyzing pesticides, including phenylureas, from large volumes of natural waters (34,35). Phenylureas together with other nonacidic pesticides were in both cases isolated from acidic ones by differential elution.

Off-line SPE with C-18 cartridges has been adopted for analyzing various sulfonylureas in water samples (36–40).

Solid-phase extraction with polymeric material (SDB) has been used for extracting some sulfonylureas from 250 ml of water samples from various sources (41). Recovery of three analytes out of four were better than 85%, while recovery of tribenuron-methyl was about 75%. In both cases, the pH of the water samples was adjusted to 3, before analyte extraction.

C-18 cartridges associated with combined GC and LC techniques have been involved in a procedure to monitor three phenylureas and one sulfonylurea in small lochs surrounded by different crops (42).

By a means of a Carbograph 4 cartridge and within a single step, seven commonly used sulfonylureas were extracted from large volumes of various types of water and isolated from coextracted nonacidic compounds and humic acids (27). Before extraction, no adjustment of the water pH was needed.

c. *Off-Line Solid-Phase Extraction with Adsorbents Imbedded in Membranes.* In recent years, commercially available filter disks containing both C-18 and PS-DVB materials with particle sizes finer than those used with cartridges (8 μm against 40 μm) imbedded in a Teflon matrix have been used for both off- and online SPE of pesticides. The specific advantages claimed for disk design over cartridge design are as follows: shorter sampling flow rate due to faster mass transfer and lack of channeling effects; decreased plugging by particulate matter due to the large cross-sectional area; cleaner background interferences. The last advantage derives from the fact that, unlike SPE with cartridges, the extraction apparatus with disks consists of glass. The use of an extraction disk is particularly simple. The membrane is placed in a filtration apparatus connected to a vacuum source by a water pump. After the disk has been washed/conditioned with 10 ml of methanol and 10 ml of distilled water, the aqueous sample is passed through the disk (Fig. 5). After eliminating part of the water by vacuum, the assembly supporting the disk is transferred to a second vacuum flask containing a vial. Then, 5–10 ml of the eluent phase, usually methanol or acetonitrile, is slowly drawn onto the membrane by moderate vacuum. The vacuum is interrupted for 2–4 min to allow the liquid to soak the membrane. Thereafter, analytes are eluted and collected into the vial. This operation is repeated by applying another 5-ml aliquot of the eluent phase to the top of the disk.

C-18 Empore extraction disks were used for the isolation and trace enrichment of linuron in 4 L of river water (43). At the 0.25-μg/L level, recovery of this phenylurea was 94% with 6% coefficient of variation.

A procedure based on C-18 extraction disks and GC after derivatization has been elaborated for analyzing traces of chlorsulfuron and metsulfuron methyl in water (44). A combination of the SPE disk technology for extracting analytes in water and supercritical fluid extraction technology for re-extracting analytes from disks has been applied to the analysis of sulfuron methyl and chlorsulfuron (45).

Online Solid-Phase Extraction. With the off-line SPE technique, a certain skill and care is required of the analyst. Moreover, rapid screening procedures of many samples for monitoring pesticides requires analysis automation. In recent years, fully automated analysis of contaminants in water by the online coupling of SPE to LC or GC instrumentation has received increasing attention. Besides allowing rapid analysis, additional positive features of online SPE are that analyte loss due to evaporation does not occur and that all of the sample is introduced into the chromatographic instrumentation, instead of a fraction as with off-line procedures. In this way, the sampled volume can be drastically reduced, thus lowering the costs of cooled sample transportation and storage. With online SPE, the water sample is pumped through a short precolumn (typically 10-mm length $\times$ 2-mm ID) filled with small particles (15–25 μm) of either C-18 or PS-DVB adsorbing media. Solutes are trapped, while water is wasted. Eventually, the precolumn can be washed with a small volume of a water/methanol mixture. By a system of switching valves, the solutes are then removed from the precolumn by the LC mobile phase itself and transported into the LC column (Fig. 6). When using a precolumn packed with an adsorbent having a larger affinity for analytes than that filling the analytical column, broad peaks for the last-eluted analytes are obtained. In this case, analyte backward-elution from the precolumn with the LC mobile phase can eliminate peak broadening.

Precolumn technology in LC has been applied to the determination of phenylureas in the presence of their anilines (46). These latter were removed from the water sample by inserting a platinum phase in a short precolumn upstream from the C-18 precolumn. A fully automated sample handling system has been developed for LC that combines the advantages of a disposable cartridge system and precolumn technology with its high automation potential. As an example, this device was applied to the analysis of some phenylureas in river water (47).

A precolumn packed with a PS-DVB copolymer or

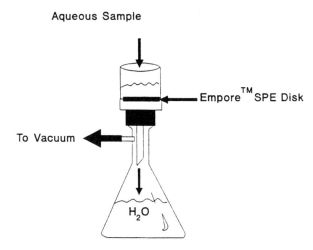

Fig. 5 Schematic of an off-line SPE device with membranes. (From Ref. 45.)

a stack of eight 4.6-mm-diameter C-18-loaded membrane extraction disks have been coupled to LC-MS for trace analysis of 15 phenylureas in surface water and drinking water (48). By sampling only 50 ml water, limits of detection ranged between 5 and 20 ng/L. For analyzing a test mixture of 17 pesticides, which included three phenylureas, in tap water and river water, the analytes were preconcentrated from 15-ml samples on a short analytical LC column and then gradient-eluted into a LC-MS apparatus (49). Automated online trace enrichment, LC analysis with diode array detection were investigated for the determination of widely used pesticides (19 phenylureas were included in this

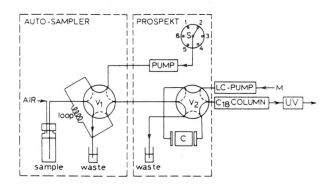

Fig. 6 Design of the automated sample handling (PROSPEKT) system. V_1, injection valve; V_2, high-pressure switching valve; S, six-port solvent selection valve; M, mobile phase; C, 10 $\times$ 2-mm ID cartridge packed with 40 μm C$_8$ silica. (From Ref. 47.)

list) in environmental waters (50). Detection limits of 0.1 μg/L were obtained using 150 ml of river waters.

For the purpose of developing a single-class MRM involving online SPE in combination with LC-MS for sulfonylureas in aqueous samples, eight sulfonylureas were selected (51).

e. Online Extraction with Liquid Membranes. Sample preparation by means of liquid membrane extraction is a technique that in essence contains two LLE extractions in one step. The setup is easily automated, and sample preparation is performed in a closed system, thus minimizing the risk for contamination and losses during the process. Because the extraction is made from an aqueous phase (donor) to a second, also aqueous phase (acceptor), further enrichment on a precolumn is possible before injection into the LC apparatus. Liquid membranes were used for enrichment of metsulfuron-methyl and chlorsulfuron from clean aqueous samples (52) and natural waters (53).

f. Solid-Phase Microextraction. A new technique for extracting analytes from water is emerging, the so-called solid-phase microextraction (SPME). A 0.5–1-mm-ID uncoated fiber or coated with suitable immobilized liquid phase (in the second case this technique should be more correctly called liquid-phase microextraction) is immersed in a continuously stirred water sample. After equilibrium is reached (a good exposure time takes 15–25 min), the fiber is introduced into the injection port of a gas chromatograph, where analytes are thermally desorbed and analyzed. Positive features of this technique are that it is rapid and very simple and does not use any solvent. In addition, like online SPE, this technique requires small sample volumes (2–5 ml), because of all the sample extract is injected into the analytical column.

Recently, SPME has been successfully used coupled to LC (Fig. 7). An automated in-tube SPME has been proposed for analyzing six phenylureas in aqueous samples (54).

Table 5 summarizes the use of various extraction procedures for assaying phenylurea and sulfonylurea herbicides in environmental waters.

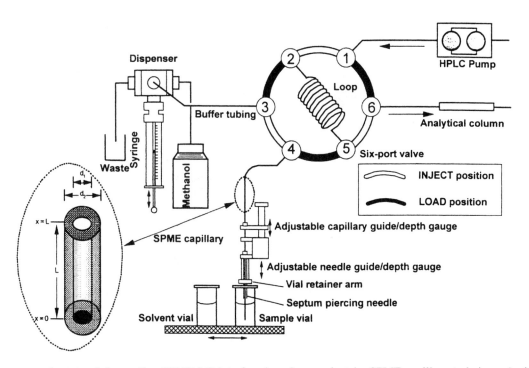

Fig. 7 Instrumental setup of the on-line SPME-LC interface based on an in-tube SPME capillary technique. A piece of GC column (in-tube SPME) hosts in the position of the former needle capillary. The aqueous sample is frequently aspirated from the sample vial through the GC column and dispensed back to the vial (INJECT position) by movement of the syringe. After the extraction step, the six-port valve is switched to the LOAD position for the desorption of the analytes from the in-tube SPME by flushing 100% methanol from another vial through the SPME capillary. The volume is transferred to the loop. After switching the Valco valve to the INJECT position, an isocratic separation using a mixture of 60:40 acetonitrile/water was performed. A detailed view of the in-tube SPME capillary is included at the left side of the figure. (From Ref. 54.)

Table 5 Solid-Phase Extraction of Phenylurea and Sulfonylurea Herbicides from Water Samples

Compound	Sample	Mode	Sorbent	Eluent phase	Ref.
14 Phenylureas	2 L DW,^{a} 2 L GW,^{b} 1 L RW^c	Off-line	Carbograph 1, 0.25 g	6 ml CH$_2$Cl$_2$/CH$_3$OH (95:5)	22
Isoproturon, chlortoluron, linuron, diuron	1 L DW	—	C-18, 1 g	6 ml CH$_3$OH	28
Diuron, linuron, monuron	0.1 L DW	—	C-18, 1 g	5 ml CH$_3$OH	29
Fenuron, metoxuron, in MRM^d	2 L DW	—	Carbograph 1, 1 g	6 ml CH$_2$Cl$_2$/CH$_3$OH (80:20)	31
7 Phenylureas in MRM	4 L DW, 2 L GW, 0.5 L RW	—	Carbograph 4, 0.5 g	1.5 ml CH$_3$OH, then 8 ml CH$_2$Cl$_2$/ CH$_3$OH (90:10)	32
9 Phenylureas and related chloroanilines	4 L DW, 2 L GW, 0.5 L RW	—	Carbograph 4, 0.5 g	1.5 ml CH$_3$OH, then 6 ml CH$_2$Cl$_2$/ CH$_3$OH (80:20) acidified with 10 mM HCl	33
5 Phenylureas in MRM	2 L DW	—	Carbograph 1, 0.25 g	6 ml CH$_2$Cl$_2$/CH$_3$OH (80:20)	34
13 Phenylureas in MRM	2 L DW, 1 L GW	—	Carbograph 1, 0.5 g	5 ml CH$_2$Cl$_2$/CH$_3$OH (90:10)	35
Linuron in MRM	4 L RW, 4 L simulated SW^e	—	C-18 disks	20 ml CH$_3$OH	43
Chlorsulfuron, metsulfuron methyl	1 L DW (pH 2)	—	C-18 disks	20 ml ethylacetate	44
Chlorsulfuron, sulfometuron methyl	1 L PW^f	—	C-18 disks	2% CH$_3$OH- modified CO$_2$	45
16 Phenylureas	0.2 L DW, 0.2 L RW	—	Anti-isoproturon antibody on silica support	4 ml CH$_3$OH/H$_2$O (70:30)	23
12 Sulfonylureas	0.5 L MW^g (pH 3)	—	RP-102, 0.5 g	10 ml CH$_3$OH	40
15 Phenylureas	10 ml RW	Online	Pt-phase + C-18	CH$_3$OH/H$_2$O (30:70)	46
5 Phenylureas	10 ml RW	—	C-18	CH$_3$OH/20 mM phosphate buffer (45:55)	47
15 Phenylureas	50 mL RW	—	PLRP-S or several C-18 disks	CH$_3$OH/0.1 M AcNH$_4$ (gradient elution in the backflush mode)	48
Monuron, diuron, neburon in MRM	0.1 L DW, 0.1 L RW	—	PLRP-S	CH$_3$CN/H$_2$O, gradient elution	49
16 Phenylureas in MRM	0.15 L RW	—	PLRP-S or C-18	CH$_3$CN/phosphate buffer, gradient elution	50
Linuron in MRM	25 ml DW, GW, SW	—	Yeast immobilized on silica	CH$_3$CN/phosphate buffer (gradient elution in the backflush mode)	26
8 Sulfonylureas	0.1 L DW (pH 3)	—	C-18	CH$_3$OH/0.1 M acetic acid, gradient elution	51
16 Phenylureas	50 ml SW	—	Antichortoluron antibody on silica support	CH$_3$CN/phosphate buffer, gradient elution	25

Table 5   Continued

Compound	Sample	Mode	Sorbent	Eluent phase	Ref.
16 Phenylureas	20 ml DW, 20 ml RW	—	Antichortoluron antibody on silica support	CH₃CN/0.5% acetic acid	24
6 Phenylureas	25 μl PW	—	SPME with a glass capillary coated with Carbowax	38 μl CH₃OH	54

[a]DW = drinking water.
[b]GW = groundwater.
[c]RW = river water.
[d]MRM = multiresidue method.
[e]SW = seawater.
[f]PW = pure water.
[g]MW = marsh water.

D.   Separation and Detection Methods

1.   Gas Chromatography

a.   Phenylurea Herbicides.   For determining phenylureas in water, the GC technique with the use of selective detectors, such as the electron capture (ECD), nitrogen phosphorous, and, chiefly, mass spectrometry (MS) detectors, has attracted the attention of many searchers. Direct GC analysis of phenylurea herbicides has been reported in the literature (55–61). However, it also has been shown that a significant number of these compounds cannot be analyzed in this way owing to their thermal instability. They partly decompose into isocyanates and amines, the main contributory factor being the NH moiety. Some methods have relied on quantification of the degradation product formed into the injection port (55,63–65).

To make phenylureas amenable to GC analysis, various derivatization procedures have been elaborated. It has to be pointed out that all these reactions consist in substitution of the free hydrogen attached to the nitrogen atom close to the aromatic moiety by different groups. These procedures can be grouped into four reaction classes:

1. *Direct acylation.* The reaction most frequently carried out is perfluoroacylation, by reacting the analyte with trifluoroacetic anhydride (TFAA) or heptafluorobutyric anhydride (HFBA) (66–70). These derivatizaion agents are chosen for exploiting sensitive electron capture (ECD) detection (Fig. 8).
2. *Indirect acylation.* Here, phenylureas are first converted to their anilines, and then the later are reacted with the aforementioned derivatization

agents (71–76). Aniline derivatives are more readily formed than the corresponding phenylureas, but derivative preparation is time consuming. Figure 9 shows a chromatogram obtained by following this derivatization procedure.

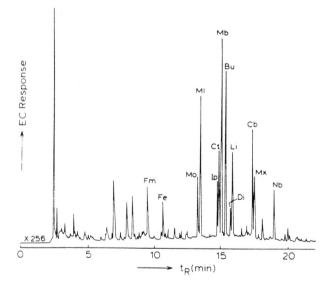

Fig. 8   Capillary GC with fused-silica column coated with CP-Sil 5 (analogous to SE 30 and OV-101) of HFB derivatives of 13 phenylureas obtained after extraction of a Bosbaan river water sample spiked at the 1-μg/L level, and direct derivatization with HFBA. Injected amount corresponds to 0.1 ng of each herbicide. *Symbol explanation*: Fm, fluometuron; Fe, fenuron; Mo, monuron; Ml, monolinuron; Ip, isoproturon; Ct, Chlorotoluron; Mb, metobromuron; Bu, buturon; Di, diuron; Li, linuron; Cb, chlorbromuron; Mx, metoxuron; Nb, neburon. (From Ref. 69.)

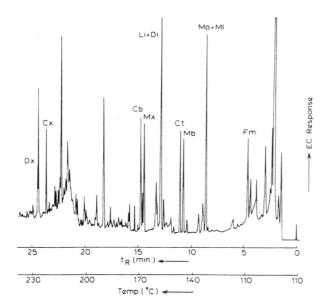

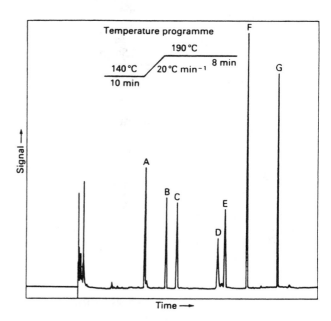

Fig. 9 Capillary GC with fused-silica column coated with CP-Sil 5 (analogous to SE 30 and OV-101) of HFB anilines obtained after extraction of a pure water sample spiked with 1 μg/L parent herbicides, and subsequent hydrolysis and derivatization. Stationary phase: CP-Sil 5. *Symbol explanation:* Fm, fluometuron; Mo, monuron; Ml, monolinuron; Mb, metobromuron; Ct, chlorotoluron; Di, diuron; Li, linuron; Mx, metoxuron; Cb, chlorbromuron; Cx, chloroxuron; Dx, difenoxuron. (From Ref. 75.)

Fig. 10 GC with a fused-silica capillary column and an NPD detector of some urea herbicides after conversion to their methylated forms. *Symbol explanation:* A, monuron; B, isoproturon; C, chlorotoluron; D, linuron; E, diuron; F, methabenzthiazuron; G, tebuthiuron. (From Ref. 28.)

3. *Alkylation.* The reagents most frequently used are trimethylaniline hydroxide (TMAH) (77,78) and alkyl iodide (28,79–85). The reaction with TMAH can be carried out on column by direct injection of a mixture of the phenylureas and the reagent in methanol. Derivatization with alkyl iodide can distinguish between derivatives of the parent phenylurea herbicide and those of the *N*-dimethyl metabolites if ethyl iodide is used in place of methyl iodide as alkylating agent (84,85). Figure 10 shows a typical GC chromatogram of some phenylureas after their conversion to alkylated species.

4. *Sylilation.* Sylilating reagents are commonly used to block polar groups. One paper has described preparation of phenylurea herbicides by various sylilating reagents (86). This procedure is not often followed because the sensitivity and selectivity of the ECD detector cannot be exploited.

b. *Sulfonylureas.* Two GC methods have been elaborated for analyzing chlorsulfuron and metsulfuron methyl in water (18,44). Sulfonylureas are even less

volatile and more thermally labile than phenylureas, so they need to be converted to more volatile compounds before GC analysis. Diazomethane has been used to convert the two sulfonylureas to their stable *N,N'*-dimethyl derivatives (Fig. 11).

Table 6 summarizes selected methods for analyzing urea herbicides in water by the GC technique.

2. Supercritical Fluid Chromatography (SFC)

The SFC technique is closely related to LC, but it is three to five times faster, allows more rapid generation of high efficiency, and can be used with both typical GC and LC detectors, simultaneously. Higher efficiency coupled with multiple detection simplify the task of resolving complex mixtures, such as in screening for the presence or absence of a large number of target compounds without the expense of a mass spectrometer.

Supercritical fluid chromatography coupled to online SPE has been proposed for monitoring four sulfonylureas in natural waters (Fig. 12) with detection limits as low as 50 ng/L (87).

3. Liquid Chromatography

Liquid chromatographic systems for environmental pesticide analysis has been extensively reviewed in two

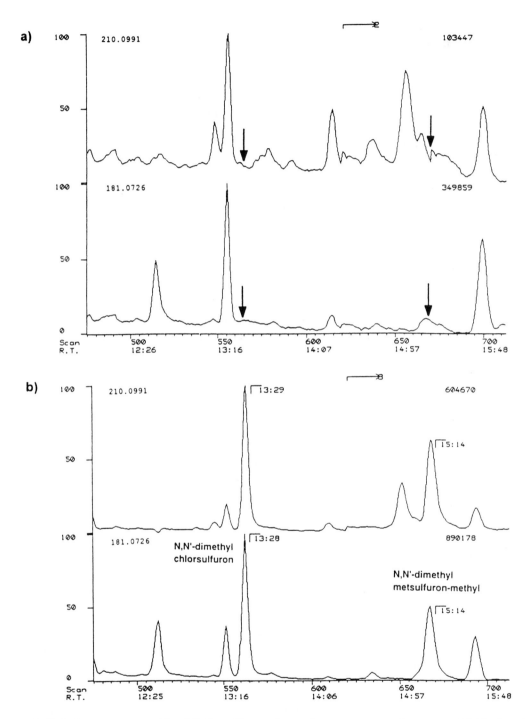

Fig. 11 GC-MS SIM chromatograms with a HP-5 capillary column of pure water sample extracts, (a) blank and (b) spiked with 0.1 μg/L of chlorsulfuron and metsulfuron methyl. (From Ref. 44.)

Table 6 Selected Capillary Column GC Methods for Determining Urea Herbicides in Water Samples

Compound	Derivatizing agent	Column characteristics	Injection device	Detector	LOD	Refs.
15 Phenylureas	HFBA[a]	CP-Sil 5 on 25 m × 0.22 mm	Splitless	ECD,[b] MS[c]	1 pg	69, 75, 76
Isoproturon, chlorotoluron, linuron, diuron	Iodomethane	BP1 on 25 m × 0.22 mm, 0.25 μm film thickness	Split	NPD,[d] MS	<0.1 μg/L	28
Metsulfuron-methyl, chlorsulfuron	Hydrolysis	DB-17 on Megabore column	On-column	ECD	0.1 μg/L	18
Metsulfuron-methyl, chlorsulfuron	Diazomethane	HP-5 on 25 m × 0.22 mm, 0.33-μm film thickness	Splitless	ECD, NPD	<0.1 μg/L	44

[a] HFBA = heptafluorobutyric anhydride.
[b] ECD = electron capture detector.
[c] MS = mass spectrometry.
[d] NPD = nitrogen-phosphorous detector.

previous papers (88,89). The increasing use of LC methods for pesticides is chiefly the result of their suitability for thermally labile and polar pesticides, such as phenylurea and sulfonylurea herbicides, which require derivatization prior to GC analysis. Such LC methods of analysis also have the important advantage over GC methods in that online pre- and postcolumn reaction systems are compatible with LC instrumentation. Furthermore, the LC apparatus can easily be coupled on line with the enrichment step using SPE on

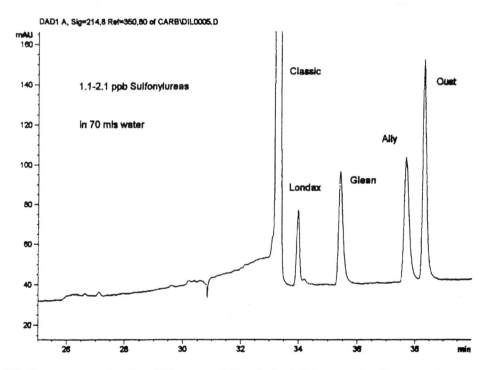

Fig. 12 SFC-UV chromatogram of online SPE extract of 70 ml of 1.1–2.1 μg/L of sulfonylureas in water. *Column*: eight standard LC columns, each 200 × 4.6-mm ID with 5-μm silica particles. The first column was C-18, while the rest were Hypersil silica. The mobile phase was 1% CH_3OH in CO_2. After a 4-min hold, CH_3OH was programmed to 16% at 0.5%/min, then held. The temperature was 60°C. (From Ref. 87.)

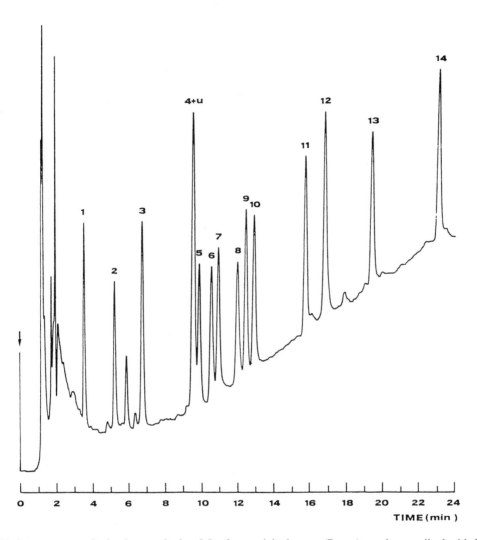

Fig. 13 LC-UV chromatogram obtained on analyzing 2 L of a municipal water (Rome) specimen spiked with 30 ng/L of each phenylurea. *Column*: 250 × 4.6-mm ID packed with 5-μm C-18. *Elution*: $CH_3OH–CH_3CN$ (85:15)/H_2O linear gradient elution from 47% organic modifier to 70% in 20 min. *Peak numbering*: 1, fenuron; 2, metoxuron; 3, monuron; 4, monolinuron; 5, fluometuron; 6, chlorotoluron; 7, metobromuron; 8, difenoxuron; 9, isoproturon; 10, diuron; 11, linuron; 12, chlorbromuron; 13, chloroxuron; 14, neburon; u = unknown compound. (From Ref. 22.)

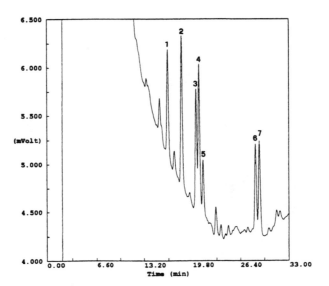

Fig. 14 LC-UV chromatogram obtained by analyzing 4 L of drinking water spiked with seven sulfonylureas at the individual level of 10 ng/L. *Column*: Alltima 250 × 4.6-mm ID containing 5-μm C-18. *Elution*: CH$_3$CN/H$_2$O (both containing 3 mmol/ L trifluoroacetic acid) linear gradient elution from 32:68 to 62:38 (v/v) in 40 min. *Peak numbering*: 1, thifensulfuron methyl; 2, metsulfuron methyl; 3, triasulfuron methyl; 4, chlorsulfuron; 5, rimsulfuron; 6, bensulfuron methyl; 7, tribenuron methyl. (From Ref. 27.)

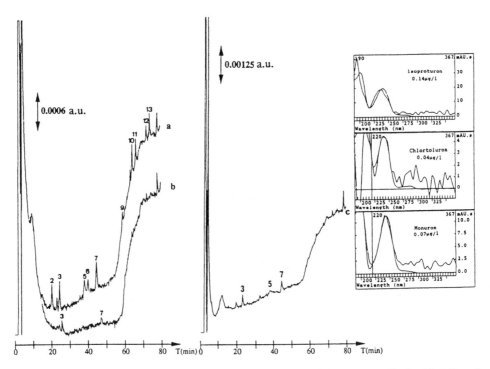

Fig. 15 LC-DAD analysis after online preconcentration of 50 mL of River Seine water spiked with 0.1 μg/L with (a) phenylureas and (b) nonspiked on a precolumn containing 0.22 g of silica bonded with antichlorotoluron antibodies, and (c) identification of three compounds in the nonspiked river water. Detection was performed at 244 nm. *Column*: 250 × 4.6-mm ID packed with Baker Narrow Pore C-18. *Peak numbering*: 1, fenuron; 2, metoxuron; 3, monuron; 4, methabenzthiazuron; 5, chlorotoluron; 6, fluometuron; 7, isoproturon; 8, difenoxuron; 9, buturon; 10, linuron; 11, chlorbromuron; 12, diflubenzuron; 13, neburon. (From Ref. 24.)

Table 7 Liquid Chromatographic Methods with UV Detection for Determining Urea Herbicides in Water Samples

Compound	Column	Mobile phase	Detector	LOD (ng/L)	Ref.
14 Phenylureas	C-18 (5-μm) in 25 cm × 4.6 mm	H$_2$O/CH$_3$OH–CH$_3$CN (85:15), gradient elution	UV—250 mm	1	22
5 Phenylureas	C-18 (5-μm) in 25 cm × 4.6 mm	H$_2$O/CH$_3$CN, gradient elution	DAD^a/UV detection at 244 nm	100	23
11 Phenylureas	C-18 (5-μm) in 25 cm × 4.6 mm	H$_2$O/CH$_3$CN, gradient elution	DAD/UV detection at 220 nm	ca. 100	24
Linuron	C-18 (5-μm) in 10 cm × 4.6 mm	2 mM KH$_2$PO$_4$/CH$_3$CN, gradient elution	DAD	20–80	26
Fenuron, metoxuron	C-18 (5-μm) in 25 cm × 4.6 mm	H$_2$O + 2% CH$_3$OH/CH$_3$CN, gradient elution	UV—210 nm	3–20	31
Metoxuron, monuron, monolinuron, linuron, chlorouxoron	C-18 (5-μm) in 25 cm × 4.6 mm	H$_2$O + 0.05% TFA/CH$_3$CN + 0.025% TFA, gradient elution	UV—225 nm	30–60	34
12 Phenylureas	C-18 (5-μm) in 25 cm × 4.6 mm	1 mM p.b.^{b} (pH 7)/CH$_3$CN, gradient elution	UV—220 nm	3–15	35
12 Phenylureas	C-18 (8 μm) in 25 cm × 4.6 mm	H$_2$O/CH$_3$OH 40:60, isocratic elution	UV—245 nm	Few μg/L	46
5 Phenylureas	C-18 (5 μm) in 20 cm × 4.6 mm	20 mM p.b. (pH 7)/CH$_3$OH (55:45), isocratic elution	UV—245 nm	n.r.^{c}	47
Methabenzthiazuron, linuron, isoproturon	C-18 (5-μm) in 25 cm × 4.6 mm	50 mM p.b. (pH 7)/CH$_3$CN, gradient elution	DAD/UV detection at 220 nm	ca. 100	50
6 Phenylureas	C-18 (4 μm) in 10 cm × 10 mm	H$_2$O/CH$_3$CN (40:60), isocratic elution	UV—245 nm	2800–4100	54
Chlorotoluron, diuron, isoproturon	C-18 (5-μm) in 15 cm × 4.6 mm	H$_2$O/CH$_3$CN, gradient elution	DAD/UV detection at 245 nm	2–23	41
7 Sulfonylureas	C-18 (5-μm) in 25 cm × 4.6 mm	H$_2$O/CH$_3$CN, both containing 3 mM TFA, gradient elution	UV—230 nm	0.6–2 in drinking water	27
Thifensulfuron, metsulfuron, chlorosulfuron, tribenuron	C-18 (5-μm) in 15 cm × 3.9 mm	20 mM p.b. (pH 3.4)/CH$_3$CN, gradient elution	DAD/UV detection at 225 and 262 nm	<100	42
Sulfometuron, chlorsulfuron	C-1 (5-μm) in 25 cm × 4.6 mm	H$_2$O/CH$_3$CN (75:25) + H$_3$PO$_4$ (pH 3), isocratic elution	UV—230 nm	n.r.	45
Thifensulfuron, metsulfuron, chlorsulfuron, tribenuron	C-18 (3-μm) in 15 cm × 2.1 mm	H$_2$O/CH$_3$CN (70:30) + ACH (pH 3), isocratic elution	UV—240 nm	50–100	53

^{a}DAD = diode array detector.
^{b}p.b. = phosphate buffer.
^{c}n.r. = not reported.

precolumns, thereby making the analysis fully automated.

Many LC methods involving the use of UV (17,18,22,27,31,34,35,39,45–47,52–54,90), diode array (23,24,26,41,42,50), electrochemical (91), and photoconductivity (38) detectors have been developed for analyzing phenylurea and sulfonylurea herbicides in water samples. As examples, Figs. 13–15 show LC chromatograms obtained by injecting extracts of real water samples spiked with trace amounts of urea herbicides.

In Table 7, selected LC methods for assaying urea pesticides in water are listed.

4. Capillary Electrophoresis

Electrophoresis is a process in which charged species are separated according to differences in their electrophoretic mobilities, and these are related to their charge densities. In the mid-1980s, instruments able to fractionate charged analytes into a capillary column were introduced. This new technique is called capillary electrophoresis (CE) or capillary zone electrophoresis (CZE). The separation is usually carried out in a short capillary fused silica filled with a buffer solution. Typically, capillary columns are 25–100 cm in length with ID ranging between 25 and 100 μm. Electrodes are usually platinum and are connected to a power supply able to provide constant voltages up to 30 kV and currents up to 100 μA. A particular characteristic of the electro-osmotic flow is that the profile of the liquid front is practically flat, instead of being parabolic, because it occurs when a liquid is forced to pass through a tube by hydrodynamic pressure. This effect, coupled to the absence of any resistance to the mass transfer, enables CE to separate compounds in 10 min with an efficiency of more than 200,000 plates. Extremely

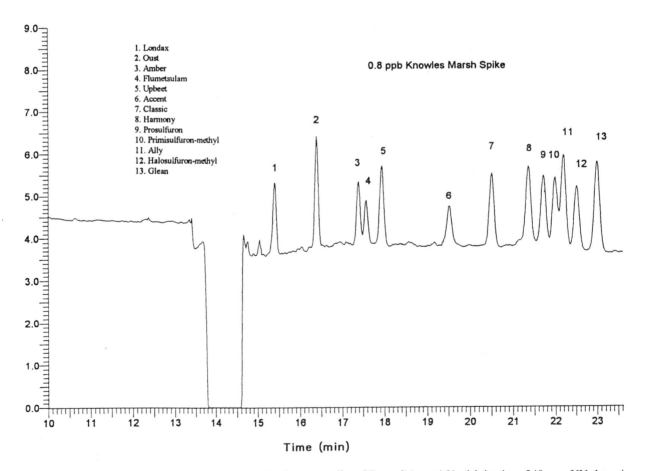

Fig. 16   Electropherogram of 0.8 μg/L Knowles Marsh water spike. *CE conditions*: 161-nl injection; 240 nm–UV detection; mobile phase: 50 mmol/L ammonium acetate at pH 4.75, with 12% acetonitrile added to inlet buffer vial; 30 kV, 30 μA, 30°C. *Capillary*: length 122 cm (100 cm effective length) × 75-μm ID bare fused silica high-sensitivity optical cell. (From Ref. 40.)

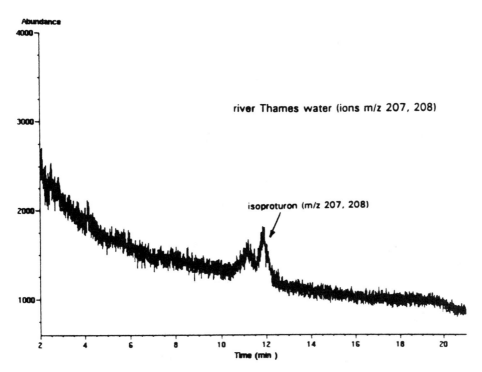

Fig. 17 Online SPE trace-enrichment LC-TS-MS chromatogram of River Thames water (ions m/z 207 and 208 monitored) using a PLRP-S precolumn. *Column*: 150 × 4.6-mm-ID column containing 5-μm Rosil C-18 bonded silica. *Elution*: CH$_3$OH/ 0.1 mol/L ammonium acetate, linear gradient elution from 40:60 to 80:20 (v/v) in 20 min. The contamination level of isoproturon in the river water was estimated to be a few nanograms per liter. (From Ref. 48.)

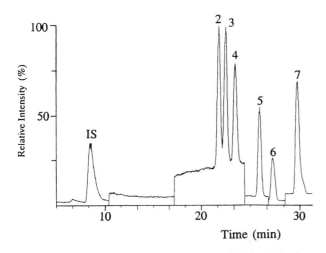

Fig. 18 Online SPE followed by LC-TS-MS SIM chromatogram for 100 ml of a drinking water sample spiked with seven sulfonylureas at the 1-μg/L level. *Column*: 125 × 3-mm-ID Lichrocart cartridge packed with 5-μm LiChrospher 60 RP. *Elution*: CH$_3$OH/0.1 mol/L acetic acid, linear gradient elution from 20:80 to 95:5 in 45 min. *Peak numbering*: 2, thifensulfuron methyl; 3, metsulfuron methyl; 4, chlorsulfuron; 5, tribenuron methyl; 6, bensulfuron methyl; 7, chlorimuron methyl. IS = caffeine (internal standard). (From Ref. 51.)

sharp peaks for the analytes also reflect that CE instruments equipped with UV detectors are able to detect analyte quantities as low as 0.2 pg. On the other hand, only a few nanoliters of a sample volume can be injected into the capillary without affecting the electrophoretic process. This results in method detection limits of several hundreds of ppb, which are too high for practical environmental applications. Several techniques have been reported for on-column concentration to enhance detection in CZE. Among these, the field-amplified technique seems to offer the best possibilities, in terms of sensitivity. By this expedient, a 10-fold analyte concentration can be reached, provided the sample volume occupies only a small section of the capillary.

Sulfonylureas are ionogenic compounds, and thus they lend themselves to analysis by CE. Some CE procedures involving UV detection have been developed for analyzing sulfonylureas extracted from natural waters (40,92–94). A typical electropherogram of sulfonylureas is shown in Fig. 16. One study has described the use of CE coupled with MS for the rapid online separation and characterization of sulfonylureas as synthetic mixtures (95).

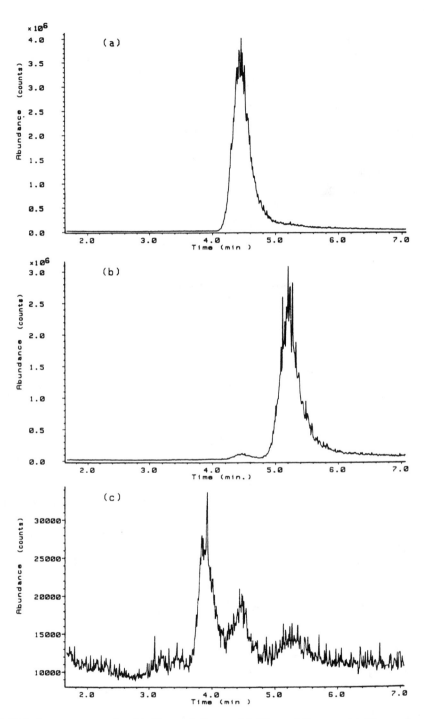

Fig. 19 LC-PB-MS SIM chromatograms for: (a) 50 ng of diuron; (b) 50 ng of linuron; (c) 70 ng of monuron. *Column*: 300 × 2.1-mm-ID μBondapak 10-μm C-18 column. *Elution*: CH_3OH/H_2O (68:32, v/v), isocratic elution. (From Ref. 29.)

5. Liquid Chromatography–Mass Spectrometry

A serious weakness of methods based on LC with conventional detectors is that they lack sufficient specificity for showing without doubt the presence of traces of target compounds in complex aqueous matrices. If a photodiode array is employed as the LC detector, UV spectra can be used to confirm peak identification. However, the UV spectra of many pesticides belonging to the same compound class are very similar, and differences between compound classes are frequently small. This limits the use of these spectra for peak confirmation. Furthermore, peak overlapping precludes quantification of target compounds, even by the use of diode array detectors. Thus, the use of MS as a detector is a key consideration for the future development of many LC methods.

Despite elaborate research efforts in LC-MS interfacing in the last 25 years, it is only in the last few years that LC-MS has become a technique that can be used routinely in analytical laboratories. Some LC-MS instrumentation is now actually being sold as an integrated detector for LC, i.e., to enter the chromatography lab rather than the mass spectrometry laboratory.

Nowadays, thermospray (TS), particle beam (PB), and electrospray (ES) are the most commonly used LC-MS interfaces for target analysis of pesticides from various matrices. The feasibility of using the TS interface for analyzing phenylureas (43,48,96–106) and sulfonylureas (51,107,108) by LC-MS has been extensively investigated. Two typical TC-TS-MS chromatograms are presented in Figs. 17 and 18.

The drawbacks of the TS source are that it does not provide structure-significant ions and does nor produce a sufficiently stable ion abundance over an 8-h period.

The lack of structural information by TS/MS has encouraged several researchers to reconsider the use of the PB interface for the detection of several classes of pesticides, because it is able to generate classical electron-impact spectra. The LC/PB/MS technique has been proposed for analyzing phenylureas in water (29). Figure 19 shows a chromatogram obtained by injecting three phenylureas into an LC-PB-MS apparatus from standard solutions. The feasibility of using LC/PB/MS for monitoring 104 pesticides, including phenylureas, and related compounds in groundwater has been investigated (109). The authors concluded that the PB/MS

Table 8 Major Ions, Among Parent and Fragment Ions, and Their Relative Abundances for Selected Urea Herbicides Resulting from the In-Source CID Process with the ES/MS System

Compound	Molecular weight	m/z of main ions[a] (relative abundance)
Chlorotoluron	212	46 (90), 72 (95), **213** (100)
Diuron	232	46 (85), 72 (100), **233** (90)
Fenuron	164	72 (90), **165** (100)
Isoproturon	206	46 (30), 72 (35), **207** (100)
Linuron	248	160 (35), 182 (35), **249** (100)
Methabenziathiazuron	221	165 (100), 222 (10)
Metobromuron	258	148 (100), 170 (80), **259** (45)
Metoxuron	230	72 (100), **231** (20)
Monolinuron	214	126 (100), 148 (70), **215** (30)
Monuron	200	72 (100), **201** (10)
Neburon	275	88 (55), 114 (25), **276** (100)
Bensulfuron-methyl	410	149 (80), 182 (60), **411** (100)
Chlorsulfuron	387	167 (100), 205 (10), **388** (30)
Metsulfuron-methyl	381	167 (100), 264 (10), **382** (30)
Rimsulfuron	431	182 (100), 326 (20), **432** (70)
Thifensulfuron-methyl	387	167 (100), 205 (10), **388** (30)
Triasulfuron	402	167 (100), 219 (10), **403** (60)
Tribenuron-methyl	395	155 (100), 364 (15), **396** (20)

Source: Refs. 27 and 32.

[a]Ion signals (adduct ions are reported in **boldface**) produced by the in-source CID process in an ES/MS single-quadrupole system Mod. Platform (Micromass, Manchester, U.K.) after setting the cone voltage to 30 V.

arrangement was not sufficiently sensitive for monitoring about half of the compounds considered at the 0.1 μg/L level. Nevertheless, a relatively recent study on the use of LC-PB-MS for target analysis of natural water samples has shown that several phenylureas could be detected at the 30–50-ng/L level, provided large enrichment factors can be achieved during the extraction step (110).

The ES interface is the newest device introduced for LC-MS coupling. It has opened new and exciting perspectives to the LC-MS technique. It is sufficient to say that the ES interface enables LC-MS analysis of compounds having up to 4 million u (unified atomic mass unit) by a conventional quadrupole, because the ES process is capable of forming multiply charged ions from polymeric compounds. The ES/MS system is a very robust, sensitive, and versatile device and is continually gaining in popularity at the expense of the TS source. As a matter of fact, several analytical methods developed in the past for analyzing pesticides by LC-TS-MS have recently been modified by replacing the TS source with the ES one (111–113). In addition to being more sensitive, an important advantage of the ES source over the TS one is that diagnostic fragment ions can easily be obtained by means of collision-induced dissociation (CID) after suitably adjusting the potential difference in the preanalyzer region. Provided that the analyte is separated from other species by the LC column, a single-mass analyzer can provide CID spectra very similar to those obtained by the much more expensive MS-MS technique. If the LC column fails to separate the target compound from other species present in the water sample, mixed CID spectra are obtained. Even in this case, however, evidence for the

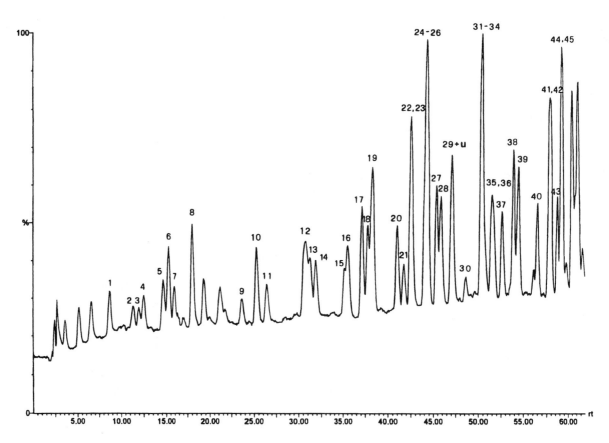

Fig. 20 Full-scan LC-ES-MS chromatogram obtained by analyzing 4 L of tap water spiked with 45 widely used pesticides pertaining to several classes. *Spike level*: 50 ng/L. *Column*: Alltima 250 × 4.6—m ID containing 5-μm C-18. *Elution*: CH₃OH/ H₂O linear gradient elution from 15:85 to 90:10 (v/v) in 60 min. *Peak numbering*: 23, monolinuron; 24 chlorotoluron; 27, isoproturon; 28, methabenzthiazuron; 29, diuron; 33, linuron; 40, neburon. u = unknown. (From Ref. 32.)

presence in water of a targeted compound can be gained by identifying on the mixed CID spectra those characteristic ion signals produced by its adduct ion and relative fragment ions. Characteristic parent and daughter ions of selected urea herbicides obtained by the in-source CID process are listed in Table 8.

Several LC methods relying on the high sensitivity and confirmation power of the ES/MS detector have been developed for analyzing traces of urea herbicides in natural waters (23,25,27,30,32,33,40,49,112,114,115). Figures 20 and 21 show LC-ES-MS chromatograms

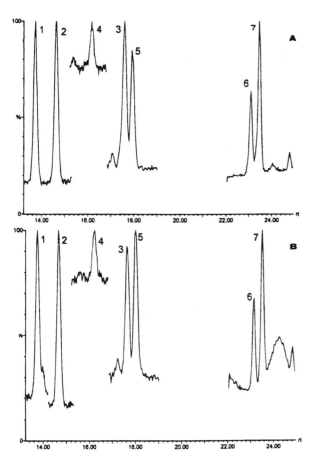

Fig. 21 Time-scheduled three-ion SIM LC-MS chromatograms obtained by analyzing (A) 4 L of drinking water spiked with seven sulfonylureas at the individual level of 3 ng/L and (B) 0.2 L Tiber River water sample spiked with the analytes at the individual level of 60 ng/L. *Column*: Alltima 250 × 4.6-mm ID containing 5-μm C-18. *Elution*: CH$_3$CN/ H$_2$O (both containing 3 mmol/L trifluoroacetic acid), linear gradient elution from 32:68 to 62:38 (v/v) in 40 min. *Peak numbering*: 1, thifensulfuron methyl; 2, metsulfuron methyl; 3, triasulfuron methyl; 4, chlorsulfuron; 5, rimsulfuron; 6, bensulfuron methyl; 7, tribenuron methyl. (From Ref. 27.)

obtained by analyzing water samples spiked with ng/L levels of selected urea herbicides.

Selected LC-MS methods for analyzing urea pesticides are summarized in Table 9.

IV. CONCLUSIONS

Liquid–liquid extraction with traditional solvents is still used for the isolation of pesticides from water samples. It tends to consume large volumes of high-purity solvents, which may have significant health hazards and disposal costs associated with their use. Furthermore, it is frequently plagued by problems, such as emulsion formation. The SPE technique with various adsorbing materials packed in cartridges or imbedded in membranes and used in the off-line or online mode, as shown by this review, is now definitely preferred to LLE.

The broad spectrum of well-established GC methods with selective detectors available today allows the identification and determination of hundreds of contaminants in aqueous environmental samples. However, several classes of pesticides, among these phenylurea and sulfonylurea herbicides, are not amenable to GC without time-consuming derivatization procedures. For such compounds, the LC technique is the most appropriate separation method. As a matter of fact, in recent years only LC applications have been of practical importance for phenylurea and sulfonylurea herbicides. Many of these applications rely on the use of conventional UV detectors. A serious weakness of these methods is that UV detection does not provide qualitative information sufficient to recognize sub-μg/L levels of target compounds in complex mixtures with a low probability of false positives. In such cases, only the use of a diode array detector can give some relief to this problem. In terms of qualitative and quantitative analysis, this discussion has shown that monitoring of urea herbicides in real water samples can greatly benefit from the use of LC-MS. In the last few years many sensitive and selective LC-MS methods that use different interfaces have been proposed. Today, the electrospray ion source is definitely considered to have the most promising future. It is expected that the recent introduction of less expensive, easy-to-use benchtop LC-ES-MS instrumentation will further stimulate practical applications of the recently developed analytical methodologies, enabling sensitive and reliable monitoring of the aforementioned compounds in a variety of aqueous environmental samples.

LIVERPOOL
JOHN MOORES UNIVERSITY
AVRIL ROBARTS LRC
TITHEBARN STREET
LIVERPOOL L2 2ER
TEL. 0151 231 4022

Table 9 Selected LC-MS Methods for Analyzing Urea Pesticides in Water

Compound	Interface	Acquisition mode	LOD (ng/L)	Ref.
Monuron, diuron, linuron	PB[a]	One-ion SIM[b]	160–500	29
Linuron, monolinuron	ES[c]	Two-ion SIM	5–9	30
7 Phenylureas	ES	Full-scan	1.4–3.1	32
Linuron	TS[d]	Full-scan	2000	43
15 Phenylureas	TS	One-ion SIM	5–120	48
Monuron, diuron, neburon	EF	Full-scan	7–3000	49
		One-ion SIM	0.1–200	
Diuron, fluometuron, linuron, monuron	TS	Full-scan	500–18,000	101
14 Phenylureas	TS	Full-scan	100–500	103
		One-ion SIM	2–20	
Metoxuron, monuron, chlorotoluron, diuron, linuron, chlorbromuron	PB	Full-scan	30–5000	110
Chlorotoluron, isoproturon, diuron, linuron, diflubenzuron	ES	Full-scan	0.6–8	112
Chlorotoluron, isoproturon, diuron, linuron, diflubenzuron	TS	Full-scan	1.5–8	112
7 Phenylureas	ES	MS-MS two of the most intense ions for Q[e] 3	10–50	115
7 Sulfonylureas	ES	Three-ion SIM	0.5–3	27
12 Sulfonylureas	ES	One-ion SIM	10–30	40
8 Sulfonylureas	TS	Sull-scan	>500	51
		Three-ion SIM	50–400	

[a] PB = particle beam.
[b] SIM = selected ion monitoring.
[c] ES = electrospray.
[d] TS = thermospray.
[e] Q = quadrupole.

The governments of some European countries, namely, Denmark and Sweden, are considering decreasing the maximum admissible concentration of an individual pesticide in drinking water from 100 to 10 ng/L and including in the list of undesired compounds those pesticide degradation products (DPs) that are toxic. It is possible in the near future that other European countries will follow this strategy. This will urge European analytical chemists to develop new analytical LC-ES-MS methodologies that, in addition to being more sensitive than most of the actual ones, will also be capable of simultaneously analyzing pesticides and their health-hazardous DPs, such as the case of phenylureas and their related chloroanilines. The latter compounds are more toxic than the parent compounds, and some of them are suspected of inducing cancer.

REFERENCES

1. Gy. Matolcsy, M. Nadasy, V. Andriska. Pesticide Chemistry. Elsevier, Amsterdam, PP 682–684 (1988).
2. R.E. Gosselin, R.P. Smith, H.C. Hodge. Clinical Toxicology of Commercial Products. 5th ed. Williams and Wilkins, Baltimore, p 11 (1984).
3. T.S. Scott. Carcinogenic and Toxic Health of Aromatic Amines. Elsevier, New York (1962).
4. C. Tomlin (ed.). The Pesticide Manual. British Crop Protection Council, Farnham, Surrey (1994).
5. D.J. Munch, R.L. Graves, R.A. Maxey, TM Engel. Environ. Sci. Technol. 24:1446 (1990).
6. US Environmental Protection Agency. National Survey of Pesticides in Drinking Water Wells. Phase II Report, EPA 570/9-91-020. National Technical Information Service, Springfield, VA (1992).

7. G. Vincent. In: G Angeletti, A. Bjørseth (Eds.). Organic Micropollutants in the Aquatic Environment. Lisbon Symposium, Kluwer, Dordrecht, pp 285–292 (1991).

8. M. Fielding, D. Barceló, A. Helweg, S. Galassi, L. Thorstensson, P. Van Zoonen, R. Wolter, G. Angeletti. In: Pesticides in Ground and Drinking Water (Warter Pollution Research Report, 27), Commission of the European Communities, Brussels, pp 1–136 (1992).

9. C.D. Watts, L. Clark, S. Hennings, K. Moore, C. Parker. In: Pesticides: Analytical Requirements for Compliance with EC Directives (Water Pollution Research Report, 11), Commission of the European Communities, Brussels, pp 16–34 (1989).

10. J.S. Smith. In: Keith, L. H. (ed.). Principles of Environmental Sampling. American Chemical Society, Washington, DC, p 225 (1988).

11. C.F. Nordin. In: Proceedings of International Symposium on River Sedimentation, 2nd. Beijing, Water Resources and Electric Power Press. Nanjing, China, p 1145 (1983).

12. R.H. Meade. Suspended Sediments in the Amazon River and Its Tributaries in Brazil During 1982–1984. Open File Rep–U.S. Geol. Surv., No. 85-492 (1985).

13. D.J. Munch, C.P. Frebis. Environ. Sci. Technol. 26:921 (1992).

14. Standing Committee of Analysts. The Determination of Carbamates, Thiocarbamates, Related Compounds and Ureas in Water. HM Stationery Ofice, London, pp 15–21 (1987).

15. US Environmental Protection Agency. Methods for the Determination of Organic Compounds in Drinking Water PB91-231480. Revised. National Technical Information Service, Springfield, VA (1991).

16. I. Ahmad. J. Ass. Off. Anal. Chem. 70:745 (1987).

17. V. Leoni, C. Cremisini, A. Casuccio, A. Gullotti. Pestic. Sci. 31:209 (1991).

18. D.G. Thompson, L.M. McDonald. J. Assoc. Off. Anal. Chem. 75:1084 (1992).

19. Method 525.1 Determination of Organic Compounds in Drinking Water by Liquid–Solid Extraction and Capillary Column Gas Chromatography/Mass Spectrometry (Revision 2.2). Environmental Monitoring Systems Laboratory, Office of Research and Development, U.S. Environmental Protection Agency, Cincinnati, OH.

20. C. Crescenzi, A. Di Corcia, M.D. Madbouly, R. Samperi. Environ. Sci. Technol. 29:2185 (1995).

21. S.A. Senseman, T.L. Lavy, J.D. Mattice, B.M. Myers, B.W. Skulman. Environ. Sci. Technol. 27:516 (1993).

22. A. Di Corcia, M. Marchetti. J. Chromatogr. 541:365 (1991).

23. V. Pichon, L. Chen, M.-C. Hennion, R. Daniel, A. Martel, F. Le Goffic, J. Abian, D. Barceló. Anal. Chem. 67:2541 (1995).

24. V. Pichon, L. Chen, N. Durand, F. Le Goffic, M.-C. Hennion. J. Chromatogr. 725:107 (1996).

25. I. Ferrer, M.-C Hennion, D. Barceló. Anal. Chem. 69:4508 (1997).

26. A. Martin-Esteban, P. Fernandez, C. Camera. Anal. Chem. 69:3267 (1997).

27. A. Di Corcia, C. Crescenzi, R. Samperi, L. Scappaticcio. Anal. Chem. 69:2819 (1997).

28. S. Scott. Analyst 118:1117 (1993).

29. M. J. Incorvia Mattina. J Chromatogr. 549:237 (1991).

30. D. Giraud, A. Ventura, V. Camel, A. Bermond, P. Arpino. J. Chromatogr. 777:115 (1997).

31. A. Di Corcia, A. Marcomini, R. Samperi, S. Stelluto. Anal. Chem. 65:907 (1993).

32. C. Crescenzi, A. Di Corcia, E. Guerriero, R. Samperi. Environ. Sci. Technol. 31:479 (1997).

33. A. Di Corcia, A. Costantino, C. Crescenzi, R. Samperi. Paper submitted to Anal. Chem.

34. A. Di Corcia, M. Marchetti. Anal. Chem. 63:580 (1991).

35. A. Di Corcia, M. Marchetti. Environ. Sci. Technol. 26:66 (1992).

36. E.W. Zhanow. J. Agric. Food Chem. 33:479 (1985).

37. M.J.M. Wells, J.L. Michael. J. Chromatogr. Sci. 25:345 (1987).

38. E.G. Cotterill. Pestic. Sci. 34:291 (1992).

39. G.C. Galletti, A. Bonetti, G. Dinelli. J. Chromatogr. 692:27 (1995).

40. A.J. Krynitsky. J. Assoc. Off. Anal. Chem. 80:1084 (1997).

41. M.S. Young, J. Assoc. Off. Anal. Chem. 80:108 (1998).

42. J.J. Jimenez, J.L. Bernal, M.J. del Nozal, J.M. Rivera. J. Chromatogr. 778:289 (1997).

43. D. Barceló, G. Durand, V. Bouvot, M. Nielen. Environ. Sci. Technol. 27:271 (1993).

44. P. Klaffenbach, P. T. Holland. J. Agric. Food Chem. 41:396 (1993).

45. A.L. Howard, L.T. Taylor. J. Chromatogr. Sci. 30:374 (1992).

46. C.E. Goewie, P. Kwakman, R.W. Frei, U.A. Th. Brinkman, W. Maasfeld, T. Sehadri, A. Kettrup. J. Chromatogr. 284:73 (1984).

47. M.W.F. Nielen, A.J. Valk, R.W. Frei, U.A. Th. Brinkman, Ph. Mussche, R. De Nijs, B. Ooms, W. Smink. J. Chromatogr. 393:69 (1987).

48. H. Bagheri, E.R. Brouwer, R.T. Ghijsen, U.A. Th. Brinkman. Analysis 20:475 (1992).

49. J. Slobodnik, A.C. Hogenboom, J.J. Vreuls, J.A. Rontree, B.L.M. van Baar, W.M.A. Nissen, U.A. Th. Brinkman. J. Chromatogr. 741:59 (1996).

50. V. Pichon, M.-C. Hennion. J. Chromatogr. 665:269 (1994).

51. D. Volmer, J.G. Wilkes, K. Levsen. Rapid Commun. Mass Spectrom. 9:767 (1995).

52. G. Nilvé, R. Stebbins. Chromatographia 32:269 (1991).

53. G. Nilvé, M. Knutsson, J.A. Jönsson. J. Chromatogr. 688:75 (1994).

54. R. Eisert, J. Pawliszyn. Anal. Chem. 69:3140 (1997).

55. C.E. McKone, R.J. Hance. J. Chromatogr. 36:234 (1968).

56. C.E. McKone. J. Chromatogr. 44:60 (1969).

57. D. Spengler, B. Hamroll. J. Chromatogr. 49:205 (1970).

58. H. Buser, K. Grolimund. J. Ass. Offic. Anal. Chem. 57:1294 (1974).

59. K.H. Bowmer, J.A. Adeney. Pestic. Sci. 9:324 (1978).

60. R. Deleu, A. Copin. J. High Resolut. Chromatogr. 3: 299 (1980).

61. K. Grob Jr. J. Chromatogr. 208:217 (1981).

62. C.E. McKone, R.J. Hance. J. Chromatogr. 36:234 (1968).

63. W.P. Cochrane, B.P. Wilson. J. Chromatogr. 63:364 (1971).

64. R. Deleu, A. Copin. J. Chromatogr. 171:263 (1979).

65. J.L. Tadeo, J.M. García-Baudín, T. Matienzo, S. Pérez, H. Sixto. Chemosphere. 18:1673 (1989).

66. D.G. Saunders, L.E. Vanatta. Anal. Chem. 46:1319 (1974).

67. J.J. Ryan, J.F. Lawrence. J. Chromatogr. 135:117 (1977).

68. A.H. Hofberg Jr, L.C. Heinrichs, V.M. Barringer, M. Tin, G.A. Gentry. J. Ass. Offic. Anal. Chem. 60:716 (1977).

69. U.A. Th. Brinkman, A de Kok, R.B. Geerdink. J. Chromatogr. 283:113 (1984).

70. S. Perez, M.T. Matienzo, J.L. Tadeo. Chromatographia 36:195 (1993).

71. W.E. Bleidner, H.M. Baker, M. Levitsky, W.K. Lowen. J. Agr. Food. Chem. 2:476 (1954).

72. I. Baunok, H. Geissbühler. Bull. Environ. Contam. Toxicol. 3:7 (1968).

73. H. Kussmaul, M. Hegazi, K. Pfeilsticker. Vom Wasser 44:31 (1975).

74. A.H.M.T. Scholten, B.J. de Vos, J.F. Lawrence, U.A. Th. Brinkman, R.W. Frei. Anal. Lett. 13A:1235 (1980).

75. A. de Kok, I.M. Roorda, R.W. Frei, U.A. Th. Brinkman. Chromatographia 14:579 (1981).

76. A. de Kok, Y.J. Vos, C. van Garderen, T. de Jong, M. van Opstal, R.B. Gerdink, R.W. Frei, U.A. Th. Brinkman. J. Chromatogr. 288:71 (1984).

77. F.S. Tanaka, R.G. Wien. J. Chromatogr. 87:85 (1973).

78. L. Ogierman. Fresenius Z. Anal. Chem. 320:365 (1985).

79. J.F. Lawrence, G.W. Laver. J. Agric. Food Chem. 23: 325 (1975).

80. R. Greenhalgh, J. Kovacikova. J. Agric. Food Chem. 23:325 (1975).

81. J.F. Lawrence. J. Agric. Food Chem. 24:1236 (1976).

82. J.F. Lawrence. J. Assoc. Offic. Anal. Chem. 59:1061 (1976).

83. G. Glad, T. Popoff, O. Heander. J. Chromatogr. Sci. 16:118 (1978).

84. J.F. Lawrence, C. Van Buuren, U.A. Th. Brinkman, R.W. Frei. J. Agric. Food Chem. 28:630 (1980).

85. S. Pérez, J.M. García-Baudín, J.L. Tadeo. Fresenius Z. Anal. Chem. 339:413 (1991).

86. L. Fishbein, W.L. Zielinski. J. Chromatogr. 20:9 (1965).

87. T.A. Berger. Chromatographia 41:133 (1995).

88. D. Barceló. Analyst 116:681 (1991).

89. D. Barceló. Chromatographia 25:928 (1988).

90. P. Jandera, J. Churacek, P. Butzke, M. Smrz. J. Chromatogr. 387:155 (1987).

91. Q.G. von Nehring, J.W. Hightower, J.L. Anderson. Anal. Chem. 58:2777 (1986).

92. G. Dinelli, A. Vicari, P. Catizone. J. Agric. Food Chem. 41:742 (1993).

93. G. Dinelli, A. Boetti, P. Catizone, G. Galletti. J. Chromatogr. 656:275 (1994).

94. G. Dinelli, A. Vicari, A. Bonetti. J. Chromatogr. 700: 195 (1995).

95. F. Garcia, J. Henion. J. Chromatogr. 606:237 (1992).

96. D. Barceló. Org. Mass Spectrom. 24:219 (1989).

97. D. Barceló, J. Albaiges. J. Chromatogr. 474:13 (1989).

98. R.D. Voyksner. In: Rosen (ed.). Application of New Mass Spectrometric Techniques in Pesticide Chemistry. Wiley, New York, p 146 (1987).

99. W.M.A. Niessen, R.A.M. Van Der Hoeven, M.A.G. De Kraa, C.E.M. Heeremans, U.R. Tjaden, J. Van Der Greef. J. Chromatogr. 478:325 (1989).

100. D. Volmer, K Levsen. J. Am. Soc. Mass. Spectrom. 655:5 (1994).

101. W.L. Bellar, T.A. Budde. Anal. Chem. 60:2076 (1988).

102. D. Volmer, K. Levsen, W. Engewald. Vom Wasser 82: 335 (1994).

103. D. Volmer, K. Levsen. J. Chromatogr. 660:231 (1994).

104. R.D. Voyksner, J.T. Bursey, E.D. Pellizzari. Anal. Chem. 56:1507 (1984).

105. R.J. Vreeken, U.A. Th. Brinkman, G.J. de Jong, D. Barceló. Biomed. Environ. Mass Spectrom. 19:481 (1990).

106. R.J. Vreeken, W.D. van Dongen, R.T. Ghijsen, U.A. Th. Brinkman. Int. J. Environ. Anal. Chem. 54:119 (1994).

107. L.M. Shalaby, S.W. George. In: M.A. Brown (ed.). Liquid Chromatography/Mass Spectrometry. Applications in Agricultural, Pharmaceutical and Environmental Chemistry (ACS Symposium Series, No 420). American Chemical Society, Washington, DC, p 62 (1990).

108. L.M. Shalaby, F.Q. Bramble Jr, P.W. Lee. J. Agric. Food Chem. 40:513 (1992).

109. C.J. Miles, D.R. Doerge, S. Bajic. Arch. Environ. Contam. Toxicol. 22:247 (1992).

110. H. Bagheri, J. Slobodnik, R.M. Marce Recasens, R.T. Ghijsens, U.A. Th. Brinkman. Chromatographia 37: 159 (1993).

111. C. Molina, M. Honing, D. Barceló. Anal. Chem. 66: 4444 (1994).

112. C. Molina, G. Durand, D. Barceló. J. Chromatogr. 712:113 (1995).

113. S. Lacorte, D. Barceló. Anal. Chem. 68:2464 (1996).

114. C. Aguilar, I. Ferrer, F. Borrul, R.M. Marcé, D. Barceló. J. Chromatogr. 794:147 (1998).

115. A.C. Hogenboom, P. Speksnijder, R.J. Vreken, W.M.A. Niessen, U.A. Th. Brinkman. J. Chromatogr. 777:81 (1997).

26

Analysis of Organochlorinated Pesticides in Water

Filippo Mangani, Michela Maione, and Pierangela Palma
University of Urbino, Urbino, Italy

I. INTRODUCTION

Several kinds of pesticides have been used over the past decades in the attempt to defeat the huge number of crop-eating insects, approximately 700 species worldwide, that caused infective and parasitic diseases to humans and loss to harvest. The use of pesticides has contributed to the drastic reduction in those diseases transmitted by insects, most of them life-threatening, while protecting crops during growth and storage.

Before World War II the selection of insecticides was more or less the same as those available for a thousand and more years before. It was in the 1940s and '50s that a new concept of pest control took hold, opening a new era of synthetic, highly effective compounds.

The extensive use of synthetic pesticides was at first greeted with enthusiasm, but in a few years it appeared clear that these compounds and their residues contaminate ground- and surface water. In most cases these substances were organochlorinated compounds, also known as chlorinated hydrocarbons, chlorinated organics, chlorinated insecticides, chlorinated synthetics and organochlorinated pesticides (OCPs) (Table 1). Their use has been superseded in most countries, but interest in these compounds is still high. In several organisms, organochlorines are found in a higher concentration than in the environment in which they live.

II. PHYSICAL AND CHEMICAL PROPERTIES

The popularity of chlorinated pesticides was based on certain important properties:

Extremely high stability
Very low solubility in water
High solubility in organic media
High toxicity to insects and low toxicity to humans

Due to their high solubility in organic media, most organochlorines present in an aquatic environment penetrate to the adipose tissue of fishes and accumulate there. This phenomenon is called *bioconcentration*. The *bioconcentration factor* (*BCF*) represents the capability of a substance to accumulate in aquatic organisms and enter the food chain, leading to dangerous or lethal concentrations when the aquatic organisms are consumed by higher organisms. It is defined as the equilibrium constant of the concentration of a particular chemical substance in a fish compared to its concentration in the surrounding water, when the single source of such substance for the fish is the mechanism of diffusion. It depends upon both the lipophylic nature of the substance and the ability of the organism to metabolize it. Since the values of the BCF can vary significantly, depending not only on the compound but also on the fish species, the BCF of these pesticides are determined in the same laboratory determining the de-

Table 1 Selected Pesticides

Pesticide	CASRN	Formula	Molecular weight
Aldrin	309-00-2	C12H8Cl6	364.9
Chlordane	57-74-9	C10H6Cl8	409.8
Chlordecone	143-50-0	C10Cl10O	490.6
Dieldrin	60-57-1	C12H8Cl6O	377.9
o,p'-DDE	3424-82-6	C14H8Cl4	315.9
p,p'-DDE	72-55-9	C14H8Cl4	315.9
o,p'-DDD	72-54-8	C14H10Cl4	318.0
p,p'-DDD	72-54-8	C14H10Cl4	318.0
o-p'-DDT	789-02-6	C14H9Cl5	354.5
p,p'-DDT	50-29-2	C14H9Cl5	354.5
Endosulfan	115-29-7	C9H6Cl6O3S	406.9
	Isomer I 959-98-8		
	Isomer II 33213-65-9		
Endosulfan sulfate	1031-07-8	C9H6Cl6O4S	422.9
Endrin	72-20-8	C12H8Cl6O	380.9
Heptachlor	76-44-8	C10H5Cl7	373.3
Heptachlor epoxide	1024-57-3	C10H5Cl7O	385.8
Hexachlorobenzene (HCB)	118-74-1	C6Cl6	284.8
Hexachlorocyclohexane	680-73-1	C6H6Cl6	290.8
Lindane	58-89-9	C6H6Cl6	290.8
Methoxychlor	72-43-5	C16H15Cl3O2	345.7
Mirex	2385-85-5	C10Cl12	545.6

gree of lipophilicity. A good model of the adipose tissues of a fish can be represented by 1-octanol. The selected compound is partitioned in a water/1-octanol binary phase system. The partition coefficient K_{ow} 1-octanol/water describes the lipophilic or hydrophobic properties of a chemical:

$$K_{ow} = \frac{[S]_{octanol}}{[S]_{water}}$$

Since the values of K_{ow} range from $<10^{-4}$ to $>10^8$, it is more often expressed as a pK_{ow}, also called log P. The higher the value of pK_{ow}, the easier a chemical binds to organics of the soil to eventually migrate to the adipose tissue of a living organism. The value of the bioconcentration factor can be estimated, up to a factor of 10, using log P correlation.

Water solubility is a very important property for predicting the fate of a chemical and its tendency to partition to soil and sediment and to bioaccumulate. There are several methods for estimating it, the most practical of which involves regression-derived correlation using log P and the use of the melting point and molecular weight with log P.

The toxicity of a chemical compound is expressed in terms of lethal dose (LD) and, in particular, as LD_{50} which represents the amount of substance necessary to kill half of the laboratory animals treated with that particular chemical. The LD_{50} values are expressed in milligrams of substance per kilogram of weight of the animal (Table 2).

A. Dichlorodiphenyltrichloroethane (DDT)

The most notorious pesticide ever used is by far dichlorodiphenyltrichloroethane (DDT). It was synthesized in 1874 by a German student but was rediscovered by Paul Mueller, a Swiss scientist who recognized it as a powerful insecticide for the control of some important insect-vectored (body lice) diseases, such as malaria, yellow fever, and thyphus. For this important discovery Paul Mueller was awarded the Nobel Prize in Medicine in 1948.

p, p'-DDT

Table 2 Some Physical and Chemical Properties of Selected Pesticides

Pesticide	Log K_{ow} (log P)	Water solubility at 25°C (ppm)	Melting point (°C)	Vapor pressure (mPa) at 25°C	LD$_{50}$ mg/kg
Aldrin	6.5	0.01–0.2	104		
Chlordane	6.16	0.1	104–107	61	Rats 457–590
Chlordecone	5.41	7.6	350		
Dieldrin	5.4	0.186 at 20°C	176	0.39	Rats 46
o,p'-DDE	6.51				
p,p'-DDE	6.51		88–90	2.09	
o,p'-DDD	6.02	0.1			
p,p'-DDD	6.02	0.05	88–90, 109–112	0.62	
o,p'-DDT					
p,p'-DDT	6.91	0.0077 at 20°C	108.5	0.025 at 20°C	Rats 115
Endrin	5.2	0.23	200	0.026	
Endosulfan	(I) 3.83	(I) 0.32 at 22°C	(I) 70–100	0.0012 at 80°C	Rats 80–110 TC[a]
		(II) 0.33 at 22°C	(II) 213.3		
Endosulfan sulfate	3.66	0.117	181		
Heptachlor	6.1	0.18	46–74	40	Rats 147–220
Heptachlor epoxide	4.98				
Hexachlorobenzene (HCB)	5.73	0.006	226	1.45 at 20°C	Rats 10,000
Hexachlorocyclohexane	4.26		112	5.6 at 20°C	
Lindane	3.72	7	112	5.6 at 20°C	Rats 88–270 Mice 59–246
Methoxychlor	5.08	0.12	89		Rats 6000 TC
Mirex		7×10^{-5}	485	0.1	

[a]TC = technical grade.

o, p'-DDT

The production of DDT, started in 1943, and DDT was heralded as a miraculous compound, becoming widely used during World War II. After World War II, DDT was used in agriculture for a pest control. The results were excellent: DDT proved itself very effective against a number of pests, and agricultural yields increased rapidly.

Unfortunately, the unrestrained use of DDT, especially in agriculture, led to a rapid increase in its concentration in the environment, and soon after its introduction the scientific community expressed concern about this insecticide. The longtime permanence of DDT in the soil resulted in a magnification through the food chain, causing the destruction of the balance of sodium and potassium ions within the axons of neurons in a way that prevents the normal transmission of nerve impulses in both insects and mammals. Fields and forests were sprayed on a wide scale, causing the elimination not only of the "bad" insects but also of those that pollinate plants.

In 1962 a book entitled *Silent Spring*, by Rachel Carson, brought the consequences of the use of DDT to public notice. The author referred to the progressive extinction of several aviary species, among them the bald eagle, explaining in powerful language the dangers of DDT and pesticides in general. The book triggered a massive and concerned reaction, after an intensive investigation of the issues of pesticide use. The first country to withdraw DDT from the market was Hungary, in 1968, followed shortly thereafter by other industrialized countries.

The World Bank recommends that DDT not be used in agricultural production. Altogether, DDT is banned in 12 countries (plus the entire European Economic Community) and severely restricted in 15.

B. Dichlorodiphenyldichloroethane (DDD) and Dichlorodiphenylchloroethane (DDE)

Environmental levels of DDT soon began to fall. But despite the ban, in the 1980s they started increasing again, leading to the cracking and thinning of eggshells. Like the other OCPs, DDT is very soluble in organic solvents and, therefore, in adipose tissue. Several species of animals can metabolize DDT, in alkaline conditions, leading to the formation of metabolites dichlorodiphenyldichloroethane (DDD) and dichlorodiphenylchloroethane (DDE).

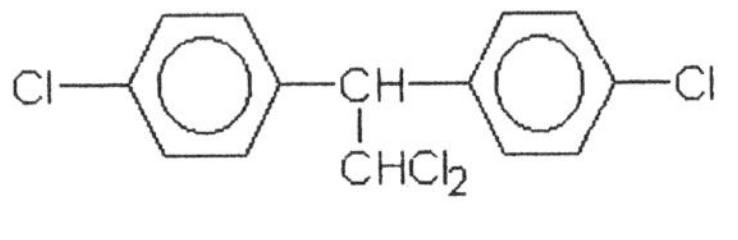

p, p'-DDD

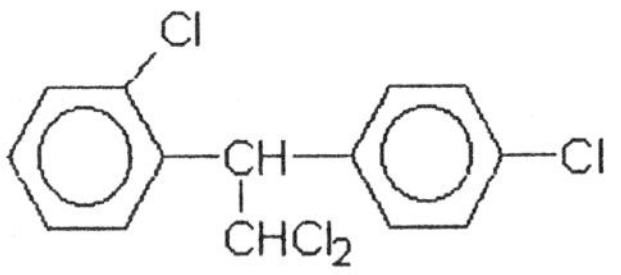

o, p'-DDD

The compound DDD was used for controlling a number of insects on vegetables and tobacco. It is also very active against mosquito larvae. It can be released to the environment through its use and as a biodegradation product of DDT. It binds very strongly to soil and does not leach appreciably to groundwater, although it can be transported there. When in water, it adsorbs to sediment.

Biodegradation of DDD is very slow, and photodegradation is not expected to occur. In air it is more likely to be found on particulate matter and can be removed mainly by fallout and washout. Due to its persistence in the environment and its stability, it bioconcentrates through the food chain.

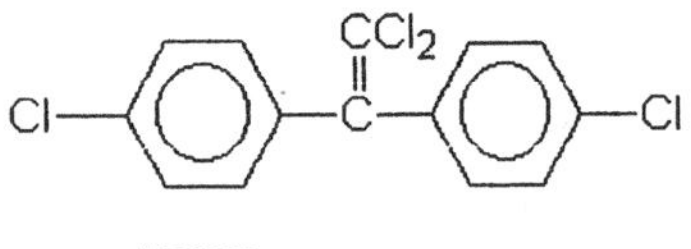

p, p'-DDE

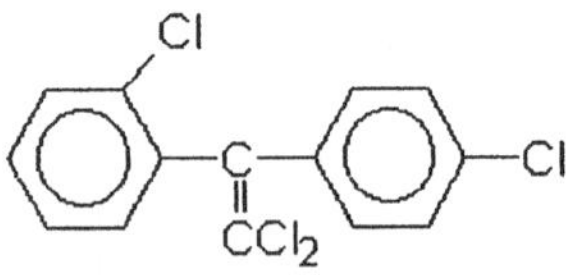

o,p'-DDE

The compound DDE is a metabolite of DDT as well as an impurity, so its presence in the environment is correlated strictly to the use of DDT. Like its precursor, it sticks very strongly to soil and, if released in water, adsorbs to the sediment, where photolysis is unlikely to occur. It does not biodegrade easily, but it does evaporate significantly from surfaces with low organic content.

DDE interferes, in some birds, with the enzyme that regulates calcium distribution, so their eggshells are very weak, unable to bear the weight of the parent during brooding. This phenomenon was caused partly by the illegal use of DDT and by windblown DDT, but the primary cause has been attributed to the use of Dicofol, a pesticide manufactured from DDT and containing it as a contaminant. Although DDT is not easily absorbed through the skin, it concentrates in the food chain, where it is very stable and accumulates in the fatty tissues of man and animals. Most of the DDT in human body fat is present in its metabolized form, DDE, which is extremely soluble in organics and remains in our organism for a long time. Several Third World countries, ignoring bans and restrictions, still use and export DDT, whose contamination is present in virtually all foods and living things.

DDT is a very toxic compound. Exposure to high doses can affect the central nervous system, provoking paralysis of the tongue, lips, and face, apprehension, irritability, dizziness, tremors, and convulsions. In moderately severe poisoning cases, cardiac and respiratory failure can occur.

C. Methoxychlor

Methoxychlor is a synthesized compound with chemical structure and properties similar to those of DDT. It is rarely phytotoxic and is used on crops, including several types of seeds. It is also effective against mosquito larvae and houseflies. Methoxychlor is also used as an insecticide on cattle. When released to soil, it remains primarily on the upper layer; however, a small percentage may reach lower layers and from there migrate to groundwater.

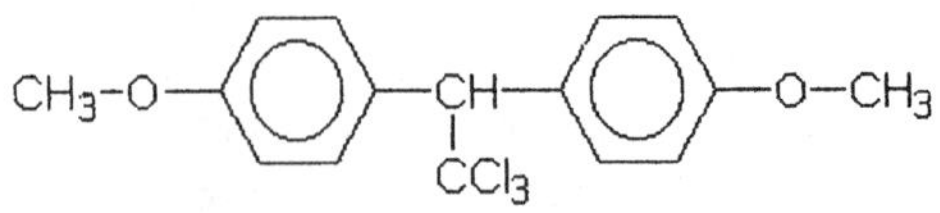

METHOXYCHLOR

Methoxychlor biodegrades more easily than DDT and, especially in anaerobic conditions, is the dominant removal mechanism. Major degradation products are dechlorinated methoxychlor (DMDD) and mono- and di-hydro derivatives. When released in water, it adsorbs to suspended particles and photolyzed quite easily. Volatilization is also an appreciable means of removal and transport of methoxychlor. Aquatic organisms metabolize it and transform it into other, less toxic substances; therefore it does not lead to significant bioaccumulation phenomena.

D. Hexchlorocyclohexane (HCH) and Lindane

The insecticide hexchlorocyclohexane (HCH) is a mixture of five isomers, alpha, beta, gamma, delta, and epsilon, but only gamma-HCH shows insecticide properties when ingested, by contact and as fumigant. Consequently, the gamma isomer is isolated and sold as the odorless insecticide Lindane.

LINDANE

Lindane is commonly used externally on animals to kill lice and ticks and internally to discourage the propagation of parasites. It is used also to control several skin diseases. In the forest, Lindane is employed to control several kinds of insects on conifers and in agriculture as seed dressing.

Lindane hydrolyzes in water only in alkaline conditions. While volatilization is very slow in water, it represents the major source of release from soil, together with a slow leaching to groundwater. It biodegrades rapidly under anaerobic conditions, more slowly under aerobic conditions. Photodegradation is not a major environmental process. Bioconcentration is low, but present, so human exposure is due mainly to food consumption.

Short-term exposure to HCH and Lindane interferes with the transmission of nerve impulses, producing effects that resemble those of DDT but that occur much more rapidly; long-term exposure leads to liver and kidney damage. Suspected of carcinogenicity and proven to be mutagenic, HCH and Lindane bioaccumulate in human fat for a long time.

The manufacturing process is very economical; for this reason HCH has been widely used in many developing countries. The World Bank recommends that HCH, Lindane, and other organochlorine pesticides not be used in agricultural production. HCH is banned in 12 countries plus the entire European Economic Community, and is severely restricted in seven countries. Lindane is banned or severely restricted in most countries.

E. Cyclodienes

After World War II, cyclodienes, a new class of pesticides, appeared on the scene, comprising chlordane (1945), aldrin and dieldrin (1948), heptachlor (1949), endrin (1951), mirex (1954), endosulfan (1956), chlordecone (1958), and some others of minor importance. Most of them are very stable in sunlight and persistent in soil, and they were used to control termites and other insects. Their effectiveness led to insect resistance and bioaccumulation in the food chain. For these reasons their use was banned between 1984 and 1988. In contrast to DDT and HCH, their toxicity increases with temperature. They all affect the central nervous system (CNS) in the same way, causing tremors, convulsions, and prostration to the maximum extent, depending on the rate and on the time of exposure.

1. Endrin

Endrin is a solid, white-to-tan, almost odorless substance, produced and sold since 1986. It has been widely used as an insecticide on cotton as well as an avicide and rodenticide. It is very persistent, even though, when exposed to light, part of it decomposes to produce endrin ketone and endrin aldehyde, whose properties are little known. It can evaporate to air or adsorb on dust, and, although it is not likely to leach into groundwater, it can reach surface water via surface runoff. In surface water it adsorbs to sediments without hydrolyzing or biodegrading. It significantly bioconcentrates in aquatic organisms.

ENDRIN

Human exposure results primarily from food. Short-term exposure can vary, depending on the amount of substance and the mode of exposure. The target of this pesticide is the nervous system. Swallowing large amounts of endrin may lead to death in a few hours, while breathing it can provoke headaches, vomiting, and convulsions. No long-term effects have been detected in workers exposed. Studies of laboratory animals show that endrin is not likely to produce cancer, and these data are confirmed in humans, since no increase in cancer has been found in exposed factory workers. For these reasons and because information is insufficient, endrin has not been classified as a human carcinogen.

2. Aldrin and Dieldrin

Aldrin and dieldrin are two pesticides widely used in agriculture, to control insects in soil, and in public health, to defeat mosquitoes and tsetse flies. Dieldrin has also been used in veterinary medicine. These two pesticides have similar structure and chemical properties and show similar toxicity. Sunlight and bacteria slowly convert aldrin to dieldrin in plants, animals and in the environment. Biodegradation of aldrin in soil is slow, and it is not expected to leach to groundwater. When in water, aldrin volatilizes from the surface at a rate directly proportional to the wind velocity and inversely proportional to the depth of the water table. In this case, significant photolysis takes place. Aldrin adsorbs strongly to sediments and bioconcentrates in aquatic organisms. The environmental impact of dieldrin is very similar to that of aldrin, as is its environmental fate, with the exception that dieldrin photodegradates very slowly or rarely.

ALDRIN

DIELDRIN

These two compounds were widely used from the 1950s through the 1970s, when most uses were banned. In 1987, the EPA banned all uses. Contamination by dieldrin is detectable in soil, where it binds tightly and evaporates slowly. It can be absorbed by plants and decomposes very slowly. Eating food products grown in soil or water treated with dieldrin or near waste sites represents a means of contamination for animals and humans, in which dieldrin is stored in the fatty tissue. Exposure may be due to inhalation and absorption through the skin, especially among factory workers, but contamination via air and water is in general less likely to occur. Short-term exposure to high levels of aldrin and dieldrin have effects on the central nervous system, such as headache, nausea, convulsions, and, to the maximum extent, death. Long-term exposure to low levels does not seem to show health effects, even though these compounds accumulate in adipose tissue. Although studies on laboratory animals show that aldrin and dieldrin may be carcinogens, there is inadequate evidence that they can induce cancer in humans. The EPA considers them probable cancer agents.

3. Heptachlor

Heptachlor is a chemical used extensively as a termiticide in homes and buildings and as a pesticide on food crops, especially corn. Heptachlor epoxide is a breakdown product of heptachlor that is more likely to be found in the environment than is heptachlor.

HEPTACHLOR

HEPTACHLOR EPOXIDE

The use of heptachlor has been regulated, and it is now used only for control of fire ants in power transformers. Its presence in the environment is more likely due its extensive use prior to 1983. Evaporation and hydrolyzation of heptachlor from moist soil surfaces is significant, but heptachlor adsorbed to soil evaporates and hydrolyzes very slowly. In soil it degrades to heptachlor epoxide, 1-hydroxychlordene, and another, unknown metabolite less hydrophilic than heptachlor epoxide. Biodegradation is less significant than hydrolysis. In water, unadsorbed heptachlor undergoes photolysis, but it is likely to bioconcentrate in fish, the rest sticking strongly to sediment.

Heptachlor epoxide is not commercially available but is a result of heptachlor degradation. The epoxide can pollute water and soil for a long time. It adsorbs strongly to soil, with almost no biodegradation, and only little photolysis and volatilization take place. It binds to aquatic and air sediments and can be washed out by rain. It bioconcentrates and builds up in the food chain. The most contaminated foods are dairy products, meat, poultry, and fish. As described for most OCPs, exposure may be due to the ingestion of contaminated food or water. Animals eating food contaminated with heptachlor can transform it into heptachlor epoxide. Since these pesticides have been used extensively in houses and buildings, the occupants of the treated structures were exposed to them. These two compounds bioaccumulate along the food chain, and in the body heptachlor is readily epoxidized.

Short-term exposure affects principally the CNS, depending, as with the other OCPs, on the concentration of the compound. Long-term exposure may bring significant risk of hepatic and renal damage, hormonal alterations, and neurological changes. Heptachlor and its epoxide are considered probable oncogenes, even though there are insufficient data to establish this clearly.

The use of heptachlor slowed down in the 1970s and stopped in 1988 except for the control of fire ants in power transformers.

4. Endosulfan and Endosulfan Sulfate

Two other cyclodienes used to control insects on crops are endusulfan, a mixture of the two isomeric forms endosulfan I and endosulfan II, and endosulfan sulfate. Their physical properties and toxicity are similar to those of the other cyclodienes. They can persist in soil for several years, evaporating and breaking down very slowly. The sulfate can be found in the environment as a result of the use of endosulfan. It adsorbs on soil and does not leach to groundwater. Biodegradation and hydrolyzation are likely to occur. It adsorbs to water sediments and bioconcentrates in living organisms. Dissolved endosulfan sulfate can be transported in water tables by evaporation. Even though no information about hydrolysis has been found, this may be an important process.

ENDOSULFAN

ENDOSULFAN SULFATE

For the foregoing reasons, these cyclodienes accumulate in aquatic organisms and enter the food chain. Exposure may result from breathing air, drinking water, or eating food that has been treated with endosulfan. They affect the CNS, causing convulsions, tremors, decreased breathing, and death. Endosulfan and endosulfan sulfate are not classified as human carcinogens.

5. Chlordane

Chlordane, widely used from 1948 to 1988 on crops to fight against termites and many other insect pests, is a thick liquid that is colorless to amber. The only permitted use is for fire ant control in power transformers.

CHLORDANE

Technical grade comprises chlordane isomers, the most abundant being the alpha, or cis, isomer and the gamma, or trans, isomer, and compounds such as heptachlor, chlordene, and other minor components. Commercial formulations contain 10% heptachlor.

Chlordane is very stable, is corrosive to iron and other metals, and attacks plastic and rubber. It is toxic and carcinogenic and bioaccumulates, persisting in the environment, where it can be found either in soil or in aquatic sediments. Its persistence in soil is very long, although, despite its low solubility in water, it can move slowly to groundwater. It usually leaves the soil by evaporation, depending on the percentage of moisture present in the ground. Biochemical and chemical degradation are quite minor, and, although it is photodegradable, it can stay in the soil for over 20 years. In water it does not undergo biological, chemical, or physical degradation, and it adsorbs strongly to sediments. It is expected to volatilize from water to the atmosphere, where direct photolysis does not occur, since chlordane is stable in UV light under normal conditions. It can be removed from the atmosphere by rainfall.

Bioconcentration is quite significant. Chlordane enters the organism through the skin, the respiratory system, and the gastrointestinal tract, where it is metabolized. It accumulates primarily in fat, but it can be also found in other organs, such as liver, kidneys, brain, and muscles.

Despite a ban on all uses, chlordane is still produced for export to Third World countries.

As with other OCPs, short-term exposure to high doses of chlordane can affect the CNS. Long-term exposure can cause anemia and different forms of leukemia and can damage the liver. It reduces fertility in rats.

6. Mirex and Chlordecone

Mirex and chlordecone are two different pesticides, but with similar chemical structure, used from the late 1950s to the early '80s. Mirex was used for fire ant control and flame retardance; chlordecone was used as an insecticide on different kind of trees and in ant and roach traps. There is evidence that mirex degrades extremely slowly to its monohydro- and dihydro-derivatives and to chlordecone. Since these compounds are not likely to undergo chemical or biological degradation, they can be considered persistent in the environment. They adsorb very strongly to organics in soil, aquatic sediments, and particles in the atmosphere. Mirex and chlordecone build up in aquatic organisms and, as a consequence, enter the food chain. It is not known how mirex affects our health; however, people exposed to high levels of chlordecone for long periods showed effects on the CNS, skin, liver, and reproductive system. Even though there is no direct evidence that these compounds cause cancer in humans, they can be anticipated to be carcinogens.

MIREX

CHLORDECONE

7. Hexachlorobenzene

Hexachlorobenzene (HCB) is another OCP used, until 1965, to protect crops against fungi. It has also been employed in the munitions and rubber industries. Because of its high toxicity, it has been withdrawn or severely restricted in most countries.

HEXACHLOROBENZENE

Hexachlorobenzene is a very stable compound and, consequently, is highly persistent in the environment. When released in soil it adsorbs strongly and does not biodegrade, and it is unlikely to contaminate groundwater. In water it partitions to sediment; however, it volatilizes rapidly. Biodegradation and hydrolysis are not significant. Like the other OCPs, it shows a very low solubility in water and remains in the sediments of water basins for a long time, stuck strongly on particle surfaces. It enters the food chain and bioaccumulates in both plants and animals. Exposure occurs from eating contaminated food or eating milk or dairy products from contaminated cattle. Drinking polluted water as well as breathing contaminated air are also sources of exposure.

HCB interferes with lipid metabolism and transport, and it affects liver enzymes. It may affect the central nervous system to differing degrees, depending on the amount and time of exposure, and it can lead to dermal irritation and liver and kidney damage. Thyroid enlargement, anemia, and pulmonary and stomach damage have also been reported. Studies on laboratory animals have demonstrated that exposure to HCB can induce cancer of the liver, kidneys, and thyroid. Although there is no evidence on humans, HCB is suspected to be a carcinogen.

III. PREANALYTICAL TECHNIQUES

The overall analytical procedure to be used in the determination of organochlorine pesticides present in water requires extraction from the aqueous matrix, cleanup, and enrichment procedures prior to the analysis by GC-ECD or GC-MS.

As previously stated, OCPs represent a toxic and ubiquitous class of compounds. Therefore, they are included among those chemical species that need to be analyzed, even though present in very low concentrations, such as parts per billion (ppb). Furthermore, when the molecular weight increases, the chromatographic separations of the compounds of interest from interfering compounds become more and more difficult, particularly when the chemical structure of the interfering compounds is similar to that of the analytes.

For these reasons and because of the need to push the analytical procedure to the minimum detection limits technologically available, the analysis of such compounds requires highly sophisticated instrumentation and, more important, careful cleanup and preconcentration procedures.

The final aim of preconcentration procedures is to obtain the sample to be analyzed in a suitable solvent within a concentration range compatible with the sensitivity and detection limits of the instrumentation used. One of the major problems related to preconcentration methods is the possibility of severe and/or nonreproducible losses of the analytes during sample manipulation. Thus, the procedure adopted should be carefully evaluated for sample losses and reproducibility. In the following sections the various preconcentration techniques formerly and currently applied to the analysis of OCPs will be examined.

A. Liquid–Liquid Extraction (LLE)

Methods for the extraction of OCPs from water-grab samples, which made use of separative funnels to perform a liquid–liquid extraction (LLE) of the organic compounds from the aqueous matrix, have been described since the early 1960s (1). The LLE technique is based on the low value of the partition coefficient for the organic compounds between water and organic solvents, and it is particularly advantageous in trace analysis when the compounds of interest have a very high solubility in the organic solvent and a very low solubility in water so that a relatively small amount of solvent is able to extract a relatively large amount of water. The EPA method 508 (2) involves the use of methylene chloride as the organic solvent to be added, in a separative funnel, to a 1-L sample of water. Threefold replicate extractions must be performed, adding 60 ml of solvent each time. The extract is then dried and exchanged to hexane during concentration to a volume of ca. 10 ml.

Liquid–liquid extraction can be also carried out almost automatically utilizing continuous liquid–liquid extractors (3), exploiting the lower boiling point and density of methylene chloride as compared to water. The vapors produced in the flask containing the organic solvent bubble in the water and extract the organic compounds. The liquid solvent is collected and reintroduced into the original flask by means of a siphon system. The process is quite similar in principle to that of Soxhlet extraction.

Using LLE as the extracting technique, several interferences may occur due to the presence of contaminants in solvents and glassware and to the presence of phthalate esters that are ubiquitous in a laboratory. Furthermore, matrix interferences caused by contaminants coextracted from the actual sample may occur as well. However, the presence of matrix interferences can vary considerably according to the nature and source of the water sample, and only when the actual sample is not particularly clean is a cleanup procedure required. In that case, a Florisil column should be used to fractionate the different compounds and to eliminate polar interferences, by elution with solvents of increasing polarity. To remove elemental sulfur, activated copper powder is used.

In recent years, a micro LLE method has been described. Extraction is performed on 400-ml water samples extracted once with 500 μl of toluene. Extracts are then analyzed directly, without any further treatment, by gas chromatography–electron capture detection (GC-ECD) (4). The use of LLE as an extraction technique has several drawbacks:

The time necessary to perform the extraction is quite long.
It uses relatively large volumes of expensive and potentially toxic solvents.
The formation of emulsions resulting in analyte losses is likely to occur.
The extensive use of glassware can contaminate the sample.
Solvent impurities are magnified as well, thus affecting the analysis of trace compounds.
Preconcentration of the sample prior to analysis is often required.
Evaporative losses of analytes may occur.
Repeatability is not always satisfactory.
Only a fraction of the extraction liquid is injected, so potential sensitivity is lost.

However, an interesting new LLE technique was recently developed by Cramers and coworkers (5). The method implies extraction of the analytes of interest from an aqueous matrix by passing the water sample through a sorption cartridge containing particles consisting of a polymeric liquid phase. As extraction phase, polydimethylsiloxane (PDMS) was used, which appears to be a solid but has sorptive characteristics similar to those of a liquid phase. Retention of analytes is not based on adsorption of the solutes onto the surface of the PDMS material; rather, the solutes are dissolved (partitioned) into the bulk of this high-viscosity liquid phase. The extraction cartridges are then thermally desorbed to ensure full transfer of all analytes onto the GC column. In this way the consumption of organic solvents is eliminated and maximum sensitivity is attained, since all solutes trapped from the sample are actually introduced into the GC column. In this respect, such a technique proved to be more advantageous than solid-phase extraction (SPE) and solid-phase microextraction (SPME), described in the following sections. In SPE, based on adsorption of analytes onto an active surface, only a fraction of the desorption liquid is injected, thus affecting the sensitivity of the method. In SPME the active part is a sorbent material very similar to that used in this "new" liquid–liquid extraction. However, the two methods differ dramatically, since SPME is an equilibrium method; meanwhile, in LLE the extraction is complete, thus enhancing the overall sensitivity. Such a novel method was shown to be particularly suitable for the analysis OCPs in water. For these relatively apolar compounds, quantitative extraction can be obtained for sample volumes up to at least 100 ml, yielding detection limits in the low ppt range. The method can therefore be considered an attractive alternative for certain LLE, SPE, and SPME applications.

B. Solid-Phase Extraction (SPE)

To overcome the drawbacks typical of classical liquid–liquid extraction, and particularly to avoid any large consumption of hazardous chlorinated organic solvents, in the 1970s solid-phase extraction methods were introduced as an alternative to LLE. The SPE technique is based on the principle of liquid–solid chromatography. A polypropylene cartridge (Fig. 1) is filled with an adsorbent in a fine mesh range (150–400). Water passed through such a cartridge should not be retained, nor should it produce any significant modification in the physicochemical properties of the adsorbent, which, instead, should completely retain the organics dissolved in the water. Moreover, water should not behave as an eluent for the organics under analysis. In this way, large amounts of polluted water can be passed through the cartridge, even in liter quantities, leaving the organic compounds fully adsorbed. The retention volumes of

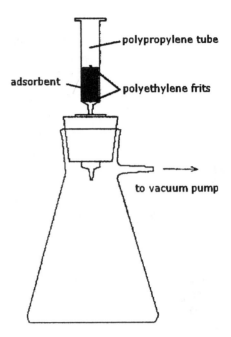

Fig. 1 SPE apparatus.

organic compounds should be nearly infinite when the mobile phase is water. Once the desired amount of water is passed through the cartridge, the latter is dried by means of a nitrogen flow and is eluted with an organic solvent. In this situation the retention volume of the organics should be nearly zero, so the pollutants can be eluted rapidly with a retention volume only a few times higher than the dead volume of the cartridge.

Several advantages are obtained by operating in this way:

The method is fast, simple, and low in cost.
Sample handling is reduced and a lower volume of glassware is necessary.
Extraction and preconcentration are performed in a single step.
The formation of emulsions do not take place.
Very small volumes of solvents are used and the solvent removal step either is eliminated or is very quick.
Extraction can be performed on line.
Sample storage space is considerably reduced.
At least a 1,000-fold enrichment is achieved in a single step.
The recovery of the organics from the adsorbent is complete and reproducible.

The working mechanism of the cartridges is strongly dependent on the mutual properties of water, of the solvent, of the adsorbent used, and, of course, of the molecules involved in the enrichment process. Therefore, a specific enrichment process for every class of analytes has to be developed. The adsorbent to be used should preferably be a nonpolar, nonspecific, nonporous solid with a substantial surface area. In the absence of these properties, irreversible adsorption of the molecules of interest may occur with either severe sample losses or the necessity of using large amounts of solvent, thereby spoiling the preconcentration method.

In the early days when SPE was introduced, porous polymers such as Tenax and polystyrenedivinylbenzene (XAD-2) were used as the adsorbents (6,7). With the important and rapid progress made in the synthesis and availability of new materials for reverse-phase HPLC (C_8–C_{18} bonded silica), these materials have become very popular for SPE (8–10). These materials have a smooth surface that is made nonpolar by the attached carbon chain. Absence of small pores is another advantage. Furthermore, unlike porous polymers and resins, once cleaned by passing water and some organic solvent in the SPE bed, they do not give rise to contamination of the water and solvents.

Another class of adsorbents, graphitized carbon blacks (GCB), exhibits the same positive characteristics as the bonded silicas, but is also more retentive and absolutely nonpolar and nonporous. With GCBs, rather good results have been obtained for classic chlorinated pesticides (11–14).

Recoveries of selected OCPs from different adsorbents are reported in Table 3.

In the first papers on the use of SPE, larger amounts of adsorbent were preferred, but today, because of the advancement of the analytical equipment and because of the higher attention given to the choice of the adsorbents and solvents, the trend is to use a small amount of water and a small amount of solvent. As an example, the already mentioned paper by Junk and Richard (9) described the determination of many pesticides present in concentrations of about 0.1 ng/ml using 100 ml of water extracted by means of a cartridge containing only 100 mg of silica C_{18} with 0.1 ml of ethyl acetate. The great advantage of this procedure is in the combination of small water volumes, fast flow rates, small columns, and small eluate volumes. Of course, the use of this method is only possible when the determination is limited to within the ppb range.

Later on, membrane disks were introduced instead of cartridges. Disks are made of a network of PTFE fibers in which C_{18}-bonded silica is enmeshed to form a strong porous membrane (15,16). They are advantageous with respect to the SPE cartridge, whose narrow internal diameter limits the flow rate and whose small cross-sectional area is easily clogged by suspended solids, thus prolonging the extraction of a large-volume sample. Using membrane disks that have smaller particles, a large diameter, and a shorter length, the extraction process is faster while maintaining its effectiveness. It has also been stated that the performance of membrane disks is enhanced by placing two or three of them in series. Furthermore, they can be used online (Fig. 2).

Solid-phase extraction was accepted as an EPA method for the determination of OCPs and other organic compounds in drinking water (17). Method 525.2 uses a solid-phase extraction based on commercial SPE cartridges or disks. The cartridges are packed with re-

Table 3 Recoveries of Some Chlorinated Pesticides from Different Adsorbents

Pesticide	Recovery (%)			
	GCB	Tenax	Porapack P	C_{18}
α-BHC	93	81	55	95
β-BHC	100	81	60	93
γ-BHC	100	77	51	93
Heptachlor	97	94	70	96
δ-BHC	97	94	50	96
Aldrin	96	88	71	88
Heptachlor epoxide	95	100	63	99
4,4′-DDE	100	87	77	93
Dieldrin	100	95	80	95
Endrin	99	89	75	94
4,4′-DDD	100	83	63	92
4,4′-DDT	100	86	55	95

Source: Ref. 14.

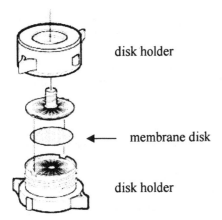

Fig. 2 In-line extraction disk holders.

verse-phase liquid chromatography packing materials. The disks are made of Teflon containing silica coated with a chemically bonded C-18 organic phase. A 1-L water sample is passed through the cartridges or disks, and the OCPs are sorbed on the solid phase. After air-drying, the organic compounds are eluted using a very small volume of an organic solvent.

One of the major problems connected with the analysis of organics in water using SPE as preanalytical procedure is that the breakthrough volume (BTV) of the compounds of interest may become much lower than forecast by spiking experiments, due to the presence of naturally occurring substances such as fulvic and humic acids, humins, and dissolved organic matter (DOM). In fact, binding interactions with such substances influence the solubility and particle adsorption of hydrophobic compounds, thus affecting the efficiencies of their extraction from water. Johnson et al. (18) have effectively shown that the SPE of water containing a commercially available humic acid failed when silica C_{18} cartridges containing 500 mg of the adsorbent were used, thus yielding inferior recoveries with respect to LLE, and also have shown that the recovery changed according to the different pesticides tested. However, such a difficulty can be overcome either by using a larger amount of adsorbent or by eliminating the interfering compounds prior to extraction by chemical oxidation. When particulate matter is present in water, which is quite normal when analyzing lowlands river or lake waters, dwell water, or marine water, most organics are adsorbed on the particulate itself, which passes undisturbed through the adsorbing bed. However, particulate can easily be eliminated by placing a fiberglass filter before the SPE cartridge or high-density glass beads to be used on top of SPE extraction disks.

In a recent paper (19), where the chemical stability of some pesticides extracted from seawater on C_{18} disks was evaluated, the conditions affecting SPE were investigated as well. Unfiltered seawater samples exhibited a lower recovery compared to filtered seawater, with recoveries ranging from 49.8 to 72.3% and from 64.8 to 110.1% for unfiltered and filtered water, respectively. Therefore, filtered seawater showed similar recoveries when compared to spiked clean water samples. Filtration appears to be necessary when the level of humic acids or particulate matter is higher than 10%. For samples heavily contaminated with particulates, these should be analyzed separately from the water for adsorbed OCPs in order to determine the pollution load for the whole sample.

In conclusion, it can be stated that SPE is an effective method for extracting OCPs from water, but problems linked to the presence of dissolved organic matter and/or particulate must be taken into account.

C. Solid-Phase Microextraction (SPME)

Solid-phase microextraction (SPME) was introduced at the end of the 1980s by Pawliszyn and coworkers as a technique for extracting organic micropollutants from aqueous matrices (20–23). This technique involves exposing a fused silica fiber that has been coated with a nonvolatile polymeric coating to a sample or its headspace (Fig. 3). The absorbed analytes are then thermally desorbed in the injector of a gas chromatograph for separation and quantitation. Therefore, no solvent extraction is required.

The fiber is mounted in a syringelike holder that protects the fiber during storage and penetration of septa in the sample vial and in the GC injector. This device is operated like an ordinary GC syringe for sampling and injection. The extraction principle can be described as an equilibrium process in which the analyte partitions between the fiber and the aqueous phase, according to the following equation:

$$n = \frac{K_{fs}V_fC_0V_s}{K_{fs}V_f + V_s}$$

where n is the mass of analyte adsorbed by the coating, V_f and V_s are the volumes of the coating and of the sample, respectively, K_{fs} is the partition coefficient of the analyte between the coating and the sample matrix, and C_0 is the initial concentration of the analyte in the sample. Because the coatings used in SPME have strong affinities for organic compounds, K_{fs} values for targeted analytes are quite large. Therefore, SPME is very effective in concentrating analytes, and it leads to

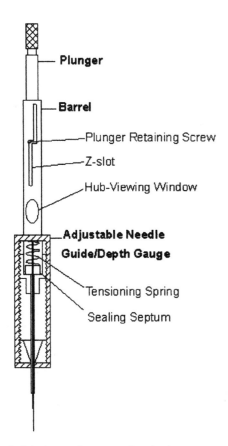

- Plunger
- Barrel
- Plunger Retaining Screw
- Z-slot
- Hub-Viewing Window
- **Adjustable Needle Guide/Depth Gauge**
- Tensioning Spring
- Sealing Septum

Fig. 3 Solid-phase microextraction device.

good sensitivity. Since SPME is a process dependent on equilibrium more than total extraction, the amount of analyte extracted at a given time is dependent upon the mass transfer of an analyte through the aqueous phase. Therefore, a shorter equilibrium time can be attained by simply agitating the solution by means of a magnetic stirrer.

The main advantages of SPME over other preanalytical methodologies can be summarized as follows:

It is fast, simple, and inexpensive.
It is solventless.
Because of its cylindrical geometry, the fiber cannot be plugged.
It can easily be automated using a conventional autosampler.
It is portable and therefore amenable to field use.
It is compatible with GC and LC.
It has a large linear dynamic range while retaining excellent detection limits.
It can be used both as a fast screening technique and in the quantitative analysis of selected compounds.

Since its introduction, SPME has found numerous applications in the analysis of different classes of compounds present in various matrices. Several analytical methods (24–27) for the determination of OCPs in water samples that make use of SPME as an extracting and preconcentrating technique have been described. Magdic and Pawliszyn (24) analyzed environmental water samples for the determination of OCPs using a polydimethylsiloxane- (PDMS-) coated fiber (film thickness 100 μm). The PDMS was preferred to the other commercially available coating, i.e., polyacrylate, because the latter is more polar. In fact, due to the relatively high octanol–water coefficient of OCPs, these analytes are expected to partition more readily into a more nonpolar fiber than a polar one. Optimization of extraction conditions by means of matrix modification was investigated as well. In fact, the more soluble the analyte is in the water, the lower the affinity of the analyte toward the fiber. Therefore, the amount of analyte extracted can be increased by decreasing its solubility in water. This can be achieved by altering the ionic strength via the addition of salt to the matrix or by adjusting the pH of the water. Eighteen OCPs were detected by using such an SPME method coupled either with ECD or MS, obtaining appreciable results. A fully automated analytical method based on in-line coupling of SPME to GC for a continuous analysis of OCPs and other organic contaminants present in surface and sewage water was described (25). Water sample is pumped continuously through a flow-through cell mounted on a commercial GC autosampler, and the fiber is dipped at regular intervals into the flowing sample, thus allowing continual monitoring of OCPs in aqueous systems (Fig. 4). Fibers different than the commercial ones are described as well. A pencil lead modified by a water stream at 800°C was used as SPME device for the determination of OCPs at sub-ppb levels in ground- and surface water (27). Since the equilibrium times quoted for OCPs fall in the range 30–180 min, nonequilibrium SPME can be used for a fast screening of such compounds (26). If a highly sensitive detection system (such as ECD) is available, a reduction in extraction time is possible; in fact, linear responses having good precision are possible even by using extraction times shorter than equilibrium times. In this paper, an extraction time of 2 min was used, thus producing a further reduction of the sample preparation time.

Because SPME is a partition process, interferences due to dissolved and nondissolved organic matter are not likely to occur.

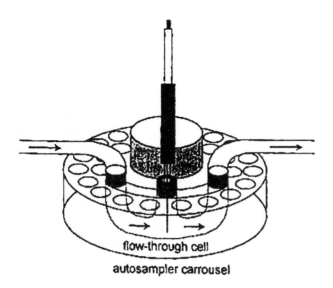

Fig. 4 Automated SPME apparatus. (From Ref. 25.)

IV. COLUMNS

The chromatographic separation of chlorinated pesticides was performed with packed columns until the late 1970s early '80s, when silica capillary columns were not fully developed and available in a wide variety of stationary phases. In most cases the columns used were 2–3 m long, with an internal diameter of 2 mm, packed with a selected stationary phase or with a mixture of stationary phases. The liquid phases most commonly used were siliconic phases, such as OV-17 (34), SP2250, SP2401 (12,13) DC QF-1, SE-30, OV-210, SF-96, OV-225 (28), trifluoromethylpropyl silicone coating a solid inert support such as diatomaceous earths. The operating isothermal temperature was in most cases around 200°C. Columns with different polarity characteristics were selected as a "working pair" to resolve the identification of coeluting peaks. A useful pair is represented by a 1.5% OV-17/1.95% OV-210 and a 5% OV-210. As an alternative, a column 4% SE-30/6% OV-210 can be validly used.

In Fig. 5, the peak elution profile of 13 chlorinated pesticides for the abovementioned columns is shown. Column A shows the coelution of β-BHC and heptachlor; column B shows two coelutions: lindane and β-BHC, and endrin and o,p'-DDT. Column C achieves the best results with the only coelution of β-BHC and heptachlor while three other pairs are separated at 30% of valley. In all cases, the time of analysis is around 18 minutes.

Rus and colleagues (29) proposed a column packed with 2.5% OV-11 + 1% QF + 0.5% XE 60 in order

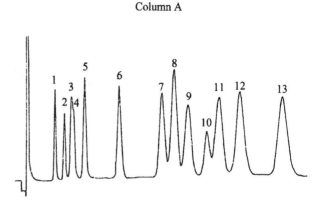

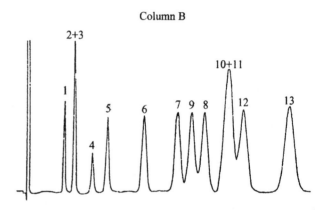

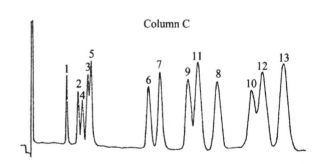

Fig. 5 Peak elution profile of 13 chlorinated pesticides for three different columns. Peak identification: (1) α-BHC; (2) lindane; (3) β-BHC; (4) heptachlor; (5) aldrin; (6) heptachlor epoxide; (7) p,p'-DDE; (8) dieldrin; (9), o,p'-DDD; (10) endrin; (11) o,p'-DDT; (12) p,p'-DDD; (13) p,p'-DDT.

to achieve the separation of hexachlorobenzene from hexachlorocyclohexane isomers (α, β, γ, δ) and of DDT isomers from its metabolization products. Table 4 compares retention times, relative to aldrin, for this column and column A. It can be clearly seen that the

Table 4 Relative Retention Time (RRT, Aldrin = 1) of OCPs

Compound	RRT (column Rus et al.)	RRT (column A)
Hexachlorobenzene	0.45	
α-BHC	0.57	0.53
γ-BHC	0.74	0.67
Heptachlor	0.84	0.83
Aldrin	1.0 (4′06″)	1.0 (4′36″)
β-BHC	1.10	0.77
δ-BHC	1.25	0.77
Heptachlor epoxide	1.45	
p,p'-DDE	2.36	2.19
Dieldrin	2.65	2.36
o,p'-DDD	2.89	2.59
o,p'-DDT	3.25	3.09
p,p'-DDD	4.09	3.41
p,p'-DDT	4.58	4.07

Source: Ref. 29.

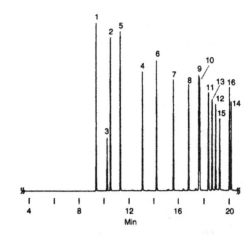

Fig. 6 Separation of 16 chlorinated pesticides obtained with a PTE-5 capillary column, 30 m long × 0.25-mm ID, and 0.25-μm film thickness. *Temperature program*: 4 min at 150°C, then raised to 290°C at 6°C/min, held 5 min. Carrier gas helium at 25 cm/s. Detector ECD. *Peak identification*: (1) α-BHC; (2) γ-BHC; (3) β-BHC; (4) heptachlor; (5) δ-BHC; (6) aldrin; (7) heptachlor epoxide; (8) endosulfan I; (9) p,p'-DDE; (10) dieldrin; (11) endrin; (12) p,p'-DDD; (13) endosulfan II; (14) $p,p,'$-DDT; (15) endrin aldehyde; (16) endosulfan sulfate.

column proposed by Rus et al. allows the separation of the 14 pesticides without coelution in 18.6 min, while the other column does not permit one to resolve β-BHC from δ-BHC in a slightly shorter run time.

Nowadays, fused silica capillary columns are used exclusively for the GC analysis of chlorinated pesticides. Due to their high efficiency and low carrier gas flow rates, they allow the separation of OCPs and the direct coupling with a mass spectrometer, with all the advantages that this powerful technique can offer.

The siliconic stationary phases (methylphenylsilicon polymer) are most frequently used, such as PTE-5. Figure 6 reports GC-ECD analysis of 16 compounds, using a PTE-5 column, 30 m × 0.25-mm ID, 0.25-μm film thickness. The programmed temperature was 150°C for 4 min and then raised to 290°C at 6°C/min and held for 5 min. The detector temperature was 310°C. In this case the time of analysis is slightly longer, but the complete separation of all the compounds of interest is achieved.

V. DETECTION METHODS

The most common device for detecting OCPs is the electron capture detector (ECD). Its high sensitivity and excellent selectivity have made it the detector of choice in pesticide analysis. However, GC-ECD analysis without prefractionation is unlikely to provide unambiguous pesticide identification in the presence of a

wide range of chlorinated organics. In this case the use of mass spectrometry (MS) as confirmation method is required.

A. Electron Capture Detector (ECD)

Since its introduction in the early 1960s, the electron capture detector has played a major role as a gas chromatographic detection technique in the analysis of OCPs and other halogenated organic pollutants. The first applications of ECD to environmental problems appeared in 1961, when two papers were published simultaneously that showed the ubiquitous distribution of chlorinated pesticides (30,31). These works, possible only because of the availability of the ECD, had an enormous impact on the scientific world and on public opinion, giving rise to the great interest in the fate of such compounds of environmental concern.

The principle on which the ECD is based is rather simple. It uses a radioactive β emitter (electrons) to ionize some of the carrier gas and produce a current between a biased pair of electrodes. When organic molecules that contain electronegative functional groups, such as halogens, phosphorous, and nitro groups, pass by the detector, they capture some of the electrons and reduce the current measured between the electrodes.

Thus a negative peak is generated whose area should be proportional to the amount injected.

The first ECDs were equipped with a tritium (^{3}H) adsorbed on palladium β-ray electron source. Later on, due to the necessity for high-temperature cell operation, tritium, which evaporates in these conditions, was substituted with a ^{63}Ni β-ray source, having similar electron energy (0.018 MeV tritium, 0.067 MeV ^{63}Ni) and a higher half-life (12.5 years tritium, 85 years ^{63}Ni). At first ECDs operated with a low constant electric field of 10–20 eV between the electrodes. Later on, pulsed voltage was used instead of constant voltage. The advantage of this mode of operation is that the slow negative ions formed in the gas after collision with thermal electrons are not collected. Moreover, the space charge, due to the formation of positive ions, is strongly reduced in the cell. The main purpose of such a modification to the original design is to increase the linear dynamic range of the detector. A limited linear dynamic range is in fact a major drawback of ECD.

Since the introduction of ECD, the radioactive electron source has remained unchanged, and considerable effort has been made to develop nonradioactive alternatives. A new version of such a detector—the pulsed-discharge electron capture detector (PDECD)—employs a pulsed discharge in helium as the primary source of electron generation (32). A modified version of PDECD that makes use of methane as dopant gas and of a sapphire-and-quartz insulation was used for detecting OCPs (26,33). The relatively low ionization potential of methane reduces interference from extraneous ionization peaks, thus enhancing sensitivity, while the highly inert sapphire-and-quartz insulation allows operation at temperatures up to 400°C. Minimum detectable quantities (MDQs) of OCPs obtained with PDECD are in the mid-femtogram range (Table 5), and

a linear dynamic range of over 3–4 orders of magnitude is attained.

In conclusion, it can be stated that chlorinated pesticide analysis must be considered one of the most important applications of the ECD, since no other detectors have a comparable sensitivity and selectivity. For this reason, ECD is still widely and successfully used for the routine analysis of OCPs (2,4,12,19,26). However, because an extremely high degree of specificity is sometimes required for the analysis of certain compounds, mass spectrometry is the accepted instrumentation for many controversies.

B. Mass Spectrometry (MS)

Mass spectrometry is a destructive technique in which the sample is consumed and removed from the analytical system during the analysis. Bassed on this characteristic, the mass spectrometer should be considered a mass flow detector. In the ion source of a mass spectrometer, energy is transferred to the molecule of interest, causing it to be ionized. Once the molecule is ionized the excess energy serves to decompose it, giving rise to several neutral and charged fragments. The resulting fragmentation pattern constitutes the mass spectrum. A mass spectrum yields two kinds of information of crucial importance: the mass of ions obtained and their abundance. The mass of the molecular ion and of fragments is expressed as mass/charge ratio, abbreviated to m/z. The mass m is equal to the sum of the atomic masses (in daltons) of all the atoms that compose the fragment, and z represents the number of charges carried by the ion. In addition to the fragment pattern, the intact ionized molecule is in most cases stable enough to be detected and its abundance reported. This particular ion is called a molecular ion. Generally, the mass spectrum is specific for each compound, and, under the same conditions, it is always reproducible and represents a sort of chemical fingerprint for sample characterization. In this respect, mass spectrometry can be considered a primary confirmation method.

The most common technique currently used for the production of positive ions is electron impact (EI). Electron impact spectra of selected OCPs are reported in Table 6. As can be seen, the major drawback of EI consists in the generation of low-mass fragments for several OCPs, which might compromise both sensitivity and selectivity. Furthermore the molecular ion is seldom generated.

The total ion current (TIC) acquisition mode is activated when all ion current contributions are summed

Table 5 Minimum Detectable Quantity (MDQ) of PDECD for Pesticides

Pesticide	MDQ (fg)	Pesticide	MDQ (fg)
α-BHC	34	Dieldrin	40
Lindane	36	o,p'-DDD	77
β-BHC	83	Endrin	84
Heptachlor	37	o,p'-DDT	80
Aldrin	38	p,p'-DDD	70
Heptachlor epoxide	40	p,p'-DDT	98
p,p'-DDE	50		

Signal-to-noise ratio = 2.
Source: Ref. 33.

Table 6 EI Mass Spectra of Selected OCPs

Compound	m/z									
Aldrin	66	79	91	163	101	261	65	154	39	92
	(100%)	(42%)	(34%)	(31%)	(28%)	(20%)	(19%)	(17%)	(16%)	(12%)
Dieldrin	79	82	81	108	39	77	27	263	80	277
	(100%)	(42%)	(35%)	(20%)	(20%)	(19%)	(17%)	(16%)	(14%)	(13%)
o,p'-DDE	246	248	318	176	316M	247	105	320	250	210
	(100%)	(68%)	(34%)	(27%)	(23%)	(20%)	(19%)	(18%)	(13%)	(12%)
p,p'-DDE	246	318	248	316M	320	176	105	247	250	281
	(100%)	(79%)	(66%)	(61%)	(40%)	(30%)	(18%)	(18%)	(16%)	(15%)
p,p'-DDT	235	237	165	236	212	239	199	238	176	28
	(100%)	(68%)	(38%)	(16%)	(13%)	(12%)	(11%)	(10%)	(10%)	(9%)
p,p'-DDD	235	237	165	236	239	199	178	238	75	88
	(100%)	(65%)	(41%)	(17%)	(11%)	(11%)	(10%)	(10%)	(9%)	(9%)
Endrin	67	66	209	39	79	27	36	149	101	235
	(100%)	(24%)	(21%)	(20%)	(17%)	(16%)	(16%)	(16%)	(16%)	(14%)
Heptachlor	100	272	274	270	102	65	276	237	135	337
	(100%)	(63%)	(50%)	(35%)	(34%)	(26%)	(22%)	(18%)	(17%)	(15%)
Heptachlor epoxide	81	353	355	351	357	27	53	151	263	51
	(100%)	(94%)	(72%)	(48%)	(32%)	(25%)	(21%)	(19%)	(19%)	(18%)
Hexachlorobenzene	284	286	282M	94	288	142	249	107	251	247
	(100%)	(81%)	(55%)	(39%)	(37%)	(33%)	(28%)	(23%)	(19%)	(19%)
Lindane	181	183	217	219	109	111	51	221	77	185
	(100%)	(86%)	(67%)	(66%)	(60%)	(58%)	(39%)	(34%)	(33%)	(31%)
Methoxychlor	227	28	228	114	15	32	212	152	18	141
	(100%)	(25%)	(17%)	(76%)	(59%)	(59%)	(53%)	(50%)	(48%)	(39%)
Endosulfan I	195	241	197	239	207	237	277	75	243	69
	(100%)	(84%)	(77%)	(76%)	(71%)	(69%)	(68%)	(65%)	(63%)	(60%)
Endosulfan II	29	39	27	31	48	30	41	15	49	63
	(100%)	(30%)	(27%)	(26%)	(22%)	(20%)	(20%)	(17%)	(13%)	(13%)
Endosulfan sulfate	272	274	229	387	270	227	239	237	389	276
	(100%)	(91%)	(77%)	(67%)	(58%)	(58%)	(56%)	(53%)	(48%)	(45%)

Relative mass abundance in parentheses ($\underline{M}$ = molecular ion).

together at the detector output to give a nonspecific chromatographic profile. Each scan is a full spectrum. For quantitative results, scans have to be very rapid, the better to follow the peak shape during its chromatographic elution. The main drawback of TIC appears when trace analysis must be performed: During each scan, fast enough to follow the peak profile, most of the time is lost in the acquisition of low-intensity ion species, since only a few ions in every mass spectrum have enough intensity to generate a noticeable current. This becomes critical below 1 ng for most organic samples. To overcome this problem, the selected ion monitoring technique (SIM) is employed. With this technique the abundance of a few selected ions is followed during the chromatographic run. This implies the detection of specific analytes with the consequent loss of all other information.

Electron impact mass spectrometry is widely used in the analysis of OCPs, employing both the total ion current (34) and the selected ion monitoring (SIM) mode (14,18,35). The SIM technique and total ion current–selected ion extraction (TIC-SIE) mass spectrometry were considered, together with electron capture detection (36). The SIM technique was used as semiquantitative detection and TIC-SIE as quantitative detection, and differences in the precision and detection limit of both methodologies were evaluated by determining their linear dynamic range and response factor values. A higher degree of linearity was obtained for SIM detection, with correlation coefficients closer to 1. Greater measurement precision was evidenced for TIC-SIE and ECD, with lower average response factor % relative standard deviations (RSDs) (5% RSD for TIC-SIE and ECD, vs. 20% RSD for SIM). As expected, the detection limit of SIM was ca. 30 times more sensitive. Table 7 reports the SIM and TIC-SIE fragment ions used in this work.

An alternative ionization technique to EI is positive

Table 7 SIM and TIC-SIE Fragment Ions and Relative Abundances

	Ion (relative abundances %)	
Compound	SIM	TIC-SIE
α-BHC, β-BHC, γ-BHC, δ-BHC	219 (74.9), 181 (100), 221 (38.8), 263 (0.0)	219, 217, 221
Heptachlor	100 (100), 274 (51.9), 272 (64.7), 261 (0.0)	100, 274, 272
Aldrin	263 (100), 261 (64.7), 101 (100), 293 (41.6)	263, 261, 101, 293
Heptachlor epoxide	353 (100), 355 (80.5), 351 (51.9), 357 (38.8)	353, 355, 351, 351
Endosulfan I, endosulfan II	195 (100), 241 (86.5), 207 (74.9), 387 (0.0)	195, 241, 207
Dieldrin	108 (100), 263 (20,1), 277 (15.0), 207 (0.0)	108, 263, 277
4,4'-DDE	246 (100), 318 (80.5), 316 (60.1), 235 (0.0)	246, 318, 248
4,4'-DDD	235 (100), 237 (64.7), 165 (44.8), 178 (11.2)	235, 237, 165
Endosulfan sulfate	272 (100), 274 (92.8), 277 (44.8), 195 (0.0)	272, 274, 277
4,4'-DDT	246 (100), 235 (100), 237 (69.6), 178 (0.0)	235, 237, 165
Endrin	317 (55.9), 315 (38.8), 345 (25.0), 343 (18.7)	317, 315, 345

Source: Ref. 36.

chemical ionization (CI), where the amount of energy involved in the ionization process is lower so that the molecular ion formed is less likely to fragment as compared with that formed using EI. Therefore, CI is very useful for obtaining the molecular ion and, consequently, the molecular weight of compounds. However, compounds with high electron affinity show higher sensitivity and selectivity when electron capture negative chemical ionization (ECNCI) is used. The ECNCI technique makes use of a moderating gas (at high pressure similar to those used for CI) that does not participate in any reaction with the analyte but, rather, slows down the energy of the electrons generated by the filament. In these conditions, the thermal electrons are captured by the predisposed compounds with high electron affinity, generating negative molecular ions. Fragmentation is usually less extended than in EI, and high-mass fragments are more likely to be formed. Advantages of ECNCI over EI and CI include selective ionization in the presence of complex matrices and greater sensitivity. Using methane as nonreactive reagent gas (37), spectra of 24 bridged polycyclic chlorinated pesticides were obtained (Fig. 7).

Ideally, the negative molecular ion is stable enough or undergoes only minor decomposition to high-mass

fragments, but unfortunately the negative molecular ions sometimes decompose to low-mass fragments that are representative of only the electrophilic moiety of the original molecule. For this reason, the use of a derivative that under ECNI conditions would produce a stable molecular ion or fragment ion consisting of a major portion of the molecule of interest is recommended. As an example, the oxygen-induced dechlorination of DDT and its metabolites (p,p'-DDE and p,p'-DDD) can be mentioned (38). Indeed, the reaction between an organochlorine compound and oxygen corresponds to the addition of one oxygen atom with the simultaneous loss of one chlorine atom, thus generating ions at 19 u lower than the molecular ion. Because the oxygen addition–induced dechlorination reactions are regioselective at the aliphatic sites, they could provide some structural information on the various DDT metabolites.

Even if gas chromatography is the separation technique of choice for OCPs, a "multianalysis" system for the automated analysis of environmental water samples was described (39). In such a system, SPE-GC analyses were performed simultaneously with SPE-LC (liquid chromatography) analyses, employing a single mass spectrometric detector operated in total ion cur-

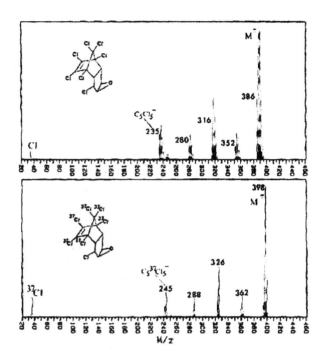

Fig. 7 Methane-enhanced negative ion mass spectra of heptachlor epoxide and ^{37}Cl-heptachlor epoxide. Ion source, 100°C. (From Ref. 37.)

rent mode, using a 45–400 amu (atomic mass unit) scan range for positive-ion EI, and 65–400 amu for NCI (negative chemical ionization) detection. For target analysis, the SIM mode was used. Prior to entering the MS, the LC eluent was allowed to pass through the flow cell of a UV diode array detector (DAD). However, because the UV spectra are not high in information content, DAD was used mainly for a first screening to find suspect samples or peaks. Furthermore, LC-DAD data provided additional means for quantitation and yield-complementary spectral information. The goal of this study was to integrate the three different techniques (GC-MS, LC-MS, and LC-DAD) in one system. The relative standard deviations (RSDs) of retention times were lower than 0.2% in all systems, while RSDs of peak areas ranged from 5 to 15%. Detection limits (DLs) in total ion current mode were below 0.1 μg/L for GC-MS (10-ml samples). For LC-MS, 0.5–7 μg/L and 0.05–1 μg/L DL values were obtained in the TIC and SIM modes, respectively. Negative chemical ionization with methane as reagent gas improved the sensitivity of halogenated compounds 3- to 30-fold and provided relevant information for the structural elucidation of unknown compounds.

REFERENCES

1. J.I. Teasley, W.S. Cox. J. Am. Water Works Assoc. 55: 1093–1098, 1963.
2. USEPA. Determination of Chlorinated Pesticides in Water by Gas Chromatography with an Electron Capture Detector, Method 508, rev 3.1. National Exposure Research Laboratory, Office of Research and Development, U.S. Environmental Protection Agency, Cincinnati, OH, EPA-600/R-95/131, August 1995.
3. L. Kahn, C.H. Wayman. Apparatus for continuous extraction of nonpolar compounds from water applied to determination of chlorinated pesticides and intermediates. Anal. Chem. 36:1340–1343, 1964.
4. A. Zapf, R. Heyer, H.J. Stan. Rapid micro liquid–liquid extraction for trace analysis of organic contaminants in drinking water. J. Chromatogr. A 694(2):453–461, 1995.
5. E. Baltussen, H.G. Janssen, P. Sandra, C.A. Cramers. A novel type of liquid/liquid extraction for the preconcentration of organic micropollutants from aqueous samples: application to the analysis of PAHs and OCPs in water. J. High Resol. Chromatogr. 20:395–399, 1997.
6. V. Leoni, G. Puccetti, A. Grella. Preliminary results on the use of Tenax for the extraction of pesticides and polynuclear aromatic hydrocarbons from surface and drinking waters for analytical purposes. J. Chromatogr. 106(1):119–124, 1975.
7. G.A. Junk, J.J. Richard, M.D. Grieser, D. Witiak, J.C. Witiak, M.D. Arguello, R. Vick, H.J. Svec, J.S. Frizt, G.V. Calder. Use of macroreticular resins in the analysis of water for trace organic contaminants. J. Chromatogr. 99(0):745–762, 1974.
8. D.A. Hinkley, T.F. Bidleman. Analysis of pesticides in seawater after enrichment onto C$_8$ bonded-phase cartridge. Anal. Chem. 23:995–1000, 1988.
9. G.A. Junk, J.J. Richard. Organics in water: solid phase extraction on a small scale. Anal. Chem. 60:451–454, 1988.
10. J.C. Moltò, C. Albeda, G. Font, J. Mañes. Solid phase extraction of organochlorine pesticides from water samples. J. Environ. Anal. Chem. 41:21–26, 1990.
11. R. Petty. Elution of adsorbed organics from graphitized carbon black. Anal. Chem. 53:1548–1551, 1981.
12. F. Mangani, G. Crescentini, F. Bruner. Sample enrichment for determination of chlorinated pesticides in water and soil by chromatographic extraction. Anal. Chem. 53:1627–1632, 1981.
13. F. Bruner, G. Crescentini, F. Mangai, R. Petty. Comments on sorption capacities of graphitized carbon black in determination of chlorinated pesticides traces in water. Anal. Chem. 55:793–795, 1983.
14. F. Mangani, G. Crescentini, P. Palma, F. Bruner. Performance of graphitized carbon black cartridges in the extraction of some organic priority pollutants from water. J. Chromatogr. 452:527–534, 1988.

15. D.F. Hagen, C.G. Markell, G.A. Schmitt, D.D. Blevins. Membrane approach to solid phase extraction. Anal. Chim. Acta 236:157–164, 1990.

16. B.A. Tomkins, R. Merriwaether, R.A. Jenkins. Determination of eight organochlorine pesticides at low ng/L concentrations in groundwater using filter disk extraction and gas chromatography. J. AOAC Int. 75: 1091–1099, 1992.

17. USEPA. Determination of Organic Compounds in Drinking Water by Liquid Solid Extraction and Capillary Column Gas Chromatography/Mass Spectrometry, Method 525.2, rev 2.0. National Exposure Research Laboratory, Office of Research and Development, U.S. Environmental Protection Agency, Cincinnati, OH, EPA-600/R-95/131, August 1995.

18. W.E. Johnson, N.J. Fendinger, J.R. Plimmer. Solid phase extraction of pesticides from water: possible interferences from dissolved organic material. Anal. Chem. 63:1510–1513, 1991.

19. K.K. Chee, M.K. Wong, H.K. Lee. Determination of organochlorine pesticides in water by membranous solid-phase extraction, and in sediment by microwave-assisted solvent extraction with gas chromatography and electron capture detector and mass spectrometric detection. J. Chromatogr. A 736:211–218, 1996.

20. R. Berlardi, J. Pawliszyn. The application of chemically modified fused silica fibers in the extraction of organics from water matrix samples and their rapid transfer to capillary columns. Water Pollution Res. J. Canada 24: 179–185, 1989.

21. C.L. Arthur, J. Pawliszyn. Solid phase microextraction with thermal desorption using fused silica optical fibers. Anal. Chem., 62:2145–2148, 1990.

22. C.L. Arthur, D. Potter, K. Buchholz, S. Motlagh, J. Pawliszyn. Solid phase microextraction for the direct analysis of water: theory and practice. LC-GC 10:656–661, 1992.

23. Z. Zhang, M. Yang, J. Pawliszyn. Solid phase microextraction: a new solvent-free alternative for sample preparation. Anal. Chem. 66:844A–853A, 1994.

24. S. Magdic, J. Pawliszyn. Analysis of organochlorine pesticides by solid phase microextraction. J. Chromatogr. A 723:111–122, 1996.

25. R. Eisert, K. Levsen. Development of a prototype system for quasi-continuous analysis of organic contaminants in surface or sewage water based on in-line coupling of solid-phase microextraction to gas chromatography. J. Chromatogr. A 737:59–65, 1996.

26. G.P. Jackson, A.R.J. Andrews. A new fast screening method for organochlorine pesticides in water by using solid-phase microextraction with fast gas chromatographic and a pulsed-discharge electron capture detector. Analyst 123:1085–1090, 1998.

27. D. Djozan, Y. Assadi. Determination of pesticides in water using solid phase microextraction and capillary chromatography. Proceedings of the 20th International Symposium on Capillary Chromatography, Riva del Garda, Italy, May 26–28, 1998.

28. T.A. Amin, R.S. Narang. Determination of volatile organics in sediment at nanogram-per-gram concentrations by gas chromatography. Anal. Chem. 57:648–651, 1985.

29. V. Rus, I. Funducm, A. Crainiceanu, S. Trestianu. Gas chromatographic column for separation of organochlorine insecticide residues. Anal. Chem. 49:2123–2124, 1977.

30. E.S. Goodwin, R. Goulden, J.G. Reynolds. Gas chromatography with electron capture ionization detector for rapid identification of pesticide residues in crops. Proceedings of the 18th International Congress on Pure and Applied Chemistry, Montreal, Canada, August 1961.

31. J.D. Watts, A.K. Klein. Determination of chlorinated pesticides residues by electron capture gas chromatography. Proceedings of the 75th Annual Meeting of the Association of Official Agricultural Chemists, Washington, DC, October 1961.

32. W.E. Wentworth, E.D. D'Sa, H. Cai, S.D. Stearns. Pulsed discharge electron capture detector. J. Chromatogr. Sci. 30:478–485, 1992.

33. H. Cai, W.E. Wentworth, S.D. Stearns. Characterization of the pulsed discharge electron capture detector. Anal. Chem. 68:1233–1244, 1996.

34. E.J. Bonelli. Gas chromatograph/mass spectrometer techniques for determination of interferences in pesticide analysis. Anal. Chem. 44:603–606, 1972.

35. C. Bourbon, K. Levsen. Analysis of pesticides and their degradation product in rainwater. Proceedings of the 20th International Symposium on Capillary Chromatography, Riva del Garda, Italy, May 26–28, 1998.

36. A. Robbat Jr., C. Liu, T.Y. Liu. Field detection of organochlorine pesticides by thermal desorption gas chromatography–mass spectrometry. J. Chromatogr. 625: 277–288, 1992.

37. E.A. Stemmler, R.A. Hites. Methane enhanced negative ion mass spectrometry of hexachlorocyclopentadiene derivatives. Anal. Chem. 57:684–692, 1985.

38. F.L. Lépine, S. Milot, O.A. Mamer. Regioselectivity of oxygen addition induced dechlorination of PCBs and DDT metabolites in electron capture mass spectrometry. J. Am. Soc. Mass Spectrom. 7:66–72, 1996.

39. J. Slobodník, A.C. Hogenboom, A.J.H. Louter, U.A.Th. Brinkman. Integrated system for on-line gas and liquid chromatography with a single mass spectrometric detector for the automated analysis of environmental samples. J. Chromatogr. A 730:353–371, 1996.

27

Residue Analysis of Carbamate Pesticides in Water

Evaristo Ballesteros Tribaldo
University of Jaén, Jaén, Spain

I. INTRODUCTION

The demand for efficient agricultural production has resulted in the increasing development and subsequent production of a large number of substances and preparations used for the destruction of pests. Traces of pesticides and some of their degradation products are regularly detected in surface water and ground water throughout Europe and North America as a consequence of their widespread use for agricultural and nonagricultural purposes. Priority lists have been published to protect the quality of drinking and surface waters (1).

The World Health Organization (WHO) recommended the classification of pesticides by hazard into five classes on the basis of LD_{50} values (p.o. for rat): extremely hazardous (Ia), highly hazardous (Ib), moderately hazardous (II), slightly hazardous (III), and unlikely to present hazard in normal use (III+). Most herbicides belong to class III+, insecticides and rodenticides fall into groups Ia, Ib, and II, while most fungicides belong to groups II, III, and III+ (2). Pesticides are also divided into distinct classes according to their chemical structure: organochlorines (OCs), organophosphates (OPs), carbamates, triazine herbicides, etc. The OCs, OPs, and carbamates exhibit a large degree of persistence in the environment, which causes various health and safety problems. Due to the persistence of the organochlorine pesticides and the toxicity of the organophosphorous pesticides and their metabolites, the carbamates offer a viable alternative. Generally the carbamate insecticides demonstrate a high insect toxicity but have a low toxicity toward warm-blooded nontarget species, are more biodegradable and less persistent than the organochlorine pesticides, and have relatively less toxic decomposition products (3).

Many gas chromatographic methods have been developed for carbamate determination. Some reasons for derivatizing carbamates include overcoming their problem of thermolability during direct gas liquid chromatographic analysis, increasing detector sensitivity, increasing volatility, and improving chromatographic separation (3). Other chromatographic techniques frequently used for the determination of carbamates include thin-layer chromatography (4) and high-performance liquid chromatography (5). In many cases, the analysis of environmental samples entails the preconcentration of the sample to improve the sensitivity of the analytical method and to separate the analytes from the matrix (6).

This chapter covers some aspects of the general chemistry, toxicology, and environmental persistence of carbamate pesticides in water, with an emphasis on analytical residue methodology. The discussions are devoted primarily to the carbamate insecticides, especially the *N*-methylcarbamates, but the other carbamate pesticides are considered, albeit in less detail.

II. CHARACTERISTICS OF CARBAMATES

A. Chemistry, Structure, and Nomenclature of Carbamates

The carbamates discussed in this chapter are those that are used mainly in agriculture, which include insecticides, fungicides, herbicides, nematocides, and sprout

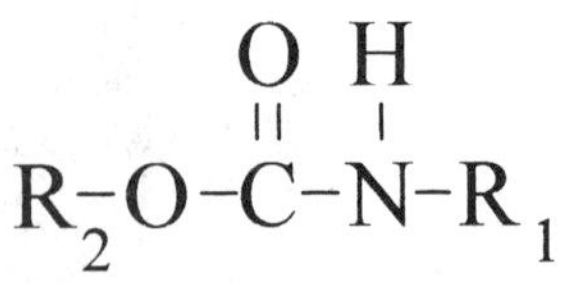

$$\underset{R_2}{} O-\overset{\overset{O}{\|}}{C}-\overset{\overset{H}{|}}{N}-R_1$$

Fig. 1 General structure of carbamate pesticides.

inhibitors. They are also used as biocides for industrial or other applications and in household products.

In most cases, carbamates are N-substituted esters of carbamic acid. Their general formula is presented in Fig. 1, where R_2 is an aromatic or aliphatic moiety. These carbamate pesticides fall into three main groups: (a) carbamate insecticides (R_1 is a methyl group); (b) carbamate herbicides (R_1 is an aromatic moiety); and (c) carbamate fungicides (R_1 is a benzimidazole moiety). There are five principal groups of carbamates as N-substituted esters of carbamic acid: N-methylcarbamates, oxime N-methylcarbamates, N,N-dimethylcarbamates, N-phenylcarbamates, and benzimidazole carbamates. The sulfur analogs of carbamic acid show analogous reactions. Derivatives of thiolcarbamic acid form herbicidal compounds (EPTC), whereas those of dithiocarbamic acid form herbicides (sulfallate) or fungicides (ferbam), and derivatives of ethylenebisdithiocarbamic acid are fungicides (maneb). Table 1 shows the chemical structures and chemical names of some carbamates. The table also gives the pesticidal activity and the dose required to kill 50% of rats (LD_{50}) for each pesticide (3,7,8).

B. Chemical and Physical Properties

The carbamate pesticides exhibit differing stabilities. Thus, N-substituted derivatives of simple esters of carbamic acid are unstable compounds, especially under alkaline conditions, with decomposition resulting in the formation of the parent alcohol, phenol, ammonia, amine, or carbon dioxide. Salts and esters of substituted carbamic acid are more stable than carbamic acid.

In general, carbamate pesticides in a pure state are almost-odorless, white, and crystalline solids of high melting point and low vapor pressure and with variable, but usually low, water solubility; they are moderately soluble in solvents such as benzene, toluene, xylene, chloroform, dichloromethane, and 1,2-dichloroethane. Generally, they are poorly soluble in nonpolar organic solvents such as n-hexane and petroleum ether but highly soluble in polar organic solvents such as methanol, ethanol, and acetone (3).

C. Degradation and Metabolic Processes of Carbamates — Environmental Persistence

Carbamates are generally nonpersistent (relative to the organochlorine pesticides); however, there is evidence that they are sufficiently persistent to have an effect on the aquatic environment.

In general, the pesticides can undergo three general types of degradation processes in the aquatic environment: chemical (hydrolysis), biological, and photochemical. Metabolism of carbamates, whether it be chemically or biochemically via enzyme catalyzed reactions, occurs through three principal routes: hydrolysis, oxidation, and conjugation (3).

In the aquatic environment, factors such as pH, temperature, ionic strength, presence of suspended solids, UV light, and microorganisms have an effect on the persistence of carbamates. There is only a small body of literature on studies of carbamate persistence (relative to other types of pesticides). Aly and El-Dib studied these factors with respect to carbaryl, propoxur, and dimetilan in water and demonstrated that the N,N-dimethylcarbamates are more stable than N-methylcarbamates in alkaline media (9). Vassilieff and Ecobichon showed that, in acid freshwater, aminocarb would be rather stable and would persist long enough to be bioaccumulated by tropic levels of food chains (10). The carbamate fungicides carbendazim, benomyl, and thiophanate are related. Carbendazim is slowly hydrolyzed by alkali to 2-aminobenzimidazole, but it is stable as acid-forming water-soluble salts (11). Benomyl is rather unstable in common solvents (12). Other studies have been conducted on the persistence of carbamates in water (13–19); they show that the carbamates have short half-lives in water but may be prone to total destruction, and small amounts may persist for long periods. Recently, the degradation kinetics of three N-methylcarbamates (carbaryl, propoxur, and carbofuran) and their corresponding phenols (1-naphthol, 2-isopropoxyphenol, and 3-hydroxycarbofuran, respectively) were monitored in river water (20). Figure 2 shows the degradation curves obtained over a 6-week period for these pesticides. As can be seen in this figure, carbaryl was stable for a shorter time (10 h) than propoxur (30 h) and carbofuran (80 h). The phenols started to degrade before the pesticide, except for 1-naphthol (10 h). Lartiges and Garrigues (21) have studied over a 6-month period the evolution of a mixture containing organophosphorous and organonitrogen pesticides at µg/L levels in different water types (ultrapure water, natural seawater, river water, filtered river water) and un-

Table 1 Chemical Structures, Names, and Pesticidal Activity of Some Carbamate Pesticides

No.	Common name Trade or other names Chemical name Activity LD_{50} for rats (oral): mg/kg	Chemical structure
	N-Methylcarbamates	
1	Bendiocarb Bendiocarbe 2,3-Isopropylidenedioxiphenyl *N*-methylcarbamate Insecticide 40–156	O H ‖ | O-C-N-CH₃ (benzene ring fused with dioxolane bearing C(CH₃)₂)
2	Bufencarb Bux, metalkamate 3:1 Mixture of 3-(1-methylbutyl)phenyl *N*-methylcarbamate and 3-(1-ethylpropyl)phenyl *N*-methylcarbamate Insecticide 87	O H ‖ | O-C-N-CH₃ (phenyl with CH(CH₃)C₃H₇) and O H ‖	O-C-N-CH₃ (phenyl with CH(C₂H₅)C₂H₅)
3	Carbanolate Banol, chlorxylam 2-Chloro-4,5-dimethyl *N*-methylcarbamate Insecticide 30–55	O H ‖ | O-C-N-CH₃ (phenyl with Cl, CH₃, CH₃)
4	Carbaryl Sevin® 1-Naphthyl *N*-methylcarbamate Insecticide 540	O H ‖ | O-C-N-CH₃ (naphthyl)
5	Carbofuran Furadan®, NIA-10242 2,3-Dihydro-2,2-dimethylbenzofuran-7-yl *N*-methylcarbamate Insecticide 8–14	O H ‖ | O-C-N-CH₃ (benzofuran with CH₃ CH₃)
6	Landrin® — 4:1 Mixture of 3,4,5-trimethylphenyl *N*-methylcarbamate and 2,3,5-trimethylphenyl *N*-methylcarbamate Insecticide 208	O H ‖ | O-C-N-CH₃ (phenyl with CH₃ CH₃ CH₃) and O H ‖	O-C-N-CH₃ (phenyl with CH₃ CH₃ CH₃)

Table 1 Continued

No.	Common name Trade or other names Chemical name Activity LD$_{50}$ for rats (oral): mg/kg	Chemical structure
7	Methiocarb Mesurol®, metmercapturon, mercaptodimethur 4-Methylthio-3,5,-xylyl N-methylcarbamate Insecticide 135	
8	Mobam — 4-Benzothienyl N-methylcarbamate Insecticide 20–125	
9	Propoxur Baygon® 2-Isopropoxyphenyl N-methylcarbamate Insecticide 128	
	Aminophenyl N-methylcarbamates	
10	Aminocarb Matacil® 4-Dimethylamino-m-tolyl N-methylcarbamate Insecticide <51	
11	Mexacarbate Zectran® 4-Dimethylamino-3,5-xylyl N-methylcarbamate Insecticide 24	

Table 1 Continued

No.	Common name Trade or other names Chemical name Activity LD_{50} for rats (oral): mg/kg	Chemical structure

<table>
<tr><td colspan="3" align="center">Oxime N-methylcarbamates</td></tr>
</table>

12	Aldicarb Temik®, UC-21149 2-Methyl-2-(methylthio)propinaladehyde O-methylcarbamoyloxime Insecticide 0.9	
13	Methomyl Lannate 1-(Methylthio)acetaldehyde O-methylcarbamoyloxime Insecticide 17–24	
14	Oxamyl Vydate, DPX-1410 2-Dimethylamino-1-(methylthio)glyoxal O-methylcarbamoyloxime Insecticide, nematicide 5.4	
15	Thiofanox Dacamox 3,3-Dimethyl-1-(methylthio)-2-butanone O-methylcarbamoyloxime Insecticide 8.5	
16	Dimetilan Snip 1-Dimethylcarbomyl-5-methylpyrazoyl-3-yl N,N-dimethylcarbamate Insecticide <50	
17	Pirimicarb Pirimor 2-Dimethylamino-5,6-dimethylpirimidin-4-yl N,N-dimethylcarbamate Insecticide 147	

<table>
<tr><td colspan="3" align="center">N-Phenylcarbamates</td></tr>
</table>

| 18 | CIPC
Chlorprophan, chloro-IPC
Isopropyl 3-chlorophenylcarbamate (isopropyl
  m-chlorocarbanilate)
Herbicide
5,000–7,000 | |

Table 1   Continued

No.	Common name Trade or other names Chemical name Activity LD_{50} for rats (oral): mg/kg	Chemical structure
19	IPC Prophan 1-Methylethyl phenylcarbamate (isopropyl carbanilate) Herbicide 5,000	
20	Swep — Methyl 3,4-dichlorophenyl carbamate (methyl 3,4-dichlorocarbanilate) Herbicide 552	

Benzimidazole carbamates		
21	Benomyl — Methyl 1-(butylcarbomoyl)benzimidazol-2-yl-carbamate Fungicide >10,000	
22	Carbendazim Carbendazime, carbendazol Methyl benzimidazol-2-yl-carbamate Fungicide >15,000	

Thiocarbamates		
23	Butylate Sutan S-Ethyl di-isobutylthiocarbamate Herbicide 5,366	
24	Cycloate Ro-Neet S-Ethyl cyclohexylathylthiocarbamate Herbicide 3,160	
25	Diallate Avadex S-2,3-Dichloroallyl di-isopropylthiocarbamate (as [E] and [Z] isomers) Herbicide 395	

Table 1 Continued

No.	Common name Trade or other names Chemical name Activity LD_{50} for rats (oral): mg/kg	Chemical structure
26	EPTC Eptam S-Ethyl dipropylthiocarbamate Herbicide 1,367	
27	Molinate Ordram S-Ethyl N,N-hexamethylenethiocarbamate Herbicide 564	
28	Pebulate Tillan S-Propyl butylethylthiocarbamate Herbicide 1,120	
29	Triallate Awadex BW S-2,3,3-Trichloroallyl di-isopropylthiocarbamate Herbicide 1,100	
30	Vernolate Vernam S-Propyl dipropylthiocarbamate Herbicide 1,800	

	Dithiocarbamates	
31	Ferbam Fermate Ferric dimethyldithiocarbamate Fungicide 17,000	
32	Thiram Thirame, Thiuram Tetramethylthiuram disulfide (TMTD) Fungicide 780	
33	Ziram Milbam, Zerlate Zinc dimethyldithiocarbamate Fungicide 320	

Table 1 Continued

No.	Common name Trade or other names Chemical name Activity LD_{50} for rats (oral): mg/kg	Chemical structure
	Ethylenebisdithiocarbamates	
34	Mancozeb Dithane M-45, Manzeb Manganese ethylenebisdithiocarbamate (polymeric) complex with zinc salt Fungicide >5,000	$\left\{\begin{array}{c}\text{H} \quad \text{S} \\ \mid \quad \parallel \\ \text{CH}_2\text{-N-C-S- Mn} \\ \mid \\ \text{CH}_2\text{-N-C-S-} \\ \mid \quad \parallel \\ \text{H} \quad \text{S}\end{array}\right\}_x \text{Zn}_y$
35	Maneb Dithane M-22, Manzate Manganese ethylenebisdithiocarbamate (polymeric) Fungicide 6,750	$\left\{\begin{array}{c}\text{H} \quad \text{S} \\ \mid \quad \parallel \\ \text{CH}_2\text{-N-C-S-Mn} \\ \mid \\ \text{CH}_2\text{-N-C-S-} \\ \mid \quad \parallel \\ \text{H} \quad \text{S}\end{array}\right\}_x$
36	Nabam Dithane D-14, parzate, nabame Disodium ethylenebisdithiocarbamate Fungicide, algicide 395	$\begin{array}{c}\text{H} \quad \text{S} \\ \mid \quad \parallel \\ \text{CH}_2\text{-N-C-S- Na} \\ \mid \\ \text{CH}_2\text{-N-C-S- Na} \\ \mid \quad \parallel \\ \text{H} \quad \text{S}\end{array}$
37	Zineb Dithane Z-78 Zinc ethylenebisdithiocarbamate (polymeric) Fungicide 5,200	$\left\{\begin{array}{c}\text{H} \quad \text{S} \\ \mid \quad \parallel \\ \text{CH}_2\text{-N-C-S-Zn} \\ \mid \\ \text{CH}_2\text{-N-C-S-} \\ \mid \quad \parallel \\ \text{H} \quad \text{S}\end{array}\right\}_x$

der various conditions (21). Table 2 lists some of the half-lives of carbamates in water, including references.

Some studies on the photodegradation of carbamate pesticides have appeared in the literature (22–24). Ultraviolet light has a marked effect on carbamates, and as a result they undergo photodissociation in aqueous media, although the degree of photodecomposition is reduced in deep water. Bertrand and Barceló have studied the photodegradation of the pesticides aldicarb, carbofuran, and carbaryl in distilled water, pond water, and artificial seawater samples with or without humic acids (23). For this study, two different light sources, a xenon arc lamp and a high-pressure mercury lamp, were used to compare the photolysis behavior. The unfiltered high-pressure mercury lamp enhanced the degradation of the carbamate insecticides in comparison with the use of filtered high-pressure mercury and/or xenon lamps, confirming the importance of wavelength on the photodegradation.

D. Toxicology

Pesticides kill by interfering with various biological compounds, and a single reaction or biochemical function may be affected by the pesticide, which produces a causal chain of events leading from chemical interaction to death of the organism. The mode of action of carbamates depends on the type of pesticide.

1. Carbamate Insecticides

This type comprises *N*-methylcarbamates, oxine *N*-methylcarbamates, and *N*,*N*-dimethylcarbamates. Their function is based on their ability to inhibit acetylcholesterase action in the transmission of impulses in the nervous system (25). The carbamate insecticides form a reversible complex with the enzyme, producing a carbylated enzyme. At this stage the transmitter builds up, disrupting normal nerve function (26). The acute toxicity of different carbamates ranges from highly toxic

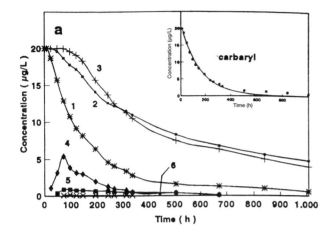

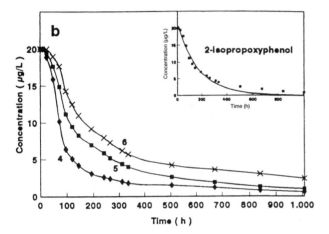

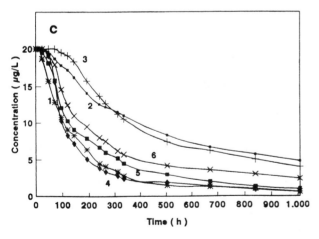

Fig. 2 Degradation curves for (a) *N*-methylcarbamates, (b) phenols, and (c) the mixture of pesticides and phenols spiked at 20 μg/L in river water (10 L) under natural conditions (pH 7.5; ambient temperature 10–30°C; sunlit, closed 25-1 PTFE bottles). (1) carbaryl, (2) propoxur, (3) carbofuran, (4) 1-naphthol, (5) 2-isopropoxyphenol, and (6) 3-hydroxycarbofuran. (Reprinted from Ref. 20, copyright 1996, with permission from the American Chemical Society.)

to only slightly toxic or practically nontoxic. The LD$_{50}$ for the rat ranges from less than 1 mg/kg to over 5000 mg/kg body weight.

2. Carbamate Herbicides

This group includes *N*-phenylcarbamates, thiocarbamates, and some dithiocarbamates. The inhibiting effect of these herbicides is on cell division in the root of the plant (8,26). Thiocarbamate herbicides are volatile liquids normally incorporated into the soil that exert their herbicidal action at the early stages of seedling growth (27).

3. Carbamate Fungicides

This type of pesticide includes some dithiocarbamates, ethylenebisdithiocarbamates, and benzimidazole carbamates. The dithiocarbamate fungicides are respiratory inhibitors and apparently are effective through the formation of heavy-enzyme complexes, such as with lipoic acid and pyravic acid dehydrogenase (3,28). The ethylenebisdithiocarbamates differ chemically from the dithiocarbamates in that they have a reactive hydrogen on the nitrogen atom; this reduces their stability and results in a different biological behavior. The mode of action of ethylenebisdithiocarbamates is not fully understood, but they appear to affect oxidation-reduction reactions (11). Benzimidazole carbamates are systemic fungicides with protective and curative action. They are adsorbed through the leaves and roots of plants, with translocation principally acropetally.

E. Regulations

Natural waters are contaminated with pesticides or their transformation products. Herbicides and fungicides are potential contaminants of natural waters because they are applied directly to the soil and are transported into groundwater or leached to the surface water. Insecticides are transported into groundwater in dust or rainwater, which are washed out by precipitation and fall onto the soil. Due to pesticidal impact on the environment, most countries have established the maximum legal limits for pesticide residues in water. In the same way, different international organizations have published priority list of pesticides. The EEC Directive 80/778, which is concerned with the quality of water designated for human consumption, has established the maximum admissible concentration of each individual pesticide at 0.1 μg/L and the total amount of pesticides at 0.5 μg/L (29). The United States, through the National Pesticide Survey (30,31) organized by the U.S.

Table 2 Half-Lives of Some *N*-Methylcarbamates and Hydrolysis Products in Water

Compound	Water type	Temperature (°C)	pH	$t_{1/2}$	Ref.
Aminocarb	Stream	Env. conditions	7.1	8.7 days	13
	Pond	Env. conditions	5.5	4.4 days	13
Carbaryl	Estuarine	20	8.0	4 days	14
	River	Env. conditions	7.5	109.5 h	20
	Buffered	20	7.0	10.5 days	9
	Buffered	20	8.0	1.8 days	9
	Buffered	20	9.0	2.5 h	9
	Deionized	25	9.0	173 min	15
Carbofuran	River	Env. conditions	7.5	324 h	20
	Paddy (field)	26–30	7.8–8.5	67 h	16
	Pond	26–30	7.8–8.5	55 h	16
	Deionized	27 ± 2	7.0	864 h	16
	Deionized	27 ± 2	8.7	19.4 h	16
	Deionized	27 ± 2	10.0	1.2 h	16
Landrin	Water	38	8.0	42 h	17
Mexacarbate	River (nonsterile)	20	8.2	9.1 days	18
	River (sterile)	20	8.2–8.4	6.2 days	18
	Buffered	20	7.0	25.7 days	18
	Buffered	20	8.4	4.6 days	18
	Buffered	12–13	9.5	ca. 2 days	19
	Buffered	12–13	7.4	ca. 2 weeks	19
Propoxur	River	Env. conditions	7.5	394 h	20
	Buffered	20	8.0	16 days	9
	Buffered	20	9.0	1.6 days	9
	Buffered	20	10.0	4.2 h	9
1-Naphthol	River	Env. conditions	7.5	63 h	20
2-Isopropoxyphenol	River	Env. conditions	7.5	121 h	20
3-Hydroxycarbofuran	River	Env. conditions	7.5	177 h	20

Environmental Protection Agency, established a list of compounds on the amount used, water solubility, and hydrolysis half-life. For more than 20 years the Joint FAO/WHO (Food and Agriculture Organization/World Health Organization) Meeting on Pesticide Residues (JMPR) and the International Agency for Research on Cancer (IARC) have been evaluating toxicity and carcinogenicity data on the different carbamates (25).

III. SAMPLE PRETREATMENT

Carbamate pesticides, unlike the organochlorine and organophosphate insecticides, are not generally amenable to multiresidue gas chromatography (GC) or other chromatographic analysis. A GC analysis of intact *N*-methylcarbamates is hindered by their thermal lability and their lack of sensitivity to electron capture detection; detection of carbamates is usually based on the

detection of a heteroatom using an element-specific detector. These detectors, however, are not specific to individual pesticides, but only to a specific element, and the presence of coextractives or artifacts often makes the job of pesticide identification difficult. Treatment of the sample prior to injection into the instrument is designed to increase the analyte concentration and/or remove substances that might interfere with the determination. The sample pretreatment consists most of the time of extracting traces of pesticides from the aqueous media, concentrating these traces, and removing from the matrix other components (cleanup) that have been coextracted and coconcentrated and that might interfere in the chromatographic analysis (32).

A. Off-Line Extraction Techniques

Although most of the official methods for pesticide analysis in water use liquid–liquid extraction (LLE) because of its simplicity and because it is a fully developed technique (33), solid-phase extraction (SPE) techniques have, however, gained in popularity, and some have already been validated by different official institutions, viz. the Environmental Protection Agency (EPA) and Ames laboratories (34).

1. Liquid–Liquid Extraction

Liquid–liquid extraction is an effective method for extracting pesticides form water samples. The selectivity of LLE is dependent on the solvent used and on the nature of the water matrix. Other parameters, such as pH, ionic strength, water:solvent ratio, number of extractions, and type and concentration of analyte, must also be taken into account (35). In most cases, the solvents used for the LLE of carbamates were: dichloromethane, chloroform, ethyl ether, hexane, and ethyl acetate (36–52). Table 3 summarizes the methods employed. A clean-up step after LLE is necessary in many cases, although in fewer cases than with nonaqueous samples (3). However, LLE methods do have some disadvantages: they are laborious, time consuming, and expensive and are subject to problems arising from the formation of emulsions, the evaporation of large solvent volumes, the disposal of toxic or flammable solvents, impure and wet extracts, nonquantitative extraction, and irreproducible results.

2. Solid-Phase Extraction

Because of recent regulatory pressures to reduce the use of organic solvents in analytical laboratories, there has been considerable development in the area of solid-phase extraction as a reliable and convenient alternative to liquid–liquid extraction. In addition, many modern pesticides and identified degradation products are fairly soluble in water and therefore are less suitable for solvent extraction. One further advantage is that cartridges or precolumns can be loaded in the field, thus avoiding the transport of voluminous water samples, and the stability of many pesticides in the adsorbed form has been demonstrated. In SPE the analytes are sorbed onto a solid while the matrix (water) passes through it. The retained analytes are then eluted with small volumes of an appropriate solvent. The experimental procedure involves the following stages: (i) activation of the sorbent with a solvent; (ii) elimination of the solvent with distilled water; (iii) retention of the analytes; (iv) elimination of interferent compounds with a suitable solvent; (v) elution of the analytes; and (vi) regeneration of the column if necessary. The potential of SPE in the analysis of pesticides in waters can be found in recent reviews on both solid-phase extraction and pesticide analysis (53–58).

The first report of SPE found in the bibliography dates from 1979, where natural water was neutralized and XAD-2 resin was used to extract carbamates, which were eluted with ethyl acetate and finally analyzed by GC (59). Some studies employing SPE in water samples have used C_{18}, C_8, graphitized carbon black, XAD-2, and XAD-7 as sorbent materials, which were placed in columns, cartridges, or disks (52,60–78). The solvents used for the elution-adsorbed carbamates included: dimethyl chloride, hexane, ethyl acetate, and acetonitrile. As can be seen in Table 3, no cleanup step was necessary.

However, problems may occur for certain pesticides stored at 4°C and for many at room temperature (79,80). The losses observed during the storage period are related to the physicochemical properties of the pesticides, e.g., high vapor pressure or high water solubility, which will favor either losses through volatilization or hydrolysis. The type of SPE material is another factor that must be taken into consideration. Problems arise from either the silanol groups present in the C_{18} material (favoring hydrolysis due to the water left after incomplete drying) or the chemical composition of the surface, as on graphitized carbon black (catalyzing reactions or chemisorption) (81).

B. Online Extraction Techniques

Sample pretreatment is still the weakest link and the time-determining step in the whole analytical procedure, accounting for about two-thirds of the total analysis time and being the primary source of errors and discrepancies between laboratories. The development of more rapid and reliable strategies requires the removal of intermediate steps such as transfer, evaporation, and derivatization. It is the current trend to emphasize automation and techniques that maximize sample throughput.

The use of SPE coupled online with liquid and gas chromatography has grown considerably in the last few years, enabling the preconcentration of smaller sample volumes and the achieving of the detection limits required by the current stringent regulations.

*1. Solid-Phase Extraction Coupled Online with
High-Performance Liquid Chromatography*

The coupling of SPE to liquid chromatography can easily be used routinely; this approach has been described by several authors for the online preconcentration of

Table 3 Off-Line Methods for Pretreatment of Water Samples and Determination of Carbamate Pesticides

Carbamate pesticides[a]	Sample	Extraction technique[b]	Cleanup technique	Derivatization	Determination[c]	Chromatographic column	Analytical figures of merit[d]	Ref.
4, 7, 9, 10, 11	River water	LLE (CHCl$_3$)H+	—	—	TLC	—	—	36
18, 19	Water	LLE (ethyl ether) pH 2	Alumina	—	TLC	—	—	37
4, 5	River water	LLE (CH$_2$Cl$_2$)H+	—	Pentafluorobenzyl bromide	GC/ECD	—	—	38
13	Water	LLE (CH$_2$Cl$_2$)	Florisil	—	GC/FPD	—	—	39
4	Distilled, river, well water	LLE (CH$_3$Cl)	XAD-8	Heptafluorobutyric anhidride	GC/ECD	OV-17, Silicone XF-1105, diethylene glycol succinate	LD: 2.5–10 ppb rec: 94–102%	40
2, 4, 5, 7, 9, 10, 11	Water	LLE (CH$_2$Cl$_2$)	Silica gel	2,4-Dinitrofluorobenzene	GC/ECD	—	—	41
5	Rice paddy water	LLE (CH$_2$Cl$_2$—ether 3:1)	—	—	GC	—	—	38
1, 2, 4, 5, 7, 8, 9	Tap, river, stream, lake water	LLE (CH$_2$Cl$_2$) pH 3–4	Silica gel	Pentafluorobenzyl bromide	GC/ECD	—	—	42
23, 24, 25, 26, 27, 28, 30	Water	LLE (iso-octane)	—	—	GC/FPD	—	—	42
5	Runoff water	LLE (CH$_2$Cl$_2$), 0.25 N (HCL)	Celite® 545-MgO-Norit® charcoal	—	GC	—	—	43
5	Rice paddy water	LLE (CH$_2$CL$_2$)	—	—	GC	—	—	44
5	Water	LLE (CHCl$_3$)	—	Hydrolysis (Na$_2$CO$_3$)-derivatized (2,4-dinitrofluorobenzene)	GC/NPD	OV-17-QF-1 (1:1) (1.8 m)	LD: 4 ppb	45
5	Aqueous solution	LLE (CH$_2$Cl$_2$)	Sep-PaK C$_{18}$ cartridge	—	GC/NPD	OV-101 (1.2 m)	LD: 33 µg/L rec: >90% RSD: 3.1%	46
4, 9, 22	River water	LLE (CH$_2$Cl$_2$)	Florisil	Trifluoroacetic anhydride or diazomethane	GC/MS	Ultra-2 (25 m)	LD: 0.014–0.18 ng/ml rec: 83–127% RSD: 2.6–22.6%	47
5	Aqueous solution	LLE (CH$_2$Cl$_2$)	Sep-PaK C$_{18}$ cartridge	—	HPLC/UV	Bondopak C$_{18}$ (30 cm)	LD: 37 µg/L rec: >90% RSD: 2.2%	46
12, 13, 14	Water	LLE (CHCl$_3$)	—	—	HPLC/FL	—	DL: 0.2 ng rec: 77–95%	48
4, 5, 9, 12, 13	Water	LLE (CH$_2$Cl$_2$)	—	—	HPLC/UV	—	DL: 5–10 ng/g rec: 62–109%	49
4	Water	LLE (CH$_2$Cl$_2$/n-hexane) SFE (CO$_2$)	Florisil	—	HPLC/UV	Lichrospher 100 C-18 (12.5 cm)	—	50
4, 5, 9, 13, 14	Water	LLE (CH$_2$Cl$_2$) or SPE (C$_{18}$)	—	—	HPLC/MS	—	DL: 1 ppb rec: 98%	51
4, 5, 7, 9, 10, 11, 12, 13, 14, 15, 19	Water	LLE (CH$_3$Cl)	—	—	HPLC/MS (ionization chemistry)	Supelco LC-18 (25 cm); thermospray interface	DL: 0.4–16 µg/L rec: 71%	52
4, 5, 7, 9, 10, 11, 12, 13, 14, 15, 19	Water	SPE (C$_{18}$), eluted with methanol	—	—	HPLC/MS (ionization chemistry)	Supelco LC-18 (25 cm); thermospray interface	DL: 17–180 µg/L rec: 76% RSD: 17%	
4, 5, 7, 9, 10, 11, 13, 17	Natural water	SPE (XAD-2), eluted with ethyl acetate	—	—	GC/NPD	—	—	59
4, 27	River water, agricultural drains	SPE (XAD-4) eluted with CH$_2$Cl$_2$/acetone/methanol	—	—	GC/NPD or ECD	DB-17 (30 m)	DL: 25 ng/L rec: 47.3–68.0%	60
4, 5	Water	SPE (XAD-2, C$_{18}$) eluted with ethyl acetate	—	—	GC/MS	—	DL: 0.1 ng/L rec: 83–92%	61
4	Groundwater	SPE (Analytichem C$_{18}$) eluted with CH$_2$Cl$_2$	—	—	GC/NPD	BP-10 (12 m)	DL: 1 ppb rec: 106%	62

Number of carbamate pesticides	Sample	Extraction[a]		Derivatization	Detection[b]	Column	DL, rec, RSD[c]	Ref.
5	Shallow well water	SPE (C$_{18}$ cartridges) eluted with ethyl acetate	—		GC/NPD	SPB-20 (60 m)	DL: 0.02–0.05 μg/L rec: 77%	63
4, 5, 23	Water	SPE (XAD-2 and XAD-7) eluted with CH$_2$Cl$_2$	—		GC/MS (ionization chemistry)	DB-5 (15 m)	DL: 0.005–1 μg/L rec: 80.4–101.1%	64
4	Pond and well water	SPE (C$_8$) eluted with CH$_2$Cl$_2$–acetonitrile–hexane (50:3:47)	—		GC/NPD	DB-5 (30 m)	rec: 84–106%	65
24, 27	Drinking water	SPE (C$_8$ silica membrane) eluted with ethyl acetate	—		GC/NPD		DL: 0.02–05 μg/L rec: 67–78%	66
4, 5, 17, 27	River, lake, irrigation channel water	SPE (C$_{18}$) eluted with ethyl acetate–n-hexane (1:1)	—		GC/NPD or ECD	BP-5 (25 m)	DL: 0.065–1.049 μg/L rec: 40–89%	67
4, 17, 27	Water	SPE (C$_8$ and C$_{18}$ disks or column) eluted with ethyl acetate–n-hexane (1:1)	—		GC/NPD or ECD	DB-5 (30 m)	rec: 51–92% RSD: 9–39%	68
4, 5, 13, 14, 18, 19	Drinking and groundwater	SPE (graphitized carbon black cartridge) eluted with methanol–CH$_2$Cl	—		HPLC/UV	LC-18 (25 cm)	DL: 0.003–0.07 μg/L rec: 86–101%	69
1, 2, 3, 4, 5, 6, 7, 9, 12, 13, 14, 15	River and lake water	SPE (C$_{18}$, C$_{18}$–OH) eluted with acetonitrile	—	Postcolumn hydrolysis derivatization with o-phthalaldehyde-2-mercaptoethanol	HPLC/FL	Superphere RP-8 (25 cm)	DL: 20–30 ng/L rec: 76–106% RSD: 0.5–8.5%	70
4, 9	River and seawater	SPE (C$_{18}$ discs) eluted with methanol	—		HPLC/UV	Lichrospher 100 RP-18 (12.5 cm)	DL: 0.01–3 μL rec: >74% RSD: 5–10%	71
5, 7, 12, 13	Ground, drinking, and sea water	SPE (Sep-Pak C$_{18}$) eluted with acetonitrile	—		HPLC/DAD	Hypersil Shandon C$_{18}$ (15 cm)	DL: 0.1–0.3 μg/L rec: 14–110% RSD: 2.2–22.4%	72
18, 20	River, ground, and tap water	SPE (Sep-Pak Light C18 ODS) eluted with acetonitrile	—		HPLC/MS (FAB MS with FRIT interface)	Bondasphere C$_8$ (15 cm)	DL: 0.01–015 μg/L rec: 92.9–98.7%	73
4, 5, 12, 13, 14, 18, 19	River water	SPE (graphitized carbon black cartridge) eluted with methanol and CH$_3$Cl	—		HPLC/MS (particle beam interface, SIM mode)	C$_{18}$ (25 cm)	DL: 0.2–30 ppb rec: 85–101%	74
4, 5, 13, 14, 22, 27	Lake water	SPE (C$_8$) eluted with acetonitrile–CH$_2$Cl$_2$ (1:1), added 10% sodium chloride	—		HPLC/UV	Spherisorb C18 (25 cm)	CL: 0.07–018 μg/L rec: 52–95% RSD: 3–9%	75
5, 29	Milli-Q water	SPE (C$_{18}$, polystyrenedivinylbenzene, discs or cartridges) eluted with THF/methanol	—		HPLC/DAD	Hypersil ODS (25 cm)	DL: 144–1057 μg/L rec: 35–107% RSD: 4.1–14.7%	76
4, 5, 7, 9	Water	SPE (C$_{18}$) eluted with ethyl acetate	—	p-nitrobenzene-diazonium fluoborate (developer reagent)	HPTLC densitometric scanning Chemiluminiscent assays	Silica gel	DL: 300 ng rec: 96.8% RSD: 7.5%	77
5	Tap water	SPE (C$_{18}$)	—			Silica gel	DL: 0.2–1 μg/L	78
1, 4, 5, 6, 7, 9, 10, 12, 13, 14, 15, 17, 18, 19, 20, 21, 22, 23, 24, 25, 26, 27, 28, 29, 30	Drinking water	SPE (RP-C$_{18}$) eluted with methanol or LLE (CH$_2$Cl$_2$)	—		AMD-HPTLC screening	Silica gel	DL: 5–250 ng	182

[a] Number of carbamate pesticides (see Table 1). [a]LLE: liquid–liquid extraction; SPE: solid-phase extraction; SFE: supercritical fluid extraction. [b]TLC: thin-layer chromatography; GC: gas chromatography; HPLC: high-performance liquid chromatography; HPTLC: high-performance thin-layer chromatography; FPD: flame photometric detection; FID: flame ionization detection; ECD: electron capture detection; AFID: alkali flame ionization detection; MS: mass spectrometry detection; NPD: nitrogen-phosphorus detection; UV: UV detection; FL: fluorescence detection; DAD: diode array detection; AMD automated multiple development. [c]DL: detection limit; RSD: relative standard deviation (%); rec: recovery (%).

organic compounds in environmental water samples (32,53–58,82–84).

The easiest way to couple SPE to liquid chromatography is shown in Fig. 3, where a precolumn, filled with an appropriate sorbent, is inserted in the loop of a six-way injection valve. After the sorbent has been conditioned, the sample is loaded by means of a pump and the analytes retained on the precolumn. Then the precolumn is connected online to the analytical column by switching the valve to the inject position. The trapped compounds are then eluted directly from the precolumn into the analytical column by an appropriate mobile phase that is also suitable for the chromatographic separation. In this way, manipulation of the sample is minimal, minimizing its contamination or the possible loss of analytes. Furthermore, the complete sample is introduced in the analytical column, which permits the use of smaller sample volumes than those used in off-line SPE (53). The addition of a second switching valve allows both direct injection onto the analytical column and preconcentration via the precolumn. In online SPE-HPLC procedures the choice of sorbent is determined not only by its efficiency in trapping the analytes of interest, but also by its compatibility with the actual chromatographic system. Generally the sorbent in the analytical column should have retention capabilities the same as or higher than those of the precolumn material, although the precolumn should be as small as possible in order to prevent additional band broadening. Table 4 summarizes the methods for determining carbamate pesticides by using solid-phase extraction online coupled to HPLC (85–95). In some cases, desorption after enrichment is assisted by increasing the temperature (thermally assisted desorption, TAD), because it is necessary to elute in very small volumes (95).

Automation is very easy and several devices are now commercially available [Gynkotek GWBH; Waters Inc.; Tracer MCS® (Kontron Ltd.); Mus®, Multiport Streamswitch (Spark Holland Inc.); Promis®, Programmable Multidimensional Injection System (Spark Holland Inc.); Prospekt® (Spark Holland Inc.); HP 1090 (Hewlett Packard Inc.); and OSP-2 (Merck, Darmstadt)] (53). Two of these commercial systems can be observed in Fig. 4. Trace enrichment was executed by means of an OSP-2 online sample preparator from Merck, combined with an L-6200 intelligent pump and an AS-4000 autosampler, installed with a 5-ml loop; or a Prospekt online sample preparation system from Spark Holland, consisting of the Prospekt apparatus, a solvent delivery unit (SDU), and a Marathon autosampler, equipped with a peristaltic pump and a 3-ml loop. The whole sequence can be programmed and can be readily performed on a sample while the online analysis of a previous sample is still being carried out. Some authors have used these commercial systems for the determination of carbamate pesticides in water samples (96–98). This full automation has been used for on-site monitoring of pesticides in surface waters as an early warning alarm system (99–102). These studies

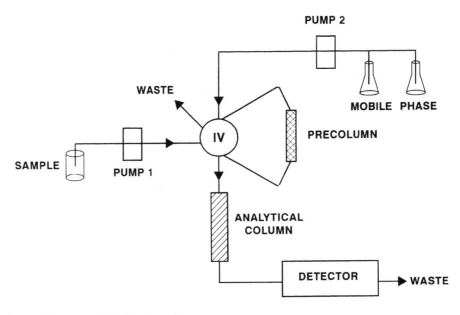

Fig. 3 SPE-HPLC coupling setup. IV: injection valve.

have largely contributed to demonstrating that online SPE-HPLC is a robust and reliable technique that can be applied routinely in the field. No major problems were encountered for over 1000 analyses (over ca. 6-month periods) except for the exchange of a deuterium lamp and the clogging (twice) of the precolumn capillary when using nonfiltered surface water samples (99,100).

2. Pretreatment Systems Coupled Online to a Gas Chromatograph

The GC technique offers the advantages of high separation and a range of very sensitive and selective detection modes. In environmental analysis, the sample pretreatment for GC is almost always an off-line procedure, with the inherent limitation of a final extract volume between 50 and 150 μl, whereas the injection volume is often only 1–5 μl. However, in recent years the increasing use of on-column injections in combination with a column of deactivated silica (retention gaps) and partial concurrent solvent evaporation has allowed a real increase in the injection volume. The main technical problem faced in connecting a liquid treatment system online with a gas chromatograph arises from the difference in the state of aggregation of the fluids crossing the integrated modules (liquid and gas, respectively).

Coupling liquid and gas chromatography online has become more important in analytical chemistry, but almost all applications consist of a normal-phase pre-separation via an LC column and a subsequent separation by means of GC (103,104). An intermediate step involving an online liquid–liquid extraction using an addition of organic solvent and a phase separation has been proposed, but the setup is rather complicated (105–107). A continuous liquid–liquid extractor connected online to a gas chromatograph was proposed for the simultaneous extraction of various N-methylcarbamates (with or without derivatization) in aqueous samples (108,109). Figure 5 depicts the combined extraction analyzer–chromatographic system used. The flow extraction system consisted of a peristaltic pump, pumping tubes, PTFE tubes, and a custom-made phase separator furnished with a fluoropore membrane (110). The extraction unit is connected to the gas chromatograph via an injection valve that permits the introduction of vaporized sample directly into the instrument injection port (111). Shutting the inlet with a stopcock allowed the instrument to be used manually with a nitrogen carrier. This online system has proven to be ef-

fective with various types of samples and analytes (112).

Much effort has been devoted to the online coupling of the SPE pretreatment of aqueous samples with GC. A totally automated online setup with a Prospekt unit for the SPE of water samples was developed (100,113–115). In essence, one is dealing with online SPE-GC, where the SPE procedures are the same as discussed earlier for SPE-HPLC (99–101). The only difference is that sample volumes of 1–10 ml are used, in contrast with the 10–100 ml quoted for LC systems, because of the much greater performance of GC detection. The water samples are preconcentrated on a precolumn, and then the precolumn is dried by a stream of nitrogen. A syringe pump is used for delivering the desorption solvent (normally ethyl acetate) and the analytes are transferred to the GC. Because traces of water can form an azeotrope with ethyl acetate, a retention gap and retaining precolumn are necessary to prevent the introduction of water into the analytical column.

Recently, an online solid-phase extractor coupled to a gas chromatograph was developed for the preconcentration and determination of N-methylcarbamates in water samples (20). The continuous system (Fig. 6) comprised a peristaltic pump furnished with pumping tubes, an injection valve, and a laboratory-made adsorption column, placed in the loop of the injection valve. The interface unit between the SPE system and the gas chromatograph was an injection valve similar to that used for coupling an extraction unit to a gas chromatograph (111). Various sorbent materials (activated carbon, RP-C$_{18}$, and XAD-2) were assayed for the retention of N-methylcarbamates, with the result that XAD-2 is a better sorbent than activated carbon or RP-C$_{18}$.

C. Other Techniques for Sample Pretreatment

Other techniques, such as solid-phase microextraction, microextraction, supercritical fluid extraction, and immunoextraction, have recently been employed for the preconcentration of pesticides in water samples.

Solid-phase microextraction (SPME) was developed and introduced by Pawliszyn and coworkers (116,117). The SPME method is based on an equilibration of the analytes between the aqueous and an immobilized liquid phase coated onto a silica fiber as a stationary phase, e.g., polydimethylsiloxane or polyacrylate polymers. It is particularly appropriate for the direct extraction of volatile and semivolatile compounds from water by dipping the fiber into the aqueous sample.

Table 4 On-Line Methods for Pretreatment of Water Samples and Determination of Carbamate Pesticides

Carbamate pesticides[a]	Sample	Extraction technique[b]	Elution	Derivatization	Determination[c]	Chromatographic column	Analytical figures of merit[d]	Ref.
4	Water	SPE (RP-C$_{18}$)	Catalytic hydrolysis, o-phthalaldehyde reagent		HPLC/FL	RP-C$_{18}$	DL: 0.4–2.0 ng rec: 104–106% RSD: 2%	85
4, 5, 9, 18, 19, 23	Tap, distilled, deionized, commercial spring water	SPE (Spherisorb C$_{18}$, Spherisorb C$_8$, Vydac Reserse Phase TP-201, Ultrasil ODS or Co-Pell ODS)	Acetonitrile	—	HPLC/UV	Supelcosil C$_8$ (25 cm) or Spherisorb C$_{18}$ (15 cm)	DL: 10–460 pg/ml	86
4, 5, 9, 10, 18, 19, 21, 22, 23	Water	SPE (Spherisorb C$_{18}$ cartridge)	Acetonitrile–phosphate buffer	—	HPLC/UV	Spherisorb C$_{18}$ (15 cm)	DK: 10–500 ng/L	87
12, 21, 22	Drinking water	SPE (C$_8$)	Acetonitrile	—	HPLC/UV	C$_8$	range: 2.5–11.0 ng/ml	88
4, 5, 12	Drinking water	SPE (C$_{18}$ Empore discs)	Acetonitrile–methanol	—	HPLC/UV	Supersphere 60 RP-8	DL: 0.01–0.03 μg/L	89
10, 13, 14	River and groundwater	SPE (ODS, PLRP-S, porous graphitic C)	Acetonitrile–phosphate buffer	—	HPLC/UV	Supelcosil C$_{18}$ (25 cm)	DL: 0.1 μg/L rec: 23–100%	90
12, 13	Groundwater	SPE (PRP-1)	Formate–acetonitrile	—	HPLC/MS (thermospray interface; time-scheduled selected-ion monitoring)	ODS-2		91
4, 5, 7, 9, 17, 18, 19	Drinking water	SPE (PLR-s, PRP-1)	Methanol	—	HPLC/MS (thermospray interface)	C$_8$-bonded silica (12.5 cm)	DL: 1–100 ng/L rec: >60% RSD: 1–15%	92
4, 7, 9, 12	River water	SPE (Bondesil-C18/OH)	Methanol–ammonium acetate	—	HPLC/MS or GC/MS	Supelco LC-18 DB (25 cm); or HP-1 (12 m)	DL: 0.1–8 μg/L RSD: 5–35%; or DL: 0.05–4 ng	93
4, 7, 9, 12, 13, 14	Drinking, river water	SPE (graphitized carbon and Empore-activated carbon disks)	Acetonitrile	—	HPLC/UV	Supelcosil C$_{18}$ (25 cm)	DL: 0.05–1 μg/L RSD: 4–14%	94

No. of carbamate pesticides[a]	Sample	Extraction[b,c]	Solvent	Reagent	Detection	Column	DL,[d] RSD, rec	Ref.
5	Water	SPE (divinylbenzene-ethylvinylbenzene copolymer and perfluorinated polyethylene)	Thermally assisted desorption	—	HPLC/DAD or UV/VIS	Supersphere 60 RP-18 (12.5 cm)	DL: 100 ng/L	95
12	Tap, river water	SPE (PLRP-S cartridges) in OSP-2 or Prospekt commercial systems	Acetonitrile, methanol	—	HPLC/DAD	Hypersil C_{18} (10 cm) or LiChrosorb RP-18 (15 cm) or Chromosphere Pesticide (25 cm)	DL: 0.1 μg/L RSD: 0.3–10%	96
1, 2, 4, 5, 7, 9, 12, 13, 14, 15	Drinking, river water	SPE (PLRP-S, C_{18}/OH) in OSP-2 or Prospekt commercial systems	Acetonitrile, methanol	o-Phthalaldehyde reagent	HPLC/FL	Supersphere RP-8 (25 cm)	DL: 30–50 ng/L RSD: 2–10% rec: 60–108.4%	97
4	Freeze-dried waters	SPE (C_{18}) in a Prospekt commercial system	Acetonitrile	—	HPLC/UV (enzymic sensor)	C_{18} (15 cm)	DL: 0.01–0.05 μg/L	98
1, 4, 5, 7, 9, 10	Water	LLE (ethyl acetate)	Acetic anhydride		GC/FID	HP-17 (10 m)	DL: 0.2–0.4 mg/L RSD: 1.9–3.9%	108
4, 5, 9	River, pond, and wastewater	SPE (XAD-2, RP-C_{18}, activated carbon)	Ethyl acetate	—	GC/FID	HP-1 (15 m)	DL: 0.7–1 μg/L RSD: 2.9–3.7% rec: 94–103.5%	20
4	Well and river water	SPE (Bond Elut Cl_{18} cartridges)	CH_2Cl_2	—	FA/FTIR		DL: 15 μg/L RSD: 8.6% rec: 97–103%	134

[a]Number of carbamate pesticides (see Table 1).

[b]LLE: liquid-liquid extraction; SPE: solid-phase extraction.

[c]TLC: thin-layer chromatography; GC: gas chromatography; HPLC: high-performance liquid chromatography; HPTLC: high-performance thin-layer chromatography; FPD: flame photometric detection; ECD: electron capture detection; FID: flame ionization detection; AFID: alkali flame ionization detection; MS: mass spectrometry detection; NPD: nitrogen-phosphorus detection; UV: UV detection; VIS: visible detection; FL: fluorescence detection; DAD: diode array detection; FA: Flow analysis; FTIR: Fourier transform infrared.

[d]DL: detection limit; RSD: relative standard deviation (%); rec: recovery (%).

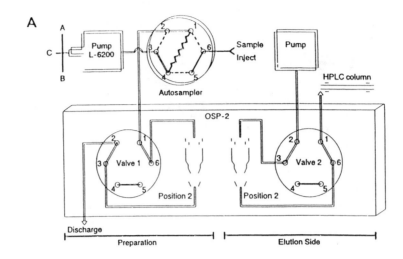

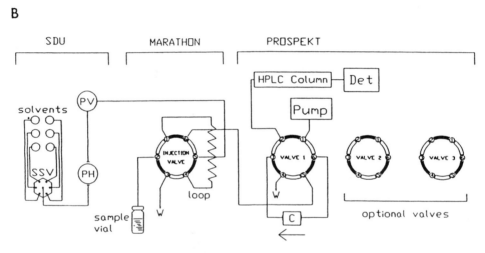

Fig. 4 Details of (A) the OSP-2 system and (B) the Prospekt system. SSV: solenoid valve; PH: purge pump; PV: pulse damper; C: cartridge; DET: detection system; W: waste. (Reprinted from Ref. 97, copyright 1994, with permission from Elsevier Science.)

Hence sampling, extraction, and concentration take place in a single step. After adsorption of the analytes, the fiber is transferred into the heated injector of the gas chromatograph and exposed for a given period of time; there the organic compounds are thermally desorbed from the polymeric phase. Routine analysis of large number of samples is facilitated by easy automation using a modified GC autosampler. The small sample volume used and the absence of organic solvents during the whole process are the main advantages. Pesticides were extracted from water samples (4 ml) onto polydimethylsiloxane- or polyacrylate-coated fiber by stirring for 50 min at room temperature. After extraction, the pesticides were thermally desorbed from the fiber for analysis by GC with N-P (118) or MS

detection (119). Recently, a prototype of an analysis system allowing the quasi-continuous monitoring of organic contaminants in surface water was developed that consists of a flow-through cell and an automated solid-phase microextraction unit, coupled in-line to a gas chromatograph (120,121).

A very interesting alternative to conventional LLE is microextraction (ME), based on the use of an aqueous phase/organic ratio on the order of 100 or 200, where the aqueous solution must be saturated with inorganic salts in order to avoid high irreproducibility due to high phase ratios. This technique is based on the physicochemical principle that the solubility of organic compounds in water is due to their capacity to make hydrogen bonds. A comparative study between

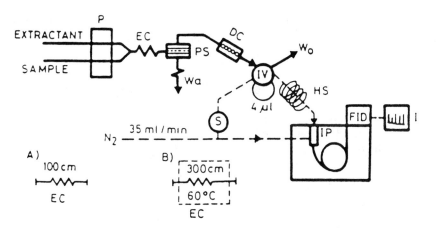

Fig. 5 Flow injection system for the extraction and extraction/derivatization of *N*-methylcarbamate pesticides. EC: extraction coil (A for extraction and B for extraction/derivatization); PS: phase separator; DC: desiccating column; IV: injection valve; HS: heating system; S: tube stopcock; IP: injection port; FID: flame ionization detector; I: integrator. (Reprinted from Ref. 108, copyright 1993, with permission from Elsevier Science.)

LLE, SPE, and ME was conducted for the extraction of N-P pesticides (e.g., primicarb) from aqueous samples (122). In the ME procedure a Kaltron organic phase (immiscible in water) and an aqueous phase saturated with inorganic salts (monohydrated sodium dihydrogen phosphate and anhydrous ammonium sulphate) were used. The best results were obtained with ME, for which preconcentration factors (15–45) were superior to those of LLE (≤13) and SPE (<10).

Supercritical fluid extraction (SFE) appears to be a powerful technique for the extraction of pesticides from environmental matrices; usually after extraction, the samples are analyzed by GC techniques. When compared with the conventional Soxlhet, LLE, or SPE procedures, SFE has various advantages: (a) less expensive in terms of solvent cost and time expended for sample extraction; (b) more selective, since its selectivity can, to a degree, be controlled by varying the fluid pressure and temperature, and by adding small amounts of modifiers; (c) amenable to automation since equipment is now available that allows multiple sample extractions in parallel or sequentially; and (d) requires

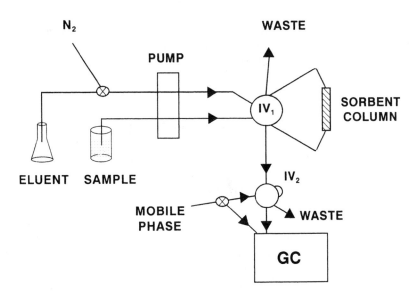

Fig. 6 Setup used for the continuous preconcentration of *N*-methylcarbamates in water samples. IV: injection valve; GC: gas chromatograph.

minimal amounts of solvent to collect the extracted material, thus furthering the concept of pollution prevention (123,124). Two reviews have been published in the last few years regarding the potential of the techniques in environmental analysis (125,126). Some applications of SFE for the extraction of N-methylcarbamates from aqueous samples or samples with high moisture content (e.g., wet soil, fruit, cereal, vegetable, meat) have been developed (50,127–130). The SFE technique was also combined with enzyme immunoassays for rapid detection of aldicarb, carbaryl, carbofuran, and other pesticides in environmental samples (131).

The development of highly selective sorbents involving antigen–antibody interactions has been widely used for the determination of different analytes in the medical field. However, the use of antibodies immobilized in an appropriate support material (immunosorbents) to preconcentrate pesticides from environmental samples has appeared in the last few years as a clear alternative to the traditional sorbents. Since antibody–antigen interactions occur over short distances, steric effects are involved in the coupling reaction. These steric effects are what make the antigen–antibody interaction so selective, and only the antigen that produced the immune response, or very closely related molecules, are capable of binding with the antibody. Therefore, in this type of application only one antibody is used for the selective preconcentration of one analyte (132). This approach has been employed in the determination of carbofuran (133).

A new study to improve the limits of detection in Fourier transform infrared (FTIR) for the determination of carbaryl has been developed. This method consists of a preconcentration step, using conventional SPE cartridges, and online elution with dichloromethane, with absorbance measurements obtained by FTIR (134).

IV. DETERMINATION

The history of carbamate methodologies parallels the development of procedures and instrumentation for other pesticide classes, and also the increase in the use of carbamate pesticides and requests for their analysis. Originally many methods were based on spectroscopic or thin-layer chromatographic (TLC) techniques, with their inherent difficulties, particularly with respect to sensitivity. The development of gas and liquid chromatography as routine analytical techniques represented a major step forward for all pesticide analysis, and the use of sensitive detectors allowed for better quantitation of small amounts of pesticides. In recent years, new methods for the determination of carbamates have been proposed that are based on the technological advances in instrumentation.

A. Classical Methods

Nowadays the classical techniques for the determination of carbamate pesticides in water samples are gas chromatography and high-performance liquid chromatography. The present situation in the field of pesticide analysis can be characterized as a coexistence between these two chromatographic techniques, which sometimes might look like a silent competition to provide maximal information flow in the analyses of particular groups of compounds (135).

1. Gas Chromatography

Gas chromatography has traditionally been the technique most commonly employed in the analysis of pesticides, due to its separation efficiency, the high analysis speed, and the great variety of highly sensitive detectors, including sample coupling with mass spectrometry (MS), which permits the unequivocal identification of unknown compounds. However, the direct CG analysis of N-methylcarbamates often leads to their breakdown in the injection port or in the column during analysis. There are two solutions available for this problem: (a) the preparation of more stable derivatives, and (b) the use of lower temperatures and shorter analysis times.

The reasons for derivatization carbamates, apart from improving their thermolability during GC analysis, include increasing detector sensitivity (particularly with electron capture detection, ECD), increasing volatility, obtaining better chromatographic separations, extending a procedure to multiresidue analysis, and enhancing stability of the compounds. There are two general approaches to the analysis of N-methylcarbamates by derivatization; derivatization of the intact pesticides and derivatization of a hydrolysis product, one of which is always the volatile methylamine. Methods for the derivatization of pesticides have been the subject of excellent reviews (3,136–138). The reactions typically used to obtain derivatives of both intact and hydrolysis products of N-methylcarbamates are methylation (139), silylation (140), halogenation (141), acylation (142), and esterification (143). For analytical purposes, acetyl derivatives (144) and methyl derivatives (145) have been prepared for use with flame ionization detection (FID), but most analysts prefer to introduce electron-capturing moities for use with the more selective and sensitive electron capture detection

(ECD) as halogen derivatives. Reagents such as heptafluoroburic anhydride (40,146), 2,4-dinitrofluorobenzene (41,45,147), trifluoroacetic anhydride (47,148), pentafluorobenzyl bromide (42,149,150), and pentafluoropropionic anhydride (151,152) have been used for this purpose. A method that combines online derivatization-extraction and gas chromatography for the determination of *N*-methylcarbamates in aqueous samples has been reported to reduce human involvement; in it, the acetate esters of the hydrolysis products of six pesticides were formed by reacting with acetic anhydride in a continuous-flow system (108).

Many of the non-*N*-methylcarbamate pesticides may be determined by direct GC of quickly and easily by other techniques, without resorting to chemical derivatization. In many cases, chemical derivatization of these compounds is used primarily as a confirmatory technique, although in a multiresidue application or in natural samples these derivatives could be formed concomitant with the *N*-methylcarbamates (153,154). Much of the attention on the non-*N*-methylcarbamates has been directed at the carbamate herbicides, particularly the *N*-phenylcarbamates, which are similar in structure to the *N*-phenylurea herbicides. Thus, Lawrence demonstrated that CIPC, IPC, and Swep as well as seven *N*-phenylureas could be determined or confirmed as their methylated derivatives (155,156).

In order to avoid the additional labor required for derivatization, another approach in GC carbamate analysis has been to adjust the separation conditions to make them suitable for direct carbamate analysis. Lower temperatures, shorter columns, and temperature-programmed vaporization injectors have usually been used (157). Cold on-column injection marks many *N*-methylcarbamate analyses, for it tends to reduce thermal degradation in the injection port (158). The use of electronic pressure programming at the GC inlet enabled rapid sweeping of the injected sample into the column to reduce thermal decomposition of pesticides on the hot injector surface (159).

Capillary GC, in conjunction with selective detectors [mainly nitrogen-phosphorous (NPD), electron capture (ECD), and flame photometric (FPD)] or mass spectrometry (MS) is still one of the most common techniques for the determination of environmental pesticide residues (Tables 3 and 4). Several reviews on the use of GC-NPD and GC-ECD (135,160,161) and GC-MS in various operational modes, such as electron impact (EI) and positive (PCI) and negative (NCI) chemical ionization, have been published (162,163).

The GC stationary liquid phases range from nonpolar through intermediate polarity to polar. Due to the thermal instability of *N*-methylcarbamates, the use of short columns and nonpolar liquid phases allows these compounds a minimum of column residence time and thus minimizes the possibility of on-column breakdown (3). However, in most cases capillary columns are used, but derivatizing reactions of carbamates are necessary. The commonly used liquid phases in GC are composed of 50% phenyl–50% methyl polysiloxane (HP-17, OV-17, DB-17) (40,45,60,108), 5% phenyl–95% methyl polysiloxane (HP-5, DB-5, Ultra-2) (47,64,65,67,68), or 100% methyl polysiloxane (HP-1, OV-101) (20,46).

2. High-Performance Liquid Chromatography

The use of liquid chromatography for carbamate analysis is another solution for problems encountered during their GC separation. Most HPLC methods for *N*-methylcarbamate pesticides have employed reversed-phase chromatography with C_{18} or C_8 columns and aqueous mobile phases (52,69,70–75,85-90,92–98). Almost all normal-phase methods were reported before 1984; thus diol and nitrile (164) have been used in the normal phase. Sparacino and Hines (165) studied retention and resolution of 14 *N*-methylcarbamates and metabolites in normal- and reversed-phase modes on a variety of columns. Silica, cyanopropyl, and propylamine were studied in normal-phase mode with two mobile-phase systems (isopropanol–heptane and dichloromethane–heptane); C_{18} and ether-phase columns were studied in reversed-phase mode with three mobile-phase systems (water–methanol, water–tetrahydrofuran, and water–acetonitrile). Although normal-phase mode was for the most part satisfactory, reversed-phase mode gave generally superior results.

The UV spectrophotometer has been the detector most widely used in the development of methods for carbamate determination (46,49,50,69,71,75,86–90, 94,95,98). However, at present the diode array detector is usually chosen to obtain the UV spectra of each compound and also to confirm the presence of target compounds (72,76,95,96).

An HPLC separation followed by postcolumn derivatization and subsequent fluorescence detection of carbamate pesticides has been widely recognized for its sensitivity (5,48,70,85,97,166–168). This method consists of a two-stage reaction that converts the carbamate into a fluorescent derivative. The first stage is the hydrolysis of the carbamate molecule (in alkali solution) to release methylamine. The second stage is the derivatization of the released methylamine with *o*-phthaldehyde and 2-mercaptoethanol or 3-mercaptopropionic acid to produce the highly fluorescent 1-hydroxyethylthio-2-methylioindole. A simplification of the post column reaction technique that eliminates the need for both an alkali solution postcolumn pump and solid-phase or photolytic reactors was reported by McGarvey (169). This method was evaluated for 11 *N*-methylcarbamate pesticides in water with a detection limit of 0.1 ng and a relative standard deviation between 0.2 and 2.3%.

Electrochemical LC detectors provide very sensitive, simple, inexpensive, and somewhat selective alternatives to conventional UV and fluorescence detectors for LC. Thus, Anderson and Chesney (170) studied oxidation electrochemical detection of aminocarb, carbaryl, and other pesticides at applied potentials of up to 1.2 V, but they successfully detected only aminocarb, which exhibited a detection limit of 140 pg. These same authors determined aminocarb in water (detection limit 53 pg) using a microarray electrochemical detector (171). Apart from their high sensitivity, the problem with the use of electrochemical detectors is the maintenance and their operational ability. One reason is the accumulation of the reaction products on the surface of the electrode, resulting in blockage of the active surface. To avoid this drawback, in situ cleaning by pulsing the electrode periodically to extreme potentials can be used (172).

Several workers have reported the use of mass spectrometry for the detection of *N*-methylcarbamates. The earliest reports (173) involved the use of a fraction collector to collect carbaryl peaks as they were eluted from a UV detector and confirmation of their identity off-line using low- and high-resolution field desorption MS. The advances in the coupling of LC to MS converts it into one of the most powerful techniques available today to confirm the presence of carbamates in water samples. The most common interfaces used between the LC and MS are thermospray (52,91,92,174) and particle beam interfaces (74,175,176). Recently, Honing et al. have studied the influence of various LC/MS interfaces on the determination of carbofuran (177). With thermospray and particle beam interfaces the quantification of this pesticide is affected by both the ion source pressure and the temperature, whereas quantification using the recently developed atmospheric pressure ionization interfaces, atmospheric pressure chemical ionization, electrospray, and ionspray is less dependent on these parameters.

B. New Trends in Analytical Technology

Other techniques, such as supercritical fluid chromatography, capillary electrophoresis, immunoassays, biosensors, spectrophotometry, and electrochemical meth-

ods have recently been employed for the determination of carbamate pesticides in water samples.

1. Supercritical Fluid Chromatography

Because supercritical fluids are gaslike in some respects and liquid in others (they are typically 10–100 times less viscous than liquids), they can be used as mobile phases, thus forming complementary aids to HPLC and GC. The main advantages of this technique are the shorter retention times involved in the analysis of moderately polar and thermally labile pesticides and the compatibility with most LC and GC detectors (UV, FID, NPD, MS, and FTIR spectrometry).

France and Voorhees (178) have used a capillary supercritical fluid chromatography SFC/MS unit for the quantification of benthiocarb and carbaryl pesticides and herbicides. Diethylcarbamates were subjected to SFC on capillary columns of DB-5, DB-17, and SB-octyl-50 with an NPD detector (179). An off-line SPE/SFC system has been proposed for the preconcentration/determination of carbamates with a detection limit of 110 ng/L (180).

2. Thin-Layer Chromatography

As a result of its low cost and simplicity, thin-layer chromatography (TLC) was frequently used for the determination of carbamate pesticides prior to development of GC. In recent years, TLC applications have been limited to the cleaning of samples, to metabolic studies, and, perhaps most importantly, as one piece of confirmatory evidence in clinical and medical-legal cases in which positive, unambiguous identification and quantitative determination are needed. Thus, Rathore (181) has proposed a spot test for the detection of carbaryl. In this method an ethanolic carbaryl solution was spotted on a glass plate coated with a slurry of silica gel G-water (1:2). After the plate was developed in hexane–acetone, it was dried and sprayed with 1% copper(II) chloride solution followed by 0.1% ammonium metavanadate reagent. A violet spot confirms the presence of carbaryl.

Recently, high-performance thin-layer chromatography with automated multiple development (AMD-HPTLC) was used to screen water samples for pesticides (182). The design of the instrument used in this method is shown in Fig. 7. This unit consists of the actual developing chamber for 20 × 10-cm plates (1), six solvent reservoir bottles, (2) connected via the seven-port motor-driven valve, (3) to the two-step gradient mixer, (4) a wash bottle, (5) for external preparation of the gas phase, the gas-phase reservoir, (6) and

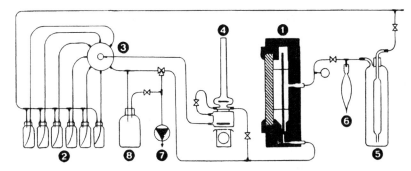

Fig. 7 Flow diagram of the AMD-HPTLC developing unit. 1: developing chamber; 2: solvent reservoir bottles; 3: 7-port motor-driven valve; 4: two-step gradient mixer; 5: wash bottle; 6: gas-phase reservoir; 7: vacuum pump; 8: waste collection bottle. (Reprinted from Ref. 182, copyright 1995, with permission from the American Chemical Society.)

a vacuum-tight tube system with control valves. The AMD-HPTLC method was applied to the determination of 265 pesticides in drinking water spiked at the 100 mg/l level.

3. Capillary Electrophoresis

Most of the published methods using capillary electrophoresis have centered on exploiting the high resolution obtained in the separation of pesticides, but the application of this technique to environmental water samples has not yet been reported.

Süsse and Müller (183) have developed a method for the simultaneous separation of triazines, carbamates, and organophosphorous pesticides by micellar electrokinetic capillary electrophoresis with UV/VIS detection. The detection limit obtained ranged from 0.08 to 0.13 mg/L, with a reproducibility of migration (1.6–2.0%) and peak areas (1.7–5.1%). Dithiocarbamates were also determined by capillary electrophoresis with diode array detection; in it all compounds gave linear calibration graphs up to 50 μg/ml, with detection limits in the 0.1–1-μg/ml range (184).

4. Immunoassays

This technique is related directly to immunosorbent techniques (see Sec. III.C). Immunoassays have been widely used for decades for the determination of compounds of clinical interest; only in the last few years have they been applied to the determination of pesticides in environmental samples. There are various types of immunoassays, but the most frequently used in the determination of pesticides is the enzyme-linked immunosorbent assay (ELISA). This new technology

provided reactions specific to a compound or a group of compounds (185). Thompson (186) has developed an immunoassay for avian serum butyrylcholinesterase and its use in assessing exposure to organophosphorous and carbamate pesticides. An indirect ELISA assay for carbofuran has recently been proposed (187). In this method a detection limit of 0.2 ng/ml was obtained, in which cross-reactivity studies showed that the immunoassay was quite specific for carbofuran, since, of the six N-methylcarbamates assayed, only benthiocarb was significantly recognized.

5. Biosensors

A chemical sensor is a device capable of indicating the concentration of an analyte in a continuous and reversible way. When the sensor involves the use of an enzyme, an antibody of any biological component, it can be called a biosensor. Most biosensors are based on the monitoring of acetyl cholesterase activity and its inhibition by one or a group of pesticides (basically carbamates and organophosphorous). Amperometric biosensors based on the inhibition of acetyl cholesterase enzyme on different substrates have been proposed for the determination of carbaryl or carbofuran (188–190). Dithiocarbamates are also determined by their inhibiting effect on the catalytic activity of a tyrosinase electrode; the amperometric inhibition measurements are carried out at −0.2 V (191) or 0.25 (192) vs. Ag/AgCl.

Biosensors based on lipid membranes have been proposed that are based on the interaction of analytes with a membrane to generate a reproducible transient current signal that varies depending on the analyte. In this way, carbofuran was electrochemically detected by using a bilayer lipid membrane supported in a KCl electrolyte solution containing HEPES buffer and Ca^{2+} (detection limit, 45 pM) (193).

Another type of biosensor involves the use of flow techniques. A flow-through sensor for the determination of carbaryl, carbofuran, and propoxur based on retention of these compounds on C_{18} bonded phase beads packed in a flow-cell or a diode array spectrophotometer was developed (194). Kumaran and Tran-Minh (195) have proposed an FIA system for the quantification of carbamate pesticides in seawater consisting of a single string reactor containing acetyl cholesterase immobilized on glass beads and a pH electrode with a wall-jet entry to assay the activity of this enzyme.

6. Spectrophotometric Methods

Prior to the development of GC, spectrophotometric methods were developed for carbaryl, and some procedures were extended to other carbamates. The general approach is to couple a chromogenic reagent with the parent compound or a hydrolysis product, such as phenol, and determination of the colored complex spectrophotometrically at the appropriate wavelength (196). In this section, new methodologies based on spectrophotometric techniques are reported.

Synchronous spectrofluorimetry in combination with derivative techniques allows the simultaneous determination of various pesticides with similar fluorescent spectral characteristics in water samples without previous separation. In this way, prohan (197) and carbaryl (198–200) have been determined in aqueous samples. A QAE-Sephadex A-25 gel has been used to increase the sensitivity in the determination of carbaryl; the carbamate is fixed in this gel, and finally the fluorescence of the gel is measured by derivative synchronous spectrofluorimetry (201).

Stopped-flow methods for the spectrophometric determination of *N*-methylcarbamates have been proposed by Quintero et al. (202–205). These methods are based on the rate of coupling between the hydrolysis product of the pesticide and diazotized sulphanilic acid (202,203) or on the enzymatic inhibition of acetyl cholesterase by the carbamate pesticide using 5,5-dithiobis(2-nitrobenzoic) acid as chromogenic reagent (204,205). Stopped flow coupled with a diode array detector, using a partial least-squares calibration method, were applied to determine the optimum time for the simultaneous determination of carbaryl and chlorpyrifos (206). Recently, a stopped-flow technique in combination with micelle-stabilized room-temperature liquid phosphorimetry was developed for the de-

termination of carbaryl in irrigation water (detection limit 0.01 μg/ml) (207).

Continuous-flow systems have also been proposed for the determination of *N*-methylcarbamates. In this way, carbaryl was determined by Fourier-transform infrared spectrometry by using a continuous system (208). On the other hand, simultaneous kinetic spectrophotometric determination of carbamate pesticides (carbaryl, propoxur, ethiofencarb, and formetanate) after continuous derivatization of hydrolysis products with *p*-aminophenol were proposed in which a partial least-squares calibration method was used (209). An indirect determination of a diethyldithiocarbamate pesticide (ziram) by atomic absorption spectrometry has been reported by Jiménez de Blas et al. (210). The method was based on the formation of the dithiocarbamate ion–Cu^{2+} complex, which was extracted into isobutyl methyl ketone in a flow injection system.

7. Electrochemical Methods

Polarographic and voltammetric techniques have often been used in the determination of Cl-, P-, and N-containing pesticides (211,212). In this way, electrochemical methods for the determination of carbaryl, after prior oxidation to 1,4-naphthoquinone in natural water, were proposed, using differential pulse polarography (I) and adsorptive stripping voltammetry (II) (213). Detection limits were found to be 0.4 mg/L (previously preconcentrated by SPE on C_{18} Sep-Pak cartridges) and 5 μg/L (without preconcentration) for methods I and II, respectively. A differential pulse voltammetric method was used for the simultaneous determination of carbaryl and carbofuran, based on the anodic peaks observed following their alkaline hydrolysis (formation of phenol derivatives) (214). The overlapped peaks of two carbamate pesticides were resolved using a partial least-squares calibration method.

V. APPLICATIONS IN WATER ANALYSIS

The physicochemical properties of pesticides as well as the point of application directly affect their persistence and distribution in the environment (see Sec. II.C). After application, a significant portion of pesticides remains associated with the soil over long periods of time. This persistence depends on the lipophobicity of the pesticide, the soil mineralogy, the organic matter content, and the soil humidity. All these parameters have a great influence on pesticide leaching to ground- and surface waters. In this way, studies concluded that a pesticide can reach groundwater if its water solubility is higher than 20 mg/L, its partition coefficient between soil organic carbon and water (K_{OC}) is lower than 300–500 ml/g, its soil half-life is longer than about 2–3 weeks, its hydrolysis half-life is longer than approximately 6 months, and its photolysis half-life is longer than 3 days (55). Thus, there are two parameters that affect the mobility of pesticides to reach groundwater via soils: K_{OC} and the soil half-life ($T_{0.5}$). The Commission of European Communities (1) has published a report of an exhaustive study based on the GUS index (Ground Water Ubicity Score), which indicated the leaching probability of a given pesticide. This index can be calculated according to the following equation: GUS = log $T_{0.5}$ × (4 − log K_{OC}). Thus, pesticides are classified as probable leachers (GUS > 2.8), transient leachers (1.8 ≤ GUS ≤ 2.8), and improbable leachers (GUS < 1.8). Only eight carbamate pesticides (aldicarb, carbaryl, carbendazim, EPTC, maneb, methiocarb, prophan, and ziram) show GUS indices higher than 1.8, which are thus potential contaminants of groundwater. Therefore, it is reasonable to assume that pesticides of this kind will also be present in surface water (river, lakes, etc.) as a result of atmospheric deposition, groundwater discharge, and nonagricultural applications, such as highway and railway maintenance close to rivers. They can also affect the coastal seawaters. Tables 3 and 4 summarize the kind of water samples normally analyzed for carbamate pesticides: drinking, tap, well, pond, river, stream, lake, lagoon, groundwater, and seawater. In the majority of cases, recovery studies have been made in samples spiked with carbamate pesticides, because none of the pesticides were detected in natural water. For example, two chromatograms obtained in the analysis of pond water (20) and lake surface water (70) by GC and HPLC, fortified with N-methylcarbamates, are shown in Figs. 8 and 9, respectively. In both cases, solid-phase extraction is used to preconcentrate pesticides prior to chromatographic determination.

VI. CONCLUSIONS

The solving of the dilemma of technique selection for the analysis of carbamate pesticides is usually based on several factors: (a) the behavior of the analyte in the separation/determination system; (b) the need of analyte derivatization and the labor required for it; (c) the personal preference of the analyst for a particular technique; and (d) the availability of the technique in the laboratory and the general purpose of the analysis. In

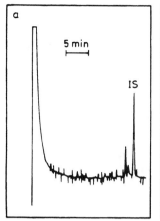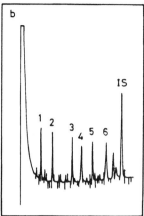

Fig. 8 Gas chromatogram obtained in the analysis of pond water: (a) unspiked and (b) spiked with 20 µg/L N-methylcarbamates and phenols. (1) 2-isopropoxyphenol, (2) 3-hydroxycarbofuran, (3) 1-naphthol, (4) propoxur, (5) carbofuran, (6) carbaryl and (IS) internal standard. (Reprinted from Ref. 20, copyright 1996, with permission from the American Chemical Society.)

the majority of cases, the analytical chemist prefers the GC or HPLC technique because either offers high resolution power and the possibility of incorporating sensitive and specific detectors.

The general concern about pesticide residues gives rise to an increasing need for rapid and reliable methods to determine these compounds at trace levels in water samples. Because the concentrations of pesticides in natural waters are generally low, an enrichment of analytes (normally by LLE or SPE) from the aqueous samples is needed prior to their determination. For this purpose, in recent years there has been an intense development of automated devices in pesticide analysis due to the growing concern about the quality of our environment. Thus, sample throughput is maximized without sacrificing data quality. The coupling of chromatographic methods to immunoassays and biosensors can be of future interest, because they are very fast methods for multiresidue analysis of contaminants in surface waters and groundwaters.

ACKNOWLEDGMENTS

The author wishes to express his thanks to Drs. M. Valcárcel, M. Gallego, and D. Bryce for carefully reading and commenting on the text.

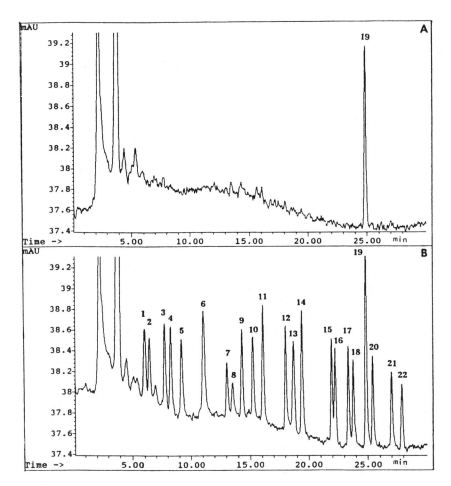

Fig. 9 HPLC separation of Lake IJsselmeer surface water samples after SPE of 50 ml of (A) a blank water sample and (B) water sample fortified with 11 *N*-methylcarbamates and 10 metabolites at the 0.1-μg/L level on 500-mg C$_{18}$, OH cartridges. (5) Oxamyl; (6) methomyl; (14) aldicarb; (15) propoxur and bendiocarb; (16) carbofuran; (17) carbaryl; (18) thiofanox; (19) landrin (internal standard); (20) carbanolate; (21) methiocarb; (22) bufencarb. The other peaks are metabolites of *N*-methylcarbamates. (Reprinted from Ref. 70, copyright 1992, with permission from Elsevier Science.)

REFERENCES

1. M Fielding, D Barceló, A Helwey, S Galassi, L Tortenson, P van Zoonen, R Wolter, G Angeletti. Pesticides in Ground and Drinking Water. Water Pollutions Research Report, No. 27. Brussels: Commission of the European Community, 1992, pp 1–136.
2. FAO/WHO. Pesticide residues in food. Report of the 1983 Joint Meeting of the FAO Panel of Experts on Pesticide Residues in Food and the Environment and the WHO Expert Group on Pesticide Residues. Rome: Food and Agriculture Organization of the United Nations (FAO Plant Production and Protection Paper 56), 1984.
3. BD Ripley, ASY Chau. Carbamate pesticides. In: JW Robinson, AS Chau, BK Afghan, eds. Analysis of Pesticides in Water. Vol. III. Nitrogen-Containing Pesticides. Boca Raton, FL: CRC Press, 1982, pp 1–182.
4. HS Rathore, T Begum. Thin-layer chromatographic behavior of carbamate pesticides and related compounds. J Chromatogr 643:321–329, 1993.
5. W Blad. Determination of methyl carbamate residues using online coupling of HPLC with post-column fluorimetric labelling technique. Fresenius J Anal Chem 339:340–343, 1991.
6. JA Leenheer. Concentration, partitioning, and isolation techniques. In: RA Minear, LH Keith, eds. Water Analysis. Vol. III: Organic Species, Orlando, FL: Academic Press, 1984, Chap 3.
7. K Packer. Nanogen Index. A Dictionary of Pesticides and Chemical Pollutants. Freedom, CA: Nanogen International, 1975.

8. C Tomlin. The Pesticide Manual. 10th ed. Surrey: British Crops Protection Council, 1994.

9. OM Aly, MA El-Dib. Studies of the persistence of some carbamate insecticides in the aquatic environmental. In: RF Could, ed. Fate of Organic Pesticides in the Aquatic Environment, Advances in Chemistry, Series 111. Washington, DC: American Chemical Society, 1972, Chap 11.

10. I Vassilieff, DJ Ecobichon. Stability of Matacil® in aqueous media as measured by changes in anticholinesterase potency. Bull Environ Contam Toxicol 29:366–370, 1982.

11. AK Sypesteyn, HM Dekhuyzen, JW Vonk. Biological conversion of fungicides in plants and microorganisms. In: MR Siegel, HD Sisler, eds. Antifungal Compounds. Vol. 2. New York: Marcel Dekker, 1977, pp 91–147.

12. M Chiba, F Doornbos. Instability of benomyl in various conditions. Bull Environ Contam Toxicol 11:273–274, 1974.

13. KMS Sundaram, Y Volpe, GG Simith, JR Duffy. A Preliminary Study on the Persistence and Distribution of Matacil in a Forest Environment, Report CC-X-116. Ottawa, Ontario: Chemical Control Research Institute, 1976.

14. JF Karinen, JG Lamberton, EN Stewart, LC Terriere. Persistence of carbaryl in the marine stuarine environment. Chemical and biological stability in aquarium system. J Agric Food Chem 15:148–153, 1967.

15. RD Wauchope, R Haque. Effects of pH, light and temperature on carbaryl in aqueous media. Bull Environ Contam Toxicol 9:257–261, 1973.

16. JN Seiber, MP Catahan, CR Barril. Loss of carbofuran from rice paddy water: chemical and physical factors. J Environ Sci Health B13:131–134, 1978.

17. RI Asai, FA Gunther, WE Westlake. Influence of some soil characteristics on the dissipation rate of Landrin insecticide. Bull Environ Contam Toxicol 11:352–358, 1974.

18. EW Mathews, SD Faust. The hydrolysis of Zectram in buffered and natural waters. J Environ Sci Health B12:129–131, 1977.

19. CF Hosler Jr. Degradation of Zectran in alkaline water. Bull Environ Contam Toxicol 12:599–603, 1974.

20. E Ballesteros, M Gallego, M Valcárcel. On-line preconcentration and gas chromatographic determination of N-methylcarbamates and their degradation products in aqueous samples. Environ Sci Technol 30:2071–2077, 1996.

21. SB Lartiges, PP Garrigues. Degradation kinetics of organophosphorous and organonitrogen pesticides in different waters under various environmental conditions. Environ Sci Technol 29:1246–1254, 1995.

22. OM Aly, MA El-Dib. Photodecomposition of some carbamate insecticides in aquatic environments. In: SJ Faust, JV Hunter, eds. Organic Compounds in Aquatic Environment. New York: Marcel Dekker, 1971, Chap 20.

23. N de Bertrand. D Barceló. Photodegradation of the carbamate pesticides aldicarb, carbaryl and carbofuran in water. Anal Chim Acta 254:235–244, 1991.

24. MJ Climent, MA Miranda. Gas-chromatographic–mass-spectrometric study of photodegradation of carbamate pesticides. J Chromatogr A 738:225–231, 1996.

25. WHO. Environmental Health Criteria 64. Carbamates Pesticides: A General Introduction. Geneva: World Health Organization, 1986.

26. JR Corbett, K Wright, AC Baillie. The Biochemical Mode of Action of Pesticides. 2nd ed. London: Academic Press, 1984.

27. PC Kearney, DD Kaufman. Herbicides. Chemistry, Degradation and Mode of Action. Vol. 1. New York: Marcel Dekker, 1975, pp 323–348.

28. WN Aldridge, L Magos. Carbamates, thiocarbamates and dithiocarbamates. Luxembourg: Commission of the European Communities, 1978.

29. Economic European Communities. Drinking Water Directive, Official Journal No. 299/11, Directive 80/778/EEC. Brussels: EEC, 1988.

30. U.S. Environmental Protection Agency. National Pesticide Survey. Phase I Report PB-91-125765. Springfield, VA: National Technical Information Service, 1990.

31. U.S. Environmental Protection Agency. National Pesticide Survey. Phase II Report EPA 570/9-91-020. Springfield, VA: National Technical Information Service, 1992.

32. D Barceló, MC Hennion. On-line sample handling strategies for the trace-level determination of pesticides and their degradation products in environmental waters. Anal Chim Acta 318:1–41, 1995.

33. IH Suffet, SD Faust. Liquid–liquid extraction of organic pesticides from water: the p-value approach to quantitative extraction. In: RF Could, ed. Fate of Organic Pesticides in the Aquatic Environment, Advances in Chemistry, Series 111. Washington, DC: American Chemical Society, 1972, Chap 2.

34. JJ Richard, GA Junk. Solid-phase versus solvent extraction of pesticides from water. Mikrochim Acta I: 387–394, 1986.

35. JA Coburn, BD Ripley, ASY Chau. Analysis of pesticide residues by chemical derivatization. II. N-Methylcarbamates in water and soil. J Assoc Off Anal Chem 59:188–195, 1976.

36. JW Eichelberger, JJ Lichtenberg. Persistence of pesticides in river water. Environ Sci Technol 5:541–545, 1975.

37. JD MacNeil, RW Frei, M Frei-Häusler, Electron donor-acceptor reagents in the analysis of pesticides. Int J Environ Anal Chem 2:323–326, 1973.

38. JN Seiber, DG Crosby, H Fouda, CJ Soderquist. Ether derivatives for the determination of phenols and phe-

nol-generating pesticides by electron capture gas chromatography. J Chromatogr 73:89–94, 1972.

39. RG Reeves, DW Woodham. Gas chromatographic analysis of methomyl residues in soil, sediment, water and tobacco utilizing the flame photometric detector. J Agric Food Chem 22:76–79, 1974.

40. K Nagasawa, H Uchiyama, A Ogamo, T Shinozuka. Gas chromatographic determination of microamounts of carbaryl and 1-naphthol in natural water as sources of water supplies. J Chromatogr 144:77–84, 1977.

41. JF Thompson, SJ Reid, EJ Kantor. A multiclass, multiresidue analytical method for pesticides in water. Arch Environ Contam Toxicol 6:143–147, 1977.

42. R Frank, HE Braun, G Sirons, MVH Holdrinet, BD Ripley, D Onn, R Coote. Stream Flow Quality—Pesticides in Eleven Agricultural Watersheds in Southern Ontario, Canada, 1974–1977. Windsor, Ontario: International Joint Commission, 1978.

43. AM Kadoum, DE Mock. Herbicide and insecticide residues in tailwater pits: water and pit bottom soil from irrigated corn and sorghum fields. J Agric Food Chem 26:45–50, 1978.

44. LE Deuel Jr, JD Price, FT Turner, KW Brown. Persistence of carbofuran and its metabolites, 3-keto and 3-hydroxy carbofuran, under flooded rice culture. J Environ Qual, 8:23–27, 1979.

45. A Bilikova, A Kuthan. Determination of carbofuran in water. Vodni Hospod B 33:215–219, 1983.

46. CF Ling, G Perez-Melian, F Jiménez-Conde, E Revilla. Determination of carbofuran in a nutrient solution by GLC-NPD (nitrogen-phosphorous detection) and HPLC. Chromatographia 30:421–423, 1990.

47. T Okumura, K Imamura, Y Nishikawa. Determination of carbamate pesticides in environmental samples as their trifluoroacetyl or methyl derivatives by using gas chromatography–mass spectrometry. Analyst 120: 2675–2681, 1995.

48. A Dekker, NWH Houx. Simple determination of oxime carbamates in soil and environmental water by high-pressure liquid chromatography. J Environ Sci Health B 18:379–392, 1983.

49. E Grou, V Radulescu, A Csuma. Direct determination of some carbamate pesticides in water and soil by high-performance liquid chromatogrraphy. J Chromatogr 260:502–506, 1983.

50. R Alzaga, G Durand, D Barceló, JM Bayona. Comparison of supercritical-fluid extraction and liquid–liquid extraction for isolation of selected pesticides stored in freeze-dried water samples. Chromatographia 38:502–508, 1994.

51. JD Rosen. Applications of New Mass Spectrometry Techniques in Pesticide Chemistry. New York: Wiley-Interscience, 1987, pp 146–160.

52. TA Bellar, WL Budde. Determination of nonvolatile organic compounds in aqueous environmental samples using liquid chromatography/mass spectrometry. Anal Chem 60:2076–2083, 1988.

53. NWF Nielen, RW Frei, UATh Brinkman. On-line sample handling and trace enrichment in liquid chromatography. The determination of organic compounds in water samples. In: RW Frei, K Zech, eds. Selective Sample Handling and Detection in High Performance Liquid Chromatography. Journal of Chromatography Library. Vol. 39A. Amsterdam: Elsevier, 1988, Chap 1.

54. J Namiesnik, T Gorecki, M Biziuk, L Torres. Isolation and preconcentration of volatile organic compounds from water. Anal Chim Acta 237:1–60, 1990.

55. D Barceló. Occurrence, handling and chromatographic determination of pesticides in the aquatic environment. Analyst 116:681–689, 1991.

56. G Font, J Mañes, JC Moltó, Y Picó. Solid-phase extraction in multi-residue pesticide analysis of water. J Chromatogr 642:135–161, 1993.

57. BD McGarvey. High-performance liquid chromatographic methods for the determination of N-methylcarbamate pesticides in water, soil, plants and air. J Chromatogr 642:89–105, 1993.

58. A Martin-Esteban, P Fernández, A. Fernández-Alba, C. Cámara. Analysis of polar pesticides in environmental waters: a review. Quim Anal 17:51–66, 1998.

59. KMS Sundaram, SY Szeto, R Hindle. Evaluation of Amberlite XAD-2 as the extractant for carbamate insecticides from natural water. J Chromatogr 177:29–32, 1979.

60. JE Woodrow, MS Majewski, JN Seiber. Accumulative sampling of trace pesticides and other organics in surface water using XAD-4 resin. J Environ Sci Health 21:143–164, 1986.

61. GA Junk, JJ Richard. Organics in water: solid-phase extraction on a small scale. Anal. Chem 60:451–454, 1988.

62. MW Brooks, D Tessier, D Soderstrom, J Jenkins, JM Clark. Rapid method for the simultaneous analysis of chlorpyrifos, isofenphos, carbaryl, iprodione and triadimefon in groundwater by solid-phase extraction. J Chromatogr Sci 28:487–489, 1990.

63. RG Nash. Solid-phase extraction of carbofuran, atrazine, simazine, alachlor and cyanazine from shallow well water. J Assoc Off Anal Chem 73:438–439, 1990.

64. CG Mattern, JB Louis, JD Rosen. Multipesticide determination in surface water by gas chromatography–chemical ionization mass spectrometry–ion-trap detection. J Assoc Off Anal Chem 74:982–986, 1991.

65. DW Beyers, CA Carlson, JD Tessari. Solid-phase extraction of carbaryl and malathion from pond and well water. Environ Toxicol Chem 10:1425–1429, 1991.

66. LM Davi, M Baldi, L Penazzi, M Liboni. Evolution of the membrane approach to solid-phase extractions or pesticide residues in drinking water. Pestic Sci 35: 63–67, 1992.

67. Y Picó, JC Moltó, MJ Redondo, E Viana, J Mañes, G Font. Monitoring of the pesticide levels in natural wa-

ters of the Valencia Community (Spain). Bull Environ Contam Toxicol 53:230–237, 1994.

68. E Viana, MJ Redondo, G Font, JC Moltó. Disks versus columns in the solid-phase extraction of pesticides from water. J Chromatogr 733:267–274, 1996.

69. A Di Corcia, M Marchetti. Multiresidue method for pesticides in drinking water using a graphitized carbon black cartridge extraction and liquid chromatographic analysis. Anal Chem 63:580–585, 1991.

70. A de Kok, M Hiemstra, UAT Brinkman. Low ng/l level determination of twenty N-methylcarbamate pesticides and twelve of their polar metabolites in surface water via off-line solid-phase extraction and high-performance liquid chromatography with post-column reaction and fluorescence detection. J Chromatogr 623: 265–276, 1992.

71. G Durand, S Chiron, V Bonvot, D Barceló. Use of extraction discs for trace enrichment of various pesticides from river and sea-water samples. Int J Environ Anal Chem 49:31–42, 1992.

72. P Parrilla, JL Martínez Vidal, M Martínez Galera, AG Frenich. Simple and rapid screening procedure for pesticides in water using SPE and HPLC/DAD detection. Fresenius J Anal Chem 350:633–637, 1994.

73. T Ohkura, T Takechi, S Deguchi, T Ishimaru, T Maki, H Inouye. Simultaneous determination of herbicides in water by FRIT-FAB LC-MS. J Toxicol Environ Health 40:266–273, 1994.

74. A Cappiello, G Famiglini, F Bruner. Determination of acid and basic/neutral pesticides in water with a new microliter flow rate LC/MS particle beam interface. Anal Chem 66:1416–1423, 1994.

75. B Jiménez, JC Moltó, G Font. Influence of dissolved humic material and ionic strength on C_8 extraction of pesticides from water. Chromatographia 41:318–324, 1995.

76. J Schülein, D Martens, P Spitzauer, A Kettrup. Comparison of different solid-phase extraction materials and techniques by application of multiresidue methods for the determination of pesticides in water by high-performance liquid chromatography. Fresenius. J Anal Chem 352:565–571, 1995.

77. SC McGinnis, J Sherma. Determination of carbamate insecticides in water by C-18 solid-phase extraction and quantitative HPTLC. J Liq Chromatogr 17:151–156, 1994.

78. P Moris, I Alexandre, M Roger, J Remache. Chemiluminescence assays of organophosphorous and carbamate pesticides. Anal Chim Acta 302:53–59, 1995.

79. Document CEN/TC 230 N 180 and CEN/TC 230/WG 1N 104. Mendmenhan, UK: Water Research Center, June 1994.

80. SA Senseman, TL Lavy, JD Mattice, EE Gbur. Influence of dissolved humic acid and calcium-montmorillinite clay on pesticide extraction efficiency from water using solid-phase extraction discs. Environ Sci Technol 29:2647–2653, 1995.

81. D Barceló, MF Alperdurada. A review of sample storage and preservation of polar pesticides in water samples. Chromatographia 42:704–712, 1996.

82. MC Hennion, P Scribe. Sample handling strategies for the analysis of organic compounds from environmental water samples. In: D Barceló, ed. Environmental Analysis: Techniques, Applications and Quality Assurance. Amsterdam: Elsevier Science, 1993.

83. UAT Brinkman. Online sample treatment for or via column liquid chromatography. J Chromatogr 665: 217–231, 1994.

84. J Sherma. Pesticides. Anal Chem 67:1R–20R, 1995.

85. LK She, UAT Brinkman, RW Frei. Liquid-chromatographic residue analysis of carbaryl based on a post-column catalytic reactor principle and fluorigenic labelling. Anal Lett 17:915–931, 1984.

86. CH Marvin, ID Brindle, CD Hall, M Chiba. Development of an automated high-performance liquid-chromatographic method for the online pre-concentration and determination of trace concentrations of pesticides in drinking water. J Chromatogr 503:167–176, 1990.

87. CH Marvin, ID Brindle, RP Singh, CD Hall, M Chiba. Simultaneous determination of trace concentrations of benomyl, carbendazim (MBC) and nine other pesticides in water using a automated online pre-concentration high-performance liquid-chromatographic method. J Chromatogr 518:242–249, 1990.

88. CH Marvin, ID Brindle, CD Hall, M Chiba. Rapid online pre-column high-performance liquid-chromatographic method for the determination of benomyl, carbendazim and aldicarb species in drinking water. J Chromatogr 555:147–154, 1991.

89. S Chiron, D Barceló. Determination of pesticides in drinking water by online solid-phase disc extraction followed by various liquid-chromatographic system. J Chromatogr 645:125–134, 1993.

90. S Guenu, MC Hennion. Online trace-enrichment of polar pesticides or degradation products in aqueous samples: analytical columns. Anal Methods Instrum 2: 247–253, 1995.

91. S Chiron, A Valverde, A Fernández-Alba, D Barceló. Automated sample preparation for monitoring groundwater pollution by carbamate insecticides and their transformation products. J Assoc Off Anal Chem Int 78:1346–1352, 1995.

92. S Sennert, D Volmer, K Levsen, G Wünsch. Multiresidue analysis of polar pesticides in surface and drinking water by on-line enrichment and thermospray LC-MS. Fresenius J Anal Chem 351:642–649, 1995.

93. J Slobodnik, SJF Hoekstra-Oussoren, ME Jager, M Honing, BLM van-Baar, UAT Brinkman. On-line solid-phase extraction–liquid chromatography–particle beam mass spectrometry and gas chromatography–mass spectrometry of carbamate pesticides. Analyst 121:1327–1334, 1996.

94. J Slobodnik, Ö Öeztezkizan, H Lingeman, UAT Brinkman. Solid-phase extraction of polar pesticides from environmental water samples on graphitized carbon and Empore-activated carbon discs and online coupling to octadecyl-bonded silica analytical columns. J Chromatogr 750:227–238, 1996.

95. T Renner, D Baumgarten, KK Unger. Analysis of organic pollutants in water at trace levels using fully automated solid-phase extraction coupled to high-performance liquid chromatography. Chromatographia 45:199–205, 1997.

96. J Slobodník, ER Brouwer, RB Geerdink, WH Mulder, H Lingeman, UAT Brinkman. Fully automated on-line liquid chromatographic separation system for polar pollutants in various types of water. Anal Chim Acta 268:55–65, 1992.

97. M Hiemstra, A de Kok. Determination of N-methylcarbamate pesticides in environmental water samples using automated on-line trace enrichment with exchangeable cartridges and high-performance liquid chromatography. J Chromatogr 667:155–166, 1994.

98. JL Marty, N Mionetto, S Lacorte, D Barceló. Validation of an enzymic biosensor with various liquid-chromatographic techniques for determining organophosphorous pesticides and carbaryl in freeze-dried waters. Anal Chim Acta 311:265–271, 1995.

99. J Slobodník, MGM Groenewegen, ER Brouwer, H Lingeman, UAT Brinkman. Fully automated multi-residue method for trace level monitoring of polar pesticides by liquid chromatography. J Chromatogr 642:359–370, 1993.

100. UAT Brinkman, J Slobodník, JJ Vreuls. Trace-level detection and identification of polar pesticides in surface water: the SAMOS approach. Trends Anal Chem 13:373–381, 1994.

101. PJM van Hoult, UAT Brinkman. The Rhine Basin Program. Trends Anal Chem 13:382–388, 1994.

102. UAT Brinkman, H Lingeman, J Slobodník. SAMOS LC: a fully automated liquid-chromatographic method for trace-level monitoring of polar pesticides. LC-GC Int 7:157–163, 1994.

103. W Bertsch, WJ Jenning, P Sandra. On-Line Coupled LC-GC. Heidelberg: Hüthig, 1991, Chap 6.

104. P van Zoonen, EA Hogendoorn, GR van der Hoff, RA Baumann. Selectivity and sensitivity in coupled chromatographic techniques as applied in pesticide residue analysis. Trends Anal Chem 11:11–17, 1992.

105. EC Goosens, MH Broekman, MH Wolters, RE Strijker, D de Jong, GJ de Jong, UAT Brinkman. Continuous two-phase reaction system coupled online with capillary chromatography for the determination of polar solutes in water. J High Resolut Chromatogr 15:242–248, 1992.

106. GR van der Hoff, RA Baumann, UAT Brinkman, P van Zoonen. Online combination of automated micro liquid–liquid extraction and capillary gas chromatog-

107. raphy for the determination of pesticides in water. J Chromatogr 644:367–373, 1993.

107. EC Goosens, D de Jong, GJ de Jong, FD Rinkema, UAT Brinkman. Continuous liquid–liquid extraction combined online with capillary gas chromatography-atomic emission detection for environmental analysis. J High Resolut Chromatogr 18:38–44, 1995.

108. E Ballesteros, M Gallego, M Valcárcel. Automatic determination of N-methylcarbamate pesticides by using a liquid–liquid extractor derivatization module coupled on-line to a gas chromatograph equipped with a flame ionization detector. J Chromatogr 633:169–176, 1993.

109. E Ballesteros, M Gallego, M Valcárcel. Automatic gas chromatographic determination of N-methylcarbamates in milk with electron capture detection. Anal Chem 65:1773–1778, 1993.

110. M Gallego, M Silva, M Valcárcel. Indirect atomic absorption determination of anionic surfactants in wastewaters by flow injection continuous liquid–liquid extraction. Anal Chem 58:2265–2269, 1986.

111. E Ballesteros, M Gallego, M Valcárcel. Online coupling of a gas chromatograph to a continuous liquid–liquid extractor. Anal Chem 62:1587–1591, 1990.

112. M Valcárcel, E Ballesteros, M Gallego. Continuous liquid–liquid extraction and derivatization module coupled on-line with gas chromatographic detection. Trends Anal Chem 13:68–73, 1994.

113. JJ Vreuls, WJGM Cuppen, GJ de Jong, UAT Brinkman. Ethyl acetate for the desorption of a liquid-chromatographic pre-column online into a gas chromatograph. J High Resolut Chromatogr 13:157–161, 1990.

114. UAT Brinkman, JJ Vreuls. Solid-phase extraction for online sample treatment in capillary gas chromatography. LC-LC Int 8:694–698, 1995.

115. D Jahr, JJ Vreuls, AJH Louter, W Loebel. Water analysis by online solid-phase extraction–gas chromatography–mass spectrometry. GIT Fachz Lab 40:178–183, 1996.

116. J Chen, JB Pawliszyn. Solid-phase microextraction coupled to high-performance liquid chromatography. Anal Chem 67:2530–2533, 1995.

117. T Gorecki, R Mindrup, J Pawliszyn. Pesticides by solid-phase microextraction: results of a round robin test. Analyst 121:1381–1386, 1996.

118. TK Choudhury, KO Gerhardt, TP Mawhinney. Solid-phase microextraction of nitrogen- and phosphorous-containing pesticides from water and gas-chromatographic analysis. Environ Sci Technol 30:3259–3265, 1996.

119. AA Boyd-Boland, S Magdic, J Pawliszyn. Simultaneous determination of 60 pesticides in water using solid-phase microextraction and gas chromatography–mass spectrometry. Analyst 121:929–938, 1996.

120. R Eisert, K Levsen. Solid-phase microextraction coupled to gas chromatography: a new method for the

analysis of organics in water. J Chromatogr A 733: 143–157, 1996.

121. R Eisert, K Levsen. Development of a prototype system for quasi-continuous analysis of organic contaminants in surface or sewage water based on in-line coupling of solid-phase microextraction to gas chromatography. J Chromatogr A 737:59–65, 1996.

122. C Saenz-Barrio, E Romero-Melgosa, J Sanz-Asensio, J Galbán-Bernal. Extraction of pesticides from aqueous samples: a comparative study. Mikrochim Acta 122:267–277, 1996.

123. SA Westwood. Supercritical Fluid Extraction and Its Use in Chromatographic Sample Preparation. Dordrecht: Backie Academic and Professional, Chapman and Hall, 1992.

124. M Valcárcel, MD Luque de Castro, MT Tena. Extracción con Fluidos Supercríticos en el Proceso Analítico. Barcelona: Reverté, 1993.

125. RW Vannoort, JP Chervet, H Lingeman, GJ de Jong, UAT Brinkman. Coupling of supercritical-fluid extraction with chromatographic techniques. J Chromatogr 505:45–77, 1990.

126. K Jinno. Hyphenated Techniques in Supercritical Fluid Chromatography and Extraction. Journal Chromatography Library. Vol. 53. Amsterdam: Elsevier, 1992.

127. R Hillmann, K Baechmann. Extraction of pesticides using supercritical trifluoromethane and carbon dioxide. J Chromatogr A 695:149–154, 1995.

128. W Wuchner, R Grob. Application of supercritical-fluid extraction in water analysis. Analysis 23:227–229, 1995.

129. RJ Argauer, KI Eller, MA Ibrahim, RT Brown. Determining propoxur and other carbamates in meat using HPLC fluorescence and gas chromatography–ion-trap mass spectrometry after supercritical-fluid extraction. J Agric Food Chem 43:2774–2778, 1995.

130. A Izquierdo, MT Tena, MD Luque de Castro, M Valcárcel. Supercritical-fluid extraction of carbamate pesticides from soils and cereals. Chromatographia 42: 206–212, 1996.

131. V Lopez-Avila, C Charan and WF Beckert. Using supercritical fluid extraction and enzyme immunoassays to determine pesticides in soils. Trends Anal Chem 13: 118–126, 1994.

132. A Martin-Esteban, P Fernández, C Cámara. Immunosorbents: a new tool for pesticide sample handling in environmental analysis. Fresenius J Anal Chem 357: 927–933, 1997.

133. GS Rule, AV Mordehai, J Henion. Determination of carbofuran by on-line immunoaffinity chromatography with coupled-column liquid chromatography/mass spectrometry. Anal Chem 66:230–235, 1994.

134. S Garriges, MT Vidal, M Gallignani, M de la Guardia. On-line preconcentration and flow analysis–Fourier transform infrared determination of carbaryl. Analyst 119:659–664, 1994.

135. I Liska, J Slobodník. Comparison of gas and liquid chromatography for analyzing polar pesticides in water samples. J Chromatogr A 733:235–258, 1996.

136. K Blau and GS King. Handbook of Derivatives for Chromatography. London: Heydon and Sons, 1977.

137. JF Lawrence. Derivatization in chromatography. J Chromatogr Sci 17:113–124, 1979.

138. WP Cochrance. Application of chemical derivatization techniques for pesticides analysis. J Chromatogr Sci 17:124–132, 1979.

139. R Greenhalgh, J Kovacicova. A chemical confirmatory test for organophosphorous and carbamate insecticides and triazine and urea herbicides with reactive NH moities. J Agric Food Chem 23:325–329, 1975.

140. L Fishbein, WL Zielinski Jr. Gas chromatography of trimethylsilyl derivatives. I. Pesticidal carbamates and ureas. J Chromatogr 20:9–13, 1965.

141. JW Ralls, A Cortes. Determination of Sevin in green beans by bromination and electron capture gas chromatography. J Gas Chromatogr 2:132–136, 1964.

142. JJK Boulton, CBC Boyce, PJ Jewess, RF Jones. Comparative properties of N-acetyl derivatives of oxime N-methylcarbamates and aryl N-methylcarbamates as insecticides and acetylcholinesterase inhibitors. Pest Sci 2:10–13, 1971.

143. V Drevenkar, B Stengl, B Tkalcevic, Z Vasilic. Occupational exposure control by simultaneous determination of N-methylcarbamates and organophosphorous pesticides residues in human urine. Int J Environ Anal Chem 14:215–230, 1983.

144. RT Coutts, EE Hargesheimer, FM Pasutto. Application of a direct aqueous acetylation technique to the gas chromatographic quantitation of nitrophenols and 1-naphthol in environmental water samples. J Chromatogr 195:105–112, 1980.

145. L Ogierman. Gas-liquid chromatographic derivatization and chromatography of N-methylcarbamate methoxy derivatives formed with trimethylanilinium hydroxide. J Assoc Off Anal Chem 65:1452–1456, 1982.

146. JF Lawrence, DA Lewis, HA McLeod. Detection of carbofuran and metabolites directly or as their heptafluorobutyryl derivatives using gas-liquid or high-pressure liquid chromatography with different detectors. J Chromatogr 138:143–150, 1977.

147. K Helrich. Official Methods of Analysis of the Association of Official Analytical Chemists. Vol. 1. 15th ed. Arlington, VA: Association of Official Analytical Chemists, 1990, p 291.

148. SC Lau, RL Marxmiller. Residue determination of Landrin insecticide by trifluoroacetylation and electron-capture gas chromatography. J Agric Food Chem 18:413–416, 1970.

149. GH Tjan, JTA Jansen. Gas-liquid chromatographic determination of thiabendazole and methyl 2-benzimidazole carbamate in fruits and crops. J Assoc Off Anal Chem 62:769–773, 1979.

150. RE Cline, GD Todd, DL Ashley, J Grainger, JM McCraw, CC Alley, RH Hill. Gas chromatographic and spectral properties of pentafluorobenzyl derivatives of 2,4-dichlorophenoxyacetic acid and phenolic pesticides and metabolites. J Chromatogr Sci 28:167–172, 1990.

151. J Sherma. Chromatographic analysis of pesticide residues. Crit Rev Anal Chem 3:299–303, 1973.

152. T Albi. Insecticide residues in olives. I. Determination of carbamates in olives. Carbaryl residues in olive-pickling processes. Grasas Aceites (Seville) 32:381–386, 1981.

153. WP Cochrane, R Purkayastha. Analysis of herbicide residues by gas chromatography. Toxicol Environ Chem Rev 1:137–140, 1973.

154. SU Khan. Chemical derivatization of herbicide residues for gas liquid chromatographic analysis. Residue Rev 59:21–25, 1975.

155. JF Lawrence. Evaluation and confirmation of an alkylation-gas-liquid chromatographic method for the determination of carbamate and urea herbicides in foods. J Assoc Off Anal Chem 59:1061–1065, 1976.

156. JF Lawrence. Confirmation of some organonitrogen herbicides and fungicides by chemical derivatization and gas chromatography. J Agric Food Chem 24:1236–1239, 1976.

157. WZ Zhong, AT Lemley, J Spalik. Quantitative determination of parts-per-billion levels of carbamate pesticides in water by capillary gas chromatography. J Chromatogr 299:269–274, 1984.

158. HM Mueller, HJ Stan. Thermal degradation observed with different injection technique: quantitative estimation by the use of thermolabile carbamate pesticides. J High Resolut Chromatogr 13:759–763, 1990.

159. PL Wylie, KJ Klein, MQ Thompson, BW Hermann. Using electronic pressure programming to reduce the decomposition of labile compounds during splitless injection. J High Resolut Chromatogr 15:763–768, 1992.

160. FI Onuska. Pesticide residue analysis by open-tubular column gas chromatography: trials, tribulations and trends. HRCCC, J High Resolut Chromatogr Chromatogr Commun 7:660–670, 1984.

161. J Tekel, S Hatrik. Pesticide residue analyses in plant material by chromatographic methods: clean-up procedures and selective detectors. J Chromatogr A 754:397–410, 1996.

162. J Gilbert. Application of Mass Spectrometry in Food Science. New York: Elsevier, 1987.

163. T Cairns, RA Baldwin. Pesticide Analysis in Food by MS. Anal Chem 67:552A–557A, 1995.

164. I Fogy, ER Schmid, JFK Huber. Determination of carbamate pesticides in fruits and vegetables by multidimensional high-pressure liquid chromatography. Z Lengsm Unters Forsch 170:194–199, 1980.

165. CM Sparacino, JW Hines. High-performance liquid chromatography of carbamate pesticides. J Chromatogr Sci 14:549–565, 1976.

166. HA Moye, SJ Scherer, PA St John. Dynamic fluorogenic labelling of pesticides for high performance liquid chromatography: detection of N-methylcarbamates with o-phthaldehide. Anal Lett 10:1049–1053, 1977.

167. RT Krause. Resolution, sensitivity and selectivity of a high-performance liquid-chromatographic post-column fluorimetric labelling technique for determination of carbamate insecticides. J Chromatogr 185:615–624, 1979.

168. W Blass. Determination of methyl carbamate residues using online coupling of HPLC with a post-column fluorimetric labelling technique. Fresenius J Anal Chem 339:340–343, 1991.

169. BD McGarvey. Liquid chromatographic determination of N-methylcarbamate pesticides using a single-stage post-column derivatization reaction and fluorescence detection. J Chromatogr 481:445–451, 1989.

170. JL Anderson, DJ Chesney. Liquid chromatographic determination of selected carbamate pesticides in water with electrochemical detection. Anal Chem 52:2156–2161, 1980.

171. JL Anderson, KK Whiten, JD Brewster, TY Ou, WK Nonidez. Micro-array electrochemical flow detectors at high applied potentials and liquid chromatography with electrochemical detection of carbamate pesticides in river water. Anal Chem 57:1366–1373, 1985.

172. MB Thomas, PE Sturrock. Determination of carbamates by high-performance liquid chromatography with electrochemical detection using pulsed-potential cleaning. J Chromatogr 357:318–324, 1986.

173. I Stoeber, HR Schulten. Combined application of high-pressure liquid chromatography and field-desorption mass spectrometry for determination of biocides of the phenylurea and carbamate type in surface water. Sci Total Environ 16:249–262, 1980.

174. JD Rosen. Applications of New Mass Spectrometry Techniques in Pesticide Chemistry. New York: Wiley Interscience, 1987, pp 161–175.

175. BL Kleintop, DM Eades, RA Yost. Operation of a quadrupole trap for particle beam LC/MS analyses. Anal Chem 65:1295–1300, 1993.

176. M Honing, D Barceló, ME Jager, J Slobodník, BLM van Baar, UAT Brinkman. Effect of ion source pressure on ion formation of carbamates in particle-beam chemical-ionization mass spectrometry. J Chromatogr A 712:21–30, 1995.

177. M Honing, D Barceló, BLM van Baar, UAT Brinkman. Limitations and perspectives in the determination of carbofuran with various liquid chromatography–mass spectrometry interfacing systems. Trends Anal Chem 14:496–504, 1995.

178. JE France, KJ Voorhees. Capillary supercritical-fluid chromatography with ultra-violet multi-channel detec-

tion of some pesticides and herbicides. HRCCC, J High Resolut Chromatogr Chromatogr Commun 11: 692–696, 1988.

179. L Mathiasson, JA Jonsson, L Karlsson. Determination of nitrogen compounds by supercritical-fluid chromatography using nitrous oxide as the mobile phase and nitrogen-sensitive detection. J Chromatogr 467:61–74, 1989.

180. TA Berger, WH Wilson, JF Deye. Analysis of carbamate pesticides by packed-column supercritical-fluid chromatography. J Chromatogr Sci 32:179–184, 1994.

181. HS Rathore. Chromatographic and related spot tests for the detection of water pollutants. J Chromatogr A 733:5–17, 1996.

182. S Butz, HJ Stan. Screening of 265 pesticides in water by thin-layer chromatography with automated multiple development. Anal Chem 67:620–630, 1995.

183. H Süsse, H Müller. Application of micellar electrokinetic capillary chromatography to the analysis of pesticides. Fresenius J Anal Chem 352:470–473, 1995.

184. AWM Lee, WF Chan, FSY Yuen, CH Lo, RCK Chan, Y Liang. Simultaneous determination of dithiocarbamates by capillary electrophoresis with diode array detection and using factor analysis. Anal Chim Acta 339: 123–129, 1997.

185. BM Kaufman, M Clower. Immunoassay of pesticides. J Assoc Off Anal Chem 74:239–247, 1991.

186. HM Thompson. Development of an immunoassay for avian serum butyrylcholinesterase and its use in assessing exposure to organophosphorous and carbamate pesticides. Bull Environ Contam Toxicol 54:237–244, 1995.

187. A Abad, MJ Moreno, A Montoya. A monoclonal immunoassay for carbofuran and its application to the analysis of fruit juices. Anal Chim Acta 347:103–110, 1997.

188. D Martorell, F Cespedes, E Martínez-Fabregas, S Alegret. Amperometric determination of pesticides using a biosensor based on a polishable graphite-epoxy biocomposite. Anal Chim Acta 290:343–348, 1994.

189. C La Rosa, F Pariente, L Hernández, E Lorenzo. Determination of organophosphorous and carbamic pesticides with an acetylcholinesterase amperometric biosensor using 4-aminophenyl acetate as substrate. Anal Chim Acta 295:273–282, 1994.

190. C La Rosa, F Pariente, L Hernández, E Lorenzo. Amperometric flow-through biosensor for the determination of pesticides. Anal Chim Acta 308:129–136, 1995.

191. JL Bescombes, S Cosnier, P Labbe, G Reverdy. A biosensor as warming device for the detection of cyanide, chlorophenols, atrazine and carbamate pesticides. Anal Chim Acta 311:255–263, 1995.

192. MT Perez-Pita, AJ Reviejo, FJ Manuel de Villena, JM Pingarron. Amperometric selective biosensing of dimethyl- and diethyldithiocarbamates based on inhibi-

tion processes in a medium of reversed micelles. Anal Chim Acta 340:89–97, 1997.

193. DP Nikolelis, UJ Krull. Direct electrochemical sensing of insecticides by bilayer lipid membranes. Anal Chim Acta 288:187–192, 1994.

194. B Fernández-Band, P Linares, MD Luque de Castro, M Valcárcel. Flow-through sensor for the direct determination of pesticide mixture without chromatographic separation. Anal Chem 63:1672–1675, 1991.

195. S Kumaran, C Tran-Minh. Determination of organophosphorous and carbamate insecticides by flow-injection analysis. Anal Biochem 200:187–194, 1992.

196. T Cyr, N Cyr, R Haque. Spectrophotometric methods. In: G Zweig, J Sherma, eds. Analytical Methods for Pesticides and Plant Growth Regulators. Vol. 9. New York: Academic Press, 1977, Chap 3.

197. F García-Sánchez, C Cruces-Blanco. Determination of the carbamate herbicide prophan by synchronous derivative spectrofluorimetry following fluorescamine fluorogenic labelling. Anal Chem 58:73–76, 1986.

198. F García-Sánchez, C Cruces-Blanco. Determination of carbaryl and its metabolite 1-naphthol in commercial formulations and biological fluids. Talanta 37:573–578, 1990.

199. JL Vilchez, J Rohand, R Avidad, A Navalón, LF Capitan-Vallvey. Simultaneous determination of carbaryl and azinphos-methyl in water by first-derivative synchronous spectrofluorimetry. Fresenius J Anal Chem 350:626–629, 1994.

200. JL Vilchez, J Rohand, LF Capitan-Vallvey, A Navalón, R Avidad. Simultaneous determination of carbaryl and thiabendazole in water by derivative synchronous spectrofluorimetry. Quim Anal 14:195–200, 1995.

201. LF Capitan-Vallvey, J Rohand, A Navalón, R Avidad, JL Vilchez. Simultaneous determination of carbaryl and o-phenylphenol residues in waters by first-derivative synchronous solid-phase spectrofluorimetry. Talanta, 40:1695–1701, 1993.

202. MC Quintero, M Silva, D Pérez-Bendito. Stopped-flow determination of carbaryl and its hydrolysis product in mixtures in environmental samples. Talanta 35: 943–948, 1988.

203. MC Quintero, M Silva, D Pérez-Bendito. Analysis of carbofuran residues in soil by stopped-flow technique. Analyst 114:497–500, 1989.

204. MC Quintero, M Silva, D Pérez-Bendito. Enzymatic stopped-flow determination of carbofuran residues at the nanomolar level in environmental waters. Int J Environ Anal Chem 39:239–243, 1990.

205. MC Quintero, M Silva, D Pérez-Bendito. Enzymic determination of N-methylcarbamate pesticides at the nanomolar level by the stopped-flow technique. Talanta 38:1273–1277, 1991.

206. A Espinosa-Mansilla, F Salinas, Z Zamoro. Simultaneous determination of chlorpyrifos and carbaryl by differential degradation using diode-array spectropho-

tometry optimized by partial least squares. Analyst 119:1183–1188, 1994.

207. S Panadero, A Gómez-Hens, D Pérez-Bendito. Usefulness of the stopped-flow mixing technique for micelle-stabilized room-temperature liquid phosphorimetry. Anal Chem 66:919–923, 1994.

208. M Gallignani, S Garrigues, A Mártinez-Vado, M de la Guardia. Determination of carbaryl in pesticide formulations by Fourier-transform infra-red spectrometry with flow analysis. Analyst 118:1043–1048, 1993.

209. JM García, AI Jiménez, JJ Arias, KD Khalaf, A Morales-Rubio, M de la Guardia. Application of the partial least-squares calibration method to the simultaneous kinetic determination of propoxur, carbaryl, ethiofencarb and formetanate. Analyst 120:313–317, 1995.

210. O Jiménez de Blas, JL Pereda de Paz, J Hernández-Méndez. Indirect determination of diethyl-dithiocarbamate by atomic-absorption spectrometry with continuous extraction: application to the determination of the fungicide ziram. J Anal At Spectrom 5:693–696, 1990.

211. International Union of Pure and Applied Chemistry. Electrochemical analysis of organic pollutants. Pure Appl Chem 59:245–256, 1987.

212. J Skopalova, M Kotoucek. Polarographic and voltammetric determination of pesticides. Chem Listy 89:270–279, 1995.

213. JA Pérez-López, A Zapardiel, E Bermejo, E Arauzo, L Hernández. Electrochemical determination of carbaryl oxidation in natural water and soil samples. Fresenius J Anal Chem 350:620–625, 1994.

214. A Guiberteau, T Galeano-Díaz, F Salinas, JM Ortiz. Indirect voltammetric determination of carbaryl and carbofuran using partial least squares calibration. Anal Chim Acta 305:219–226, 1995.

28

Organophosphates

Monica Culea

NATEX s.r.l., Cluj-Napoca, Romania

Simion Gocan

University "Babes-Bolyai," Cluj-Napoca, Romania

I. PHYSICOCHEMICAL PROPERTIES

Pollution monitoring in the water environment involves either the analysis of a compound not naturally found in the environment or the determination of the increase in concentration of a compound above its "natural" level. The widespread agricultural use of several hundred pesticides results in their appearance in surface or drinking water. The transport of pesticides by ground- or surface water and atmospheric processes contributes to environmental contamination too (1). The analytical interest in organophosphorus pesticides (OPPs) in recent years is explained by the progressive replacement of the highly persistent organochlorine pesticides with organophosphates. Organochlorines were banned for their toxicity, persistence, and bioaccumulation in environmental matrices. Organophosphates are less persistent due to their low stability under different environmental conditions such as light, temperature, and humidity. The required level for pesticides in drinking water, groundwater and surface water specified in EEC Directive 80/778/EEC explains the need for straightforward, selective, and sensitive methods for trace-level determination of organophosphates in water and the recent development of techniques in this area.

A. Classification

Most organophosphorus insecticides have the following general structure:

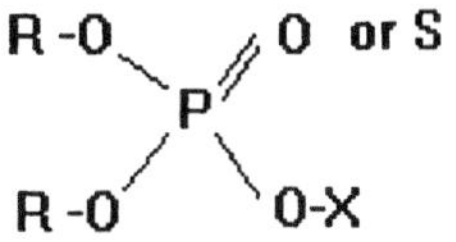

The two R groups are usually methyl or ethyl and are the same, while X is frequently an aliphatic, homocyclic, or heterocyclic group.

Organophosphorus pesticides are classified according to chemical:

Phosphates (e.g., chlorfenvinphos, dichlorvos, mevinphos) having four oxygen atoms arranged around the phosphorus.

Thionophosphates (phosphorothionates): (e.g., bromophos, diazinon, fenitrothion, parathion, pyrimiphos) if sulphur replaces the oxygen in the double-bond part.

Thiolphosphates (phosphorothiolates): (e.g., demeton-*S*-methyl, vamidothion) if sulphur replaces the oxygen in the thiol position.

Dithiophosphates (phosphorothiolothionates): (e.g., azinphos-methyl, dimethoate, disulfoton, malathion, phorate, etion) if two oxygen atoms are replaced by sulphur.

Phosphonates: (e.g., trichlorfon) if a carbon atom of the X group is attached directly to the phosphorus.

Phosphoramides: (e.g., methamidophos) containing NH_2 as the X group.

Pyrophosphates: (e.g., sulphotep) having two phosphorothionate groups bonded to the oxygen atom.

B. Toxicity

Organophosphorus pesticides form a large group with widely different structures and biological activities: insecticides, acaricides, and nematicides. The structural variability is reflected not only in physical properties but also in the diversity of mechanisms by which they are attacked by enzymes.

The toxic effects of organophosphorus compounds are due to the inactivation of acetylcholinesterase at muscarinic receptors for acetylcoline, causing symptoms such as cold sweating, salivation, nausea, bronchoconstriction, and a decrease in blood pressure and, at nicotinic receptors, causing a twitching of muscles, muscular cramp, or paralysis. Exposure could be by inhalation, skin absorption, or ingestion (accidental or deliberate). Special tests involve the measurement of cholinesterase levels in serum and in red blood cells. Mutagenic and carcinogenic potential have been signaled (2).

C. Physical Parameters Predicting the Environmental Fate of Pesticides

Octanol–water partition coefficient (K_{OW}) describes the partitioning of a pesticide between octanol and water.

K_D is the constant describing the distribution between the organic phase and the water phase.

Henry's low constant (H) describes the ratio of the concentration of a chemical in the atmosphere and in solution.

Vapor pressure (P_V) describes the contribution to the pressure of a chemical in the gas phase at a given temperature. Higher vapor pressure increases the tendency to evaporate.

Bioconcentration factor (BCF) describes the affinity of the chemical for aquatic organisms, indicating the accumulation of the chemical in fish as compared with the water in which they swim.

Water solubility (S_W) describes the maximum amount of chemical that could be dissolved in water at a given temperature.

A list of values for these properties for pesticides has been compiled by the U.S. Environmental Protection Agency (EPA) indicating high groundwater contamination: $S_W > 30$ mg/L; $K_D < 5$, usually < 1; $K_{OW} < 300$; $H < 0.01$ atm·mol/m^3, hydrolysis $t_{1/2} > 25$ weeks, pho-

tolysis > 1 week, and field dissipation half-life > 3 weeks (3–7).

D. Degradation of Organophosphorus Pesticides

Photodegradation studies on organophosphates have shown oxidation of P═S to P═O (2). Aqueous hydrolysis at the P atom or in the alkyl chain, favored in alkaline pH, is the most important reaction of organophosphates, greater with the change of P═S to P═O. Light and temperature induce the thion–thiol isomerization, transforming *O*-phosphorothioates into the more toxic *S*-phosphorothioates. The conversion of phosphorodithioates into phosphates and metabolites also can increase their toxicity (4–7). Malathion and parathion hydrolysis in aqueous solutions of pH 2, 8, 9, and 10 was reported (8). Degradation of fenthion in water explains the lower than 60% recoveries always found. Various transformation products (TPs) from water samples spiked with fenthion were identified by mass spectrometry (MS) (5). Water degradation of OPPs is studied under different pH and times by measuring their residual concentration (5).

The regulations of the European Economic Community (EEC) and the U.S. Environmental Protection Agency (EPA) regarding water also include the TPs, which could be sometimes more toxic than the parent compounds. A lot of degradation studies of OPPs at different pH, temperature, or light exposures have been reported (5,9–14). Mass spectrometry serves an important function, especially in the identification of TPs of OPPs.

E. Regulations

The EEC and EPA have issued priority lists of pesticides in water (15,16). The chromatographic protocols of the EPA employed in the National Pesticide Survey (NPS) include more than 100 pesticides and tens of TPs. The OPPs are included in NPS Method 1 (EPA 507). EEC Directive 80/778/EEC (17,18) sets a maximum admissible concentration (MAC) of 0.1 μg/L^{-1} for individual pesticides and related products, and of 0.5 μg/L^{-1} for total pesticides in drinking water (17) and 1–3 μg/L^{-1} in surface water (3,6,18–21). The majority of pesticides discharged in the aquatic environment of the EEC listed in the Council, Directive 76/464/EEC (the so-called black list) are organophosphorus pesticides (3,19,21). The chemical structures of some of these pesticides is given in Table 1. The huge quantities of pesticides used, some of them

Table 1 Chemical Structures of Some OPPs

Acephate, M = 183 ($C_4H_{10}NO_3PS$), m/z: 42, 136, 94

Azinphos methyl, M = 317 ($C_{10}H_{12}N_3O_3PS_3$), m/z: 160, 77, 132

Azinphos ethyl, M = 345 ($C_{12}H_{16}N_3O_3PS_2$), m/z: 132, 77, 160

Chlorfenvinphos, M = 358 ($C_{12}H_{14}Cl_3O_4P$), m/z: 267, 323, 81

Coumaphos, M = 362 ($C_{14}H_{16}ClO_5PS$), m/z: 362, 109, 226

Demeton-O, M = 258 ($C_8H_{19}O_3PS_2$), m/z: 88, 60, 115

Demeton-S, M = 258 ($C_8H_{19}O_3PS_2$), m/z: 88, 89, 171

Diazinon, M = 304 ($C_{12}H_{21}N_2O_3PS$), m/z: 179, 137, 152, 304

Table 1   Continued

Dichlorvos, M = 220 ($C_4H_7Cl_2O_4P$), m/z: 109, 185, 79

Dimethoate, M = 229 ($C_5H_{12}NO_3PS_2$), m/z: 87, 93, 125

Disulfoton, M = 274 ($C_8H_{19}O_2PS_3$), m/z: 88, 89, 274

EPN, M = 323 ($C_{14}H_{14}NO_4PS$), m/z: 157, 63, 323

Fenitrothion, M = 277 ($C_9H_{12}NO_5PS$), m/z: 277, 125, 109

Fenthion, M = 278 ($C_{10}H_{15}O_3PS_2$), m/z: 278, 125, 109

Leptophos, M = 410 ($C_{13}H_{10}BrCl_2O_2PS$), m/z: 171, 377, 77

Table 1 Continued

Malathion, M = 330 ($C_{10}H_{19}O_6PS_2$), m/z: 173, 125, 127, 93

Methamidophos, M = 141 ($C_2H_8NO_2PS$), m/z: 15, 94, 47, 141

Mevinphos, M = 224 ($C_7H_{13}O_6P$), m/z: 127, 109, 43

Paraoxon, M = 275 ($C_{10}H_{14}NO_6P$), m/z: 109, 149, 81, 275

Parathion, M = 291 ($C_{10}H_{14}NO_5PS$), m/z: 97, 291, 109

Parathion-methyl, M = 303 ($C_8H_{10}NO_8PS$), m/z: 109, 125, 163

Phorate, M = 260 ($C_7H_{17}O_2PS_3$), m/z: 75, 121, 260

Tetrachlorvinphos, M = 364 ($C_{10}H_9Cl_4O_4P$), m/z: 329, 109, 331

Table 1 Continued

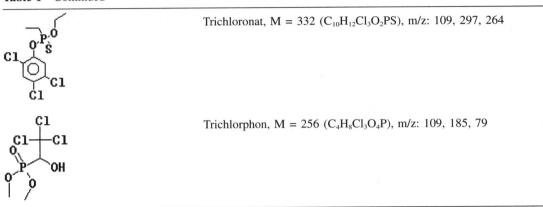

Trichloronat, M = 332 ($C_{10}H_{12}Cl_3O_2PS$), m/z: 109, 297, 264

Trichlorphon, M = 256 ($C_4H_8Cl_3O_4P$), m/z: 109, 185, 79

at a rate of 50,000–500,000 kg per annum, increase the probability of leaching in ground- and drinking waters. Methods with a very low limit of detection (LOD of 0.02 μg/L^{-1}) are needed for this complicated task, and they are very difficult to achieve for all pesticides.

Most of the EPA methods have the following general guidelines:

1. Acceptance of recoveries in the range 70–130%, with maximum relative standard deviation (RSD) < 30% each.
2. Preservation and storage at 4°C and for a maximum of 14 days. For organonitrogen- and organophosphorus-containing pesticides, special attention is given to storage because some of them present 100% loss (disulphoton sulphoxide, diazinon, fenitrothion, terbufos) when stored under usual conditions at 4°C for 14 days; the water sample is better analyzed immediately.
3. Description of apparatus and equipment, safety considerations, reagents, standards, consumable materials.
4. Use of two columns of different polarity, with a DB-5 as primary column and a confirmatory column such as DB-1701; in liquid chromatography (LC) the primary column is a C_{18} bonded silica and the second a cyano type.
5. A way to proceed with blank samples, internal standards, surrogate solutions, interferences, calibration, standardization, and quality control. The internal standard must be an analyte that is not a sample component (21). A continuing revision of the methods for the determination of organic components in water is necessary (22–24).

In order to avoid "false positives" in the determination of pesticides in water samples, confirmatory techniques are needed; the second column of different polarity in the EPA methods has this goal. So-called two-dimensional capillary gas chromatography (GC) can also be used. A third possibility is chemical derivatization, when the original pesticide peaks disappear and the derivatives appear at a different retention times. The most widely used confirmation technique is GC mass spectrometry (GC-MS). The EPA and UK Standing Committee of Analysis (SCA) methods have already implemented GC-MS in some of their protocols (EPA Method 525).

II. SAMPLE PREPARATION

Trace pollutants in water involve pretreatment—extraction and concentration for analysis at these concentrations, to increase the analyte concentration to within the instrument's sensitivity. Determination of residual pesticides at the levels required for drinking water and surface water needs a preconcentration step before chromatographic analysis. Most analytical methods for organics involve the extraction of the compounds from the water before the chromatographic analysis, because direct injection of the water sample into the chromatograph is counter recommended. Gas chromatographic columns are usual incompatible with water, and the deposition of nonvolatile solids on the column will shorten column lifetime. The extraction technique must be selective for the analyte, simplifying the chromatogram and enabling a lower limit of detection.

The preconcentration step could be: (1) solvent extraction, (2) headspace analysis, (3) purge and trap, (4) solid-phase extraction (column and discs), or (5) solid-phase microextraction (SPME).

A typical pretreatment would go as follows:

1. Extraction of the organic components into a suitable solvent
2. Drying the solvent using a column containing 5 g of sodium sulphate
3. Further concentration of the extract to 1 ml via partial evaporation by bubbling dry nitrogen or argon through the sample
4. Cleanup of the extract by means of column chromatography

Here are guidelines for the storage of samples and their subsequent analysis: (a) *Microbial degradation:* storage below 0°C will lower microbial activity. (b) *Photolitic decomposition:* aqueous solutions should be stored in the dark. (c) *Contamination from the container:* glass bottles should be used; bottles made of organic polymers will leach potentially interfering monomers and additives into the sample. (d) *Loss of the analyte on the container walls:* low-solubility organic compounds can be adsorbed on the container walls. The best way of minimizing this effect is to proceed with the analysis as quickly as possible, (e) Glassware should be scrupulously cleaned or new.

A.   Liquid–Liquid Extraction

For water sample liquid–liquid extraction (LLE), certain steps are followed. The sample is shaken with an immiscible organic solvent in which the components are soluble. The most common extraction solvents are hexane and light petroleum, oxygenated and chlorinated solvents. The organic layer is separated and, after concentration, is injected into the chromatograph. By altering the pH of the aqueous layer, the extraction can be made selective toward acidic and basic components. The solvent must be distilled to avoid interferences with peaks from trace impurities and the analyte in the chromatogram.

The partitioning of the different substrates between the aqueous sample and the organic extracting solvent is in accord with Nernst's law: Any neutral species will distribute between two immiscible solvents such that the ratio of the concentrations remain constant. The distribution constant K_D is given by:

$$K_D = \frac{C_o}{C_{aq}} \qquad (1)$$

were C_o is the concentration of compound A in the organic phase and C_{aq} is the concentration of A in the aqueous phase.

The fraction of an analyte removed from the aqueous sample by extraction is given by:

$$E = 1 - \left(\frac{1}{(1 + K_D V)}\right)^n \qquad (2)$$

where $V = V_o/V_{aq}$ is the ratio between the organic- and aqueous-phase volumes and n is the number of extractions. The aliquot of aqueous sample is shaken with an equal volume of the immiscible organic solvent in a separator funnel. It is more efficient to use several small portions, rather than all of the solvent at once. The selectivity of LLE is dependent on the nature of the organic solvent and the aqueous matrix, pH (25–27), ionic strength (5,25,28–32); E depends on K_D, the water/solvent ratio, and the number of extractions.

The LLE technique appears simple, but it is laborious, time consuming, expensive, and subject to emulsification, the use of large volumes of highly pure and possible toxic solvents, and the risk of loss and contamination.

Water samples are still widely processed by classical LLE. Dichloromethane is especially popular, because it is sufficiently polar to extract efficiently the most OPPs, and its high density is convenient for extraction in a separating funnel (5,26,27,29,33–42). Less polar solvents, such as *n*-pentane (28), *n*-hexane (36,37,29,35,42–44), *n*-heptane (45), and *n*-hexane/dichloromethane (3:1) (35) or (1:3) (46), give qualitative extraction for a limited number of OPPs. Water extraction of OPPs was also done with cyclohexane (36), toluene (25), benzene (44), chloroform (36,47–49), light petroleum/ethyl acetate (1:1) (32), and xylene (31) (Table 2). For polar OPPs, polar solvents such as acetonitrile, methanol, nitromethane, dimethylacetamide, dimethylformamide (50), and dichloromethane/ethyl acetate (1:0) (46) were used.

Greven and Goewie (51) show that in the case of surface water and industrial effluents, OPPs could remain adsorbed in silt. The best solution is to separate the aqueous phase from the solid phase and to extract separately with a low-polarity solvent and a high-polarity one (40,52). Preparation of standard solution in water (53,54) and general stock solutions of OPPs prepared in ethyl acetate (55) is also a very important step.

B.   Solid-Phase Extraction

Solid-phase extraction (SPE) is an alternative to LLE. Known as liquid–solid extraction, the technique has been applied to cleanup, to fractionation, solvent changeovers, and compound concentration. The two major mechanisms of analyte retention on solid support are adsorption and partitioning.

Table 2 Solvent Extraction Recoveries (%) and RSD (%, in Parentheses) for LLE of OPPs from Spiked NaCl Bidistilled Water and Spiked Ultrapure Water

Pesticide	Spiked (μg/L)	Solvent						Ref.
		n-Hexane	n-Hexane/ CH$_2$Cl$_2$(3:1)	CH$_2$Cl$_2$	Ethyl acetate	CHCl$_3$	C$_6$H$_6$	
Azinphos-methyl	1.594	50 (16)	72 (5)	74 (27)				35
	200	85 (12)		92 (7)	105 (7)	95 (13)	89 (5)	36
Azinphos-ethyl	1.626	81 (6)	95 (5)	107 (13)				35
	200	78 (9)		82 (1)	82 (2)	85 (3)	82 (2)	36
Bromophos	200	104 (10)		79 (2)	86 (4)	82 (7)		35
Chlorpyriphos	0.400	108 (10)	85 (11)	95 (4)				36
Chlorpyriphos-methyl	200	94 (6)		77 (4)	75 (5)	76 (4)	80 (3)	35
Chlorthion	0.398	98 (9)	93 (7)	101 (5)				35
Coumaphos	1.590	91 (3)	99 (10)	97 (15)				35
Diazinon	0.360	107 (8)	86 (17)	89 (7)				35
Dichlorvos	0.300	21 (39)	72 (57)	60 (17)				35
	200	82 (7)		79 (5)	81 (6)	77 (6)	86 (3)	36
Dimethoate	0.400	nd	51 (11)	110 (17)				35
	200	nd		83 (2)	52 (4)	91 (6)	nd	36
EPN	0.830	107 (22)	106 (6)	109 (10)				35
Ethion	0.400	103 (13)	95 (3)	105 (5.)				35
Fenitrothion	0.400	99 (6)	87 (11)	100 (7)				35
Isofenphos	200	80 (3)		92 (7)	51 (5)	95 (13)	81 (5)	36
Leptophos	0.816	105 (16)	97 (7)	111 (16)				35
Malathion	0.400	95 (10)	89 (8)	103 (4)				35
Methidathion	0.720	87 (4)	90 (6)	105 (6)				35
Parathion-methyl	200	91 (12)		82 (5)	78 (7)	76 (1)	83 (8)	36
Parathion-ethyl	0.360	90 (8)	87 (17)	94 (16)				35
Sulphotep	0.360	91 (19)	74 (29)	73 (12)				35
TBP (tributyl phosphate)	0.380	84 (15)	89 (24)	90 (10)				35
TiBP (tri-isobutyl phosphate)	0.360	94 (17)	88 (32)	81 (10)				35
Tetrachlorvinphos	0.720	74 (4)	94 (4)	106 (5)				35
Tributoxy- ethylphosphate	2.380	85 (19)	97 (6)	104 (12)				35
Tris(2-ethyl- hexyl)phosphate	1.980	85 (13)	97 (6)	104 (12)				35

nd = not detected.
Sources: Refs. 35 and 36.

Adsorption: Common sorbents using the adsorption mechanism for trace enrichment are charcoal and porous polymers (Amberlite XAD-2 (35,56–59), XAD-4 (37,59–61), XAD-7 (37), Tenax GC (2,6-diphenyl-*p*-phenylene oxide), and Wolfatit Y77). Desorption of the compounds of interest from the concentration column is performed with a small quantity of a suitable solvent.

Partitioning: The bonded silica developed for LC are the solid phases used for extraction. The one most commonly used contains the octadecylsilane (C18) (33,55,62–80) or octylsilane (C8) (67,81) group chemically bonded onto a silica support (C18 and C8). Two different configurations of SPE devices uses either (a) a column in the form of a packed cartridge containing suitable sorbent (100–500 mg) to trap the analyte, a plastic syringe, and a filter of 0.4 μm to avoid impurities and to stop the flux of sample by cartridge or (b) a disc configuration of membrane filters. The SPE discs are currently used in EPA methods 506, 513, 525, and 550.1 (63). The sample and solvent flow through the sorbent by gravity or by positive (syringe) or negative (vacuum) pressure.

By suitable choice of the extraction column material, selective retention and elution of the analyte can be achieved, providing both sample cleanup and analyte concentration. A pretreatment of the column by passing a small quantity of a conditioning solvent is needed before extraction. Sampling automation by laboratory robots using precolumns with bonded silica cartridges or polymers coupled with GC or HPLC is of special use for pesticide analysis (82–87).

A further development of SPE devices is the use of extraction discs, where the adsorbent material is held within a filter disc. The disc is conditioned with a layer of methanol, and the flux is preserved using a vacuum container. The extracted compounds are then eluted using a suitable solvent.

The extraction recovery of pesticides from water depends on a number of factors: (a) the type of water samples (presence of particulate matter, ionic strength of the water), (b) pH, (c) sample volume, and (d) sorbent treatment (activation, washing, desorption).

A typical SPE sequence involves the following steps.

1. Activation (conditioning) of the sorbent (wetting with the appropriate solvent)
2. Sample introduction
3. Washing of the bonded phases
4. Elution of the retained pesticides
5. Regeneration of the column

Ethyl acetate is the most frequently used solvent for desorption (11,53,55,72,58). Also used are methanol (65,71), methanol/0.1 M ammonium acetate (10:90, v/v) (80), acetone and hexane (72), ethyl acetate and hexane (11), acetonitrile (11), and methyl ether (MTBE) (67). Good recoveries were obtained by a concentration of samples before elution on a microcolumn (82–131%; RSD 3–14%) or rotary evaporation (87–137%; RSD 1–6%) (88).

The bed volume is formatted from the interstitial and pore volumes, being 1.2 μl/mg for a particle size of 40 μm and a pore diameter of 60 Å. A five-bed volume is needed for an optimal elution (89).

Recovery of OPPs by SPE is shown in Table 3. A lower flow of sample and solvent (acetone) have led to high recovery for dimethoate (33), usual very low due to its high solubility in water. Good validation parameters were obtained (M. Culea, personal communication, 1996). The validation of the method for each compound of interest is the first step needed in a

Table 3 SPE Recovery of OPPs

Pesticide	Recovery (%)	RSD (%)	Ref.
Azinphos-methyl	106	8	93
	97.2	15	67
Bromophos-ethyl	61	12	55
Chlorpyryphos	83	6	93
Coumaphos	80	7	55
Diazinon	89.3	9	67
	87	4	55
	85	8	93
Dichlorvos	56.9	10	67
Dimethoate	6.4–10.2	9	62
	35	9	93
	49	14.3	33
EPN	86.3	13	67
Ethion	72	5	55
Ethoprop	92	7	67
Famphur	99.9	13	67
Fenchlorfos	78	7	55
Fenitrothion	89	5	93
Fenthion	15	19	93
Fonofos	66	7	93
Formothion	12	50	93
Malathion	92	3	93
Methidatthin	95	4	93
Mevinphos	70	10	55
Parathion	94.3	11	67
Parathion methyl	96.3	11	67
Parathion ethyl	84	4	55
Phorate	54.6	9	67
Phosalone	94	2	93
Phosmet	88	9	93
Pyrazophos	82	8	55
Tokuthion	70.4	12	67
Triazophos	78	8	55

Solvents used: MTBE (Ref. 67); ethyl acetate (Refs. 55 and 93); acetone (Ref. 33).

quantitative determination (90). Good reproducibility and recoveries for phorate, parathion-methyl, and parathion-ethyl were obtained by two SPE procedures (72). A rapid SPE isolation on a C-18 cartridge for OPP determination is achieved by GC separation on a widebore capillary column (91).

The type of water used is a very significant loss factor in recovery. Experiments are usually carried out on aqueous samples with low ionic strength, free from colloidal particles, such as distilled, deionized, tap, or finished water, different from natural water or seawater. Recovery of OPPs from different types of water is presented in Table 4 (66,69–71,83,92,93). Similar recov-

Table 4 Recovery of OPPs from Several Types of Water

Pesticide	Mili-Q water	Tap water	River water	Lake water	Seawater	Groundwater	Ref.
Azinphos-methyl	106					127	93
	97	92					70
Carbaryl	98	64					70
Chlorpyryfos	83					79	93
Cumaphos		82.4		98.4	90		92
Diazinon	85					87	93
	100	81					70
		90.4		89.1	89.1		92
Dimethoate	35					44	93
		9.0		10.2	6.4		92
Disulfaton		95		72	93		66
Ethion		93		91	90		66
Fenitrothion			100				71
		83.8	86.3				69
	89					94	93
	98	90					70
Fenthion	15					15	93
Fonofos		94		78	96		66
	66					74	93
Formothion		61.0		64.2	60.9		88
Guthion			103				71
		87.2	84.4				69
Heptenophos		87		83	88		66
Malathion	92					90	93
		103		92	99		66
Methidathion	95					114	93
Parathion-ethyl		95		92	97		66
			98				71
		79.3	81.7				69
	100	92					70
Parathion-methyl		108		90	104		66
	98	88					70
			109				71
Paraoxon			105				71
		84.4	91.2				69
Phenthoate		89		75	91		66
Phorate		51		61.9	46.0		92
Phosalone	94					100	93
Phosmet	97	92					70
Piridafenthion		87.6		81.2	80.7		92
Pyrazophos		97.2		96.3	92.0		92
Quinalphos		81.3		90.0	89.0		92
Sumithion		99		85	102		66
Tetrachlorvinphos		88.7		84.7	86.2		92
Triazophos		87.0		86.9	84.0		92
Trichlorfon		5.5		5.7	6.3		92
Trithion		100		79	101		66

eries were obtained for 1 L and 10 L water with the same amount of OPPs, except for trichlorfon, phorate, dimethoate, fermothion, piridafenthion, pyrazophos (68), fonophos, and disulphoton (66). Maximum recovery values were obtained at a pH of 4.8–8 (66,92).

Recoveries obtained for 10 OPPs in water, pH 7, for ethyl acetate, n-hexane, and light petroleum were 75–100%, 36–94%, and 29–84%, respectively. Better recoveries were obtained for heptenophos, fonofos, and ethion by increasing the solvent volume to 5 ml.

The effect of sample volume in SPE is of crucial importance for water samples. A sample of 200 ml to 1 L is needed in order to obtain low levels of OPPs. The breakthrough volume is the maximum sample volume from which 100% recovery can be achieved. For reverse-phase sorbents, the breakthrough volume is a function of the hydrophobicity of the solute and the mass of the sorbed used. Lacorte and Barcelo (74) have determined the breakthrough volume for fenamiphos, pyridafenthion, and temephos to be 200 ml.

The characteristics of styrene-divinylbenzene copolymers XAD-2, XAD-4, and XAD-7 are: specific surface area of 330, 780, and 450 m^2/g, respectively, and average pore diameter of 9.5 or 8 nm (35,89). PLRS-S has the same structure as XAD-2 (19,74,94–97). Polystyrene-divinylbenzene copolymer sorbents PRP-1 produce the highest efficiency in the enrichment of nonpolar pesticides such as OPPs (70). A new sorbent is ethylvinylbenzene-divinylbenzene copolymer LiChrolute EN (98). This copolymer has a specific surface area of ca. 1200 m^2/g, due to its microporous structure. LiChrolute EN copolymer shows hydrophilic properties without any surface modification and consequently has the highest efficiency toward polar molecules (98).

The solvents frequently used for OPP desorption from polymers are: methanol (57), acetonitrile (35), ethyl acetate (37,59,94,97), acetone (35,58,61), and dichloromethane (60). Table 5 shows recovery values obtained for OPPs on different polymers. Recovery values of 85–100% for OPPs (99) and for 11 OPPs on XAD-4 (60) and XAD-7 (37) have been reported. On-line system -S cartridges have been used with a gas chromatography–flame photometric detector (GC-FPD) (19) or with a gas chromatography–atomic emission detector (GC-AED) (95). Polyurethane foam (PUF) is also an effective sorbent for water pollutant removal (16). The recoveries obtained in a comparative study of six different sorbents for OPPs from 200 ml groundwater spiked at 0.2 μg/L by liquid chromatography ion spray mass spectrometry showed good values

for all OPPs on Lichrolut EN. The RSD values for $n = 6$ varied from 10 to 15% for recoveries higher than 70%, and up to 30% for recoveries of 8–50% (6).

Poor recoveries for dichlorvos (75) were attributed to its high vapor pressure, with losses during preparation steps. When the extract is not evaporated to dryness and the drying of the cartridge was improved, less hydrolysis was produced and recovery was increased (6).

Discs represent the new generation of SPE devices. The extraction disc consists of a PTFE fibril network (10%) that holds bonded silica (90%) of 8-μm particle diameter and 60 Å pore size. The small size (4.6-cm diameter) corresponds to 500 mg bonded silica C18 and the larger diameter (9 cm) to 2000 mg. The SDB disc has a 6.8-μm particle diameter, an 80-Å pore size, and a 350-m^2/g surface area. Recoveries obtained on discs (C18 and SDB) using ethyl acetate as solvent were between 70 and 99% (6), except for dichlorvos and monocrotophos. High recovery (70–100%) was obtained for dimethoate on C18 discs, using acetone solvent (M. Culea, personal communication, 1996).

Trace OPPs from natural water were determined on an Empore disc 1 (4.7-cm diameter, 0.4-mm thickness), using 1 L Water spiked with 0.1–50 μg/L OPPs and 5 ml methanol. The elution solvent was methanol/ethyl acetate (1:1, v/v). Surface water (rivers, lake, sea) showed lower recovery values than distilled and underground water (100).

Tap water from the rivers Rhine and Garone was analyzed with a simple online technique using disc extraction (C18 and XAD) coupled to capillary GC-NPD (nitrogen phosphorus detector) and with ethyl acetate as solvent (88). Recoveries of 89–91% were reported with a computerized procedure for OPP determination from aqueous samples by membrane discs (4.7-cm Empore WAD-2 discs, 0.5-mm thickness, ethyl acetate solvent) online with capillary GC (53). Recoveries of 70% were reported for oxydemethon-methyl, demethon-methyl, dimethoate, and fenamiphos obtained for 0.1 μl/L spiked OPPs, using Empore C18 and SDB discs and C18 cartridges (ASPEC XL), by liquid chromatography electrospray mass spectrometry (LC/ESP-MS) (75). Eighteen OPP recoveries on Empore discs and by LLE were compared (71). Supercritical fluid extraction (SFE) for elution of OPPs from a C18 disc PTFE membrane was investigated. After adsorption of OPPs from water, the disc was placed in an SFE cell and eluted with methanol-modified CO_2. Recoveries and RSD were: dichlorvos 95.6% (6.6%); diazinon 93% (6.1%); and malathion 84.6% (2.1%) (4).

Table 5   Mean Recovery (%) of OPPs and RSD (%, in Parentheses)

Pesticide	XAD-2	XAD-2 + Teflon	XAD-2 + XAD-7	PLRP-S	XAD-4	Ref.
Azinphos-ethyl	91 (10)	95 (12)	90 (9)			35
Azinphos-methyl	74 (13)	81 (26)	71 (16)			35
Bromophos-ethyl				42 (22)		95
Chlorpyryphos	92 (13)	82 (15)	84 (9)			35
Chlorthion	71 (50)	60 (28)	54 (54)			35
Coumaphos	91 (13)	100 (13)	91 (9)			35
Diazinon	86 (15)	91 (16)	85 (10)			35
				110 (9)		95
				91 (7)		19
					75	59
Dichlorvos	60 (37)	71 (43)	54 (24)			35
EPN	78 (19)	66 (22)	61 (38)			35
Ethion				72 (11)		95
Fenchlorfos				85 (4)		95
Fenitrothion	98 (18)	87 (16)	88 (9)			35
				97 (4)		19
Leptophos	86 (7)	89 (12)	88 (9)			35
Malathion	57 (35)	72 (29)	44 (87)			35
					97	59
					95	61
Methidathin	105 (11)	98 (13)	94 (5)			35
Methychlorpyrifos	88 (21)	79 (12)	76 (11)			35
Mevinphos				74 (9)		19
				87 (7)		95
Monocrotophos	96 (36)	97 (40)	79 (8)			35
Parathion-ethyl				95 (10)		95
					72	59
					90	61
Parathion-methyl	75 (39)	63 (24)	60 (47)			35
Phosalone					105	61
Pyrazophos				96 (4)		95
Sulphotep				95 (6)		95
	80 (16)	86 (9)	85 (10)			35
TBP	87 (16)	89 (11)	86 (6)			35
Tetrachlorvinphos				86 (9)		95
TiBP	76 (22)	77 (21)	73 (15)			35
Triazophos				104 (5)		95
Tributoxyethylpohosphate	88 (16)	85 (15)	82 (4)			35
Tris(2-ethylhexyl)phosphate	59 (10)	61 (23)	58 (3)			35

C. Solid-Phase Microextraction

Pawliszyn and coworkers (7,101–113) have introduced solid-phase microextraction (SPME). The SPME technique is based on an equilibration of the analytes between the aqueous and an immobilized liquid phase into a silica fiber as a stationary phase. It is an equilibrium extraction and can be described by Nernst's law (102): The amount of analyte adsorbed by fiber at equilibrium is given by:

$$n = \frac{K_o V_f C_o V_s}{K_o V_f + V_s} \tag{3}$$

where n = the mass adsorbed by the fiber coating, V_f is the fiber coating the stationary phase, V_s is the volume of the sample aqueous phase, K_o is the distribution coefficient constant of an analyte partitioning between the stationary and aqueous phases, and C_o is the initial concentration of the analyte in the aqueous phase. When K_o is very large, $K_o V_f \gg V_s$, and

$$n = C_o V_s \qquad (4)$$

expresses the case when the analyte is totally extracted by the fiber coating.

The dynamic of the adsorption process has been mathematically modeled (103). A rapid extraction (less than 1 min) is predicted for the fully stirred solution.

The SPME technique involves a Hamilton syringe. A fused silica fiber coated with stationary phase (1 cm) is glued into a stainless steel tubing that enters the syringe needle. Stationary phases investigated were: polydimethylsiloxane (PDMS) (7,111–116), polyacrylate (PACRY) (7,79,112,114,116,117), carbowax-divinylbenzene (CX-DVD) (114), and PDMS-DVB (114); thickness was 7–100 μm. The polar coatings CX and PACRY are suitable for polar compounds. For SPME-GC the desorption is made via direct injection into the injector port. For SPME-HPLC a solvent desorption in a minimum amount is used before analysis (110). Table 6 presents the equilibrium times and K_o values for several OPPs. An interlaboratory study was carried out to demonstrate the efficacy of SPME. Eleven laboratories have made validation tests for 12 pesticides at ppb levels (dichlorvos, ethoprofos, and chlorpyriphos-methyl) by measuring repeatability, reproducibility, and accuracy. Good results were obtained. The SPME method is fast, and inexpensive, with complete elimination of the solvent and a limit of detection (LOD) of ppt when an ion trap detector was used.

III. ANALYTICAL METHODS FOR THE DETERMINATION OF ORGANOPHOSPHORUS PESTICIDES IN WATER

The primary methods used for pesticide residue analysis are GC, HPLC, and SFC, which have been increasing steadily. The development of immunoassay methods for pesticide analysis is the research area with the most likelihood of significant progress. Most available residue methods have been devised for the analysis of only a single pesticide or pesticide class.

A. Gas Chromatography

The most widely used technique for determining organic pollutants is gas chromatography (GC). A variety of detectors can be used. Liquid chromatography is preferred for highly polar and nonvolatile compounds. The hyphenated techniques, such as gas chromatography–mass spectrometry (GC-MS) and liquid chromatography–mass spectrometry (LC-MS), bring together the ideal techniques for separation with the ideal technique for identification.

The most common GC detectors used for environmental trace analysis are:

Flame ionization detector (FID): sensitive universal detector for organic compounds

Flame photometric detector (FPD): element-specific detector for compounds containing sulphur and phosphorus

Nitrogen phosphorus detector (NPD)

Electron capture detector with ^{63}Ni (ECD)

Thermionic ionization detector (TID): element-specific detector for compounds containing nitrogen and phosphorus

Plasma emission detector (PED)

Atomic emission detector (AED)

Infrared detector (IRD)

Mass spectrometric detector (MSD): highly specific and sensitive detector for all organic compounds, used for peak identification.

Table 7 presents the detectors used most frequently in the last 15 years. The ECD and EI-MS are similar in sensitivity. LOD, and linearity, but selectivity is better for MS. The NPD and PCI-MS also show similar values. The S-FPD (sulphur flame photometric detector) and P-FPD (phosphorus flame photometric detector) have similar values, poor sensitivity, and LOD (34).

The range of GC columns available is extensive. For an efficient separation, it's very important to consider the most appropriate station phase and column dimension. A classification in order of decreasing separation efficiency would be:

1. Narrow-bore capillary columns: 30–60-m length, 0.2-mm ID, flow rate 0.4 ml/min of He
2. Wide-bore capillary columns: 15–30-m length, 0.53-mm ID, flow rate 2.5 ml/min of He
3. Packed columns: 2-m length, 2-mm ID, flow rate 20 ml/min of He

Most recent analytical methods for water analysis use the first two types of column. A splitting device is necessary for the introduction of the sample.

The organophosphorus compounds of environmental interest are of high relative molecular mass and have low volatility. Consequently, high oven temperatures are necessary for them. Silicone polymers are often the favored stationary phase. The best separation efficiencies are achieved when the stationary phase has a polarity similar to those of the components of the analyte. Pesticides are often separated on medium-polarity col-

Table 6 Equilibrium Time and K Values for Some OPPs

Pesticide	Equilibrium time (min)		K_o values		Ref.
	PDMS	PACRY	PDMS	PACRY	
Azinphos-methyl	30	15	350	2,400	7
	30 (13)	60 (10)			112
Chlorpyryphos	120 (8)	90 (8)			112
	90	45	4,500	150,000	7
Diazinon	60	40	1,500	6,000	7
	60 (10)	90 (15)			112
Dichlorvos	30	30	130	300	7
	30 (12)	30 (16)			112
Dimethoate	30	25	350	800	7
	30 (13)	30 (17)			112
Disulfoton	60	45	1,600	22,000	7
	60 (6)	90 (8)			112
EPN	90	25	4,000	4,100	1
	30 (4)	30 (6)			112
Etroprophos	15	60	180	300	7
	30 (13)	90 (11)			112
Famphur	60	30	150	700	7
	30 (10)	30 (10)			112
Fenchlorfos	120	45	3,500	22,000	7
	120 (10)	90 (15)			112
Fenitrothion	60	30	1,000	9,100	7
	30 (10)	30 (12)			112
Isoxathion	90	75	1,800	37,400	7
	60 (7)	90 (10)			112
O,O,O-Triethylphosphorothioate	30	40	180	500	7
	30 (11)	90 (3)			112
Parathion-ethyl	60	45	1,100	45,000	7
	60 (12)	90 (4)			112
Parathion-methyl	30	40	130	4,000	7
	30 (7)	90 (15)			112
Phorate	90	40	1,200	27,000	7
	60 (10)	90 (2)			112
Prothiophos	60	45	10,000	9,800	7
	90 (7)	60 (17)			112
Sulphotep	90	40	1,600	20,000	7
	90 (8)	60 (5)			112
Thionazin	15	40	50	400	7
	30 (12)	60 (14)			112

Table 7 Frequency of Use of Several Types of Detectors for OPP Analysis

Type of detector	Packed column					Capillary column							
	FPD	NPD	ECD	TID	FID	FPD	NPD	ECD	FID	PED	AED	MSD	IRD
Number of works	6	1	4	3	3	13	24	6	6	1	3	18	1

FPD: flame photometric detector; NPD: element-specific detector for compounds containing nitrogen and phosphorus; ECD: electron capture detector; TID: thermionic ionization detector; FID: flame ionization detector; PED: plasma emission detector; AED: atomic emission detector; MSD: mass spectrometric detector; IRD: infrared detector.

umns, e.g., diphenyldimethylsilicone. The stationary phase may be adsorbed or chemically bonded on the column walls of capillary and wide columns or on the support material in packed columns. For analysis close to the limit of detection and at high oven temperatures, column bleeding may become a significant factor. The use of low-loaded columns (0.1–0.25-μm film thickness) of chemically bonded phases may reduce this effect.

B. Limit of Detection and Retention Time

The limit of detection (LOD) of an individual analytical procedure is the lowest amount of an analyte in a sample that can be detected but not necessarily quantified as an exact value. It is achieved by repeatedly analyzing a blank matrix and one with the analyte present at a concentration that produces a response in the chromatographic system equivalent to the mean blank response plus three standard deviations. The limit of quantification (LOQ) is the lowest amount of an analyte in a sample that can be measured with an acceptable level of precision and accuracy. Usually it is taken as 10 times the signal-to-noise ration. The LOD estimated for several detectors for OPPs was discussed (34) (Table 8).

Gas chromatography qualitative information is obtained by using retention time (t_R) or relative retention when referred to a properly resolved and quantitatively added internal standard. Tables 9 and 10 present, respectively, t_R and relative retention for OPPs on different GC columns (OV-5 and DB-1701 are recommended by the EPA for monitoring OPPs (53)). The retention index, the Kovats index I, calculated for hydrocarbons, allows one to avoid chromatographic variations (carrier gas flow, temperature program, column length, etc.) in the identification of OPPs compounds (Table 11):

$$I = 100z + \frac{10[t_R - t_R(Cz)]}{t_R(Cz + 1) - t_R(Cz)} \tag{5}$$

where t_R is the retention time to be determined and $t_R(Cz)$ and $t_R(Cz + 1)$ are the retention times for the internal standard, with z being the number of rings in the organonitrogen internal standard. The I values of 100, 200, 300, and 400 are arbitrary, attributed to pyridine, quinoline, benzo, and naphtoquinoline.

C. High-Performance Liquid Chromatography

The most common form of high-performance liquid chromatography (HPLC) uses ultraviolet absorption as its method of detection. Few pollutants have sufficient ultraviolet absorption for direct detection, so analytes would have to be derivatized before analysis. There are few advantages over gas chromatography for trace organics in water samples for specific classes of compounds.

In order to monitor low concentrations, sample preconcentration is needed. Sensitivity can be maximized if the detector is capable of changing the excitation and detection wavelengths throughout the chromatographic run, since each compound has different optimum settings. The range of wavelengths used is 270–300 nm for excitation and 330–500 nm for detection. High-performance liquid chromatography offers an alternative for analyzing low-volatility and thermally labile OPPs. The separation of OPPs from water are usually done on C18 reversed-phase (RP) (45,51,54,57,64,69,70,71,74,94,122–124) or C8 (54,77,78,81,125) columns and with chemical siloxane bonded phase with cyanopropyl (75,6). The packed columns Partisil PXS-10/25 silica (38) and Spherisorb S5W silica (120) were used in normal-phase (NP)-HPLC. The common detection of OPPs was by ultraviolet absorption (UV) (38,45,54,57,69,70,124). Simultaneous detection at several wavelengths by photodiode array UV detector (DAD) gave suitable wavelengths free from interference (51,64,74,78,81,122). Dual electrochemical (ED) (reductive-oxidative) detection was also reported (71,123). Table 12 shows LOD values obtained by LC with different detectors. Retention values of OPPs analyzed by LC with a UV detector are presented in Table 13. Capacity factors k and UV absorption data are also given (24).

D. Supercritical Fluid Chromatography

Supercritical fluid chromatography (SFC) is used today more as a separation technique, complementary to GC and HPLC, suitable for thermally labile OPPs. Detectors similar to those used in GC and LC are employed. Mass spectrometry and Fourier transformed infrared spectroscopy (FTIR) are more easily interfaced to SFC. The SFC technique could be a better choice for complex mixture analysis than LC because it is three to five times faster and more efficient. The liquid chromatography–supercritical fluid chromatography–thermionic ionization detector (LC-SFC-TID) or SPE-SFC-TID (solid-phase extraction–supercritical fluid chromatography–thermionic ionization detector) on different-polarity OPPs, Lichrosorb RP-18 fused-silica column (13.5 cm × 0.32-mm ID) at 50°C, pressure programming from 150 bar (1 min) to 200 bar over 3

Table 8 Limit of Detection (ng/L) for OPPs

						Detector[a]						
	FID	TID	NPD			FPD		MSD			AED	
Pesticide	(7)	(100)	(7)	(97)	(116)	(93)	(19)	(7)	(79)	(5)	(96)	(115)
Azinphos-methyl	5000		280			1000		3	100			
Bromophos-ethyl					11		2				150	500
Bromophos-methyl					13							500
Chlorfenvinphos-cis				10								
Chlorthion					9							
Chlorpyrifos	780		17			3000		2	100			
Coumaphos									550		275	
Cyanophos					8							
Diazinon	900	1	26		10	3000	1	2	100	13	30	1000
Dichlorvos	1350		500	10				6	110			
Dichrotophos					90							
Dimethoate	5200		50			9000		73				
Disulfoton	9000		50									
EDDP										24		
EPN	2400		15					8				
Ethion		0.8					2				40	
Ethoprophos	280		161	10	50			100				500
Etrimphos					15							
Famphur	1400	310						3				
Fenchlorphos	940		15				2	2			30	
Fenthion						3000			210			
Fenithrothion	240		32			3000		4				
Fonofos						2000						
Formothion						8000						
Iodfenphos					40							
Iprofenfos	1780		130					2		27		
Isoxanthion	460		40					2				
Malathion		1				3000				44		
MEP										32		
Methidathioin					7					72		
Mevinphos							20		100		50	
MPP										120		
Parathion-ethyl	1960	1	9	10	16	1.2	2	5			30	1000
Parathion-methyl	1030		114		13			11	560			500
Phorate	690		11					2	330			
Phosalone						3000				73		
Phosmet						5000				48		
Prothiophos	540		23		5			2				
Pyrazophos							5				130	
Salithion										13		
Sulfotep	470		16		1	1		2			15	
Tetrachlorvinphos							2				70	
Thionazin	430	180						2				
Triazophos							3				75	
Trichloronate									100			
Triethylphosphorothiate	4900		2600					9				

[a]Relevant reference numbers are shown in parentheses. LOD $S/N = 3$.

FID: flame ionization detector; TID: thermionic ionization detector; NPD: element-specific detector for compounds containing nitrogen and phosphorus; FPD: flame photometric detector; MSD: mass spectrometric detector; AED: atomic emission detector.

Table 9 Retention Times of OPPs on Different Columns[a]

Pesticide	DB-5 (53)	DB-1701 (53)	RSL-300 (53)	DB-5 (88)	DB-1 (100)	HP-5 (100)	SPB-5 (112)	DB-17 (118)	SP-22 (119)
					t_R (min)				
Acephate									2.14
Azinphos-ethyl	27.12	34.13	33.12	16.08			41.55		
Azinphos-methyl									18.99
Chlorpyryfos	20.07	25.92	21.67				31.44	7.76	24.75
Coumaphos	28.05	36.11	33.64	16.93			19.93		
Demeton	23.41	25.86	22.56	10.28					
Demeton-*S* methyl									3.44
Diazinon	17.39	19.28	17.68	11.64	34.40	29.28		4.79	25.07
Dichlorvos	8.59	11.18	7.59	7.24			14.32	0.71	19.22
Dimethoate				11.16			26.02	5.66	
Disulfoton	19.22	17.53	17.92	11.17			27.54	5.43	25.25
EPN							40.34	16.22	
Ethion					60.06	41.61			31.31
Ethoprop	15.10		14.78					3.01	
Ethoprofos							23.53		
Famphur							38.05		
Fenamiiphos			29.29	25.53					
Fenchlorphos	19.13	24.32	24.69				30.18		
Fenitrothion	19.34	26.44	21.68	12.63			30.40		
Fenthion				12.88				8.77	
Fenitrooxon	24.22	25.70							
Fensulfothion	23.18	29.61	27.02						
Fonofos	17.19	19.42	17.74					5.01	
Foxim				13.60					
Iprofenfos (IBP)							28.36		
Isoxathion							31.15		
Malaoxon	24.22	25.59	21.15						
Malathion	19.53	26.17	21.89	12.75	39.54	32.79		8.18	27.57
Methamidophos				7.09				0.95	
Mevinfos				8.85					21.42
Monocrotophos					30.56	26.82			
Omethoate				10.10					
O,O,O-Triethylphosphorthiate							12.40		
Paraoxon-ethyl	19.12	26.44							
Paraoxon-methyl	17.45	24.91	21.91						
Parathion-ethyl	20.05	27.08	21.98	12.91	40.74	33.20	31.46		27.45
Parathion-methyl					36.80	31.16	29.41	7.04	
Phosalone								7.24	
Phorate	16.04	17.50	15.86					25.18	
Phosalone									
Phosmet	25.26	35.14	31.03					35.15	
Pyridafenthion	25.26	34.85	30.45						
Ronnel								5.67	
RPA400629	22.44	29.00	20.40						
Stirofos	21.49	30.26	26.09						
Sulfotep							25.09		
Terbufos								4.29	
Thianozin							23.14		
Triazofos				14.78					
Trichlorfon				9.06					3.09
Vamidothion		30.99		13.73					

[a]The relevant reference numbers are shown in parentheses.

Table 10 Relative Retenton for Some OPPs*

				Relative retention			
Pesticide	SPB-5 (121)[a]	OV-101 (121)[b]	SP-2250, SP-2401 (121)[c]	HP Ultra 2 (29)[e]	SPB-20 (29)[f]	SPB-20 (29)[g]	CBP-10 (120)[h]
Acephate	0.26	[d]	[d]				
Azinphos-ethyl	2.29	3.14	2.91				
Azinphos-methyl	2.16	2.83	2.66				
Bromophos	1.24	1.41	1.07				
Chlorfenvinphos E	1.31	1.47	1.24				
Chlorfenvinphos Z	1.35	1.55	1.32				1.08
Chlorpyryfos	1.18	1.30	0.97				0.81
Chlorpyryfos-methyl	0.99	1.00	0.82	1.000	0.980	0.974	
Cyanofenphos							1.39
Cyanophos							0.63
Demeton-O	0.48	0.45	0.60				
Demeton-S methyl	0.49	0.44	0.56	0.661			
Demeton-S methylsulfone	1.09	[d]	0.130.56				
Diazinon	0.84	0.81	0.55				0.48
Dichlorvos							0.09
Dimethoate	0.69	0.64	0.89	0.784	1.090	1.136	0.66
Disulfoton							0.52
Edifenphos							1.34
EPN							1.57
Ethion	1.73	2.18	1.69				
Fenitrothion	1.11	1.16	1.09	1.078	1.100	1.136	0.92
Fenthion							0.89
Fonofos				0.855	0.825	0.764	
Heotenphos	0.42	0.39	0.46				
Iprofenfos (IBP)							0.59
Leptophos							1.54
Malaoxon	1.03	1.04	1.19				
Malathion	1.16	1.22	1.09	1.103	1.179	1.258	
Methamidophos	0.09	[d]	[d]	0.370			
Methidathion	1.40	1.69	1.46	1.334	1.320	1.428	
Monocrotophos	0.64	0.66	1.12	0.712			
Omethoate	0.48	0.52	0.90				
Paraoxon-ethyl	1.06	0.82	1.21				
Paraoxon-methyl	0.85	0.84	1.08				
Parathion-ethyl	1.19	1.30	1.16				1.000
Parathion-methyl	1.000	1.000	1.000	1.000	1.000	1.000	
Phenthoate							1.08
Phorate				0.741	0.695	0.605	
Phosalone	2.20	2.90	2.45	2.045	1.850	2.146	1.79
Phosmet				1.904	1.800	2.068	1.63
Prothiophos							1.11
Pyridafenthion							1.63
Pyrimiphos-methyl	1.11	1.58	0.92				
Salithion							0.42
Tetrachlorfenvinphos	1.41	1.19	1.46				
Vamidothion	1.45	[d]	1.87				

*The relevant reference numbers are shown in parantheses.

[a]SPB-5 fused-silica capillary column (30 m × 0.53-mm ID); temperature programmed from 140°C (2 min) to 240°C at 5°C/min and hold 2/min; N$_2$ carrier gas, 15 ml/min. Parathion-methyl t_R: 9.20 min.

[b]Glass column (2 m × 3-mm ID) packed with 4% OV-101 on 80–100-mesh Supelcoport; temperature programmed from 170°C to 250°C at 5°C/min and hold 2 min; N$_2$ carrier gas, 45 ml/min. Parathion-methyl t_R: 5.15 min.

[c]Glass column (2 m × 3-mm ID) packed with 4% OV-101 on 80–100-mesh Supelcoport; temperature programmed from 170°C to 250°C at 5°C/min and hold 2 min; N$_2$ carrier gas, 45 ml/min. Parathion-methyl t_R: 8.25 min.

[d]Compounds not revealed at these operation conditions.

[e]HP Ultra 2 column (25 m × 0.2-mm ID), 0.33-μm film thickness; operating conditions: 90°C (1 min) programmed to 180°C at 30°C/min, then at 4°C/min to 270°C (hold 15 min). Parathion-methyl t_R: 12.2 min.

[f]Supelco SPB-20 column (30 m × 0.53-mm ID), 0.50-μm film thickness; operating conditions: 150°C (1 min) programmed with 4°C/min to 270°C (hold 25 min). Parathion-methyl t_R: 16.5 min.

[g]Supelco SPB-20 column (30 m × 0.53-mm ID), 0.50-μm film thickness; operating conditions: 150°C (1 min) programmed from 2°C to 270°C (hold 25 min). Parathion methyl t_R: 23.9 min.

[h]CBP 10 fused-silica capillary column, 12 × 0.53-mm ID, coated with 1.0-μm film (Shimadzu Corp.). Operating conditions: 140°C for 1 min, programmed to 160°C at 10°C/min, followed by 2°C/min to 185°C and then 20°C/min to 225°C, 15 min. Parathion-ethyl T_R: 15.6 min.

Table 11 Retention Index of OPPs on Several Stationary Phases[a]

	Retention index I				
Pesticide	SB-30 (124)	OV-7 (124)	OV-17- (124)	CPSIL-8-CB (36)	CPSIL-13-CB (36)
Azinphos-ethyl				386.8	385.9
Azinphos-methyl				378.7	376.1
Bromophos				323.3	316.7
Carbophenothion	231.0	249.4	270.8		
Chlormephos Z				236.1	233.5
Chlorpyryfos-ethyl				319.3	312.5
Chlorpyryfos-methyl	186.6	206.0	223.4	307.6	303.9
Coumaphos				396.8	397.2
Diazinon	176.9	189.9	203.2	295.2	288.8
Dichlorvos	122.0	133.4	145.4	201.2	199.3
Disulfoton	177.6	190.6	208.0		
Ethoprophos				269.5	265.9
Fenthion	194.4	211.2	227.8	318.7	313.8
Fonofos				291.5	286.2
Iodofenphos				328.9	320.1
Malathion	191.7	210.7	228.7	316.4	310.8
Methidathion	209.2	226.8	247.5		
Mevinphos	139.4	153.2	168.0	236.1	233.5
Parathion-ethyl				319.3	313.4
Parathion-methyl				307.6	305.1
Phorate				278.8	285.3
Phosalone	248.8	273.6	296.0		
Phosmet				369.8	363.5
Phosphamidon	185.0	201.6	221.6		
Terbuphos				291.5	286.2
Tetrchlorfenvinphos				335.2	326.9
Trazophos				354.5	345.5

[a]The relevant reference numbers are shown in parentheses.

min, showed an LOD of 50 pg/L (73). In an extensive paper, Berger demonstrated (76) the wide applicability of SFC to water pesticide analysis. Recoveries were of 80–90%, with LODs of 100–500 ppt for 5-ml samples, detectors being DAD, ECD, and NPD simultaneously, and with series columns (1.6 m × 4.6-mm ID), the first Hypersil C18 and the rest Hypersil silica (5 μm).

E. Thin-Layer Chromatography

Thin-layer chromatography (TLC) is a type of LC in which the sorbent is a thin layer on a flat surface. New layers prepared from small-diameter particles (5–15 μm) introduced HPTLC. The identification of OPP compounds from water is made by comparing retention factor R_f values, color and UV-visible spectrum, with those of a known compound (126). The LODs are presented in Table 14. Table 15 shows R_f values for some

OPPs. The plates in the S_1 and S_2 systems were examined under UV radiation or by spraying with diphenylamine and zinc chloride in acetone, heated at 200°C, or bromide vapor for 1 min, prior to sequential spraying with two reagents: fluoresceine and silver nitrate (124), obtaining colored spots. In the S_3 system the plates were dipped in hexane solution 0.3% N,2,6-trichlorobenzoquinoneimine and heated at 110°C for 10–15 min, yielding red-orange spots that were scanned and quantitatively determined (68).

F. Flow Injection Analysis

Flow injection analysis (FIA) can be achieved with a system consisting of a peristaltic pump, an injection valve, a reactor chamber, and a detector. An FIA system incorporating an acetylcholinesterase (AChE) single bead string reactor (SBSR) for OPP determination was

Table 12 LOD in LC with Several Detector Systems

Pesticide	DAD (78)	ED (71)	ED (123)	APCI/MS (125)	TSP/MS (SIM) (80)	(77)	APCI/MS Full-scan	APCI/MS SIM (94)	MS/MS
Azinphos-ethyl	50			14					
Azinphos-methyl	59			12					
Bromophos-ethyl							n.d.	n.m.	n.m.
Chlorfenvinphos	90			1					
Chlorpyryfos						88	2500	150	n.m.
Coumaphos							350	15	75
Diazinon				2	30				
Dichlorvos	99			6					
Dimethoate					10		20	1	7
Disulfoton					10	2			
Fenamiphos						12	10	0.8	1
Fenchlorphos							n.d.	n.d.	n.d.
Fenitrothion	40		180	9					
Fenthion	65	20					3000	n.m.	n.m.
Guthion		20							
Isofenphos						11			
Malathion	190			3	15	29			
Metidathion						16			
Mevinfos-*cis*	111			2					
Mevinfos-*trans*	39			2					
Paraoxon		20	350						
Paraoxon-ethyl	48	30	330	15	90				
Parathion-methyl	48	20	210	9					
Pyridafenthion						11			
Temephos						38			

[a]The relevant references are shown in parentheses.
n.m., not monitored.
n.d., not detected at the high concentration analyzed.
DAD: photodiode array UV detector; ED: dual electrochemical detector; APCI/MS: atmospheric pressure chemical ionization–mass spectrometry; TSP/MS: thermospray–mass spectrometry; SIM: selected ion-monitoring mode; MS/MS: mass spectrometry–mass spectrometry.

reported (129). The concentration of OPPs is determined by means of enzyme inhibition. The LOD of 0.5 ppb for malathion and 275 ppb for bromophos-methyl was found. The analysis of OPPs by FIA-HPLC-UV has been reported (130,131).

G. Spectrophotometric Analysis

Quantitative determinations of OPPs, at ppm levels, by a spectrophotometric method have been described (132–134). Oxidation of indole to fluorescent indoxyl with sodium perborate in the presence of organophosphorus esters was followed by excitation at 320 nm and measurements at 488 nm. The optimal pH was 9.3–9.6. Removal of ions was accomplished with hexamethaphosphate. The reaction is specific for organophosphorus esters but cannot be applied to thionates and dithionates (132). Water OPP determination (61,135) has been published. Total OPPs, adsorbed on Amberlyte XAD-4, were converted into orthophosphoric acid by oxygen-flask combustion, and measured by spectrophotometry at 830 nm as phosphonomolybdenum blue (61).

Table 13   Retention Data for LC Analysis of Pesticides[a]

Pesticide	t_R (min) RP-HPLC				Adjusted retention value NP-HPLC	Relative retention volume NP-HPLC	
	(77)	(80)	(64)	(54)	(38)	[38][b]	[38][c]
Azinphos-methyl		35.1		65	27.86	1.244	1.198
Azinphos-methyl oxon				57.68	2.576	2.480	
Carbotuthion		47.8					
Chlorfenvinphos			14.9				
Chlorpyrifos-ethyl			18.7	3.72	0.166	0.160	
Chlorpyrifos-methyl	33.9		16.4				
Chlorpyrifos-methyl oxon	28.5						
Crufomate				6.2			
Coumaphos				7.7			
Diazinon	35.8	42.2		92	12.62	0.564	0.543
Diazinon oxon	29.1						
Dicapton				6.7			
Dichlorvos			5.0				
Dimethoate		19.3	3.1	52.48		2.637	
Disulfoton	30.1	43.8					
Disulfoton sulfoxide	24.6						
Fenamiphos	28.7						
Fenamiphos sulfoxide	20.9						
Fenchlorphos				10.9			
Fensulfothion				5.1			
Fenitrothion				6.4			
Fenthion	34.1						
Fenthion sulfoxide	24.9						
Fonofos				8.9			
Gophacide				7.1			
Isofenphos	30.3						
Isofenphos oxon	25.3						
Malathion	31.7	37.8			17.03	0.761	0.732
Metidathion	31.0						
Metidathion oxon					59.48	2.657	2.557
Paraoxon					33.51	1.497	1.441
Parathion-ethyl		41.2			6.85	0.306	0.294
Parathion-methyl				5.9	10.83		0.454
Phosmet				5.6			
Pyridafenthion	29.3						
Pyridafenthion oxon	23.5						
Ronnel					7.84		0.256
Temephos	31.3			10.4	16.94		0.541
Temephos oxon	31.6						
Triazophos			13.7				
Zytron				10.2			

[a]The relevant reference numbers are shown in parentheses.

[b]Adjusted retention volume relative to benzyl alcohol.

[c]Adjusted retention volume relative to carbaryl.

Table 14 Limits of Detection[a]

Pesticide	LOD (μg) S_1 (127)	S_2 (128)	S_3 (128)	S_4 (128)
Azinphos-ethyl	0.010			
Azinphos-methyl	0.010			
Chlorfenvinphos	0.010		0.1	
Chlorfenvinphos-ethyl	0.020			
Chlorfenvinphos-methyl	0.020			
Diazinon	0.020			
Dimethoate	0.030	0.02	0.05	0.10
Fenthion			0.05	0.10
Mevinphos		0.05	0.05	
Paraoxon-ethyl	0.040			
Paraoxon-methyl	0.040		0.05	
Parathion-ethyl	0.005			
Parathion-methyl	0.040		0.05	
Pyrimiphos-ethyl	0.005			
Propenfenphos	0.010			
Trichlorfon		0.05		

[a]The relevent references are shown in parentheses.
S_1—HPTLC with silica 60F254/TLC scanner with $S/N = 3$.
S_2—Polygram cel 300/silver nitrate-2-phenoxyethanol.
S_3—Polygram SIL G/4(4'-nitrobenzyl)pyridimine.
S_4—Kiselgel 60/2,6-dichloro- or 2,6-dibromo-monobenzo-quinone-N-chloroimine.

H. Immunoassay Analysis

Enzymatic determination of OPPs is based on the decrease of enzyme activity correlated with the pesticide's concentration, because of the pesticide's reaction with the enzyme (136–154).

Immunoassay methods have been developed as quantitative analytical methods (155–158). They are used for rapid tests, in direct analysis, and in unconcentrated water samples, showing a good LOD. Chlorpyrifos was determined in surface water at 0.2–0.07-ppb levels. A good correlation with GC-MS and HPLC determinations was found (157). Official guidelines for antibody kits used in pesticide detection in water have been presented (158).

I. Mass Spectrometry and Hyphenated Methods

The identification of separated compounds based on retention times is uncertain. Mass spectrometry is a highly sensitive and specific technique for use in environmental organic analysis. The compound is ionized under high vacuum, often using electron impact, in-

Table 15 hR_f Values for OPPs

Pesticide	hR_f[a] S_1 (124)	S_2 (124)	S_3 (68)
Azinphos-ethyl			31
Carbophenothion	58	37	63
Diazinon	30	20	54
Dichlorvos	0	0	
Dimethoate	3	1	
Disulfoton	52	35	
Fenitrothion	50	17	
Fonofos			62
Iodofenphos	62	31	
Malathion			33
Meditathion	25	13	
Mevinphos	3	2	
Parathion-ethyl			47
Parathion-methyl			38
Phosalone	39	17	
Phosphamidon	4	2	
Trichlorfon	2	1	

[a]$hR_f = 100\ R_f$.

ducing molecular fragmentation. The ions produced are focused into a beam, accelerated, and then separated according to their mass-to-charge ratios, m/z. High-resolution mass spectrometers separate the ions using both magnetic and electrostatic fields. Low-resolution spectrometers, which are commonly found in gas chromat-

ograph–mass spectrometer (GC/MS) systems, tend to use quadrupole separation.

The peak with the highest mass/charge ratio is usually, but not always, the molecular ion and can confirm the molecular mass of the compound. The fragmentation pattern can give an indication of the chemical

Table 16 Fragment Ions and Relative Abundances (%) in Mass Spectra of OPPs by GC/MS Analysis

| Compound | MW | m/z Relative intensity (%) of main ions | |
		EI	NCI
Azinphos-ethyl	345	104 (19), 132 (100), 160 (82)	185 (100)
Chlorpyryfos	349	97 (59), 125 (22), 197 (100), 258 (37), 314 (33)	169 (23), 212 (40), 313 (100)
Coumaphos	362	89 (310, 97 (94), 109 (100), 210 (56), 226 (49) 362 (79)	169 (6), 362 (100)
Demeton	258	109 (47), 125 (39), 169 (100)	169 (15), 187 (100)
Diazinon	304	93 (29), 137 (100), 152 (74), 179 (98), 199 (59) 304 (22)	169 (100)
Dichlorvos	220	79 (18), 109 (100), 185 (30)	125 (100), 134 (58), 170 (80)
Disulfoton	274	88 (100), 97 (31), 125 (26), 147 (27), 153 (27), 185 (25)	185 (100)
Ethoprophos	242	97 (57), 126 (48), 139 (46), 158 (100), 200 (31)	199 (100)
Fenamiphos	303	122 (29), 154 (100), 217 (54), 260 (30), 288 (31), 303 (97)	153 (100)
Fenchlorphos	340	79 (17), 109 (28), 125 (65), 285 (100)	141 (20), 211 (100), 270 (54)
Fenitrooxon	261	109 (97), 127 (20), 244 (100), 261 (12)	261 (100)
Fenitrothion	308	95 (58), 109 (48), 125 (81), 139 (39), 140 (82), 141 (64), 153 (34), 156 (100), 293 (84)	169 (100), 293 (8)
Fensulfothion	308	97 (58), 109 (48), 125 (81), 139 (39), 140 (82), 141 (64), 156 (100), 293 (84)	169 (100), 293 (8)
Fenthion	278	125 (78), 278 (100)	
Fonofos	246	109 (100), 137 (57), 246 (36)	109 (39), 169 (100)
Isofenphos	345	232 (88)	
Malaoxon	314	99 (23), 109 (19), 125 (25), 127 (100), 142 (16), 173 (12), 195 (17)	141 (100)
Malathion	330	93 (40), 125 (85), 127 (93), 158 (43), 173 (100)	157 (100)
Mevinphos	224	192 (23), 224 (4)	
Monocrotophos	223	127 (100), 223 (3)	
Paraoxon-ethyl	275	81 (38), 109 (100), 127 (27), 139 (51), 149 (66), 220 (26), 247 (25), 275 (27)	275 (100)
Paraoxon-methyl	247	79 (15), 96 (41), 109 (100), 230 (36), 247 (25)	247 (100)
Parathion-ethyl	291	97 (83), 109 (100), 125 (49), 139 (67), 155 (54), 186 (29)	154 (100)
Phorate	260	75 (100), 97 (27), 121 (58), 260 (20)	185 (100)
Phosmet	317	77 (4), 93 (4), 160 (100), 317 (2)	157 (100)
Pyridafenthion	340	77 (30), 97 (68), 125 (47), 188 (89), 199 (100), 204 (58), 349 (79)	169 (40), 340 (100)
RPA-400629	368	97 (25), 109 (24), 121 (31), 153 (14), 171 (100), 215 (30), 233 (11)	169 (49), 185 (100)
Stirofos	366	79 (11), 109 (100), 331 (66),	125 (100)
Vamidothion	287	58 (13), 60 (15), 87 (100), 109 (39), 125 (20), 142 (39), 145 (74), 169 (26)	141 (100)

Sources: Refs. 53 and 58.

Table 17 Mass Fragments and Relative Abundances (%) in PCI Mode

Compound	M	[M + H]⁺	[M + Na]⁺	[M + NH₄]⁺	[M + 64]⁺	[M + 42]⁺	Fragments
			LC-TSP-MS (*Source:* Refs. 77 and 80)				
Azinphos-methyl	317	318 (100)		335 (78)			
Chloropyrifos-methyl	322	323 (58)		340 (100)			
Diazinon	304	305 (100)					
Disulfoton	274	275 (100)		292 (26)			153 (75)
Disulfoton sulfone	306	307 (100)		324 (44)			
Disulfoton sulfoxide	290	291 (100)		308 (15)			159 (73)
Fenamifos	303	304 (100)		321 (17)			
Fenamifos sulfone	335	336 (30)		353 (100)		377 (11)	
Fenamifos sulfoxide	319	320 (93)		337 (100)		361 (17)	
Fenthion	278	279 (25)		296 (100)			183 (53)
Fenthion oxon	262	263 (24)		280 (100)		304 (8)	171 (13)
Isofenphos	345	346 (89)					287 (100)
Malathion	330	331 (12)		348 (100)			175 (20)
							192 (20)
Malaoxon	314	315 (54)		332 (100)			
Metidation	302	303 (18)		320 (100)			
Pyridafenthion	340	341 (100)					
Temephos	466	M + (53)		484 (100)			
Temephos sulfoxide	482	M + (45)					348 (44)
		483 (100)					
			LC-APCI-MS (*Source:* Refs. 125 and 94)				
Azinphos-ethyl	345	346 (15)	368 (20)				132 (50)
							160 (100)
Azinphos-methyl	317	318 (15)	340 (10)				132 (45)
							160 (100)
Chlorfenvinphos	358	359 (45)	381 (20)				155 (100)
Chlorpyriphos	349	350 (100)					322 (13)
Coumaphos	362	363 (100)				404 (12)	177 (41)
							211 (4)
							329 (7)
Diazinon	304	305 (100)					197 (60)
Dichlorvos	220	221 (100)					109 (10)
Dimethoate	229	230 (37)					
Fenamiphos	303	304 (100)				345 (12)	262 (4)
							276 (3)
Fenitrothion	277						97 (100)
							248 (100)
							262 (15)
Fenthion	278	279 (100)					97 (80)
							247 (20)
Malation	330	331 (20)	353 (20)				127 (100)
							285 (25)
Mevinphos	224	225 (50)	247 (20)				127 (20)
							193 (100)
Parathion-ethyl	291	292 (10)					110 (100)
							262 (90)
Parathion-methyl	263						97 (20)
							110 (100)
							234 (95)

Table 17 Continued

Compound	M	[M + H]⁺	[M + Na]⁺	[M + NH₄]⁺	[M + 64]⁺	[M + 42]⁺	Fragments
			LC-PA-ESP-MS (*Source:* Refs. 94 and 75)				
Chlorpyriphos	349	nd					
Coumaphos	362	363 (100)	385 (21)				
Dimethoate	229	230 (10)	252 (8)				157 (5)
							171 (58)
							199 (100)
Fenamophos	303	304 (100)	326 (31)		367 (19)		217 (3)
							234 (7)
							262 (10)
							276 (13)
Fenthion	278	nd					
Demethon-*S*	230		253 (50)				89 (100)
							109 (20)
							169 (10)
Dichlorvos	220		243 (100)				109 (55)
Fenamiphos sulfoxide	335		358 (100)				
Fenamiphos sulfoxide	319	320 (10)	342 (100)	358 (20)	[M + K]⁺		
Fenitrooxon	261	262 (21)	234 (21)				109 (10)
Oxydemeton-methyl	246		269 (100)				164 (30)
							169 (19)
							191 (34)
Trichlorfon	256		279 (100)				109 (50)
							133 (89)

LC-TSP-MS: liquid chromatography–thermospray–mass spectrometry; LC-APCI-MS: liquid chromatography–atmospheric pressure chemical ionization–mass spectrometry; LC-PA-ESP-MS: liquid chromatography–pneumatic pressure–electrospray–mass spectrometry.

groups in the molecule and the molecular structure. However, a simple method of identification is comparison with the pure sample or a reference library.

When a mass spectrometer is used as a GC detector, the total ion current is monitored. After separation, the chromatographic peaks may be identified from the mass spectrum. The purity of the compounds can be confirmed from the mass spectra.

By using high-resolution mass spectrometry, the exact mass, to four decimal places, of an ion of the analyte could be monitored. This is an obvious advantage, for the pretreatment of the sample may be reduced.

A further development is tandem mass spectrometry (MS/MS), where a single ion is subjected to a second fragmentation to confirm the identity of the ion.

The most common coupled methods used in environmental organic analysis include GC, LC or SFC, and MS. Hyphenated techniques such as GC-MS, HPLC-MS, SFC-MS, Fourier transform infrared spectrometry (GC-FTIR), and GC-MS-MS are solutions for identification (159). Multidimensional GC/IR/MS quantitative analyses for environmental extracts have been reported (27). Mass spectrometry has found use

as a sensitive and highly selective detection method. The development of inexpensive benchtop integrated GC-MS explains its widespread application. In the last decade a lot of ionization techniques have been developed, including electron impact (EI), chemical ionization (CI), positive chemical ionization (PCI), negative chemical ionization (NCI), fast-atom bombardment (FAB), atmospheric pressure ionization (API), thermospray ionization (TSI), and electrospray ionization (ESI). The fused-silica capillary GC columns are coupled directly to the ion source of MS, but the packed columns need an interface, such as a molecular separator, jet separator, or membrane separator, to reduce the high pressure of the carrier gas when it enters the high-vacuum ion source of the MS (10^{-6}–10^{-7} torr). Advances have been made via cheap computer data processing and storage to handle the massive amount of information produced, even from a single chromatographic separation. Library search procedures that compare more than 275,000 spectra with the unknown and report a degree of fit are available for any GC-MS.

In the selective ion monitoring (SIM) mode, using a few selected ions, the sensitivity is increased by one

Table 18 GC Applications to OPP Analysis from Water

Apparatus	Injector	Column	Detector	OPPs analyzed	Ref.
1. Varian 3700 GC		3% OV-225 (1 m × 4 mm) on Gas-ChromQ (100–120 mesh); 170°C (10'), 20°C/'–240°C (7')	FPD N$_2$, 40 ml''	5	161
2. HP 5880	Split/splitless 250°C	Ultra-2 (25 m × 0.32-mm ID, 0.33 μm) 160°C, 1', 4°C/'–230°C, 2', 20°C/'–280°C, 6'	ECD	12	93
HP 5890 GC	1-μl sample	HP 1 (12.5 m × 0.22-mm ID, 0.33 μm); 45°C (1'), 30°C/'–170°C (2'), then 20°C/'–270°C (2')	FPD		
3. Varian 3400 GC/Varian Saturn MS	SPME 250°C	SPB-5 capillary (30 m × 0.25-mm ID, 0.25-μm film); 40°C (5'), 30°C–100°C, then 5°C/'–275°C, then 30°C/'–300°C (2')	Ion-trap MS 250°C	21	112
4. HP 5890 GC		DB-1 capillary (30 m × 0.3-mm ID), 100% methyl silicon, 0.25 μm	Plasma emission detector (PED)	6	162
5. HP 5890 GC	Automatic 220°C	(1) DB-5, (2) DB-1701 (30 m × 0.54-mm ID)	(1) TID, (2) FPD-P	43	163
6. Varian 3700 GC		(a) Capillary SPB-5 (30 m × 0.53-mm ID); 140°C (2'), 5°C/'–240°C (2'), N$_2$: 15 ml/'; (b) glass, (2 m × 3-mm ID), 4% OV-101 on Supelcoport (80–100 mesh); 170°C, 5°C/'–250°C (2'), (c) glass, (2 m × 3-mm ID), 1.5% SP2250 + 1.95% SP-2401 on Supelcoport (100–120 mesh); 175°C (2'), 5°C/'–240°C (10'); ULTRA-2 (HP), (25 m × 0.2-mm ID, 0.11 μm), 80° (2'), 20°C/'–140°C, 3°C/'–250°C	FPD-P 250°C	30	121
7. Perkin Elmer Sigma 3B GC	Splitless 220°C	3% SE-30 (2 m × 4-mm ID) on Chromosorb WHP. N$_2$: 50 ml'' 3% OV-7 and OV-17	MSD 5970 B SIM mode, EI 70 eV; FID; FID; NPD	14	124
8. HP 5987		DB-5 (25 m)	MS (EI and CI)	13	164
9. GC	250°C	1.5% OV-17 + 1.95% OV-202 (2 m × 3-mm ID, on Chromosorb WHP (80–100 mesh). N$_2$: 30 ml/'. 190°C (2'), 15°C/'–260°C (15')	FPD-P 260°C	7	165
10. HRGC Mega 5160	Split/splitless 250°C 2-μl, 60-s splitless	DB 1701 (30 m × 0.25-mm ID), 14% cyanopropyl methylpolysiloxane, 0.25 μm, 90°C (1'), 15°C/'–180°C, 5°C/'–280°C (13')	NPD 280°C	13	166
11. HP 5890 II GC	SPME	HP-5 (25 m × 0.25 mm, 0.33 μm) 40°C, 10°C/'–250°C	FID	2	114
12. HP 5713A GC	250°C	3% OV-1 (1.8 m × 2-mm ID) (1) on Gas Chrom Q (100–120 mesh), 165°C (2), 3% OV-1 on Gas Chrom Q (100–140 mesh) 180°C,	ECD (^{63}Ni), 300°C	9	44
13. HP 5730A GC	Splitless 200°C	50 m SE-54 chemically bonded glass capillary 0.15 μm, 25°C (4')–100°C, 4°C/'	ECD (^{63}Ni), 350°C	3	167
14. Perkin Elmer Sigma 3B	Splitter (1:20) TCFI	A wide-bore SE-30	FPD; FID	13	59
15. GC		Capillary, Carbowax-20M	FPD-P, TID		51

#	Instrument	Injector	Column	Detector		
16.	Varian Aerograph Series 1400 and 2800	200°C / 200°C	(1) 20% Triton X-305 (1.5 m × 2-mm ID) on Chromosorb WAW DMCS (0.16–0.20 mm) (2) 4% SE-30 + 6% OV-210 (1.8 m × 2-mm ID) on Gas Chrom Q (0.16–0.20 mm), 10-cm 10% Carbowax 20M on Chromosorb W NAW 155°C	TID 235°C 210°C	6	168
17.	HP 5710 A	Split 250°C; 1:20	FSOT DB-1 (30 m × 0.31-mm ID, 0.25 μm) 190°C	NPD 250°C	2	60
18.	Konik 2000-C GC	260°C splitless 5° (6') 30°C/'–140°C	BP-1 (25 m × 0.22-mm ID, 0.25 μm), He as carrier gas, 140°C (2') 5°C/'–260°C (5')	TID 280°C	14	66
19.	Varian 3400	Wide-bore splitter	100°C (5'), 8°C/'270°C (5'), DB-5 (30 m × 0.53 mm)	FID 300°C FPD 300°C	12	67
20.	GC	Split 1:20, 220°C	DB-608 (30 m × 0.53 mm) SE 54 QK (25 m × 0.32 mm, 0.25 μm) DB 1701 QK (30 m × 0.32 mm, 0.25 μm) 50°C (5'), 5°C/'–220°C (21'), Ar/methane (90:10)	NPD 240°C Makeup Ar/methane 30 ml/'	11	43
21.	GC 6000 Vega series HP 5988A	300°C	SPB-5 (30 m × 0.25-mm ID, 0.12 μm) 60°C, 6°C/'–300°C (15')	NPD 320°C Makeup He 30 ml/' MS in PCI and NCI	3	40 52
22.	Konik 2000-C GC	280°C; 50°C (0.8') 30°C/'–140°C	BP-5 (25 m × 0.22-mm ID, 0.25 μm), DB-17 (30 m × 0.24 mm, 0.25 μm), 140°C (2'), 5°C/'–280°C	Alkali FID 300°C	11	92
23.	GC	250°C	1.5% OV-17 + 1.95% OV-202 (2 m × 3-mm ID) on Chromosorb WHP (80–100 mesh), N₂ 30 ml/', 190°C (2') 15°C/'–260°C	FPD-P 265°C	6	169
24.	HP 5840 GC HP 5890A GC	250°C 250°C	Supelcowax 10 (15 m × 0.53 mm ID, 1.5 μm) He 5 ml/'; 190°C (8') 15°C/'–210°C OV-17, (10 m × 0.53-mm ID, 2.0 μm), He 5 ml/'; 120°C (2') 30°C/'–180°C, 3°C/'–270°C (5')	NPD 265°C FPD 280°C	18	26
25.	HP 5890 Sereis II GC	On-column injector 5 m × 0.32-mm ID	DB-5 precolumn (3 m × 0.32-mm ID, 0.25 μm), (25 m × 0.32-mm ID, 0.1 μm), 80°C, 15°C/'–280°C	NPD	19 6	88 159
26.	Shimadzu GC-14A HP 5890 Series II	On-column injector 45°C, 15°C/'–300°C	(1) CPSIL-8-CB Chrompack (50 m × 0.25-mm ID, 0.12 μm), (2) CPSIL-13-CB Chrompack (50 m × 0.25-mm ID, 0.40 μm), He 2 ml/'; 40°C, 5°C/'–280°C (20'), He 0.5 ml/'	NPD 300°C MSD 5971 70 Ev, 2000 V	23	36
27.	Perkin Elmer 3920 GC Tracor 560 GC	On-column inlet 230°C	Glass (1.8 m × 6-mm ID), 3.6 OV-101 and 5% OV-210 on ChromosorbW (80–100 mesh), 190°C–220°C; SPB-5 (30 m × 0.75-mm ID, 10 μm)	FPD 250°C NPD	13	31
28.	HP 5890 Series II GC Perkin Elmer Sigma		HP Ultra 2 (25 m × 0.2-mm ID, 0.33 μm), 90°C (1'), 30°C/'–180°C; 4°C/'–270°C (15'); 150°C (1'), 4°C/'–270°C (25')	NPD ECD	18	29
29.	HRGC 5300 Mega Series GC	290°C	(1) DB-1701, (2) FSQT RSL-300, (3) DB-5 (30 m × 0.25-mm ID), He (1) 60°C, 10°C/'–90°C; 6°C/'–280°C (3'), (2) 60°C, 10°C/'–90°C; 6°C/'–310°C	NPD-40 Makeup He air and H₂ 290°C EI-MS Ts: 200°C 70 eV, NCI-MS	12	53
	HP 5995 GC HP 5983A GC	Splitless (35 s)	DB-17 fused silica, 90°C, 6°C/'–280°C DB-5, same conditions as GC-EI-MS			

Table 18 Continued

Apparatus	Injector	Column	Detector	OPPs analyzed	Ref.
30. HP 5880 GC	Splitless 250°C	HP-1 (12 m × 0.2-mm ID, 0.32 μm), He 1 ml/', 45°C (2'), 30°C/'–190°C (2'), 20°C/'–250°C (3')	NPD 280°C	11	87
31. Konik 2000-C GC	Splitless 285°C Splitless at 50°C	DB-5 (25 m × 0.22-mm ID, 0.25 μm), DB-17 (30 m × 0.24-mm ID, 0.25 μm), 50°C (0.8'), 30°C/'–140°C (2'), 5°C/'–280°C (12')	TID ECD (63Ni) 300°C	10	11
32. Varian aerograph 3700 GC	220°C	(2 m × 2-mm ID), (1) 10% OV-101 + 15%), V-210 on Chromosorb WHP (80–100 mesh), (2) 10% OV-101 on Chromosorb WAW (80–100 mesh)	ECD (63Ni) 300°C NPD 250°C	3	72
33. HP 5890 GC	Split/splitless 290°C	SE-54 (25 m × 0.25-mm ID), SP-2100 (30 m × 0.25-mm ID), 50°C (1'), 4°C/'–290°C (10')	SIM, MS NPD 290°C	6	58
34. HP 5890 GC	Split, 85°C 0.75'; 250°C	HP-1 (25 m × 0.20-mm ID), He, transfer line, 280°C, 85°C, 6°C/'–285°C (2')	MSD 5971 A SIM	3	4
35. Carlo Erba Mega GC		DB-1 (5 m × 0.32-mm ID) deactivated connected (15 m × 0.32-mm ID, 0.14 μm), He: 70°C–300°C	FID; NPD; FPD	5	19
36. HP 5890 II GC	Large-volume loop-type interface	Precolumn (3 m × 0.32-mm ID, 0.25 μm), 100°C (5'), 20°C/'–270°C (5')	HP 5981A AED	11	95
37. HP 5890 II	SPME, 205°C	DB5.625 (30 m × 0.31-mm ID, 0.25 μm), 50°C (4'), 30°C/'–180°C (1'), 3°C/'–202°C (5')	HP 5981A AED	6	115
38. HP 5890 II	275°C	HP Ultra-2 (25 m × 0.20-mm ID, 0.33 μm), 90°C (1'), 30°C/'–180°C, 4°C/'–270°C (5')	ECD 300°C NPD 270°C	3	30
39. HP 5890 II	275°C	SPE-GC, 30 cm × 50 μm ID; 5 m × 32-mm ID, then SPB-5 (15 m × 0.32-mm ID, 0.25 μm) Desorbtion: 100 μl ethyl acetate SVE between precolumn–column 80°C (5'), 20°C/'–280°C (2')	Make-up He 40 ml/'.	11	96
40. HP 5890 II	Split/splitless (1 m × 0.32 mm)	DB-17 (30 m × 0.53 mm, 1 μm), EPP at 20 psi, 57 ml/', 150°C, 5°C/'–250°C	FPD-P	29	118
41. HP 5890 GC	Split/splitless SPME (5')	PTE-5 (30 m × 0.32 mm, 0.25 μm, 60°C, 30°C/'–180°C (2'), 2°C/'–200°C, 6°C/'–220°C	NPD	14	116
42. HP 5890 GC	Split/splitless 210°C	SE-54 (25 m × 0.32-mm ID, 0.17 μm), 100°C (1'), 30°C/'–150°C (2'), 3°C/'–205°C, 10°C/'–260°C		25	25
43. Varian 3700 GC	220°C	10% OV-101 and 15% OV-210 on Chromosorb WHP (80–100 mesh)	ECD (63Ni) NPD	2	39
44. Shimadzu 14A GC	220°C	DB-1 (30 m × 0.32-mm ID), 55°C (2'), 5°C/'–210°C (20'), 20°C/'–270°C (4')	TID 250°C	8	100
HP 5890 B	240°C	HP-5 (25 m × 0.25-mm ID, 55°C (2'), 5°C/'–210°C, 20°C/'–270°C (4')	5971A MSD; EI, 70 eV		

Instrument	Injection	Column	Detector		
45. HP 5790 GC	250°C splitless 1.5'	Ultra-2 crosslinked 5% phenylmethylsilicone (25 m × 0.25-mm ID), 70°C, 3°C/'–250°C	JEOL-DX303 MS IS 250°C EI 70 eV, SIM	14	5
46. HP 5890 GC Varian 3400 GC	250°C	CP-Sil 19CB; CP-Sil 5CB (20 m × 0.32-mm ID, 0.25 μm) (Chrompak); 90°C (1'), 30°C/'–180°C, 4°C/'–260°C (12')	ECD (^{63}Ni) NPD 300°C	11	170
47. Carlo Erba 8000 GC	Large-volume 100 μl PCSE;	(1.0–2.5 m × 0.53-mm ID)–(2 m × 0.32 mm, 1 μm) DB-1 (10 m × 0.32-mm ID, 1 μm), HP-5-MS (25 m × 0.25-mm ID, 0.25 μm), 77°C, 10°C/'–300°C	FID MSD	9	55
48. Carlo Erba Mega GC	Online SPE	CPSIL-19CB (50 m × 0.32-mm ID, 0.2 μm) (Chrompak)–10 m × 0.32-mm ID; cosolvent: 125°C 3'), 4°C/'–245°C (20'), 10°C/'–275°C (20'); no cosolvent: 90°C (3'), 4°C/'–215°C (10')	NPD 280°C	7	97
49. HP 58901 GC	280°C splitless 0.8'	Ultra-2 (25 m × 0.32-mm ID, 0.52 μm), 50°C (0.8'), 30°C/'–140°C (2'), 10°C/'–280°C (2'); SGE-1701 (25 m × 0.25-mm ID, 0.25 μm), 50°C (0.8'), 20°C/'–150°C (2'), 10°C/'–240°C (9')	NPD 300°C	7	171
50. Shimatzu-14A GC	On-column 45°C, 30°C/'–300°C	CPSIL-8-CB (50 m × 0.25-mm ID, 0.12 μm), carrier gas He 2–4 ml/', 40°C, 5°C/'–280°C (20')	TID 300°C	18	34
HP 5890II GC	Splitless 1' 250°C on-column inj. 45°C, 30°C/'–300°C (15')	CPSIL-8-CB (50 m × 0.25-mm ID, 0.25 μm) 50°C, 5°C/'–290°C (15') CPSIL-8-CB (50 m × 0.25-mm ID, 0.12 μm) 40°C, 5°C/'–280°C (20')	ECD (^{63}NI) 300°C N$_2$ 60 ml/' MSD 5971 300°C EI, 70 eV; 2 kV PCI, methane		
51. HP 5890 GC	250°C	OV-1701 (25 m × 0.25-mm ID, 0.20 μm) (Chrompak); 60°C (1'), 25°C/'–190°C, 2°C/'–225°C, 5°C/'–280°C (10')	FPD 225°C	24	35
52. HP 5890II GC	1-μl splitless, after injection 55-s purge	Restek Rt$_x$-5 (30 m × 0.32-mm ID, 1 μm) (5% diphenyl–95% dimethyl polysiloxane) + Reastek Rt$_x$-1701 (30 m × 0.32-mm ID, 1 μm) (14% cyanopropyl phenyl–86% dimethyl polysiloxane), 45°C, 4°C/'–260°C (30')	Parallel HP 5965B IRD and HP 5970B MSD EI	6	27
53. GC	SPME 5' 310°C	5% Phenyl methyl silicon (30 m × 2.5-mm ID, 0.25 μm), 35°C (15'), 10°C/'–300°C (5')	MSD 70-eV SIM	20	79
54. Perkin Elmer 990 GC		Glass, 2.45% OV-17 (1 m × 4-mm ID) Chromosorb G, 100–120 mesh, 140–275°C, 8°C/'; capillary OV-17 (25 m × 0.25-mm ID, 0.53 mm) 150°C, 2', 6°C/'–300°C, ID-MS	SIM-EI-MS	5	33
55. Varian 6000 GC Varian 3400 CX	Splitless 250°C for SPME; direct injection	PTE-5 (30 m × 0.25-mm ID, 0.25 μm), 50°C (3'), 20°C/'–170°C (1.5'), 10°C/'–190°C (2'), 10°C/'–290°C (1')	FID 300°C IT-MSD, IT 250°C	20	7
56. GC		2% OV-17 (2.1 m × 3.2 mm) on Chromosorb WAW DMCS (69–80 mesh), 170°C (13'), 3°C/'–240°C (15'), N$_2$: 42 ml/'	NPD		32

Table 18 Continued

Apparatus	Injector	Column	Detector	OPPs analyzed	Ref.
57. GC		PTE-5 (30 m × 0.25-mm ID, 0.25 μm), 100°C, 4°C/′–300°C; He 20–30 cm/s	NPD		172
58. GC		JWDB (30 m × 0.32-mm ID, 0.25 μm), 100°C (5′), 5°C/′–260°C	NPD		173
59. GC		ChromosorbW-HP (80–100 mesh) (2 m × 2-mm ID), 230°C N2, 30 ml/′	FID	3	48
60. GC		SE-54 capillary, fused silica	NPD	7	28
61. GC		3% OV-17 (1.5 m × 2-mm ID) on Chromosorb W (100–120 mesh), 180°C, 2.5°C/′–225°C	ECD		47
62. GC		SPB-608; SPB-5; SPB-35 capillary column	MSD; NPD	20	174
63. GC		Capillary	FPD; TID		42
64. GC		3% PPE-6R (1 m × 3 mm) on Chromosorb WAW DMCS 180°C; 3% PPE-7R, Chromosorb W-HP	FID	5	175
65. GC		GLC	TID		176
66. HP 5995		DB-1701 (30 m × 0.25-mm ID), 90°C, 6°C/′–280°C	SIM-EI-MS	11	177
67. GC	Splitless	DB-5 (30 m × 0.25-mm ID, 0.25 μm), 50°C, 15°C/′–300°C	SIM-ID-GC-MS	1	178
68. GC	Split/splitless	OV-1 WCOT (14.5 m × 0.25-mm ID, 0.4 μm), 115°C, 4°C/′–190°C, 8°C/′–260°C	TID	26	179

′ = min; °C/′ = °C/min

GC: gas chromatograph; FPD: flame photometric detector; ECD: electron capture detector; MS: mass spectrometer; TID: thermoionic ionization detector; MSD: mass spectrometric detector; SIM: selected ion monitoring; EI: electron impact; NPD: nitrogen and phosphorus detector; CI: chemical ionization; TCFI: thermal desorption cold trap flash; PSCE: partially concurrent solvent evaporation; SPME: solid-phase microextraction; IRD: infrared detector; PCI: positive chemical ionization; SIM-EI-MS: selected ion monitoring–electron impact–mass spectrometry; IT-MSD: ion trap–mass spectrometry; SPE: solid-phase extraction; SIM-ID-GC-MS: selected ion monitoring–isotopic dilution–gas chromatography–mass spectrometry.

or two orders of magnitude. The SIM-MS technique with high-resolution double-focusing mass spectrometer and SIM-GC/MS are very useful in quantitative work and are usually accomplished by means of isotopic dilution (ID). Quantification is performed by the addition of known amounts of internal standards to the sample before extraction. The method will compensate for sample losses in the cleanup stage, assuming that the losses of the standard are identical with those of the analyte. Stable isotopically labeled homologous compounds, ^{13}C or ^{15}N, are the best choice (33,159).

Typical fragment ions obtained in EI mode for OPPs have been studied (53,80). A high sensitivity could be obtained for OPP identification in NCI mode (53,58) (Table 16).

Interfacing LC to a mass spectrometer includes: direct LC/CI; continuous-flow FAB/FAB; moving belt/EI/CI/FAB; thermospray-CI (TSP-CI); electrospray (ESP); atmospheric pressure chemical ionization (APCI); monodisperse aerosol-generation interface (Magic)/EI/CI. In recent papers OPPs and TPs have been analyzed by LC/MS (77, 125,80,94) using ESP-MS, TSP-MS (77), and APCI-MS (125). The PCI (Ta-

ble 17) and NCI modes show characteristic ions for OPPs and TPs. The parent molecules can be recognized.

The coupling of LC with pneumatic-assisted electrospray LC-PA-ESP/MS/MS was used for trace determinations of OPPs in water (94). The LODs in the range 26–57 ppb (69) and 0.01–0.20 μg/L by MS were determined.

Online SPE-LC-TSP-MS results for determining TPs of OPPs in the SIM mode are shown in Table 17 (77,80).

Many GC-MS determinations, in EI, PCI, or NCI mode, of OPPs from water have been reported (5,34,36,100,112). Many automated SPE-GC-MS procedures for water analysis have been published (97,159,160), the LC-MS method gave LODs at the levels required by EU regulations. Some GC applications to OPP analysis from aqueous samples are listed in Table 18.

Water analysis by the LC/MS technique is also used a lot (6,45,60,75,77,94,125). Automated methods have been developed using either SPE-GC-MS or SPE-LC/MS for continuous water sample analysis (82,94). A

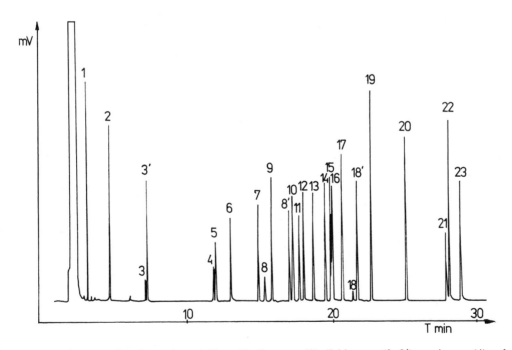

Fig. 1 GC separation of organophosphorus insecticides: (1) dimetox; (2) dichlorvos; (3–3′) mevinvos; (4) naled; (5) dicrotophos; (6) dimethoate; (7) fonofos; (8–8′) phosphamidon; (9) diazinon; (10) parathon-methyl; (11) malaxon; (12) paraoxon; (13) fenitrothion; (14) malathion; (15) parathion-ethyl; (18) chlorpyrifos; (17) bromophos; (18–18′) chlorfenvinphos; (19) tetrachlorvinphos; (20) ethion; (21) phosalone; (22) azinphos-methyl; (23) azinphos-ethyl. Carlo Erba Fractovap 2150 with thermionic detector in P mode (NPSD), GC column: glass WCOT 14.5 × 0.25 mm, 0.4-μm film of OV-1; 115°C, 4°C/min to 190°C, then 8°C/min to 260°C. (Reproduced with permission of HÜTHIG-Fachverlag from the *Journal of HRC&CC*, Ref. 179.)

"multianalysis" system, combining LC and GC separation with a single MS, was used to analyze OPPs (tetrachlorvinphos, coumaphos, fenchlorphos, promophos, chlorpyritophos) in surface water and river sediment extracts. Three detection systems—GC-MS, LC-DAD, and LC-MS—with particle beam interface (LC-PB-MS) showed good LODs 5–50 ng/L for the first two and 0.5–7 μg/L for the last one.

Sensitive HPLC determinations (51,54,64,69,75, 81,122,123) of OPPs in water using spectrophotometry (51), UV detection (54,69), DAD (81,118), ED (123), or ESP-MS (75) have been published. An online extraction and concentration system with SPE-HPLC by continuous flow and a UV detector was developed for the determination of OPPs in water (57,70).

Also developed have been online systems with UV and MS (SIM mode) detection for OPPs in spiked river water (45), SPE-LC-TSP-MS (liquid chromatography–thermospray–mass spectrometry) (80), LC-PA-ESP (liquid chromatography–pneumatic-assisted–electrospray), or LC-APCI (liquid chromatography–atmospheric pressure chemical ionization) followed by MS-MS detection for tap water monitoring (94). Off-line SPE-LC automated systems were also developed (71) in drinking and natural water (74) and in groundwater and wastewater (78). An online SPE-LC-DAD system was validated by interlaboratory studies with LODs at 0.1-μg/L levels (77). Online systems SPE-LC-TSP-MS and SPE-LC-APCI-MS in PCI and NCI modes have been used for OPP analyses in groundwater at ppt levels (125). A very sensitive automated technique (LOD 0.01–0.20 μg/L) was developed by Barcelo and co-workers (6) for polar and thermally labile OPPs by using SIM mode by SPE-LC-ISP-MS.

Compounds of high toxicity need to be analyzed at very low concentrations. Ultratrace analysis includes pollutants concentration ranges of nanograms per kilogram and below. The analysis requires not only highly sensitive and selective instrumentation but also a high degree of analytical skill and expertise. It is important to measure such a low concentration because the concentration in an organism can easily be increased over that of the environment in which it is living (157).

A GC separation of organophosphorus insecticides on a capillary column is shown in Fig. 1.

REFERENCES

1. JM Zabik, JN Seiber. Atmospheric transport of organophosphate pesticides from California's Central Valley to the Sierra Nevada Mountains. J Environmental Quality 22:80–90, 1993.

2. KA Hassall. Organophosphorus pesticides. In: The Biochemistry and Uses of Pesticides. Weinheim: VCH, 1990, pp 81–123.

3. S Hatrik, J Tekel. Extraction methodology and chromatography for the determination of residual pesticides in water. J Chromatogr A 733:217–233, 1996.

4. IJ Barnabas, JR Dean, SM Hitchen, SP Qwen. Selective extraction of organophosphorus pesticides using a combined solid-phase extraction–supercritical fluid extraction approach. Anal Chim Acta 291:261–267, 1994.

5. T Okumura, Y Nishikawa. Determination of organophosphorus pesticides in environmental samples by capillary gas chromatography–mass spectrometry. J Chromatogr A 709:319–331, 1995.

6. C Molina, P Grasso, E Benfenati, D Barcelo. Automated sample preparation with extraction columns followed by liquid chromatography–ionspray mass spectrometry. Interferences, determination and degradation of polar organophosphorus pesticides in water samples. J Chromatogr A 737:47–58, 1996.

7. S Magdic, A Boyd-Boland, K Jinno, JB Pawliszyn. Analysis of organophosphorus insecticides from environmental samples using solid-phase microextraction. J Chromatogr A 736:219–228, 1996.

8. J Manes, G Font, Y Pico, JC Molto. Residues of organophosphates in food. In: LML Nollet, ed. Handbook of Food Analysis. New York: Marcel Dekker, 1996, pp 1437–1459.

9. YL Loukas, E Antoniadou-Vyza, A Papadaki-Valiraki. High-performance liquid chromatography in stability studies of an organophosphorus insecticide in free form and after formulation as emulsifiable concentrate. Response surface design correlation of kinetic data and parameters. J Chromatogr A 677:53–61, 1994.

10. N Ohashi, Y Tsuchiya, H Sasano, A Hamada. Ozonation products of organophosphorus in water (abstr). Jpn J Toxicol Environ Health 40:185–192, 1994.

11. E Viana, JC Molto, J Manes, G Font. Clean up and confirmation procedures for gas chromatographic determination of pesticide residues in contaminated waters. I. J Chromatogr A 655:285–292, 1993.

12. HO Esser. Terminal residues of organophosphorus insecticides in animals. Pesticide terminal residues. Pure Appl Chem (Suppl) 1971, pp 33–56.

13. D Barcelo, G Durand, N de Bertrand, J Albaiges. Determination of aquatic photodegradation products of selected pesticides by gas chromatography–mass spectrometry and liquid chromatography–mass spectrometry. Sci Total Environ 132:283–296, 1993.

14. CG Daughton, DG Crosby, RL Garnas, DPH Hsieh. Analysis of phosphorus-containing hydrolytic products of organophosphorus insecticides in water. J Agric Food Chem 24:236–241, 1976.

15. U.S. Environmental Protection Agency. Agricultural Chemicals in Ground Water: Proposed Pesticide Strategy. Washington, DC: EPA pp 1–50.

16. CD Watts, L Clark, S Hennings, K Moore, C Parker. In: Pesticides: Analytical Requirements for Compliance with EC Directives (Water pollution research report, 11). Commission of the European Communities, Brussels, 1989, pp 16–34.

17. EEC Drinking Water Guidelines, 80/779/EEC No. L229/11-29. EEC, Brussels, 1980, pp 1–20.

18. M Fielding, D Barcelo, A Helweg, S Calassi, L Torstensson, P Van Zoonen, R Wolter, G Angeletti. Water Pollution Research. Report 27. Commission of the European Communities, Brussels, 1992.

19. Y Pico, AJH Louter, JJ Vreuls, UAT Brinkman. Online trace-level enrichment gas chromatography of triazine herbicides, organophosphorus pesticides and organosulfur compounds from drinking and surface waters. Analyst 119:2025–2031, 1994.

20. D Barcelo. Applications of gas chromatography–mass spectrometry in monitoring environmentally important compounds. Trends Anal Chem 10:323–329, 1991.

21. D Barcelo. Environmental Protection Agency and other methods for determination of priority pesticides and their transformation products in water. J Chromatogr 643:117–143, 1993.

22. EPA Procedures for Water Pollution Analyses. GC Bulletin 775 C. Supelco 1988.

23. US Environmental Protection Agency. Methods for the determination of organic compounds in drinking water. Supplement 1, PB91-140627. National Technique Information Service, Springfield, VA, 1990.

24. U.S. Environmental Protection Agency. Methods for the determination of organic compounds in drinking water. PB91-231480. National Technique Information Service, Springfield, VA, revised, 1991.

25. A Zapf, R Heyer, HJ Stan. Rapid micro liquid–liquid extraction method for trace analysis of organic contaminants in drinking water. J Chromatogr A 694:453–461, 1995.

26. LM Davi, M Baldi, L Panazzi M Liboni. Evaluation of the membrane approach to solid-phase extractions of pesticide residues in drinking water. Pestic Sci 35:63–67, 1992.

27. KA Krock, CL Wilkins. Quantitative analysis of contaminates environmental extracts by multidimensional gas chromatography with infrared and mass spectral detection (MDGC-IR-MS). J Chromatogr A 726:167–178, 1996.

28. J Brodesser, HF Schoeler. Improved extraction method for quantitative analysis of pesticides in water (abstr). Zentralbl Bakteriol Mikrobiol Hyg Ser B 185:183–185, 1987.

29. F Hernandez, I Morell, J Beltran, FJ Lopez. Multiresidue procedure for the analysis of pesticides in ground water: application to samples from the Comunidad Valenciana, Spain. Chromatographia 37:303–312, 1993.

30. F Hernandez, J Beltran, FJ Lopez. Study of sample cleanup in organochlorine and organophosphorus pesticide residue analysis. Application to soil and water samples from the vadose zone. Quim Anal 13:115–120, 1994.

31. D Bourgeois, J Gaudet, P Deveau, VN Mallet. Micro-extraction of organophosphorus pesticides from environmental water and analysis by gas chromatography. Bull Environ Contam Toxicol 50:433–440, 1993.

32. Y Ji, S Wang, Q Zhang, W Li, J Lu. Determination of multiresidues of organochlorine, organophosphorus and carbamate pesticides in water by gas chromatography (abstr). Sepu 10:94–96, 1992.

33. M Culea, I: Fenesan, S Cobzac, S Gocan, M. Chiriac, N Palibroda. Trace analyses of triazines and organophosphorus pesticides in water. Fresenius J Anal Chem 335:748–749, 1996.

34. SB Lartiges, P Garrigues. Gas-chromatographic analysis of organophosphorus and organonitrogen pesticides with different detectors. Analysis 23:418–421, 1995.

35. I Tolosa, JW Readman, LD Mee. Comparison of the performance of solid-phase extraction techniques in recovering organophosphorus and organochlorine compounds from water. J Chromatogr A 725:93–106, 1996.

36. S Lartiges, P Garrigues. Determination of organophosphorus and organonitrogen pesticides in water and sediments by GC-NPD (nitrogen-phosphorus detection) and GC-MS. Analysis 21:157–165, 1993.

37. C Mallet, VN Mallet. Conversion of a conventional packed-column gas chromatograph to accommodate megabore column. II. Determination of organophosphorus pesticides in environmental water. J Chromatogr 481:37–44, 1989.

38. JN Seiber, DE Glotfelty, AD Lucas, MM McChesney, J Sagebiel, TA Wehner. Multiresidue by high-performance liquid chromatography–based fractionation and gas-chromatographic determination of trace levels of pesticides in air and in water. Arch Environ Contam Toxicol 19:583–592, 1990.

39. GE Miliadis. Organochlorine and organophosphorus pesticides residues in the water of the Pinios river, Greece. Bull Environ Contam Toxicol 54:837–840, 1995.

40. D Barcelo, C Porte, J Cid, J Albaiges. Determination of organophosphorus compounds in Mediterranean coastal waters and biota samples using chromatography with nitrogen-phosphorus and chemical-ionization mass-spectrometric detection. Int J Environ Anal Chem 38:199–209, 1990.

41. P Herzsprung, L Weil, KE Quentin, I: Zombola. Determination of organophosphorus compounds and carbamates by their inhibition of cholinesterase. II. Esti-

mation of detection limits for insecticide determination by concentration, oxidation and inhibition values. Vom Wasser 74:339–350, 1990.

42. U.S. Department of the Environment. Organophosphorus pesticides in sewage sludge; organophosphorus pesticides in river and drinking water, an addition (abstr). 1985. Methods Exam Waters Assoc Mater 1986, pp 20.

43. J Alberti, W Stock. Determination of insecticides in water by gas chromatography with special reference to alternative isolation and concentration methods. Gewasserschutz, Wasswer, Abwasser 106:204–218, 1989.

44. HB Lee, LD Weng, ASY Chau. Confirmation of pesticide residue identity. XI. Organophosphorus pesticides. J Assoc Off Anal Chem 67:553–556, 1984.

45. A Farran, JL Cortina, J De Pabloo, D Barcelo. Online continuous-flow extraction system in liquid chromatography with ultra-violet and mass spectrometric detection for the determination of selected organic pollutants. Anal Chim Acta 234:119–126, 1990.

46. G Durand, R Alonso, D Barcelo. Extraction and analysis of organophosphorus and chlorotriazine pesticides in water and soil samples. Quim Anal 10:157–168, 1991.

47. A Neicheva, E Kovacheva, G Marudov. Determination of organophosphorus pesticides in apples and water by gas-liquid chromatography with electron-capture detection. J Chromatogr 437:249–253, 1988.

48. IV Pershina, DB Popov, EK Ivanova, TV Polenova. Gas-chromatographic determination of some organophosphorus pesticides in waters in presence of fulvic acids (abstr). Zh Anal Khim 44:1475–1479, 1989.

49. G Schmaland, H Schulert. An efficient variant of the .DELTA.pH method for the determination of cholinesterase inhibitors in water samples. Acta Hydrochim Hydrobiol 14:19–25, 1986.

50. VA Frankovskii, VI Kofanov, YV Lushnikova. Liquid–liquid separation and chromatographic determination of some organochlorine and organophosphorus pesticides in three-phase systems (abstr). Zh Anal Khim 47:1058–1065, 1992.

51. PA Greve, CE Goewie. Developments in the determination of organophosphorus pesticides. Int J Environ Anal Chem 20:29–39, 1985.

53. S Lacorte, C Molina, D Barcelo. Screening of organophosphorus pesticides in environmental matrices by various gas-chromatographic techniques. Anal Chim Acta 281:71–84, 1993.

54. VN Mallet, M Duguay, M Bernier, N Trottier. Evaluation of high-performance liquid chromatography–UV detection for the multi-residue analysis of organophosphorus pesticides I environmental water. Int J Environ Anal Chem 39:271–279, 1990.

55. S Ramalho, T Hankemeier, M de Jong, UAT Brinkman, RJJ Vreuls. Large volume oncolumn injections for gas chromatography. J Microcolumn 7:383–394, 1995.

56. GAV Rees, L Au. Use of XAD-2 macroreticular resin for recovery of ambient trace levels of pesticides and industrial organic pollutants from water. Bull Environ Contam Toxicol 21:561–566, 1979.

57. A Farran, J De Pablo, S Hernandez. Continuous-flow determination of organophosphorus pesticides using solid-phase extraction coupled online with high-performance liquid chromatography. Int J Environ Anal Chem 46:245–253, 1992.

58. M Psathaki, E Manoussaridou, EG Stephanou. Determination of organophosphorus and triazine pesticides in ground and drinking water by solid-phase extraction and gas chromatography with nitrogen-phosphorus and mass spectrometric detection. J Chromatogr A 667:241–248, 1994.

59. A Verweji, MA Van Liempt Van Houten, HL Boter. Isolation, concentration and subsequent analysis by capillary gas chromatography of trace amounts of organophosphorus compounds from aqueous samples. Int J Environ Anal Chem 21:63–77, 1985.

60. JE Woodrow, MS Majewski, JN Seiber. Accumulative sampling of trace pesticides and other organics in surface water using XAD-4 resin. J Environ Sci Health Part B B21:143–164, 1986.

61. Z Froebe, V Drevenkar, B Stengl, Z Stefanac. Oxygen-flask combustion of accumulated organophosphorus pesticides for monitoring water pollution. Anal Chem Acta 206:299–312, 1988.

62. G Font, J Manes, JC Molto, Y Pico. Solid-phase extraction in multi-residue pesticide analysis of water. J Chromatogr. 642:135–161, 1993

63. KK Chee, MK Wong, HK Lee. Sample preparation technique for the analysis of organophosphorus pesticides in environmental samples: a review (abstr). Bull. Sing N I Chem 21:81–99, 1993.

64. JL Martinez Vidal, P Parrilla, M Martinez Galera, A Garrido. Cross-section of spectrochromatograms for the resolution of folpet, procymidone and triazophos pesticides in high-performance liquid-chromatography with diode-array detection. Analyst 121:1367–1382, 1996.

65. V Drevenkar, Z Froebe, B Stengl, B Tkalcevic. [Octadecyl] C-18 reversed-phase trace enrichment of organophosphorus pesticides and residues in water. Mikrochim Acta I:143–156, 1985.

66. J Manes, JC Molto, C Igualada, G Font. Isolation and concentration of organophosphorus pesticides from water using a C-18 reversed phase. J Chromatogr 472:365–370, 1989.

67. PR Loconto, AK Gaind. Isolation and recovery of organophosphorus pesticides from water by solid-phase extraction with dual wide-bore capillary gas chromatography. J Chromatogr 27:569–573, 1989.

68. J Sherma, W Bretchneider. Determination of organophosphorus insecticides in water by C-18 solid-phase

extraction and quantitative TLC. J Liq Chromatogr 13: 1983–1989, 1990.

69. R Carabias-Martinez, E Rodriguez-Gonzalo, MJ Amigo-Moran, J Hernandez-Mendez. Sensitive method for the determination of organophosphorus pesticides in fruits and surface water by high-performance liquid chromatography with ultraviolet detection. J Chromatogr 607:37–45, 1992.

70. MR Driss, MC Hennion, ML Bouguerra. Determination of carbaryl and some organophosphorus pesticides in drinking water using online liquid-chromatographic pre-concentration techniques. J Chromatogr 639:352–358, 1993.

71. R Carabias-Martinez, E Rodriguez-Gonzalo, F Garay-Garcia, J Hernandez-Mendez. Automated high-performance liquid-chromatographic method for the determination of organophorphorus pesticides in waters with dual electrochemical (reductive-oxidative) detection. J Chromatogr 644:49–58, 1993.

72. GE Miliadis. Gas chromatographic determination of pesticides in natural waters of Greece. Bull Environ Contam Toxicol 50:247–252, 1993.

73. BN Zegers, H de Geus, SHJ Wildenburg, H Lingeman, UAT Brinkman. Large-volume injection in packed-capillary supercritical-fluid chromatography. J ChromatogrA 677:141–150, 1994.

74. S Lacorte, D Barcelo. Validation of an automated pre-column exchange system (PROSPEKT) coupled to liquid chromatography with diode array detection. Application to the determination of pesticides in natural waters. Anal Chim Acta 296:223–234, 1994.

75. C Molina, M Honing, D Barcelo. Determination of organophosphorus pesticides in water by solid-phase extraction followed by liquid chromatography–high flow pneumatically assisted electrospray mass spectrometry. Anal Chem 66:4444–4449, 1994.

76. TA Berger. Feasibility of screening large aqueous samples of thermally unstable pesticides using high-efficiency packed-column supercritical-fluid chromatography with multiple detectors. Chromatographia 41: 471–484, 1995.

77. S Lacorte, D Barcelo. Determination of organophosphorus pesticides and their transformation products in river waters by automated online solid-phase extraction followed by thermospray liquid chromatography–mass spectrometry. J Chromatogr A 712:103–112, 1995.

78. S Lacorte, D Barcelo. Improvements in the determination of organophosphorus pesticides in ground and wastewater samples from interlaboratory studies by automated online liquid-solid extraction followed by liquid chromatography–diode array detection. J Chromatogr A 725:85–92, 1996.

79. Solid-phase micro-extraction of organophosphate insecticides and analysis by capillary GC-MS. Supelco Appl Note 94:2, 1996.

80. H Bagheri, ER Brouwer, RT Ghijsen, UAT Brinkman. Online low-level screening of polar pesticides in drinking and surface waters by liquid chromatography–thermospray mass spectrometry. J Chromatogr 647:121–129, 1993.

81. R Tauler, S Lacorte, D Barcelo. Application of multivariate self-modelling curve resolution to the quantitation of trace levels of organophosphorus pesticides in natural waters from interlaboratory studies. J Chromatogr A 730:177–183, 1996.

82. J Slobodnik, AC Hogenboom, AJH Louter, UAT Brinkman. Integrated system for on-line gas and liquid chromatography with a single mass spectrometric detection for the automated analysis of environmental samples. J Chromatogr A 730:353–371, 1996.

83. D Barcelo, MC Hennion. Sampling of polar pesticides from water matrices. Anal Chim Acta 338:3–18, 1997.

84. C Aguilar, F Borrull, RM Marcé. On-line and off-line solid-phase extraction with styrene-divinylbenzene-membrane extraction discs for determining pesticides in water by reversed-phase liquid chromatography-diode-array detection. J Chromatogr A. 754:77–84, 1996.

85. E Pocurull, RM Marce, F Borull. Improvement of on-line solid-phase extraction for determining phenolic compounds in water. Chromatographia 41:521–526, 1995.

86. C Aguilar, F Borrull, RM Marcé. Determination of pesticides by on-line trace enrichment–reversed-phase liquid chromatography–diode-array detection and confirmation by particle-beam mass spectrometry. Chromatographia 43:592–598, 1996.

87. C De la Colina, A Pena-Heras, G Dios-Cancela, F Sanchez-Rasero. Determination of organophosphorus and nitrogen-containing pesticides in water samples by solid-phase extraction with gas chromatography and nitrogen-phosphorus detection. J Chromatogr A 655: 127–132, 1993.

88. PJM Kwakman, JJ Vreuls, UAT Brinkman, RT Ghijsen. Determination of organophosphorus pasticides in aqueous samples by online membrane disc extraction and capillary gas chromatography. Chromatographia 34:41–47, 1992.

89. KC van Horne. Handbook of Sorbent Extraction Technology. Analytchem International Inc., Harbor City, CA, 1985, pp 4–66.

90. M Sargent, G MacKay. Guidelines for Achieving Quality in Trace Analysis. Cambridge: The Royal Society of Chemistry, 1995, pp 12–36.

91. J Liu, O Suzuki, T Kumazova, T Seno. Rapid isolation with Sep-Pak C18 cartridge and wide-bore capillary GC of organophosphate pesticide (abstr). Forensic Sci Int 41:67–72, 1989.

92. C Molto, Y Pico, G Font, J Manes. Determination of triazines and organophosphorus pesticides in water samples using solid-phase extraction. J Chromatogr 555:137–145, 1991.

93. C de la Colina, F Sanchez-Rasero, GD Cancela, ER Taboada, A Pena. Use of a solid phase extraction method for the analysis of pesticides in groundwater by gas chromatography with electron capture and flame photometric detectors. Analyst 120:1723–1728, 1995.

94. J Slobodnik, AC Hogenboom, JJ Vreuls, JA Rontree, BLM van Baar, WMA Niessen, UATh Brinkman. Trace-level determination of pesticide residues using on-line solid-phase extraction-column liquid chromatography with atmospheric pressure ionization mass spectrometric and tandem mass spectrometric detection. J Chromatogr A 741:59–74, 1996.

95. FD Rinkema, AJH Louter, UAT Brinkman. Large-volume injections in gas chromatography–atomic-emission detection: an approach for trace-level detection in water analysis. J Chromatogr A 678:289–297, 1994.

96. T Hankemeier, AJH Louter, FD Rinkema, UAT Brinkman. Online coupling of solid-phase extraction and gas chromatography with atomic-emission detection for analysis of trace pollutants in aqueous samples. Chromatographia 40:119–124, 1995.

97. THM Noij, MME van der Kool. Automated analysis of polar pesticides in water by on-line solid phase extraction and gas chromatography using the co-solvent effect. J High Resol Chromatogr 18:535–539, 1995.

98. A Junker-Buchheit, M Witzenbacher. Pesticide monitoring of drinking water with the help of solid-phase extraction and high-performance liquid chromatography. J Chromatogr A 737:67–74, 1996.

99. WA Steller, NR Pasarela. Gas-liquid chromatographic method for the determination of dimethoate and dimethoxon in plant and animal tissues, milk and eggs. J Assoc Off Anal Chem 55:1280–1287, 1972.

100. TA Albanis, DG Hela. Multi-residue pesticide analysis in environmental water samples using solid-phase extraction discs and gas chromatography with flame thermionic and mass-selective detection. J Chromatogr A 707:283–292, 1995.

101. CL Arthur, R Belardi, K Pratt, S Matlagh, J Pawliszyn. Environmental analyses of organic compounds in water using solid phase microextraction. J High Res Chrom 15:741–744, 1992.

102. CL Arthur, J Pawliszyn. Solid phase microextraction with termal desorption using fused silica optical fibers. Anal Chem 62:2145–2148, 1990

103. D Louch, S Motlaghand, J Pawliszyn. Dynamics of organic compound extraction from water using liquid-coated fused silica fibers. Anal Chem 64:1187–1199, 1992.

104. CL Arthur, IM Killam, KD Bucholz, J Pawliszyn. Automation and optimization of solid-phase microextraction. Anal Chem 64:1960–1966, 1992.

105. CL Arthur, DW Potter, KD Buchholz, S Motlagh, J Pawliszyn. Solid-phase microextraction for the direct analysis of water: theory and practice. LC.GC 10:656–661, 1992.

106. Z Zhang, J Pawliszyn. Headspace solid-phase microextraction. Anal Chem 65:1843–1852, 1993.

107. Z Zhang, MJ Yang, JB Pawliszyn. Solid-phase microextraction. Anal Chem 66::844A–853A, 1994.

108. Z Zhang, JB Pawliszyn. Quantitative extraction using an internally cooled solid-phase microextraction device. Anal Chem 67:34–43, 1995.

109. T Goewcki, J Pawliszyn. Sample introduction approaches for solid-phase microextraction/rapid GC. Anal Chem 67:3265–3274, 1995.

110. J Chen, JB Pawliszyn. Solid-phase microextraction coupled to high-performance liquid chromatography. Anal Chem 67:2530–2533, 1995.

111. S Magdic, JB Pawliszyn. Analysis of organochlorine pesticides using solid-phase microextraction. J Chromatogr A 723:111–122, 1996.

112. AA Boyd-Boland, S Magdic, JB Pawliszyn. Simultaneous determination of 60 pesticides in water using solid-phase microextraction and gas chromatography–mass spectrometry. Analyst 121:929–938, 1996.

113. T Gorecki, R Mindrup, J Pawliszyn. Pesticides by solid-phase microextraction. Results of a round robin test. Analyst 121:1381–1386, 1996.

114. MT Sng, FK Lee, HA Lasko. Solid-phase microextraction of organophosphorus pesticides from water. J Chromatogr A 759:225–230, 1997.

115. R Eisert, K Levsen, G Wuensch. Element-selective detection of pesticides by gas chromatography–atomic emission detection and solid-phase microextraction. J Chromatogr A 683:175–183, 1994.

116. R Eisert, K Levsen. Determination of organophosphorus, triazine and 2,6-dinitroaniline pesticides in aqueous samples via solid-phase microextraction (SPME) and gas chromatography with nitrogen-phosphorus detection. Fresenius J Anal Chem 351:555–562, 1995.

117. R Eisert, K Levsen. Determination of pesticides in aqueous environmental samples via solid-phase microextraction (abstr). GIT-Fachz Lab 39:25–26, 28–30, 32, 1995.

118. DR Erney. Determination of organophosphorus pesticides in whole/chocolate/skim-milk and infant formula using solid-phase extraction with capillary gas chromatography–flame photometric detection. J High Resolut Chromatogr 18:59–62, 1995.

119. NE Spingarn, DJ Norhington, T Pressley. Analysis of non-volatile organic hazardous substances by GC/MS. J Chromatographic Sci 20:571–574, 1982.

120. K Sasaki, T Suzuki, Y Saito. Simplified cleanup and gas chromatographic determination of organophosphorus pesticides in crops. J Assoc Off Chem 70:460–464, 1987.

121. V Leoni, AM Caricchia, S Chiavarini. Multiresidue method for quantitation of organophosphorus pesticides in vegetable and animal foods. J Assoc Off Anal Chem 75:511–518, 1992.

122. WA Minnaard, J Slobodnik, JJ Vreuls, KP Hupe, UAT Brinkman. Rapid liquid-chromatographic screening of

organic micropollutants in aqueous samples using a single short column for trace enrichment and separation. J Chromatogr A 696:333–340, 1995.

123. C Garcia-Pinto, JL Perez-Pavon, B Moreno-Cordero, Cloud point preconcentration and high-performance liquid-chromatographic determination of organophosphorus pesticides with dual electrochemical detection. 67:2606–2612, 1995.

124. MD Osselton, RD Snelling. Chromatographic identification of pesticides. J Chromatogr A 368:265–271, 1986.

125. S Lacorte, D Barcelo. Determination of parts per trillion levels of organophosphorus pesticides in groundwater by automated on-line liquid–solid extraction followed by liquid chromatography/atmospheric pressure chemical ionization mass spectrometry using positive and negative ion modes of operation. Anal Chem 68:2464–2470, 1996.

126. S Gocan, G Câmpan. Compound identification in thin layer chromatography using spectrometric methods. Rev Anal Chem 16:1–26, 1997.

127. S Butz, HJ Stan. Screening of 265 pesticides in water by thin-layer chromatography with automated multiple development. Anal Chem 67:620–630, 1995.

128. K Fodor-Csarba, F Dutka. Selectivity and sensitivity of some thin-layer chromatographic detection systems. J Chromatogr A 365:309–314, 1986.

129. S Kumaran, C Tran-Minh. Determination of organophosphorus and carbamate insecticides by flow-injection analysis. Anal Biochem 200:187–194, 1992.

130. A Farran, J De Pablo. Evaluation of combined flow injection-high-performance liquid chromatography for the determination of three organophosphorus pesticides in liquid wastes. Int J Environ Anal Chem 30: 59–68, 1987.

131. A Farran, E Figuerola, J dePablo, S Hernandez. Treatment and determination of organophosphorus pesticides in waste water by using continuous flow methodologies coupled on line with high performance liquid chromatography. Int J Environ Anal Chem 33: 245–256, 1988.

132. I Hornyak, L Kozma, E Brisching. Spectrophotometric determination of organophosphorus insecticides. Microchem J 51:187–190, 1995.

133. ZK Gou. Schoenemann reaction for organophosphorus pesticides (abstr). Fenxi Huaxue 22:41–43, 1994.

134. ZH Chohan, AI Shah. Simple spectrophotometric screening method for the detection of organophosphate pesticides on plant material. Analyst 117:1379–1380, 1992.

135. HS Rathore, I Ali, A Kumar. Capillary spot test for detection of trace levels of organophosphoric insecticides in water, soil and vegetation. Int J Environ Anal Chem 35:199–206, 1989.

136. JL Marty, N Mionetto, S Lacorte, D Barcelo. Validation of an enzymic biosensor with various liquid-chro-matographic techniques for determining organophosphorus pesticides and carbaryl in freeze-dried waters. Anal Chim Acta 311:265–271, 1995.

137. C Ristori, C Del-Carlo, M Martini, A Barbaro, A Ancarani. Potentiometric detection of pesticides in water samples. Anal Chim Acta 325:151–160, 1996.

138. VLF Cuhna-Bastos, J Cuhna-Bastos, JS Lima, MV Castro-Faria. Brain acetylcholinesterase as an in vitro detector of organophosphorus and carbamate insecticides in water. Water Res 25:835–849, 1992.

139. P Durand, JM Nicaud, J Mallevialle. Detection of organophosphorus pesticides in water with an immobilized cholinesterase electrode. J Anal Toxicol 8:112–117, 1984.

140. P Durand, D Thomas. Use of immobilized enzyme coupled with an electrochemical sensor for the detection of organophosphate and carbamate pesticides. J Environ Pathol Toxicol 5:51–57, 1984.

141. J Manem, J Mallevialle, P Durand, E Chabert. Detection of organophosphorus and carbamate pesticides with a butyrylcholinesterase electrode (abstr). Eau Ind Nuisances 74:31–34, 1983.

142. H Lay, U Draeger. Determination of pesticides using a reactivatable enzyme electrode (abstr). Dtsch Lebensm Rundsch 88:349–352, 1992.

143. D Barcelo, S Lacorte, JL Marty. Validation of an enzymatic biosensor with liquid chromatography for pesticide monitoring. Trends Anal Chem 14:334–340, 1995.

144. G Palleschi, M Bernabei, C Cremisini, M Mascini. Determination of organophosphorus insecticides with a choline electrochemical biosensor. Sens Actuators B B7:513–517, 1992.

145. GL Ellman, KD Courtney, Z Andres, RM Featherstone. A new and rapid determination of acetylcholinesterase activity. Biochem Pharmacol 7:88–95, 1961.

146. S Uk. Time-controlled cholinesterase reactions as a means of estimating water-soluble residues of organophosphorus-insecticide spray deposits on plant surfaces. Pestic Sci 10:308–312, 1979.

147. C Alsen, O Christensen, H Kruse. Detection of acetylcholinesterase inhibitors in potable and surface water. Vom Wasser 58, 1–58, 1982.

148. F Galgani, G Bocquene. Method for routine detection of organophosphates and carbamates in sea-water. Environ Technol Lett 10:311–322, 1989.

149. P Herzsprung, L Weil, KE Quentin, I: Zombola. Determination of organophosphorus compounds and carbamates by their inhibition of cholinesterase. II. Estimation of detection limits for insecticide determination by concentration, oxidation and inhibition values. Vom Wasser 74:339–350, 1990.

150. T Du, S Zhou. Enzymatic analytical method for the direct determination of trace organophosphorus compounds in chlorine-containing water (abstr). Fenxi Huaxue 15:109–113, 1988.

151. EB Nikol'skaya, GA Evtyugin, AV Svyatkovskii, RR Iskanderov, EV Suntsov, AA Prokopov, SN Moralev, BN Kormilitsin, VZ Latypova. Test unit for the detection of trace amounts of organophosphorus pesticides and pharmaceutical preparations of anticholinesterase action (abstr). Zh Anal Khim 49:374–380, 1994.

152. P Moris, I Alexandre, M Roger, J Remacle. Chemiluminescence assays of organophosphorus and carbamate pesticides. Anal Chim Acta. 302:53–59, 1995.

153. L Campanella, M Achilli, MP Sammartino, M Tomassetti. Butylcholine enzyme sensor for determining organophosphorus inhibitors (abstr). Bioelectrochem Bioenerg 26:237–249, 1991.

154. C La Rose, F Pariente, L Hernandez, E Lorenzo. Amperometric flowthrough biosensor for the determination of pesticides (abstr). Anal Chim Acta 308:129–136, 1995.

155. A Montiel. Determination of pesticide residues in water: possibilities for enzyme immunoassays. Analysis 19:i34–i37, 1991.

156. MH Erhard, A Juengling, S Brendgen, J Kellner, U Loesch. Development of a direct and indirect chemiluminescence immunoassay for the detection of an organophosphorus compound. J Immunoassay 13:273:287, 1992.

157. A S Hill, J H Skerritt, R. J Bushway, W Pask, K A Larkin, M Thomas, W Korth, K Bowmer. Development and application of laboratory and field immunoassays for chlorpyryfos in water and soil matrices. J Agric Food Chem 42:2051–2058, 1994.

158. BM Kaufman, M Clower,jun. Immunoassay of pesticides. J Assoc Off Anal Chem 74:239–247, 1991.

159. AJH Louter, UAT Brinkman, RT Ghijsen. Fully automated water analyzer based on online solid-phase extraction–gas chromatography. J Microcolumn Sep 5:303–315, 1993.

160. PJM van Hout, UAT Brinkman. The Rhine Basin Program. Trends Anal Chem 13:382–388, 1994.

161. RV Martindale. Determination of residues of a range of fungicides, anti-sprouting agents and (organochlorine and organophosphorus) insecticides in potatoes by gas-liquid and high-performance liquid chromatography. Analyst 113:1229–1233, 1988.

162. G Knapp, E Leitner, M Michaelis, B Platzer, A Schalk. Element-specific GC detection by plasma atomic emission spectroscopy—a powerful tool in environmental analysis. Int J Environ Anal Chem 38:369–378, 1990.

163. HP Hsu, HJ Schtenberg III, MM Garza. Fast turnaround multiresidue screen for pesticides in produce. J Assoc Off Anal Chem 74:886–892, 1991.

164. AK Singh, DW Hewetson, KJ Jordon, M Ashraf. Analysis of organophosphorus insecticides in biological samples by selective ion monitoring as chromatography-mass spectrometry. J Chromatogr A 369:83–96, 1986.

165. H Steinwandter. A collaborative study for intermethod comparison I:. Micro and macro extraction methods for the determination of organophosphorus pesticides in fruits and vegetables. Fresenius J Anal Chem 348:688–691, 1994.

166. P Cabras, A Angioni, M Melis, EV Minelli, FM Pirisi. Simplified multiresidue method for the determination of organophosphorus insecticides in olive oil. J Chromatogr A 761:327–331, 1997.

167. S Sportstoel, K Urdal, H Drangsholt, N Gjoes. Description of a method for automated determination of organic pollutants in water. Int J Environ Anal Chem 21:129–138, 1985.

168. V Drevenkar, Z Froebe, B Stengl, B Tkalcevic. [Octadecyl] C-18 reversed-phase trace enrichment of organophosphorus pesticides and residues in water. Mikrochim Acta 1:143–156, 1985.

169. H Steinwandter. Contributions to the ethyl acetate application in residue analysis. II. Micro online method for extracting organophosphorus pesticides from fruits, vegetables and feedstuffs. Fresenius J Anal Chem 343:887–889, 1992.

170. S Bengtsson, A Ramberg. Solid-phase extraction of pesticides from surface water using bulk sorbents. J Chromatografic Sci. 33:554–556, 1995.

171. C Gomez-Gomez, MI Arufe-Martinez, JL Romero-Palanco, JJ Gamero-Lucas, MA Vizcaya-Rojas. Monitoring of organophosphorus insecticides in the Guadalete river (Southern Spain). Bull Environ Contam Toxicol 55:431–438, 1995.

172. Analyse nytrogen-, phosphorus- or halogen-containing pesticides in water using a PTE-5 capillary GC column. Supelco Rep 9:6–7, 1990.

173. P Branca, P Quaglino. Determination of organophosphorus pesticides and triazine herbicides in food and water (abstr). Boll Chim Ig Parte Sci 40:71–78, 1989.

174. M Kuehni. Reliable determination of pesticides and herbicides. Chem Rundsch 40:5, 1987.

175. M Horiba. Gas chromatographic determination of fenitrothion and some other organophosphorus pesticides in technical materials and formulations. J Chromatogr 287:189–191, 1984.

176. H Takehara, Y Hiratsuka. Rapid determination of organophosphorus pesticides in waste waters by gas chromatography with a flame thermionic detector. Bunseki Kagaku 31:529–531, 1982.

177. G Durand, M Mansour, D Barcelo. Identification and determination of fenitrothion photolysis products in water–methanol by gas chromatography–mass spectrometry. Anal Chim Acta 262:167–178, 1992.

178. V Lopez-Avila, P Hirata, S Kraska, M Flanagan, JH Taylor, Jr. Determination of atrazine, lindane, pentachlorophenol, and diazinon in water and soil by isotope dilution gas chromatography/mass spectrometry. Anal Chem 57:2797–2801, 1985.

179. M Wolf, R Deleu, A Copin. Separation of pesticides by capillary gas chromatography. J HRC&CC 4:346–347, 1981.

29

Fungicide and Herbicide Residues in Water

H. S. Rathore and A. A. Khan
Aligarh Muslim University, Aligarh, India

I. INTRODUCTION

Pesticides are chemical substances used to kill or control pests. Pests should include insects and mites that damage crops; weeds that compete with field crops for nutrients and moisture; diseases of plants caused by fungi, bacteria, and viruses; nematodes, snails, and slugs; and rodents that feed on grain, young plants, and the bark of fruit trees.

Pesticides have saved millions of lives by controlling human disease vectors, by greatly increasing the yields of agricultural crops, and by protecting foodstuffs. However, in recent years humanity has become ever more conscious of the way in which the environment is becoming increasingly polluted by chemicals that may harm plants, animals, or even humankind itself. Amongst these chemicals, the insecticides, particularly organochlorines have been found to be a major cause of anxiety to ecologists, while the chemicals used to protect crops and foodstuffs, such as fungicides and herbicides, have been of great concern to people and animals.

The extent and seriousness of the potential hazards due to these chemicals still remain to be fully evaluated. The available information shows that the occurrence of residues in various parts of the environment is very uneven and localized. Water is one of the most important commodities of the environment, so much research work has been carried out all over the world, and a great deal of data on pesticide residues in water is available. This chapter presents the methods of analysis and the levels of residues of fungicides and herbicides in water.

II. FUNGICIDES

The chemicals used to kill or halt the growth of fungi are known as fungicides. However, for agricultural purposes they are considered chemicals used to control bacterial as well as fungal plant pathogens, the causal agents of most plant diseases. Thus fungicides fall into two general classifications: agricultural and industrial. Fungicides for plants act by direct contact and often injure the host as well as the fungus. Fungi are parasitic plants comprising the molds, mildews, rusts, smuts, mushrooms, and allied forms that are capable of destroying higher plants. They may attack seeds, the growing plant, and plant materials.

Some fungicides are poisonous to people and to cattle and should be used with caution. Spraying and dusting schedules are now being evolved for controlling certain types of insect pests and plant diseases in one operation. We should, however, become familiar with the compatibility of insecticides with fungicides of different types.

Thus the chemical control of plant diseases represents a profit-induced poisoning of the environment. It is moderately expensive; to achieve good results, the operation should be carried out in accordance with the advice of technical experts, using an approved fungicide at the right time to the right part of the plant and in correct dosage. The operation should be repeated if rain or wind or both are likely to interfere with the action of the fungicides.

The concentration of fungicide residues is increasing day by day in our flora and fauna due to repeated and continuous application of fungicides in plant protection and food preservation. Therefore, the present knowledge of fungicide residues in air, water, plants, animals, soils, sediments, and foodstuffs will be of great interest to scientists, engineers, farmers, regulators, journalists, bureaucrats, and politicians who are concerned with chemical crop protection as well as health programs.

The use of fungicides containing copper, sulphur, or mercury as their active ingredient started as early as the late sixteenth century and has continued. Even today most of our plant diseases could be controlled by these chemical groups. However, many of these compounds are so persistent in the soil that crops have been damaged by their soil residues; therefore, organic fungicides were developed. These sometimes have greater fungicidal activity and usually have lower phytotoxicity. General-purpose fungicides for agriculture still include inorganic forms of copper, sulphur, metallic complexes of cadmium, chromium, and zinc along with a variety of organic compounds. The general-purpose lawn-and-garden fungicides are few in number and are usually organic compounds. Fungicides may be classified in a number of chemical groups (Table 1). These will be discussed next.

A. Inorganic Fungicides

1. Sulphur Fungicides

Sulphur is probably the oldest effective fungicide known, and it is still a very useful garden fungicide. Three formulations of sulphur exist: finely ground sulphur dust, floating or colloidal sulphur, and sulphur with a wetting agent or as lime-sulphur solution. These are specific against rusts and powdery mildews. Sulphur also kills a kind of pest known as mites. It acts on fungi both by contact and by producing gases (fumigant effect) that have fungicidal action. Acting at a distance, its fumigant effect is very important in killing spores of powdery mildews. The mechanism of action of sulphur is to interfere in electron transport along the cytochromes and to reduce to hydrogen sulphide, which is toxic to most cellular proteins. Sulphur has little residual action. In using sulphur fungicides, a good coverage is desirable, though not essential.

2. Copper Fungicides

Copper fungicides commonly used are Bordeaux mixture, burgundy mixture, proprietary compounds containing copper oxychloride and cuprous oxide, col-loidal copper, and copper carbonate. The various formulations include Bordeaux mixture, named after the Bordeaux region in France, where it originated. Bordeaux is a chemically undefined mixture of copper sulphate and hydrated lime that was discovered accidentally. When the copper mixture was sprayed, downy mildew, a disease of grapes, disappeared from the treated plants. Bordeaux mixture and burgundy mixture are used exclusively for spraying. The proprietary compounds are available both for spraying and dusting. Chestnut compound is used for soil disinfection for controlling "damping-off" disease. Copper carbonate is now used to a limited extent for seed treatment. A few of the copper fungicides used over the years are given in Table 1.

The copper ion, which becomes available from both the highly soluble and the relatively insoluble copper salts, provides the fungicidal as well as phytotoxic and poisonous properties. It is interesting to note that in all forms in which copper is used for spraying, it is insoluble, by the action of weather and/or germinating spores, copper in a soluble form is produced in traces that kill the spores, and the sprayed surface is kept protected as long as copper is present on it. This is the basis for the use of relatively insoluble, or "fixed" copper fungicides, which release only very low levels of copper, adequate for fungicidal activity but not enough to affect the host plant.

Although copper is an essential element for plants, there is some danger in the accumulation of copper in agricultural soils due to repeated and continuous use. In fact, certain citrus growers in Florida have experienced a serious problem of copper toxicity after using fixed copper disease control. The mechanism of action of these fungicides is nonspecific denaturation of protein. The copper ion (Cu^{++}) reacts with sulphhydryl groups of the enzyme of the vulnerable fungal cells.

3. Mercury Fungicides

Mercurial fungicides were used for disinfection or disinfestation of seeds and other propagating stocks. The divalent mercury ions are toxic to all forms of life. Therefore materials treated with mercurial fungicides should not be consumed or fed to cattle. Ceresan is a typical example of organo-mercurial compounds used as seed treatments. Another mercurial, phenylmercury acetate (PMA), was useful as a seed treatment for turf diseases and as dormant sprays for fruit trees. The mechanism of action of these fungicides is the nonselective inhibition of enzymes, especially those containing iron and sulphhydryl sites. Quite recently, mercurial

Table 1 Chemical Classification of Fungicides

Functional group and examples	Toxicity to mammals, LD_{50} (acute) (mg/kg)		Action	Formula
	Oral (rat)	Dermal (rabbit)		
Inorganic Fungicides				
Sulphur Finely ground sulphur dust Floating or colloidal sulphur Wettable sulphur	Relatively nontoxic		Contact and fumigant fungicide	
Copper				
Cupric sulphate	470	—	Contact fungicide for seed treatment and to prepare Bordeaux mixture	$CuSO_4 \cdot 5H_2O$
Copper hydrazine sulphate	—	—	Contact fungicide for powdery mildews (black spot of roses)	$CuSO_4(C_2H_5)_2SO_4$
Copper oxychloride	700–800	—	Contact fungicide for powdery mildews	$3Cu(OH)_2CuCl_2$
Copper oxychloride sulphate	—	—	Contact fungicide for many fungal diseases	$3Cu(OH)_2CuCl_2 3Cu(OH)_2CuSO_4$
Copper zinc chromates	—	—	Contact fungicide for protecting potato, tomato, cucurbits, peanuts, and citrus	$15CuO \cdot 10ZnO \cdot 6CrO_3 \cdot 25H_2O$
Cuprous oxide	—	—	Contact fungicide for powdery mildews	Cu_2O
Basic copper sulphate	1,000	—	Contact fungicide for seed treatment and to prepare Bordeaux mixture	$CuSO_4 \cdot Cu(OH)_2 \cdot H_2O$
Cupric carbonate	—	—	Contact fungicide for many fungal diseases	$3Cu(OH)_2CuCO_3$
Mercury				
Various ethylmercury salts: ethylmercury chloride, ethylmercury phosphate, ethylmercury sulphonamide, etc. (Ceresan)	—	—	Seed treatment fungicides	C_2H_5HgCl
Phenylmercury acetate (PMA)	100	22–44	Eradicant fungicide with little protective value and selective herbicide	⬡—HgO—C(=O)—CH₃
Calcium hypochlorite (chloride of lime, bleaching powder)	—	—	Contact fungicide and bactericide	$CaOCl_2$
Sodium chlorate and magnesium chlorate	1,200	—	Contact fungicide	$NaOCl$ $Mg(OCl)_2$
Organic Fungicides				
Dithiocarbamates Tetramethylthiuram disulphide (Thiuram)	780	—	Seed protectant fungicide and animal repellent	(CH₃)₂N—C(=S)-S-S-C(=S)—N(CH₃)₂

Table 1   Continued

Functional group and examples	Toxicity to mammals, LD$_{50}$ (acute) (mg/kg)		Action	Formula
	Oral (rat)	Dermal (rabbit)		
Manganese ethylene-bisdithiocarbamate (Maneb)	6,750	—	Turf fungicide	
Ferric dimethyl-dithiocarbamate (Ferbam)	>17,000	—	Protective fungicide	
Zinc dimethyl-dithiocarbamate (Ziram)	14,000	—	Protective fungicide	
Sodium N-methyldithio-carbamate (Vapam)	820	—	Soil fungicide	
Zinc ethylenebisdithio-carbamate (Zineb)	>5,200	—	Protective fungicide	
Quinones				
2,3,5,6-Tetrachloro-1,4-benzoquinone (Chloranil, Spergon)	4,000	—	Seed treatment fungicide	

Table 1 Continued

Functional group and examples	Toxicity to mammals, LD_{50} (acute) (mg/kg)		Action	Formula
	Oral (rat)	Dermal (rabbit)		
2,3-Dichloro-1,4-naphthoquinone (Dichlone, Phygon)	1,300	5,000	Fungicide	
5,10-Dihydro-5,10-dioxonaphtho(2,3b)-*p*-dithin-2,3-dicarbonitrile (Dithianon, Delan)	638	—	Protective fungicide	
Substituted benzene				
Hexachlorobenzene	—	—	Seed protectant fungicide	
Pentachloronitrobenzene (PCNB)	1,700	—	Selected foliage fungicide	
1,4-Dichloro-2,5-dimethyl-benzene (Chloroneb)	11,000	—	Cotton seedling fungicide	

Table 1 Continued

Functional group and examples	Toxicity to mammals, LD$_{50}$ (acute) (mg/kg)		Action	Formula
	Oral (rat)	Dermal (rabbit)		
Pentachlorophenol (PCP)	50–140	—	Wood preservative fungicide	
Tetrachloroisophthalonitrile (Chlorothalonil)	>10,000	>10,000	Foliage protectant fungicide	
Nitrophenols 2,4-Dinitro-6-(2-octylphenyl-crotonate) (Dinocap)	980	—	Home fungicide	
Thiazoles 5-Ethoxy-3-trichloromethyl-1,2,4-thiadiazole (Terrazole)	1,077	—	Soil fungicide	
Triazines 2,4-Dichloro-6-(o-chloro-anilino)-s-triazine (Anilazine)	>5,000	>9,400	Turf fungicide	

Table 1 Continued

Functional group and examples	Toxicity to mammals, LD$_{50}$ (acute) (mg/kg)		Action	Formula
	Oral (rat)	Dermal (rabbit)		
Dicarboximides				
N-(Trichloromethyl)thio-4-cyclohexene-1,2-dicarboximide) (Captan)	10,000	—	Nonspecific fungicide	
N-(Trichloromethylthio-thalimide) (Folpet)	>10,000	—	Nonspecific fungicide	
Tetrachloroethylmercapto-cyclohexenedicarboximide (Captafol, Difolatan)	62,000	—	Nonspecific fungicide	
Methanesulfonyl-N-trichloro-methylsulphenyl-4-chloroanilide (Misulfan)	—	—	Nonspecific fungicide	
Organotins				
Triphenyltin hydroxide (Duter, Fentin hydroxide)	108	—	Protective fungicide	

Table 1 Continued

Functional group and examples	Toxicity to mammals, LD$_{50}$ (acute) (mg/kg)		Action	Formula
	Oral (rat)	Dermal (rabbit)		
Triphenyl tin acetate (Brestan)	140	—	Protective fungicides	
Aliphatic nitrogen compounds n-Dodecylguanidine acetate (Dodine)	1,000	1,500	Orchard fungicide	
Systemic Fungicides				
Oxathiins 2,3-Dihydro-5-carboxanilido-6-methyl-1,4-oxathiin (Carboxin)	3,820	8,000	Systemic fungicide	
2,3-Dihydro-5-carboxanilido-6-methyl-1,4-oxathiin-4,4-dioxide (Oxycarboxin)	2,000	16,000	Systemic fungicide	
Benzimidazoles Methyl 1-(butylcarbamoyl)-2-benzimidazole carbamate (Benomyl)	>10,000	—	Systemic fungicide	

Table 1 Continued

Functional group and examples	Toxicity to mammals, LD$_{50}$ (acute) (mg/kg)		Action	Formula
	Oral (rat)	Dermal (rabbit)		
2-(4'-thiazolyl)benzimid-azole (Thiabendazole)	3,100	—	Systemic fungicide	
1,2-Bis(3-ethoxycarbonyl-2-thioureido)benzene (Thiophanate)	15,000	—	Systemic fungicide	
5-Ethoxy-3-trichloromethyl-1,2,4-thiadiazole (Terrazole)	1,077	—	Systemic fungicide	
5-*n*-Butyl-2-dimethylamino-4-hydroxy-6-methylpy-rimidine (Dimethirimol)	2,350	—	Systemic fungicide	
1,4-Dichloro-2,5-dimethoxy-benzene (Chloronibe)	197	—	Systemic fungicide	
Fumigants				
Sodium *N*-methyldithio-carbamate (Metam-sodium, SMDC)	820	—	Soil fungicide	

Table 1 Continued

Functional group and examples	Toxicity to mammals, LD$_{50}$ (acute) (mg/kg)		Action	Formula
	Oral (rat)	Dermal (rabbit)		
Trichloronitromethane (Chloropicrin)	250	—	Soil fungicide	CCl_3NO_2
Bromomethane (Methyl bromide)	0.20	—	Soil fungicide	CH_3Br
Methylisothiocyanate (MIT)	350	—	Soil fungicide	CH_3NCS
Antibiotics				
2,4-Diguanidino-3,5,6-trihydroxycyclohexyl-5-deoxy-2-o(2-deoxy-2-methylamino-α-glucopyranosyl)-3-formyl pentofuranoside (Streptomycin)	—	—	Antibiotic fungicide	
3-{2-(3,5-Dimethyl-2-oxocyclohexyl)-2-hydroxyethyl-}glutarimine (Actidione)	2	—	Antibiotic fungicide	

fungicides, either inorganic or organic or both, have been banned, with one or two exceptions, by the U.S. Environmental Protection Agency on the basis of their toxicity to warm-blooded animals and the accumulation of mercury in the environment.

4. Bleaching Powder Fungicides

Bleaching powder has been used for controlling storage diseases of apples (fly speak). The fruits are dipped in a 5% bleaching powder suspension for 1 minute, exposed to air for 10 minutes, washed with plain water, and wiped until dry with a cloth.

5. Chlorate Fungicide

Aqueous solutions of sodium chlorate and magnesium chlorate (1%) are used as fungicides. In fact, they act as contact poisons that are translocated and may be absorbed from the soil to kill both plant roots and tops. Chlorates cause chlorosis of leaves and a starch depletion in stems and roots when applied in less than lethal doses.

B. Organic Fungicides

Inorganic fungicides were found to be harsher and less selective, so they were replaced by extremely efficient, safer, and readily degradable organic fungicides of very low phytotoxicity. The latter have been developed over the past 50 years. The first of the organic sulphur fungicides, thiram, was discovered in 1931; this was followed by many others, e.g., Zineb and Captan. At present, more than 200 fungicides containing different functional groups are in use or are in various stages of development. Some of them are discussed next.

1. Dithiocarbamate Fungicides

The Du Pont Company (U.S.A.) found that some of the derivatives of dithiocarbamic acid had insecticide and fungicide properties. In the early 1930s and 1940s, Thiram, Maneb, Ferbam, Ziram, Vapam, and Zineb were developed. The dithiocarbamates probably act by being metabolized to the isothiocyanate radical (—N=C=S), which inactivates the sulphhydryl group in amino acids of the individual pathogen cells. Chelation is another important function in plants that may explain the mode of action for dithiocarbamates as fungicides. Enzymes require some metal ions in traces. The chelates, in turn, disrupt protein synthesis and metabolism. When the metals required in traces appear in abundance or excessive quantities, this is equivalent to the introduction of potent poisons in the cells. Thus chelation of heavy metals plays an important role in both the life and death of cells.

2. Quinone Fungicides

Quinones are potential fungicides, and their number is very large in comparison to any other functional group. Generally, quinones affect respiration in many fungi and act by coupling to the —SH groups in enzymes, thus inhibiting their action and indirectly uncoupling oxidative phosphorylation.

3. Substituted Benzene Fungicides

The substituted benzenes have been in use for the last four decades. PCP has been used since 1936, and hexachlorobenzene, PCNB, chlorothalonil, and chloroneb were introduced in 1945, 1930, 1964, and 1965, respectively. Their modes of action are different. They reduce growth rates and sporulation of fungi, probably by combining with —NH$_2$ or —SH groups of essential metabolic compounds.

4. Nitrophenol Fungicides

Nitrophenols such as Dinocap are excellent substitutes for sulphur for controlling certain species of mites on sulphur-sensitive plants like certain varieties of cucur-

bits and apples. Dinocap has been used since the late 1930s as both an acaricide and a fungicide. It acts in the vapor phase, and it is quite effective against powdery mildews, whose spores germinate in the absence of water. The mode of action of nitrophenols is uncoupling oxidative phosphorylation in cells, with an attendant upset of the energy systems within the cells.

5. Thiazole Fungicides

Terrazole is one of the fungicides belonging to the thiazoles. The five-membered ring of the thiazoles is cleaved under soil conditions to give either fungicidal isothiocyanate (—N=C=S) or dithiocarbamate. Their mode of action is similar to that of the dithiocarbamates.

6. Triazine Fungicides

Anilazine was introduced in 1955 and has been used widely to control potato and tomato leaf spots and turf grass diseases. Most triazins are used as herbicides; only anilazine is used as a fungicide.

7. Dicarboximide Fungicides

The dicarboximides are the safest pesticides recommended for lawn-and-garden use as seed treatment and as protectants for mildews, late blight, and other diseases. Their lethal effect on disease organisms is probably due to the inhibition of the syntheses of amino compounds and enzymes containing the sulphhydryl (—SH) radical. Fungicides of this group, such as Captan, Folpet, and Captafol, appeared in 1949, 1962, and 1961, respectively.

8. Organotin Fungicides

The organotins were first introduced in 1960 by American Cyanamide Co. The triaryl compounds are suitable for protective use and also have acaricidal properties. They probably block oxidative phosphorylation.

9. Aliphatic Nitrogen Fungicide

Dodine, an aliphatic nitrogen compound, was introduced in 1959 by the American Cyanamide Co. as a fungicide for the control of fruit diseases. It is taken up rapidly by fungal cells, causing leakage in these cells, probably by alteration in membrane permeability. It is also known that the guanidine nucleus of Dodine inhibits the synthesis of RNA.

C. Systemic Fungicides

Initially, fungicidal treatment was based on only the external application of the fungicides. Recently, after successful internal protection with 8-hydroxy quinoline against Dutch elm disease, systemic fungicides have been marketed.

Systemics are applied to one part of the plant, say, the root or leaf, and carried by translocation through the cuticle and across leaves to the growing point. Some systemics can be applied as soil treatments and are slowly absorbed through the roots to give prolonged disease control. The systemics are therapeutic and possess eradicant properties, so they can be used to cure plant diseases and to stop the progress of existing infections.

Systemics offer much better control of diseases than is possible with protectant fungicides. The latter require uniform application and physical presence essentially on the plant surfaces. Systemics containing different functional groups are described next.

1. Oxathiin Fungicides

Carboxin and oxycarboxin, of the oxathiins group, were introduced in 1966. These are used in seed treatments because they are selectively toxic to embryo-infecting smut fungi, smuts, and rusts and to *Rhizoctonia*. The oxathiins concentrate selectively in the fungal cells, followed by the inhibition of succinic dehydrogenase.

2. Benzimidazole Fungicides

The benzimidazoles, such as benomyl and thiabendazole, were introduced in 1968 and have received wide acceptance as systemic fungicides against a broad spectrum of diseases (*Sclerolinia*, *Botrytis*, and *Rhizoctonia* species, powdery mildews, and apple scab). Another compound, thiophanate, of similar spectrum of activity, was introduced in 1969. Benzimidazole is converted to this group by the host plant and the fungus through their metabolism. The mode of action of this group of fungicides seems to be induction of abnormalities in spore germination, cellular multiplication, and growth. With frequent and continuous fungicidal treatment, the systemic fungicides can reduce the risk of environmental pollution.

D. Fumigants

Certain poisons are transmitted to plants in the gaseous state. Some of them are applied directly as gases; other fumigants, such as liquid nicotine or solid naphthalene, give fumes or vapors after application. Chloropicrin, methyl bromide, and MIT are the pertinent examples of this class. They are multipurpose pesticides, for they control fungi, insects, nematodes, and weed seeds in the soil. MIT acts like dithiocarbamates. Other members are hydrocyanic acid, carbon disulphide, methyl allyl chloride, ethylene oxide, calcium cyanide, aluminium phosphide, and SMDC. They penetrate plant or organism by different routes simultaneously.

E. Antibiotics

The successful use of antibiotics against animal disease led to their use for the control of plant diseases. The first antibiotic, prepared from streptomyces and named antimycin A, was used as a protective fungicide for the control of apple scab and tomato early blight, *Alternaria solani*. The mode of action of streptomycin is not fully understood, but it probably interferes with the synthesis of proteins. Most antibiotics are of complex structure and are generally highly unstable, so their use has been restricted. Several antibiotics, such as penicillin, gliotoxin, streptomycin, blastomycin, blastocidin S, cycloheximide, auerofungin, and griseofulvin have been tested for the control of diseases. However, only streptomycin and cycloheximide have been used commercially for plant protection. Cycloheximide gives stable derivatives, such as acetate, oxime, and semicarbazone, that are active as fungicides. It probably interferes with glutamine and protein synthesis. Actidione is used in the control of powdery mildew, rusts, turf diseases, and certain blight.

III. HERBICIDES

An herbicide is any chemical substance that kills or inhibits the growth of plants. Herbicides provide a more effective and economical means of weed control than cultivation, hoeing, and hand-pulling. They are very important due to the shortage of agricultural labor in the production of cotton, sugar beets, grains, potatoes, and corn. They are used extensively in areas such as industrial sites, roadsides, ditch banks, irrigation canals, fence lines, recreational areas, railroad embankments, and power lines. Their residues have been found in water resources, especially in surface water, such as rivers, lakes, ponds, and springs, due to rainwater, spray drift, runoff from the site of application, and industrial effluents.

Herbicides can be classified in several ways based on selectivity, contact versus translocation, timing, area covered, and chemical functional group. Simply, herbicides are classed as *selective* when they are used to kill weeds without harming the crop and as *nonselective* when the purpose is to kill all vegetation. The selectivity is not absolute, but is governed by the amount of the chemical applied, the way it is applied, the degree of wetting of the foliage, the amount of rainfall following, the tolerance of different plants to a specific chemical, and the difference in the growth habits of the crop and the weed.

Herbicides can also be classified on the basis of whether their effect on the plants occurs by contact or by translocation. *Contact* herbicides kill the plant parts to which the chemical is applied and are most effective against annuals, those weeds that germinate from seeds and grow to maturity each year. *Translocated* herbicides are absorbed by either roots or above-ground parts of plants and then moved within the plant system to distant tissues. Translocated herbicides are effective against all types of weeds; however, they have been found to be of extra advantage when used to control established perennials.

The timing of herbicide application with respect to the stage of crop or weed development has also been used in classifying the herbicides. The three categories of timing are preplanting, pre-emergence, and post-emergence. Herbicide application based on area covered involves four categories: band, broadcast, spot treatments, and directed spray.

Herbicides may be inorganic or organic in nature.

A. Inorganic Herbicides

The ancient Romans used brine and a mixture of salt and ashes to sterilize the land, as early as in Biblical times. In 1896, copper sulphate was used selectively to kill weeds in grain fields. In the period 1906–1960, sodium arsenite solutions were the standard herbicides of commerce. Arsenic trioxide has been used for years. In 1942, ammonium sulphamate ($NH_4SO_3NH_2$) was introduced for brush control. Other salts, such as ammonium thiocyanate, ammonium nitrate, ammonium sulphate, iron sulphate, and copper sulphate, were applied heavily as a foliar spray. The mechanism of action is either desiccation and plasmolysis.

Borate herbicides, such as sodium tetraborate ($Na_2B_4O_7 \cdot 5H_2O$), sodium metaborate ($Na_2B_2O_4 \cdot 4H_2O$), and amorphous sodium borate ($Na_2B_8O_{13} \cdot 4H_2O$), were used to give a semipermanent form of sterility to areas where no vegetation of any sort is wanted. Borates are absorbed by plant roots, are translocated to above-ground parts, and are nonselective and persistent herbicides. Boron accumulates in the reproductive structures of plants, but its mechanism of toxicity is not clear.

Sodium chlorate, a nonselective herbicide, was used for the last 40 years. It acts as a soil sterilant at 200–1000 pounds per acre, and it can be used as a foliar spray at 5 pounds per acre as a defoliant of cotton. Sulphuric acid has also been used as a foliar herbicide. Its corrosiveness to metal spray rings makes it of limited use. Its mechanism of action is also desiccation and plasmolysis.

Inorganic herbicides have been found to be persistent, so they have been replaced by organic substances.

B. Organic Herbicides

Some organic herbicides are recorded in Table 2 and are discussed in the following subsections.

1. Petroleum Oil Herbicides

Petroleum oils are in use for spot treatment. Their effect is temporary, but they are fast-acting and very safe to use around the home. These herbicides exert their lethal effect by penetrating and disrupting plasma membranes.

2. Organic Arsenical Herbicides

Cacodylic acid and its sodium salt are used as contact herbicides. DSMA and MSMA are translocated to underground tubers and rhizomes, making them extremely useful against Johnson grass and nut sedges. They are usually applied as spot treatments.

3. Phenoxyaliphatic Acid Herbicides

The first of the phenoxy herbicides (phenoxyacetic acid derivatives), or hormone weed killers, 2,4-D, was introduced in 1944. 2,4-DB, MCPA, Silvex, 2,4-D, and 2,4,5-T are highly selective for broadleaf weeds and are translocated throughout the plant. Their mechanism is similar to that of auxins (growth hormones). They affect cellular division, activate phosphate metabolism, and modify nucleic acid metabolism.

4. Substituted Amide Herbicides

Allidochlor is one of the selective herbicides for grasses. It inhibits germination or early seedling growth of most annual grasses, probably by alkylation of the —SH groups of proteins. Diphenamid is used as a pre-

Table 2 Chemical Classification of Organic Herbicides

Functional group and examples	Toxicity to mammals, LD_{50} (acute) (mg/kg)		Action	Formula
	Oral (rat)	Dermal (rabbit)		
Petroleum oils	Mixtures		Effective contact herbicides for all vegetation	
Crank case oil	Mixtures of alkanes,			
Gasoline	alkenes, alicyclics, and			
Kerosene	aromatics containing			
Diesel	traces of nitrogen and			
	sulphur			
Organic arsenicals				
Arsonic acid	Highly poisonous		Translocation herbicide	
Arsinic acid	Highly poisonous		Translocation herbicide	
Monosodium methane arsonate (MSMA)	700	—	Translocation herbicide	
Hydroxydimethylarsine oxide (Cacodylic acid)	700	—	Contact herbicide	
Disodium methanearsonate (DSMA)	1,000	—	Translocation herbicide	
Sodium salt of cacodylic acid	—	—	Contact herbicide	
Phenoxyaliphatic acids				
2,4-Dichlorophenoxyacetic acid (2,4-D)	370	—	Translocation herbicide	

Table 2 Continued

Functional group and examples	Toxicity to mammals, LD$_{50}$ (acute) (mg/kg)		Action	Formula
	Oral (rat)	Dermal (rabbit)		
2,4,5-Trichlorophenoxyacetic acid (2,4,5-T)	500	—	Translocation herbicide	
4-(2,4-Dichlorophenoxy)-butyric acid (2,4-DB)	1,960	—	Translocation herbicide	
4-Chloro-2-methylphenoxy-acetic acid (MCPA)	700–800	—	Translocation herbicide	
2-(2,4,5-Trichlorophenoxy)-propionic acid (2,4,5-TP, Silvex)	650	—	Translocation herbicide	
Substituted amides				
N,N-Diallyl-2-chloro-acetamide (Allidochlor CDAA)	750	—	Pre-emergence herbicide	
N,N-Dimethyl-2,2-diphenylacetamide (Diphenamid)	10,000	—	Pre-emergence herbicide	
3',4'-Dichloropropionanilide (Propanil)	1,384	—	Postemergence herbicide	

Table 2 Continued

Functional group and examples	Toxicity to mammals, LD_{50} (acute) (mg/kg)		Action	Formula
	Oral (rat)	Dermal (rabbit)		
Nitroanilines				
N-Butyl-N-ethyl-α,α,α-trifluoro-2,6-dinitro-p-toluidine (Benefin)	>10,000	—	Pre-emergence herbicide	
α,α,α-Trifluoro-2,6-dinitro-N,N-dipropyl-p-toluidine (Trifluralin)	>10,000	—	Pre-emergence herbicide	
Substituted ureas				
3-(p-Chlorophenyl)-1,1-dimethylurea (Monuron)	3,600	—	Pre-emergence herbicide	
3-(3,4-Dichlorophenyl)-1,1-dimethyl urea (Diuron)	3,400	—	Pre-emergence herbicide	
Carbamates				
Isopropylcarbanilate (Propham)	5,000	—	Selective pre-emergence and postemergence herbicide	
4-Chloro-2-butynyl m-chlorocarbanilate (Barbane)	600	—	Selective pre-emergence and postemergence herbicide	

Table 2 Continued

Functional group and examples	Toxicity to mammals, LD$_{50}$ (acute) (mg/kg)		Action	Formula
	Oral (rat)	Dermal (rabbit)		
Isopropyl-*N*-(3-chloro-phenyl)carbamate (Chlorpropham)	3,800	—	Selective pre-emergence and postemergence herbicide	
2,6-Di-*tert*-butyl-*p*-tolyl methylcarbamate (Terbutol)	34,000	—	Selective pre-emergence and postemergence herbicide	
Thiocarbamates S-Ethyldipropylthio-carbamate (EPTC)	1,630	—	Selective weedicidal herbicide	
S-Propylbutylethylthio-carbamate (Pebulate)	1,120	2,936	Selective weedicidal herbicide	
Heterocyclic nitrozines Triazines 2-Chloro-4-(ethylamino)-6-(isopropylamino)-*s*-triazine (Atrazine)	3,080	—	Postemergence herbicide	
2-Chloro-4,6-bis(ethyl-amino)-*S*-triazine (Simazine)	10,000	3,100	Postemergence herbicide	

Table 2   Continued

Functional group and examples	Toxicity to mammals, LD$_{50}$ (acute) (mg/kg)		Action	Formula
	Oral (rat)	Dermal (rabbit)		

Triazoles
  3-Amino-S-triazole
    (Amitrole)
	25,000	10,000	Postemergence herbicide	

Pyridine derivatives
  4-Amino-3,5,6-trichloro-
    picolinic acid (Picloram)
	8,200	—	Translocated herbicide	

Uracils
  3-tert-Butyl-5-chloro-6-
    methyluracil (Terbacil)
	5,000–7,500	—	Translocated herbicide	

Aliphatic acids Trichloroacetic acid (TCA)	5,000	—	Translocated herbicide	
2,2-Dichloropropionic acid (Dalapon)	970	—	Translocated herbicide	

Arylaliphatic acids
  2-Methoxy-3,6-dichloro-
    benzoic acid (Dicamba)
	800–2,900	—	Pre-emergence herbicide	

Table 2 Continued

Functional group and examples	Toxicity to mammals, LD_{50} (acute) (mg/kg)		Action	Formula
	Oral (rat)	Dermal (rabbit)		
(2,3,6-Trichlorophenyl)acetic acid (Fenac)	1,780	3,160	Pre-emergence herbicide	
Dimethyltetrachloro-terephthalate (DCPA)	3,000	10,000	Pre-emergence herbicide	
Amino-2,5-dichlorobenzoic acid (Chloramben)	5,620	—	Pre-emergence herbicide	
Phenol derivatives 3-*sec*-Butyl-4,6-dinitrophenol (Dinoseb)	40–60	—	Nonselective herbicide (preharvest defoliant)	
Pentachlorophenol (PCP)	50–140	—	Nonselective herbicide (preharvest defoliant)	

Table 2 Continued

Functional group and examples	Toxicity to mammals, LD_{50} (acute) (mg/kg)		Action	Formula
	Oral (rat)	Dermal (rabbit)		
4,6-Dinitro-*o*-cresol (DNOC)	20–50	—	Nonselective herbicide (preharvest defoliant)	
Substituted nitriles 2,6-Dichlorobenzonitrile (Dichlobenil)	3,160	—	Wide-spectrum herbicide	
3,5-Dibromo-4-hydroxy-benzonitrile (Bromoxynil)	190	—	Wide-spectrum herbicide	
Bipyridyliums 6,7-Dihydrodipyrido(1,2-α: 2′,1′-*c*)pyrazidinium (Diquat)	400–440	—	Contact herbicide	
1,1′-Dimethyl-4,4′-bipyridylium ion (Paraquat)	150	—	Contact herbicide	
Miscellaneous compounds Methyl bromide	0.20	—	Fumigant herbicide	
Allyl alcohol	60	89	Fumigant herbicide	

Table 2 Continued

Functional group and examples	Toxicity to mammals, LD_{50} (acute) (mg/kg)		Action	Formula
	Oral (rat)	Dermal (rabbit)		
7-Oxabicyclo(2,2,1)heptane-2,3-dicarboxylic acid, sodium salt (Endothall)	51	70	Pre- and postemergence herbicide (an outstanding example of environmental protection through pesticide selectivity)	
O,O-Diisopropyl-phosphorodithioate S-ester with N-(-2-mercaptoethyl)-benzenesulphonamide (Bensulphide)	1,082	—	Pre-emergence herbicide (one of the better turf herbicides)	
Arolein	—	—	Aquatic herbicide (another example of environmental protection through pesticide selectivity)	$CH_2 = CH - CHO$

emergence soil treatment and has little contact effect. Propanil has been used extensively on rice fields as a selective postemergence control for a broad spectrum of weeds. These herbicides show multimodes and multiple sites of action. One of the important modes of action is the inhibition of the Hill reaction. Some of them are applied only to the soil and are active through the root system or seeds; others are applied only to foliage. The amide herbicides are simple molecules that have diverse biological properties and are easily degraded by plants and soil.

5. Nitroaniline Herbicides

Trifluralin shows minimum leaching and movement away from the target. This may be due to its very low water solubility. The nitroanilines inhibit the growth of both roots and shoots when absorbed by roots. They show biochemical effects, which include inhibiting the development of several enzymes and the uncoupling of oxidative phosphorylation.

6. Substituted Urea Herbicides

The substituted ureas are a utilitarian group of compounds that are strongly adsorbed by the soil and then absorbed by roots. Their mechanism of action is the inhibition of photosynthetic production of plant sugars and, indirectly, through the inhibition of the Hill reaction.

7. Carbamate and Thiocarbamate Herbicides

Carbamate herbicides were discovered in 1945. They are physiologically quite active, so some carbamates are insecticides, while others are fungicides. They kill plants by stopping cell division and plant tissue growth, i.e., the cessation of protein production and the shortening of chromosomes undergoing mitosis (duplication).

Another group, the thiocarbamates, are selective herbicides marketed for weed control. These herbicides are quite volatile and must be incorporated in the soil after application.

8. Triazine, Triazole, Pyridine, and Uracil Herbicides

The well-known triazines are strong inhibitors of photosynthesis. Triazoles act in the same way as the triazines. Pyridine derivatives and substituted uracils also belong to this group.

9. Aliphatic Acid Herbicides

Aliphatic acids, such as TCA and Dalafon, are heavily used herbicides. Their mechanism of action is the precipitation of protein within the cells. They are used against grasses. Dalafon is popularly known to control Bermuda grass around homes.

10. Arylaliphatic Acid Herbicides

The arylaliphatic acids are applied to the soil to kill germinating seeds and seedlings. The mechanism of action of these herbicides is not fully understood. It is supposed that Dicamba and Fenac interfere with nucleic acid metabolism (similar to 2,4-D), while DCPA and Chloramben produce auxinlike growth effects in plants.

11. Phenol-Derivative Herbicides

Phenol derivatives are familiar compounds. They have been used as insecticides, ovicides, fungicides, and blossom-thinning agents. They are highly toxic to humans by every route of entry into the body and are nonselective foliar herbicides that are most effective in hot weather. The nitrophenols, DNOC, were first introduced in 1932. Dinoseb acts by uncoupling oxidative phosphorylation. PCP is destructive to all living cells due to its wide effectiveness and multiple routes of action, such as plasmolysis, protein precipitation, and desiccation.

12. Substituted Nitrile Herbicides

The substituted nitriles have a wide spectrum of uses. Their action is slow, due to rapid permeation and the release of cyanide ions. Their mechanism of action is also broad; it involves seedling growth inhibition, potato sprout inhibition, and gross disruption of tissues by inhibiting oxidative phosphorylation and preventing the fixation of CO_2.

13. Bipyridylium Herbicides

Bipyridyliums reduce photosynthesis and are more effective in presence of light than in the dark. They are not active in soil and are used by professional weed control specialists to achieve spectacular results. They damage plant tissues quickly via cell membrane destruction.

14. Miscellaneous Herbicides

The miscellaneous herbicides include methyl bromide, which is used as a fumigant against organisms present in soil or plant parts. Allyl alcohol is a volatile, water-soluble fumigant also used for this purpose. Endothall is used to control aquatic weeds as well as selective field crops. It interferes with RNA synthesis and has low toxicity to fish. Bensulphide, the least toxic and nontranslocative organophosphate, is used to control certain grasses and broadleaf weeds by inhibiting cell division in root tips. Acrolein, a general plant toxicant, is used to control aquatic weeds without harming fish populations. It produces a frightening tear-gas effect. It destroys plant cell membranes and reacts with various enzyme systems.

For further information on fungicides and herbicides, see Refs. 1–17.

IV. ANALYSIS OF PESTICIDE RESIDUES IN WATER

The analysis of pesticides residues involves four steps: detection, sample preparation, estimation, and confirmation.

A. Detection: Preliminary Characterization of Test Material

Generally, a preliminary characterization of the residue is required before undertaking sophisticated and costly instrumental analysis. Spot tests have been found to be extremely useful for the preliminary field detection of the test material. Different pesticide characteristics have been utilized to maximize the sensitivity, selectivity, and specificity of the tests. Recently, several types of spot tests, such as the capillary spot test, the thin-layer chromatographic spot test, the paper chromatographic spot test, the ion-exchange spot test, and the enzymatic spot test, have been developed for the detection of fungicide and herbicide residues in water at trace levels.

B. Sample Preparation

Sample preparation involves two steps: enrichment and cleanup of the test material. The concentration of pesticide residues in water is generally low, below the lower limit of estimation of the analytical methods. Therefore, a suitable preconcentration/enrichment method is coupled with the analytical technique. The choice of preconcentration method depends on factors such as the volatility and solubility of the fungicide or herbicide, the degree of concentration required, and the nature of the analytical technique to be used. The following methods of preconcentration have been used.

1. Concentration Methods

In concentration methods, water is removed in order to enrich the dissolved residues. The following techniques are being used.

a. Freeze Concentration. This technique removes water as a solid phase, and it concentrates the fungicide or herbicide residues in the unfrozen portion. It is applicable at the level of milliliter volumes of water sample. It has been observed that the recovery of the analyte decreases with the increasing concentration of the inorganics dissolved in the water. This technique has been used to enrich residues of phenols in water, with a 20-fold volume reduction and 80% recovery. The carbamates in acetonitrile or acetone solutions have also been enriched by using this technique.

b. Lyophilization. In this technique, water is frozen and then removed by sublimation under vacuum. The concentration of the analyte can be raised several thousand times, but the recovery of the organics is poor from water samples containing inorganics. This method can be used for the preconcentration of nonvolatile or less volatile fungicide or herbicide residues in water. It has been successfully used in the analysis of organochlorines.

c. Evaporation. This is a common method that has been very successful in inorganic analysis and in the analysis of nonvolatile fungicide and herbicide residues of higher melting point in water.

d. Distillation. Steam distillation is useful in collecting water-soluble organics from water samples containing large quantities of nonvolatile organics and inorganics. Vacuum distillation is more advantageous but more laborious than lyophilization. The former has been used to enrich pesticides in water, sediments, and tissues; the latter is used to enrich ethylene chlorohydrin and acrylonitrile in water.

e. Reverse Osmosis. This is used to purify water in many places. Commercial units that can treat several liters of water per day are available in the marketplace. When coupled with solvent extraction and dialysis, about 30–40% of the organics in water can be recov-

ered. Cellulose acetate membranes reject 90–97% of inorganics and organics with molecular weight more than 200. The drawback of the process lies in the fact that the membrane may either absorb the analyte irreversibly or release the contaminants into the analyte.

f. Ultrafiltration. This process is more selective than reverse osmosis, because it concentrates only species of higher molecular weight (1000). In fact, it is a filtration under pressure through a membrane, and it gives better results than lyophilization for enriching 82–85% humic materials and endosulfan residues in water.

2. Isolation Methods

These are the methods that take out pesticide residues in water. The following techniques have been used for this purpose.

a. Liquid–Liquid Extraction. This is a very old technique of considerable significance to the chemical industry. Countercurrent-flow systems and Teflon helix continuous liquid–liquid extractors have become the routine tools of analysis. Numerous pesticide residues, including 2,4-D, Silvex, and 2,4,5-T, at ppm levels in water have been enriched and cleaned up by using toluene, *n*-heptane, cyclohexane, methylene chloride, petroleum ether, and the like as the extractants. The average recovery found was 82–115%. It is advisable to tailor organic solvent and extraction parameters such as pH, ionic strength, and temperature before using the procedure under certain circumstances. The solvents to be used must be sufficiently pure, and great care must be taken to avoid contamination by contact with solid support. The solvent may form an emulsion on shaking with water at room temperature, and the emulsion behaves differently from the pure solvent. The emulsion formation tendency decreases with increasing temperature; thus, extraction has been found to be more effective at relatively higher temperatures. It is claimed that the *p*-value approach can be considered a guide to optimizing the liquid–liquid extraction. It has been found that the *p*-value determined in distilled water is also applicable to other types of water, including river water, seawater, and wastewater.

b. Solid–Liquid Extraction. The solid phases used with different procedures are discussed in the following subsections.

Activated Carbons: Activated carbon columns and filters have been developed and widely used in European countries for removing organics from drinking water. Parathion, malathion, β-naphthoxyacetic acid, β-naphthaleneacetic acid, and trichloroacetic acid present in water have been enriched by using carbon columns. However, the average recovery has been found to be poor, so the method is highly useful for removing pesticide residues from drinking water but has limited use in pesticide residue analysis.

Organic Polymeric Adsorbents: Synthetic organic adsorbents, such as XAD-2, XAD-4, XAD-7, and XAD-8, are available in the form of hard, nondusting spheres with a composition between that of activated carbon and that of polymeric adsorbents. The residues of fungicides and herbicides such as organonitrogen, chlorophenol, chlorophenoxy acids, and humic acids present in water have been enriched by the use of this technique. The average recovery was found to be 50–90%. However, total recovery is not possible.

In Situ Polymerized Resin: A new in situ polymerized resin process using open-pore polyurethane foam has been developed. These resins have a relatively high adsorption capacity that is nearly equivalent to that of activated charcoal. The resins have a high affinity for phenols, polycyclic aromatic hydrocarbons (PAHs), primary amines, and nitrogen-containing hetrocyclic fungicides and herbicides and their residues in water. In situ polymerization resin is readily adaptable to large-scale use, and at this stage it seems of promising potential for pesticide residues in water.

Polyurethane Foam: This is a nonhygroscopic, semisolid, nonvolatile, and relatively nonreactive adsorbent. It has been used for concentrating phthalate esters, PAHs, toxaphene, indole-3-acetic acid, β-naphthaleneacetic acid, β-naphthoxyacetic acid, TCA, etc. in large volumes of finished and raw waters at a flow rate of 250 ± 10 ml/min and $60 \pm 2°C$. The average recovery was 62–88%. This technique cannot be used reliably to enrich polychlorinated biphenyls (PCBs) from turbid natural water and wastewater. Foam impregnated with GC liquid phases has also been used for this purpose. For example, DC-200-coated foam was used to enrich organochlorine pesticides at ppb levels in water with 90% recovery.

Inorganic Adsorbents: Activated alumina, calcium phosphate, florisil, hydroxylapatite, magnesia, and silica gel have been used for this purpose. It is a rapid and selective process. Bonded-phase silica columns give good recovery for chlorinated fungicides and herbicides in water.

Ion Exchangers: These represent a simple, inexpensive, and widely used technique. The extra advantage of ion-exchange resins/beads lies in the fact that

either the preconcentrated pesticide residues can be detected and determined on the beads or the analyte may be eluted with a suitable solvent and then analyzed. Aliphatic as well as aromatic acids and chloro- and nitrophenols in water have been preconcentrated using strong anion-exchange resins. This technique has also been utilized to enrich 2,4-dichlorophenoxyacetic acid (2,4-D) and tetrachloroterphthalic acid in wastewater at ppm levels. The average recovery efficiency was found to be 70–100%. Inorganic ion exchangers—e.g., zirconium phosphate, ammonium molybdophosphate, and titanium hydroxide—have been tried for this purpose without any appreciable results. Ion-exclusion chromatography for preconcentration of low-molecular-weight carboxylic acid fungicides and herbicides have been reported. Ion-retardation resins and immobilized ion-exchange site resins have also been developed in order to improve the simplicity and selectivity of the already-known adsorbents or ion exchangers. However, these developments have been found of limited utility.

Precipitation: It is well known that humic acids are soluble in bases but insoluble in alcohols. Therefore, humic materials may be precipitated from aqueous solutions by acidification with glacial acetic acid in the presence of isoamyl alcohol. This method has been applied in a cursory manner to fresh underground water and seawater for some organics.

3. Cleanup or Separation Methods

The removal of interfering substances from the extract is usually referred to as the cleanup procedure. The nature of impurities present in the enriched sample depends upon the impurities present in the water sample. For example, a water sample from agricultural runoff often contains many biocides, such as insecticides, fungicides, herbicides, humic materials, and several inorganics. However, lake or river water samples in industrial areas contain sulphur, organosulphur, PCBs, phthalic esters, inorganics, etc., depending upon the nature of the industry. These impurities can be removed partly or completely by a proper combination of the preceding techniques, including gel permeation chromatography (GPC), paper chromatography (PC), thin-layer chromatography (TLC), electrophoresis (EL), high-performance liquid chromatography (HPLC), and gas–liquid chromatography (GLC).

C. Estimation

The quantity of pesticide residues, together with the metabolites and breakdown products in the cleaned-up extract or concentrate, is estimated by using the following methods.

Functional group analysis
Biological test methods
Chromatographic methods
    PC
    TLC
    GLC
    GC-MS
    HPLC
    Liquid–liquid chromatography
Spectroscopic methods
    Ultraviolet and visible
    Infrared
    Fluorescence and phosphorescence
Radiochemical methods
Electrochemical methods
    Voltametry
    Amperometry
    Potentiometry
    Polarography
Mass spectrometry
Electrophoresis
Hyphenated techniques

D. Confirmation

The confirmation of the presence of a residue is carried out by using a different method or by the formation and identification of a derivative. It is important that the identity and level of pesticide residues determined be confirmed by a different method from that used in the determination. When the end determination method is TLC, the pesticide can be removed from the chromatoplate with a suitable solvent and injected directly onto a gas chromatograph. A simple confirmation technique for GLC is afforded by TLC when the residue amounts present are large enough.

For more information, the reader is directed to Refs. 1–17.

V. FUNGICIDE AND HERBICIDE RESIDUE DATA

A. Experimental Details

The experimental details, such as the technique used, the methodology, sample preparation, the sensitivity, and the level of residues of fungicides and herbicides are given in Table 3.

B. Preliminary Characterization Tests

Some of the tests used for the preliminary characterization of pesticide residues are described next.

1. Nitrogen-Containing Pesticides

A simple and inexpensive spot test for the detection of nitrogen-containing pesticide residues in water was developed (18). Citric-acid-impregnated paper in the presence of acetic anhydride was used as the coloring reagent. The limit of detection for amitrol, azobenzene, bavistin, calixin, 2,4-lutidine, 2,6-lutidine, nicotine acid, β-picoline, and quinoline was found to be 0.04, 40, 2, 3.2, 4, 100, 0.04, 0.4, and 100 μg, respectively. The test was successfully applied in acidic, basic, and saline water. To perform the test, the pesticide residue was enriched by extracting river water with chloroform. A drop of the chloroform extract was applied with the help of a micropipette at the center of a piece of filter paper impregnated with citric acid and dried to remove the solvent completely. Acetic anhydride (2–3 drops) was applied on the spot of the test solution, and it was then heated at 165 $\pm$ 2°C for 2 min. The color developed was subsequently noted.

2. TCA

A fluorescence spot test (31) was developed to detect TCA at ppm levels in water and soil. A dosed sample (TCA 0.40 ppm) of deep-well water was extracted with diethyl ether. The organic layer was collected separately and dried to remove solvent, and the residue was dissolved in 1 ml of methanol. The methanolic test solution (0.1 ml) was treated with NaOH and phenol solution in a micro test tube at 100°C for 2 min and cooled, and one drop of the reagent (aqueous solution of hydrazine soluphate and sodium sulphate) was added to it. The reaction mixture was neutralized with acetic acid and then shaken with diethyl ether (0.5 ml). The ethereal layer was spotted on a strip of filter paper, dried, and exposed to UV light of 366 nm. A greenish-yellow stain on a purple background indicated the presence of TCA.

3. Phenoxy Herbicides

A sensitive and selective novel technique (32) was developed for the field detection and semiquantitative determination of phenoxy herbicides. The water sample containing traces of 2,4-D acid or 2,4-D ethyl ester or 2,4-D sodium salt was taken in a test tube with an arm, evaporated to dryness, and cooled to room temperature. A pinch of solid $ZnSO_4 \cdot 7H_2O$ was added. The reagent

Table 3 Analysis of Fungicide and Herbicide Residues in Water

Fungicides/herbicides	Experimental conditions	Inferences	Refs.
i. Indole-3-acetic acid ii. β-Naphthaleneacetic acid iii. β-Naphthoxyacetic acid	*Technique:* Column chromatography *Column:* 90-mm × 15-mm-ID flexible polyurethane foam plug in 1-m × 17-m-ID glass column *Flow rate:* 23 ± 5 ml/min *Sample:* Spiked deep-well water	*Average recovery:* (i) 62%, (ii) 65%, (iii) 70%	19
i. Indole-3-acetic acid ii. β-Naphthaleneacetic acid iii. β-Naphthoxyacetic acid iv. Phenoxyacetic acid v. Trichoroacetic acid	*Technique:* Adsorption studies by batch equilibrium process *Adsorbent:* Charcoal and alumina *Sample:* Spiked deep-well water	*Sequence of adsorption:* (iii), > (ii), > (i), > (iv), > (v) on charcoal; (i), > (iii), > (v), > (iv), > (ii) on alumina	20, 21
Indole-3-acetic acid β-Naphthaleneacetic acid β-Naphthoxyacetic acid Phenoxyacetic acid Trichloroacetic acid Nitrophenols	*Technique:* TLC *Stationary phase:* The following adsorbents and their admixtures: calcium sulphate, ammonium molybdate, alumina, calcium carbonate, copper sulphate, ammonium phosphate, ferric chloride, magnesium chloride, nickel sulphate, sodium molybdate, titanium oxide, zinc oxide, barium sulphate, calcium phosphate *Mobile phase:* Ethanol, distilled water, ethylacetate, propanol, benzene, carbon tetrachloride, acetone *Detection:* Ethanolic alkaline bromophenol blue solution (1%) *Sample:* Spiked deep-well water	Trichloroacetic acid can be separated quantitatively from other herbicides under study. This technique can be used for cleanup of herbicides. It can also be used for confirmation and identification on the basis of R_f values.	22–29
Chlorophenoxyacetic acid 2,4-Dichlorophenoxyacetic acid Indole-3-acetic acid Indole-3-propionic acid α-Naphthaleneacetic acid β-Naphthaleneacetic acid β-Naphthoxyacetic acid Trichloroacetic acid 2,4,5-Trichlorophenoxyacetic acid Dalapon	*Technique:* Paper chromotography *Stationary phase:* Whatman chromatographic paper strips and strips impregnated with cetrimide, paraffin oil, and cetrimide–paraffin oil *Mobile phase:* Benzene, chloroform, ethanol, methanol, distilled water, ethylacetate, chlorobenzene, acetone, acetophenone, buta-2-ol, 1,4-dioxan, nitrobenzene, propan-2-ol, pyridine *Detection:* Ethanolic alkaline bromophenol blue solution (0.1%) *Sample:* Spiked deep-well water	This is an inexpensive method for the cleanup of fungicides and herbicides.	30
Dimethyl tetrachloroterephthalate (DCPA) Monomethyl tetrachloroterephthalate (MM) Tetrachloroterephthalate (TCPA)	*Technique:* HPLC and GC-MS **For HLPC:** *Column:* 15.0-cm × 4.6-mm-ID × 15-µm Supelco LC-PAH-modified C18 *Guard column:* 2-cm × 2-mm-ID × 30–40-µm pellicular Perisorb RP 18; precolumn filter with a 2-µm replaceable frit *Flow cell:* 18 µl *Mobile phase:* m_1:m_2:m_3 = 20:70:10 and 40:50:10 m_1: 15 ml of 0.5 M TBAP in 1 L of CH_3CN m_2: 15 ml of 0.5 M TBAP + 50 ml of CH_3CN in 1 L of H_2O	More than 8800 samples of potable groundwater taken from over 4000 different potable-water wells over a period of 5 years were analyzed. MM was not detected; TCPA was found in 1107 samples at concentration of more than 10 µg/L. Additionally, 262 individual wells were found to have TCPA concentrations of more than 50 µg/L; 280 wells had TCPA concentrations in the range 10–50 µg/L. The highest concentration of TCPA found in a well water sample to date was 1750 µg/L.	38

m_3: 15 ml of 0.5 M TBAP + 25 ml of CH$_3$CN in 1 L of 0.01 M K$_2$HPO$_4$3H$_2$O

TBAP: Tetrabutylammonium phosphate

Flow rate: 1 ml/min

Detection: LC-480 autoscan diode array

For GC-MS:

Column: 12-m × 0.20-mm-ID × 0.33-μm HP-1, crosslinked methyl silicone

Mobile phase: He

Flow rate: 1 ml/min

Detector: Series mass selective detector

Sample: S_1 and S_2

S_1: For HPLC/UV analysis, samples were collected in 4-oz high-density polyethylene bottles; unless analyzed immediately, samples were stored in a 10°C freezer. Fresh or completely thawed samples were transferred into 4-ml disposable autosampler vials for loading into the HPLC autosampler. A 500-μl sample volume was injected sequentially every 23 min.

S_2: For GC/MS analysis, a 50-ml aliquot of a well water sample was taken in a separatory funnel; 15 ml of 10 N sulphuric acid was added to it and the mixture was then extracted with 16 ml of diethyl ether:petroleum ether (50:50). The ether layer was centrifuged to remove aqueous acid and evaporated to near dryness. The residue was cooled, and sequentially 100 μl of isooctane and 50 μl of neat BTFA were added to it. The contents were heated and cooled, and 0.3 ml of hexane/heptachlor epoxide solution was added to it and then thoroughly mixed. A 3-μl volume of the derivatized sample extract after mixing with internal standard was injected into GC/MS.

BTFA: *N,O*-Bis(trimethylsilyl)trifluoroacetamide.

Chemicals	Method	Results	Ref.
Simazine Thiram Thiobencarb Propyzamide Diazinon Chlorothalonil Iprobenfos Thiobencarb Isoprothiolane Isoxanthion Asulam Oxine-copper Thiram	*Technique:* Solid-phase extraction and high-performance liquid chromatography *Enrichment column:* SEP-PAK PS-2 cartridge **For HPLC:** *Column:* 15-cm × 4.6-mm-ID ODS-80 TM *Mobile phase:* Acetonitrile *Flow rate:* 1.0 ml/min *Detection:* Photodiode array *Sample:* The analyte was enriched by solid-phase extraction and was eluted with acetonitrile and analyzed by HPLC/photodiode array.	The recoveries and coefficients of variations of the chemicals were within the range 83–107% and 6.4–9.3%, respectively. The lower limit of detection was 0.2 μg/L. This method could be used as a screening test before GC/MS analysis to determine the agricultural chemicals in water.	29–43
Simazine Thiobencarb Iprodione Bensulide	*Technique:* GC-MS *Sample:* water Other conditions not available.	Preconcentration of the analyte was carried out by solid-phase microextraction on fiber mode of polydimethylsilonaxe/divinylbenzene. The average recoveries ranged from 1.03% to 17.49%, and coefficients of variance ranged from 5.04% to 17.43%.	44–48

Table 3　Continued

Fungicides/herbicides	Experimental conditions	Inferences	Refs.
Atrazine Phenylurea Isoproturon Chlortoluron Diuron	Not available	Analysis is being carried out on the three main rivers in the Paris area to monitor a number of products from the triazine and urea families. Atrazine is the most important contaminant, and its concentration exceeds the value of 100 ng/L most of the time. For a period of several months every year, concentrations approaching 1000 ng/L are observed in all of the catchment areas being studied. These are the results of the rapid transfer of atrazine in runoff water.	49
Vamidothion 4-NP MCPA Mecoprop Dinoseb Dinoterb Isoproturon Ametryn Fenitrothion Prometryn Terbutryn Parathion-ethyl	*Technique:* Reversed-phase liquid chromatography with diode array detection (RPLC-DAD) and particle beam (PB) mass spectrometer *Enrichment column:* solid-phase extraction (SPE) 10×2.0-mm cartridge packed with PLRP-S styrene-divinylbenzene copolymer **For liquid chromatography:** *Column:* 200×4.6-mm-ID $\times$ 5-μm Spherisorb ODSZ *Mobile phase:* A gradient methanol:0.1 M ammonium acetate (30:70) to (70:30) in 20 min, then 20 min isocratic elution at (70:30); after 2 min at 90% methanol *Flow rate:* 0.4 ml/min *Detection:* DAD **For MS:** A Hewlett-Packard 5989A MS Engine equipped with a dual electron impact/chemical ionization (EI/CI) source was connected to the outlet of the DAD via a Hewlett-Packard PB interface. This connection was made with a 50-cm length of 0.12-mm-ID stainless steel capillary tubing. Methane was used as reagent gas. *Sample:* Tap and river water samples were filtered; pH was adjusted and then enriched.	SPE with a PLRP-S precolumn coupled online to RPLC and DAD enables a group of pesticides to be detected in water at levels between 0.05 and 0.5 μg/L with RSDS ($n = 3$) below 13%.	50
Ametryn Dinoseb Mecoprop Terbutryn Bentazone Isoproturon MCPA Molinate Prometryn Vamidothion	*Technique:* Liquid chromatography—mass spectrometry (LC-MS) *Enrichment column:* 10×2.0-mm cartridge packed with PLRP-S sytrene-divinylbenzene copolymer **For LC:** *Column:* 200×4.6-mm-ID $\times$ 5-μm Spherisorb ODSZ *Mobile phase:* The linear gradient was from 30% to 60% methanol in 20 min, then 70% at 40 min. After 5 min at 90% methanol, the mobile phase was returned to initial conditions. *Flow rate:* 0.4 ml/min *Detection:* Diode array detection (DAD) **For MS:** An HP 5989 A mass spectrometer equipped with electron ionization (EI)/chemical ionization (CI) source was connected to the liquid chromatograph through a 50-cm $\times$ 0.12-mm stainless steel capillary. Methane was used as reagent gas.	EI was more sensitive than positive and negative chemical ionization (PCI and NCI). Of the latter two modes, NCI gave higher responses, especially for organophosphorus compounds. When online solid-phase extraction–LC-PB-MS was applied to real samples, limits of detection in full-scan mode were in the range of 0.5–10 μg/L for EI. The present results of poor analyte detectability versus good studies devoted to improve the particle beam interface.	34

Compounds	Method	Results	Ref.
Ametryn Atrazine Prometryn Terbutryn	*Sample:* Prior to analysis, river water samples were filtered through a 0.45-μm filter, pH was adjusted to 2, and then the samples were concentrated. *Technique:* Microextraction—GC-MS *Microextraction:* Polyacrylate fiber system from Supelco **For GC:** *Column:* 30-m × 0.25-mm × 0.25-μm crosslinked methylsilicone film *Mobile phase:* He *Flow rate:* not available *Detector:* not available **For MS:** An HP 5972 mass spectrometer equipped with selected-ion monitoring (SIM) acquisition. *Sample:* Tap and river water samples were filtered through a 0.45-μm membrane filter and preconcentrated.	Limits of detection at the sub-μg/L level were achieved with GC and MS under the full-scan acquisition mode and at the ng/L level for MS under SIM acquisition. The repeatability of the method for tap water spiked at a level of 0.5 μg/L ($n = 5$) was below 25% (RSD).	52
Ametryn Atrazine Bentazone Dinoseb Isoproturon MCPA Mecoprop Molinate 4-NP Prometryn Terbutryn Vomidothion Dinoterb Fluometuron Terbuthylazine	*Technique:* Solid-phase extraction (SPE), liquid chromatography (LC)–atmospheric pressure chemical ionization–mass spectrometry (APCI-MS), and particle beam–mass spectrometry (PB-MS) *Enrichment column:* 10 × 2.0-mm cartridge packed with C18 and Lichrospher Si 100 RP-18 **For LC-APCI-MS:** *Column:* 200 × 4.6-mm-ID × 5-μm Spherisorb ODS2 stainless steel column *Mobile phase:* Methanol:0.1 M ammonium acetate (acidified to pH 4.5 with acetic acid) with gradient elution *Flow rate:* 0.8 ml/min *MS:* The system was equipped with an APCI interface, a nitrogen flow rate of 10 L/h was used for the nebulization, and the flow rate for the drying nitrogen was 200 L/h. **For LC-PB-MS:** *LC column:* as for LC-APCI-MS *Flow rate:* 0.4 ml/min *MS:* The system was equipped with a PB interface through a 50-cm × 0.12-mm stainless steel capillary tubing. The helium nebulizer pressure was 60 psi under CI conditions, and methane was used as the reagent gas for PCI and NCI. *Sample:* Tap water volumes of 200 ml spiked with the pesticides and the corresponding internal standard were preconcentrated on disposable precolumns at a flow rate of 5 ml/min. Water samples were acidified to pH 3 with hydrochloric acid, filtered through a 0.45-μm filter, and then concentrated.	For APCI under positive-ion (PI) mode, limits of detection (LODs) were between 0.8 and 4 ng/L, and under negative-ion (NI) mode of operation, between 4 and 20 ng/L; for PB the LODs were between 0.05 and 0.2 μg/L under EI conditions and from 0.02 to 0.1 μg/L under chemical ionization in both positive and negative acquisition modes. The study demonstrates the higher sensitivity of HPLC-APCI-MS over HPLC-PB-MS, and both are potential techniques for confirming the presence of fungicide and herbicide residues in water. The methods developed were validated by participating in aquacheck interlaboratory exercises.	53

Table 3   Continued

Fungicides/herbicides	Experimental conditions	Inferences	Refs.
Captan Ferbam Nebam 2,4-D 2,4,5-TP (Silvex) Dalapon Diuron Simazine	Not available	Fundamentals of the subject in terms of the importance of the pesticide issue, the different types of major pesticides, and the analysis of their persistence, removal, distribution and detection in water. It is expected that such information can be useful in understanding, visualizing, synthesizing, and perhaps handling more complex aspects of pesticide technology. A total of 40 pesticides have been detected in surface water in monitoring reports received from 12 governmental groups that included federal, state, regional, and local agencies in Florida. These monitoring results indicate the presence of 32 pesticides in the sediment samples. All agency reports generally indicate that the impact of the sporadic and low level of pesticide detection is below adverse levels. The concentration frequency distributions of the majority of pesticides detected in surface water and in sediment samples of Florida followed a decreasing exponential curve. It is claimed that the pesticide concentration distributions obtained can provide some assessments for identifying pesticides for further monitoring in sensitive surface waters of Florida. The water quality assessments, based on the sampling of surface water, shallow and deep drinking water wells, and irrigation wells at 14 locations of this citrus grove for about 7 years (January 1989 to October 1995), indicated that residues of Bromacil, Diuron, and simazine were observed sporadically and were contained largely in the middle part of the surface water system of the site. As a net result of dilution, degradation, best management practices (BMPs) and detection areas, the citrus conversion process had no significant off-site adverse water quality impact during the first three phases of the citrus grove development.	54–57
Molinate Simazine Atrazine Ametryn Prometryn Terbutryn Bentazone	*Technique:* SPE-GC with ECD-MS *Enrichment:* SPE cartridges with 200 mg of porous-column ethylvinylbenzene-divinylbenzene copolymer (Li Chrolut EN) **For GC:** *Capillary column:* 30-m × 0.25-mm-ID × 0.25-μm crosslinked methylsilicone *Mobile phase:* Helium and nitrogen *Flow rate:* 2.00 ml/min *Detection:* Electron Capture detector (^{63}Ni) (ECD) **For MS-SIM:**	The detection limit of the fungicides and herbicides in tap water was 0.02–0.1 μg/L when GC-MS-SIM acquisition mode was used and 0.2–1 ng/L for ECD. The method was applied for analyzing drinking water and surface water. In Ebro River water and water from the Ebro Delta, some pesticides could be determined by GC-MS-SIM and GC-ECD, some of them being confirmed by GC-MS under full-scan acquisition.	58

The column was inserted directly into the ion source of the mass spectrometer.

Sample: Off-line trace enrichment was carried out using the Bond Elut-Vac Elut system. The cartridge was activated by passing 5 ml of hexane, 5 ml of ethyl acetate, and 5 ml of Milli-Q water in sequence through it with a low vacuum. The spiked water sample (500 ml) containing 15 g/L of NaCl was passing through the cartridge at a flow rate of approximately 20 ml/min using a vacuum system. The cartridge was dried under vacuum, and the elution was carried out by sequentially adding 5 ml of hexane and 10 ml of ethyl acetate under vacuum. The eluate was collected in a tube, and the internal standard was added (15 mg/L) for 1-chlorooctadecane (MS) and (50 mg/L) for bromophos-ethyl (ECD). The eluate was evaporated to 1 ml, and a 1-μl aliquot was injected into the capillary column. River water samples were filtered using a 0.45-μm membrane filter before being preconcentrated.

Simazine
Cyanazine
Chlortoluron
Atrazine
Isoproturon
Ametryn
Prometryn
Terbutryn

Technique: SPE-RPLC

Enrichment column: 4.6-mm-ID S-DVB Empore extraction disks for online SPE were placed in the conventional millipore apparatus connected to a vacuum system.

For RPLC:

Column: 25 × 0.46-cm-ID × 5-μm Spherisorb ODS-2

Mobile phase: A gradient profile of a phosphate buffer (pH 7) containing 5% acetonitrile (solvent A) and acetonitrile (solvent B), going from 15% of solvent B to 30% in 15 min and 100% of solvent B at 35 min and then back to the initial conditions in 5 min. All mobile phases were degassed with helium.

Flow rate: 1 ml/min

Detection: Diode array detection (DAD)

Sample: Water samples from the Ebro River were filtered through a 0.45-μm PTFE (Millipore). Samples were spiked with pesticides and 10 g/L NaCl and extracted using off-line or online SPE at a flow rate of 5 ml/min. The disks were flushed with 15 ml of methanol and injected in the LC column.

Lower limit of detection of online method was found to be 0.1 μg/L for most pesticides to be determined in tap water, as required by European Community (EC) rules, while the off-line mode determines pesticides at 1–3 μg/L in surface water, as recommended by EC rules.

59

Molinate
Simazine
Atrazine
Ametryn
Prometryn

Technique: SPE-GC-MS

Enrichment column: 10 × 2-mm-ID × 20-μm hand-packed styrene-divinylbenzene copolymer

For GC-MS:

Column: The SVE kit consisted of a 5-m × 530-μm-ID retention gap, a 2-m × 250-μm-ID retaining precolumn, and a 30-m × 250-μm-ID, 0.25-μm analytical column, both columns HP-5 MS, and a solvent vent valve.

Mobile phase: Helium

Flow rate: 1.2 ml/min

Detection: Electron ionization (EI)

MS: Selected ion monitoring (SIM) acquisition

The limit of detection was found to be 2–20 mg/L for tap and river waters. Molinate was identified and quantified by SPE-GC-MS under full-scan acquisition in Ebro River water at a concentration of 0.1 μg/L.

60

Table 3 Continued

Fungicides/herbicides	Experimental condititions	Inferences	Refs.
	Sample: River and tap water samples were filtered using 0.45-μm filter. The precolumn was conditioned with 3 ml of methanol, activated with 3 ml of water; 10 ml of the sample containing 30% methanol was passed through the precolumn. The precolumn was dried with N_2 for 30 min, and the analyte was eluted with ethyl acetate (100 μl, 47 μl/min) and then injected onto the GC column.		
Simazine Atrazine Methomyl Oxamyl MCPA Bentazone	*Technique:* SPE-LC *Enrichment column:* 10 × 3-mm ID packed with carbon black carbopack B 120/400, highly crosslinked styrene–divinyl benzene copolymer and functionalized polymeric resin **For LC:** *Column:* 250 × 4.6-mm-ID × 5-μm Spherisorb ODS-2 *Mobile phase:* HPLC–gradient-grade acetonitrile and Milli-Q quality water adjusted to pH 3 with sulphuric acid *Flow rate:* 1 ml/min *Detection:* Shimadza SPD-10A UV spectrophotometric detector *Sample:* The water samples were acidified with hydrochloric acid to pH 2.5, passed through the preconcentration column, eluted in the backflush mode with acetonitrile, and analyzed by LC.	The recovery from tap water was 86%, and the detection limit was 0.03–0.17 μg/L. Interferences due to high concentrations of fulvic and humic acids were contolled by adding 10% Na_2SO_4 solution to the water samples.	61
Oxamyl Methomyl 2,4-Dinitrophenol 2-Chlorophenol Bentazone Simazine MCPA Atrazine	*Technique:* SPE-LC *Enrichment column:* 10 × 3-mm ID packed with chemically modified polystyrene-divinylbenzene resin with an *O*-carboxybenzoyl moiety **For LC:** *Column:* 250 × 4.6-mm-ID × 2.5-μm Spherisorb ODS *Mobile phase:* The gradient solution was carried out with Milli-Q water adjusted to pH 3 with sulphuric acid (solvent A) and methanol as organic modifier (solvent B). The solvent program was initially 20% B, 50% B after 25 min, 100% at 32 min, isocratic for 2 min, and the mobile phase returned to initial conditions in 2 min for subsequent analysis run. *Flow rate:* 1 ml/min *Detection:* UV detector at 240–280 nm *Sample:* Water samples from the tap and the Ebro River were filtered through 0.45-μm nylon membranes, acidified with hydrochloric acid to pH 2.5; 10% sodium sulphide solution was added. Samples were preconcentrated, eluted with methanol in the backflush mode, and analyzed by LC.	The newly synthesized sorbent enables higher volumes of sample to be concentrated for determining pesticides. Better results were obtained when 0.5–1 ml of 10% Na_2SO_3 solution for every 100 ml of sample was added to tap and river water, respectively.	62
Phenol 2-Nitrophenol 4-Nitrophenol 2,4-Dinitrophenol 2-Methyl-4,6-dinitrophenol	*Technique:* SPE–supercritical fluid chromatography (SFC) *Enrichment column:* 10 × 3-mm-ID × 10-μm Spherisorb ODS-2, 20-μm PLRP-S, and 80–160-μm highly crosslinked styrene-divinylbenzene copolymer ENVI Chrom P **For SFC:** *Column:* 150 × 4.6-mm-ID × 5-μm HP Spherisorb ODS-2 and 250 × 4.6-mm-ID × 5-μm HP LiChrospher Diol, and HP Hypersil silica	SFC separated phenolic compounds in less than 6 min with good resolution. The lower limit of detection was found to be 0.2–1 ppb for tap water and 2–6 ppb for river water. The RSDs ($n = 4$) were less than 10%. It is a better technique than SPE-RPLC with UV detection because of no matrix interference.	63

Analytes	Method	Comments	Ref.
	Mobile phase: Carbon dioxide and HPLC-grade methanol *Flow rate:* 2 ml/min *Detection:* HP 1050 diode array detector *Sample:* The water sample was preconcentrated and the tubes were dried with helium. The analyte was eluted in the backflush mode and then transferred to the analytical column. Samples of river or tap water were filtered; 0.3 ml of 10% Na_2SO_3 solution was added to it, and the pH was adjusted to 9.0 with 0.1 M NaOH. Tetrabutylammonium bromide was added, preconcentrated, eluted, and then analyzed by SFC.	This study showed that sodium sulphite is the best reagent for removing humic substances from tap or river water.	64
Resorcinol Oxamyl Methomyl 4-Nitrophenol 4-Chlorophenol Bentazone Phenol	*Technique:* SPE-LC *Enrichment column:* 10 × 3-mm ID packed with chemically modified polystyrene-divinylbenzene resin with an *O*-carboxybenzoyl moiety **For LC:** *Column:* 25 × 0.46-cm-ID × 5-μm Kromasil 100 C18 *Mobile phase:* Milli-Q quality water (solvent A) and acetonitrile as organic modifier (solvent B) *Flow rate:* 1 ml/min *Detection:* Shimadza SPD-10A UV spectrophotometric detector at 240–280 nm *Sample:* Water samples are filtered. Sodium sulphate solution was added and adjusted to pH 3. The sample was preconcentrated and the analyte was eluted with acetonitrile (solvent B) in the backflush mode.		
Oxamyl Methomyl Phenol 4-Nitrophenol 2,4-Dinitrophenol 2-Chlorophenol Bentazone Simazine MCPA Atrazine	*Technique:* SPE-LC *Enrichment column:* 10 × 3-mm-ID × 20-μm PLRP-S, Envi-chrom P, and Lichrolut EN **For LC:** *Column:* 250 × 4-mm-ID × 2.5-μm Spherisorb ODS *Mobile phase:* Milli-Q water adjusted to pH 3 with sulphuric acid (solvent A) and methanol (solvent B) *Flow rate:* 1 ml/min *Detection:* Shimadza SPD-10A UV spectrophotometric detector *Sample:* Real samples were filtered through 0.45-μm nylon membranes and adjusted to pH 2.5 with hydrochloric acid.	The newly synthesized stationary phase used in the enrichment column had higher recoveries of pesticides in surface and tap water than the commercially available sorbents, such as PLRP-S and Envi-chrom P. The matrix effect was found to be similar to that of Envi-Chrom P and Lichrolut EN.	65
Simazine Atrazine Diazinon	*Technique:* Membrane extraction discs–GC-MS *Enrichment disc:* 47-mm-ID × 0.5-mm thick × 500-mg of C18-bonded silica or stylene-divinylbenzene copolymer (SDB) for GC *Column:* 12-m × 0.2-mm-ID × 0.33-μm crosslinked methyl silicone gum *Mobile phase:* Helium as the carrier gas and methane as the reagent gas in the PCI and NCI modes *Flow rate:* 1.8 and 1.7 torr *Detection:* A Hewlett-Packard 5989 A MS Engine equipped with a dual EI-CI source in conjunction with an HP 5890 and an HP-UX 59944 C data system and electron impact ionization	The use of membrane extraction discs and GC-MS with EI ionization can determine pesticides at 0.06–0.2 μg/L. SDB extraction discs gave better recoveries than C18 extraction discs.	66

Table 3 Continued

Fungicides/herbicides	Experimental conditions	Inferences	Refs.
Oxamyl Methomyl Aldicarb Cyanazine Monuron Propoxur Carbofuran Simazine Carbaryl Fluometuron Atrazine Diuron Linuron Barban	Sample: The water sample was filtered, mixed with 1% NaCl, and passed through the discs. The trapped pesticides were eluted with 2 × 15 ml of ethyl acetate. The internal standard was added into the extract. Ethyl acetate was evaporated to 100 μl, and 1 μl was injected into the GC-MS system. Technique: Reversed-phase liquid chromatography (RPLC)–diode array detection–particle beam mass spectrometry (DAD-PB-MS) Enrichment column: 10 × 2.0-mm-ID cartridge packed with 15–25 μm RLRP-S styrene-divinylbenzene copolymer For RPLC: Column: 200 × 4.0-mm-ID × 5-μm Spherisorb ODSZ Mobile phase: A linear gradient of methanol–0.1 M ammonium acetate (pH 5.0) from 30:70 to 88:12 in 34 min Flow rate: 0.4 ml/min Detection: DAD detector For MS: MS: A Hewlett-Packard 5989 A MS Engine with a dual EI/chemical ionization source was connected to the DAD outlet via a Hewlett-Packard PB interface. Interface tuning and signal optimization were conducted by injecting solutions of 500 ng monuron by flow injection analysis using methanol–0.1 M ammonium acetate (60:40) as the carrier stream. Typical operating pressures were 0.5 torr at the second-stage momentum separator and 1.5×10^{-5} torr in the ion-source chamber. Sample: A 100-ml sample was preconcentrated (4 ml/min) on the enrichment column. The analyte was eluted in the backflush mode with methanol–0.1 M ammonium acetate (pH 5.0) (30:70) and transferred to the analytical column.	The present study demonstrates that online trace enrichment–RPLC-DAD-PB-MS presents no experimental problems and can be used for the trace-level determination of fungicides and herbicides in surface water (0.2–5 $\mu g/L$) and drinking waters were analyzed and some pesticides were detected at sub-$\mu g/L$ levels.	67
Atrazine Propazine Terbuthylazine	Technique: SPE-GC-MS Enrichment column: Alumina column (Accu bond—J & W) For GC: Column: 30-m × 0.25-mm-ID × 0.25-μm HP-5 Mobile phase: Not available Pressure: 50–99–5 psi Detection: mass detector HP 5972 Sample: The sample was extracted with dichloromethane, evaporated, and then diluted with acetonitrile and cleaned by using a column packed with alumina, copper, and Na_2SO_4. The eluate was collected, evaporated, redissolved in 1 ml dichloromethane, and analyzed.	The analysis of pesticides in soil and sediments is important, because various types of compounds are present in water and sediments. This procedure can be applied to the determination of nitrogen-containing fungicides and herbicides.	68–70
Prometrine Terbuthylazine Simazine Atrazine	Technique: SPE-GC–thermoionic detector (NPD) or electron capture detector (ECD) Enrichment column: The sorptive commercial cartridges—SDBI (styrene-divinyl-benzene copolymer)	These studies included optimization of bed drying (lyophilization, air, or nitrogen drying) and eluate drying (solvent exchange by evaporation to dryness or adding anhydrous sodium sulphate). Good results were obtained	71–72

For GC:

Column: 30-m × 0.25-mm-ID × 0.25-mm (5% diphenyl polysiloxane, 95% dimethyl polysiloxane)

Mobile phase: Hydrogen with makeup N_2

Flow rate: 120 kPa

Sample: The model mixture was prepared by injecting 100 μl of the standard solution of the pesticide into 1 L of deionized water. The pesticide was concentrated from water sample by SPE, eluted with dichloromethane, evaporated, redissolved in methanol, and then analyzed.

using the chemical drier anhydrous sodium sulphate. The optimal procedure with air drying (0.5 L) was chosen for the determination of the selected pesticides in natural (sea, river, lake, tap) waters, with good results.

Atrazine

Technique: Enzyme inhibition–amperometry

Biosensor: A thin slice of potato (solanum tuberosum) + tissue rich in the enzyme polyphenoloxidase (PPO) or a commercial O_2-selective Clark electrode

Bioelectrode: A commercial O_2-sensing electrode was covered by three different membranes: First, a poly(tetrafluoroethylene) (PTFE) gas-permeable membrane to eliminate interferences from electroactive substances; second, a thin slice of potato tissue; third, a dialysis membrane to prevent microbial attack and/or contamination of the enzyme and leaking of the enzyme itself from the membrane; and, finally, a rubber O-ring to fix the three layers on the tip of the oxygen sensor.

Cell: The bioelectrode was connected to an amperometric detector. A constant potential of −650 mV was applied between the platinum cathode and the Ag/AgCl anode of the oxygen electrode. The electrode jacket was filled with an internal filling solution of KH_2PO_4 and KCl, both 0.1 M, pH 7.4. The experiment was carried out in 5 ml of 0.1 M phosphate buffer, pH 6.6, in a glass cell at 25°C with uniform magnetic stirring.

Determination: The determination of atrazine was performed in 1.0 M phosphate buffer, pH 6.6, at 25°C. Atrazine was incubated to maintain the contact in the thermostated cell in order to allow the inhibition of the enzyme. The inhibition power of atrazine was measured by using catechol at different concentrations.

Sample: Atrazine samples were prepared by dissolving 21.5 mg in methanol (1.6 ml); then it was added to a buffer solution of phosphate 0.1 M, pH 6.6, to yield a final volume of 100 ml and a final concentration 1 mM of atrazine.

This is a new, simple, and inexpensive method of determination. The low initial and operating costs of this device and its good analytical potential suggest its application in the field of environmental analysis, especially in the continuous monitoring of atrazine in water in risk areas.

73–75

Carbendazim
Thiabendazole
Propiconazole
Vincolozolin
Iprodion
Triadimefon
Triadimenol
Captan
Imazalil

Technique: SPE-bioautography-TLC

Enrichment column: 500-mg C_{18} Sep-Pak cartridges

For TLC:

Stationary phase: 20 × 20-cm × 250-μm × 250-μm layer coated with Kieselgel G

Mobile phase: A freshly prepared mixture of hexane:acetone (75:25)

Mode of development: Ascending for a distance of 15 cm in a saturated chamber

Trichoderma viride could be recommended as one of the most universal detection agents in a bioautography. *Fusarium* sp. or *Botrytis cinerea* are suitable as complementary agents. For semiquantitative determination of some fungicides in water samples, C18 SPE followed by TLC with bioautograph detection proved to be a useful procedure.

76

Table 3 Continued

Fungicides/herbicides	Experimental conditions	Inferences	Refs.
Flusilazole Fenasimol Myclo-butanil Flutnafol Procymidone Hexaconazole Tebuconazole Dichloftuanid	*Detection:* Spore suspensions of *Penicillium* sp., *Botrytis cinerea*, *Trichoderma viride*, *Fusarium* sp., and *Alternaria* sp. were prepared just before use in sterilized distilled water. The spore suspension was mixed 2:1 with liquid nutrient medium of sterile filtered malt extract. *Sample:* 500-ml water samples were passed through enrichment cartridges under vacuum at a rate of 10 ml/min. The analyte was eluted from the dried cartridges with 1 ml of methanol. The solvent was evaporated, redissolved in 20 μl of dichloromethane, and spotted on TLC plates for analysis.		
Bentazone 2,4-D MCPA Metoxuron Fluazifop acid Simazine Atrazine Monoliuron Metobromuron Diuron Linuron	*Technique:* SPE-HPLC *Enrichment column:* 5-mm-ID × 0.5-g C18 *Guard column:* 40-mm × 4.6-mm-ID cartridge × 5-μm RP-18 **For HPLC:** *Column:* 100-mm × 4.6-mm-ID × 5-μm RP-18 Spheri and 250-mm × 4.6-mm-ID × 5-μm Lichrosorb-CN *Mobile phase:* Methanol:0.1 M acetic acid–sodium acetate buffer (pH 3.8) (50:50) and methanol:water (2:8) for urea and triazine herbicide in second column *Flow rate:* 1 ml/min *Detection:* Variable-wavelength UV detector *Sample:* Water samples were fortified with a known volume of a standard solution, adjusted to pH 2 by adding 2.5 M sulphuric acid and 10 g of sodium chloride. The samples were passed through an SPE column. The analyte was eluted with 1 ml of methanol from a washed-and-dried column. The solvent was evaporated, redissolved in 0.5 ml of methanol:0.1 M acid–sodium acetate buffer of pH 3.8 (1:1) and analyzed by HPLC.	This is a simple, reliable, and inexpensive multiresidue method for different classes of herbicides in water. The enrichment factor is about 2000. The sensitivity of the method is 0.1 μg/L.	77
Dicamba Bentazone Benazolin 2,4-D MCPA	*Technique:* SPE-RP HPLC *Enrichment column:* Sep-Pak cartridges **For RP-HPLC:** *Column:* 250-mm × 4.6-mm-ID × 5-μm Lichrosorb RP-18 *Mobile phase:* Aqueous solution of TEA (0.01 M) adjusted to pH 6.9 with 1 M phosphoric acid:methanol (80:20) for isocratic elution. For programmed elution, the conditions were as follows: 100% A 0.01 M TEA (pH 6.9)–methanol (80:20) for 4 min, then 100% B 0.01 M TEA (pH 6.9)–methanol (70:30) in 1 min, at 100% B for 16 min, then back to A in 12 min (TEA = triethylamine). *Flow rate:* 1 ml/min *Detection:* Variable-wavelength UV detector *Sample:* Samples of drinking water or groundwater (1 L) were adjusted to pH 6.9 with TEA (0.03 M) and phosphoric acid (1 M) and spiked with the herbicides to 0.05–1 μg/L. The mixture was passed through the enrichment column. The analyte was eluted with 2 ml of methanol. The solvent was evaporated, redissolved in 1 ml of HPLC mobile phase, and then analyzed.	This is a reproducible and sufficiently sensitive method that is particularly useful for increasing the retention capacity of weakly retained compounds, such as Dicamba.	78

Fenoxaprop-ethyl Fenoxaprop	*Technique:* SPE-IP-HPLC *Enrichment column:* Sep-Pak cartridges **For IP-HPLC:** *Column:* 250-mm × 4.6-mm-ID × 5-μm Lichrosorb RP-18 *Mobile phase:* (A) Methanol:triethylamine buffer (9:1) and (B) methanol:triethylamine buffer (4:6). A for fenoxaprop-ethyl and A:B (6:4) for fenoxaprop. *Flow rate:* 1 ml/min *Detection:* Variable-wavelength UV detector	This is an effective, sensitive, and reproducible method for studies monitoring drinking water. Triethylamine as ion-pairing reagent provides milder conditions of analysis and prevents hydrolysis of fenoxaprop-ethyl.	79, 80
Atrazine Simazine 2,4-D MCPA	*Technique:* SPE-HPLC-GC *Enrichment column:* Lichrolut RP-18 encapped cartridges (500 mg) **For HPLC:** *Column:* Not available *Mobile phase:* Not available *Flow rate:* Not available *Detection:* UV detector **For GC:** *Column:* 1.8-m × 2-mm-ID × 5% Carbowax 20 M, 20-m × 0.32-mm-ID × 0.25-μm HP-5 and 15-m × 0.32-mm-ID × 1-μm HP-1701 *Mobile phase:* Not available *Flow rate:* Not available *Detection:* ECD and NPD *Sample:* All samples were collected in brown glass 1-L bottles, filtered through a 0.45-μm membrane filter, refrigerated at 4°C up to a maximum of 1 week, enriched, and then analyzed.	Atrazine was the ingredient most frequently found, followed by Alachlor and 2,4-D. The contamination was associated mostly with continuous agricultural application. The main strategies for preventing groundwater contamination have to do with the best management of agricultural practice and of the spectrum of active ingredients.	81
Picloram Atrazine Simazine 2,4,5-T Procymidone 2,4-D Vindozolin Terbuthylzine Metiram Metalaxyl Maneb Mancozeb MCPA Linuron Iprodione Loxynil Glyphosate Diquat Dinocap Diazinon Chloridazon Chloramben Amitrole	Not available	The results of the groundwater monitoring show levels of pesticides below 0.1–1 mg/m^3. Atrazine and Simazine have been most frequently detected in groundwater. There is a need for continued monitoring of groundwater systems, particularly those considered to be at risk of contamination. The unconfined groundwater in New Zealand can be contaminated by pesticides, and in some situations there can be contamination to levels greater than health standards would allow. Contaminated wells were significantly shallower, had higher water tables, had less depth between the water table and the well screen, and had slightly lower temperatures than uncontaminated wells.	82–86

Table 3 Continued

Fungicides/herbicides	Experimental conditions	Inferences	Refs.
Alachlor Acephate Chlorbufam Propazine Mectolachlor			
Metalachlor Alachlor Terbuthylazine Linuron Isoproturon Simazine Mecoprop MCPA Chloridazon Chlorpyrifos Bensulfuron Bonthiocarb Dalapon sodium Flurenol Molinate Propanil Oxadiazon TCA sodium Triclopyre TCA Carbofuran Dimethoate Isoproturon Trifluralin	*Technique*: SPE-HPLC-GC *Enrichment*: C$_8$ membrane **For HPLC:** *Column*: 15-cm × 4.6-mm-ID × 5-μm Supelcosil LC-18 *Mobile phase*: Acetonitrile:water (50:50) *Flow rate*: 1.6 ml/min *Detection*: UV detector **For GC:** *Column*: 12-m × 0.32-mm-ID × 12-μm Chrompack *Mobile phase*: Helium and makeup gas orgon:methon (95:5) *Flow rate*: Head pressure 31 kPa *Detection*: EC detector *Sample*: The water samples were generally collected within 24 hr and stored in the dark at below 4°C for a maximum of 10 days prior to extraction and analysis.	Real field data for pesticide release to surface water were compared with data obtained by simulations of a fugacity-derived model. A fugacity model was developed to take into account the unsteady-state condition of the actual field treatments, since pesticides are typically applied once or twice on the same area. The comparison between predicted and measured concentrations indicated that the model is a useful tool for the prediction of surface water concentration. The authors have also described a validation exercise of the soil fugacity model using field runoff data from Rosemaund Farm (UK). The soil fugacity model follows a middle path because, while simple in terms of data requirements and ease of use, it attempts to predict average pesticide concentrations in stream water following particular rainfall events in given scenarios.	87–91
Alachlor Atrazine Chlorfenvinphos Chlortoluron Cyanazine 2,4-D MCPA Metazachlor Metabrouron Metolachlor Metoxuron Monuron Parathion Pendimethalin Propazine	*Technique*: SPE-HPTLC *Enrichment column*: RP 18 **For HPTLC:** *Stationary phase*: 18 × 20-cm RP 18 F$_{254s}$ Plates *Mobile phase*: Methanol:water (70:30) *Mode of development*: Ascending *Detection*: UV detector *Sample*: Standard solution (10 ng/μl) of herbicides	SPE is a very useful tool in sample preparation particularly if an automated system is used.	92

Compounds	Technique/Conditions	Comments	Ref.
Sebuthylazine Simazine 2,4,5-T Terbuthylazine Trifluralin Vinclozolin			
1-Naphtalin-1-suphonic acid Procloraz Triazoxid Ethidimuron Simazine Bromazil Metribuzin Dinoseb 1-Hydroxyatrazin Formetanat Triadimenol Metalaxyl Isoproturon Diuron Dimethylaminosulphanilide Methidalthion 2,4-D-Isobutyl ester Ethalfluralin 2,2-Bis-(4-chloro-phenyl)-1,1-dichloroethane 2,4-DB 2,4,5-TP-methyl ester Oxamyl Cyanazin Propazin Sebuthylazin Terbuthylazin Chlotoluron Metobromuron Metoxuron Monuron MCPA 2,4,5-T 2,4-D Alachlor Metazachlor Metalachlor	*Technique:* SPE-TLC/HPTLC *Enrichment column:* Nonend-capped RP 18 cartridges **For TLC/HPTLC:** *Stationary phase:* Silica gel *Mobile phase:* Methanol-dichloromethane and *n*-hexane *Mode of development:* Ascending stepwise-gradient development *Detection:* UV multiwavelength detection *Sample:* The water sample was adjusted to pH 2 and passed through an SPE column. Pesticides were eluted with dichloromethane or methanol. The solvent was evaporated and then analyzed by TLC.	The automated multiple-development (AMD) technique has been accepted as a German standard for identification and quantification of active ingredients of plant-protecting chemicals present in ground-, raw, drinking, and mineral water. The AMD technique brings the advantages of stepwise-gradient development into the field of thin-layer chromatography, leading to an enormous increase in the selectivity of the chromatographic separation.	83, 96
2,4-D MCPA Oxamyl Chloroxuron Linuron	*Technique:* GC-HPLC Other conditions not available.	Herbicide concentrations measured in lake water samples do not immediately demonstrate worries about water drinkability, but, in any event, they show evidence of pollution risk that must be taken into consideration and prevented in time.	94, 95

Table 3 Continued

Fungicides/herbicides	Experimental conditions	Inferences	Refs.
Metobromuron Mitoxuron Monuron Azilazine Atrazine Propazine Simazine Terbutrine Terbutilazine Prometrine Ethion Propham			
Dichlorprop Bentazon	Not available	In this study, leaching tests conducted in field lysimeters for the purpose of pesticide registration are evaluated, particularly in terms of factors such as the effects of soil type, variability in leaching between replicate lysimeters, and simulation of worst-case scenarios. Dichlorprop and Bentazon, fairly mobile, were chosen as test compounds.	97
Atrazine	*Technique:* Real-time biospecific interaction analysis (BIA)–surface plasmon resonance (SPR) *Biosensor:* The sensor chip is a glass slide coated on one side with a thin gold film. A matrix of carboxymethylated dextran is covalently attached to it. The surface was then washed for 1 min with distilled water. The ester formed by activation reacts readily with unchanged amino groups present in the binding molecule. Atrazine derivative (carboamido-ethylamine-4-ethylamino-6-chloro-1,3,5-triazine) was spread on the sensor chip, incubated, washed with borate–ethanol solution, and stored in 20% aqueous ethanol at 4°C. *Sensogram:* BIA core from Pharmacia Biosensor AB, Uppsala, Sweden. *Sample:* A stock solution of 100 ppb in 10% aqueous ethanol was prepared and frozen at −20°C. Monoclonal antibodies (Mabs)	This method seems to be a promising approach for the detection of pesticide residues in drinking water. The lower limit of detection is 0.05 ppb for Atrazine, with a precision of less than 5% over the measuring range employing monoclonal antibodies (Mab clone 1). These authors have also developed a cholinesterase-disposable biosensor (99), cholinesterase-cobalt phthalocyanine–modified composite electrodes (100), thick-film carbon-based sensors and biosensors (101), disposable screen-printed electrodes (102), and ruthenized screen-printed choline oxidase-based biosensors for the detection and determination of pesticides in water.	90–103

against atrazine were mixed with the sample containing herbicide, and then it was flown over the surface when the interaction was completed (after 7 min). The concentration of the antibody in the surface layer gives a surface. The plasmon resonance (SPR) response is inversely related to the atrazine concentration in the sample.

Atrazine

Technique: Piezoelectric crystal biosensor

Crystal: Quartz piezoelectric crystals (AT-Cut, resonance frequency of 10 MHz) with gold electrodes deposited on their sides (Universal Sensor, New Orleans, LA)

Detection: Microprocessor-controlled piezoelectric crystal detector for gas-phase analysis

Immobilization of Atrazine derivative: The crystal surface was treated with 5% α-aminopropyltriethoxysilane (APTES) solution in acetone for 2 hr. It was dried at 100°C for 1 hr and immersed in a 2.5% glutaraldehyde (GA)–100 mM phosphate solution of pH 7 for 1 hr. Next it was air-dried. 500 nmoles of modified derivative of Atrazine (ethylaminoexanoic–dl acid–derivative) were deposited on the surface, incubated overnight, washed, and stored at 4°C.

Immobilization of Anti-Atrazine (MAbs): The gold electrodes were treated sequentially with the following solutions: 1.2 N NaOH for 20 min, washed with water, 1.2 N HCl for 5 min, concentrated HCl for 2 min, air-dried for 30 min. Then 3 μg/side Protein A were placed on the electrodes in a buffer, 50 mM acetate/phosphate of pH 5.5. After 30 min, 30 μg/side Protein A was deposited again. 20 min later 1.5 μg/side MAbs were placed on the surface, incubated for 2 hr, washed with water, dried, and stored at 4°C.

Sample: In the direct assay, the crystals with the immobilized Mabs were dipped in 25-ml Atrazine standard solutions prepared in distilled water for 30 min, washed, and air-dried for 30 min, and then the resonance was recorded.

In the indirect assay, the crystals with the Atrazine derivative immobilized were incubated in the presence of anti-Atrazine antibodies (4 μg) and samples of different concentrations of Atrazine. The competition between free and bound Atrazine for a limited number of IgG binding sites occurs, and the resulting frequency decrease is measured.

The piezoelectric crystal technology seems to be a promising tool for the determination of pesticides in water. The competitive assay gives more reliable results than the direct assay. The method is slow—the operation time for measurement is 1 hr. Most of the time is consumed in crystal drying. The operation time may be reduced by using flow system analysis.

104

(one of two tiny crystals of chromotropic acid sodium salt in 0.20 ml of concentrated sulphuric acid) was taken in the arm of the test tube. The test tube was stoppered, and the arm was connected to the suction pump. The test tube was heated at 180°C for 2 min. Violet color formation in the arm confirmed the presence of 2,4-D. This technique was successfully used to detect 2,4-D in formulations, leaves, soil, and water at ppm levels.

4. Phenolic/Carboxylic/Nitrogen Pesticides

A new and versatile technique (33,34), the pressure capillary spot test, was developed for the detection and semiquantitative determination of fungicides and herbicides containing a phenolic group, a carboxylic group, and nitrogen in river water. A capillary containing a cotton plug impregnated with p-dimethylamino-benzaldehyde and trichloroacetic acid was used as detector. The river water sample was dosed with the test solution. This sample (0.1 ml) was taken in a test tube, and the contents were heated to remove the solvent and all moisture. Then the detector was fitted in the mouth of the test tube, and it was heated at $180° \pm 2°C$ for 5 min. The vapors/fumes so produced were passed through the detector under pressure. The lower limit of detection (color) for indole-3-acetic acid, pyrogallol, resorcinol, and orcinol was found to be 0.1 (dark violet), 10 (dark red), 10 (dark red), and 20 μg (dark red), respectively.

5. Mancozeb

An already-known test (31) for dithiocarbamates was extended for the selective detection of Mancozeb (35) at microgram levels in water. A water sample dosed with the traces of Mancozeb was taken in a micro test tube and neutralized with 1 M acetic acid; then buffer solution of pH 6 (2 ml), the reagent (1 ml of 1% copper chloride in acetic acid), and chloroform (0.5 ml) were added to it. After thorough mixing, the colors of both the aqueous and the organic layers at room temperature (25°C) were recorded. A red-brown color in the organic layer indicated the presence of Mancozeb.

A new color reaction (36) based on the formation of black lead sulphide on treating Mancozeb with alkaline plumbite solution was developed. The lower limit of detection was found to be 0.45 μg. A drop of deep-well water containing Mancozeb was mixed with a drop of plumbite reagent in a micro test tube. At room temperature (250°C) a yellow color developed, which slowly changed to black. If the mixture was boiled, the

black color appeared quickly. The test was successfully applied for the detection of Mancozeb in river water, wastewater, pond water, and soil extracts.

REFERENCES

1. Royal Society of Chemistry. The Argochemicals Handbook. Unwin, London, 1983.
2. SL Chopra, JS Kanwar. Analytical Agricultural Chemistry. Kalyani Publications, New Delhi, India, 1991.
3. B Nath, L Candela, L Hens, JP Robinson, eds. Proceedings of the International Conference on Environmental Pollution. Vols. 1 and 2. European Centre for Pollution Research, London, 1993.
4. B Nath, L Lang, E Meszaros, JP Robinson, eds. Proceedings of the International Conference on Environmental Pollution. Vols. 1 and 2. European Centre for Pollution Research, London, 1996.
5. RK Trivedy, PK Goel, eds. Current Pollution Research in India. Environmental Publications, Karad, India, 1985.
6. AK Sharma, A Sharma, eds. Impact of Development of Science and Technology on Environment. Indian Science Congress Association, Calcutta, India, 1981.
7. CCS Haryana Agricultural University, Hisar, India: Indo-German Conference on the Impact of Modern Agriculture on the Environment. Abstracts, 1993.
8. SK Arora, M Singh, RP Agarwal. Recent Advances in Environmental Pollution and Management. Haryana Pollution Control Board and Haryana Agricultural University, Hisar, India, 1990.
9. RG Lewis, RF Moseman, DW Hodgson, eds. Analysis of Pesticide Residues in Human and Environmental Samples. U.S. Environmental Protection Agency, Washington, DC, 1979.
10. ASY Chau, BK Afghan, JW Robinson, eds. Analysis of Pesticides in Water. Vols. I, II, III. CRC Press, Boca Raton, FL, 1982.
11. RE Clement, PW Yang. Environmental analysis. Anal. Chem. 69:251R–287R, 1997.
12. ML Nollet, ed. Handbook of Food Analysis. Vols. 1, 2. Marcel Dekker, New York, 1996.
13. RT Meister, GL Berg, C Sine, S Meister, H Shepard, eds. Farm Chemicals Handbook. Meister, Willoughby, 1990.
14. KC Dhingra. Handbook of Pesticides. Small Industry Research Institute, Roop Nagar, Delhi, India, 1996.
15. CA Edwards, ed. Environmental Pollution by Pesticides. Plenum Press, New York, 1973.
16. HS Rathore. The synthesis, ion-exchange properties and analytical applications of titanium molybdates and stannic arsenates. Ph.D. dissertation, Aligarh Muslim University, Aligarh, India, 1971.

17. SK Saxena. Studies on the Analysis of Some Organic Pollutants in Water. Ph.D. dissertation, Aligarh Muslim University, Aligarh, India, 1989.

18. M Qureshi, HS Rathore, AM Sulaiman. A new sensitive and simple spot test for the detection of tertiary amine fungicides and herbicides in water. Water Res. 16:435–440, 1982.

19. SR Ahmad, HS Rathore, I Ali, SK Sharma. Extraction and recovery of some carboxylic acids by flexible polyurethane foam from water. J. Indian Chem. Soc. LXII:786–787, 1985.

20. SK Sharma, HS Rathore, I Ali, SR Ahmad. Adsorption of some carboxylic pollutants from saline water on alumina. IAWPC Tech. Annual XII:79–83, 1985.

21. SK Sharma, HS Rathore, I Ali, SR Ahmad. Adsorption of agrochemicals from saline water by charcoal. Indian J. Environ. Health 27(2):130–139, 1985.

22. HS Rathore, K Kumari, M Agarwal. Quantitative separation of citric acid from trichloroacetic acid on plates coated with calcium sulphate containing zinc oxide. J Liquid Chromatog. 8(7):1299–1317, 1985.

23. HS Rathore, HA Khan. Characterization of barium sulphate as TLC material for the separation of plant carboxylic acids. Chromatographia 23(6):432–434, 1987.

24. HS Rathore, I Ali, HA Khan. Quantitative separation of trichloroacetic acid from some carboxylic herbicides on BaSO$_4$–CaSO$_4$ coatings impregnated with coconut oil. J Planar Chromatog. 1:252–254, 1988.

25. HS Rathore, I Ali, S Gupta, T Begum. Chromatographic characteristics of calcium sulphate. J. Planar Chromatog. 2:119–127, 1989.

26. HS Rathore, SK Saxena. A comparative study of some carboxylic acid herbicides by normal-phase, sequential, and ion-pair reversed-phase thin-layer chromatography on calcium sulphate. J. Planar Chromatog. 2:387–390, 1989.

27. HS Rathore, SK Saxena, R Sharma. Chromatographic behavior of some herbicides on a mixture of silica gel and calcium sulphate. J. Planar Chromatog. 3:251–256, 1990.

28. HS Rathore, T Begum. Thin-layer chromatographic behavior of carbamate and related compounds. J. Chromatog. 643:321–329, 1993.

29. HS Rathore, T Begum. Thin-layer chromatographic methods for use in pesticide residue analysis. J. Chromatog. 643:271–290, 1993.

30. HS Rathore, T Begum. Chromatographic behavior of some carboxylic acid herbicides and plant growth regulators on impregnated papers. J. Planar Chromatog. 4:451–455, 1991.

31. HS Rathore, SK Saxena, T Begum. A selective fluorescence spot test for the detection of TCA in soil and water. J. Indian Chem. Soc. 69:798–799, 1992.

32. HS Rathore, HA Khan. A novel method for the detection and semiquantitative determination of trace level of 2,4-D and related compounds. Water Res. 23(7):899–905, 1989.

33. HS Rathore, I Ali, S Gupta, HA Khan. Pressure capillary spot test for the detection of nitrogen-containing pollutants in crops, vegetation and environment. Water Supply 6:343–346, 1988.

34. HS Rathore, S Gupta, HA Khan. Pressure capillary spot-test for the detection of pollutants in crops, vegetation and environment. Analytical Lett. 19(15,16):1545–1560, 1986.

35. HS Rathore, R Sharma, S Mital. Spot test analysis of pesticides: detection of carbaryl and mancozeb in water. Water, Air Soil Pollution 97:431–441, 1997.

36. HS Rathore, S Mital. Spot test analysis of pesticides: TLC detection of mancozeb. J. Planar Chromatog. 10:124–127, 1997.

37. F Feigl, V Anger. Spot test in Organic Analysis. 7th ed. Elsevier, Amsterdam, 1966, pp. 48–49.

38. RA Carpenter, RH Hollowell, KM Hill. Determination of the metabolites of the herbicide dimethyl tetrachloroterephthalate in drinking water by high-performance liquid chromatography with gas chromatography/mass spectrophotometry confirmation. Anal. Chem. 69:3314–3320, 1997.

39. SI Itoh, S Setsuda, S Naito. On the simultaneous determination of simazine, thiram and thiobencarb in water by solid phase extraction/high performance liquid chromatography. J. Japan Water Works Asso. 63(5):22–26, 1994.

40. SI Itoh, S Setsuda. On the stability of the mixture of 13 agricultural chemicals in solvent. . . . J. Japan Water Works Asso. 65(1):24–29, 1996.

41. S Setsuda, SI Itoh, S Naito. Determination of asulam and oxine-copper in water by using solid-phase extraction. Bull. Kanagawa P. H. Lab. 23:39–42, 1993.

42. S Setsuda, S Itoh, S Naito. Decomposition of thiram by chlorination. Bull. Kanagawa P. H. Lab. 22:41–44, 1992.

43. S Setsuda, SI Itoh, S Naito. On a pre-treatment for analysis of organo-chloric pesticides in water by using solid phase extraction. Bull. Kanagawa P. H. Lab. 22:25–28, 1990.

44. SI Itoh, S Setsuda. Determination of trichlorofon (DEP) in water by using solid phase-extraction/GC-MS technique. J. Japan Water Works Asso. 64(4):9–13, 1995.

45. H Uemura, SI Itoh, S Setsuda. Analysis of DDTs with solid phase extraction/GC-MS method. J. Japan Water Works Asso. 65(11):47–51, 1996.

46. SI Ihoh, H Uemura, S Setsuda. Determination of Agricultural Chemicals in water by using solid phase microextraction (SPME)/GC-MS Technique. J. Japan Water Works Asso. 65(12):10–17, 1996.

47. S Setsuda, SI Itoh, H Uemura. Simultaneous analysis of five pesticides in water. Bull. Kanagawa P. H. Lab. 26:39–42, 1996.

48. SI Itoh, H Uemura, S Setsuda. On the stability of the mixture of 13 agricultural chemicals in solvent (II). J. Japan Water Works Asso. 66(5):22–23, 1997.

49. MA Tisseu, N Fauchon, J Cavard, T Vandevelde. Pesticide contamination of water resources: a case study —the rivers in the Paris region. Water Sci. Tech. 34(7–8):147–152, 1996.

50. C Aquilar, F Borrull, RM Marce. Determination of pesticides by on-line trace enrichment–reversed-phase liquid chromatography–diode-array detection and confirmation by particle-beam mass spectrometry. Chromatographia 43(11/12):592–598, 1996.

51. C Aguilar, F Borrull, RM Marce. Identification of pesticides by liquid chromatography–particle beam mass spectrometry using electron ionization and chemical ionization. J. Chromatog. A 805:127–135, 1998.

52. C Aguilar, S Penalver, E Pocurull, F Borrull, RM Marce. Solid-phase microextraction and gas chromatography with mass spectrometric detection for the determination of pesticides in aqueous samples. J. Chromatog. A 795:105–115, 1998.

53. C Aguilar, I Ferrer, F Borrull, RM Marce, D Barcelo. Comparison of automated on-line solid-phase extraction followed by liquid chromatography–mass spectrometry with atmospheric pressure chemical ionization and particle beam mass spectrometry for the determination of a priority group of pesticides in environmental waters. J. Chromatog. A 794:147–163, 1998.

54. AN Shahane. Pesticides in the water. J. I. W. W. A. IX(3):209–219, 1977.

55. AN Shahane. Pesticide detection in surface waters of Florida. In: Toxic Substances and the Hydrologic Sciences. American Institute of Hydrology, Denver, 1994, pp. 408–416.

56. AN Shahane. Pesticide concentration frequencies in surface water and sediments of Florida. In: Water Resources at Risk. American Institute of Hydrology, Denver, 1995, pp. LL-95-LL-111.

57. AN Shahane. Water assessments at a citrus grove in South Florida. In: Hydrology and Hydrogeology of Urban and Urbanizing Areas. American Institute of Hydrology, Denver, 1996, p. 12.

58. C Aguilar, F Borull, RM Marce. Determination of pesticides in environmental waters by solid-phase extraction and gas chromatography with electron-capture and mass spectrometry detection. J. Chromatog. A 771:221–231, 1997.

59. C Aguilar, F Borull, RM Marce. On-line and off-line solid-phase extraction with styrene-divinylbenzene membrane extraction disks for determining pesticides in water by reversed-phase liquid chromatography diode-array detection. J. Chromatog. A 754:77–84, 1996.

60. E Pocurull, C Aguilar, F Borrull, RM Marce. On-line coupling of solid-phase extraction to gas chromatography with mass spectrometric detection to determine pesticides in water. J. Chromatog. A 818:85–93, 1998.

61. N Masque, RM Marce, F Borrull. Comparison of different sorbents for on-line solid-phase extraction of pesticides and phenolic compounds from natural water followed by liquid chromatography. J. Chromatog. A 793:257–263, 1998.

62. N Masque, M Galia, RM Marce, F Borull. New chemically modified polymeric resins for solid-phase extraction of pesticides and phenolic compounds from water. J. Chromatog. A 803:147–155, 1998.

63. E Pocurull, RM Marce, F Borrull, JL Bernal, L Toribo, ML Serna. On line solid-phase extraction coupled to supercritical fluid chromatography to determine phenol and nitrophenols in water. J. Chromatog. A 755:67–74, 1996.

64. N Masque, RM Marce, F Borrull. Chemical removal of humic substances interfering with the on-line solid-phase extraction–liquid chromatographic determination of polar water pollutants. Chromatographia 48(3/4):231–236, 1998.

65. N Masque, M Galia, RM Marce, F Borrull. Solid-phase extraction of phenols and pesticides in water with a modified polymeric resin. Analyst 122:425–428, 1997.

66. C Crespo, RM Marce, F Borull. Determination of various pesticides using membrane extraction discs and gas chromatography–mass spectrometry. J. Chromatog. A 670:135–144, 1994.

67. RM Marce, H Prosen, C Crespo, M Calull, F Borrull, UATh Brinkman. On-line trace enrichment of polar pesticides in environmental waters by reversed-phase liquid chromatography–diode array detection–particle beam mass spectrometry. J. Chromatog. A 696:63–74, 1995.

68. L Dabrowski, B Truchanowicz, A Zwir, M Biziuk, J Gaca. Analysis of some organic compounds in soil and sediments using alumina as a clean up medium. Adv. Chromatogr. Electrophor. Related Sep. Methods. In press, 1998.

69. M Wiergowski, L Dabrowski, K Galer, B Makuch, M Biziuk. Determination of pesticides and phenols in post-flood sediments and water. Acta Hydrochimica Hydrobiologia. In press, 1998.

70. M Biziuk, A Przyjazny, J Czerwinski, M Wiergowski. Review: occurrence and determination of pesticides in natural and treated waters. J. Chromatog. A 754:103–123, 1996.

71. M Biziuk, A Kot, B Makuch, Z Polkowska, L Wolska, D Gorlo, W Janieki, M Wiergowski, A Wasik, B Zygmunt, M Turska, J Namiesnik. New approach for determination of selected organic compounds in sea water of the southern Baltic and Vistula River. J. AOAC Int. In press, 1998.

72. M Wiergowski, M Biziuk, J Zywicka. Different meth-

ods of drying in procedure of solid phase extraction for determination of selected pesticides in water. Adv. Chromatogr. Electrophor. Related Sep. Methods. In press, 1998.

73. F Mazzei, F Botre, G Lorenti, G Simonetti, F Porcelli, G Scibona, C Botre. Plant tissue electrode for the determination of atrazine. Analytica Chemica Acta 361: 79–82, 1995.

74. F Mazzei, F Botre, C Botre. Acid phosphatase/glucose oxidase-based biosensors for the determination of pesticides. Analytica Chimica Acta 336:67–75, 1966.

75. F Botre, G Lorenti, F Mazzei, G Simonetti, F Porcelli, C Botre, G Scibona. Cholinesterase based bioreactor for determination of pesticides. Sensors Actuators B 18–19:689–693, 1994.

76. A Balinova. Extension of the bioautograph technique for multiresidue determination of fungicide residues in plants and water. Analytica Chimica Acta 311:423–427, 1995.

77. A Balinova. Solid-phase extraction followed by high-performance liquid chromatographic analysis for monitoring herbicides in drinking water. J. Chromatog. A 643:203–207, 1993.

78. A Balinova. Ion-pairing mechanism in the solid-phase extraction and reversed-phase high-performance liquid chromatographic determination of acidic herbicides in water. J. Chromatog. A 728:319–324, 1996.

79. A Balinova. Strategies for chromatographic analysis of pesticide residues in water. Review. J. Chromatog. 754:125–135, 1996.

80. AM Balinova. Analysis of fenoxaprop-ethyl and fenoxaprop in drinking water using solid-phase extraction and ion-pair HPLC. Pestic. Sci. 48:219–223, 1996.

81. AM Balinova, M Mondesky. Pesticide contamination of ground and surface water in Bulgarian Danube Plain. J. Environ. Sci. Health, in press, 1998.

82. ME Close, L Pang, JPC Watt, KW Vincent. Leaching of picloram, atrazine, and simazine through two New Zealand soils. Gee Geoderma 84:45–63, 1998.

83. M Close. Pesticides in New Zealand's groundwater. Proceedings of the First Ag Research/Landcare Research Pesticide Residue Workshop. Hamilton, 1995, pp. 47–52.

84. ME Close. Assessment of pesticide contamination of groundwater in New Zealand. 2. Results of groundwater sampling. New Zealand J. Marine Freshwater Res. 27:267–273, 1993.

85. ME Close. Assessment of pesticide contamination of groundwater in New Zealand. Ranking of regions for potential contamination. New Zealand J. Marine Freshwater Res. 27:257–266, 1993.

86. ME Close. Survey of pesticides in New Zealand groundwaters. New Zealand J. Marine Freshwater Res. 30:455–461, 1996.

87. A Di Guardo, D Calamari, G Zanin, A Consalter, D Mackay. A fugacity model of pesticide runoff to surface water: development and validation. Chemosphere 28(3):511–531, 1994.

88. A Di Guardo, RJ Williams, P Matthiessen, DN Brooke, D Calamari. Simulation of pesticide runoff at Rosemaund Farm (UK) using the soil fugacity model. Environ. Sci. Pollution Res. 1(3):151–160, 1994.

89. G Zanin, M Borin, L Altissimo, D Calasnari. Simulation of herbicide contamination of the aquifer north of Vicenza (north coast Italy). Chemosphere 26(5): 929–940, 1993.

90. R Barra, M Vighi, A Di Guardo. Prediction of surface water input of chloridazon and chlorpyrifos from an agricultural watershed in Chile. Chemosphere 30(3): 485–500, 1995.

91. M Vighi, D Sandroni, CS Floretti. Modelling herbicide fate in paddy fields. Proceedings of the X Symposium Pesticide Chemistry-Modelling Systems, 1998, pp. 449–456.

92. G Pfaab, H Jork. Application of AMD to the determination of crop-protection agents in drinking water. Part III: solid phase extraction and affecting factors. Acta hydrochim. hydrobiol. 22(5):216–223, 1994.

93. GE Morlock. Analysis of pesticide residues in drinking water by planar chromatography. J. Chromatog. A 754:423–430, 1996.

94. V Caffarelli, MR Rapagnani, A Correnti, G Cecchini, L Cirilli, R Ercoli, A Frugis, C Turco, P Di Luzio, G Ciampi, S Rizzo. Impacts of pesticides on water quality in the Braccciano Lake Basin: preliminary results. Proceedings of the X Symposium Pesticide Chemistry —Pesticide Residues in the Environment, Piacenza, Italy, 1996, pp. 491–498.

95. G Chemello, C Gaggi, V Caffarelli, E Bacci. Hazard assessment of carbaryl and endosulfan in the Lake Vico ecosystem (VT—Italy). Proceedings of the 80th Simposio Chimica degli Antiparassitari. Universita Calloica del Sacro Cuore, Italy, 1991, pp. 75–95.

96. GE Morlock. Organische Pflanzenbehandlungsmittel Prakische Erfahrung mit der DC—Bestimmung nach DIN 38407, Teil 11, in Chromatographic: Chronologie einer Analysentechnik, Praxis-status-Trends. Darmstadt, 1996, pp. 222–230.

97. LF Bergstrom, NJ Jarvis. Leaching of dichlorprop bentazon, ^{36}Cl in undistorbed field lysimeters of different agricultural soils. Weed Science 41:251–261, 1993.

98. M Minunni, M Mascini. Detection of pesticides in drinking water using real-time biospecific interaction analysis (BIA). Analyt. Lett. 26(7):1441–1460, 1993.

99. I Palchetti, A Cagnini, MD Carlo, C Coppi, M Mascini, APF Turner. Determination of anticholinesterase pesticides in real samples using a disposable biosensor. Analytica Chimica Acta 337:315–321, 1997.

100. P Skladal, M Mascini. Sensitive detection of pesticides using amperometric sensors based on cobalt phthalocyanine-modified composite electrodes and immobi-

lized cholinesterases. Biosensors Bioelectronics 7:
335–343, 1992.

101. A Cagnini, I Palchetti, MD Carlo, M Mascini. Thick-
film carbon-based sensors and biosensors. Conference
Proceedings. SIF, Bologna, 1996, 54:91–98.

102. MD Carlo, I Lionti, M Taccini, A Cagnini, M Mascini.
Disposable screen-printed electrodes for the immu-

nochemical detection of polychlorinated biphenyls.
Analytica Chimica Acta 342:189–197, 1997.

103. A Cagnini, I Palchetti, M Mascini, APF Turner. Mikrochim. Acta 121:155–166, 1995.

104. M Mununni, P Skladal, M Mascini. A piezoelectric
crystal biosensor for detection of atrazine. Life Chemistry Rep. 11:391–398, 1994.

30

Methods for the Determination of Polychlorobiphenyls (PCBs) in Water

Roger Fuoco and Alessio Ceccarini
University of Pisa, Pisa, Italy

I. INTRODUCTION

Polychlorobiphenyls (PCBs) are a class of nonpolar semivolatile organic compounds that includes 209 congeners divided into ten congener classes (Fig. 1) and named according to the IUPAC numbering from PCB1 to PCB209. They were synthesized at the end of the last century and commercialized under various trade names (e.g., Aroclor, Clophen, Phenoclor, Kanechlor, Fenclor) from around 1930. The PCBs have been massively used as hydraulic and heat transfer fluids, dielectric fluids in capacitors, plasticizers, adhesives, etc. (1–3). It has been estimated that about 1.2×10^6 tons were produced worldwide between 1930 and 1974. Of this, about one-third has been released into the environment, more than 60% is still in use or deposited in landfills, and only 4% has been destroyed or incinerated (3). Such PCB mixtures as Aroclor 1260 are also currently being released to the environment, both from landfills containing PCB waste materials and products and from open areas due to the illegal disposal of PCB materials, such as waste transformer fluid. This has led to the widespread occurrence of PCBs in the terrestrial ecosystem (1,4), even in remote areas (5). The PCBs are chemically and thermally very stable, and they are one of the most persistent widespread classes of environmental pollutants. The half-life of PCBs ranges from a few days to about 10 years for mono- to pentachlorobiphenyls, and it can be as high as 20 years for higher substituted congeners (1). Moreover, they are soluble in fatty and lipid-rich tissues and organs of biota, where they accumulate and may act as cancer initiators (1,2,6–8). The toxicity of a PCB mixture is associated mainly with the presence of nonortho and mono-ortho substituted coplanar congeners (9–12). Biological assay, based on PCB interaction with Ah-receptors, has shown that the most toxic congeners are PCB15, PCB37, PCB77, PCB81, PCB126, and PCB169 (8,13). There is also evidence supporting the hypothesis that PCBs may cause reproductive failure in animals (1). Many different analytical procedures have consequently been developed, and their diversity relates to the nature of the matrix to be analyzed (1,14,15).

In this chapter an overview of the most widely used analytical procedures for the determination of PCBs in water samples is given.

II. WATER REGULATION

There are very few regulations and laws dealing with the presence of PCBs in water, either at an international or a national level. In these few cases the total PCB concentration is always indicated, and no mention is made about specific congeners. In particular, the U.S. Environmental Protection Agency (USEPA) has fixed the maximum concentration level of PCBs in drinking water at 0.5 μg L^{-1} (16). Moreover, a recent Italian regulation fixed the PCB water quality criteria to be

LIVERPOOL JOHN MOORES UNIVERSITY
LEARNING SERVICES

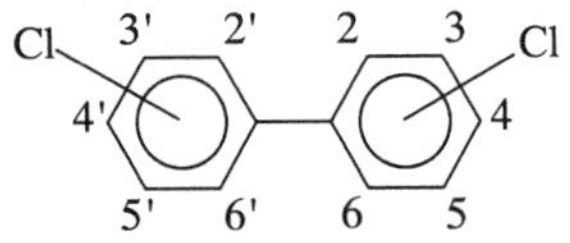

Congener class	Number of isomers	(IUPAC Nos.)
monochloro	3	PCB1 - PCB3
dichloro	12	PCB4 - PCB15
trichloro	24	PCB16 - PCB39
tetrachloro	42	PCB40 - PCB81
pentachloro	46	PCB82 - PCB127
hexachloro	42	PCB128 - PCB169
heptachloro	24	PCB170 - PCB193
octachloro	12	PCB194 - PCB205
nonachloro	3	PCB206 - PCB208
decachloro	1	PCB209
Total	209	

Fig. 1 Congener classes, number of isomers, and IUPAC numbering of PCBs.

reached in the Venetian Lagoon ecosystem at 0.04 ng L^{-1} (17). The Ministry of Agriculture, Fisheries and Food of the UK (MAFF) reported that there was a large decline in the estimated average UK dietary intake of PCBs from 1.0 mg/person/day in 1982 to 0.34 mg/person/day in 1992 (18). A tolerable daily intake level of 1 μg kg^{-1} body weight per day of total PCBs is also currently under review in Canada (19).

III. PHYSICAL AND CHEMICAL PROPERTIES

Polychlorobiphenyls are noninflammable and water-insoluble compounds. Their solubility decreases from about 6 to 0.08 mg L^{-1} for mono- and dichloro congener classes, respectively, and it ranges from 0.175 to 0.007 mg L^{-1} for all other classes (Table 1). They are chemically inert under acid and alkali conditions and very stable to oxidation (the thermal decomposition rises up to 1000°C) (2). They have high boiling points and low electrical conductivity. The boiling point of PCBs and their vapor pressures vary with the degree of chlorination and with the position of chlorine atoms in the biphenyl structure (Table 1). Both vapor pressure and octanol–water partition coefficients (K_{ow}) are used largely in the diffusion model of PCBs in the environment. The mean value of log K_{ow} varies quite linearly with the number of chlorine atoms from 4.1 to 9.6 (Table 1). Density and viscosity also increase with the degree of chlorination. Congeners with 1–4 chlorine atoms are oily fluids, pentachlorobiphenyls are honeylike oils, and the higher chlorinated PCBs are greases and waxy substances. Table 2 shows apparent color, distil-

lation range, molecular weight average, density, and viscosity for various Aroclor mixtures.

IV. ANALYTICAL METHODS

Whatever chemical species have to be determined, the first step is to define correctly the information needed. Then the analytical methods to obtain it should be chosen and tested. As for the latter, the following steps should be considered:

Sample collection and storage
Sample preparation
Instrumental analysis
Data evaluation, including analytical quality control

A. Sample Collection and Storage

Before planning a sampling program, all the data available on the system in question should be collected, including those relevant to chemical, physical, and biological parameters that may affect the concentration level of analytes in the sample. Water samples are generally collected by a Teflon or stainless steel pumping system, without any lubricant or oil in order to avoid any contamination. The sample volume may vary from a few to a thousand liters, depending on the expected concentration level (20–24). For large-volume sampling, a filtering system containing a suitable amount of adsorbing resin (20,21,24) is generally used. If required, the water sample is filtered on a 0.45-μm-pore-size membrane filter, and particulate matter is analyzed separately. If it is not possible to extract the samples immediately after sampling, they are generally stored at −20°C in stainless steel containers.

B. Sample Preparation

1. Extraction of Polychlorobiphenyls

Table 3 shows a selection of extraction procedures for PCB determination in water samples, including snow and ice. The PCBs are nonpolar compounds and consequently can be extracted from water samples with nonpolar immiscible solvents. N-Hexane (5) and dichloromethane (25–27) or pentane (22) are the most widely used solvents in liquid–liquid extraction techniques. Solid-phase extraction (SPE) and elution with different mixtures of solvents is also very common (28–30). XAD-2 (21,20,24,31–33), polyurethane foam (34), C_8-bonded silica (3), and C_{18}-bonded silica (32) are the most widely used adsorbing resins. The adsorb-

Table 1 Physical and Chemical Properties of PCBs

IUPAC no.	Compound	Boiling point (°C)	Vapor pressure (mmHg, 25°C)	Log K_{ow}	Solubility (mg L^{-1})
—	Biphenyl	255	9.5×10^{-3}	4.10	7.2
	Monochlorobiphenyls				
1	2	274	8.4×10^{-3}	4.56	5.9
2	3	284	1.5×10^{-3}	4.72	3.5
3	4	291	4.6×10^{-3}	4.69	1.19
	Dichlorobiphenyls				
4	2,2'		1×10^{-3}	5.02	1.50
5	2,3	172 (30)			
7	2,4		1.8×10^{-3}	5.15	1.40
8	2,4'			<5.32	1.88
9	2,5	171 (15)	1.4×10^{-3}	5.18	0.59
11	3,3'	322–324	6.8×10^{-4}	5.34	
12	3,4	195–200 (15)			
14	3,5	166 (10)			
15	4,4'	315–319	1.9×10^{-5}	5.28	0.08
	Trichlorobiphenyls				
18	2,2',5		9×10^{-5}	5.65	0.14
33	2',3,4		7.7×10^{-5}	6.1	0.078
28	2,4,4'			5.74	0.085
29	2,4,5		3.3×10^{-4}	5.77	0.092
30	2,4,6		8.8×10^{-4}		
31	2,4',5		3.0×10^{-4}	5.77	
33	2',3,4				0.078
37	3,4,4'			5.90	0.015
	Tetrachlorobiphenyls				
40	2,2',3,3'		7.3×10^{-5}	6.67	0.034
44	2,2',3,5'			6.67	0.170
47	2,2',4,4'		8.6×10^{-5}	6.44	0.068
52	2,2',5,5'		3.7×10^{-5}	6.26	0.046
53	2,2',5,6		2.1×10^{-4}		
54	2,2',6,6'			5.94	
60	2,3,4,4'				0.058
61	2,3,4,5			6.39	0.019
66	2,3',4,4'		4.6×10^{-5}	6.67	0.058
70	2,3',4',5		4.4×10^{-6}	6.39	0.041
77	3,3',4,4'		2.3×10^{-6}	6.52	0.175
80	3,3',5,5'			6.58	
	Pentachlorobiphenyls				
86	2,2',3,4,5		5.8×10^{-7}	6.38	0.0098
87	2,2',3,4,5'		1.6×10^{-5}	6.58	0.022
88	2,2',3,4,6			7.51	0.012
99	2,2',4,4',5		2.1×10^{-5}		
101	2,2',4,5,5'		9.0×10^{-6}	6.85	0.031
105	2,3,3',4,4'		6.8×10^{-6}		
116	2,3,4,5,6			6.85	0.0068
118	2,3',4,4',5	195–220 (10)	9.0×10^{-6}		
	Hexachlorobiphenyls				
128	2,2',3,3',4,4'		2.6×10^{-6}	7.44	0.00044
129	2,2',3,3',4,5			8.26	

Table 1 Continued

IUPAC no.	Compound	Boiling point (°C)	Vapor pressure (mmHg, 25°C)	Log K_{OW}	Solubility (mg L^{-1})
134	2,2',3,3',5,6			8.18	0.00091
138	2,2',3,4,4',5'		4.0×10^{-6}		
149	2,2',3,4',5',6		1.1×10^{-5}		
153	2,2',4,4',5,5'		5.2×10^{-6}	7.44	0.0088
					0.0013
155	2,2',4,4',6,6'		1.3×10^{-5}	7.12	0.00091
156	2,3,3',4,4',5		1.6×10^{-6}		
	Eptachlorobiphenyls				
170	2,2',3,3',4,4',5		6.3×10^{-7}		
171	2,2',3,3',4,4',6		1.8×10^{-6}		
180	2,2',3,4,4',5,5'	240–280	9.7×10^{-7}		
185	2,2',3,4,5,5',6			7.93	0.00048
187	2,2',3,4',5,5',6		2.3×10^{-6}		
	Octachlorobiphenyls				
194	2,2',3,3',4,4',5,5'			8.68	0.0070
					0.0014
202	2,2',3,3',5,5',6,6'			8.42	0.00018
	Nonachlorobiphenyls				
206	2,2',3,3',4,4',5,5'6			9.14	0.00011
209	Decachlorobiphenyl			9.60	0.015
					0.00049

ing material is generally supported inside a column or fixed on a membrane disk (32). Solid-phase extraction has several advantages: use in field applications, easy automation, low solvent consumption (35,36), and less critical cleanup of the eluate. Solid-phase microextraction (SPME) is a modified SPE procedure based on the use of a coated fiber, usually made of fused silica, which in many cases avoids the use of organic solvents, since it allows the direct transfer of the analytes in the chromatographic system to be performed (30,36). Chromatographic stationary phases, such as poly(methylsiloxane), are generally used as chemically bonded coatings of the fiber (37,38). However, solid-phase extraction has some drawbacks that might limit its application to water samples, such as low and irreproducible recoveries due to matrix effects, low capacity for samples that have a high content of organic matter, and the need for critical calibration procedures for

Table 2 Characteristics of Aroclor Mixtures

	Apparent color	Distillation range (°C)	Molecular weight average	Density (g mL^{-1} at 20°C)	Viscosity (Saybolt universal sec.) at 98.9°C
Aroclor 1260	Light yellow, soft, sticky resin	385–420	366–372	1.62	72–78
Aroclor 1254	Light yellow, viscous liquid	365–390	326.4–327	1.54	44–58
Aroclor 1248	Colorless mobile oil	340–375	291.9–288	1.44	36–37
Aroclor 1242	Colorless mobile oil	325–366	257.5–261	1.38	34–35
Aroclor 1016	Colorless mobile oil	323–356	—	1.37	—

Table 3 Extraction of PCBs from Water Samples

PCB congeners	Matrix	Extraction	Comments	Refs.
101, 136, 151, 118, 153, 138	Seawater	Liquid–solid	The water was passed through Amberlite XAD-2 and extracted from the resin with methanol followed by dichloromethane and hexane. The extracts were then transferred to hexane.	21
Congener classes	Seawater and snow	Liquid–solid	The water was passed through Amberlite XAD-2 resin columns. The PCBs were extracted from resin with ethanol.	20
Total	Seawater	Liquid–liquid	Seawater was filtered and extracted with pentane.	22
8, 18, 22, 26, 28, 31, 44, 49, 52, 70, 77, 101, 105, 110, 118, 126, 128, 138, 149, 153, 156, 157, 167, 169, 170, 180, 183, 187, 189, 194, 199	Seawater	Liquid–solid	Water was filtered on GF/C filters and the filtrate was passed through a 120-cm^3 column filled with Amberlite XAD-2 resin at a flow rate of 30 L h^{-1}. The analytes were extracted with 150 ml acetonitrile containing 15% water for 6 h in a modified Soxhlet apparatus.	24, 31
Total	Snow	Liquid–liquid	Extraction was performed in a continuous-flow, large-volume, liquid-phase extractor with about 150 ml of dichloromethane. The organic extracts were stored frozen and sent to the laboratory in 500-ml bottles. In laboratory the solutions were dried, and then iso-octane was added. The dichloromethane/iso-octane solutions were reduced to 1 ml.	25
Total	Melted snow, ice, and lake water	Liquid–solid	Suitable volume of water was passed on a polyuretane foam. Analytes were recovered with an hexane/methylene chloride mixture in a separatory funnel.	34
Total	Snow and seawater	Liquid–solid	2–3 L of seawater or melted snow were pulled at 0.8–1.3 L h^{-1} through 500 mg C$_8$-bonded silica cartridges. Analytes were eluted with 3 ml of 1:1 ethyl ether–hexane. Ethyl ether was eliminated by nitrogen stream and replaced with hexane.	3
Total	Wastewater	Liquid–liquid	About 1 L of water was extracted with three aliquots (60 ml) of methylene chloride in a separatory funnel. The extracts were passed through a column containing anhydrous sodium sulphate and collected into a Kuderna-Danish concentrator. 20–30 ml of methylene chloride were added into the column to complete the quantitative transfer. The solution was concentrated to about 1 ml.	26
Total	Water	Solid-phase microextraction	6.1 × 10^{-4} cm^3 of 100-μm poly(dimethilsiloxane) fiber was immersed in 100 ml of water. After 30 min of adsorption time, the fiber was inserted into a GC injector. The analytes were desorbed in 5 min at 260°C.	38
28, 52, 44, 70, 101, 118, 153, 138, 187, 128, 180, 170	Seawater	Liquid–solid	300 L of seawater was filtered on a kiln-fired GF/F filter (293 mm in diameter) at a flow rate of 500 L h^{-1}. 50 to 100 L of filtered seawater was passed on a column packed with 50 g of XAD-2 resin. PCBs were eluted from the XAD-2 resin with 200 ml of methanol followed by 200 ml of dichloromethane.	33

Table 3 Continued

PCB congeners	Matrix	Extraction	Comments	Refs.
28, 52, 101, 118, 153, 138, 180	Seawater	Liquid–solid	50–100 L of seawater were passed through a column filled with XAD-2 resin at a flow rate of 400 ml/min for 2–4 h. PCBs were eluted from the XAD-2 column with 200 ml of MeOH followed by 200 ml of dichloromethane. The methanol fraction was concentrated to half volume followed by extraction with 25 ml of n-hexane. The hexane extract was dried, combined with the dichloromethane one, and finally concentrated at 1–2 ml.	32
28, 52, 101, 118, 153, 138, 180	Seawater	Liquid–solid	5–15 L of seawater were passed through a C_{18} disk at a flow rate of 50 ml/min for 2–5 h. The C_{18} disk was then dried for 10 min and eluted with 40 ml of MeOH, spiked with octachloronaphthalene and perdeuterated pyrene, and concentrated to dryness in two steps: by rotary vacuum solvent evaporation and by means of a gentle stream of dry nitrogen. Dried extracts were diluted with 1 ml of n-hexane for fractionation purposes.	32
Congener classes	Lake water	Liquid–liquid	PCBs were extracted from filtered water samples using a continuous-flow liquid–liquid Goulden large-sample extractor with dichloromethane.	27
Total	Seawater	Liquid–liquid	20 L of seawater was extracted with two 30-ml aliquots of n-hexane.	5

MeOH = methanol.

quantitative determinations (36), particularly for SPME methods.

Ice and snow samples are generally allowed to melt in a clean laboratory, and extraction is undertaken as soon as they are melted, following the same procedures as those applied to water samples (25).

2. Cleanup of the Extract

Major matrix components or other trace organic compounds that are coextracted with PCBs might cause interference on the instrumental response; thus they should be eliminated by suitable cleanup procedures. Table 4 shows a selection of cleanup procedures for the determination of PCBs in organic extracts of water samples. The cleanup is generally performed by column chromatography on suitably activated silica (25), allumina (22,27,32,33), or Florisil (synthetic magnesium silicate) (5,26,34). The retention of analytes in the column should be checked by standard solutions in order to find out both the best solvent or mixture of solvents and the optimum volume to be used for selectively eluting PCBs and leaving interferents in the

column (35). N-Hexane and dichloromethane are the most widely used solvents. Moreover, specific treatments are very often used for eliminating specific interfering substances. For instance, activated copper powder with (16) or without mercury (21,32), and fuming sulphuric acid (3,20) are generally used for removing elemental sulphur and lipids, respectively.

C. Instrumental Analysis

Physical-chemical characteristics of the analytes, along with both the detection limit and the chromatographic resolution required, should be carefully considered in the choice of the analytical instrumentation. Then extraction and cleanup procedures should be optimized accordingly. In the case of complex matrices, high-resolution chromatographic techniques, equipped with a low detection limit, high selectivity, and a high-identification-power detector, should be used whenever possible. Capillary high-resolution gas chromatography (HRGC) coupled with a mass spectrometric detector (MS) has been widely applied to the determination of

Table 4 Cleanup of Water Sample Organic Extracts for PCB Analysis

PCB congeners		Comments	Refs.
Total	Seawater	The organic extract was dried with anhydrous Na_2SO_4, treated with activated Cu powder to remove sulphur, and then cleaned up by alumina and silica column chromatography for the separation of organochlorine pesticides.	21
Congener classes	Seawater	The ethanol eluate was transferred to hexane and the hexane was purified with 5% fuming H_2SO_4.	20
52, 101, 118, 153, 138, 180	Seawater	The organic extract was reduced in volume and treated with 0.01 N NaOH. PCBs in the organic phase were separated from pesticides by alumina column chromatography (1% Et_2O in hexane as eluent) and then on silica column (3% water in hexane).	22
8, 18, 22, 26, 28, 31, 44, 49, 52, 70, 77, 101, 105, 110, 118, 126, 128, 138, 149, 153, 156, 157, 167, 169, 170, 180, 183, 187, 189, 194, 199	Seawater	The organic extract (15% H_2O in acetonitrile) was concentrated and extracted three times with 10 ml portions of n-hexane. The hexane were dried on anhydrous Na_2SO_4 and cleaned up by HPLC (column: Nucleosil 100-5; eluent: 20% dichloromethane–80% pentane). The fraction containing PCBs was concentrated at 20–50 ml by a gentle N_2 stream at room temperature.	24, 31
Total	Seawater	The organic extract was cleaned up by normal-phase chromatography on activated silica gel. PCBs were eluted with hexane followed by a 1:1 dichloromethane/hexane rinse.	25
Total	Seawater	The organic extract (1:1 ethyl ether–hexane) was transferred in n-hexane, and the hexane solution was shaken with 18 M sulphuric acid.	3
Total	Wastewater	The organic extract was cleaned up on a column containing 20 g of Florisil. PCBs were eluted with 200 ml of 6% ethyl ether in hexane. Elemental sulphur was eliminated by shaking the eluate with mercury or activated copper powder.	26
28, 52, 44, 70, 101, 118, 153, 138, 187, 128, 180, 170	Seawater	The organic extract was dried with anhydrous Na_2SO_4 and reduced to 1–2 ml by rotary vacuum evaporation. The extract was then fractionated by column chromatography on 2 g of alumina, previously activated at 70°C. PCBs were separated from DDE and DDT by elution with 6 ml of n-hexane. Elemental sulphur was removed by shaking the eluate with activated copper.	33
28, 52, 44, 70, 101, 118, 153, 138, 187, 128, 180, 170	Seawater	The organic extract was concentrated by rotary evaporation to half volume, and it was extracted three times with 25 ml of n-hexane. The hexane extract was dried with anhydrous Na_2SO_4 and reduced by rotary vacuum evaporation to 1–2 ml. Then it was fractionated by column chromatography using 2 g of alumina, previously activated at 120°C. PCBs were eluted with 6 ml of n-hexane. The solution eluted was concentrated to dryness and reconstituted with 50 μl of iso-octane.	32, 33
Homolog concentrations	Lake water	The organic extract was passed on a column containing 2 g of anhydrous Na_2SO_4 (first layer), 5 g of 100% activated neutral alumina (second layer), and 1 g of anhydrous Na_2SO_4 (third layer). Analytes were eluted with 4 × 25 ml of 2% (v/v) dichloromethane in hexane and reduced to 3 ml.	27
Total	Seawater	The organic extracts were dried with anhydrous Na_2SO_4 and reduced to about 6 ml. The extracts were loaded on a Florisil column (about 1.5 g), and PCBs were eluted with 8 ml of n-hexane/dichlorometane (85:15) mixture. The eluate (2 ml) was concentrated at 100 μl for instrumental analysis.	5

PCBs in organic extracts of water samples. In fact, it allows the extremely high resolution of HRGC to be combined with the very high sensitivity and identification power of mass spectrometry, which makes it possible to determine an analyte at low pg/μl levels in the final organic extract. The electron capture detector (ECD) is also very often used for PCB determination, since it is the least expensive high-sensitivity detector for chlorinated compounds.

1. High-Resolution Gas Chromatography (HRGC)

The PCBs can be separated on a 30–50-meter fused-silica capillary column with various chemically bonded stationary phases. The most widely used one is 95% dimethyl–5% phenyl polysiloxane, although 100% dimethyl polysiloxane (39,40,2), 14% cyanopropyl–1% vinyl–85% methyl-polysiloxane (41–43), bis-cyanopropyl-phenyl-polysiloxane (43), 1,2-dicarba-closo-dodecarborane dimethyl-polysiloxane (43), 50% cyanopropyl-methyl, 50% phenylmehyl-polysiloxane (24,44), and 50% dimethyl–50% phenyl-polysiloxane are also utilized (45). On-column injection is very often used, while several oven temperature programs have been applied to PCB analysis. The initial temperature is generally 10–15°C lower than the boiling point of the solvent if a cold on-column injector is used, and the final temperature does not exceed 290–300°C. Table 5 shows numerous combinations of column lengths, stationary phases, oven temperature programs, and detectors. For low-polarity stationary phases, the boiling point is the major retention factor, so the retention time of PCBs generally increases with increasing chlorine content, in agreement with basic gas chromatography theory. For stationary phases with a higher polarity, several low-chlorinated PCBs are retained in the column stronger than high-chlorinated ones. The effect of the stationary phase polarity is more evident for compounds with none or one ortho-chlorine substituted. No single column is able to separate all of the approximately 140 congeners present in technical mixtures, so at least two columns should be used (43). A complete PCB analysis has been proposed that entails the simultaneous injection of the sample (a glass t-split was used) into two chromatographic columns with different polarities (42). Multidimensional GC techniques have also been suggested as a tool for the effective separation of PCBs that cannot be separated on a single column. Two capillary columns are arranged in series such that the second column receives only small preselected fractions eluting from the first one (44).

2. Electron Capture Detector (ECD)

The electron capture detector has a very high sensitivity and selectivity toward halogenated compounds, so it has been the most widely used detector for gas chromatographic analysis of PCBs. However, it is subject to positive interference from many organic compounds that might be present in a water sample, such as phthalate esters, organochlorine pesticides, and chloroaromatics (46). It is also subject to other types of interferences that do not give specific signals, such as elemental sulphur (1). Other disadvantages of this detector are the high variability of the response factor, also within the same PCB congener class, and the large variability of the sensitivity within the same day (47,48). A detection limit of 0.05–0.5 pg of each congener injected in the GC can be obtained.

3. Mass Spectrometric Detector (MS)

The mass spectrometric detector is very useful to detect PCBs, since they generally have a very intense molecular ion, along with a typical chorine cluster associated with the two naturally occurring chlorine isotopes (^{35}Cl and ^{37}Cl). It does not have many of the drawbacks of ECD, and can very easily be interfaced to a GC. At present, due to their relatively low cost, robustness, and ease of operation, quadrupole and ion-trap-based instruments are the most popular and are included in the basic instrumentation of most public and private analytical laboratories. Mass spectrometry may be divided into the following categories, according to the ionization process and the polarity of the detected ions:

Electron impact ionization (positive-ion detection) (EIMS)
Chemical ionization (positive-ion detection) (CIMS)
Chemical ionization (negative-ion detection) (NCIMS)

Since PCBs produce a high abundance of molecular ions by electron impact, CIMS does not give any further improvement, so only EIMS and NCIMS will be discussed.

Electron impact mass spectrometry is the most popular MS ionization technique, and it is routinely applied for PCB analysis (49,50). It can be used either in the full-scan mode (observation of the full mass range), from which specific ions can be extracted, or in the selected-ion monitoring (SIM) mode (observation of a few selected ions).

The EI mass spectra of PCBs are characterized by the cluster of the chorine isotopic distribution (i.e., 75.8% ^{35}Cl and 24.2% ^{37}Cl), which is very useful for

Table 5 HRGC Experimental Conditions for PCB Analysis

Stationary-phase composition (nomenclature)	Column length × internal diameter (film thickness)	Temperature program	Detector	Refs.
Bis-cyanopropyl phenyl polysiloxane (Sil-88)	50 m × 0.25 mm (0.20 μm)	2 min at 90°C, then 20°C/min to 150°C, isothermal 7.5 min, then at 3°C/min to 240°C	ECD/MS	43
95% Dimethyl–5% phenyl polysiloxane (DB-5)	60 m × 0.25 mm (0.25 μm)	15°C/min from 60°C to 100°C, then 6°C/min to 300°C, hold for 10 min	FID/MS	32, 33, 65
95% Dimethyl–5% phenyl polysiloxane (SE-54)	25 m × 0.32 mm (0.25 μm)	4°C/min from 140 to 250°C	FID	24, 44
50% Cyanopropyl–methyl 50% phenylmethyl polysiloxane (OV210) (multidimention GC)	30 m × 0.32 mm (0.25 μm)	20 min at 160°C, then 4°C/min to 230°C		
95% Dimethyl–5% phenyl polysiloxane (CP-Sil 8)	50 m × 0.22 mm (0.2 μm)	1 min at 80°C, then 3°C/min to 270°C	ECD	22, 66
95% Dimethyl–5% phenyl polysiloxane (DB-5)	60 m × 0.25 mm (0.11 μm)	1 min at 90°C, then 25°C/min to 180°C, hold for 2 min, then 1.5°C/min to 220°C, hold for 2 min, then 3°C/min to 275°C	ECD	42
95% Dimethyl–5% phenyl polysiloxane (SE-54)	25 m × 0.2 mm (0.11 μm)	2 min at 35°C, then 8°C/min to 300°C, hold 5 min	FPD	67
95% Dimethyl–5% phenyl polysiloxane (SE-54)	30 m × 0.25 mm (0.15 μm)	2 min at 75°C, then 40°C/min to 120°C and 2°C/min to 240°C		68
95% Dimethyl–5% phenyl polysiloxane (SE-54)	60 m × 0.25 mm (0.15 μm)	2 min at 110°C, then 10°C/min to 180°C, hold 8 min, and 4°C/min to 220°C, hold for 5 min, and 4°C/min to 270°C		21
95% Dimethyl–5% phenyl polysiloxane (BP-5)	25 m × 0.32 mm (0.25 μm)	0.8 min at 50°C, then 30°C/min to 140°C, hold 2 min, then at 5°C/min to 280°C		69
95% Dimethyl–5% phenyl polysiloxane (CP-Sil 8CB)	50 m × 0.25 mm (0.25 μm)	2 min at 60°C, then 15°C/min to 180°C, hold for 6 min, then 4°C/min to 220°C, hold for 2 min, then 5°C/min to 280°C, hold for 25 min	MS	5, 12, 70
95% Dimethyl–5% phenyl polysiloxane (CP-Sil 8CB)	50 m × 0.15 mm (0.25 μm)	3 min at 90°C, then 30°C/min to 215°C		71
14% Cyanopropyl–1% vinyl–85% methyl polysiloxane (CP-Sil 19CB)	50 m × 0.15 mm (0.20 μm)	5°C/min to 270°C, hold for 24 min		
95% Dimethyl–5% phenyl polysiloxane (DB-5)	30 m × 0.32 mm (0.25 μm)	1 min at 50°C, then 4°C/min to 150°C, then 2°C/min to 280°C	ECD	72
95% Dimethyl–5% phenyl polysiloxane (DB-5)	60 m × 0.32 mm (0.25 μm)	2 min at 180°C, then 15°C/min to 205°C, then 2°C/min to 300°C	LRMS-SIR	73
95% Dimethyl–5% phenyl polysiloxane (DB-5)	30 m × 0.25 mm (0.25 μm)	1 min at 100°C, then 5°C/min to 140°C, hold for 1 min, then 1.5°C/min to 250°C, hold for 1 min, and then 10°C/min to 300°C	MS	74

Table 5 Continued

Stationary-phase composition (nomenclature)	Column length × internal diameter (film thickness)	Temperature program	Detector	Refs.
95% Dimethyl–5% phenyl polysiloxane (DB-5)	60 m × 0.25 mm (0.25 μm)	1 min at 60°C, then 50°C/min to 170°C, then 3°C/min to 200, then 4°C/min to 250°C	ECD	75
95% Dimethyl–5% phenyl polysiloxane (DB-5)	30 m × 0.25 mm (0.25 μm)	100°C to 280°C over 170 min	ECD	27
95% Dimethyl–5% phenyl polysiloxane (DB-5)	30 m × 0.25 mm (0.25 μm)	1 min at 120°C, 23°C/min to 200°C, 3°C/min to 270°C		76
95% Dimethyl–5% phenyl polysiloxane (DB-5)	30 m × 0.25 mm (0.25 μm)	1 min at 80°C, 10°C/min to 150°C, 2°C/min to 250°C		77
95% Dimethyl–5% phenyl polysiloxane (DB-5)	30 m × 0.25 mm (0.25 μm)	0.5 min at 100°C, ramp to 180°C at 20°C/min, 3°C/min to 250°C, 20°C/min to 300°C		78
95% Dimethyl–5% phenyl polysiloxane (DB-5)	30 m × 0.25 mm (0.25 μm)	15°C/min from 60°C to 100°C, 6°C/min to 300°C	ECD/MS	65
95% Dimethyl–5% phenyl polysiloxane (PTE-5)	60 m × 0.32 mm (0.25 μm)	2 min at 180°C, then 15°C/min to 205°C and at 2°C/min to 300°C	LRMS-SIR	73
95% Dimethyl–5% diphenyl polysiloxane (SPB-5)	30 m × 0.2 mm (0.25 μm)	10 min at 100°C, then 5°C/min to 280°C	ECD/MS	79
95% Dimethyl–5% diphenyl polysiloxane (Sil-8)	50 m × 0.25 mm (0.26 μm)	2 min at 90°C, then 20°C/min to 170°C, isothermal 7.5 min, then at 3°C/min to 280°C	ECD/MS	43
95% Dimethyl–5% phenyl polysiloxane (HP-5)	25 m × 0.2 mm (0.33 μm)	1 min at 90°C, then 15°C/min to 220°C and 5°C/min to 300°C	ECD	3
50% Dimethyl–50% phenyl polysiloxane (DB-17)	30 m × 0.32 mm (0.25 μm)	1 min at 50°C, then 4°C/min to 150°C, then 2°C/min to 280°C	ECD	80
14% Cyanopropylphenyl–1% vinyl dimethyl polysiloxane (Sil-19)	50 m × 0.25 mm (0.26 μm)	2 min at 90°C, then 20°C/min to 170°C (only 150°C for Sil-88), hold for 7.5 min, then at 3°C/min to 280°C	ECD/MS	43
14% Cyanopropyl–1% vinyl–85% methyl polysiloxane (DB-1701)	60 m × 0.25 mm (0.15 μm)	1 min at 90°C, then 25°C/min to 180°C, hold for 2 min, then 1.5°C/min to 220°C, hold for 2 min, then 3°C/min to 275°C	ECD	42
100% Dimethyl polysiloxane (Sil-5)	50 m × 0.25 mm (0.15 μm)	2 min at 90°C, then 20°C/min to 170°C, hold for 7.5 min, then at 3°C/min to 280°C	ECD/MS	43
100% Dimethyl polysiloxane (OV-1)	30 m × 0.25 mm (0.1 μm)	1 min at 50°C, then ballistically to 120°C and at 2°C/min to 240°C		81
100% Dimethyl polysiloxane (CP-Sil 5)	25 m × 0.32 mm (0.12 μm)	From 150°C to 200°C at 3°C/min		39
100% Dimethyl polysiloxane (DB-1)	50 m × 0.22 mm (0.25 μm)	1 min at 80°C, then 15/min to 300°C, hold for 15 min		82
100% Dimethyl polysiloxane (HP-1)	50 m × 0.25 mm (0.25 μm)	2 min at 75°C, then 30°C/min to 120°C and 10°C/min to 275°C		83
1,2-Dicarba-cloro-dodecarborane dimethylpolysiloxane (HT-5)	25 m × 0.25 mm (0.1 μm)	2 min at 90°C, then 20°C/min to 200°C, hold for 7.5 min and 3°C/min to 280°C, hold for 20 min	ECD/MS	43

LRMS-SIR: low-resolution mass spectrometer–selected ion recording; ECD: electron capture detector; MS: mass spectrometer; FT-IR: Fourier transform infrared spectroscopy; FID: flame ionization detector; FPD: flame photometric detector.

the identification of chlorinated species. The most abundant fragments are obtained by chlorine elimination; the odd-electron species are favored (i.e., $[M]^+$, $[M - Cl_2]^+$, $[M - Cl_4]^+$) (2). Asymmetrically substituted ortho-chloro PCBs only exhibit the $[M - 35]^+$ fragment ion. For less chlorinated isomers, the loss of HCl is also observed. Tables 6 and 7 show the mean relative response of every congener class and the relative abundance of $[M]^+$, $[M - 35]^+$, and $[M - 70]^+$ ions for many PCB congeners, and the relative intensity of $[M]^+$, $[M + 2]^+$, and $[M + 4]^+$ ions for every PCB congener class, respectively (1,2). The detection limit of EIMS in the SIM mode is in the range of a few picograms of each congener injected in the GC.

Negative chemical ionization is one of the soft ionization techniques, which produce fewer fragments, and favors the molecular ion. The NCI technique generates relatively simple mass spectra that may be affected by the physical and geometrical parameters of the ion source (51–54), including the temperature and the reagent pressure of the ion source. The presence of water and oxygen might also affect the ionization process (55,56). The NCI mass spectra of mono-, di-, and trichlorobiphenyls are dominated by m/z 35 and 37, whereas the molecular ion $M^{\cdot-}$ is the most abundant one in those of PCBs with more than four chlorine atoms. This is attributed to the stabilization effect of a negative charge due to the higher number of chlorine atoms (57). The detection limit of NCI-MS in the selected-ion monitoring (SIM) is in the range of 0.05–0.1 pg of each congener with more than four chlorines injected into the GC.

V. DATA EVALUATION

A. Identification and Quantification of Polychlorobiphenyls

The major drawback to comparing analytical data from different bibliographic sources is related mainly to the method of PCB quantification, since there is no widely accepted standard procedure. The following have been the most commonly used:

1. Individual concentration of all identified congeners.
2. Individual concentration of only a limited number of selected congeners. The following seven congeners are very often measured: PCB28, PCB52, PCB101, PCB118, PCB138, PCB153, and PCB180. A good estimate of the total PCB content in natural water samples may be obtained by multiplying the sum of these seven congeners by a factor of 4 (58,59).
3. Total PCB concentration expressed as an equivalent quantity of a commercial mixture, such as Aroclor or Clophen, taken as a reference. The concentration is calculated on the basis of the area of the most representative chromatographic peaks.

In the first case, PCB congeners are initially identified by GC/MS on a standard solution of several Aroclors (i.e., 1221, 1232, 1248, and 1260). The relative retention time (RRT) for each identified congener is then calculated by using one or more internal standards (ISs). The RRTs are then applied for the chromatographic peak assignment of real samples, which can be analyzed by either GC/MS or GC/ECD. Experimental response factors (RFs) are generally obtained for a limited number of selected PCB congeners (i.e., IUPAC Nos. 13, 28, 35, 52, 81, 101, 118, 127, 138, 153 169, 180) and for the IS, in a suitably selected concentration range. The relative response factors (RRFs) to the IS are then calculated and used in turn to calculate the RRFs for all congeners by extrapolating the values reported in the literature (60). The final extracts of real samples are analyzed after adding a known amount of an IS, and quantification can be performed by using commercial computer programs, which automatically assign chromatographic peaks on the basis of RRTs and calculate the concentration of each congener on the basis of the RRFs and the IS concentration/peak area ratio. If an MS detector in the SIM mode is used, deuterated PCBs are generally used as internal standards, and at least two ions should be selected for each analyte: one as target and one as qualifier. Finally, it is also worth mentioning that the determination of nonortho and mono-ortho substituted PCBs is often performed, since they are the most toxic congeners.

B. Determination of Coplanar Nonortho Substituted Polychlorobiphenyls

The toxicity of a PCB mixture is associated principally with the presence of isomers having a planar configuration, which facilitates the process of penetration of cellular membranes of living organisms. This configuration is typical for those isomers that don't have any chlorine atoms in the four ortho positions available on the biphenyl structure. Biological assays (8) based on PCB interactions with Ah-receptors, showed that the

Table 6 Mean Relative Response of PCB Homologue Classes and Relative Abundance of Molecule and Fragment Ions of PCB Congeners in EI-SIM Mass Spectra

IUPAC no.	Compound	Relative abundance			Mean relative response[a]
		$[M]^+$	$[M-35]^+$	$[M-70]^+$	
	Monochlorobiphenyls				3.331 (0.643)
1	2-	100	12	—	
2	3-	100	10	—	
3	4-	100	12	—	
	Dichlorobiphenyls				2.027 (0.447)
4	2,2'-	100	2.8	79	
7	2,4-	100	1.5	39	
10	2,6-	100	2.4	35	
11	3,3'-	100	6.1	38	
12	3,4-	100	6.7	35	
15	4,4'-	100	1.4	38	
	Trichlorobiphenyls				1.573 (0.341)
28	2,4,4'-	100	1	35	
30	2,4,6-	100	1	36	
	Tetrachlorobiphenyls				0.951 (0.175)
47	2,2',4,4'-	100	4.0	56	
52	2,2',5,5'-	100	13	71	
54	2,2',6,6'-	100	2.0	75	
61	2,3,4,5-	100	1.0	38	
65	2,3,5,6-	100	1.5	44	
66	2,3',4,4'-	100	0.5	35	
77	3,3',4,4'-	100	0.5	30	
80	3,3',5,5'-	100	1.0	33	
	Pentachlorobiphenyls				0.720 (0.120)
	Hexachlorobiphenyls				0.514 (0.079)
133	2,2',3,3',5,5'-	100	2	54	
138	2,2',3,4,4',5-	100	5	36	
153	2,2',4,4',5,5'-	100	2	52	
155	2,2',4,4',6,6'-	100	1	56	
169	3,3',4,4',5,5'-	100	1	31	
	Eptachlorobiphenyls				0.361 (0.024)
	Octachlorobiphenyls				0.253 (0.030)
194	2,2',3,3',4,4',5,5'-	100	6	53	
197	2,2',3,3',4,4',6,6'-	100	3	38	
202	2,2',3,3',5,5',6,6'	100	8	45	
	Nonachlorobiphenyls				0.230 (0.034)
209	Decachlorobiphenyl	100	1	65	0.213

[a]Relative to d_6-3,3',4,4'-tetrachlorobiphenyls.

Table 7 Relative Intensity of Molecular Ions of PCB Congener Classes According to ^{35}Cl and ^{37}Cl Isotope Abundance in Mass Spectra Obtained by EI-MS-SIM

Congener class	Primary	Relative intensity	Secondary	Relative intensity	Tertiary	Relative intensity
Mono	188	100	190	33	—	—
Di	222	100	224	66	226	11
Tri	256	100	258	99	260	33
Tetra	292	100	290	76	294	49
Penta	326	100	328	66	324	61
Hexa	360	100	362	82	364	36
Epta	394	100	396	98	398	54
Octa	430	100	432	66	428	87
Nona	464	100	466	76	462	76
Deca	498	100	500	87	496	68

most toxic congeners are PCB15, PCB37, PCB77, PCB81, PCB126, and PCB169. Table 8 shows the IUPAC number and the name of these PCBs, along with their relative toxicity factor (RTF). The RTFs were evaluated against 2,3,7,8-tetrachlorodibenzo-*p*-dioxin, whose value is assumed to be 100. If we consider that the concentration ratio of PCBs to dioxins in biota is generally higher than 50 and can reach values as high as 10,000, it follows that the global toxic effect of PCBs may be much higher than that of dioxins, even for those congeners whose RTF is 0.1. All these considerations support the need for reliable analytical procedures to determine the contents of planar and nonplanar congeners. Since the ratio between ortho-PCBs and nonortho-PCBs is typically 100 in commercial mixtures as well as in environmental matrices, and may be as high as 10,000, the concentration of nonortho-PCBs in the final extract is generally lower than the detection limit of GC/MS. On the other hand, peak overlapping between ortho and nonortho-PCB isomers may occur. This makes the determination of nonortho-PCBs impossible without a preseparation step. High-performance LC on a porous graphitic carbon (PGC) column with hexane as the eluent has been suggested for this purpose (12). In particular, an aliquot of the eluate obtained after the cleanup was injected into the HPLC for fractioning. The first fraction, which was collected from 0.8 to 6.4 min, contained all the ortho-PCBs, whereas the second one was collected from 6.5 to 23 min and contained all the nonortho-PCBs. Both fractions were reduced in volume and analyzed by either GS-MS or GC-ECD.

C. Analytical Quality Control

Whenever complex analytical methods are used, reliable data can be obtained only if a suitable program for analytical quality control and quality assurance is run in the laboratory. The main goal of any analytical quality control procedure is to obtain data within assigned values of accuracy and precision. Analytical quality control is achieved primarily by the use of certified reference materials and participation in intercomparison exercises, though calibration solutions and spiked samples can also be used (61,62).

Certified reference materials (CRMs) are the most useful tool for analytical quality control (61,63). However, spiked samples might be an alternative in many cases, though it must be remembered that spiked analytes generally behave differently from native ones (64). For a correct use of CRMs, their analysis should be scheduled within the time sequence of the analysis of real samples, and the results should be reported, for example, on a working analytical control chart. Participation in intercomparison exercises is also very useful. In fact, it is a unique opportunity for a laboratory to assess the quality of its analytical capability. Finally, calibration solutions are very useful to test an analytical procedure, but their preparation and storage are still among the main sources of error in these analyses (61). Certified analytes in neat form (purity higher than 99%) should be preferred for preparing calibration solutions following suitable procedures (61), although commercial standard solutions in various solvents are in fact often used. For the sake of comparison, Table 9 shows the relative response factors to PCB209 of a few PCBs:

Table 8   Relative Toxicity Factors (RTFs) to 2,3,7,8-Tetrachlorodibenzo-*p*-dioxin (RTF = 100) of Nonortho Chlorine Substituted PCBs and Coeluted Ortho-PCBs in GC on a 95% Dimethyl–5% Diphenyl Polysiloxane

IUPAC no. (nonortho PCBs)	Compound	RTF	IUPAC no. (coeluted ortho PCBs)
PCB 15	4,4'-dichlorobiphenyl	0.1	PCB 17
PCB 37	3,4,4'- trichlorobiphenyl	0.1	PCB 59
PCB 77	3,3',4,4'-tetrachlorobiphenyl	1.0	PCB 110
PCB 81	3,4,4',5- tetrachlorobiphenyl	0.1	PCB 87
PCB 126	3,3',4,4',5- pentachlorobiphenyl	10	PCB 178
PCB 169	3,3',4,4',5,5'- hexachlorobiphenyl	5.0	—

PCB28, PCB52, PCB101, PCB118, PCB138, PCB153, PCB156, and PCB180. The analysis was performed by GC-ECD on three commercial standard solutions and another one was prepared from analytes in neat form. The following experimental conditions were used: fused-silica capillary column (stationary phase: 95% dimethyl–5% phenyl polysiloxane, length 50 m, internal diameter 0.25 mm, film thickness 0.25 μm), temperature program: 2 min at 60°C, then 15°C/min to 180°C, hold for 6 min, then 4°C/min to 220°C, hold for 2 min, then 5°C/min to 280°C, hold for 25 min. The results obtained showed that the response factor varied

Table 9 Comparison of Relative Response Factors (RRFs) (Relative to PCB 209) of PCB Congeners in Different Standard Solutions

IUPAC no.	A	B	C	D	RRF	s.d.
PCB 28	0.43	0.42	0.40	—	0.42	0.02
PCB 52	0.37	0.35	0.34	—	0.35	0.02
PCB 101	0.56	0.51	0.54	—	0.54	0.03
PCB 105	0.81	—	—	0.78	—	—
PCB 118	0.68	0.66	0.65	0.67	0.67	0.02
PCB 138	0.79	0.75	0.76	—	0.77	0.02
PCB 153	0.69	0.66	0.71	—	0.69	0.03
PCB 156	—	—	1.00	0.98	—	—
PCB 180	1.07	1.10	1.04	—	1.07	0.03

Concentrations were always in the range $7-10$ ng g^{-1}; standard deviation (s.d.) was calculated on five replicate measurements.
A = Supelco (USA), CRADA PCBs mixture in isooctane; B = BCR (EC), CRM365 PCBs mixture in isooctane; C = Accustandard, Inc. (USA); D = Accustandard, Inc. (USA).

by a factor of about 3, and the relative standard deviation was typically better than 5%.

REFERENCES

1. Erickson M.D. Analytical Chemistry of PCBs (Butterworth, Stoneham, MA, 1986).
2. Hutzinger O., Safe S., Zitko V. The Chemistry of PCBs (CRC Press, Cleveland, OH, 1974).
3. Bidleman T.F., Patton G.W., Hinckley D.A., Walla M.D., Cotham W.E., Hargrave B.T. In: Long-Range Transport of Pesticides, chap. 23, p. 347 (Kurtz D.A., ed., Lewis, Chelsea, MI, 1990).
4. Fowler S.W. In: PCBs and the Environment, chap. 8, p. 209 (Waid, J.S., ed., CRC Press, Boca Raton, FL, 1986, vol. III).
5. Fuoco R., Colombini M.P. Microchemical J. 51 (1995) 106.
6. Lang V. J. Chromatogr. 595 (1992) 1.
7. Borlakoglu J.T., Dils R.R. Chem. Brit. (Sept. 1991) 815.
8. Tillitt D.E., Giesy J.P., Ankley G.T. Environ. Sci. Technol. 25 (1991) 87.
9. Mes J., Weber D. Chemosphere 19 (1989) 1357.
10. Hong C.S., Bush B. Chemosphere 21 (1990) 173.
11. Haglund P., Asplund L., Janberg U., Jansson B. Chemosphere 20 (1990) 887.
12. Fuoco R., Colombini M.P., Samcova E. Chromatographia 36 (1993) 65.
13. De Voogt P., Wells D.E., Reutergårdh L., Brinkman U.A.Th. In: Environmental Analytical Chemistry of PCBs/Current Topics in Environmental and Toxicological Chemistry, p. 151 (Albaiges J., ed., Gordon and Breach Science, Amsterdam, 1993).
14. Fuoco R., Griffiths P.R. Ann. Chim. (Rome) 82 (1992) 235.
15. Fuoco R., Colombini M.P., Abete C. Int J. Environ. Anal. Chem. 55 (1994) 15.
16. EPA National Primary and Secondary Drinking water regulations, 40 CFR part 141.
17. Law approved on April 23, 1998.
18. Ministry of Agriculture Fisheries and Food. Polychlorinated Biphenyls in Food—UK Dietary Intakes. Food Surveillance Information Sheet No. 89, MAFF, London.
19. Feeley M.M., Grant D.L. Reg. Toxicol. Pharmacol. 18 (1993) 428.
20. Tanabe S., Hidaka H., Tatsukawa R. Chemosphere 12 (1983) 277.
21. Gupta S.R., Sarkar A., Kureishey T.W. Deep-Sea Res. II 43 (1996) 119.
22. Kelly A.G., Cruz I., Wells D.E. Anal. Chim. Acta 276 (1993) 3.
23. Ewen A., Lafontaine H.J., Hidesheim K.-T., Stuurmen W.H. Int. Chrom. Laboratory 14 (1993) 4.
24. Schulz-Bull D.E., Petrick G., Duinker J.C. Marine Chemistry 36 (1991) 365.
25. Gregor D.J., Gummer W.D. Environ. Sci. Tech. 23 (1989) 561.
26. EPA National Primary and Secondary Drinking Water Regulations, 40 CFR, appendix to part 136, method 608.
27. Pearson R.F., Hornbuckle K.C., Eisenreich S.J., Swackhamer D.L. Environ. Sci. Technol. 30 (1996) 1429.

28. Vo-Dinh T. In: Chemical Analysis of Polycyclic Aromatic Compounds, p. 494 (Vo-Dinh T., ed., Wiley, New York, 1989).

29. Lintelmann J.W., Sauerbrey R., Kettrup A. Fres. J. Anal. Chem. 352(78) (1995) 735.

30. Poole G., Thibert B., Lemaire H., Sheridan B., Chiu C. Organohalogen Compd. 24 (1995) 47.

31. Schulz-Bull D.E., Petrick G., Kannan N., Duinker J.C. Mar. Chem. 48 (1995) 245.

32. Dachs J., Bayona J.M. Chemosphere 35 (1997) 1669–1679.

33. Dachs J., Bayona J.M., Albaigés J. Mar. Chem. 57 (1997) 313.

34. Murphy T.J., Schinsky A.W. J. Great Lakes Res. 9(1) (1983) 92.

35. Peltonen K., Kuljukka T. J. Chromatogr. 710 (1995) 93.

36. Zhang Z., Yang M.J., Pawliszyn. J. Anal. Chem. 66 (1994) 844A.

37. Zhang Z., Pawliszyn J. Anal. Chem. 65 (1993) 1843.

38. Llompart M., Li K., Fingas M. Anal. Chem 70 (1998) 2510.

39. Driss M.R., Sabbah S., Bouguerra M.L. J. Chromatogr. 552 (1991) 213.

40. Smith L.M., Stalling D.L., Johnson J.L. Anal. Chem. 56 (1984) 1830.

41. De Boer J., Dao Q.T. J. High Resolu. Chromatogr. 12 (1989) 755.

42. Storr-Hansen E. In: Environmental Analytical Chemistry of PCBs/Current Topics in Environmental and Toxicological Chemistry, p. 25 (Albaigés J., ed., Gordon and Breach Science, Amsterdam, 1993).

43. Arsen B., Bøwadt S., Tilio R. In: Environmental Analytical Chemistry of PCBs/Current Topics in Environmental and Toxicological Chemistry, p. 3 (Albaigés J., ed., Gordon and Breach Science, Amsterdam, 1993).

44. Duinker J.C., Schulz D.E., Petrick G. Anal. Chem. 60 (1988) 478.

45. Durel G.S., Sauer T.C. Anal. Chem. 62 (1990) 1867.

46. Hanneman L.F. The New PCB Issue—Analysis of Specific Congeneres Produced by Unintentional By-Product Chemistry. Presented at Capillary Chromatography '82—An International Symposium, Tarrytown, NY, October 4–6, 1982.

47. Zitco V. Bull. Environ. Contam. Toxicol. 6(5) (1971) 464.

48. Hattori Y., Kuge Y., Nakamoto M. Bull. Chem. Soc. Jpn. 54(9) (1981) 2807.

49. Safe S. Overview of Analytical Identification and Spectroscopic Properties. In: National Conference on Polychlorinated Biphenyls (November 19–21, 1975 Chicago), Ayer, F.A., ed., EPA-560/6-75-004; NTIS No. PB 253-248, March 1976.

50. Stan H.J. In: Pesticide Analysis (Das K.G., ed., Marcel Dekker, New York, Chap. 9, 1981).

51. Stemmler E.A., Hites R.A., Aborgast B., Budde W.L., Deinzer M.L., Dougherty R.C., Eichelberger J.W., Foltz R.L., Grimm C., Grimsrud E.P., Sakashita C., Sears L.J. Anal. Chem. 60 (1988) 781.

52. Oehme M., Stocki D., Knoppei H. Anal. Chem. 58 (1986) 554.

53. Stemmler E.A., Hites R.A. Biomed. Environ. Mass. Spectrom. 15 (1988) 659.

54. Stemmler E.A., Hites R.A. Biomed. Environ. Mass. Spectrom. 17 (1988) 311.

55. Stemmler E.A., Hites R.A. Anal. Chem. 57 (1985) 684.

56. Budzikiewicz H. Mass Spectrom. Rev. 5 (1986) 354.

57. Bagheri H., Leonards P.E.G., Ghijsen R.T., Brinkman U.A.TH. In: Environmental Analytical Chemistry of PCBs/Current Topics in Environmental and Toxicological Chemistry, p. 221 (Albaigés J., ed., Gordon and Breach Science, Amsterdam, 1993).

58. Screitmüller J., Ballschmiter K. Fresenius J. Anal. Chem. 348 (1994) 226. Screitmüller J., Vigneron M., Bacher R., Ballschmiter K. Int. J. Environ. Anal. Chem. 57 (1994) 33.

59. Mössner S., Barudio I., Spraker T.S., Antonelis G., Early G., Geraci J.R., Beker P.R., Ballchmiter K. Fresenius J. Anal. Chem. 349 (1994) 708.

60. Mullin M.D., Pochini C.M., McCrindle S., Romkes M., Safe S.H., Safe L.M. Environ. Sci. Technol. 18 (1984) 468–476.

61. Wells D.E., Maier E.A., Griepink B. Int. J. Environ. Anal. Chem. 46 (1992) 255.

62. Jacob J. Quality Assurance for Environmental Analysis, p. 649 (Quevauviller P., Maier E.A., Griepink B., eds., Elsevier Science, Amsterdam, 1995, Vol. 17).

63. Porte C., Barcelo D., Albaigés J. J. Chromatogr. 442 (1988) 386.

64. Langenfeld, J.J., Hawthorne S.B., Miller D.J. Anal. Chem. 67 (1995) 1727.

65. Tolosa I., Bayona J.M., Albaigés J. Environ. Sci. Technol. 29 (1995) 2519.

66. Cruz I., Wells D.E., Marr I.L. Anal. Chim. Acta 283 (1993) 280.

67. Tolosa I., Merlini L., De Bertrand N., Bayoina J.M. Environ. Toxicol. Chem. 11 (1992) 145.

68. Kalas L., Onuska F.I., Mudroch A. Chemosphere 18 (1989) 2141.

69. Hong C.-S., Bush B. Chemosphere 21 (1990) 434.

70. Fuoco R., Colombini M.P., Abete C., Carignani S. Int. J. Environ. Anal. Chem. 61 (1995) 309.

71. De Boer J., Dao Q.T. J. High Res. Chrom. 12 (1989) 755.

72. Durel G.S., Sauer T.C. Anal. Chem. 62 (1990) 1867–1871.

73. Bavel B.N.C., Bergqvist P.A., Broman D., Lundgren K., Papakosta O., Rolff C., Strandberg B., Zebòhr Y., Zook D., Rappe C. Mar. Pollution Bull. 32 (1995) 210.

74. Ennicutt M.C., McDonald S.J., Sericano L.J., Boothe P., Oliver J., Safe S., Presley B.J., Liu H., Douglas W., Wade T.L., Crockett A., Bockus D. Environ. Sci. Technol. 29 (1995) 1279.

75. Annenberger D., Lerz A. Deutsche Hydrographische Zeitschrift/German J. Hydrography 47(4) (1995) 301.

76. Kuehl D.W., Butterworth B.C., Libal J., Marquis P. Chemosphere 22 (1991) 849.

77. Schmidt L.J., Hesselberg R.J. Arch. Environ. Contam. Toxicol. 23 (1992) 37.

78. Guevremont K., Yosr R.A., Jamieson W.D. Biomed. Environ. Mass Spectrom. 14 (1987) 434.

79. Focardi S., Bargagli R., Corsolini S. Antarctic Science 7(1) (1995) 31.

80. Durel G.S., Sauer T.C. Anal. Chem. 62 (1990) 1867.

81. Kominar R.J., Onuska F.I., Terry K.A. J. High Resol. Chromatogr. 8 (1985) 585.

82. Hong C.-S., Bush B. Chemosphere 21 (1990) 434.

83. Mills A.G., Jefferies T.M. J. Chromatogr. 643 (1993) 409.

31

Determination of PCDDs and PCDFs in Water

Anna Laura Iamiceli, Luigi Turrio-Baldassarri, and Alessandro di Domenico
National Institute of Health, Rome, Italy

I. INTRODUCTION

A. Physicochemical Properties and Environmental Relevance

The polychlorinated dibenzo-*p*-dioxins (PCDDs) and dibenzofurans (PCDFs) (1,2) are tricyclic aromatic compounds with two and one oxygen atoms, respectively (Fig. 1). The hydrogen atoms at the 1,2,3,4- and 6,7,8,9-positions are partially or totally replaced by chlorine atoms, thereby yielding 210 congeners (75 PCDDs and 135 PCDFs), different in the number and/or position of chlorine atoms (Table 1). These notes will focus only on the PCDDs and PCDFs having four to eight chlorine atoms: these congeners are chemically and thermally stable and known to be very persistent in the environment. In addition, those with the 2,3,7,8-chlorosubstitution pattern are considered to be potent carcinogenic compounds [in particular, 2,3,7,8-T_4CDD has been classified a Class 1 carcinogen by IARC (3)] and endocrine disrupting agents (4).

PCDDs and PCDFs are substantially insoluble in water (Table 2) and exhibit a strong lipophilic character, as shown by the high values of *n*-octanol–water partition coefficients K_{ow} (log K_{ow} ranges from 6.8 for 2,3,7,8-T_4CDD to 8.2 for O_8CDD). Therefore, these compounds tend to accumulate in fat and fatty matrices and display strong affinity for soil, sediment, and sludges. PCDDs and PCDFs are considered semivolatile substances: their vapor pressure ranges from 10^{-7}–10^{-6} Pa for T_4CDDs and T_4CDFs to 10^{-10} Pa for O_8CDD and O_8CDF (Table 2).

PCDDs and PCDFs are not produced intentionally by industry. In fact their formation and eventual release into the environment occur primarily in thermal or combustion processes, particularly when wastes are incinerated. In addition to this, PCDDs and PCDFs can be obtained as unwanted by-products in industrial processes involving chlorine.

B. Analysis

The standard methods developed to determine PCDDs and PCDFs generally include the assessment of 2,3,7,8-T_4CCD and the other 16 2,3,7,8-chlorosubstituted congeners. This selection is based principally on toxicological considerations: as anticipated, these compounds are considered to be the most toxic representatives of both families. However, the number of target analytes may be higher when the analysis aims at the identification of specific pollution sources.

The choice of an analytical method for the determination of contaminants in environmental samples strongly depends on the reliability (accuracy and precision) requested. For PCDD and PCDF analysis, an increasing need for accurate, congener-specific determinations is observed, and complex analytical procedures are required. The very low levels these compounds generally exhibit in the environment demand the use of highly sensitive procedures capable of detecting absolute amounts in the 0.1–0.01-pg range. Besides, in cases where such low levels have to be measured, the detection or quantification limits are often

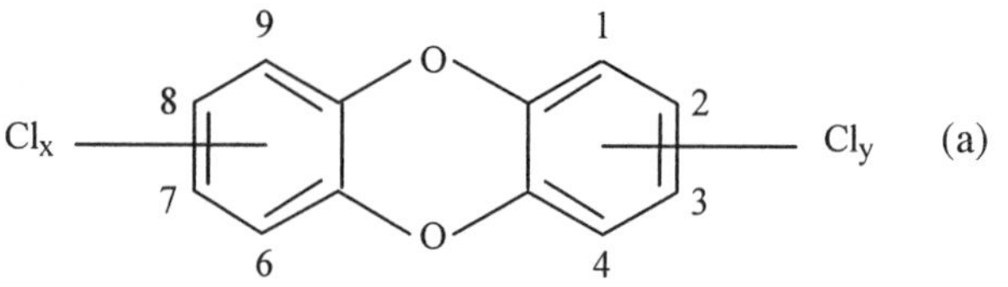

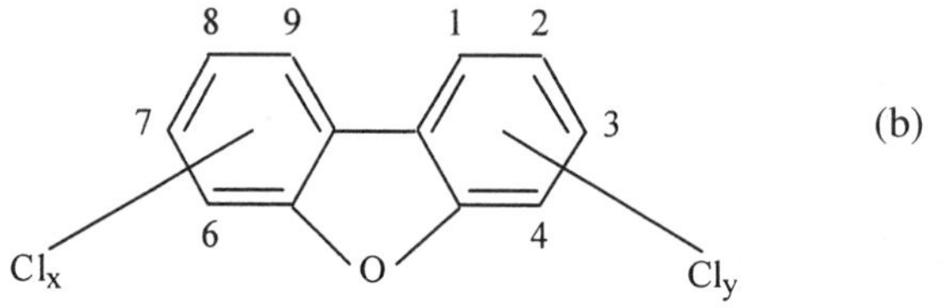

Fig. 1 Molecular structure of (a) PCDDs and (b) PCDFs. x and y = 1–8.

influenced by the apparent response observed in reagent blanks rather than by the instrumental sensitivity itself; an exhaustive definition of limit of detection (LOD) is given by Skoog and co-workers (5). Moreover, the number of possible interfering species arising from sample matrix or from ambient sources can increase remarkably when such low analytical limits must be reached. Most interferences are usually removed by cleanup procedures; others should be separated as effectively as possible by gas chromatography (GC) coupled to a specific detection system. The difficulty in resolving and quantifying the large number of PCDD and PCDF congeners, kin analytes, and interfering substances present at low levels in water samples is such that a combination of mass spectrometry (MS) and high-resolution gas chromatography (HRGC) is the only instrumental technique actually employed for the determination of PCDDs and PCDFs.

Table 1 PCDDs and PCDFs According to Positional Isomer Groups

Chlorination degree	Number of isomers and acronyms			
	PCDD		PCDF	
Monochloro-	2	M_1CDD	4	M_1CDF
Dichloro-	10	D_2CDD	16	D_2CDF
Trichloro-	14	T_3CDD	28	T_3CDF
Tetrachloro-	22	T_4CDD	38	T_4CDF
Pentachloro-	14	P_5CDD	28	P_5CDF
Hexachloro-	10	H_6CDD	16	H_6CDF
Heptachloro-	2	H_7CDD	4	H_7CDF
Octachloro-	1	O_8CDD	1	O_8CDF
Total	75		135	

In order to give an overall view of the analytical methods in use for the quantification of tetra- through octachlorosubstituted PCDDs and PCDFs in water samples, the most recent literature has been reviewed. In spite of the large number of publications, few papers were found explicitly devoted to the determination of interest (Table 3). The analytical procedures described in these papers are applicable only by expert laboratories, equipped with sophisticated instrumentation and good infrastructural facilities.

Reference is also made to the method elaborated by the U.S. Environmental Protection Agency (US EPA) for determination of the tetra- through octachlorosubstituted PCDD and PCDF toxic congeners in aqueous, solid, and tissue matrices by HRGC in tandem with high-resolution mass spectrometry (HRMS) (7). It is worth noting that this reference method is particularly suitable for routine measurements and for those cases where samples with high levels of contamination have to be assayed.

C. Internal Standard Quantification Method

Because some amounts of the substances under investigation may be lost during the complex preparative procedures, the internal standard technique is generally adopted to provide proper correction for analyte losses. To this aim, known quantities of "internal quantification standards" (commonly referred to as *internal standards*, IS's) are added to the samples "at the earliest possible stage of extraction" (8) or to "the main collecting parts" (9) before sampling, in the case of high-efficiency collecting systems. The most suitable IS's are the isotopically labeled analytes ("tracers"), where

Table 2 Selected Chemicophysical Properties of Most Toxic PCDDs and PCDFs[a]

	PCDDs				PCDFs		
	Water solubility (ng/L)	Vapor pressure (Pa)	$\log K_{ow}$		Water solubility (ng/L)	Vapor pressure[b] (Pa)	$\log K_{ow}$
$2,3,7,8\text{-}T_4CDD$	1.93×10^1	2.0×10^{-7}	6.80	$2,3,7,8\text{-}T_4CDF$	4.19×10^2	2×10^{-6}	6.53
$1,2,3,7,8\text{-}P_5CDD$	—	5.8×10^{-8}	6.64	$1,2,3,7,8\text{-}P_5CDF$	—	2.3×10^{-7}	6.79
				$2,3,4,7,8\text{-}P_5CDF$	2.36×10^2	3.5×10^{-7}	6.92
$1,2,3,4,7,8\text{-}H_6CDD$	4.42	5.1×10^{-9}	7.80	$1,2,3,4,7,8\text{-}H_6CDF$	8.25	3.2×10^{-8}	—
$1,2,3,6,7,8\text{-}H_6CDD$	—	4.8×10^{-9}	—	$1,2,3,6,7,8\text{-}H_6CDF$	1.77×10^1	2.9×10^{-8}	—
$1,2,3,7,8,9\text{-}H_6CDD$	—	6.5×10^{-9}	—	$1,2,3,7,8,9\text{-}H_6CDF$	—	2.4×10^{-8}	—
				$2,3,4,6,7,8\text{-}H_6CDF$	—	2.6×10^{-8}	—
$1,2,3,4,6,7,8\text{-}H_7CDD$	2.40	7.5×10^{-10}	8.00	$1,2,3,4,6,7,8\text{-}H_7CDF$	1.35	4.7×10^{-9}	7.92
				$1,2,3,4,7,8,9\text{-}H_7CDF$	—	6.2×10^{-9}	—
$1,2,3,4,6,7,8,9\text{-}O_8CDD$	7.4×10^{-2}	1.1×10^{-10}	8.20	$1,2,3,4,6,7,8,9\text{-}O_8CDF$	1.16^b	5×10^{-10}	8.78

Source: Ref. 3. Chemicophysical properties determined at 25°C for PCDDs and 22.7°C for PCDFs, except where noted.
[b] At 25°C.

Table 3 Examples of PCDD and PCDF Determination in Water Matrices

Ref.	Sample size	Sampling technique	Extraction	Cleanup	Instrumental analysis	TEQ values (LOD for 2,3,7,8-T_4CDD)[a]
			River water			
16	500–1000 L	Centrifugation[b]; sorption on PUF plugs[c]	Soxhlet extraction[b,c]	Neutral, basic, and acid silica; basic alumina; Amoco PX-21; Carbopack C	HRGC-HRMS	73–41 ng/kg[b] 4.0–17 fg/L[c]
20	—	Semipermeable-membrane device[c]	Dialysis[c]	Basic and acid silica; GPC; Hypercarb	HRGC-LRMS	Results expressed in relative units
29	150–250 L	Filtration[b]; sorption on PUF plugs[c]	Soxhlet extraction[b,c]	Carbopack C	HRGC-HRMS	77 fg/L[d] (24 fg/L)[d]
30	430 L	Filtration[b]; sorption on PUF plugs[c]	Soxhlet extraction[b,c]	Carbopack C	HRGC-HRMS	14 fg/L[d]
			Seawater			
14	1500–2000 L	Tangential-flow filtration[b]; sorption on PUF plugs[c]	Soxhlet extraction[b,c]	HPLC fractionation; Smith's procedure (Ref. 28)	HRGC-HRMS	0.8–3.3 fg/L[b] (0.05–0.2 fg/L)[b] 0.4–3.6 fg/L[c] (0.2–1.9 fg/L)[c]
19	2000 L	Filtration through glass fiber filters[b]; sorption on XAD-2[c]	Soxhlet extraction[b,c]	H_2SO_4; silica; alumina	HRGC-HRMS	92 fg/L[d] (5 fg/L)[d]
			Water from Municipal Treatment Plants (Wastewater)			
22	4 L	Direct sampling	Filtration; Soxhlet extraction[b]	Silica; acid silica; alumina	HRGC-LRMS	(1–2 pg/L)[b]
6	4 L	Direct sampling	PUF trapping; Soxhlet extraction[c]	—	HRGC-HRMS	0.26–3.8 pg/L[c] (0.05–1.3 pg/L)[c]
29	1–5 L	Direct sampling	Filtration; Soxhlet extraction[b]; LLE[c]	Silica; basic and acid silica; silica; Alox; Carbopack C	HRGC-HRMS	1.37 pg/L[d] (1.2 pg/L)[d], ingoing water; 0.98 pg/L[d] (0.26 pg/L), outgoing water

[a]Values rounded off to two figures.
[b]Particulate matter.
[c]Dissolved fraction.
[d]Particulate matter and dissolved fraction.

one or more atoms of the isotope ^{2}H, ^{13}C, or ^{37}Cl have substantially replaced an equal number of ^{1}H, ^{12}C, or ^{35}Cl.

The rationale for using isotopically labeled congeners as IS's is that they normally exhibit physicochemical properties (and thereby an analytical behavior) very similar to the corresponding natural analogues, the difference in molecular weight making them suitable for a specific MS measurement regardless of the natural PCDDs and PCDFs present. At least one IS should be added for each homologous group. If one labeled congener is added for each analyte, the method is commonly known as the *isotope dilution technique* (7) (however, the terminology may also be encountered where methods employing a lesser number of tracers are described). Before the cleanup step, other IS's may be added to evaluate the efficiency of the cleanup process. Moreover, *internal sensitivity standards* (known also as *injection* or *syringe* standards) (8) are added to the final extracts immediately before GC-MS determination to provide a check on instrumental response sensitivity and to allow for increased reliability in the measurement of IS recoveries. Of course, a given labeled congener cannot be used simultaneously as internal quantification standard, cleanup standard, and sensitivity standard; therefore, it is important to have different labeled (i.e., with ^{13}C, ^{2}H, and ^{37}Cl) standards of the same congener.

D. Laboratory Safety

Chemical compounds and reagents used for the determination of PCCDs and PCDFs should be treated as potential health hazard. A strict safety program for handling these substances should be developed by each laboratory. Most of the precautionary measures are suggested by the literature (10–12). Waste handling and decontamination of glassware, towels, laboratory coats, etc., is described elsewhere (7).

II. PRINCIPLES BEHIND THE METHOD

Due to their strong lipophilic properties and poor solubility, the PCDDs and PCDFs dissolved in water were generally measured at levels lower than 1 pg/L. If particulate matter is present, most PCDDs PCDFs are sorbed onto the suspended solid material. In this case, the determination should be performed on the aqueous phase and particulate matter as well. In general, the analysis of PCDDs and PCDFs dissolved in water requires large sample volumes, possibly up to hundreds of liters. The use of systems to collect and concentrate the analytes is also recommended (13). To this aim, solid hydrophobic adsorbents or nonpolar organic solvents are used.

Many analytical methods follow the general scheme of Fig. 2. During sampling, or immediately after, known amounts of isotopically labeled analytes are introduced into the sample. Care is taken to ensure a complete homogenization of the labeled compounds in sample matrix. This is generally achieved by preparing the spiking solutions in a solvent completely miscible with water, such as methanol or acetone.

A separation of the suspended particulate present in the sample could be necessary. This operation is carried out by either centrifugation or filtration.

Extraction of the analytes can be performed essentially in two ways: by liquid extraction with organic solvents directly from water, or by elution with organic solvents after sorption of the organic contaminants on a solid sorbent.

Purification of the raw extract is always necessary in order to remove interfering compounds and prepare the sample for the chromatographic separation and final

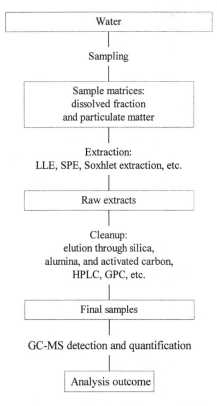

Fig. 2 General flowchart of PCDD and PCDF quantification procedures.

instrumental assessment. Many of the cleanup procedures include a step where PCDDs and PCDFs are treated with concentrated sulfuric acid; otherwise, the planarity and aromaticity of these molecules are often used to selectively adsorb them on the surface of a carbonaceous material such as activated or graphitized carbon. As anticipated, before cleanup the raw extract may be spiked with some other labeled compound(s) in order to check the efficiency of purification steps. The instrumental analysis is carried out by HRGC-MS (preferably, HRMS).

III. SAMPLING

A. Preliminary Remarks

The collection of samples representative of the environment to be studied is not a simple task. In the case of the analysis of PCDDs and PCDFs in water, difficulties in sampling are connected with the very low concentrations generally exhibited by the dissolved substances and their high affinity toward the surface of particulate matter. As a consequence, large volumes of water are normally sampled. Two main approaches have been suggested (13). For small to medium-size sample volumes (up to 5 L), direct collection of water may be adequate; in these cases, filtration of suspended material and extraction of the analytes dissolved (and possibly of those sorbed on the solid fraction) are both performed in the laboratory. For larger sample volumes, sampling techniques specifically developed for each circumstance appear to have been employed: what follows is the description of the techniques that, in our view, seem to be the most significant.

Generally, large-volume samplers consist of two units: the first one for removing the suspended particulate matter, the second one for collecting the dissolved analytes.

B. Suspended Particulate Matter

The size threshold for particles is often set at 0.45 μm, a cutoff value normally adopted for removal of suspended material from aqueous matrices. However, the threshold distinguishing particulate matter from dissolved material in water is selected as a compromise between the actual practicability to remove smaller particles and their negligibility. The presence of particulate matter, its origin, and particle size distribution strongly depend on the kind of water under consideration. Sea- or river water, for example, certainly contains higher amounts of particulate matter than drinking water.

The choice of the system to be used for collecting particulate matter depends on the water volume to be sampled and the expected particle concentration therein. Borosilicate glass microfiber filters have proved to be appropriate to separate the suspended material from the aqueous solution for a sample size up to 300 L (14). A limiting factor may be represented by filter plugging, as demonstrated in a validation study conducted by Uza and coworkers to test the efficiency of an automated preconcentration water sampler (APS) for PCDDs and PCDFs (15). The APS system is composed of a two-stage particulate filtering unit, with coarse and fine filters, and an Amberlite XAD-2 resin column to trap dissolved hydrophobic compounds. Various filter combinations were tested and, in all runs, filter plugging was observed.

This problem seems to be overcome by the tangential-flow filtration system, specifically developed for a large sample size (1–2 m^3) by Broman and coworkers (14) in order to collect Baltic Sea water where the suspended particulate generally occurs at concentrations of 0.2–0.3 mg/L. The system is made of a microporous poly(vinyliden diflouride) membrane, where the filtrate is separated from the retentate with a constant cutoff of 0.45 μm. A tangential flow of water sweeps the surface of the membrane so that the particles are kept in suspension, avoiding membrane plugging. Each 3–5-L volume of retentate (liquid) is collected and centrifuged; the supernatant and pellet are thereafter extracted and analyzed.

An alternative method to collect suspended material is centrifugation (16,17). The nature of the particulate separated from the aqueous matrix depends on the operational parameters. For example, a flow-through centrifuge, operating at 0.7 L/min and at 10,000 rpm, was employed by Götz and coworkers for the centrifugation of about 1000 L of river water containing a particularly high level of suspended material (20–50 mg/L) (16).

C. Dissolved Fraction

If the collection of particulate matter requires great efforts, the sampling of PCDD and PCDF dissolved fraction is probably an even greater challenge. In fact, as already mentioned, the PCDD and PCDF levels present in the aqueous phase may be several orders of magnitude lower than those present in the particulate fraction, so, in most cases, the determination of particle-bound PCDDs and PCDFs is a good approximation of the total content (16). The most widely adopted system to sample dissolved material is the sorption on hydrophobic solid sorbents. In fact, in comparison with the traditional liquid–liquid extraction approach, such a system displays good collection efficiency, overcomes the difficult problem of transporting large volumes of liquid, and allows the use of smaller amounts of solvents during the extraction step. Polyurethane foam (PUF) plugs, and the aforementioned XAD-2 resin proved to be particularly suitable to collect organic compounds from water samples. Before sampling, the sorbents are often spiked with known amounts of IS's in order to check the efficiency of the process. This can vary greatly with the different kinds of water: sea-, lake-, and river waters are rather complex and heterogeneous matrices when compared to drinking water. The presence of numerous and different organic contaminants can result in significant analyte losses during sampling owing to a possible competition between the chlorinated compounds and the other hydrophobic species present in the sample. In these cases, care should be taken to choose the most appropriate sampling system and set the most suitable sampling parameters (sample flow rate, volume of sorbent material, etc). In the recent literature, several examples are reported of PUF use to trap dissolved PCDDs and PCDFs from water. One of them is represented by the already-mentioned sampling equipment developed by Broman and coworkers for Baltic Sea waters that allows PCDD and PCDF determination at mean levels of 230 and 120 fg/L for the particle-associated fraction and the dissolved material, respectively (14). The system consists of two independent units: a tangential-flow filtration unit for the collection of particulate matter (see earlier) and an online filtration unit for sampling dissolved material. This system has proved to be particularly suitable for sampling water volumes up to 300 L. Larger sample volumes containing a high level of suspended particulate matter would result in a rapid plugging of the glass filter cartridges (16).

To efficiently overcome this problem, Götz and co-workers designed a FILtration/ADsorption (FILAD 1) system to sample the PCDDs and PCDFs dissolved in the River Elbe waters containing high levels (20–50 mg/L) of particulate matter (16). As shown in Fig. 3, the equipment consisted of three modules: two filtering units for the separation of suspended particulate, and a sorbing unit mounting PUF plugs for collection of dissolved analytes. The good separating efficiency of the system was demonstrated by the different congener profiles found in the particulate matter and for dissolved analytes. In fact, if a relevant portion of particulate had been collected by the PUF plugs, the same pattern would have been observed in the two congener profiles.

The parameters to be used with the XAD-2 resin in an automated water sampling unit were studied in Ottawa (Canada) by Lebel and coworkers (18). In this case, 200-L tap water was passed at different flow rates through columns packed with XAD-2 resin. To evaluate recoveries, ^{13}C-PCDD IS's were added at the pg/L level directly to the sample stream through an HPLC pump. This spiking technique proved to be more representative of analyte sampling in which solutions of contaminants continuously flow through the resin bed. Results demonstrated that flow rate was of great importance to obtain quantitative recoveries (>75%) and

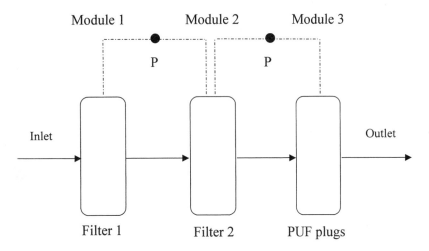

Fig. 3 Sampling device developed by Götz and coworkers: filter 1: propylene filter (∅ 254 mm, pore size 5 μm); filter 2: nylon 66 membrane filter (∅ 254 mm, pore size 0.45 μm); PUF plugs (∅ 60 mm, length 200 mm); P: difference pressure manometer. (Adapted from Ref. 16.)

that a flow rate of 70 mL/min, corresponding to three bed-volumes/min, is adequate. In fact, at higher flow rates, a breakthrough of congeners into the backup columns was observed.

An application of the XAD-2 resin to sampling seawaters was reported by Matsumura and coworkers (19). The equipment was made of a series of filters of decreasing porosity (from 50 down to 1 μm), followed by a glass fiber filter with a pore size of 0.5 μm for the removal of suspended material, and finally by an XAD-2 resin column for collecting the dissolved matter. The resin volume was 5000 ml, and the flow rate was approximately 10 L/min, equivalent to two bed-volumes/min. The system was employed in the determination of PCDDs and PCDFs in water at a level of fg/L, with recoveries greater than 90%.

In addition to solid–liquid collection techniques, in situ sampling methods were developed that contemporarily performed analyte collection and extraction. These systems utilize either sequential or continuous liquid–liquid extraction. An example is provided by the aqueous-phase liquid extractor (APLE) of Clement and coworkers, designed to sample surface water near dump sites where PCDDs and PCDFs were present at μg/L levels (17). The APLE was capable of extracting up to 200 L of water in a single-batch process. A spray bar on the top dispersed a heavier-than-water solvent (methylene chloride) as a fine spray across the surface of the water sample, pushed into the system by a submersible pump. Efficiency in extraction is ensured by continuous recirculation of the solvent. Devices like the one described avoid the problem of transporting large volumes of sample to the laboratory, but they often remain cumbersome and difficult to transport.

An alternative sampling method was used by Lebo and coworkers to collect dissolved PCCDs and PCDFs at lower than the pg/L range in Bayou Meto (20). In this case a semipermeable-membrane device, functioning as a passive in situ aquatic sampler, was developed that consisted of polyethylene tubing containing a neutral lipid with a large molecular mass, such as triolein, where hydrophobic contaminants can be solubilized. This system has the advantage of working unattended and collecting large volumes of water for a very long time (a sampling period of 28 days was reported by the authors), thus attaining very low quantification levels. In addition, if only the truly dissolved and bioavailable contaminants have to be determined, the system automatically excludes particle-associated pollutants, thereby rendering a filtration step unnecessary. However, results are expressed in relative units rather than as conventional concentrations.

IV. EXTRACTION

A. Preliminary Remarks

Solid and liquid samples not spiked during sampling should be added with IS's before extraction. The aim of every extraction is to bring quantitatively the compounds of interest into a stable solution. This can be achieved in different ways, depending on both the nature of matrix and the sampling technique adopted. When only a few liters of water are directly sampled and taken to the laboratory, a separation of particulate matter from aqueous phase might be appropriate. If "visible" particles are present, US EPA Method 1613 recommends filtering the water through a glass fiber filter. This operation, in our opinion, should be performed as soon as the sample is delivered to the laboratory, to avoid precipitation or adsorption of solid material onto the container walls, subsequently resulting in analyte losses that later may not be accounted for.

B. Extraction from Liquid Matrices

In the case of a liquid sample, two procedures are generally adopted to carry out the extraction, as described next. The liquid–liquid extraction (LLE), usually performed in separatory funnels, involves treatment of the aqueous sample with a virtually immiscible solvent. Despite the simplicity of this technique, the use of large volumes of solvents, generally toxic and difficult to dispose of (e.g., methylene chloride, toluene, hexane), and the long time necessary to perform the extraction when stable emulsions form are serious disadvantages, particularly when several liters of waters have to be treated.

A good alternative to the foregoing procedure is solid-phase extraction (SPE). This is a two-step process involving first the passage of the aqueous sample through a sorbent selectively retaining the analytes and then their elution from the sorbent with a proper solvent. XAD-2 resins and PUF plugs, whose properties have already been illustrated in the preceding pages, are the most widely used sorbents in PCDD and PCDF assessment. Solid-phase extraction is often carried out in ready-to-use packed cartridges, although the use of an octadecyl (C18) membrane sorbent was reported as an alternative sample preparation technique to capture hydrophobic contaminants (7,21). Compared to conventional LLE, SPE requires a shorter sample preparation time and reduced solvent volumes, resulting in a reduction of blank contamination levels.

C. Extraction from Solid Matrices

For solid samples, Soxhlet extraction is the technique most commonly used to quantitatively remove PCDDs and PCDFs from environmental matrices (such as sediments, fly ash, particulate matter) and solid sorbents employed in air or water sampling. The efficiency of Soxhlet extraction is ensured by the continuous contact of the solid matrix with freshly distilled solvent, usually toluene. A Dean–Stark apparatus is often used in combination with the Soxhlet extractor to remove traces of water (22). Besides the classical Soxhlet apparatus, other innovative techniques have been developed for the extraction of organic substances from solid matrices. Among these techniques, supercritical fluid extraction (SFE) and pressurized fluid extraction (PFE) are destined to find wide acceptance in the scientific community. For a thorough examination of these subjects and their application in the field of environmental samples, consultation of specialized texts and/or articles is recommended (23–27).

V. CLEANUP

In every extraction procedure, interfering chemical species are coextracted with the analytes. As a result, the raw extract has to be purified in order to isolate the compounds of interest. This is generally accomplished with a multistep process whose efficiency may be checked by adding "cleanup standards" (also isotopically labeled) to the raw sample extract.

Generally, procedures devised for the purification of extracts from biological or other environmental matrices were eventually adapted to water matrices. For this reason we will describe the most frequently used cleanup procedures to assess PCDDs and PCDFs, with particular attention to those actually employed in water analysis. It should be stressed that contaminant determination at concentrations lower than the ng/kg level requires the use of both highly selective cleanup procedures and very specific and sensitive detection systems.

Many of the analytical procedures actually used for environmental assessment refer to the method developed by Smith and coworkers for the analysis of PCDDs and PCDFs at ng/kg levels in freshwater fish species (28). The selectivity of this method is based on the resistance of PCDDs and PCDFs to concentrated sulfuric acid or on their poor chemical sensitivity to properly calibrated basic solutions, as well as on their affinity for the activated carbon surface. According to

the cleanup scheme, the sample extract is first eluted through a series of alternate potassium silicate and silica layers and then on an Amoco PX-21 activated carbon column to retain by adsorption a restricted number of organic compounds having a (pseudo-)planar multiring (hetero)aromatic structure such as PAHs, some PCBs, PCDDs, and PCDFs. Subsequent desorption is carried out by backflushing with an aromatic solvent, after the organic molecules with a nonplanar geometry, have been eluted from the carbon column with non-aromatic solvents. After an additional passage on cesium silicate and silica gel impregnated with sulfuric acid, the enriched aromatic phase is further fractionated by elution through an alumina column.

A wide range of analytes in water was determined by Broman and coworkers (14) using straight-phase high-performance liquid chromatography (HPLC) for the fractionation of specific groups of compounds. In particular, a μBondpak NH_2 semipreparative column, with n-hexane as mobile phase, allowed the isolation of three fractions containing, respectively, aliphatic and monoaromatic (e.g., hexachlorobenzene), diaromatic (e.g., PCBs, PCDDs, and PCDFs), and polyaromatic (e.g., PAHs) compounds. After separation, the fraction containing PCDDs and PCDFs was further purified according to Smith's procedure. Other cleanup procedures use different adsorbents to remove interferences: besides alumina, Florisil and Carbopack have been widely used for cleanup of environmental and, in particular, water extracts (29–31). Finally, the use of gel permeation chromatography (GPC) has also been reported for samples where high-molecular-weight interferences (e.g., polimeric materials, humic acids) have to be removed.

VI. INSTRUMENTAL ANALYSIS

A. Gas Chromatography

Fused-silica capillary columns are regularly employed for the determination of PCDDs and PCDFs, with stationary phases of varying polarity, according to the isomer specificity to be reached. For such samples as biological matrices, containing mainly the 2,3,7,8-chlorosubstituted congeners, low-polarity 5% phenyl–95% dimethylsiloxane columns (e.g., J & W DB-5, Hewlett Packard HP Ultra-2) are generally considered to be sufficiently selective (13). In fact, this kind of column can separate all homologue groups and 2,3,7,8-substituted congeners from each other but not all the non-2,3,7,8-chlorosubstituted congeners (32). Thus, 2,3,7,8-T_4CDD can be separated from 1,2,3,7,8-P_5CDD but not from 1,3,7,8-T$_4$CDD. On the other hand, for the assay of environmental samples, where very complex PCDD and PCDF mixtures are generally encountered, the use of the more polar cyanopropyl silicon columns (e.g., Supelco SP-2330, SP-2331) is recommended (13). In fact, an almost complete resolution of the 2,3,7,8-chlorosubstituted congeners in the presence of all the remaining ones can be achieved only with polar columns, even if incomplete separations are observed in the case of 2,3,7,8-T$_4$CDF, 1,2,3,7,8-P$_5$CDF, and 1,2,3,4,7,8-H$_6$CDF. In spite of its relatively low thermal stability, the SP-2330 column has often been used for the determination of PCDDs and PCDFs up to the octachlorosubstituted congeners (29). Its use was reported by Götz and coworkers for the determination of 23 PCDDs and 46 PCDFs in the analysis of water samples (16).

Stationary-phase thermal stability is of great importance for a successful use of GC-MS analysis: a higher stability implies lower bleeding and noise and a longer life. For this reason, technological research in the last decade focused on the development of columns with better thermal stability and resolution. In this context, in our laboratory the use of the low-polarity SGE BPX5 column, which can operate at temperatures up to 370°C, proved to be suitable for the determination of 2,3,7,8-chlorosubstituted congeners. In addition, a resolution better than that provided by a conventional 5% phenyl siloxane column can be achieved.

B. Mass Spectrometry

Electron impact (EI) ion sources are normally used in the MS determination of PCDDs and PCDFs, with conventional electron energies of 30–35 eV. Selected-ion monitoring (SIM) is canonically employed to improve specificity and sensitivity. In the SIM mode, the two or three most abundant ion masses of the predominant ion cluster of each analyte are those generally monitored.

Both low-resolution (LR) and high-resolution (HR) MS can be virtually used for PCDD and PCDF determination. However, HRMS is actually the most widely used instrumental technique, providing the best sensitivity and highest selectivity. Instrumental quantification limits in the range of 0.5–5 pg (injected) for the tetra- to heptachlorosubstituted congeners and 5–20 pg (injected) for the octachloroderivatives can be achieved with quadrupol instruments (LRMS) equipped with an EI source and operating in the SIM mode. The use of high-resolution instruments at a resolving power of 5000–10,000 improves sensitivity by one to two orders of magnitude.

Despite the extensive cleanup procedures used for PCDDs and PCDFs, compounds with gas chromatographic properties similar to those of the analytes may interfere with the assessment of the latter when LRMS is employed. In some cases, the difference between the molecular ion exact mass of the analyte and that of the interfering species is so small that it is impossible to achieve their separation even when HRMS is used. For instance, the polychlorodibenzothiophenes (PCDTs) are sulfur analogues of PCDFs detected in some environmental samples; because these compounds have molecular masses differing only by 0.0177 mass units from those of PCDDs with corresponding chlorosubstitution degree, a resolving power of at least 18,000 is required for a selective GC-MS determination of PCDDs in presence of PCDTs. To this purpose, the MS-MS technique was successfully applied to single out PCDDs from interfering compounds (33).

C. Identification and Quantification of Congeners

Performance and calibration of the GC-MS system should be verified periodically, possibly daily, for all PCDDs and PCDFs and labeled compounds. Only after all performance criteria are met, injections (typically, 1 μL) of samples, blanks (at least one for each batch of samples), and calibrant solutions may be performed. Internal quantification standards, cleanup standards, and injection standards should be present in samples, blanks, and calibrant solutions in comparable amounts. The calibrant solutions should also contain all the congeners to be analyzed at accurate concentrations ("external standards"). Identification of each congener is carried out by comparing its relative retention time [retention time of each congener vs. retention time of appropriate internal standard(s)] in the sample chromatogram with the corresponding one in the calibrant solution. An additional criterion to consolidate congener identity recognition is the intensity ratio of the two (or three) most abundant components of the molecular ion cluster of each congener: such a ratio is accepted as good if it is within ±15% of that expected for that chlorosubstitution degree.

For the quantification of the analytes, the "isotope dilution" technique is always adopted. In the unknown samples, the ratio of each analyte response vs. that of the pertinent IS must be determined and compared with those found for the calibrant solution (relative response factor, RRF). Further details can be found in EPA Method 1613, including the mathematical equations

Table 4 I-TEFs for the 2,3,7,8-Substituted PCDDs and PCDFs

PCDD			PCDF		
Congener	I-TEF[a]	WHO-TEF[b]	Congener	I-TEF[a]	WHO-TEF[b]
2,3,7,8-T$_4$CDD	1	1	2,3,7,8-T$_4$CDF	0.1	0.1
1,2,3,7,8-P$_5$CDD	0.5	1	1,2,3,7,8-P$_5$CDF	0.05	0.05
			2,3,4,7,8-P$_5$CDF	0.5	0.5
1,2,3,4,7,8-H$_6$CDD	0.1	0.1	1,2,3,4,7,8-H$_6$CDF	0.1	0.1
1,2,3,6,7,8-H$_6$CDD	0.1	0.1	1,2,3,6,7,8-H$_6$CDF	0.1	0.1
1,2,3,7,8,9-H$_6$CDD	0.1	0.1	1,2,3,7,8,9-H$_6$CDF	0.1	0.1
			2,3,4,6,7,8-H$_6$CDF	0.1	0.1
1,2,3,4,6,7,8-H$_7$CDD	0.01	0.01	1,2,3,4,6,7,8-H$_7$CDF	0.01	0.01
			1,2,3,4,7,8,9-H$_7$CDF	0.01	0.01
1,2,3,4,6,7,8,9-O$_8$CDD	0.001	0.0001	1,2,3,4,6,7,8,9-O$_8$CDF	0.001	0.0001
All other PCDDs	0	0	All other PCDFs	0	0

[a]*Source*: Ref. 3.
[b]*Source*: Ref. 34.

applied for RRF calculation and to estimate the final concentration of each analyte.

VII. DATA REPORTING

Analytical data are generally expressed in concentrations: as an amount per water volume (e.g., fg/L) for the assessment of the dissolved fraction, and as an amount per dry weight of particulate (e.g., ng/kg) for the assessment of the particle-associated fraction; however, cumulative (combined) concentration levels will be given as an amount per water volume or weight. For each analyte, efficiency of recovery is evaluated on the basis of that measured for the pertinent or corresponding IS. Acceptable recovery variation range for each labeled congener is reported in the EPA Method 1613.

Congener-specific or cumulative analytical levels may be expressed as 2,3,7,8-T$_4$CDD toxicity equivalents (TEQs), multiplying each assessed congener level by the pertinent toxicity equivalency factor (TEF) (3). Each factor represents an estimate of the toxic potency of a given congener relative to the congener traditionally considered the most toxic, that is, 2,3,7,8-T$_4$CDD (TEF = 1). Different TEF systems are available, with the most used one being the "international" system (I-TEFs) (Table 4), developed within the framework of North Atlantic Treaty Organization Committee on Challenges to Modern Society (NATO/CCMS) activities in the late 1980s and later adopted for risk management by several countries.

NOMENCLATURE

congener	Any one particular member of the same chemical family (e.g., the PCDD family has 75 congeners)
fg	10^{-15} g
GC	Gas chromatography
GC-MS	Gas chromatography combined with mass spectrometry
GPC	Gel permeation chromatography
homologue	A group of structurally related chemicals with the same degree of chlorination (e.g., there are eight homologues of PCDDs and PCDFs)
HPLC	High-performance liquid chromatography
HRGC	High-resolution gas chromatography
HRMS	High-resolution mass spectrometry
IARC	International Agency for Research on Cancer
internal standard	A ^{13}C−, ^{2}H−, or ^{37}Cl-labeled PCDD or PCDF congener
isomer	Any one member of the same homologous group (e.g., the 22 T$_4$CDDs)
I-TEF	International toxicity equivalency factor
L	Liter
LLE	Liquid–liquid extraction
LOD	Limit of detection
MS	Mass spectrometry
ng	10^{-9} g
PAH	Polycyclic aromatic hydrocarbon
PCB	Polychlorinated biphenyl

PCDD	Polychlorinated dibenzo-*p*-dioxin
PCDF	Polychlorinated dibenzofuran
PCDT	Polychlorodibenzothiophene
PFE	Pressurized fluid extraction
pg	10^{-12} g
PUF	Polyurethane foam
RRF	Relative response factor
SFE	Supercritical fluid extraction
SPE	Solid-phase extraction
spiking	Addition of ^{13}C−, ^{2}H−, or ^{37}Cl-labeled PCDD and PCDF congener(s)
TEF	Toxicity equivalency factor
TEQ	Toxicity equivalent
US EPA	U.S. Environmental Protection Agency

REFERENCES

1. AKD Liem, RMC Theelen. Dioxin: Chemical Analysis, Exposure and Risk Assessment. PhD dissertation, Utrecht University, Utrecht, Netherlands, 1997.

2. C Näf. Some Biotic and Abiotic Aspects of the Environmental Chemistry of PAHs (polycyclic aromatic hydrocarbons) and PCDD/Fs (polychlorinated dibenzodioxins and dibenzofurans). PhD dissertation, University of Stockholm, Stockholm, 1991.

3. Polychloritated dibenzo-*para*-dioxins and dibenzofurans. IARC Monographs on the Evaluation of Carcinogenic Risks to Humans. Vol. 69, IARC, Lyon, France, 1997.

4. U.S. Environmental Protection Agency. Special Report on Environmental Endocrine Disruption: An Effects Assessment and Analysis. Risk Assessment Forum. EPA/630/R-96/012, 1997.

5. DA Skoog, FJ Holler, TA Nieman. Principles of Instrumental Analysis. 5th ed. Saunders College, 1998, pp 13–14.

6. C Rappe, R Andersson, M Bonner, K Cooper, H Fielder, F Howell. PCDDs and PCDFs in municipal sewage sludge and effluent from Potw in the State of Mississippi, USA. Chemosphere 36(suppl 2):315–328, 1998.

7. USEPA Method 1613, Tetra- through Octa-Chlorinated Dioxins and Furans by Isotope Dilution HRGC/HRMS. EPA 821-B-94-005, 1994.

8. PF Ambidge, EA Cox, CS Creaser, M Greenberg, MG de M Gem, J Gilbert, PW Jones, MG Kibblewhite, J Levey, SG Lisseter, TJ Meredith, L Smith, P Smith, JR Startin, I Stenhouse, M Whitworth. Acceptance criteria for analytical data on polychlorinated dibenzo-*p*-dioxins and polychlorinated dibenzofurans. Chemosphere 21(suppl 8):999–1006, 1990.

9. European Standard. Stationary source emissions—determination of the mass concentration of PCDDs/

PCDFs—Part 1: Sampling. European Standard EN 1948-1, 1996.

10. Working with carcinogens—Department of Health, Education & Welfare, Public Health Service, Centers for Disease Control, NIOSH, Publication 77-206, August 1977, NTIS PB-277256.

11. Appendix A to part 136—Method 613-2,3,7,8-Tetrachlorodibenzo-*p*-dioxin. 40 CFR 136 (49 FR 43234), 1984, Section 4.1.

12. APHS. Standard Methods for the Examination of Water and Wastewater. 18th ed. and later revisions. American Public Health Association, Washington, DC, 1–35:Section 1090 (Safety), 1992.

13. EA Maier, B Griepink, U Fortunati. Round table discussions. Outcome and recommendations. Fresenius J Anal Chem 348:171–179, 1994.

14. D Broman, C Näf, C Rolff, Y Zebühr. Occurrence and dynamics of polychlorinated dibenzo-*p*-dioxins and dibenzofurans and polycyclic aromatic hydrocarbons in the mixed surface layer of remote coastal and offshore waters of the Baltic. Environ Sci Technol 25(suppl 11):1850–1864, 1991.

15. M Uza, R Hunsiger, T Thomson, RE Clement. Validation studies of an automated preconcentration water sampler (APS) for chlorinated dibenzo-*p*-dioxins and dibenzofurans. Chemosphere 20(suppl 10–12):1349–1354, 1990.

16. R Götz, P Enge, P Friesel, K Roch, LO Kjeller, SE Kulp, C Rappe. Sampling and analysis of water and suspended particulate matter of the River Elbe for polychlorinated dibenzo-*p*-dioxins (PCDDs) and dibenzofurans (PCDFs). Chemosphere 28(suppl 1):63–74, 1994.

17. RE Clement, SA Suter, HM Tosine. Analysis of large-volume water samples near chemical dump sites using the aqueous phase liquid extractor (APLE). Chemosphere 18(suppl 1–6):133–140, 1989.

18. G Lebel, T Williams, BR Hollebone, C Shewchuk, J Brownlee, H Tosine, R Clement, S Suter. Analysis of polychlorinated dibenzo-*p*-dioxins in raw and treated waters. Part 2: optimization of an XAD-2 resin column methodology. Int J Environ Anal Chem 38:21–29, 1990.

19. T Matsumura, H Ito, T Yamamoto, M Morita. Development of pre-concentration system for PCDDs and PCDFs in seawater. Organohalogen Compounds 19:109–112, 1994.

20. JA Lebo, RW Gale, JD Petty, DE Tillit, JN Huckins, JC Meadows, CE Orazio, KR Echols, DJ Schroeder. Use of the semipermeable membrane device as an in situ sampler of waterborne bioavailable PCDD and PCDF residues at sub-parts-per-quadrillion concentrations. Environ Sci Technol 29:2886–2892, 1995.

21. SM Price. Validation of Empore™ solid-phase sorbent disks for the extraction of PCDDs and PCDFs from wastewaters—EPA Method 1613B. Organohalogen Compounds 23:19–22, 1995.

22. NH Mahle, LL Lamparski, TJ Nestrick. A method for determination of 2,3,7,8-tetrachlorodibenzo-*p*-dioxin in processed wastewater at parts per quadrillion level. Chemosphere 18(suppl 11–12):2257–2261, 1989.

23. MD Luque de Castro, M Valcárel, MT Tena. Analytical Supercritical Fluid Extraction. Springer-Verlag, Berlin, Heidelberg, 1994.

24. ML Lee, KE Markides. Analytical Supercritical Fluid Chromatography and Extraction. Chromatography Conferences, Provo, Utah, 1990.

25. H Wagenaar, N Pronk, H Olthof. Evaluation of accelerated solvent extraction of PCDDs and PCDFs from native contaminated samples. Organohalogen Compounds 27:265–268, 1996.

26. B van Bavel, H Wingfors, S Lundstedt, G Lindström. Extraction under pressure: the sustainable alternative in environmental analysis. Organohalogen Compounds 35:71–74, 1998.

27. I Windal, G Eppe, AC Gridelet, LE Garcia-Ayuso, MD Luque de Castro, E De Pauw. Fast extraction of dioxins from fly ash: comparison of Soxhlet, supercritical fluid extraction and microwave-assisted Soxhlet extraction. Organohalogen Compounds 35:199–203, 1998.

28. LM Smith, DL Stalling, JL Johnson. Determination of part-per-trillion levels of polychlorinated dibenzofurans and dioxins in environmental samples. Anal Chem 56:1830–1842, 1984.

29. C Rappe, LO Kjeller, R Anderssen. Analyses of PCDDs and PCDFs in sludge and water samples. Chemosphere 19(suppl 1–6):13–20, 1989.

30. C Rappe, LO Kjeller, SE Kulp. Sampling and analysis of PCDDs and PCDFs in surface water and drinking water at 0.001 ppq levels. Organohalogen Compounds 2:207–210, 1990.

31. S Marklund, LO Kjeller, M Hansson, M Tysklind, C Rappe, C Ryan, H Collazo, R Dougherty. Evaluation of XAD-2 resin cartridge for concentration/isolation of chlorinated dibenzo-*p*-dioxin and furans from drinking water at the parts-per quadrillion level. In: C Rappe, G Choudhary, LH Keith, eds. Chlorinated Dioxins and Dibenzofurans in Perspective. Lewis, Chelsea, MI, 1986, pp 329–341.

32. APJM de Jong, AKD Liem, R Hoogerbrugge. Study of polychlorinated dibenzodioxins and furans from municipal waste incinerator emissions in the Netherlands: analytical methods and levels in the environment and human food chain. J Chromatogr 643:91–106, 1993.

33. HR Buser, C Rappe. Determination of polychlorodibenzothiophenes, the sulfur analogues of polychlorodibenzofurans, using various gas chromatographic/mass spectrometric techniques. Anal Chem 63:1210–1217, 1991.

34. M Van der Berg, L Birnbaum, BTC Bosvelt, B Brunström, P Cook, M Feeley, J Giesy, A Hangerg, R Hasegawa, SW Kennedy, T Kubiak, JC Larsen, FXR Leeuwen, AKD Liem, C Nolt, RE Peterson, L Poellinger, S Safe, D Schrenk, D Tillitt, M Tysklind, M Younes, F Waern, T Zacharewski. Environ. Health Persp. 106:775–792, 1998.

32

Polycyclic Aromatic Hydrocarbons

Miren López de Alda-Villaizán

Spanish Council for Scientific Research, Barcelona, Spain

I. INTRODUCTION

A. General Remarks

Polycyclic aromatic hydrocarbons (PAHs) are ubiquitous environmental pollutants that represent the largest class of suspected chemical carcinogens. In fact, PAHs were the first class of compounds demonstrated to be carcinogenic in experimental animals (1). PAHs are composed of two or more benzene rings, with adjacent rings sharing two carbon atoms; nonaromatic rings may also be present. The molecular structures and the carcinogenic evaluation (2,3) of the PAHs of primary environmental concern (4) are shown in Table 1.

Physicochemical characteristics of PAHs are highly determined by molecular weight (MW). Thus, resistance to oxidation and reduction, vapor pressure, and aqueous solubility, for instance, decrease with increasing MW (5). Several compilations of the physicochemical and spectral properties, occurrence, toxicity, methods of analysis, etc., of a broad range of PAHs have been published through the years (2,6–13). Some other reviews have dealt specifically with the sources, fate, occurrence, and significance of PAHs in the aquatic environment (5,14–19).

B. Sources

The PAHs can be formed from both natural and anthropogenic sources. With the exception of anthracene, which was used in the dye industry, PAHs have no commercial use except for research purposes (2). Apart from small amounts of geochemical and biosynthetic origin (they are synthesized by some algae, bacteria, fungi, and plants) (15,20,21), PAHs are mainly anthropogenic. Heating, power plants using fossil fuels (22–26), industrial processes (27), incineration of industrial and domestic wastes (28), forest fires, volcanic activity, vehicle exhausts (29), asphalt pavements, and, in general, all incomplete combustion at high temperature and pyrolytic processes involving fossil fuels (peat, coal, petroleum), or more generally, materials containing C and H, are major sources of PAHs (5,17).

C. Occurrence in Water

The PAHs enter the aquatic environment through several mechanisms, such as natural biological processes, atmospheric deposition [dry fallout (30), rainfall (31), vapor-phase deposition], surface runoff (32), leachate from solid waste disposal sites (33,34), petroleum spillage (35), and industrial and domestic wastewaters.

Because of their nonpolar hydrophobic nature and relatively low aqueous solubility (in the range ppm to ppb), which in grossly polluted waters can be slightly enhanced by tensoactive agents and other organic compounds (5,14,15,18,36), PAHs entering the aquatic environment quickly become adsorbed on suspended particular matter, sediments, and biota (37–43). The concentration of PAHs in the water column is usually orders of magnitude lower than in the sediments and in the tissues of the organisms where they accumulate, and can be expected to decrease approximately logarithmically with distance from the source (5,22). Thus,

Table 1 Molecular Structure and Carcinogenic Evaluation of the PAHs of Primary Environmental Concern

Structure	IUPAC systematic name	M.W.	Carcinogenic evaluation IARC (Ref. 2)	USEPA (Ref. 3)	Abbreviation
	Naphthalene	128	Nonevaluated	D[a]	Na
	Acenaphthene	152	Nonevaluated	—	Ace
	Acenaphthylene	154	Nonevaluated		Acy
	Fluorene	166	Inadequate data	D	Flu
	Anthracene	178	No evidence	D	An
	Phenanthrene	178	Inadequate data	—	Ph
	Fluoranthene[c]	202	No evidence of being carcinogenic per se		Fl
	Pyrene	202	No evidence	D	Pyr
	Benz[a]anthracene	228	Sufficient evidence	B2[b]	BaA
	Chrysene	228	Limited evidence	B2	Chrys
	Benzo[b]fluoranthene[c]	252	Sufficient evidence	B2	BbF
	Benzo[k]fluoranthene[c]	252	Sufficient evidence	B2	BkF
	Benzo[a]pyrene[c]	252	Sufficient evidence	B2	BaP
	Benzo[ghi]perylene[c]	276	Inadequate data	D	BghiP
	Indeno[1,2,3-cd]pyrene[c]	276	Sufficient evidence	B2	IP
	Dibenz[a,h]anthracene	278	Sufficient evidence		dBahA

[a]Group D: Not classifiable. Inadequate or no human and animal evidence of carcinogenicity.
[b]Group B2: Probable human carcinogen. Sufficient evidence from studies.
[c]Regulated in water by the European Union Community.

water bodies near industrialized and populated areas are primary repositories of aquatic PAHs.

Natural routes of removal of PAHs from the aquatic environment include volatilization (44), adsorption, and a variety of degradative processes (45), such as photo-oxidation (46), chemical oxidation, and biological transformation by aquatic bacteria, fungi, and animals, most of which are in general impeded in groundwater (47), oxygen-poor water basins, and deep oceans, where PAHs can therefore be more persistent.

Conventional water treatment procedures combine both mechanical methods, such as flocculation, sedimentation, and filtration, which remove PAHs bound to particulate matter (about two-thirds of the PAHs found in raw water), and chemical methods, such as oxidation, which further remove the remaining part, constituted of dissolved PAHs (48,65). Activated carbon filtration has been shown to be the most effective chemical method, achieving removal efficiencies of 99–100% (49) and providing water from chemical wastewater capable to fulfill even the official requirements for drinking water (50). The widely used chlorination method is currently being questioned, owing to the potential toxicity of the by-products formed in the process (18,19,51–53). For site remediation, biotreatment processes are receiving increasing attention, and some recently proposed methods for processing hazardous waste disposal are ultrasounds (or sonolytic decomposition) (54) and cloud-point approaches (55).

Literature values for PAH concentrations in water run as low as 1 pg/L in seawater (56) and up to several mg/L in heavily polluted wastewaters (25–57). Of the various natural waters, groundwater has the lowest concentration of PAHs, due to the natural filtration experienced by the organic contaminants through several solid matrices (58), though many current studies indicate the increasing pollution of this important drinking water resource by leachates from solid waste dumps (47) and landfill sites (25,33,34) and by infiltration of runoff waters (32).

According to the World Health Organization, the average concentrations of the six PAHs formerly selected as indicators of the whole group (59) are typically 10–50 ng/L in groundwaters, 50–250 ng/L in relatively unpolluted river waters, and higher in polluted rivers and effluents. Rainwater may contain significant concentrations of PAHs, sometimes much higher than the receiving water body (60), whereas fog and cloud water have been shown to be much more polluted than precipitation water (61).

Effluents and wastewaters represent the most polluted water bodies. Commonly cited industrial sources of PAH-contaminated wastewater effluents are oil refineries, industries utilizing solid and liquid hydrocarbon feedstocks for the manufacture of chemical by-products, the plastics and dyestuffs industries, high-temperature furnaces (especially those using anthracite electrodes), the lime industry, metal smelting industries, and many others (5,15,62). Domestic wastewaters, including raw sewage and storm sewer runoff, may contain significant quantities of PAHs. Household sewage contains PAHs derived from cooking and washing activities and from urine and feces. During heavy rainstorms, sewer runoff may increase raw sewage PAH concentrations by up to 100-fold over dry-weather conditions, probably owing to wear and leaching of tires and road surface materials and to condensation from vehicular exhaust (29,60,62,63).

In water supplies PAHs can be found, even after water treatment (49,64), at concentrations similar to, or slightly higher than, that of groundwater. On occasion, PAH concentration increases through the distribution system, as a consequence of their penetration from the surrounding contaminated soil through relatively permeable pipework into the water (47) and/or contact with the tar or bitumen with which the pipes and water storage tanks are coated (52,65). In the last case, an increase in the level of fluoranthene is particularly marked (59,62,65,66).

D. Exposure

Human exposure to PAHs occurs principally by inhalation of tobacco smoke and polluted air, ingestion of contaminated and processed food and water, or by dermal contact with soot, tars, and oils. The greatest exposure is likely to take place in the workplace of certain facilities (tar-production plants, coking plants, asphalt-production plants, coal-gasification sites, smoke houses, municipal trash incinerators) via air. The first observation of work-related cancer, reported by the English surgeon Percival Pott in 1775 in soot workers, was related to PAHs.

The relative contribution of drinking water to PAHs is minor, having been estimated as 0.1–0.3% of the total exposure (21). However, water pollution by PAHs represents a potential human health hazard not only from the direct consumption of drinking water but also because of the possible introduction of PAHs in the food chain from environmental water (14,67).

E. Legislation

Based on the available toxicological data (2,68–72) and on some other considerations, such as frequency of occurrence and analytical method availability, a number of organizations have promulgated certain regulations with regard to the presence and monitoring of PAHs in water.

In 1984 the World Health Organization proposed a guideline value of 0.01 μg/L for BaP in drinking water (21,59), thus changing a previously proposed limit of 200 ng/L for the sum of six PAHs (Fl, BbF, BkF, BaP, BghiP, and IP) that were relatively easy to detect and could serve as indicators for the whole group but a limit that had not been selected based on toxicological considerations. The same value of 200 ng/L is the current maximum admissible concentration established for the same six-PAH indicator in Council Directive 80/778 of the Commission of the European Communities (73) for water intended for human consumption.

Limiting values for PAHs are also set by the European Union (EU) in surface waters intended for the abstraction of drinking water. According to Council Directive 75/440/EEC (74), sources of drinking water are divided into three quality categories, A1, A2, and A3, that, with regard to PAH content, should not contain more than 0.2, 0.2, and 1 μg/L, respectively, of the six-PAH indicator, and for which certain treatment methods are specified. This directive and some of the other directives to be mentioned shortly will be integrated by the year 2007 into the Water Quality Framework Directive, which is currently in preparation and will provide the basis of all future Community water legislation.

Reference to PAHs, specifically as "PAHs with carcinogenic effects" (75), or inferred among "the group of substances which has been proved to possess carcinogenic or mutagenic properties in or via the aquatic environment" (76–78), is also made in other European directives (75–78) on water pollution, and in some other provisions for marine pollution control of specific areas, but no quality or emission standard values are set.

A survey of the directives of the European Community (EC) Council related to the monitoring of organic contaminants in general in water has been developed by Bedding et al. (19) and more recently by Hennion et al. (79).

In the United States the principal legal framework for the prevention and control of water pollution is contained in the Safe Drinking Water Act (SDWA), the Federal Water Pollution Control Act, commonly known as the Clean Water Act (CWA), the Resource Conservation and Recovery Act (RCRA), and the Superfund Amendments and Reauthorization Act [formerly Comprehensive Environmental Response, Compensation, and Liability Act (CERCLA)], also called the Super-

fund Act. Under these acts, the U.S. Environmental Protection Agency (EPA) has issued the following lists of organic compounds, which include PAHs, to be monitored in various types of water:

National Primary Drinking Water List (80), with application to public water systems. It specifies a maximum contaminant level of 0.2 μg/l of BaP.

EPA's Priority Toxic Pollutant List (4), with application to municipal and industrial wastewaters. The U.S. EPA classifies 16 PAHs as priority pollutants, subsequently referred to as "16 EPA priority PAHs," and mandates their routine monitoring for regulatory administration, but no water quality standards applicable to freshwater and saltwater have been established for them.

Hazardous Constituents List (81) and Ground-Water Monitoring List (82), with application to wastewater and solids and to groundwater, respectively, at active hazardous treatment, storage, and disposal sites. Both lists include 19 PAHs; the former contains 10 of the 16 EPA priority PAHs, and the latter all 16 EPA priority PAHs plus 7,12-dimethylbenz[a]anthracene, 3-methylcholanthrene, and 2-methylnaphthalene.

Contract Laboratory Program (CLP) Target Compound List, of application to groundwater, sediment, and soil samples at abandoned hazardous waste sites. It includes the 16 EPA priority pollutant PAHs plus 2-methylnaphthalene.

Under the National Pollutant Discharge Elimination System (NPDES), effluent limitations guidelines and standards regulating the discharge of PAHs and other pollutants from a number of sources have been established (83).

II. ANALYSIS

Because several reviews concerning the analytical determination of PAHs (6,10–12,84), some of them in the water environment (14,85,86), have been published throughout the years, this chapter will focus mainly on literature published since 1980. A selection of the most interesting and best-validated recent methods, with regard to detection limits, accuracy, and precision, using liquid and gas chromatographic techniques are summarized in Tables 2 and 3, respectively.

A. Reference Methods

Table 4 lists some standard and reference methods published by various official organizations for the analysis

Table 2 HPLC Methods for Analysis of PAHs in Water

Analyte	Water sample type	Sample volume	Sample preparation	Stationary phase	Detection	LD	RSD (%)	Recovery (%)	Levels (individual PAHs)	Publication year (Ref.)
16 EPA PAHs	Purified, drinking, surface	50 ml	Online SPE (Zorbax ODS1, 5 μm, 10 × 2 mm)	LiChrospher PAH (4 μm, 125 × 2 mm)	FL (program) UV	0.1–2 ng/L 100 ng/L (Acy)	—	—	—	1997 (258)
Na, Ace, Flu, Ph, An, Fl, Pyr, BaA, Chrys, BeP, BaP, dBahA, BghiP	Tap, reservoir	1.5 L	SPE (Sep-Pak vac tC-18)	Hypersil Green PAH (5 μm, 100 × 4.6 mm)	FL (program)	ng/L	0.4–10	60–96	0 (tap) Low ng/L (reservoir)	1996 (40)
16 EPA PAHs except Acy	River, seawater	10 ml	Online SPE (Pelliguard Guard C18)	Spherisorb S5 PAH (5 μm, 150 × 4.6 mm)	FL (program)	0.04–1.98 ng/L	3.28–9.76	>57, except for Na (3.89)	10–30 ng/L (river); 5–60 ng/L (seawater)	1996 (259)
16 EPA PAHs except Acy and Na	Bottled, network supply, river	25 ml	Cloud-point extraction (Triton X-114)	Lichrospher 100 RP-18 (5 μm, 125 × 4.0 mm)	FL (program)	0.3–6.7 ng/L	0.8–6.0	>70, except for An (32.9)	1.6–26.8 ng/L (river)	1996 (260)
Na, Ace, An, Fl, Pyr, BaA, BbF, BkF, BaP, BghiP, IP	Raw, finished, drinking	1 L	(a) LLE (DCM) (b) SPE (Sep-Pak tC18)	Superspher 100 C18 (4 μm, 250 × 4.6 mm)	FL (program)	(a) 0.007–1.1 μg/L (b) 0.011–1.3 μg/L	(a) 6–14 (b) 8–15	(a) >75 (b) >72	ND–53 ng/L (raw) ND–43 ng/L (finished)	1996 (261)
17 PAHs (from Na to dBahA)	Seawater	5 L	LLE (n-hexane)	Waters Nova Pak C18 (4 μm, 150 × 3.9 mm), micellar LC	FL (program)	0.10–1.94 ng/L	0.6–12.1	71–91	ND–0.27 μg/L	1996 (262)
16 EPA PAHs	Rainwater	180 ml	SPE (Sep-Pak C18)	LiChrospher PAH (5 μm, 250 × 4 mm)	UV (for Acy at 225 nm) FL (program)	0.03 (BkF)–36 ng/L (Na)	3–32	10 (low MW)–98 (high MW)	3 ng/L–1 μg/L	1996 (31)
16 EPA PAHs except Acy plus BeP	Wastewater	1 L	LLE (c-hexane)	Hypersil Green PAH (5 μm, 100 × 4.6 mm)	FL (program)	0.17–30 ng/L	5–21	70–120	1.2–900 ng/L	1996 (263)
Ace, Acy, 6 EU PAHs	Ground, surface, rain, drinking, river	0.5 L	SPE (C18)	C₁₈-RP (3 μm, 100 × 3.2 mm)	UV (300 nm) Amperometric	36.4–435.7 ng/L (UV) 3.1–64.4 ng/L (amperometric)	<10%	77–130	Low ng/L (surface, rain, and river), <QL (ground)	1996 (264)
16 EPA PAHs	Aqueous media	1 L	SPE (EnvirElute/PAH SPE column)	PE Chrompher-3 PAH column (3 μm, 100 × 4.6 mm)	UV DAD	—	—	Average: 83	—	1995 (265)
14 PAHs with MW >178 (from Ph to IP)	Drinking, surface (made 0.3 mM in Brij-35)	50 ml	Online SPE (Boos silica, 20 μm, 10 × 4 mm)	Supelcosil LC-PAH C18 (5 μm, 250 × 4.6 mm)	PB-MS (EI or CI)	0.003–4 μg/L (potable) (1.5–2-fold higher in surface water)	—	—	ND	1995 (266)

Table 2 Continued

Analyte	Water sample type	Sample volume	Sample preparation	Stationary phase	Detection	LD	RSD (%)	Recovery (%)	Levels (individual PAHs)	Publication year (Ref.)
16 EPA PAHs	HPLC-grade water	1.5 ml	Online SPE (Aquapore RP-18 cartridge)	Keystone PAH (150 × 4.6 mm)	UV and FL (program)	0.007–0.29 μg/L, except for Acy (4.44 μg/L)	6.9–32.5	26–85	—	1995 (267)
16 EPA PAHs	Drinking, spring	—	SPE (ENVI-18)	Supelcosil LC-PAH (5 μm, 150 × 4.6 mm)	UV (254 nm) and FL	—	2.9–7.0	96–103	ND	1995 (268)
6 EU PAHs	Tap, lake, bog, sewage treatment plant outlet (STPO)	50 ml	Onsite SPE (BioPAcK, 30–40 μm, 10 × 4 mm), coupled-column HPLC	Bakerbond Wide-pore RP-18 (5 μm, 250 × 4.6 mm)	FL (ex 365, em 470 nm)	0.06–0.74 ng/L	1.1–6.5 (tap); 5.4–56.7 (lake); 2.2–15.6 (bog); 5.8–32.2 (STPO)	>60 (except for IP in bog water)	2.84–40.83 ng/L (lake); 4.82–22.04 ng/L (bog); 3.1–15.09 ng/L (STPO)	1995 (269)
16 EPA PAHs	Surface (0.45 μm filtered)	10 ml	Online micelle-mediated SPE (ODS or Boos silica, 10 or 20 × 3 mm)	Supelcosil LC-PAH ODS (5 μm, 250 × 4.6 mm)	UV-DAD and FL (program)	0.5–70 ng/L (150 ng/L for Acy)	1.0–8.5	>90 (lower recovery for nonfiltered water)	Na = 20 ng/L Phe = 10 ng/L	1994 (270)
16 EPA PAHs	Drinking, groundwater	—	Filtration through a 0.2-μm membrane filter	PE ChromSpher-3 PAH (3 μm, 100 × 4.6 mm)	UV-DAD (280–335–360 nm) FL (program)	UV: 0.1–3 μg/L FL: 1–30 ng/L	0.5–1.2 (for standards)	—	—	1993 (271)
11 EPA PAHs (from An to IP, including 6 EU PAHs)	Surface, groundwater	1043 ml	SPE (NH2 + C18 cartridges in series)	Bakerbond Wide Pore C18 (5 μm, 250 × 4.6 mm)	FL (program)	0.15–2.73 ng/L	6.83–11.27	69.2–92.1	Low ng/L	1993 (272)

Analytes	Matrix	Volume	Extraction	Column	Detection	LOD	Linear range	Recovery (%)	Concentration	Year (Ref.)
9 PAHs (from Fl to BghiP)	Cloud, rain	8–1000 ml	LLE (DCM), silica gel column cleanup	Lichrospher 60 RP-Select B (5 μm, 250 × 4 mm)	FL (program)	—	1.0 (for BaP)	100.5 (for BaP)	Average = 371 ng/L (Σ PAHs)	1993 (61)
16 PAHs	Drinking	1 L	SPE (NH2 + C18)	Bakerbond Wide Pore PAH (5 μm, 250 × 4.6 mm)	UV-DAD (254, 280 nm) and FL (program)	—	—	70–85 (13 of 16 PAHs) Lower for Na, Ace, Acy	—	1989 (273)
An, Fl, BaA, Peryl, BaP	Lake	1 L	Cloud-point extraction (Triton X-100)	Supelco C18 (5 μm, 150 × 4.0 mm)	FL (program)	—	—	86–102	0.3, 1.6, 10.6 ng/L for BaP, Per, Fl, respectively)	1988 (274)
6 EU PAHs	River (unfiltered)	2 L	Continuous LLE (c-hexane)	Erbasil C18 (5 μm, 165 × 4.6 mm)	FL (ex 290 nm, em 440–420–500 nm)	—	5–23	75–95	0.11–7.22 ng/L	1986 (275)
13 PAHs from Na to dBahA	Refinery effluent	600 ml	SPE (Sep Pak C18 cartridge)	Rad Pak C-18 (5 μm, 100 × 8 mm)	UV and FL	0.005–25 ng	2–10	85–96	0.11–45 μg/L	1983 (276)
16 EPA PAHs	Wastewater	1 L	LLE (DCM), silica gel column cleanup	Perkin Elmer ODS (10 μm, 250 × 2.6 mm)	UV (254 nm) and FL	0.03–20 μg/L	2–37	76–115	ng/L–low μg/L	1982 (277)
6 EU PAHs	River (filtered)	2 L	LLE (2,2,4-trimethylpentane)	LiChrosorb ODS or Partisil 5 PPS (150 × 4.6 mm)	FL	0.3 ng/L (BaP)	—	74–96	ND–15.3 ng/L	1981 (278)
6 EU PAHs	Reservoir	2 L	SPE (C18 Sep-Pak cartridge)	LiChrosorb RP-18 (5 μm, 250 × 4.0 mm)	FL (ex 360 nm, em = 460 nm)	0.1–0.5 ng/L	—	48–78	0.1–6.7 ng/L	1981 (66)

Abbreviations (not included in Table 1 or in the text): BeP, benzo[e]pyrene; Peryl, perylene; QL, quantitation limit; ND, not detected.

Table 3 GC Methods for Analysis of PAHs in Water

Analyte	Water sample type	Sample volume	Sample preparation	Stationary phase	Detection	LD	RSD (%)	Recovery (%)	Levels (individual PAHs)	Publication year (Ref.)
23 PAHs, including 16 EPA PAHs	Lake, river, groundwater	1 L	SPE (Empore disk)	RTX-5 (0.25 μm, 60 m × 0.25 mm)	MS	9–56 ng/L	1–15	26.4–125.8	Low ng/L	1996 (279)
Flu, Pyr, BaA, Chrys, BbF, BkF, BeP, BaP, and Peryl	River	1 L	Microextraction (n-hexane)	DB-17 (0.15 μm, 15 m × 0.25 mm)	FT-IR	0.1–0.4 μg/L	—	80–90	ND	1996 (280)
16 EPA PAHs	Drinking, surface	1 L	SPE (C18 cartridges)	DB-5 (0.25 μm, 30 m × 0.32 mm)	MS (SIM)	0.5 ng/L	—	—	Low ng/L	1996 (281)
16 EPA PAHs	Seawater	500 ml	SPE (C-18 disks), MASE (microwave-assisted solvent elution)	HP-5 (0.25 μm, 25 m × 0.32 mm)	MS	—	2.4–9.1	67.9–82.4	—	1996 (39)
Na, Ph, An, Fl, Pyr, Chrys, BaP	Cresosote-contaminated wetland water	38 ml	SPME (polydimethylsiloxane fiber)	DB-5 (0.25 μm, 60 m × 0.32 mm)	MS or FID	30–420 μg/L	<10	>87	2–507 μg/L	1996 (282)
16 EPA PAHs	Contaminated river	—	SPMD, dialysis, GPC, silica gel column cleanup	DB-5 (0.25 μm, 30 m × 0.32 mm)	MS (SIM)	—	<10	—	ng/ml triolein	1996 (283)
Acy, Flu, Ph, An, Pyr, BaA, Chrys, BaF, BkF, BaP, IP, BghiP, dBahA	Drinking	250 ml	SPE (ENVI-18 cartridge)	PTE-5 (0.25 μm, 30 m × 0.25 mm)	FID and ECD	—	7.1–11, except Acy (28)	>90, except for Acy (66)	—	1995 (284)
16 EPA Pahs	Biologically purified oil wastewater	250 ml	SPE (ENVI-18)	ULTRA 2 (0.52 μm, 30 m × 0.32 mm)	MS (SIM)	0.08–1 μg/L	13–20	60–90	low μg/L	1995 (285)

Analytes	Sample	Volume	Extraction/cleanup	Column	Detector	LOD	RSD (%)	Recovery (%)	Concentration	Year (Ref.)
16 EPA PAHs	Distilled water	1 L	SPE (C18 disk), supercritical fluid elution	DB-5 MS (0.25 µm, 30 m × 0.25 mm)	MS	—	0.7–11	91–106	—	1995 (286)
Na, An, BaA, BaP	Ground, treated sewage water	In situ	SPME (polydimethylsiloxane or Carbopack B)	SPB-5 (0.25 µm, 30 m × 0.25 mm)	MS	1–20 ng/L	8–11	8–120	mg/L (ground)	1994 (287)
12 PAHs (from Ph to IP)	Drinking, drainage	800 ml	Microextraction (toluene), large-volume injection GC	OV-1 (0.4 µm, 20 m × 0.32 mm)	MS (SIM)	0.2–1 ng/L	0.9–136	43.0–186	0.7–3.5 ng/L (drinking); 4–15,000 ng/L (drainage)	1993 (32)
Na, Ace, Flu, Ph	Industrial plant process water	20 ml	LLE (hexane)	Rtx-5 (0.25 µm, 30 m × 0.32 mm)	FID	0.31–50 µg/L	—	72–90 (for 16 EPA PAHs)	1–590 µg/L	1993 (288)
16 EPA PAHs	Raw sewage, primary effluent, primary sludge	1 L (50 ml for primary sludge)	LLE (c-hexane), aluminum oxide, and silica gel column cleanup	SE-34 liquid phase, 0.11 µm, 25 m × 0.2 mm	MS (SIM)	0.01–0.05 µg/L	10.3–21.3	69–127	0.1–2 µg/L	1988 (289)
Low-MW PAHs (up to Fl)	Surface, groundwater	50 or 100 ml	SPE (C-18)	DB-5 or OV-1701 (30 m × 0.25 mm)	FID	—	<12	>80	2–433 µg/L (ground)	1988 (290)
16 PAHs	Wastewater	1 or 0.1 L	LLE (DCM)	DB-5 (0.25 µm, 30 m)	Automated MS (EI)	1–2 µg/L (= QL)	1–29	—	—	1985 (291)
16 PAHs	Seawater	1 L	LLE (hexane), silica gel, and aluminum oxide microcolumn fractionation	(a) SE-52 (0.15 µm, 25 m)	FID	24–40 ng/L	—	44–85	—	1984 (292)
Na, Acy, Flu, Fl, Ph, An, Pyr, BaA	Wastewater	90 ml	Microextraction (hexane)	SP2100 WCOT (10.6 m × 0.25 mm)	FID	—	14–25	87–97	—	1980 (293)

Abbreviations (not included in Table 1 or in the text): BeP, benzo[e]pyrene; Peryl, perylene; BaF, benzo[a]fluoranthene; QL, quantitation limit; ND, not detected.

Table 4 Standard and Reference Methods Proposed for Analysis of PAHs in Water

Organization (Ref.)	Method no.	Method focus	Procedure
EPA (294)	550	PAHs in drinking water	LLE, HPLC-UV-FL (16 PAHs)
EPA (295)	550.1	PAHs in drinking water	LSE, HPLC-UV-FL (16 PAHs)
EPA (296, 297)	525.1, 525.2	Organic compounds in drinking water	LSE, GC-MS (13 PAHs)
EPA (298)	610	PAHs in municipal and industrial wastewater	(LLE, optional silica gel column cleanup, HPLC-UV-FL, and GC-FID) (16 PAHs)
EPA (299)	625	Base/neutrals and acids in municipal and industrial wastewater	LLE, GC-MS (16 PAHs)
EPA (300)	1625	Semivolatile organic compounds in municipal and industrial wastewater	LLE, GC-MS (16 PAHs)
EPA (301)	8310	PAHs in groundwater and wastewater at active hazardous treatment, storage, and disposal sites	LLE, optional silica gel column cleanup, HPLC-UV-FL (16 PAHs)
EPA (302)	8100	PAHs in groundwater and wastewater at active hazardous treatment, storage, and disposal sites	LLE, optional silica gel column cleanup, GC-FID (24 PAHs)
ASTM (303)	4657	PAHs in water and wastewater	LLE, silica gel column chromatography, HPLC-UV-FL (16 PAHs)
APHA, AWWA, WEF (304)	6440	PAHs in municipal and industrial wastewater	LLE, optional silica gel column cleanup, HPLC-UV-FL, and GC-FID (16 PAHs)
APHA, AWWA, WEF (305)	6410	Extractable base/neutrals and acids in municipal and industrial wastewater	LLE, GC-MS (16 PAHs)
EU (87)	Directive 79/ 869/EEC	PAHs in surface water intended for the abstraction of drinking water	TLC-FL (6 PAHs)
EU (73)	Directive 80/ 778/EEC	PAHs in water intended for human consumption	LLE-FL or TLC-FL or TLC-GC (6 PAHs)

of PAHs in water. The European reference methods provide only succinct descriptions of analytical procedures (87,73), which focus on the measurement of the six PAHs selected as standard substances (Table 1), hereinafter referred to as the "6 EU reference PAHs." The remaining methods are detailed procedures for the determination, in most cases, of the 16 EPA priority pollutant PAHs.

B. Certified Reference Materials

Table 5 lists certified reference materials (CRMs) available for PAHs. Unfortunately, none of these materials reproduce actual water samples containing PAHs. Only calibration solutions and related reference materials (column selectivity mixtures, isotopically labeled compound solutions, etc.), which are used mainly to cali-

brate the measurement system and to spike or fortify samples, are available (88).

In addition to the materials listed in Table 5, a great number of CRMs for individual PAH standards (authentic reference compounds certified for purity) are prepared by the Community Bureau of Reference (BCR), European Commission (Belgium). Other sources of standard materials for individual PAHs or mixtures of them are chemical standards companies, such as Supelco (Bellefonte, PA) and Promochem (Sweden).

C. Sample Collection, Preservation, and Handling

Since sampling is the subject of another chapter, only a few considerations will be presented here. Several

Table 5 Reference Materials for Determination of PAHs

Laboratory	Catalog no.	Material	Certified constituent
NIST	SRM 869	Polycyclic aromatic hydrocarbon mixture	Column selectivity
NIST	SRM 1491	Aromatic hydrocarbons in hexane–toluene	23 PAHs, including the 16 EPA priority PAHs (4–8 g/ml)
NIST	SRM 1586	Isotopically labeled priority pollutants in methanol	10 organics, including An and BaP as PAHs (deuterated and nondeuterated in two separated solutions)
NIST	SRM 1647d	Priority pollutant PAHs in acetonitrile	16 EPA priority PAHs
NIST	SRM 2260	Aromatic hydrocarbons in toluene	23 PAHs, including the 16 EPA priority PAHs (50–70 g/ml)
NIST	SRM 2269	Perdeuterated PAH-I	5 perdeuterated PAHs
NIST	SRM 2270	Perdeuterated PAH-II	6 perdeuterated PAHs
NRCCRM	GBW08701	Benzo(*a*)pyrene in methanol	BaP (5.75 μg/ml)
NRCCRM	GBW08702	Benzo(*a*)pyrene in methanol	BaP (10.0 μg/ml)

Abbreviations: NIST, National Institute of Standards and Technology, Gaithersburg, MD; NRCCRM, National Research Center for Certified Reference Materials, Beijing, China.

factors determine the type of sampling procedure to follow: the aim of the study or program, the nature of the water to be sampled, the characteristics of the analytes of interest, and the subsequent method of analysis, among others (89,90).

Grab samples or composite samples will be considered according to the stated goal. *Grab sampling* is the procedure required in most instances and for most types of water (including drinking water, surface water, groundwater, and wastewater), but continuous-collection strategies may be necessary when monitoring water systems for occasional discharges.

For rainwater collection, wet-only samplers are the devices commonly employed (29,31,91,92). Cloud water can be collected with the help of passive or active samplers of the impactor type, the latter having been reported as more reliable (61). Seawater is probably the type of water sample that presents the most difficulties for collection (56,90,93). Besides the large sample volumes required (up to 4000 L), due to the extremely low concentrations at which PAHs are found in the ocean, the depth at which samples need to be collected, potential contamination, as well as handling and cleaning of the collection devices are additional inconveniences.

Conventional sampling and preservation practices should be followed (94–97) except for the following recommendations (295,303–305).

Grab samples must be collected in glass containers. The materials recommended in the European legislation are glass or aluminum (87). On the other hand, López García et al. (1992) (98) compared a number of different materials and found glass and also PTFE to be appropriate for sampling and storage of PAH-containing waters. Stainless steel vessels are seldom used (56,269).

Glassware must be thoroughly cleaned. Washing the glassware with water and detergent followed by rinsing with distilled water, acetone, and hexane is the procedure recommended in EPA method 550.1 (295). The ASTM standard test method (303) recommends subsequent heating of the glassware (if the type and size permits) in a muffle furnace to approximately 400°C for 15–30 min. In the APHA standard methods (304,305) and in some laboratories (289,89), rinsing with organic solvents is substituted for oven-drying of the glassware.

The bottle must not be prerinsed with the sample before collection.

All samples must be iced or refrigerated at 4°C from the time of collection until extraction.

PAHs are known to be light sensitive; therefore, samples, as well as extracts and standards, should be protected from the light.

Samples must be extracted within 7 days of collection.

Extracts must be analyzed within 30 days of extraction.

Samples known or suspected to contain residual chlorine must be preserved with a reducing agent. Sodium sulfite has been used for this purpose

(99), but sodium thiosulfate (80–100 mg/L) is most commonly used (48) and is recommended in the standard methods (295,304,305). However, the time at which the sample must be preserved varies with the standard method. According to the APHA methods, chemical preservatives, when used, should be added to the sample bottles as soon as the samples are collected (304,305). According to the ASTM standard method, chemical preservatives should be added to the samples only if extraction does not take place within 48 h after collection; in such a case, sodium thiosulfate needs to be added at 35 mg per part per million of free chlorine per liter, and the pH should be adjusted in the range 6.0–8.0 (303).

The addition of 1% v/v formaldehyde solution as a preserving agent to wastewater samples has been reported (263,289), though, according to the ASTM (142), its use should, in general, be avoided.

One-liter sample volume is used mostly because it usually allows sufficient overall method sensitivity. Larger samples are costly to ship and difficult to process and store in medium-sized refrigerators (100).

On the other hand, due to the already-mentioned strong hydrophobicity of the PAHs, adsorption on suspended solids and onto glass and metal surfaces of containers and connection tubes (258,278,98,101–103) is, along with contamination (89) and losses by evaporation (104), the most common cause of inaccurate results for PAH analyses. Low-MW PAHs are more likely to undergo volatilization, whereas losses due to adsorption are more pronounced for high-MW compounds (265,103). To avoid sorption problems during sampling and storage of PAHs, the following different approaches have been proposed.

Previous treatment of glassware with a silanizing reagent such as dimethyldichlorosilane (in toluene) (286,101).

Addition of an organic modifier, such as isopropanol (10–20%) (66,264,272,273,105), ACN (acetonitrile) (265), methanol (20%) (37,101), acetone (37), or tetrahydrofuran (THF) (258) prior, during, or right after sampling.

Addition of the extraction solvent to the sampling bottle prior to the addition of the water sample, which Crosby et al. (278) demonstrated improved recovery after 3 days of sample storage

Addition of "micelle-mediating" surfactants (270), such as SDS (sodium dodecyl sulphate), cetyl trimethylammonium chloride (CTACl) or polyoxyethylene lauryl ether (Brij-35) (266,270,98),

which according to López García et al. (1992) (98) produces the same effect as the addition of 40% ACN to the water sample, with the additional advantages of safety (surfactant solutions are less toxic, less volatile and less flammable than organic solvents), low cost, ability to achieve larger preconcentration factors because there is no dilution of the sample, and improvement of further specific operations

Rinsing the glass bottle with an appropriate solvent and combining the rinse with the extract (265,303–305)

Other alternatives to overcome or reduce sorption problems are solid-phase microextraction (SPME) (287), a new technique that allows sampling directly from the source (lake, fountain, etc.), and on-site extraction on conventional solid-surfaces (104), although in this last case the addition of an organic modifier or a solubilizing agent is also sometimes recommended (66,269,98,106–108). The fundamentals of SPME will be discussed with more detail in the extraction section.

On-site solid-phase extraction has also been pointed out as an especially convenient method for seawater sampling as a means of reducing the risks of contamination so often associated with sampling of this type of water sample (56,93). In addition, this approach offers some other advantages, as compared to classical grab sampling in containers, such as reduction of shipping and handling expenses and easy storage. However, the stability of PAHs stored on solid supports is not clear yet, as indicated by the often-contradictory results obtained in various studies describing this subject (269,279,109,110). In general, it appears that PAHs are more stable when stored on solid supports than when stored in the whole water samples, although both types of storage yield lower recoveries than samples extracted and analyzed shortly after collection (279). The immediate extraction and analysis of the sample after collection is always desirable.

Another sampling method that has attracted attention since its introduction by Huckins et al. in 1990 (111) is the use of the semipermeable-membrane device (SPMD) for passive in situ monitoring of organic contaminants in water (283,112–116). The device consists of a thin film of neutral lipid, such as triolein, enclosed in thin-walled lay-flat tubing made of low-density polyethylene or another nonporous polymer. Unlike other sampling methods, the SPMD samples "truly" dissolved or bioavailable contaminants, and it has been proposed as an advantageous alternative to biomonitoring with sentinel organisms. It has been described as a valuable tool for detecting spatial trends or temporal changes in PAH contamination, for identifying point sources of contamination in aquatic ecosystems, for predicting contaminant uptake by aquatic organisms, and for calculating contaminant half-lives or persistence in an aqueous compartment (113). However, since the estimation of average concentrations of dissolved organic contaminants in water need to be done through the application of mathematical models, its ability to function as quantitative monitoring device has yet to be fully demonstrated (112).

D. Sample Preparation

1. Filtration

Prior to extraction, samples are often filtered. Filtration is usually required for waters containing high levels of suspended solids or turbidity and especially when extraction is going to be carried out by adsorption on solid supports (263,278,90,103). When filtered, both the filtered sample and the filtrate should be analyzed separately, because high-MW PAHs will be adsorbed onto the particles (31,40,79,110,117).

2. Extraction

Because PAHs are in general present in water along with a variety of other pollutants and at very low concentrations—that is, at nanograms-per-liter levels—their analytical determination normally involves isolation and preconcentration before the final analysis. Direct analysis of the water samples has been attempted only on a few occasions, and in these cases the approaches used included capillary LC (118), and classical LC (271,119) and GC analysis of high sample volumes (up to 5 ml) (120–122). Fluorescence (FL) was used for detection with the liquid chromatographic methods and flame ionization (FID) with gas chromatography, and the reported detection limits were at ppb (118,119,121,122) and ppt levels (271,120).

The procedures most often used for PAH isolation and preconcentration of PAHs from water are liquid–liquid extraction (LLE) and solid-phase extraction (SPE). In the past, sample preparation was dominated by conventional LLE, but in recent years SPE has gained in popularity. Both methods present both advantages and disadvantages (104).

LLE advantages

No plugging
PAHs both dissolved in water and adsorbed on suspended matter are extracted (79)

LLE disadvantages

Use of large volumes of expensive, high-purity, potentially hazardous organic solvents
Emulsion formation
Needs more intensive further concentration stage
Lack of sensitivity for more volatile analytes
Time consuming
Not very amenable to automation

SPE advantages

Reduced or no solvent requirements (287)
Selectivity derived from the sorbent choice (106) and the elution procedure followed
Large sample volumes can be extracted with certain sorbents, such as XAD-2 resins
Little or no concentration step
Fully automatable, though it requires sophisticated, expensive technology
Sampling in the field or on-site sampling (123), which avoids possible breakage in transit, facilitates transport and storage, and reduces the possibility of changes of the sample

SPE disadvantages

SPE cartridges and disks susceptible to plugging
Only dissolved PAHs extracted
Batch-to-batch variation (104)
Breakthrough (259,124)

From the recovery and the standard deviation values reported by various authors in the literature, the assessment of whether LLE, SPE, or any other approach performs best for the extraction of PAHs from water is hard to determine, given that the range of PAHs studied as well as certain procedural steps, such as spiking and sample prefiltration (if performed at all), vary widely among the studies. Spiking levels are often unrealistically high, and the type of water used for the recovery tests frequently does not reproduce appropriately the nature of the sample analyzed. Several authors have compared the extraction efficiency of both LLE and SPE after their application to the same samples and analytes (38,42,261,263,275,103,110,123,125), and in general it has been concluded that either both techniques are comparable or LLE is superior as a result of its comparably higher recoveries and lower dependence of the extraction efficiency on the analyte MW and the suspended matter content of the water sample. As reported by many authors, one of the major drawbacks of the reversed-phase SPE approach for the extraction of PAHs from water is the considerably variable water solubility, polarity, and, as a result, breakthrough volume of the commonly investigated 16 EPA priority PAHs. With this in mind, and in view of the mentioned advantages of SPE compared to LLE, SPE could be the method of choice for the analysis of the six PAHs defined as standard by the EU or of BaP, defined as the standard by the WHO, in waters with relatively small organic matter content. For a satisfactory overall recovery of the 16 EPA priority PAHs by SPE, two different solid-phase materials or two different protocols, each optimized for low- and high-MW compounds would be advisable.

Currently, the European reference methods (73) recommend LLE with hexane; dichloromethane (DCM) is the solvent selected for extraction in the EPA methods. However, the EPA and other environmental bodies are attempting to find alternative extraction methods that minimize the use of solvents (295,297,126).

a. Liquid–Liquid Extraction. Several LLE techniques [including stirring, manual and automated shaking, and continuous LLE (275,288,127,128)], solvents (44,263,277,127,129), and sample pHs (41,277,128) have been compared in the literature with regard to extraction efficiency. Even though results do not always agree, manual shaking (288,127) and continuous extraction (275,128) as techniques, DCM (277,129), cyclohexane (263,127), n-hexane (292), and carbon tetrachloride (292,127) as solvents, and an acidic pH (~2) (41,277,128) appear to be the most favorable extraction conditions.

According to Pörschmann and Stottmeister (41), a change of pH from neutral to acidic leads to nearly complete release of the pollutants bound to humic material. Emulsion formation can be either prevented by the addition of a salt, such as sodium sulphate (278), or broken by means of centrifugation, stirring, filtration of the emulsion through glass wool, or the addition of more extraction solvent (130), or, when these techniques fail, by continuous extraction (131,305).

Conventional LLE is usually performed in two to three steps, consuming a total volume of 100–300 ml of solvent. However, reasonable recoveries have also been achieved through microextraction, a one-step extraction in which the solvent:water ratio normally ranges from 1:40 to 1:100 (31,280,293).

b. Extraction Procedures Using Solid Sorbents. Extraction of PAHs from water matrices using solid sorbents is dominated by but not limited to solid-phase extraction in which the analytes are accumulated directly from a liquid phase, the water matrix in this case, and subsequently eluted with a proper liquid phase (104). Other procedures have been described for trapping volatile PAHs, stripped from the water matrix by means of a stream of inert gas from the headspace gas above the sample, with subsequent desorption on the gas or liquid phase of gas chromatographs (126,131) or liquid chromatographs (132), respectively. These methods employed the relatively new solid-phase microextraction (SPME) procedure, which will be discussed in more detail later, designated as headspace SPME when the adsorption process takes place from a gas phase.

A very interesting review on the use of solid sorbents in general, and on solid-phase extraction techniques in particular, for direct accumulation of organic compounds from water matrices was published by Liska et al. in 1989 (104). In recent years, solid-phase extraction (SPE), also known as liquid-solid adsorption (LSA), liquid-solid extraction (LSE), or sorbent extraction, of PAHs from water has been performed mostly by means of octadecyl (C18) silica bonded phases.

Graphitized carbon black sorbents have rarely been used for this purpose (133–135), and according to Liska et al. (104) they are not suitable adsorbents for preconcentration of PAHs from water.

Polymer sorbents, on the contrary, have been employed to a greater extent. Among these, Tenax (27,64,104,136,137), polyurethane foams (104,109,110), and polytetrafluoroethylene (PTFE, or Teflon) (124) have occasionally been mentioned in the literature, while styrene-divinylbenzene copolymers (ST-DVB) (138), in particular, Amberlite XAD-2 (52,56,93,109 and references therein, 139–141), are the most popular. The main advantage of this type of resin over bonded-silica adsorbents is the large sample volumes that can be processed (up to several thousand liters) and at high flow rates (400 ml/min), which is especially desirable for the analysis of water with very low PAH levels, such as seawater (52,56,93).

Recently, phthalocyanine-based sorbents, such as "blue cotton," "blue pearls," "Boos silica," and "Boos glass," have been described for the selective preconcentration from water of PAHs and other mutagens and carcinogens having three or more fused rings (269,270,142,143). Conventional C18-bonded silica displays better trace-enrichment characteristics for the smaller PAHs, but these sorbents are more selective for the larger PAHs with which they exhibit specific hydrophobic and steric interactions. Another advantage is that larger sample volumes can be extracted (e.g., 35–50 L), and they have been proposed as an alternative to XAD resins.

Capitán-Vallvey et al. (144,145) have also reported the use of various Sephadex gels, which are common materials in size exclusion or gel permeation chromatography, for the preconcentration of PAHs from water and posterior solid-state fluorimetric determination.

As mentioned previously, SPE of PAHs from aqueous media is usually carried out on bonded-phase silica adsorbents (269,123), and in particular, in octadecyl-silica (C18) stationary phases. This type of SPE offers the advantage, as compared to the earlier-mentioned sorbents, of being suitable for online or coupled-column LC analysis. In coupled-column and online procedures, the C18 sorbent is contained in precolumns, also called extraction columns. For off-line procedures, cartridges, and recently also disks (39,279,286,103), packed with the C18 phases are used. Both types of devices are equally effective, but cartridges are more easily clogged by suspended solids, thus slowing the extraction, and disks require larger eluting volumes (a minimum of 15 ml) and are more expensive.

Critical factors affecting the extraction efficiency, regardless of the type of dispositive used (cartridges, disks, etc.), are sample volume, concentration of PAHs in the sample, addition of a modifier to the sample, flow rate of the sample through the solid sorbent, drying of the sorbent after sample loading, and the solvent, volume, and number of steps used for elution of PAHs.

Prior to sample loading, the sorbent is usually cleaned with an apolar solvent, e.g., DCM, or with the solvent that will be later used for elution. Then it is activated with a water-miscible, more polar solvent, commonly methanol or 2-propanol, and finally, the sorbent is conditioned with reagent water or a suitable buffer. For sorbent amounts up to 1 g, the volume used in every step usually varies from 4 to 12 ml.

The selection of the sample volume is limited by breakthrough values, which in the majority of situations are determined by analyte capacity factors and not by overloading of the sorbent, since PAHs are at relatively low levels in water. Sample volumes from 10 ml (259,270) in online enrichment procedures to 2 L (66) have been used, but the critical and most often-used sample volume is 1 L.

The addition of a modifier has two contradictory effects. By keeping the sorbent octadecyl chains in an activated form and by reducing or preventing sorption on suspended particulates and surfaces, it improves the overall efficiency of extraction of the PAHs, particularly of the high-MW PAHs, and by increasing the eluotropic strength of the sample it lowers the breakthrough volumes and consequently the recovery of the lower-MW PAHs. An acceptable compromise is rather difficult to reach when the target PAHs widely differ in MW, as happens with the 16 EPA priority PAHs. In this case, the best results have been obtained by addition of 15% 2-propanol (66,264,107) or 20% methanol (31,276) to sample volumes close to 0.5 L.

High flow rates can lead to lower recoveries of the compounds, with retention volumes close to the sample volume, due to nonequilibrium processes (31,104). In the literature, flow rates usually vary from 2.5 (264) to 25 ml/min (290); manufacturers recommend flow rates close to 10 ml/min (284).

In off-line procedures, drying of the solid phase after sample loading has been reported to improve both recovery and precision (31) by simplifying the subsequent desorption step. However, total drying of the support should be avoided, because it can yield lower recoveries of the more volatile PAHs due to evaporation (284,290,104). A stream of air (with negative or positive pressures) (264,269,272,279,284,290,108) or nitrogen (265) for about 5–15 minutes, centrifugation (273), or combination of both (31,40) are the most often-used means to perform this step.

In off-line procedures, the analytes can be eluted with solvent mixtures (281,284), via elution steps (265), or most commonly with a single solvent. Several authors have compared the elution properties of various solvents, including ACN, methanol, 2-propanol, *n*-hexane, *n*-pentane, acetone, ethyl ether, ethyl acetate, DCM, tetrahydrofuran, cyclohexane, and benzene (31,40,279,146,147). Tetrahydrofuran (31,66,146,148) and DCM (31,48,264,272,273,107,108) are among the most appropriate and frequently used elution solvents. The total volume of eluting solvent ranges from as little as 100 μl (147) to about 15 ml (48,279), with the number of elution steps generally varying between one and four.

Variations of the conventional procedure are the microwave-assisted solvent elution procedure described by Chee et al. (39) and supercritical fluid elution (286,149). In the latter case, pure or modified carbon dioxide is used as the eluting phase, which presents the advantages over common eluting solvents of being inexpensive, nontoxic, noncombustible, chemically inert, and easy to discard. In online procedures, the eluting phase is equal to the mobile phase in the LC analytical column.

At the present time, an increasing interest is devoted to online methods, with automated sample preparation interfaced to LC (258,260,266,267,270,150,151) or GC (152) instruments. Several working schemes have been proposed for LC procedures, with either several LC pumps and several valves (258,266,270,151,152) or just one LC pump and one (260) or several valves (267,150). In GC a microprocessor-controlled loop interface (Autoloop 2000, Interchro) was used to elute the adsorbed compounds with ethyl acetate for GC-MS analysis (152).

The main advantages of online systems lie in the fact that all of the analytes present in the sample are injected into the chromatographic system and that analysis time is significantly shortened, because manipulation of the sample is reduced to a minimum.

Certain drawbacks of the online approach are the possibility of memory effects when the precolumn is reused and less flexibility with respect to the choice of eluting solvents. A major problem, however, is the adsorption of the analytes to suspended solids and to the inner walls of the tubing of the LC system. As has been observed, this is usually diminished by the addition of an organic modifier to the sample, which has an adverse effect on the breakthrough of the low-MW PAHs. A promising alternative to this approach is the method proposed by Slobodnik et al. (266), which involves the addition of the surfactant Brij-35 to the sample, the disruption of the micelles immediately before the preconcentration column through the addition of water, and the deviation of the first portion of effluent from the mass detector, all of which avoids the influence of the modifier on the breakthrough and the contamination of the detector with Brij-35.

Another interesting modification of classic online procedures is the thermally assisted desorption (TAD) introduced by Renner et al. (258). Desorption of PAHs from the enrichment C-18 precolumn is thermally assisted at 95°C, which reduces the peak broadening resulting from online enrichment by a factor of about 10% and increases the efficiency of the subsequent online separation as well as analyte recovery.

The already-mentioned solid-phase microextraction (SPME) is a recently introduced extraction technique that has been reported to offer some advantages over LLE or SPE for the screening of PAHs from relatively clean water samples. The SPME consists of a fused-silica fiber coated with a polymer, generally polydimethylsiloxane, and housed inside a syringe for ease of handling. The fiber is either dipped into the aqueous sample or placed in the headspace above the sample matrix for adsorption of the corresponding analytes, and subsequently transferred into a GC (126,131, 153,154) or LC (287,132,155) system for analysis. The technique is simple, fast, inexpensive, solventless, and easily automated and requires small sample sizes (e.g., 1–5 ml). Under controlled pH and ionic strength conditions, the technique is relatively insensitive to matrix effects and when coupled to hyphenated techniques is very sensitive (156).

c. Other Extraction Procedures. An extraction and preconcentration technique that has not received much application in spite of presenting some advantages (cost, simplicity, sensitivity, safety hazard considerations, sorption effects prevention) over LLE and SPE is surfactant-mediated extraction, also referred to as cloud-point or micelle-mediated extraction or flotation enrichment (157). The reason probably lies in the fact that the surfactants often interfere in the subsequent UV or fluorimetric detection. The entire topic of this technique has been the focus of a review (55). Triton X-100 (274) and Triton X-114 (260,158) are among the nonionic surfactants (159) most commonly used for PAH extraction from water.

Only two references on supercritical fluid extraction (SFE) of PAHs from water have been found (125,160). In principle SFE is not applicable to water samples (286); however, SFE and accelerated solvent extraction (ASE) have been judged the best techniques for the extraction of PAHs from suspended particulate matter and related materials obtained from natural water filtration (161).

3. Purification

Extracts from heavily contaminated waters may require further cleanup prior to analytical determination, which is carried out by thin-layer chromatography (TLC) (27,52,275,289,136,162) or column chromatography (61,277,91,99,100,103,163). Because only an isolated PAH fraction is desired, TLC cleanup generally involves one-dimensional development (164). Alumina and, preferably, silica gel (277), are the common stationary phases in both approaches, while solvents such as hexane, cyclohexane, and benzene and mixtures of them are the common mobile phases. Most standard methods for the determination of PAHs in wastewaters (see Table 4) include silica gel column cleanup (sometimes it is optional) for subsequent GC or LC analysis.

4. Concentration

To achieve sufficient overall method sensitivity or for solvent exchanging for further analysis, extracts usually need to be concentrated. The concentration is a critical step and can result in considerable losses of the more volatile compounds (52,273,103). Constable (165) evaluated several solvents and solvent reduction methods for concentration of PAH solutions. The volume reduction methods tested included rotary evaporation, nitrogen evaporation, and four different distillation columns (Vigreaux, Snyder, stainless steel gauze, and a spinning-band column). Rotary evaporation and nitrogen evaporation are, according to this author, the simplest methods; but due to the comparably longer time needed for solvent reduction, these methods are less desirable than some of the distillation techniques. Among the distillation techniques, the spinning-band column performs the best with regard to overall recoveries, but it is also very time consuming, whereas almost-comparable recoveries, but in a shorter time, can be reached with the stainless steel gauze column when low-boiling-temperature solvents are used. The standard methods (303–305) recommend the use of a Kuderna–Danish apparatus attached to a Snyder column. Valkenburg et al. (128) found that concentration of the sample (in DCM) without significant loss of analytes can be achieved either by macro-Snyder/Kuderna–Danish or under nitrogen stream. Among the solvents tested (DCM, *n*-pentane, ethyl ether, acetone, and ACN), DCM is suggested as the most advantageous, and, in general, the more volatile solvents are reported to give higher recovery factors.

To obtain the best recoveries, a few precautions should be followed:

Control flow rate and temperature when nitrogen evaporation is used (265,268).

Carry out distillations at the boiling point of the solvent used.

Protect sample solutions from light, to prevent losses by photolytic degradation.

Prevent samples from being reduced to dryness unless proper precautions, such as downstream trapping of volatile components or the addition of keepers, such as long-chain alcohols (e.g., 1-hexanol or 1-octanol), are taken (129,165). Allowing the sample to go to dryness results in serious losses of the low-MW PAH (29,292,165).

E. Separation Techniques

Up until the 1980s the most popular analytical techniques for the determination of PAHs in water were paper chromatography with ultraviolet detection (PC-UV), gas liquid chromatography–mass spectrometry (GLC-MS), and, especially, thin-layer chromatography (TLC) (14,85).

Thin-layer chromatography with fluorescence detection in the UV is currently one of the methods recommended by European legislation for the analysis of PAHs in surface and drinking waters (73,87). However, the number of recent papers describing its use for this purpose is rather limited (37,49,64,164,166–178). The number of manual steps and the associated unstan-

dardized errors, as well as the lack of resolution reported for some PAHs (175), have probably favored the extended use of other techniques, such as LC and GC, which are more powerful and already constitute common analytical tools in most laboratories. Nowadays, TLC can be valuable as a simple, inexpensive screening technique (164,167,177).

In the last two decades, PAHs in water have been analyzed mostly by reversed-phase liquid chromatography (RP-LC) with UV and/or fluorescence detection or by GC in combination with a flame ionization or mass spectrometric detector.

At present, LC can be considered more appropriate than GC for the analysis of PAHs from the point of view of resolution and detection limits, since a number of commercial LC columns (275,179,180), especially those specifically manufactured for PAH monitoring, allow complete resolution of the 16 EPA priority PAHs (Fig. 1) and of isomers that are often difficult to separate by capillary column GC (270,277,304,305,180), such as chrysene and triphenylene; BbF, BjF, (benzo[*j*]fluoranthene) and BkF; and dBacA (dibenz[*a,c*]anthracene) and dBahA. A comprehensive review on the determination of PAHs by liquid chromatography has recently been published (180); it includes a useful list of over 40 commercial C_{18} columns classified into polymeric, intermediate, and monomeric-like columns according to the corresponding value of the selectivity factor $\alpha_{TBN/BaP}$, which is indicative of the separation efficiency of the column toward PAHs.

The LC analysis of PAHs is carried out in reverse-phase mode using mostly octadecyl silica stationary phases (C18) (181,182). Mixtures of water with organic solvents such as methanol and ACN with gradient elutions from 20–50% to 100% organic solvent are generally used as mobile phases, though micellar mobile phases for the so-called micellar liquid chromatography (MLC) have also been employed (262,270). Temperature is an important parameter affecting efficiency and capacity factors (183,184), and its control is an absolute prerequisite for programmed multiple-wavelength fluorescence detection (185). Best separations are reported at slightly below ambient temperatures (186,187).

On the other hand, GC has superior separation efficiency to LC and, combined with MS, is more advantageous for the complete characterization of sample PAH contents, i.e., when the aim of the study is not evaluating specific target compounds, such as the 16 EPA priority PAHs or the 6 EU reference PAHs. The common GC limiting factor related to the analyte volatility is not relevant in this case unless PAHs with MW

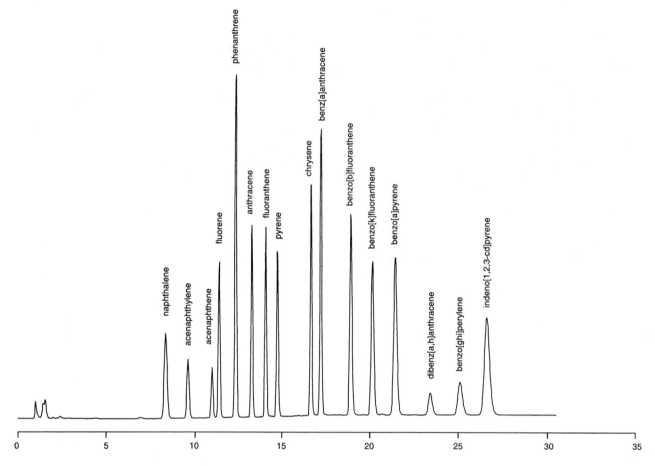

Fig. 1 Reversed-phase LC separation of the 16 EPA priority PAHs in standard reference material (SRM) 1647c. *Column*: Hypersil Green PAH. *Mobile phase*: 3 min, hold at 50% water:50% acetonitrile; 15 min linear gradient to 100% acetonitrile; 15 min hold at 100% acetonitrile. UV detection at 254 nm.

greater than 300, and therefore larger than the 16 EPA priority PAHs, are to be determined.

Gas chromatographic determinations of PAHs are usually carried out with high-resolution capillary columns, which came to replace the formerly widely used packed columns (164). Glass or fused-silica capillary columns with a variety of coated or chemically bonded stationary phases, such as SE-30 (27,188), OV-101 (26), DB1 (92), DB-5 (286,288,291,91,103), Pte-5 QTM (189), OV-1 (25,32), SE-54 (121,122), SPB-5 (287,190), and HP-5 (130), and some others already specified in Table 3, have been used for PAH separations from water extracts based on boiling points. PS-089 (120) and SE-54 (121,122) were used as stationary phases by Kaiser et al. (120) and Goretti et al. (121,122), respectively, for the analysis of PAHs from the direct injection of relatively high volumes of water sample (1–5 ml) into the chromatographic column.

In general these GC capillary columns are not able to completely separate certain PAH isomers, namely, chrysene/triphenylene and BjF/BbF/BkF, which therefore have to be quantified together. Burrows et al. (191) described the preparation of a Poly 179 capillary column, a polyphenyl sulphone phase, for separation of PAHs with five or more rings. This column could achieve a resolution of about 1 for the pairs of isomers BaP/BeP and BjF/BbF, making it therefore interesting as a complement to other stationary phases that can be more selective for lower-MW PAHs. Other stationary phases that allow the separation of certain isomers (BbF/BjF/BkF and dBacA/dBahA) are DB-5ms and DB-17 (192).

The use of liquid-crystal phases is also suitable for the separation of isomeric PAHs (193), since retention in this type of stationary phase is governed by shape selectivity in addition to boiling point. However, these phases suffer from limited useful temperature ranges (120–300°C) and higher bleeding rates and present shorter lifetimes (2–3 months).

Temperature programming from approximately 45 to 300°C is in general satisfactory for the elution of PAHs from naphthalene to coronene (164). Helium, hydrogen and nitrogen are employed equally as carrier gases. Split/splitless injection is the most extensively used injection technique, although oncolumn injection (288), and a number of other injection systems, including programmed temperature vaporizing (PTV) injectors (194) and loop-type interfaces (280,152), have been described for large-volume injections, with the aim of increasing chromatographic sensitivity (280,288,194) or for interfacing liquid systems to gas chromatographs (152).

The GC experimental conditions for the analysis of a wider range of PAHs commonly present in other types of environmental matrices, including alkylated derivatives and polycyclic aromatic compounds containing heteroatoms such as sulfur and nitrogen, have been described by other authors (195–197).

A few attempts to apply capillary liquid chromatography with a variety of detectors [UV (198), fluorescence (118), and PB-MS (199)] to PAH analysis can be found in the literature. This technique employs low mobile-phase rates, therefore consuming little solvent, improving sensitivity (and efficacy), and offering the possibility of coupling to MS. But to date it has not received much application in this field yet.

Immunoassays based on the principle of enzyme-linked immunosorbent assay (ELISA) for the semi-quantitative determination of PAHs in soil and water (200), micellar electrokinetic capillary chromatography (201), and supercritical fluid chromatography (SFC) hyphenated to UV detection (57) are among the techniques that have been rarely applied to this type of analysis. Supercritical fluid chromatography was used for the analysis of PAHs in waters from a canal marina site and from a stream running close to an abandoned coal mine (57). Except for Ace and Acy, which showed very similar retention times, the resolution achieved for the 16 EPA priority PAHs with SFC was better than that obtained with the same column under classical LC conditions, and in a quarter of the time. A complete analysis could be performed every 7 minutes.

F. Detectors

1. Liquid Chromatography Detectors

Ultraviolet absorption and fluorescence detection are the most widely used LC detectors for the measurement of PAHs. They are usually used in series, and when the 16 EPA priority PAHs are the target analytes, UV is generally employed for the determination of Acy, which does not fluoresce; the remaining PAHs are determined by fluorescence. The standard test method for PAHs in water (303) recommends the UV detector for the determination of Na, Ace, Acy, and fluorene at 254 nm and the fluorescence detector for the remaining PAHs.

Ultraviolet detection of PAHs, by means of fixed- or variable-wavelength UV detectors, or by means of the diode array detector (DAD), is usually performed at 254 nm, a wavelength at which most PAHs exhibit some absorption. However, other wavelengths can be used to improve sensitivity and selectivity (271,138). An optimized 280-nm, 335-nm, 360-nm UV program for the analysis of the 16 EPA priority PAHs eluting in positions 1–5, 6–10, and 11–16, respectively, is proposed by Dong (271). In general, the UV absorption maximum characteristic of the PAHs increases with the number of condensed aromatic rings.

Fluorescence is by far more sensitive and selective toward PAHs than is UV detection. Fluorescence detection, with detection limits (DLs) generally at picogram and even subpicogram levels, is typically two to three orders of magnitude more sensitive than UV (DL ~ng) (271,185). As a result, the need for concentration techniques is reduced. Since almost all PAHs fluoresce, whereas most other groups of organic pollutants do not, fluorescence detection hyphenated to LC requires less extensive purification than GC (181). Additionally, the selection of the appropriate excitation and emission wavelengths for a particular PAH permits its individual determination in the presence of coeluting compounds (181). Table 6 lists the excitation and emission wavelengths currently employed at the National Institute of Standards and Technology (NIST) for the analysis of PAHs in environmental samples (180). Such wavelengths represent a compromise between maximum sensitivity of the target compounds and minimum interference from possible unresolved compounds.

Current UV and fluorescence detectors allow the recording of the corresponding UV or fluorescence spectra of the chromatographic peaks without stopping the flow. However, neither type of spectra is usually characteristic enough of the specific PAH to permit unambiguous identification (164). Ratioing between peak intensities recorded at different wavelengths is more useful to aid in identification and peak purity assessment (271,273).

Another detector that has been used in LC is the amperometric detector. This detector has been reported

Table 6 Fluorescence Wavelength Programming Conditions for LC Determination of the 16 EPA Priority PAHs

Wavelength change	Excitation wavelength (nm)	Emission wavelength (nm)	PAHs determined
1	280	340	Naphthalene
2	289	321	Acenaphthylene,[a] acenaphthene, fluorene
3	249	362	Phenanthrene
4	250	400	Anthracene
5	285	450	Fluoranthene
6	333	390	Pyrene
7	285	385	Benz[a]anthracene
8	260	381	Chrysene
9	295	420	Benzo[b]fluoranthene
10	296	405	Benzo[k]fluoranthene, benzo[a]pyrene, dibenz[a,h]anthracene
11	380	405	Benzo[ghi]perylene
12	300	500	Indeno[1,2,3-cd]pyrene

[a]Acenaphthylene has a low fluorescence quantum yield, and it is usually measured by UV detection at 254 nm.

to be more sensitive (202), about 5–10 times more sensitive (264), than UV for the determination of PAHs. However, its use for this purpose has been very limited.

The relatively new LC-hyphenated-with-mass-detection (LC-MS) has been described as an interesting alternative to GC analysis for the determination of higher-molecular-weight PAHs. The interface most commonly used for analysis of PAHs from water (266,199,203) and other environmental samples (204) has been the particle beam (PB) in combination with electron impact (EI) ionization mode, which provides "classical" mass spectra for extremely useful peak identity confirmation and for automation purposes by comparison with available libraries. However, this technique has been found unsuitable for the determination of the volatile low-molecular-mass PAHs (MW 178), and it is often preferred to carry out quantification by other means, due to the nonlinear responses, high standard deviation, and low sensitivity associated with this technique.

The GC-MS of PAHs provides detection limits for PAHs that are about two orders of magnitude better than those obtained with LC-PB-MS (266). However, in a study conducted by Anacleto et al. (205) to evaluate the suitability of three LC-MS interfaces—moving belt (MB) and particle beam (PB) with EI, and heated pneumatic nebulizer (HPN) with atmospheric pressure chemical ionization (APCI)—for the analysis of complex mixtures of PAHs, the HPN interface with APCI provided the best overall performance, because it is compatible with a wide range of mobile-phase compositions, exhibits excellent transmission efficiencies and detection limits (in the picogram range) for both low- and high-MW PAHs, and provides excellent linearity of response.

2. Gas Chromatography Detectors

Although several types of detectors, such as gas phase fluorescence (206), infrared (130,207), and electron capture (208) detectors have been applied to the GC determination of PAHs, flame ionization and mass spectrometry are by far the detection systems most commonly used. Because of its universality of response, FID is usually employed in tandem with MS, which provides positive identification of known PAHs and MW information for unknown PAHs. Several reviews on the mass spectrometric determination of PAHs have been published (209,210). For most isomeric PAHs, the fragmentation patterns are too similar; in the case of coelution, their separate quantification is usually impossible (197,210). Regardless of the ionization mode used—typically, electron impact (EI) and chemical ionization (CI)—PAHs exhibit little fragmentation, and their mass spectra are characterized by a very intense molecular ion and small fragment ions. Selected-ion monitoring (SIM) of the molecular ion allows detection of PAHs at subnanogram levels. A recent study carried out by Landrock et al. (211) reveals that chemical ionization with water as reactant offers greater selectivity and lower detection limits (10 to 50 times higher response) than 70 eV EIMS and methanol chemical ionization.

The direct analysis of residues from liophilized aqueous solutions for the determination of PAHs after derivatization by nitration has been carried out by tandem mass spectrometry (212), but individual detection of isomeric PAHs was not possible.

Several authors have compared the suitability of LC and GC for the analysis of PAHs in environmental samples (213,214). According to Wise et al. (213) LC-FL and GC-MS are complementary methods, because, depending on the particular PAH, one method or another can be more advantageous than the other, in terms of resolution and sensitivity. Sim et al. (214), on the contrary, thinks that both LC and GC are suitable, but only as long as mass detection, and not optical detection, is used. In our opinion, if either the 16 EPA priority PAHs or the 6 EU reference PAHs are to be determined, LC is the method of choice, since all the target compounds, including the isomeric ones, are completely resolved in a single chromatographic run. As for the detector, we consider FL to be more adequate, not only because of its sensitivity but also because of its selectivity, which would be necessary as coupled to LC, to distinguish, for instance, the target PAH BbF from the nontarget, but commonly coeluting, isomeric PAH perylene. This differentiation would not be possible using LC-MS.

G. Optical Spectrometric Techniques

The analysis of PAHs by optical spectrometric techniques has been described in detail elsewhere (12,215). Although the use of Fourier-transform infrared spectroscopy (FTIR) (216) and phosphorescence (217,218) for the analysis of PAHs from water has been reported, fluorescence spectrometry has been by far the most widely used optical technique, as a result of its comparatively much higher sensitivity and selectivity toward PAHs.

In practice, however, the application of conventional spectrofluorimetry (219,220) to multicomponent analysis is limited by the overlapping and featureless fluorescence spectra (36). Hence, several techniques or modifications of conventional spectrofluorimetry have been proposed to improve selectivity and sensitivity further.

Among these, Shpol'skii fluorimetry (221–224) offers similar sensitivity but higher selectivity than conventional spectrofluorimetry (225). Thus, full widths at half maximum (fwhm) of Shpol'skii spectra are typically 1–10 cm^{-1} compared to fwhm of the order of 300 cm^{-1} for conventional fluorescence and phosphorescence spectra of PAHs (226). The selectivity of the Shpol'skii effect lies in the inherent sharp absorption

bandwidths exhibited by PAHs in appropriate frozen *n*-alkane hosts. Because of its high specificity and sensitivity, which can be further improved by the application of selective laser excitation (225,226), and its capability for measurements in "dirty" samples (227–229), it has been proposed as a useful independent method to check individual steps (e.g., cleanup) as well as final determinations by other procedures (225,230,231). However, a number of disadvantages (requirement of rather specialized and expensive apparatus, not easy automation or hyphenation with other techniques, etc.) have precluded its application as a method in routine analysis (225,230).

Synchronous fluorescence spectroscopy in its various modes—constant-wavelength (144,159,232–240), constant-energy, and variable-angle synchronous spectrofluorimetry (241)—has also been shown to yield narrower spectral bands for PAHs through the simultaneous scanning of both monochromators, and the selectivity can also be enhanced by using the derivative spectra (239–241).

Both conventional (242) and synchronous spectrofluorimetry (144,145,236,240,243) have been applied to the measurement of PAHs sorbed from water on different solid-phase extraction (SPE) materials, such as C18-bonded silica beads (240), C18 membranes (242,243), and Sephadex gels (144,145,236), by the socalled solid-state or solid-phase luminescence spectroscopy (SSLS). Because of its simplicity and relatively high sensitivity (DS in the range μg/L to ng/L), combined SPE-SSLS is a promising technique for field and laboratory screening analyses of PAHs from water. However, to date its applicability to the individual quantification of PAHs in complex matrices or mixtures does not seem possible unless higher selectivity is achieved by the luminescence methods and certain other difficulties (residual fluorescence, scatter of the solid supports, etc.) are overcome (243). Further selectivity and sensitivity can often be achieved with reagents that selectively quench or enhance fluorescence (e.g., metal cations) (244), as well as with organized media, such as micelle (159,233,234,244).

Several authors have studied and compared the effect of various surfactants on the fluorescence signal of certain mixtures of PAHs (159,233,234,244). In general, PAHs present larger and narrower peaks in micellar media than in aqueous solutions. However, some surfactants also exhibit fluorescence, which may interfere with the fluorescent signal of some PAHs. Bockelen and Niessner (159) tested various nonionic surfactants for their suitability as media not only for the fluorescence measurement of PAHs but also for their extraction from water by the previously mentioned micellar or cloud-point extraction. This author found Genapol X-80 and APC 600 CS to be the only surfactants, of the eight tested, capable of fulfilling the properties (clear and transparent in fluorescence, low critical micellar concentration and temperature point, etc.) required for both purposes.

Finally, the spectrofluorimetric technique that is attracting most attention nowadays is laser-induced time-resolved fluorescence over fiber-optic cables for on-line in situ determination of PAHs in water (36,245–253). The application of the time-resolved fluorescence technique to the determination of PAHs has been reviewed by McGown and Nithipatikom (254). This technique, particularly in combination with sophisticated chemometric methods, provides enhanced specificity as a result of the combined measurement of both decay times and fluorescence spectra. The existence of fiber-optic sensor systems with fiber lengths of 50 m and the size of a briefcase (249), the speed and sensitivity of the fluorescence technique (DL in the ng/l range), and the inherent advantages of the direct, real-time analysis, absent any sample manipulation, would make this technique ideal for the analysis of PAHs in aquatic systems in general, but in particular for in-field monitoring of PAHs in remote systems such as groundwaters and wells. However, none of the fluorescence techniques described offer sufficient specificity to resolve mixtures of PAHs, for they can be found in real samples, or to eliminate interferences derived from the water matrix (e.g., humic acids, quenching effects) or from the technique itself (e.g., chromatic dispersion). Until now, most of the research done in this field has been confined to the characterization of mixtures with a limited number of PAHs. A great deal of research will be devoted in the future to the development of improved, more selective fluorescence techniques to be used in combination with fiber-optic systems.

H. Calibration

Both internal and external standard calibration procedures are commonly used for quantification, and both are suggested in the standards methods (304). The use of the internal standard approach, however, renders a very advantageous alternative to the often-laborious external standard or standard addition procedures, since perdeuterated PAHs are excellent internal standards. They react in the same way as the parent compound to the sample preparation, are generally baseline-resolved from the nondeuterated PAHs in the subsequent chromatographic analysis, and present nearly the same op-

tical characteristics for detection (180,225–227, 229,255).

I. Standards

Due to the low solubility of PAHs in water, standards for fortification are prepared in water-miscible organic solvents capable of readily dissolving PAHs, such as acetone, DMSO, methanol, and ACN. For standard mixtures containing the 16 EPA priority PAHs, ACN is preferred over methanol, due to the low solubility of dBahA and BghiP in the latter (270). For the preparation of calibration standard solutions, the most commonly used solvents, selected according to the final analytical technique, are hexane, DCM, ACN, and methanol.

All standards should be stored in the dark and preferably at temperatures of at most 4°C (304,305), although some authors keep them at room temperature (276,289). In some cases, for complete dissolution, the standards solutions have been treated in a sonic bath (274,276). This step, however, should be avoided or performed for a very short time, since the destruction of PAHs by ultrasound has been reported (54).

J. Applications in Water Analysis

Of the various types of water, seawater is probably the least commonly investigated, due to the extremely low concentrations at which PAHs are usually present in it, to the difficulties associated with its sampling, and to the possibility of using alternative, more advantageous matrices, such as sediments (35) or marine organisms (256) (commonly bivalve molluscs) for monitoring seawater pollution. Drinking water and wastewater, on the other hand, are the most frequently investigated water samples and the focus of most official methods (Table 4). Basically, similar analytical schemes can be applied to the analysis of PAHs in the various types of water as long as a cleanup step, when needed, is included in the procedure. However, as it has already been pointed out throughout the chapter, a number of matrix-related factors can strongly determine the method of choice. Thus, samples with low levels of PAHs, such as groundwater and seawater, call for methods allowing the use of relatively large sample volumes and involving the use of sensitive detection techniques, such as fluorescence; samples with high contents of particulate matter, such as some surface waters and wastewaters, require either prefiltration of the sample and subsequent separate analysis of both the filtrate and the filtrated sample, or application of the extraction procedures

(LLE) and the precautions (addition of organic modifiers or surfactants, pH adjustment, etc.) that have been shown to overcome best the losses of PAHs due to adsorption processes, complex matrices, such as some wastewaters, demand a cleanup step and selective detection (FL or MS).

K. Future

The next few years will no doubt see the general application of multidimensional techniques to help identify analytes and prevent coelution and of coupled-column (257) and online procedures, which improve analytical performance (analyte traceability, reliability, and repeatability), increase sample throughput, and diminish operating costs and contamination risks. Great advances are also expected in the field of the fiber-optic sensor systems. New procedures using the most recent technologies will be developed in an effort to fulfill the requirements of continually updated environmental legislation.

REFERENCES

1. EL Kennaway, I Hieger. Carcinogenic substances and their fluorescence spectrum. Br Med J 1:1044, 1930.
2. International Agency for Research on Cancer. IARC Monographs on the Evaluation of the Carcinogenic Risk of Chemicals to Humans. Vol. 32. Polynuclear Aromatic Compounds, Part 1, Chemical, Environmental and Experimental Data. Lyon, France: IARC, 1983.
3. EPA. Drinking Water Regulations and Health Advisories, EPA-822-B96-002. Washington, DC: U.S. Environmental Protection Agency, Office of Water, 1996.
4. EPA. Toxics criteria for those states not complying with Clean Water Act section 303(c)(2)(B). 40 CFR 131.36, 1995, pp. 531–548.
5. JM Neff. Polycyclic Aromatic Hydrocarbons in the Aquatic Environment. Sources, Fates and Biological Effects. London: Applied Science, 1979.
6. ML Lee, MV Novotny, KD Bartle. Analytical Chemistry of Polycyclic Aromatic Compounds. New York: Academic Press, 1981.
7. W Karcher, RJ Fordham, JJ Dubois, PGJM Glaude, JAM Ligthart. Spectral Atlas of Polycyclic Aromatic Compounds. Dordrecht: Reidel, 1983.
8. W Karcher. Spectral Atlas of Polycyclic Aromatic Compounds. Vol. 2. Dordrecht: Kluwer Academic, 1988.
9. W Karcher, J Devillers, Ph Garrigues, J Jacob. Spectral Atlas of Polycyclic Aromatic Compounds. Vol. 3. Dordrecht: Kluwer Academic, 1991.
10. A Bjørseth. Handbook of Polycyclic Aromatic Hydrocarbons. New York: Marcel Dekker, 1983.
11. A Bjørseth, T Ramdahl. Handbook of Polycyclic Aromatic Hydrocarbons. Vol. 2. Emission, Sources and Recent Progress in Analytical Chemistry. New York: Marcel Dekker, 1985.
12. T Vo-Dinh. Chemical Analysis of Polycyclic Aromatic Hydrocarbons. New York: Wiley, 1989.
13. RG Harvey. Polycyclic Aromatic Hydrocarbons. New York: Wiley, 1997.
14. JB Andelman, JE Snodgrass. Incidence and significance of polynuclear aromatic hydrocarbons in the water environment. CRC Crit Rev Environ Control 4:69–83, 1974.
15. RM Harrison, R Perry, RA Wellings. Polynuclear aromatic hydrocarbons in raw, potable and waste waters. Water Res 9:331–346, 1975.
16. DW Connell, GJ Miller. Petroleum hydrocarbons in aquatic ecosystems—behavior and effects of sublethal concentrations: Part 1. CRC Crit Rev Environ Control 11:37–162, 1980.
17. ND Bedding, AE McIntyre, R Perry, JN Lester. Organic contaminants in the aquatic environment. I. Sources and occurrence. Sci Total Environ 25:143–167, 1982.
18. ND Bedding, AE McIntyre, R Perry, JN Lester. Organic contaminants in the aquatic environment. II. Behavior and fate in the hydrological cycle. Sci Total Environ 26:255–312, 1983.
19. ND Bedding, AE McIntyre, JN Lester. Organic contaminants in the aquatic environment. III. Public health aspects, quality standards and legislation. Sci Total Environ 27:163–200, 1983.
20. PM Niaussat, J Trichet, M Heros, NT Luong, JP Ehrhardt. Mise en évidence de benzo-3-4-pyrène dans l'eau et les sédiments organiques de mares saumâtres d'atolls polynésiens. Etude de certains facteurs biotiques et abiotiques associés. CR Acad Sci (Paris) Ser D 281:1031–1034, 1975.
21. WHO (World Health Organization). Guidelines for drinking-water quality. Vol. 2. Health criteria and other supporting information. Geneva: World Health Organization, 1984, pp. 182–189.
22. HW Armstrong, K Fucik, JW Anderson, JM Neff. Effects of oil-field brine effluent on benthic organisms in Trinity Bay, Texas. Mar Environ Res 2:55–69, 1979.
23. A López-García, E Blanco-González, JI García-Alonso, A Sanz-Medel. Determination of some selected polycyclic aromatic hydrocarbons in environmental samples by high-performance liquid chromatography with fluorescence detection. Chromatographia 33:225–230, 1992.
24. RA Pandey, PL Muthal, NM Parhad, P Kumaran. PAH in LTC [low-temperature carbonization] waste water. Indian J Environ Health 33:40–44, 1991.

25. B Davani, JL Gardea, JA Dodson, GA Eiceman. Hazardous organic compounds in liquid wastes from disposal pits for production of natural gas. Int J Environ Anal Chem 20:205–223, 1985.

26. SK Gangwal. Gas chromatographic investigation of raw wastewater from coal gasification. J Chromatogr 204:439–444, 1981.

27. R Kadar, K Nagy, D Fremstad. Determination of polycyclic aromatic hydrocarbons in industrial waste water at the ng ml^{-1} level. Talanta 27:227–230, 1980.

28. FW Karasek, GM Charbonneau, GJ Reuel, HY Tong. Determination of organic compounds leached from municipal incinerator fly ash by water at different pH levels. Anal Chem 59:1027–1031, 1987.

29. G Kiss, A Gelencser, Z Krivacsy, J Hlavay. Occurrence and determination of organic pollutants in aerosol, precipitation and sediment samples collected at Lake Balaton. J Chromatogr A 774:349–361, 1997.

30. AR Fernandes, BR Bushby, JE Faulkner, DS Wallace, P Clayton, BJ Davis. The analysis of toxic organic micro-pollutants (PCDDs, PCDFs, PCBs, and PAHs) in ambient air and atmospheric deposition. Chemosphere 25:1311–1316, 1992.

31. G Kiss, Z Varga-Puchony, J Hlavay. Determination of polycyclic aromatic hydrocarbons in precipitation using solid-phase extraction and column liquid chromatography. J Chromatogr A 725:261–272, 1996.

32. R Kubinec, P Kuráň, I Ostrovský, L Soják. Determination of polycyclic aromatic hydrocarbons from bitumen concrete roads in drainage water by micro-extraction, large-volume sampling and gas chromatography–mass spectrometry with selected-ion monitoring. J Chromatogr A 653:363–368, 1993.

33. FB DeWalle, ESK Chian. Detection of trace organics in well water near a solid waste landfill. J Am Water Works Assoc 73:206–211, 1981.

34. R Dickin, L Eastcott. Environmental assessment and remediation of historical creosote contamination. 37th Annual Pacific North West International Section Conference (PNWIS), British Columbia, 1997, abstract No. 22.1.

35. F Mangani, G Crescentini, E Sisti, F Bruner, S Cannarsa. PAHs, PCBs and chlorinated pesticides in Mediterranean coastal sediments. Int J Environ Anal Chem 45:89–100, 1991.

36. U Panne, F Lewitzka, R Niessner. Fibre-optical sensors for detection of atmospheric and hydrospheric polycyclic aromatic hydrocarbons. Analusis 20:533–542, 1992.

37. J Zerbe, D Baralkiewicz, H Gramowska, J Zminkowska. Extraction efficiency of polynuclear aromatic hydrocarbons from waters containing high suspended-solids content. Chem Anal (Warsaw) 29:455–459, 1984.

38. GE Carlberg, K Martinsen. Influence of humus with time on organic pollutants and comparison of two analytical methods for analyzing organic pollutants in humus water. In: A Bjørseth, G Angeletti, eds. Analysis of Organic Micropollutants in Water. Boston: Reidel, 1981, pp 42–44.

39. KK Chee, MK Wong, HK Lee. Microwave-assisted solvent elution technique for the extraction of organic pollutants in water. Anal Chim Acta 330:217–227, 1996.

40. MN Kayali-Sayadi, S Rubio-Barroso, C Beceiro-Roldan, LM Polo-Diez. Rapid determination of PAH in drinking water samples using solid-phase extraction and HPLC with programmed fluorescence detection. J Liq Chromatogr Relat Technol 19:3135–3146, 1996.

41. J Pörschmann, U Stottmeister. Methodical investigation of interactions between organic pollutants and humic organic material in coal waste waters. Chromatographia 36:207–211, 1993.

42. ET Gjessing, L Berglind. Adsorption of PAH to aquatic humus. Arch Hydrobiol 92:24–30, 1981.

43. A Nelson, N Auffret, J Readman. Initial applications of phospholipid-coated mercury electrodes to determination of polynuclear aromatic hydrocarbons and other organic micro-pollutants in aqueous systems. Anal Chim Acta 207:47–57, 1988.

44. R Cini, P Desideri, L Lepri. Transport of organic compounds across the air–sea interface of artificial and natural marine aerosols. Anal Chim Acta 291:329–340, 1994.

45. ND Bedding, AE McIntyre, JN Lester, R Perry. Analysis of waste waters for polynuclear aromatic hydrocarbons. II. Errors, sampling and storage. J Chromatogr Sci 26:606–615, 1988.

46. F Berthou, J Ducreux, G Bodennec. Analysis of water-soluble acidic compounds derived from spilled oil in a controlled marine enclosure. Int J Environ Anal Chem 21:267–282, 1985.

47. BCJ Zoeteman, E de Greef, FJJ Brinkman. Persistence of organic contaminants in groundwater, lessons from soil pollution incidents in The Netherlands. Sci Total Environ 21:187–202, 1981.

48. MJ Fernández, C. García, RJ García-Villanova, JA Gómez. Evaluation of liquid–liquid extraction and liquid–solid extraction with a new sorbent for the determination of polycyclic aromatic hydrocarbons in raw and finished drinking waters. J Agric Food Chem 44: 1785–1789, 1996.

49. DK Basu, J Saxena. Polynuclear aromatic hydrocarbons in selected U.S. drinking waters and their raw water sources. Environ Sci Technol 12:795–798, 1978.

50. J Kalman, P Siklos, I Szebenyi, L Majerusz, R Hajos. Removal of PAH from petrochemical waste water by use of activated carbon powder. Magy Kem Lapja 39: 402–406, 1984.

51. Y Mori, S Naito, H Matsushita. Simultaneous determination of residual pyrene and monochloropyrene

produced by chlorination of dissolved pyrene in water using derivative spectrofluorimetry. Bunseki-Kagaku 35:513–517, 1986.

52. S Onodera. Characterization and determination of organic compounds in the mutagenic XAD-2 extracts of drinking water. J Chromatogr 557:413–327, 1991.

53. National Academy of Sciences. Drinking Water and Health. Vol. 2. Washington, DC: National Academy Press, 1980, pp 163–164.

54. AP D'Silva, SK Laughlin, SJ Weeks, WH Buttermore. Destruction of polycyclic aromatic hydrocarbons with ultrasound. Polycyclic Aromatic Compounds 1:125–135, 1990.

55. WL Hinze, E Pramauro. A critical review of surfactant-mediated phase separations (cloud-point extraction): theory and applications. Crit Rev Anal Chem 24:133–177, 1993.

56. G Petrick, DE Schulz-Bull, V Martens, K Scholz, JC Duinker. An in situ filtration/extraction system for the recovery of trace organics in solution and on particles tested in deep ocean water. Marine Chemistry 54:97–105, 1996.

57. DM Heaton, KD Bartle, AA Clifford, P Myers, BW King. Rapid separation of polycyclic aromatic hydrocarbons by packed-column supercritical fluid chromatography. Chromatographia 39:607–611, 1994.

58. PL McCarty, M Reinhard, BE Rittmann. Trace organics in groundwater. Environ Sci Technol 15:40–51, 1981.

59. WHO (World Health Organization). Guidelines for drinking-water quality. Vol. 1. Recommendations. Geneva: World Health Organization, 1984, pp 47–102.

60. M Fielding, RF Packham. Organic compounds in drinking water and public health. J Inst Water Eng Sci 31:353–375, 1977.

61. K Levsen, S Behnert, P Mussmann, M Raabe, B Priess. Organic compounds in cloud and rain water. Int J Environ Anal Chem 52:87–97, 1993.

62. J Borneff, H Kunte. Carinogenic substances in water and soil—XVII: about the origin and evaluation of the PAH in water. Arch Hyg Bakteriol 149:226–243, 1965.

63. H Hagenmaier, H Kaut, P Krauss. Analysis of polycyclic aromatic hydrocarbons in sediments, sewage sludges and composts form municipal refuse by HPLC. Int J Environ Anal Chem 23:331–345, 1986.

64. D Baralkiewicz, H Gramowska, G Lesnierowski. Isolation and concentration of polycyclic aromatic hydrocarbons from water samples by adsorption on Tenax GC. Chem Anal (Warsaw) 34:149–153, 1989.

65. RI Crane, M Fielding, TM Gibson, CP Steel. A survey of polycyclic aromatic hydrocarbon levels in British waters. Technical Report TR 158, Water Research Center, Medmenham, U.K., 1981.

66. N Van de Hoed, MTH Halmans, JS Dits. Determination of polycyclic aromatic hydrocarbons (PAHs) at the low ng/L level in the Biesbosch water storage reservoirs (Neth.) for the study of the degradation of chemicals in surface waters. In: A Bjørseth, G Angeletti, eds. Analysis of Organic Micropollutants in Water. Boston: Reidel, 1981, pp 188–192.

67. JT Coates, AW Elzerman, AW Garrison. Extraction and determination of selected polycyclic aromatic hydrocarbons in plant tissues. J Assoc Off Anal Chem 69:110–114, 1986.

68. JK Fawell, S Hunt. Environmental Toxicology: Organic Pollutants. Chichester: Ellis Horwood, 1988, pp 241–269.

69. SK Katiyar, R Agarwal, H Mukhtar. Introduction: sources, occurrence, nomenclature, and carcinogenicity of polycyclic aromatic hydrocarbons. In: HS Rathore, ed. CRC Handbook of Chromatography: Liquid Chromatography of Polycyclic Aromatic Hydrocarbons. Boca Raton, FL: CRC Press, 1993, pp 1–17.

70. GR Shaw, DW Connell. Prediction and monitoring of the carcinogenicity of polycyclic aromatic compounds (PACs). Rev Environ Contam T 135:1–62, 1994.

71. National Academy of Sciences. Drinking Water and Health. Washington, DC: National Academy of Sciences, 1977, pp 691–694, 794.

72. PG Wislocki, AYH Lu. Carcinogenicity and mutagenicity of proximate and ultimate carcinogens of polycyclic aromatic hydrocarbons. In: SK Yang, BD Silverman, eds. Polycyclic Aromatic Hydrocarbon Carcinogenesis: Structure–Activity Relationships. Vol. 1. Boca Raton, FL: CRC Press, 1988, pp 1–30.

73. Council of the European Communities. Council Directive of 15 July 1980 relating to the quality of water intended for human consumption. Directive 80/778/EEC. Off J Eur Comm L229:11–29, 1980.

74. Council of the European Communities. Council Directive of 16 June 1975 concerning the quality required of surface water intended for the abstraction of drinking water in the Member States, Directive 75/440/EEC. Off J Eur Comm L194:26–31, 1975.

75. Council of the European Communities. Council Directive of 20 March 1978 on toxic and dangerous waste, Directive 78/319/EEC. Off J Eur Comm L84:43–48, 1978.

76. Council of the European Communities. Council Directive of 4 May 1976 on pollution caused by certain substances discharged into the aquatic environment of the Community, Directive 76/464/EEC. Off J Eur Comm L129:23–29, 1976.

77. Council of the European Communities. Council Directive of 24 September 1996 concerning integrated pollution prevention and control, Directive 96/61/EC. Off J Eur Comm L257:26–40, 1996.

78. Council of the European Communities. Council Directive of 17 December 1979 on the protection of groundwater against pollution caused by certain dangerous substances, Directive 80/68/EEC. Off J Eur Comm L20:43–50, 1980.

79. MC Hennion, V Pichon, D Barcelo. Surface water analysis (trace-organic contaminants) and EC regulations. TrAC, Trends Anal Chem (Pers-Ed) 13:361–372, 1994.

80. Environmental Protection Agency. National Primary Drinking Water Regulations, EPA 600/4-88-039. Washington, DC: U.S. Environmental Protection Agency, Office of Water, 1996.

81. Environmental Protection Agency. Hazardous constituents. 40 CFR Pt 261, App VIII, 1996, pp 77–85.

82. Environmental Protecton Agency. Ground-water monitoring list. 40 CFR Pt 264, App IX, 1996, pp 341–347.

83. Environmental Protection Agency. Effluent guidelines and standards. 40 CFR Pt 400 to 471, 1995–1996.

84. HS Rathore. CRC Handbook of Chromatography: Liquid Chromatography of Polycyclic Aromatic Hydrocarbons. Boca Raton, FL: CRC Press, 1993.

85. COST Project 64B Management Committee. A comprehensive list of polluting substances which have been identified in various fresh waters, effluent discharges, aquatic animals and plants and bottom sediments. 2nd ed. Brussels: EUROCOP-COST Secretariat, The Commission of the European Communities, 1976.

86. DJ Futoma, SR Smith, TE Smith, J Tanaka. Polycyclic Aromatic Hydrocarbons in Water Systems. Boca Raton, FL: CRC Press, 1981.

87. Council of the European Communities. Council Directive of 9 October 1979 concerning the methods of measurement and frequencies of sampling and analysis of surface water intended for the abstraction of drinking water in the Member States, Directive 79/869/EEC. Off J Eur Comm L271:44–53, 1979.

88. SA Wise. Standard reference materials for the determination of trace organic constituents in environmental samples. In: D Barceló, ed. Environmental Analysis: Techniques, Applications and Quality Assurance. Amsterdam: Elsevier Science, 1993, pp 403–446.

89. K Ballschmiter. Sample-treatment techniques for organic trace analysis. Pure Appl Chem 55:1943–1956, 1983.

90. B Josefsson. Recent concepts in sampling methodology. In: A Bjørseth, G Angeletti, eds. Analysis of Organic Micropollutants in Water. Boston: Reidel, 1981, pp 7–15.

91. H Prast, R Niehaus, B Scheulen, HD Narres, HW Duerbeck. New concept for the determination of PAH in air and in rain water. Fresenius Z Anal Chem 333:709–710, 1989.

92. DI Welch, CD Watts. Collection and identification of trace organic compounds in atmospheric deposition from a semi-rural site in the UK. Int J Environ Anal Chem 38:185–198, 1990.

93. MB Yunker, FA McLaughlin, RW Macdonald, WJ Cretney, BR Fowler, TA Smyth. Measurement of natural trace dissolved hydrocarbons [in water] by in situ column extraction: an inter-comparison of two adsorption resins. Anal Chem 61:1333–1343, 1989.

94. American Society for Testing and Materials (ASTM). Standard practices for preparation of sample containers and for preservation of organic constituents. ASTM Standard, D 3694–96, 1998.

95. American Society for Testing and Materials (ASTM). Standard practices for sampling water from closed conduits. ASTM Standard, D 3370–95a, 1998.

96. American Society for Testing and Materials (ASTM). Standard specification for equipment for sampling water and steam in closed conduits. ASTM Standard, D 1192–95, 1998.

97. AE Greenberg, LS Clesceri, AD Eaton. Standard Methods for the Analysis of Water and Waste Water. 18th ed. Washington, DC: American Public Health Association, American Water Works Association, Water Environment Federation, 1992, Method 1060 Collection and preservation of samples, pp 1-18–1-23.

98. A López-García, E Blanco-González, JI García-Alonso, A Sanz-Medel. Potential of micelle-mediated procedures in the sample preparation steps for the determination of polynuclear aromatic hydrocarbons in waters. Anal Chim Acta 264:241–248, 1992.

99. RK Sorrell, R Reding. Analysis of polynuclear aromatic hydrocarbons [PAH] in environmental waters by high-pressure liquid chromatography. J Chromatogr 185:655–670, 1979.

100. M Reinhard, JE Schreiner, T Everhart, J Graydon. Specific analysis of trace organics in water using gas chromatography and mass spectroscopy. J Environ Pathol Toxicol Oncol 7:417–435, 1987.

101. K Ogan, E Katz, W Slavin. Concentration and determination of trace amounts of several polycyclic aromatic hydrocarbons in aqueous samples. J Chromatogr Sci 16:517–522, 1978.

102. DW Schults, SP Ferraro, LM Smith, FA Roberts, CK Poindexter. Comparison of methods for collecting interstitial water for trace organic compounds and metals analyses. Water Res 26:989–995, 1992.

103. EY Zeng, AR Khan. Extraction of municipal waste water effluent using 90-mm C18 bonded disks. J Microcolumn Sep 7:529–539, 1995.

104. I Liška, J Krupčik, PA Leclercq. The use of solid sorbents for direct accumulation of organic compounds from water matrices—a review of solid-phase extraction techniques. J High Resol Chromatogr 12:577–590, 1989.

105. F Eisenbeiss, H Hein, R Joester, G Naundorf. The separation by LC and determination of polycyclic aromatic hydrocarbons in water using an integrated enrichment step. Chromatogr Newsletter 6:8–12, 1978.

106. HG Kicinski, A Kettrup. Solid-phase extraction and HPLC analysis of polycyclic aromatic hydrocarbons in drinking water. Vom Wasser 71245–254, 1988.

107. HG Kicinski. Solid-phase extraction and HPLC analysis of PAH in drinking water, soil and used [cooking] oil. SLZ, Schweiz Lab Z 47:152–154, 157–158, 1990.

108. HG Kicinski. Solid-phase extraction of PAH from drinking water and slightly polluted surface water by addition of Hyamine. Z Wasser Abwasser Forsch 25:289–294, 1992.

109. M Dressler. Extraction of trace amounts of organic compounds from water with porous organic polymers. J Chromatogr 165:167–206, 1979.

110. BK Afghan, RJ Wilkinson, A Chow, TW Findley, HD Gesser, KI Srikameswaran. Comparative study of the concentration of polynuclear aromatic hydrocarbons by open-cell polyurethan foams. Water Res 18:9–16, 1984.

111. JN Huckins, MW Tubergen, GK Manuweera. Semipermeable membrane devices containing model lipid: a new approach to monitoring the bioavailability of lipophilic contaminants and estimating their bioconcentration potential. Chemosphere 20, 533–552, 1990.

112. JN Huckins, GK Manuweera, JD Petty, D Mackay, JA Lebo. Lipid-containing semipermeable membrane devices for monitoring organic contaminants in water. Environ Sci Technol 27:2489–2496, 1993.

113. HF Prest, JN Huckins, JD Petty, S Herve, J Paasivirta, P Heinonen. A survey of recent results in passive sampling of water and air by semipermeable membrane devices. Marine Pollution Bull 31:306–312, 1995.

114. JA Lebo, JL Zajicek, JN Huckins, JD Petty, PH Peterman. Use of semipermeable membrane devices for in situ monitoring of polycyclic aromatic hydrocarbons in aquatic environments. Chemosphere 25:697–718, 1992.

115. JA Lebo, JL Zajicek, CE Orazio, JD Petty, JN Huckins, EH Douglas. Use of the semipermeable membrane device (SPMD) to sample polycyclic aromatic hydrocarbon pollution in a lotic system. Polycyclic Aromatic Compounds 8:53–65, 1996.

116. JB Moring, DR Rose. Occurrence and concentration of polycyclic aromatic hydrocarbons in semipermeable membrane devices and clams in three urban streams of the Dallas–Fort Worth Metropolitan Area, Texas. Chemosphere 34:551–566, 1997.

117. PCM Van-Noort, E Wondergem. Isolation of some polynuclear aromatic hydrocarbons from aqueous samples by means of reversed-phase concentrator columns. Anal Chim Acta 172:335–340, 1985.

118. VF Ruban, IA Anisimova. Determination of polycyclic aromatic hydrocarbons in water by capillary HPLC with fluorimetric detection. Zh Anal Khim 46:2035–2040, 1991.

119. KA Pinkerton. Direct LC analysis of selected priority pollutants in water at ppb levels. J High Resol Chromatogr—Chromatogr Commun 4:33–34, 1981.

120. RE Kaiser, R Rieder. High-boiling organic traces in drinking water. Quantitative analysis by liquid–liquid enrichment within the analytical glass capillary. J Chromatogr 477:49–52, 1989.

121. G Goretti, MV Russo, E Veschetti. Use of the same capillary column for both sampling and gas-chromatographic analysis of aqueous organic pollutants. J High Resolut Chromatogr 15:51–54, 1992.

122. G Goretti, MV Russo, E Veschetti. Trapping efficiency of aqueous micropollutants sampled and analysed using the same capillary column. Theoretical aspects and experimental results. Chromatographia 35:653–660, 1993.

123. E Chladek, RS Marano. Use of bonded-phase silica sorbents for sampling of priority pollutants in waste waters. J Chromatogr Sci 22:313–320, 1984.

124. CM Josefson, JB Johnston, R Trubey. Adsorption of organic compounds from water with porous poly(tetrafluoroethylene). Anal Chem 56:764–768, 1984.

125. SR Sargenti, HM McNair. Comparison of solid-phase extraction and supercritical fluid extraction for extraction of polycyclic aromatic hydrocarbons from drinking water. J Microcolumn Sep 10:125–131, 1998.

126. RE Shirey, GD Wachob. New thin film fiber for solid phase microextraction of semivolatiles. Supelco Rep 13:6–7, 1994.

127. L González-Bravo, L Rejthar. Quantitative determination of tract concentration of organics in water by solvent extraction and fused-silica capillary gas chromatography: aliphatic and polynuclear hydrocarbons. Int J Environ Anal Chem 24:305–318, 1986.

128. CA Valkenburg, WD Munslow, LC Butler. Evaluation of modifications to extraction procedures used in analysis of environmental samples [water] from Superfund sites. J Assoc Off Anal Chem 72:602–608, 1989.

129. A Matthiessen. Use of a keeper to enhance the recovery of volatile polycyclic aromatic hydrocarbons in HPLC analysis. Chromatographia 45:190–194, 1997.

130. J Poerschmann, FD Kopinke, M Remmler, K Mackenzie, W Geyer, S Mothes. Hyphenated techniques for characterizing coal waste waters and associated sediments. J Chromatogr A 750:287–301, 1996.

131. Z Zhang, J Pawliszyn. Analysis of organic compounds in environmental samples by headspace solid-phase micro-extraction. J High Resolut Chromatogr 16:689–692, 1993.

132. I Haag. Potential applications of coupled solid-phase microextraction-HPLC. LaborPraxis 20:66–68, 71, 1996.

133. F Mangani, G Crescentini, P Palma, F Bruner. Performance of graphitized carbon black cartridges in the extraction of some organic priority pollutants from water. J Chromatogr 452:527–534, 1988.

134. A Lagana, BM Petronio, M Rotatori. Concentration and determination of polycyclic aromatic hydrocarbons in aqueous samples on graphitized carbon black. J Chromatogr 198:143–149, 1980.

135. F Bruner, G Furlani, F Mangani. Sample enrichment for gas chromatographic–mass spectrometric analysis of polynuclear aromatic hydrocarbons in water and in organic mixtures. J Chromatogr 302:167–172, 1984.

136. S Monarca, G Scassellati-Sforzolini, A Savino. Determination of polycyclic aromatic hydrocarbons in water. Comparison between different methods of extraction. Inquinamento 20:41–45, 1978.

137. JF Pankow, MP Ligocki, ME Rosen, LM Isabelle, KM Hart. Adsorption–thermal desorption with small cartridges for the determination of trace aqueous semivolatile organic compounds. Anal Chem 60:40–47, 1988.

138. X Xu, Z Jin. High-performance liquid-chromatographic studies of environmental carcinogens in China. J Chromatogr 317:545–555, 1984.

139. M Kilarska, R Rajtar, W Solarski, E Zieliński. Application of XAD-2 resin for benzo[a]pyrene adsorption from water eluate obtained from used molding sands. Chem Anal (Warsaw) 37:279–284, 1992.

140. AI Krylov, IO Kostyuk, NF Volynets. Determination of polycyclic aromatic hydrocarbons in water by high-performance liquid chromatography with preconcentration and fractionation on XAD-2. Zh Anal Khim 50:543–551, 1995.

141. C O'Donnell. Routine HPLC of polynuclear aromatic hydrocarbons. In: A Bjørseth, G Angeletti, eds. Analysis of Organic Micropollutants in Water. Boston: Reidel, 1981, pp 174–177.

142. M Geisert, T Rose, RK Zahn. Extraction and trace enrichment of genotoxicants from environmental samples by solid-phase adsorption on "Blue Pearls." Fresenius Z Anal Chem 330:437–438, 1988.

143. H Hayatsu. Cellulose bearing covalently linked copper phthalocyanine trisulphonate as an adsorbent selective for polycyclic compounds and its use in studies of environmental mutagens and carcinogens. J Chromatogr 597:37–56, 1992.

144. LF Capitán-Vallvey, M Del-Olmo-Iruela, R Avidad-Castaneda, JL Vilchez-Quero. Determination of benzo[a]pyrene in water by synchronous fluorimetry following pre-concentration of Sephadex gels. Anal Lett 26:2443–2454, 1993.

145. LF Capitán-Vallvey, M del Olmo, R Avidad, N Navalón, I De Orbe, JL Vilchez. Close overlapping discrimination of polycyclic aromatic hydrocarbons by synchronous scanning at variable-angle solid-phase spectrofluorimetry. Anal Chim Acta 302:193–200, 1995.

146. W Gerlich, G Martin, H Panning. Extraction process for PAH determination in drinking and ground water. LaborPraxis 15:942–944, 1991.

147. GA Junk, JJ Richard. Solid-phase extraction on a small scale. J Res Natl Bur Stand (US) 93:274–276, 1988.

148. C Cavelier. Determination of polycyclic aromatic hydrocarbons in water by high-performance liquid chromatography with fluorimetric detection. Analusis 8:46–48, 1980.

149. PH Tang, JS Ho, JW Eichelberger. Determination of organic pollutants in reagent water by liquid–solid extraction followed by supercritical-fluid elution. J AOAC Int 76:72–82, 1993.

150. P Dolezel, M Krejci, V Kahle. Enrichment technique in an automated liquid microchromatograph with a capillary mixer. J Chromatogr A 675:47–54, 1994.

151. H Zobel, H Panning, S Nowak. Determination of PAH by online coupling of SPE with HPLC. LaborPraxis 19:28–31, 1995.

152. D Jahr, JJ Vreuls, AJH Louter, W Loebel. Water analysis by online solid-phase extraction–gas chromatography–mass spectrometry. GIT Fachz Lab 40:178, 180–183, 1996.

153. JJ Langenfeld, SB Hawthorne, DJ Miller. Quantitative analysis of fuel-related hydrocarbons in surface water and waste water samples by solid-phase microextraction. Anal Chem 68:144–155, 1996.

154. J Poerschmann, Z Zhang, FD Kopinke, J Pawliszyn. Solid-phase microextraction for determining the distribution of chemicals in aqueous matrices. Anal Chem 69:597–600, 1997.

155. C Woolley, R Mindrup. Solid-phase microextraction: rapid and versatile extraction for GC or HPLC applications. Supelco Rep 15:5–7, 9, 1996.

156. R Eisert, K Levsen. Solid-phase microextraction coupled to gas chromatography: a new method for the analysis of organics in water. J Chromatogr A 733:143–157, 1996.

157. RP Frankewich, WL Hinze. Evaluation and optimization of the factors affecting non-ionic surfactant-mediated phase separations. Anal Chem 66:944–954, 1994.

158. C García-Pinto, JL Pérez-Pavón, B Moreno-Cordero. Cloud point preconcentration and high-performance liquid-chromatographic determination of polycyclic aromatic hydrocarbons with fluorescence detection. Anal Chem 66:874–881, 1994.

159. A Bockelen, R Niessner. Combination of micellar extraction of polycyclic aromatic hydrocarbons from aqueous media with detection by synchronous fluorescence. Fresenius J Anal Chem 346:435–440, 1993.

160. V Janda, J Fanta, J Vejrosta. Factors affecting SFE of PAHs from water samples. J High Resolut Chromatogr 19:588–590, 1996.

161. OP Heemken, N Theobald, BW Wenclawiak. Comparison of ASE and SFE with Soxhlet, sonication, and methanolic saponification extractions for the determination of organic micropollutants in marine particulate matter. Anal Chem 69:2171–2180, 1997.

162. JW Hofstraat, S Griffioen, RJ van de Nesse, UATh Brinkman, C Gooijer, NH Velthorst. Coupling of narrow-bore column liquid chromatography and thin layer chromatography. Interface optimization and

characteristics for normal-phase liquid chromatography. J Planar Chromatogr 1:220–226, 1988.

163. DF Bishop. GC/MS methodology for priority organics in municipal wastewater treatment. Report No. EPA-600/2-80-196. Environmental Protection Agency, municipal Environmental Research Laboratory, Cincinnati, OH, 1980.

164. DJ Futoma, SR Smith, J Tanaka, TE Smith. Chromatographic methods for analysis of polycyclic aromatic hydrocarbons in water systems. CRC Crit Rev Anal Chem 12:69–153, 1981.

165. DJC Constable. Comparison of solvent reduction methods for concentration of PAH solutions. Environ Sci Technol 18:975–978, 1984.

166. H Kunte. Determination of the polycyclic aromatic hydrocarbons (PAH) specified in the drinking-water regulations: investigation of the possible influence of other PAH. Fresenius Z Anal Chem 301:287–289, 1980.

167. L Weil, G Grimmer, H Hellmann, B De-Jong, H Kunte, M Sonneborn, I Stoeber. Semi-quantitative test for detection of polycyclic hydrocarbons in drinking water. Z Wasser Abwasser Forsch 13:108–111, 1980.

168. L Weil, HE Hauck. New method for semi-quantitative determination of polycyclic hydrocarbons [PAH] in drinking water. GIT Fachz Lab 24:538–540, 1980.

169. D Grange, P Clement. Application of liquid phase and thin layer chromatography to the analysis of organic water pollutants. A bibliography. Rapp Rech LPC 103: 41, 1981.

170. H Hellmann. Comparison of group and single determination methods in the analysis of polycyclic aromatic compounds by fluorescence detection. Fresenius Z Anal Chem 314:125–128, 1983.

171. VL Olsanski, T Yatzus. Method for determination of [total] polycyclic aromatic hydrocarbons by fluorodensitometry. Microchem J 28:151–154, 1983.

172. G Matysik, E Soczewinski. Pre-concentration thin-layer chromatography of polyaromatic hydrocarbons on octadecyl silica. Chem Anal (Warsaw) 28:521–525, 1983.

173. H Schoessner, W Falkenberg, H Althaus. One-dimensional thin-layer chromatographic separation and fluorimetric quantification of polycyclic aromatic hydrocarbons in water samples. Z Wasser Abwasser Forsch 16:132–135, 1983.

174. N Schmidt. Pre-coated reversed-phase high-performance TLC plates in daily routine work. GIT Suppl 4:78–81, 1985.

175. SK Poole, TA Dean, CF Poole. Preparation of environmental samples for the determination of polycyclic aromatic hydrocarbons by thin-layer chromatography. J Chromatogr 400:323–341, 1987.

176. A Colmsjö, M Ericsson. Synthesis and performance of HPTLC [high-performance TLC] plates modified with cyanopropyl- and octadecyl-trichlorosilane. J High

Resolut Chromatogr Chromatogr Commun 10:177–180, 1987.

177. Department of the Environment (UK). The determination of six specific polynuclear aromatic hydrocarbons in waters (with notes on the determination of other PAH) 1985. Methods Exam Waters Assoc Mater, 1988, pp 45.

178. W Funk, G Donnevert, B Schuch, V Gluck, J Becker. Quantitative HPTLC determination of six polynuclear aromatic hydrocarbons (PAH) in water. J Planar Chromatogr Mod TLC 2:317–320, 1989.

179. R Amos. Evaluation of bonded phases for high-performance liquid-chromatographic determination of polycyclic aromatic hydrocarbons in effluent waters. J Chromatogr 204:469–478, 1981.

180. SA Wise, LC Sander, WE May. Determination of polycyclic aromatic hydrocarbons by liquid chromatography. J Chromatogr 642:329–349, 1993.

181. SA Wise. High-performance liquid chromatography for the determination of polycyclic aromatic hydrocarbons. In: A Bjørseth, ed. Handbook of Polycyclic Aromatic Hydrocarbons. New York: Marcel Dekker, 1983, pp 183–256.

182. SA Wise. Recent progress in the determination of PAH by high performance liquid chromatography. In: A Bjørseth, T Ramdahl, eds. Handbook of Polycyclic Aromatic Hydrocarbons. Vol. 2. Emission Sources and Recent Progress in Analytical Chemistry. New York: Marcel Dekker, 1985, pp 113–191.

183. JA Schmit, RA Henry, RC Williams, JF Dieckman. Applications of high speed reversed-phase liquid chromatography. J Chromatogr Sci 9:645–651, 1971.

184. R Fischer. Determination of the six polycyclic aromatic hydrocarbons according to drinking-water regulations on a special C18 reversed-phase column: dependence of retention on temperature, detecton limit by UV absorption, and selection of mobile phases. Fresenius Z Anal Chem 311:109–111, 1982.

185. M Makela, L Pyy. Effect of temperature on retention time reproducibility and on the use of programmable fluorescence detection of fifteen polycyclic aromatic hydrocarbons. J Chromatogr A 699:49–57, 1995.

186. LC Sander, SA Wise. Subambient temperature modification of selectivity in reversed-phase liquid chromatography. Anal Chem 61:1749–1754, 1989.

187. A Gratzfeld-Huesgen. New HPLC offers reliable control of column parameters. Retention-time stability and PNA analysis. Hewlett Packard Peak 3:2–4, 1995.

188. VM Pozhidaev, VG Berezkin, AA Korolev, TP Popova, KA Pozhidaeva,-KA. Capillary chromatography of polycyclic aromatic hydrocarbons on a home-produced capillary column with immobilized stationary liquid phase SE-30. Zh Anal Khim 42:2222–2226, 1987.

189. Anonymous. New capillary column for rapid screening of hazardous waste samples. Supelco Rep 9:2–5, 1990.

190. M Geissler, H Hohmann, HW Stuurman, A Weimar. Different techniques for water analysis. LaborPraxis 17:30, 33–37, 1993.

191. R Burrows, DG Gillespie, M Cooke, O Eddib. Simple preparation of a Poly-S 179 capillary column for the analysis of polycyclic aromatic hydrocarbons and triglycerides. J Chromatogr 404:248–253, 1987.

192. Anonymous. High resolution, low bleed GC analysis of polynuclear aromatic hydrocarbons. J and W Scientific GC Environ Appl Note E2, 1994.

193. DL Poster, MJ López de Alda, MM Schantz, LC Sander, MG Vangel, SA Wise. Certification of a diesel particulate related standard reference material (SRM 1975) for PAHs. Polycyclic Aromatic Compounds (in press).

194. HGJ Mol, M Althuizen, HG Janssen, CA Cramers, UAT Brinkman. Environmental applications of large volume injection in capillary GC using programmed temperature vaporizing injectors. J High Resolut Chromatogr 19:69–79, 1996.

195. B Sortland. Analysis of polycyclic aromatic hydrocarbons by gas chromatography. In: A Bjørseth, ed. Handbook of Polycyclic Aromatic Hydrocarbons. New York: Marcel Dekker, 1983, pp 257–300.

196. KD Bartle. Recent advances in the analysis of polycyclic aromatic compounds by gas chromatography. In: A Bjørseth, T Randahl, eds. Handbook of Polycyclic Aromatic Hydrocarbons. Vol. 2. Emission Sources and Recent Progress in Analytical Chemistry. New York: Marcel Dekker, 1985, pp 193–236.

197. JC Fetzer. Gas- and liquid-chromatographic techniques. In: T Vo-Dinh, ed. Chemical Analysis of Polycyclic Aromatic Hydrocarbons. New York: Wiley, 1989, pp 59–109.

198. T Takeuchi, D Ishii. Application of ultra-micro high-performance liquid chromatography to trace analysis. J Chromatogr 218, 199–208, 1981.

199. TJ Gremm, FH Frimmel. Application of liquid chromatography–particle beam mass spectrometry and gas chromatography–mass spectrometry for the identification of metabolites of polycyclic aromatic hydrocarbons. Chromatographia 38:781–788, 1994.

200. DTSC–OPPTD PAH RaPID ASSAY®, Certification No: 96-01-022. PAH RaPID ASSAY® (Immunoassay for Polynucleated Aromatic Hydrocarbons in Soil and Water).

201. X Fu, J Lü, A Zhu. Micellar electrokinetic capillary chromatography with mixed ethanol–water solvent. Fenxi Huaxue 18:791–795, 1990.

202. JL Peschet, C Tinet. Analysis of mineral and organic pollution in various waters by high-performance ion and liquid chromatography. Eau Ind Nuisances 114: 47–50, 1987.

203. DR Doerge, J Clayton, PP Fu, DA Wolfe. Analysis of polycyclic aromatic hydrocarbons using liquid chromatography-particle beam mass spectrometry. Biol Mass Spectrom 22:654–660, 1993.

204. MA Brown, RD Stephens, IS Kim. Liquid chromatography–mass spectrometry—a new window for environmental analysis. Trends Anal Chem 10:330–336, 1991.

205. JF Anacleto, L Ramaley, FM Benoit, RK Boyd, MA Quilliam. Comparison of liquid chromatography/mass spectrometry interfaces for the analysis of polycyclic aromatic compounds. Anal Chem 67:4145–4154, 1995.

206. B Galle, P Grennfelt. Instrument for polycyclic aromatic hydrocarbon analysis of air-borne particles by capillary gas chromatography with laser-induced fluorescence detection. J Chromatogr 279:643–648, 1983.

207. JF Schneider, KR Schneider, SE Spiro, DR Bierma, LF Sytsma. Evaluation of gas chromatography–matrix-isolation infrared spectroscopy for the quantitative analysis of environmental samples. Appl Spectrosc 45: 566–571, 1991.

208. MF Mehran, WJ Cooper, M Mehran, R Diaz. Effluent stream splitting to two different detectors. J High Resolut Chromatogr Chromatogr Commun 7:639–640, 1984.

209. B Josefsson. Mass spectrometric analysis of polycyclic aromatic hydrocarbons. In: A Bjørseth, ed. Handbook of Polycyclic Aromatic Hydrocarbons. New York: Marcel Dekker, 1983, pp 301–321.

210. RA Hites. Mass spectrometry of polycyclic aromatic compounds. In: T Vo-Dinh, ed. Chemical Analysis of Polycyclic Aromatic Hydrocarbons. New York: Wiley, 1989, pp 219–261.

211. A Landrock, H Richter, H Merten. Water CI+, a new selective and highly sensitive method for the detection of environmental components using ion trap mass spectrometers. Fresenius J Anal Chem 351:536–543, 1995.

212. DF Hunt, J Shabanowitz, TM Harvey, M Coates. Scheme for the direct analysis of organic[compound]s in the environment by tandem mass spectrometry. Anal Chem 57:525–537, 1985.

213. SA Wise, LR Hilpert, GD Byrd, WE May. Comparison of liquid chromatography with fluorescence detection and gas chromatography/mass spectrometry for the determination of polycyclic aromatic hydrocarbons in environmental samples. Polycyclic Aromatic Compounds 1:81–98, 1990.

214. PG Sim, RK Boyd, RM Gershey, R Guevremont, WD Jamieson, MA Quilliam, RJ Gergely. A comparison of chromatographic and chromatographic/mass spectrometric techniques for the determination of polycyclic aromatic hydrocarbons in marine sediments. Biomed Environ Mass Spectrom 14:375–381, 1987.

215. EL Wehry. Optical spectrometric techniques for determination of polycyclic aromatic hydrocarbons. In: A Bjørseth, ed. Handbook of Polycyclic Aromatic Hydrocarbons. New York: Marcel Dekker, 1983, pp 323–396.

216. G Mille, M Guiliano, H Reymond, H Dou. Analysis of hydrocarbons by Fourier-transform infrared spectroscopy. Int J Environ Anal Chem 21:239–260, 1985.

217. AV Karyakin, YuN Smirnov, VI Vershinin. Phosphorimetric determination of polycyclic aromatic hydrocarbons in aqueous saline matrices. Zh Anal Khim 43: 728–732, 1988.

218. AD Campiglia, JP Alarie, T Vo-Dinh. Development of a room-temperature phosphorescence fiber-optic sensor. Anal Chem 68:1599–1604, 1996.

219. YaI Korenman, VA Vorob'ev, KI Zhilinskaya. Determination of polycyclic aromatic hydrocarbons in mineral waters. Zh Prikl Khim (Leningrad) 63:910–912, 1990.

220. GI Romanovskaya, NA Lebedeva. Luminescence determination of polycyclic aromatic hydrocarbons against the background of self-fluorescence of natural, drinking and waste waters. Zh Anal Khim 48:1983–1990, 1993.

221. L Paturel, J Jarosz, C Fachinger, J Suptil. Realisation d'un dispositif experimental pour le dosage des hydrocarbures aromatiques polycycliques par spectrofluorimetry Shpol'skii a 10 K. Anal Chim Acta 147: 293–302, 1983.

222. AV Karyakin, TS Sorokina, NF Efimova. Rapid luminescence determination of polycyclic aromatic substances in water. Zh Anal Khim 39:1697–1699, 1984.

223. M Wittenberg, J Jarosz, L Paturel, M Vial, M Martin-Bouyer. Analysis by Shpol'skii spectrofluorimetry at 10 K of polynuclear aromatic hydrocarbons in the environment: sampling and extraction procedures, quantitative results. Analysis 13:249–260, 1985.

224. B Santoni, C Mandon. Determination of polycyclic aromatic hydrocarbon (PAH) in water and sediments by low-temperature fluorimetry. Analysis 9:259–264, 1981.

225. JW Hofstraat, WJM Van-Zeijl, F Ariese, JWG Mastenbroek, C Gooijer, NH Velthorst. Laser-excited Shpol'skii fluorimetry: applications in marine environmental analysis of polynuclear aromatic hydrocarbons. Mar Chem 33:301–320, 1991.

226. Y Yang, AP D'Silva, VA Fassel. Laser-excited Shpol'skii spectroscopy for the selective excitation and determination of polynuclear aromatic hydrocarbons. Anal Chem 53:894–899, 1981.

227. JW Hofstraat, HJM Jansen, GPH Hoornweg, C Gooijer, NH Velnthorst. Quantitative determination of polycyclic aromatic hydrocarbons in harbor sediment via high-resolution Shpol'skii spectroscopy. Int J Environ Anal Chem 21:299–332, 1985.

228. Y Yang, AP D'Silva, VA Fassel, M Iles. Direct determination of polynuclear aromatic hydrocarbons in coal liquids and shale oil by laser excited Shpol'skii spectrometry. Anal Chem 52:1350–1351, 1980.

229. Y Yang, AP D'Silva, VA Fassel. Deuterated analogues as internal reference compounds for the direct determination of benzo[a]pyrene and perylene in liquid fuels by laser-excited Shpol'skii spectrometry. Anal Chem 53:2107–2109, 1981.

230. JWG Mastenbroek, F Ariese, C Gooijer, NH Velthorst. Shpol'skii fluorimetry as an independent identification method to upgrade routine HPLC analysis of polycyclic aromatic hydrocarbons. Chemosphere 21:377–386, 1990.

231. Ph Garrigues, MP Marniesse, SA Wise, J Bellocq, M Ewald. Identification of mutagenic methylbenz[a]anthracene and methylchrysene isomers in natural samples by liquid chromatography and Shpol'skii spectrometry. Anal Chem 59:1695–1700, 1987.

232. J Simal-Lozano, MA Lage-Yusty, L Vázquez-Oderiz, MJ López de Alda-Villaizán. Mise au point d'une method de fluorescence synchrone pour le dosage de HAP dans l'eau potable. J Francais d'Hydrologie 23: 37–55, 1992.

233. JJ Santana-Rodríguez, Z Sosa-Ferrera, A Afonso-Perera, V González-Diaz. Sensitive simultaneous determination of benzo[a]pyrene, perylene and chrysene by synchronous spectrofluorometry in nonionic micellar media. Talanta 39:1611–1617, 1992.

234. JJ Santana-Rodríguez, J Hernandez-García, MM Bernal-Suarez, AB Martin-Lazaro. Analysis of mixtures of polycyclic aromatic hydrocarbons in seawater by synchronous fluorescence spectrometry in organized media. Analyst (London) 118:917–921, 1993.

235. M López de Alda-Villaizán, E Alvárez-Pineiro, S García-Falcón, A Lage-Yusty, J Simal-Lozano. Hexane as a solvent in the determination of PAH by spectrofluorimetry. Analysis 22:495–498, 1994.

236. JL Vilchez, M Del-Olmo, R Avidad, LF Capitán-Vallvey. Determination of polycyclic aromatic hydrocarbon residues in water by synchronous solid-phase spectrofluorimetry. Analyst (London) 119:1211–1214, 1994.

237. MJ López de Alda-Villaizán, J Simal-Lozano, MA Lage-Yusty. Determination of polycyclic aromatic hydrocarbons in drinking and surface waters from Galacia (N.W. Spain) by constant-wavelength synchronous spectrofluorimetry. Talanta 42:967–970, 1995.

238. MJ López de Alda-Villaizán, S García-Falcón, MA Lage-Yusty, J Simal-Lozano. Synchronous spectrofluorimetric determination of total amounts of the six polycyclic aromatic hydrocarbons officially designated as indicators of drinking water quality. J AOAC Int 78:402–406, 1995.

239. MJ López de Alda-Villaizán, MS García-Falcón, S González-Amigo, J Simal-Lozano, MA Lage-Yusty. Second-derivative constant-wavelength synchronous-scan spectrofluorimetry for determination of benzo[b]fluoranthene, benzo[a]pyrene and indeno[1,2,3-cd]pyrene in drinking water. Talanta 43:1405–1412, 1996.

240. P Canizares, MD Luque-de-Castro. Flow-through sensor based on derivative synchronous fluorescence

spectrometry for the simultaneous determination of pyrene, benzo[*e*]pyrene and benzo[*ghi*]perylene in water. Fresenius J Anal Chem 354:291–295, 1996.

241. YQ Li, XZ Huang. Rapid resolution of five polynuclear aromatic compounds in a mixture by derivative non-linear variable-angle synchronous fluorescence spectrometry. Fresenius J Anal Chem 357:1072–1075, 1997.

242. EJ Poziomek, D Eastwood, RL Lidberg, G Gibson. Solid-phase extraction and solid-state spectroscopy for monitoring water pollution. Anal Lett 24:1913–1921, 1991.

243. D Eastwood, ME Dominguez, RL Lidberg, EJ Poziomek. Solid-phase extraction–solid-state luminescence approach for monitoring PAH in water. Analysis 22:305–310, 1994.

244. K Nithipatikom, LB McGown. Effects of metal cations on the fluorescence intensity of polycyclic aromatic hydrocarbons in sodium taurocholate micellar solutions. Anal Chem 60:1043–1045, 1988.

245. SH Lieberman, SM Inman, GA Theriault. Use of time-resolved spectral fluorimetry for improving specificity of fiber optic-based chemical sensors. Proc SPIE Int Soc Opt Eng 1172:94–98, 1989.

246. R Niessner, W Robers, A Krupp. New analytical concept: remote laser-induced and time-resolved fluorescence (LIF) of PAH in aerosols or water. Fresenius Z Anal Chem 333:708–709, 1989.

247. R Niessner, W Robers, A Krupp. Fiber optical sensor system using a tunable laser for detection of PAH [polycyclic aromatic hydrocarbons] on particles and in water. Proc SPIE Int Soc Opt Eng 1172 (Chem Biochem Environ Sens):145–156, 1989.

248. SM Inman, P Thibado, GA Theriault, SH Lieberman. Development of a pulsed-laser, fiber-optic-based fluorimeter: determination of fluorescence decay times of polycyclic aromatic hydrocarbons in seawater. Anal Chim Acta 239:45–51, 1990.

249. R Niessner, U Panne, H Schroeder. Fibre-optic sensor for the determination of polynuclear aromatic hydrocarbons with time-resolved, laser-induced fluorescence. Anal Chim Acta 255:231–243, 1991.

250. U Panne, R Niessner. Fiber-optical sensor for polynuclear aromatic hydrocarbons based on multidimensional fluorescence. Sens Actuators-B B13:288–292, 1993.

251. E Jaeger, H Lucht, A Weissbach. Determination of polyaromatic hydrocarbons by laser fluorescence. Part 1. LaborPraxis 18:42–44, 51–53, 1994.

252. J Bublitz, W Schade. Laser spectrometry in environmental analysis. GIT Fachz Lab 39:117–118, 121–123, 1995.

253. R Kotzick, R Niessner. Application of time-resolved, laser-induced and fiber-optically guided fluorescence for monitoring of a PAH-contaminated remediation site. Fresenius J Anal Chem 354:72–76, 1996.

254. LB McGown, K Nithipatikom. Phase-resolved fluorescence spectroscopy. In: T Vo-Dinh, ed. Chemical Analysis of Polycyclic Aromatic Hydrocarbons. New York: Wiley, 1989, pp 201–218.

255. WF Kline, SA Wise, WE May. The application of perdeuterated polycyclic aromatic hydrocarbons (PAH) as internal standards for the liquid chromatographic determination of PAH in a petroleum crude oil and other complex mixtures. J Liq Chromatogr 8:223–237, 1985.

256. M Ogata, K Fujisawa. Organic sulphur compounds and polycyclic hydrocarbons transferred to oyster and mussel from petroleum suspension. Water Res 19:107–118, 1985.

257. UAT Brinkman. Multidimensionality: a hot topic. Analysis 24:M12–M13, 1996.

258. T Renner, D Baumgarten, KK Unger. Analysis of organic pollutants in water at trace levels using fully automated solid-phase extraction coupled to high-performance liquid chromatography. Chromatographia 45:199–205, 1997.

259. R Ferrer, J Guiteras, JL Beltrán. Optimization of an online precolumn preconcentration method for the determination of polycyclic aromatic hydrocarbons (PAHs) in water samples (river and sea water). Anal Lett 29:2201–2219, 1996.

260. R Ferrer, J Guiteras, JL Beltrán. Use of cloud-point extraction methodology for the determination of PAH priority pollutants in water samples by high-performance liquid chromatography with fluorescence detection and wavelength programming. Anal Chim Acta 330:199–206, 1996.

261. MJ Fernández, C García, RJ García-Villanova, JA Gómez. Evaluation of liquid–liquid extraction and liquid–solid extraction with a new sorbent for the determination of polycyclic aromatic hydrocarbons in raw and finished drinking waters. J Agric Food Chem 44:1785–1789, 1996.

262. MA Rodríguez-Delgado, MJ Sánchez, V González, F García-Montelongo. Determination of polynuclear aromatic hydrocarbons in environmental samples by micellar liquid chromatography. J High Resolut Chromatogr 19:111–116, 1996.

263. E Manoli, C Samara. Polycyclic aromatic hydrocarbons in waste waters and sewage sludge: extraction and clean up for HPLC analysis with fluorescence detection. Chromatographic 43:135–142, 1996.

264. HP Nirmaier, E Fischer, A Meyer, G Henze. Determination of polycyclic aromatic hydrocarbons in water samples using high-performance liquid chromatography with amperometric detection. J Chromatogr A 730:169–175, 1996.

265. NC Fladung. Optimization of automated solid-phase extraction for quantitation of polycyclic aromatic hydrocarbons in aqueous media by high-performance liquid chromatography–UV detection. J Chromatogr 692:21–26, 1995.

266. J Slobodnik, SJF Hoekstra-Oussoren, UAT Brinkman. Automated determination of polycyclic aromatic hydrocarbons by online solid-phase extraction–liquid chromatography–particle beam–mass spectrometry. Anal Methods Instrum 2:227–235, 1995.

267. F Lai, L White. Automated precolumn concentration and high-performance liquid-chromatographic analysis of polynuclear aromatic hydrocarbons in water using a single pump and a single valve. J Chromatogr A 692:11–20, 1995.

268. S Moret, S Amici, R Bortolomeazzi, G Lercker. Determination of polycyclic aromatic hydrocarbons in water and water-based alcoholic beverages. Z Lebensm Unters Forsch 201:322–326, 1995.

269. J Lintelmann, Kwaadt, R Sauerbrey, A Kettrup. On-site solid-phase extraction and coupled-column high-performance liquid-chromatographic determination of six polycyclic aromatic hydrocarbons in water. Fresenius J Anal Chem 352:735–742, 1995.

270. ER Brouwer, ANJ Hermans, H Lingeman, UAT Brinkman. Determination of polycyclic aromatic hydrocarbons in surface water by column liquid chromatography with fluorescence detection, using online micelle-mediated sample preparation. J Chromatogr A 669:45–57, 1994.

271. MW Dong. An improved HPLC method for polynuclear aromatic hydrocarbons (PAHs). Int Labmate 18: 21–24, 1993.

272. J Lintelmann, WJ Guenther, E Rose, A Kettrup. Behavior of polycyclic aromatic hydrocarbons (PAH) and triazine herbicides in water and aquifer material of a drinking-water recharge plant. I. The area of investigation and the determination methods for PAH and triazine herbicides in the aqueous matrix. Fresenius J Anal Chem 346:988–994, 1993.

273. HG Kicinski, S Adamek, A Kettrup. Trace enrichment and HPLC analysis of polycyclic aromatic hydrocarbons in environmental samples using solid-phase extraction in connection with UV-visible diode-array and fluorescence detection. Chromatographia 28:203–208, 1989.

274. X Bo-Xing, F Yu-Zhi. Determination of polynuclear aromatic hydrocarbons in water by flotation enrichment and HPLC. Talanta 35:891–894, 1988.

275. G Cartoni, F Coccioli, M Ronchetti, L Simonetti, L Zoccolillo. Determination of polycyclic aromatic hydrocarbons in natural waters by thin-layer chromatography and high-performance liquid chromatography. J Chromatogr 370:157–163, 1986.

276. RK Symons, I Crick. Determination of polynuclear aromatic hydrocarbons in refinery effluent by high-performance liquid chromatography. Anal Chim Acta, 151:237–243, 1983.

277. PE Strup. Determination of Polynuclear Aromatic Hydrocarbons in Industrial and Municipal Wastewaters. Report No. EPA-600/4-82-025. USEPA, Environmental Monitoring and Support Laboratory, Cincinnati, 1982.

278. NT Crosby, DC Hunt, LA Philp, I Patel. Determination of polynuclear [polycyclic] aromatic hydrocarbons in food, water and smoke using high-performance liquid chromatography. Analyst (London) 106: 135–145, 1981.

279. G Michor, J Carron, S Bruce, DA Cancilla. Analysis of 23 polynuclear aromatic hydrocarbons from natural water at the sub-ng/L level using solid-phase disc extraction and mass-selective detection. J Chromatogr A 732:85–99, 1996.

280. T Hankemeier, HTC van-der-Laan, JJ Vreuls, MJ Vredenbregt, T Visser, UAT Brinkman. Detectability enhancement by the use of large-volume injections in gas chromatography—cryotrapping Fourier transform infrared spectrometry. J Chromatogr A 732:75–84, 1996.

281. M Biziuk, J Namiesnik, J Czerwinski, D Gorlo, B Makuch, W Janicki, Z Polkowska, L Wolska. Occurrence and determination of organic pollutants in tap and surface waters of the Gdansk district. J Chromatogr A 733:171–183, 1996.

282. JJ Langenfeld, SB Hawthorne, DJ Miller. Quantitative analysis of fuel-related hydrocarbons in surface water and waste water samples by solid-phase microextraction. Anal Chem 68:144–155, 1996.

283. ER Bennett, TL Metcalfe, CD Metcalfe. Semi-permeable devices (SPMDS) for monitoring organic contaminants in the Otonabee River, Ontario. Chemosphere, 33:363–375, 1996.

284. Anonymous. ENVI-18 SPE tube ensures low background for monitoring organic compounds in drinking water by EPA Method 525. Supelco Appl Note 65, p 2, 1995.

285. D Papazova, A Pavlova. Determination of polycyclic aromatic hydrocarbons in purified waste waters from Neftochim. Anal Lab 4:197–202, 1995.

286. DC Messer, LT Taylor. Method development for the quantitation of trace polyaromatic hydrocarbons from water via solid-phase extraction with supercritical fluid elution. J Chromatogr Sci 33:290–296, 1995.

287. DW Potter, J Pawliszyn. Rapid determination of polyaromatic hydrocarbons and polychlorinated biphenyls in water using solid-phase microextraction and GC-MS. Environ Sci Technol 28:298–305, 1994.

288. PL Morabito, T McCabe, JF Hiller, D Zakett, D. Determination of polynuclear aromatic hydrocarbons in water samples using large volume on-column injection capillary gas chromatography. J High Resolut Chromatogr 16:90–94, 1993.

289. ND Bedding, AE McIntyre, JN Lester, R Perry. Analysis of waste waters for polynuclear aromatic hydrocarbons. I. Method development and validation. J Chromatogr Sci 26:597–605, 1988.

290. GA Junk, JJ Richard. Organics in water: solid-phase

extraction on a small scale. Anal Chem 60:451–454, 1988.

291. S Sporstoel, K Urdal, H Drangsholt, N Gjoes. Description of a method for automated determination of organic pollutants in water. Int J Environ Anal Chem 21:129–138, 1985.

292. PG Desideri, L Lepri, D Heimler, S Giannessi, L Checchini. Concentration, separation and determination of hydrocarbons in sea-water. J Chromatogr 284: 167–178, 1984.

293. JW Rhoades, CP Nulton. Priority-pollutant analyses of industrial waste waters using a microextraction approach. J Environ Sci Health, Part A, 15:467–484, 1980.

294. U.S. Environmental Protection Agency. Methods for the Determination of Organic Compounds in Drinking Water—Supplement I, EPA-600/4-90/020. Cincinnati, OH: U.S. Environmental Protection Agency, Office of Research and Development, 1990, pp 121–142.

295. U.S. Environmental Protection Agency. Methods for the Determination of Organic Compounds in Drinking Water—Supplement I, EPA-600/4-90/020. Cincinnati, OH: U.S. Environmental Protection Agency, Office of Research and Development, 1990, pp 143–168.

296. U.S. Environmental Protection Agency. Methods for the Determination of Organic Compounds in Drinking Water, EPA-600/4-88/039. Cincinnati, OH: U.S. Environmental Protection Agency, Office of Research and Development, 1988, pp 323–360.

297. U.S. Environmental Protection Agency. Methods for the Determination of Organic Compounds in Drinking Water. Supplement III, EPA-600/R-95/131. Cincinnati, OH: U.S. Environmental Protection Agency, Office of Research and Development, 1995.

298. Environmental Protection Agency. Method 610–Polynuclear aromatic hydrocarbons. 40 CFR Pt 136, App A, 1995, pp 761–774.

299. Environmental Protection Agency. Method 625–Base/neutrals and acids. 40 CFR Pt 136, App A, 1995, pp 821–848.

300. Environmental Protection Agency. Method 1625 Revision B–Semivolatile organic compounds by isotope dilution GC/MS. 40 CFR Pt 136, App A, 1995, pp 862–881.

301. U.S. Environmental Protection Agency. Test Methods for Evaluation of Solid Waste. 3rd ed. SW-846 Ch 4.3.3. Washington, DC: U.S. Environmental Protection Agency, Office of Solid Waste, 1988.

302. U.S. Environmental Protection Agency. Test Methods for Evaluation of Solid Waste. 3rd ed. SW-846 Ch 4.3.1. Washington, DC: U.S. Environmental Protection Agency, Office of Solid Waste, 1988.

303. American Society for Testing and Materials (ASTM). Standard test method for polynuclear aromatic hydrocarbons in water. ASTM Standard, D 4657–92, 1998.

304. AE Greenberg, LS Clesceri, AD Eaton. Standard Methods for the Analysis of Water and Waste Water. 18th ed. Washington, DC: American Public Health Association, American Water Works Association, Water Environment Federation, 1992, Method 6440 Polynuclear aromatic hydrocarbons, pp 6-96–6-101.

305. AE Greenberg, LS Clesceri, AD Eaton. Standard Methods for the Analysis of Water and Waste Water. 18th ed. Washington, DC: American Public Health Association, American Water Works Association, Water Environment Federation, 1992, Method 6410 Extractable base/neutrals and acids, pp 6-76–6-89.

33

Analysis of BTEX Compounds in Water

Ignacio Valor, Mónica Pérez, Carol Cortada, and David Apraiz
LABAQUA, Alicante, Spain

Juan Carlos Moltó
University of Valencia, Valencia, Spain

I. INTRODUCTION

The BTEX group of contaminants consists of benzene, toluene, ethylbenzene, and the three isomers *o*-, *m*- and *p*-xylene. Because of the high concentration of BTEX compounds in petroleum and the massive use of petroleum products as an energy source, as solvents, and in the production of other organic chemicals, their presence in water creates a hazard to public health and the environment. The BTEX chemicals are present in a standard gasoline blend in approximately 18% (w/w), and the group is considered to be the largest one that is related to any health hazards (1). Naphthalenes, which are also a health risk, make up only 1% (w/w) of gasoline. Benzene, which is recognized as the most toxic compound among BTEX, represents 11% of the total BTEX fraction in gasoline (26% toluene, 11% ethylbenzene, and 52% total xylenes).

Concentrations of BTEX have been found in surface water, groundwater, and drinking water at a few micrograms per liter (2–5). Accidental emissions can lead to higher concentrations in groundwater. The BTEX compounds are introduced into water via industrial effluents and atmospheric pollution caused mainly by vehicular emissions. One of the most common sources of BTEX contamination of soil and groundwater are spills involving the release of petroleum products, such as gasoline, diesel fuel, and lubricating and heating oil

from leaking tanks (6,7). Compared to the other main group of hydrocarbons present in gasoline, such as aliphatics, BTEX are very soluble in water, permitting their transfer to the groundwater (8).

The BTEX compounds are included in the EPA list of 129 priority pollutants (9) and also in the corresponding European Community Priority Pollutant List, List I (10), as compounds to be monitored in water due to their health and environmental significance. Sensitive and accurate analytical methods need to be developed in order to detect concentrations under the maximum permitted levels.

A. Physicochemical Properties and Toxicology

The BTEX compounds represent a homogeneous group of aromatic volatile hydrocarbons with similar physicochemical properties. Nevertheless, for a better understanding of analytical questions, and because of the different behavior of a given BTEX in the aquatic environment, it is of interest to know the particular physicochemical properties summarized in Table 1.

The World Health Organization has issued guidelines for drinking water quality (11) on the basis of risk assessments carried out by experts in each area over a period of 4 years. With regard to BTEX, the following toxicology information is given.

Table 1 Physicochemical Properties of BTEX Compounds

Property	Benzene	Toluene	Ethylbenzene	*m*-Xylene	*p*-Xylene	*o*-Xylene
Chemical structure		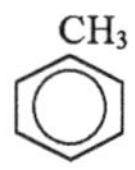	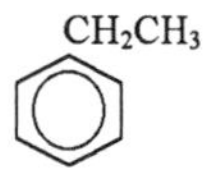	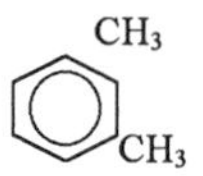		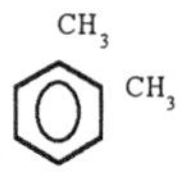
CAS number	71-43-2	108-88-3	100-41-4	108-38-3	106-42-3	95-47-6
Molecular weight (g/mol)	78.11	92.13	106.16	106.16	106.16	106.16
Water solubility (g/L)	0.7	0.5	0.2	0.2	0.2	0.2
Vapor pressure (hPa 20°C)	101	29	9.3	8	8.2	6.7
Density (20°C g/ml)	0.8786	0.8669	0.8670	0.8642	0.8611	0.8802
Octanol–water partition coeff. K_{ow} 20°C	135	489	1412	—	1510	589
Henry's law constant (atm × cm^3/mol)	126	340	528	—	831	654
Melting point (°C)	−11	−95	−95.01	−47.4	13–14	−25
Boiling point (°C)	80.1	110.6	136.2	139.1	138.35	144.4

1. Benzene

Acute exposure of humans to high concentrations of benzene affects primarily the central nervous system. At lower concentrations benzene is toxic to the hematopoietic system, causing a continuum of hematological changes, including leukemia. Because benzene is carcinogenic to humans, the International Agency for Research on Cancer (IARC) has classified it in Group 1. Based on a risk estimate using data on leukemia from epidemiological studies involving inhalation exposure, it was calculated that a drinking water concentration of 10 μg/L level was associated with an increased lifetime cancer risk of 10^{-5}. A 10 μg/L guideline value was then established.

2. Toluene

The acute oral toxicity is low. There is no evidence for carcinogenicity of toluene, and genotoxicity tests in vitro were negative. In vivo genotoxicity assays showed conflicting results with respect to chromosomal aberrations. The guideline value for drinking water is 700 μg/L.

3. Ethylbenzene

The acute oral toxicity is low. No definite conclusions can be drawn from teratogenicity data. Ethylbenzene has shown no evidence of genotoxicity in in vitro or in vivo systems. A guideline value for drinking water of 300 μg/L was stated.

4. Xylenes

Acute oral toxicity is low. No convincing evidence for teratogenicity has been found. In vitro and in vivo mutagenicity tests have proved negative. The proposed guideline value for drinking water was 500 μg/L.

B. Regulations

In the last decade important progress has been made by several regulatory agencies that have put great emphasis on organic chemical regulations regarding water pollution. In the U.S. EPA Primary Drinking Water Regulations, maximum contamination levels (MCLs) were established for BTEX compounds. The lowest stated MCL was 5 μg/L for the carcinogen benzene.

The other MCLs were 2 mg/L for toluene, 0.7 mg/L for ethylbenzene, and 10 mg/L for total xylenes.

In Europe, on the other hand, a proposed addition to the existing 80/778/CEE Directive relating to water quality for human consumption was presented in 1995 (12); it reviewed the existing parameters to be controlled in water, including benzene for the first time and proposing a maximum concentration level of 1 μg/L. In 1997 this proposal was finally accepted by the European Community Council (13), obliging EC members to establish a maximum concentration level of 1 μg/L for benzene within 2 years of its official publication.

C. Analytical Methods

The technique of choice for the analysis of BTEX compounds in water is gas chromatography. Nevertheless, as generally occurs in chromatographic methods, two separate steps can be distinguished in the analytical scheme. In the first step an extraction/concentration process is carried out; in it the analytes are extracted from the matrix and are generally concentrated before injection in the gas chromatograph for analysis. Thus gas chromatographic methods for the analysis of BTEX can be classified according to the nature of the extracting agent. The extraction can be made by a gas, a liquid, or a solid. Organic compounds with high volatilities are generally extracted by means of a gas. This is the case with the most popular and established gas–liquid extraction methods for volatile compounds in water, such as static headspace, purge-and-trap, and close-loop stripping analysis, which are discussed later. On the other hand, when a solid adsorbent is responsible for the extraction, we are talking about solid–liquid extraction, which is the mechanism behind the novel technique of solid-phase microextraction proposed in this chapter as a good choice for BTEX analysis in water.

The second step, gas chromatographic analysis, is the same in the four methods discussed in this chapter. The reader is referred to the next section (I.D) for all the information related to columns, detectors, and operational conditions of the gas chromatograph for BTEX analysis.

D. Detectors, Columns, and Gas Chromatographic Conditions for BTEX Analysis

1. Detectors

Three different detectors could be used for BTEX analysis: the flame ionization detector (FID), the photoion-

ization detector (PID), and the mass spectrometer detector (MSD). Throughout the chapter we will refer to quadrupole detectors as MSD and MS-ITD for the case of ion-trap mass spectrometer detectors. Temperatures in the range of 260–300°C are recommended for both FID and PID. Gas flow conditions (air, nitrogen, and helium) are defined depending on the manufacturer. A typical electron ionization energy of 70 eV must be used for MSD and MS-ITD scanning in the range of 35–150 for the analysis of BTEX. Recommended quantification ions are: 78 for benzene, 91 for toluene and ethylbenzene, and 106 for o-, m-, and p-xylene.

2. Columns

Nonpolar stationary phases especially designed for volatiles with 1.8–3-μm-thick films and 30–60-m $\times$ 0.32–0.50-mm-ID capillary columns are recommended for a good resolution for these compounds. Megabore as well as packed columns could be used for BTEX separation. A 50°C and a 200°C initial and final temperature, respectively, could be programmed at the appropriate temperature and carrier gas flow rate, depending on the selected column.

II. STATIC HEADSPACE (SHS)

In static headspace the sample is introduced in a closed system (generally a septum-sealed ampoule) at a given temperature, for a period of time in which volatiles are transferred from the condensed phase to the headspace until the system reaches equilibrium. After equilibrium is reached, an aliquot of the gaseous phase is injected in the gas chromatograph with either a gas-tight syringe or a similar device and analyzed. Static headspace has been used successfully since the 1960s (14–20) for the analysis of volatile organic compounds in aqueous and solid samples by gas chromatography and is still being used today (21–30).

For quantitative headspace gas analysis, parameters affecting the equilibrium in the system must be taken into consideration. Theoretical aspects of the thermodynamic equilibrium have been studied by different authors (31–34). The distribution of the solute in the gas-condensed-phase system, the partition coefficient K_H according the Henry's Law, is defined as the ratio of the concentration in the gaseous phase (C_g) to that in the condensed phase (C_c): $K_H = C_g/C_c$. This constant is dependent upon the analyte, the composition of the phases, and the pressure and the temperature of the system. Nevertheless, the pressure and gas-phase composition are parameters with no practical interest in optimizing the static headspace analysis, because the sample is loaded in a sealed bottle with ambient air filling the headspace, and the pressure in the system is generally a parameter fixed by the selected temperature.

A. Factors Affecting the Technique

The concentration of the analytes in the gaseous phase can be increased by raising the temperature of the system or by increasing the activity coefficients of the analytes, that is, by adding a salt to the sample. This is known as the "salting-out" effect. Agitation of the sample reduces the time needed to reach equilibrium. Increasing the absolute amount of the sample in a headspace bottle does not improve its sensitivity, because it is the concentration of a compound in the gas phase that is in fact analyzed (35), not the total amount of that compound in the bottle. Hence the volume of the sample is determined mainly from practical considerations, such as the ease of handling the sample with syringes or pipettes. Generally, 20–50-ml glass hypovials sealed with silicon or rubber septums are used, leaving a headspace over the water sample of about half of the total volume of the bottle. Parameters affecting this equilibrium are discussed next.

1. Temperature

Improvements in sensitivity can be achieved by increasing the temperature of the sample. As a general rule, a temperature increase of about 30°C doubles the peak height (35). Nevertheless, this increase is limited by the boiling point of the condensed phase (100°C). The sample is heated during the equilibration time by immersing the sealed sample bottle in a water bath, placing the vial on a heated plate, or placing it in an oven. Microwave-assisted systems have also been used for this purpose (28). In order to avoid sample condensation over the syringe walls it is necessary to maintain the syringe at at least 10°C over the sample equilibration temperature by placing the syringe in a heated device between injections. For all of these reasons, a sample equilibration temperature of 60–80°C is a good compromise between sensitivity and practical considerations. When automated headspace autosamplers systems are used it is easier to work with temperatures around 80–90°C, maintaining the sample loop and transfer lines at higher temperatures to avoid condensation before injection.

2. "Salting-Out" Effect

Saturation of the condensed phase by an inorganic salt increases the concentration of analytes in the vapor phase and is a practical way of increasing sensitivity. Bassett et al. (14) proved that the addition of 1.2 g of sodium sulphate to 2 ml of the sample increased peak heights about 4–7 times when analyzing aqueous solutions. Saturation of the samples with an inorganic salt (sodium carbonate, sodium chloride, etc.) can be used for this purpose, not only permitting the enhancement in sensitivity but also buffering the matrix effect caused when analyzing samples with a different content of salts. Salt addition should not exceed the saturation point, in order to avoid absorption of the analytes over the precipitated salts. It is of course assumed that calibration solutions must be added with the same amount of salt used in the sample analysis.

3. Agitation and Equilibrium Time

Agitation of the sample favors the mass transfer of the analytes from the condensed phase to the gas phase, thus reducing the time needed to reach equilibrium. A magnetic stirrer is a simple and effective way to agitate the sample, allowing equilibration times of about 10 min in the case of BTEX, depending on the selected temperature. Manual agitation (36) or the most effective ultrasonic agitation can be also used for this purpose.

In conclusion, sample preparation in static headspace analysis can be optimized by saturating the sample with an inorganic salt and heating the sealed sample vial while agitating the condensed phase during the time required to reach equilibrium. Although analysis can be carried out in nonequilibrium conditions, the maximum sensitivity and also the highest precision for headspace analysis occur when sample and headspace are in equilibrium. Once the analyst has decided on the sample conditions to be used (temperature, inorganic salt, agitation, etc.), invariably in the analytical sequence he must calculate the equilibration time in those conditions. The nature of the sample matrix affects the rate of diffusion of volatile components from the sample to the headspace. Therefore, the analyst should consider the sample matrix when deciding on the length of the equilibration period. Low-viscosity fluids equilibrate faster than high-viscosity ones. To determine the equilibration time, perform a series of analyses by increasing the equilibrium period. After that, record the signals (area counts, peak height, etc.) obtained in the different analyses for the peaks of interest. Finally, plot the area versus equilibration time. When the curve levels out, equilibrium has been reached. This equilibrium is maintained even though this time exceeds the equilibration time, permitting the analysis of the heated vials after this equilibrium is reached. Figure 1 shows typical equilibration under specific conditions.

4. Sample Introduction

Due to the fact that a gas instead of a liquid is injected in the gas chromatograph for analysis, special devices must be used. There are mainly two systems for gas sample injection: the common gas-tight syringes (Fig. 2) generally used for manual injection, and the electropneumatic systems installed in full, automated headspace autosamplers based on a sample loop for sample injection.

Manual injection with gas-tight syringes is an inexpensive method that can give good results with careful handling. The syringe temperature must always be higher than that of the sample to avoid losses by condensation in the inner walls of the syringes. Once the vial septum is pierced, the syringe should be filled slowly and emptied back into the sample flask at least four times to minimize losses resulting from adsorption. Finally, in order to avoid memory effects between injections, it is important to clean the syringe by removing the plunger and passing a current of nitrogen through the interior while maintaining it at the operation temperature.

Pauchman and Göke (37) described a simple arrangement in which the excess pressure of the headspace gas within the equilibration container was utilized to introduce the sample in the gas chromatograph. Drozd and coworkers (37) developed a novel device for charging headspace gas samples into the gas chromatograph that eliminates the use of injection syringes. On the basis of these considerations, modern, fully automated headspace autosamplers are commercially available today. The electropneumatic dosing diagram and sample steps for this equipment are given in Fig. 3. These autosamplers allow full programming of the different parameters, such as equilibration time, equilibrium temperature, mixing power, and gas sample size. The main advantages of these systems are better precision, the minimization of memory effects, and the reduction of time consumed.

B. Analytical Procedure Proposed for Static Headspace

Reagents.

Ultrapure reagent water

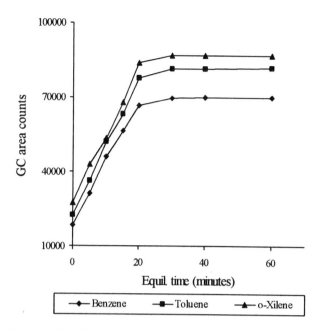

Fig. 1 Equilibrium-time profile using static headspace, 60°C, and 20% NaCl with magnetic agitation.

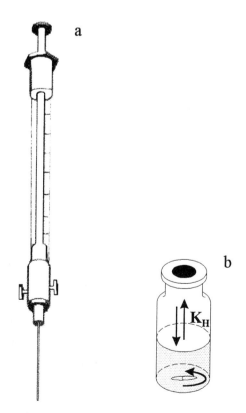

Fig. 2 Static headspace: (a) 1-ml gas-tight syringe; (b) 50-ml glass hypovial with high-temperature septum and crimp cap.

1. SAMPLE EQUILIBRATION, MIXING AND STABILIZATION.

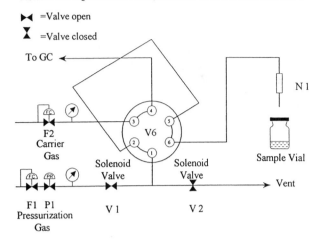

2. VIAL PRESSURIZATION AND LOOP FILL

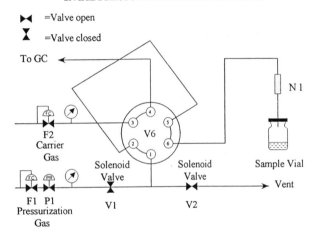

3. INJECT

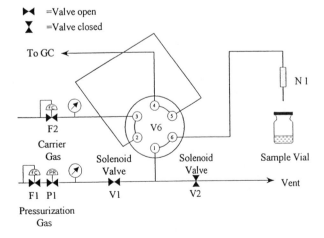

Fig. 3 Static headspace autosampler, sampling steps.

BTEX mix standard working solution of 20 mg/L of each component in methanol (pesticide grade)

Internal standard working solution of fluorobenzene at 5 mg/L in methanol

Sodium chloride (A.R.)

Materials.

1–2-mL gas-tight syringe

50-ml glass hypovials with high-temperature septa and crimp caps (or similar)

Thermostatic water bath or oven

5–25-μl syringes

20 × 4-mm magnetic stirrer bars

Magnetic agitator

Instrumentation.

Capillary GC equipped with a split–splitless injector and FID, PID, or MSD detector

Capillary column

Procedure. Fill a sealed 50-ml hypovial with 25 ml of the sample spiked with 25 μl of the internal standard solution corresponding to a 5-μg/L concentration and 5 g NaCl. To agitate the sample, a stirrer bar is introduced in the vial. Equilibrium is achieved by placing the vial in a water bath at 60°C for 30 min. Finally, 1 ml of the headspace is taken, following the recommendations explained earlier, and injected with the split mode (split ratio 1:10) into the gas chromatograph for analysis. See Sec. VI for the quality control procedure.

The typical limits of detection and precision (expressed as relative standard deviation, RSD) are shown in Table 2. These results were obtained by capillary gas chromatography (in the specific static headspace conditions described earlier) attached to different detectors (FID, PID, and MSD) in the analysis of potable waters. Highly contaminated wastewater samples can change the partition coefficient of the BTEX and also generate noisy baselines, giving higher detection limits. Figure 4 shows a typical analysis following this proposed method by GC-FID.

III. PURGE-AND-TRAP

Purge-and-trap could be defined as a headspace gas analysis in which volatiles are stripped from the sample with an inert gas, trapped into a solid sorbent, and thermally desorbed into a gas chromatograph for analysis. Since the first attempts by Swinnerton and Linnenbom in 1967 and the development of the popular system pioneered by Bellar and Lichtenberg (38), purge-and-

Table 2 Typical LODs and Precision Obtained in Analysis of BTEX in Potable Waters by Static Headspace with Capillary GC Coupled to FID and MS-ITD

	FID[a]		MS-ITD[b]	
	RSD (%)	LOD (μg/L)	RSD (%)	LOD (μg/L)
Benzene	3	0.15	6	0.15
Toluene	8	0.13	4	0.07
Ethylbenzene	6	0.08	4	0.03
m- + *p*-Xylene	6	0.21	3	0.04
o-Xylene	1	0.41	9	0.12

[a]*Source*: Ref. 160.
[b]Data obtained by I. Valor et al. by calculating the LOD from concentration generating *s/n* = 3 and RSD (%) for *n* = 5.

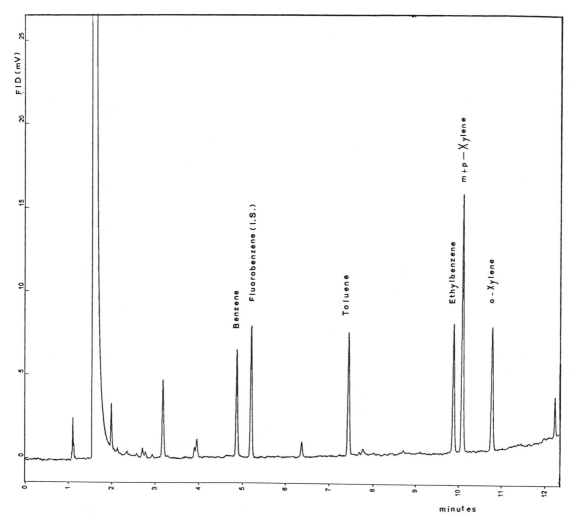

Fig. 4 High-resolution gas chromatogram of a spiked water sample at a 10-μg/L level for each BTEX compound (I.S. = internal standard) by static headspace (SHS). *Column*: DB-624 (30 m × 0.32 × 1.8-μm film). *Temp. program*: 45°C (2 min), 8°C/min to 150°C (5 min). *Carrier gas*: helium 4-ml/min flow. *Injection temp.*: 260°C in split mode (split ratio 1/10). *Detector*: FID.

trap has been widely used in environmental analysis for volatile organic pollutants in water (39–55) and has been extended to the analysis of foods (56), clinical applications (57), and other matrices (58,59). Purge-and-trap is the most widely used technique for the routine quality control of organic volatiles in any kind of water and is the official method required in many countries. The U.S. Environmental Protection Agency (EPA) has standardized this method (60), proposing different protocols for the analysis of volatiles in water using purge-and-trap. Those in which BTEX are included, such as the 502.2 (61), 503.1 (62), and 524.2 63) methods for drinking waters and the 602 (64) and 624 (65) methods for wastewaters.

The purge-and-trap method consists of three separate processes carried out by three basic pieces of equipment. In the first process, an aliquot of the sample (generally 5 ml or 25 ml in order to achieve better detection limits) is loaded into a glass sparging container. The sample is purged with either ultrapure helium or nitrogen at specified flow rate, temperature, and time. In this way, volatile analytes from the aqueous phase are transferred into the gas phase. In the second process, volatiles are retained on a solid sorbent trap that is packed into a column at ambient temperature. Finally, this trap is desorbed by a rapid increase in temperature, and a small volume of inert gas transfers the concentrated analytes from the sample to the gas chromatograph for further analysis.

A schematic drawing of a modern purge-and-trap system is shown in Fig. 5. In the purge mode, the inert gas passes through the water sample to the adsorbent trap and is vented out. In the desorption mode, a six-port valve opens and closes automatically. Flow is in a way that the heated trap is backflushed by the inert gas to the gas chromatograph. Although this system is usually performed in the online mode (66,67), it can also be carried out in an off-line mode (42,43,68,69), in which adsorption and desorption steps take place in different apparatuses, thereby permitting an easy design of home-made systems. However, the advantages of the online systems are generally associated with the possibility of fully automated systems, in which it is even possible to program a sequence of several sample extractions and analyses. Variables affecting each of the three processes are discussed next.

A. Factors Affecting the Technique

1. Purge

The glass sparging container must be designed to accept the sample volume (5–25 ml generally), where the

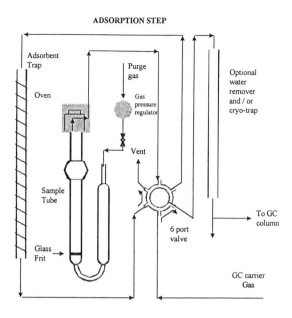

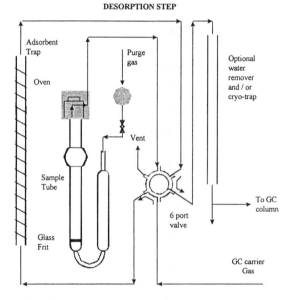

Fig. 5 Schematic drawing of a modern online purge-and-trap system.

volatiles are stripped by an inert gas helium or nitrogen. Volatiles can be removed by a current of the inert gas passing over the surface of the liquid phase (with or without agitation), which is called "sweeping" (70); but generally the gas purges the sample with a glass frit installed at the base of the sample chamber so that the extracting gas passes through the water column as finely divided bubbles. With this purge system the foaming of wastewater samples is sometimes a serious problem. The foam can climb through the apparatus to the sorbent trap, causing several problems, such as deactivation of the trap and the introduction of thermal decomposition products from labile, nonvolatile materials. For samples that do not form a highly persistent foam it is possible to reduce the foam by decreasing the purge flow or by inserting a mechanical barrier to the foam. When this is insufficient, the alternatives are (1) to apply heat to dissipate the foam, and (2) to add silicone-based commercial antifoam emulsions (71,72).

The kinetics of purging for volatiles in water have been studied in depth by Lin et al. (73). The two main variables that account for the extraction efficiency of an analyte using the purge-and-trap technique are the total volume of gas through the sample and sample temperature. Since the earliest purge-and-trap methodology, purge time has been specified as 11 minutes at a flow rate of 40 ml/min, that is, a total extraction volume of 440 ml. Extraction of analytes with high boiling points can generally be improved by increasing the total extraction volume (74–76), but care must be taken with this assumption because losses of the most volatile organics are possible due to breakthrough problems, as reported by many authors (42,74,77,78).

On the other hand, heating the sample during the purge period favors the extraction power (79). Extractions obtained at 40°C versus 25°C increase by a factor of 1.5–2.5 for high-boiling-point polar compounds included in EPA method 524.2 (80). In the past, heating the sample during extraction had the drawback of transferring excessive moisture to the trap and finally to the detector, causing quenching of the detector response (45). Improvements in integrated water-removal systems reduce this problem drastically (81–83).

The BTEX compounds are in a good intermediate volatility position, not enough to cause breakthrough problems but with high purging recoveries. Then, as was concluded in the earlier discussion of variables affecting the purge step, the standard accepted period—room-temperature helium or nitrogen purge of a 5-ml sample for 11 min at a flow rate of 40 ml/min—seems to be a good compromise for purging BTEX compounds.

2. Trap

Trapping efficiency is affected by several factors, including the vapor pressure of the compound, the surface area of the adsorbent, and thermodynamic interactions between the analyte and the adsorbent (80). Trapping adsorbents generally improve in efficiency at lower temperatures. To minimize breakthrough, the concentrator trap should be near 25°C or within 1–2°C

of ambient temperature. The choice of the proper adsorbent should be based mainly on breakthrough volumes for a target analyte list, but thermal stability, pure chromatographic blanks, the presence of irreversible active sites, and low affinity to water are also important parameters that should be taken into account. Different solid adsorbents applied in purge-and-trap are reviewed in depth by Núñez et al. (84). Their classification is as follows.

a. Activated Carbon and Graphitized Sorbents. Activated carbon was used in the first applications of trapping volatiles with solid sorbents (85–87) due to its high specific surface and thermal stability (up to 700°C). Graphitized sorbents, or carbon blacks (88–90) (Carbopack, Carbotrap, etc.), are very homogeneous sorbents obtained by the pyrolysis of different hydrocarbons. In spite of their high sorbent capacity and thermal stability, activated carbon and graphitized sorbents have a strong affinity to water, excessive surface activity (activated carbon), or active sites for polar compounds (graphitized sorbents) and the need for too high a temperature for desorption. Both sorbents may be used when porous polymers have insufficient sampling capacity and when liquid desorption is needed to extract very high-boiling-point compounds, which are not readily desorbed.

b. Porous Polymers. The use of porous polymers has gained in popularity in recent years because it eliminates problems associated with water retention, irreversible adsorption, or thermal decomposition phenomena. Porous polymers in spite of their smaller specific active surface compared to carbon and graphitized sorbents are excellent adsorbents for nonpolar compounds, and the development of technologies in this field has led to a wide range of possibilities for the selection of a suitable trapping medium for a given determination of headspace volatiles. General precautions with any of these polymeric materials must be observed in order to avoid various detrimental effects on the performance of the polymer (84): first, oxidizing atmospheres when working at high temperatures; second, heavy organic molecules deposited on the surface of the polymeric sorbent that could modify either its chemical structure or its adsorptive properties, and finally, not heating the polymer over its maximum permitted temperature, generally specified by the manufacturer. Table 3 summarizes the porous polymers most commonly used as trap packings.

Although Chromosorb, Porapak, and Amberlite XAD series are very common adsorbents for the preconcentration of headspace volatiles, Tenax is by far the most widely used organic polymer in different me-

Table 3 Physical Properties of Porous Polymeric Adsorbents Commonly Used for Preconcentration of Trace Organic Volatiles in Purge-and-Trap

Sorbent	Composition	Specific surface area (m²/g)	Temperature limit (°C)	Refs.
Tenax	Poly(2,6-diphenyl-p-phenylene oxide)	19–30	450	38,42,70,104,107,108,111–123
Chromosorb	Depends on the type	15–800	250–300	70,87,113,124,125
Porapak	Depends on the type	225–840	200–300	126–134
Amberlite	DVB copolymers	100–750	150–200	91,108,135–138
Other	—	70–620	200–340	139–141

DVB = divinylbenzene.

dia (94–97, 110) due to its high thermal stability (up to 450°C), low affinity to water, and high recoveries on thermal elutions in spite of its limited specific surface area (19–30 m²/g). Tenax GC permits relatively low background levels, which can be decreased easily by a prior conditioning at 200–250°C overnight under a stream of purified inert gas. Nevertheless, certain ghost peaks in the blank gas chromatograms have been reported (91). Even after thermal conditioning, aliphatic and aromatic hydrocarbons are observed in the blanks (92). A new modified Tenax polymer appeared in 1982. MacLeod and Ames (93) compared the background produced on heat-desorbing preconditioned Tenax GC with Tenax TA. Heating the new polymer under purified nitrogen for 4 h at 340°C and then at 300°C for two separate periods of 15 min just before use resulted in better blank chromatograms with Tenax TA than with Tenax GC.

3. Trap Packing and Design

The adsorbent system consists of a packed column made either of deactivated glass (93) or, more generally, of stainless steel, with different lengths and internal diameters, varying from 5 to 25 cm and 2 to 5 mm, respectively, and packed with a single or multiple solid sorbents (94). Two or more adsorbent tubes can also be used that can be connected in series or in parallel (42). The standard trap proposed in EPA method 502.2 consists of a column 25 cm long and with at least a 2.7-mm ID. Starting from the inlet, the trap must contain the following amounts of adsorbents: 1/3 of 2,6-diphenyl-p-phenylene oxide polymer, 1/3 of silica gel, and 1/3 of coconut charcoal. It is recommended to insert 1.0 cm of methyl silicone–coated packing (OV-1) at the inlet to extend the life of the trap. Nevertheless, if it is unnecessary to analyze for dichlorodifluoromethane (Freon 12, boiling point −29°C), the charcoal can be

eliminated and the polymer increased to fill 2/3 of the trap. If only compounds boiling above 35°C are to be analyzed, both the silica gel and charcoal can be eliminated and the polymer increased to fill the entire trap.

Therefore for the analysis of BTEX, in which the lowest boiling point is benzene's (80°C), the proposed trap described in EPA method 503.1 for the analysis of volatile aromatic and unsaturated organic compounds is shown in Fig. 6. Before initial use, the trap should be conditioned overnight at 180°C by backflushing with an inert gas at at least 20 ml/min.

4. New Types of Concentrator Traps

The basic requirement to be fulfilled in gas chromatographic techniques is that the sampling procedure concentrate the sample as a sufficiently narrow band in the inlet section of the column. As will be discussed shortly, many attempts have been made to achieve this end. One of them consists in reducing the dimensions of the concentration trap to such an extent that it can be connected directly to the capillary column with a sufficiently high flow rate to desorb analytes from the trap. Wall-coated traps, as described by Grob and Habich (98), are one of the solutions used in purge-and-trap capillary gas chromatography (99,102).

5. Desorption

Once the volatiles have been purged and trapped from the sample, they must be desorbed by a rapid increase in temperature and a convenient volume of an inert gas, in order to transport the analytes to the gas chromatograph through a narrow band to avoid tailing chromatographic peaks. Again, as in the extraction, desorption is dependent upon flow rate and temperature. When the purge-and-trap technique was developed, packed-column flow rates greater than 30 ml/min were sufficient for rapid analyte transfer from the trap, and trap heating

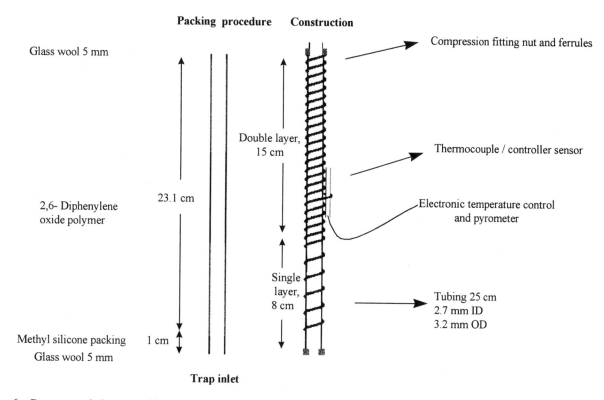

Fig. 6 Recommended trap packing and trap construction for BTEX analysis by purge-and-trap (recommendation proposed in EPA method 503.1).

rates of 200°C/min where high enough for this objective. The transition to the higher-resolution-power capillary columns introduces an additional problem to purge-and-trap systems when a packed column (the sorbent trap) desorbed with a high flow rate, around 40 ml/min, must be connected to a 0.20–0.53-mm-ID capillary column with typical flow rates ranging from 1–10 ml/min. For a better understanding of the problem let's consider the following example. An efficient trap designed for the concentration of dilute headspace volatiles may consist of a 3-mm-ID tube packed with a solid adsorbent, which may be connected directly to a 0.3-mm-ID capillary column. The carrier gas may pass through the column at 50 cm/s. Because the cross section of the trap is 100 times greater than that of the column, the linear flow rate in the trap is only 5 mm/s. Strategies to solve this problem have been developed in recent years and are described next.

6. High Desorption Speeds

The trend in trap desorption is to reduce the time required to transfer the analytes to the GC. The faster the trap heating rate, the quicker the analyte is released from the adsorbent. There are many advantages associated with the increase in the heating rate. The volume of inert gas needed to transport the analytes is reduced when the desorption rate increases, but this reduction of transfer time also reduces the water and carbon dioxide transferred to the GC, improving the peak shape, narrowing the peak width, and increasing sensitivity (80). Recently the use of microwave-assisted systems instead of electrical ovens has improved the desorption efficiency, permitting an increase in the heating rate of up to 1000°C/min. Thermal desorption devices can be situated before the GC injector (104–106). The equipment required for this is generally complex, requiring hardware associated with the instrument, switching valves, and heating transfer lines. Therefore homemade systems are rarely found with this scheme, but commercially available purge-and-trap equipment incorporates this design. A second possibility is to introduce the trap tube (5–10 cm long) directly into the GC injector port, resulting in an off-line mode widely used by many workers (107,108).

7. Interface Techniques

In the past the most common technique to connect the trap to the column in the online mode was to use a low-dead-volume stainless union connected in the carrier gas inlet just a few centimeters before the injector body, via a heated transfer line. This is very simple to install and allows the analyst to make injections with other non–purge-and-trap systems. However, this is not possible when capillary columns are used. In this case, two main interface techniques have been developed: the cryofocusing injector interface and the split injection interface (open split interface). In the cryofocusing injector interface (99,100,103,109), once trap desorption has begun, a short section of capillary column (uncoated fused-silica column) is connected via a low-dead-volume joint (press-fit union) before the analytical column, or the proper analytical column is cooled at liquid nitrogen temperatures ($-160°C$). When trap desorption is completed, the interface is heated rapidly (1000°C/min) under a stream of carrier gas, transferring the analytes to the analytical column in a narrow band. This interface is proposed in EPA method 524.2 (63). The main advantage of this interface is that all of the volatiles from the sample are transferred to the column offering the highest sensitivity. Problems with detector saturation or blockage of the interface by frozen water from the trap should be controlled. In the split injector interface, the gas used for desorption of the trap passes through a heated transfer line directly into the carrier gas inlet of the split injector. The main advantages is that a trap desorbed at a normal flow could be connected to very-small-diameter columns (0.20-mm ID) by controlling the split ratio flow to the column. Obviously only around 1/20 (depending on the split ratio) is introduced on the column, with a corresponding decrease in sensitivity.

B. Analytical Procedure Proposed for Purge and Trap

Reagents.

Ultrapure reagent water
BTEX mix standard working solution of 2 mg/L of each component in methanol (pesticide grade)
Internal standard working solution of fluorobenzene at 5-mg/L level in methanol

Materials.

5–25-μl syringes
5-ml glass syringes with Luer-Lok tip
5-ml frit sparger vessels for the purge-and-trap autosampler

Instrumentation.

Gas chromatograph with purge-and-trap autosampler and FID, PID, or MSD detectors
Capillary column

Procedure. Spike 5 ml of the sample with 5 μl internal standard solution, corresponding to a 5-μg/L concentration, and introduce this to the sparger sampler tube of the purge-and-trap system with a glass syringe. The sample is purged for 11 min at ambient temperature. After the 11 min of purge, the system is placed in the desorption mode and the trap preheats to 180°C without flowing the desorption gas. Then, simultaneously, the flow of the desorption gas for 4 min and the program of the gas chromatographic analysis starts. With the cryogenic interface, after the trap desorption cycle the cryogenic trap is heated rapidly and the chromatographic run started simultaneously. After desorbing of the sample, a trap-reconditioning step is recommended, by maintaining the trap temperature at 180°C for 7 min. When the trap is cool and the gas chromatographic run has finished, the next sample can be analyzed. Table 4 summarizes analytical conditions in BTEX analysis by purge-and-trap. See Sec. VI for a quality control procedure.

Typical limits of detection and precision (expressed as RSD) are shown in Table 5. These results were obtained by capillary gas chromatography coupled to different detectors (FID, PID, and MSD) in the analysis of BTEX in potable waters. Matrix effects are not very strong in headspace techniques; nevertheless, the limits of detection are generally higher than those obtained in clean waters, due to noisy baselines caused by the matrices. A typical analysis by the purge-and-trap technique is shown in Fig. 7.

IV. CLOSED-LOOP STRIPPING ANALYSIS (CLSA)

In 1973 Grob introduced the closed-loop stripping analysis (CLSA) system for the analysis of trace organics in water (85). In CLSA, volatiles from the water sample are stripped by a recirculating stream of air and trapped in a small activated-carbon filter. The adsorbed analytes are extracted from the carbon filter by a small volume (10–50 μl) of a organic solvent, generally carbon disulphide, and analyzed without solvent concentration by gas chromatography. A schematic representation of a CLSA apparatus is given in Fig. 8. A 1–5-L sample flask immersed in a water bath is stripped

Table 4 Analytical Conditions Proposed for BTEX Analysis by Online Purge-and-Trap Capillary GC

Purge-and-trap parameters	
Trap	OV-1/Tenax
Trap temperature	Ambient
Purge flow	40 ml/min
Purge time	11 min
Desorption temperature	180°C
Desorption time	4 min
Trap reconditioning after desorption	180°C (7 min)
Transfer line temperature	150°C
GC parameters	
Injector interface	Split injector interface or cryofocusing interface
Column	30–75-m × 0.32–0.53-mm ID × 1.8–3-μm film
Column flow rate	2–10 ml/min
Column program	35°C for 2 min, 35°C to 200°C at 8°C/min
Detectors	PID/FID or MSD

with air generated by a pump. An activated-carbon filter containing 1–2 mg of carbon is connected with stainless steel tubing of 2-mm ID in the closed system. The gas stream leaving the water is always heated about 10°C above the bath temperature in order to prevent water condensation. The water bath and the filter holder temperatures are monitored independently. The parts adjacent to the sample flask are spirally rolled in order to protect the glass–metal connections during manipulation. The selection of the appropriate materials for the equipment is of primary importance. Any rubber or plastic material must not be used, in order to avoid adsorptions. Glass and metal equipment can be used without restriction. The system must be absolutely

gas-tight to avoid any leak. A few liters of ambient air entering the stripping circuit will seriously disturb the results when analyzing ng/L levels. A modified system consisting of an open circuit, called open-loop stripping analysis (OLSA), has also been investigated. The advantages and disadvantages of both systems are discussed by different authors (143–145).

Comparing CLSA to purge-and-trap, we find that CLSA gives better results for high-molecular-weight and nonpolar compounds (achieving detection limits at the ng/L level) than purge-and-trap systems because of the higher recoveries obtained with the solvent elution instead of thermal desorption. On the other hand, recoveries of polar compounds such as phenols are very

Table 5 Typical LODs and Precision Obtained in the Analysis of BTEX in Potable Waters by an Automated Online Purge-and-Trap with Capillary GC Coupled to PID, FID, and MSD

	FID[a]		PID[b]		MSD[c]	
	RSD (%)	LOD (μg/L)	RSD (%)	LOD (μg/L)	RSD (%)	LOD (μg/L)
Benzene	3	0.02	1	0.01	6	0.04
Toluene	3	0.03	2	0.02	7	0.11
Ethylbenzene	6	0.03	1	0.03	7	0.05
m-Xylene	n.r.[d]	0.01	1	0.02	6	0.04
p-Xylene	5	0.03	1	0.02	7	0.10
o-Xylene	2	0.01	1	0.02	7	0.09

[a]*Source*: Ref. 99.
[b]Data obtained from EPA method 502.2. Average of the results obtained with two different columns.
[c]Data obtained from EPA method 524.2. Average of the results obtained with three different columns.
[d]n.r. = not reported.

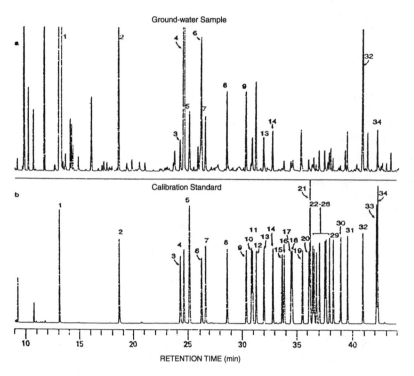

Fig. 7 High-resolution gas chromatograms resulting from purge-and-trap HRGC-FID showing (a) volatile hydrocarbons in petroleum-contaminated groundwater, (b) calibration standard used for benzene and C_7–C_{10} aromatic hydrocarbons. 1: benzene; 2: toluene; 4: ethylbenzene; 5: *m-* + *p*-xylene; 3, 6, and 7: internal standards. (From Ref. 50.)

low using CLSA. This is why both techniques are complementary for the analysis of a wide range of volatiles. For the analysis of BTEX, similar recoveries could be obtained with both CLSA and purge-and-trap systems. In contrast to its low acceptance in the United States,

this versatile technique has become very popular in European laboratories, where it can be performed on a commercially available system. And CLSA has been widely used in the analysis of taste and odor problems in water (146–149), which requires analytical techniques at nanogram-per-liter levels, and in the analysis of different hydrocarbons, including BTEX, in water (150–153).

A. Factors Affecting the Technique

The recovery rates of volatile organic compounds in water by the CLSA technique are influenced by the efficiency of stripping from the water, the efficiency of adsorption on the filter, and the efficiency of extraction from the filter.

1. Stripping

Among the parameters that govern the efficiency of stripping, both the water temperature and the stripping time are the most relevant. The following discussion assumes the most common operational conditions, in which ambient air is used as the purging gas and the

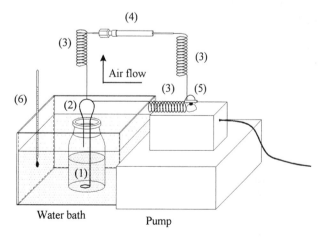

Fig. 8 Schematic drawing of a CLSA apparatus: (1) 1-L water sample stripped with a coarse gas frit; (2) fused glass–metal connections; (3) rolled steel tubing; (4) thermostated filter holder; (5) stainless steel tubing; (6) thermometer.

pump gives a 2–3-L/min air flow through the closed system.

Grob (85) investigated the stripping efficiency of a list of hydrocarbons from C_{10} to C_{24} spiked in 5-L sample and stripped with air at room temperature. The 44-h period didn't improve the recoveries up to C_{20} compared to the 20-h stripping time. In a different work, Grob et al. (154) optimized the stripping procedure by reducing the sample volume to 1 L and by immersing the whole flask, including the glass joint, in a water bath. In this work they investigated the extraction of n-alkanes, from octane to dodecane, added in concentrations of 100 ng/L to fresh tap water. After 10 hours of stripping, more than half of the added material had disappeared. Stripping for 5 h caused a reduction of about one-third, whereas 1, 2, and 3 h gave virtually identical recoveries. This was attributed to biological degradation, and the effect is then more important when more biologically active samples are analyzed. The reduced stripping time nevertheless caused losses in the recovery of heavier substances, e.g., for the alkanes beyond eicosane.

Curvers et al. developed a theoretical model of the stripping process (155), concluding that to permit the determination of low concentrations, sample, volume, and stripping time should be increased. For polar compounds, large sample volumes will result in very long stripping times. For nonpolar compounds, however, the stripping time increases less, in proportion, than does the sample volume. For nonpolar compounds this offers the possibility of processing large volumes of water within a reasonable time. On the other hand, stripping recoveries are expected to improve at higher temperatures, owing to the increased vapor pressures. This assumption was experimentally proved, obtaining a 103% recovery in the stripping process at 35°C water temperature, compared to 93% at 25°C.

The effect of stripping temperature was also evaluated by Thomason and Bertsch (156) in the extraction of EPA base/neutral priority pollutants. When raising the stripping temperature from 25 to 40°C, an increase of recoveries was observed. On increasing the stripping temperature to 60°C, the expected increase of base neutrals was not observed. Water condensed on the charcoal filter and the recoveries reduced to zero. Heating of the stripping gas before contact with the filter did not eliminate this problem. Gómez Belinchón and Albaigés (157) optimized the operational parameters affecting the CLSA technique, comparing the general consensus in performing the stripping at 30–35°C for 2 h with 45° during 0.5 h in the extraction of n-C_8 to n-C_{19} and aromatic hydrocarbons. The results obtained

demonstrated that the proposed 45°C stripping temperature for 0.5 h provided a better performance, with the great advantage of reducing the analysis time. Our experience in the analysis of BTEX with CLSA agrees with these results. A stripping period of 0.5 h, maintaining the water sample at 45°C, gives recoveries in the stripping step of around 100%. Higher-stripping periods must be used only in the analysis of high-molecular-weight or polar compounds

The "salting-out" effect—raising the ionic strength of the water sample with an inorganic salt such as sodium chloride or sodium sulphate—increases the stripping rate of many organic compounds (146).

2. Adsorption

The efficiency of the adsorption depends on the selected adsorbent, the amount of trapping material, particle size, geometry and the preparation and conditioning of the adsorbent filters. Due to the fact that a small volume of solvent must be used in the extraction, generally 10–50 μl (depending on the desired sensitivity), a small amount of adsorbent media must also be used. That is why only charcoal with a very high specific surface allows good recoveries. Once one has selected the ideal adsorbent, the parameters that have the most influence on trapping and desorption efficiencies are the size and the geometry of the adsorbent filter. The filter disc should be as small as possible, to contain carbon particles of the smallest possible diameter while still giving a low flow resistance to the stripping gas. It should be mechanically stable so as to allow a virtually unlimited number of runs with unchanged characteristics, and should be easy to prepare. All these partly contradictory requirements lead to a compromise proposed initially by Grob (85,154) that is still accepted today: A 1–1.5-mg amount of activated carbon with particles that are 0.05–0.10 mm thick is assembled in a cylindrical disc 2.5 mm in diameter and 0.8–1.2 mm in depth.

It is important to use a set of filters matched in resistance for each group of samples and calibration standards. It is possible to determine the filter resistance by filling the longer glass tube above the charcoal with a solvent and, after several rinses, to wet the filter, measuring the time needed to pass through a known volume of the solvent from the top of the filter tube to the surface of carbon. Rates for new, commercially prepared filters vary significantly and decrease with use. For optimal recoveries, filters with flow greater than 0.9 ml/min should be used. A filter holder such as shown in Fig. 9 is used to connect the filter in the

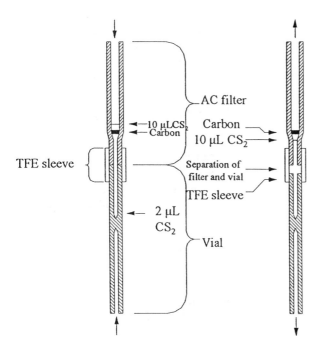

Fig. 9 Carbon filter extraction in CLSA with CS$_2$.

closed system, and a heating system is installed to prevent condensation. It is advisable to reconditioning the filter immediately after use, when it is still wetted by the extracting solvent.

3. Filter Extraction

In CLSA the extraction efficiency of the solvent has great importance, because quantitative extraction must be obtained with the smallest possible volume in order to obtain the biggest concentration factor. Different authors have investigated the extraction efficiency of charcoal filters using different solvents. Grob and Zürcher (158) proved that a complete extraction for ethylbenzene and *m*-xylene (99.3% and 99.4%, respectively) from charcoal filters is achieved with 15 μl of CS$_2$, and it is necessary to use 30 μl of CH$_2$Cl$_2$ to obtain the same efficiency. Curvers et al. (155) tested the extraction efficiency of five different solvents—carbon disulphide, methylene chloride, methanol, ethyl acetate, and *n*-pentane—in the extraction of *n*-undecane, *o*-dichlorobenzene, 3-heptanone, and 1-heptanol, concluding that carbon disulphide can't be used for the extraction of alcohols and ketones but that on the other hand it is highly efficient for aliphatic and aromatic compounds; methylene chloride appears to be the best for alcohols and ketones, but aromatic compounds are not well extracted. Borén et al. (143) investigated the extraction of a mixture of a variety of compounds by

using seven different solvents and four solvent mixtures, concluding that dichloromethane and carbon disulphide were both efficient but that carbon disulphide was more successful in extracting aromatic compounds. For all of these reasons, CS$_2$ is by far the most common solvent used in CLSA extractions, giving excellent results for BTEX compounds.

B. Analytical Procedure Proposed for CLSA

Reagents.

Ultrapure reagent water
BTEX mix standard working solution of 2 mg/L of each component in methanol (pesticide grade)
Surrogate standard working solution of 1-cloropentane and 1-clorohexane 0.5-g/L level in methanol
Carbon disulphide, acetone, and methylene chloride (pesticide grade)

Materials.

Carbon filters with 1.5 mg activated carbon
Syringes of 5–25-L capacity
Glass vials, 50-μl capacity, with gas-tight stoppers (as described in Fig. 9)
TFE sleeve, 5-mm-ID flexible tubing approximately 2 cm long

Instrumentation.

CLSA apparatus (as described in Fig. 6)
Gas chromatograph and FID, PID, or MSD detectors
Capillary column

Procedure. One liter of the sample water is rinsed in the stripping bottle by adding 10 μl of the surrogate working solution corresponding to a 5-μg/L concentration. Then the sample flask is tightly stopped and introduced to a water bath with a glass joint below the water level. The hermetic stripping circuit is then established at the operating temperature with an auxiliary adsorbent filter mounted in the filter holder. The pump is turned on for 1 min in order to purify the air remaining over the water sample. The auxiliary filter is then replaced with the cleaned analytical filter, and the stripping period is started. The stripping conditions are 45°C for 0.5 h. Salting-out could also be applied by adding 200 g sodium chloride to the 1-L water sample. Nevertheless, due to the high stripping efficiency for BTEX, a good sensitivity is obtained without the need of salt addition, also avoiding salt precipitations in the pneumatic system.

Extracting the Filter. Once the stripping period is finished, remove the filter from the filter holder and extract with CS_2, as indicated in Fig. 9. The filter is connected to the sample tube using a suitable piece of PTFE tubing in such a way that the two restricted ends of the filter and sample tube are in direct contact. Then with a syringe place 5 μl of carbon disulphide above the carbon, and force the small volume of solvent to move up and down through the charcoal layer by separating and reconnecting the filter and vial while still within the PTFE sleeve. Repeat this operation five times and then let it stand for 1 min so as to permit diffusion of the solvent within the carbon particles. Again move up and down five times. After this, cool the vial with a piece of ice, taking care not to freeze the CS_2. This cooling draws the solvent to the vial. This procedure is repeated three times with 5 μl of carbon disulphide each time, obtaining a final volume, owing to evaporation, of about 15 μl. Finally, an aliquot of 2 μl of the extract is injected in the gas chromatograph in the splitless of on-column mode for analysis. Vials are stoppered and stored at 20°C until analysis. See Sec. VI for the quality control procedure. Figure 10 shows a typical analysis of BTEX in water by the proposed analytical procedure.

Reconditioning the Adsorbent Filter. Fill the glass tube with CS_2 and let the solvent flow through the filter. Repeat for at least three times with acetone and three times with methylene chloride. If the flow is too slow due to deposited salts, immerse the filter in 1 N HNO_3 and empty three to five times so that the acid penetrates the carbon completely. After washing with acid, rinse with distilled water and acetone, and continue with cleaning as before. After the final rinse, connect the filter to the vacuum for approximately 5 min. For stored filters, a short rinse and evacuation immediately is recommended before an analysis. Clean stoppers with methylene chloride, immersing at least overnight. Rinse the TFE sleeve with methylene chloride and acetone and store it in acetone until ready to use.

Typical limits of detection and precision (expressed as RSD) are shown in Table 6. These results were obtained with capillary gas chromatography coupled to different detectors (FID and MSD) in the analysis of BTEX in potable waters.

V. SOLID-PHASE MICROEXTRACTION (SPME)

Solid-phase microextraction (SPME) consists of a small, cylindrical polymeric-coated fused-silica fiber

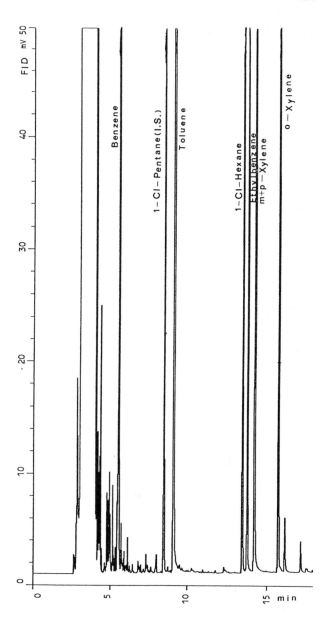

Fig. 10 High-resolution gas chromatogram of a spiked water sample at a 5-μg/L level of each BTEX and the corresponding internal standards (I.S.) analyzed by the proposed CLSA method. Chromatogram courtesy of J. Romero and F. Ventura from Aguas de Barcelona (Laboratori de Sant Joan d'Espí, Barcelona, Spain).

mounted in a modified gas chromatography syringe so that the fiber can be handled during extraction and injection processes. The analytes are extracted by absorption over the fiber, which is immersed in the water sample, generally agitated to favor diffusion (direct SPME) or in the headspace above the sample (headspace SPME), for a predetermined time. For each an-

Table 6 Typical LODs and Precision Obtained in Analysis of BTEX in Potable Water by CLSA with Capillary GC Coupled to FID and MS-ITD

	FID[a]		MS-ITD[a]	
	LOD (ng/L)	RSD (%)	LOD (ng/L)	RSD (%)
Benzene	2.4	7	1.3	5
Toluene	1.2	5	0.6	3
Ethylbenzene	1.5	8	0.8	6
m- + *p*-Xylene	1.2	9	0.4	8
o-Xylene	1.2	6	0.6	7

[a]Data provided by J. Romero and F. Ventura. Sociedad general de Aguas de Barcelona S.A. (Lab. Sant Joan de Despí, Spain).

alyte, once equilibrium between the water and the polymeric fiber is reached, the fiber is removed from the sample and withdrawn into the needle of the microsyringe. Then the needle is used to pierce the injector septum of the gas chromatograph to introduce the fiber into the glass liner, where the analytes are thermally desorbed and analyzed. The SPME method is a solvent-free technique that allows analysis at ng/L levels of volatile and semivolatile analytes in liquid, gases, and solid matrices with very low-volume samples, generally 2 ml. In SPME, sample manipulation is low, with very short sampling times, and because of its simplicity it is easily automated. The specificity of this extraction technique depends on the affinity of the analytes to be extracted for the selected type of polymeric fiber. The SPME technique has been widely used for the analysis of BTEX in all kinds of water (159–165) and solid matrices. The technique was initially applied to the analysis of volatile compounds (166–170) but has been extended to semivolatile compounds such as pesticides (171–178), phenols (179–182), PAHs, and PCBs (183) in the environmental field as well as to food analysis (184–187), clinical applications (188), and other fields (188–191).

The first attempts at the SPME technique were introduced by Pawliszyn and Liu in 1987 (192) when investigating rapid techniques for sample introduction in capillary gas chromatography in order to prevent peak broadening. They used a modified optical fiber and tested the laser desorption of three different compounds (triethylene glycol, polyethylene glycol, and carbowax) by immersing about 1 mm of the fiber tip into a 10% solution of the compound in chloroform. After that the solvent was removed by evaporation. The optical fiber was then introduced carefully into the cap-

illary column, and the compounds were laser-desorbed and analyzed by FID. In 1989, on the basis of these experiences and prior studies carried out by Pankow in 1982 (116) with solid-phase extraction (SPE) coupled to thermal desorption, R. Belardi and J. Pawliszyn (193) first saw SPME as a promising alternative to the existing sample preparation methods for water analysis. In 1990 Arthur and Pawliszyn (194) analyzed volatile chlorinated organic compounds, polychlorinated byphenils, and BTEX compounds in water at µg/L levels, using either an uncoated, fused silica or the polyimide film that the optical fibers are shipped with. The SPME was constructed by attaching a 22-cm fused-silica fiber inside a Hamilton 7000 series syringe (Fig. 11a). The metal plunger wire assembly is removed and replaced by the fiber, which is glued, using a high-temperature-resistant epoxy glue, into a length of stainless steel tubing. Only 1 cm of the fiber end is used to absorb the analytes.

Earlier fibers were purchased from optical fiber manufactures with a polyimide coating or with a polydimethylsiloxane film coating (40,42), and modified syringes were constructed in the laboratory as explained earlier. The increasing interest in the revolutionary SPME technique raised interest in the analytical companies, and in 1993 Supelco Inc. commercialized an SPME holder, as shown in Fig. 11b, and Varian developed and patented the first SPME autosampler, permitting automation of the analytical process, with the possibility of programming the parameters affecting the extraction. This equipment was further improved in 1996 in a new version that included the possibility of agitation during extraction. At the same time, efforts in

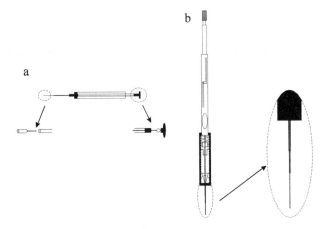

Fig. 11 Modified syringes for SPME: (a) homemade syringe constructed from a Hamilton 7000 series; (b) commercially available SPME syringe.

the development of new fibers permitted the commercialization of different thick films of the existing polydimethylsiloxane polymer and a new polyacrilate polymer very useful in the extraction of polar compounds (195).

The analysis of BTEX compounds in water by SPME could be carried out by immersion of the fiber into the water sample, known as direct solid-phase microextraction. But sometimes, for the analysis of complex matrices such as wastewater, it is preferable to analyze by headspace SPME (HS-SPME), in which the absorption takes place in the headspace above the water sample (196–198).

A. Factors Affecting the Technique

Two different processes take place in an SPME analysis: absorption of analytes over the polymeric fiber, and desorption of the fiber in the injector body of the gas chromatograph. The absorption process is nevertheless the most important step influencing the effectiveness of the extraction, and factors affecting this process, such as fiber selection, extraction time, agitation of the sample, and temperature, should be taken into consideration. In the next few sections, factors affecting the extraction of BTEX by direct SPME and HS-SPME are discussed.

B. Absorption

1. Direct Solid-Phase Microextraction

a. Fiber Selection. The most popular and most investigated polymeric coatings are poly-dimethylsiloxane (PDMS), commercialized with different-thickness films (7, 30, and 100 μm), and polyacrilate (PA), commercialized with 85-μm-thick film. Whereas PDMS is a nonpolar polymer with liquid properties, PA is more polar and with solid properties. This makes PDMS the best choice for the analysis of the nonpolar BTEX compounds. The film thickness determines the stationary-phase volume and, consequently, the amount of analyte absorbed. Comparing both 7 and 100 μm, 100-μm film produces, on average, a 20-fold increase in the absorption of BTEX (160). Although larger film thicknesses require longer equilibrium times, BTEX equilibrium is reached in less than 10 min with both fibers; therefore 100-μm PDMS film seems to be the best choice for BTEX analysis when low detection limits are needed. In recent years other, different polymeric fibers (carbowax/divinylbenzene CW/DVB, poly-dimethylsiloxane/divinylbenzene PDMS/DVB, carboxen

polydimethylsiloxane, and carbowax/templated resin CW/TPR) have been brought onto the market, to cover a higher range of polarities for the analysis of different compounds with even the possibility of HPLC analysis.

b. Factors Affecting the Diffusion.

Agitation and Temperature. In SPME an exhaustive extraction of analytes from the sample does not occur, but an equilibrium is developed between the aqueous and the polymeric organic phases. The distribution constant K of each compound between both systems governs the process. The extraction in an SPME analysis is completed when the system reaches equilibrium. During this process, in the first step analytes in the water sample should migrate to the thin layer around the fiber, where the absorption takes place. There then exists a concentration gradient near the surface of the fiber. In a second step, a migration inside of the viscous liquid polymer takes place in order to maintain a constant concentration of the analyte along the polymer. Dynamics in the extraction process in SPME have been studied in depth by Louch et al. (199), who obtained mathematical models of the process. They concluded that when the polymer coating is a liquid, which is the case for PDMS, diffusion is the only mass transport mechanism that determines the migration of the analyte molecules in the system, and so diffusion is the limiting step in the absorption process. For this reason, factors that increase this diffusion, such as agitation of the sample and temperature, will speed up the process. These factors determine the time needed to reach equilibrium.

For practical reasons, magnetic agitation is generally used, by placing a magnetic stirbar inside the 2-ml vials. As can be seen in Fig. 12, increasing the stirring rate reduces the equilibrium time. In the case of *o*-xylene, equilibrium in a well-agitated sample is reached in 10 min; without agitation it takes about 120 min. High-molecular-weight compounds are expected to have longer equilibration times than low-molecular-weight compounds because of their smaller diffusion coefficients. This equilibrium is reached in about 2 min for benzene and toluene and in 10 min for ethylbenzene and xylenes. More effective agitation systems have been used in SPME, such as intrusive mixing and sonication (200), in the analysis of heavier compounds (pesticides, PCBs, etc.), in which the mixing power is of more interest in order to reduce the equilibrium times.

Diffusion of the analytes in the aqueous phase increases as temperature rises. However, the absorption is an exothermic process, and increasing temperature

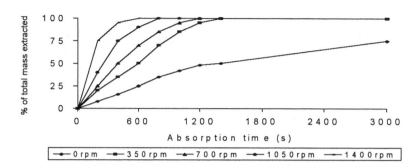

Fig. 12 Effect of stirring rate on *o*-xylene equilibrium-time profile by direct SPME.

has a negative effect in such processes. These effects are clearly shown in Fig. 13, in which the equilibrium time for *o*-xylene is reduced from 10 min at 25°C to 5 min at 50°C. Nevertheless, the total amount extracted at 50°C is reduced, because the distribution constant decreases when temperature increases. It has been reported that better conditions can be obtained by heating the sample and internally cooling the fiber (201). Nevertheless, 10 min is a reasonable extraction time, and, due to the decrease in absorption efficiency with higher temperatures, the ambient temperature with magnetic stirring is the best extraction condition for BTEX compounds.

"Salting-Out" Effect. The addition of salt to the water sample increases the partition coefficient K between the water sample and the coating. As shown in Fig. 14, a 100-g/L sodium chloride addition increases the total amount of BTEX absorbed by the fiber. In the best case, that of benzene, the amount of analyte extracted is double that of a water sample extracted without salt addition. Nevertheless, detection limits obtained without salt addition are very low, permitting the simplification of the procedure to the minimum.

Sample Volume. Due to the fact that SPME is an extraction governed by an equilibrium (Fig. 15a), the volume of the sample does not affect the amount of analyte extracted. The total amount of analyte adsorbed in direct SPME by a liquid coating such as PDMS can be calculated from the following equation:

$$n = \frac{C_0 V_1 V_2 K}{K V_{1+} V_2} \quad (1)$$

where n is the mass adsorbed by the coating, V_1 and V_2 are the volumes of the coating and the aqueous sample, respectively, K is the partition coefficient of the analyte between the coating and the water, and C_0 is the initial concentration of the analyte in the water sample. Because the volume of the coating is very small compared to $V_2(V_1 \ll V_2)$ and K is small, this results in

$$n = C_0 V_1 K \quad (2)$$

The total amount absorbed is dependent on the initial

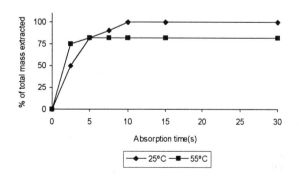

Fig. 13 Effect of temperature on *o*-xylene extraction by direct SPME.

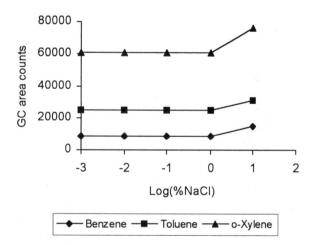

Fig. 14 Effect of salting-out on BTEX extraction by direct SPME.

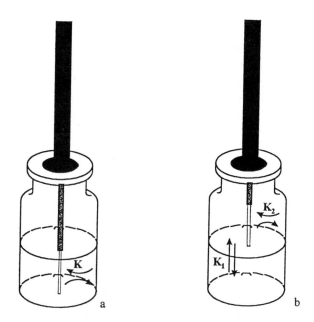

Fig. 15 (a) Direct SPME; (b) headspace SPME.

concentration, the volume of the coating (film thickness), and the partition coefficient. The volume of the sample does not affect the total amount extracted, and then it is decided by practical purposes. Generally, typical 2-ml GC autosampler vials are used for direct SPME.

2. Headspace Solid-Phase Microextraction (HS-SPME)

In headspace SPME, equilibrium is governed by two constants (Fig. 15b): the equilibrium between the liquid and the gaseous phase, K_1, which corresponds to Henry's constant, the limiting step of the process, and the equilibrium between the gaseous phase and the polymeric coating K_2. Because diffusion in HS-SPME occurs in the gaseous phase and, in the case of benzene, for example, the diffusion coefficient in the headspace is 0.077 cm^2/s compared with that in the aqueous phase of 1.8×10^{-5} cm^2/s, it is evident that an increase in diffusion with temperature or agitation is of no interest because equilibrium between the gaseous phase and the coating occurs (once reached, equilibrium between the water and the gaseous phase) in 40 s for benzene, in 1 min for ethylbenzene and o-xylene, and in 2 min for m- and p-xylene (196). The parameters of interest for HS-SPME are those that increase the concentration of the analyte in the vapor phase and hence affect the total amount of analyte extracted; the "salting out" effect; temperature; and the headspace/sample volume ratio.

a. "Salting-Out" Effect. The addition of salt to the water sample increases the partition coefficient between the aqueous and the gaseous phase, as discussed in the section on static headspace. C_0 will increase, as shown in Fig. 1, and then the total amount of BTEX absorbed is as according to Eq. (2).

b. Temperature. Although an increase in temperature reduces the time to reach the first equilibrium, K_1, and also increases the concentration of analyte in the gaseous phase, the total amount absorbed at 25°C is higher than that at 40°C (18). This could be explained by the fact that absorption by the fiber at lower temperatures is more important than the increase in analyte gaseous-phase concentration, because absorption is an exothermic process.

c. Headspace/Sample Volume Ratio. The total amount of analyte absorbed in HS-SPME is determined by the following equation:

$$n = \frac{C_0 V_1 V_2 K_1 K_2}{K_1 K_2 V_1 + K_2 V_3 + V_2} \quad (3)$$

in which the new term, V_3, is the volume of the headspace. Nevertheless $K = K_1 K_2$ and $V_1 \ll V_2$. Then Eq. (3) can be rewritten as:

$$n = \frac{C_0 V_1 V_2 K}{K_2 V_3 + V_2} \quad (4)$$

This equation has an extra term, $K_2 V_3$, compared to the amount extracted by direct SPME (Eq. 1). This supposes that the mass extracted by HS-SPME is less than that in direct SPME and that the sensitivities will be lower. Nevertheless, by reducing the headspace/sample volume ratio to the minimum value and considering the small K_2 values for BTEX (see Table 7), very similar

Table 7 Partition Coefficients of BTEX

	K (atm cm^3/mol)	K_1 (atm cm^3/mol)	K_2 (atm cm^3/mol)
Benzene	126	0.26	493
Toluene	340	0.26	1322
Ethylbenzene	528	0.16	3266
m-Xylene	—	—	—
p-Xylene	831	0.24	3507
o-Xylene	654	0.15	4417

Source: Ref. 196.
K_1 is the gas/water partition coefficient, K_2 is the coating/gas partition coefficient, and K is the coating/water partition coefficient.

recoveries could be obtained with HS-SPME and direct SPME. Taking into consideration that the length of the fiber is 1 cm and must be above the surface of the water sample to reduce the headspace/sample volume ratio, it is necessary to use vials with small diameters and high heights. In our laboratory we have tested 20-ml commercially available hypovials with 8-cm height and 2-cm diameter, permitting us to reduce this ratio to 0.25 and thus obtaining similar recoveries with HS-SPME and direct SPME.

C. Desorption

Once analytes have been extracted, either by direct SPME or HS-SPME, the polymeric fiber is desorbed in the GC injector. The fiber is withdrawn into the needle, and, after piercing the GC septum, the fiber is released inside the glass insert, where thermal desorption occurs. There are two parameters affecting this process: temperature and desorption time. Polydimethylsiloxane is stable up to 300°C; nevertheless, thermal desorptions at 150°C for 2 min have been found to give a rapid and complete desorption (27,171), demonstrated by the fact that no tailing chromatographic peaks were observed and carryovers after highly contaminated water samples were less than 0.2% for all BTEX compounds.

D. Analytical Procedure Proposed for Direct and Headspace Solid-Phase Microextraction

As mentioned earlier, in 1996 Varian commercialized an SPME autosampler that permits full automation of the process. Absorption time, desorption time, temperature of the sample, and agitation can be programmed. Agitation of the sample is carried out by an intense vibratory movement of the fiber, instead of by a magnetic stirring of the sample. With this autosampler, the mixing power is not controlled, but a constant intensity is applied. And SPME extractions can be done with or without agitation with this equipment. Our experiences with this autosampler have demonstrated that the mixing efficiency obtained with this system permits 10-min equilibrium times for BTEX, as observed with magnetic stirring. A carousel with 48-position 2-mL vials could be programmed for direct SPME, or with 12-position 10-mL vials for HS-SPME. The analytical procedure, described next, applies to manual SPME and also to automatic SPME. Special devices besides the modified syringe (Fig. 11) are commercially available for easy handling during the extraction.

Reagents.

Polydimethylsiloxane 100-μm film thickness for manual or autosampler use
Modified SPME syringe
BTEX mix standard working solution 2 mg/L of each component in methanol (pesticide grade)
Internal standard working solution of fluorobenzene at 5-mg/L level in methanol

Materials.

5–25-μl syringes
2-ml autosampler vials for direct SPME, or 20-ml hypovials for HS-SPME with high-temperature septa and crimp caps
7-mm × 2-mm magnetic stirrer bars

Procedure. For direct SPME, 1.5 mL sample is placed in the 2-ml autosampler vials. To agitate the samples, a magnetic stirrer bar is placed in the sample vial. The fiber is placed in direct contact with the stirred sample for 10 min at room temperature. Finally, the fiber is desorbed in the GC injector at 260°C for 2 min. Splitless or on-column injection modes can be used. For HS-SPME, 15 mL sample is filled in the 20-ml hypovials. After 1 h at room temperature, during which the water/gaseous equilibrium is reached, the vials are analyzed. The fiber is placed in the headspace above the water sample for 5 min, during which the gas/coating equilibrium is reached. After this, the fiber is desorbed in the GC injector at 260°C for 2 min. See Sec. VI for the quality control procedure.

Typical limits of detection and precision (expressed as RSD) are shown in Table 8. These results are obtained with capillary gas chromatography coupled to

Table 8 Typical LODs and Precision Obtained in Analysis of BTEX in Potable Waters by Direct SPME with Capillary GC Coupled to FID and MS-ITD

	FID[a]		MS-ITD[b]	
	LOD (ng/L)	RSD (%)	LOD (ng/L)	RSD (%)
Benzene	0.22	4	0.11	6
Toluene	0.11	3	0.05	4
Ethylbenzene	0.04	9	0.03	7
m- + *p*-Xylene	0.09	9	0.05	8
o-Xylene	0.03	9	0.03	7

[a]*Source*: Ref. 160.
[b]Data experimentally obtained by I. Valor et al. by calculating the LOD from concentration generating *s/n* = 3 and RSD (%) for *n* = 5.

FID and MS-ITD in the analysis of BTEX in potable water. Figures 16 and 17 represent the typical chromatorgams obtained by means of direct SPME and HS-SPME, respectively.

VI. QUALITY CONTROL

A. Calibration

In static headspace, an aliquot of the headspace gas is analyzed, the resultant signal from a given compound (e.g., benzene) represents the concentration of this compound in the gas space above the sample. But the concentration of interest, that of the benzene in the condensed phase, could be calculated by knowing the partition coefficient of benzene in the gas-condensed system. In the purge-and-trap technique and CLSA, BTEX compounds from the water sample are stripped with a gas, trapped into a sorbent, and desorbed either with a gas in purge-and-trap or with a solvent in CLSA, and again an aliquot of the gas or solvent is analyzed. The concentration of the water sample could be calculated from the chromatographic signal obtained, by knowing the recovery of that compound during the extraction process. In the last case of direct SPME and HS-SPME it would be necessary to know constants affecting each process in order to apply Eqs. (2) and (3) from Sec. V.B.1.b. Nevertheless, for practical analysis, to avoid the determination of these data, model reference systems are analyzed to make calibration. Ultrapure water is spiked at known concentration levels and analyzed following the same procedure as for unknown samples. Signals obtained with these analyses are used to generate the calibration curve.

1. Calibration Procedure by Means of Model Reference Systems

A multipoint curve is recommended for method calibration. The number of calibration solutions needed depends on the calibration range desired. A minimum of three calibration solutions is required to calibrate a 20-factor concentration range. For a factor of 100, use at least five standards. The first calibration solution should contain each BTEX at a concentration 2–10 times greater than the method detection limit. To prepare a calibration standard, dilute the appropriate volume of the working standard solution to ultrapure water in a volumetric container. If the salting-out technique is used, each calibration solution should be added with the appropriate amount of the selected salt. Every calibration solution must be spiked with the internal standard solution at the same concentration. The concentration of the internal standard must give a good, quantifiable chromatographic signal. Because BTEX standard solutions are not stable, they should be prepared daily.

Starting with the standard of the lowest concentration, analyze each calibration standard according to the proposed analytical procedure. Peak-area-relative ratios are used to generate the calibration curve. At the beginning of a series of analyses, and once every 10 samples, a calibration solution should be analyzed to check the calibration curve. The result should be within the expected result $\pm$ 3 times the standard deviation of the method for this analyte.

2. Calibration by Means of the Standard Addition Method

The effect caused by the matrix when analyzing very highly contaminated wastewater samples, viscous water samples, etc. affects the intermolecular interaction between the analyte and the water. Subsequently, calibration has to be carried out in the same matrix. Nevertheless, this requirement can sometimes cause serious problems, because the matrix of the sample is not available in a pure form or cannot be simulated to a sufficient extent. This matrix effect applies especially to the static headspace technique and to the direct and HS-SPME techniques. When this occurs, the standard addition method has demonstrated good results for static headspace (202–205) and could also be applied to direct and HS-SPME.

In the standard addition method, the sample of interest is analyzed twice, with the addition of a known amount of the compound to one of the two replicates. The increase in the peak area corresponds to the amount added and permits the calculation of the original concentration.

Procedure. Analyze the sample following the normal procedure. Let the area peak of a detected compound in this analysis be A_0 and let C_0 be the unknown concentration of, for example, toluene in this sample. In a second step, a defined volume of a standard solution of toluene in methanol is introduced into the closed system with a liquid syringe, by piercing the rubber septum (for static headspace or direct and HS-SPME). This results in a known concentration added to the water sample C_1. After this the sample is again analyzed, giving a peak area of A_1. The initial concentration of toluene in the water sample, C_0, is calculated by the following formula:

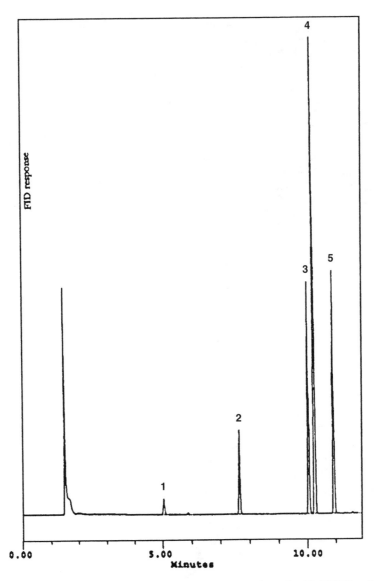

Fig. 16 High-resolution gas chromatogram of industrial wastewater contaminated with BTEX using direct SPME (100-μm PDMS fiber). *Column*: VOCOL Rtx-624 (30 m × 0.32-mm ID × 1.8-μm film). *Temp. program*: 40°C (2 min), 8°C/min to 120°C (1 min). *Carrier gas*: helium 4-ml/min flow. *Injector temp.*: 260°C, splitless injection mode. *Detector*: FID. 1: benzene; 2: toluene; 3: ethylbenzene; 4: *m-* + *p*-xylene; 5: *o*-xylene. (From Ref. 160.)

$$C_0 = \frac{A_0 C_1}{A_1 - A_0} \tag{5}$$

Background Control. Before any samples are analyzed, a laboratory reagent blank must be quantified in order to check that a reasonably low level is obtained. In a sequence of analyses, a blank after a highly contaminated water sample should be analyzed to avoid carryover. The blank concentration must always be below the detection limit for each analyte. Sources of high background levels are:

Static headspace: Contamination of the syringe, glassware, or equipment; ambient air contamination

Purge-and-trap: glassware, purity of the purge gas, a badly reconditioned trap, equipment, or a carryover problem

CLSA: Leaks in the system, with air entering in the circuit; badly reconditioned filters, glassware, or a carryover problem

Direct and HS-SPME: Badly desorbed fiber, glassware

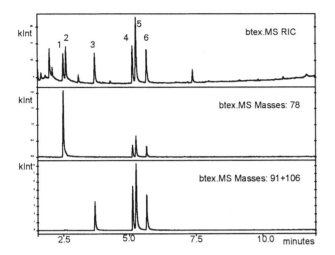

Fig. 17 High-resolution gas chromatogram of a spiked water sample at 10-μg/L level of each BTEX analyzed by the proposed HS-SPME method. 1: benzene; 2: fluorobenzene (internal standard); 3: toluene; 4: ethylbenzene; 5: *m-* + *p-*xylene; 6: *o*-xylene. *Column:* VOCOL Rtx-624 (30 m $\times$ 0.32-mm-ID $\times$ 1.8-μm film). *Temp. program:* 40°C (2 min), 8°C/min to 120°C (1 min). *Carrier gas:* helium 4-ml/min flow. *Injector temp.:* 260°C, splitless injection mode. *Detector:* MS-ITD.

Method Validation. The laboratory must validate the method in order to know the method detection limits, accuracy, precision, linearity, and matrix effects.

VII. OTHER METHODS FOR BTEX ANALYSIS IN WATER

Nowadays, purge-and-trap, headspace, CLSA, and SPME represent probably more than 95% of analytical technique used in routine laboratory analysis of BTEX in water. Nevertheless, other analytical techniques can be used for this purpose. In the direct injection technique, the water sample is injected directly into the GC without any pretreatment. The main problem with this technique in the analysis of BTEX is the low sensitivity of PID, FID, or MSD. Detection limits are in the range of 0.1–1 mg/L for the flame ionization detector (206). The same problem exists with the simple liquid–liquid extraction technique, due to the fact that evaporation to obtain a concentrated extract is not possible for BTEX because of the high volatility of the analytes. Steam distillation and vacuum distillation (207) have also been applied to the analysis of volatiles in water.

Recently, a novel technique called *membrane extraction* has gained popularity for the analysis of vol-atile organic compounds (VOCs) in water samples (208–211). Membrane extraction involves a semiper-meable membrane. Volatiles migrate from the aqueous phase to the surface of the membrane and dissolve in the inside surface layer of the membrane. The next step is either evaporation or stripping of the analytes into the stripping gas. Finally, enzyme immunoassay using antibodies is a powerful technique with very high selectivity and sensitivity. Initially such techniques were limited to clinical applications, but more and more they are being introduced in environmental analysis. The EPA has incorporated these techniques in EPA method 4031 (212) for BTEX analysis in soil. Enzyme immunoassay kits are commercially available for the analysis of BTEX in water (213).

ACKNOWLEDGMENTS

The authors want to thank Mr. A. Pérez Martín for his great contribution to all the schematic drawings of equipment.

REFERENCES

1. T. Walden, L. Spence. Risk Based BTEX screening criteria for a groundwater irrigation scenario. Human Ecological Risk Assessment 4:699–722, 1997.
2. R. C. Borden, C. M. Kao. Evaluation of groundwater extraction for remediation of petroleum-contaminated aquifers. Water Environ. Res. 64:28–36, 1992.
3. A. I. Stubin, T. M. Brosnan, K. D. Porter, L. Jiménez, H. Lochan. Organic priority pollutants in New York City municipal waste waters: 1989–1993. Water Environ. Res. 68:1037–1044, 1996.
4. A. J. Gregory, L. A. Viorica, A. H. Ronald. Organic compounds in an industrial wastewater: a case of study of their environmental impact. Environ. Sci. Technol. 12:88–96, 1978.
5. D. F. LaBranche, M. Robin Collins. Stripping volatile organic compounds and petroleum hydrocarbons from water. Water Environ. Res. 68:348–358, 1996.
6. C. M. Aelion. Impact of aquifer sediment grain size on petroleum hydrocarbon distribution and biodegradation. J. Contaminant Hydrol. 22:109–121, 1996.
7. P. J. Squillace, J. S. Zogorski, W. G. Wilber, C. V. Price. Preliminary assessment of the occurrence and possible sources of MTBE in groundwater in the United States. Environ. Sci. Technol. 5:1721–1730, 1996.
8. P. T. Katsumata, W. E. Kastenberg. Fate and transport of methanol fuel from spills and leaks. Hazardous waste and hazardous materials. 13:485–498, 1996.

9. L. H. Keith, W. A. Teilliard. Priority pollutants I—a perspective view. Environ. Sci. Technol. 13:416–423, 1976.

10. Liste I de la Directive 76/464/CEE du Conseil. JO. No. 1129 du 18/5 1976, 23.

11. Guidelines for drinking-water quality. 2nd ed. Vol. 1. Recommendations. World Health Organization, Geneva, 1993.

12. Propuesta de la Directiva del Consejo relativa a la calidad de las aguas destinadas al consumo humano. Diario Oficial de las Comunidades Europeas. No. C 131/5 de 30.05.1995.

13. Posición Común (CE) No. 13/98. Diario Oficial de las Comunidades Europeas. De 26.03. 1998.

14. R. Bassette, S. Özeris, C. H. Whitnah. Gas chromatographic analysis of head space gas of dilute aqueous solutions. Anal. Chem. 34:1540–1543, 1962.

15. S. Ozeril, R. Bassette. Quantitative study of gas chromatographic analysis of headspace gas of dilute aqueous solutions. Anal. Chem. 35:1091–1093, 1963.

16. Edward S. K. Chian and Powell P. K. Kuo. Distillation/headspace/gas chromatographic analysis for volatile polar organics at ppb level. Environ. Sci. Technol. 11:282–285, 1977.

17. R. E. Kepner, H. Maarse, J. Strating. Gas chromatographic head space technique for the quantitative determination of volatile components in multicomponent aqueous solutions. Anal. Chem. 36:77–82, 1964.

18. K. L. E. Kaiser, B. G. Oliver. Determination of volatile halogenated hydrocarbons in water by gas chromatography. Anal. Chem. 48:2207–2209, 1976.

19. R. Bassette, C. H. Whitnah. Removal and identification of organic compounds by chemical reaction in gas chromatography analysis. Anal. Chem. 32:1098–1200, 1960.

20. E. Valero. Determinación de compuestos volátiles mediante técnicas de espacio de cabeza. Cromatografía y técnicas afines. 16:3–9, 1995.

21. J. L. Guinamant. Extraction of volatile and/or semivolatile compounds. Analusis 20:43–45, 1992.

22. A. G. Vitenberg. Efficiency of different variants of headspace analysis. Zh. Anal. Khim. 46:764–769, 1991.

23. B. D. Page, H. B. S. Conacher, T. Salminen, G. R. Nixon, G. Riedel, B. Mori, J. Gagnon, R. Brousseau. Survey of bottled drinking water sold in Canada. Part I. Select volatile organic compounds. J. AOAC. Int. 76:26–31, 1993.

24. B. D. Page, H. B. S. Conacher, T. Salminen, G. R. Nixon, G. Riedel, B. Mori, J. Gagnon, R. Brousseau. Survey of bottled drinking water sold in Canada. Part II. Select volatile organic compounds. J. AOAC. Int. 76:26–31, 1993.

25. B. Kolb. Quantitative trace analysis of volatile organic compounds in air, water and soil using equilibrium headspace gas chromatography. LC-GC 14:44–54, 1996.

26. M. T. Mehran, N. Golkar, W. T. Cooper, A. K. Vickers. Headspace analysis of some typical organic pollutants in drinking water using differential detectors: effects of columns and operational parameters. J. Chromatogr. Sci. 34:122–129, 1996.

27. J. R. J. Pare, J. M. R. Belanger, K. Li. Microwave-assisted process (MAP): application to the headspace analysis of VOC's in water. J. Microcolumn Sep. 7: 37–40, 1995.

28. H. Ohno, J. Aoyama, H. Kishimoto. Simultaneous determination of volatile organic compounds by head space gas-chromatographic analysis with dual detection. Jpn. J. Toxicol. Health 38:84–92, 1992.

29. N. P. Zinov'era. Determination of benzene, styrene and toluene in water. Gig. Sanit. 7:76–77, 1993.

30. M. Denaro, S. Giammarioli, M. R. Milana, M. Mosca, R. Marcoaldi. Analysis of volatile substances in bottled minerals waters. I. Determination of benzene. Rapp. ISTISIAN, 1991, p. 34.

31. E. Hala, J. Pick, V. Fried, O. Vilím. Vapor–Liquid Equilibrium. 2nd ed. Pergamon Press, Oxford, 1967, p. 117.

32. A. G. Vitenberg, B. V. Ioffe, V. N. Borisov. Application of phase equilibria to gas chromatographic trace analysis. Chromatographia 7:610–619, 1974.

33. D. Mackay, A. T. K. Yeun. Mass transfer coefficient correlations for volatilization of organic solutes from water. Environ. Sci. Technol. 17:211–217, 1983.

34. O. V. Rodinkov, L. N. Moskvin. Computational prediction of the distribution ratios of volatile organic compounds in gas–liquid system. Zh. Anal. Khim. 50: 164–166, 1995.

35. B. Kolb. Application of an automated headspace procedure for trace analysis by gas chromatography. J. Chromatogr. 122:553–568, 1976.

36. A. Montiel. Méthode de dosage rapide et sensible des composés organovolatils et de leurs précurseurs dans les eaux. La tribune du CEBEDEAU 422:23–30, 1979.

37. J. Drozd, J. Novak. Headspace gas analysis by gas chromatography. J. Chromatogr. 165:141–165, 1979.

38. A. T. Bellar, J. J. Lichtenberg. Determining volatile organics at the μg/L level by gas chromatography. J. Am. Water Works Assoc. 66:739–794, 1974.

39. A. T. Bellar, J. J. Lichtenberg. Semi-automated headspace analysis of drinking waters and industrial waters for purgeable volatile organic compounds. In: C. E. Van Hall, ed. Measurement of organic pollutants in water and wastewater. STP 686, American Soc. Testing and Materials, Philadelphia.

40. S. A. Hazard, J. L. Brown, W. R. Bertz. Extraction and analysis of hydrocarbons associated with leaking underground storage tanks. LC-GC 9:40–42, 1991.

41. T. N. Barnung, O. Grahl-Nielsen. Determination of benzenes and naphthalenes in water by purge and trap isolation and capillary column chromatography. J. Chromatogr. 466:271–278, 1989.

42. W. Bertsch, E. Anderson, G. Holzer. Trace analysis of organic volatiles in water by gas chromatography–mass spectrometry with glass capillary columns. J. Chromatogr. 112:701–718, 1975.

43. E. Woollfenden, J. Barbeiro. Purge and trap analysis of volatile organic compounds (VOCs) in water. Int. Lab. 22:25–30, 1992.

44. V. López-Avila, R. Wood, M. Flanagan, R. Scott. Determination of volatile priority pollutants in water by purge and trap and capillary column gas chromatography/mass spectrometry. J. Chromatogr. Sci. 25:286–291, 1987.

45. M. Duffy, J. N. Driscoll, S. Pappas, W. Sanford. Analysis of ppb levels of organics in water by means of purge-and-trap, capillary gas chromatography and selective detectors. J. Chromatogr. 441:73–80, 1988.

46. B. D. Page, H. B. S. Conacher, J. Salminen, G. R. Nixon, G. Riedel, B. Mori, J. Gagnon, R. Brousseau. Survey of bottled drinking water sold in Canada. Part II. Selected volatile organic compounds. J. Assoc. Off. Anal. Chem. Int. 76:26–31, 1993.

47. X. Yan, K. R. Carney, E. B. Overton. Application of purge-and-trap method for fast, convenient field analysis of water and soil samples. J. Chromatogr. Sci. 30:491–496, 1992.

48. M. Biziuk, Z. Polkowska, D. Gorko, W. Janick, J. Namiesnik. Determination of volatile organic compounds in water intakes and trap water by purge and trap and direct aqueous injection–electron capture detection techniques. Chem. Anal. 40:299–307, 1995.

49. Perkin-Elmer. Analysis of volatile organic compounds (VOCs) in water by USEPA Method 624 using a programmable split/splitless injector. Appl. Note GCA-63, 1994, pp. 8–12.

50. R. P. Eganhouse, T. F. Dorsey, C. S. Phinney, A. M. Westcott. Determination of C_6 to C_{10} aromatic hydrocarbons in water by purge and trap capillary gas chromatography. J. Chromatogr. 628:81–92, 1993.

51. N. K. Kristiansen, E. Lundanes, M. Froshaug, H. Utkilen. Determination and identification of volatile organic compounds in drinking water produced offshore. Chemosphere 25:1631–1642, 1992.

52. D. K. Nguyen, A. Bruchet, P. Arpino. High-resolution capillary GC-MS analysis of low-molecular-weight organic compounds in municipal waste water. J. High Res. Chromatogr. 17:153–159, 1994.

53. Improved purge-trap-GC analysis of volatiles in drinking water by USEPA Method 524.2, Supelco Appl. Note, p. 12, 1994.

54. A. K. Vickers, L. M. Wright. An automated GC/MS system for the analysis of volatile and semi-volatile organic compounds in water. J. Automatic Chem. 15:133–139, 1993.

55. J. S. Ho, P. Hodakievic, T. A. Bellar. A fully automated purge-and-trap system for analyzing volatile organics in drinking water. Am. Lab. July:40–51, 1989.

56. V. B. Stain, R. S. Naraung. Simplified method for the determination of volatile compounds in eggs using headspace analysis with photoionization detector. Arch. Environ. Contam. Toxicol. 19:593–596, 1990.

57. R. E. Hurts. A method of collecting and concentrating headspace volatiles for gas chromatographic analysis. Analyst 99:302–305, 1974.

58. J. L. Esteban, I. Martínez-Castro, J. Sanz. Evaluation and optimization of the automatic thermal desorption method in the gas chromatographic determination of plant volatile compounds. J. Chromatogr. 657:155–164, 1993.

59. Y. Xinwei, R. C. Kenneth, E. B. Overton. Application of purge and trap method to fast and convenient field analysis of water and soil samples. J. Chromatogr. Sci. 30:491–496, 1992.

60. U.S. Environmental Protection Agency. Contact laboratory program protocol, September 1984, exhibit D, analytical methods, revised July 1985.

61. EPA 502.2. Volatile organic compounds in water by purge and trap capillary column gas chromatography with photoionization and electrolytic conductivity detectors in series. Revision 2.0, 1989.

62. EPA Method 503.1.

63. EPA 524.2 Measurement of purgeable organic compounds in water by capillary column gas chromatography/mass spectrometry. Revision 3.0, 1989.

64. EPA Method 602.

65. Guidelines establishing test procedures for analysis of pollutants under the Clean Water Act; final rule and interim final rule and proposed rule. Method 624, purgeables. U.S. Environmental Protection Agency, 40 CFR Part 136. Fed. Regist. 1984, pp 141–152.

66. J. G. Schnable, M. B. Capangpangan, I. H. Suffet. Continuous automatic monitoring of volatile organic compounds in aqueous streams by a modified purge and trap system. J. Chromatogr. 549:335–344, 1991.

67. J. G. Schnable, B. Dussert, J. H. Suffet, C. D. Hertz. Comparison of quarter-hourly on-line dynamic headspace analysis to purge-and-trap analysis of varying volatile organic compounds in drinking water sources. J. Chromatogr. 513:47–54, 1990.

68. W. E. Hammers, H. F. P. M. Bosman. Quantitative analysis of a dynamic headspace analysis technique for non-polar pollutants in aqueous samples at the ng/kg level. J. Chromatogr. 360:425–432, 1986.

69. A. B. Robinson, D. Partridge, M. Turner, R. Teranishi, L. Pauling. An apparatus for the quantitative analysis of volatile compounds in urine. J. Chromatogr. 85:19–29, 1973.

70. K. E. Murray. Concentration of headspace, airborne and aqueous volatiles on Chromosorb 105 for examination by gas chromatography and gas chromatography–mass spectrometry. J. Chromatogr. 135:49–60, 1977.

71. M. E. Rose, B. N. Colby. Reduction in sample foaming in purge and trap gas chromatography/mass spectrometry analyses. Anal. Chem. 51:2176–2180, 1979.

72. M. D. Erickson, M. K. Alsup, P. A. S. Hyldburg. Foam prevention in purge and trap analysis. Anal. Chem. 53: 1265–1269, 1981.

73. D. P. Lin, C. Falkenberg, D. A. Payne, J. Thakkar, C. Tang, C. Elly. Kinetics of purging for the priority volatile organic compounds in water. Anal. Chem. 65: 999–1002, 1993.

74. S. M. Abeel. The effect of various parameters on extraction efficiencies of volatile organic compounds during purge and trap analysis. Presented at the Pittsburgh Conference and Exposition, Chicago, February 28–March 3, 1994.

75. R. Kostiainen. Effect of operating parameters in purge-and-trap GC-MS of polar and non-polar organic compounds. Chromatographia 38:709–714, 1994.

76. D. Djzan and Y. Assadi. Optimization of the gas stripping and cryogenic trapping method for capillary gas-chromatographic analysis of traces of volatile halogenated compounds in drinking water. J. Chromatogr. A. 697:525–532, 1995.

77. R. H. Brown, C. J. Purnell. Collection and analysis of trace organic vapor pollutants in ambient atmospheres. The performance of a Tenax-GC adsorbent tube. J. Chromatogr. 178:79–90, 1979.

78. E. D. Pellizzari, J. E. Bunch, R. E. Berkley, J. McRae. Collection and analysis of trace organic vapor pollutants in ambient atmospheres. The performance of a Tenax GC cartridge sampler for hazardous vapors. Anal. Lett. 9:45–63, 1976.

79. A. Bianchi, M. S. Varney. Modified analytical technique for the determination of trace organics in water using dynamic headspace and gas chromatography–mass spectrometry. J. Chromatogr. 467:111–128, 1989.

80. S. M. Abeel, A. K. Vickers, D. Decker. Trends in purge and trap. J. Chromatogr. Sci. 32:328–338, 1994.

81. R. Westendorf. Managing water in purge and trap GC/MS. Environ. Lab. Aug–Sep:36–39, 1992.

82. J. F. Pankow. Technique for removing water from moist headspace and purge gas containing volatile organic compounds. Application in the purge with whole-column cryotrapping (PIWGC) method. Environ. Sci. Technol. 25:123–126, 1991.

83. W. Janicki, L. Wolska, W. Wardencki, T. Namiesnik. Simple device for permeation removal of water vapor from purge gases in the determination of volatile organic compounds in aqueous samples. J. Chromatogr. A 654:279–285, 1993.

84. A. J. Núñez, L. F. González, J. Jának. Preconcentration of headspace volatiles for trace organic analysis by gas chromatography. J. Chromatogr. 300:127–162, 1984.

85. K. Grob. Organic substances in potable water and its precursor. Part I. Method for their determination by gas–liquid chromatography. J. Chromatogr. 84:255–273, 1973.

86. B. Levadie, S. M. MacAskill. Analysis of organic solvents taking on charcoal tube sample by a simplified technique. Anal. Chem. 48:76–78, 1976.

87. R. G. Clark, D. A. Cronin. The use of activated charcoal for the concentration and analysis of headspace vapors containing food aroma volatiles. J. Sci. Fd. Agic. 26:1615–1624, 1975.

88. M. T. Mehran, M. G. Nickelsen, N. Gokar, W. T. Cooper. Improvement in the purge and trap technique for the rapid analysis of volatile organic pollutants in water. J. High Res. Chromatogr. 13:429–433, 1990.

89. M. Marchand, M. Termonia, J. C. Caprais, M. Wybaw. Purge-and-trap GC-MS analysis of volatile organic compounds from the Guayamas Basin hydrothermal site (gull of California). Analusis. 22:326–331, 1994.

90. R. Takeuchi, H. Ueno, K. Naito, Y. Fujiki. Simplified determination of volatile organic compounds in water with use of active carbon. Bunseki Kagaku 44:651–657, 1995.

91. G. Holzer, H. Shanfield, A. Zlatkis, W. Bertsch, P. Juarez, H. Mayfield, H. M. Liebich. Collection analysis of trace organic emissions from natural sources. J. Chromatogr. 142:755–764, 1977.

92. M. J. Lewis, A. A. Williams. Potential artifacts from using porous polymers for collecting aroma components. J. Sci. Food Agric. 31:1017–1026, 1980.

93. G. MacLeod, J. M. Ames. Comparative assessment of the artifact background on thermal desorption of Tenax GC and Tenax TA. J. Chromatogr. 335:393–398, 1986.

94. A. Bianchi, M. S. Varney, I. Phillips. Modified analytical technique for the determination of trace organics in water using headspace and gas chromatography mass spectrometry. J. Chromatogr. 467:111–128, 1989.

95. Y. Gao, G. Zhang, Y. Ma, B. Xin. Purge-and-trap apparatus for the analysis of trace volatile organics in water. Fenxi Haxue 17:944–948, 1989.

96. M. R. Driss, M. L. Bouguerra. Analysis of volatile organic compounds in water by purge and trap and gas chromatographic techniques. Operational parameters optimization of the purge step. Int. J. Environ. Anal. Chem. 45:193–204, 1991.

97. K. Ventura, M. Dostal, J. Churacek. Retention characteristics of some volatile compounds on Tenax GP. J. Chromatogr. 642:379–382, 1993.

98. K. Grob, A. Habich. Headspace gas analysis: the role and the design of concentrator traps specifically suitable for capillary gas chromatography. J. Chromatogr. 321:45–58, 1985.

99. H. Kessels, W. Hoogerwerf, J. Lips. The determination of volatile organic compounds from EPA method 524.2 using purge-and-trap capillary gas chromatography, ECD and FID. J. Chromatogr. Sci. 30:247–255, 1992.

100. A. Gamble. Concentrating volatile organics for gas chromatography. Lab. Equip. 29:31–35, 1991.

101. H. Kessels, W. Hoogerwerf, J. Lips. Determination of volatile organic compounds with EPA 524.2 using purge and trap gas chromatography, ECD and FID detection. Analusis 20:55–60, 1992.

102. B. V. Burger, M. LeRoux. Trace determination of volatile organic compounds in water by enrichment in ultra-thick film capillary traps and gas chromatography. J. Chromatogr. 642:117–122, 1993.

103. T. Jursik, K. Stransky, K. Ubik. Trapping system for trace organic substances. J. Chromatogr. 586:315–322, 1991.

104. P. P. K. Kuo, E. S. K. Chian, F. B. DeWalle, J. H. Kim. Gas-stripping sorption and thermal-desorption procedures for pre-concentrating volatile polar water-soluble organics from water samples for analysis by gas chromatography. Anal. Chem. 49:1023–1029, 1977.

105. W. E. May, S. N. Chesler, S. P. Cram, B. H. Gump, H. S. Hertz, D. P. Enagorio, S. M. Dyszel. Chromatographic analysis of hydrocarbons in marine sediments and seawater. J. Chromatogr. Sci. 13:535–540, 1975.

106. J. A. Settlage, W. G. Jennings. Inexpensive method for rapid heating of cold traps. J. High Res. Chromatogr. Commun. 3:146–150, 1980.

107. H. Peterson, G. A. Eiceman, L. R. Field, R. E. Sievers. Gas-chromatographic injector attachment for direct insertion and removal of a porous-polymer sorption trap. Anal. Chem. 50:2152–2154, 1978.

108. B. J. Dowty, L. E. Green, J. L. Laseter. Automated gas-chromatographic procedure to analyze volatile organics in water and biological fluids. Anal. Chem. 48:946–949, 1976.

109. J. F. Pankow. Cold-trapping of volatile organic compounds on fused-silica capillary columns. HRC&CC 4:156–163, 1981.

110. P. M. Buszka, D. L. Rose, G. B. Ozuna, G. Z. Groschen. Determination of nanogram per liter concentration of volatile organic compounds in water by capillary gas chromatography and selected ion monitoring mass spectrometry and its use to define ground water flow directions in Edwards Aquifer, Texas. Anal. Chem. 67:3659–3669, 1995.

111. H. J. Brass. Analysis of trihalomethanes in drinking water by purge and trap and liquid–liquid extraction. Int. Lab. 10:17–18, 1980.

112. W. N. Billings, T. E. Bidleman. Field comparison of polyurethane foam and Tenax-GC resin for high-volume air sampling of chlorinated hydrocarbons. Environ. Sci. Technol. 14:679–683, 1980.

113. T. Tanaka. Chromatographic characterization of porous polymer adsorbents in a trapping column for trace organic vapor pollutants in air. J. Chromatogr. 153:7–13, 1978.

114. R. G. Melcher, V. J. C. Caldecourt. Delayed injection–preconcentration gas-chromatographic technique for parts-per-billion determination of organic compounds in air and water. Anal. Chem. 52:875–881, 1980.

115. H. G. Eaton. Gradient heat-desorption technique of preconcentrated Tenax-GC tubes for use in GC-MS analysis. J. Chromatogr. Sci. 18:580–582, 1980.

116. J. F. Pankow, L. M. Isabelle. Adsorption-thermal desorption as a method for determination of low levels of aqueous organics. J. Chromatogr. 237:25–39, 1982.

117. B. V. Ioffe, V. A. Isidorov, I. G. Zenkevich. Gas chromatographic–mass spectrometric determination of volatile organic compounds in an urban atmosphere. J. Chromatogr. 142:787–795, 1977.

118. E. D. Pellizari, B. H. Carpenter, J. E. Bunch, E. Sawicki. Collection and analysis of trace organic vapor pollutants in ambient atmospheres. I. Technique for evaluating concentration of vapors by sorbent media. II. Thermal desorption of organic vapors from sorbent media. Environ. Sci. Technol. 9:552–560, 1975.

119. K. Grob, G. Grob. Gas–liquid chromatographic–gas spectrometric investigation of C6 to C20 organic compounds in an urban atmosphere. Application of ultra-trace analysis on capillary column. J. Chromatogr. 62:1–13, 1971.

120. A. Raymond, G. Guiochon. Gas-chromatographic analysis of C_8 to C_{18} hydrocarbons in Paris air. Environ. Sci. Technol. 8:143–148, 1974.

121. R. E. Sievers, R. M. Barkley, G. A. Eiceman, R. H. Shapiro, H. F. Walton, K. J. Kolonko, L. R. Field. Environmental trace analysis of organic in water by gas-capillary column chromatography and ancillary technique. Products of ozonolysis. J. Chromatogr. 142:745–754, 1977.

122. V. Leoni, G. Pucceti, A. Grella. Preliminary results on the use of Tenax (a porous polymer based on 2,6,diphenyl-p-diphenylene oxide) for the extraction of pesticides and polynuclear aromatic hydrocarbons from surface and drinking waters for analytical purposes. J. Chromatogr. 106:119–124, 1975.

123. A. C. Hu, P. H. Weiner. Modification of methods for volatile organic analysis at trace levels. J. Chromatogr. Sci. 18:333–342, 1980.

124. R. F. Simpson. Influence of gas volume sampled on wine on headspace analysis using pre-concentration on Chromosorb 105. Chromatographia 12:733–736, 1979.

125. J. P. Mieure. Determining volatile organics in water. Environ. Sci. Technol. 14:930–935, 1980.

126. N. R. Rakshieva, S. Wicar, J. Nanák, J. Janák. Effect of carrier-gas velocity on retention volume in gas chromatography on Porapak sorbents. J. Chromatogr. 91:59–66, 1974.

127. O. L. Hollis. Porous polymers used in gas and liquid chromatography. J. Chromatogr. Sci. 11:335–342, 1973.

128. J. F. Johnson, E. M. Barrall. I. Study of micro-pore structure of porous polymer columns. J. Chromatogr. 40:209–212, 1969.

129. F. Onusaka, J. Janák, S. Duras, M. Krcmárová. Analysis of formaldehyde (solutions) by gas chromatography using Porapak N. J. Chromatogr. 40:209–212, 1969.

130. M. Dresser, O. K. Guha, J. Janák. Chromatographic behavior of isomeric compounds on Poropak P. J. Chromatogr. 65:261–269, 1972.

131. F. M. Zado, J. Fabecic. Physico-chemical fundamentals of gas-chromatographic retention on porous polymer columns: Porapaks Q and T. J. Chromatogr. 51:37–44, 1970.

132. G. Castello, G. Dámato. Gas-chromatographic analysis of light hydrocarbons for characterization of Porapak columns. J. Chromatogr. 196:245–254, 1980.

133. G. Castello, G. Dámato. Characterization of Chromosorb porous polymer-bead columns by gas chromatographic retention values of light hydrocarbons and carbon dioxide. J. Chromatogr. 212:261–269, 1981.

134. J. DeGreef, M. DeProft, F. DeWinter. Gas chromatographic determination of ethylene in large air volumes at the fractional parts per billion level. Anal. Chem. 48:38–41, 1976.

135. V. Niederschulte, K. Ballschmitter. Isolation and preconcentration of chlorophenoxyacetic acids from water with macroreticular resins XAD-2 and XAD-7. Z. Anal. Chem. 269:360–363, 1974.

136. G. Hunt, N. Pangaro. Potential contamination from the use of sinthetycal adsorbents in air-sampling procedures. Anal. Chem. 54:369–372, 1982.

137. P. Van Rossum, N. J. Webb. Isolation of organic water pollutants by XAD resins and carbon. J. Chromatogr. 150:381–392, 1978.

138. G. A. Junk, J. J. Richard, M. D. Gresser, D. Witiak, J. L. Witiac, M. D. Arguello, R. Vick, H. J. Svec, J. S. Fritz, G. V. Calder. Use of macro-reticular resins in analysis water for trace organic contaminants. J. Chromatogr. 99:745–762, 1974.

139. O. Dufka, J. Malinsky, J. Churacek, K. Kamarck. Use of "Synachrom," a macroporous polymer, as soption material for GSC. J. Chromatogr. 54:111–115, 1970.

140. L. Wennrich, W. Egenwald, T. Welsch, B. Wenzel. Application of porous polymers of the Ceka.chrom series as enrichment materials in organic trace analysis. Chem. Tech. (Leipzig) 33:203–206, 1981.

141. K. I. Sakodinskii, N. S. Klinskaya, L. I. Panina. In: A. Zlatkis, ed. Polymer sorbents based on polyimides. (Thermally stable packings for gas chromatography). Anal. Chem. 45:1369–1374, 1973.

142. B. A. Collenut, S. Thorburn. Novel technique for sampling and gas chromatographic analysis of vapors. Chromatographia 12:519–522, 1979.

143. H. Borén, A. Grimvall, J. Palmborg, R. Sävenhed, B. Wigilius. Optimization of the open stripping system for the analysis of trace organics in water. J. Chromatogr. 348:67–78, 1985.

144. S. Skrabakova, E. Matisova, I. Novak, D. Berek. Use of a novel carbon sorbent for the desorption of organic compounds from water. J. Chromatogr. A 665:27–32, 1994.

145. N. K. Kristiansen, E. Lundanes, M. Froshaug, G. Bechner. Evaluation of the open-loop stripping technique used for the determination of volatile organic compounds in water. Anal. Chim. Acta. 280:111–117, 1993.

146. C. J. Hwang, S. W. Krasner, M. J. McGuire, M. S. Moylan, M. S. Dale. Determination of subnanogram per liter levels of earthy-musty odorants in water by the salted closed-loop stripping method. Environ. Sci. Technol. 18:535–541, 1984.

147. M. J. McGuire, S. W. Krasner, C. J. Hwang, G. Izaguirre. Closed-loop stripping analysis as a tool for solving taste and odor problems. Res. Technol. Oct: 530–537, 1981.

148. S. W. Krasner, C. J. Hwang, M. J. McGuire. Development of a closed-loop stripping technique for the analysis of taste- and odor-causing substances in drinking water. In: L. H. Keith, ed. Advances in the Identification and Analysis of Organic Pollutants in Water. Vol. 2. Ann Arbor Science Publishers. Ann Arbor, MI, 1981.

149. S. W. Krasner, C. J. Hwang, M. J. McGuire. A standard method for quantification of earthy-musty odorants in water. Water Sci. Technol. 18:535–542, 1983.

150. K. Grob, G. Grob. Organic substances in potable water and in its precursor. Part II. Application in the area of Zürich. J. Chromatogr. 90:303–313, 1974.

151. J. W. Graydon, K. Grob, F. Zuercher, W. Giger. Determination of highly volatile organic contaminants in water by the closed-loop gaseous stripping technique followed by thermal desorption of the activated carbon filters. J. Chromatogr. 285:307–318, 1984.

152. J. I. Gómez Belinchón. Analysis of volatile organic compounds. The closed-loop stripping procedure. Chromatogr. Anal. April:9–12, 1989.

153. F. Ventura, J. Romero, M. R. Boleda, I. Martí. Identificación de contaminantes orgánicos volátiles en agua. Tecnología del agua. 131:17–28, 1994.

154. K. Grob, K. Grob, G. Grob. Organic substances in potable water and in its precursor. Part III. The closed-loop stripping procedure compared with rapid liquid extraction. J. Chromatogr. 106:299–315, 1975.

155. J. Curvers, T. Noy, C. Cramers, J. Rijks. Possibilities and limitations of dynamic headspace sampling as a pre-concentration technique for trace analysis of organics by capillary gas chromatography. J. Chromatogr. 289:171–182, 1984.

156. M. M. Thomason, W. Bertsch. Evaluation of sampling methods for the determination of trace organics in water. J. Chromatogr. 279:383–393, 1983.

157. J. I. Gómez Belinchón, J. Albaigés. Organic pollutants in water. I. Optimization of operational parameters of the CLSA technique. Int. J. Environ. Anal. 30:183–195, 1987.

158. K. Grob, F. Zürcher. Stripping of trace organic substances from water. Equipment and procedure. J. Chromatogr. 117:285–294, 1976.

159. S. P. Thomas, R. S. Rjan, G. R. B. Webster, L. P. Sarna. Protocol for the analysis of high concentrations of benzene, toluene, ethylbenzene and xylene isomers in water using automated solid phase microextraction–GC-FID. Environ. Sci. Technol. 30:1521–1526, 1996.

160. I. Valor, C. Cortada. Direct solid phase microextraction for the determination of BTEX in water and wastewater. J. High Res. Chromatogr. 19:472–474, 1996.

161. F. Mangani, R. Cendciarini. Solid phase microextraction using fused silica fibers coated with graphitized carbon black. Chromatographia 41:678–683, 1995.

162. L. P. Sarna, G. R. B. Webster, M. R. Friesen-Fischer. Analysis of the petroleum components benzene, toluene, ethylbenzene and the xylenes in water by commercially available solid phase microextraction and carbon-layer open tubular capillary columns gas chromatography. J. Chromatogr. A 677:201–205, 1994.

163. C. L. Arthur, L. M. Killam, S. Motlagh, M. Lim, D. V. Potter, J. Pawliszyn. Analysis of substituted benzene compounds in groundwater using solid phase microextraction. Environ. Sci. Technol. 26:979–983, 1992.

164. B. L. Wittkamp, S. B. Hawthorne. Determination of aromatics compounds in water by solid phase microextraction and ultraviolet absorption spectroscopy. 1. Methodology. Anal. Chem. 69:1197–1203, 1997.

165. D. W. Potter, J. Pawliszyn. Detection of substituted benzenes in water at the pg/ml level using solid phase microextraction and gas chromatography–ion trap mass spectrometry. J. Chromatogr. 625:247–255, 1992.

166. M. Chai, C. L. Arthur, J. Pawliszyn. Determination of volatile chlorinated hydrocarbons in air and water with solid phase microextraction. Analyst 118:1501–1505, 1993.

167. C. L. Arthur, K. Pratt, S. Motlagh, J. Pawliszyn. Environmental analysis of organic compounds in water using solid phase microextraction. J. High Res. Chromatogr. 15:741–744, 1992.

168. R. E. Shirey. Rapid analysis of environmental samples using solid phase microextraction (SPME) and narrow bore capillary columns. J. High Res. Chromatogr. 66: 689–692, 1994.

169. T. Nilsson, F. Pelusino, L. Montanarella, B. Larsen, S. Facchetti, J. Madsen. An evaluation of solid phase microextraction for analysis of volatile organic compounds in drinking water. J. Agric. Food Chem. 43: 2138–2143, 1995.

170. N. Xu, S. Vandegrift, G. W. Sewell. Determination of chloroethenes in environmental biological samples using gas chromatography coupled with solid phase microextraction. Chromatographia 42:313–317, 1996.

171. I. Valor, J. C. Moltó, D. Apraiz, G. Font. Matrix effects on solid phase microextraction of organophophorous pesticides from water. J. Chromatogr. 767:195–203, 1997.

172. I. J. Barnabas, J. R. Dean, I. A. Fowlis, S. P. Owen. Automated determination of s-triazines herbicides using solid phase microextraction. J. Chromatogr. A 705: 305–312, 1995.

173. R. Eiser, K. Levsen. Determination of organophosphorous, triazine and 2,6-dinitroaniline pesticides in aqueous samples via solid phase microextraction (SPME) and gas chromatography with nitrogen-phosphorous detection. Fresenius J. Anal. Chem. 351:555–562, 1995.

174. R. Eisert, K. Levsen. Determination of pesticides in aqueous samples by solid phase microextraction on-line couple to gas chromatography–mass spectrometry. J. Am. Soc. Mass Spectrom. 6:1119–1130, 1995.

175. A. A. Boyd-Boland, J. B. Pawliszyn. Solid phase microextraction of nitrogen-containing herbicides. J. Chromatogr. A 704:163–172, 1995.

176. X. P. Lee, T. Kumazawa, K. Sato, O. Suzuki. Detection of organophosphate pesticides in human body fluids by headspace solid phase microextraction (SPME) and capillary gas chromatography with nitrogen-phosphorous detection. Chromatographia 42:135–140, 1995.

177. R. Young, V. López Avila. On-line determination of organochlorine pesticides in water by solid phase microextraction and gas chromatography with electron capture detection. J. High Res. Chromatogr. 19:247–256, 1996.

178. H. B. Wan, H. Chi, M. K. Wong, C. Y. Mok. Solid phase microextraction using pencil lead as sorbent for analysis of organic pollutants in water. Anal. Chim. Acta 298:219–223, 1994.

179. K. D. Bulchholz, J. Pawliszyn. Optimization of solid phase microextraction conditions for determination of phenols. Anal. Chem. 66:160–167, 1994.

180. K. D. Buchholz, J. Pawliszyn. Determination of phenols by solid phase microextraction and gas chromatographic analysis. Environ. Sci. Technol. 27:2844–2848, 1993.

181. M. Möder, S. Schrader, U. Frank, P. Popp. Determination of phenolic compounds in waste water by solid phase microextraction. Fresenius J. Anal. Chem. 357: 326–332, 1997.

182. B. Schäfer, W. Engewald. Enrichment of nitrophenols from water by means of solid phase microextraction. Fresenius J. Anal. Chem. 352:535–536, 1995.

183. D. W. Potter, J. Pawliszyn. Rapid determination of polyaromatic hydrocarbons and polychlorinated bi-

phenyls in water using solid phase microextraction and GC/MS. Environ. Sci. Technol. 28:298–305, 1994.

184. S. B. Hawthorne, D. J. Miller, J. Pawliszyn, C. L. Arthur. Solventless determination of caffeine in beverages using solid phase microextraction with fused-silica fibers. J. Chromatogr. 603:185–191, 1992.

185. B. D. Page, G. Lacroix. Application of solid-phase microextraction to the headspace gas chromatographic analysis of halogenated volatiles in selected foods.

186. J. Wang, M. Bonilla, H. M. McNair. Solid phase microextraction associated with microwave assisted extraction of food products. High Res. Chromatogr. 20: 213–216, 1997.

187. D. de le Calle García, S. Magnaghi, M. Reichenbächer, K. Danzer. Systematic optimization of the analysis of wine bouquet components by solid phase microextraction. J. High Res. Chromatogr. 19:257–262, 1996.

188. T. Kumazawa, K. Sato, H. Seno, A. Ishii, O. Suzuki. Extraction of local Anaesthetics from human blood by direct immersion solid phase microextraction (SPME). Chromatographia 43:331–333, 1996.

189. K. G. Miller, C. F. Poole, T. M. P. Pawloski. Classification of the botanical origin of cinnamon by solid-phase microextraction and gas chromatography. Chromatographia 42:639–646, 1996.

190. K. G. Furton, G. Bruna, J. R. Almirall. A simple, inexpensive, rapid, sensitive and solventless technique for the analysis of accelerants in free debris based on SPME. J. High Res. Chromatogr. 18:625–629, 1995.

191. Z. Zhang, J. Pawliszyn. Sampling volatile organic compounds using a modified solid phase microextraction device. Anal. Commun. 33:219–221, 1996.

192. J. Pawliszyn, S. Liu. Sample introduction for capillary gas chromatography with laser desorption and optical fibers. Anal. Chem. 59:1475–1478, 1987.

193. R. Belardi, J. Pawliszyn. The application of chemically modified fused silica fibers in the extraction of organics from water matrix samples and their rapid transfer to capillary columns. Water Pollution Res. J. Canada. 24:179–185, 1989.

194. C. L. Arthur, J. Pawliszyn. Solid phase microextraction with thermal desorption using fused silica optical fibers. Anal. Chem. 62:2145–2148, 1990.

195. Polyacrilate film fiber for solid phase microextraction of polar semivolatiles from water. Supelco report 17. T-394011.

196. Z. Zhang, J. Pawliszyn. Headspace solid phase microextraction. Anal. Chem. 65:1843–1852, 1993.

197. Z. Zhang, M. J. Yang, J. Pawliszyn. Solid phase microextraction. Anal. Chem. 66:844A–853A, 1994.

198. A. Zhang, J. Pawliszyn. Analysis of organic compounds in environmental samples using headspace solid phase microextraction. J. High Res. Chromatogr. 16:689–692, 1993.

199. D. Louch, S. Motlagh, J. Pawliszyn. Dynamics of organic compound extraction from water using liquid-coated fused silica fibers. Anal. Chem. 64:1187–1199, 1992.

200. S. Motlagh, J. Pawliszyn. On-line monitoring of flowing samples using solid phase microextraction–gas chromatography. Anal. Chim. Acta. 284:265–273, 1993.

201. Z. Zhang, J. Pawliszyn. Quantitative extraction using an internally cooled solid phase microextraction device. Anal. Chem. 67:34–43, 1995.

202. J. Drozd, J. Novák. Quantitative head space gas analysis by the standard additions method. J. Chromatogr. 136:37–44, 1977.

203. W. J. Khazal, J. Vejroseta, J. Novák. Comparison of headspace gas and liquid extraction determination of hydrocarbons in water by the standard addition method. J. Chromatogr. 157:125–131, 1978.

204. J. Drozd, J. Novák. Quantitative and qualitative head space gas analysis of parts per billion amounts of hydrocarbons in water. J. Chromatogr. 158:471–482, 1978.

205. J. Drozd, J. Novák. Headspace determination of benzene in gas aqueous liquid systems by the standard additions method. J. Chromatogr. 152:55–61, 1978.

206. P. Kurán, L. Soják. Environmental analysis of volatile organic compounds in water and sediment by gas chromatography. J. Chromatogr. A 733:119–141, 1996.

207. M. H. Hiatt, C. M. Farr. Volatile organic compound determination using surrogate-based correction for method and matrix effects. Anal. Chem. 67:426–433, 1995.

208. P. S. H. Wong, R. G. Cooks, M. E. Cisper, P. H. Hemberger. On-line, in situ analysis with membrane introduction MS. Environ. Sci. Technol. 29:215–218, 1995.

209. M. A. Menden, R. S. Pimpim, J. Kotiaho, M. N. Eberlin. A cryotrap membrane introduction mass spectrometry system for analysis of volatile organic compounds in water at the low parts-per-trillion level. Anal. Chem. 68:3502–3506, 1996.

210. Y. H. Xu, S. Mitra. Continuous monitoring of volatile organic compounds using on-line membrane extraction and microtrap gas chromatography system. J. Chromatogr. A 688:171–180, 1994.

211. C. S. Creaser, S. W. Stygall. Membrane introduction–ion trap mass spectrometry for the determination of volatile organics in aqueous effluents. Anal. Proc. 32: 7–9, 1995.

212. EPA Method 4031. Soil screening for BTEX by immunoassay. May 1994.

213. R. L. Allen, W. B. Manning, K. D. McKenzie, T. A. Withers, J. P. Mapes, S. B. Friedman. A rapid and sensitive immunoassay for the detection of gasoline and diesel fuel in contaminated soil. J. Soil Contamination 1:227–237, 1992.

34

Oil and Greases and Petroleum Hydrocarbon Analysis

Rossiza Belcheva

Refinery and Petrochemical Research Institute, Bourgas, Bulgaria

I. INTRODUCTION

The levels of total oil and grease (TOG) and total petroleum hydrocarbons (TPH) in water have attracted environmental researchers' attention for a long time. Contamination with crude oil and its derivatives, used in almost every economy, are everywhere. Marine waters, surface waters, groundwaters, and wastewaters are all polluted. Countries all over the world take measures to prevent, limit, and reduce pollution. The protection of water as an essential part of human existence is a question of ecological culture and corresponding resources. Unfortunately, few of our planet's inhabitants or our respective institutions and governments can be proud of this. That is why TOG and TPH nowadays remain two of the most important indexes of water quality. They are part of a control-and-examination system. The success of the determination of these indexes depends on careful planning and on performance of all manipulations as well as on a knowledge of this subject. This chapter is an overview and a directory for sampling and analyzing TOG and TPH in water.

II. DEFINITIONS AND INTERFERENCES

TOG is defined as any material, soluble in the solvent, recovered and extracted from water. In most standards the solvent is specified (1–6). It is important to realize that, unlike constituents that represent distinct chemical elements, ions, compounds, or groups of compounds,

TOG is defined by the method used for their determination. Moreover, an absolute quantity of a specific substance is not measured. The term TPH represents the nonpolar and weakly polar hydrocarbons extracted from water, not adsorbed from one of the following sorbents: aluminium oxide, silica gel, or florisil (4,7–12).

Since TOG and TPH contents are defined as a result of the test procedure, interferences are included. Groups of substances with similar physical characteristics and chemical structures as a result of their common solubility in an organic extracting solvent can be extracted and determined quantitatively. Those compounds are generally organic solvents, sulphur compounds, hydrocarbon derivatives of chlorine, sulphur, and nitrogen, certain organic dyes, and chlorophyll. Some interfering compounds may be adsorbed by the sorbent; others may not. What is not adsorbed is measured as TPH. Data interpretation on the basis of chemical structure must be carried out carefully because of the variety of measured compounds.

III. PHYSICAL AND CHEMICAL PROPERTIES

The origin of petroleum water pollution is quite diverse. Their boiling points vary greatly. Pollutants can be gaseous, liquid, and solid. Gaseous water pollutants do not last long. When liquid pollutants first enter water, the water loses the low-boiling part because of evaporation into the atmosphere as a result of solar warmth. The carbon contents of the pollutants in a 24-hour period begins from five carbon atoms upwards.

TOG and TPH in water have different migration forms: dissolved, emulsified, suspended, as colloids, sorbed on solid parts and like spots over the surface. As a result of evaporation, sorption, biochemical, and chemical oxidation processes, the component content changes, and important differences in chemical composition can develop (9). The part passed to solution, colloids, emulsion, or suspension after the initial evaporation is rich in branched-chain hydrocarbons and low-molecular-weight aromatic compounds, which can be assimilated by living organisms and can fall to the bottom layers and sediments (9,13).

The chemical behavior of these compounds is dependent on chemical and biological oxidation processes. Those processes run very slowly for TPH. They remain uninfluenced for the longest time. Chemical oxidation represents about 15% of all processes, but the basic contribution to oxidation, especially for hydrocarbons, is caused by microorganisms (9). Unsaturated straight chains and aromatic compounds may be amenable to their influence. The biochemical stability of the different hydrocarbons increases from normal alkanes to their isomers and then to cyclic and aromatic hydrocarbons. On this basis, the accumulation of aromatic hydrocarbons in water is possible. The last ones are the most soluble constituents of crude oils or their refined products (14). Thus the solubility of liquid hydrocarbons in water decreases exponentially with increasing molar volume. Benzene and $C_{1,2}$-benzenes tend to be the dominant monoaromatic hydrocarbons in contaminated waters. The aromatic part of TPH is toxic to living organisms. Several aromatic compounds are on the U.S. EPA list of Priority Pollutants and are carcinogens or suspected to be carcinogens. TPH accumulate in any sort of aquatic flora and fauna and exert a negative influence on the uppermost aquatic vegetation. They disturb many of the photosynthesic processes, the dynamic of the phytoplankton, and the live activity in water, as well as the gas exchange with the atmosphere and warmth and radiation-exchange. Furthermore TPH alter the organoleptic quality of water, which is why they are an object of interest.

IV. DISTRIBUTION IN WATERS

Distribution of TOG and TPH in water is a result of different local conditions: speed and character of water movement, presence of sediments and silt, physical and

chemical nature of the available compounds, and their stability in water. Usually there is a dynamic equilibrium among these factors. Distribution of TPH in the water body is not uniform. If we were to split the water vertically, it would be divided in three layers: surface (with a thickness of $50-500$ μm), intermediate, and bottom. The surface layer is an area of active reactions for a lot of compounds. There the equilibrium process of mass exchange between air and water takes place. Biological activity is very intense (13). The intermediate layer can also be divided into three parts: upper, intermediate, and bottom. Its properties depend on the mixing of the water body. Regarding the bottom layer, the accumulation of the pollutants is of interest. Here the process of substance transport from water to sediment and from sediment to water takes place, and insoluble parts can accumulate. The last process is a consequence of an increased number of molecules from the oxidation processes. Areas with stronger pollution are the surface and the bottom water layers and sediments.

V. SAMPLING WATER

To acquire reliable results regardless of the method used, each analysis must go through the following steps: preparing containers for sampling water, sampling of the water, conservation, concentration, and measurement. Each step must be executed correctly to minimize errors, given the extent of TOG/TPH pollution.

Collecting samples is a substantial and important analysis step, since the sample must fully characterize the examined object. It depends on the TOG and TPH distribution and is related to the already-mentioned local conditions: speed and character of the water movement, silt and sediment availability, physical and chemical properties of the composite compounds, and their stability in water. World standards give different ways for collecting samples, depending on the aim of the analysis and the forms of TOG and TPH. If TOG/TPH are like a spot, a filter paper or, in recent years, PTFE sheets are used, and the result is given in mg/m^2. A sample taken from the surface layer is defined as "general" and the result is given in mg/L. Because of the wide variety of conditions found in streams, lakes, reservoirs, and other water bodies, it is not possible to point out the exact spot for collecting the sample. Where the water in a stream is mixed enough to approach uniformity, a sample taken at any point in the cross section is satisfactory. For large rivers or streams, which certainly not mixed uniformly, more samples are desirable, and they are usually taken at a number of points at the surface across the entire width and at a number of depths at each point. Ordinarily, samples are taken on the whole water column height for getting data on the pollution of the water body (15). The observations of many authors of different TOG/TPH analyses show a gradient in their distribution (Table 1) (16). Taking samples at a 20–30-cm depth is a general practice, and it gives an idea about the trends in pollution (7,8,12). As seen in Table 1, data from the 30-cm level are averages of those taken from the surface and from the bottom layers at the same point of collection.

The choice of location of the sampling point depends on the desired information and local conditions. It is good practice to choose the sampling points with extreme care, usually before mixing pipes, where the best water mixing is assumed, or where the current is the strongest and most turbulent (7,13,15). It is best to take samples by dipping the container itself. Poured samples are not representative because of the TOG/TPH tendency to line the container walls. By immersing the glass bottle to a depth of 20–30 cm, a part of the surface and the upper part of the intermediate layers runs into it. In this way the sample is a mixture of a "general" sample and one from 20–30-cm depth.

The needed sample volume can be determined approximately by organoleptic and other physical water characteristics (appearance, color, opalescence). An important condition for getting representative data is to analyze the whole sample volume. A 1-L sample should be taken for water containing up to 10 mg/L TOG/TPH, and 2–3 L or more for those containing less than 0.01 mg/L. When high quantities are expected, the volume can be minimized to 0.5 L. In case of a spill, the sample volume is reduced to 250 or 100 ml.

VI. CONTAINER PREPARATION FOR SAMPLING WATER

Sampling of water needs special devices. Contact with different materials may contaminate the samples and change the composition. TOG and TPH are water admixtures that can be held on the container walls; for that reason, the container can't be rinsed with analysis water (7,12,15,17–20). Sampling the water depends on the aims of the analysis, on the site, on the dissolving model, and on the applied analytical technique. General practice is to collect samples from lakes, wells, streams, reservoirs, processing tanks, ponds, oceans, and rivers for TOG/TPH determination for systematic observations. Such samples are called *grab*. Standard methods for sampling water from different sources have been developed (15,18,19), but the conditions for grab sampling are given precisely in ASTM D-3370 (15), BDS 16714 (21), and ISO-5661 (16). When one has to take samples from a spill, ASTM D-4489 (20) and NT CHEM:001 (21) are recommended. The last two are divided into the corresponding levels (A, B and A, B, C) according to the thickness of the polluted surface.

The best devices for collecting grab samples are those known as *bailers*, *kemmerers*, and other kinds of samplers usually made of metal or plastic. Samples from different levels can easily be taken. Plastic or metal samplers with polymer lining on the sample-contact side have certain inconveniences for TOG and TPH analysis. Polymer materials are known accumulators of organic impurities (13). When waters are examined for organic pollutants such as TPH, individual hydrocarbons, pesticides, herbicides, and polychlorinated biphenyls, such samplers turn out to be inconvenient (9). Bottles and dry wide-mouth vials with glass stoppers are most convenient for TOG/TPH examination. Transparent, colorless, and chemically stable glass is the best material. Besides glass stoppers, screw plugs with PTFE liners on the side of sample contact can be used (7,8). PTFE-lined samplers are the most acceptable for collecting samples from different layers.

The bottle volume must correspond to the collected sample volume, because in the following analysis steps all samples must be treated. Generally this volume varies from 3 l to 100 ml. For spills, PTFE sheets are used (22). The preparation of the containers takes some steps and aims at the removal of all pollution: rinsing by acetone (sometimes), rinsing by water, cleaning with a $K_2Cr_2O_7$–H_2SO_4 mixture (detergents and synthetic washing agents may not be used), washing once again with water and distilled water, heating in neck-up po-

Table 1 TPH Distribution, by Water Column Height, in Purified Wastewater[a]

		TPH, mg/L		TOG, mg/L	
Item	Water column height, cm	Sample 1	Sample 2	Sample 1	Sample 2
1	Surface	0.032	0.022	0.076	0.203
2	30	0.046	0.034	0.160	0.203
3	100	0.076	0.048	0.160	0.231

[a]The purification process includes the following steps: mechanical, physicochemical, and biological purification, as well as natural oxidation for about 40 days.

sition in a drying oven, and, after cooling, finally rinsing with a solvent used for the extraction step to remove any residue that might interfere with the analysis.

It is clear that, for TOG and TPH analyses, the container material and its special preparation are critical.

VII. CONSERVATION

This step aims to minimize the water processes connected with TOG/TPH changes. Conservation is necessary if the time between sampling and the analysis is more than 4 hours (7,9). Whatever the preservative, it cannot fully preserve TOG and TPH from alteration, which is why the analysis of a conserved sample must not be put off for long. A week (12), or 28 days according to the U.S. EPA and ASTM, at a temperature of 4°C is the maximum (3,10,17,23). When the analytical object is wastewater, its conservation is complex, especially when it contains insoluble parts. The preservative cannot influence them. There are a few preservatives:

  The extraction solvent, with the aim to extract the principal part of TOG/TPH at the sampling moment (7)
  Decreasing pH to below 2 (with HCl or H_2SO_4), with the aim to suppress oxidative processes (3,10–12)
  Precipitating with an $Al_2(SO_4)_3$ solution, to separate the solution from the sediment, to dissolve it, and to determine TOG/TPH in the dissolved sediment and in solution (7,8)

The best choice is a pH change, because of continuing oxidative processes.

VIII. SAMPLE PREPARATION

Preparatory work for TOG/TPH analysis consists of their isolation and concentration. This is an important part of the procedure. The techniques used include extraction processes: liquid–liquid (LLE), gas (headspace–HSE) and solid-phase extraction (SPE). The extraction grade has importance; it depends on the distribution coefficient and defines the choice of extraction solvent.

A. Liquid–Liquid Extraction

Liquid–liquid extraction and concentration of TOG/TPH was the first extraction type. Nowadays it is widely practiced in all standardized methods. It is used mainly for the extraction of semivolatile and nonvolatile pollutants. Organic solvents in large volumes are used in this process. The most frequently used solvents are chlorinated hydrocarbons (1–7,10–12,24–27) (dichlorometane, chloroform, carbon tetrachloride, tetrachloroetylene, 1,1,2-trichlorotrifluoroethane—freon 113), carbon disulphide, toluene, and n-hexane (28–30). A rise in the distribution coefficient is achieved in a few ways: a salting-out process, mixed solvents, an increase in temperature, successive extractions. Of the neutral salts (LiCl, NaCl, KCl, KBr) in water or substances (tris-butylphosphate) in organic solvent for decreasing the dissolution of the components in water, adding NaCl is used most (1–7). The synergistic action of the individual constituents determines the effect of mixed-solvent use (31,32). The effect significance is almost the same as with the salting-out process. A temperature increase is also applied, but it is not put into practice as much as is successive extractions. The number of extractions depends on the water pollution level. For bigger quantities (100–1000 mg/L) the practice is a three-way extraction (1–6,10), for smaller quantities [0.05–1 mg/L (11) and 0.3–10 mg/L (32)], a two-way extraction and a one-way extraction (5,6), respectively. A higher extraction is in general related to how long the two phases remain in contact. The constituent quantity and the saturation of the solvent are also important. This contact time for two- and three-way extraction is 2 and 3 min (1–3,11,32), for a one-way extraction 10 min (5) or 15 min (6,12). For a concentration range up to 1 mg/L, a two-way extraction is quite satisfactory (Table 2) (16). The average recovery is 98%, calculated on an additionally added TPH quantity. The results are estimated for constant (t_a) and linearly variable (t_b) systematic errors: $t_a = 0.36 < t_{tabl.} = 2.45 > t_b = 0.42$.

Some characteristics of the solvents most often used for TOG/TPH extraction are shown in Table 3. They may be classified by degree and speed of extraction as follows: CCl_4, Freon 113, CH_2Cl_2, n-C_6H_{14}, C_2Cl_4, $CHCl_3$. The first two are the best, although Freon 113 requires a little more time. Besides the mentioned solvents, many solvent mixtures are experimented with (31,32). Extraction of all organic pollutants isn't possible. No solvent can selectively extract TOG/TPH. Heavier crude oil can contain a sufficient portion of materials that are not soluble in any solvent (4).

Because CCl_4 is very harmful to the human organism, it is replaced by petroleum ether, n-hexane, Freon 113, and the mix of n-hexane/methyl-tert-butyl-ether. We have to keep in mind that TOG is defined by the solvents used in the determination method. The low-

Table 2 Recovery via TPH Extraction from Water, Determined by Standard Addition Method

Item	Quantity, mg Added	Quantity, mg Found	Recovery, %
1	0.032[a]	0.033	103.1
2	0.054[a]	0.053	98.1
3	0.067[a]	0.065	97.0
4	0.073[a]	0.075	102.7
5	0.1[b]	0.089	89.0
6	0.32[a]	0.327	102.2
7	0.33[b]	0.324	97.1
8	0.35[b]	0.330	94.3
9	0.35[a]	0.348	99.4
10	0.35[b]	0.376	107.7
11	0.385[b]	0.347	97.1
12	0.44[b]	0.432	98.2
13	0.45[a]	0.424	94.2
14	0.51[b]	0.468	91.8
		Mean:	98.9

[a]Mixture is with marine water.
[b]Mixture is with purified wastewater.

boiling-point fractions that volatilize at temperatures below 70°C or 85°C when the mentioned solvent mixture is used are lost (4).

The problems for the environment with chlorofluorocarbons are well known the world over. The Montreal Protocol, accepted by most of the civilized nations, is an attempt to stop the depletion of the ozone layer in the upper atmosphere. According to this protocol, the manufacture of volatile chlorinated hydrocarbons including Freon 113 has been banned since January 1995. The protocol does not make exceptions. As a consequence, the attention of analysts was directed to a new method and an alternative solvent for extraction

(4,29). AWWA Standard Methods (4) put into practice alternative solvents or mixtures; but despite good results, they cannot fully replace Freon 113. The analyst has a choice of solvents in the extraction process when the concern is high TPH content. For quantities of 1 mg/L or more, one can use *n*-hexane and the mixed solvent (*n*-hexane/methyl-tert-butyl-ether) instead of Freon 113 or CCl₄. World practice in such cases goes in this direction. But for quantities under 1 mg/L and especially under 0.2 mg/L, the choice is very limited. In this case, CCl₄ can almost not be replaced by any alternative solvent (11,21,33). A lower degree of extraction cannot be compensated by increasing the sample volume.

The techniques for LLE application vary: classic, microwave-assisted, and supercritical fluid extraction (SFE). The two last techniques have significant advantages as compared to the classic one: a drastic reduction in the volume of the toxic solvent, a simplified procedure, reduction of the environmental hazard, safety, speed, elimination of the analytical concentration step, and cost saving. The SFE technique is suitable for low-boiling-point and temperature-unstable parts of the analytical object. The highest recovery has been noted for microwave-assisted extraction. The new techniques need specific appliances.

B. Gas Extraction

Headspace extraction is a process of obtaining volatile compounds from a solution by increasing the temperature (static version) or by the passing of a gas (dynamic version) (34). It is an indirect extraction method mainly for GC determination (34–41). Depending on the instrument, this technique is for the detection and determination of compounds with boiling points under 250°C; HSE needs no solvents. There is a possibility

Table 3 Solvents Most Often Used

Item	Solvent	Formula	BP, °C	Ref. index/20°C (n_D^{20})	Density (d_4^{20})	Solubility in water, 20°C, g/L[a]
1	1,1,2-Trichlorotrifluoroethane (Freon 113)	$C_2Cl_3F_3$	48	1.3596	1.5745	Insoluble
2	Carbon tetrachloride	CCl_4	76.75	1.4607	1.5935	0.8
3	Chloroform	$CHCl_3$	61	1.4459	1.4817	8.0
4	Dichloromethane	CH_2Cl_2	39.8	1.4250	1.3231	20.0
5	Tetrachloroethylene	C_2Cl_4	121.2	1.5054	1.6230	0.15
6	*n*-Hexane	C_6H_{14}	68.7	1.3748	0.6592	0.014[b]

[a]Merck, E., Catalogue *Reagents, Chemicals, Diagnostics*, 1996, Darmstadt, Merck, pp. 374, 501, 1170, 1223.
[b]In volumetric %, *Short Handbook of Practical Data*, 1965, Kiev, Naukova dumka, p. 249.

for analysis automation and increased sensitivity. Because of the limitation of the GC temperature, the whole quantity of TOG/TPH cannot be blown out. Comparing European and American methods that use HSE techniques, one observes a preference of the dynamic HSE and purge-and-trap in the United States, Scandinavian countries, and the Netherlands, while in Great Britain and the West European countries the static HSE is chosen (41).

C. Solid-Phase Extraction (SPE)

Solid-phase extraction is a method of adsorption of organic pollutants on the sorbent in a column or on a disk (42–46). This technique is a potential substitute for LLE, with comparable extraction effectiveness. It is faster and cheaper than LLE, allowing the analyst to concentrate and purify a sample in one step. Based upon specific molecular interactions it can provide a better cleanup than other techniques and yield highly reproducible results without the LLE drawbacks, such as emulsion formation, hazardous concentration procedures, and generation of large volumes of hazardous wastes. At the same time, SPE can be easily automated, so operator technique will not affect data accuracy or precision. It is carried out by cartridges filled by an adsorbent or disks with surfaces modified by different phases and configured as membrane filters. Cartridges may have some inconveniences: a limited speed of passing water because of the small sizes, and often blocking up the adsorbent surface by fine solid parts. The diameters of the larger disks (47 and 90 mm) and the smaller thickness (0.6–1 mm) hasten the passing of water with an equivalent effectiveness. The sorbents used are: tenaks (poly-2,6-diphenyl-phenilene oxide) (36,47) and polymer sorbents usually having a silicium base with additives (42,48). The desorption is carried out by increasing the temperature and passing gas (in GC analysis the carrying gas) or via organic solvents (for spectral and GC measurements). By this technique, TOG/TPH are eluted in small volumes and can easily be concentrated for spectral or GC determinations.

Solid-phase microextraction (SPME) is a relatively new technique. There is no use of solvents for the isolation of impurities, but concentration takes place with the help of quartz fiber filled with an organic layer and installed in a headspace device (35,36,38,49,50). The impurities are isolated and concentrated on the fiber and transported to the analytical instrument. This technique reaches detection limits in the ppt (ng/L) region. It is used for volatile compounds.

The ban of the manufacture and use of volatile fluorocarbons by the Montreal Protocol gives alternatives for TOG/TPH determination. For several reasons, the best replacement solvent appears to be SPE by n-hexane (29). However, because of its IR absorption at the same frequencies where oil products absorb, it must be evaporated from the sample before IR analysis can be carried out. Maybe this method will be standardized in the future. In its favor are speed, use of minimal solvent quantities, and compatibility for IR and GC aims. The questions of regeneration of the used filters for the next use and their price have no answers at the present moment. Moreover, the separation of TPH from TOG by the same technique is not resolved yet.

Sludges, substances between waters and soils are usually analyzed like waters because of their high water content. Any oils or solid or viscous greases are separated from the liquid sample phase. Such objects are analyzed after Soxhlet extraction (4). In this way, the short-chain hydrocarbons and simple aromatics disappear because of their volatilization.

The question of a separate estimation of petroleum hydrocarbons and natural hydrocarbons remains without answer. It is known that in each water body as a result of development, decay, conversion, and death of the plant and animal worlds, there are some hydrocarbons of natural origin. Their quantity in many cases is important for an estimation of the degree of pollution. Natural and petroleum hydrocarbons were distinguished using pond water from the last degree of wastewater cleaning. For that sample the TPH level was 0.4 mg/L and the ratio of natural to petroleum hydrocarbons was 1:10 (unpublished results). The IR, GC, and GC-MS methods were applied.

IX. DETECTION AND DETERMINATION METHODS

The methods for detection of TOG and TPH in water are nowadays the same as their determination methods. Detection and determination methods of individual or groups of compounds in the TOG/TPH composition exit, but no detection methods exist for the two entire groups. This depends on the object specificity. Initially the presence of TPH was found out visually and organoleptically. At present this is done by the following determination methods: spectroscopic [infrared (IR)], fluorescent (Fl), ultraviolet and visible spectroscopy (UV-VIS), and chromatographic [gas chromatography (GC)]. The distinction between qualitative and quantitative determination is in the calibration. Laser methods in this analysis sphere has a lot more applications. These methods can be divided into two groups: the first is not influenced by wavelength (λ) in water; the second is connected with λ changes (fluorescence and Raman spectra). New techniques, i.e., photoacoustic sensors (51), Fl together with fiber optics (52), and others (53–56) reveal specific characteristics, but none of them reaches wide popularity in practice. The future instruments for the detection and determination of TOG and TPH involve a coupling of techniques.

The wide variety of TOG and TPH migration forms causes difficulties in their total determination, because these factors influence the quantity. The separate determination of dissolved, emulsified, colloidal, and sorbed TOG and TPH is connected with a lot of experimental difficulties. Moreover, in routine analysis they are determined as a sum.

In recent years, no advances were made in the analysis of these indexes. Two reviews of determination methods nowadays are published (57,58). The quantitative determination of TOG and TPH depends roughly on the amount of pollution. Gravimetric analysis (4—method B,10) was applied a long time ago. In API 75153 and U.S. EPA 413.1 the same technique is applied. Success of this technique depends precisely on the performance of each manipulation, on all reagents and the cleanliness of the laboratory glassware, and on not allowing external contamination. Since the number of process steps is not small, there is a greater chance of erroneous results. This is why this technique is not recommended for the determination of concentrations near to or under 0.3 mg/L.

The instrumental methods of analysis for quantities of about 10 and even 100–1000 mg/L have wide application. The optical methods most applied are: IR (1–9,11,12,16,21,24,28–30,42,59–61), Fl (33,52,54,55,62–66), UV-VIS (67,68), GC and GC-MS coupling methods (14,25–27,34–41,43–50,69–81), as well as coulometry (82).

Today, enough spectroscopic methods exist that in comparison with other techniques (chemical, biological, and so on) have some advantages:

The speed of transformation of the light signals of the information studied in electronic signals
Remote control as a result of identical diffusion of light in water and in the atmosphere and its transition through their borderline
High sensitivity of determination, with the possibility for hyperfaint-signal registration

Universality (by one or another optical spectroscopic method, identification of all substances is possible)
Great selectivity.

Infrared spectroscopy is the most used application among the optical techniques (1–9,11,12,16,21,24,28–30,42,59–61). It is the method of TOG and TPH determination applied in almost all standard techniques. The common stretch vibration $\nu_{c\text{-}h}$ of CH_3^- and CH_2^- groups and CH^- from aromatic rings for most organic compounds in the region 3100–2700 cm^{-1} is used. The quantitation can be made by absorption of the most intensive band (ν_{2927}) by the sum of band absorptions at 3030, 2957, 2927, 2872, and 2855 cm^{-1}, or by an integral absorption of the area c between the baseline and the spectral curve among two endpoints (around 3050 and 2885 cm^{-1}).

Flame ionization (FI) is another method used but not standardized (33,52,54,55,62–66). This technique is applied as a standard for the identification of petroleum oil spill (33) and is accepted from the International UNESCO Oceanographic Commission standard (IOC/UNESCO, Manuals and Guides No. 3, Paris, 1984, p. 35). The applicability of this method depends on the fact that the primary aromatic hydrocarbons from the water pollutants fluoresce in the UV spectral region with considerable displacement to longer radiation waves. Information on the other TPH is lacking, but it is known that a part of the fluorescence is caused not by hydrocarbons but by polar compounds of unknown origin. The advantage of the method is a great selectivity. A detection limit of 0.5 ng/L can be attained by laser-induced FI (52), and the possibility of increasing sensitivity with laser multiwavelength excitation in the 240–360-nm region is possible. A series of fluorescent characteristic parameters (fluorescent band format and position, the excited status lifetime, and so on) can be used individually or in combination for the quantitative estimation of TPH and TOG.

The UV-VIS methods aren't used very much (67,68). The incomplete TOG and TPH evaluation in this method is a disadvantage. The predominant part of crude oils and their refined products, aliphatic hydrocarbons, are transmitted in this region.

Gas chromatographic and GC-MS coupling are the other applied instrumental techniques (14,22,25–27,34–41,43–50,69–81). Most of these methods are used for TPH identification and individual determination, for TOG, or for the part of them extracted by a fixed solvent. Their specific possibilities for separation of compounds with boiling points in a wide temperature range onto an appropriate phase and other advantages make them valuable tools, especially for identification aims and for the detection of the origin of pollution, for volatile and semivolatile organic compound determination, and for those known as "purgeables" (41,44,50,77–79). Division of the volatile and semivolatile pollutants is random. It is based on the technique used for their determination. The pollutants are considered volatile if they can be analyzed by the purge-and-trap technique and semivolatile if they are not so effectively purged because of their lower partial pressure. This analytical technique lacks a standardized method for a quantitative estimation of general TOG and TPH. Some chloro- and oxygen-containing compounds—phenols, phthalates, nitrosamines, halogenated hydrocarbons and esters, polycyclic aromatic hydrocarbons, polychlorinated biphenyls, pesticides, and others—are usually joined with the TPH analysis. Capillary GC is most successfully applied among different GC work techniques. The detection can be carried out by a flame ionization detector—FID (35,36,69), by flame photometric detectors —PID (73), by electron capture detectors—ECD (34,71), and by mass spectrometric methods—MS (21,27,44,50,76–80). The last have greater possibilities because of simultaneous separation and determination. Since the analytical object is an exceptionally complex mixture with boiling points in a range of a few to a hundred degrees, the full separation usually needs preliminary enrichment and separation into classes of a similar structure.

The work techniques applied nowadays are used for a fast TOG/TPH detection and determination in the field in most cases. These instruments are small, quite portable, and usable. The distinction between TOG and TPH can be discerned only by a few of them. They work predominantly on the principle of water extraction by solvent. The detection is accomplished in different ways. The most popular is via the NDIR (nondispersive infrared) analyzer (Ocma 220–Horiba, OMS-2—GSA, HC-404—BUCK Scientific, Infracal —Wilks Enterprise Inc.). In other instruments it is based on a photometric principle in the visible region (DR 2000—Hach), or a detection on a reflected-light basis (EE 440–012—ELE International). With the last instrument, the analyst can work without an extraction step. In almost all instruments, calibrations with some kind of crude oil or its refined products are memorized. On this basis among the data obtained by such instruments and from laboratory-analyzed samples from the same source, serious distinctions can often be achieved. Advantages are speed, possibilities

to measure TPH concentrations in a water flow, and the easy manipulation in field conditions.

X. PROBLEM ESTIMATION

The last 10 years' practice of TOG/TPH determination in almost all kinds of waters by standardized IR methods (1,3,5,11,21) show some difficulties and error sources. For the determination of small concentrations these difficulties and errors are obvious. The calibration is one of the crucial moments. Using different reagent batches, deviations in blank samples can be observed. Also, in analyzing drinking and distilled waters, an unchangeable "background" contamination can be seen. Reagent purity is necessary, especially when analyzing small concentrations.

A. Calibration

The preferred calibration practice involves a sample of TOG/TPH present in the sample of water or wastewater awaiting analysis. This is the case of a known and constant source of pollution. Unfortunately, that happens in only a few cases. When the pollution source is unknown or has a variable composition, one uses a mixture for calibration. Its composition reflects (qualitatively and partly quantitatively) the presence of straight- and branched-chain n-alkanes and monoaromatic hydrocarbons. This is a mixture from cetane (n-hexadecane), iso-octane (2,2,4-trimethyl-pentane), and benzene (4—methods C and F, 5,12,20), a mixture from cetane, iso-octane, and chloro-benzene (2,3—in older versions), or a mixture of only the first two compounds. Replacement of benzene by chloro-benzene is done to create a less hazardous work process. After this, the benzene component is eliminated. Because of the hazards of benzene exposure, which acts here only as a diluent that gives no contribution to the A_{2927} (very low absorptivity), this chemical is eliminated as a component for calibration. To maintain relevance between current and future analytical data and the data of the past it is necessary to compensate for the difference in concentration and density between the former and the present calibration standards. A factor of 1.4 is used because the wright ratio between iso-octane and cetane in the new two-component mixture and the older three-component mixture is 1.000 to 0.717, or 1.4. In some of the instruments in the field, the calibration is read by the so-called "squalan" value ($C_{30}H_{62}$). This value is used in DIN (5), but it is not so close to the true estimation. For an unknown pollutant in three- or five-wavelength absorptions or with integral absorption, the three-component mixture is the best (5,12,20).

B. Reagents

In the analytical procedure, NaCl, Na_2SO_4, Al_2O_3 (SiO_2), and a solvent are used.

1. Hard Reagents

The value of TPH pollution working with "p.a." and "suprapure" reagents is a minimum 0.1 mg/L. Regardless of the reagent grade used, some differences can be noticed using one or another package. The last is of more significance when waters near the regulated value are analyzed. The regulated values in different countries vary between small limits. With 0.05 mg/L regulation (the Bulgarian one) and working with commercially accessible reagents, it is not difficult to reach the "absence" result. Washing the hard reagents with a solvent (for spectroscopy) is a possible way out. The solvent consumption naturally depends on the specific package pollution, but 100–150 ml solvent is usually sufficient for 100 g "p.a." reagent. An absence of absorption bands in the IR region 3100–2770 cm^{-1} (especially near 2927 cm^{-1}) of the last washing portions is a measure of washing degree.

The sorbent quantity to be used for the separation of hydrocarbons from other compounds can be changed depending on the sample charge. The recommended quantities in the different methods are from 2 to 8 g, but every analyst is supposed to remember that the sorption power of 1 g sorbent is 0.65 mg polar compounds.

2. Solvents

TOG/TPH extraction is carried out with one or another solvent, and in all standards it is necessary to be "for spectroscopy." In the case of mass work, it makes the analysis more expensive. Using "p.a." solvents combined with a dispersive spectrometer increases the detection limit. If the analyst wants to be sure of result correctness and to lower the detection limit, the solvent used must be for spectroscopy or preliminary purified. This last can be attained with activated charcoal (16 hours at 200–220°C or 30 min at 400–450°C), used only once for a contact time of 48 hours, or in countercurrent with the solvent at 2-ml/min speed. There are special instruments for solvent cleaning on the market, working on the same principle (GSA, Germany). By means of such instruments, or in a way mentioned ear-

lier, a solvent with up to 50 mg/L TPH can be cleaned. The cleaning process can be carried out repeatedly and the consumption is considerably less.

XI. RECOMMENDED METHODS

The methods recommended for TOG/TPH determination are listed in the following sections.

A. For Total Determination

1. *For Quantities of 100 mg/L and More*

Gravimetric analysis:	ASTM D-4281–83
	U.S. EPA-9070
	AWWA 5520.B
	U.S. EPA 413.1
	API 75153
IR analysis	U.S. EPA 413.2

2. *For Quantities up to 100 mg/L*

IR analysis	ASTM D-3921–85
	DIN 38409–18
	AWWA 5520.C and F

3. *For Quantities up to 10 mg/L*

IR analysis	BDS 16714–96

4. *For Quantities up to 1 mg/L*

IR analysis	BDS 16714–96
	ASTM D—Proposal 162
	U.S. EPA 418.1

B. For Sludges

Gravimetric analysis	AWWA 5520.E

C. For Determination of Individual Hydrocarbons and Other Organic Compounds in TOG Composition Via GC Analysis

Purgeables	U.S. EPA 624
	U.S. EPA 524.2–4
	U.S. EPA 602
Semivolatiles	U.S. EPA 1624
	U.S. EPA 502.1
	U.S. EPA 502.2
	U.S. EPA 503.1
	ASTM D-2908–87
	AWWA 6210
Volatile aromatics	AWWA 6220

D. For Detection of Spill Pollution Sources and for Comparison of Waterborne Petroleum Oils

Sampling	ASTM D-4489 (methods A and B) NT CHEM 001 (methods A, B and C)
Preservation	ASTM D-3325–85
Preparation	ASTM D-3326–84
Identification	ASTM D-3415–79 (GC, IR, Elemental)
IR analysis	ASTM D-3414–80
FI analysis	ASTM D-3650–78
GC analysis	ASTM D-3328–78
GC/MS analysis	NT CHEM: 001

The experience of water analysis permits us to conclude that the applicability of any of the recommended and standardized methods must be estimated while taking into consideration the nature of the sample nature, the aims of the analysis, and the instrumental technique available.

REFERENCES

1. U.S. EPA Method 413.2. Oil and grease. Total recoverable (spectrophotometric, infrared), issued 1974, editorial revision 1978. Methods for Chemical Analysis of Water and Wastes. 3rd ed. Cincinnati, OH: U.S. Environmental Protection Agency, 1983.
2. U.S. EPA Method 418.1. Petroleum hydrocarbons. Total recoverable (spectrophotometric, infrared), issued 1978. Methods for Chemical Analysis of Water and Wastes. 3rd ed. Cincinnati, OH: U.S. Environmental Protection Agency, 1983.
3. ASTM D-3921-85. Standard test method for oil and grease and petroleum hydrocarbons in water. Annual Book of ASTM Standards. Philadelphia, 1990, 11.02: 62–66.
4. AWWA Method 5520. Oil and grease. Standard methods for the examination of water and wastewater. 18th ed. Washington, DC: 1992, pp 5:24–5:29.
5. DIN 38409 teil 18. Summarische wirkungs—und stoffkenngrossen bestimmung von Kohlenwasserstoffen. Deutsche Industrialische Normen. V. Berlin, 1981, pp 1–10.
6. NF T 90-203. Essais des aux. Effluents aqueux des raffineries de petrole. Dosage des hydrocarbones totaux. Norme francaise, Paris: AFNOR, 1973, pp 1–3.
7. COMECON. Petroleum hydrocarbons. Standardized methods for examination of water quality. Chemical Methods for Water Analysis. 3rd ed. Moscow, 1977, pp 1:359–388.
8. Collection of methods for examination of wastewaters and sediments. 1st ed. Sofia, Bulgaria, 1987, pp 13–21.
9. Methods for Sea Water Analysis. 1st ed. Leningrad: Gidrometeo, 1981, pp 8–15.
10. ASTM D-4281-89. Test method for oil and grease (fluorocarbon extractable substances) by gravimetric determination. Annual Book of ASTM Standards. Philadelphia, 1990, 11.02:55–56.
11. ASTM D-19 Proposal 162. Proposed method of quantification for petroleum oil in water. Annual Book of ASTM Standards. Philadelphia, 1985, 11.02:1077–1083.
12. BDS 16714-87. Hydrosphere. Determination of Petroleum Products in Water by Infrared Spectrometry. 1st ed. Sofia, Bulgaria, 1988, pp 1–6.
13. MT Dimitriev, NM Kasmina. Sanitary-Chemical Analysis of Environmental Pollutant Substances. 1st ed. Moscow: Chemistry, 1989, pp 13–18, 21–27.
14. RP Eganhouse, TF Dorsey, CS Phynney. Processes affecting the fate of monoaromatic hydrocarbons in an aquifer contaminated by crude oil. Environ Sci Technol 30:3304–3312, 1966.
15. ASTM D-3370-82. Standard practices for sampling water. Annual Book of ASTM Standards. Philadelphia, 1990, 11.01:104–110.
16. R Belcheva, K Kovacheva. Determination of petroleum hydrocarbons in water—assessment of the problems. Anal Labor 5:200–205, 1996.
17. ASTM D-3694-89. Standard practices for preparation of sample containers and for preservation of organic constituents. Annual Book of ASTM Standards. Philadelphia, 1990, 11.02:12–18.
18. ISO 5661. Water quality—sampling. Parts 2–7, 9–11. Guidance on sampling techniques, preservation and handling of samples, sampling from: lakes—natural and man-made, drinking water, rivers and streams, water and stream in boiler plants, marine waters, waste waters, groundwaters. 1987, 1991–1994, 1996, pp 1–9, 1–31, 1–5, 1–8, 1–9, 1–16, 1–8, 1–10, 1–10.
19. LD Johnson, RH James. Sampling and analysis techniques for hazardous waste. In: HM Freeman, ed. Standard Handbook of Hazardous Waste Treatment and Disposal. Chap. 13. New York: McGraw-Hill, 1989, pp 13.3–13.15.
20. ASTM D-4489-85. Standard practices for sampling of waterborne oils. Annual Book of ASTM Standards. Philadelphia, 1990, 11.02:295–297.
21. BDS 16714-96. Nature protection. Hydrosphere. Quality indexes of water. Method for Infrared Determination of Oil and Grease and Petroleum Hydrocarbons. 2nd ed. Sofia, Bulgaria, 1996, pp 1–9.
22. NT CHEM 001. Oil spill identification. Nordtest Method. 2nd ed. Finland: Nordtest, 1991, pp 1–24.
23. U.S. EPA. Sample preservation. Methods for Chemical Analysis of Water and Waste. 3rd ed. Cincinnati, OH: U.S. Environmental Protection Agency, 1983, Table 1, p XVIII.

24. AJ Gaches, PS Wilson. Measurement of oil, grease and other hydrocarbons in water. Spectroscopy World 3:22, 1991.

25. I Harrison, RU Leader, JJW Higgo, JC Tjell. Determination of organic pollutants in small samples of groundwaters by liquid–liquid extraction and capillary gas chromatography. J Chrom A 688:181–188, 1994.

26. N Theobald, W Lange, W Gahlert, F Renner. Mass spectrometric investigations of water extracts of the River Elbe for the determination of potential inputs of pollutants into the North Sea. Fresenias J Anal Chem 353:50–56, 1995.

27. J Bartulewicz, E Bartulewicz, J Gawlowski, J Niedzilski. Simple and rapid method for the determination of petroleum products in water. Chem Anal (Warsaw) 39: 167–177, 1994.

28. H Nguyen. Solid phase extraction for the isolation of oil and grease. Am Environm Lab 6:16–19, 1992.

29. PA Wilks Jr. Replacing Freon 113 as an IR solvent: how the attempted solution of one environmental problem is crating another. Spectroscopy (Eugene, Ore.) 12:47–48, 1997.

30. J Daghbouche, S Garriques, A Morales-Rubio, M dela Guardia. Evaluation of extraction alternative for Fourier-transform infrared spectrometric determination of oil and grease in water. Anal Chim Acta 345:161–177, 1997.

31. ASTM D-2778-74. Standard practices for solvent extraction of organic matter. Annual Book of ASTM Standards. Philadelphia, 1981, 31:639–644.

32. ASTM D-2910-85. Standard practice for concentration and recovery of organic matter from water by activated carbon. Annual Book of ASTM Standards. Philadelphia, 1990, 11.02:8–11.

33. ASTM D-3650-78. Standard test method for comparison of waterborne petroleum oils by fluorescence analysis. Annual Book of ASTM Standards. Philadelphia, 1990, 11.02:263–267.

34. AG Vitenberg, BV Ioffe. Gas Extraction in Chromatographic Analysis: Headspace Analysis and Related Methods. 1st ed. Leningrad: Chemistry, 1982, pp 116–117.

35. JJ Langenfeld, SB Hawthorne, DJ Miller. Quantitative analysis of fuel-related hydrocarbons in surface water and waste water samples by solid-phase microextraction. Anal Chem 68:144–155, 1996.

36. A Saraullo, PA Martos, J Pawliszyn. Water analysis by solid-phase microextraction based on physical-chemical properties of the coating. Anal Chem 68:1992–1998, 1997.

37. ASTM D-3871-84. Standard test method for purgeable organic compounds in water using headspace sampling. Annual Book of ASTM Standards. Philadelphia, 1990, 11.02:186–193.

38. Z Zhang, J Pawliszyn. Headspace solid-phase microextraction. Anal Chem 65:1843–1852, 1993.

39. B Rothweiler. The use of static and dynamic headspace sampling for the trace analysis of volatile organic pollutants in environmental samples. Hewlett Packard Appl. Note 228-261:1–13, 1994.

40. B Kolb. Quantitative trace analysis of volatile organic compounds in air, water and soil using equilibrium headspace gas chromatography. LC-GC Int 8:512–524, 1995.

41. TC Voice, B Kolb. Comparison of European and American techniques for the analysis of volatile organic compounds in environmental matrices. J Chrom Sci 32: 306–311, 1994.

42. DL Heglund, DC Tillota. Determination of volatile organic compounds in water by solid phase microextraction and IR spectroscopy. Environ Sci Technol 30: 1212–1219, 1996.

43. N Theobald. Rapid preliminary separation of petroleum hydrocarbons by solid-phase extraction cartridges. Anal Chim Acta 204:135–144, 1988.

44. EY Zeng, AR Khan. Extraction of municipal wastewater effluent using 90-mm C-18 bonded disks. J Microcolumn Separ 7:529–539, 1995.

45. AI Krylov, NF Volynets, IO Kostyuk, VV Buevets. Concentration of organic compounds from aqueous solutions on SKS, SKN and Carbochrom carbon sorbents. Ah Anal Khim 50:924–930, 1995.

46. AI Krylov, NF Volynets, IO Kostyuk. Concentration of organic compounds from aqueous solutions by polymeric sorbents of polisorb and polichrom types. Zh Anal Khim 48:1462–1468, 1993.

47. JM Warner, RK Beasly. Purge and trap chromatographic method for the determination of acrilonitrile, chlorobenzene, 1,2-dichloroethane and ethylbenzene in aqueous samples. Anal Chem 56:1953–1956, 1984.

48. TV Silkina, MF Prokop'eva. Gas-chromatographic determination of organic compounds in wastewaters. Express information. Series: Methods for Output Analysis and Quality Checking 1:4–6, 1988.

49. Anon. Monitor BTEX compounds and fuels in water, using solid-phase microextraction and capillary gas chromatography. Supelco Appl, 1995, 81, p 2.

50. SS Stafford, RH Kolloff, JR Thorp, JS Hollis, LC Doherty. Analysis of volatiles using purge and trap with GC/MS—methods and instrument configurations. Hewlett Packard Appl. Note 228-278:1–7, 1994.

51. P Hodgson, KM Quan, HA Mackenzie, SS Freeborn, J Hannigan, EM Johnson, F Greig, TD Binnie. Application of pulsed laser photoacoustic sensors in monitoring oil contamination of water. Sens Actuators B 329:339–344, 1995.

52. W Shade, J Bublitz. On-site laser probe for the detection of petroleum products in water and soil. Environ Sci Technol 30:1451–1458, 1996.

53. J Pianosi. Methods to detect, measure NAPLs vary. Int Groundwater Technol 2:31–32, 34, 1996.

54. EM Filipova, VV Chubarov, VV Fadeev. New possibilities of laser fluorescence spectroscopy for diagnostics of petroleum hydrocarbons in natural water. Can J Appl Spectrosc 38:139–144, 1993.

55. MF Quinn, AS Al-Otaibi, A Abdullah, PS Setki, F Al-Bahrani, O Alanelddine. Determination of intrinsic fluorescence lifetime parameters of crude oils using a laser fluorosensor with a streak camera detection system. Instrum Sci Technol 23:201–215, 1995.

56. JD Hanby. New method for the detection and measurement of aromatic compounds in water. Int Labmate 16:11–16, 1991.

57. M Ehrhard, J Klungsoeyr, RJ Law. Hydrocarbons: review of methods for analysis of seawater, biota and sediments. Tech Mar Environ Sci 12:14, 1991.

58. JN Discoll. Review of field screening methodology for analysis of hydrocarbons in soils and groundwater. Int Labmate 17:27–32, 1992-1993.

59. M Boeck. Quantitative determination of hydrocarbons in water and soil. Erdoel, Khole, Erdgas, Petroleum 44:182–183, 1991.

60. E Sensfelder, J Buerck, H Ache. Determination of hydrocarbons in water by evanescent-wave absorption spectroscopy in the near-infrared region. Fresenius J Anal Chem 354:848–851, 1996.

61. B Minty, ED Ramsey, R Lewis. Hydrocarbons in water: analysis using on-line aqueous supercritical-fluid-extraction-Fourier-transform infra-red spectroscopy. Anal Commun 33:203–204, 1996.

62. Y Fang. Determination of oil in water by extraction-fluorescence photometry. Shanghai Huanjing Kexue 9:19–20, 1990.

63. M Picer, N Picer. Evaluation of modifications of the simple spectrofluorimetry method for estimating petroleum hydrocarbons levels in fresh and waste water samples. Chemosphere 24:1825–1834, 1992.

64. J Liu, Z Xie, X Wang, ZR Wang. Study of synchronous high order derivative spectrofluorimetry identification of crude oil and heavy fuel oil pollutants. Fenxi Shiyanshi 12:28–30, 1993.

65. M Picer, N Picer. Evaluation of modifications of the simple spectrofluorimetry method for estimating petroleum hydrocarbon levels in sea water. Bull Environ Contam 50:802–810, 1993.

66. AV Karyakin, AV Galunin. The fluorescence of water-soluble compounds of oils and petroleum products that constitute the petrogenic pollution of waters. Zh Anal Khim 50:1178–1180, 1995.

67. H Dong, H Jin, M Jiang, J Lechua. Solvent flotation spectrophotometric determination of oil content in sewage water. Huaxue Fence 28:98–99, 1992.

68. T Bastov, WH Durnie, A Jeferson, J Pang. UV spectroscopy for the analysis of oil-in-water effluent using isopropanol as co-solvent. Appl Spectroscopy (Austral) 51:318–322, 1997.

69. D Papazova, A Pavlova. Determination of individual hydrocarbons in waste waters from petrochemical plant near Burgas. Isv Khim 24:337–344, 1991.

70. N Theobald, A Rave, K Jerzycki-Brandes. Input of hydrocarbons into the North Sea by the River Elbe. Frenius J Anal Chem 353:83–87, 1995.

71. U.S. EPA Method 1624. Volatile organic compounds by isotope dilution GC/MS. Methods for Chemical Analysis of Water and Waste. 3rd ed. Cincinnati, OH: U.S. Environmental Protection Agency, 1983.

72. U.S. EPA Method 1625. Semivolatile organic compounds by isotope dilution GC/MS. Methods for Chemical Analysis of Water and Waste. 3rd ed. Cincinnati, OH: U.S. Environmental Protection Agency, 1983.

73. ASTM D-2908-87. Standard practice for measuring volatile organic matter in water by aqueous-injection gas chromatography. Annual Book of ASTM Standards. Philadelphia, 1990, 11.02:194–199.

74. AP O'Brien, G McTaggart. The analysis of light aromatics in aqueous effluent by gas chromatography. Chromatographia 16:301–303, 1982.

75. ZD Wang, M Fingas. Differentiation of the source of spilled oil and monitoring of the oil weathering process using gas chromatography—mass spectrometry. J Chrom 712:321–343, 1995.

76. FA Medvedev, LSh Vorob'eva, ON Chernisheva. GC-MS analysis of oil contaminants of water and marine organisms. Zh Anal Khim 51:1181–1185, 1996.

77. AWWA Method 6210. Volatile Organics. Standard methods for the examination of water and wastewater. 18th ed. Washington, DC: 1992, pp 6:17–6:35.

78. AWWA Method 6220. Volatile aromatic organics. Standard Methods for the Examination of Water and Wastewater. 18th ed. Washington, DC: 1992, pp 6:38–6:46.

79. U.S. EPA Method 624. Purgeables. Methods for Chemical Analysis of Water and Wastes. 3rd ed. Cincinnati, OH: U.S. Environmental Protection Agency, 1983.

80. Z Luo, Y Hsia, K Xie, Y Zhang. Analysis of diesel compounds in soil and water contaminated by semivolatile synthetic organic compounds. J Chrom Sci 33:263–267, 1995.

81. S Skrabakova, E Matisova, E Benicka, I Novak, D Berek. Use of a novel carbon sorbent for the adsorption of organic compounds from water. J Chrom A 665:27–32, 1994.

82. BK Zuev, OK Timonina, VD Podrugina. Rapid method for determining the total concentration of organic impurities in water. Zh Anal Khim 50:663–668, 1995.

35

Asbestos in Water

Leo M. L. Nollet

Hogeschool Gent, Ghent, Belgium

I. ASBESTOS MATERIALS AND HEALTH EFFECTS

Asbestos is a mineral with a fibrous form. It usually occurs as veins in rocks. Two main types exist: chrysotile and amphibole. Chrysotile, or "white asbestos," is composed of microscopic glossy tubes of magnesium silicate. Amphibole is made up of bundles of needles of complex silicates containing iron, calcium, and other elements. The various types are found all over the world. There are many industrial and commercial uses of asbestos: materials for insulation, as fire retardant, pipe materials, brake and clutch linings, and roofing materials.

The small needles are able to cause cancer. Fibers longer than 8–10 μm are more carcinogenic (length/ diameter greater than 3:1) (1). Possible cancer types are lung and throat cancers and cancers of the gastro-intestinal and urinary systems by coughing up and swallowing. Other possible diseases are mesothelioma and asbestosis (2–12). The fibers can remain and accumulate in the lungs for years. Symptoms can show up many years after exposure. Falchi and Paoletti give a review of the diseases, sampling, and analysis techniques in human samples (13).

Drinking water is considered to be a minor source of asbestos. Asbestos reaches water via leaching from rocks, mining operations for other minerals, and piping consisting of asbestos–cement compounds. Most fibers in water are shorter than 10 μm.

II. ANALYTICAL METHODS FOR ASBESTOS

Samples are taken at a tap served by asbestos–cement pipe or at a representative sampling point of the groundwater or surface water system.

Samples are preserved cool at 4°C in plastic or glass containers.

Water samples are filtered through a 37-mm, 0.8-μm mixed cellulose esters (MCE) filter. With phase contrast microscopy (PCM) the sample can be analyzed qualitatively. With that type of analysis no distinguishability between asbestos fibers and nonasbestos fibers is possible. Nevertheless it is a common analysis technique for total fiber content.

Concentrations and types of fibers can be determined with transmission electron microscopy (TEM) (14,15).

The EPA Public Water Drinking maximum concentration level (MCL) for asbestos is 7 million fibers/liter (fibers longer than 10 μm). The detection limit is 0.01 MFL (million fibers per liter).

The European Union (16) stipulates that the disposal of liquid waste containing asbestos shall not result in pollution of the aquatic environment. A limit value of 30 grams of total suspended matter per cubic meter of aqueous effluent shall not be exceeded. The reference method of analysis to determine total suspended matter (filterable matter from the nonprecipitated sample), expressed as mg/L, shall be filtering through a 0.45-mm filter membrane, drying at 105°C, and weighing. Samples must be taken in such a way as to be representative of the discharge over a 24-hour period. This determination must be conducted to a precision of ±5% and an accuracy of ±10%.

REFERENCES

1. TM Pal, J Schaaphok, J Coenraads. Cah. Notes Doc. 138:254–257, 1990.
2. IARC Monographs on the Evaluation of Carcinogenic Risk to Humans. Supplement 7. International Agency for Research on Cancer, Lyon, 1987.
3. JC McDonald, AD McDonald. Ann. Occup. Hyg. 41: 699–705, 1997.
4. FDK Liddell, AD McDonald, JC McDonald. Ann. Occup. Hyg. 41:13–36, 1997.
5. AD McDonald, BW Case, A Churg. Ann. Occup. Hyg. 41:707–719, 1997.
6. FDK Liddell, AD McDonald, JC McDonald. Ann. Occup. Hyg. 42:7–20, 1998.
7. M Camus, J Siemiatycki, and B. Meek. New Engl. J. Med. 338:1565–1571, 1998.
8. SQ Zou, YX Wu, FS Ma, HS Ma, WZ Sueng, ZH Jjiang. Retrospective mortality study of asbestos workers in Laiyuan. Proc. Int. Pneumocon. Conf. - Part II, Aug 23–26, 1988, Pittsburgh, PA, pp 1242–1244.
9. P Landrigan. New Engl. J. Med. 338:1618–1619, 1998.
10. D Egilman, A Reinert. Am. J. Ind. Med. 30:398–406, 1996.
11. M Meldrum. Review of Fibre Toxicology. Health and Safety Executive, London, 1996.
12. Scientific Committee on Toxicity, Ecotoxicity and the Environment (CSTEE)—Opinion on chrysotile asbestos and candidate substitutes expressed at the 5th CSTEE plenary meeting, Brussels, 15 September 1998.
13. M Falchi, L Paoletti. Ann. Ist. Super. Sanita. 34:139–149, 1994.
14. Determination of Asbestos Structures over 10 μm in Length in Drinking Water. EPA 600/R 94-134, 1994.
15. Analytical Method for Determination of Asbestos Fibers in Water. EPA 600/4-83-043, 1983.
16. Council Directive 87/217/EEC of 19 March 1987 on the prevention and reduction of environmental pollution by asbestos, 28/03/1987.

36

Analysis of Surfactants

Bieluonwu Augustus Uzoukwu
University of Port Harcourt, Port Harcourt, Nigeria

Leo M. L. Nollet
Hogeschool Gent, Ghent, Belgium

I. INTRODUCTION

Detergents are complex formulations used for cleansing purposes industrially and domestically, and many of them consist of mixtures of different detergents intended to achieve specific surface-active purposes. The active ingredients in these detergents are called *surface-active agents*, or *surfactants*. A surfactant is, therefore, a surface-active reagent consisting of a long alkyl chain, or hydrophobic portion, attached to a water-soluble functional group, or hydrophilic portion. Although some surfactants may not have detergent properties, they are, however, present in all types of detergents and constitute the most important group of detergent components, on account of their mode of action. Components of detergents for domestic and industrial uses can be categorized into surfactants, builders, bleaching agents, and auxiliary agents, and each of the listed components of a detergent has its own specific functions in the washing process.

In order to improve the quality and appeal of the product, certain substances are added where necessary during the production process. Wetting power, adsorption, and wash effectiveness (detergency) of surfactants have been associated with carbon chain branch and length. Subsequently, surfactants with few branched carbon chains in their hydrophobic portion very often show poor wetting tendencies and good wash effec-

tiveness. On the other hand, those with highly branched carbon chains show good wetting agents but poor wash effectiveness. Generally, surfactants that are used for detergent formulations are expected to possess the following characteristics (1): specific adsorption, soil removal, low sensitivity to water hardness, dispersion properties, soil antiredeposition capability, high solubility, wetting power, desirable foaming characteristics, neutral odor, low intrinsic color, storage stability, favorable handling characteristics, minimal toxicity to humans, favorable environmental behavior, assured raw material supply, and economy.

The presence of surfactants in our environment raises ecological, health, and conservation concerns, particularly those surfactants that contain branched-chain hydrocarbons and aromatic rings that cannot be biodegraded effectively by microorganisms. The health concerns stem from the ability of surfactants to interact with proteins and enzymes, in some cases resulting in the formation of complexes with these body structures, leading to protein denaturation and cell damage. The other concern is the ability of surfactants to penetrate an organism through the skin, although the amount involved tends to be quite low and can therefore be regarded as nonhazardous (2–4). The consequences of the interaction with the body components are more severe with charged surfactants, such as the anionic and cationic surfactants. This is because these interactions give rise to complex formations, attributed to polar interactions between the hydrophilic functional groups of the cationic and anionic surfactants and charged sites on the protein molecule. Interactions with body components involving nonionic surfactants are insignificant. This is due to the absence of charged sites in nonionic surfactants. One of the obvious potential consequences of this interaction is interference with material transport due to changes in membrane permeability. Another important health concern also is the ability of surfactants to emulsify lipids, leading to damage of the lipid film layer that covers the skin surface (3). This subsequently interferes with the basic barrier function of the lipids, leading to increased permeability and loss of moisture.

On ingestion, both anionic and nonionic surfactants can be readily absorbed through the gastrointestinal tract (2,5) leading to vomiting and possible diarrhea. Damage to the mucous membranes of the gastrointestinal cell membranes can occur. The eye is so sensitive to surfactants that low concentrations of anionic surfactant can cause eye irritation (6). Surfactants can cause skin irritation, and the level of skin irritability can be related to the length of the alkyl chain of anionic surfactants. Hence, in a given homologous series, saturated alkyl chains of 10–12 carbon atoms are reported to cause the most severe skin irritability (6). However, neither oral ingestion nor skin exposure during long-term study has ever shown that any of the groups of surfactants possesses carcinogenic (7), mutagenic, or teratogenic (7) properties. It is pertinent to point out that all surfactants used for detergent formulation are well tolerated at the concentration level at which they are used.

Depending on the type of charge present in the hydrophobic portion of the surfactant after dissociation in an aqueous medium, a surfactant can be placed in any one of the following four groups: anionic surfactants, cationic surfactants, nonionic surfactants, and amphoteric surfactants.

A. Anionic Surfactants

These are surfactants in which the hydrophilic group is an anion, consisting mostly of carboxy group (soap), sulphonates, sulphates, and, in rare cases, phosphates, and the hydrophobic group consists of a length of carbon chain. The anionic surfactants are the agents most commonly used for the formulation of detergents for washing, laundry, and other domestic cleansing activities. Due to sensitivity to water hardness and scale formation with hard water, soap ($RCH_2CO_3^-Na^+$, R = C_{10-16}) has gradually ceased playing a major role as an anionic surfactant in cleansing formulations (8). Some of the commonly used anionic surfactants include alkyl aryl sulphonates (para-$RC_6H_4SO_3^-Na^+$, R = C_{10-13}); alkanesulphonates ($R^1CH(R^2)SO_3^-Na^+$), $R^1 + R^2 = C_{11-17}$); α-olefine sulphonates (mixture of $CH_3(CH_2)_mCH$=$CH(CH_2)_nSO_3^-Na^+$, $n + m = 9$–15, and $RCH(OH)$-$(CH_2)_nSO_3^-Na^+$, R = C_{8-16}, $n = 1, 2, 3$); α-sulpho fatty acid methyl esters ($RCH(SO_3^-Na^+)CO_2CH_3$ and $RCH(SO_3^-Na^+)CO_2H$, R = C_{14-16}); alkyl sulphates ($RCH_2-O-SO_3^-Na^+$, R = C_{11-17}); fatty alcohol ether sulphates ($R^1CH_2CH_2-O-(CH_2CH_2-O)_nSO_3^-Na^+$, $n = 1$–4; $R^1 = C_{10-12}$); oxo alcohol ether sulphates ($R^1CH(R^2)CH_2-O-(CH_2CH_2-O)_nSO_3^-Na^+$, $n = 1$–4; $R^1 = H; C_{1,2,...} R^1 + R^2 = C_{11-13}$); sulpho succinate esters ($CO_2NaCH(SO_3^-Na^+)CH_2CO_2R$ and $CO_2RCH(SO_3^-Na^+)CH_2CO_2R$). When sodium is replaced by other types of cations, the properties of the surfactant changes, making it suitable for other uses, such as hand-cleaning gel, dry-cleaning detergents, cosmetic detergents, and emulsifier in pesticide formulations, among others. The most widely used anionic surfactants in detergent formulation are the alkyl aryl sulphonates, because of their excellent physical properties, interesting foaming characteristics, and low cost of production (they can easily be made from available materials).

More detailed information on the analytical chemistry of anionic surfactants can be found in Ref. 71.

B. Cationic Surfactants

Cationic surfactants have a cationic hydrophilic group that consists of a quaternary ammonium group and an alkyl chain, which imparts hydrophobicity to the surfactant. The hydrophobic group could consist of 10 carbon atoms or more, of which the following are some of the common surfactants (where X^- is chloride, sulphate, or bromide):

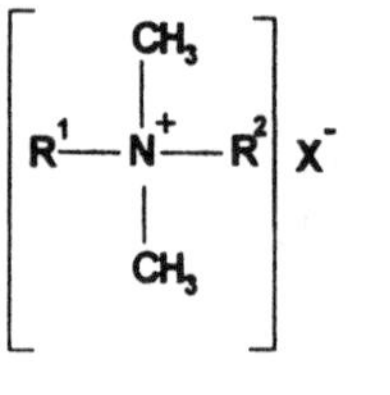

R^1, R^2 = C_1, C_{16-18} Alkyltrimethylammonium salt
R^1, R^2 = C_{16-18} Dialkydiamethylammonium salt

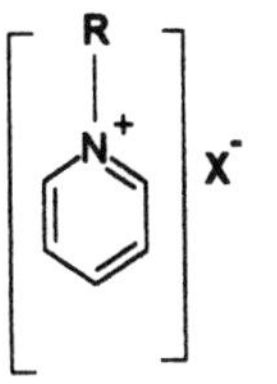

R = C$_{16-18}$ Alkylpyridinium salt

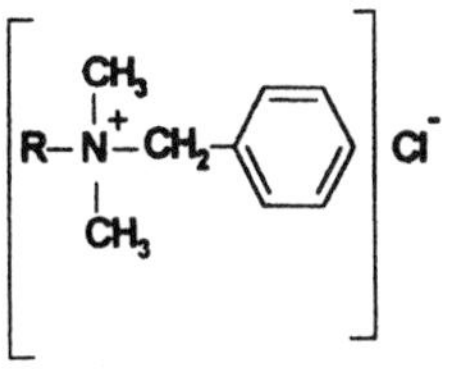

R = C$_{8-18}$ Alkyldimethylbenzylammonium chloride

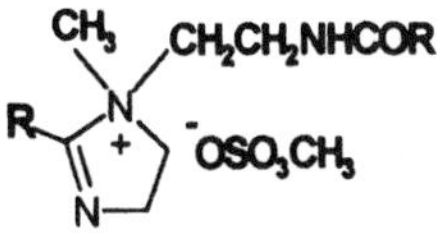

R = C$_{16-18}$ Imidazoliumquaternary-ammonium methylsulphate

Distearyldimethylammonium chloride, known for its extraordinarily high sorption power with respect to a wide variety of surfaces (9,10), was one of the earliest cationic surfactants developed and was used as a fabric softener. Another cationic surfactant known to have high adsorption capacity is alkyldimethylbenzylammonium chloride, used mainly as a disinfecting agent. Cationic surfactants have charges that are opposite to those of anionic surfactants; thus, a solution mixture of anionic and cationic surfactants produces neutral salts with extremely low water solubility and no washing effect. Hence, cationic surfactants are employed in laundry and cleansing agents only for certain special purposes, such as fabric softeners and disinfecting agents.

C. Nonionic Surfactants

These are surfactants that have properties that show distinct hydrophobic and hydrophilic regions in the molecule, but they have no overall electronic charges in the molecule. Hence, hydrophobic interaction can be used to explain the adsorption properties (11) shown by nonionic surfactants, even though this is obviously

not due to ion-exchange or ion-pair interactions. The following are some common nonionic surfactants:

Alkylphenol poly(ethyleneglycol) ethers

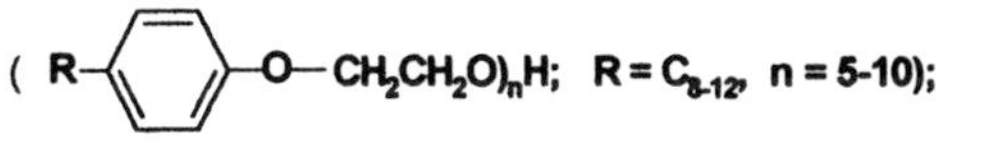

Fatty acid alkanolamides

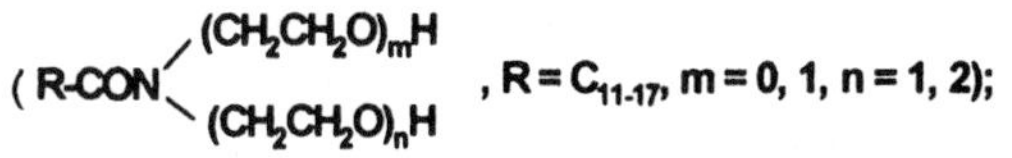

Alkyl glycosides

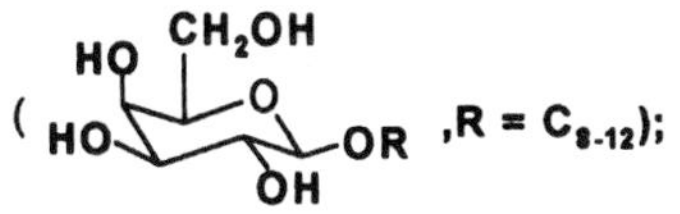

alcohol ethoxylates (R(OCH$_2$CH$_2$)$_n$OH, R = C$_{9-18}$, n = 1–40); fatty alcohol polyglycol ethers (RO-(CH$_2$CH$_2$O)$_m$–(CH$_2$CH(CH$_3$)O)$_n$H, R = C$_{8-18}$, m = 3–6, n = 3–6); ethylene oxide–propylene oxide block polymers (H(OCH$_2$CH$_2$)$_m$–O–CH(CH$_3$)CH$_2$O—[CH$_2$-CH(CH$_3$)–O–(CH$_2$CH$_2$O)$_m$H)]$_n$, m = 15–80, n = 2–60); fatty acid polyol and sorbitol esters (RCOO-(CH$_2$CH$_2$OH)$_4$–CH$_2$OH, R = C$_{12-18}$).

Generally, nonionic surfactants do not dissociate in aqueous solution because of the absence of electrostatic interactions. Nonionic surfactants have, therefore, found major application in a variety of products where their interfacial effects of wetting, dispersion, and solubilization can enhance product quality. These days, many products, including cosmetics, paints, pesticides and pharmaceuticals, contain nonionic surfactants.

D. Amphoteric Surfactants

These are surfactants that possess both anionic and cationic groups in the same molecule, even in aqueous solutions. This type of surfactant is employed only rarely, in the formulation of detergents for special applications; hence, they are not widely used for domestic or industrial application in comparison with the other types of surfactants. They include alkylbetaines (RN$^+$(CH$_3$)$_2$–CH$_2$–COO$^-$, R = C$_{12-18}$) and alkylsulphobetaines (RN$^+$(CH$_3$)$_2$–(CH$_2$)$_3$SO$_3^-$, R = C$_{12-18}$).

II. QUANTITATIVE ANALYSIS OF SURFACTANTS IN WATER

A. Anionic Surfactants

1. Determination of Anionic Surfactants of Water Using Methylene Blue Method [methylene blue active substances (MBAS)]

The analysis is based on the formation of ion-pair complexes from a reaction between anionic surfactants and methylene blue dye (12), which acts as the cationic species. The ion-pair complexes are extracted into an organic phase while methylene blue is not, so this forms the analytical basis for the application of this technique in the analysis of anionic surfactants. The organic extract is analyzed colorimetrically, because the intensity of the color of the organic extract is proportional to the concentration of extracted anionic surfactant. Methylene blue has the property of a cationic surfactant; hence, several anionic surfactants could react with it to form blue complexes that are extractable into organic solvents. Subsequently, the test is designated as that of methylene blue active substances (MBAS).

　　a.　Reagents and Apparatus.

1. Prepare a standard stock solution of sodium lauryl sulphate (100 mg·L^{-1} sodium lauryl sulphate) by dissolving 0.1 g of sodium lauryl sulphate (reagent pure) in a small quantity of water in a 1-L volumetric flask before diluting to mark with water. Prepare a range of standard solutions of sodium lauryl sulphate (0–5 mg·L^{-1}) in 100-ml volumetric flasks by pipetting appropriate volumes of standard stock solution into 100-ml volumetric flasks and diluting to mark with water. One of the volumetric flasks serving as blank should have 0 mg of sodium lauryl sulphate and be subjected to the same analytical procedure as the rest. Transfer the standard solutions and blank into 250-ml separation funnels.

2. Prepare: (a) 0.05% neutral methylene blue solution by dissolving 0.05 g of methylene blue in a small quantity of water in a 100-ml volumetric flask before diluting to mark with water. Prepare fresh and leave the solution at 20–25°C for 24 hours before use; (b) 0.05% acidified methylene blue solution by dissolving 0.05 g of methylene blue in a small quantity of water in a 100-ml volumetric flask, add 0.6 ml of concentrated H$_2$SO$_4$ before diluting to mark with water. Prepare fresh and leave the solution at 20–25°C for 24 hours before use.

3. Prepare a phosphate buffer (pH 10) by dissolving 1.252 g of Na$_2$HPO$_4$·2H$_2$O in 50 ml of water in a 100-ml volumetric flask, and adjust to pH 10 with 0.2 M NaOH solution before diluting to mark with distilled water.

4. Obtain 200 ml of chloroform (reagent pure), a set of 50-ml separation funnels, and a set of 50-ml volumetric flasks, a spectrophotometer or colorimeter, and 1-cm cuvette.

　　b.　Procedure.　Introduce a known quantity v ml of water sample (ca. 100–200 ml) into a 250-ml separation funnel. From this point treat the standard solutions, blank, and water sample in 250-ml separation funnels the same way, as follows: Add 10 ml of phosphate buffer, 5 ml of neutral methylene blue solution, and 15 ml of CHCl$_3$ to each of the 250-ml separation funnels. Carry out the extraction process by shaking the separation funnels vigorously for about 1 minute. Allow the phases to settle, and drain each CHCl$_3$ extract off into another 250-ml separation funnel containing 100 ml of water and 5 ml of acidified methylene blue solution. Shake each of the second set of 250-ml separation funnels for about 1 minute. Allow the phases to settle before draining off the CHCl$_3$ extract in each separation funnel through a chloroform-moistened cotton wool on a funnel into 50-ml volumetric flasks. Twice, repeat the two stages of extraction processes just described on all solutions and samples with fresh 10-ml portions of CHCl$_3$, and collect all three extracts from solution with the same 50-ml volumetric flask before topping to mark with CHCl$_3$. Addition of fresh quantities of phosphate buffer and methylene blue solution are not necessary when repeating the extractions. Obtain the absorbances of the extracts at 650 nm against CHCl$_3$ blank. Prepare a calibration curve by plotting the absorbances of the extracts from standard solutions against their corresponding concentrations. Compare the absorbance of the extract from water sample with those of the calibration curve to obtain the concentration c_1 mg·L^{-1} of anionic surfactants in the water sample extract. Compare the absorbance of the extract from blank with those of the calibration curve to obtain the concentration c_2 mg·L^{-1} of baseline correction for water sample extract.

$$\text{Concentration of anionic surfactants (mg·L}^{-1}$$
$$\text{sodium lauryl sulphate) in water sample}$$
$$= \frac{50 \text{ mg} \times (c_1 - c_2) \text{ mg·L}^{-1}}{v \text{ ml}}$$

Note: Other suitable reagents for preparing standard solutions for the analysis of anionic detergents include sodium salt of dodecane-1-sulphonic acid and sodium tetrapropylene benzene sulphonate. The result in each case should be stated in $mg \cdot L^{-1}$ of the relevant standard anionic surfactant used.

Zanette et al. (106) discuss investigation of the MBAS method for wastewater and surface water in ISO 7875/1 and IRSA E-012.

2. Determination of Anionic Surfactants of Water Using the Ethyl Violet Method [Ethyl Violet Active Substances (EVAS)]

Ethyl violet (13) forms ion-pair complexes with various anionic surfactants that are extractable into organic solvents, in the same way as does methylene blue. Its ion-pair complexes are, however, more soluble in many organic solvents than those of methylene blue due to their higher distribution ratios. In this respect ethyl violet is more efficient and more sensitive in the analysis of anionic surfactants than is methylene blue. Extraction of the colored ion-pair complexes into an organic phase forms the analytical basis for the extraction–spectrophotometric analysis of anionic surfactants.

  a. Reagents and Apparatus.

1. Prepare a standard stock solution of sodium lauryl sulphate (100 $mg \cdot L^{-1}$ sodium lauryl sulphate) by dissolving 0.1 g of sodium lauryl sulphate (reagent pure) as described earlier. Prepare a range of standard solutions of sodium lauryl sulphate (0–2 $mg \cdot L^{-1}$) and blank in 100-ml volumetric flasks as described earlier, and transfer the standard solutions into 250-ml separation funnels.
2. Prepare 0.001 M ethyl violet solution by dissolving 0.049 g of ethyl violet in water in a 100-ml volumetric flask before diluting to mark.
3. Prepare 0.2 M acetic acid solution by diluting 11.56 ml of concentrated CH_3COOH (99% w/w; d, 1.049) in a 1-L volumetric flask to mark with water. Prepare 0.2 M sodium acetate solution by dissolving 27.216 g of $CH_3COONa \cdot 3H_2O$ in a quantity of water in a 1-L volumetric flask before diluting to mark. Prepare acetate buffer (pH 5) solution by mixing 12 ml of acetic acid solution and 28 ml of sodium acetate solution in a 50-ml flask that contains 5.68 g of Na_2SO_4 and 0.3 g of sodium EDTA.
4. Obtain 200 ml of toluene (reagent pure), a set of 50-ml separation funnels, a set of 50-ml volumetric flasks, a spectrophotometer or colorimeter, and 1-cm cuvette.

  b. Procedure. Transfer a known quantity v ml of water sample (ca. 50–200 ml) into a 250-ml separation funnel. Treat the standard solutions, blank, and water sample in 250-ml separation funnels the same way, as follows: Add 5 ml of the acetate buffer solution, 2 ml of the ethyl violet solution, and 5 ml of toluene to each of the 250-ml separation funnels. Carry out the extraction process by shaking the separation funnels vigorously for about 10 minutes. Allow the phases to settle, and drain each toluene extract off into a 10-ml volumetric flask and dilute to mark with toluene. Obtain the absorbances of the extracts at 615 nm against toluene blank. Prepare a calibration curve by plotting the absorbances of the extracts from standard solutions against their corresponding concentrations. Compare the absorbance of the extract from water sample with those of the calibration curve to obtain the concentration c $mg \cdot L^{-1}$ of anionic surfactants in water sample extract after taking the baseline value from the blank into consideration.

Concentration of anionic surfactants ($mg \cdot L^{-1}$ sodium lauryl sulphate) in water sample

$$= \frac{10 \text{ ml} \times c \text{ mg} \cdot^{-1}}{v \text{ ml}}$$

3. Determination of Anionic Surfactants of Water by Two-Phase Titration Using Benzethonium Chloride

Titrimetric methods are possible for the analysis of anionic surfactants because anionic surfactants are moderately acidic. Hence, using a suitable base, such as benzethonium chloride (14), and solvent, medium anionic surfactants can be analyzed. The major drawback in this mode of analyzing anion surfactants is interference from other acidic species in the sample that are not anionic surfactants and that could equally react directly with a basic titrant. This can be overcome by titrating with a cationic surfactant and the use of two nonmixable solvent phases. Thus, in this two-phase titration, anionic surfactants react with cationic surfactants in the aqueous phase to form a neutral species that is extractable into the organic phase. With the aid of a cationic and anionic dye added to the aqueous phase, the continued extraction of the neutral species into the organic phase can be monitored toward the endpoint. This is because before the endpoint, the anionic surfactant forms a colored species with the cationic dye, which is extracted into the organic phase, thereby coloring the organic phase. At the end point, the cationic surfactant (titrant) will displace the cationic dye from its association with the anionic surfactant so that the dye leaves the organic layer along with its color. Slight addition of the cationic surfactant beyond the endpoint will react with the anionic dye in the aqueous phase to form species of a different color, which is also extractable into the organic phase. Hence, a change in color of the organic layer is observed, indicating the endpoint. The two indicator dyes used in this type of titration are dimidium bromide (cationic) and disulfine blue (anionic), which impart a pink and a blue color to the organic layer, respectively.

a. Reagents and Apparatus.

1. Prepare a standard stock solution of sodium lauryl sulphate (2000 mg·$^{-1}$ sodium lauryl sulphate) by dissolving 2 g of sodium lauryl sulphate (reagent pure) in a small quantity of water in a 1-L volumetric flask before diluting to mark with water. Prepare a standard solution of sodium lauryl sulphate (1000 mg·L^{-1}) in 100-ml volumetric flasks by pipetting appropriate volumes of standard stock solution into 100-ml volumetric flasks and diluting to mark with water. Transfer the standard solution into a 250-ml conical flask.

2. Prepare 1000 mg·L^{-1} solution of benzethonium chloride by dissolving 0.1 g of benzethonium chloride in a small quantity of water in a 100-ml volumetric flask before diluting to mark with water. Standardize the benzethonium chloride solution against standard 1000 mg·L^{-1} sodium lauryl sulphate solution. Calculate the concentration of standard benzethonium chloride using the equation $M_1V_1 = M_2V_2$, expressing M in mg·L^{-1}.

3. Prepare a mixed indicator solution by dissolving 0.5 g of dimidium bromide and 0.25 g of disulfine blue in 100 ml of warm 10% aqueous ethanol in a 250-ml flask and diluting to mark with 10% aqueous ethanol. Introduce 4 ml of the solution in a 100-ml volumetric flask, add 10 ml of 1 M sulphuric acid solution, and dilute to mark with water.

4. Obtain 200 ml of dichloromethane (reagent pure), a set of 250-ml conical flasks, and a set of 100-ml graduated cylinders with glass stoppers.

b. Procedure. Reduce a known quantity v ml of water sample (ca. 0.5–2 L) to about 20–50 ml on a steam bath. Transfer to a 100-ml graduated cylinder, and neutralize to phenolphthalein with drops of dilute NaOH or H$_2$SO$_4$ solutions. Add 10 ml of dichloromethane and 10 ml of mixed indicator solution. The organic layer will become pink if the sample contains

anionic surfactant. Titrate by adding drops of the standard solution of benzethonium chloride from a burette, shaking the immiscible phases well each time a drop is added. The endpoint occurs when the pink color of the organic layer is discharged and a pale blue color is obtained. Calculate the concentration (M$_2$) of anionic surfactants (mg·L^{-1} sodium lauryl sulphate) in water sample using $M_1V_1 = M_2v$.

B. Cationic Surfactants

1. Determination of Cationic Surfactants of Water Using the Disulfine Blue Method [Disulfine Blue Active Substances (DBAS)]

Cationic surfactants are determined by this method (15) through formation of ion-pair complex species with anionic species. Water-soluble disulfine blue (15) dye is the anionic species used in this analysis. The dye is extracted from the aqueous phase into a chloroform solution of a cationic surfactant and the organic extract analyzed spectrophotometrically. The method suffers from negative interference from anionic surfactants and positive interference from cationic species that may also be present in the water sample. Interference from these substances can be removed by the inclusion of an ion-exchange step to isolate the cationic surfactants (16) from these species.

a. Reagents and Apparatus.

1. Prepare an anion-exchange column by putting a 50–100 mesh Cl$^-$ form of Bio-Rad AG1-X2 to 10-cm depth in a 1.3 × 17-cm glass column equipped with stopcock and reservoir.

2. Introduce a range of cetyltrimethylammonium bromide (0–1 mg) into a set of 100-ml volumetric flasks as the standard. One of the flasks, serving as the blank, should have 0 mg of cetyltrimethylammonium bromide and be subjected to the same analytical procedure as the rest.

3. Prepare disufine blue reagent by dissolving 0.064 g of disulfine blue in 100 ml of 10% aqueous ethanol solution.

4. Prepare acetate buffer by dissolving 11.5 g of anhydrous sodium acetate (NaCH$_3$COO) in 50 ml of water in a 100-ml volumetric flask, adjusting to pH 5 with glacial acetic acid before diluting to mark with water.

5. Obtain 1 L methanol, 200 ml chloroform, a set of 100-ml volumetric flasks, a set of 50-ml conical centrifuge tubes with stopcock, a centrifuge, a spectrophotometer or colorimeter, and 1-cm cuvette.

b. Procedure. The ion-exchange process is performed as follows: Precondition the anion exchanger by passing through the column 10 ml of 1 mg·L^{-1} of the cetyltrimethylammonium bromide at a rate of 2 ml·min^{-1} before washing by passing through the column 600 ml of methanol at the same rate. Obtain a known quantity v ml of water sample (ca. 1–5 L), reduce it to a manageable level through evaporation on a steam bath, and add about 150 ml methanol to it. Pass the solution through the glass column containing the anion exchanger at a flow rate of about 1–2 ml·min^{-1}, and collect the eluate in a beaker of suitable size until all the eluate has been collected. Elute the column at the same rate with 150 ml of methanol and collect the eluate with the same beaker. Evaporate the aqueous methanol in the eluate to dryness on a steam bath. Add 20 ml of chloroform to the residue to dissolve it, and transfer the solution to a 100-ml volumetric flask. Rinse the beaker with fresh quantity of chloroform and transfer rinsing to the same volumetric flask. Add about 20 ml of chloroform to each of the standard samples obtained in the preceding section on Reagents and Apparatus to make a solution. From this point, treat both the chloroform solutions of the standards and sample the same way. Dilute the sample and standard solutions to the 100-ml mark with chloroform, and shake to get a uniform solution. Introduce 2.5 ml of the acetate buffer, 1 ml of disulfine reagent, and 20 ml of water into a 50-ml centrifuge tube with stopcock. Add 10 ml of the chloroform solutions to each of the centrifuge tubes, cock, and shake vigorously for about 5 minutes to ensure complete extraction into the organic phase. Centrifuge the two phases to ensure proper separation of the phases. Allow the phases to stand for about 30 second before transferring some quantity of the lower layer of organic extract into a 1-cm cuvette using a pipette. Obtain the absorbance of each organic extract at 628 nm against chloroform. Prepare a calibration curve by plotting the absorbances of the standard solutions against the corresponding mass of surfactant. Compare the absorbance of organic extract from water sample with those of the calibration curve to obtain the mass c mg of cationic surfactants in the water sample after taking the baseline value from blank into consideration.

Concentration of cationic surfactants (mg·L^{-1}

cetyltrimethylammonium bromide) in water sample

$$= \frac{10^3 \text{ ml} \times c \text{ mg·L}^{-1}}{v \text{ ml}}$$

2. Determination of Cationic Surfactants of Water by Titration with Sodium Tetraphenylborate Solution

Most common cationic surfactants can be determined by the titration method, except that in most cases the pH of solution is critical and, thus, has to be controlled carefully to get good results. The basis of the two-phase titration method for cationic surfactants is that the cationic surfactants react with anionic surfactants (17) in the aqueous phase to form a species that is extractable into the organic phase. With the aid of visual indicators added to the system the continued extraction of the species into the organic phase can be monitored toward the endpoint. The endpoint is indicated by a change in color of the organic phase as the titration proceeds. The method is similar to the two-phase titration method described earlier for anionic surfactants; hence, anionic surfactants interfere seriously in this determination. Subsequently, an anion-exchange step may be incorporated to remove the interference of anionic surfactants.

a. Reagents and Apparatus.

1. Prepare an anion-exchange column as described in Sec. II.B.1.
2. Prepare 10 mg·L^{-1} solution of cetyltrimethylammonium bromide by dissolving 1 mg of dry cetyltrimethylammonium bromide in a quantity of water in a 100-ml volumetric flask before diluting to mark with water.
3. Prepare 15 mg·L^{-1} solution of sodium tetraphenylborate by dissolving 15 mg of dry sodium tetraphenylborate in a quantity of 0.001 M NaOH solution in a 1-L volumetric flask before diluting to mark with 0.001 M NaOH. Standardize the sodium tetraphenylborate solution against standard 10 mg·L^{-1} cetyltrimethylammonium bromide solution. Calculate the concentration of standardized sodium tetraphenylborate using the equation $M_1V_1 = M_2V_2$, expressing M in mg·L^{-1}.
4. Prepare an indicator solution by dissolving 0.2 g of potassium tetrabromophenolphthalein ethyl ester in 100 ml ethanol.
5. Prepare a phosphate buffer by bringing a 50-ml quantity of 0.3 M solution of Na$_2$HPO$_4$ to pH 6 with drops of 5 M sulphuric acid in a 100-ml flask.
6. Obtain 1 L of methanol, 200 ml of dichloromethane (reagent pure), and a set of 200-ml conical flasks.

b. Procedure. The ion-exchange process is performed as described in Sec. II.B.1, and a known quantity *v* ml of water sample (ca. 1–5 L) is treated as described in Sec. II.B.1. Add 20 ml of the phosphate buffer and a drop of indicator solution to the sample residue and shake. Introduce 2 ml of dichloromethane to the sample and shake. Titrate by adding drops of the standardized solution of sodium tetraphenylborate from a burette, shaking the immiscible phases properly each time a drop is added. Before the endpoint, the dichloromethane will appear blue, but it will turn green in the vicinity of the endpoint. Beyond the endpoint, it will turn yellow.

Calculate concentration (M_2) of cationic surfactants (mg·L^{-1} cetyltrimethylammonium bromide) in water sample using $M_1V_1 = M_2v$.

C. Nonionic Surfactants

1. Determination of Nonionic Surfactants of Water Using the Cobalt Thiocyanate Method (cobalt thiocyanate active substances (CTAS))

When nonionic surfactants react with ammonium cobaltothiocyanate (12) in the aqueous phase, colored ion association complex species are formed. The complexes are readily extractable into dichloromethane and analyzed spectrophotometrically. The method, however, suffers from interference of other forms of surfactants, particularly the cationic surfactant, which must be removed from the sample. Because it is not very sensitive when the surfactant is of low concentration, preconcentration of the sample may be carried out using an ion exchanger to increase the sensitivity and specificity. The ion-exchange process may also reduce the interference from other surfactants.

a. Reagents and Apparatus.

1. Prepare an ion-exchange column consisting of anion-exchange and cation-exchange resins by putting a 50–100-mesh OH$^-$ form of Bio-Rad AG1-X2 anion-exchange resin and 50–100 mesh-H$^+$ form of Bio-Rad 50W-X8 cation-exchange resin, each 10 cm depth, in a 1 × 30-cm glass column.
2. Introduce a range of nonylphenol decaglycol ether (0–2 mg) into a set of 100-ml extraction flasks as standards and evaporate to dryness on a steam bath. One of the beakers, serving as blank, should have 0 mg of nonylphenol decaglycol ether and be subjected to the same analytical procedure as the rest. Any other ethoxylate used as standard must have 6–50 ethylene oxide groups

for it to respond to analysis. Ethoxylates with four or fewer ethylene oxide groups do not react and, hence, cannot be analyzed.

3. Prepare the cobalt thiocyanate reagent by dissolving 30 g of $Co(NO_3)_2 \cdot 6H_2O$ and 200 g of NH_4SCN in a quantity of water in a 1-L volumetric flask before diluting to mark.
4. Obtain 600 ml of methanol, 200 ml of dichloromethane (CH_2Cl_2), a set of 100-ml separation funnels, 2 g of glass wool, a spectrophotometer or colorimeter, and 2-cm cuvette.

b. Procedure. The ion-exchange process is performed as follows: Pass a known quantity *v* ml of water sample (ca. 1–5 L) containing about 150 ml methanol through the glass column containing the mixed-bed ion exchanger at the rate of a few drops per second on the column, and collect the eluate in a beaker of suitable size until all the eluate has been collected. Elute the column with 150 ml of methanol and collect the eluate with the same beaker. Evaporate the aqueous methanol in the eluate to dryness on a steam bath. From this point, treat both the sample residue and those of the standards obtained in the preceding section on Reagents and Apparatus the same way. Add 5 ml of the cobalt thiocyanate reagent to each of a set of 100-ml separation funnels. Introduce 10 ml of CH_2Cl_2 to each of the extraction flasks or beaker containing the standards and sample residue, and swirl carefully so that the CH_2Cl_2 can pick up the residues. Avoid vigorous swirling so that CH_2Cl_2 is not lost through evaporation. Transfer the CH_2Cl_2 solutions, without rinsing, into the separation funnels and shake properly for about 2 minutes. Allow the two immiscible liquid phases to separate before running off the lower portion of the CH_2Cl_2 extract through a plug of wool moistened with CH_2Cl_2 into a 2-cm cuvette. Obtain the absorbances of the CH_2Cl_2 extracts at 620 nm against CH_2Cl_2 blank. Prepare a calibration curve by plotting the absorbances of the extracts from standard solutions against the corresponding mass of surfactant. Compare the absorbance of the extract from the water sample with those of the calibration curve to obtain the mass *c* mg of nonionic surfactants in the water sample, after taking the baseline value from blank into consideration.

Concentration of nonionic surfactants (mg·L^{-1} nonylphenol decaglycol ether) in water sample
$$= \frac{10^3 \text{ ml} \times c \text{ mg·L}^{-1}}{v \text{ ml}}$$

2. Determination of Nonionic Surfactants of Water Using the Barium Iodobismuthate Method [Bismuth Active Substances (BiAS)]

In this method, nonionic surfactants react with tetraiodobismuthate ion (modified Dragendorff reagent) (18) to form orange precipitates. The precipitate is isolated and then dissolved with ammonium tartrate solution, and the liberated bismuth ion in the complex is determined spectrophotometrically in the aqueous phase. Extraction of the complex from the aqueous solution to an organic phase is not necessary. The method suffers from interference of the presence of cationic surfactants.

a. Reagents and Apparatus.

1. Prepare an ion-exchange column as described in Sec. II.C.1.

2. Introduce a range of nonylphenol decaglycol ether (0–2 mg) into a set of 200-ml beakers as standards, and evaporate to dryness on a steam bath. One of the beakers should have 0 mg of nonylphenol decaglycol ether and be subjected to the same analytical procedure as the rest. Any other ethoxylate used as standard must have 6–50 ethylene oxide groups for it to respond to analysis. Ethoxylates with four or fewer ethylene oxide groups may not react, hence, may not be analyzed.

3. Prepare 1.7% barium iodobismuthate solution by dissolving 1.7 g of bismuth nitrate ($BiNO_3 \cdot H_2O$) in 100 ml of 20% acetic acid solution. Obtain 32.5% potassium iodate solution by dissolving 65 g of KIO_3 in 200 ml of water. Introduce the 100 ml of 1.7% barium iodobismuthate solution and the 200 ml of 32.5% potassium iodate solution into a 1-L volumetric flask containing 200 ml acetic acid, and dilute to mark with water to get solution (a).

4. Prepare 29% barium chloride solutin by dissolving 29 g of $BaCl_2 \cdot 2H_2O$ in a small amount of water in a 100-ml volumetric flask before diluting to mark to get solution (b).

5. Prepare the precipitation reagent by mixing 200 ml of solution (a) and 100 ml of solution (b) in an amber flask and store in a dark place. Solution should be prepared weekly.

6. Prepare ammonium tartrate solution by dissolving 1.8 g of ammonium tartrate (($NH_4)_2C_4H_4O_6$) in a quantity of water in a 100-ml volumetric flask before diluting to mark with water.

7. Prepare EDTA solution (0.02 M) by dissolving 0.744 g of ethylenediaminetetraacetic acid disodium dihydrate (($NaO_2CCH_2N(CH_2CO_2H) CH_2$-$CH_2N(CH_2CO_2Na)CH_2CO_2H.2H_2O$) in a small quantity of water in a 100-ml volumetric flask before diluting to mark.

8. Prepare 0.1% bromcresol purple indicator by dissolving 0.1 g of bromcresol purple in a small quantity of methanol in a 100-ml volumetric flask before diluting to mark with methanol.

9. Obtain a set of 100-ml volumetric flasks, a set of 50-ml filtering crucibles, a spectrophotometer, and a 2-cm cuvette.

b. Procedure. The ion-exchange process is performed as described in Sec. II.C.1. A known quantity *v* ml of water sample (ca. 1–5 L) is similarly treated, as described under Sec. II.C.1. From this point, treat the sample residue and those of the standards obtained in the preceding section on Reagents and Apparatus the same way. Add about 50 ml of water to each of the standards and sample and add three drops of bromcresol purple indicator solution. Stir the solutions before adjusting to get a yellow color with 0.2 M HCl. Introduce 30 ml of the precipitation reagent to each of the solutions with stirring, and allow the solutions to stand for about 10 minutes for the precipitates to separate from solution. Arrange a sintered-glass filtering crucible and filter the solutions through the crucible under suction. Wash each residue with 10 ml of glacial acetic acid. Dissolve the precipitate from each solution by pouring 50 ml of hot ammonium tartrate solution through the crucible in aliquots of 5 ml, followed by 20 ml of water poured in aliquots of 5 ml using suction, and collect the resultant solution with the original 200-ml beaker. Add 5 ml of EDTA solution to each of the solutions, stir to ensure proper mixture before transferring to a 100-ml volumetric flask. Bring each of the 100-ml volumetric flasks to mark with water, and measure the absorbance of each solution with a 2-cm cuvette at 264 nm against water. Prepare a calibration curve by plotting the absorbances of the standard solutions against the corresponding mass of surfactant. Compare the absorbance of that of the water sample with those of the calibration curve to obtain the mass *c* mg of nonionic surfactants in the water sample after taking the baseline value from blank into consideration.

Concentration of nonionic surfactants ($mg \cdot L^{-1}$ nonylphenol decaglycol ether) in water sample

$$= \frac{10^3 \text{ ml} \times c \text{ mg} \cdot L^{-1}}{v \text{ ml}}$$

Zanette et al. (105) discuss the investigation of the BiAS method for wastewater and surface water in ISO 7875/2 and IRSA E-013.

3. Determination of Nonionic Surfactants of Water by Titration with Sodium Pyrrolidinecarbodithioate Solution

The basis of this method of analysis is that the precipitate that is formed from the interaction between nonionic surfactants and tetraiodobismuthate ion is dissolved with ammonium tartrate solution and the liberated bismuth ion in the complex determined tetrimetrically (19) rather than spectrophotometrically. The method can be used to analyze samples containing $mg \cdot L^{-1}$ levels of nonionic surfactants and has been found suitable for the analysis of most types of nonionic surfactants.

a.   Reagents and Apparatus.

1. Prepare an ion-exchange column as described in Sec. II.C.1.
2. Introduce a range of nonylphenol decaglycol ether (0–1 mg) into a set of 200-ml beakers as standards, and evaporate to dryness on a steam bath. One of the beakers should have 0 mg of nonylphenol decaglycol ether and be subjected to the same analytical procedure as the rest. Any other ethoxylate used as standard must have 6–50 ethylene oxide groups for it to respond to analysis. Ethoxylates with four or fewer ethylene oxide groups may not react; hence, they may not be analyzed.
3. Prepare the precipitation reagent as described in Sec II.C.2.
4. Prepare ammonium tartrate solution by dissolving 1.24 g of tartaric acid ($C_4H_6O_6$) in 1.8 ml of concentrated ammonium hydroxide in a 100-ml volumetric flask before diluting to mark with water.
5. Prepare acetate buffer solution by carefully adding 4 g of sodium hydroxide to 12 ml of glacial acetic acid in a 100-ml volumetric flask before diluting to mark with water.
6. Prepare 0.001 M sodium pyrrolidinecarbodithoate solution by dissolving 0.206 g of sodium pyrrolidinecarbodithoate in a small quantity of distilled water in a 1-L volumetric flask, add 10 ml n-amyl alcohol and 0.5 g of sodium bicarbonate before diluting to mark with distilled water. Solution should be prepared weekly.
7. Prepare 0.1% bromcresol purple indicator by dissolving 0.1 g of bromcresol purple in a small quantity of methanol in a 100-ml volumetric flask before diluting to mark with methanol.
8. Obtain a potentiometer with platinum/calomel or platinum/silver chloride electrodes with a 250-mV measuring range, a 50-ml burette flasks, and a magnetic stirrer.

b.   Procedure.   The ion-exchange process and treatment of known quantity v ml of water sample (ca. 1–5 L) are performed as described in Sec. II.C.1. From this point, treat both the sample residue and those of the standards obtained in the preceding section on Reagents and Apparatus the same way. Add about 50 ml of water to each of the standards and sample, and add 3–5 drops of bromcresol purple indicator solution. Stir the solutions before adjusting to get a yellow color with 0.2 M HCl. Introduce 30 ml of the precipitation reagent to each of the solutions with stirring, and allow the solutions to stand for about 10 minutes for the precipitates to separate from solution. Arrange a sintered-glass filtering crucible and filter the solutions through the crucible under suction. Wash each residue with 10 ml of glacial acetic acid. Dissolve the precipitate from each solution by pouring 50 ml of hot ammonium tartrate solution through the crucible in aliquots of 5 ml. Rinse with aliquots of ammonium tartrate solution followed by 20 ml of water poured in aliquots of 5 ml using suction, and collect the resultant solution with the original beaker. Ensure that the total volume of solution is between 50 ml and 200 ml. Add 3–5 drops of bromcresol purple indicator solution to the solutions and stir. Add drops of 1% ammonium hydroxide solution until the color of solution changes to violet. Bring the solution to pH 5 by adding 10 ml of the acetate buffer. Set the potentiometer, immerse the electrodes into the solution, and titrate with the sodium pyrrolidinecarbodithoate solution at a rate of 2 $ml \cdot min^{-1}$ beyond the potential jump. Determine the volume of titrant at the equivalence point of the potential curve of each titration using the extrapolation method. Prepare a calibration curve by plotting the volume at the equivalence point against the corresponding standard mass of surfactant. Compare the volume at the equivalence point for the water sample with those of the calibration curve to obtain the mass c mg of nonionic surfactants in water sample after taking the baseline value from blank into consideration.

Table 1 Photometric Methods for Surfactant Analysis

Analyte	Type of water	Method	Detection limit	Ref.
Anionic surfactants		Spectroscopy, N-alkylnaphtylazopyridium salts	$\pm 10^{-6}$ M	29
Nonionic surfactants	Seawater	Spectroscopy, 380 nm	0.1 mg/L	31
Anionic SDS		Spectroscopy, flow injection extraction	0.03 ppm (with a 65-μl injection)	30
Anionic surfactants		Spectrophotometry, spectrofluorometry, FIA coupled with LLE, Rhodamine B and 4-[[4-(dimethylamino) phenyl]azo]-2-methylquinoline		33
Anionic surfactants		Spectrophotometry, ethyl violet		34
Anionic surfactants		Spectrophotometry, 4-[[4-(dimethylamino) phenyl]azol]-2-methylquinoline		35
Anionic surfactants	Water	Adsorptive preconcentration, photometric detection	2 μg/L	66
Anionic surfactants	Wastewater	Spectrophotometry at 650 nm	0.02 mg/L	68
Cationic surfactants	River water, seawater	In water samples containing anionic surfactants, spectrophotometry at 610 nm; tetrabromophenolphtalein ethyl ester dye		72
Anionic surfactants	River water	Spectrophotometry at 590 nm, bromophenol blue, and hexadecylpyridinium chloride		77
Nonionic surfactants	Natural water, wastewater	Photometry at 680 nm, molybdophosphoric acid, pyrocatechol violet 0.05%, dodecylpyridinium bromide 0.15%	10–20 μg/L	79
Anionic surfactants	Water	FIA, continuous LLE with online monitoring, spectrophotometry, methylene blue	8 ng/ml	78
Anionic surfactants, SDS, sodium dodecylbenzene sulfonate (SDBS)	Water	Spectrophotometry at 555 nm, complex of iron(II)-3-(2-pyridyl)-5,6-diphenyl-1,2,4-triazine	SDS: 4.3 μg/L SDBS: 5.2 μg/L	83
Anionic surfactants	Tap water, river water, pond water, wastewater	FIA, fluorimetry at 550 nm, rhodamine 6G		88
Anionic surfactants	Wastewater	Dual-wavelength β-correction spectrophotometry (550 nm/640 nm), ethyl violet	0.01 mg/L	94
Anionic surfactants	Seawater	Spectrometry at 560 nm, ion pair with bis[2-(5-diethylamino)-phenolato]cobalt (III) counter ion	0.005 mg/L	104
Anionic surfactants	River water, pond water	Spectrophotometry at 614 nm, dodecyldimethylbenzylammonium bromide, bromothymol blue		112
Anionic surfactants	Water	Spectrophotometry at 630 nm, (p-nitrophenylazo)resorcinol (PNBAR)-cetyltrimethyl-ammonium bromide (CTMAB)		113
Anionic surfactants	River water, groundwater, wastewater	FIA, detection at 625 nm, 0.002% brilliant green in 0.04% ethanol	0.1 ppm	117
Anionic surfactants	Wastewater	FIA, methylene blue method, detection at 652 nm		119

Table 2   Chromatographic Methods for Surfactant Analysis

Analyte	Type of water	Method	Detection limit	Ref
Heptaethylene glycol, dodecyl ether	Wastewater	HPLC, fluorescence detection		37
Aliphatic ionic	Wastewater	HPLC, fluorescence detection		38
Nonylphenol ethoxylate	Wastewater	HPLC, fluorescence detection		40
Alkylbenzenesulfonates	Tap (1) and river (2) water	Anion-exchange HPLC, UV detection	± 01 μg/L (1), 100 μg/L (2)	39
Alcohol ethoxylates		TSP LC/MS		41
Nonionic surfactants	Wastewater	FIA MS/MS and TSP LC MS/MS SIM		42
Alkyl sulfates, alkyl ethoxysulfates	Wastewater, river water	Ion spray LC/MS	<1 ppb	43
Anionic and nonionic tensides		FIA TSP LC/MS		44
Secondary alkanesulfonates		GC/MS, concentration on SPE disk, injection port, derivatization		46
LAS		HPLC, SPE, fluorometric detection	0.8 ppb	50
LAS		RP HPLC, UV detection		51
Cationic surfactants		HPLC, conductivity detection	16 μg/L for ditallowimidazolimium methosulfonate, 16 μg/L for distearyldimethylammonium chloride, 6 μg/L dodecylpyridinium chloride	52
Alkyl sulfate surfactants	Wastewater and water	GC-FID, RP extraction, anion and cation exchange Derivatization: N,O-bis(trimethylsilyl)-trifluoramide (+1% trichloromethylsilane)		53
LAS		GC-ECD Derivatization: N-methylanilide	0.1 ng	54
Surfactants	River water	SPE LC-MS, fluorescence, MS CI interference	Alkylphenol ethoxylates: 0.05 μg/L; alkylbenzene sulfonate: 0.005 μg/L	64
Alkylphenol ethoxylate surfactants	Water	SPME HPLC, detection at 220 nm, GC (after derivatization), ion trap MS	1.57 μg/L	67
Nonionic polyethoxylate, aliphatic ethoxylate alcohols (AEO), nonphenol polyethoxylates (NPEO), LAS	Raw water, treated water, river water, drinking water	LC-electrospray MS; MS: positive-ion mode	River water: 0.002 μg/L; drinking water: 0.0002 μg/L	74
Alcohol ethoxylate surfactants	Wastewater	HPLC-electrospray MS	10 ppb	84

Table 2 Continued

Analyte	Type of water	Method	Detection limit	Ref
Nonionic surfactants	Waste water, river water	HPLC-APCI MS, UV detection, tandem MS	1 μg/L	87
Nonionic aliphatic and aromatic polyethoxylated surfactants	Natural waters, wastewater	RP-HPLC, fluorometric detection	0.1 μg/L	89
Nonionic surfactants		High-temperature CGC		91
Alkyl ethoxylate surfactants	River water, wastewater	GC, MS single-ion monitoring mode with EI ionization		92
Aromatic surfactants, biodegradation, intermediates, nonylphenolpoly-ethoxylates (NPEO), LAS	Sewage	SPE, LC, fluorimetry at 295 nm		93
Nonylphenol ethoxylate surfactants	Wastewater	HPLC, detection at 229 nm	Total surfactants: 40 μg/L	97
Nonionic surfactants		HPLC, fluorescence, ELSD		101
Nonionic surfactants	Trade effluents	HPLC, fluorometric detection	0.1 mg/L	109
Ethoxylated nonionic surfactants	Wastewater	HPLC	3 μg/L	110
Nonionic surfactants	Water	HPLC, ELSD, UV fluorescence		120
Propoxylated alcoholic surfactants	Wastewater	TSP HPLC/MS		57

Table 3 Various Methods for Determination of Surfactants in Water

Analyte	Type of water	Method	Detection limit	Ref.
Corexit 9527	Natural waters	Flame AAS, extraction in methyl isobutyl ketone and bis(ethylenediamine)copper(II)		32
Poly(ethylene)glycol residues	Surface water	^{1}H and ^{13}C NMR		36
Nonionic, alcohol ethoxylates		^{1}H NMR	5 μg/L	48
Surfactants	Raw and drinking waters	FAB MS, FAB MS/MS		45
Nonionic	Tap water	Desorption chemical ionization MS		47
Alcoholic phenol ethoxylates		SFE of SPE disks, SFC		49
LAS		FAB MS/MS	0.5 μg/L	55

Table 3 Continued

Analyte	Type of water	Method	Detection limit	Ref.
[(Alkyloxy)polyethoxy]-carboxylates	Raw and finished drinking waters	FAB MS/MS		56
Surfactants		Flow injection system		58
Nonionic	Wastewater	Polarography, molybdophosphoric acid complexes		59
Surfactants		Polarography	0.01 mg/dm^3	60
Nonionic surfactants	Effluent	Solid-state electrodes		69
Nonionic surfactants	River water, lake water, underground water	AAS, reaction with Mo-complex	0.01 mg/L	70
Surfactants	River water	SPE, Amberlit XAD-4/RP 18 extraction		73
Anionic surfactants	Standard methods for the analysis of anionic surfactants as methylene-blue-active substances modified method			75
Nonionic surfactants	Effluent	Tensommetry, BiAS separation procedure		76
LAS	Wastewater	CE	1 mg/l	81
Anionic surfactants	River water	FIA, detection at 662 nm, malachite green	18 ppb	82
Nonionic surfactants	Water	FTIR spectrometry	0.25 mg/L	85
Anionic surfactants	Standard method	Methylene blue index MBAS		96
Surfactants, fluorotensides		ISE, fluoride determination	1 μM fluoride	98
Nonionic surfactant	Surface water	Tensammetry		99
Surfactants		Tensammetry, mercury electrode		100
Surfactants	SFE extraction			102
Anionic surfactants		FIA, ISE		103
Nonionic and anionic surfactants	Water	Alternating-current voltammetry	0.1–1 μg/L	107
Surfactants	Seawater	Voltammetry, calibration with model substances		114
SDS, total surfactant concentration, equivalent concentration of SDS	Seawater	Voltammetry		115
Anionic surfactants	Natural water, tap water, wastewater	Removal of dyes by SDS from the silica gel, spectrophotometry, diffuse reflectance spectrometry, luminescence	10–50 ppb SDS	116
Nonionic surfactants	Surface waters	Indirect tensammetry	Nonfiltered water: 6 μg/L	118
Cationic surfactants	Lake water, river water, seawater	ISFET devices		123

Concentration of nonionic surfactants ($mg \cdot L^{-1}$)

nonylphenol decaglycol ether) in water sample

$$= \frac{10^3 \text{ ml} \times c \text{ mg} \cdot L^{-1}}{v \text{ ml}}$$

III. RECENT METHODS

In recent years, many reviews on surfactants have been written (23–28). Aboul-Kassim and Simoneit (20) review the chemical composition and determination of surfactants in seawater. Gonzalez-Mora and Gomez-Para (122) discuss anionic surfactants, LAS, and degradation products in the same water type.

Preconcentration and extraction by SPE are evaluated by Kloster et al. (22). Chromatographic analysis methods (HPLC, GC, and SFC) are treated by Marcomini and Zanette (62), Kiewiet and de Voogt (63), and Miszkiewicz and Szymanowski (111).

Reviews have been published on different analysis techniques: MS characterization (21), electrochemical detection (61), LC-MS (95), FTIR spectroscopy (90), ISE (108), and FIA (121).

Lukaszewski and Szymanski (65) evaluate sources of error in the analysis methods for nonionic surfactants in water. Szymanski et al. (80) study preservation parameters of nonionic surfactants. Of chloroform, formaldehyde, Cu(II), and Hg(II), formaldehyde is the most effective in a concentration of 1% for long-term storage or a concentration of 0.1% for short-term storage (≤ 6 days).

For recent photometric methods, see Table 1; for chromatographic methods, see Table 2. In Table 3, various methods are given.

ABBREVIATIONS

AAS	atomic absorption spectroscopy
AEO	aliphatic ethoxylate alcohols
APCI	atmospheric pressure chemical ionization
BiAS	bismuth active substance
CE	capillary electrophoresis
CGC	capillary gas chromatography
CTAS	cobalt thiocyanate active substances
DBAS	disulfine blue active substances
ECD	electron capture detector
EI	electron impact
ELSD	evaporative light-scattering detector
EVAS	ethyl violet active substances
FAB	fast atom bombardment

FIA	flow injection analysis
FID	flame ionization detection
FTIR	Fourier transform infrared
GC	gas chromatography
HPLC	high-performance liquid chromatography
IRSA	Italian CNR Water Research Institute
ISE	ion-selective electrode
ISFET	ion-sensitive field-effect transistors
ISO	International Organization for Standardization
LAS	linear alkylbenzenesulfonates
LC	liquid chromatography
LLE	liquid–liquid extraction
MBAS	methylene blue active substances
MS	mass spectrometry
NMR	nuclear magnetic resonance
NPEO	nonphenol polyethoxylates
RP	reversed phase
SDBS	sodiumdodecylbenzenesulfonate
SDS	sodiumdodecylsulfate
SFC	supercritical fluid chromatography
SFE	supercritical fluid extraction
SIM	single ion monitoring
SPE	solid-phase extraction
SPME	solid-phase microextraction
TSP	thermospray
UV	ultraviolet

REFERENCES

1. M. F. Cox, T. P. Matson, J. L. Berna, A. Moreno, S. Kawakami, M. Suzuki, J. Am. Oil Chem. Soc. 61:330, 1984.
2. R. B. Drotman. Toxicol. Appl. Pharmacol. 52:38, 1977.
3. J. G. Black, D. Howes. Anionic Surfactant Biochemistry, Toxicology, Dermatology, Surfactant Science Series, 10:51. Marcel Dekker, New York, 1981.
4. R. B. Drotman. Cutaneous Toxicity. Academic Press, New York, 1977, p. 96.
5. B. Isomaa. Food Cosmet. Toxicol. 13:231, 1975.
6. W. Kästner. Anionic Surfactant Biochemistry. Toxicology, Dermatology, Surfactant Science Series, 10: 139. Marcel Dekker, New York 1981.
7. K. Oba. Anionic Surfactant Biochemistry, Toxicology, Dermatology, Surfactant Science Series, 10:327. Marcel Dekker, New York, 1981.
8. H. Andree, P. Krings. Chem. Ztg. 99:168, 1975.
9. W. M. Linefield. Surfactant Science Series, Cationic Surfactants, 4:49. Marcel Dekker, New York, 1970.
10. K. Bräuer, H. Fehr, R. Puchta. Tenside Deterg. 17:281, 1980.
11. M. J. Schwuger. Fette Seifen Anstrichm. 72:565, 1970.

12. American Public Health Association. Standard Methods for the Examination of Water and Wastewater. 17th ed. APHA, Washington, DC, 1989, Section 5540.

13. S. Montomizu, S. Fujiwara, A. Fujiwara, K. Toei. Anal. Chem. 54:392, 1982.

14. V. M. Reid, G. F. Longman, E. Heinerth. Tenside 4: 292, 1967; 5:90, 1968.

15. J. Waters, W. Kupfer. Anal. Chim. Acta 85:241, 1976.

16. Q. W. Osburn. J. Am. Oil Chem. Soc. 59:453, 1982.

17. M. Tsubouchi, H. Mitsushio, N. Yamasaki. Anal. Chem. 53:1957, 1981.

18. J. Waters, G. F. Longman. Anal. Chim. Acta 93:341, 1977.

19. R. Wickbold. Tenside 9:173, 1972; 10:179, 1973.

20. T. A. Aboul-Kassim, B. R. T. Simoneit. Crit. Rev. Environm. Sci. Technol. 23:325–376, 1993.

21. F. Ventura. Tech. Instrum. Anal. Chem. 13:481–520, 1993.

22. G. Kloster, M. Schoester, H. Prast. Tenside, Surfactants, Deterg. 31:23–28, 1994.

23. A. Marcomini. Riv. Ital. Sostanze Grasse. 68:339–344, 1991.

24. W. Huber. Tenside, Surfactants, Deterg. 28:106–110, 1991.

25. W. Kalbfus. Muench. Beitr. Abwasser-, Fisch.-Flussbiol. 44:195–204, 1990.

26. E. Matthijs, E. C. Hennes. Tenside, Surfactants, Deterg. 28:22–27, 1991.

27. T. Sakomoto. Jpn. J. Toxicol. 3:223–230, 1990.

28. L. Huber, H. Wagner. Beitr. Abwasser-, Fisch.-Flussbiol. 45:336–353, 1991.

29. Y. Shimoishi, H. Miyata. Fresenius J. Anal. Chem. 345:456–461, 1993.

30. H. Liu, P. K. Dasgupta. Anal. Chim. Acta 288:237–245, 1994.

31. A. A. Boyd-Boland, J. M. Eckert. Anal. Chim. Acta 27:311–314, 1993.

32. G. M. Scelfo, R. S. Tjeerdema. Mar. Environ. Res. 31: 69–78, 1991.

33. S. Motomizu, M. Kobayashi. Anal. Chim. Acta 261: 471–475, 1992.

34. D. Gorenc, B. Adam, B. Gorenc. Mikrochim. Acta 1: 311–315, 1991.

35. H. Kubota, M. Katsuki, S. Motomizu. Anal. Sci. 6: 705–709, 1990.

36. J. A. Leenheer, R. L. Wershaw, P. A. Brown, T. I. Noyes. Environ. Sci. Technol. 25:161–168, 1991.

37. I. Fujita, K. Nishiyama, Y. Nagano, K. Harada, M. Nakayama, A. Sugii. Int. J. Environ. Anal. Chem. 56: 57–62, 1994.

38. G. Kloster, M. Schoester, M. J. Schwuger. Commun. Jorn. Com. Esp. Deterg. 24:25–33, 1993.

39. Y. Yokoyama, M. Kondo, H. Sato. J. Chromatogr. 643: 169–172, 1993.

40. M. J. Scarlett, J. A. Fisher, H. Zhang, M. Ronan. Water Res. 28:2109–2116, 1994.

41. K. A. Evans, S. T. Dubey, L. Kravetz, I. Dzidic, J. Gumulka, R. Mueller, J. R. Stork. Anal. Chem. 66: 699–705, 1994.

42. H. F. Schroeder. J. Chromatogr. 647:219–234, 1993.

43. D. D. Popenoe, S. J. Moris, P. S. Horn, K. T. Norwood. Anal. Chem. 66:1620–1629, 1994.

44. H. F. Schroeder. Vom Wasser 79:193–209. 1992.

45. F. Ventura, I. Caixach, J. Romero, I. Espadaler, J. Rivera. Water Sci. Technol. 25:257–264, 1992.

46. J. A. Filed, T. M. Filed, T. Poiger, W. Giger. Environ. Sci. Technol. 28:497–503, 1994.

47. M. Vincenti, C. Minero, E. Pelizzetti. Ann. Chim. 83: 381–396, 1993.

48. L. Cavalli, A. Gellera, G. Cassani, M. Lazzrin, C. Maraschin, G. Nucci. Riv. Ital. Sostanze Grasse 70: 447–452, 1993.

49. M. Kane, J. R. Dean, S. M. Hitchen, C. J. Dwole, R. Tranter. Anal. Proc. 30:399–400, 1993.

50. A. Di Corcia, M. Marchetti, R. Samperi, A. Marcomini. Anal. Chem. 63:1179–1182, 1991.

51. Y. Yokoyama, H. Sato. J. Chromatogr. 555:155–162, 1991.

52. L. Nitschke, R. Mueller, G. Mezner, L. Huber. Fresenius J. Anal. Chem. 342:711–713, 1992.

53. N. J. Fendlinger, W. M. Begley, D. C. McAvoy, W. S. Eckhoff. Environ. Sci. Technol. 26:2493–2498, 1992.

54. H. Kataoka, N. Muroi, M. Makita. Anal. Sci. 7:585–588, 1991.

55. A. J. Borgerding, R. A. Hites. Anal. Chem. 64:1449–1454, 1992.

56. F. Ventura, D. Fraisse, J. Caixach, J. Rivera. Anal. Chem. 63:2095–2099, 1991.

57. H. F. Schroeder. DVGW—Schrifleur., Wasser. 108: 121–144, 1990.

58. H. Sawamoto, K. Gamoh. Anal. Sci. 7:1707–1709, 1991.

59. I. Zjawiony, Z. Klima. Z. Chem. Anal. 36:741–747, 1991.

60. E. Bednarkiewicz. Electroanalysis 3:839–845, 1991.

61. M. Gerlache, J. M. Kauffman, G. Quarin, J. C. Vire, G. A. Brijant, J. M. Talbot. Talanta 43:507–519, 1996.

62. A. Marcomini, M. Zanette. J. Chromatogr. 733:193–206, 1996.

63. A. T. Kiewiet, P. de Voogt. J. Chromatogr. 733:185–192, 1996.

64. S. D. Scullion, M. R. Clench, M. Cooke, A. E. Ashcroft. J. Chromatogr. 733:207–216, 1996.

65. Z. Lukaszewski, A. Szymanski, Mikrochim. Acta 123: 185–196, 1996.

66. L. N. Moskvin, D. N. Nikolaeva, N. V. Mikhailova. Zh. Anal. Khim. 51:304–307, 1996.

67. A. A. Boyd-Boland, J. B. Pawliszyn. Anal. Chem. 68: 1521–1529, 1996.

68. L. N. Moskvin, J. Simon, P. Loeffer, N. V. Michailova, D. N. Nicolaevna. Talanta 43:819–824, 1996.

69. R. K. Chernova, A. I. Kulapin, L. A. Yurova. Zh. Anal. Khim. 50:855–858, 1995.

70. Z. G. Chen, X. Chen. Fenxi Ceshi Xuebao 17:75–77, 1998.
71. J. Cross (ed.). Anionic Surfactants: Analytical Chemistry. 2nd ed. Marcel Dekker, New York, 1998.
72. E. Nakamura, A. Inove, M. Okubo, H. Namiki. Bunseki Kagaku 47:141–144, 1998.
73. A. Kreisselmeier, M. Schoester, G. Kloster. Fresenius J. Anal. Chem. 353:109–111, 1995.
74. C. Crescenzi, A. Di Corcia, R. Samperi, A. Marcomini. Anal. Chem. 67:1797–1804, 1995.
75. S. Chitikela, S. K. Dentel, H. E. Allen. Analyst 20:2001–2004, 1995.
76. B. Wyrwas, A. Szymanski, Z. Lukaszewski. Talanta 41:1529–1535, 1994.
77. H. J. Fan, Y. Xiong, G. C. Hu. Fenxi Huaxue 22:1051–1053, 1994.
78. M. Agudo, A. Rios, M. Valcarcel. Analyst 119:2097–2100, 1994.
79. L. V. Bolva, I. N. Urazova, Y. Y. Vinnikov. Zh. Anal. Khim. 49:381–384, 1994.
80. A. Szymanski, Z. Swit, Z. Lukaszewski. Anal. Chim. Acta 311:31–36, 1995.
81. K. Heinig, C. Vogt, G. Werner. Analyst 123:349–353, 1998.
82. T. Sakai, H. Harada, X. Q. Liu, N. Ura, K. Takeyoshi, K. Sugimoto. Talanta 45:543–548, 1998.
83. Y. Chen, S. Y. Wang, R. F. Wu, D. Y. Qi, T. Z. Zhou. Anal. Lett. 31:691–701, 1998.
84. K. A. Evans, S. T. Dubey, L. Kravetz, S. W. Evetts, I. Dzidic, C. C. Dooyema. J. Am. Oil Chem. Soc. 74:765–773, 1997.
85. A. Jelev, M. F. Ciobanu, L. Frunza. Spectosc. Lett. 30:1149–1154, 1997.
86. A. Marcomini, G. Pojana. Analusis 25:M35–M37, 1997.
87. T. Yamagishi, S. Hashimoto, M. Kanai, A. Otsuki. Bunseki Kagaku 46:537–547, 1997.
88. J. Y. Gao, Z. M. Zhang, Z. H. Zhu. Fenxi Huaxue 26:568–570, 1998.
89. A. Marcomini, G. Pojana, L. Patrolecco, S. Capri. Analusis 26:64–69, 1998.
90. W. Wang, L. M. Li, S. Q. Xi. Fenxi Huaxue 22:1273–1281, 1994.
91. W. C. Brumbey, W. J. Jones, A. H. Grange. LC-GC 13:228–238, 1995.
92. N. J. Fendenger, W. M. Begley, D. C. McAvory, W. S. Eckhoff. Environ. Sci. Technol. 29:856–863, 1995.
93. A. DiCorcia, R. Samperi, A. Marcomini. Environ. Sci. Technol. 28:850–858, 1994.
94. H. W. Gao. Anal. Proc. 32:197–200, 1995.
95. M. Carei, P. Manini, M. Maspero. Ann. Chim. 84:475–508, 1994.
96. Water Quality—Determination of anionic surfactants by measurement of the methylene blue index MBAS (methylene blue active substances). British Standards Institution. BS 6068: Section 2.23:1994, 1994.
97. J. A. Fisher, H. Zhang, M. Roman. Water Res. 28:2109–2116, 1994.
98. H. Fritsche, S. H. Huettenhain. Chemosphere 29:1797–1801, 1994.
99. A. Szymanski, B. Wyrwas, Z. Lukaszewski. Anal. Chim. Acta 305:256–264, 1995.
100. H. Lohse. Anal. Chim. Acta 305:269–272, 1995.
101. S. T. Dubey, L. Kravetz, J. P. Salanitro. J. Am. Oil Chem. Soc. 72:23–30, 1995.
102. M. Kane, J. R. Dean, S. H. Hitchen, C. J. Dowbe, R. L. Trantes. Analyst 20:355–359, 1995.
103. J. Alonso, J. Baro, J. Bartrolli, J. Sanchez, M. del Volle. Anal. Chim. Acta 308:115–121, 1995.
104. I. Kasahara, K. Hashimoto, T. Kawabe, A. Kunita, K. Magawa, N. Hataz, S. Taguchi, K. Goto. Analyst 120:1803–1807, 1995.
105. M. Zanette, A. Marcomini, S. Capri, A. Liberatori. Ann. Chim. 85:201–220, 1995.
106. M. Zanette, A. Marcomini, S. Capri, A. Liberatori. Ann. Chim. 85:221–233, 1995.
107. S. Sander, G. Henze. Electroanalysis 9:243–246, 1997.
108. Laborpraxis 19:35, 1995.
109. N. H. A. Ibrahim, B. B. Wheals. Analyst 121:239–142, 1996.
110. A. T. Kiewiet, J. M. D. van der Steen, J. R. Parsons. Anal. Chem. 67:4409–4415, 1995.
111. W. Miszkiewicz, J. Szymanowski. Crit. Rev. Anal. Chem. 25:203–246, 1996.
112. Y. S. Wang, G. R. Li, C. X. Liu, C. Y. Lu. Henxi Shiyanshi 15:53–55, 1996.
113. W. C. Yang, M. Zhang. Fenxi Shiyanshi 15:67–69, 1996.
114. B. Cosovic, V. Vojvodic. Electroanalysis 10:429–434, 1998.
115. L. Novotny, T. Navratil. Electroanalysis 10:557–561, 1998.
116. O. A. Zaporozhets, O. Y. Nadzhafova, V. V. Verba, S. A. Dolenko, T. Y. Keda, V. V. Sukhan. Analyst 123:1583–1586, 1998.
117. R. Patel, K. S. Patel. Analyst 123:1691–1695, 1998.
118. B. Wyrwas, A. Szymanski, Z. Lukaszewski. Anal. Chim. Acta 331:131–139, 1996.
119. S. H. Fan, Z. L. Fang. Fresenius J. Anal. Chem. 357:416–419, 1997.
120. T. C. G. Kibbey, T. P. Yavaski, K. F. Hayes. J. Chromatogr. 752:155–165, 1996.
121. T. Masadone, T. Imato. J. Flow Injection Anal. 13:120–136, 1996.
122. E. Gonzalez-Maro, A. Gomez-Parra. Trends Anal. Chem. 15:375–380, 1996.
123. L. Campanella, L. Arello, C. Colapicchioni, M. Tomassetti. Anal. Lett. 30:1611–1629, 1997.

37

Instruments and Techniques

A-M Siouffi

University of Aix-Marseille, Marseille, France

From a glance at textbooks on water analysis and more generally on analysis it is obvious that techniques are more and more instrumentalized and more and more automated. Simple, reliable instruments that are easy to handle are readily available from manufacturers. There is a clear trend towards miniaturized analytical techniques driven by the increasing needs for higher sensitivity and efficiency in the analysis of samples that keep getting smaller in volume and lower in concentration.

The attention of analytical chemists has been focused on the necessity to demonstrate the validity of analytical data and to develop systems that will enable those data to be compared with data generated in another laboratory or in another country.

The International Organization for Standardization (ISO) checks reference methods: "thoroughly investigated method, clearly and exactly describing the necessary conditions and procedures for measurement of one or more property values that has been shown to have accuracy and precision commensurate with its extended use and that can therefore be used to assess the accuracy of the other methods for the same measurement, particularly in permitting the characterization of reference material" (39).

The ISO 14,000 series of standards has been developed by ISO technical committee 207 for guidance in environmental management systems and tools. This series has similar structure and philosophy to the ISO 9,000 series.

Among the different methods, we can distinguish:

Multielement methods, such as chromatography, electrophoresis, and, to a lesser extent, atomic absorption spectrometry

Single-element determination methods, such as ion-selective electrode and dissolved oxygen measurement

I. CONCEPTS, CHARACTERISTICS, TESTS, AND TECHNIQUES

A. Acid-Consuming Capacity

This is the number of hydrochloric acid milliequivalents consumed on titration of 1000 ml water until the color of certain indicators change (or until the attainment of certain pH values in the case of electrometric measurement).

The p-value is the number of ml 1N hydrochloric acid consumed for 1000 ml water on using phenolphthalein as indicator (pH range 8.2–10.0) (or until attainment of pH 8.2 with electrometric measurement):

$$p = \frac{a \cdot f \cdot 1000}{c}$$

where

a = ml HCl 0.1 N consumed until change of indicator

c = number of ml water sample used

f = normality of HCl.

The m-value is the number of ml 1N hydrochloric

acid consumed for 1000 ml water on using methyl orange as indicator (pH range 4.3–6.0) (or attainment of pH 4.3 with electrometric measurement):

$$m = \frac{b \cdot f \cdot 1000}{c}$$

where:

   b = ml HCl 0.1 N consumed until change of indicator
   c = number of ml water sample used
   f = normality of HCl
   0.05 = m-value of neutral water without buffering constituents, in meq/L

Titration is not applicable when the water contains substances other than carbonic acid and its anions, which have a buffering effect in the pH ranges around 4.3 or 8.2.

B. Base-Consuming Capacity

This is the number of milliequivalents of sodium hydroxide solution consumed on titration of 1000 ml of water until the color change of certain indicators (or in the case of electrometric measurement until attainment of certain pH values). A negative p-value represents the number of ml 1N sodium hydroxide consumed for 1000 ml water, using phenolphthalein as indicator (pH range 8.2–10.0).

Alkalinity is the capacity of an aqueous solution to react with H^+ ions (ISO 6107-2). Alkalinity determined with methyl red (pH range 4.3–6.0) yields HCO_3^- (methyl orange) CO_3^{2-}, and OH^- (total alkalinity). TAC is alkalinity determined with phenolphthalein:

$$\text{negative } p\text{-value} = \frac{a \cdot f \cdot 1000}{c}$$

where

   a = ml 0.1 N sodium hydroxide consumed until change of indicator (phenolphthalein)
   c = ml water sample used
   f = normality of sodium hydroxide

A negative m-value represents that number of ml 1 N sodium hydroxide consumed for 1000 ml water, using methyl orange as indicator:

$$\text{negative } m\text{-value} = \frac{b \cdot f \cdot 1000}{c}$$

where

   b = mL 0.1 N sodium hydroxide consumed until color change of indicator (methyl orange)

   c = ml water sample used
   f = normality of sodium hydroxide

The titration is not applicable when the water contains substances other than carbonic acid and its ions, which effect a buffering in the pH ranges around 4.3 or 8.2.

C. pH Value

By definition, pH = $\log[1/(H^+)]$ = $-\log(H^+)$ where (H^+) is the hydrogen ion activity. The relationship between the activity of a species and its molar concentration [M] is given by $(a)_i = f_i[M]$, where f is a dimensionless quantity called the *activity coefficient*. The activity coefficient, and thus the activity of M, varies with the ionic strength of a solution.

A pH of 2 is equivalent to a hydrogen ion concentration of 10^{-2} M. The activity range for the hydrogen ion as defined by the dissociation product $K_w = (H^+)$ (OH^-) is 10^0–10^{-14}. The pH scale is 0–14. Each unit on the pH scale represents a tenfold change in activity.

The pH is measured:

With nonbleeding pH indicator strips and indicator papers. Special indicator strips and papers (see, for example, Merckoquant) are accurate up to 0.2 pH units.
Glass electrode (Fig. 1).

pH electrodes measure the pH of a solution potentiometrically. When a pH-sensing electrode comes in

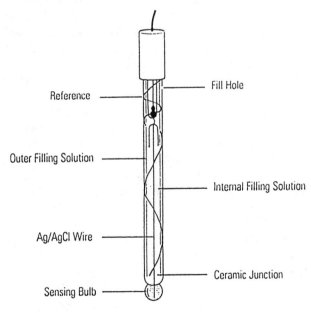

Fig. 1 Glass pH electrode. (Courtesy of Orion Research, Beverly, MA.)

contact with a sample, a potential difference is created across the sensing membrane surface. The membrane potential varies with the pH. Measurement requires a second, unvarying potential to quantitatively compare the changes of the sensing membrane potential. A reference electrode serves this function.

Electrode behavior is described by the Nernst equation:

$$E = E_0 + 2.3 \frac{RT}{nF} \log(H^+)$$

where

E = measured potential
E_0 = potential of electrode (related to reference electrode)
$2.3(RT/nF)$ = Nernst factor (also called *electrode slope*)
n = ion charge
F = Faraday's constant (96,500 coulombs)
R = gas law constant
T = temperature in K

When the temperature is 25°C, the Nernst factor is 59.16 mV/pH unit.

D. Toxicity Test

The principle here is that harmless bacteria produce light by bioluminescence when in contact with a toxic sample; the intensity of light decreases and is measured via luminometer. In operation, we use *Photobacterium fischerii* NRRLB-11177.

E. Color Test

The Pt/Co method utilizes a comparison with a reference solution, which consists of:

K_2PtCl_6	1.245 g
$CoCL_2, 6H_2O$	1.0 g
HCl	100 ml
H_2O_{qsp}	100 ml

By definition this solution exhibits a 500-Hazen-units color. Three wavelengths are checked:

$\lambda_1 = 436$ nm

$\lambda_2 = 525$ nm

$\lambda_3 = 620$ nm

One Hazen unit corresponds to the color of a solution containing 1 mg of Pt (as chloroplatinic acid) and 2

mg cobalt chloride hexahydrate per liter. A calibration range is executed via dilutions.

F. Hardness

The total hardness of a water sample is the sum of the alkaline earths (magnesium, calcium, strontium, and barium ions) bound as carbonates, sulphates, chlorides, nitrates, and phosphates, expressed in mmol/L. It may also be expressed in degrees:

French°: hardness of a solution containing 10 mg $CaCO_3$/L or 4 mg Ca
German°: hardness of a solution containing 10 mg CaO/L
UK° or Clark°: hardness of a solution containing 1.43 mg $CaCO_3$/L
US°: expressed in ppm $CaCO_3$

In Table 1 the correspondence of the different degrees is shown.

Carbonate hardness means the hardness induced by the proportion of the alkaline earth ions, which is equivalent to the carbonate and hydrogen carbonate ions contained in the water and to the hydroxyl ions resulting from their hydrolysis.

1. Method

The method involves dosage with an EDTA (ethylenediaminetetraacetic acid) solution 0.02 N with eriochrome black. The eriochrome color changes from red to grey and then to green.

$$\text{Hardness (meq/L)} = 100 \times \frac{CV_1}{V_2}$$

where

C = concentration of EDTA solution in meq/L
V_1 = volume (in ml) of the EDTA solution
V_2 = sample volume

Table 1 Correspondence of the Different Degrees of Hardness

Degree					
French	1.0	0.70	0.56	0.58	0.2
German	1.43	1.0	0.80	0.83	0.286
British	1.79	1.25	1.0	1.04	0.358
American	1.72	1.2	0.96	1.0	0.34
milliequivalents	5	3.5	2.8	2.9	1.0

2. Reagents

  Buffer solution pH 10
  Na_2EDTA solution 0.01 mole/L
  Standard Ca solution $(CaCO_3)$ 0.01 mole/L
  Eriochrome black

G. Turbidity

This is a measure of the transparency of water, which is important for drinking water. Light passing through matter may be scattered by inhomogeneities, with subsequent reduction of transmitted light. Light scattering depends upon the size, shape, and surface characterization of particles and the wavelength of the light. The characteristic of the solutions that causes light scattering is called *turbidity*.

To overcome the difficulty of differentiating small changes against a large background, measurement is performed at right angles to the incident light and is called *nephelometry*. ISO 7027 has set a detection angle of 90° and the light wavelength at 860 nm. The intensity I_0 of a parallel incident beam is attenuated in its passage through the scattering sample according to $I = I_0 \exp(-\tau l)$, where τ is the extinction of the incident beam and l is path length.

$$I = I_0 K \frac{N V^2}{\lambda^4} \sin \phi$$

where

  I = intensity of Tyndall light in a direction separated
     from initial beam direction by angle ϕ
  I_0 = intensity of incident light
  N = number of particles that cause deviation
  V = particle volume
  λ = wavelength of incident radiation

The units are turbidity units (TU), nephelometric turbidity units (NTU), and formazine turbidity units (FTU). There is a limit of 4 NTU for drinking water.

Formazine is a suspension that is not commercially available and has to be made from the reaction between hexamethylenetetramine and hydrazine sulphate. The instrument used is a probe immersed directly into an open channel. A wiper keeps the optical faces clear.

H. Suspended Solids

Suspended solids analysis gives an excellent indication of bulk water quality. The standard test methods are: filtration on glass microfiber filters, filtration on membranes, or centrifuging the sample and weighing the residue after drying (1).

I. Conductivity

The conductance of water is highly dependent on the measurement conditions. It is the reciprocal of the resistance (in ohms) measured between the opposite faces of a 1-cm cube of liquid at a specific temperature. The basic unit of conductance is the siemens (S) (formerly called the mho).

From Ohm's law, conductance = current/voltage. A measurement yields the conductance, so it is necessary to convert the measured value to the conductivity. This is achieved by measuring a cell constant for each setup by means of solutions of known conductivity.

$$\text{cell conductance} \times K = \text{conductivity}(\kappa)$$

The cell constant is related to the physical characteristics of the measuring cell. For two flat, parallel measuring electrodes, K is defined as the electrode separation distance (d) divided by the electrode area (A). For a 1-cm cube of liquid, $K = d/A = 1$ cm^{-1}. With pure water, conductivity is very low and resistivity is measured. A four-electrode cell for conductivity measurement is displayed in Fig. 2. A four-electrode conductivity cell contains two drive (current) electrodes and two sense (voltage) electrodes. The drive electrodes are powered by an alternating voltage to overcome polarization. The amplitude of the alternating voltage applied to the drive electrodes is controlled by the voltage measured at the sense electrodes. The alternating current that flows is measured to determine the conductivity.

Solution conductivity is due to ion mobility. The conductivity depends on the number of ions present. If the concentration is C (in moles per unit volume), then the molar conductivity is $\Lambda_m = \kappa/C$. Since the resistance is measured in ohms (Ω), the units are $\Omega^{-1} \cdot cm^{-1}$. Conductivity is expressed in siemens/cm, or S/cm.

Total conductivity is sometimes expressed in terms of the salt content, mg/L NaCl. The equivalent conductivity is sometimes used. It is expressed in terms of the number of individual charges that are being carried. The equivalent conductance of H^+ is 349.8 cm^2/Ω equiv (higher than that of all other ions).

J. Resistivity

Resistivity is the reciprocal of conductivity. The conductivity C $(\mu S\text{-cm})$ is given by the sum of the contribution of the individual ions:

$$\sum \frac{Ic_i \cdot Ec}{Ew} \cdot 10^{-3}$$

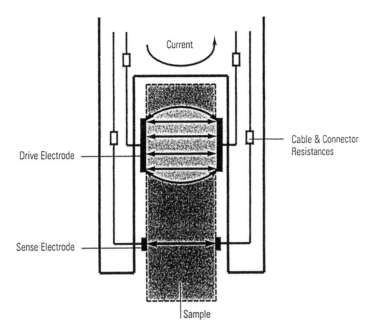

Fig. 2 Scheme of a four-electrode cell for conductivity measurements. (Courtesy of Orion Research.)

where

Ic_i = ion concentration in $\mu g/L$
Ec = equivalent conductance of ions in S/cm^2
Ew = equivalent mass (mg/equivalent)

Since the resistance is measured in ohms (Ω), the resistivity is measured in Ω-cm. Here are some examples:

Resistivity of ultrapure water $\cong$ 18 MΩ-cm at 25°
Resistivity of drinking water $\cong$ 50 MΩ-cm

K. Determination of CO_2

Free CO_2 is determined by neutralization with a slight excess of NaOH. NaOH not consumed is then dosed with acidic solution until a change of the phenolphthalein color.

Total CO_2 is determined via Van Slyke's method, whose principle of the operation is that O_2, N_2, and CO_2 gases evolved are trapped in vacuum and volume and pressure are measured. By treatment with NaOH, CO_2 reacts, and the remaining gases are left to expand to the same volume as before treatment. The pressure difference yields the CO_2 partial pressure:

$$V_{0/760} = P \times \frac{ai}{760(1 + 0.00384t)} \left(1 + \frac{S\alpha}{A - S}\right)$$

where

$V_{0/760}$ = CO_2 volume under 0°C and 760 mm of Hg
$P = P_1 - P_2$ = partial pressure of CO_2
a = gas volume (typically 2 ml)
i = coefficient of gas reabsorption in liquid of reacting flask ($i = 1.014$ for CO_2, $i = 1$ for O_2, H_2, N_2)
t = temperature in °C
S = volume of sample and reagents
A = volume of reaction chamber
α = distribution coefficient of CO_2 between liquid and gas phases

L. Biochemical Oxygen Demand (BOD)

The BOD is a test used to assess the relative oxygen requirements of wastewater, effluent, and polluted water samples. The measure BOD 5 is the quantity of oxygen (in mg) consumed in the test (incubation for 5 days at 20°C under darkness) by degrading organic species through biological pathways. The principle is to measure the oxygen depletion that occurs during the test.

1. Dilution Method

The dilution method is as follows:

1. Prepare a solution obtained by diluting the water sample with diluting water.

2. Incubate for 5 days at 20°C ± 1°C.
3. Measure the consumed oxygen.

X_0 = oxygen content (mg/L) of the water to analyze at start of the experiment

X_5 = oxygen content (mg/L) at the end (5 days) of the test

V_e = volume (mL) of water to analyze in the gauge flask

V_t = volume of the gauge flask

V_d = dilution volume $V_d = V_t - V_e$

$$\text{BOD} = X_0 - X_5 = \frac{V_t}{V_e(T_0 - T_5)} - \frac{V_d}{V_e(D_0 - D_5)}$$

where

D_0 = oxygen content of diluting water when filling flask

D_5 = oxygen content of diluting water 5 days later

T_0 = oxygen content of one dilution when filling flask

T_5 = oxygen content of one dilution 5 days later

Oxygen depletion can be measured by the Winkler method, an iodometric titration based on the oxidizing property of dissolved oxygen. $Mn(OH)_2$ precipitates in water and traps oxygen by formation of $Mn(OH)_4^{2-}$. Adding HCl forms $MnCl_2$

$$MnCl_2 + IK \rightarrow I_2$$

I_2 dosage with thiosulphate is carried out:

$$2S_2O_3Na_2 + I_2 \rightarrow 2NaI + S_4O_6Na_2$$

Current technique, an electrometric method uses a membrane covered electrode.

The Clark electrode (Fig. 3) consists of a thin organic membrane covering a layer of electrolyte and two metallic electrodes. The membrane allows the influx of

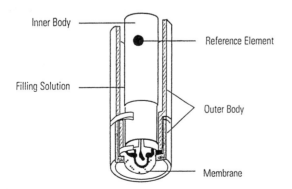

Fig. 3 Gas-sensing electrode (Clark type). (Courtesy of Orion Research.)

molecular oxygen to the cathode surface, at which it is reduced. The cathode potential is fixed such that only oxygen is reduced to water.

$$O_2 + 4H^+ + 4e^- \rightarrow 2H_2O$$

The reduction of O_2 creates a cathodic current, which is directly proportional to the partial pressure of the oxygen. Electrodes measure the partial pressure of the oxygen. For a given partial pressure, the concentration present in saturated pure water is fixed at any temperature. The presence of dissolved oxygen lowers the amount of the oxygen that can dissolve. A salinity correction is suggested for use with any commercial dissolved-oxygen meter.

Commercially available meters use cathode current, sample temperature, membrane temperature, barometric pressure, and salinity correction to calculate the dissolved oxygen content of the sample. No stirring is required. Data transfer is carried out via infrared signal transmission.

2. Respirometric BOD Measurement

The measuring bottle contains a premeasured amount of sample and a sealed reserve of air. The sample is incubated for 5 days at 20°C. Microorganisms degrade the organic species and consume oxygen. Stirring causes the oxygen in the sample to be replaced with oxygen from the air reserve. A pressure sensor monitors the resulting drop in pressure. Carbon dioxide produced is absorbed by potassium hydroxide and does not contribute any error.

M. Dissolved Oxygen Demand

Dissolved oxygen demand is measured with same meters as those utilized for BOD determinations. Meters are derived from the Clark electrode, which measures the partial pressure of oxygen. Improvements have been made to membranes (two types are in use: loose membranes and membrane cap assemblies), on steady-state measurements, on electrode surface (the larger the electrode, the higher the signal). Calibration is performed in water-saturated air. The concentration of oxygen is expressed in mg/L of water or in parts per million (ppm).

N. Ion-Selective Electrode (ISE)

Chemical sensors are devices that can quantitatively and reversibly measure an analyte. There is an analyte

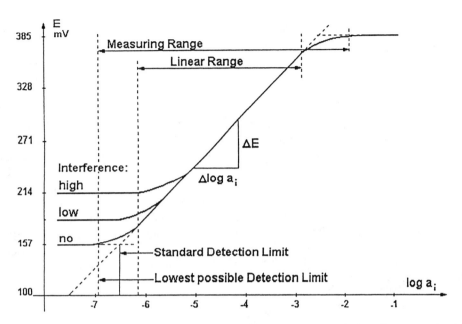

Fig. 4 Ion-selective measurements: Nernst equation, measuring range, detection limit, interferences, selectivity coefficients.

recognition process and a transduction translating this process to an electrical signal.

In ion selective electrodes (ISE) the transduction is potentiometric. An ISE is an electrochemical half-cell with a membrane for the analyte recognition process, an internal filling solution (electrolyte), and an internal reference electrode. The potential difference developed between the sensing and reference electrodes is a measure of the activity of the reactive species. As the activity varies, so does the potential measured between the two electrodes: From the Nernst equation: $E = E_0 + S \log a$. The Nicolsky Eisemann formalism is

$$E(mV) = E_0 + S \log \left[a_i + \sum_{j \neq 1} K_{ij}^{pot}(a_j)^{Zi/Zj} \right]$$

where E is the measured voltage, E_0 is a combination of several constants, including the liquid junction potential difference generated between the electrolyte of the reference electrode and the sample solution, Zi and Zj are the charge numbers of the primary ion i and the interfering ion j, respectively, a_i and a_j are concentrations (mol/L), K_{ij}^{pot} is the selectivity factor, and S is the slope of the electrode response (Fig. 4).

By measuring the electrode potential in both a standardizing solution and in a sample solution, it is possible to calculate the concentration of the unknown solution:

$$C_\alpha = C_i \times 10^{\Delta E/S}$$

where

C_α = concentration of unknown solution
C_i = concentration of standardizing solution
ΔE = observed difference potential
S = electrode slope

Response time and selectivity are important parameters. *Response time* is the time taken for the potential to reach a value of 1 mV from the final equilibrium potential after an instantaneous change in determined activity. *Selectivity* is related to interferences from other ions, it is characterized by the selectivity coefficient. Direct measurement is the simplest way. The millivoltmeter has a very high input impedance. Calibration is performed in a series of standards.

Measurements from ISE are gathered in five groups according to the nature of the membrane:

1. Glass membrane (Na^+, K^+)
2. Liquid membrane (NH_4^+, Ca^{2+}, fluoroborate, nitrate, nitrite, K^+)
3. Crystalline (for example, with an insoluble salt of the element) (Cl^-, Br^-, CN^-, Cu^{2+})
4. Enzymatic membrane (glucose)
5. Permselective to gases (NH_3, CO_2).

In titration experiments, a reagent is incrementally added that reacts with the sample species. An ISE is used for determination of the endpoint. Examples are: Ca with EDTA (electrode pCa), Ag with Ag/S electrode, SO_4^{2-} with pPb^{2+}.

O. Chemical Oxygen Demand (COD)

This is the oxygen concentration in mg/L, which is equivalent to the quantity of potassium dichromate consumed by organic species when a liquid sample is oxidized.

1. Principle

The experiment is conducted in concentrated acidic media, with silver sulphate as a catalyst and Hg as chloride complexing agent. Excess dichromate is dosed with a titrated ammonium and iron sulphate (Mohr salt) solution, with forroine as indicator.

2. Reagents

Distilled water

Mercury sulphate

Silver sulphate solution [6.6 g in 1000 ml H_2SO_4 ($d = 1.84$)]

Iron ammonium sulphate 0.25 N [98 g Mohr salt, 20 ml H_2SO_4 ($d = 1.84$) in 1000 mL water]

Potassium dichromate solution 0.25 N (12.2588 g in 1000 ml water)

Ferroine solution (1.485 g 1.10-phenanthroline + 0.685 g Fe sulphate in 100 ml water)

$$COD = 8000 \frac{V_o - V_i}{V_c} f \quad \text{in mg } O_2/L$$

where

V_o = volume of Mohr solution in ml in sample dosage

V_i = volume of Mohr solution in ml in blank experiment

V_c = volume of sample

f = normality of solution

P. Coulometric Titration

Coulometry permits determination of chemical substances by measuring the quantity of electricity required for their conversion to a different oxidation state. The quantity of electricity or charge is measured in coulombs. (The coulomb is the quantity of charge that is transported in 1 second by a constant current of 1 ampere). For a constant current of I amperes, the number of coulombs Q is $Q = It$.

The Faraday's law states that $Q = Fnz$, where F is the Faraday's constant, Q is the number of coulombs required to convert n moles of reactant to product by a reaction involving z electrons per ion or molecule of reactant. For a variable current, $Q = \int_0^t I \, dt$.

Two techniques are in current use: potentiostatic and amperostatic. The first involves maintaining the potential of the working electrode (the electrode at which the reaction takes place) at a constant level. The second method involves a constant current, which is maintained until completion of the analytical reaction. The quantity of electricity required to attain the endpoint is calculated from the magnitude of the current and the time of its passage. It is called a *coulometric titration*, since a titrant is electrolytically generated by a constant current. Ions to be evaluated react with another ion, which in turn is produced by an electrode reaction.

Coulometric titrations are performed with a constant-current generator (amperostat), which detects small variations of intensity and answers by changing the voltage to the cell in such a way that the current is set back to its initial value. Applications: chloride, As(III), Fe(II), Ce(II), Cr(VI) determinations.

Q. Measurement of Total Carbon (TC)

Total carbon is the sum of the TOC (total organic carbon) and the IC (inorganic carbon). A scheme of an instrument is displayed in Fig. 5. The carrier gas is flow regulated to 150 ml/min and allowed to flow through the TC combustion tube, which is packed with catalyst and kept at 680°C. When the sample enters the TC combustion tube, the total carbon in the sample is oxidized to carbon dioxide. The carrier gas containing the combustion products from the TC combustion tube flows through the IC reaction vessel, dehumidifier, and halogen scrubber and finally reaches the sample cell of the nondispersive infrared (NDIR) detector, which measures the carbon dioxide content. The output signal (analog) of the NDIR detector is displayed as peaks. The peak areas are measured and processed by a data-processing unit. Since the peak areas are proportional to the total carbon concentrations, the total carbon in a sample may be easily determined from the calibration curve prepared using standard solutions of known carbon content.

R. Measurement of Inorganic Carbon (IC)

The sample is injected into the IC reaction vessel, which has carrier gas bubbling through the IC reaction

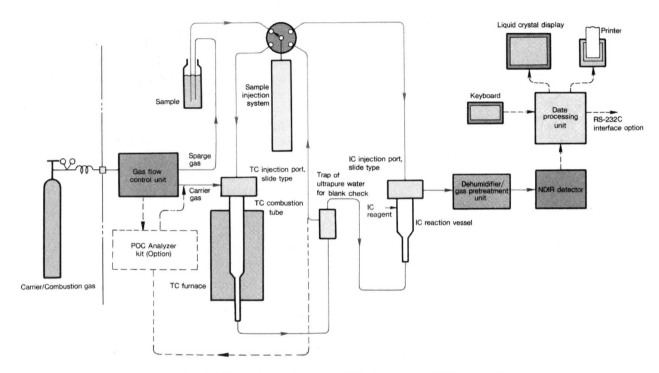

Fig. 5 Measurement of total carbon (TC) and inorganic carbon (IC). (Courtesy of Shimadzu, Tokyo.)

solution. Only the inorganic carbon in the sample is converted into carbon dioxide, which is then detected via NDIR detector. The concentration of inorganic carbon is the sum of carbonate and hydrogen carbonate.

S. Measurement of Total Organic Carbon (TOC)

The TOC concentration is obtained by subtracting the IC concentration from the TC concentration.

T. Nitrogen Determination

1. Kjeldahl's Method

In this procedure a weighted sample is digested in sulphuric acid in the presence of a catalyst. The excess acid is then neutralized and the solution made strongly basic. At this point the solution is distilled, to liberate ammonia for collection in neutral water, followed by titration:

$$R-N + H_2SO_4 \xrightarrow{T_0} (NH_4)_2SO_4 + H_2O + CO_2 + SO_2$$

$$(NH_4)_2SO_4 + NaOH \xrightarrow{T_0} NH_3 + Na_2SO_4 + H_2O$$

$$NH_3 + H_2O \rightarrow NH_4OH$$

Accuracy is dependent on apparatus, reagents, and

technique. The method is time consuming, and low-level nitrogen determination is tedious.

Amines and amides are quantitatively transformed to ammonium ions by H_2SO_4. Conversely, nitro, azo, and azoxy groups may yield nitrogen or nitrogen oxides, which may evolve. Reduction is carried out prior to acidic treatment.

A Kjeldahl system is a complete set of modular units for digestion, distillation, tube racking, and automatic titration. Colorimetry can be used for determination of free and fixed ammonia levels.

Elemental analysis of nitrogen is performed in an automatic nitrogen analyzer. Sample (liquid or solid) is submitted to flash combustion in the presence of NiO, followed by a reduction section. After removal of water, CO_2 and N_2 are analyzed via GC.

2. Chemiluminescene Method

Oxidative combustion of nitrogen-containing compounds produces nitric oxide. Nitric acid when in contact with ozone produces a metastable nitrogen dioxide molecule:

$$NO + O_3 \rightarrow NO_2^* + O_2$$

which relaxes to a stable state by emitting at a wavelength of 700–900 nm:

$$NO_2^* \rightarrow NO_2 + h\nu$$

As the light emission occurs, the light intensity is measured via a photomultiplier tube through a band-pass filter.

To convert chemically bound nitrogen to nitric oxide, the sample is subjected to oxidative pyrolysis:

$$p = \frac{N_1 - N_0}{b} \times f$$

where

p = concentration, in mg/L, of bound nitrogen in the sample
N_1 = sample response
N_0 = blank response
b = slope of calibration curve
f = dilution factor

3. Nitrogen by Elemental Analyzer

The substance is weighed in a special tin container and placed in an automatic sampling system. At preset times the sample container is introduced into the combustion chamber, which is maintained at 1020°C. At the same time, oxygen is introduced. In the presence of oxygen the tin initiates a strongly exothermic reaction (1800°C) lasting a few seconds (fast combustion). NiO particles (a good oxidation catalyst) fill the combustion chamber. The combustion chamber outlet is connected directly to the reduction section, which consists of a quartz reactor filled with reduced Cu at 700°C. At the outlet of the reduction section, a mixture of CO_2, N_2, and H_2O passes through an absorber packed with magnesium perchlorate or through a molecular sieve, where water is retained. A second absorber retains CO_2. Nitrogen and helium carrier enter the chromatographic column packed with Porapak Q and then reach the conductivity detector.

U. Biological Assay

The enzyme-linked immunosorbent assay (ELISA) (Fig. 6) exhibits high specificity and sensitivity. The use of 96 well format allows the quantification of a high number of samples.

The binding of antibodies to antigens in an ELISA is detected via the cleavage of a chromogenic substrate by an enzyme conjugate, for instance, alkaline phosphatase or horseradish peroxidase, coupled either to an antibody or to streptavidin. The colored reaction product is quantified by measuring the absorbance.

Assays with higher sensitivities use chemiluminescence. Electroluminescence immunoassay (ECLIA) is a promising approach to eliminate background signals. Unlike conventional chemiluminescence assay, the signal generated is a result of a reaction (typically oxido reduction) under a well-controlled electric potential. In this way, reproductibility does not depend on the substrate concentration and reaction time.

V. Immunosensing Devices

Immunosensors take advantage of antibody (Ab) as binding proteins for the analyte. To produce measurable signal from the Ab–analyte binding reaction, one can use a tracer, which provides the signal, or use a tracer-free variant. In the first approach, Ab is immobilized onto a solid phase and an enzyme analyte conjugate is used as a tracer. If Ab and tracer concentration are kept constant, tracer binding to the Ab depends on the analyte concentration in the sample.

In tracer-free mode, the Ab is immobilized in a matrix on the sensor chip, which forms one wall of a microflow cell. The sample is injected over the surface in a controlled flow. Any change in surface concentration resulting from the interaction is detected.

Immunosensors can be operated at a quasi-continuous mode.

Complexation between antibody and antigen is written

$$Ab + H \leftrightarrow AbH$$

$$K = \frac{(AbH)}{(Ab)(H)}$$

Antibodies generally possess enormous selectivity, targeting unique structural elements (epitopes) that consist of 3–12 amino acids in polypeptides or a few carbohydrate residues in oligosaccharides. Immunological recognition is based on the spatial complementarity of groups in the epitope of the antigen with those in the paratope of the antibody.

A biosensor can be thought of as comprising a bioactive substance (typically an enzyme, multienzyme system, membrane component, or microorganism) that can specifically recognize species of interest in intimate contact with a suitable transducing system. The purpose of a transducer is to convert the biochemical signal into an electronic signal that can be suitably processed. Enzyme electrodes result from combining any type of electrochemical sensor with a layer of enzyme, such as 10–200 μm, in close proximity to the active

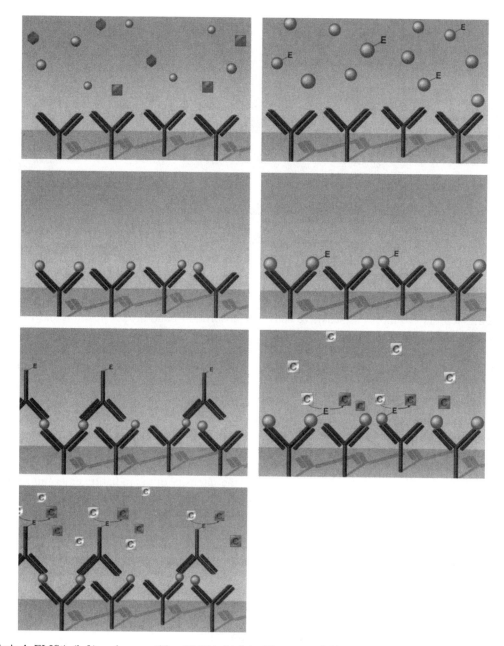

Fig. 6 Sandwisch ELISA (left) and competitive ELISA (right). (Courtesy of Riedel-De Haen, Seelze, Germany).

surface of the transducer. Amperometric enzyme electrodes combine anion-selective electrodes with an immobilized enzyme. The signal is a current flowing through the working electrode.

The difficulties in achieving direct electron transfer between enzymes and electrodes has prompted the use of small-molecule electroactive indicators, to enhance the rate at which the transfer of electrons occurs. The role of the mediator is to shuttle electrons efficiently between electrode and enzyme. The material consists of salts comprising two planar organic molecules with extorted π-electron systems formed into segregated stacks. Under suitable conditions, overlap of the π-orbitals can lead to the delocalization of charge through the stack, thus giving rise to electrical conductivity.

The potentiometric sensor relies on the relationship between the emf of an electrochemical cell and the concentration of the chemical species in the sample. Most measurements are performed as "zero-current potentiometry." In asymmetrical devices, the selective mem-

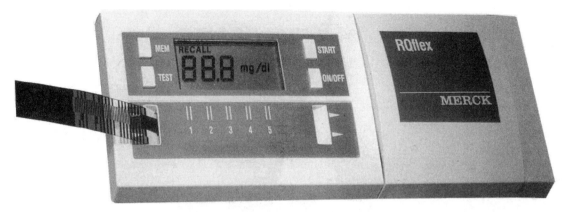

Fig. 7 Merck reflectometer for color test. (Courtesy of Merck, Darmstadt, Germany.)

brane is in an asymmetrical arrangement with respect to the sample.

Conductometric sensors are most gas-sensitive sensors. The interaction takes place at the surface, while the modulation and the measurement of the conductance is performed along the surface of the device.

For more information on environmental immunoassays and biosensors, see Refs. 2 and 3.

W. Color Tests for Elements

These rely upon the selective and sensitive reaction of the elements with a complexing or chelating agent. According to the sensitivity, we can distinguish:

> *Semi-quantitative strips*: These quick tests are able to monitor a parameter.
> *Quantitative strips*: These provide precise data with the use of a reflectometer (Fig. 7).
> Sensitivity is around 1 mg/L.
> *Tube test*: Sample is added to the reagent in a tube and reading is carried out with a photometer.

Spot tests can be performed on paper or TLC plates. For example, carbonyl compounds can be detected by the Ehrlich diazo test or by Millon's test. A review of chromatographic and related spot tests for the detection of water pollutants at the microgram level has been published (4). A list of colorimetric determinations of different elements can be found in Table 2.

II. DETECTION METHODS

A. UV-VIS Absorption

UV-VIS spectra arise from electronic transitions within molecules. Broad absorption bands are usually ob-

served due to the contribution of vibrational and rotational energy levels. The principal characteristics of an absorption band are its position and intensity. The position of maximum wavelength (λ_{max}) corresponds to the wavelength of radiation whose energy is equal to that required for an electronic transition.

The intensity of an absorption is expressed by the transmittance:

$$T = \frac{I_o}{I}$$

where I_o is the intensity of the radiant energy and I is the intensity of the radiation emerging from the sample.

The Beer–Lambert law is expressed as

$$\log_{10} T = A = \varepsilon l c$$

where

> ε = molar absorptivity of solute
> l = path length through sample
> c = concentration of solute
> A = absorbance

The spectrum of a mixture $A(\lambda)$ is given by the sum of the products of concentration c, molar absorptivities, ε, and path length l of each component:

$$A = \varepsilon_1 l c_1 + \varepsilon_2 l c_2 + \varepsilon_3 l c_3 + \cdots$$

written in matrix form this is $\mathbf{A}/\mathbf{l} = \mathbf{E} \cdot \mathbf{C}$. where $\mathbf{A}$ is the measured spectrum, $\mathbf{E}$ is the matrix of the component spectra ordered in columns, and $\mathbf{C}$ is the matrix of the concentration of the component ordered in a column.

Spectrophotometers are either of single-beam (Fig. 8) or double-beam design. Resolution in spectrophotometry is the ratio of natural bandwidth to spectral

Table 2 Colorimetric Determination of Elements (Courtesy of Merck)

Element	Method	Measurement range	λ (nm)
Aluminum	Colorimetry of color lacquer from chromazurol S and aluminum in an acetate-buffered solution	0.07–0.8 mg/L	545
Ammonia	Colorimetry with indophenol blue (Berthelot's reaction)	0.025–0.4 mg/L	690
Ammonia	Potassium tetraiodomercurate forms a yellow-brown reaction product with ammonia ions in an alkaline medium (Nessler method)	0.05–0.8 mg/L	
Chloride	Choride reacts in presence of iron(III) with mercury(II)-thiocyanate to mercury(II) chloride, chloromercurate(II) anions, and orange-red iron(III) thiocyanate	5–300 mg/L	525
Chlorine	Oxidation of N,N-diethyl-1,4-phenyldiamine by chlorine, hypochlorite, and HOCl to a red-violet, semi-quinoid product	0.01–0.3 mg/L	557
Chrome	Reduction from chrome(VI) to chrome(III) during simultaneous oxidation to diphenylcarbazide and subsequent formation of a red-violet chrome(III) diphenylcarbazone complex	0.005–0.1 mg/L	540
Copper	Reaction of copper(II) with cuprizone in an alkaline medium with the formation of a blue color complex	0.05–0.5 mg/L	595 585
Cyanide	Cyanide is converted to chlorine cyanide. Subsequently there is a reaction with pyridine to glutacondialdehyde and condensation with 1,3-dimethyl-barbituric acid to a violet polymethine dye	0.002–0.03 mg/L	585
Hydrogen sulphide	Reaction with N,N'-dimethyl-1,4-phenylene-diammoniumchloride and oxidation with iron(III) to methylene blue	0.02–0.25 mg/L	665
Iron	Reaction of iron with 1,10-phenanthroline in a thioglycolate buffer with the formation of an orange color complex	0.25–15 mg/L	492
Iron	Reduction of Fe(II) with ferrospectral in a thioglycolate buffer with the formation of a violet color complex	0.01–0.2 mg/L	565
Manganese	Colorimetry of the red-brown color complex of manganese with formaldoxime	0.03–0.5 mg/L	445
Nickel	Colorimetry of red-brown dimethylglyoxime complex of nickel in an alkaline oxidizing medium	0.02–0.5 mg/L	445
Nitrite	Reaction of nitrite with sulfanilic acid and N-(1-naphtyl)-ethylenediamide to a red azo dye (Griess reaction)	0.005–0.1 mg/L	
Phosphate, determined as PMB	Orthophosphate ions form, with molybdate ions in sulphuric acid solution, molybdatophosphoric acid, which in turn is reduced to phosphomolybdic blue (PMB) with ascorbic acid	0.015–0.14 mg/L P, 0.046–0.43 mg/L PO$_4$	690
Phosphate, determined as VM	Colorimetry of the orange-yellow color complex of molybdovanadate phosphoric acid (VM)	1.0–40 mg/L	405
Silicium	Colorimetry of the reduced β-silicon molybdic acid	0.01–0.25 mg/L	650
Sulphate	Reaction with barium iodate to barium sulphate and iodate; iodate forms a slightly brown-red color complex with tannic acid in a slightly acidic medium	25–300 mg/L	525 820
Zinc	Colorimetry of blue-green ternary complex of zinc with thiocyanate and brillant green in an acidic medium	0.1–5.0 mg/L	565

bandwidth; it must be $5 \geq 1$. In modern spectrophotometers (for colorimetric procedures), tests are programmed and procedures are displayed on a liquid crystal display (LCD) screen. Test parameters are automatically set when any test is selected. Results are automatically displayed, and they are printed if necessary. Some spectrophotometers include a six-filter diode array system.

Wavelength reproducibility is more important than absolute accuracy, especially if measurement is being

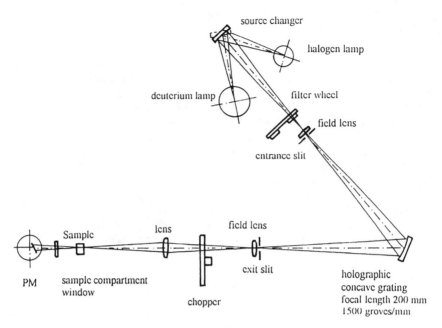

Fig. 8   A single-beam UV spectophotometer (from Uvikon.)

made on the side of a peak. With microprocessor-controlled stepping motors, a reproducibility of ±0.002 nm is achieved.

Derivative spectra are obtained by differentiating the absorbance (A) spectra of the sample with respect to wavelength. First-, second-, or higher-order derivatives may be generated (Fig. 9). All the derivatives emphasize the features of the original spectra by enhancing small changes in slope. Derivative spectra are gener-

ated optically, electronically, or mathematically. The usual optical method utilizes the wavelength modulation technique, where the wavelength of incident radiation is rapidly modulated over a narrow wavelength range. Electronically generated derivatives are produced by analog resistance capacitance (*RC*) devices. Mathematical methods are obviously the best performing. The Savitsky–Golay algorithm generates derivatives with a variable degree of smoothing.

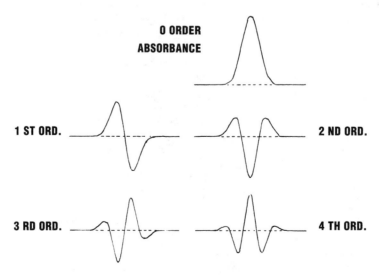

Fig. 9   Gaussian absorption band and its first- to fourth-order derivative.

Derivatives are used for:

Enhancement of spectral differences
Enhancement of resolution
Selective substraction
Selective discrimination

For two coincident bands of equal intensity, the nth-order derivative amplitude of a sharp band A is greater than that of a broad band B by a factor that increases with derivative order.

B. Fourier Transform Spectrometry

In the Michelson interferometer, a collimated light beam is divided at a beam splitter into two coherent beams of equal amplitude that are incident normally on two plane mirrors. The reflected beams recombine coherently at the beam splitter to give circular interference fringes at infinity focused by a lens at the plane of the detector (see figure 47 on GC-FTIR).

For monochromatic light of wavelength λ_0 and intensity $B(\lambda_0)$, the intensity at the center of the fringe pattern as a function of the optical path difference x between the two beams is given by

$$I_0 = B(\lambda_0) \left[1 + \cos \frac{2\pi x}{\lambda_0} \right]$$

$$= B(\sigma_0) \left[1 + \cos 2\pi\sigma_0 x \right]$$

where

$\sigma = \nu/c$ (σ is the wavenumber)
ν = frequency of the light in s^{-1}
c = speed of light in cm-s^{-1}

If x is changed by scanning one of the mirrors, the recorded intensity (the interferogram) is a cosine of spatial frequency σ_0. Its temporal frequency is given by $f_0 = v\sigma_0$, where v is the rate of change of the optical path. If the source contains more than one frequency, the detector sees a superposition of such cosines (Fig. 10):

$$I_0(x) = \int_0^\infty B(\sigma) \cdot (1 + \cos 2\pi\sigma x) \, d\sigma$$

Substracting the constant intensity $\int_0^\infty B(\sigma) \, d\sigma$ corresponding to the mean value of the interferogram $\langle I(x) \rangle$ yields:

$$I(x) = I_0(x) - \langle I(x) \rangle = \int_0^\infty B(\sigma) \cos(2\pi\sigma x) \, d\sigma$$

The right-hand side contains all the spectral information in the source and is the cosine Fourier transformed of the source distribution $B(\sigma)$.

The latter can be recovered by the inverse Fourier transform:

$$B(\sigma) = \int_0^\infty I(x) \cos(2\pi\sigma x) \, dx$$

Actually, the Fourier transform reproduces $B(\sigma)$ and adds a mirror image $B(-\sigma)$ at negative frequencies.

In fact, the interferogram is never totally symmetric about $x = 0$. To recover the full spectral information, it is necessary to take the complex rather than the cosine Fourier transform. The interferogram is recorded to a finite path difference L rather than infinity. It is actually recorded by sampling it at discrete intervals Δt.

C. Neutron Activation Analysis

The sample and the standard are irradiated with neutrons in a nuclear reactor to form radioactive isotopes. The decay of some radioisotopes is accompanied by emission of γ rays with energies characteristic of the nuclide emitting them. Comparison of intensity of the appropriate γ radiation of the sample and the known standard provides quantitative measurement of the abundance of the element.

D. Atomic Emission Spectrometry (AES)

Excited atoms are able to produce radiation at a specific wavelength when returning to ground state. A simple way to excite atoms is to burn the sample in a flame, although excitation is limited by the low temperature a flame can achieve. An electric arc or a spark works better. Inductively coupled plasma is the usual source. The sample solution is nebulized, and the mist is transformed through a spray chamber into a high-temperature plasma in which all the elements are atomized and excited to emit radioactive lines, with a specific wavelength for each element. Excitation in a plasma system in thermal equilibrium can be treated in terms of the Saha relationship and the dissociation in terms of the general equation governing chemical equilibrium. The ratio of the number of atoms (n_e) is given by Boltzmann's law:

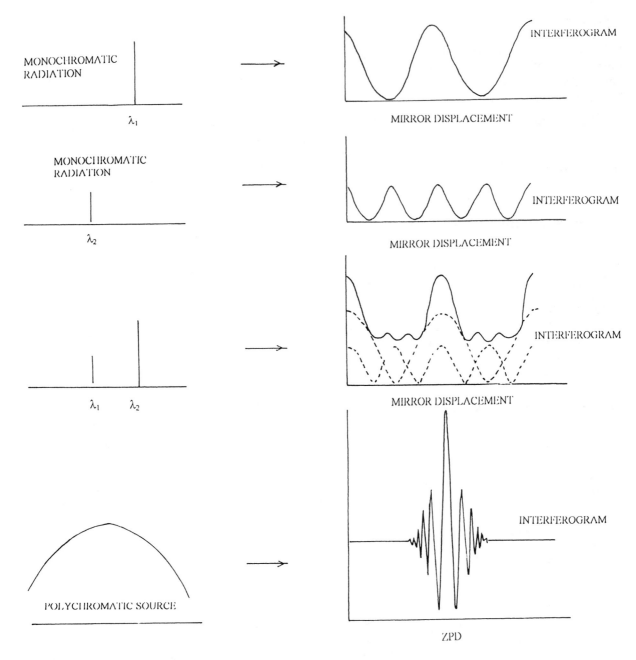

Fig. 10 Fourier transform signal.

$$\frac{n_e}{n_0} = \frac{W_e}{W_0} \exp\left[-\frac{E_e}{kT}\right]$$

where

    n = number of atoms per cubic centimeter
    W = statistical weight
    E_c = excitation energy in eV

k = Boltzmann's constant
T = absolute temperature

Thermal excitation is of extremely low efficiency.

The absolute intensity (photons-cm^{-3}-s^{-1}) of a spectral line associated with a given spontaneous transition is obtained by multiplying the number of statistically excited atoms by the transition probability. Microwave-

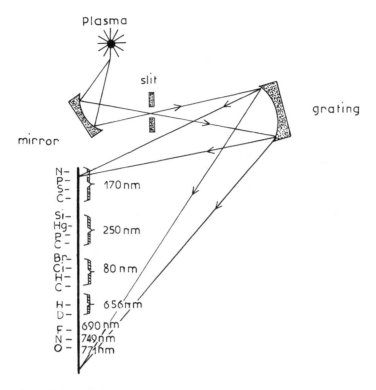

Fig. 11 Scheme of an atomic emission spectrometer.

induced plasmas (MIP) are operated at 2450–2500 MHz. The design is simple, and the operating cost is low (a scheme is displayed in Fig. 11). The instrument consists of a microwave generator, a capacitor, and a plasma tube.

E. Atomic Absorption Spectroscopy

This method has become very popular for the elemental analysis of solutions. It is based on selective absorption of line radiation by the elemental atomic species in vapor phase. The element of interest is elevated to high temperature, where is it dissociated from its chemical bonds. In this atomized state it is capable of absorbing radiation in discrete lines of narrow bandwidth.

Only atoms at the ground state (n_0) (see atomic emission) are capable of absorbing the radiation line. An atomic absorption spectrum consists predominantly of resonance lines. Measurement of light intensity before and after the atomic vapor path yields the absorption percentage. The principle of operation is diagrammed as follows (see bottom of the page).

1. Instrument

Production of the radiation line must fulfill certain requirements:

The line must be a resonance line.
It must be a sharp line.
I_0 should be high enough.
It must be stable.

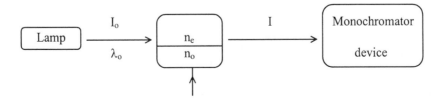

Hollow cathode lamps are the most popular. This type of lamp consists of a tungsten anode and a cylindrical cathode coated with a metal film of the selected element. The cathode is sealed in a glass tube that is filled with neon or argon at a pressure of 1–5 torr. From the potential difference between electrodes, ionization of gas occurs, with the production of electrons $(e)^-$ and positively charged ions that are directed at the cathode. If the kinetic energy is high enough, sputtering of metal atoms from the cathode occurs. Some of the sputtered metal atoms are in excited states and thus emit their characteristic radiation as they return to the ground state. Atoms collide with e^- and positive ions, which induce transition to an excited state and light emission. A monochromator selects the desired radiation line. Available hollow lamps are either single-element lamps or multielement lamps. The lamps are mounted on a turret.

Electrodeless discharge lamps produce radiation lines by submitting atoms to high-frequency electromagnetic fields. The element (a salt) is placed in a quartz bulb under low-pressure inert gas. A coil surrounds the bulb, which can produce a radio-frequency field. Electrodeless discharge lamps are commercially available for 15 or more elements. They are often considered less reliable than hollow cathode lamps.

2. Atomic Vapor Production

a. Flame. The solution to analyze is nebulized with a pneumatic nebulizer that produces a spray. Air is often replaced by N_2O. Pneumatic nebulizers do not produce a homogeneous spray. An ultrasonic nebulizer (800–3500 kHz) produces good droplets of fine, homogeneous diameter with good yield. Flames are produced via combustion of acetylene, propane, or hydrogen with either air, oxygen, or N_2O. An acetylene-rich flame is a reductive flame; an oxygen- or nitrogen protoxide–rich flame is an oxidative flame. Conventional flame technique requires approximately 1 ml of sample. Laminar-flow burners provide a quiet flame and a long path length, which increases sensitivity and reproducibility.

b. Flameless or Electrothermal Vaporization (Fig. 12). Samples are placed in a graphite furnace or glassy carbon furnace. Atomization is carried out by joule heating. The usual commercial atomizers are resistively heated tubular furnace devices placed in the sample beam of the atomic absorption unit. Temperature increase is very fast (1000°C/s). The result is an absorption peak of a brief duration, the height or area of which is used for quantitation. The furnace is swept by a very pure inert gas. The lifetime of the furnace is not very high (about 100 measurements). Electrothermal analyzers provide high sensitivity for small volumes of sample (0.5–10 μL).

c. Hydride Generation. This technique is used to enhance sensitivity with the analysis of arsenic, antimony, tin, selenium, bismuth, or mercury. Sample is mineralized and then acidified. Reaction with sodium borohydride ($NaBH_4$) at elevated temperature under inert gas produces a volatile hydride that is more easily volatilized (Fig. 14). Mercury hydride is decomposed at low temperatures. High-speed instruments with an autosampler can determine 10 elements in 20 samples within 1 hour.

Background measurements in atomic absorption are extremely important, especially when low concentrations must be determined. Nonspecific absorption of light at the wavelength of the absorption line of the element causes a systematic error in the quantitative determination of the element. The Zeeman effect is of-

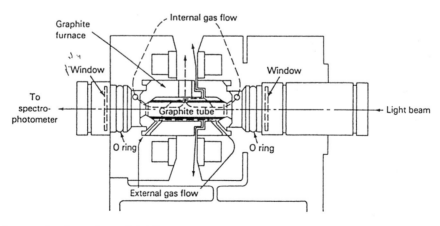

Fig. 12   Graphite furnace for electrothermal AAS.

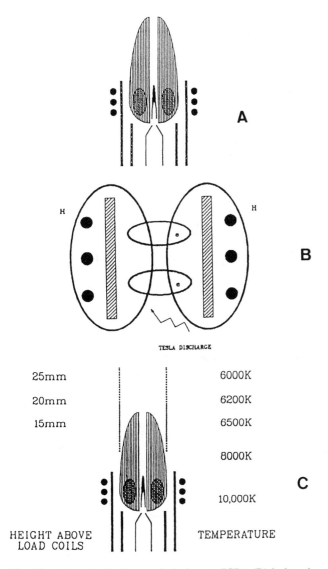

25mm   6000K

20mm   6200K

15mm   6500K

                8000K

               10,000K

HEIGHT ABOVE   TEMPERATURE
LOAD COILS

Fig. 13 (A) Inductively coupled plasma (ICP); (B) induced electric and magnetic fields; (C) temperature profile in the ICP.

F. Inductively Coupled Plasma (ICP)

Inductively coupled optical emission spectroscopy (ICP-OES) has emerged as a major analytical technique for trace metal determinations. Inductively coupled plasma mass spectrometry (ICP-MS) has revolutionized the mass spectroscopy of inorganic species. Detection limits in the sub-ppm range are possible for many elements, and the technique offers the ability to determine isotope ratios.

1. Principle of Operation

A plasma is a very hot gas in which a significant fraction of the atoms or molecules are ionized. Because of its ionic nature, a plasma is able to react when it is irradiated with electromagnetic beams. A plasma surrounded by a time-varying magnetic field is inductively coupled, i.e., currents flows are induced in the ionized medium. These current flows cause ohmic (resistive) heating of the plasma gas, enabling the plasma to be self-sustaining. At room temperature, argon gas does not contain any ion to initiate the plasma formation. To create a small number of ions, a high-voltage discharge is required. When species are entering the plasma, a nearly complete atomization occurs. In optical spectroscopy the plasma serves as excitation source and is combined with an optical spectrometer for selective elemental detection by observation of characteristic atomic emission line spectra. In the plasma, positively charged gas ions Ar^+ and electrons e^- are produced. Singly charged ions are generated, with a degree of ionization of over 80% for the majority of all elements.

2. Instruments

In practice the source arrangement commonly used consists of a quartz tube surrounded by a multiturn copper induction coil connected to a radio-frequency (RF) generator. The generator operates in the 20–50-MHz range, with a variable output power of up to 2 kW. For hot plasmas, RF power is 1200 W, for cool plasmas, it is 600–800 W. A Tesla coil provides the starting of operation. The plasma is prevented from touching the walls of the quartz tube by a thin screen of cool gas. The RF is tuned in such a way that a toroidal shape of plasma is obtained (Fig. 13). In this way the axial zone in the center is relativity cool in comparison to the periphery. A gas stream containing sample aerosol is injected into the center of the toroid without disturbing plasma stability. A pneumatic nebulizer is utilized. A microconcentric pneumatic nebulizer operates at 30 μl/min. It is of primary importance to yield ideal electro-

ten utilized by applying a magnetic field. Atomic lines are decomposed in three parts: the two components σ^+ and σ^- and a component π. The σ and π components are polarized in two perpendicular planes. The magnetic field is implemented around the flame or around the furnace.

Calibration curves are required. Matrix effects are minimized by the method of standard addition. The stabilized-temperature platform furnace, coupled with signal integration, has largely eliminated matrix problems.

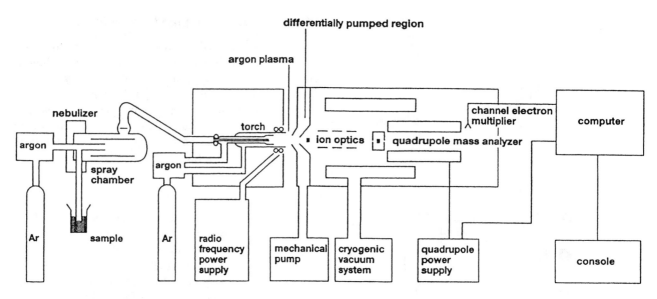

Fig. 14 Schematic diagram of a hydride generator and an ICP. (Courtesy of Perkin Elmer, Norwalk, CT.)

static conditions for ion production while keeping a stable plasma. Automatic positioning of the torch and a knitted induction coil increase the reproducibility of ion production. Physical interferences are often caused by samples that contain high levels of dissolved soils, such as seawater. Chemical interferences result from charges in vaporization or ionization (due, for example, to high amounts of sodium). Spectral interferences in atomic emission are caused by a continuum emission or overlapping emission light. The purchaser can find sequential and simultaneous multichannel instruments. An ICP-OES (optical emission spectrometer) is displayed in Fig. 15.

The ICP-MS technique uses ICP as an atmospheric-pressure ionization source. The ions are transported through successive pumping stages into the mass spectrometer at low pressure. The plasma–mass spectrometer interface is an ion-lens system. In the HP 4500, the lens system bends the ion beam into off-axis quadrupole rather than defocusing the ion around a conventional photon stopper. High ion transmission is the requirement. The charge on the element's ions from the plasma source is usually +1. Given a mass resolution of 1 atomic mass unit, one might expect minimal spectral overlap or complexity of the mass spectra.

G. Fluorescence

Fluorescence is a three-stage process. The first step is excitation via the absorption of radiation: a photon of energy $h\nu$ is supplied by a source (lamp or laser). This process distinguishes fluorescence from chemiluminescence, in which the excited state is created by a chemical reaction. Excitation occurs when the energy of the incident radiation corresponds to the energy spacing between the ground and one of the excited singlet states S_1, S_2, S_n and T_1, T_2, T_n for the excited triplet states, which by Hund's rule are lower in energy than the corresponding singlet states. A singlet state is one in which all the electrons in a molecule have a paired electron with opposite sign; a triplet state exists when two unpaired electrons have the same spin.

The excited state exists for a very short time (typically $1-10 \times 10^{-9}$ s); this is the second step.

The third step is emission. Emission is seldom observed from the higher singlet state because a radiationless process known as internal conversion results in an S_n-to-S_1 transition. From the excited singlet stage, S_1, a variety of transitions may occur. The most important are:

1. Radiationless internal conversion to S_0
2. Radiationless intersystem crossing to the triplet state, T_1
3. Radiative transition to S_0

These three transitions are all competing processes, and only the third one leads to fluorescence. Process 2 under specialized conditions can lead to phosphorescence (Fig. 16).

The quantum yield ϕ is a fundamental molecular property that describes the ratio of the number of photons emitted to the number of photons absorbed:

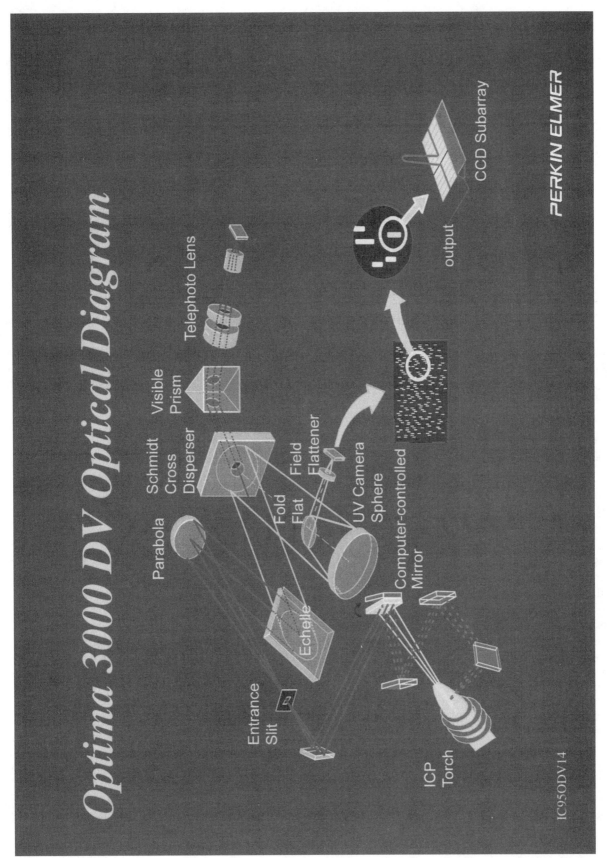

Fig. 15 An ICP-OES (optical emission spectrometer). (Courtesy of Perkin Elmer, Norwalk, CT.)

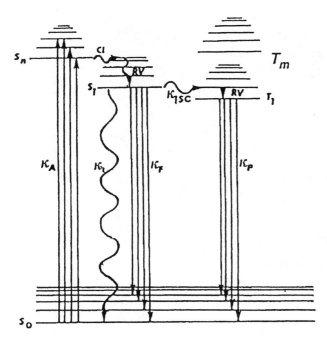

Fig. 16 Jablonski diagram: K_a = excitation rate constant; K_1 = de-excitation rate constant (pathway No. 1); K_f = fluorescence rate constant; K_p = phosphorescence rate constant; S_0 is for ground state; S_n, S_1 are for singlet state; T is for triplet state.

$$\phi = \frac{k_f}{k_f + \sum k_d}$$

where k_f is the rate constant for fluorescence emission and $\sum k_d$ is the sum of the rate constants for all the nonradiative processes that can depopulate S_1. The rate constant for radiative transition should be large relative to those for nonradiative transitions.

Numerous factors can affect molecular fluorescence, e.g., the type of solvent, the pH. A molecule that exhibits high fluorescence does not contain functional groups, which enhances the rates of radiationless transitions. In addition, such a molecule should possess a high molar absorptivity (ε).

The signal intensity I_f is given by Beer's law:

$$I_f = I_0(1 - e^{\varepsilon l C})\phi k$$

When sample absorbance is small, this expression reduces to

$$I_f = I_0 \cdot 2.3 \cdot \varepsilon \cdot l \cdot C \cdot \phi k$$

where k is the instrumental efficiency for collecting the fluorescence emission and I_0 is the intensity of the incident radiation.

1. Excitation Sources

Gas discharge lamps containing deuterium, mercury, zinc, cadmium, or xenon are the most common sources. They exhibit high spectral radiance and good stability.

> Xenon provides continuous emission in the 250–300-nm range.
> Deuterium provides continuous emission in the 200–300-nm range.
> Mercury exhibits lines at 254, 365, and 405 nm.
> Zinc exhibits lines at 214, 308, and 335 nm.
> > Laser radiation is monochromatic, and the output beams of lasers are highly collimated. With laser as excitation source, larger fluorescence signal levels are observed and nonlinear excitation is possible.

2. Excitation Wavelength Selection

Filters or monochromators are generally used. Filters are less expensive, but monochromators provide greater versatility and selectivity for excitation. Grating monochromators have a constant bandpass, regardless of wavelength selection. Gratings are either ruled or holographic.

3. Emission Wavelength Selection

There are several sources of radiation that must be selectively prevented from reaching the photomultiplicator. These included Rayleigh scattering, Raman scattering, second-order radiation, and solvent impurity emission. Appropriate emission wavelengths can be selected with monochromators or filters. Filters generally offer greater sensitivity than monochromators.

H. Chemiluminescence (CL)

Chemiluminescence is the production of electromagnetic radiation (UV, VIS, or IR) via a chemical reaction between at least two reagent, A and B, in which an electronically excited intermediate or product C* is obtained and subsequently relaxes to the ground state, with emission of a photon or by donating its energy to another molecule that then luminesces. The intensity of light emission depends on the rate of the chemical reaction, the yield of excited state, and the efficiency of light emission from the excited states. For example: Nitric oxide when in contact with ozone, produces a metastable nitrogen dioxide molecule that relaxes to a stable state by emitting at a wavelength of 700–900 nm:

$$NO + O_3 \rightarrow NO_2^* + O_2$$
$$NO_2^* \rightarrow NO_2 + h\nu$$

To convert chemically bound nitrogen to nitric oxide, the sample is submitted to oxidative pyrolysis. As the light emission occurs, light intensity is measured by a photomultiplier tube through a bandpass filter. One application is the thermal energy analyzer (TEA) for nitrosamines.

Most of chemiluminescence systems are those using:

Peroxyoxalate: reaction of hydrogen peroxide with an aryloxalate ester produces a high-energy intermediate (1,2-dioxetane-3,4-dione). In the presence of a fluorophore, the intermediate forms a charge transfer complex that dissociates to yield an excited-state fluorophore, which then emits a photon. Applications include the determination of hydrogen peroxide, polycyclic aromatic hydrocarbons, dansyl derivatives, and nonfluorescers (sulfite, nitrite) that quench the emission.

Acridinium esters: oxidation of an acridinium ester by hydrogen peroxide in alkaline medium.

Luminol: luminol (5-amino-2,3-dihydro-1,4 phthalazinedione) reacts with an oxidant (in the presence of a catalyst) to produce 3-aminophthalate, which emits at 425–435 nm in alkaline medium. One application is Co(II), Cu(II), Ni(II), Fe(III).

Firefly luciferase: luciferin reacts with adenine triphosphate (ATP) to form adenylluciferine, which oxidizes to form oxyluciferin, adenine monophosphate (AMP), CO_2, and light.

I. Mass Spectrometry

In a mass spectrometry experiment the material in the gas phase is introduced to the high-vacuum region of the ion source of the instrument. Here the molecules are ionized, usually by allowing them to interact with a beam of electrons, typically in the energy region of 70–10 eV, but there are many other ionization sources. From the ion source a mixture of molecular ions is produced, which gives molecular weight information and fragment ions, which contain the structural information.

Ions are separated according to their mass-to-charge ratio prior to detection. This is achieved by means of an external electric or magnetic field on the ion beam. It is the mass analyzer that yields a spectrum of abundance of ions (ion current) versus mass-to-charge (m/z) ratio.

A mass spectrometer consists of an introduction device, an ion source, and a analyzer (Fig. 17). The resolving power of a mass spectrometer is a measure of its ability to distinguish between two neighboring masses. Resolution is Δamu/amu (atomic mass unit). Spectrometers easily perform resolutions of 50,000 (i.e., distinguish Δamu = 0.01 when M = 500). Resolution is often written in ppm Δamu × 10^6/M.

High-resolution MS is of the double-focusing type, since a primary electrostatic analyzer lowers the dispersion of the ion beam and then a magnetic analyzer provides dispersion of the ion beam according to the mass-to-charge ratio.

The quadrupole mass filter consists of four parallel hyperbolic rods in a square array. The inside radius (field radius) is equal to the smallest radius curvature of the hyperbola. Diagonally opposite rods are electrically connected to radio frequency/direct current voltages. For a given radio frequency/direct current voltage ratio, only ions of a dedicated m/z value are transmitted to the filter and reach the detector. Ions with a different m/z ratio are deflected away from the principal axis and strike the rods. To scan the mass spectrum, the frequency of the radio frequency voltage and the ratio of the ac/dc voltages are held constant while the magnitude of ac and dc voltages are varied. The transmitted ions of m/z are then linearly dependent on the voltage applied to the quadrupole, producing an m/z scale that is linear with time.

For quadrupole mass spectrometers, selected ion monitoring (SIM) yields significantly enhanced detection limits, compared with scanning MS operation, because of the greater dwell time for signal acquisition at each selected m/z value.

Mass spectrometers are more and more miniaturized. They are routinely hyphenated to separation instruments (GC-MS, LC-MS, SFC-MS, CE-MS) (see further sections).

1. Electron Impact (EI)

Sources of EI consist of a heated, evacuated chamber in which a beam of electrons is generated from a heated metal filament. The energy of ionizing electrons is controlled by the voltage established between the cathode and the electron source filament. A voltage of 70 eV is standard practice; this is large enough to cause ionization and fragmentation of organic moieties.

Bombardment of a neutral molecule with an electron beam provides a molecular ion (or parent ion), but in many cases this ion is too unstable to be present in the spectrum.

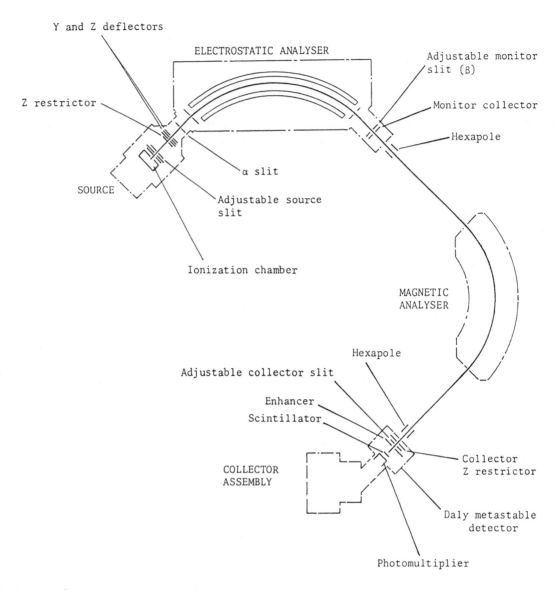

Fig. 17 Scheme of a mass spectrometer.

2. *Chemical Ionization (CI)*

In CI, mass spectra are produced by reaction between neutral organic molecules and reagent gas ion plasma. The concentration of reagent exceeds that of the sample by several orders of magnitude.

Chemical ionization sources are operated at high energies (200–500 eV), which favors the production of thermal electrons; CI produces stable molecular ions with little fragmentation.

Several gases are used in CI ionization: methane, propane, isobutane, hydrogen, ammonia, water, tetramethyl silane, and dimethyl amine.

3. *Fast Atom Bombardment (FAB)*

If a solid is bombarded by high-velocity particles, e.g., rare gas ions of about 8-keV energy, the material will be removed into the gas phase. Some of the sputtered material will be in the form of positively or negatively charged ions. An FAB source consists of

Atom gun
Atom beam
Sample holder
Lens system leading to the mass analyzer

In the mass spectrum one can even find electron molecular ions.

4. Electrospray (ES)

The electrospray process is initiated by applying an electrical potential of several kilovolts to a liquid in a narrow-bore capillary or electrospray needle.

There are three major processes in ES-MS:

1. *Production of charged droplets at the ES capillary tip*: A voltage of 2–3 kV is applied to the metal capillary. When the capillary is the positive electrode, source positive ions in the liquid will drift toward the liquid surface, and some negative ions drift away from it, until the imposed field inside the liquid is essentially removed by this charge redistribution. However, the accumulated positive charge at the surface leads to destabilization of the surface, because the positive ions are drawn downfield but cannot escape from the liquid. A liquid cone is produced. At a sufficiently high field E, the liquid cone vanishes and a fine mist of small droplets is generated. The droplets' surfaces are enriched with positive ions for which there are no negative counterions.

2. *Shrinkage of charged ES droplets*: The charge and size of the droplets depend on the spray conditions. When good conditions are provided, the droplets are small and exhibit a narrow distribution of sizes. The droplets shrink by evaporation of solvent molecules until they come close to the Rayleigh limit, which gives the condition in which the charges becomes sufficient enough to overcome the surface tension γ that holds each droplet together. They undergo fission into smaller droplets.

3. *Production of gas-phase ions*: Highly charged droplets are capable of producing gas-phase ions. Droplet evaporation is stimulated by the use of a current heated gas or heated sampling capillary. Two different mechanisms have been proposed to account for the formation of gas-phase ions from the small charged droplets. Extremely small droplets containing a single ion will give rise to a gas-phase ion. The other mechanism assumes emission or ion evaporation. Under some conditions the droplets do not undergo fission but emit gas-phase ions. The gas-phase ions are modified in the atmospheric and ion-sampling regions of the spectrometer. Analyte ion intensity depends on the analyte concentration and the pressure of the other electrolytes.

The ES technique is very popular due to the absence of critical temperature. (See Fig. 18.)

5. Thermospray (TS)

Thermospray was originally developed by Vestal and coworkers (5). The interface is as follows: A resistively heated capillary generates an aerosol from the effluent. The aerosol is sprayed into a desolvation chamber, which is maintained at fore vacuum pressure by a high-capacity pump. The aerosol spray is directed perpendicular to the MS entrance. Due to the combination of

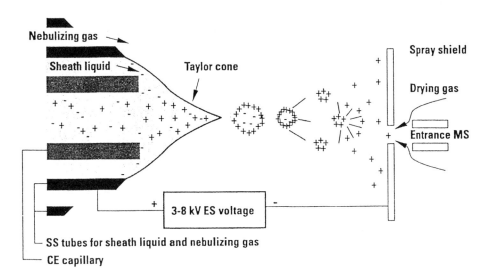

Fig. 18 Electrospray ionization/mass spectrometry.

heat and low pressure, the column effluent breaks up into large droplets, which in turn evaporate volatile solvent molecules.

Inside the desolvation chamber evaporation occurs that produces desolvated analyte molecules and primary ions, which are formed by ion evaporation. Ion evaporation is achieved by the addition of a volatile buffer salt, typically ammonium acetate. The desolvation chamber and the mass spectrometer are separated by one or more skimmers, and they are differentially pumped. Any ions in the aerosol may be forced into the mass spectrometer by a repeller electrode, which is commonly placed opposite the skimmer.

Use of thermospray is declining.

6. *Fourier Transform MS (FTMS)*

In an FTMS instrument, the ions of interest are detected by applying a very fast frequency sweep voltage to the transmitter plates following the ionization process. The frequency of this cyclotron motion is mass dependent. The coherent motion of the excited ions induces image currents in the receiver circuit. Positive ions approaching one receiver plate attract electrons. As they continue to move in their orbits they approach the opposite receiver plate and attract electrons on this surface. When the receiver plates are connected in a circuit, the induced image current of ions can be detected in the form of a time domain signal, which results from the superposition of a number of individual frequencies produced by the different ion species coherently orbiting at the same time. A mass spectrum is obtained by amplification, digitization, and conversion of this time domain signal to a frequency domain spectrum using Fourier transformation.

7. *Matrix-Assisted Laser Desorption Ionization (MALDI)*

In a MALDI/MS instrument, short-duration pulses of laser light are directed at a prepared mixture of sample plus a matrix polymer sample to be volatilized and ionized. These individual ions are accelerated to a fixed energy in an electrostatic field and directed into a field-free flight tube. The ions impact an ion detector, and the time intervals between the pulse of laser light and impact on the detector are measured.

8. *Time of Flight (TOF)*

A scheme follows (see bottom of the page).

A small number of ions is extracted from the source in a few microseconds and accelerated with few kilovolts; they are then directed to a tube. The process can be repeated 100,000 times per second. Kinetic energy is similar for every ion, and ions with higher velocities (light ions) will reach the end of the tube before heavy ions. Instruments have two tubes with a mirror in the middle; resolution may reach 5000. A schematic diagram of an instrument is displayed in Fig. 19. See also Ref. 6.

III. SAMPLE PREPARATION

A sample preparation step is often necessary to isolate the components of interest from a sample matrix. Sample preparation is often the major source of error in analytical procedures, for practitioners are often required by law to use traditional methods.

The ideal sample preparation technique should be simple, solvent free, efficient, and inexpensive. EPA method 3600C gives general guidance on the selection of cleanup methods that are appropriate for various target analytes.

A sample pretreatment will:

Improve accuracy.
Improve detectability.
Improve selectivity by removal of interfering matrix.

A result is reproducible if the sample to be analyzed is fully representative of the material to be tested. That

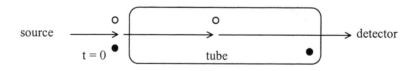

TOF - MS

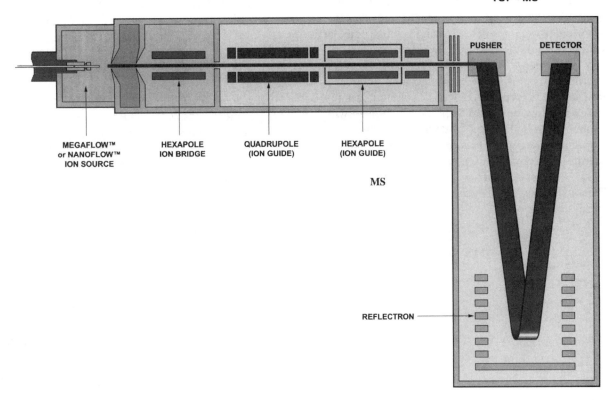

Fig. 19   Time-of-flight–mass spectrometer (TOF-MS) instrument. (Courtesy of Micromass.)

means that the sample taken can be equated with the entire batch.

We can distinguish off-line and online procedures.

A.   Lyophilization

Relatively large samples containing water-soluble analytes are frozen in a dry ice–acetone bath or in liquid nitrogen. Subsequently the frozen samples are placed in the freeze-dryer, where water is removed by vacuum sublimation. After freeze-drying, the residues can be dissolved in a suitable organic solvent.

B.   Ultrafiltration

Ultrafiltration involves the use of specialized membranes that allow rapid and gentle concentration or removal of molecules based on their molecular weight. Ultrafiltration membranes consist of a very thin and dense layer on top of a macroporous support that has progressively larger open spaces on the downstream side of the membrane. Substances that pass through the membrane will also pass easily through the macroporous support. Concentration can be accomplished be-

cause molecules smaller than the molecular-weight cutoff the membrane will flow through. Molecules larger than the membrane cutoff will be retained on the sample side of the membrane. Performances are affected by pressure. As the pressure increases, so does the flow rate, but there is a resistance from the concentrated layer on the surface of the membrane. This phenomenon is called *concentration polarization*.

Membranes are usually made of cellulose or polysulfone.

The porosity of the membrane determines the size of the molecules concentrated. In the supported liquid membrane device, analytes are extracted in a flow system from an aqueous sample through a hydrophobic membrane liquid into a second aqueous solution. The impregnated membrane is clamped between two circular PTFE blocks. Two types of transport mechanisms across the membrane are in use: pH gradient and ion-pairing formation.

C.   Dialysis

Dialysis is a method used to separate molecules through a semipermeable membrane. The concentration gradient of the components across the membrane drives the separation. Dialysis is used for removal of excess low-molecular-weight solutes. Analytes small enough to diffuse through the pores of the membrane are collected.

Microdialysis probes are implanted into the area of interest and slowly perfused with a solution, usually matching the fluid outside. The probe is equipped with a membrane through which substances pass due to the concentration gradient. It can be coupled online to liquid chromatography (HPLC), capillary electrophoresis (CE), or mass spectrometry.

The flux through a dialysis membrane is described by Fick's law:

$$I = -\frac{DA}{\tau}\frac{dc}{dx}$$

where

I = flux (mol/s)
D = solute diffusion coefficient (m^2/s)
A = membrane area (m^2)
τ = tortuosity of membrane
dc/dx = concentration gradient across membrane (mol/m^4)
$D = kT/6\pi\eta r$ (obtained from the Stokes Einstein equation)
k = Boltzmann's constant
T = absolute temperature (°K)
η = viscosity of medium
r = radius of molecule

The smaller the molecule, the larger the diffusion coefficient and the higher the flux. The relative recovery, or the dialysis factor, is the ratio of the dialysate (analyte in the outgoing liquid) to the concentration outside the membrane.

There are three modes of conducting a microdialysis experiment:

Perfusion and stirring are continuous over the monitoring period.
Perfusion is carried out only during sampling, but the stirring is maintained.
Perfusion and stirring are carried out during sampling only.

An automated sequential trace enrichment of dialysates (ASTED) instrument (Gilson) can be modified by exchanging the dialysis membrane holder for a membrane holder.

D. Methods for Volatile Organic Compounds (VOCs)

1. Headspace

a. Static Headspace. The sample is equilibrated in a closed system (usually a glass vial closed with a septum). The gas phase is manually sampled by a gas-tight syringe or with an electropneumatic sampling system in automated headspace analyzers, which can easily be interfaced with a GC chromatograph. The samples are injected directly into the GC column.

The method is simple and rather inexpensive, requires small volumes, and permits accurate quantification of major components. Conversely, low sensitivity and poor detectability of low-pressure compounds are the observed drawbacks.

b. Dynamic Headspace. Carrier gas is bubbled through an aqueous sample to purge VOCs from the matrix. These solutes are then collected using a cold trap or a sorbent trap.

2. Purge-and-Trap

Volatile organic compounds are purged from an aqueous sample by an inert gas at a moderate temperature and carried to a trap. Sorbent traps are packed with numerous sorbents (Tenax, activated carbon, Chromosorb, Carbosieve, etc.). Cold-trapping is usually performed. The temperature is brought to volatilization in a very short time (1000°C/s). Purge-and-trap systems are usually coupled online with GC. The method is the official method at U.S. water analysis laboratories. (See Fig. 20.)

3. Closed-Loop Stripping Analysis

Volatile components of the liquid phase are trapped in a sorbent by pumping the stripping gas in a closed circuit via the trap and the aqueous phase. When the purge is over, the trap is removed and organic compounds are eluted with carbon disulphide. The method is very sensitive and rapid, but recoveries for highly volatile compounds are low and CS$_2$ is very toxic.

4. Gas Stripping

An inert purge gas sparges through the aqueous sample, thus transporting volatile constituents from the liquid phase to the gas phase. Analytes are cryogenically trapped.

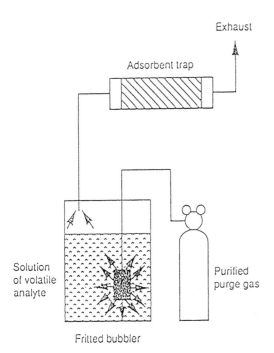

Fig. 20 Principle of purge-and-trap prior to GC. (Reproduced with permission from LC-GC Int.)

E. Liquid/Liquid Extraction (LLE)

This sample pretreatment is devoted mainly to organic compounds that can be removed from an aqueous solution by extracting them into a water-immiscible solvent (Fig. 21). Chelation or ion pairing between large and poorly hydrated ions and chelating agent may form neutral compounds that can be extracted by organic solvents.

When two immiscible solvents are placed in contact, any substance soluble in both of them will distribute or partition between the two phases in a definite proportion. According to the Nernst partition isotherm, the following relationship for a solute partitioning between two phases a and b holds:

$$\frac{(A)_a}{(A)_b} = K_d \qquad K_d = \text{partition coefficient}$$

$$(A)_a \leftrightarrow (A)_b \qquad \text{very often} \qquad (A)_w \leftrightarrow (A)_{org}$$

This assumes that no significant solute–solute interactions or strong specific solute–solvent interaction occur. The K_d value is constant when the distributing substance does not chemically react in either phase and the temperature is kept constant.

The fraction extracted R is related to K_d by

$$\frac{C_o V_o}{C_o V_o + C_w V_w} = \frac{K_d V}{1 + K_d V}$$

where C_o and C_w represent the solute concentration in organic (o) and water (w) phases, respectively, V_o and V_w are the volumes of organic and aqueous phases, and $V = V_o/V_w$.

It is possible to increase the extent of extraction with a given K_d by increasing the phase–volume ratio. When performing micro-LLE, the analyst works with an extreme ratio of extracting solvent/extracted liquid (for example, 1/800). Another way is to carry out a second and a third extraction. After n extractions, the final concentration of the compound in the aqueous phase is

$$Cw_n = C_w \left[\frac{V_w}{V_w + K_d V_o} \right]^n$$

An extraction process is more efficient if it is performed with several small portions of solvents.

Solubility increases as the values of solubility parameter δ of the solute and of the solvent get close:

$$\ln K_d = \frac{\bar{V}_s}{RT} [(\delta_s - \delta_i)^2 - (\delta_j - \delta_s)^2]$$

where $\bar{V}_s$ is the molar volume of the distributing solute, δ_s is its solubility parameter, and δ_i and δ_j are the solubility parameters of the pair of immiscible solvents.

It must be kept in mind that some species may exist under different forms in aqueous media. An acid, for example, must be written in this form:

$$HA \leftrightarrow A^- + H^+$$

$$K_d = \frac{(HA)_o}{(HA)_w} = \frac{(HA)_o}{(HA)_w + (A^-)_w}$$

Chelating agents such as dithizone or diphenylthiocarbazone are widely used for metallic cation extraction:

$$C_6H_5-N=N-C=S$$
$$\quad | $$
$$HN-NH-C_6H_5$$

may be written HQ. With metal it gives a ligand complex written MQ_n.

If we assume that no side reaction occurs in either the organic (o) or the aqueous (w) phase, then

$$M^{n+} + n(HQ)_{(o)} \leftrightarrow MQ_{n(o)} + nH^+$$

where M_{n+} is the metal ion, HQ is the chelate forming agent, and MQ_n is the chelate.

We can write

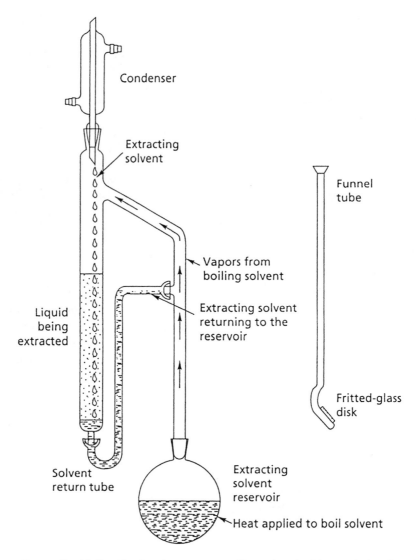

Fig. 21 Design of a continuous liquid–liquid extraction apparatus. (Reproduced with permission from LC-GC Int.)

$$K_{\mathrm{ex}} = \frac{(\mathrm{MQ}_n)_o (\mathrm{H}^+)^n}{(\mathrm{M}^{n+})(\mathrm{HQ})_o^n}$$

This equation describes the overall equilibrium. There are a number of intermediate equilibria:

HQ$_o$ ↔ HQ$_w$ Distribution of chelating agent between phases

HQ ↔ Q$^-$ + H$^+$ Dissociation of HQ in aqueous phase

M_n^+ + $n\mathrm{Q}^-$ ↔ MQ$_n$ Reaction of metal ions with ions of organic reagent in aqueous phase

MQ$_{n(o)}$ ↔ MQ$_{n(w)}$ Distribution of chelate between phases

$$K_{\mathrm{ex}} = K_d \frac{(\mathrm{H}+)^n}{(\mathrm{HQ})_o^n}$$

$$\ln K_d = \ln K_{\mathrm{ex}} + n\mathrm{pH} + n \ln(\mathrm{HQ})_o$$

The plot of this equation is sigmoid in shape and the position of the curves relative to the pH axis depends on the value of K_{ex}, the slope depends on the value of n.

Selective extractions can be carried out by careful selection of pH. When extracting organic species, it must be remembered that compounds may exist as different species, depending on the pH. For example, 8-hydroxyquinolinol(oxine) may be involved in two equilibria:

$$HQ \leftrightarrow Q^- + H^+ \quad \text{with the constant } K_{a1} = 10^{-5.5}$$

$$HQ + H^+ \leftrightarrow H_2 + Q \quad \text{with the constant } K_{a2} = 10^{-9.0}$$

The neutral HQ is very soluble in chloroform:

$$K_d = \frac{(HQ)_o}{(H_2Q^+)_w + (HQ)_w + (Q^-)_w}$$

$$K_d = \frac{(HQ)_o}{\left[1 + \dfrac{H^+}{K_{a1}} + \dfrac{K_{a2}}{(H^+)^2}\right]}$$

When the pH lies between 5.5 and 9.0, oxine is extracted by chloroform, beyond this range, recovery of oxine is nil.

P_{ow} is the partition coefficient of a solute between octanol and water. It is a common measure of hydrophobicity. The usual measurement for log P_{ow} is the shake flask method, where a compound is shaken in an octanol–water mixture. After equilibrium, the concentration is measured in one or both phases. Reversed-phase HPLC has also been extensively used. Correlation plots of log P_{ow} against retention time or capacity factors are drawn. High values of log P_{ow} give guidelines for extraction. See Table 3 for the partition coefficients of pesticides. A typical example of liquid–liquid extraction is displayed in Fig. 22.

F. Solid-Phase Extraction (SPE)

This is the most widely used method. Analytes (mainly organics) are trapped by a suitable sorbent by passing through a plastic cartridge containing an appropriate support. A selective organic solvent is used to wash out the target analytes. The SPE technique is rapid and relies upon chromatographic retention and log P_{ow}. It can easily be automated. Off-line procedures are inexpensive. Online devices are readily available from many companies.

1. Off-Line

A typical SPE cartridge is displayed in Fig. 23.

Sorbents are very similar to liquid chromatography stationary phase. The analyst can take advantage of:

Nonpolar interactions (hydrophobic): typical octadecyl modified silica, polystyrene, divinylbenzene copolymers, carbon-based solvent.

Polar interactions through hydrogen bond, for example: In this mode sorbents are bare silica, polar bonded silica, polyamide.

Ion exchange: benzene sulphonic acid (cation exchange) quaternary amine (anion exchange).

Immunosorbents: the lack of selective sorbents to trap organic analytes in water is certainly the most important weakness of the SPE technique. Selective interactions are involved with immunoaffinity sorbents.

An alternative to SPE columns is discs (Figs. 24 and 25). Discs do not exhibit bed channeling. Samples can be applied to the discs using either a syringe or a vacuum manifold. Flow rates can be made faster by pushing samples through the discs with increasing pressure on the syringe. Conditioning is carried out with methanol and water.

The procedure for the SPE technique is as follows: conditioning, sample application, washing, and elution. The adsorbent must be wetted. With bare silica, no problem generally occurs, but it does with hydrophobic sorbents. In this case, adsorbent must be treated with a suitable solvent. Methanol is preferred for most applications; however, other solvents that are miscible with water, such as isopropanol, THF, and acetonitrile, are convenient as well. Conditioning is achieved with about 2–3 column volumes of solvent. Then 1–2 column volumes of the sample solvent are poured through the column. After this step, the phase must not run dry. After adsorption of the sample molecules, the phase may run dry through either a water jet pump or flushing with inert gas. Elution is then performed with a strong solvent to elute with the lowest possible volume (Fig. 26).

The capacity is the quantity of sample molecules retained per unit quantity of adsorbent. Capacity depends on solute size. It lies in the range of 4–60 mg/g of packing. Breakthrough of solutes occurs when they are no longer retained by the sorbent. Overloading beyond the sorbent capacity may also lead to breakthrough of analytes. The breakthrough volume can be measured from the breakthrough curve obtained by monitoring the signal of the effluent from the extraction column (Fig. 27). V_b is usually defined as 1% of the initial absorbance and corresponds to the sample volume that can be handled without breakthrough. V_r is the retention volume of the analyte. V_m is defined as 99% of the initial absorbance.

The method is time consuming. Another method is proposed: A small volume spiked with a trace concentration (μg/L) level of all the analytes is percolated

Table 3 Partition Coefficients (*n*-Octanol–Water) of Pesticides

Compound	log P_{ow}	Derivation
Acrolein	0.90	Cited
Aldicarb	1.08	Measured, inversion
	1.57	Measured, inversion
	1.13	Measured, shake flask
Aldicarb sulfone	−0.57	Measured, inversion
Aldicarb sulphoxide	−1.0	?
Aldoxycarb	−0.57	Measured, RP-TLC
Aldrin	5.66	Calculated
	7.4	Measured, RP-TLC
	6.50	Measured, shake flask
Allethrin	5.0	?
Ametryn	3.07	Measured, HPLC
	3.07	Calculated
Aminocarb	1.73	Measured, shake flask
Amitrole	−0.87	Measured, inversion
Atraton	2.69	Measured, HPLC
	2.69	Calculated
Atrazine	2.40	Measured, HPLC
	2.21	Measured, HPLC
	2.64	Measured, shake flask
	2.68	or
	2.75	Cited
	2.61	Measured, HPLC
	2.61	Calculated
	2.75	Measured, RP-TLC
	2.47	Measured, HPLC
Azinphos ethyl	3.40	Measured, shake flask
Azinphos ethyl O-analogue	1.63	Measured, shake flask
Azinphos methyl	2.69	Measured, shake flask
Azinphos methyl O-analog	0.78	Measured, shake flask
Benalaxyl	3.4	Cited
Benomyl	2.12	Cited
α-BHC	3.81	Measured, shake flask
	3.78	Measured, slow stirring
β-BHC	3.80	Measured, shake flask
	3.84	Measured, slow stirring
γ-BHC (lindane)	3.72	Measured, shake flask
	3.72	Measured, shake flask
	3.66	Measured, shake flask
δ-BHC	4.14	Measured, shake flask
Bifenthrin	6.00	Cited
Bromophos	4.88	Measured, shake flask
	5.21	Measured, slow stirring
Bromophos ethyl	5.68	Measured, shake flask
	6.15	Measured, slow stirring
Camphechlor	5.50	Cited
Captafol	3.83	Measured, inversion
Captan	2.54	Measured, inversion
	2.35	Measured, inversion
Carbanolate	2.3	Measured, inversion
Carbaryl	2.34	Measured, shake flask
	2.32	Measured, inversion
	2.36	Measured, inversion

Table 3 Continued

Compound	log P_{ow}	Derivation
	2.31	Measured, shake flask
	2.29	Measured, shake flask
Carbendazim	1.40	Measured, inversion
	1.52	Measured, inversion
Carbofuran	1.63	Measured, shake flask
Carbophenothion	5.12	Measured, shake flask
	5.66	Measured, slow stirring
Carbophenothion methyl	4.82	Measured, shake flask
Carboxin	2.14	Cited
Chloramben methyl ester	2.8	Measured, inversion
Chlorbromuron	3.09	Measured, inversion
Chlordane	5.16	Calculated
	6.00	Measured, HPLC
α-Chlordane	6.00	Cited
	6.0	Cited
γ-Chlordane	6.0	Cited
Chlordimeform	2.89	Measured, shake flask
Chlorfenac methyl	3.8	Measured, inversion
Chlorfenvinphos	3.10	Measured, inversion
	3.23	Measured, inversion
	3.81	Measured, shake flask
	3.80	Measured, shake flask
	3.82	Measured, shake flask
Chloridazon	1.14	Measured, shake flask
	1.50	Measured, inversion
Chlornitrofen	3.67	Measured, shake flask
Chlorotoluron	2.41	Measured, inversion
Chloroxon	1.83	Measured, shaking
Chloroxuron	3.7	Measured, inversion
Chlorpyrifos	5.11	Measured, shake flask
	4.96	Measured, shake flask
	5.2	Measured, shaking
	5.27	Measured, slow stirring
Chlorpyrifos methyl	4.31	Measured, shake flask
	4.30	Measured, shake flask
Chlorsulfuron	1.09 to −0.41 at pH 4.5–12.0	Measured, stirring
Chlorthion	3.45	Measured, shaking
	3.63	Measured, slow stirring
Clofentezine	2.18	Cited
	3.1	Cited
Clopyralid	1.76	Calculated
Cyanazine	1.8	Measured, HPLC
	1.66	Calculated
Cyanophos	2.71	Measured, slow stirring
Cycloheximide	0.55	Measured, inversion
Cyhexatin	5.39	Cited
Cypermethrin	4.47	Measured, shake flask
2,4 Dichlorophenoxyacetic acid (2,4-D)	2.90 (undissociated)	Measured, inversion
	−0.24 (dissociated)	Measured, inversion
	2.81	Measured, shake flask
2,4-D dimethylamine	0.65	Measured, HPLC

Table 3 Continued

Compound	log P_{ow}	Derivation
2,4-D octyl ester	5.86	Measured, HPLC
	6.71	Calculated
	6.89	Calculated
DDE	5.63	Measured, HPLC
p,p'-DDE	5.69	Measured, HPLC
	5.89	Measured, HPLC
	6.96	Measured, slow stirring
	5.69	Measured, HPLC
	6.09	Measured, HPLC
DDT	4.64	Measured, HPLC
	3.98	Measured, shake flask
	5.90	Measured, shake flask
	6.12	Measured, HPLC
	5.89	Measured, HPLC
o,p'-DDT	5.75	Measured, HPLC
p,p'-DDT	6.38	Measured, shake flask
	6.2	Measured, slow stirring
	6.19	Measured, shake flask
	6.91	Measured, slow stirring
Deet	2.02	Measured, HPLC
Demetonthiol	1.93	Measured, shake flask
Desethylatrazine	1.53	Calculated
	1.51	Measured, HPLC
Desisopropylatrazine	1.12	Calculated
	1.15	Measured, HPLC
Dialifos	4.69	Measured, shake flask
Diazinon	3.11	Measured, inversion
	3.81	Measured, shake flask
	3.14	Measured, shake flask
Diazoxon	2.07	Measured, shake flask
Dicapthon	3.44	Measured, shaking
	3.58	Measured, shake flask
	3.62	Measured, shake flask
	3.72	Measured, slow stirring
Dicapthoxon	1.84	Mesured, shaking
Dicholofenthion	5.14	Measured, shake flask
3,4 Dichloroaniline	2.78	Measured, inversion
p-Dichlorobenzene	3.42	Measured, shake flask
	3.38	Measured, generator column
	3.44	Measured, slow stirring
2,4-Dichlorophenol	2.8	Measured, inversion
Dichlorvos	1.47	Measured, shake flask
Dieldrin	4.54	Measured, slow stirring
	4.32	Measured, shake flask
	5.40	Measured, slow stirring
Dimethoate	0.50	Measured, shake flask
	0.79	Measured, inversion
	0.78	Measured, shake flask
Diphenyl	3.63	Measured, shake flask
	4.00	Measured, slow stirring
	4.00	Measured, shake flask
	3.83	Measued, generator column
	4.01	Measured, slow stirring
	3.91	Measured, shake flask
Diquat dichloride	−3.55	Cited

Table 3 Continued

Compound	log P_{ow}	Derivation
Disulfoton	4.02	Measured, shake flask
Disulfoton sulfone	1.87	Measured, shake flask
Disulfoton sulfoxide	1.73	Measured, shake flask
Diuron	2.68	Measured, inversion
Dowco 275	3.51	Measured, inversion
2,2-DPA (Dalapon, 2.2-dichloropropionic acid)	0.78	Calculated or cited
Endrin	4.56	Measured, shake flask
	5.20	Measured, slow stirring
EPN (O-ethyl O-4-nitrophenyl phenylphosphonothioate)	3.85	Measured, shake flask
Ethion	5.07	Measured, shake flask
ETU (ethylene thiourea)	−0.66	Cited
Fenamiphos	3.18	Measured, inversion
	3.23	Measured, shake flask
Fenchlorphos	4.88	Measured, shake flask
	4.81	Measured, shake flask
	5.07	Measured, slow stirring
Fenitrooxon	1.69	Measured, shaking
Fenitrothion	3.38	Measured, shake flask
	3.30	Measured, shaking
	3.40	Measured, shake flask
	3.47	Measured, slow stirring
	3.44	Measured, shake flask
Fenobucarb	3.18	Measured, shake flask
Fenoprop	2.44	Cited
	3.86	Cited
Fenpropathrin	3.03	Measured, shake flask
Fensulfothion	2.23	Measured, shake flask
Fensulfothion sulfide	4.16	Measured, shake flask
Fensulfothion sulfone	2.56	Measured, shake flask
Fenthion	4.09	Measured, shake flask
	4.17	Measured, slow stirring
Fenuron	0.96	Measured, inversion
Fenvalerate	4.42	Measured, shake flask
	6.2	Measured, shaking
Flamprop	2.90 (undissociated)	Measured, inversion
	−0.40 (dissociated)	Measured, inversion
Flucythrinate	6.2	Mesured, shaking
Fluometuron	2.42	Measured, inversion
Fluorodifen	4.4	Measured, inversion
Fluvalinate	>3.85	Cited
Fonofos	3.89	Measured, shake flask
Fonofos O-analogue	2.11	Measured, shake flask
Guazatine	−1.15 at pH 3	Cited
Haloxyfop	4.47	Cited
	3.52	Cited
Haloxyfop methyl ester	4.07	Cited
Hexachlorobenzene (HCB)	5.50	Measured, shake flask
	5.44	Measured, inversion
	6.18	Measured, shake flask
	5.47	Measured, shake flask
	5.47	Measured, generator column
	5.73	Measured, slow stirring
	5.66	Measured, shake flask

Table 3 Continued

Compound	log P_{ow}	Derivation
Heptachlor	5.38	Calculated
	5.27	Measured, HPLC
	6.06	Calculated
	5.44	Measured, HPLC
	5.5	Calculated
	5.58	Measured, HPLC
Heptachlor epoxide	5.40	Measured, HPLC
Hexythiazox	2.53	Cited
Hydramethylnone	2.31	Cited
Imazapyr	0.11	Cited
Imazaquin	0.34	Cited
3-Indoleacetic acid	1.41	Cited
Iodofenphos	5.16	Measured, shake flask
Iprobenfos	3.21	Measured, shake flask
Isazofos	3.82	Calculated from solubilities
	3.82	Measured, RP-TLC and HPLC
Isofenphos	4.12	Measured, shake flask
Leptophos	6.31	Measured, shake flask
	5.88	Measured, shake flask
	4.32	Measured, shake flask
Leptophos O-analogue	4.58	Measured, shake flask
Linuron	2.76	Measured, inversion
Malathion	2.89	Measured, shake flask
	2.84	Measured, shake flask
	2.94	Measured, slow stirring
Maleic hydrazide	−0.63	Measured, inversion
Metalaxyl	1.27	Calculated from solubilities
	1.65	Measured, RP-TLC and HPLC
Metflurazon	2.67	Measured, shake flask
Methidathion	2.42	Measured, shake flask
Methiocarb	2.92	Measured, inversion
Methomyl	0.13	Measured, shake flask
Methoxychlor	3.31	Measured, shake flask
Metobromuron	2.38	Measured, inversion
Metolachlor	3.28	Calculated from solubilities
	3.13	Measured, RP-TLC and HPLC
Metoxuron	1.64	Measured, inversion
Metribuzin	1.70	Cited
Mirex	6.89	Measured, HPLC
Molinate	3.21	Measured, shake flask
Monolinuron	2.30	Measured, inversion
Monuron	1.98	Measured, inversion
Naled	1.38	Measured, shake flask
Naphthalene	3.36	Measured, inversion
	3.25	Measured, slow stirring
	3.28	Measured, shake flask
	3.31	Measured, shake flask
NIA 24 110 (5-benzylfur-3-yl-methyl-*trans*-(+)-3-cyclo-pentylidenemethyl-2,2-di-methylcyclopropanecarboxylate) (RU 11679)	7.14	?
Nitrapyrin	3.02	Measured, inversion
Norflurazon	2.30	Measured, shake flask
Oxamyl	−0.47	Measured, inversion
Oxycarboxin	0.9	Measured, inversion
Paclobutrazol	3.2	Cited

Table 3 Continued

Compound	log P_{ow}	Derivation
Paraquat di-iodide	−5.00	Cited
Paraoxon	1.59	Measured, shake flask
	1.98	Measured, shake flask
Paraoxon methyl	1.28	Measured, shake flask
	1.21	Measured, shaking
Parathion	2.15	Measured, shake flask
	3.93	Measured, inversion
	3.81	Measured, shake flask
	3.76	Measured, shake flask
Parathion amino	2.60	Measured, shake flask
Parathion methyl	2.04	Measured, shake flask
	2.99	Measured, shaking
	2.94	Measured, shake flask
	1.8	Measured, shaking
	3.04	Measured, slow stirring
Pentachlorophenol	5.01	Measured, HPLC
	3.69	Measured, shake flask or cited
	5.01	Measured, HPLC
Pentanochlor	3.7	Measured, inversion
Permethrin	6.6	Measured, calculated
	3.49	Measured, shake flask
	6.5	Measured, shaking
	5.84 (trans)	Measured, HPLC
	6.24 (cis)	Measured, HPLC
Phenothiazine	4.15	Cited
Phenoxyacetic acid	1.47	Calculated from solubilities
	1.52	Measured, HPLC
Phenthoate	3.96	Measured, slow stirring
	2.89	Measured, shake flask
o-Phenyl phenol	3.09	Cited
Phorate	4.26	Measured, inversion
	3.83	Measured, shake flask
Phorate sulfone	1.99	Measured, shake flask
Phorate sulfoxide	1.78	Measured, shake flask
Phosalone	4.30	Measured, shake flask
	4.38	Measured, shake flask
Phosmet	2.83	Measured, shake flask
	2.78	Measured, shake flask
	2.81	Measured, slow stirring
Phoxim	4.39	Measured, shake flask
Picloram	0.30	Calculated
Picloram methyl ester	2.3	Measured, inversion
Pirimiphos ethyl	4.85	Measured, shake flask
Pirimiphos methyl	4.20	Measured, shake flask
PP450 (Flutriafol)	2.29	Measured, inversion
Profenofos	4.70	Calculated from solubilities
	4.70	Measured, RP-TLC and HPLC
Profluralin	6.34	Calculated from solubilities
	5.58	Measured, RP-TLC and HPLC
Prometon	3.1	Calculated
	2.99	Measured, HPLC
Prometryn	3.48	Calculated
	3.34	Measured, HPLC
Propanil	2.8	Measured, inversion
Propazine	3.02	Calculated
	2.91	Measured, HPLC

Table 3 Continued

Compound	log P_{ow}	Derivation
Propham	2.60	Measured, inversion
Propoxur	1.58	Measured, shake flask
	1.55	Measured, shake flask
Quintozene	4.22	Measured, shake flask
Resmethrin	6.14	?
Simazine	1.51	Measured, inversion
	2.06	Measured, HPLC
	1.96	Measured, HPLC
	2.2	Calculated
	2.26	Measured, HPLC
Simetryn	2.66	Calculated
	2.8	Measured, HPLC
Strychnine	1.93	Cited
Swep	2.80	Cited
2,4,5-Trichlorophenoxyacetic acid (2,4,5-T)	0.60	Calculated
Trichloroacetic acid	1.33	Cited
	0.10	Cited
	1.96	Calculated
p,p-TDE	6.22	Measured, slow stirring
Temephos	5.96	Measured, shake flask
Terbufos	4.48	Measured, shake flask
Terbufos sulfone	2.48	Measured, shake flask
Terbufos sulfoxide	2.21	Measured, shake flask
Terbumeton	3.1	Calculated
	3.1	Measured, HPLC
Terbuthylazine	3.02	Calculated
	3.06	Measured, HPLC
Terbutryn	3.72	Calculated from solubilities
	3.74	Measured, RP-TLC and HPLC
	3.48	Calculated
	3.43	Measured, HPLC
Tetrachlorvinphos	3.53	Measured, shake flask
Tetramethrin	4.7	?
Thiazfluron	1.46	Calculated from solubilities
	1.85	Measured, RP-TLC and HPLC
Thiobencarb	3.4	Measured, shaking
	3.42	Measured, shake flask
Tolylfluanid	390	Cited
Triazophos	3.55	Measured, shake flask
Triadimefon	2.77	Measured, inversion
Trichlorfon	0.43	Measured, shake flask
Trichloronate	5.23	Measured, shake flask
2,4,6-Trichlorophenol	2.97	Measured, shake flask or calculated
Tridiphane	4.34	Cited
Trietazine	3.15	Calculated
	3.07	Measured, HPLC
Trifluralin	3.97	Measured, shake flask
Vinclozolin	3.0	Cited
Warfarin	0.05	Cited
	2.72	Cited

Source: A. Noble. Partition coefficients (*n*-octanol–water) for pesticides. J. Chromatogr. 642:3–14 (1993).

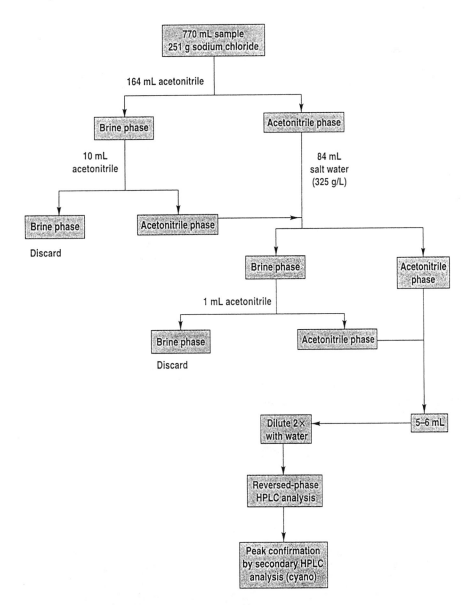

Fig. 22 Sample preparation flowchart for EPA method 8330.

through the cartridge, and peak areas are recorded. The first volume is selected so that breakthrough does not occur for any solute. The sample volume is then increased and the concentration decreased in order to have a constant amount of analytes in the percolated samples. In this mode, peak areas remain constant. The breakthrough volume of an analyte is calculated when the peak area begins to decrease, and the corresponding recovery can be also calculated by dividing the peak area obtained for the sample volume by the constant peak area obtained for sample volumes before breakthrough.

Prediction of breakthrough volume is important for selecting a convenient sorbent and consequently the amount of sorbent.

Hydrophobic sorbents: n-Alkyl silicas are by far the most utilized. A large number of applications on such sorbents has been published. The drawback (as with every bonded silica) is the poor stability in very acidic or basic media.

From liquid chromatography we know that in reversed-phase mode,

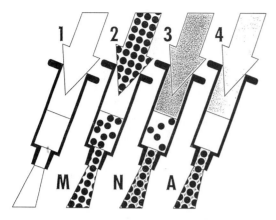

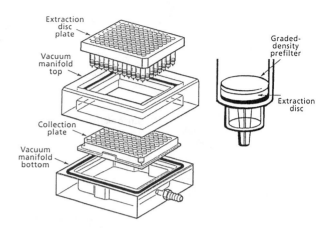

Fig. 23 Solid-phase extraction (SPE) cartridge and the four steps of operation. (Courtesy of Merck.)

Fig. 25 Schematic diagram of a 96-well SPE extraction disc microtiter plate. (Reproduced with permission from LC-GC Int. and by courtesy of Empore.)

$$\ln k = \ln k_w - S\phi \text{ or } \ln k = \ln k_w - b\phi - a\phi^2$$

where ϕ is the organic modifier volume percent in the binary mobile phase (water/modifier). One can estimate $\ln k_w$ by a graphical extrapolation to zero modifier content; $\ln k_w$ represents the hypothetical capacity factor of the solute with pure water as eluent. Since $\ln k_w$ is very often correlated with $\log P$ octanol/water, it is often taken as a hydrophobicity constant. Values of $\ln k_w$ may be as high as 3–4, which means that large sample volumes with trace amount of solutes can be handled.

Styrene divinyl benzene copolymers (PS-DVB), either porous or rigid, are stable over the whole pH range. Calculated $\ln k_w$ values on these sorbents are higher than those on C_{18}. Consequently, moderately polar compounds, which are not retained by C_{18} silica, are more readily concentrated on these sorbents. Carbon-based sorbents are gaining acceptance, with the growing availability of porous graphitic carbon. Data obtained with this support demonstrate that it exhibits high retention for apolar compounds, but some differences with PS-DVB and C_{18}-bonded

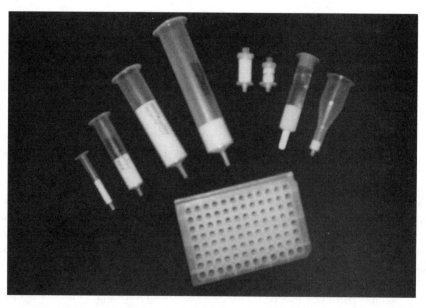

Fig. 24 Solid-phase extraction (SPE) cartridges.

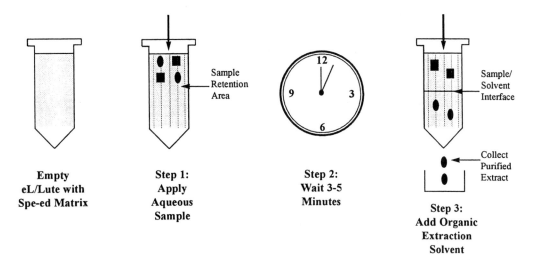

Fig. 26 Elution protocol for SPE cartridges.

silica are observed. Selection of the eluting solvent is performed through knowledge of the eluting strength ε^o of the solvent in the chromatographic mode.

Use of polar adsorbents is less advocated. The behavior of the solute is well understood since a lot of chromatographic data can be retrieved from thin-layer chromatography (TLC) experiments.

By adjusting pH, many solutes can be ionized (e.g., carboxylic acids) and trapped on ion exchangers. Owing to pH stability, resins are widely used. The drawback comes from high amounts of inorganic ions (e.g., seawater), which easily overload the capacity of the sorbent.

Immunosorbents: It is now possible to produce antibodies against some target compounds, including some small molecules such as pesticides. For example, an immunosorbent made with polyclonal anti-isoproturon antibodies covalently bound to a silica sorbent is able to concentrate several phenylureas. Due to the high selectivity, phenylureas can be detected at the 0.1-μg/L level in wastewaters.

Solid-phase microextraction uses a 1-cm length of focused silica fiber, coated on the outer surface with a stationary phase and bonded to a stainless steel plunger holder that looks like a modified microliter syringe. The fused silica fiber can be drawn into a hollow needle by using the plunger. In the first process, the coated fiber is exposed to the sample and the target analytes are extracted from the sample matrix into the coating.

The fiber is then transferred to an instrument for desorption (Fig. 28).

2. Online

Online coupling of SPE to either LC or GC is easily performed. In the simplest way, a precolumn is placed in the sample loop position of a six-port switching valve. After conditioning, sample application, and cleaning via a low-cost pump, the precolumn is coupled to an analytical column by switching the valve into the inject position. The solutes of interest are eluted directly from the precolumn to the analytical column by an appropriate mobile phase. The sequence can be fully automated, for example, in the Prospekt system (Spark Holland, Emmen NL) (Fig. 29).

A few commercial robotic systems are available, notably the Aspec (Fig. 30) (Gilson, Villiers le Bel, F) Millilab (Millipore, Bedford, MA), and Benchmate (Zymark, Hopkinton, MA). The SAMOS (System for Automated Monitoring of Organic Compounds in Surface Water) combines a Prospekt sample treatment module and an HPLC separation diode array detection unit.

3. Cleanup

Extracts obtained from either LLE or SPE contain analytes and other compounds that may interfere in the chromatographic separation. A cleanup is required. The most widely used is fractionation by LC. Extract is loaded onto a chromatographic column packed with an appropriate sorbent (silica, alumina, florisil, bonded sil-

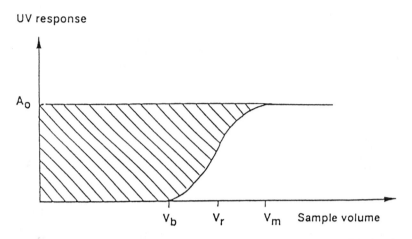

Fig. 27 Breakthrough curve obtained by percolation of a spiked water sample with a UV absorbance A_o through an SPE cartridge. Breakthrough occurs for a sample volume V_b, usually defined as 1% of the initial absorbance. V_r is the retention volume of the analyte, and V_m is defined as 99% of the initial absorbance. (Reproduced by permission of the *Journal of Chromatography*, Elsevier, Amsterdam.)

ica) and step elution with solvents is carried out. Each fraction is collected and submitted to chromatography. Derivatization prior to fractionation is sometimes performed.

Coupling two sorbents in the SPE procedure, for example, hydrophobic sorbent and ion exchange in series, is efficient.

A chart on sample preparation is available on request from LC-GC International. The reader can find further information on sample preparation in Refs. 8, 9.

IV. Flow Injection Analysis (FIA)

Principle of operation: A definite volume of a liquid sample solution is injected into a moving, nonsegmented, continuous carrier stream of a suitable liquid. The injected sample forms a zone that begins dispersing (and reacting if necessary) with the carrier stream as it is transported toward a detector. The sample is introduced through an injection valve with a loop of an accurately known volume. Injection volumes lie be-

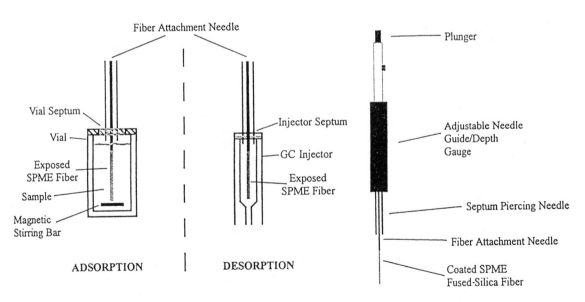

Fig. 28 Schematic of an SPME unit with adsorption and desorption steps. (Reproduced by permission of the *Journal of Chromatography*.)

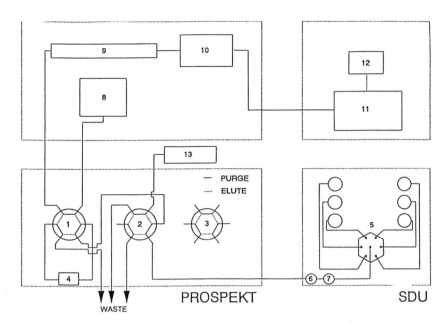

Fig. 29 Prospekt system, an automated online trace enrichment LC system for water samples: 1, 2, 3 = high-pressure valves of the Prospekt; 4 = trace enrichment cartridge of the Prospekt; 5 = solenoid valve; 6 = pulse damper; 7 = purge pump; 8 = solvent delivery system of the HPLC; 9 = analytical column; 10 = diode array detector; 11 = computer; 12 = printer; 13 = preparative pump for sample loading; SDU = solvent delivery unit. (Reproduced by permission of the *Journal of Chromatography.*)

tween 1 and 200 μl (typically 25 μl). A continuous record is carried out during passage of the sample material through the flow cell of the detector. A typical recorder output has the shape of a Poisson peak, the height of which is related to the concentration of the analyte (Fig. 31). Narrow-bore polytetrafluoroethylene (PTFE) tubing (as coils to aid mixing) is used for sample and reagent transport (Fig. 32).

Any physical parameter change can be monitored, which enables a wide range of possible detectors (absorbance, electrode potential, biosensor, etc.). Residence time is around 30 s, which permits about 100 determinations per hour.

Letting C^o be the original concentration of the injected sample solution and C be the concentration of any element of fluid along the gradient of the dispersed sample zone, the dispersion coefficient is $D = C^o/C$, which can be related to a fixed delay time.

A. Techniques

1. Gradient Dilution

The principle is based on selecting for the analytical readout any point other than the peak maximum C_{max}. Each element along the dispersed sample zone will be characterized by a fixed dispersion coefficient D-value corresponding to a fixed delay time t.

2. Gradient Calibration

To avoid calibration by means of serially diluted solutions, a series of sequential C-values spaced along the gradient are identified through increasing delay times.

3. Titrations

If a sample zone, for example, an acid, is injected into a carrier stream of a base, the dispersed zone will become gradually neutralized by the base that is penetrating through the interfaces with the carrier stream at the leading and tailing sections. Therefore an element of fluid exists both at the front and at the tail of the zone within which the acid is exactly neutralized by the base. The two equivalence points form a pair having the same response value on the recording chart, and elapsed time Δt will increase (or decrease) with increasing concentration of acid (decrease with increasing base).

4. Stopped Flow

One takes advantage of the fact that when the carrier flow stops moving, the dispersion of the sample zone will stop and D will become independent of time.

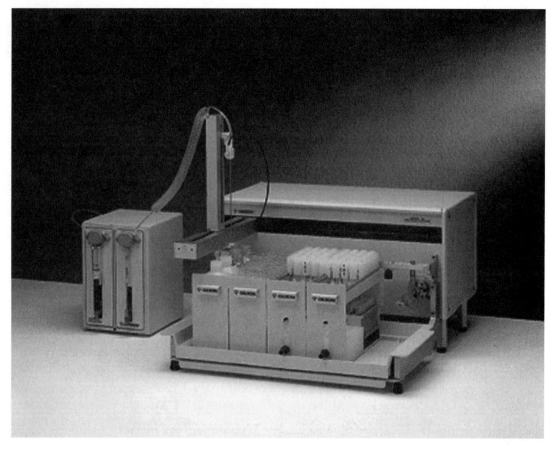

Fig. 30 Fully automated chromatographic system incorporating robotic sample handling (Aspec XL) with SPE, separation, and biosensor detection for the determination of phenols in surface waters. (Courtesy of Gilson medical electronics, Villiers le Bel, France.)

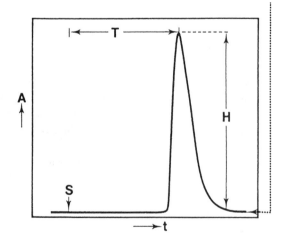

Fig. 31 Typical FIA recorder output. *S*: sample injection; *T*: residence time; *H*: peak height; *A*: absorbance; *t*: time.

5. Reagent Addition

A manifold provides the means of bringing together the fluid lines that allow chemical reaction to take place. Manifolds with several lines can be assembled (Fig. 33).

B. Detection

A lot of detecting devices can be used. Most of them are typically liquid chromatography detectors.

> *Spectrophotometry with or without diode array*: for example, nitrite detection. A reaction coil is packed with Cd granules to reduce nitrate to nitrite. Nitrite is allowed to react with sulfanilamide and *N*-naphthylethylene diamine to form a diazo compound (λ_{max} 565 nm).
>
> *Fluorimetry*.
>
> *Atomic absorption*: Reaction with a reducing agent ($NaBH_4$ or $SnCl_2$) produces the gaseous hydride

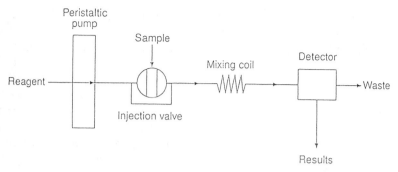

Fig. 32 Simple single-channel FIA manifold.

of the analyte element or reduces mercury to its elemental form.

ICP atomic emission spectrophotometry.

Microwave plasma torch atomic emission spectrophotometry.

Immunoanalysis, e.g., detection of atrazine: An atrazine peroxidase conjugate is used as a tracer. The immuno reaction takes place in a membrane reactor with an immobilized atrazine (Ab). The following components are consecutively pumped through the membrane reactor: sample, enzyme tracer, peroxide, and the second substrate hydroxyphenyl propionic acid. The enzyme activity of the Ab bound tracer is determined fluorimetrically. Use of monoclonal antibodies specific for atrazine and for atrazine labeled with alkaline phosphatase is possible.

Chemiluminescence: inhibition of luminol/H_2O_2. Co^{2+} and Mn^{2+} catalyze the reaction of luminol with potassium periodate. Reaction of luminol, KCN, and Cu^{2+}; H_2O_2 oxidation of luminol.

Coupling with ion-selective electrode.

Ruzicka and Hansen and Karlberg and Pacey wrote interesting works on FIA (11,12).

V. CHROMATOGRAPHY

A. General

According to the IUPAC definition, chromatography is a physical method of separation in which the components to be separated are distributed between two phases, one of which is stationary (the stationary phase) and the other of which (the mobile phase) moves in a definite direction. The mobile phase is a fluid that percolates through or along the stationary bed. Three types of fluid are in current use: liquid, gas, and supercritical. Chromatography is named principally by the nature of

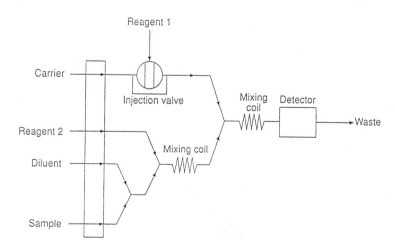

Fig. 33 Scheme of an FIA system with derivatization reagent.

the fluid. We can distinguish liquid, gas or supercritical fluid chromatography. The stationary phase may be a solid, a liquid, or a gel. The liquid may be coated onto a solid surface. A special type consists of bonding chemical moieties onto a solid surface by covalent bond. It is a bonded phase.

The combination of mobile and stationary phases unambiguously names the chromatographic mode:

Mobile phase	Gas	Liquid	Supercritical
Stationary phase	Liquid Solid	Liquid Solid	Liquid Solid

Gas–liquid chromatography and liquid–solid chromatography are by far the most popular. Liquid chromatography with alkyl-bonded phases is considered liquid–solid.

According to the nature of the stationary phase, a chromatographic mode (especially in liquid chromatography) is often named by the chemical species that governs the retention mechanism. We can thus distinguish in liquid chromatography:

Adsorption chromatography, often referred to as *normal-phase mode*
Reversed-phase chromatography (on alkyl-bonded silica)
Ion chromatography
Affinity chromatography
Size-exclusion chromatography
Chiral stationary phases
Micellar liquid chromatography

1. Chromatographic Instrumentation

The heart of the separation is the column. Solutes are injected onto the column, where they partition between the mobile and the stationary phases. When a solute is flowing off the column it is eluted.

Typically a chromatograph is depicted as follows: Solutes to separate are placed in an injector; they are driven to the separation column; when they elute, they are monitored by a detector.

There are two ways to perform chromatography: analytical mode and preparative mode. We shall consider only the analytical mode, in which solutes are infinitely diluted.

Molecules of solute that do not interact with stationary phase are unretained. Molecules of solute that can partition between both phases are retained. Detection and recording of separated solutes yields a chromatogram (Fig. 34). t_M is the retention time of the unretained solute, often written t_o. In fact, t_o and t_M are not equal (especially in packed columns), and the nature of the unretained (or inert) solute should be given in any reported experimental conditions. t_{Ri} is the retention time of the solute i. The volume of the mobile phase required to elute an unretained solute is V_M.

$$t_M = \frac{V_M}{F}$$

where F is the mobile-phase flow rate. Similarly, V_{Ri} is the retention volume of solute i. The retention factor is

$$k_i = \frac{t_{Ri} - t_M}{t_M} = \frac{V_{Ri} - V_M}{V_M} = \frac{d_{Ri} - d_M}{d_M}$$

where d_M, d_R are retention distance measured on the recording chart.

The retention factor is also equal to the ratio of the amounts of a sample component in the stationary and mobile phases, respectively, at equilibrium:

$$k = \frac{\text{amount of component in stationary phase}}{\text{amount of component in mobile phase}}$$

The adjusted retention time is

$$t'_{Ri} = t_{Ri} - t_M \quad \text{and} \quad k = \frac{t'_{Ri}}{t_M}$$

Similarly,

$$V'_{Ri} = V_{Ri} - V_M$$
$$d'_{Ri} = d_{Ri} - d_M$$

In analytical mode, peaks are presumed to be Gaussian, and retention parameters are taken at maximum peak heights. When peaks are not truly Gaussian, it is necessary to use statistical moments. The zero moment $M_0 = \int_a^b h(t)\, dt$, where $h(t)$ is the peak height at time t.

The first moment

$$M_1 = \frac{1}{M_0} \int_a^b t \cdot h(t)\, dt$$

expresses the true retention time, because it corre-

Mobile phase   solute

injector — column — detector

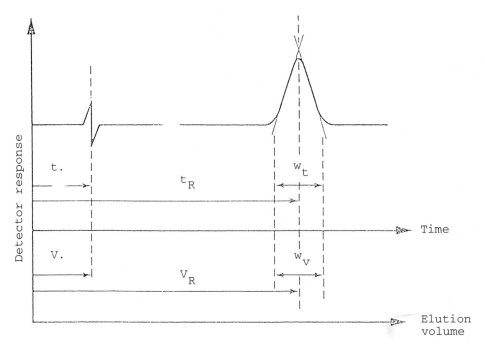

Fig. 34 A chromatogram.

sponds to the elution time of the center of the gravity of the peak.

M_2 = peak variance

M_3 = peak skew

When two consecutive (i, j) Gaussian peaks are close, the resolution is

$$R_s = \frac{t_{rj} - t_{ri}}{1/2(\omega_i + \omega_j)}$$

where ω is the peak width in the time units (see Fig. 35). When peaks are Gaussian, $\omega = 4\sigma$ (σ is the standard deviation). When two peaks are well resolved, then $R_s \geq 1.25$ (see Fig. 35). Asymmetry is determined by the A/B ratio (Fig. 36).

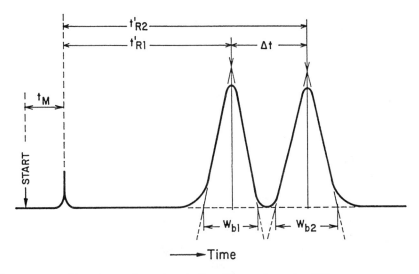

Fig. 35 Resolution between two Gaussian peaks. t_r is retention time; ω is peak width.

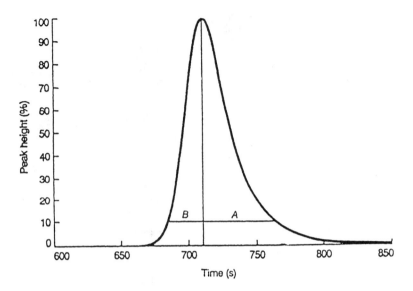

Fig. 36 Asymmetry factor by determination of the *A/B* ratio at 10% peak height.

Selectivity α is defined as $\alpha = k_j/k_i$. In this way, $\alpha \geq 1$; when $\alpha = 1$, no separation occurs.

In gas chromatography, the separation number (SN) is often referred to:

$$SN = \frac{t_{R(n+1)} - t_{Rn}}{\omega_n + \omega_{n+1}}$$

where t_{Rn} and $t_{R(n+1)}$ are the retention times of two consecutive normal paraffins. The value thus obtained represents the theoretical number of peaks separated by approximately twice the width at half the peak height that will fit between the two standards under the same optimized conditions. In the German literature, the separation number is called *Trennzahl*.

2. *Efficiency*

Sharp peaks are indicative of the efficiency of the chromatographic process. Efficiency is also called *plate number*.

$$N = \left[\frac{t_R}{\sigma_t}\right]^2 = \left[\frac{V_R}{\sigma_V}\right]^2 = \left[\frac{d_R}{\sigma_d}\right]^2$$

When peaks are Gaussian,

$$N = 16\left[\frac{t_R}{\omega}\right]^2 \quad \text{or} \quad N = 5.54\left[\frac{t_R}{\delta}\right]^2$$

where δ is the peak width at mid-height.

The area of a Gaussian peak is a function of its standard deviation and peak height, according to the following equation:

$$A = \sqrt{2\pi}\sigma h_p$$

where h_p is the peak height. From

$$N = \left[\frac{t_r}{\sigma}\right]^2$$

and rearranging, we get

$$N = \frac{2\pi(h_p t_r)^2}{A^2}$$

The plate height is the column length divided by the plate number:

$$H = \frac{L}{N}$$

the number of effective plates is

$$N_{\text{eff}} = 16\left[\frac{t'_R}{\omega}\right]^2$$

The foregoing equations can be combined. If $\bar{k} = (k_i + k_j)/2$ and assuming peaks are close enough, then

$$R_s = \frac{\alpha - 1}{\alpha + 1}\frac{\bar{k}}{1 + \bar{k}}\frac{\sqrt{N}}{4}$$

A slightly different equation is written as

$$N = 16R_s^2\left[\frac{\alpha}{\alpha - 1}\right]^2\left[\frac{1 + \bar{k}}{\bar{k}}\right]^2$$

This last equation permits us to determine the number of plates required to achieve the separation between two solutes of retention factors k_i and k_j, respectively, with a given R_s.

The peak capacity is the number of peaks which can be observed on a chromatogram with baseline resolution (sometimes with unit resolution):

$$n_p = 1 + \frac{\sqrt{N}}{4} \log \frac{t_{Rz}}{t_{r1}}$$

where t_{Rz} is the retention time of the last eluted solute and t_{r1} is the retention time of the first eluted solute.

The capacity of a column is the maximum amount of sample that can be injected into a column before significant peak distortion occurs. Peak distortion is measured either by peak skew or by the asymmetry factor.

B. Gas Chromatography (GC)

In this mode, the mobile phase is a gas. Nitrogen, helium, and hydrogen are used. The gas only acts as a carrier.

1. Instrument

A GC instrument consists of

A carrier gas delivery system
An oven in which the column is placed
An injector
A detector

Gas chromatography can be equipped with an accurate digital pressure and flow control system for both carrier and detector gases. The complete digital control of the carrier gas allows easy setup of all parameters.

By creating electronic control systems, bulky mechanical flow and pressure regulators are eliminated from the GC mainframe. Electronic pressure control provides constant or programmable flow conditions throughout temperature-programmed runs.

2. Gas Chromatography Injection Techniques

a. Packed Columns. Injection is performed simply by a syringe through a septum. The entire injected volume goes to the column inlet.

b. Open Tubular Columns.

Split Injection: Liquid samples are vaporized in a glass tube. Vapors are driven to the column entrance by the carrier gas. Column inlet is placed in the injection chamber in such a way that only one fraction (1/100) of the carrier gas can enter the column. The remainder is directed to a waste gate called the *split*

outlet (Fig. 37). Split injection can provide sharp initial bands. The technique can be used in two ways: heated chamber and programmed temperature vaporization (PTV) (Fig. 38). In PTV, the sample is introduced into a cool chamber, which is rapidly heated after the syringe needle is withdrawn.

Splitless: The design here is the same as for the split mode, but the split outlet remains closed during a period of time that permits the transfer of vapors into the column (about 30–60 s). The split outlet is then opened to flush the remaining vapors. The initial bands are broad, and reconcentration by cold-trapping or solvent effects is required. In the classical hot split/splitless injector, the injector block is heated to a constant elevated temperature prior to injection, in the programmable split/splitless system, the injector is at a lower temperature during sample injection. After the withdrawal of the syringe needle, the injector is rapidly heated up to evaporate the sample.

Direct Injection: Basically this is the same as splitless, but there is no split outlet and by consequence no purge by the carrier gas. This can be utilized with packed columns.

Solvent Split Injection: This is a combination of split, splitless, and programmed-temperature vaporization. Sample (usually a large volume) is introduced into the chamber at low temperature with an open split valve. Solvent evaporation occurs then the split outlet

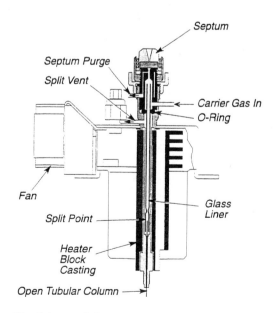

Fig. 37 Scheme of the programmable split/splitless injector system. (Courtesy of Perkin Elmer.)

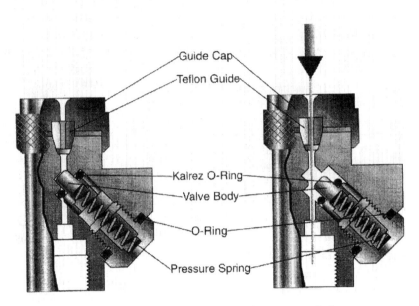

Fig. 38 PTV injector. (Reproduced by permission of the *Journal of Chromatography*.)

valve is closed and the chamber is heated. Solute is transferred onto the column. This cannot be used with very volatile solutes.

On Column (Fig. 39): The sample is deposited directly into the oven-thermostated column inlet with a syringe or into an uncoated precolumn (the retention gap). The injector is kept cool to prevent solvent evaporation inside the syringe needle. On-column injection eliminates discrimination and degradation that can result from a vaporization technique. The syringe needle passes completely through the injector and enters the column or the precolumn. Upon injection, the flow of carrier gas redistributes the inserted plug of liquid into a film on the tubing wall. Sample vaporization occurs from this film.

In the large-volume injection (LVI) technique, the liquid sample is injected through the cold on-column injector into a desolvation precolumn. The solvent vapors generated during the injection event are vented through a solvent-vapor exit system. Residual solvent and target compounds are then transferred to the analytical column. A discussion on LVI can be found in Ref. (13).

Megabore Injector: This consists of a glass injector liner held in place with a metal injector fitting. The sample is injected and rapidly vaporized in the liner. The carrier gas sweeps all of the vaporized sample into the megabore column.

3. Columns

Two types of columns are available: packed columns and open tubular columns. Packed columns are rather short (1–6 m) and exhibit average 6-mm ID. Open tubular (OT) columns are 15–100 m long. They fall into two categories:

Wide bore or megabore (ID ≥ 0.25 mm) (0.53, 0.32, 0.25 mm)
Narrow bore or true capillary columns (ID ≤ 0.25 mm) (0.25, 0.10, 0.05 mm)

Tubing is either fused silica deactivated or undeactivated or prosteel (Utimetal, for example, is virtually unbreakable).

From the outside, an open tubular column is composed of three parts (Fig. 40):

Polyimide coating
Fused silica tubing
Stationary phase

Fused silica contains less than 1 ppm metallic impurities. The inner wall of the fused silica tubing is deactivated to eliminate (or at least minimize) undesired interactions between the tubing and the sample. Stationary phase is either coated or bonded onto the inner wall (WCOT column). Crosslinked, chemically bonded, or immobilized phases are more stable, have lower bleed, and can operate at higher temperatures than coated phases.

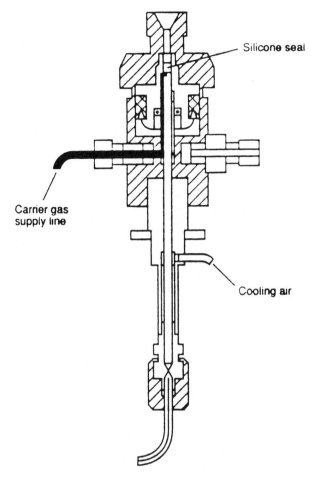

Silicone seal

Carrier gas
supply line

Cooling air

Fig. 39 On-column injector. (Courtesy of Chrompack, Middelburgh, the Netherlands.)

Average film thickness is 0.25 μm. Thick films increase retention and sample capacity. With a standard 0.25-μm-ID column and standard 0.25-μm film thickness, sample capacity is 40–350 ng.

Column bleed is essentially the breakdown of the column stationary phase induced by the combinations of elevated temperatures and oxidative agent such as water and oxygen. Silicon–oxygen bonds in the polymer backbone are broken to produce small cyclic groups (typically 3–4 siloxane units), which are very volatile. They elute from the column, producing a detector response.

Porous-layer open tubular (PLOT) columns have become very popular due to their high retention and the high selectivity for volatile compounds. Because the separation process with PLOT columns is based on adsorption, the actual elution temperatures are much higher, which avoids the use of subambient conditions. Such PLOT columns are well suited for fast analyses of gases and volatiles.

Table 4 lists some characteristics of OT columns.

4. Gas Chromatography Detectors

a. Olfactometry.   This is surprisingly effective with some solutes that exhibit intense odor. It allows 0.2 ppm to be detected.

b. Thermal Conductivity Detector.   This is a universal detector but suffers from lack of sensitivity as compared to the other types of detectors.

Principle of operation: A resistor is heated by a current and cooled by the gas stream from the carrier gas. The equilibrium temperature depends on the composition of the gas. The resistance of the resistor, in turn, depends on its temperature. In the detector device, resistors are connected to a Wheatstone bridge. Cells in one diagonal are swept by pure carrier gas, cells of the other diagonal by column effluent. When solutes are eluted, the bridge experiences a disequilibrium, which is amplified and recorded. Its highest sensitivity is ob-

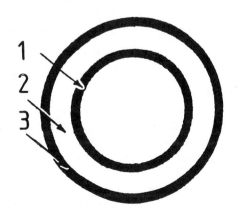

Fig. 40 Open tubular column. 1: stationary phase; 2: fused silica; 3: polyimide coating.

Table 4   Characteristics of OT Columns

Internal diameter (mm)	0.10	0.18	0.25	0.32	0.53
N/m	8,600	5,300	3,300	2,700	1,600
Sample capacity (ng)	5–10	10–20	50–100	400	1,000–2,000

tained with carrier gases that have a high thermal conductivity, e.g., hydrogen and helium.

c. Ionization Detector. All ionization detectors have the same base body. They all are miniaturized. They are not universal, with the exception of the helium ionization detector.

d. Flame Ionization Detector (FID). This detects C and H (Fig. 41). However, a response is observed for some other elements.

Principle of operation: A small hydrogen–air flame burns at a capillary jet. In the hottest part of a flame at high temperature, a certain amount of radicals are created (a few ions per million molecules). It generates a current between two electrodes. A collector electrode is located a few millimeters above the flame, and the ion current is measured by establishing a potential between the jet tip and the collector electrode. When organic compounds are burnt, an ion-producing reaction occurs:

$$CH^0 + O^* \rightarrow CHO^+ + e^-$$

This allows 0.1 pg hydrocarbons to be detected.

e. Electron Capture Detector (ECD). It is very sensitive to any electrophilic compounds and particularly well suited for organochlorine species. Very widely used in pollution control (Fig. 42).

Principle of operation: A ^{63}Ni source is emitting a β electron beam. A current between two electrodes is generated. When electrophilic species enter the detector, a decrease in the detector background current is observed due to the capture of the electrons by the electrophilic species.

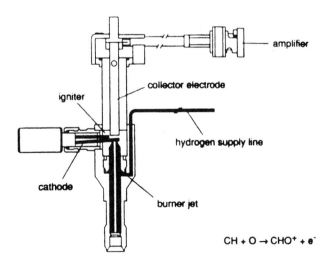

amplifier

collector electrode

igniter

hydrogen supply line

cathode

burner jet

$$CH + O \rightarrow CHO^+ + e^-$$

Fig. 41 Flame ionization detector (FID). (Courtesy of Chrompack.)

Constant-current ECD is the most common mode of operation, but fixed frequency is gaining acceptance. The thermal electron concentration in the detector cell is measured discontinuously by a pulsed voltage. In the pulsed-discharge ECD, a radioactive source is not required. The detection limit $\cong 10^{-15}$ g Cl/s or 8 fg lindane/s.

f. Pulsed-Discharge Helium Ionization Detector (PDHID). This is one of the most sensitive detectors available for GC.

Principle of operation: Photon emission in pure helium arises from excited states of He_2 and consists of a continuum extending from 11.6 eV to 21.7 eV. Since these energies are greater than the ionization potentials of all atoms and molecules, the photonization detector is a universal detector. The photon emission distribution can be characterized by energies at half maximum (14.1 eV to 16.7 eV). This range contains 66% of the photon emission.

The detection limit is 1–20 pg. A fiber-optic multiphoton ionization detection is able to detect 0.12 ng PAH.

g. Flame Photometric Detector. *Principle of operation*: Flame breaks down large molecules. Atoms and species are brought to an excited state (S_2^* or PO*) and relax, emitting a light of characteristic wavelength.

This detector is well adapted for sulfur and phosphorus determination. Two flames are often used to separate the region of sample decomposition to sample emission. Response is dependent on the environment of the sulfur atoms (thiols, sulfides, disulfides, thiophenes). (See Fig. 43.) Detection limit is around 10–12 pg P/s and 10–10 pg S/s.

h. Thermoionic Detector. *Principle of operation*: Adding an alkali metal salt to a flame enhances the response to compounds containing N_2, P, or S. The mechanism is not fully understood. Gas-phase reactions involve free alkali metal atoms in the flame that are ionized by collision with carrier gas molecules (Fig. 44).

$$A + M \rightarrow A^+ + e^- + M$$

Free radicals resulting from the pyrolysis of organic compounds containing P or N react with alkali metal atoms. In the instrument a ceramic or glass matrix is doped with an alkali metal salt that can be electrically heated. The usual salt is rubidium silicate.

The detection limit is 10^{-13} g of N.

i. Chemiluminescence Detectors. Sulfur chemiluminescence detection takes advantage of the fact that SO is produced during FID operation. When SO reacts with ozone, a strong blue chemiluminescence signal is

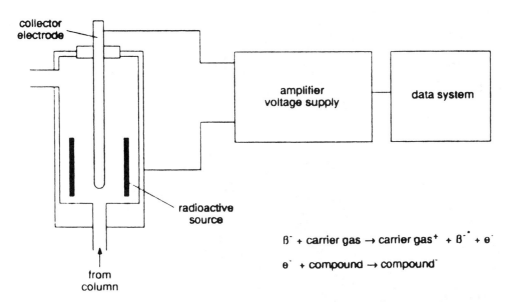

collector
electrode

amplifier
voltage supply

data system

radioactive
source

from
column

$\beta^- + \text{carrier gas} \rightarrow \text{carrier gas}^+ + \beta^{-\cdot} + e^-$

$e^- + \text{compound} \rightarrow \text{compound}^-$

Fig. 42 Electron capture detector (ECD). (Courtesy of Chrompack.)

emitted by the resulting SO_2^*. The signal is isolated from other radiations and detected by a photomultiplier tube. The detection limit is around 10 pg sulfur.

CLND pyrochemiluminescent nitrogen detection: Components eluting from the column undergo high-temperature (1000° C) oxidation. All nitrogen-containing compounds are converted into nitric oxide, NO. The resulting gases are dried and mixed with ozone in a reaction chamber. This results in the formation of nitrogen dioxide, NO_2^*, in the excited state. Light is emitted by the chemical reaction and detected by a photomultiplier tube.

j. Electrochemical Detector. The electrolytic conductivity detector (Hall detector) relies on the absorption of ionizable gases into liquid for conductivity measurement. These detectors are rarely advocated in EPA methods, probably because the electrolyte must be kept extremely clean. The limit of detection with sulfur is 1 pg sulfur.

k. Gas Chromatography/Atomic Emission Detector (GC/AED). The AED detector is a multielement detector capable of detecting elements with atomic emission lines in the vacuum UV, UV-VIS, and near-IR portions of the electromagnetic spectrum. It allows multielement measurement.

Plasma sources are capable of producing intense emission from the elements. The types of plasma used in chromatographic detection are microwave-induced plasmas (MIP) and ICP. An argon plasma is sustained in a microwave cavity that focus into a capillary discharge cell. The most widely used cavities are cylin-

drical resonance cavities and the "surfatron," which operates by surface microwave propagation along a plasma column. Atmospheric pressure cavities are very simple to interface with capillary GC columns. Figure 45 displays a GC-AED interface.

Other plasmas are glow-discharge plasmas and direct current plasmas with a continuous DC arc. Radiations from the plasma are dispersed on a diode array spectrometer.

Table 5 displays some detection limits with helium microwave-induced plasma.

l. Gas Chromatography/Mass Spectrometry. Mass spectrometers can easily be interfaced to GC (Fig. 46). The GC/MS method is now the most widely used detection technique.

Gas Chromatography/Quadrupole MS with Electron Impact: The most common design is a single-capillary column directly coupled to an EI quadrupole mass spectrometer. The sensitivity is very high, which allows detection of extremely small quantities of contaminants in water. The electron impact mode is 70 eV. The typical scan range is 65–400 amu, scan time is 0.5 s, with a delay time of 0.2 s between individual scans. Library spectra are capable of producing more than 130,000 spectra.

Gas Chromatography/Quadrupole MS with Chemical Ionization: Chemical ionization (CI) is a soft ionization technique that produces molecular ions (M^+ or M^-), adduct ions $(M + CI \text{ reagent})^{\pm}$, and fragment

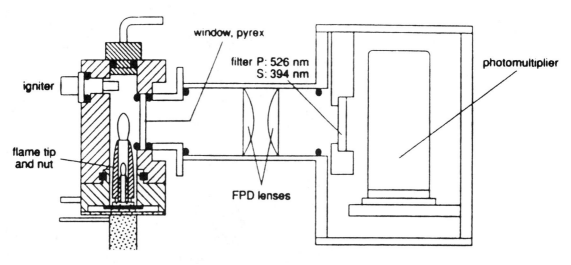

Fig. 43 Flame photometric detector. (Courtesy of Chrompack.)

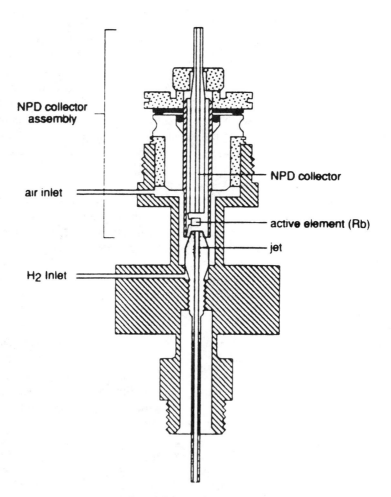

Fig. 44 Thermoionic detector (NPD). (Courtesy of Chrompack.)

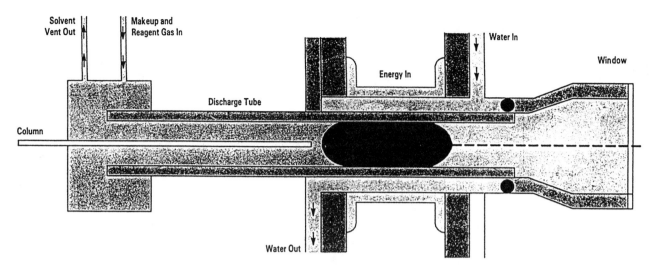

Fig. 45 Diagram of helium discharge tube for GC/AED. (Courtesy of Hewlett-Packard.)

ions. Instrumentation is more expensive, but CI permits isomer differentiation. Methane is used for CI.

Gas Chromatography/Ion-Trap MS: Sample entering the detector is ionized by thermal electrons and accelerated through an electron gate that can be opened and closed by application of the appropriate voltage. The ion trap is a quadrupole detector, but the radio frequency signal is applied to a central circular ring electrode situated between two endcaps held at the ground potential. Quadrupole ion traps use an automatic gain-control scan function. Ions are trapped and subsequently ejected by ramping the radio frequency (RF) voltage applied to the ring electrode. Ions with low m/z sequentially eject before higher m/z ions as the radio frequency voltage increases.

Before a spectrum scan occurs, the system uses a fixed ionization time and a rapid RF scan to get a gross measurement of the sample size. Ionization time is adjusted from the total ion current and automatic gain-control target value. Long ionization times produce more ions, thus increasing sensitivity.

Gas Chromatography Ion-Trap Tandem MS and Gas Chromatography Tandem MS: In this technique, an ion of interest is selected by ejecting all unwanted ions, and the selected ion is subsequently fragmented by collision with a neutral gas. The resulting mass spectrum is called a daughter or product ion spectrum, and it is characteristic of the secondary fragmentation process.

In GC tandem MS, a first quadrupole acts as a mass selective filter, and the second quadrupole is used as the collision cell with the addition of a collision gas such as helium or argon. In a third quadrupole, the full

mode is performed to obtain the full mass spectrum of the product ions.

Gas Chromatography/Time-of-Flight (TOF) MS: In GC-TOF-MS, a full spectrum can be collected in 100 μs. Several spectra must be arrayed to improve the signal-to-noise (S/N) ratio. In addition to fast data-acquisition rates, a TOF-MS also provides simultaneous sampling of all the ions produced in the ion source. Since TOF-MS has no temporal-based spectral skewing with simultaneous sampling, high data rates can be used to improve quantification, analyte determination, and deconvolution of unresolved chromatographic peaks.

Gas Chromatography Fourier Transform Infrared: Mass spectrometry cannot distinguish structural isomers. Infrared (IR) spectroscopy provides information on the intact molecule. The most common GC/FTIR instrument is the light-pipe instrument. A light pipe is a narrow-bore (100–200-μm ID) borosilicate capillary with a smooth, thin layer of gold coated on the inside surface. Reflection occurs with the gold coating, thus increasing the path length of the cell by a factor of 10 or more. A schematic of GC/FTIR instrumentation is displayed on Fig. 47.

A sensitive technique used for real-time reconstruction of chromatograms from the interferogram is the Gram Schmidt vector orthogonalization method. The Gram Schmidt method relies on the fact that the interferogram contains information on absorbing samples at all optical retardations less than the reciprocal of the width of each band in the spectrum.

Table 5 GC Detection with Helium Microwave-Induced Plasmas (MIP)

Element	Wavelength (nm)	Detection limit [pg/s(pg)]	Selectivity vs. C	LDR
Carbon (a)	247.9	2.7 (12)	1	>1,000
Carbon (b)	193.1	2.6	1	21,000
Hydrogen (a)	656.3	7.5 (22)	160	500
Hydrogen (b)	486.1	2.2	variable	6,000
Deuterium (a)	656.1	7.4 (20)	194	500
Boron (a)	249.8	3.6 (27)	9,300	500
Chlorine (b)	479.5	39	25,000	20,000
Bromine (b)	470.5	10	11,400	>1,000
Fluorine (b)	685.6	40	30,000	2,000
Sulfur (b)	180.7	1.7	150,000	20,000
Phosphorus (b)	177.5	1	5,000	1,000
Silicon (b)	251.6	7.0	90,000	40,000
Oxygen (b)	777.2	75	25,000	4,000
Nitrogen (b)	174.2	7.0	6,000	43,000
Aluminum (b)	396.2	5.0	>10,000	>1,000
Antimony	217.6	5.0	19,000	>1,000
Gallium (b)	294.3	ca. 200	>10,000	>500
Germanium (a)	265.1	1.3 (3.9)	7,600	>1,000
Tin (a)	284.0	1.6 (6.1)	36,000	>1,000
Tin (b)	303.1	(0.5)	30,000	>1,000
Arsenic (b)	189.0	3.0	47,000	500
Selenium (b)	196.1	4.0	50,000	>1,000
Chromium (b)	267.7	7.5	108,000	>1,000
Iron (b)	302.1	0.05	3,500,000	>1,000
Lead (a)	283.3	0.17 (0.71)	25,000	>1,000
Mercury (b)	253.7	0.1	3,000,000	>1,000
Vanadium (b)	292.4	4.0	36,000	>1,000
Titanium (b)	338.4	1.0	50,000	>1,000
Nickel (b)	301.2	1.0	200,000	>1,000
Palladium	340.4	5.0	>10,000	>1,000
Manganese (b)	257.6	1.6 (7.7)	110,000	>1,000

Detection limit: 3 times the signal-to-noise ratio. LDR = linear dynamic range.
(a): Conventional TM_{010} MIP (University of Massachusetts).
(b): Hewlett-Packard 5Y21A (Hewlett-Packard or University of Massachusetts).
Source: P.C. Uden. Element-specific chromatographic detection by atomic absorption, plasma atomic emission and plasma mass spectrometry. J. Chromatogr. 703:393–416 (1995). Reproduced with permission.

In off-line systems, analytes eluting from the GC column are frozen as pure substances onto the surface of an IR transparent Zn Se window. Immediately after deposition, peaks are passed under a transmittance IR microscope and scanned. This off-line procedure permits one to rescan.

5. Derivatization

When a solute is not volatile enough, a derivatization reaction is performed prior to injection to make the solute more volatilable. Each class of reactions replaces the active hydrogens of OH, NH, or SH. Alkyl-ation, silylation, and acylation are very easily performed. Reagents are sold in vials to carry out the reaction. Depending on the type of solute, different reagents are available. The Pierce Company provides a useful directory to help the analyst in selecting.

$$R-OH + Cl-Si-R_3 \rightarrow R-O-Si-R_3$$

Amines and alcohols whose boiling point is high from the hydrogen bond association are easily transformed into silylated derivatives, with the consequence of a lowered boiling point. Table 6 gives an example of a possible derivatization reaction with amines.

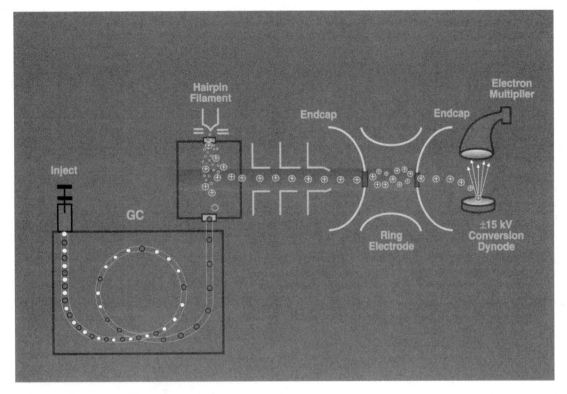

Fig. 46 GC-MS. (Courtesy of Finnigan.)

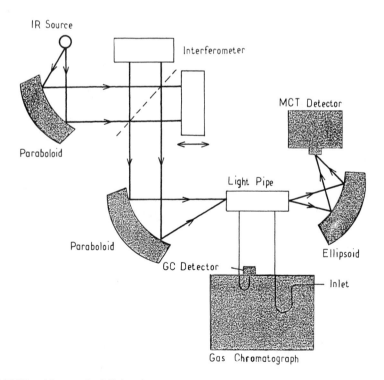

Fig. 47 Scheme of GC-FTIR with a typical light pipe.

Table 6 Derivatization Reactions for GC Determination of Amines

Reagent	Amine type[a]	Detection[b] (detection limit)
Acylation		
Trifluoroacetic anhydride	P, S, A	I (4–5 ng), E (30 pg), M (1 ng)
Pentafluoropropionic anhydride	P, S	N (4–20 pg), E, M (1 ng)
Hexafluorobutyric anhydride	P, S	N (4–20 pg), E (0.1–6 pg), M (1 ng)
Chloro- or dichloroacetic anhydride	P	E (0.2–1.5 ng), M (1 ng)
Trichloroacetyl chloride	P, S	E (0.2–2 ng)
Pentafluorobenzoyl chloride	P, S	N (0.1 ng), E (10 pg)
Heptafluorobutylimidazole	P, S	E (0.3–0.4 ng)
N-Methylbis(trifluoroacetamide)	P, S	I (80 ng), M (1 ng)
Silylation		
N,O-Bis(trimethylsilyl)trifluoroacetamide	P, S	M (4–20 pg)
N-Methyl-N-(*tert.*-butyldimethylsilyl)acetamide	P, S	M
Pentafluorophenyldimethylsilyl reagents	P, S	I (5 ng), E (5 pg)
Dinitrophenylation		
2,4-Dinitrofluorobenzene	P, S	I (1–2 ng), E (20 pg), M (0.3–4 ng)
2,4-Dinitrobenzenesulphonic acid	P, S, A	I (50–100 ng)
Permethylation		
Formamide–sodium borohydride	P, S	N (0.1 ng)
Schiff base formation		
Benzaldehyde	P	I (60 ng)
Furfural	P	I
2-Thiophenealdehyde	P	S
Pentafluorobenzaldehyde	P	E (20 pg), M (5 ng)
Dimethylformamide dimethyl acetal	P, A	S (25 pg)
Carbamate formation		
Diethylpyrocarbonate	P, A	I
Ethyl chloroformate	P, S, T	I, N (3–10 pg), M (1.5 pg)
Isobutyl chloroformate	P, S	I, N (3–20 pg), S (40 pg), M (2 pg)
Amyl chloroformate	P, S	I
2,2,2-Trifluoroethyl chloroformate	P, S	N (10 pg), M (10 pg)
Pentafluorobenzyl chloroformate	P, S, T	E (3 pg)
Sulphonamide formation		
Benzenesulphonyl chloride	P, S, A	S (6–25 pg), C (60 pg), M (5–30 pg)
p-Toluenesulphonyl chloride	S	I (10 ng), M (45 pg)
Phosphoamide formation		
Dimethylthiophosphinic chloride	P	N (0.5 pg)
Dimethylthiophosphoryl chloride	P, S	I, P
Diethylthiophosphoryl chloride	P, S, N	P (3–15 pg)

[a]P = primary amine; S = secondary amine; T = tertiary amine; A = ammonia; N = nitrosamine.
[b]I = FID; N = NPD; S = FPD (S mode); P = FPD (P mode); E = ECD; C = CLD; M = GC–MS–SIM.
Source: H. Kataoka. Derivitazation reactions for the determination of amines by gas chromatography and their applications in environmental analysis. J. Chromatogr. 733:19–34 (1996). Reproduced with permission.

6. Portable Gas Chromatography: High-Speed Gas Chromatography

A major trend in GC has been the development of portable (or at least transportable) GC instruments. Gas chromatography excels over all other techniques for the analysis of complex mixtures with high resolution, high speed, dynamic concentration range, and precision. The instrument is complete.

The GC/MS instrument is designed for site seeing. It contains the gas chromatograph and mass spectrometer as well as the sampling inlet, battery, carrier gas, internal standards, vacuum pump, control electronics, and analysis software. The mass spectrometer is self-tuning. Representatives of these instruments are Hapsite from Inficon (Fig. 48). Voyager from Perkin Elmer is not a GC/MS instrument but includes a photoionization detector and ECD (optional). An LCD screen displays the compounds and concentration detected. A portable GC instrument does not mean high-speed GC; some can do this, others not.

Relatively simple mixtures containing a few components can be separated in a few seconds, more complex mixtures containing several tens of components can be separated in less than 1 minute (1–2 minutes to separate 11 components in a pesticide mixture, for example). For an analysis time of a few seconds, programmed inlet systems that provide injection bandwidths of a few milliseconds are required. Cryofocusing inlet systems can produce injection bandwidths of 10 ms. The injector can be programmed at 100°C/s. Detectors and associated electronics are forced to respond to higher-frequency constituents than encountered in more conventional chromatographic systems. A flash GC is displayed in Fig. 49.

7. Multidimensional Gas Chromatography

Single-stage separation of a moderately complex mixture is very often unsuccessful, since only a small number relative to the peak capacity is observed. In simplest form, two-dimensional GC involves separation on a first column and trapping a segment of the chromatogram, which is directed to a second column of different polarity or different separation mechanism. It relies on column switching.

To overcome the problems of a valve leakage, Deans (14) developed the Live system. The special coupling piece is a part of a pneumatic bridge circuit similar to

Fig. 48 Picture of a portable GC-MS.

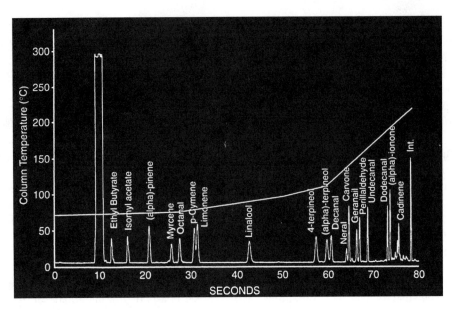

Fig. 49 Typical flash GC chromatogram. (Courtesy of Thermedics.)

a Wheastone bridge. The system (Fig. 50) consists of four flow resistors, needle valves, and throttles connected to a pressure source known as "heart cutting." In multiple parallel trap multidimensional GC, cryogenic trapping is performed. Solutes of interest are trapped before analysis on the second analytical column.

More and more multiseparation systems are under development. Coupling of SPE (solid-phase extraction) with LC followed by GC-MS allows very sensitive detection of pollutants. The system can be fully automated.

8. Flow Velocity Through the Column

The integrated Poiseuille law relates the outlet column flow velocity u_o to the column length L and to the inlet and outlet pressures P_i and P_o, respectively:

$$u_o = \frac{B_o}{2\eta L P_o}(P_i^2 - P_o^2)$$

where η is the gas viscosity and B_o is the column permeability. With an open tubular column

$$B_o = \frac{d_c^2}{32}$$

where d_c is the column diameter. With a packed column,

$$B_o = \frac{d_p^2}{1000}$$

where d_p is the mean particle diameter.

Due to gas compressibility, gas velocity varies along the column. The average velocity $\bar{u}$ is given by

$$\bar{u} = \frac{L}{t_M} = ju_o$$

where

$$j = \frac{3}{2}\frac{P^2 - 1}{P^3 - 1}$$

$$P = \frac{P_i}{P_o}$$

t_M = retention time of unretained solute

The corrected retention time is:

$$t_R^o = t_R j$$

$$t_R^o = \frac{jV_R^o}{F}$$

where V_R^o is the corrected retention volume. The net retention time is

$$t_N = j\frac{V_R'}{F}$$

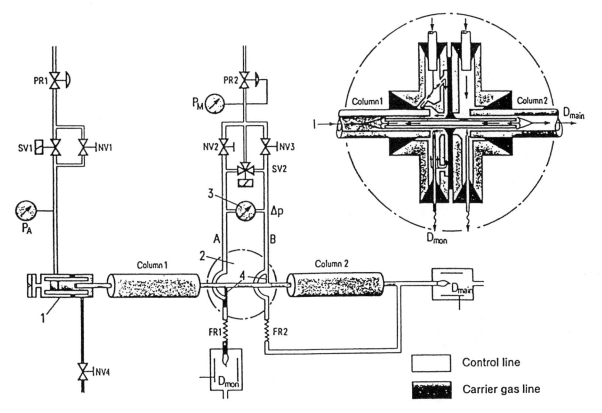

Fig. 50 "Live" column switching system: 1, injector with split valve and septum purge; 2, live T piece; 3, differential-pressure manometer; 4, ring slot; 5, needle valve (NV); solenoid valve (SV); pressure regulator (PR); flow restrictor (FR); D_{mon}, monitor detector; D_{main}, main detector; P_A, inlet pressure; P_M, mean pressure; ΔP, differential pressure; A, control line A; B, control line B.

and

$$V_N = V_R' j$$

with $V_R' = V_R - V_M$.

The net retention volume at the column temperature is the net retention volume per gram of stationary phase W_s:

$$V_g^o = \frac{V_N}{W_S}$$

The specific retention volume at 0°C is the value of V_g^o corrected to 0°C:

$$V_g = V_g^o \frac{273.15}{T_c}$$

where T_c is the column temperature.

Retention time locking, introduced by Hewlett Packard, is a new technique for obtaining nearly identical retention times column to column and instrument to instrument.

The technique is based on computing and implementing an adjustment in the inlet pressure to compensate for small differences in column and instrument characteristics.

9. Plate Height and Gas Velocity

The Van Deemter equation holds true for a packed column:

$$H = A + \frac{B}{u} + Cu$$

where A is the eddy diffusion, B is the molecular diffusion term, and C is the mass transfer term. For an open tubular column, the eddy diffusion terms (A) does not exist; the plate height is given by the Golay equation:

$$H = \frac{B}{u_o} + C_G \bar{u} + C_L \bar{u}$$

where

$$B = 2D_G$$

D_G = diffusion coefficient of solute in gas phase

$$C_G = 1 + 6k + 11k^2/(1 + k)^2 \cdot d_c^2/96D_G$$

D_L = diffusion coefficient in liquid stationary phase

$$C_L = \frac{k}{6(1 + k)^2} \frac{e_f^2}{D_L}$$

In both cases (packed and open tubular), plots of H versus u exhibit a minimum that determines the optimum conditions for achieving the best efficiency (Fig. 51).

If C_L is negligible as compared to C_G, then

$$H = \frac{B}{\bar{u}} + C_G\bar{u}$$

The theoretical minimum value of the plate height is

$$H_{\min} = 2\sqrt{BC_G} = \frac{d_c}{2(1 + k)} \sqrt{\frac{1 + 6k + 11k^2}{3}}$$

The coating efficiency is

$$\frac{H_{\text{theoretical}}}{H_{\text{actual}}} \times 100$$

It is a measure of the uniformity of the stationary phase. The optimum velocity is

$$u_{\text{opt}} = \sqrt{\frac{B}{C_G}}$$

To compare columns of different geometry, reduced variables are utilized.

Reduced Plate Height.

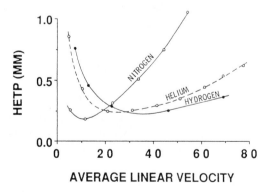

Fig. 51 Van Deemter curves of HETP versus linear velocity of different carrier gases in GC.

$$h = \begin{cases} \dfrac{H}{d_p} & \text{with packed columns} \\[2mm] \dfrac{H}{d_c} & \text{with open tubular columns} \end{cases}$$

Reduced Velocity.

$$\nu = \frac{\bar{u}dp}{D_G} \quad \text{or} \quad \frac{\bar{u}dc}{D_G}$$

Relative Retention.

$$\alpha = \frac{Vg_j}{Vg_i} = \frac{(\gamma^\infty p^o)_i}{(\gamma^\infty p^o)_j}$$

where

γ^∞ = activity coefficient at infinite dilution

p_i^o = vapor pressure of solute i

The relative retention of two compounds is the ratio of their Henry's law constants.

10. Retention Indices

The original Kovats' index is determined at constant temperature by using a homologous series of n-alkanes for the calibration. At constant temperature, n-alkanes are logarithmically spaced on the recording:

$$I_X = 100z + 100 \frac{\log t'_{Rx} - \log t'_{Rz}}{\log t'_{Rz+1} - \log t'_{Rz}}$$

where x refers to the substance of interest and z and $z + 1$ are the carbon numbers of two consecutive n-alkanes.

11. Programmed Temperature

This mode is required when the elution of strongly retained solutes is too long. When the temperature is increasing, the retention time follows the so-called Van't Hoff plot (Fig. 52):

$$\log k = \frac{A}{T} + B$$

Most utilized is the linear heating rate. In this mode, neighboring homologous peaks appear at equal intervals.

A temperature-programmed index is obtained by

$$Ip_{(X)} = 100 \left[\frac{T_{R(a)} - T_{R(z)}}{T_{R(z+1)} - T_{R(z)}} \right] + 100z$$

where T_R is the retention temperature (°K).

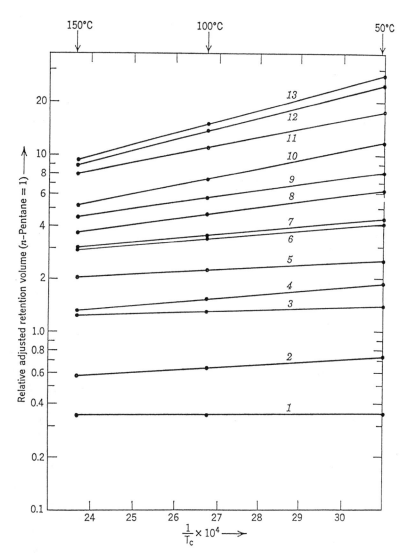

Fig. 52   So-called Van't Hoff plot in GC.

The rate of the temperature program is an important parameter. One can select for a shorter column, with a thin film and a higher program rate than for a longer column, with a thicker film.

12. Polarity

The strength of interaction between a solute and a liquid phase depends on the properties of both the solute and the phase. Following McReynolds, one can define the strength of the interaction by

$$\Delta I = I_{X(i)} - I_{Ref}$$

where $I_{X(i)}$ is the Kovats index of the solute on liquid phase i and I_{Ref} is the Kovats index of the solute on a reference liquid phase.

Ten solutes are used as probes to characterize the dominant interactions: benzene, butanol, 2-pentanone, nitropropane, pyridine, 2-methyl-2 pentanol, iodobutane, 2-octyne, dioxane, and cis-hydrindane.

McReynolds (15) utilized squalane as the reference stationary phase. It is now questionable, since GC is currently performed at higher and higher temperatures and squalane is not thermally stable. Apolane (16) is often advocated as reference. McReynolds suggested defining average polarity to be equal to the sum of the ΔI values for the test solutes. Some manufacturers utilize their own polarity scale (see, for example, the Chrompack scale, which is not very different from the McReynolds scale). McReynolds indices are the means to select a stationary phase to perform retention tuning according to the type of solutes to analyze. For further information, see Refs. 17–23. Table 7 displays EPA methods by GC. Table 8 displays EPA method 525. Fig. 53 displays a GC chromatogram according to EPA method 8270.

C. Liquid Chromatography

1. Instrument

A scheme of an LC instrument is displayed in Fig. 54. It consists of

Solvent reservoirs
A solvent delivery system
An injection device
The column
A detector
A data-acquisition system.

a. Solvent Reservoirs. Usually made of glass, they should be equipped with a degassing device. Degassing with helium is usually performed, but it can form slugs of helium in the tubing. Solvents are mixed in a mixing chamber. To account for the desired composition, solenoid valves are actuated.

b. Solvent Delivery System. Reciprocating pumps are utilized almost exclusively with conventional columns. They exhibit large column back-pressure compensation abilities and show flow rates in the 0.1–10.0-ml range with high precision. In some cases, titanium or ceramic head pumps are used to ensure biocompatibility. Syringe pumps can be used with micro- or minicolumns, but eluent compressibility and mixing of microflows is a serious drawback. With open tubular columns, split-flow techniques are probably best. Flow-splitting devices are based on the microflow processor concept.

Precise control of the flow rate is of primary importance to ensure reproducibility of retention times. At the present time, most commercially available pumping units for conventional HPLC are the single or multihead reciprocating-piston type. With capillary LC, the syringe type of pump is well suited.

The requirements for solvent delivery systems are as follows:

Flow rate stability
Flow rate accuracy (usually $\pm 0.3\%$)
Flow rate reproducibility
Large flow rate range (from 0.1–10 ml/min in HPLC)
Compatibility with any liquid
Standing with high pressures

Table 7 EPA Methods by GC

504	GC/MS	
551.		
1618	Organophosphorus insecticides	
501.3	Trihalomethanes in drinking water	GC/MS
502.2	Volatile halogenated organic compounds in water	purge-and-trap GC
503.1	Volatile aromatics and unsaturated organic compounds	purge-and-trap GC
504	1-2-dibromoethane and 1-2-dibromo-3-chloropropane	GC/ECD
505	Organohalide pesticides, aroclors	GC/ECD
507	Nitrogen- and phosphorous-containing pesticides	GC/NPD
508	Chlorinated pesticides	GC/ECD
513	2,3,7,8-tetrachlorodibenzo-p-dioxine	
515.1	Chlorinated acid	GC/ECD
515.2		
524.2	Purgeable organic compounds	GC/MS
525	Organic compounds in drinking water	GC/MS
531.1	Aldicarb and related compounds	GC/MS
548.1	Endothal in drinking water	GC/MS
551	Chlorination disaffection by-products	
552	Haloacetic acids	GC/ECD
552.1		
8271		
including	304	
	605	
	606	
	609	
	610	
	611	
	612	
	625	
	8040	
	8080	

Source: Taken in part from U.S. EPA methods. Reproduced with permission.

Table 8 EPA Method 525: Determination of Organic Compounds in Drinking Water by Liquid–Solid Extraction and Capillary Column GC-MS

Summary of the method	Organic compounds are extracted from a water sample by passing 1 L of sample through a cartridge or a disk containing a solid inorganic matrix coated with a chemically bonded C-18 organic phase; the organic compounds are eluted from the LSE cartridge or disk with a small quantity of dichloromethane and concentrated further by evaporating some of the solvent; final concentration of the extract is between 0.5 ml and 1 ml; an aliquot of $1-2$ μl is injected onto the GC-MS, and compounds are identified by their retention times and mass spectra.
GC column used	A 30-m $\times$ 0.25-mm-ID capillary column coated with DB-5 or equivalent is recommended.
Surrogate	Perylene-D_{12}.
Internal standards	Acenaphthene D_{10}, phenanthrene D_{10}, and chrysene D_{10}.

Estimated detection limits (EDLS) (μg/L) (only for modern pesticides):

Alachlor	0.09	Metoxychlor	0.08
Atrazine	0.14	Simazine	0.12

Source: U.S. EPA methods. Reproduced with permission.

1. Pyridine
2. 2-Picoline
3. Methyl methanesulfonate
4. Ethyl methanesulfonate
5. Aniline
6. 1,4-Dichlorobenzene-d_4 (Int. std.)
7. Benzyl alcohol
8. Acetophenone
9. o-Toluidine
10. Nitrobenzene-d_5 (surrogate)
11. Dimethylphenethylamine
12. Naphthalene-d_8 (Int. std.)
13. 4-Chloroaniline

14. Hexachloropropylene
15. 1,4-Phenylenediamine
16. Safrole
17. 2-Methylnaphthalene
18. cis-Isosafrole
19. 2-Fluorobiphenyl (surrogate)
20. trans-Isosafrole
21. 2-Nitroaniline
22. Acenaphthene-d_{10} (Int. std.)
23. 3-Nitroaniline
24. Dibenzofuran
25. Pentachlorobenzene
26. 1-Naphthylamine

27. 2-Naphthylamine
28. 5-Nitro-o-toluidine
29. 4-Nitroaniline
30. Diphenylamine
31. 1,3,5-Trinitrobenzene
32. Diallate (Isomer)
33. Phenacetin
34. Diallate (Isomer)
35. 4-Aminobiphenyl
36. Pentachloronitrobenzene
37. Phenanthrene-d_{10} (Int. std.)
38. Pronamide
39. 4-Nitroquinoline-N-oxide

40. Methapyrilene HCl
41. Isodrin
42. Benzidine
43. 4-Terphenyl-d_{14} (surrogate)
44. 4-Dimethylaminoazobenzene
45. Chlorobenzilate
46. Kepone
47. 3,3'-Dichlorobenzidine
48. 2-Acetylaminofluorene
49. Chrysene-d_{12} (Int. std.)
50. 3,3'-Dimethylbenzidine
51. 7,12-Dimethylbenzanthracene
52. Perylene-d_{12} (Int. std.)
53. 3-Methylcholanthrene

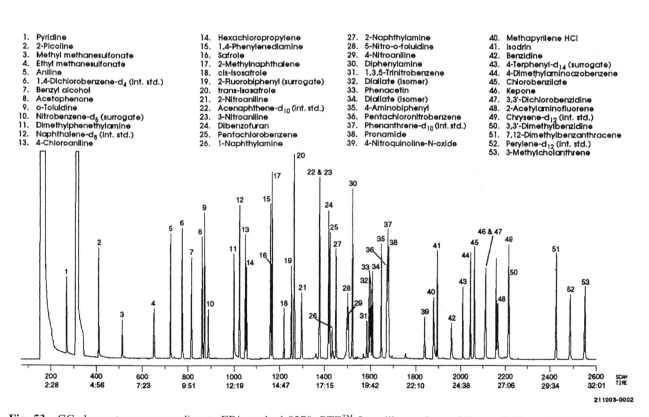

Fig. 53 GC chromatogram according to EPA method 8270. PTE™-5 capillary column, 30 m $\times$ 0.25 mm ID, 0.25 μm film, Col. Temp.: 35°C for 4 min., then to 300°C at 10°C/min., Linear Velocity: 35 cm/sec., He, Det.: MS (scan mass range M/Z = 33–550 at 0.740 sec./scan), Sample: mixture of Supelco chemical standards for US EPA 8000 series methods (50 ng each compound).

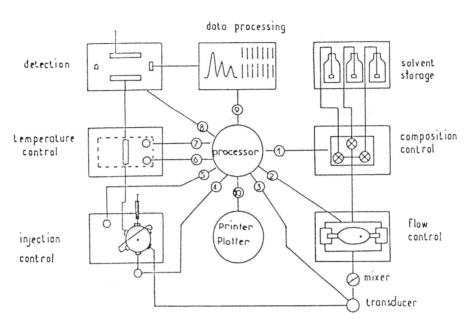

Fig. 54 Scheme of a liquid chromatography instrument.

c. Gradient Formation. Increasing the solvent strength permits one to achieve elution of highly retained solutes. Moreover, optimization procedures such as Drylab G make use of two gradients.

Two methods can generate binary solvent gradients: low-pressure mixing and high-pressure mixing. Some manufacturers display both. In the low-pressure mixing method, two solvent reservoirs and a single pump are used (Fig. 55). In high-pressure mixing mode, two pumps separately deliver the required volumes of solvents. The most critical points are the electronic control units and the mixing chamber.

Linear or curved gradients are possible, but linear gradients are used by the vast majority of chromatographers. There is obviously a time delay between the solvent composition programmed and the actual com-

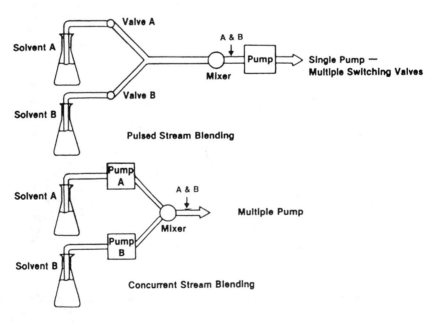

Fig. 55 Gradient Formation in LC.

position at the column inlet, due to the dwell volumes. Gradient delay and rounding are observed. Blank runs are carried out with the injector connected directly to the detector. Mobile phase A is UV transparent, and mobile phase B contains a slight proportion of UV absorbing (1% acetone in methanol, for example). The gradient is performed and the delay time is determined. This is of primary importance to obtain reliable retention times.

Gradient delay volume is the volume between the start of gradient and the top of the column. Gradient delay time is the time elapsed between the start of the gradient and the time it reaches the top of the column.

2. Column

Most of the commercially available columns have a 4.6-mm ID. They are made of stainless steel, and the inner wall of the tubing is electropolished. Variable lengths are available (10-cm-long columns are most popular). Columns or cartridges can be serially connected to increase the plate number. Guard columns (1 cm long) must be connected to the analytical column to increase the lifetime, especially when dealing with environmental samples. Columns are packed with 3-μm or 5-μm particles, either spherical or irregularly shaped. Particles are either porous or nonporous. Manufacturers provide a test chromatogram and ensure reproducibility.

To reduce solvent consumption, there is a trend towards microcolumns, minicolumns, or true capillary columns. Columns that have an inner diameter in the range 0.5–1.0 mm are called *microcolumns* (they were formally called *microbore*). Good, efficient microcolumns can be produced at the present time, since the packing procedure has been optimized.

Capillary LC columns have 10–100-μm ID. It seems that 50-μm-ID columns packed with 8-μm particles exhibit the best performances.

According to theoretical papers published in the 1980s, open tubular columns in LC can match performances of packed columns only if the diameter of the column is of the same order of magnitude as the particle diameter in conventional columns. This conclusion has led to nanoscale LC, with 5–11-μm-ID open tubular columns, which are for research laboratories but not yet for routine use.

3. Injection

For manual and automated injection, the majority of the injection systems consist of injection valves. Sample loops are usually 10 μl for injection with conven-

tional columns. In LC, the analyst can increase the volume of injection without disturbing the separation efficiency. Conversely, he must prevent any mass overloading. If so, partition of the solute is not performed in the linear portion of the isotherm, with the consequence of peak tailing.

4. Detectors

a. UV-VIS. The most popular LC detectors are the UV absorbance detectors. These detectors measure changes in the absorbance of light in the 190–350-nm region or 350–700-nm region. Basic instrumentation includes a mercury lamp with strong emission lines at 254, 313, and 365 nm, cadmium at 229 and 326 nm, and zinc at 308 nm. Deuterium and xenon lamps exhibit a continuum in the 190–360-nm region, which requires the use of a monochromator.

A filter or grating is used to select a specific wavelength for measurement. Cutoff filters pass all wavelengths of light above or below a given wavelength. Bandpass filters pass light in a narrow range (e.g., 5 nm). The flow cell is typically 8 μl with a 10-mm path length. The photodiode array (see next section) is now the best sensor.

According to the Beer's law, the greater the path length, the higher the transmitted light. Most cells are Z-shaped. With capillaries such as LC capillaries or CE capillaries, there are only limited path lengths. A free portion of capillary is brought into the light path of a UV absorbance detector. When the aperture of the source is adjusted to the inside diameter of the capillary, the effective light path is

$$l_{\text{eff}} = \frac{1}{2}\,\pi r$$

where r is the radius of the capillary. A U cell design provides a longer longitudinal light path and a substantial increase in the signal-to-noise ratio.

The limits of detection are highly dependent on the molar absorbtivity of the solute ε (see Beer's law).

b. Photodiode Array (PDA). The detection of structurally similar impurities eluting simultaneously with the analytes of interest is a problem. The analyst must detect the existence of peaks of interest, determine the extent of their purity, and confirm their identity.

Photodiode operation (Fig. 56) relies on the photovoltaic effect. In the typical photodiode there are two components of semiconductor called P and N. P is a very pure silicon with low levels of trivalent impurities, such as boron or gallium. Each impurity atom can ac-

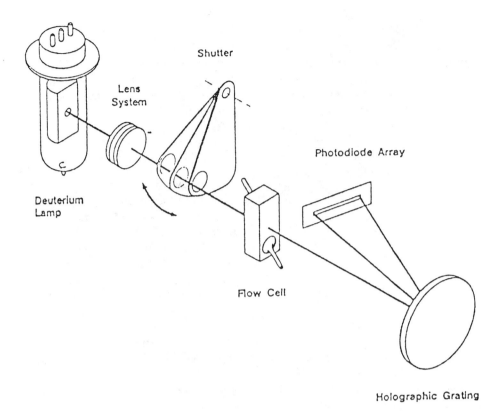

Fig. 56 Photodiode array.

cept an electron from the valence bonds, giving rise to a hole that can take part in the electrical conduction process and an immobile negatively charged impurity ion. Since the hole is positively charged, such a material is a P (positive) silicon crystal. If the impurity added is a pentavalent atom (As), the atoms behave as donors of electrons that can move through the entire silicon crystal. It is thus an N (negative) type. A photon of wavelength less than 1.1 nm is able to break a covalent bond between the silicon atoms. The free electron formed is free to move with the missing electron in the broken valence bond, which induces electrical conduction by repeated replacement. The PDA detector passes the total light through the flow cell and disperses it via a diffraction grating. The dispersed light is measured by an array of photosensitive diodes. The array of photodiodes is scanned by the microprocessor (16 times a second is usual). The reading for each diode is summed and averaged. The PDA detector can simultaneously measure the absorbance of all wavelengths versus time. The amount of data storage is a key feature in PDA. A run can easily require several megabytes for data storage. The dynamic range is usually 0.5 mAU– 2.0 AU.

Peak purity is based on the proprietary spectral-contrast algorithm, which converts spectral data into vectors that are used to compare spectra mathematically. This comparison is expressed as a purity angle. The purity angle is derived from the combined spectral-contrast angles between the peak apex spectrum and all other spectra within that peak. To determine peak purity, the purity angle is compared to the purity threshold. For a pure peak, the purity angle will be less than the purity threshold. Spectral deconvolution techniques are used when two peaks coelute. Peaks are identified by comparison with spectra contained in a library of standards.

c. Fluorescence Detection. Fluorescence emission provides more selectivity and increased sensitivity as compared to UV absorption. Laser-induced fluorescence is in current use. Various lasers are utilized (He Ne, diode, argon ion). The diode laser seems the best choice. Due to the highly collimating nature of lasers, most scattering sources are eliminated. Detection is increased with pre- or postcolumn derivatization (see, for example, organotin species complexed with fluorescent tags). A molecule detection is possible. Derivatization reactions for fluorescence are listed in Table 9.

Table 9 Derivatizing Reagents for Fluorescence Labeling of Functional Groups

Reagent	Abbreviation	Functional group
Aminoethyl-4-dimethylaminonaphthalene	DANE	Carboxyl
4-(Aminosulfonyl)-7-fluoro-2,1,3-benzoxadiazole	ABD-F	Thiol
Ammonium-7-fluorobenzo-2-oxa-1,3-diazole-4-sulfonate	SBD-F	Thiol
Anthracene isocyanate	AIC	(Amine), hydroxyl
9-Anthryldiazomethane	ADAM	Caboxyl (and other acidic groups)
Bimane, monobromo-	mBBr	Thiol
Bimane, dibromo-	bBBr	Thiol
Bimane, monobromotrimethylammonio-	qBBr	Thiol
4-Bromo-methyl-7-acetoxycoumarin	Br-Mac	See Br-Mmc
4-Bromo-methyl-7-methoxycoumarin	Br-Mmc	Carboxyl, imide, phenol, thiol
N-Chloro-5-dimethyaminonaphthalene-1-sulfonamide	NCDA	Amine (prim., sec.), thiol
9-(Chlormethyl)anthracene	9-CIMA	See Br-Mmc
7-Chloro-4-nitrobenzo-2-oxa-1,3-diazole	NBD-Cl	Amine (prim., sec.), phenol
2-*p*-Chlorosulfophenyl-3-phenylindone	DIS-Cl	Amino acids, amino sugars
9,10-Diaminophenanthrene	9,10-DAP	Caboxyl
2,6-Diaminopyridine-Cu^{2+}	2,6-DAP-Cu	Amines (prim. aromatic)
4-Diazomethyl-7-methoxycoumarin	DMC	See ADAM
5-Di-*n*-butylaminonaphthalene-1-sulfonyl chloride	Bns-Cl	See Dns-Cl
Dicyclohexylcarbodiimide	DCC	Carboxyl
N,*N*′-Dicyclohexyl-*O*-(7-methoxycoumarin-4-yl)methylisourea	DCCl	Carboxyl
N,*N*′-Diisopropyl-*O*-(7-methoxycoumarin-4-yl)methylisourea	DlCl	Carboxyl
4-Dimethylaminoazobenzene-4′-sulfonylchloride	Dbs-Cl	See Dns-Cl
N-(7-Dimethyl)amino-4-methyl-3-coumarinylmaleimide	DACM	Thiol
5-Dimethylaminonaphthalene-1-sulfonyl-aziridine	Dns-aziridine	Thiol
5-Dimethylaminonaphthalene-1-sulfonylchloride	Dns-Cl	Amine (prim., sec., tert.), (hydroxyl), imidazole, phenol, thiol
5-Dimethylaminonaphthalene-1-sulfonyl-hydrazine	Dns-hydrazine	Carbonyl
4-Dimethylamino-1-naphthoylnitrile	DMA-NN	Hydroxyl
9,10-Dimethoxyanthracene-2-sulfonate	DAS	Amine (sec., tert.)
2,2′-Dithiobis (1-aminonaphthalene)	DTAN	Aromatic aldehydes
1-Ethoxy-4-(dichloro-s-triazinyl)naphthalene	EDTN	Amine, hydroxyl (prim.)
9-Fluorenyl-methylchloroformate	FMOCCl	Amine (prim., sec.)
7-Fluoro-4-nitrobenzo-2-oxa-1,3-diazole	NBD-F	Amine (prim., sec.), phenol, thiol
4′-Hydrazino-2-stilbazole	—	α-Oxo acids
4-Hydroxymethyl-7-methoxycoumarin	Hy-Mmc	Carboxyl
4-(6-Methylbenzothiazol-2-yl)-phenyl-isocyanate	Mbp	Amine (prim., sec.), hydroxyl
N-Methyl-1-naphthalenemethylamine	—	Isocyanates (aliphatic, aromatic)
1,2-Naphthoylenebenzimidazole-6-sulfonyl chloride	NBl-SO$_2$Cl	See Dns-Cl
2-Naphthylchloroformate	NCF	Amine (tert.)
Naphthyl isocyanate	NIC	(Amine), hydroxyl
Ninhydrin	—	Amine (prim.)
4-Phenylspiro(furan-2(3*H*), 1′-phthalan)-3,3′-dione (fluorescamine)	Flur	Amine (prim., sec.), hydroxyl, (thiol)
o-Phthaldialdehyde (*o*-phthalaldehyde)	OPA	Amine (prim., sec.), thiol
N-(1-Pyrene)maleimide	PM	Thiol

Source: Reproduced from W.R.G. Bacyens and B.L. Ling. J. Planar Chromatogr 1:198.213 (1988).

d. Derivatization. Many solutes do not exhibit UV absorption; they can be converted in UV-absorbing derivatives by pre- or postcolumn derivatization. This procedure has a wider range than GC, since reaction can be performed following separation. When precolumn derivatization is carried out, the chromatographic system is obviously different from the one selected for the nonderivatized solutes. A large volume of literature deals with postcolumn reactions. These can be carried out in coils, in packed-bed reactors, or via photolysis.

The main requirement is not the completion of the reaction but the reproducibility. Reaction vessels should not produce excessive band-broadening. Table 9 displays some derivatizing reagents for the fluorescence labeling of functional groups.

  e. Electrochemical Detection

Conductivity: This detection method is based on the application of an alternative voltage E to the cell electrodes. The cell current is directly proportional to the conductance G of the solution between the electrodes, per Ohm's law:

$$G = \frac{1}{k} = \frac{i}{E}$$

The measured conductivity is the sum of individual contributions to the total conductivity of all the ions in solution. Kohlrausch's law states that

$$k = \frac{\sum\limits_{i} \lambda_i^o C_i}{1000}$$

where C_i is the concentration of each ion i and λ_i^o is the limiting equivalent conductivity, which is the contribution of an ion to the total conductivity divided by its concentration extrapolated to infinite dilution. Kohlrausch's law is valid only in dilute solutions (chromatography or electrophoresis). The magnitude of the signal is greatest for small high-mobility ions with multiple charge, such as sulphate.

Amperometric: Electrochemical detection is a concentration-sensitive technique. In amperometric mode, compounds undergo oxidation or reduction reaction through the loss or gain, respectively, of electrons at the working electrode surface. The working electrode is kept at constant potential against a reference electrode. Electrical current from the electrons passed to or from the electrode is recorded and is proportional to the concentration of the analyte present.

A thin-layer cell is displayed in Fig. 57. A thin gasket with a slot cut in the middle is sandwiched between two blocks: one contains the working electrode, the other contains the counterelectrode. The slot in the gasket forms the thin-layer channel. The reference electrode is placed downstream of the working electrode. The thin-layer design produces a high mobile-phase linear velocity, which in turn produces a high signal magnitude.

The intensity of the current is

$$i = \phi n F u^{1/2} \cdot C D^{2/3} \cdot A$$

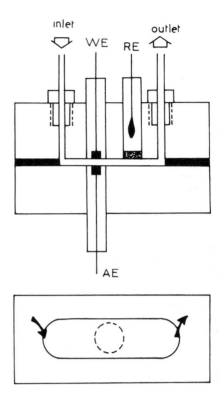

Fig. 57 Electrochemical detection in LC (thin-layer cell). WE: working electrode; AE: auxiliary electrode; RE: reference electrode.

where

  n = number of electrons
  F = Faraday's constant
  u = linear velocity of mobile phase
  C = analyte concentration
  D = diffusion coefficient of solute
  A = electrode surface
  ϕ = geometrical constant of cell

The quality of the sample cleanup procedure often determines the detection limits. The instability of the reference electrode is the source of voltage noise.

Parallel dual electrodes may be used for a number of reasons:

1. With one electrode at a positive potential and one electrode at a negative potential, oxidizable and reducible compounds can be detected in a single chromatographic run.
2. When two solutes with different redox potentials coelute from the column, the potential of one electrode can be selected such that only the most easily oxidized (or reduced) compound is detected while on other electrode both compounds

are converted. The concentration of the second compound is evaluated by substraction of the signal.

Series dual electrodes are set up such that one electrode is in oxidative mode and the other is in reductive mode. The downstream electrode measures the products of the upstream electrode. The second electrode responds only to compounds that are converted reversibly. The redox product is more selectively detected.

Voltammetric analysis is performed by scanning the potential or by applying a triangular potential wave form to the electrode. Coeluting peaks are distinguished if their voltagrams are significantly different.

f. Refractive Index (RI) Detector. This is one of the very few universal detectors available. The RI detector monitors both the eluent and the analyte. The output reflects the difference in refractive index between a sample flow cell and a reference flow cell. The measured RI response is determined by the volume fraction of the analyte in the flow cell (x) and the volume fraction of the eluent in the other flow cell ($1 - x$):

$$\eta - \eta_2 = v_1(\eta_1 - \eta_2)$$

where

v_1 = volume fraction of analyte
η_1 = refractive index of pure analyte
η_2 = refractive index of pure solvent (contained in a reference cell)
η = refractive index of solution in sample cell

There are four types of RI detectors:

The *deflection type* is by far the most popular. It relies on Snell's law, which governs the angles of incidence and refraction at an interface:

$$\eta_1 \sin \theta_1 = \eta_2 \sin \theta_2$$

where θ_1 is the angle of the beam with respect to the normal of the interface in the medium with RI of η_1.
Reflection type, according to Fresnel's law of reflection. Measurement of $\Delta\eta$ is a measure of change in reflectivity.
Interference type (utilized in capillary LC).
Christiansen-effect type.

The refractive index is very sensitive to temperature and pressure:

$$\frac{d\eta}{dt} \times 10^{-4} = \begin{cases} 0.67 & \text{for water} \\ 6.84 & \text{for dichloromethane} \end{cases}$$

$$\frac{d\eta}{dP} \times 10^5 = \begin{cases} 1.53 & \text{for water} \\ 5.56 & \text{for dichloromethane} \end{cases}$$

Detection by He/Ne laser–based RI has been developed for nanoscale LC.

g. Light-Scattering Detector. *Principle of operation:* The effluent of the LC column is vaporized in a nebulizer by means of a gas. The droplets pass through a drift tube at a temperature of 40–50°C, and the only particles left are the analyte and the solvent impurities. A laser (typically 1-mV He/Ne) irradiates the particles, and the scattered light is collected by a glass rod and transmitted to a photomultiplier tube (Fig. 58). The light measured is proportional to the amount of sample in the light-scattering chamber.

Parameters affecting the response are the particle size, the degree of nebulization (most critical), and the nature of the solvent. The amount of scattered light depends strongly on the molar absorptivity of the solute. The light-scattering detector is a universal detector but not a mass detector. Its response is nonlinear, and the calibration curve is log-log. It can easily be used with a gradient.

h. Gas Chromatography Detectors in Liquid Chromatography. Thermoionic, flame photometric, and electron capture detectors can be connected to an LC column. The LC eluent can be either transported into the GC detector or introduced directly.

D. Liquid Chromatography–Mass Spectrometry

Interfaces have been developed to solve the problem of handling high LC flow rates (1 mL/min) and the high vacuum required by the mass spectrometer. The mass spectrometer is mass flow sensitive. The enrichment factor is the ratio of the analyte concentration in the MS flow to that in the LC flow. The transfer yield is $Y = Q_m/Q_l$, where Q_m is the amount of solute transferred in MS and Q_l is the amount of solute from the LC column.

LC is not nearly as compatible with MS as is GC. Hyphenating LC and MS requires overcoming major difficulties:

Conventional packed columns are operated at 1 ml/min.
LC separations make use of nonvolatile mobile phases and very often buffer solutions.

856

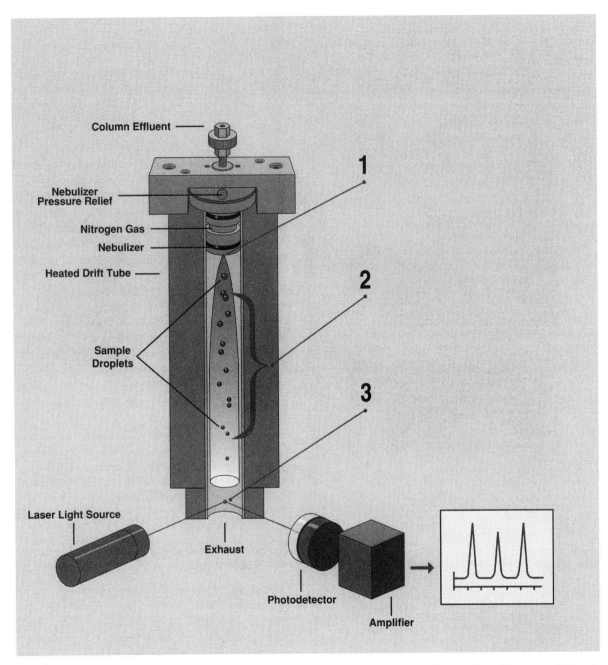

Fig. 58 Light-scattering detector in LC: 1, nebulization; 2, mobile-phase evaporation; 3, detection. (Courtesy of Alltech.)

Ionization of nonvolatile or thermally labile solutes is difficult.

However, the difficulties have been overcome, and LC-MS has become a robust and routinely applicable tool in environmental laboratories.

The first successful commercially available LC-MS interface was the transport, or moving-belt, system. The

operating principle can be separated into three main steps:

1. The liquid from the HPLC column is deposited onto a moving support—a wire or a plastic belt.
2. The solvent is removed by applying a vacuum and heat while the belt passes through vacuum locks in series, and then the neutral sample mol-

ecules are flash-heated and thermally desorbed from the belt.

3. Sample ionization immediately follows thermal desorption, using either electron impact (EI) or chemical ionization (CI).

A major advantage of the moving belt is its ability to provide the analyst with more than one ionization mode. Conversely, the belt is often fragile, due to temperature differences, and proper cleaning is sometimes difficult to achieve. The moving-belt interface is hardly used at present.

In direct liquid introduction (DLI), the effluent from the LC column is introduced directly into the MS source region. The column outlet is connected to a narrow-bore capillary. The liquid flowing through the orifice forms a stable jet of droplets. The unit is directed toward a pinhole (5–25-μm diameter). The DLI interface has a limited flow-rate capacity. It was used in the 1980s and has disappeared.

The thermospray interface was a real improvement in the 1980s. It can stand flow rates as high as 2 ml/min with reversed-phase eluents and volatile buffers. However, thermospray delivers only poor structural information.

Continuous-flow FAB is easy to implement. It can be linked to capillary or narrow-bore columns. The flow rate must be low (5–15 ml/min) to achieve mixing with the FAB matrix (e.g., glycerol). Evaporation of the solvent on a stainless steel frit yields a film subsequently bombarded by atoms.

j. Particle Beam Interface. In the particle beam interface (Fig. 59), the column effluent is pneumatically nebulized in a heated desolvation chamber at nearly atmospheric pressure. The solutes are selectively separated from the solvent vapor molecules in a two-stage momentum separator where the high-mass analytes are preferentially transferred to the MS source while the low-mass solvent molecules are pumped away. The analyte molecules are transferred as small particles to a conventional ion source, where they disintegrate upon collision at the heated source walls. The released gaseous molecules are ionized by EI or CI. The degree of desolvation of the droplets depends on the heat capacity of the solvent and the desolvation temperature of the chamber.

k. Atmospheric Pressure Ionization (API) Interface. To date, LC-thermospray MS and LC-FAB have largely been replaced by electrospray and ion spray (ESI). Atmospheric pressure ionization (API) sources are an ideal interface to MS. Since 1993, new ESI interfaces have been developed with pneumatic nebulization (Fig. 60) and the advent of hexapole or octapole devices for ion collection (Fig. 61). High flow rates (1 ml/min) can be handled. Much effort is being focused on miniaturization of the system. An API system containing an orthogonally positioned spray device has recently been introduced by Hewlett-Packard (Fig. 62).

In atmospheric pressure chemical ionization, the reagent ions for ionization of the analyte species are created by means of a corona discharge. The interface consists of a concentric pneumatic nebulizer and a large-diameter heated quartz tube. The nebulized liquid effluent is swept through the heated tube by an additional circumventing gas flow. The gases are introduced in the APCI source, where a corona discharge initiates APCI.

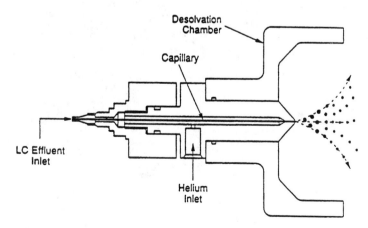

Fig. 59 Schematic diagram of a particle bean interface. (Courtesy of Hewlett-Packard.)

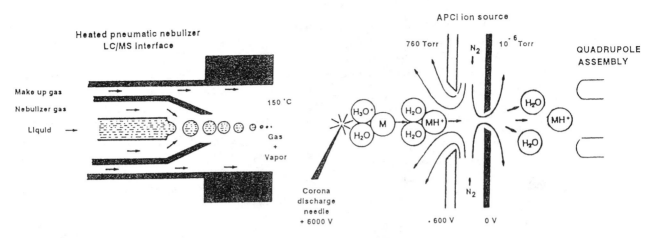

Fig. 60 Schematic diagram of a heated pneumatic nebulizer interface. (Reproduced with permission from the *Journal of Chromatography*.)

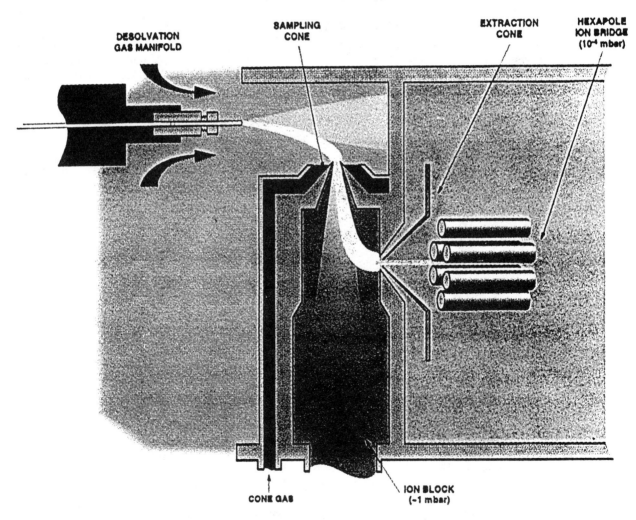

Fig. 61 Schematic diagram of the Micromass Z spray electrospray source. (Courtesy of Micromass.)

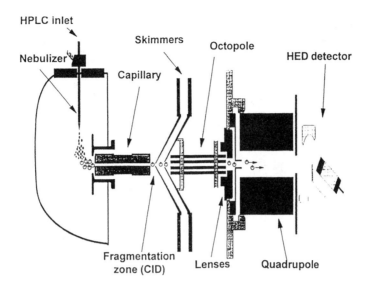

HPLC inlet

Skimmers

Octopole

HED detector

Nebulizer

Capillary

Fragmentation
zone (CID)   Lenses   Quadrupole

Fig. 62 Schematic of the LC-MS electrospray interface with orthogonal electrospray system. (Reproduced with permission from Hewlett-Packard.)

5. Different Modes of Liquid Chromatography

a. Adsorption, or Normal Phase Liquid Chromatography (NPLC).   In this mode, the stationary phase is polar. Since silicagel and alumina were utilized only in the beginning of liquid chromatography and hydrophobic alkyl-bonded phases were developed later, this mode is called *normal phase*.

Stationary Phases.   Silicagel and alumina are declining in use. Polar-bonded phases are now widely used.

Diol phase.   The following is very similar to bare silica but less retentive:

$$\text{Si—O—Si—(CH}_2)_3\text{—O—CH}_2\text{—CH—CH}_2\text{OH}$$
$$\text{OH}$$

Amino phase.   The following acts by the lone pair of electrons on the nitrogen atom:

$$\text{Si—O—Si—(CH}_2)_3\text{—NH}_2$$

Cyano phase.   The following is less polar and has been advocated in some optimization procedures as the unique phase, since it can be used with any solvent:

$$\text{Si—O—Si—(CH}_2)_3\text{—CN}$$

Other polar phases are available but are less often advocated.

Mobile Phases.   Theoretically any solvent is convenient. Extensive work has been carried out on adsorption theory and the retention mechanism since the pioneering work of Snyder (24). In this chromatographic mode, physical adsorption of solute occurs. That means weak interactions with the stationary phase. Adsorption of the solutes occurs via a donor acceptor mechanism (hydrogen bonding, charge transfer, etc.). According to Snyder's displacement model, a solute S takes the place of n molecules of previously adsorbed mobile-phase solvent molecules M.

$$n\text{M}_s + \text{S}_m \overset{K}{\leftrightarrow} \text{S}_s + n\text{M}_m$$

where the subscripts s and m refer to stationary phase and mobile phase, respectively.

$$K = \frac{(\text{S})_s}{(\text{S})_m} \times \frac{(\text{M}_m)^n}{(\text{M}_s)^n}$$

From this starting equilibrium, the Snyder–Soczewinski equation yields the retention equation:

$$\ln k = \ln \frac{V_a W}{V_m} + \alpha(\varepsilon^o - S^o A_s) + \sum \text{secondary effects}$$

where $V_a W$ is the volume of solvent forming a monolayer coverage on the adsorbent surface. When a solute is strongly adsorbed, eluent must exhibit a strong affinity toward the stationary phase to displace the solute. ε^o represents the dimensionless energy of solvent ad-

Table 10 Eluotropic Series for Different Adsorbents

Solvent	Solvent strength parameter					
	Alumina	Silica	Carbon	Aminopropyl	Cyanopropyl	Diol
Pentane	0.00	0.00				
Hexane	0.01	0.01	0.13–0.17			
Carbon tetrachloride	0.17	0.11		0.069		
1-Chlorobutane	0.26	0.20	0.09–0.14			
Benzene	0.32	0.25	0.20–0.22			
Methyl-tert. butyl ether	0.48			0.11–0.124	0.049–0.085	0.071
Chloroform	0.36	0.26	0.12–0.20	0.13–0.14	0.106	0.097
Dichloro-methane	0.40	0.30	0.14–0.17	0.13	0.120	0.096
Acetone	0.58	0.53		0.14		
Tetrahydrofuran	0.51	0.53	0.09–0.14	0.11		
Dioxane	0.61	0.51	0.14–0.17			
Ethyl acetate	0.60	0.48	0.04–0.09	0.113		
Acetonitrile	0.55	0.52	0.01–0.04			
Pyridine	0.70					
Methanol	0.95	0.70	0.00	0.24		

Source: L.R. Snyder. Mobile-phase effects. In: C.S. Horvath, ed. Liquid–Solid Chromatography in High-Performance Liquid Chromatography. Vol. 3. Academic Press, New York, 1983, p. 157. Reproduced with permission.

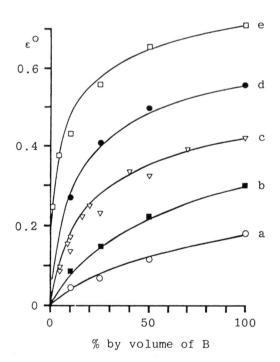

Fig. 63 Plot of eluent strength of a binary mixture of solvents (A = diluent; B = modifier) in NPLC. (a) carbon tetrachloride; (b) propyl chloride; (c) methylene chloride; (d) acetone; (e) pyridine. (Reproduced with permission from Edinburgh University Press.)

sorption. Ranking ε^o gives a scale of adsorption strength. It is obvious that *n*-alkanes exhibit weak affinity toward polar stationary phases. Conversely, silicagel and alumina are hygroscopic and exhibit strong affinity toward water. Ranking ε^o constitutes the eluotropic strength scale (Table 10).

Fine retention tuning is somewhat difficult with a single solvent. Mixtures of solvents most often constitute the eluent. Binary mixtures are a blend of apolar diluent (A) and polar modifier (B). Direct calculation of the eluting strength is possible through a formula derived by Snyder (25). Plots of ε^o versus percentage of B are displayed in Fig. 63. Such plots are readily available in most manufacturers' software. Three (or more) solvent mixtures as eluent are difficult to handle and reproduce. With a binary mixture retention is given by

$$\ln K = Ct - n \ln(X_B)$$

where X_B is the molar fraction of modifier B and Ct is a constant. Continuous increase of X_B is gradient elution (see further).

Normal-phase chromatography is well adapted to the separation of polar or moderately polar compounds. The NPLC technique is well suited to the separation of structural isomers.

b. Reversed-Phase Liquid Chromatography (RPLC). In this mode, the stationary phase is hydrophobic (apolar); to maximize the difference between the nature of both stationary and mobile phases, the latter is highly polar.

Stationary phases: more than 600 reversed-phase columns are commercially available. Most are silica bonded of the C_8 or C_{18} type:

$$\text{Si—O—Si—C}_{18} \quad \text{or} \quad C_8$$

These phases should be tested for:

Efficiency, expressed by N/m (plates per meter)
Hydrophobic properties
Steric selectivity
Silanophilic properties
Metal content

Hydrophobic properties depend on the level of the surface coverage. Manufacturers who provide carbon content should also provide the specific surface area. Nevertheless, this type of information is not sufficient, and the best way to compare reversed-phase bonded silica columns is to use a test mixture and check the chromatographic separation obtained.

An impressive amount of literature has been published on preparation of these phases. To avoid the presence of remaining silanols, an end-capping reaction with small silane is performed. Most C_{18} phases commercially available are end-capped.

Silica bonded phases are not stable over the whole pH range; low pH (≤ 2) or high pH (≥ 9) may damage the siloxane bond. To overcome this drawback, polymeric phases of the polystyrene divinylbenzene (PS-DVB) type have been developed. With a high degree of crosslinking, they are mechanically stable and can stand high pressures. Porous glassy carbons (PGC) are pH stable as well, but the number of published separations with this support is small as compared to silica bonded supports.

On these hydrophobic supports, hydrophilic solutes are not retained and by consequence water is the weakest eluent. The eluotropic strength is thus exactly the reverse of the one observed in NPLC.

Mobile phases are typically water plus organic modifier mixtures. Methanol, acetonitrile, and tetrahydrofuran (THF) are the usual organic modifiers.

The mechanism of retention has been a matter of dispute. The volume of "definitive" papers on the topic is impressive. The partition mechanism is generally accepted, but some deviations from this mechanism may be observed.

Two features are important in RPLC. In a homologous series (*n*-alkanols, saturated fatty acids, etc.), lin-

ear plots of ln k versus carbon number of the solutes are observed. Increasing the volume percentage of organic modifier in the mobile phases decreases retention according to

$$\ln k = a - b\phi - c\phi^2$$

where ϕ is the volume percentage of the modifier in a binary mixture water–organic modifier. The curvature of the quadratic plot of ln k versus ϕ is highly dependent on the nature of the solute (Fig. 64). In a more or less limited range of the volume percentage, the preceding equation can be written as

$$\ln k = \ln k_w - S\phi$$

where ln k_w is the retention with pure water as mobile phase, values of k_w are obtained by extrapolation to $\phi = 0$, and S is the slope of the regression and is characteristic of the solute.

A discrepancy can occur when methanol, acetonitrile, or tetrahydrofurane is used as modifier. k_w has often been considered as the hydrophobicity of the solute. It comes from the correlation plot of ln k_w versus log P octanol–water. As with correlation plots, some

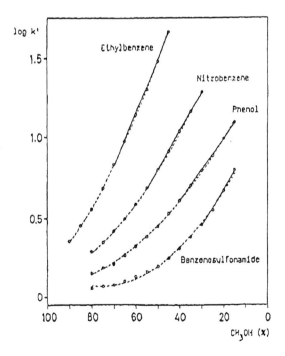

Fig. 64 Plots of ln k versus volume % of organic modifier (methanol) in RPLC with C18 column. The solid line shows the linear region of plot of ln k versus %; the dotted line shows the quadratic region. (Reproduced with permission from *Chromatographia*.)

discrepancies can occur, depending on the number of solutes in the plot and the correlation coefficient values.

The slope (S) of the plot of ln k_w versus ϕ is typical of the solute considered. When considering, for example, two solutes, one of two situations can occur: either the two solutes have the same S (in this case selectivity remains constant whatever the ϕ value) or the two solutes exhibit S_1 and S_2 values, which means that the slopes are different and by consequence there exists a ϕ value where no separation occurs. Beyond this value the order of retention is reversed. Peak crossover occurs when scanning the range of ϕ values.

The selection of organic modifier is often made by trial and error. Experienced chromatographers are well aware of solute–solvent interactions. Snyder (26) has provided a useful selectivity triangle of solvents in which dipole–dipole, proton–donor, proton–acceptor interactions are the three apices. Solvents are gathered in eight groups. Solvents in a single group exhibit similar interactions. From considering the triangle, it is obvious that acetonitrile and methanol, for example, will interact differently. When solutes are eluted according to the sole hydrophobicity (for example, a homologous n-alkanol series), the use of either methanol or acetonitrile does not change the order of retention. Conversely, when very different chemical species are separated, the use of methanol or acetonitrile will yield different retentions and selectivities. In the same way, chloroform and dichloromethane (two modifiers very often advocated in NPLC) may yield different retentions.

Gradient elution is performed when solutes are strongly retained. To decrease the t_r of these late-eluting peaks an increase of modifier (stronger eluent) is performed. From the preceding linear equation it is clear that a linear increase in ϕ will result in a linear decrease in ln k. This is called *linear solvent strength* (LSS).

Software and method development in LC separations are readily and widely available. Such products rely on the input of two or more pilot runs to calibrate a retention model (see, for example, Drylab). Once the retention model is calibrated, the software allows the chromatographer to simulate the separation under a wide variety of separation conditions. Gradient runs can be used to predict either isocratic or gradient separations. At first, software was developed for RP separations. The model has been extended to other types. The selection of initial instrument parameters and separation mode is left to the user. Binary systems were first proposed. Extension to ternary mobile phase in RPLC permits a fine selectivity tuning.

Reversed-phase phases allow a wide range of separation. They are well suited to polar organic solutes soluble in water. A rule of thumb is that retention order is similar to solubility in water or in a water–modifier mixture. Apolar solutes such as polycyclic aromatic hydrocarbons are well separated on C_{18} phases according to their hydrophobicity. Analysis of metals can be also done when a metal chelation is performed. In this mode, dithiocarbamates are often used due to the strong chelating ability of their sulfur group. Other selected chelating agents include azodyes, thiazolylazo reagents, acetylpyridine-4-ethyl-3-thiosemicarbazone, 8-quinolilol (calcium does not react), and EDTA.

In micellar liquid chromatography the mobile phase consists of surfactants at concentrations above their critical micelle concentration (CMC) in an aqueous solvent with an alkyl-bonded phase. Micelles act as a mobile-phase modifier. Retention behavior is controlled by solute partitioning from the bulk solvent into micelles and into stationary phase, as well as on direct transfer from the micelles in the mobile phase into the stationary phase. Reference (27) gives a broad overview of the method.

For more details on HPLC, see Katz et al. (28).

E. Ion Chromatography

Separation of ionic or inorganic solutes can be performed with three fundamental methods: ion-pair formation, ion exchange, and ion exclusion.

Suppose a weak acid AH. The dissociation equilibrium is

$$AH \leftrightarrow A^- + H^+$$

with the dissociation constant k_A. At low pH in a buffered solution, only AH exists. Separation with an RP system is thus possible, provided that the acid is not too strong. Alkyl sulfonic acids, for example, may exhibit very low pK_A. Separation of acids in RP occurs according to the degree of hydrophobicity. Conversely, if the pH of the solution is raised, A^- and H^+ are the only existing species. A^- can be chromatographed on an anion exchanger. Another method is ion pairing or ion interaction.

In the ion-pairing technique, a reversed-phase column (RP8 most often) is utilized, with a mobile phase consisting of an aqueous organic mixture to which an ion-pairing agent is added. The ion-pairing agent may form, with the analyte, an ion pair with increased lipophilicity, which by consequence is retained on the RP stationary phase. According to the most popular hypothesis, the ion-pairing agent induces a dynamic modification of the surface of the RP stationary phase. The lipophilic part of the ion-pairing agent coats the surface, leaving the ionic part to interact with counterions.

Retention can occur via several methods:

1. It can be retention of the ion pair formed between the cations (or the anions) of the analyte and the cation (or anion) of the ion-pairing agent. The ion pair itself is adsorbed onto the stationary phase.
2. The analyte is retained through an ion-pair complex formed with an amphiphilic ion previously adsorbed onto the surface of the hydrophobic material.
3. The analyte is retained through ion-exchange reactions with the adsorbed pair reagent.

Retention increases as the concentration of the ion-pairing agent increases, but decreases beyond a certain concentration; pH is a key parameter.

Ion-pair reagents are most often either nitrogen (+) or sulfonate (−) species: cetyl-trimethyl-ammonium bromide, diamino-dodecane, tributyl-ammonium chloride, cetyl-pyridinium chloride, CHAP (3-[(3-cholamidoproyl)-dimethyl-ammonio]-1 propane sulfonate), or tris(tris-(hydroxymethyl)aminomethane). For additional information, see Refs. 29–32.

Ion chromatography is a new version of the well-known ion-exchange chromatography. It is widely used in water analysis. The technique is specified in many standard methods. It does not suffer from interference problems that can occur in AAS. It is able to separate bromide, chloride, fluoride, nitrite, phosphate, and sulphate. It is able to speciate oxidation states of several metals, such as Fe(II) and Fe(III). Challenge to IC comes from capillary electrophoresis.

The instrumentation required for IC simply comprises a metal-free HPLC system and conductivity detection. Since conductivity is a common property of ions, this method is universal and quite sensitive.

In IC, both mobile and stationary phases are ionic.

Ion Exchange. Principle of operation: Ion exchangers are insoluble solid materials that contain exchangeable cations or anions. These ions can be exchanged for a stoichiometrically equivalent amount of other ions present in an electrolyte solution.

Exchange reactions are written as (for example, in cation exchange):

$$R{-}H^+ + Na^+ \rightleftharpoons R{-}Na^+ + H^+$$

$R{-}$ is the matrix of ion exchanger.

The thermodynamic equilibrium constant is

$$K_{H^+}^{Na^+} = \frac{(RNa^+)\cdot(H^+)}{(Na^+)\cdot(RH^+)}$$

Suppose E^+ is a monovalent element cation and M^{n+} is a sample metal ion:

$$nE^+{-}R + M^{n+} \leftrightarrow M^{n+}Rn + nE^+$$

$$K_{E/M} = \frac{(M^{n+}Rn)\cdot(E^+)^n}{(E^+{-}R)^n\cdot(M^{n+})}$$

At low loading of sample ions, the term $(E^+{-}R)$ is equal to the exchanger capacity Q.

$$K_D = \frac{(M^{n+}Rn)}{(M^{n+})}$$

is proportional to k, the retention factor, and

$$k = K_{E/M}\cdot\frac{(Q)^n}{(E^+)^n}$$

More generally,

$$\ln k_A = \frac{1}{y}\ln K_{A/E} + \frac{x}{y}\ln\frac{Q}{y} + \ln\frac{W}{V_M} - \frac{x}{y}\ln(E_m^y)$$

$$\ln k = Ct - \frac{x}{y}\log(E_m^y)$$

where

k_A = retention factor for a solute A^x
$K_{A/E}$ = ion-exchange coefficient for solute A and eluent E
Q = ion-exchange capacity of stationary phase
W = weight of stationary phase used in column
V_M = volume of mobile phase
x = charge of solute anion
E_m^y = concentration of element ion in mobile phase.

In IC, low-capacity ion-exchange columns are utilized. Eluents are in the range 1–10 mmol.

Stationary Phases: In ion exchange we can distinguish:

SCX: strong cation exchangers (functional group SO_3^-); in the H^+ form they represent "solid" acids.
WCX: weak cation exchangers (functional group COO^-).
SAX: strong anion exchangers or strong base (functional group alkanol quaternary amine)
WAX: weak anion exchangers (functional group amino).

Amphoteric ion exchangers contain anionic and cationic exchange sites.

Chelating ion-exchange groups are: iminodiacetate, 8-hydroxyquinolinol, β diketone, triphenyl methane dyes, carbamates, EDTA, and PAR [(4-2-pyridylazo)resorcinol]. New packings are compiled in the annual review that appears in LC/GC magazine.

Base Materials. The majority of stationary phases are agglomerated or pellicular materials consisting of a monolayer of charged latex particles that are electrostatically attached to a functionalized internal core particle. Polystyrene-divinylbenzene (PS-DVB) with different degrees of crosslinking is the substrate. Particle diameters are typically in the range 5–25 μm, while adsorbed latex particles are 0.1 μm. This material exhibits very good pH stability and excellent mass transfer properties. For example, analysis of oligosaccharides from the ozone bleaching of kraft pulp is performed at pH 11–12. Analyte ions interact with the functional groups on the porous latex beads and do not diffuse into the substrate, owing to the Donnan exclusion, with the result of faster kinetics. Conversely, this material is hydrophobic.

Microporous methacrylate–based materials are used mainly for anion exchange because of their resistance to high-pH eluents.

Silica-based materials are produced by grafting organic moieties via the procedure utilized in producing reversed-phase silica; aminopropyl-bonded silica or functionalized silicas such as sulphonate are available. An advantage of silica-based materials is the low probability of secondary interactions between the solute ions and the silica substrate.

Another route to produce base material is the polymer coating of silica material. A layer of polybutadiene-maleic acid (PBDMA) is deposited on silica and then crosslinked by peroxide-initiated radical chain reaction. This is used mainly in the analysis of transition metals.

Retention: Factors influencing retention and selectivity are: hydration enthalpy, hydration entropy, polarizability, charge, size, and structure of both eluent and solute ions. The concentration of eluent ion (or ionic strength) and pH play an important role too.

A secondary equilibrium occurs with the nature of the sorbent matrix. Hydrophobic interaction markedly influences retention (a typical example is the well-known amino acid sequence on the cation exchanger, which does not follow the pKa sequence). With inorganic ions, the perchlorate (ClO_4^-) effect is also well established. Nonionic eluent modifiers are often used to change the ion-exchange affinity of hydrophobic ions.

Gradient elution starts with an eluent of low ionic strength for the resolution of the most weakly retained species. The dependence of analyte retention on eluent concentration is a straight line with slope given by the ratio of the charges of the analyte and eluent. Column switching permits the separation of both inorganic and organic anions in one run. The simultaneous analysis of ions is possible; column switching is configured for anion analysis in column A and cation analysis in column B.

Separation selectivity depends on the degree of electrostatic forces. With an SCX phase, the binding force decreases with decreasing diameter:

$$H^+ < Li^+ < Na^+ < NH_4^+ < K^+ < Ag^+$$
$$< Be^{2+} < Mg^{2+} < Ca^{2+} < Sr^{2+} < Ba^{2+}$$

With SAX material, a selectivity series with respect to binding force also exists:

$$F^- < OH^- < CH_3COO^- < Cl^- NO_2^-$$
$$< CN^- < NO_3^- < SO_4^{2-}$$

Ion Exclusion. Ion exclusion relies upon Donnan equilibrium. The main parameter in this mode is the electrostatic interaction of the solute with the charged functional groups on the surface of the stationary phase. The stationary phase can be considered as a charged membrane separating the flowing mobile phase from the static, occluded mobile phase trapped in the pores of the exchanger. Ionic solutes are rejected because of their inability to penetrate the pores and so are eluted at the void volume of the column. Conversely, nonionic substances may partition between the occluded liquid phase and the flowing mobile phase. The degree of partition determines the extent of the retention. Sample anions are excluded from the resin phase by the fixed charges of the sulfonate groups of the cation-exchange resin.

Separations by ion exclusion are usually performed on a cation-exchange column with a PS-DVB resin. The solute retention volume is

$$V_R = V_o + K_D V_i$$

where V_o is the interstitial volume of eluent (eluent outside the resin beads), V_i is the occluded volume within the resin beads, and K_D is the distribution constant of the solute. When a very large solute or ion cannot enter the stationary phase, $K_D = 0$; when the solute is free to enter, $K_D = 1$.

The range of retention volumes is rather short; by consequence, ion-exclusion columns are rather long. Acids are found to elute in decreasing order of their acid dissociation constants; the stronger acids elute first.

Complexation Ion Chromatography: This mode is devoted mainly to the analysis of transition metal ions. A complexation reaction can take place in either the mobile or the stationary phase. Transition metals are firmly bound by strong electrostatic forces to the sulphonic acid groups of a cation exchanger. When using an eluent complexing ion, its effect is to reduce the effective charge of the metal ion, which in turn reduces the retention times of bivalent ions. The distribution coefficients K_D of bivalent metal ions are close. Differences in stability constants of metal complexes are larger. Stability constants decrease with pH. Lower pH values will be required for metal complexes with high stability constants.

When a basic solution contains an excess of strong complexing anions of high charge, such as EDTA ions, most metal ions will occur as anionic complexes, which can be separated by ion exchange.

Detection in Ion Chromatography. The detection methods most often quoted are:

Conductivity (both suppressed and nonsuppressed)

Amperometry
Spectrophotometry
Postcolumn reaction
Hyphenated instruments ICP/MS and MS.

Conductivity. Early IC systems detected ions eluted by strong eluents from high-capacity ion-exchange columns by measuring changes in conductivity. To achieve reasonable sensitivity, it was necessary to suppress the conductivity of the eluent prior to detection in order to enhance the overall conductance of the analyte and lower the background conductance of the eluent. This was achieved via a "suppressor" column in which counterions were exchanged with H^+ or OH^-.

Due to excessive band-broadening, column suppressors are no longer in use. Membrane-based devices are utilized. The membrane suppressor incorporates two semipermeable ion-exchange membranes sandwiched between sets of screens. The eluent passes through a central chamber. Regenerant flows in a countercurrent direction over the outer surfaces of the membranes, providing constant regeneration. Electrolysis of water produces the hydrogen or hydroxide ions required for regeneration. There is no contamination with carbonate (Fig. 65).

Solid-phase suppressors use disposable packed-bed cartridges as the suppression device. A small suppressor column is used upstream of the conductivity detector; a second suppressor column that requires regeneration is placed after the detector, where the suppressed eluent undergoes an electrolysis reaction so that protons or hydroxyl ions are generated.

In conductivity detectors, the change in conductivity Δk depends on the concentration of the injected ion (A) and its equivalent ionic conductivity λ_A compared with that of the eluent ion λ_E:

$$\Delta k = (A) \cdot (\lambda_A - \lambda_E)$$

Conductivity detectors used previously were range dependent, which is a disadvantage when analyzing environmental samples in which small amounts of one analyte are present together with a large amount of others. A single range digital conductivity detector eliminates the need for dilution.

Amperometric. This may be operated in either of two modes: constant potential or pulsed amperometric. The second mode is advantageous: a triple-pulse waveform is applied, with successive applications of a measuring potential, a cleaning potential, and a conditioning potential. Amperometric detection is utilized to determine bromide, iodide, and thiocyanate. New elec-

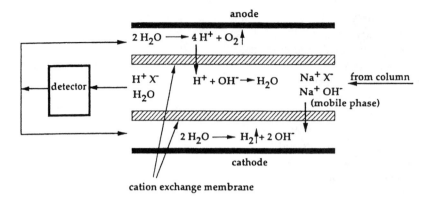

Fig. 65   Anion self-regenerating suppressor (ASRS) for ion chromatography. (Courtesy of Dionex Co.)

trode materials, such as conductive polymers (polypyrrole or polyaniline), have been developed for the detection of electrochemically inactive ions. Ion-selective electrodes for univalent cations or anions or the incorporation of ionophores in a polymer membrane are gaining in use.

Spectrophotometry (UV-VIS Fluorescence).   Indirect UV absorbance uses light-absorbing species to detect nonabsorbing ions. Sulfosalicylic acid is generally utilized with $\lambda = 298$ nm at pH 2.50.

Postcolumn Reaction.   Some ions (nitrate, nitrite) absorb in the 210-nm region. Postcolumn reaction of lanthanides with chlorophosphonazo(III) in HNO_3 yields color at 660 nm. Luminol reaction permits the trace analysis of Cr(III), Co, Cu, or Ag.

Hyphenation.   Hyphenation with AAS is possible. Nebulization may be difficult due to the nature of the ionic mobile phase.

As detector from an IC column, IPC-AES is well suited for speciation. One of the major problems that can arise upon coupling IC to MS is the possible incompatibility of the electrolyte of the mobile phase with the interface and the ion source of the MS. A micromembrane suppressor must be incorporated. Table 11 displays some applications of HPLC coupled with ICP. Table 12 displays some EPA HLPC methods.

F.   Thin-Layer Chromatography (TLC)

Thin-layer chromatography can be used for detection of inorganic species for qualitative purposes. The main use of TLC is the analysis of organic species. In TLC the chromatographic column is a plate covered with a layer of sorbent (0.20–0.25 mm thick). The inert support is either a glass plate or aluminum foil. The dimensions are 10×10 cm, 5×10 cm, 5×20 cm, or 20×20 cm. The TLC plates are covered with particles

of 11-μm average size, and high-performance thin-layer chromatography (HPTLC) plates are covered with particles of 5–7-μm average size. Most commercially available precoated layers are prepared from silicagel with a nominal pore diameter of 60 Å. The use of bonded silica is increasing, especially diol-bonded silica, which is less sensitive to moisture than is bare silica.

In normal TLC the plate is placed inside a tank in which a few milliliters of developing solvent (or solvent mixture) were previously poured. The liquid ascends the layer due to capillary forces. According to TLC optimization, the length of development is different depending on whether a TLC or HPTLC plate is used and on the diffusion coefficients of the solutes.

Retention of solutes is expressed as:

$$\text{observed Rf} = \frac{\text{migration of the solute}}{\text{migration of the eluent}}$$

In this mode, $0 \leq \text{Rf} \leq 1$, so the true Rf is

$$\frac{\text{velocity of the solute}}{\text{velocity of the eluent}}$$

since the solvent front migrates faster than the bulk of the eluent. Rf values above 0.85 are not reproducible and should be avoided for any identification purposes.

With concentration zone plates, TLC provides simultaneous sample cleanup and separation. Rf values are strongly dependent on experimental conditions. Binary mixtures of solvents (diluent plus modifier) are rather simple to handle. Complex mobile phases do not permit reproducible Rf values.

Two-dimensional TLC is a very powerful tool for complex sample separation. In this mode the plate is submitted to a development in one direction, then (after evaporation) the plate is turned 90° and a second sol-

Table 11 References Related to Water Analysis

Analyte	Chromatography	Detector	Comments
Se species	GC	ICP-ID-MS	Detection limit of 0.02 ng ml^{-1} for selenite
Species of Fe, Mn, P, and Pt	IC	ICP-AES	ICP-AES was found to suffer very few interferences
V speciation	IC, carbonate buffered 1,2-cyclohexylenedinitrilo tetraacetic acid eluent	AES	Detection limits of 145 and 70 ng ml^{-1} for V(IV) and V(V), respectively
V^{4+}, V^{5+}, NI^{2+}, Cr^{3+}, Cr^{6+}	Dionex HPLC-CS5 column with lithium hydroxide-2,6-pyridine dicarboxylic acid as the eluent	ICP-MS	Effect of varying pH, mobile phase, and ionic strength was examined
Cr^{3+} and Cr^{6+}	IC, Excelpak ICS-A23 column; EDTA–oxalic acid used as the mobile phase	ICP-MS	Detection limits of $8.1 \cdot 10^{-5}$ and $8.8 \cdot 10^{-5}$ ppm, respectively
Cr^{3+} and Cr^{6+}	Anion exchange with phthalate as the mobile phase	PC reaction with diphenylcarbizide	Sensitivity not as great as with chemiluminescence or ICP-AES/MS detection
As^{3+} and As^{5+}	IC	Electrochemical and spectrophotometric detection; reaction between heteropolymolybdoarsenic and bismuth	Detection limits of 2.9 and 13 μl l^{-1}, respectively
Cu^{2+}, Ni^{2+}, Zn^{2+}, and Mn^{2+}	Ion-PAC CS5 Column with PDCA derivatization	PC reaction with PAR	Examined the interference effect of chelating agents on recovery
Species of Cu and Pb	Silica and C_{18}-bonded columns	ETAAS	A heated microcolumn manifold was employed
Species of Cu, Cd, Pb, and Mn	C_{18} RP, Dowex anion-exchange and Chelamine columns used	ETAAS	The system showed very good reproducibility, even in complex media
Heavy-metal complexes	HPLC	ICP-ID-MS	A reliable, sensitive technique that is free from organic interference
Pb speciation	Cellex 100, Spheron oxin, Amberlite XAD-2, C_{18}, and cellulose sorbents modified with phosphoric acid and carboxymethyl groups	FAAS	Cellulose sorbents were found to have the best retention characteristics
Sb^{3+}, Sb^{5+}	HPLC, Hamilton PRP-X100 with phthalic acid as the mobile phase	HGAAS-ICP-MS	Very efficient sample transfer to the plasma; largely free from interferences, the method was applied to wastewater samples
Organotin and organo-germanium	On-column cGC	FPD	Detection limits ranged from 0.2 to 2.3 pg for ethylated, butyl- and phenyltin and from 50 to 100 pg for germanium complexes
Tributyltin	On-column cGC	FPD	TBT recovery of 90% for aquatic matrices
As^{3+} and As^{5+}	Extraction chromatography	FAAS and ETAAS	Wastewaters from a metallurgical plant were studied

Source: Taken in part from L.A. Ellis, D.J. Roberts. Chromatographic and hyphenated methods for elemental speciation analysis in environmental media. J. Chromatogr. 774:3–19 (1997).

Table 12 U.S. EPA HPLC Methods

Method	Wastewater	Drinking water	Air	RCRA	CERCLA
HPLC-UV	610 PAHs 631 Benomyl and carbendazim 632 Carbamates and urea pesticides 604.1 Hexachlorophene and dichlorophene 629 Cyanazine 635 Rotenone 636 Bensulide 637 2,2'-Dithiobis-(benzothiazole) 639 Bendiocarb 640 Mercaptobenzo-thiazole 642 Biphenyl and orthophenyl phenol 643 Bentazon 644 Picloram	549 Diquat and paraquat 550 PAHs 550.1 PAHs	TO-5 Aldehydes and ketones TO-6 Phosgene TO-8 Phenols and cresols TO-11 Formaldehyde TO-13 PAHs	8310 PAHs 8315A Aldehydes and ketones (2,4-DNPH derivatized) 8316 Acrylamide, acrylonitrile and acrolein 8317 4,4'-Methylene bis(2-chloro-aniline) (MOCA) 8321 Azo dyes, amines, organophosphorus compounds by HPLC-TSP-MS[a] 8330 Nitroaromatics and nitramine explosives 8331 Tetrazene 8332 Nitroglycerine 8333 Nitro compounds 8350 Aromatic sulfonic acids by ion-exchange chromatography	Pesticides/Aroclors, Routine Analytical Services Statement of Work, OLMO1.8 8/91, GPC cleanup
HPLC-FL	610 PAHs 641 Thiabendazole	550 PAHs 550.1 PAHs 531.1 Carbamates (postcolumn reaction) 547 Glyphosate (postcolumn reaction)	TO-8 Phenols and cresols TO-13 PAHs	8310 PAHs 8318 N-Methylcarbamates (postcolumn reaction)	
HPLC-EC[a] HPLC-MS	605 Benzidines			8321 Azo dyes, amines, organophosphorus compounds by HPLC-TSP-MS 8325 Benzidines and nitrogen-containing pesticides by HPLC-PB-MS[a] 8350 Aromatic sulfonic acids by ion-exchange chromatography	

[a]Abbreviations: EC = electrochemical detection; PB = particle beams; TSP = thermospray.

Source: Taken in part from U.S. EPA methods.

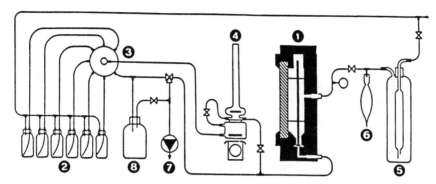

Fig. 66 Schematic diagram of an automated multiple development apparatus (AMD). (1) enclosed developing chamber; (2) solvent reservoir bottles; (3) switching valve for selecting the solvent composition; (4) gradient mixing chamber; (5) wash bottle for gas phase; (6) reservoir for gas phase; (7) vacuum pump; (8) solvent waste bottle. (Reproduced by courtesy of Camag.)

vent is utilized. It is obvious that if the same solvent is used, spots are distributed along the diagonal (full correlation). To obtain good separations the two chromatographic systems should be the least correlated. Multiple development is a fruitful approach for increasing the separation performances of TLC. See also Guiochon et al. (33).

1. Automated Multiple Development (AMD)

In an AMD separation, the chromatogram is developed repeatedly in the same direction, and successive development takes place over increasing migration distances. Increments are between 2 and 5 mm; a complete program usually comprises 20–30 cycles. Contrary to HPLC gradients, the stepwise solvent gradient starts

with the strongest eluent (e.g., methanol with bare silica as stationary phase) and proceeds through less and less eluting solvents. A schematic diagram of the instrument is displayed on Fig. 66.

2. Detection in Thin-Layer Chromatography

a. Densitometry. Densitometry is the mode for quantitative detection; optical measurement of a layer is difficult. Three types of measurements are in common use: transmission, reflection, and simultaneous transmission/reflection. Reflection mode is the most popular. A typical detecting device is shown in Fig. 67. Lamps are continuous-spectrum, halogen, or tungsten for visible spectrum, and deuterium or xenon for UV.

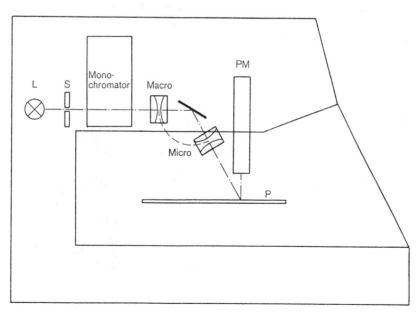

Fig. 67 Scheme of a TLC scanner. (Courtesy of Camag, Muttenz, Switzerland.)

b. Videodensitometry. Unlike scanning densitometers, videodensitometers exhibit no moving parts. A video camera permits illumination by UV light at selected wavelengths (254 nm, 356 nm). The camera focuses on the media to be scanned, and a video signal is sent to the digitizer board in the computer. The signal is also sent to a black-and-white video monitor that displays a real-time image of the media. This helps the user to position a cursor to establish the boundaries. Parameters are set and computer scans all lanes automatically. A chromatogram is produced for each lane scanned. The charge coupling device (CCD) has a number of features as an imaging detector: high sensitivity, spectral range, and two-dimensional imaging ability.

c. Blotting. Blotting is a method in which compounds are blotted from a TLC plate to a membrane (polyvinylidene difluoride). Immunostaining with TLC is less sensitive than with ELISA.

G. Overpressured-Layer Chromatography (OPLC)

This can be considered a capillary liquid chromatography with a flat column. To overcome the drawback of nonuniform mobile-phase velocity, the plate is embedded. The sorbent layer is completely covered with a flexible membrane. A scheme of an OPLC instrument is displayed in Fig. 68. Eluent is delivered by an HPLC pump. A trough is placed on the layer to ensure a linear eluent front. Chromatoplate (cassette format) is inserted into the holding unit. A special chromatoplate that is sealed at the edges is utilized. External pressure is applied (Fig. 69). A typical optimum linear velocity is 0.8 ml/min.

VI. SUPERCRITICAL FLUID CHROMATOGRAPHY

A fluid is in supercritical state when both its temperature and its pressure are above their critical values. Supercritical fluids exhibit unique physicochemical properties. Their viscosity is 5–20 times lower than that of ordinary liquids. By consequence, diffusion coefficients of solutes are greater. They thus provide a means for faster and more efficient extractions. Densities are 100–1000 times greater than those of gases, which gives them a solvating power close to that of liquids. The Hildebrand solubility parameter is

$$\delta = 1.25 P_c^{1/2} \left(\frac{\rho_g}{\rho_l} \right)$$

where P_c is the critical pressure of the fluid, ρ_g is the density of the supercritical fluid, and ρ_l is the density of the fluid in liquid state.

Selective extraction of compounds may be achieved by varying the extraction pressure. The solvation strength of a supercritical fluid may be increased by increasing its density.

Many fluids are readily available in supercritical state: N_2O, SF_6, CH_3ClF_2 (freon 22), CO_2. Carbon dioxide is by far the most widely used. We see from the pressure–density diagram (Fig. 70) that the critical point at 31.1°C and 72.8 bar permits its use under mild thermal conditions. Carbon dioxide is available in pure form, is not poisonous, and is neither flammable nor

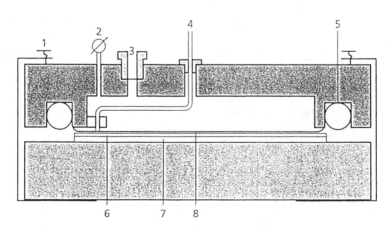

Fig. 68 Cross section of a forced-flow TLC chamber. (1) screw fasteners, (2) pressure gauge, (3) gas/liquid inlet, (4) mobile-phase inlet and tubing, (5) rubber O-ring, (6) stationary phase, (7) glass/metal plate, (8) membrane.

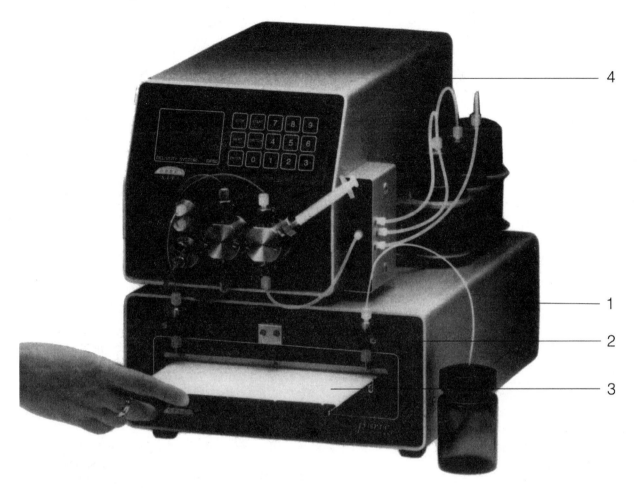

Fig. 69 Personal OPLC BS-50 system. 1: separation chamber; 2: holding unit; 3: layer cassette; 4: liquid-delivery system. (Courtesy of NIT Instrument, Budapest.)

explosive. In the supercritical state, CO_2 has a polarity comparable to that of liquid pentane. The main drawback is its lack of polarity. N_2O looks to be better suited for polar compounds, but it may cause violent explosions.

Supercritical water at 400°C and 350 bar can be used for extraction of polyaromatic hydrocarbons (PAHs) or polychlorobiphenyls.

Experimental SFE is rather simple: A pump is utilized to supply a known pressure of the extraction fluid to an extraction cell held at a temperature above the critical temperature of the fluid. When a modifier (usually methanol) is required, it is introduced via a second pump. When dealing with water solutions, some problems occur. The main problem is the solubility of water in supercritical CO_2 (0.3%), which can cause plugging of the restrictor by ice.

The SFE of water samples can be performed using the closed-loop stripping principle. The supercritical fluid (after pressurizing of the system) is recycled by a pump from the outlet of the extraction cell back into the water sample. After equilibrium, which may last more than 1 h, an aliquot of the supercritical phase is taken by means of a valve with a loop (Fig. 71).

Factors affecting SFE are:

Pressure: the higher the extraction pressure, the higher the solubility and the smaller the volume of the fluid necessary for extraction

Temperature: the density of CO_2 decreases when the temperature rises. The effect of temperature heavily depends on the sample. It is argued that increasing temperature is a useful alternative to adding organic modifiers.

Addition of a modifier: CO_2 is slightly polar, and modifiers (methanol, acetonitrile) are required to extract polar solutes. Organic pollutants that are so polar and nonvolatile that they cannot be an-

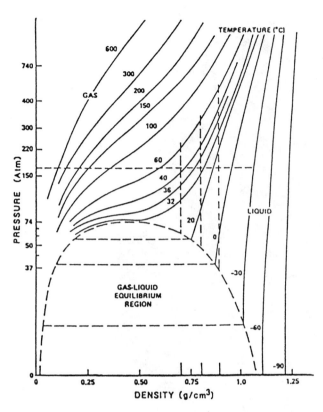

Fig. 70 Pressure/density diagram for carbon dioxide. The shaded area corresponds to the experimental domain of supercritical-phase extraction and chromatography.

alyzed by GC are not extracted very efficiently by SFE.
Fluid velocity.
Cell geometry.

Usually a small volume of water is loaded onto the cartridge packed with a suitable inert powdered material. In this mode, SFE extraction is similar to that for solids. The solubility of water in supercritical carbon dioxide is several grams per liter. Supercritical fluid enters the extraction cell at its bottom, and a restrictor is placed at its top to prevent water breakthrough.

Metal ions in aqueous samples can be extracted by supercritical CO_2 containing organic ligands. Since metals can be found as ionic species and inorganic compounds, SFE is more complicated, with the proper selection of ligands. The SFE technique may provide a means of metal speciation in natural samples.

Chelating agents are

Dithiocarbamates
β diketones

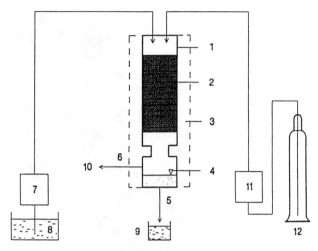

Fig. 71 Apparatus used for the continuous extraction (concurrent mode) of dilute aqueous solutions by supercritical CO_2. 1—extraction column; 2—packing; 3—thermostated aluminum block; 4—phase separator; 5 and 6—capillary restrictor; 7—reciprocal pump; 8—water sample reservoir; 9 —water output; 10—CO_2 output; 11—piston pump; 12— CO_2 cylinder. (Reproduced from the *Journal of Chromatography* with permission.)

Organophosphorous reagents
Macrocyclic ligands.

The solubility parameter of the ligands may provide a simple method for predicting the solubility of the chelates.

Coupling chemical derivatization with SFE decreases the polarity of polar analytes, which in turn increases their solubility in supercritical fluids. The goal is to replace active hydrogens such as OH, NH, and SH groups. Chlorophenoxy acid herbicides, thiouracils, and benzimidazoles are alkylated; phenols are acylated.

Organotin compounds were determined in seawater by first extracting the di- and tributyltin onto a C_{18} disc and then performing a Grignard reaction just prior to extraction with supercritical CO_2 with hexylmagnesium bromide.

A. Instrument

1. Pumping

Syringe pumps are used mainly with open tubular columns. With flow splitting problems occur with sample introduction and when a modifier must be added to CO_2. Without flow splitting, a 10–20-ml syringe pump can operate many hours without refilling.

An HPLC pump can readily pump liquid CO_2. A second pump will deliver the modifier. Pumps are cooled to pump liquid CO_2. The liquid CO_2 will be directed into a heated capillary to reach the supercritical state.

2. Injection

Flow-splitting injection is no longer in use with OTC. Solvent-venting injection, developed by Greibokk (34), allows for automated injection of microliter volumes. Injection on packed columns is straightforward. Injection is performed through average 4.6-mm columns or SPE cartridges. Large columns (up to 40 ml!) can be introduced.

3. Columns

Packed capillary columns (1–10 m long) with small particles (<5 μm) exhibit high efficiencies (>200,000 plates). They can be coated with thick films (2 μm) on 25–50-μm ID. Tubular columns increase both retention and sample capacity.

4. Detectors

Hyphenation with MS is performed with a direct-fluid-introduction interface (Fig. 72) where the effluent is admitted directly into an EI or CI ion source.

Nanoelectrospray is very promising. Droplets emitted from a 1–3-μm capillary contain only one (or two) analyte molecule when the concentration is 1 pmol/ml. The low flow rate increases the efficiency of the electrospray by a factor of 100.

Plasma emission and plasma mass detectors are used for the detection of organometallics. The problems associated with the introduction of CO_2 in plasmas have been overcome.

Coupling with FTIR is performed in two modes: direct desorption (with elimination of solvent) and flow cell detection. Xenon is better suited for the latter mode. But it is expensive and it is a poorer mobile phase than CO_2 with polar analytes.

VII. CAPILLARY ELECTROPHORESIS

Capillary electrophoresis was originally called *capillary zone electrophoresis*. Early attempts were carried out, but many authors refer to the work of Jorgenson and Luckas (35), who described separations performed in a 75–100-μm-ID glass capillary at a voltage of 30 kV. Sophisticated instruments were first introduced in 1989, and the technique very rapidly expanded.

Following J. H. Knox (36) we can call capillary electroseparation methods those techniques characterized by:

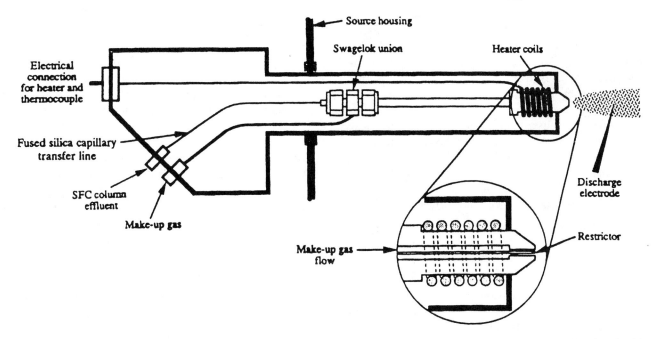

Fig. 72 SFC/MS interface for atmospheric pressure chemical ionization induced by a corona discharge. (Reproduced with permission from the *Journal of Chromatography*.)

Table 13 Methods in Capillary Electrophoresis

Technique	Open tube	Packed tube
Electrophoresis	CE	CGE
Chromatography	CMEC	CEC

Electrophoresis methods: CE = capillary electrophoresis (ions only); CGE = capillary gel electrophoresis (ions only).
Chromatographic methods: CMEC = capillary micellar electrochromatography (neutrals, ion pairs, ions); CEC = capillary electrochromatography (neutrals, ion pairs, ions).

A fine capillary within which separation occurs
A high-voltage power supply capable of delivering 50 kV at 100 μA
Two electrolyte reservoirs into which the ends of capillary dip, one connected to the high-voltage supply and the other grounded on column injection and detection
A suitable Faraday cage to ensure safe operation of the high-voltage section of the equipment
Thermostating of the capillary
Suitable electronics.

Available methods are summarized in Table 13.

A. Electro-Osmotic Flow (EOF)

Imagine a buffer inside a silica capillary tube. The surface silanol Si–OH groups are ionized at pH above about 3. The negatively charged silanoate groups attract positively charged cations from the buffer, which form a layer of cations close to the capillary wall. Close to the surface is a layer of ions that are tightly adsorbed by electrostatic interaction. This layer is often called the *Stern* or *Helmholtz layer*. An outer layer is not tightly held, because it is further away from the silanoate groups. This layer is a mobile layer that may diffuse. It is called the *Gouy Chapman layer*. These two layers make up the double layer.

The charge density σ of the excess ions falls exponentially with distance x from the surface:

$$\sigma = \sigma_o \exp\left[\frac{-x}{\delta}\right]$$

where δ, the so-called thickness of the double layer because it has the dimension of length, is inversely proportional to the concentration of electrolyte. Typically, δ lies between 1.0 and 10.0 nm

Between the two layers is a plane of shear, and an electrical imbalance is created at that plane, which is the potential difference across the layers. This is termed the *zeta potential*, ζ, which is given by

$$\zeta = 4\pi \frac{\delta e}{\varepsilon}$$

where e is the charge per unit surface area and ε is the dielectric constant of the buffer. The process is called *electro-osmosis*. The velocity of the electro-osmotic flow is

$$u_{eo} = \left[\frac{\varepsilon_o \varepsilon_r \zeta}{\eta}\right] E$$

where

ε_o = permittivity of vacuum = 8.85 $\times$ 10^{-12} $C^2 N^{-1} m^{-2}$
ε_r = dielectric constant of buffer
E = applied electric field in V/cm
η = viscosity of buffer

The electro-osmotic mobility is

$$\mu_{eo} = \left[\frac{\varepsilon_o \varepsilon_r \zeta}{\eta}\right] \quad (m^2\text{-}s^{-1}\text{-}V^{-1})$$

The elution order is cations, neutrals, anions. Charged solutes are separated according to their electrophoretic mobilities. Neutral solutes are not separated from each other. Electro-osmotic flow exhibits a flat profile (Fig. 73).

Parameters that will affect electro-osmotic flow velocity are:

Applied voltage: High voltage will increase electro-osmotic flow but will also increase Joule heating. Excessive heat, which cannot be dissipated, is produced when too high a voltage is applied. The maximum voltage depends on the capillary length and diameter. A smaller inner diameter permits a high voltage.
pH of the buffer: The zeta potential is proportional to the surface charge on the capillary wall.

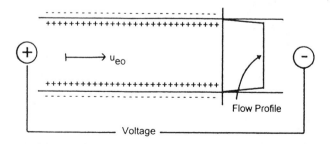

Fig. 73 Electro-osmotic flow and migration of ions.

Buffer concentration (ionic strength): Increasing the ionic strength of the buffer decreases the electro-osmotic flow.

Temperature.

Organic solvent: Adding organic solvent to the buffer will affect viscosity, dielectric constant, and zeta potential.

Chemicals.

Under the influence of an electric field (E), an electrically charged solute of radius r will migrate through a buffer with an electrophoretic velocity of $u_{ep} = \mu_{ep}E$, where u_{ep} is the electrophoretic mobility:

$$u_{eo} = \left[\frac{\varepsilon_o\varepsilon_r\zeta}{\eta}\right]f\left[\frac{r}{\delta}\right]E$$

where

$$r = \text{radius of particle or ion}$$

$$f\left(\frac{r}{\delta}\right) = \begin{cases} 1 & \text{when } r/\delta \gg 1 \\ \dfrac{2}{3} & \text{when } r/\delta \ll 1 \end{cases}$$

$$\mu_{eo} = \frac{q}{6\pi\eta r} \quad \text{assuming a spherical solute}$$

$$q = \text{charge of ionized solute}$$

The migration time t_m is the time it takes to migrate from the injection to the detector. Different from chromatography, where L is the full length of the column, we must define l, the effective capillary length:

$$u_{obs} = \frac{l}{t_m} = \mu_{obs}E$$

$$u_{obs} = u_{ep} + u_{eo}$$

where u_{eo} and u_{ep} are taken as positive when the movement is toward the cathode and negative when in the opposite direction.

$$t_m = \frac{l}{\mu_{obs}E}$$

since $E = V/L$,

$$t_m = \frac{lL}{\mu_{obs}V}$$

$$t_m = \frac{lL}{(\mu_{ep} + \mu_{eo})V}$$

From the foregoing equations, it is obvious that higher voltages, shorter capillaries, and high electro-osmotic flows give shorter migration times. The overall velocity of an ion decreases as the concentration of the background electrolyte increases.

Similar to chromatography, the number of theoretical plates is given by

$$N = \frac{l^2}{\sigma^2}$$

Since a plug flow profile is observed in CE, zone-broadening is caused by diffusion and, according to Einstein's law,

$$\sigma^2 = 2Dt$$

$$N = \frac{1}{L}\frac{1}{2D}(\mu_{ep} + \mu_{eo})V$$

Efficiency increases by increasing the applied voltage. With a Gaussian peak profile,

$$N = 16\left[\frac{t_m}{\omega}\right]^2$$

where ω is the peak width. The selectivity α is expressed in the usual way:

$$\alpha = \frac{t_{m2} - t_{nm}}{t_{m1} - t_{nm}}$$

where t_{nm} is the migration time of a neutral marker. From the preceding equation we can write

$$\alpha = \frac{\mu_1}{\mu_2}\cdot\text{constant}$$

The resolution is

$$\text{Rs} = \frac{2(t_{m2} - t_{m1})}{\omega_1 + \omega_2}$$

$$\text{Rs} = \frac{\sqrt{N}}{4}\left[\frac{\Delta u}{\bar{u}}\right]$$

where Δu is the difference in velocity between the two solutes and $\bar{u}$ is their average velocity.

$$\text{Rs} = \frac{1}{4}\left[\frac{(\mu_{ep} + \mu_{eo})V}{2D}\right]^{1/2}\left[\frac{\mu_{ep2} + \mu_{ep1}}{(\bar{\mu} + \mu_{eo})}\right]$$

In CE separations that utilize additives in the background electrolyte, analyte mobility is determined by three parameters: the electrophoretic mobilities of the free and complexed analytes and the equilibrium constant of the analyte–additive interaction. In general, CE exhibits much higher plate numbers, shorter analysis times, and different selectivities as compared to LC systems.

B. Isotachophoresis

Principle of operation: A sample is separated into zones in a supporting electrolyte. Two electrolytes are used: one contains the leading ion, which has the highest mobility of all the sample ions, and another contains the terminating ion, which has the lowest mobility. The sample is injected between the two electrolytes. Ions from the sample are not overtaken by the terminating ion. When current is applied between the electrodes, ions migrate, and separate zones are observed with the following properties:

Each zone has a single component ion.
The order of the zones is determined by the mobilities.
The concentration of the component ion is uniform in each zone.
The moving speeds of all zones are equal (origin of the name *isotachophoresis*).
The boundary between two zones is sharp.

One of the important parameters in selecting the electrolyte system is the pH value of the leading electrolyte. Most ions, except certain strong electrolytes, change their effective mobilities according to the degree of dissociation. The isotachophoregram consists of a series of steps, in which the step heights and the step lengths are indicative of the identity and the quantity of the analytes.

C. Micellar Electrokinetic Capillary Chromatography (MEKC or MECC)

Micellar electrokinetic capillary chromatography was introduced by Terabe (37). Analytes are partitioned between the background electrolyte and the micelles. Usually the micelles are charged and the analytes are uncharged, but there are other possibilities (uncharged micelles with charged analytes). The electrolyte moves at a velocity u_{eo}, while the micelles migrate at their electrophoretic velocity $u_{ep(mic)}$. The velocity of a band of analyte is

$$u = \left[\frac{k}{1 + k}\right] u_{mic} + \left[\frac{1}{1 + k}\right] u_{eo}$$

where

$$k = \frac{\text{amount of component in stationary phase}}{\text{amount of component in mobile phase}} = \frac{n_{mic}}{n_{aq}}$$

n = number of molecules

so

$$k = \frac{1/t_n - 1/t_r}{1/t_r - 1/t_{mic}} = \frac{t_r - t_o}{t_o[1 - t_r/t_{mic}]}$$

where

$$t_r = \frac{1}{u_{solute}}$$

$$t_o = \frac{1}{u_o}$$

$$t_{mic} = \frac{1}{u_{mic}}$$

$$t_n = \frac{1}{u_o}$$

u_o = velocity of a solute that does not partition

$u_o = u_{eo}$

and t_n is the time of elution of an analyte that stays in the electrolyte, t_{mic} is the time of elution of an analyte that is confined in the micelles, t_r is the retention time, and k ranges from 0 when $t_r = t_n$ to ∞ when $t_r = t_{mic}$.

D. Capillary Gel Electrophoresis (CGE)

In capillary gel electrophoresis, the capillary is filled with a crosslinked gel. Separations of molecules are based primarily on differences in their size. In packed-bed capillary, particles of silica may be porous or nonporous. In this mode the electro-osmosis phenomenon occurs with both capillary walls and silica particles. Packed-bed can be considered as a set of interconnected capillaries.

E. Capillary Electrochromatography (CEC)

In this technique, an electric field and electro-osmotic flow drive the mobile phase through capillary liquid chromatography columns. Separations are based on partitioning between a mobile phase and a stationary phase. The stationary phase is either wall coated in the capillary or in the packing of a packed-bed capillary.

F. Capillary Isoelectric Focusing

Capillary isoelectric focusing separates charged solutes on the basis of differences in their isoelectric points. Isoelectric focusing (IEF) can be used in slab and tube gels. It can be run in capillary tubes with or without a supporting gel. The surface of the capillary wall must be treated to eliminate electro-osmotic flow. A pH gra-

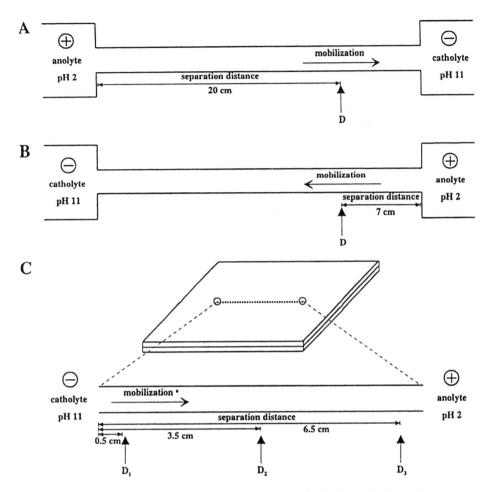

Fig. 74 Polarity mode used for capillary isoelectric focusing. (A) Normal polarity mode for capillaries, separation distance 20 cm. (B) Reversed polarity mode for capillaries, separation distance 7 cm; due to instrumental constraints, the point of detection is fixed. (C) Polarity mode for microchannels on a chip; the point of detection can be moved; separation distances of 0.5, 3.5, and 6.5 cm were chosen. *D* represents the point of detection. (Reproduced from *Analytical Chemistry* with permission.)

dient is generated with the cathode at the high-pH side of the gradient and the anode at the low-pH side. To create the pH gradient, the acid ampholyte is placed in the anode reservoir and the basic ampholyte in the cathode reservoir. Applying the elemental potential creates the pH gradient. The sample is included with the ampholytes that generate the pH gradient. Zwitterionic analytes migrate to the position where they have a net zero charge, their isoelectric point (pI). Isoelectric focusing of a water sample is displayed in Fig. 74.

G. Capillary Electrophoresis Instrumentation

Power supply: should be able to produce voltages up to 30 kV and more and currents of about 100 mA.

Electrodes are usually made of platinum.
Buffer vials.
Capillaries: capillaries are 24–100 cm long. The smaller the diameter, the better the dissipation of Joule heating. Inner diameters of 50 μm are in current use, but trends are obviously toward IDs of 12 μm or less. The smallest capillaries are achieved by channeling a silicon chip. Increasing the length of the capillary increases the resolution of the separation but increases the time of analysis and the resistance to fluid flow. A short capillary is preferred. Increasing the inner diameter of the capillary increases the path length of the detector and increases detectability, but it is detrimental to separation. The internal surface of the capillary is a parameter of paramount importance, since it controls EOF. Via coating (for example,

with a linear polyacrylamide bonded to silica), EOF can be virtually eliminated, which allows lower voltages.

Detection device: A small part of the capillary is free from any coating (inside and outside), which permits a beam to pass through. In other detecting devices a small reactor is designed (see the laser-induced fluorescence device, for example).

Buffers can be modified with surfactants (anionic, cationic, neutral, or zwitterionic), organic modifiers, or chiral selectors. Gels are polymer networks: linear polyacrylamide, dextrans, cellulose, or polyethylene oxide (Fig. 75).

H. Injection

There are two types of injection: electrokinetic and pressure differential.

1. Electrokinetic Injection

In the electrokinetic injection (or electromigration), the anodic end of the capillary is introduced along with the electrode in the sample vial. A voltage is applied for a short period (a few milliseconds), which causes migration of the sample into the tip of the capillary. The injected volume is a combination of both the electrophoretic mobility of solute molecules and the bulk electro-osmotic flow of the buffer into the capillary. The drawback of the technique is discrimination as a result of the different electrophoretic mobilities of ionic solutes in the sample, which leads to the injection of an unrepresentative sample. The injected quantity is

$$Q_{inj} = \frac{E_{inj}}{L} \pi C_{inj} t r^2 (\mu_{ep} + \mu_{eo})$$

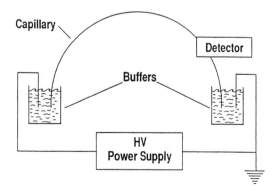

Fig. 75 Scheme of CE instrumentation.

where E_{inj} is the electric field applied during injection (typically less than 10 kV); C_{inj} is the sample concentration, t is the amount of time the voltage is applied, r is the radius of the capillary, and L is the capillary length. If the capillary has been treated to eliminate electro-osmotic flow, neutral solutes are not injected.

2. Pressure Differential, or Hydrodynamic, Injection

Sample is introduced into the capillary by placing the capillary into the vial containing the sample and generating a pressure difference across the capillary. This forces the sample to flow into the tip of the capillary.

Three different types of pressure differential injection can be performed:

Hydrostatic: The capillary is inserted into the sample vial, and the pressure difference is created by raising the sample vial to a height above the outlet vial. Reproducibility of the time taken to raise the inlet vial is the major requirement.

Vacuum: The pressure differential is generated by applying a vacuum at the outlet end of the capillary. Injection precision and reproducibility rely upon the reproducibility of the vacuum applied to the capillary.

Pressure: The capillary is placed in a sample vial and a pressure is applied to the vial, thereby creating a pressure difference across the capillary and forcing the sample to flow into the capillary. Reproducibility and duration of the applied pressure are the major requirements. Applied pressure lies in the vicinity of 250 mbar.

Example: Dynamic Compression Injection   The injected volume is

$$V_{inj} = \frac{\Delta P \pi r^4 t}{8 \eta L}$$

where ΔP is the pressure across the capillary, t is the time the pressure is applied, and L the total capillary length. The length of the sample plug in the tube is

$$L_{plug} = \frac{V_{inj}}{\pi r^2}$$

When gravity or siphoning is utilized, then

$$V_{inj} = 2.84 \times 10^{-8} \frac{H t r^8}{L}$$

where H is the height the sample is raised and L is the total capillary length.

I. Sample Preconcentration in Capillary Electrophoresis

Field amplification injection technique: When an analyte is dissolved in a sample matrix having a lower conductivity than the background electrolyte (BGE) in the capillary, then this analyte will experience locally an increased field strength and will migrate with a higher velocity that is proportional to the ratio of the conductivities in the BGE and the sample matrix. When the analyte reaches the boundary between the sample matrix zone and the background electrolyte, it will slow down again and stack in a zone much shorter than the original sample zone.

Solid-phase extraction inside the capillary: A hydrophobic phase can be covalently bonded to the inner wall of the capillary. The disadvantages are more numerous than the advantages: raw material is difficult to introduce directly, for it may plug the capillary. The use of some organic solvents is tedious.

Online coupling of isotachophoresis and CE.

J. Detection in Capillary Electrophoresis

UV-VIS: Reduction of band-broadening is of primary importance. Fast UV scanning detectors are currently available. Problems of small path length are similar to those with HPLC capillary. Optimized optical design for on-column diode array and laser-based UV detectors are in progress.

Fluorescence: This is by far the most sensitive detector. Detection of zeptomole (10^{-21} mole) is readily obtained with suitable fluorophores. Indirect fluorescence detection is possible. Laser-induced fluorescence is very popular as a detection strategy for low-level monitoring of organic pollutants (Fig. 76). The lasers are He-Cd and He-Ne, and the cell is of the sheath flow design (Fig. 77). In this design, a flat detection window reduces light scattering and background noise. A promising field is the development of multichannel detection for capillary arrays. Chemiluminescence detection will develop through electrogenerated luminescence. Detection of a single molecule has been achieved via laser-induced fluorescence.

Electrochemical: Amperometric, conductometric, and potentiometric detection are obviously possible. The electrode material is the key parameter

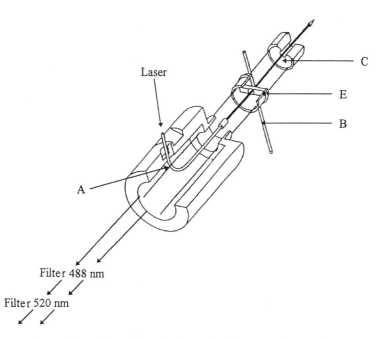

Fig. 76 Fluorimetric detector in capillary electrophoresis. A: optic fiber carrying the beam; B: electrophoresis separation capillary; C: ellipsoidal mirror; E: beam-alignment device. (Courtesy of Beckman.)

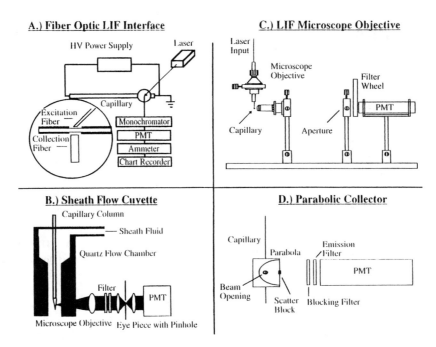

Fig. 77 Various optical arrangements reported for capillary electrophoresis/laser-induced fluorescence (CE/LIF). A: the first reported CE/LIF interface, utilizing optical fibers; B: the sheath flow cuvette interface; C: a typical CE/LIF arrangement using a microscope objective and scatter-eliminating aperture for collection; D: parabolic reflector with scatter-blocking mask. (Reproduced from Ref. 38 with permission.) (Reproduced with permission of CRC Press, Boca Raton.)

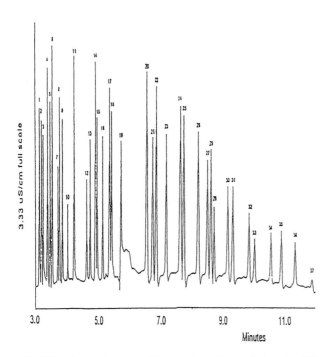

Fig. 78 Determination of inorganic and organic ions by CE with conductivity detection. (Reproduced with permission from the *Journal of Chromatography*.)

to lower the background noise of the detector. Amperometric detection should be performed with care to electrically isolate the high electric field in CE from the detector. An example is displayed in Fig. 78.

Mass spectrometry detectors: An example of CE electrospray–MS is displayed in Fig. 79.

Table 14 gives some typical operational parameters of IC and CE together with their strengths and weaknesses. Table 15 gives a comparison of different methods of nitrite determination. Further data on CE can be found in Ref. 38.

VIII. VALIDATION

The EPA defines *quality control* as follows: "The quality control process includes those activities required during data collection to produce the data quality desired and to document the quality of the collected data." The EPA defines *quality assurance* as follows: "The quality assurance process consists of management review and oversight at the planning, implementation and completion stages of the environmental data collection activity to ensure that data provided are of

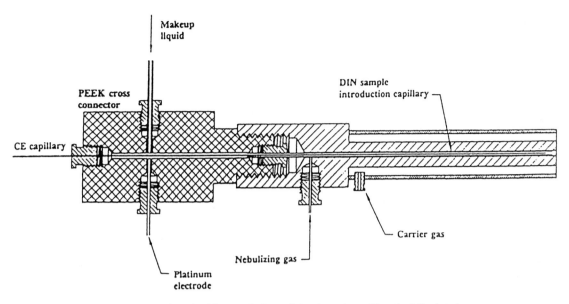

Fig. 79 CE/ICP-MS interface. (Reproduced with permission of the American Chemical Society.)

the quality required." Typical guidelines for managing quality assurance in environmental laboratories are adapted from good laboratory practice (GLP) and the ISO 9000 standards. Calibration is a "set of operations that establish, under specified conditions, the relationship between values of quantities indicated by a measuring instrument or measuring system, or values represented by a material measure or a reference material and corresponding values realized by standards. Cali-

bration constitutes the link between materials and analyzed samples that is necessary for traceability of analytical results."

A. Detection

Dynamic range: that range of concentrations of the test substance over which a change in concentra-

Table 14 Strengths and Weaknesses of IC and CE

Technique	Strengths	Weaknesses
IC	Broad range of applications Well-developed hardware Many detection options Reliability (good accuracy, precision) Accepted as standard methodology Manipulation of separation selectivity is simple High sensitivity	Moderate speed Moderate separation efficiency Intolerance to some sample matrices (e.g., high ionic strength) High cost of consumables
CE	High speed High separation efficiency Good tolerance to sample matrices (especially high pH) Low cost of consumables	Instability and irreproducibility of migration times and peak areas Moderate sensitivity Manipulation of separation selectivity is difficult Detection options are limited Routine applications are limited

Source: P.R. Haddad. Comparison of Ion chromatography and capillary electrophoresis for the determination of inorganic ions. J. Chromatogr. 770:281–290 (1997).

Table 15 Comparisons of Different Methods of
Nitrite Determination

Technique	Limit of detection[a]	Linear dynamic range[a]
Present study[b]	0.5	5–1000
Voltammetry	0.28	1.4–1400
Potentiometry	70	$0.9–1.4 \cdot 10^6$
Fluorimetry	0.3	3–60
Chemiluminescence	0.1	3–1000
Capillary electrophoresis	0.3	2.4–23.7
Spectrophotometry	0.2	0–30
Ion chromatography	0.06	14–1400
HPLC	0.15	0.3–9
GC	2.4	2.4–760

[a]All concentrations are in $\mu g/L$ NO_2-N.
[b]Precolumn formation of 2-phenylphenol with FID detection.
Source: A. Jain, R.M. Smith, K.K. Verma. Gas chromatographic determination of nitrite in water by precolumn formation of 2-phenyl phenol with flame ionization detection. J. Chromatogr. 760:319–325 (1997). Reproduced with permission.

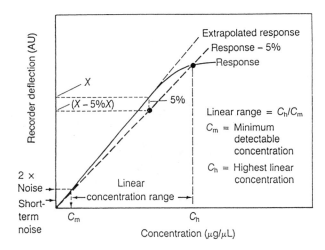

Fig. 80 Example of plot to determine the linear range of a photometric detector. (Reproduced with permission of ASTM.)

tion produces a change in detector signal. The lower limit of the dynamic range is defined as the concentration producing a detector output signal equal to a specified multiple of the detector short-term noise level. The upper limit of the dynamic range is the concentration at the point where the slope of the curve obtained by plotting detector response as a function of concentration becomes zero.

Linear range: the range of concentrations over which the sensitivity (*S*) is constant to within a defined tolerance (Fig. 80).

Limit of detection: the concentration below which the analytical method cannot reliably detect a response. A widely used detection limit technique is the 3σ approach, which is mandated for EPA testing. The standard deviation in concentration units is calculated by computing the standard deviation of blank replicates (≥ 7) and dividing by the slope of the calibration curve. This number is multiplied by the appropriate value of Student's *t* for the chosen α and for $n - 1$ degrees of freedom.

Limit of quantification: the smallest quantity of compound to be determined in given experimental conditions with defined reliability and accuracy. A signal-to-noise ratio of 10 is adequate.

Detector's noise: short-term noise in the maximum amplitude of response for all random variations of the detector signal of a frequency greater than

1 cycle per minute. Long-term noise is similar to short-term noise except that the frequency range is between 6 and 60 cycles per hour. *Drift* is the measure of the amplitude of the deviation of detector response within 1 hour.

B. Repeatability (ISO 3534)

Qualitative: the closeness of agreement between the results obtained by the same method on identical test materials under the same condition (same operator), same laboratory, same apparatus, and short interval of time.

Quantitative: the value below which the absolute difference between two single test results obtained under the foregoing conditions may be accepted to lie with a specified probability (usually 95%).

C. Reproducibility (ISO 3531)

Qualitative: the closeness of agreement between individual results obtained with the same method on identical test materials but under different conditions (different operators, different apparatus, different times).

Quantitative: the value below which the absolute difference between two single test results on individual materials obtained by operators in different laboratories using the standardized test method may be expected to lie with a specific probability (usually 95%).

D. Selectivity

This is a measure of the extent to which the method is able to determine a particular compound in the matrices without interference from matrix components.

> *Linearity of analytical procedure*: the capacity within a given interval to produce results that are directly proportional to concentration (or mass) of the compound to be determined in the sample.
>
> *Analyst must determine*:
> Linear slope
> > y-intercept
> > Correlation coefficient
> > Relative standard deviation
> > Normalized intercept/slope
>
> *Accuracy*: the difference between the true value and the mean value obtained from repeated analysis. Analysis of variance (ANOVA) estimates the within-run precision and the between-run precision.
> *Reliability*: the probability that the results lie in the interval defined by two selected limits. It gives a rigorous method for evaluating the correctness of a method of analysis in relation to two limits for error.
> *Ruggedness*: the capacity of an analytical method to produce accurate data in spite of small changes in experimental conditions.

E. Internal Standard

Used mostly in chromatography and capillary electrophoresis, this monitors the behavior of sample solutes to be analyzed and quantitatively determined.

The internal standard must fulfill certain requirements.

> It must exhibit retention behavior similar to that of the solutes.
> It must exhibit chemical functionalities and structure similar to those of the solute.
> If a derivatization step is involved in the method, the same reaction must be applied to the internal standard.
> If a sample pretreatment is required, it is better to submit the internal standard to the sample pretreatment and to check recovery.

Analysis Procedure: A standard solution contains the sample and the internal standard at concentrations C_T and C_E, respectively.

$$m_T = C_T V_T = K_T A_T$$
$$m_E = C_E V_E = K_E A_E$$

where V_T, V_E are the injected volume of the sample and of the internal standard, respectively, K_T is the response coefficient of the sample, A_T is the peak area, K_E is the response coefficient, and A_E is the peak area of the internal standard.

$$\frac{C_T V_T}{C_E V_E} = \frac{K_T A_T}{K_E A_E}$$

Usually, $V_T = V_E$.

The concentration ratio is kept constant whatever the injection volume. A sample solution contains the substance to quantitate at concentration C_X. Internal standard is added at the same C_E concentration as in the previous standard solution. We thus can write

$$m_X = C_X V_{inj} = K_X A_X$$
$$m_E = C_E V_{inj} = K_E A_E'$$

$A_E' \neq A_E$, since two injections are performed. V_{inj} is constant.

$$\frac{C_X}{C_E} = \frac{K_X}{K_E} = \frac{A_X}{A_E'}$$
$$\frac{K_X}{K_E} = \frac{K_T}{K_E}$$

so

$$\frac{C_X}{C_E} = \frac{C_T A_E}{C_E A_T} = \frac{K_X}{K_E}$$
$$\frac{C_X}{C_E} = \frac{C_T}{C_E} \frac{A_X}{A_T} = \frac{K_X}{K_E}$$

If C_E is kept constant, then

$$C_X = C_T \frac{A_E}{A_T} \frac{A_X}{A_E'}$$

It is necessary to check the detector response.

Standard solutions are prepared:

$$C_{1T} + C_E \rightarrow A_{1T} + A'_E$$

$$C_{2T} + C_E \rightarrow A_{2T} + A''_E$$

$$C_{3T} + C_E \rightarrow A_{3T} + A'''_E$$

Sample solution $\quad C_X + C_E \rightarrow A_E + A_E$

A plot of A_T/A_E versus C_T/C_E yields a regression line whose K_T/K_E is the slope.

ABBREVIATIONS

%	percent (parts per hundred)
A	absorbance ($A = \log[1/T]$)
AAS	atomic absorption spectrometry
ACN	acetonitrile
AED	atomic emission detection
AES	atomic emission spectroscopy
AMD	automated multiple development
amu	atomic mass unit
API	atmospheric pressure ionization
AU	absorbance units
AUFS	absorbance units full scale
BOD	biochemical oxygen demand
CCD	charge-couple device
CE	capillary electrophoresis
CEC	capillary electrochromatography
CGE	capillary gel electrophoresis
CI	chemical ionization
CIEF	capillary isoelectric focusing
CITP	capillary isotachophoresis
CL	chemiluminescence
CLND	chemiluminescent nitrogen detection
CMEC	capillary micellar electrochromatography
COD	chemical oxygen demand
CSFC	capillary supercritical fluid chromatography
CSP	chiral stationary phase
CZE	capillary zone electrophoresis
DAD	diode array detection
DLI	direct liquid introduction
ECLIA	electroluminescence assay
ECD	electron capture detector
EDTA	ethylenediaminetetraacetic acid
ELISA	enzyme-linked immunosorbent assay
EOF	electro-osmotic flow
EPA	Environment Protection Agency
ES	electrospay
FAB	fast atom bombardment

FIA	flow injection analysis
FID	flame ionization detector
FPD	flame photometric detection
FTIR	Fourier transform infrared
FTMS	Fourier transform mass spectrometry
GC-MS	gas chromatography—mass spectrometry
GLC	gas–liquid chromatography
GLP	good laboratory practice
HETP	height equivalent to a theoretical plate
HID	helium ionization detection
HPLC	high-performance liquid chromatography
HPTLC	high-performance thin-layer chromatography
HTCGC	high-temperature capillary gas chromatography
HTGC	high-temperature gas chromatography
IC	ion chromatography
ICP	inductively coupled plasma
ICP-OES	inductively coupled plasma–optical emission
IE	ionization energy
IEC	ion-exchange chromatography
IEF	isoelectric focusing
IPLC	ion-pairing liquid chromatography
IR	infrared
IS	internal standard
ISE	ion-selective electrode
ISO	International Organization for Standardization
ITP	isotachophoresis
LCD	liquid crystal display
LCEC	liquid chromatography with electrochemical
LDR	linear dynamic range
LIF	laser-induced fluorescence
LIFD	laser-induced fluorescence detection
LLC	liquid–liquid chromatography
LLE	liquid–liquid extraction
LSC	liquid–solid chromatography
LSS	linear solvent strength
LVI	large volume injection
MALDI	matrix-assisted laser desorption ionization
MECC	micellar electrokinetic capillary chromatography
MEKC	micellar electrokinetic chromatography
MIRS	multiple internal reflection spectroscopy

mp	melting point
MS	mass spectrometer; mass spectroscopy
MSD	mass-selective detection
MW	molecular weight (also mol wt, Mr)
NMR	nuclear magnetic resonance
NP	normal phase
NPC	normal-phase chromatography
NPD	nitrogen-phosphorus detector (see also TSD)
OPLC	overpressured-layer chromatography; overpressured-liquid chromatography
OTC	open-tubular chromatography
PAH	polycyclic aromatic hydrocarbon
PDA-UV	photodiode array UV
PES	photoelectron spectroscopy
pHs	plural of pH
PLOT	porous-layer open-tubular
PLS	porous silica layered
PS-DVB	polystyrene-divinylbenzene
PTFE	polytetrafluoroethylene
PTV	programmed temperature vaporizer/ vaporization
RF	radio frequency
RI	refractive index
RP	reversed phase
RP-HPLC	reversed-phase high-performance liquid chromatography
RPLC	reversed-phase liquid chromatography
RPTLC	reversed-phase thin-layer chromatography
SAX	strong anion exchanger
SCOT	support-coated open-tubular
SCX	strong cation exchanger
SFC	supercritical fluid chromatography
SFE	supercritical fluid extraction
SIM	selected-ion monitoring
SPE	solid-phase extraction
TC	total carbon
TCD	thermal conductivity detector
TID	thermoionic detection
TLC	thin-layer chromatography
TOC	total organic carbon
TOF	time of flight
TSD	thermionic-specific detector (see also NPD)
TSP	thermospray
TU	turbidity unit
UV	ultraviolet
UV-VIS	ultraviolet-visible

VOC	volatile organic compounds
WCOT	wall-coated open-tubular

REFERENCES

1. W.S. Holden. Water Treatment and Examination. J.A. Churchill, London (1979).
2. D.S. Aga, E.M. Thurmann, eds. Environmental Immunochemical Methods: Perspectives and Applications. ACS Symposium Series, Vol. 657, American Chemical Society, Washington, DC (1997).
3. P.R. Mathewson, J.W. Finley, eds. Biosensor Design and Applications. ACS Symposium Series, Vol. 511, American Chemical Society, Washington, DC (1966).
4. Organic Reagents for Trace Analysis. E. Merck, Darmstadt (1977).
5. C.R. Blakley, M.L. Vestal. Anal. Chem. 55:750 (1983).
6. C.M. Whitehouse, R.N. Dreyer, M. Yamashita, J.B. Fenn. Anal. Chem. 57:675 (1985).
7. B.G. Kratochvil, J.K. Taylor. NBS Technical Note, U.S. No. 1153 (1982).
8. M.C. Hennion, P. Scribe. In: D. Barcelo, ed., Environmental Analysis Techniques, Applications and Quality Assurance. Elsevier, Amsterdam (1993).
9. Special Issue on Sample Handling of Environmental and Biological Samples in Chromatography. J. Chromatogr. 665 (1994).
10. Special Issue on Chromatography and Electrophoresis in Environmental Analysis. J. Chromatogr. 733 (1996).
11. Ruzicka, E.H. Hansen. Flow Injection Analysis. Wiley Interscience, New York (1981).
12. B. Karlberg, G.E. Pacey. Flow Injection Analysis: A Practical Guide. Elsevier, Amsterdam (1989).
13. K. Grob. J. Chromatogr. 703:265 (1995).
14. D.R. Deans. Chromatographia 1:18 (1968).
15. W.O. McReynolds. J. Chromatogr. Sci. 8:685 (1970).
16. F. Riedo, D. Fritz, G. Tarjan, E.S. Kovats. J. Chromatogr. 126:63 (1976).
17. J.J. Van Deemter, H.J. Zinderweg, A. Klinkenberg. Chem. Eng. Sci. 55:271 (1956).
18. E.S. Kovats. Helv. Chim. Acta 41:1915 (1958).
19. M. Golay. In: Desty, ed. Gas Chromatography. Amsterdam Symposium. Butterworth, London (1958).
20. J. Tranchant, ed. Manuel Pratique de Chromatographie en Phase Gazeuse. Masson, Paris (1995).
21. G. Guiochon, C.L. Guillemin. Quantitative Gas Chromatography. Elsevier, Amsterdam (1988).
22. R.L. Grob, ed. Modern Practice of Gas Chromatography. 3d ed. Wiley, New York (1995).
23. W. Jennings. Analytical Gas Chromatography. Academic Press.

24. L.R. Snyder. Principles of Adsorption Chromatography. Marcel Dekker, New York (1986).

25. L.R. Snyder, Principles of Adsorption Chromatography, Appendix 3, Marcel Dekker, New York, (1986), p. 375.

26. L.R. Snyder, J. Chromatogr. Sci. 16:223 (1978).

27. Micelles as Separation Media. J. Chromatogr. 780 (1997).

28. E. Katz, R. Eksteen, P. Schoenmakers, N. Miller, eds. Handbook of HPLC. Marcel Dekker, New York (1998).

29. P.R. Haddad, P.E. Jackson. Ion chromatography: principles and applications. J. Chromatogr. Library 46 (1990).

30. J. Weiss. Ion Chromatography. 2nd ed. VCH, Weinheim (1995).

31. H. Small. Ion Chromatography. Plenum Press, New York (1989).

32. Shpigun, Y.A. Zolotov. Ion Chromatography in Water Analysis. Ellis Horwood, Chichester (1988).

33. G. Guiochon, F. Bressolle, A.M. Siouffi. J. Chromatogr. Sci. 17:365 (1979).

34. T. Greibrokk. J. Chromatogr. 703:255 (1995).

35. J.W. Jorgenson, K.D. Lukacs. J. Chromatogr. 218:209 (1981).

36. J.H. Knox. J. Chromatogr. 680:3 (1994).

37. S. Terabe, K. Otsuka, T. Ando. Anal. Chem. 57:834 (1985).

38. J.P. Landers, ed. Handbook of Capillary Electrophoresis. CRC Press, Boca Raton, FL (1994).

39. ISO5275:2 Geneva, Switzrland.

Index

4-AAP (*see* 4-aminoantipyrine)
AAS (*see* Atomic adsorption spectrometry)
AAS (*see* Atomic spectroscopic methods)
Absolute result (*see* True result)
Absorbance, 32, 79
Absorption-elution, 471
^{228}Ac, 112
Accelerated solvent extraction, 702
Accelerator mass spectrometer, 106
Accuracy, 29, 883
Ace (*see* Acenaphtene)
Acenaphtene, 705
Acenaphtylene, 705
Acetic acid, 313–345, 403, 405
Acetobacter, 300
Acid mine drainage, 51
Acidic rain, 51, 315
Acidification, 4, 262
Acidified Lugol's solution, 147–148, 152
Acidity, 51–53
  interferences, 52
  method, 52–53
    principle, 51
  sampling procedure, 52
Acridinium esters, 807
Acrolein, 630
Actidione, 620
Actinides, 106, 109–110
Actinomycetes, 75
Active chlorine, 171–177
  detection, 171
  iodometric determination, 172–173
  methyl orange test, 171
  o-tolidine test, 171

[Active chlorine]
  photometric method, 173, 175
  volumetric method, 174–175
Activity coefficient, 786
Acy (*see* Acenaphtylene)
ADDA, 145
Addition of complexing substances, 4
Adenine, 265–266
Adsorbable organic halide, 308
Adsorption chromatography, 830, 859–860
Adsorption indicators, 180–181
Adsorptive stripping voltammetry, 560
AED (*see* Atomic emission detector)
Aeration, 79
Aeromonas, 118–141
AES (*see* Atomic emission spectrometry)
Affinity chromatography, 830
AFS (*see* Atomic fluorescence spectrometry)
Ag_2S, 465
$AgCl_2^-$, 465
$AgCl_3^{2-}$, 465
Al(III), 25, 420, 430, 464
$Al(OH)_2^+$, 464
$Al(OH)^{2+}$, 464
$Al(OH)_4^-$, 464
Alanine, 264, 266–267
Alcohol ethoxylates, 769
Alcohols, 393, 840
Aldehydes, 59, 79, 313, 315, 402, 404
Aldicarb, 544, 561
Aldrin, 521–522, 530
Algae, 7, 51, 75, 115, 202, 269, 279
Algal analysis, 143–167
Algal bioassay, 278

Algal bloom, 261
Algal toxins:
   analysis, 155–165
   bioassays, 156–158
   biochemical effect measurements, 164
   chemical assays, 158–161
   immunoassays, 162–164
   sampling, 156
   toxicity testing using cell cultures, 158
Aliphatic acid herbicides, 630
Aliphatic ethers, 400
Aliphatic hydrocarbons, 404
Aliphatic nitrogen fungicide, 619
Alkaline persulfate oxidation, 268–269
Alkaline phosphatase hydrolyzable phosphorus, 280
Alkalinity, 52–56, 411, 483
Alkalinity criteria, 56
Alkalinity method, 54–55
Alkalinity method principle, 53
Alkanesulphonates, 768
Alkenes, 79
Alkyl aryl sulphonates, 768
Alkyl glycosides, 769
Alkyl sulphates, 768
Alkylbetaines, 769
Alkyldimethylbenzylammonium chloride, 769
Alkylphenol poly(ethyleneglycol) ethers, 769
Alkylphenols, 347–366
Alkylpyridinium salt, 769
Alkylsulphobetaines, 769
Alkyltrimethylammonium salt, 768
Allidochlor, 621
All-solid-state electrodes, 206
Allyl alcohol, 630
α-BHC, 5
α-ketoacids, 324, 328
α-olefine sulphonates, 768
α-sulpho fatty acid methyl esters, 768
Alpha detectors, 105
Alpha spectrometry, 105, 110
Alternaria solani, 620
Aluminium, 27, 51, 464, 467, 471
Aluminium phosphide, 620
AMD (see Acid mine drainage and Automated multiple
   development)
AMD-HPLC (see Automated multiple development, high-
   performance thin-layer chromatography)
American Public Health Association, 125–127, 129, 131, 134,
   230, 234–235, 241, 367–369, 697–698
American Society for Testing and Materials, 197, 697–698,
   755–756
Americium, 109–110
Amines, 223, 261, 264, 401, 632, 840
Amino acids, 223, 226, 228, 261, 264, 266, 387, 393, 401,
   405, 488
4-aminoantipyrine, 348

2-aminobenzimidazole, 538
Aminocarb, 538, 558
Aminophenols, 360
Aminophosphonic acids, 273
Amitrol, 633
Ammonia, 201, 223–259
   absorbance techniques, 246–247
   analysis, 229–231
   analytical method selection, 230–231
   aqueous-phase chemistry, 223–226
     effect of pH, 224–225
     effect of salinity and temperature, 225–226
   capillary zone electrophoresis, 255
   chromatographic techniques, 248–255
   conductimetric techniques, 246
   conductivity detector, 251
   derivatization methods, 249
   electroanalytical techniques, 246
   emission techniques, 247–248
   environmental effects, 229
   fluorescence detector, 251
   gas diffusion fluorescence, 244–246
   gas-sensing electrodes, 264
   human health effects, 229
   indophenol blue colorimetry, 236–240, 246
   indophenol blue colorimetry interferences, 237
   ion-selective electrode, 242–244
     interferences, 243
   Nesslerization, 240–242
   occurrence in (waste)waters, 226–229
   optodes, 247–248
   phenate and salicylate chemistries, 236–240
   photometric detector, 251
   planar chromatography, 251–255
   preparation of ammonia-free water, 231–234
   preparation of apparatus, 231
   rubazoic acid technique, 246–247
   sample collection, 234
   sample preservation, 234
   sources, 226–227
   spectrophotometric detector, 251
   spectroscopic techniques, 246–248
   titrimetric method, 235
   voltammetric techniques, 246
   why measure?, 227–229
Ammonium, 261–271
Ammonium cation, 224
Ammonium chloride, 265, 267
Ammonium hydroxide, 267
Ammonium nitrate, 621
Ammonium sulphamate, 621
Ammonium sulphate, 264, 621
Ammonium tiocyanate, 621
Amperometric enzyme electrodes, 795
Amperometric methods, 865–866
Amperometric sensors, 207, 211

Amperometry, 197, 211
Amperostat, 792
Amphibole, 765
Amphoteric surfactants, 769
AMS (*see* Accelerator mass spectrometer)
Amu (*see* Atomic mass unit)
Anabaena, 143–144
Anabaena circinalis, 278
Anabaena circularis, 145, 154, 156
Analysis of variance, 883
Anatoxin-a, 145, 159
Anatoxin-a(s), 145, 159
Anilazine, 619
Anion content, 28, 32
Anionic surfactants, 767–773, 781
Anodic stripping voltammetry, 427, 441–446
ANOVA (*see* Analysis of variance)
Anthracene, 687
Antibiotics, 620
Antichlorotoluron, 494
Anti-isoproturon, 494
Antimony, 4, 465, 467
Antimycin A, 620
Antipyrine, 268
AOBr (*see* Adsorbable organic halide)
AOCl (*see* Adsorbable organic halide)
AOI (*see* Adsorbable organic halide)
AOX (*see* Adsorbable organic halide)
APCI (*see* Atmospheric pressure chemical ionization)
APCI-MS, 601
APHA (*see* American Public Health Association)
Aphanizomenon flos-aquae, 143–147
Aphanizomenon, 144–147
Aphanizomenon ovalisporum, 143–147, 157
APHP (*see* Alkaline phosphatase hydrolyzable phosphorus)
API (*see* Atmospheric pressure ionization)
APLE (*see* Aqueous-phase liquid extractor)
Apolane, 847
Apparent color, 79
Appearance, 755
APS (*see* Automated preconcentration water sampler)
Apteronotus albifrons, 382
Aquatic humus, 388
Aqueous-phase liquid extractor, 680
Argentite, 465
Arochlor 1260, 655
Arochlors, 655, 665
Aromatic acids, 328
Aromatic amines, 487
Aromatic compounds, 266
Aromatic ethers, 401
Aromatics, 405
Arsenic, 4, 442–443
    anodic stripping voltammetry, 443
    capillary electrophoresis, 443
    cathodic stripping voltammetry, 443

[Arsenic]
    ETAAS, 443
    FIA-hydride generation, 443
    flow injection analysis, 443
    HG-AAS, 443
    HPLC-HG-AAS, 443
    hydride generation, 443
    ICP, 443
    IEC-HG-AAS, 443
    ion chromatography, 443
    nonflame AAS, 443
    spectrofluorimetric methods, 442–443
    trioxide, 621
Artemia salina, 158
Arthromyces ramosus, 380
Arylaliphatic acid herbicides, 630
As(III), 440, 443
As(V), 440, 443
Asbestos, 765–766
    analytical methods, 765–766
    health effects, 765
    maximum concentration level, 765
Aspartic acid, 266
ASPEC, 581, 825
Astatine, 169
ASTED (*See* Automated sequential trace enrichment of dialysates)
ASTM (*see* American Society for Testing and Materials)
ASV (*see* Anodic stripping voltammetry)
Atmospheric pressure chemical ionization, 558, 601, 706, 759, 857
Atmospheric pressure ionization, 352–353, 558, 595–602, 857
Atomic absorption, 828–829
Atomic absorption spectrometry, 183, 197, 370, 430, 444, 467, 472, 474, 801–803, 863
Atomic emission detector, 583
Atomic emission spectrometry, 430, 799–801
Atomic fluorescence spectrometry, 442
Atomic mass unit, 807
Atomic spectoscopic methods, 284
Atomic vapor production, 802
Atrazine, 381, 406
Auerofungin, 620
Automated multiple development, 869
    high-performance thin-layer chromatography, 558–559
Automated preconcentration water sampler, 678
Automated sequential trace enrichment of dialysates, 812
Automated titrators, 415
Auxins, 621
Averages chart, 33
AWWA Standard Methods, 757
Azide glucose broth, 129–141
Azinphos-methyl, 571
Azobenzene, 633

$B(OH)_3$, 461
$B(OH)_4$, 461

Ba^{2+}, 465

Bacillus, 129–141

Bacillus subtilis, 373

Background electrolyte, 879

Bacteria, 51, 115–141, 201–202, 224, 226, 261–262, 269, 279, 297–298, 300, 313–314, 376

Bacteriological analysis, 115–141

Bailers, 755

Baird-Parker medium, 133–134

Baker's yeast cells, 494

Bandgap, 103

BaP (*see* Benzo[a]pyrene)

BAP (*see* Bioavailable phosphorus)

Barbuturic acid, 266

Barite, 465

Barium, 27, 104, 106, 465, 787

Base-consuming capacity, 786

Bathocuproine disulfate, 419

Bavistin, 633

BbF (*see also* Benzo[b]fluoranthene), 690, 703–704, 707

BCF (*see* Bioconcentration factor)

BEAA (*see* Bile-esculin-azide agar)

Beer-Lambert law, 35–36, 796

Beer's law, 851

Benchmate, 825

Benomyl, 538, 620

Bensulphide, 630

Benthiocarb, 558–559

Benzene (*see also* BTEX compounds), 94, 721–752, 754, 760

Benzene carboxylic acids, 404

Benzene, toluene, and xylene, 94

Benzethonium chloride, 771–772

Benzimidazole carbamates, 538, 545

Benzimidazole fungicides, 620

Benzimidazoles, 872

Benzo[a]pyrene, 690, 700, 704

Benzo[b]fluoranthene, 690, 703–704, 707

Benzo[ghi]perylene, 690, 708

Benzo[j]fluorosillene, 703–704

Benzo[k]fluoranthene, 690, 703–704

Benzoquinoline, 585

Benzotriazole, 268

BeP, 704

Berthelot method, 264

Beryllium, 461, 472

β-BHC, 530–531

Beta detectors, 105–106

β-naphthaleneacetic acid, 631

β-naphtoxyacetic acid, 631

β-picoline, 633

BGE (*see* Background electrolyte)

BghiP (*see* Benzo[ghi]perylene)

Bi(II), 427

BiAS (*see* Bismuth active substances)

Bicarbonates, 70, 206, 216–217, 396

Bienzyme sensor, 380, 383

Bile-esculin-azide agar, 128–141

Bioavailable phosphorus, 273, 278–295

measures, 278–280

Biochemical oxygen demand (*see also* Biological oxygen demand), 297–307, 319, 371, 374, 379, 381, 789–790

dilution method, 298, 789–790

direct measurement of absorbance, 300

instrumental methods, 298–300

respirometric methods, 299, 790

supplied by sensors, 299

Bioconcentration, 517

factor, 517, 572

Biofilms, 115

Biological assay, 794

Biological oxygen demand, 16, 25, 62

Bioluminescence, 787

Bioprobe, 14–15

Biosensors, 351, 371–382, 559–560, 794

2,2'-bipyridine, 265

Bipyridylium herbicides, 630

Bismuth active substances, 775–776

Bismuth sulfite agar, 136

BjF (*see* Benzo[j]fluorosillene)

BkF (*see* Benzo[k]fluoranthene)

Blank value, 35

Blastocidin S, 620

Blastomycin, 620

Bleaching powder fungicides, 618

Blooms, 144, 158, 160

Blossom-thinning agents, 630

Blotting, 870

Blue baby syndrome, 202

Blue-green algae, 143–167, 201

BOD 5, 298–307, 789

BOD (*see* Biochemical oxygen demand and Biological oxygen demand)

BODS (*see* Biochemical oxygen demand, supplied by sensors)

Boltzmann's law, 799–800

Borate, 334

herbicides, 621

Bordeaux mixture, 610

Boron, 461–464, 467

Botrytis, 620

Bound active chlorine, 170

Breakthrough volume, 528

Brilliant-green lactose broth, 126–141

Brilliant-green phenol red lactose agar, 136

Bromate, 60

Bromide, 171, 181, 187–189, 192, 197, 214, 216–217, 266, 332–333, 863

electrometric method, 188

and iodide determination, 189–190

ion-selective electrodes, 188

spectrophotometric method, 187–188

Bromine, 59, 170

producing, 59

Bromoform, 80
Bromophenols, 354, 356
Bromophos, 571
BTEX compounds, 721–752
    analytical methods, 723
    calibration, 743–745
    closed-loop stripping, 732–737
    columns, 723
    columns chromatographic conditions, 723
    detectors, 723
    enzyme immunoassay, 745
    membrane extraction, 745
    physicochemical properties, 721–722
    purge-and-trap, 726–732
    quality control, 743–745
    regulations, 722–723
    salting-out effect, 724
    sample introduction, 725
    solid-phase microextraction, 737–745
    static headspace, 723–726
    toxicology, 721–722
    why detect?, 721
BTV (*see* Breakthrough volume)
BTX (*see* Benzene, toluene, and xylene)
Bubble chambers, 103
Buffer capacity, 69
Burgundy mixture, 610
Butyric acid, 334, 341

$C_{1,2}$-benzenes, 754
Ca^{2+}, 25
Cacodylic acid, 621
Cadmium, 27, 51, 420–422, 424, 433, 440–441
    anodic stripping voltammetry, 441
    differential pulse polarography, 441
    electrothermal atomic absorption spectrometry, 441
    flame atomic absorption spectrometry, 441
    fluorescence determination, 441
    inductively coupled plasma-atomic absorption spectrometry, 441
    ion-selective electrodes, 441
    neutron activation analysis, 441
    spectrophotometric methods, 441
Calcium, 27, 409–438, 787
Calcium carbonate, 409, 411, 415
Calcium cyanide, 620
Calcium fluoride, 64
Calcium-selective electrodes, 426
Calibration, 34–37
    curve, 32
Calixin, 633
Capillary column gas chromatography, 557, 703
Capillary electrochromatography, 876–877
Capillary electrophoresis, 218, 285, 336–338, 358, 443, 445, 506–509, 559, 812, 851, 873–880
    detection, 879–880

[Capillary electrophoresis]
    injection, 878–879
    sample preconcentration, 879
    mass spectrometry, 807
Capillary fused-silica columns, 328
Capillary gas-liquid chromatography, 85
Capillary GC-MIP-AES, 445
Capillary gel electrophoresis, 876
Capillary high-resolution gas chromatography, 660–662
Capillary isotachophoresis, 446
Capillary liquid chromatography, 699, 705, 851
Capillary zone electrophoresis (*see also* Capillary electrophoresis), 255, 336–338, 354–355, 433, 444, 446, 506–507
Captabol, 619
Captan, 619
Carbamate and thiocarbamate herbicides, 629
Carbamate fungicides, 538
    toxicology, 545
Carbamate herbicides, 538
    toxicology, 545
Carbamate insecticides, 538
    toxicology 544–545
Carbamate pesticides, 537–570
    gas chromatography, 537
    why detect?, 537
Carbamates, 631
    biosensors, 559–560
    capillary electrophoresis, 559
    characteristics, 537–546
    chemical and physical properties, 538
    chemistry, structure, and nomenclature, 537–538
    degradation and metabolic processes, 538–544
    derivatization, 556–558
    determination, 556–560
    electrochemical detection, 558
    electrochemical methods, 560
    environmental persistence, 538–544
    gas chromatography, 546, 556–557
    high-performance liquid chromatography, 557–558
    immunoassays, 559
    immunoextraction, 556
    liquid-liquid extraction, 546–547
    mass spectrometry, 557–558
    microextraction, 554
    offline extraction techniques, 546–547
    online extraction techniques, 547–551
    pretreatment systems, coupled online to a gas chromatograph, 551
    regulations, 545–546
    sample pretreatment, 546–556
    selective sorbents, 556
    solid-phase extraction, 547
        coupled online with high-performance liquid chromatography, 547–551
    solid-phase microextraction, 551–556

[Carbamates]
spectrophotometric methods, 560
supercritical fluid chromatography, 558
supercritical fluid extraction, 555–556
thin-layer chromatography, 556, 558–559
toxicology, 544–545
Carbaryl, 538–544, 556–561
Carbendazim, 538, 561
Carbofuran, 538–544, 558–560
Carbohydrates, 387, 401
Carbon, 265, 297
Carbon blacks, 729
Carbon cycling, 387
Carbon dioxide, 51, 411
Carbon disulphide, 620
Carbonaceous BOD, 16, 298
Carbonate hardness, 56, 787
Carbonates, 70, 216–217, 334, 338, 787
Carbonic acid equilibrium, 70–71
Carboxin, 620
Carboxyl, 397
Carboxylate, 206
Carboxylic acid spot test, 650
Carboxylic acids, 266, 324, 404
Casein, 265
CASS-3, 476
Catalytic fluorometry, 446
Catalytic spectrophotometry, 471
Cathodic stripping voltammetry, 246, 425, 442–446
Cation content, 28, 32
Cationic surfactants, 767–774
CBOD (see Carbonaceous BOD)
CBs (see Chlorinated biphenyls)
CCD (see Charge coupling device)
^{113}Cd, 102
Cd^{2+}, 378
^{137}Ce, 102, 104
CE (see Capillary electrophoresis)
CEC (see Capillary electrochromatography)
CERCLA (see Superfund Amendments and Reauthorization
Act)
Cerenkov counter, 103
Ceresan, 610
Certified reference materials, 669, 696
Certified reference solution, 476
Cetyltrimethylammonium bromide, 772–774
CGE (see Capillary gel electrophoresis)
Charge coupling device, 870
Chemical chlorine demand, 170, 176
Chemical ionization, 324, 405, 534, 595–602, 706, 808, 857,
873
mass spectrometry, 662
Chemical oxygen demand, 25, 172, 297–303, 319, 792
classic method, 302
flow injection analysis, 303
semimicro method, 302–303

Chemical parameters, 51–74
Chemically modified electrodes, 206, 425–426
Chemiluminescence, 247, 267–268, 352, 380, 430–432, 443–
446, 794, 806–807, 829, 879–880
Chemiluminescent detection, 211–212, 249, 836–837
Chemiluminescent nitrogen detection, 837
Chestnut compound, 610
Chiral stationary phases, 830
Chitopearl, 374
Chloramben, 630
Chloramines, 80, 177
Chlorate, 67
Chlorate fungicide, 618
Chlordane, 521, 523–524
Chlordecone, 521, 524
Chlorfenvinphos, 571
Chloride, 67–68, 169–171, 178–184, 192, 206, 214–217, 266,
302, 307, 315, 321, 338, 341, 396, 416, 787, 863
argentimetric methods, 179–181
atomic absorption spectrometry, 183
content, 27
detection, 179
determination, 179–184
electrochemical methods, 181
gravimetric methods, 179
mercurimetric titration, 181–182
nefelometric and turbidimetric techniques, 183–184
photometric methods, 183–184
potentiometry, 183–184
spectrometric determination, 182–183
voltammetry, 184
Chlorinated biphenyls, 10, 19
Chlorinated hydrocarbons (see also Organochlorinated
pesticides), 754, 756
Chlorinated organics (see Organochlorinated pesticides)
Chlorinated phenoxyacid herbicides, 348
Chlorinated synthetics (see Organochlorinated pesticides)
Chlorination, 59, 80, 689
schemes, 229
water disinfection, 75
Chlorine, 59, 169–170, 197, 229, 289, 298, 313, 403
binding capacity, 176
colorimetric test kits, 175–176
constant-current potentiometry, 176–177
flow injection analysis, 175
iodometric method, 59–61
producing, 59
reactive indicator paper, 176
simultaneous determination of chlorine dioxide, chlorine,
chlorate, and chlorite, 177–178
Chlorine dioxide, 60, 177–178
determination, 177–178
photometric determination, 177
residues leucomethylene blue reagent, 178
Chlorite, 60, 67
Chloroanilines, 512

Chloroaromatics, 662

Chlorofluorocarbons, 757

Chloroform, 51, 80

Chloroneb, 619

4-chlorophenol, 363

Chlorophenols, 18, 347–366, 632

Chlorophenoxy acids, 632
  herbicides, 872
  pesticides, 18

Chlorophyll, 753

Chlorophyll a, 10

Chloropicrin, 620

Chlorosis, 618

Chlorothalonil, 619

Chlorotoluron, 487–488

2-chlorophenol, 360

Chlorpyrifos, 560, 592

Chlorpyriphos-methyl, 583

Chlorpyritophos, 602

Chlorsulfuron, 487, 496

Cholera, 116

Chromatography, 829–870
  efficiency, 832–833

Chromium, 4, 27, 418, 420, 432, 443–444, 471
  ASV, 444
  chemiluminescence, 443–444
  CSV, 444
  CZE, 444
  ETAAS, 443
  FAAS, 443
  flow injection analysis, 443–444
  gas chromatography, 444
  HPLC, 444
  ICP, 444
  ion-exchange, 443
  isotope dilution-mass spectrometry, 444
  NAA, 444
  reversed-phase ion-pair HPLC, 444
  spectrophotometric methods, 443
  thermal lens spectrometry, 444

Chrysene, 703–704

Chrysotile, 765

CHX_3 (*see* Trihalomethanes)

CI (*see* Chemical ionization)

CID (*see* Collision-induced dissociation)

CI-MIMS/MS, 94

CIMS (*see* Chemical ionization mass spectrometry)

CIPC, 557

Citrobacter, 118–141

CI^-, 374, 379, 461

Cl (*see* Chemiluminescence)

Clark electrode, 790

Clark° (*see* UK°)

CLD (*see* Chemiluminescence detector)

Clean Water Act, 690

CLND (*see* Chemiluminescent nitrogen detection)

Clophen, 655, 665

Closed-loop stripping, 84–87

Closed-loop stripping analysis, 723, 732–737, 743–745, 812
  adsorption, 735–736
  affecting factors, 734–736
  analytical procedure, 736–737
  filter extraction, 736

Clostridia chromogenic or fluorogenic substrates, 132–133

Clostridia enumeration, 132–133

Clostridium, 119–141

Clostridium butyricum, 371

Clostridium perfringens, 119–141

Cloud chambers, 103

Cloud-point extraction, 702

CLP (*see* Contract Laboratory Program)

CLS (*see* Closed-loop stripping)

CLSA (*see* Closed-loop stripping analysis)

CMC (*see* Critical micelle concentration)

CMEs (*see* Chemically modified electrodes)

^{13}C-NMR, 398–401

Co(II), 426, 464

Co(III), 464

$Co(OH)_3^0$, 464

CO_2 determination, 789

CO_3^{2-}, 461

Coating efficiency, 846

Cobalt, 27, 418, 429–430, 433, 444, 464, 467, 471, 866
  AAS, 444
  adsorptive voltammetry, 444
  flow injection analysis, 444
  laser-excited AFS, 444

Cobalt thiocyanate active substances, 774–775

$CoCO_3^0$, 464

COD (*see* Chemical oxygen demand)

Coliforms, 59, 118–141
  chromogenic-fluorogenic substrates, 128–129
  enumeration, 124–129

Coliphages, 119

Collision-induced dissociation, 510–511

Colloidal state, 43

Color, 75–100, 202, 755

Color tests, 787, 796

Colorimetric closed reflux method, 302

Colorimetric protein phosphatase analysis, 164

Column chromatography, 702

Comparator, 79

Complexation ion chromatography, 865

Concentration of contaminants in suspensions and sediment, 21–22

Concentration polarization, 811

Concentrator columns, 331–332, 334

Conductimetry, 197

Conductivity, 15, 25, 28, 788, 865

Conductometric sensor, 796

Constant current stripping chronopotentiometry, 442

Constant-current electron capture detector, 836

Constant-current potentiometry, 176–177, 181
Contact herbicides, 621
Container preparation for sampling water, 755–756
Continuous-flow analyzers, 210
Continuous-flow FAB-FAB, 601
Continuous-flow fast atom bombardment, 857
Contract Laboratory Program, 690
Control chart, 33
Controlled pore glass, 373–374
Conventional membrane introduction mass spectrometry, 93–94
Cooling, 4
Copper, 409–438, 445, 866
    ASV, 445
    chemically modified electrode, 445
    chemiluminescence, 445
    ETAAS, 445
    FAAS, 445
    flow injection analysis, 445
    fungicides, 610
    ion-selective electrode, 445
    microwave plasma torch AES, 445
    oscillatory flow injection stripping potentiometry, 445
    spectrophotometric methods, 445
    voltammetric methods, 445
Copper-selective electrodes, 426
Copper sulphate, 621
Coronene, 705
Correlation, 36–37
CoS, 464
Coulometric reagent, 181
Coulometric titration, 792
Coumaphos, 602
Counting chamber, 149–152
^{13}C-PCDD, 679
CPG (see Controlled pore glass)
CPMAS (see Cross-polarization magic angle spinning)
^{134}Cs, 102
Cr(III), 440, 443–444, 464, 866
Cr(VI), 440, 443–444, 464
$Cr_2O_7^{2-}$, 464
Creatinine, 264, 268
Cresols, 354, 356
Critical micelle concentration, 862
CRMs (see Certified reference materials)
CrO_4^{2-}, 464
$Cr(OH)_2^+$, 464
$Cr(OH)^{2+}$, 464
$Cr(OH)_4^-$, 464
Cross-polarization magic angle spinning, 398
Crown ethers, 108
Cryofocusing, 843
Cryogenic desolvation-ICP-MS, 473
Cryolite, 169
Cryotrap-membrane introduction mass spectrometry, 96–98

Cryptosporidiosis, 59
Cryptosporidium, 59, 118–119
CSV (see Cathodic stripping voltammetry)
CTAS (see Cobalt thiocyanate active substances)
CT-MIMS (see Cryotrap-membrane introduction mass spectrometry)
C-toxins, 158–159
Cu(I), 419
Cu(II), 370, 378, 419–432, 445
Cupferron, 430
Cusum chart, 33
CV-AAS (see also Cold vapor-atomic absorption, spectrometry), 442
CWA (see Clean Water Act)
Cyanides, 367–385
    amperometric biosensor, 382
    analysis methods, 368–383
    atomic adsorption spectrometry, 370
    batch-and-membrane type of sensor, 371–373, 376–377
        using a gas-permeable membrane, 377–379
    biosensors, 371–382
        using enzyme-decomposing cyanides, 380–382
        using inhibition of microbial respiration, 371–376
        using microbial degradation of cyanide, 376–380
    chromatographic methods, 370
    detection with Apteronotus albifrons, 382
    electrochemical methods, 369–370
    flow injection analysis, 370–371, 380
    flow-and-reactor type of sensor, 373
        for real samples, 373–374, 379–380
    fluorometric methods, 369–370
    gas chromatography, 370
    HPLC, 370
    integrated monitoring systems, 382–383
    ion chromatography, 370
    ISFET-based peroxidase biosensor, 382
    JIS method, 369, 378, 381
    pretreatment, 367–368
    regulations, 374
    sensor with methemoglobin, 382
    spectrophotometric methods, 367–369
    standard methods, 367
    titrimetric methods, 367–368
    why detect?, 367
Cyanide-selective electrode methods, 367, 369–370
Cyanobacteria, 143–167
    calculations and results, 154
    centrifugation, 148
    counting chamber, 149–152
    enumeration, 152–153
    hepatotoxins, 145
    identification and enumeration, 147–155
    irritants, 147
    membrane filtration, 148
    microscopy, 148–149
    molecular techniques, 155

[Cyanobacteria]
  neurotoxins, 144–145
  nonspecific toxicants, 145–147
  preservation of samples, 147
  quality assurance techniques, 154–155
  recording of unidentified species, 153–154
  sample concentration technqiues, 147–148
  sampling, 147
  sedimentation, 148
  taxonomic identification, 153
  toxic genera, 143
  toxin production, 144
  toxin types, 144–147
Cyclic voltammetry, 446
Cyclodienes, 521–524
Cycloheximide, 620
Cylindrospermopsin, 143–147, 156–165
Cylindrospermopsis, 145–147
Cylindrospermopsis raciborskii, 143–147, 152, 157–161
Cyn (*see* Cylindrospermopsin)
Cypermethrin, 5
Cytosine, 266
CZE (*see* Capillary zone electrophoresis)

2,4-D (*see* 2,4-dichlorophenoxyacetic acid)
Dabsyl chloride, 249
2,4-D acid, 633–650
DAD (*see* Diode array detection)
Dalafon, 630
Dansyl chloride, 352
Data interpretation and assessment, 37–40
Data reliability, 37
Davis equation, 225
2,4-DB, 621
dBacA (*see* Dibenzo[a,c]anthracene)
dBahA (*see* Dibenzo[a,h]anthracene)
DBAS (*see* Disulfine blue substances)
DBPs (*see* Disinfection by-products)
DCPA, 630
DDD (*see also* Dichlorophenyldichloroethane), 520
DDE (*see also* Dichlorodiphenylchloroethane), 5, 520
DDT (*see also* Dichlorodiphenyltrichloroethane), 5, 518–521, 530, 534
Debye-Huckel equation, 225
Decay counting, 107–108
Dechlorination, 234
Deep water sampler, 12–13
  for trace elements, 13–14
Degrees of freedom, 30
δ-BHC, 531
Demethon-methyl, 581
Demeton-S-methyl, 571
Densitometry, 869
Density, 41–42
Derivatization, 840–841, 852–854
Detection limit, 35, 67, 476, 705

Detection methods, 796–810
Detergents, 767
Determination limit, 35
2,4-D ethyl ester, 633–650
Devarda alloy, 211
Deviation, 30
DGT (*see* Diffusive gradients in thin films)
Dialkydiamethylammonium salt, 768
Dialysis, 204–205, 811–812
Diarrheal diseases, 116
Diazinon, 571, 581
Dibenzo[a,c]anthracene, 703–704
Dibenzo[a,h]anthracene, 703–704, 708
Dibenzofurans (*see* PCDFs)
Dicamba, 630
Dicarboximide fungicides, 619
Dicarboxylic acids, 321, 324, 328, 336
Dichloramine, 229
Dichlorobiphenyls, 665
2,4-dichlorophenol, 348
2,4-dichlorophenoxyacetic acid, 621, 630–632
2,5-dichlorophenol, 363
Dichlorvos, 571, 581, 583
Dicupral, 419
Dieldrin, 5, 521–522
Dielectric track detectors, 103
Diethylamine, 266
Diethylcarbamates, 558
Diethyldithiocarbamates, 560
Differential pulse:
  anodic stripping voltammetry, 425, 446
  cathodic stripping voltammetry, 445
  chromatography, 106
  polarography, 178, 189–190, 211, 441, 560
  voltammetry, 446
Differential reinforced clostridial medium, 132–133
Diffusion-limited aggregation, 46
Diffusive gradients in thin films, 279
Dimethoate, 571, 579–581
Dimethyl diselenide, 445
Dimethyl glyoxime, 267
2,4-dimethylphenol, 352
Dimethyl selenide, 445
1,3-dimethyl-2-nitrobenzene, 491
Dimetilan, 538
DIN (*see* Direct injection nebulizer and Dissolved inorganic nitrogen)
2,3-dinitrophenol, 363
2,4-dinitrophenol, 353, 358, 363
Dinitrophenols, 356
Dinocap, 619
Dinoflagellate toxins, 156
Dinoflagellates, 143
Dinoseb, 630
Diode array detection, 205, 350, 496, 535, 557, 585, 589, 602, 705, 825–828

Diphenamid, 621–629
Dipstick, 136
Direct aqueous injection, 328
Direct competitive ELISA, 162–164
Direct desorption, 873
Direct injection, 833
Direct injection nebulizer, 444
Direct LC-CI, 601
Direct liquid introduction, 857
Direct solid-phase microextraction, 737, 739–744
Direct speciation, 461
Direct-fluid-introduction, 873
Disinfection by-products, 59
Dissolved gases, 28
Dissolved inorganic nitrogen, 223, 261
Dissolved organic carbon, 305–307, 388–408
Dissolved organic halide, 267, 307–308
Dissolved organic matter, 387–408, 528
Dissolved organic nitrogen, 261
Dissolved organic nitrogenous compounds, 223
Dissolved oxygen, 7–8, 12, 32, 34, 61–64, 195, 202, 298–
       299
    criteria, 63
    modified Winkler method, 62–63
       interferences, 62–64
          principle, 63
          sampling procedure, 62
    oxygen-membrane electrode method, 63–64
Dissolved state, 43
Distearyldimethylammonium chloride, 769
Distillation, 631
Disulfides, 836
Disulfine blue active substances, 772–773
Disulfoton, 571, 581
Dithiocarbamate fungicides, 619
Dithiocarbamates, 545, 559, 619–620, 650, 862
Dithiocarbamic acid, 538
Dithiophosphates, 571
Ditizone, 418
Diuron, 487–490
DL (*see* Detection limit)
DLA (*see* Diffusion-limited aggregation)
DLI (*see* Direct liquid introduction)
DLs (*see* Detection limits)
DNA, 155, 266
DNOC, 630
DO (*see* Dissolved oxygen)
DOC (*see* Dissolved organic carbon)
Dodecane-1-sulphonic acid, 771
Dodine, 619
DOM (*see* Dissolved organic matter)
DON (*see* Dissolved organic nitrogen and Dissolved organic
       nitrogenous compounds)
Donnan equilibrium, 864
DOX (*see* Dissolved organic halide)
DP (*see* Differential pulse)

DP-ASV (*see* Differential pulse, anodic stripping voltammetry)
DP-CSV (*see* Differential pulse, cathodic stripping
       voltammetry)
DPP (*see* Differential pulse, polarography)
DPs (*see* Pesticide degradation products)
DPV (*see* Differential pulse, voltammetry)
Dragendorff reagent, 775
DRCM (*see* Differential reinforced clostridial medium)
Drift, 882
Drift correction, 468
Drylab G, 850
2,4-D sodium salt, 633–650
DSMA, 621
Dual coulometric detection, 351
Dyes, 753
Dynamic compression injection, 878–879
Dynamic headspace, 83–84, 812
    extraction, 758
Dynamic range, 882

E. coli (*see Escherichia coli*)
EC (*see* European Community)
ECD (*see* Electron capture detection)
ECLIA (*see* Electroluminescence immunoassay)
EC medium, 126–141
ECNCI (*see* Electron capture negative chemical ionization)
ED (*see* Electrochemical detection)
EDTA, 266, 268
EI (*see* Electron impact)
EI-MS (*see also* Electron impact mass spectrometry), 583, 706
Einstein's law, 875
Elasticity, 41–42
Electric sensing zone method, 43–44
Electrical conductivity, 8
Electrochemical detection, 211, 558, 585, 602, 837, 854–855
    amperometric, 854–855
    conductivity, 854
Electrokinetic injection, 336, 878
Electroluminescence immunoassay, 794
Electrometric titrations, 178
Electrometry, 197
Electromigration injection, 336
Electron capture:
    detection, 92, 324, 354, 356, 445, 529–533, 556–557, 583,
       589, 662, 706, 760, 836, 843
    negative chemical ionization, 534–535
Electron impact, 532–535, 557, 595–602, 682, 706, 807, 857,
    873
    ionization, 324
    mass spectrometry, 662–665
Electron ionization, 404–405
Electron spectroscopy, 446
Electron spin resonance spectroscopy, 401–402
Electro-osmosis, 874
Electro-osmotic flow, 336, 506, 874–875, 878
Electrophoresis, 471

Electrospray, 352–353, 509–510, 558, 601, 809, 857
    ion chromatography-tandem mass spectrometry, 188
    ionization, 595–602
    mass spectrometry, 809
Electrothermal atomic absorption spectrometry, 441–446, 466–470
Electrothermal atomic absorption spectroscopy, 412, 422–423, 433, 443
Electrothermal vaporization, 107, 472, 802
    ICP-MS, 111, 443
ELISA (*see* Enzyme-linked immunosorbent assay)
Elution purification trap, 332
Endosulfan, 521, 523, 631
Endosulfan sulfate, 523
Endothall, 630
Endrin, 5, 521–522, 530
Endrin aldehyde, 521
Endrin ketone, 521
Enterobacter, 118–141
Enterococcus, 119–141
Enterococcus durans, 119
Enterococcus faecalis, 119
Enterococcus faecium, 119
Enterococcus hirae, 119
Enteroviruses, 119
Enzyme biosensor, 371
Enzyme electrodes, 794–795
Enzyme immunoassay, 745
Enzyme-linked immunosorbent assay, 136, 162–165, 705, 794, 805, 870
EOF (*see* Electro-osmotic flow)
EOX (*see* Extractable organic halide)
EPA (*see also* U.S. Environmental Protection Agency), 5, 178, 191, 243, 522, 525, 528, 572–576, 578, 585, 697–708, 721, 735, 765, 810, 837, 848, 866, 880–881
    Priority Toxic Pollutant List, 690
Epilimnion, 7
EPTC, 538, 561
Error, 29
    gross errors, 30–31
    random errors, 30–32, 35
    systematic errors, 30–31
ES (*see* Electrospray)
Escherichia coli, 118–141, 373, 442
ESI (*see* Electrospray ionization)
ESP (*see* Electrospray)
ESP-MS, 353–354, 511, 601–602
ESR (*see* Electron spin resonance spectroscopy)
ETAAS (*see* Electrothermal atomic absorption spectrometry)
Ether, 401
Ethiofencarb, 560
Ethion, 571, 581
Ethoprofos, 583
Ethyl violet active substances, 771
Ethylbenzene, 721–752
Ethylbenzene (*see also* BTEX compounds)

Ethylene oxide, 620
Ethylene oxide-propylene oxide block polymers, 769
Ethylenebisdithiocarbamates, 545
Ethylenebisdithiocarbamic acid, 538
ETV (*see* Electrothermal vaporization)
ETV-ICP-MS, 107
EU (*see* European Union)
European Commission, 696
European Community, 348, 459, 488, 519, 521, 545, 561, 571–576, 690
European Union, 700, 703, 707, 765–766
Eutrophication, 202, 229, 261
Evaporation, 631
EVAS (*see* Ethyl violet active substances)
External drift correction, 476
Extractable organic halide, 308
Extraction chromatography, 109–110
Extraction techniques, 17–21

FA (*see* Fulvic acids)
FAAS (*see* Flame atomic absorption spectrometry and Flame atomic absorption spectroscopy)
FAB (*see* Fast-atom bombardment)
FAES (*see* Flame atomic emission spectroscopy)
Faraday's law, 792
Fast-atom bombardment, 595–602, 808–809, 857
Fats, 387
Fatty acid alkanolamide, 769
Fatty acid polyol and sorbitol esters, 769
Fatty acids, 393
Fatty alcohol ether sulphates, 768
Fatty alcohol polyglycol ethers, 769
Fatty carboxylic acids, 79
FCF (*see* Field conversion factor)
FCP (*see* Filterable condensed phosphates)
FD (*see* Field desorption)
Fe(II), 25, 370, 374, 378, 381, 418–419, 425–427, 432, 440, 445, 863
Fe(III), 25, 410, 418–420, 426–427, 430, 440, 446, 863
Fecal streptococci, 119–141
    chromogenic-fluorogenic substrates, 131
    enumeration, 129–132
Fenac, 630
Fenamiphos, 581
Fenchlorphos, 602
Fenclor, 655
Fenitrothion, 571
Fenthion, 572
Fenvalerate, 5, 9
Ferbam, 538, 619
Fermothion, 581
Ferrocene tagging agents, 249
Ferrozine, 418
FI (*see* Field ionization, Flame ionization, Flow injection)
FIA (*see* Flow injection analysis)
FIA-HPLC-UV, 590

FIA-ICP-AES, 424
FIAstar unit, 420
Fiber optics, 759
Fiber-optic sensor systems, 708
Fiber-optic sensors (*see* Optodes)
Fiber-optic-based residual chlorine monitor, 175
Fick's law, 44, 48, 279, 812
FID (*see* Flame ionization detection)
Field conversion factor, 151, 154
Field desorption, 404
Field ionization, 404–405
FILAD (*see* FILtration/ADsorption system)
Filled amplification injection technique, 879
Filterable condensed phosphates, 273–295
Filterable organic phosphorus, 273–295
Filterable reactive phosphorus, 273–295
FILtration/ADsorption system, 679
Firefly luciferase, 807
Fl (*see* Fluoranthene and Fluorescent)
Flame atomic absorption spectrometry, 303, 441–446
Flame atomic absorption spectroscopy, 414, 420–424, 433
Flame emission spectrometry, 444
Flame ionization, 759
    detection, 92, 248, 328, 340–341, 354–360, 556, 583, 699,
        706, 723, 726, 732, 736–737, 743, 760, 836
Flame photometric detection, 557, 583, 836
Flame photometry, 197
Flameless vaporization, 802
Flavobacteria, 128
Flavor profile analysis, 81
Flavor wheel, 81
Floc morphology, 46
Flow cell detection, 873
Flow cytometric detection, 136
Flow cytometry, 155
Flow injection, 471–472
    analysis, 175, 178, 184, 196–197, 205–210, 212, 247, 263–
        269, 280–289, 303, 348, 369–371, 380, 414, 416–420,
        424, 426–427, 442, 445, 589–590, 826–829
    detection, 828–829
    techniques, 827–828
Flow injection-HG-AAS, 444
Flow meter, 14
Flow rate, 8
Flow-splitting injection, 873
Fluometuron, 488, 490
Fluoranthene, 689–690
Fluorene, 705
Fluorescence, 699, 708, 804–806, 879
Fluorescence detection, 852
Fluorescence emission wavelength selection, 806
Fluorescence excitation sources, 806
Fluorescence excitation wavelength selection, 806
Fluorescent spectroscopy, 758–760
Fluoride, 64–69, 169–170, 184–187, 192, 216, 341, 863
    criteria, 65

[Fluoride]
    detection, 185
    electrometric methods, 185–187
    ion chromatography method, 67–69
    ion-selective electrodes, 185–187
        method, 65–67
            interferences, 66
            principle, 65
            sampling procedure, 66
        preliminary treatment, 65
        sampling, 65
        spectrophotometric determination, 185–187
        potentiometric methods, 186–187
Fluoride-selective electrode, 426
Fluorimetry, 828
Fluorine, 64, 169–170
Fluorite, 169
Fluorosis, 64–65
Fluorspar, 169
Folpet, 619
Fonophos, 581
Food and Agriculture Organization/World Health Organization
        Joint Meeting on Pesticide Residues, 546
FOP (*see* Filterable organic phosphorus)
Formamide, 265
Formazine turbidity units, 788
Formetanate, 560
Formic acid, 313–345, 403
Fourier transform infrared, 556
    mass spectrometry, 799, 810
    spectrometry, 404, 560
    spectroscopy, 354, 585, 707, 873
FPA (*see also* Flavor profile analysis), 81
FPD (*see* Flame photometric detection)
Fractal objects, 46
Fractal theory, 46
Free active chlorine, 170, 173–174
Free chlorine:
    linear potential sweep voltammetry, 176
    quantitative determination, 171–177
    volumetric determination, 171–172
Free radicals, 401–402
Freeze concentration, 631
Freezing, 4
French°, 787
Freon, 113, 756–757
Fresnel's law, 855
FRP (*see* Filterable reactive phosphorus)
F-test, 38
FTIR (*see* Fourier transform, infrared spectrometry, infrared
        spectroscopy)
FTMS (*see* Fourier transform mass spectrometry)
FTU (*see* Formazine turbidity units)
Fulvic acids, 75–81, 210, 387–408, 461, 528
Fulvic matter, 43
Fumigants, 620

Fungicide residue, 609–654
  analysis, 630–633
  cleanup or separation methods, 632
  concentration methods, 631
  confirmation, 633
  estimation, 632–633
  isolation methods, 631–632
  liquid-liquid extraction, 631
  preliminary characterization tests, 633–650
  sample preparation, 630–632
  solid-liquid extraction, 631–632
  why detect?, 610
Fungicides, 545, 609–620, 630
Furans, 405
Furfurals, 405

Gamma detectors, 103–105
Gas chromatography, 19, 83, 91–93, 189, 248–249, 299, 315,
    328–329, 350–362, 472, 494–502, 511, 528–529, 555–
    557, 560, 579, 583–585, 632, 674, 681–682, 701–707,
    723, 730, 737, 758–760, 812, 830, 833–848, 853
  columns, 834–835, 838
  derivatization, 840–841
  detection, 835–840
  flow velocity through the column, 844–845
  injection techniques, 838–839
  instrument, 833
  portable systems, 842–843
  tandem MS, 839
Gas chromatography-atomic emission detector, 581, 837
Gas chromatography-electron capture detection, 159, 525–526
Gas chromatography-flame ionization detection, 336
Gas chromatography-flame photometric detector, 581
Gas chromatography-Fourier transform infrared spectroscopy,
    354, 595, 839–840
Gas chromatography-ion-trap mass spectrometry, 839
Gas chromatography-mass spectrometry, 93, 96, 159, 324, 360,
    364, 807, 837–840, 842–844
Gas chromatography-nitrogen phosphorus detector, 554, 557,
    581
Gas chromatography-quadrupole MS with chemical ionization,
    837–839
Gas chromatography-quadrupole MS with electron impact,
    837
Gas chromatography-time-of-flight MS, 839
Gas diffusion fluorescence, 244–246
Gas extraction, 757–758
Gas liquid chromatography-mass spectrometry, 703
Gas stripping, 812–813
Gas transfer across a water-gas interface, 48
Gas velocity, 845–846
Gas-sensing electrode, 242–243
Gas-tight syringes, 725
Gaussian distribution, 31
Gaussian peaks, 830–832
GC (see Gas chromatography)

GC-AED (see Gas Chromatography-atomic emission detector)
GCB (see Graphitized carbon black)
GC-ECD, 524, 557, 666–669
GC-FID, 703
GC-FPD (see Gas chromatography-flame photometric detector)
GC-FTIR (see Gas chromatography-Fourier transform infrared)
GC-ICP-MS, 445
GC-IR-MS, 595
GC-MIP-AES (see Gas chromatography-microwave-induced
    plasma-atomic emission spectrometry)
GC-MS (see also Gas chromatography-mass spectrometry),
    403, 524, 535, 554, 557, 576, 592–595, 601–602, 665,
    669, 682, 702–707
GC-MS-MS, 360, 595
GC-NPD (see Gas chromatography-nitrogen phosphorus
    detector)
GC-TOF-MS (see Gas chromatography/time-of-flight MS)
^{152}Gd, 102
Geiger-Müller gas-flow counter, 105
Gel filtration chromatography, 430
Gel permeation chromatography, 681, 700–701
Gel-phase photometry, 286–287
Geosmin, 81, 313
German°, 787
Germanium detector, 103
GFAAS (see Graphite furnace atomic absorption spectrometry)
Ghosting, 329
Giardia, 119
Giardia lamblia, 59
Gibbs' rule, 42
Glassy carbon, 425
GLC-MS (see Gas liquid chromatography-mass spectrometry)
Gliotoxin, 620
GLP (see Good laboratory practice)
Glucose, 267, 378
Glutamate, 378
Glutamic acid, 228
Glycine, 265, 267–268
Glycolic acid, 341
Glycoxylic acid, 313–345
Golay equation, 845–846
Good laboratory practice, 881
Gouy Chapman layer, 874
GPC (see Gel permeation chromatography)
Grab, 755
Grab sampling, 697
Gradient:
  delay, 851
    volume, 851
  elution, 862
  formation, 850–851
Gram Schmidt vector orthogonalization method, 840
Graphite electrothermal atomizer, 444
Graphite furnace, 444
Graphite furnace atomic absorption spectrometry, 466–476
Graphitized carbon black, 360 493–495, 527, 547, 700

Graphitized sorbents, 729
Graphs, 32–34
Graticule, 149
Greases, 753–764
Griess reaction, 265
Griseofulvin, 620
Ground-Water Monitoring List, 690
Growth hormones, 621
GTX toxins, 158
Guanidine, 265, 268
Guard columns, 851
Guntelberg equation, 225
Gus index, 561
GUS (*see* Ground Water Ubicity Score)
Gynkoteck GWBH, 550

H (*see* Henry's low constant)
$H_2BO_3^-$, 461
1,2,3,4,7,8-H_6CDF, 682
H_2S, 79
$H_2VO_4^{2-}$, 464
HA (*see* Humic acids)
Haber process, 223
Hall detector, 837
Halogenated hydrocarbons, 760
Halogenides, 4
Halogenonitrobenzofurazans, 249
Halogens, 169–194
    determination methods, 170–190
    ion chromatography, 190–192
    physical and chemical properties, 169
    why detect?, 170–171
Hanging mercury drop, 425
Hapsite, 843
Hardness, 56–59, 787–788
    interferences, 57
    method, 57–59
    method principle, 56–57
    sampling procedure, 57
Hazardous Constituents List, 690
Hazen unit, 787
HBOD (*see* Headspace biochemical oxygen demand)
HCB (*see* Hexachlorobenzene)
HCH (*see* Hexachlorocyclohexane)
HCO_3^-, 461
$HCrO_4^-$, 464
HCT (*see* High combustion temperature)
Headspace, 83–87, 700, 812
    analysis, 576
    biochemical oxygen demand, 299
    extraction, 757–758
    solid-phase microextraction, 737, 739, 741–744
Heated pneumatic nebulizer, 706
Heavy metals, 5, 51, 70, 137, 377–378, 418, 439–457, 619
    analytical methods, 441–448
    ion-exchange, 441

[Heavy metals]
    microorganism immobilization procedures, 441
    microwave digestion, 441
    non-immobilized substrates, 441
    pretreatment, 440–441
    sampling, 439–440
    solid-phase extraction, 441
    storage and preservation, 440
    ultrasound digestion, 441
    ultraviolet photolysis, 441
    why detect?, 439
Helmholtz layer, 874
Hemoglobin, 202
Henry's low constant, 572
Heptachlor, 521–523, 530
Heptachlor epoxide, 522–523
1-heptanol, 736
3-heptanone, 736
Heptenophos, 581
Herbicide, 374, 383, 387, 487–515, 545, 620–630, 755
Herbicide residue, 609–654
    analysis, 630–633
    cleanup or separation methods, 632
    concentration methods, 631
    confirmation, 633
    estimation, 632–633
    isolation methods, 631–632
    liquid-liquid extraction, 631
    preliminary characterization tests, 633–650
    sample preparation, 630–632
    solid-liquid extraction, 631–632
    why detect?, 610, 620–621
Heterotrophic plate count, 119, 122–141
Hexachlorobenzene, 524, 530, 619
Hexachlorocyclohexane, 521, 530
^{174}Hf, 102
HG (*see* Hydride generation)
Hg(I), 442
Hg(II), 427, 442
HG-AAS (*see also* Hydride generation-atomic absorption spectrometry), 441–444
HG-direct current plasma-AES, 444
HG-ICP-AES, 443
HG-ICP-MS, 443–445
High-frequency titration, 198
High-performance liquid chromatography, 19, 93, 158, 160, 164, 165, 249, 328, 336, 370, 404, 430, 527, 557–558, 579, 585, 592, 681, 812, 825
High-performance thin-layer chromatography, 589, 866
High-resolution gas chromatography, 660–662, 674, 678
High-resolution mass spectrometry, 674, 678, 682, 807
High-speed gas chromatography, 842–843
High-temperature combustion, 267–268
HMD (*see* Hanging mercury drop)
^{1}H-NMR, 398–401
HP 1090, 550

HPC (*see* Heterotrophic plate count)
HPLC (*see* High-performance liquid chromatography)
HPLC-HG-AAS, 443
HPLC-MS, 159–160, 595
HPLC-MS-MS, 155, 159–160, 165
HPN (*see* Heated pneumatic nebulizer)
HPTLC (*see* High-performance thin-layer chromatography)
HRGC (*see* High-resolution gas chromatography)
HRMS (*see* High-resolution mass spectrometry)
HS (*see* Humic substances)
HSE (*see* Headspace extraction)
HS-SPME (*see* Headspace solid-phase microextraction)
HTC (*see* High-temperature combustion)
Humic acids, 75–81, 210, 217, 387–408, 461, 494, 528, 544, 632, 681, 708
Humic material, 631–632, 700
Humic matter, 43
Humic solutes, 387–388
Humic substances, 223, 387–408
    acidification, 391
    analytical methods, 389–391
    chemical degradation methods, 402–404
    column chromatography, 390
    concentration, 389–390
    extraction, 390–391
    Fourier transform infrared spectrometry, 404
    functional groups of the dissolved organic carbon, 396
    gas chromatography, 403–404
    HPLC, 404
    hydrolysis, 402
    ion-exchange chromatography, 394–396
    isolation by chromatographic methods, 391–396
    mass spectrometry, 403–406
    nonionic macroporous sorbents, 391–394
    nuclear magnetic resonance, 398–401
    oxidation techniques, 402
    pretreatment, 389
    purification, 391
    Py-GC-FTIR, 404–406
    Py-GC-MS, 404–406
    Py-MS, 404–406
    pyrolysis, 404–406
    quantities of organic solutes obtained by the XAD technique, 396–397
    structural characterization, 396–406
    structural characterization degradative methods, 402–406
    structural characterization nondegradative methods, 398–402
    thermal degradation methods, 404–406
    verification, 391
Humins, 388, 528
Hybridization assays, 136
Hydrazine, 268
Hydride generation, 443, 472, 802
Hydride generation-atomic absorption spectrometry, 442
Hydrides, 4
Hydrocarbons, 387

Hydrocyanic acid, 620
Hydrodynamic injection, 336, 878–879
Hydrogen peroxide, 265
Hydrophobic properties, 861
Hydrostatic injection, 336
Hydroxy acids, 393
3-hydroxycarbofuran, 538
1-hydroxychlordene, 522
Hydroxy derivatives, 404
Hydroxyl radicals, 265
Hydroxyphenols, 347–366
8-hydroxy quinoline, 620
Hydroxy-quinoxilinol, 328
Hypochlorous acid, 176
Hypolimnion, 7

^{131}I, 102, 104
IARC (*see* International Agency for Research on Cancer)
IC (*see also* Inorganic carbon, Ion chromatography, Ion-exchange chromatography)
ICP (*see also* Inductively coupled plasma), 443–444, 475
ICP-AES (*see also* Inductively coupled plasma atomic emission spectrometry), 423–424, 427, 433, 442–443, 446
ICP-MS (*see also* Inductively coupled plasma mass spectrometry), 284, 289, 423–424, 430, 433, 441–445
ICP-OES (*see* Inductively coupled plasma optical emission spectrometry)
ID (*see* Isotopic dilution)
IDA (*see* Inisodiacetic acid)
IDA-Novarose, 424
ID-MS (*see* Isotope dilution-mass spectrometry)
IEC (*see* Ion-exclusion chromatography)
IEC-HG-AAS, 443
IEF (*see* Isoelectric focusing)
Imidazoliumquaternaryammonium methylsulphate, 769
Immunoanalysis, 829
Immunoassay analysis, 355, 583, 559, 592, 705
Immunoelectrochemical detection, 136
Immunofluorescence microscopy, 136
Immunosensing devices, 794–796
Immunosensors, 794
Immunosorbents, 494, 815–825
Implemented techniques, 472
Improved formate lactose glutamate medium, 124–141
In, 468
^{115}In, 102
INAA (*see* Instrumental neutron activation analysis)
Indeno[1,2,3-cd]pyrene, 690
Indicator organisms, 117–120
Indirect competitive ELISA, 162–164
Indirect speciation, 461
Indole-3-acetic acid, 632–650
Indophenol blue, 236–238, 246
Indoxyl-E-D-glucuronide, 129–141
Inductively coupled mass spectrometry, 803–804

Inductively coupled plasma, 414, 423–424, 799, 803–804, 837, 866
   atomic absorption spectrometry, 441
   atomic emission spectrometry, 284, 466–476, 829
   mass spectrometry, 106–108, 110–112, 466–476
   optical emission spectrometry, 466–474, 803–804
Inficon, 843
Infrared detector, 583
Infrared spectroscopy, 198, 402, 758–760, 762
Inisodiacetic acid, 424
Inorganic ammonium salts, 268
Inorganic carbon, 304–305, 792–793
Inorganic fungicides, 610–618
Inorganic herbicides, 621
Inositol phosphates, 273
Insecticides, 9, 15–16, 545, 630
Instrumental neutron activation analysis, 466–470, 476
Instruments, 785–886
Interference effects, 469
Interferogram, 799
Internal sensitivity standards, 677
Internal standards, 665, 674–677, 883–884
International Agency for Research on Cancer, 546, 673, 722
International Atomic Energy Agency, 105
International Organization for Standardization, 785, 788, 882–883
International UNESCO Oceanographic Commission standard, 759
Inverted compound microscope, 149
IOC/UNESCO (see International UNESCO Oceanographic Commission)
Iodate, 60
Iodide, 171, 181, 189–190
Iodide argentimetric titrations, 189
Iodine, 27, 169–170
   measurements, 190
Ion chromatography, 60, 67–69, 106–107, 190–192, 197–198, 212–218, 248–251, 427–430, 443–445, 471, 830, 862–866
   detection methods, 865–866
   interferences, 67–68
Ion exchange, 111, 231, 471, 769, 815, 863–865
Ion-exchange chromatography, 212–215, 218, 329–333, 394–396, 863
Ion exclusion, 864–865
Ion-exclusion chromatography, 216, 251, 333–336
Ion-interaction chromatography, 215–217, 251, 430
Ionization chambers, 103
Ionization detector, 836
Ion-pair, 769
Ion-pairing chromatography, 251, 430, 557
Ion-pairing chromatography-AES, 866
Ion-pairing technique, 862
Ion-selective electrodes, 185–188, 206, 242–244, 370–371, 426, 441–446, 790–792, 829
Ion-selective field-effect transistors, 206

Ion spray, 352–353, 558, 857
Ion-trap mass spectrometer detector, 723, 743
IP (see Indeno[1,2,3-cd]pyrene)
IPB (see Indophenol blue)
IPC (see Ion-pairing chromatography)
IR (see Infrared)
IRD (see Infrared detector)
Iron, 8, 27, 51, 79, 409–438, 445–446
   adsorption methods, 279
   capillary isotachophoresis, 446
   chemical sensors, 446
   chemiluminescence, 445
   CZE, 446
   electrochemical methods, 446
   electron spectroscopy, 446
   FAAS, 445
   flow injection analysis, 445
   hanging drop electrode, 446
   laser-induced breakdown spectroscopy, 446
   reversed-phase HPLC, 445
   sequential injection analysis, 445
   spectrophotometric methods, 445
Iron sulphate, 621
IS (see Internal standard)
ISEs (see Ion-selective electrodes)
ISFETs (see Ion-selective field-effect transitors)
ISO (see also International Organization for Standardization), 3, 5, 755, 771
Isoelectric focusing, 876
Isoprene, 328
2-isopropoxyphenol, 538
Isoproturon, 487–488
Isotachophoresis, 285, 876, 879
Isothiocyanate, 619
Isotope dilution, 442
   GC-MS, 442
   technique, 677
   mass spectrometry, 444
Isotopes, 101–114
Isotopic dilution, 601
ISP (see Ion spray)
IUPAC, 655

^{40}K, 102
Kanamycin-esculin-azide agar, 131–141
Kanechlor, 655
KEAA (see Kanamycin-esculin-azide agar)
Kemmerers, 755
Kentucky Water Quality Standards, 65
Ketoacids, 59, 321, 338–340
Ketomalonic acid, 313–345
Ketones, 79, 393
KF-streptococcus agar, 131–141
Kjeldahl method, 263–268, 793
Klebsiella, 119–141
Klett-Summerson colorimeter cell, 196

K$_{OC}$ (*see* Partition coefficient between soil organic carbon and water)
Kohlrausch's law, 854
König reaction, 368–369
Korolef method, 268
Kovats index, 847
K$_{OW}$ (*see* Octanol-water partition coefficient)
K$_p$, 572

^{138}La, 102
^{140}La, 104
Lactic acid, 341
Lactococcus, 119–141
Lactose broth, 124–141
Lanthanides, 110
Large-volume injection, 834
LAS (*see* Linear alkylbenzenesulfonates and Linear alkylate sulfonate)
Laser-enhanced ionization spectrometry, 446
Laser-excited atomic fluorescence spectrometry, 442, 444
Laser-induced breakdown spectroscopy, 446
Laser-induced flame ionization, 759
Laser-induced fluorescence, 852
Laser-induced time-resolved fluorescence, 708
Latex agglutination, 136
Lauryl tryptose lactose broth, 124–141
Lauryl tryptose manitol broth, 126–141
LC (*see also* Liquid chromatography), 442, 529, 535, 699–707
LC-APCI, 602
LC-APCI-MS, 353
LC-DAD, 535, 602
LC-ESP-MS, 511, 581
LC-MS (*see also* Liquid chromatography-mass spectrometry), 494, 496, 535, 601–602, 706
LC-PA-ESP, 602
LC-PA-ESP-MS-MS, 601
LC-PB-MS, 510, 602, 706
LC-SFC-TID, 585
LD (*see* Lethal dose)
LD$_{50}$, 518, 537–538
Lead, 27, 410, 420–432, 441–442
  anodic stripping voltammetry, 442
  cathodic stripping voltammetry, 442
  flame atomic absorption spectrometry, 442
  flow injection analysis, 442
  hydride generation-atomic absorption spectrometry, 442
  ICP-AES, 442
  ICP-MS, 442
  isotope-dilution GC-MS, 442
  liquid scintillation counting, 442
  spectrophotometric methods, 442
Legionella, 59
LES Endo agar, 127–141
Lethal dose, 518
Levoglucosan, 405

Liebig reaction, 368
Light blockage, 44
Light scattering, 43–44
Light-scattering detector, 855
Lignins, 405
Limit of detection, 214, 576, 585, 674, 882
Limit of quantification, 476, 585, 882
Lindane, 5, 9, 521
Linear alkylate sulfonate, 377, 381
Linear alkylbenzenesulfonates, 781
Linear potential sweep voltammetry, 176
Linear range, 882
Linear solvent strength, 862
Linearity, 883
Linuron, 487–488, 490, 494
Lipids, 405, 768
Liquid chromatography, 248–251, 356, 358–360, 471–472, 494, 496, 500–506, 511, 356, 826–827, 830, 844, 848–870
  columns, 851
  detectors, 851–855
  different modes, 859–862
  gas chromatography detectors, 855
  injection, 851
  instrument, 848–851
  mass spectrometry, 509–511, 807, 855–858
Liquid-liquid extraction, 17, 19, 87–88, 355–356, 358, 418, 491–493, 497, 525–528, 546–547, 554–555, 577, 581, 631, 656, 680, 699–702, 708–709, 756–757, 813–815, 826
Liquid scintillation, 112
  counting, 442, 444
Liquid-solid adsorption, 700
Liquid-solid extraction, 700
Lithium, 461
Lithium-drifted germanium detector, 103
Live system, 843–844
LLE (*see* Liquid-liquid extraction)
LOD (*see* Limit of detection)
Longitudinal gradient description, 8–9
Long-path capillary detectors, 286
Loop-type interfaces, 705
LOQ (*see* Limit of quantification)
Low-resolution MS, 682
LRMS (*see* Low-resolution mass spectrometry)
LSA (*see* Liquid-solid adsorption)
LSE (*see* Liquid-solid extraction)
^{176}Lu, 102
Luminescence techniques, 430–432
Luminol, 380, 432, 444, 807, 866
Lund cell, 151–154
2,4-lutidine, 633
2,6-lutidine, 633
LVI (*see* Large-volume injection)
Lyngbya, 143
Lyngbya majuscula, 147

Lyngbya toxins, 147
Lyophilization, 631, 811

MAC (*see* Maximum admissible concentration)
MacConkey broth, 124–141
Macro-Kjeldahl, 264
MAFF (*see* Ministry of Agriculture, Fisheries and Food)
MAGIC (*see* Magnesium hydroxide-induced coprecipitation
    and Monodisperse aerosol-generation ionization)
MAGIC-EI-CI, 601
Magnesium, 409–438, 787
Magnesium hydroxide-induced coprecipitation, 277
Major metals, 409–438
   acid digestion, 411
   analysis methods, 414–433
   capillary zone electrophoresis, 433
   chelation by solid-phase sorbents, 412–414
   coprecipitation, 412
   electrochemical methods, 424–427
   electrothermal atomic absorption spectroscopy, 422–423
   filtration and digestion, 411–412
   flame absorption spectroscopy, 420–423
   flame atomic emisson spectroscopy, 420–423
   flow injection analysis, 416–420
   gel filtration chromatography, 430
   HPLC, 430
   inductively coupled plasma, 423–424
   ion chromatography, 427–430
   ion-interaction chromatography, 430
   ion-selective electrodes, 426
   liquid-liquid extraction, 418
   luminescence techniques, 430–432
   neutron activation analysis, 432–433
   paper chromatography, 430
   potentiometric stripping analysis, 426–427
   potentiometric techniques, 426–427
   preconcentration, 412–414
   properties and importance, 409–410
   sample preparation, 410–414
   sampling and storage, 410–411
   sequential injection analysis, 416
   solvent extraction, 412
   spectrophotometric method, 416–420
   thin-layer chromatography, 430
   ultraviolet digestion, 411–412
   voltammetric methods, 424–426
   volumetric methods, 415
   x-ray fluorescence spectrometry, 432
Malathion, 571–572, 581, 631
MALDI (*see* Matrix-assisted laser desorption ionization)
Malonic acid, 403
Mammalian bioassays, 155
Mancozeb spot test, 650
Maneb, 538, 561, 619
Manganese, 8, 27, 51, 79, 409–438, 446
   catalytic fluorometry, 446

[Manganese]
   chemiluminescence, 446
   colorimetric methods, 446
   DP-CSV, 446
   electron spectroscopy, 446
   FAAS, 446
   flow injection analysis, 446
   laser-enhanced ionization spectrometry, 446
   stripping potentiometry, 446
Mass spectrometry, 85, 90, 92, 103, 106–107, 354, 358, 430,
    499, 509–511, 529, 531–535, 556–558, 572, 583, 585–
    589, 592–602, 660–662, 674–678, 682, 703, 706, 723,
    726, 732, 736–737, 760, 807–810, 812, 866, 880
Mass-to-charge ratio, 807
Matrix-assisted laser desorption ionization, 810
Matrix effects, 469
Maximum admissible concentration, 572
Maximum concentration level, 765
Maximum contamination levels, 722
MB (*see* Moving belt)
MBAS (*see* Methylene blue active substances)
MBTH (*see* 3-methyl-2-benzo-thiazolinone hydrazone)
MCL (*see* Maximum concentration level)
MCLs (*see* Maximum contamination levels)
MCPA, 621
mCP medium (*see* Membrane *Clostridium perfingens* medium)
McReynolds indices, 847–848
MDL (*see* Method detection limit)
MDQs (*see* Minimum detectable quantities)
ME (*see* Microextraction)
mE agar (*see* Membrane enterococcus agar)
Mean, 30–40
MECA (*see* Molecular emission cavity analysis)
MECC (*see* Micellar electrokinetic capillary chromatography)
Median, 30–40
Megabore injector, 834
MEKC (*see* Micellar electrokinetic chromatography)
Membrane *Clostridium perfingens* medium, 133
Membrane enterococcus agar, 131–141
Membrane extraction, 745
Membrane fecal coliform, 127
Membrane filtration method, 121–141
Membrane heterotrophic plate count agar, 121–141
Membrane introduction mass spectrometry, 90, 93–98
Membrane-lactose glucuronide agar, 129–141
Memory effects, 329
Mercury, 410, 433, 442, 444
   atomic fluorescence spectrometry, 442
   biological substrates, 442
   constant current stripping chronopotentiometry, 442
   CV-AAS, 442
   ETAAS, 442
   FAAS, 442
   fungicides, 610–618
   gas chromatography, 442
   ICP-AES, 442

[Mercury]
ICP-MS, 442
isotope dilution, 442
LC, 442
photoacoustic spectroscopy, 442
reversed-phase liquid chromatography, 442
spectrophotometric methods, 442
Mercury chloride, 27
Metalimnion, 7
Metals, 409–438
acid digestion, 411
analysis methods, 414–433
capillary zone electrophoresis, 433
chelation by solid-phase sorbents, 412–414
coprecipitation, 412
electrochemical methods, 424–427
electrothermal atomic absorption spectroscopy, 422–423
filtration and digestion, 411–412
flame atomic absorption spectroscopy, 420–423
flame atomic emission spectroscopy, 420–423
flow injection analysis, 416–420
gel filtration chromatography, 430
HPLC, 430
inductively coupled plasma, 423–424
ion chromatography, 427–430
ion-interaction chromatography, 430
ion-selective electrodes, 426
liquid-liquid extraction, 418
luminescence techniques, 430–432
multiple metals detection, 446–448
neutron activation analysis, 432–433
paper chromatography, 430
potentiometric stripping analysis, 426–427
potentiometric techniques, 426–427
preconcentration, 412–414
properties and importance, 409–410
sample preparation, 410–414
sampling and storage, 410–411
sequential injection analysis, 416
solvent extraction, 412
spectrophotometric method, 416–420
thin-layer chromatography, 430
ultraviolet digestion, 411–412
voltammetric techniques, 424–426
volumetric methods, 415
x-ray fluorescence spectrometry, 432
Methabenzthiazuron, 487–488
Methamidophos, 571
Methemoglobin, 202
Methemoglobinemia, 202
Methiocarb, 561
Method detection limit, 348
Methoxychlor, 520–521
Methoxyl, 397
Methoxyphenols, 401, 405
Methyl allyl chloride, 620

Methyl bromide, 620, 630
4-methyl-4,6-dichlorophenol, 348
Methyl mercury, 442
Methyl orange, 268
3-methyl-2-benzo-thiazolinone hydrazone, 348
Methylene blue active substances, 770–771
Methylene blue reduction method, 196, 198
Methylisoborneol, 79, 93
2-methylisoborneol, 81, 313
2-methylphenol, 360
4-methylphenol, 352
Metobromuron, 487
Metsulfuron methyl, 496, 500
Mevinphos, 571
mFC (*see* Membrane fecal coliform)
MFL (*see* Million fibers per liter)
m-HPO (*see* Membrane heterotrophic plate count)
m-hydroxy aromatic compounds, 403
Micellar electrokinetic capillary chromatography, 559, 705, 876
Micellar electrokinetic chromatography, 355
Micellar liquid chromatography, 703, 830
Micelle-mediated extraction, 702
Michelson interferometer, 799
Micro- liquid-liquid extraction, 525
Microbial biosensor, 371
Micrococcus luteus, 373
Microcolumns, 851
Microcystin, 143–147
Microcystins immunoassays, 162–163
Microcystis, 143
Microcystis aeruginosa, 152–153
Microdialysis, 812
Microextraction, 554–555
Micro-Kjeldahl, 264
Micro-liquid-liquid extraction, 813
Microorganisms, 75, 115–141
analysis methods, 120–136
indicator organisms, 117–120
safety of drinking and recreational water, 115–117
and water quality, 115
Microscope calibration, 149
Microtox, 158
Microwave digestion, 441
Microwave plasma torch:
AES, 445
atomic emission spectrophotometry, 829
Microwave-assisted digestion, 263–264
Microwave-assisted LLE, 757
Microwave-assisted systems, 724, 731
Microwave-induced plasmas, 800–801, 837
atomic emission detector, 360
Mie theory, 43–44
Millerite, 464
Millilab, 825
Million fibers per liter, 765

MIMS (*see* Membrane introduction mass spectrometry)
Mineral acidity, 51
Mineralization, 262
Minimum detectable quantities, 532
Ministry of Agriculture, Fisheries and Food, 656
MIP (*see* Microwave-induced plasmas)
Mirex, 521, 524
MIT, 620
MLC (*see* Micellar liquid chromatography)
m-LGA (*see* Membrane-lactose glucuronide agar)
Mn(II), 374, 381, 409, 418–425, 430, 432, 446
Mn(III), 409
Mn(IV), 425
Mn(VII), 446
MnO_4^-, 409
MOB (*see also* 2-methylisoborneol), 313
Mobile-phase ion chromatography, 251
Mode, 30–40
Mohr indicator, 179–180
Mohr titration, 180
Molds, 115
Molecular emission cavity analysis, 247
Molecular structure, 41
Molybdate-reactive phosphorus, 274–295
Molybdene, 4, 27, 32, 443
Monocarboxylic acids, 321, 328, 333
Monochloramine, 229
Monochlorobiphenyls, 665
Monoclonal antibodies, 162–164
Monocrotophos, 581
Monodisperse aerosol-generation ionization, 601
Monolinuron, 487–488
Montreal Protocol, 757–758
Most probable number, 120–141
Mouse bioassay:
    for cylindrospermopsin, 157
    for mycrocystins, 157
    for paralytic shellfish poisons, 156–157
Moving belt, 706
    EI-CI-FAB, 601
    system, 856
MPIC (*see* Mobile-phase ion chromatography)
MPN (*see* Most probable number)
MRMs (*see* Multiresidue methods)
MRP (*see* Molybdate-reactive phosphorus)
MS (*see* Mass spectrometry)
MSD (*see* Mass spectrometric detector)
m-7h FC medium, 127
MS-ITD (*see* Ion-trap mass spectrometer detector)
MSMA, 621
MS-MS, 94, 511, 595, 682
m-T7 agar, 137
MTA (*see* m-tolouoyl chloride)
MTBSTFA (*see* n-(ter-butyldimethylsilyl)-n-methyl-
    fluoracetamide)
m-tolouoyl chloride, 249

MU-GAL (*see* 4-methylumbelliferyl-E-D-galactopyranoside)
MU-GLC (*see* 4-methylumbelliferyl-E-D-glucoside)
MU-GLU medium, 128–141
Multidimensional gas chromatography, 843–844
Multielectrode electrochemical detection, 351
Multielement methods, 785
Multiple-tube test, 121
Multiple-wavelength fluorescence detection, 703
Multiport streamswitch, 550
Multiresidue methods, 495–496
MW (*see* Microwave)
m-xylene, 721–752
Mycrocystin LR, 159
Mycrocystins, 156–165
Mycrocystis aeruginosa, 159
M/z ratio (*see* Mass-to-charge ration)

N_2, 262
Na^+, 374
NAA (*see* Neutron activation analysis)
NaCl, 381
Nanoelectrospray, 873
Naphthalene-2,3-dialdehyde, 369–371
Naphthalenes, 620, 705, 721
1-naphthol, 538
Naphtoquinoline, 585
Narrow bore columns, 583, 834
NASS-4, 476
National Institute of Standards and Technology, 705
National Pesticide Survey, 545–546, 572
National Pollutant Discharge Elimination System, 690
National Primary Drinking Water List, 690
National Primary Drinking Water Regulations, 120
NATO/CCMS (*see* North Atlantic Treaty Organization
    Committee on Challenges to Modern Society)
Natural organic matter, 387–408
NCI (*see* Negative chemical ionization)
NCIMS (*see* Negative-ion chemical ionization mass
    spectrometry)
^{144}Nd, 102
NDA (*see* Naphthalene-2,3-dialdehyde)
NDIR (*see* Nondispersive infrared)
Neburon, 488, 490
Negative chemical ionization, 557, 595–602
Negative-ion chemical ionization mass spectrometry, 662–665
Neocuproin, 419
Neosaxitoxin, 145, 158, 164
Nephelometric turbidity units, 788
Nephelometry, 788
Neptunium, 110
Nernst equation, 787, 791
Nernst partition isotherm, 813
Nessler method, 264
Nessleriser, 79
Nesslerization, 240–242
Neutron activation analysis, 412, 432–433, 441, 444, 799

Neutron detectors, 103

NH_4, 8

NH_x (*see* Total ammoniacal nitrogen)

$Ni(OH)_2$, 464

Ni^{2+}, 464

^{63}Ni, 444

Nickel, 27, 418, 429–430, 444, 464, 467, 471
    chemiluminescence, 444
    electrothermal AAS, 444
    liquid scintillation counting, 444
    spectrophotometric detection, 444
    voltammetric methods, 444

Nicolsky Eisemann formalism, 791

Nicotinamide, 264

Nicotine, 620

Nicotinic acid, 264, 633

NiS, 464

NIST (*see* National Institute of Standards and Technology)

Nitrate online direct UV measurement, 205

Nitrate pollution, 203

Nitrate reductase, 210–211

Nitrates, 51, 67, 201–223, 226–228, 261–271, 297, 315, 341, 787, 866
    amperometry, 214–215
    capillary electrophoresis, 218
    chromatographic methods, 212–218
    conductivity, 214–216
    direct detection methods, 204–207
    direct electrochemical methods, 205–207
    fluorescence, 215
    formation of nitrosil chloride, 211
    indirect detection methods, 207–212
    ion chromatography, 212–218
        sampling handling, 216–218
    and nitrites reduction to ammonium, 211
    potentiometric methods, 205–206
    reduction to hydroxylamine, 211
    reduction to nitrogen oxide, 211–212
    sample handling and preservation, 203–204
    spectrophotometry, 214–215
    UV spectroscopy, 204–205
    voltammetric and amperometric methods, 206–207

Nitrate-selective electrode, 206

Nitrifying bacteria, 16

Nitrate reductase, 211

Nitrites, 62, 201–223, 226–228, 261–271, 297, 341, 863, 866
    amperometry, 214–217
    capillary electrophoresis, 218
    chromatographic methods, 212–218
    conductivity, 214–216
    direct detection methods, 204–207
    direct electrochemical methods, 205–207
    fluorescence, 215
    formation of nitrosil chloride, 211
    indirect detection methods, 207–212
    ion chromatography, 212–218

[Nitrites]
    sample handling, 216–218
    kinetic methods, 212
    photometric detection, 207
    potentiometric mehods, 205–206
    reduction of nitrate to nitrite, 207–210
    reduction to hydroxylamine, 211
    reduction to nitrogen oxide, 211–212
    sample handling and preservation, 203–204
    spectrophotometry, 214–215
    UV spectroscopy, 204–205
    voltammetric and amperometric methods, 206–207

Nitroaniline herbicides, 629

Nitrobacteria, 297

Nitroderivatives, 264

Nitrogen, 8, 25, 201, 224, 261–271, 793–794
    chemiluminescence method, 793–794
    cycle, 201, 223, 226–227
    elemental analyzer, 794
    fixation, 201
    inorganic, 25
    Kjeldahl's method, 793
    pesticides spot test, 650

Nitrogen oxides, 70

Nitrogen phosphorus detector, 248–249, 356, 499, 583, 589

Nitrogen trichloride, 229

Nitrogen-containing pesticides spot tests, 633

Nitrogenous oxygen demand, 16

3-nitrophenol, 363

4-nitrophenol, 355–356, 358, 360, 363

Nitrophenol fungicides, 619

2-nitrophenol, 363

Nitrophenols, 261, 266, 347–366, 630, 632

Nitrosamines, 203, 760

N-methylcarbamates, 538–561

NMR (*see* Nuclear magnetic resonance)

N,N-dimethylcarbamates, 538, 544

N-nitroso compounds, 203

NO_2^-, 197, 201–222, 369

NO_3^-, 197, 201–222, 369

Nodularia, 143–147

Nodularia spumigena, 145

Nodularin, 145, 164

Noice, 882

NOM (*see* Natural organic matter)

Nondispersive infrared analyzer, 307, 760

Nondispersive infrared detector, 792–793

Nonflame AAS, 443

Nonionic macroporous sorbents, 391–394

Nonionic surfactants, 768–781

Nonortho-PCB isomers, 665–669

Nonporous-membrane ion-selective electrode, 206

Nonpurgeable organic carbon, 305–307

Nonpurgeable organic halide, 308

Nonselective herbicides, 621

Nonsuppressed conductometric detection, 197

Nonsuppressed ion chromatography, 213, 216–217
Nonvolatile organic carbon, 305–307
Nonylphenol decaglycol ether, 774–781
Normal error curve, 31
Normal-phase chromatography, 830
Normal-phase liquid chromatography, 585, 859
North Atlantic Treaty Organization Committee on Challenges
    to Modern Society, 683
Nostoc, 143–147
No-toxic-effect-level, 117
NO$_x$, 262
N-P pesticides, 555
NPD (*see* Nitrogen phosphorus detector)
NPDES (*see* National Pollutant Discharge Elimination System)
N-phenylcarbamates, 538, 545, 557
N-phenyl-N′-alkyl-N′-methoxyureas, 487
N-phenyl-N′N′-dialkylureas, 487
N-phenylurea herbicides, 557
NPLC (*see* Normal-phase liquid chromatography)
NPOC (*see* Nonpurgeable organic carbon)
NPOX (*see* Nonpurgeable organic halide)
NPS (*see* National Pesticide Survey)
NTP (*see* U.S. National Toxicology Program)
Nephelometric turbidity units, 277
n-(*tert*-butyldimethylsilyl)-*N*-methyl-fluoracetamide, 324
NTU (*see* Nephelometric turbidity units)
Nuclear magnetic resonance, 398–401
Nucleic acids, 223, 273
Number-size distribution, 43
n-undecane, 736
NVOC (*see* Nonvolatile organic carbon)
NWRI agar, 123

O-(2,3,4,5,6-pentafluorobenzyl)hydroxylamine, 319, 324
o,p′-DDT, 530
O$_8$CDD, 673
O$_3$CDF, 673
OBSs (*see* Optical backscatter sensor)
OCPs (*see* Organochlorinated pesticides)
OCs (*see* Organochlorine pesticides and Organochlorines)
Octanol-water partition coefficient, 5, 518, 572, 656, 673
o-dichlorobenzene, 736
Odor, 28, 75–100, 202, 229, 313, 347
Odor and taste, 79
   threshold concentrations, 79
OEAA (*see* Oxolinic acid-esculin-azide agar)
OFISP (*see* Oscillatory flow injection stripping potentiometry)
OH$^-$, 461
Ohm's law, 854
Oils, 753–764
Olfactometry, 835
OLSA (*see* Open-loop stripping analysis)
Oncolumn injection, 705, 834
o-nitrophenol, 355
Online biomonitors, 8
ONP-GAL (*see* Ortho-nitrophenyl-E-D-galactopyranoside)

OPA (*see* *o*-phthalaldehyde and Orthophtaldialdehyde)
Opalescence, 755
Open tubular columns, 833–834
Open-loop stripping analysis, 733
Open split interface, 732
o-phthalaldehyde, 249, 369–370, 557–558
OPLC (*see* Overpressured-layer chromatography)
OPPs (*see* Organophosphates)
OPs (*see* Organophosphates)
Optical backscatter sensor, 14
Optical chemical sensor, 355
Optodes, 247–248
Orcinol, 650
Organic acids, 59, 313–345, 392–393, 396
   analysis methods, 319–338
   in atmospheric precipitation, 315, 341
   capillary electrophoresis, 336–338, 341
   column chromatography, 338
   derivatization, 319–328, 336
   distillation-titration, 338
   in drinking water, 313–314, 338–340
   extraction/concentration, 321
   filtration/centrifugation, 319
   gas chromatography, 319–321, 324–329, 338–340
   in groundwater, 315
   high-performance liquid chromatography, 328, 336
   ion-chromatography, 338–340
   ion-exchange chromatography, 319–321, 329–333
   ion-exclusion chromatography, 319, 333–336, 338, 341–
     342
   in landfill leachates, 315–319
   liquid chromatography, 329–338
   liquid-liquid extraction, 321, 324
   p-bromophenacyl esters, 324, 328
   in seawater, 315
   sample preparation, 319–328
   sample preservation, 319
   separation techniques, 328–338
   solid-phase extraction, 321, 324
   in wastewater, 314–315, 340–341
   why detect?, 313–319
Organic arsenical herbicides, 621
Organic bases, 392
Organic carbon, 266–267, 278
Organic condensed phosphates, 273
Organic contaminants:
   analysis, 81–98
   gas chromatography, 91–93
   gas chromatography-mass spectrometry, 93
   headspace methods, 83–87
   high-performance liquid chromatography, 93
   liquid-liquid extraction, 87–88
   membrane introduction mass spectrometry, 93–98
   preconcentration methods, 82–91
   preliminary sample treatments, 82–83
   sensory methods, 81–82

[Organic contaminants]
  solid-phase extraction, 88–90
  solid-phase microextraction, 90
  vacuum distillation, 90–91
Organic content, 28
Organic fungicides, 618–619
Organic herbicides, 621–630
Organic matter, 59, 80, 159, 205, 265, 297–311, 319, 393, 440
  oxidation methods, 306
Organic modifier, 862
Organic nitrogen, 261–271
  alkaline persulfate oxidation, 268–269
  analytical methods, 262–269
  AWWA-APHA procedure, 263
  compounds, 211
  conductimetric detection, 264
  filtration, 262
  flow injection analysis, 263, 266–267, 269
  gas-diffusion techniques, 264
  high-temperature combustion, 267–268
  Kjeldahl method, 263–264
  microwave-assisted digestion, 263–264
  photometric detection, 264
  photo-oxidation, 265–267
  potentiometry, 264
  sample collection and preservation, 262
  segmented flow systems, 264–269
Organic solvents, 753
Organochlorines, 571, 631, 836
  columns, 530–531
  detection methods, 531–535
  electron capture detector, 531–532
  liquid-liquid extraction, 525–526
  mass spectrometry, 531–535
  pesticides, 5, 10, 18, 517–537, 546, 571, 632, 662
  physical and chemical properties, 517–524
  preanalytical techniques, 524–529
  solid-phase extraction, 526–527
Organoleptic quality, 754–755
Organoleptical properties, 75–100
  why detect?, 75
Organonitrogen, 18, 632
  pesticides, 538
Organophosphates, 537, 546, 571–608, 630
  analytical methods, 583–602
  classification, 571–572
  degradation, 572
  flow injection analysis, 589–590
  gas chromatography, 583–585
  headspace analysis, 576
  high-performance liquid chromatography, 585
  immunoassay analysis, 583, 592
  liquid-liquid extraction, 577
  mass spectrometry and hyphenated methods, 592–602
  physical parameters predicting environmental fate, 572

[Organophosphates]
  physicochemical properties, 571–576
  purge and trap, 576
  regulations, 572–576
  sample preparation, 576–583
  solid-phase extraction, 576–581
  solid-phase microextraction, 576, 582–583
  solvent extraction, 576
  spectrophotometric analysis, 590
  storage of samples, 577
  supercritical fluid chromatography, 585–589
  thin-layer chromatography, 589
  toxicity, 572
Organophosphorous pesticides, 538, 559
Organophosphorus, 18
Organophosphorus pesticides (*see* Organophosphates)
Organosulphur, 632
Organotin, 852
Organotin fungicides, 619
Ortho-nitrophenyl-E-D-galactopyranoside, 129
Orthophosphate, 273–295
Orthophtaldialdehyde, 269
Oscillatoria, 143–147
Oscillatoria flow injection stripping potentiometry, 445
OSP-2, 550
OT (*see* Open tubular)
OTCs (*see* Odor and taste threshold concentrations)
Overpressured-layer chromatography, 870
Ovicides, 630
Oxalic acid, 313–345, 403
Oxamic acid, 265
Oxathin fungicides, 620
Oxidant demand, 59–61
  criteria, 61
  method, 60–61
  method principle, 60
Oxides, 4
Oxidized nitrogen, 8
Oxime *N*-methylcarbamates, 538, 544
Oximes, 400
Oxo alcohol ether sulphates, 768
Oxohalides, 171
Oxolinic acid-esculin-azide agar, 131–141
Oxycarboxin, 620
Oxydemethon-methyl, 581
Oxygen, 59
Oxygen-membrane electrode method, 63–64
o-xylene, 721–752
Ozonation, 59, 80–81, 314
  by-products, 338–340
Ozone, 59, 171, 188, 313
Ozone layer, 757

P/A (*see* Presence-absence)
Packed columns, 583, 833
PA-ESP (*see* Pneumatic-assisted electrospray)

PAHs (*see also* Polycyclic aromatic hydrocarbons and
    Polynuclear aromatic hydrocarbons), 681, 687–720, 738
  analysis, 690–709
  calibration, 708
  certified reference materials, 696
  concentration, 702–703
  detectors, 705–707
  exposure, 689
  extraction, 699–702
  gas chromatography detectors, 706–707
  legislation, 689–690
  liquid chromatography detectors, 705–706
  liquid-liquid extraction, 699–702
  occurrence in water, 687–689
  optical spectrometric techniques, 707–708
  preservation, 696–699
  purification, 702
  reference methods, 690–696
  sample collection, 696–699
  sample preparation, 699–703
  separation techniques, 703–705
  solid-phase extraction, 699–702
  solid-phase microextraction, 698
  sources, 687
  standards, 708
p-aminophenol, 560
P&T (*see* Purge and trap)
Paper chromatography, 251–255, 430, 703
Paralytic shellfish poisons, 143–145, 156, 159
Parathion, 571–572, 631
Parathion-ethyl, 9, 579
Parathion-methyl, 579
Particle beam, 509–510, 558, 706
  interface, 857
Particle counting, 43
Particle size distributions, 43–45
Particulate organic carbon, 388–408
Particulate organic matter, 388–408
Particulate state, 42–43
Particulates, 43
Partition coefficient between octanol and water, 815
Partition coefficient between soil organic carbon and water,
    561
PB (*see* Particle beam)
Pb^{2+}, 25, 378
^{206}Pb, 102
^{207}Pb, 101
^{208}Pb, 101
^{210}Pb, 442
^{214}Pb, 112
p-benzoquinone, 369
PB-MS, 705
PC (*see* Paper chromatography)
PCB101, 669
PCB118, 669
PCB126, 668

PCB138, 669
PCB15, 668
PCB153, 669
PCB156, 669
PCB169, 668
PCB180, 669
PCB209, 669
PCB28, 669
PCB37, 668
PCB52, 669
PCB77, 668
PCB81, 668
PCBs (*see also* Polychlorobiphenyls and Trihalomethane
    polychlorinated biphenils), 632, 681, 738–739
PCDDs, 673–685
  analysis, 673–674
  cleanup, 681
  data reporting, 683
  in dissolved fraction, 678–680
  environmental relevance, 673
  extraction, 680–681
    from liquid matrices, 680
    from solid matrices, 681
  gas chromatography, 681–682
  identification and quantification of congeners, 682–683
  instrumental analysis, 681–683
  internal standard quantification method, 674–677
  laboratory safety, 677
  mass spectrometry, 682
  physicochemical properties, 673
  principles behind the method, 677–678
  sampling, 678–680
  in suspended particulate matter, 678
PCDFs, 673–685
  analysis, 673–674
  cleanup, 681
  data reporting, 683
  in dissolved fraction, 678–680
  environmental relevance, 673
  extraction, 680–681
    from liquid matrices, 680
    from solid matrices, 681
  gas chromatography, 681–682
  identification and quantification of congeners, 682–683
  instrumental analysis, 681–683
  internal standard quantification method, 674–677
  laboratory safety, 677
  mass spectrometry, 682
  physicochemical properties, 673
  principles behind the method, 677–678
  sampling, 678–680
  in suspended particulate matter, 678
PCDTs (*see* Polychlorodibenzothiophenes)
PCI (*see* Positive chemical ionization)
PCI-MS, 583
PCM (*see* Phase contrast microscopy)

PCNB, 619
PCP (see Pentachlorophenol)
PCR (see Polymerase chain reaction)
PDA (see Photodiode array detection)
PDECD (see Pulsed-discharge electron capture detector)
PDHID (see Pulsed-discharge helium ionization detector)
PDMS (see Polydimethylsiloxane)
PEARLS process, 109–110
PED (see Plasma emission detector)
Penicillin, 620
Pentachlorobiphenyls, 655–656
Pentachlorophenol, 10, 347–366, 619, 630
Peptides, 401
Permanent hardness, 56
Permethrin, 5
Peroxidase, 380
Peroxyoxalate, 807
Perylene, 707
Pesticide degradation products, 512
Pesticides, 4, 9–10, 15, 22, 201, 353, 356, 358, 383, 387, 383, 387, 517–654, 738–739, 755, 760, 768
Pesticides hazard classes, 537
Petroleum hydrocarbon analysis, 753–764
Petroleum oil herbicides, 621
PFBHA (see o-(2,3,4,5,6-pentafluorobenzyl)hydroxylamine)
PFE (see Pressurized solvent extraction)
1,2,3,7,8-P$_5$CDD, 681–682
1,2,3,7,8-P$_5$CDF, 682
Pfizer selective enterococcus agar, 129–141
P-FPD (see Phosphorus flame photometric detector)
PGC (see Porous glassy carbons, Porous graphitic carbon)
pH, 8, 25, 28, 32, 34, 69–72, 319, 411
    criteria, 72
    electrodes, 786–787
    interferences, 71
    method, 71–72
    method principle, 71
    sampling procedure, 71
    value, 786–787
Phase contrast microscopy, 765
Phenoclor, 655
Phenol index, 349
Phenol-derivative herbicides, 630
Phenolic acid spot test, 650
Phenolic acids, 402
Phenolic compounds, 347–366, 405
    amperometric detection, 350
    biosensors, 351
    capillary zone electrophoresis, 354–355
    chemiluminescence, 352
    classification and chemical characterization, 347–348
    colorimetric methods, 348–349
    coulometric detection, 350
    derivatization, 352
    EPA and other official methods, 348–349
    flow injection analysis, 351

[Phenolic compounds]
    gas chromatography, 354, 356, 360–362
        electron capture detection, 348
        flame ionization detection, 348
        mass spectrometry, 348
    immunoassay, 355
    liquid chromatography, 349–353, 356, 358–360
        with electrochemical detection, 350–351
        with fluorescence detection, 352
        mass spectrometry, 352–353
        with UV and diode array detection, 350
    liquid-liquid extraction, 348, 355–356
    liquid-solid extraction, 348
    micellar electrokinetic chromatography, 355
    optical chemical sensor, 355
    reversed-phase liquid chromatography, 350
    sample preparation, 355–364
    solid-phase extraction, 356–363
    supercritical fluid chromatography, 355, 362–363
    supercritical fluid extraction, 363
    supported liquid membrane, 363
    why detect?, 347
Phenolphthalein alkalinity, 53
Phenols, 80, 87, 347–366, 397, 401, 403, 631–632, 738, 760, 872
Phenoxy herbicides spot test, 633–650
Phenoxyaliphatic acid herbicides, 621
Phenyl isocyanate, 249
Phenylmercury acetate, 610
Phenylurea herbicides, 487–515
Phenylureas, 487–515
PHI (see Phenyl isocyanate)
Phorate, 571, 579–581
Phosphate-ion-selective electrodes, 287
Phosphates, 28, 202, 206, 209, 214–217, 273–295, 332–334, 338, 341, 571, 787, 863
    algal bioassay, 278
    analysis, 280–286
    APHA-AWWA-WEF Standard Methods, 275
    atomic spectroscopic methods, 284
    automated batch analyzers, 286
    automated methods, 285–286
    in brackish and estuarine waters, 288–289
    capillary electrophoresis, 285
    chromatographic methods, 284–285
    colorimetry/spectrometry, 280–284
    detection methods, 286–288
    digestion, 277–278
    direct photometry, 280–284
    electrochemical detection, 287
    enzymatic detection, 287
    enzymatic methods, 279–280
    filtration, 276
    flow injection analysis, 286
    gel filtration chromatography, 284–285
    gravimetric methods, 280

[Phosphates]
  high-temperature combustion and fusion, 277
  HPLC, 284
  inductively coupled plasma atomic emission spectrometry,
      284
  indirect photometry, 284
  ion chromatography, 284, 287
  iron oxide adsorption methods, 279
  membrane filtration, 276
  microwave digestion, 277
  photoluminescent detection, 287
  photometric detection, 286–287
  physical and chemical properties, 273–274
  portable and in situ analysis systems, 287–288
  in potable water, 288
  preconcentration, 276–277
  sample preservation, storage, and pretreatment, 274–278
  in seawater, 289
  segmented-flow analysis, 286
  thermal digestion methods, 277–278
  ultrafiltration, 276
  ultraviolet photo-oxidation, 277–278
  voltammetric techniques, 287
  volumetry, 280
  in wastewaters, 288
  wet chemical digestion, 277
Phosphate-selective enzyme electrodes, 289
Phosphoamides, 273
Phospholipids, 273
Phosphomolybdenum blue, 280–284
Phosphonates, 277, 571
Phosphonomolybdenum blue, 590
Phosphoproteins, 273
Phosphoramides, 571
Phosphorescence, 707
Phosphorothiolates, 571
Phosphorothiolothionates, 571
Phosphorothionates, 571
Phosphorus compounds, 268–269
Phosphorus, 8, 25–27, 169, 265
Phosphorus flame photometric detector, 583
Phosphorus-containing pesticides, 273
Photoacoustic sensors, 759
Photoacoustic spectroscopy, 442
Photobacterium fischerii, 787
Photobacterium phosphoreum, 158
Photodiode array detection, 160, 851–852
Photographic film, 103
Photoionization detector, 723, 726, 732, 736
Photoluminescence, 430–432
Photometry, 107
Photon correlation spectroscopy, 44
Photosynthesis, 61
Phtalates, 760
Phthalates, 79, 406
Phthalic esters, 632

Phthalate esters, 632, 662
Physical properties, 41–50
  of impure water, 42–47
  of pure water, 42–42
Phytoplankton, 261–262
  growth, 9
PID (see Photoionization detector)
Piridafenthion, 581
Planar chromatography, 251–255
Plankton, 79
Plasma emission detector, 583, 873
Plasma mass detector, 873
Plate count agar, 121–141
Plate height, 845–846
Plate number, 831
PLOT (see Porous-layer open tubular columns)
Plutonium, 105, 110
PMA (see Phenylmercury acetate)
PMB (see Phosphomolybdenum blue)
P-n junction detector, 105
PN (see Pneumatic nebulization, 107
Pneumatic-assisted electrospray, 601
PN-ICP-MS, 111
PO_4^{3-}, 197, 420
POC (see Particulate organic carbon and Purgeable organic
      carbon)
POD (see Peroxidase)
Poiseuille law, 844
Polarity, 847–848
Polarography, 198, 424
Polyaromatic hydrocarbons, 871
Polycarboxylic acids, 405
Polychlorinated biphenyls (see also Polychlorobiphenyls), 632,
      755, 760
Polychlorinated dibenzo-p-dioxins (see PCDDs)
Polychlorobiphenyls, 655–671
  analytical methods, 656–665
  analytical quality control, 669
  biological assays, 665–668
  cleanup, 660
  data evaluation, 665–669
  determination of coplanar nonortho substituted
      polychlorobiphenyls, 665–669
  extraction, 656–660
  HRGC, 662
  identification and quantification, 665
  instrumental analysis, 660–665
  liquid-liquid extraction, 656
  mass spectrometry, 662–665
  physical and chemical properties, 656
  sample collection and storage, 656
  sample preparation, 656–660
  solid-phase extraction, 656
  solid-phase microextraction, 658–660
  water regulation, 655–656
  why detect?, 655

Polychlorodibenzothiophenes, 682
Polyclonal antibodies, 162–164
Polycyclic aromatic hydrocarbons (*see also* PAHs), 632, 760, 836, 862
Polydimethylsiloxane, 526
Polymerase chain reaction, 136, 155
Polynuclear aromatic hydrocarbons, 93
Polypeptides, 402, 405
Polysaccharides, 402, 405
POM (*see* Particulate organic matter)
Porous glassy carbons, 861
Porous graphitic carbon, 360
Porous membrane electrodes, 206
Porous-layer open tubular columns, 835
Positive chemical ionization, 557, 595–602
Potassium, 409–438
Potassium-40, 102
Potential bioavailable phosphorus, 278
Potentiometric sensor, 795
Potentiometric stripping analysis, 426–427
Potentiometric stripping mode, 445
Pott Percival, 689
Pour plate technique, 120–141
P_{ow} (*see* Partition coefficient between octanol and water)
POX (*see* Purgeable organic halide)
p,p'-DDD, 534
p,p'-DDE, 534
Precision, 30
Presence-absence test, 121
Preservation chemical methods, 27–28
   physical methods, 25–27
   of water, 25–28
Pressure differential injection, 878–879
Pressurized solvent extraction, 681
Primicarb, 555
Programmed temperature, 846–847
   vaporization, 705, 833
Prohan, 560
Promis, 550
Promophos, 602
Propanil, 629
Propanoic acid, 403
Prophan, 561
Propionic acid, 333, 341
Proportional counters, 105
Propoxur, 538, 560
Prospekt, 550–551, 825
Protein phosphatase inhibition assays, 155
Proteins, 226, 261, 264, 393, 405
Proteus, 133
Protozoa, 115, 119
PSA (*see* Potentiometric stripping analysis)
PSE (*see* Bile-esculin-azide agar)
Pseudomonas, 376
Pseudomonas aeruginosa, 120, 373
Pseudomonas fluorescens, 376–379, 383

Pseudomonas putida, 442
Pseudomonas stutzeri, 383
PSP toxins immunoassays, 164
PSPs (*see* Paralytic shellfish poisons)
Pt, 468
^{190}Pt, 102
PTV (*see* Programmed temperature vaporization)
^{239}Pu, 110
^{240}Pu, 110
PUHs (*see* Phenylurea herbicides)
Pulsed amperometric detection, 350
Pulsed-discharge electron capture detector, 532, 836
Pulsed-discharge helium ionization detector, 836
Purex process, 109–110
Purge, 728–729
Purgeable organic carbon, 305–307
Purgeable organic halide, 308
Purgeables, 760–761
Purge-and-trap, 84, 90, 178, 445, 576, 723, 726–734, 743–745, 758, 760, 812
   affecting factors, 728–732
   analytical procedure, 732
   desorption, 730–731
   interface techniques, 731–732
P_v (*see* Vapor pressure)
p-xylene, 721–752
Py (*see* Pyrolysis)
Py-GC-FTIR, 404–406
Py-GC-MS, 404–406
Py-MS, 404–406
Pyrazophos, 581
Pyrethroids, 5
Pyridine, 265, 267, 585, 630
Pyridine-barbituric acid, 368–369
Pyridine-pyrazolone, 369
Pyridoxal, 369
Pyrimiphos, 571
Pyrogallol, 650
Pyrolysis, 404–406
Pyrophosphates, 572
Pyrroles, 405
Pyruvic acid, 313–345

Quadrupole mass spectrometer, 807
Quality assurance, 881
Quality control, 880–881
Quin-2, 431
Quinoline, 585, 633
Quinone fungicides, 619
Quinoxilinols, 328

^{226}Ra, 105, 112
^{228}Ra, 105, 112
R2A agar, 123
Rad-disk technology, 112
Radio frequency, 839

Radioanalytical methodology, 101–114
Radiochemical neutron activation, 106
Radiometry, 198
Radionuclides, 101–114
  alpha detectors, 105
  beta detectors, 105–106
  detectors, 102–108
  extraction chromatography, 109–110
  gamma detectors, 103–105
  in the environment, 101–102
  ion exchange, 108
  liquid-liquid extraction, 108
  mass spectrometers, 106–107
  separation techniques, 108–109
  solid-phase extraction, 109
Radiostrontium, 109
Radium Rad-disk, 112
Radon, 105
Rainwater acidity, 229
Raman spectroscopy, 402
Randomly amplified polymorphic PCR, 155
Ranges chart, 33
RAPD (see Randomly amplified polymorphic PCR)
Rapid fecal coliform test, 127
Rapid thermonuclease test, 134
Rappaport Vassiliadis medium, 134
Rayleigh scattering, 806
^{87}Rb, 102
RC (see Resistance capacity)
RCRA (see Resource Conservation and Recovery Act)
^{187}Re, 102
Reactive phosphorus, 273
Ready bioavailable phosphorus, 278
Recovery of stressed organisms, 136–137
Reduced plate height, 847
Reduced velocity, 846
Refractive index detector, 855
  Christiansen-effect type, 855
  deflection type, 855
  interference type, 855
  reflection type, 855
Relative humidity, 42
Relative response factor, 665, 682–683
Relative retention, 846
Relative retention time, 665
Relative toxicity factor, 668
Reliability, 883
Repeatability, 476, 822
Reproducibility, 476, 882–883
Residual chlorine, 52–53
Resins, 387
Resistance capacitance, 798
Resistivity, 788–789
Resorcinol, 650
Resource Conservation and Recovery Act, 690
Response time, 791

Restriction fragment length polymorphism, 155
Retention temperature, 846
Reverse osmosis, 59, 277, 389–390, 631
Reversed-phase liquid chromatography, 268, 350, 585, 703, 830, 860–862
RF (see Radio frequency)
RFLP (see Restriction fragment length polymorphism)
Rh, 468
Rhizoctonia, 620
Rhodanese, 380–381
RI (see Refractive index)
Ribosomal RNA, 155
Rotating disk electrode, 446
Rotating voltammetric electrode, 178
RPLC (see Reversed-phase liquid chromatography)
RRF (see Relative response factor)
rRNA (see Ribosomal RNA)
RRT (see Relative retention time)
RTF (see Relative toxicity factor)
Rubazoic acid technique, 246–247
Rubidium-87, 102
Ruggedness, 883

Saccharomyces cerevisiae, 373–374, 442, 494
Safe Drinking Water Act, 690
Saha relationship, 799
Salinity, 10, 288–289
Salmonella, 118–141
Salmonella alternative methods, 136
Salmonella presence-absence test, 134–136
Salting-out, 724, 740–741
SAMOS (see also System for Automated Monitoring of Organic Compounds in Surface Water), 825
Sample number, 3–4
Sample preparation, 810–826
Sample volume, 4
Sampler for deep water, 12
Sampler for large quantities in water with little depth, 12
Sampler for little depth, 12
Samples:
  contamination, 4
  loss, 4
  recommended storage, 5
  sorption, 5
  storage and conservation, 4–5
Sampling, 754–755
  average concentrations, 15
  buoy, 15
  continuous, 3
  discharge-proportional, 3
  discontinuous, 3
  equipment, 11–22
  event-controlled, 3, 15–16
  in estuarine and marine environments, 10
  in lakes and reservoirs, 7
  in streams and rivers, 8–10

[Sampling]
in urban areas, 10–11
location within the stream, 8
methods, 1–24
methods general aspects, 2–5
methods temporal aspects, 3
of suspended particulates, 1
quantity-proportional, 3
site, 3
spatial aspects, 2–3
standard methods, 755
strategies, 5–11
systems automatic, 15–17
systems for the benthic boundary layer, 14–15
systems manual, 12–14
time-proportional, 3
Savitsky-Golay algorithm, 798
SAX (*see* Strong anion exchangers)
Saxitoxin, 145, 158, 164
$Sb_2S_{4(aq)}^{2-}$, 465
Sb_2S_3, 465
$SbO_{3(aq)}^-$, 465
SBSR (*see* Single bead string reactor)
Sc, 468
SCA (*see* UK Standing Committee of Analysis)
Schistocera gregaria, 158
SCIC (*see* Single-colimn ion chromatography)
Scintillation counters, 103, 105–106
Scintillation detection systems, 103–104
Sclerolinia, 620
SCN^-, 369
Scoop bottle according to Meyer, 12
SCX (*see* Strong cation exchangers)
SDU (*see* Solvent delivery unit)
SDWA (*see* Safe Drinking Water Act)
Se(IV), 444–445
Se(VI), 444
Secondary amines, 255
Sedgwick-Rafter chamber, 148–154
Sedimentation, 47
Segmented continuous-flow analysis, 289
Segmented flow analysis, 209
Segmented flow systems, 264–269, 286
Selected ion monitoring, 93, 533–535, 595–602, 662–665,
    682, 706, 807
Selective electrodes, 467
Selective flotation, 471
Selective herbicides, 621
Selective precipitation, 471
Selective sorbents, 556
Selectivity, 791, 832, 883
Selenastrum capricornutum, 279
Selenate, 464
Selenium, 4, 444–445, 464–465, 467, 471
    AAS, 444
    capillary electrophoresis, 445

[Selenium]
capillary GC-MIP-AES, 445
CSV, 445
differential pulse CSV, 445
ETAAS, 444
FAAS, 444
FI-HG-AAS, 444
flow injection analysis, 444
fluorometric methods, 444
gas chromatography, 445
HG-AAS, 444
ICP-MS, 444–445
ion chromatography, 444–445
single-wavelength-dispersive XRF, 444
spectrophotometry, 444
Semicarbazones, 264
Semiconductor detector, 103
Semipermeable-membrane device, 17–19, 698
Semivolatile organic compounds, 98
Sensors based on fiber optics, 205
Sensory defects, 115
Sensory gas chromatography, 81–82
Sensory methods, 81–82
SeO_3^{2-}, 464
SeO_4^{2-}, 464
Separation number, 832
Sequential injection analysis, 218, 416, 445
SFC (*see* Supercritical fluid chromatography)
SFC-MS, 558, 595
S-FPD (*see* Sulphur flame photometric detector)
Shigella, 118–141
Shinn-Griess reaction, 269
Shpol'skii fluorimetry, 707
SHS (*see* Static headspace)
SIA (*see* Sequential injection analysis)
Significant digits, 31
Silent Spring, 519
Silica content, 28
Silicates, 483–486
    colorimetry, 483–485
    gravimetric method, 483, 485–486
    recent methods, 486
    why detect?, 483
Silicon, 483–486
Silver, 465–866
Silver disc microelectrode, 190
Silvex, 621, 631
SIM (*see* Selection ion monitoring)
Simazine, 381
SIM-GC-MS, 601
SIM-MS, 601
Sinapis alba L, 158
Single bead string reactor, 589
Single-column IC, 251
Single-element methods, 785
Single-wavelength-dispersive XRF, 444

SiO_2^-, 420
Size exclusion chromatography (*see also* Gel filtration chromatography), 830
Skewchart chart, 33
Skewed distributions, 32
SLM (*see* Supported liquid membrane)
SLRS-3, 476
Sludges, 758
^{147}Sm, 102
^{148}Sm, 102
SMDC, 620
Smoluchowski's equation, 47
SN (*see* Separation number)
Sn^{2+}, 465
Sn^{4+}, 465
Snyder's displacement model, 859
Snyder-Soczewinski equation, 859
SO_3^{2-}, 369
SO_4^{2-}, 197, 420, 461
Soap curd, 56
Sodium, 409–438, 705
Sodium 2-(parasulfophenlazo)-1,8-dihydroxy-3,6-naphthalene disulfonate method, 65
Sodium borate, 621
Sodium chlorate, 621
Sodium fluoride, 64
Sodium lauryl sulphate, 770–772
Sodium metaborate, 621
Sodium nitrate, 267
Sodium nitrite, 267
Sodium silicofluoride, 64
Sodium tetraborate, 621
Sodium tetrapropylene benzene sulphonate, 771
Soil half-life, 561
Solid-liquid extraction, 723
Solid-phase extraction, 17, 19, 88–90, 156, 159–161, 321, 324, 350–364, 492–498, 500–502, 526–527, 547, 576–581, 656, 680, 699–702, 707, 738, 758, 781, 815–826, 844
  adsorption, 578
  cleanup, 825–826
  inside the capillary, 879
  membrane disks, 527–528
  off-line, 815–825
  online, 825
  partitioning, 578
Solid-phase luminescence spectroscopy, 707
Solid-phase microextraction, 90, 358, 497, 528–529, 551–556, 576, 582–583, 658–660, 698, 700, 702, 723, 737–745, 758, 825
  absorption, 739–742
  affecting factors, 739
  analytical procedure, 742–743
  desorption, 742
  fiber selection, 739
Solid-phase suppressors, 865
Solid-state electrodes chemically modified, 207

Soluble impurities, 47–48
Solvent delivery system, 848
Solvent delivery unit, 550
Solvent reservoirs, 848
Solvent split injection, 833–834
Solvent-extraction, 471
Solvent-venting injection, 873
Sonolytic decomposition, 689
SPADNS (*see* Sodium 2-(parasulfophenylazo)-1,8-dihydroxy-3,6-naphthalene disulfonate)
Spark chambers, 103
SPE (*see* Solid-phase extraction)
Speciation, 461
Spectrofluorimetry, 247, 560, 707
Spectrophotometry, 796–798, 828
SPE-GC, 534
SPE-GC-MS, 601
SPE-HPLC, 602
SPE-LC, 534, 602
SPE-LC-APCI-MS, 602
SPE-LC-DAD, 359, 602
SPE-LC-ISP-MS, 602
SPE-LC-MS, 601
SPE-LC-TSP-MS, 601–602
SPE-SFC-TID, 585
Split injection, 329, 732, 833
Split outlet, 833
Splitless injection, 833
Split/splitless injection, 705
SPMD (*see* Semipermeable membrane device)
SPME (*see* Solid-phase microextraction)
Spot tests, 630, 633–650
Spray and trap, 85
Spread, 30
Spread plate technique, 120–141
SPS (*see* Suspended-particle sampler)
Squalan value, 760–761
Squalane, 847
^{89}Sr, 106, 109
^{90}Sr, 106, 109
S-R (*see* Sedgwick-Rafter)
SRM1643d, 476
$SrSO_4^0$, 465
SSLS (*see* Solid-phase luminescence spectroscopy)
Stabilized temperature platform furnace, 442
Standard compound microscope, 148–149
Standard deviation, 30–40
Staphylococcus aureus enumeration, 133–134
Static headspace, 83–84, 723–726, 744, 812
  analytical procedure, 725–726
  temperature, 724
Statistical assessment, 37–38
Statistical treatment:
  definitions, 29–31
  of data, 29–40
  why?, 29

Sterilization with radiation, 25
Stern layer, 874
Stoke's law, 47
Stokes-Einstein equation, 44
Storm water monitoring systems, 11
STPF (*see* Stabilized temperature platform furnace)
Streptococcus, 119–141
Streptococcus bovus, 119, 129
Streptococcus equinus, 119, 129
Streptomyces, 620
Streptomycin, 620
Stripping, 734–735
Stripping potentiometry, 446
Stripping voltammetry, 467
Strong anion exchangers, 863
Strong cation exchangers, 863
Strontium, 27, 106, 465, 787
Strontium Rad-disk, 112
Substituted amide herbicides, 621–629
Substituted benzene fungicides, 619
Substituted nitrile herbicides, 630
Substituted urea herbicides, 629
Succinic acid, 403
Sugar phosphates, 273
Sugars, 393
Sulfallate, 538
Sulfate, 67, 195–198, 206, 214–217, 297, 315, 321, 332–333,
   338, 341, 787, 863
   colorimetric methods, 196–197
   gravimetric method, 195–196
   ion chromatography, 197
   turbidimetric method, 196
Sulfate oxidase, 380–381
Sulfide, 28, 195, 198, 209, 369, 836
   HMSO analytical method, 198
   ion chromatography, 198
   methylene blue method, 198
Sulfite, 195, 198, 341
   ion chromatography, 198
   titration method, 198
Sulfonylurea herbicides, 487–500
Sulfonylureas, 487–500
Sulfur, 52, 169, 195
   compounds, 25, 753
   flame photometric detector, 583
   fungicides, 610
Sulfur dioxides, 70
Sulfuric acid, 621
Sulfur-oxidizing bacteria, 115
Sulpho succinate esters, 768
Sulphotep, 572
Supercritical fluid chromatography, 268, 355, 357, 362–363,
   558, 585–589, 705, 830, 870–873
   detectors, 873
   instrument, 872–873
   MS, 807

Supercritical fluid extraction, 363, 496, 500, 555–556, 581,
   702, 757
Superfund Amendments and Reauthorization Act, 690
Supported liquid membrane, 363
Suppressed conductivity, 331, 334
   IC, 251
Suppressed conductometric detection, 197
Suppressed ion chromatography, 214–216, 427–428
Suppressed-mode anion chromatography, 191
Suppressor, 190
Suppressor column, 865
Surface energy, 42
Surface tension, 42
Surface-active agents (*see* Surfactants)
Surface-barrier detector, 105
Surface-grab procedures, 8
Surfactant-mediated extraction, 702
Surfactants, 84, 483, 767–783
   barium iodobismuthate method, 775–776
   cobalt thiocyanate method, 774–775
   disulfine blue method, 772–773
   ethyl violet method, 771
   methylene blue method, 770–771
   quantitative analysis, 770–781
   recent methods, 781
   titrimetric method, 771–781
   why detect?, 767
Suspended material, 234, 440
Suspended matter, 15, 52, 54, 57, 410, 755–756
Suspended particles, 203–205
Suspended-particle sampler, 22
Suspended particulate matter, 10, 678
Suspended particulates, 1
Suspended solids, 8–9, 788
Suspensions coagulation and flocculation, 45–47
SVOCs (*see* Semivolatile organic compounds)
S_w (*see* Water solubility)
Sweeping, 728
Swep, 557
Synchronous fluorescence spectroscopy, 707
Systemic fungicides, 620

2,4,5-T, 631
T&R-MIMS (*see* Trap and release-MIMS)
T_4CDD, 673
T_4CDF, 673
Tables, 32–34
TAD (*see* Thermally assisted desorption)
Tangential flow filtration systems, 276, 678–679
Tangential-flow process, 390
Taste, 75–100, 229, 313, 347
TC (*see* Total carbon)
Tc^{4+}, 112
Tc^{7+}, 112
^{99}Tc, 111–112
TCA, 630, 632

TCA spot test, 633
TCD (*see* Thermal conductivity detection)
1,3,7,8-T$_4$CDD, 682
2,3,7,8-T$_4$CDD, 673, 681–683
2,3,7,8-T$_4$CDF, 682
TC-TS-MS, 509
TDE, 5
t-distribution, 38
TDN (*see* Total dissolved nitrogen)
TDS (*see* Total dissolved solids)
TEA (*see* Thermal energy analyzer)
Technetium, 111
Technetium-99, 107, 109
Technetium-99 membrane, 112
Technicon Auto-Analyzer, 196
Techniques, 785–886
TEF (*see* Toxicity equivalency factor)
TEM (*see* Transmission electron microscopy)
Temephos, 581
Temperature, 7, 8, 25, 32
    reduction, 25–27
Temporal changes of water quality, 9
Temporary hardness, 56
TEQs (*see* Toxicity equivalents)
Terrazole, 619
2,3,7,8-tetrachlorodibenzo-p-dioxin, 668
2,3,4,6-tetrachlorophenol, 355
2,3,5,6-tetrachlorophenol, 363
Tetrachlorvinphos, 602
Tetraethyl lead, 442
TEVA-Spec™ resin, 110
TFP (*see* Total filterable phosphorus)
Thallium, 465, 467
^{230}Th, 110–111
^{232}Th, 110
Theoretical plates, 875
Thermal conductivity, detection, 92, 248, 328, 340, 835–836
Thermal energy analyzer, 807
Thermal ionization mass spectrometer, 106
Thermal lens spectromery, 444
Thermal lensing techniques, 286
Thermal stratification, 7
Thermally assisted desorption, 550, 702
Thermionic ionization detector, 583
Thermogravimetry, 198
Thermoionic detector, 836
Thermospray, 352–353, 509, 558, 809–810, 857
    ionization, 595–602
Thermospray-CI, 601
Thermotolerant coliforms, 119–141
Thiabendazole, 620
Thiazole fungicides, 619
Thin mercury film, 425
Thin-layer chromatography, 251, 430, 556, 558–559, 589,
        702–703, 825, 866–870
    detection, 869–870

Thiocarbamates, 545
Thiolcarbamic acid, 538
Thiolphosphates, 571
Thiols, 836
Thiophanate, 538, 620
Thiophenes, 836
Thiophosphates, 571
Thiouracils, 872
Thiourea, 265, 267
Thiram, 619
THMS (*see* Trihalomethanes)
Thorium, 102, 110–111
Thorium-232, 101
TIC (*see* Total ion current)
TIC-SIE (*see* Total ion current-selected ion extraction)
TID (*see* Thermionic ionization detector)
Time of flight, 810
    mass spectrometers, 106
TIMS (*see* Thermal ionization mass spectrometer)
Tin, 465, 467
TISAB (*see* Total ionic strength adjusting buffer)
Titanium, 4
TKN (*see* Total Kjeldahl nitrogen)
Tl^0, 465
Tl^{3+}, 465
TLC (*see* Thin-layer chromatography)
TMF (*see* Thin mercury film)
TN (*see also* Total nitrogen), 264
TOC (*see* Total organic carbon)
TOF (*see* Time of flight)
TOG, 753–764
    calibration, 760–761
    conservation, 756
    container preparation, 755–756
    definitions and interferences, 753
    detection and determination methods, 758–760
    detection in the field, 760
    distribution in waters, 754
    gas extraction, 757–758
    GC-MS, 759–760
    gravimetric analysis, 759, 761
    IR analysis, 761
    laser methods, 759
    liquid-liquid extraction, 756–757
    optical methods, 759
    physical and chemical properties, 753–754
    problem estimation, 760–762
    reagents, 761–762
    recommended methods, 761
    sample preparation, 756–758
    sampling, 754–755
    solid-phase extraction, 758
    spectroscopic methods, 759
    UV-VIS methods, 759
TOG/TPH methods:
    for detection of spill pollution sources and for comparison
        of waterborne petroleum oils, 762

[TOG/TPH methods]
    for determination of individual hydrocarbons, 761
    for sludges, 761
    for total determination, 761
Toluene (*see also* BTEX compounds), 94, 721–752
Toluidine blue O-DNA agar, 134
TOM (*see* Total organic matter)
Total active chlorine, 170, 173–175
Total alkalinity, 53
Total ammoniacal nitrogen, 226–227
Total carbon, 304, 792
Total dissolved nitrogen, 267
Total dissolved solids, 8, 475
Total filterable phosphorus, 274–295
Total hardness, 56
Total ion current, 532
    selected ion extraction, 533
Total ionic strength adjusting buffer, 186
Total Kjeldahl nitrogen, 263
Total nitrogen, 9, 264–269
Total oil and grease (*see* TOG)
Total organic carbon, 60, 75, 83, 297, 303–307, 388–408,
    792–793
    conductivity, 307
    flow injection anaylsis, 307
    high combustion temperature methods, 305–307
    oxidation methods, 306
    photodecomposition, 307
    radiative methods, 307
Total organic halide, 297, 307–308
    adsorption-pyrolysis-titrimetric method, 307–308
    direct measurement of absorbance, 308
    why detect?, 307
Total organic matter, 388–408
Total petroleum hydrocarbons (*see* TPH)
Total phosphorus, 8–9, 269, 273–295
Total reactive phosphorus, 274–295
Total suspended solids, 31
TOX (*see* Total organic halide)
Toxaphene, 632
Toxicity equivalency factor, 683
Toxicity equivalents, 683
Toxicity test, 787
Toxicity testing using cell cultures, 158
TP (*see* Total phosphorus)
TPH, 753–764
    calibration, 760–761
    conservation, 756
    container preparation, 755–756
    definitions and interferences, 753
    detection and determination methods, 758–760
    detection in the field, 760
    distribution in waters, 754
    gas extraction, 757–758
    GC-MS, 759–760
    gravimetric analysis, 759, 761

[TPH]
    IR analysis, 761
    laser methods, 759
    liquid-liquid extraction, 756–757
    optical methods, 759
    physical and chemical properties, 753–754
    problem estimation, 760–762
    reagents, 761–762
    recommended methods, 761
    sample preparation, 756–758
    sampling, 754–755
    solid-phase extraction, 758
    spectroscopic methods, 759
    UV-VIS methods, 759
TPs (*see* Transformation products)
Trace analysis, 4
Trace elements, 459–482
    analytical control, 476
    analytical methods, 466–473
    analytical protocol, 474–476
    applications in different waters, 472–473
    data quality, 474–476
    guideline values, 459
    implemented techniques, 472
    instrument operating conditions, 467–469
    interference effects, 469–471
    physical and chemical characteristics, 460–465
    sample preparation, 465–466
    sampling, 465–466
    separation and preconcentration methods, 471–472
    stabilization, 466
    techniques, 466–467
    why detect?, 459
Track detector, 103
Transformation products, 572
Translocated herbicides, 621
Transmissiometer, 14
Transmission electron microscopy, 765
Trap, 729–730
Trap GC-MS, 94–96
Trap membrane introduction mass spectrometry, 90, 94–98
Trap packing and design, 730
Trap and release-MIMS, 98
Trennzahl (*see* Separation number)
Triazine fungicides, 619
Triazines, 5, 537, 559, 630
Triazole, 630
Tribenuron-methyl, 495
Tricarboxylic acids, 336
Trichlorfon, 571, 581
Trichloroacetic acid, 631
Trichlorobiphenyls, 665
2,4,5-trichlorophenol, 356, 363
2,4,6-trichlorophenol, 348
Trichodesmium, 143–144
Trichodesmium erythraeum, 145–147

Trichome, 143, 152

Trichosporon cutaneum, 373

Trifluralin, 629

Trihalomethane polychlorinated biphenils, 307

Trihalomethanes, 59, 75, 80, 177, 229, 387

Triolein, 17

Triphenylene, 703–704

Tritium, 106

True result, 29

TRU-Resin™ method, 109–110

Tryptophan, 264, 266

Tryptose sulfite cycloserine agar, 132–133

TS (*see* Thermospray)

TSC (*see* Tryptose sulfite cycloserine agar)

TSI (*see* Thermospray ionization)

TSP (*see* Thermospray)

TSP-MS, 601

TSS (*see* Total suspended solids)

t-test, 38

TU (*see* Turbidity units)

Turbidimeter, 43

Turbidity, 8, 43, 79, 204, 277, 410, 788

Turbidity units, 788

Turex process, 109–110

Two-dimensional TLC, 866–869

U.S. Environmental Protection Agency, 105–106, 109, 197, 230, 239, 348–349, 459, 476, 488, 490–491, 494, 545–546, 618, 655, 674, 680, 682–683, 690, 728–732, 754, 756, 759

U.S. Federal Storm Water Regulations, 10

U.S. National Toxicology Program, 65

UK Standing Committee of Analysis, 576

UK°, 787

Ultrafiltration, 390, 631, 811

Ultrasound digestion, 441

Ultraviolet, 160, 164

 irradiation, 4

 photolysis, 441

 radiation, 59

 and visible spectroscopy, 758–760, 851

Umezakia, 145

Umezakia natans, 143–147, 157

Unreactive phosphorus, 279

Uracil, 266, 630

Uranium, 102, 105, 110–112

^{226}U, 111

$^{233-236}$U, 110

^{234}U, 111

^{235}U, 101

^{238}U, 102, 110

Urea, 261, 264, 266–268

Urea herbicides, 487–515

 analytical methods, 488–511

 capillary electrophoresis, 506–509

 derivatization procedures, 499–500

[Urea herbicides]

 extraction, 490–498

 gas chromatography, 499–500

 liquid chromatography, 500–506

 liquid chromatography-mass spectrometry, 509–511

 liquid-liquid extraction, 491–493

 online extraction with liquid membranes, 497

 physicochemical properties, 487–488

 potable water sampling, 490

 regulations, 488

 sample storage, 490

 sampling, 488–490

 solid-phase extraction, 492–498

  off-line with adsorbents imbedded in membranes, 495–496

  off-line with cartridges, 494–495

  online, 496–497

 solid-phase microextraction, 497

 supercritical fluid extraction, 500

 surface water sampling, 490

 well water sampling, 489

Uric acid, 264

US°, 787

Utermöhl counting chamber, 149–154

UV-VIS absorption, 796–799

UV-VIS (*see* Ultraviolet and visible spectroscopy)

$V(OH)_2^+$, 464

V^{5+}, 464

Valeric acid, 341

Validation, 880–884

Validation of detection, 882

Vamidothion, 571

Van Slyke's method, 789

Vanadium, 27, 433, 464, 471

Van't Hoff plot, 846

Vapam, 619

Vapor pressure, 42, 572

VFA (*see* Volatile fatty acids)

Vibrio alginolyticus, 131

Videodensitometry, 870

Viscosity, 42

Viscous shear, 42

VOC (*see* Volatile organic carbon)

VOCs (*see* Volatile organic compounds)

Volatile fatty acids, 314–345

Volatile organic acid (*see* Volatile fatty acids)

Volatile organic carbon, 305–307

Volatile organic compounds, 83–98, 745, 812–813

 closed-loop stripping analysis, 812

 gas stripping, 812–813

 headspace, 812

 purge-and-trap, 812

Volhard method, 180

Voltammetry, 424–426

Volume-size distributions, 44

Voyager, 843

Wall-coated open-tubular columns, 834–835
Water hardness, 415, 426
Water microbiological standards, 120
Water physical properties, 41–50
Water Pollution Control Law Japan, 374
Water preservation, 25–28
Water purification, 79–81
Water sampler according to Ruttner, 12–13
Water solubility, 572
WAX (*see* Weak anion exchangers)
Waxes, 387
WCOT (*see* Wall-coated open-tubular columns)
WCX (*see* Weak cation exchangers)
Weak anion exchangers, 863
Weak cation exchangers, 863
Weeds, 79
Wet-only samplers, 697
Whipple grid, 149–153
White asbestos, 765
WHO (*see* World Health Organization)
Wide-bore capillary columns, 583, 834
Winkler method, 790
Winkler titration, 298
World Bank, 519, 521
World Health Organization, 179, 459, 537, 689–690, 721

X-ray fluorescence spectrometry, 412, 432–433
X-ray photoelectron spectroscopy, 402
XRFS (*see* X-ray fluorescence spectrometry)

Xylene (*see also* BTEX compounds), 94, 721–752
Xylose lysine brilliant-green agar, 136
Xylose lysine desoxycholate agar, 136

Yeast extract agar, 121–141
Yeasts, 115
^{90}Y, 106
Y-GLU (*see* Indoxyl-E-D-glucuronide)

Zeeman effect, 802–803
Zero-point titration, 181
Zero-tolerance principle, 118
Zeta potential, 874
Zinc, 409–438, 446
    ASV, 446
    DP-ASV, 446
    ETAAS, 446
    FAAS, 446
    flow injection analysis, 446
    ICP-AES, 446
    ion-selective electrode, 446
    spectrofluorometric method, 446
Zincon, 418
Zineb, 619
Ziram, 560–561, 619
Zn(II), 374, 378, 381, 410, 432, 446
Zn(OH)$_2$, 410
ZnCO$_3$, 410